Student Study Guide/
Solutions Manual

to accompany

Organic Chemistry

Third Edition

Janice Gorzynski Smith
University of Hawai'i at Mānoa

and

Erin R. Smith Berk

The McGraw-Hill Companies

Student Study Guide/Solutions Manual to accompany
ORGANIC CHEMISTRY, THIRD EDITION
JANICE GORZYNSKI SMITH, ERIN R. SMITH BERK

Published by McGraw-Hill Higher Education, an imprint of The McGraw-Hill Companies, Inc., 1221 Avenue of the Americas, New York, NY 10020. Copyright © 2011, 2008, and 2005 by The McGraw-Hill Companies, Inc. All rights reserved.

Printed in the United States of America.

3 4 5 6 7 8 9 0 QDB 10 9 8 7 6 5 4 3 2 1

ISBN: 978-0-07-729665-0
MHID: 0-07-729665-6

www.mhhe.com

Contents

Chapter 1: Structure and Bonding

♦ **Important facts**

- **The general rule of bonding:** Atoms strive to attain a complete outer shell of valence electrons (Section 1.2). H "wants" 2 electrons. Second-row elements "want" 8 electrons.

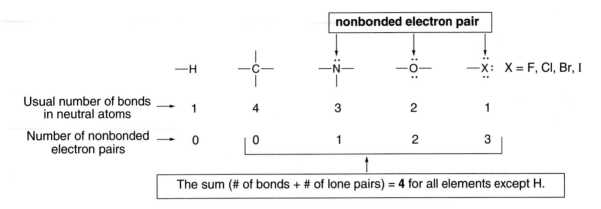

		nonbonded electron pair		
—H	—C̶—	—N̈—	—Ö—	—Ẍ: X = F, Cl, Br, I

Usual number of bonds in neutral atoms →	1	4	3	2	1
Number of nonbonded electron pairs →	0	0	1	2	3

The sum (# of bonds + # of lone pairs) = **4** for all elements except H.

- **Formal charge** (FC) is the difference between the number of valence electrons of an atom and the number of electrons it "owns" (Section 1.3C). See Sample Problem 1.4 for a stepwise example.

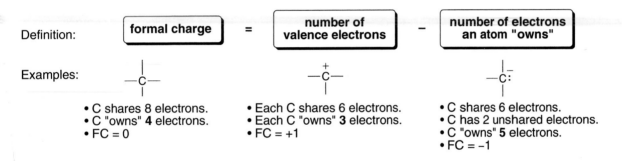

Definition: **formal charge** = **number of valence electrons** − **number of electrons an atom "owns"**

Examples:

—C̶—
- C shares 8 electrons.
- C "owns" **4** electrons.
- FC = 0

—C⁺—
- Each C shares 6 electrons.
- Each C "owns" **3** electrons.
- FC = +1

—C̈:⁻
- C shares 6 electrons.
- C has 2 unshared electrons.
- C "owns" **5** electrons.
- FC = −1

- **Curved arrow notation** shows the movement of an electron pair. The tail of the arrow always begins at an electron pair, either in a bond or a lone pair. The head points to where the electron pair "moves" (Section 1.5).

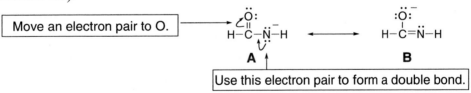

Move an electron pair to O.

A ⟷ B

Use this electron pair to form a double bond.

- **Electrostatic potential plots** are color-coded maps of electron density, indicating electron rich and electron deficient regions (Section 1.11).

Chapter 1–2

♦ The importance of Lewis structures (Sections 1.3, 1.4)

A properly drawn Lewis structure shows the number of bonds and lone pairs present around each atom in a molecule. In a valid Lewis structure, each H has two electrons, and each second-row element has no more than eight. This is the first step needed to determine many properties of a molecule.

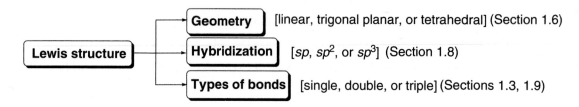

♦ Resonance (Section 1.5)

The basic principles:
- Resonance occurs when a compound cannot be represented by a single Lewis structure.
- Two resonance structures differ *only* in the position of nonbonded electrons and π bonds.
- The resonance hybrid is the only accurate representation for a resonance-stabilized compound. A hybrid is more stable than any single resonance structure because electron density is delocalized.

The difference between resonance structures and isomers:
- Two **isomers** differ in the arrangement of *both* atoms and electrons.
- **Resonance structures** differ *only* in the *arrangement of electrons*.

♦ Geometry and hybridization

The number of groups around an atom determines both its geometry (Section 1.6) and hybridization (Section 1.8).

Number of groups	Geometry	Bond angle (°)	Hybridization	Examples
2	linear	180	sp	BeH_2, HC≡CH
3	trigonal planar	120	sp^2	BF_3, CH_2=CH_2
4	tetrahedral	109.5	sp^3	CH_4, NH_3, H_2O

♦ **Drawing organic molecules (Section 1.7)**

• Shorthand methods are used to abbreviate the structure of organic molecules.

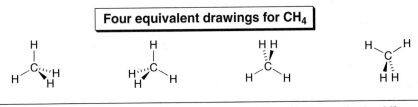

skeletal structure isooctane condensed structure

• A carbon bonded to four atoms is tetrahedral in shape. The best way to represent a tetrahedron is to draw two bonds in the plane, one in front, and one behind.

Four equivalent drawings for CH_4

Each drawing has two solid lines, one wedge, and one dashed line.

♦ **Bond length**

• Bond length decreases across a row and increases down a column of the periodic table (Section 1.6A).

$$-\overset{|}{\underset{|}{C}}-H \quad > \quad -\overset{|}{N}-H \quad > \quad -O-H \qquad H-F \quad < \quad H-Cl \quad < \quad H-Br$$

Increasing bond length   Increasing bond length

• Bond length decreases as the number of electrons between two nuclei increases (Section 1.10A).

$$CH_3-CH_3 \quad < \quad CH_2{=}CH_2 \quad < \quad H-C{\equiv}C-H$$

Increasing bond length

• Bond length increases as the percent s-character decreases (Section 1.10B).

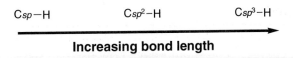

$C_{sp}-H \qquad\qquad C_{sp^2}-H \qquad\qquad C_{sp^3}-H$

Increasing bond length

• Bond length and bond strength are inversely related. Shorter bonds are stronger bonds (Section 1.10).

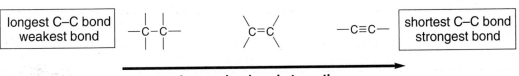

longest C–C bond weakest bond			shortest C–C bond strongest bond

Increasing bond strength

- Sigma (σ) bonds are generally stronger than π bonds (Section 1.9).

| 1 strong σ bond | 1 stronger σ bond
1 weaker π bond | 1 stronger σ bond
2 weaker π bonds |

◆ Electronegativity and polarity (Sections 1.11, 1.12)

- Electronegativity increases across a row and decreases down a column of the periodic table.
- A polar bond results when two atoms of different electronegativity are bonded together. Whenever C or H is bonded to N, O, or any halogen, the bond is polar.
- A polar molecule has either one polar bond, or two or more bond dipoles that reinforce.

◆ Drawing Lewis structures: A shortcut

Chapter 1 devotes a great deal of time to drawing valid Lewis structures. For molecules with many bonds, it may take quite awhile to find acceptable Lewis structures by using trial-and-error to place electrons. Fortunately, a shortcut can be used to figure out how many bonds are present in a molecule.

Shortcut on drawing Lewis structures—Determining the number of bonds:
   [1] Count up the number of valence electrons.
   [2] Calculate how many electrons are needed if there were no bonds between atoms and every atom has a filled shell of valence electrons; i.e., hydrogen gets two electrons, and second-row elements get eight.
   [3] Subtract the number obtained in Step [2] from the sum obtained in Step [1]. **This difference tells how many electrons must be shared** to give every H two electrons and every second-row element eight. Since there are two electrons per bond, dividing this difference by two tells how many bonds are needed.

To draw the Lewis structure:
   [1] Arrange the atoms as usual.
   [2] Count up the number of valence electrons.
   [3] Use the shortcut to determine how many bonds are present.
   [4] Draw in the two-electron bonds to all the H's first. Then, draw the remaining bonds between other atoms making sure that no second-row element gets more than eight electrons and that you use the total number of bonds determined previously.
   [5] Finally, place unshared electron pairs on all atoms that do not have an octet of electrons, and calculate formal charge. You should have now used all the valence electrons determined in the first step.

Example: Draw all valid Lewis structures for CH_3NCO using the shortcut procedure.

[1] Arrange the atoms.

```
     H
  H  C  N  C  O
     H
```

- In this case the arrangement of atoms is implied by the way the structure is drawn.

[2] Count up the number of valence electrons.

3H's	x	1 electron per H	=	3 electrons
2C's	x	4 electrons per C	=	8 electrons
1N	x	5 electrons per N	=	5 electrons
1O	x	6 electrons per O	=	+ 6 electrons

22 electrons total

[3] Use the shortcut to figure out how many bonds are needed.

- Number of electrons needed if there were no bonds:

3 H's	x	2 electrons per H	=	6 electrons
4 second-row elements	x	8 electrons per element	=	+ 32 electrons

38 electrons needed if there were no bonds

- Number of electrons that must be shared:

    38 electrons
    − 22 electrons

    16 electrons must be shared

- Since every bond takes two electrons, 16/2 = **8 bonds are needed.**

[4] Draw all possible Lewis structures.

- Draw the bonds to the H's first (three bonds). Then add five more bonds. Arrange them between the C's, N, and O, making sure that no atom gets more than eight electrons. There are three possible arrangements of bonds; i.e., there are three resonance structures.
- Add additional electron pairs to give each atom an octet and check that all 22 electrons are used.

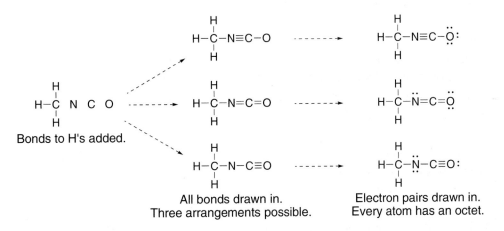

Bonds to H's added.

All bonds drawn in.
Three arrangements possible.

Electron pairs drawn in.
Every atom has an octet.

- Calculate the formal charge on each atom.

$$H-\overset{\underset{|}{H}}{\underset{|}{C}}-N\equiv C-\overset{..}{\underset{..}{O}}: \quad \longleftrightarrow \quad H-\overset{\underset{|}{H}}{\underset{|}{C}}-\overset{..}{N}=C=\overset{..}{\underset{..}{O}} \quad \longleftrightarrow \quad H-\overset{\underset{|}{H}}{\underset{|}{C}}-\overset{..}{N}-C\equiv O:$$

                    +1   −1   −1   +1

- You can evaluate the Lewis structures you have drawn. The middle structure is the best resonance structure, since it has no charged atoms.

Note: This method works for compounds that contain second-row elements in which every element gets an octet of electrons. It does NOT necessarily work for compounds with an atom that does not have an octet (such as BF_3), or compounds that have elements located in the third row and later in the periodic table.

Chapter 1: Answers to Problems

1.1 The **mass number** is the number of protons and neutrons. The **atomic number** is the number of protons and is the same for all isotopes.

	Nitrogen-14	Nitrogen-13
a. number of protons = atomic number for N = 7	7	7
b. number of neutrons = mass number – atomic number	7	6
c. number of electrons = number of protons	7	7
d. The group number is the same for all isotopes.	5A	5A

1.2 The **atomic number** is the number of protons. The **total number of electrons** in the neutral atom is equal to the number of protons. The number of **valence electrons** is equal to the group number for second-row elements. The **group number** is located above each column in the periodic table.

	a. atomic number	b. total number of e^-	c. valence e^-	d. group number
[1] $^{31}_{15}P$	15	15	5	5A
[2] $^{19}_{9}F$	9	9	7	7A
[3] $^{2}_{1}H$	1	1	1	1A

1.3 **Ionic bonds** form when an element on the far left side of the periodic table transfers an electron to an element on the far right side of the periodic table. **Covalent bonds** result when two atoms *share* electrons.

a. F—F
covalent

b. Li$^+$ Br$^-$
ionic

c. H—C—C—H (with H H on top and H H on bottom)
All C–H and C–C bonds are covalent.

d. Na$^+$:N—H (with H below)
ionic
Both N–H bonds are covalent.

1.4 a. Ionic bonding is observed in NaF since Na is in group 1A and has only one valence electron, and F is in group 7A and has seven valence electrons. When F gains one electron from Na, they form an ionic bond.

b. Covalent bonding is observed in $CFCl_3$ since carbon is a nonmetal in the middle of the periodic table and does not readily transfer electrons.

1.5 Atoms with one, two, three, or four valence electrons form one, two, three, or four bonds, respectively. Atoms with five or more valence electrons form [8 – (number of valence electrons)] bonds.

a. O 8 – 6 valence e^- = 2 bonds

b. Al 3 valence e^- = 3 bonds

c. Br 8 – 7 valence e^- = 1 bond

d. Si 4 valence e^- = 4 bonds

1.6 [1] Arrange the atoms with the H's on the periphery.
[2] Count the valence electrons.
[3] Arrange the electrons around the atoms. Give the H's 2 electrons first, and then fill the octets of the other atoms.
[4] Assign formal charges (Section 1.3C).

a.

[1]
```
    H  H
 H  C  C  H
    H  H
```

[2] Count valence e⁻.
$2C \times 4\ e^- = 8$
$6H \times 1\ e^- = 6$
total e⁻ = 14

[3]
```
    H  H
H–C–C–H
    H  H
```
All 14 e⁻ used.
All second-row elements have an octet.

b.

[1]
```
    H
 H  C  N  H
    H  H
```

[2] Count valence e⁻.
$1C \times 4\ e^- = 4$
$5H \times 1\ e^- = 5$
$1N \times 5\ e^- = 5$
total e⁻ = 14

[3]
```
    H
H–C–N–H
    H  H
```
⟶
```
    H
H–C–N̈–H
    H  H
```
12 e⁻ used.
N needs 2 more electrons for an octet.

c.

[1]
```
    H
 H  C
    H
```

[2] Count valence e⁻.
$1C \times 4\ e^- = 4$
$3H \times 1\ e^- = 3$
negative charge = 1
total e⁻ = 8

[3]
```
    H
H–C
    H
```
⟶
```
    H
H–C̈:⁻
    H
```
6 e⁻ used.
C needs 2 more electrons for an octet.
[The –1 charge on C is explained in Section 1.3C.]

d.

[1]
```
    H
 H  C  Cl
    H
```

[2] Count valence e⁻.
$1C \times 4\ e^- = 4$
$3H \times 1\ e^- = 3$
$1Cl \times 7\ e^- = 7$
total e⁻ = 14

[3]
```
    H
H–C–Cl
    H
```
⟶
```
    H
H–C–Cl̈:
    H
```
8 e⁻ used.
Cl needs 6 more electrons for an octet.
Complete octet.

1.7 Follow the directions from Answer 1.6.

a. HCN

H C N

Count valence e⁻.
$1C \times 4\ e^- = 4$
$1H \times 1\ e^- = 1$
$1N \times 5\ e^- = 5$
total e⁻ = 10

H–C–N
4 e⁻ used.

H–C≡N:
Complete N and C octets.

b. H_2CO

```
H C O
  H
```

Count valence e⁻.
$1C \times 4\ e^- = 4$
$2H \times 1\ e^- = 2$
$1O \times 6\ e^- = 6$
total e⁻ = 12

```
H–C–O
   H
```
6 e⁻ used.

```
H–C=Ö:
   H
```
Complete O and C octets.

c. $HOCH_2CO_2H$

```
        H  O
 H  O  C  C  O  H
        H
```

Count valence e⁻.
$2C \times 4\ e^- = 8$
$4H \times 1\ e^- = 4$
$3O \times 6\ e^- = 18$
total e⁻ = 30

```
    H  O
H–O–C–C–O–H
    H
```
16 e⁻ used.

```
    H  :O:
H–Ö–C–C–Ö–H
    H
```
Complete octets.

1.8 **Formal charge** (FC) = number of valence electrons – [number of unshared electrons + 1/2 (number of shared electrons)]

a.
$$\left[\begin{array}{c} H \\ H-N-H \\ H \end{array} \right]^+$$

$5 - [0 + 1/2(8)] = +1$

$5 - [0 + 1/2(8)] = +1$

b. $CH_3-N\equiv C:$

$5 - [0 + 1/2(8)] = +1$
$4 - [0 + 1/2(8)] = 0$

$4 - [2 + 1/2(6)] = -1$

c. $:\ddot{O}=\ddot{O}-\ddot{O}:$

$6 - [2 + 1/2(6)] = +1$

$6 - [4 + 1/2(4)] = 0$ $6 - [6 + 1/2(2)] = -1$

1.9

a. CH_3O^-

[1]   H C O
           H

[2] Count valence e⁻.
1C x 4 e⁻ = 4
3H x 1 e⁻ = 3
1O x 6 e⁻ = 6
―――――――――――
total e⁻   = 13
Add 1 for (–) charge = 14

[3]
$$H-\underset{\underset{H}{|}}{\overset{\overset{H}{|}}{C}}-O \longrightarrow H-\underset{\underset{H}{|}}{\overset{\overset{H}{|}}{C}}-\ddot{O}:$$

8 e⁻ used.

[4]
$$H-\underset{\underset{H}{|}}{\overset{\overset{H}{|}}{C}}-\ddot{\ddot{O}}:^-$$

Assign charge.

b. HC_2^-

[1]   H C C

[2] Count valence e⁻.
2C x 4 e⁻ = 8
1H x 1 e⁻ = 1
―――――――――――
total e⁻   = 9
Add 1 for (–) charge = 10

[3]
$H-C-C \longrightarrow H-C\equiv C:$

4 e⁻ used.

[4]
$H-C\equiv C:^-$

Assign charge.

c. $(CH_3NH_3)^+$

[1]   H H
      H C N H
         H H

[2] Count valence e⁻.
1C x 4 e⁻ = 4
6H x 1 e⁻ = 6
1N x 5 e⁻ = 5
―――――――――――
total e⁻   = 15
Subtract 1 for (+) charge = 14

[3]
$$H-\underset{\underset{H}{|}}{\overset{\overset{H}{|}}{C}}-\underset{\underset{H}{|}}{\overset{\overset{H}{|}}{N}}-H$$

14 e⁻ used.

[4]
$$H-\underset{\underset{H}{|}}{\overset{\overset{H}{|}}{C}}-\overset{+}{\underset{\underset{H}{|}}{\overset{\overset{H}{|}}{N}}}-H$$

Assign charge.

d. $(CH_3NH)^-$

[1]   H
      H C N H
         H

[2] Count valence e⁻.
1C x 4 e⁻ = 4
4H x 1 e⁻ = 4
1N x 5 e⁻ = 5
―――――――――――
total e⁻   = 13
Add 1 for (–) charge = 14

[3]
$$H-\underset{\underset{H}{|}}{\overset{\overset{H}{|}}{C}}-\underset{}{\overset{}{N}}-H$$

10 e⁻ used.

[4]
$$H-\underset{\underset{H}{|}}{\overset{\overset{H}{|}}{C}}-\ddot{\underset{}{\overset{}{N}}}^--H$$

Complete octet and assign charge.

1.10

a. $C_2H_4Cl_2$ (two isomers)

Count valence e⁻.
2C x 4 e⁻ = 8
4H x 1 e⁻ = 4
2Cl x 7 e⁻ = 14
―――――――――――
total e⁻   = 26

$$H-\underset{\underset{H}{|}}{\overset{\overset{H}{|}}{C}}-\underset{\underset{:\ddot{Cl}:}{|}}{\overset{\overset{H}{|}}{C}}-\ddot{Cl}:$$

$$H-\underset{\underset{:\ddot{Cl}:}{|}}{\overset{\overset{H}{|}}{C}}-\underset{\underset{H}{|}}{\overset{\overset{H}{|}}{C}}-\ddot{Cl}:$$

b. C_3H_8O (three isomers)

Count valence e⁻.
3C x 4 e⁻ = 12
8H x 1 e⁻ = 8
1O x 6 e⁻ = 6
―――――――――――
total e⁻   = 26

$$H-\underset{\underset{H}{|}}{\overset{\overset{H}{|}}{C}}-\underset{\underset{H}{|}}{\overset{\overset{H}{|}}{C}}-\underset{\underset{H}{|}}{\overset{\overset{H}{|}}{C}}-\ddot{O}-H$$

$$H-\underset{\underset{H}{|}}{\overset{\overset{H}{|}}{C}}-\underset{\underset{H}{|}}{\overset{\overset{:\ddot{O}:}{|}}{C}}-\underset{\underset{H}{|}}{\overset{\overset{H}{|}}{C}}-H$$

$$H-\underset{\underset{H}{|}}{\overset{\overset{H}{|}}{C}}-\underset{\underset{H}{|}}{\overset{\overset{H}{|}}{C}}-\ddot{O}-\underset{\underset{H}{|}}{\overset{\overset{H}{|}}{C}}-H$$

c. C₃H₆ (two isomers)

Count valence e⁻.
3C x 4 e⁻ = 12
6H x 1 e⁻ = 6
total e⁻ = 18

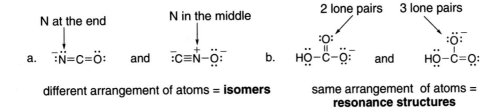

1.11 Two different definitions:
- **Isomers** have the same molecular formula and a *different* arrangement of atoms.
- **Resonance structures** have the same molecular formula and the *same* arrangement of atoms.

a. N at the end   N in the middle

:N̈=C=Ö: and   :C≡N⁺–Ö:⁻

different arrangement of atoms = **isomers**

b. 2 lone pairs   3 lone pairs

HÖ–C–Ö:⁻   and   HÖ–C=O:⁻

same arrangement of atoms = **resonance structures**

1.12 Isomers have the same molecular formula and a *different* arrangement of atoms.
Resonance structures have the same molecular formula and the *same* arrangement of atoms.

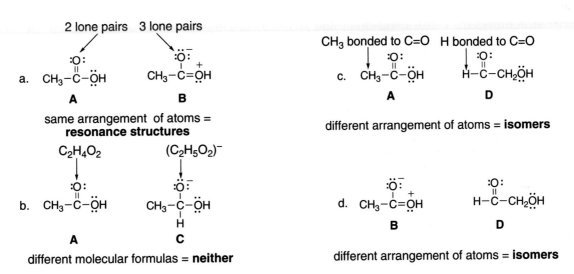

a.   2 lone pairs   3 lone pairs

CH₃–C–ÖH   CH₃–C=ÖH⁺
A   **B**

same arrangement of atoms =
resonance structures

b.   C₂H₄O₂   (C₂H₅O₂)⁻

CH₃–C–ÖH   CH₃–C–ÖH
A   **C**

different molecular formulas = **neither**

c.   CH₃ bonded to C=O   H bonded to C=O

CH₃–C–ÖH   H–C–CH₂ÖH
A   **D**

different arrangement of atoms = **isomers**

d.   CH₃–C=ÖH⁺   H–C–CH₂ÖH
B   **D**

different arrangement of atoms = **isomers**

1.13 Curved arrow notation shows the movement of an electron pair. The tail begins at an electron pair (a bond or a lone pair) and the head points to where the electron pair moves.

a.   H–C=Ö: ⟷   H–C⁺–Ö:⁻
           | |
           H   H

The net charge is the same
in both resonance structures.

b.   CH₃–C–C=CH₂   ⟷   CH₃–C=C–C̈H₂
              | | | |
              H H   H H

The net charge is the same
in both resonance structures.

1.14 Compare the resonance structures to see what electrons have "moved." **Use one curved arrow to show the movement of each electron pair.**

a. $\overset{+}{CH_2}-C=C-CH_3$ ⟷ $CH_2=C-\overset{+}{C}-CH_3$
   with H H below

One electron pair moves: one curved arrow.

b. :Ö=C–Ö:⁻ ⟷ :Ö–C–Ö:⁻

Two electron pairs move: two curved arrows.

1.15 To draw another resonance structure, **move electrons only in multiple bonds and lone pairs** and keep the number of unpaired electrons constant.

a. $CH_3-C=C-\overset{+}{C}-CH_3$ ⟷ $CH_3-\overset{+}{C}-C=C-CH_3$

c. $H-C=C-\overset{..}{Cl}:$ ⟷ $H-\overset{-}{C}-C=\overset{+}{Cl}:$

b. $CH_3-\overset{+}{C}-CH_3$ ⟷ CH_3-C-CH_3
   :Cl: :Cl+

1.16 **A "better" resonance structure is one that has more bonds and fewer charges.** The better structure is the major contributor and all others are minor contributors. To draw the resonance hybrid, use dashed lines for bonds that are in only one resonance structure, and use partial charges when the charge is on different atoms in the resonance structures.

a. $CH_3-\overset{+}{C}-\overset{..}{N}-CH_3$ ⟷ $CH_3-C=\overset{+}{N}-CH_3$
   H H   H H

   hybrid:

   $\overset{\delta+}{CH_3}-C\overset{\delta+}{=}N-CH_3$
   H H

All atoms have octets. one more bond **major contributor**

b. $CH_2=C-CH_2$ ⟷ $^-CH_2-C=CH_2$
   H   H

These two resonance structures are equivalent. They both have one charge and the same number of bonds. They are **equal contributors** to the hybrid.

hybrid:

   $\overset{\delta-}{CH_2}=C=\overset{\delta-}{CH_2}$
   H

1.17 Draw a second resonance structure for nitrous acid.

   $H-\overset{..}{O}-\overset{..}{N}=\overset{..}{O}:$ ⟷ $H-\overset{..}{O}=\overset{+}{N}-\overset{..}{O}:^-$

   $H-\overset{\delta+}{O}=N=\overset{\delta-}{O}:$

major contributor   minor contributor   hybrid
fewer charges

1.18 All representations have a carbon with two bonds in the plane of the page, one in front of the page (solid wedge) and one behind the page (dashed line). Four possibilities:

1.19 To predict the geometry around an atom, **count the number of groups (atoms + lone pairs),** making sure to draw in any needed lone pairs or hydrogens: 2 groups = linear, 3 groups = trigonal planar, 4 groups = tetrahedral.

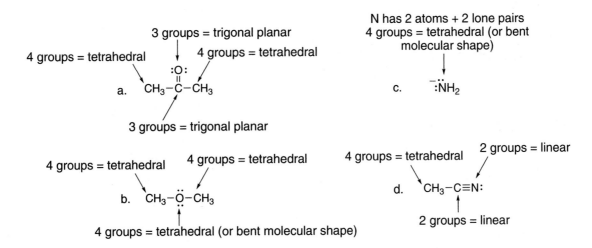

1.20 To predict the bond angle around an atom, **count the number of groups (atoms + lone pairs),** making sure to draw in any needed lone pairs or hydrogens: 2 groups = 180°, 3 groups = 120°, 4 groups = 109.5°.

2 groups = 180°

a. $CH_3-C\equiv C-Cl$

2 groups = 180°

This C has 3 groups, so both angles are 120°.

b. $CH_2=C-Cl$ (with H)

This C has 4 groups, so both angles are 109.5°.

c. CH_3-C-Cl (with H, H)

1.21 To predict the geometry around an atom, use the rules in Answer 1.19.

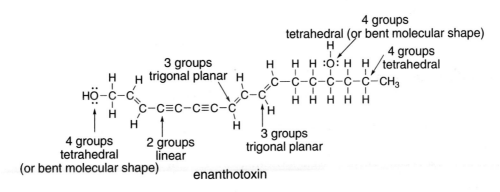

enanthotoxin

1.22 Reading from left to right, draw the molecule as a Lewis structure. Always check that carbon has four bonds and all heteroatoms have an octet by adding any needed lone pairs.

$(CH_3)_2CHCH(CH_2CH_3)_2$

a.

b.

c. $(CH_3)_3CCH(OH)CH_2CH_3$

$(CH_3)_2CHCHO$

d.

double bond needed to give C an octet

$CH_3(CH_2)_4CH(CH_3)_2$

1.23 Simplify each condensed structure using parentheses.

a. $CH_3CH_2CH_2CH_2CH_2Cl$

$CH_3(CH_2)_4Cl$

b. $CH_3CH_2CH_2-\overset{\overset{\displaystyle CH_2CH_3}{|}}{\underset{\underset{\displaystyle H}{|}}{C}}-CH_2CH_3$

$CH_3(CH_2)_2CH(CH_2CH_3)_2$

c. $HOCH_2-\overset{\overset{\displaystyle CH_2OH}{|}}{\underset{\underset{\displaystyle H}{|}}{C}}-CH_2CH_2CH_2-\overset{\overset{\displaystyle CH_3}{|}}{\underset{\underset{\displaystyle CH_3}{|}}{C}}-CH_2-\overset{\overset{\displaystyle CH_3}{|}}{\underset{\underset{\displaystyle CH_3}{|}}{C}}-CH_3$

$(HOCH_2)_2CH(CH_2)_3C(CH_3)_2CH_2C(CH_3)_3$

1.24 Draw the Lewis structure of lactic acid.

$CH_3CH(OH)CO_2H \longrightarrow$

1.25 In shorthand or skeletal drawings, **all line junctions or ends of lines represent carbon atoms**. The carbons are all tetravalent.

a.

octinoxate
(2-ethylhexyl 4-methoxycinnamate)

b.

avobenzone

1.26 In shorthand or skeletal drawings, **all line junctions or ends of lines represent carbon atoms**. Convert by writing in all carbons, and then adding hydrogen atoms to make the carbons tetravalent.

a.

b.

c.

d.

1.27 A charge on a carbon atom takes the place of one hydrogen atom. **A negatively charged C has one lone pair, and a positively charged C has none.**

a.

b.

c.

d.

positive charge	negative charge	positive charge	negative charge
no lone pairs	one lone pair	no lone pairs	one lone pair
no H's needed	one H needed	one H needed	one H needed

1.28 Draw each indicated structure. Recall that in the skeletal drawings, a carbon atom is located at the intersection of any two lines and at the end of any line.

a. $(CH_3)_2C=CH(CH_2)_4CH_3$ =

c. = $(CH_3)_2CH(CH_2)_2CONHCH_3$

b.
$CH_3-\overset{\overset{H}{|}}{C}-\overset{\overset{H}{|}}{C}-CH_2CH_2Cl$
$H_2N-\overset{\underset{H}{|}}{C}-\overset{\underset{H}{|}}{C}-H$
=

d. HO = $HO(CH_2)_2CH=CHCO_2CH(CH_3)_2$

1.29 To determine the orbitals used in bonding, **count the number of groups** (atoms + lone pairs):
4 groups = sp^3, 3 groups = sp^2, 2 groups = sp, H atom = $1s$ (no hybridization).
All covalent single bonds are σ, and all double bonds contain one σ and one π bond.

Each H uses a → H H H ← All single bonds are
1s orbital. H–C–C–C–H σ bonds.
H H H

Each C–C bond is Csp^3–Csp^3.
Each C–H bond is Csp^3–$H1s$.

Total of 10 σ bonds.

Each C has 4 groups and is
sp^3 hybridized.

1.30 [1] Draw a valid Lewis structure for each molecule.
[2] **Count the number of groups** around each atom: 4 groups = sp^3, 3 groups = sp^2, 2 groups = sp,
H atom = $1s$ (no hybridization).

Note: **Be and B** (Groups 2A and 3A) do not have enough valence e^- to form an octet, **and do not form an octet in neutral molecules.**

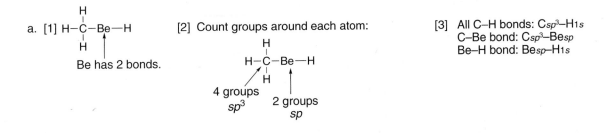

a. [1] H–C–Be–H (with H above and below C)
Be has 2 bonds.

[2] Count groups around each atom:
H–C–Be–H
4 groups sp^3
2 groups sp

[3] All C–H bonds: Csp^3–$H1s$
C–Be bond: Csp^3–$Besp$
Be–H bond: $Besp$–$H1s$

b. [1]
$$CH_3-\overset{\overset{\displaystyle CH_3}{|}}{\underset{|}{B}}-CH_3$$

B forms 3 bonds.

[2] Count groups around each atom:

$$CH_3-\overset{\overset{\displaystyle CH_3}{|}}{B}-CH_3$$

4 groups   3 groups
sp^3 sp^2

[3] All C–H bonds: C_{sp^3}–H_{1s}
C–B bonds: C_{sp^3}–B_{sp^2}

c. [1]
$$H-\overset{\overset{\displaystyle H}{|}}{\underset{\underset{\displaystyle H}{|}}{C}}-\overset{..}{\underset{..}{O}}-\overset{\overset{\displaystyle H}{|}}{\underset{\underset{\displaystyle H}{|}}{C}}-H$$

[2] Count groups around each atom:

$$H-\overset{\overset{\displaystyle H}{|}}{C}-\overset{..}{\underset{..}{O}}-\overset{\overset{\displaystyle H}{|}}{\underset{\underset{\displaystyle H}{|}}{C}}-H$$

4 groups
sp^3

4 groups
sp^3

[3] All C–H bonds: C_{sp^3}–H_{1s}
C–O bonds: C_{sp^3}–O_{sp^3}

1.31 To determine the hybridization, **count the number of groups** around each atom: 4 groups = sp^3, 3 groups = sp^2, 2 groups = sp, H atom = $1s$ (no hybridization).

a. $CH_3-C\equiv CH$

4 groups   2 groups
sp^3 sp

b. $=\overset{..}{N}-CH_3$

3 groups   3 groups
sp^2 sp^2

c. $CH_2=C=CH_2$

3 groups   2 groups
sp^2 sp

1.32 All single bonds are σ. Multiple bonds contain one σ bond, and all others are π bonds.

All C–H bonds are σ bonds.

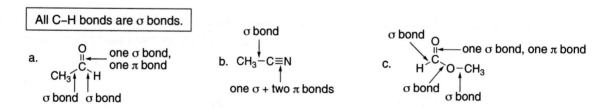

a.   one σ bond, one π bond
σ bond σ bond

b. $CH_3-C\equiv N$
σ bond
one σ + two π bonds

c.   σ bond — one σ bond, one π bond
σ bond   σ bond

1.33 Bond length and bond strength are inversely related: **longer bonds are weaker bonds**. Single bonds are weaker and longer than double bonds, which are weaker and longer than triple bonds.

a.
bond 1:
single bond

$-C\equiv C-$

bond 3:
double bond

bond 2:
triple bond

increasing bond strength: 1 < 3 < 2
increasing bond length: 2 < 3 < 1

b.
bond 1:
single bond

$CH_3\overset{H}{\underset{}{N}}$ N bond 2:
double bond

$-C\equiv N$

bond 3:
triple bond

increasing bond strength: 1 < 2 < 3
increasing bond length: 3 < 2 < 1

1.34 Bond length and bond strength are inversely related: **longer bonds are weaker bonds**. Single bonds are weaker and longer than double bonds, which are weaker and longer than triple bonds. Increasing percent *s*-character increases bond strength and decreases bond length.

a. $CH_3-C\equiv C-H$ or

   C_{sp}–H1*s*
   50% *s*-character
   shorter bond

   $\underset{H}{\overset{CH_3}{C}}=CH_2$

   C_{sp^2}–H1*s*
   33% *s*-character

c. $CH_2=\overset{..}{N}-H$ or $CH_3-\overset{H}{\underset{..}{N}}-H$

   N_{sp^2}–H1*s*
   33% *s*-character
   shorter bond

   N_{sp^3}–H1*s*
   25% *s*-character

b. $\overset{H}{\underset{H}{C}}=\overset{..}{\overset{..}{O}}$ or $H-\overset{H}{\underset{H}{C}}-\overset{..}{\underset{..}{O}}H$

   C_{sp^2}–H1*s*
   33% *s*-character
   shorter bond

   C_{sp^3}–H1*s*
   25% *s*-character

1.35 **Electronegativity increases across a row of the periodic table and decreases down a column.** Look at the relative position of the atoms to determine their relative electronegativity.

a.
most electropositive
   most electronegative
   Se < S < O
   increasing
   electronegativity

b.
most electropositive
   most electronegative
   Na < P < Cl
   increasing
   electronegativity

c.
most electropositive
   most electronegative
   S < Cl < F
   increasing
   electronegativity

d.
most electropositive
   most electronegative
   P < N < O
   increasing
   electronegativity

1.36 Dipoles result from unequal sharing of electrons in covalent bonds. More electronegative atoms "pull" electron density towards them, making a dipole. **Dipole arrows point towards the atom of higher electron density.**

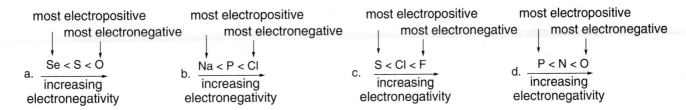

a. $\overset{\delta^+}{H}-\overset{\delta^-}{F}$

b. $\overset{\delta^+}{B}-\overset{\delta^-}{C}$

c. $\overset{\delta^-}{C}-\overset{\delta^+}{Li}$

d. $\overset{\delta^+}{C}-\overset{\delta^-}{Cl}$

1.37 Polar molecules result from a net dipole. To determine polarity, draw the molecule in three dimensions around any polar bonds, draw in the dipoles, and look to see whether the dipoles cancel or reinforce.

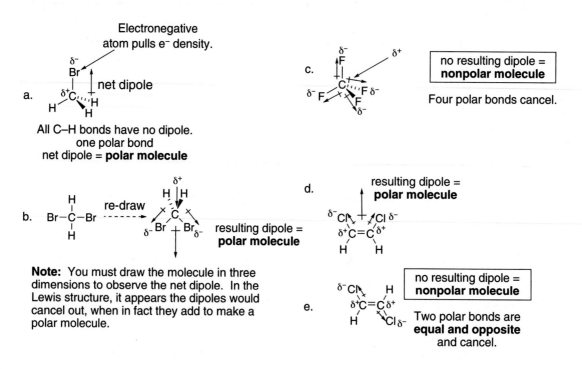

a.
All C–H bonds have no dipole.
one polar bond
net dipole = **polar molecule**

b.
Note: You must draw the molecule in three dimensions to observe the net dipole. In the Lewis structure, it appears the dipoles would cancel out, when in fact they add to make a polar molecule.

c.
no resulting dipole =
nonpolar molecule
Four polar bonds cancel.

d.
resulting dipole =
polar molecule

e.
no resulting dipole =
nonpolar molecule
Two polar bonds are
equal and opposite
and cancel.

1.38

The C–O and O–H bonds are polar.

The two circled C's are sp^3 hybridized.
All the C–H bonds are nonpolar.
All H's bonded to O and N bear a partial positive charge (δ^+).

one possibility

1.39 Use the definitions in Answer 1.1.

	Iodine-123	Iodine-131
a. number of protons = atomic number for I = 53	53	53
b. number of neutrons = mass number – atomic number	70	78
c. number of electrons = number of protons	53	53
d. The group number is the same for all isotopes.	7A	7A

1.40 Use bonding rules in Answer 1.3.

a. Na$^+$ I$^-$ b. Br—Cl   c. H—Cl   d. H—C—N—H   e. Na$^+$ $^-$O—C—H

ionic   covalent   covalent   all covalent bonds   ionic   All other bonds are covalent.

1.41 Formal charge (FC) = number of valence electrons – [number of unshared electrons + 1/2 (number of shared electrons)]. C is in group 4A.

a. CH$_2$=CH

4 – [0 + 1/2(8)] = 0   4 – [2 + 1/2(6)] = –1

b. H—C—H

4 – [2 + 1/2(4)] = 0

c. H—C—H

4 – [1 + 1/2(6)] = 0

d. H—C—C

4 – [0 + 1/2(8)] = 0   4 – [0 + 1/2(6)] = +1

1.42 Formal charge (FC) = number of valence electrons – [number of unshared electrons + 1/2 (number of shared electrons)]. N is in group 5A and O is in group 6A.

a.   CH$_3$—N—CH$_3$

5 – [4 + 1/2(4)] = –1

5 – [0 + 1/2(8)] = +1

b.   :N=N=N:

5 – [4 + 1/2(4)] = –1   5 – [4 + 1/2(4)] = –1

c.   CH$_3$—N≡N:

5 – [0 + 1/2(8)] = +1   5 – [2 + 1/2(6)] = 0

6 – [2 + 1/2(6)] = +1

:OH

d.   CH$_3$—C—CH$_3$

e.   CH$_3$—O ·

6 – [5 + 1/2(2)] = 0

5 – [2 + 1/2(6)] = 0

f.   CH$_3$—N=O

6 – [4 + 1/2(4)] = 0

1.43 Follow the steps in Answer 1.6 to draw Lewis structures.

a. CH_2N_2

$$H-\overset{+}{C}=N=\overset{\cdot\cdot}{\underset{\cdot\cdot}{N}}:^{-}$$
$$\underset{H}{|}$$
or

$$H-\overset{-}{\underset{\cdot\cdot}{C}}-\overset{+}{N}\equiv N:$$
$$\underset{H}{|}$$

valence e⁻	
1C x 4 e⁻ =	4
2H x 1 e⁻ =	2
2N x 5 e⁻ =	10
total e⁻ =	16

b. CH_3NO_2

$$\overset{\textstyle H}{\underset{\textstyle H}{|}}$$
$$H-\overset{|}{C}-\overset{+}{N}-\overset{\cdot\cdot}{\underset{\cdot\cdot}{O}}:^{-}$$
$$\underset{:O:}{||}$$
or
$$H-\overset{|}{\underset{|}{C}}-\overset{+}{N}=\overset{\cdot\cdot}{O}$$
$$\underset{:O:^{-}}{|}$$

valence e⁻	
1C x 4 e⁻ =	4
3H x 1 e⁻ =	3
1N x 5 e⁻ =	5
2O x 6 e⁻ =	12
total e⁻ =	24

c. CH_3CNO

$$\overset{\textstyle H}{\underset{\textstyle H}{|}}$$
$$H-\overset{|}{C}-C\equiv \overset{+}{N}-\overset{\cdot\cdot}{\underset{\cdot\cdot}{O}}:^{-} \quad \text{or} \quad H-\overset{|}{\underset{|}{C}}-\overset{-}{C}=\overset{+}{N}=O:$$

valence e⁻	
2C x 4 e⁻ =	8
3H x 1 e⁻ =	3
1N x 5 e⁻ =	5
1O x 6 e⁻ =	6
total e⁻ =	22

or
$$H-\overset{|}{\underset{|}{C}}-\overset{\cdot\cdot}{\underset{}{C}}{}^{2-}-\overset{+}{N}\equiv\overset{+}{O}:$$
$$\underset{H}{}$$

d. HCO_2^{-}

valence e⁻	
1C x 4 e⁻	= 4
1H x 1 e⁻	= 1
2O x 6 e⁻	= 12
1 for (–) charge	= 1
total e⁻	= 18

$$H-C=\overset{\cdot\cdot}{\underset{\cdot\cdot}{O}}: \quad \text{or} \quad H-C-\overset{\cdot\cdot}{\underset{\cdot\cdot}{O}}:^{-}$$

e. HCO_3^{-}

valence e⁻	
1C x 4 e⁻	= 4
1H x 1 e⁻	= 1
3O x 6 e⁻	= 18
1 for (–) charge	= 1
total e⁻	= 24

$$H-\overset{\cdot\cdot}{\underset{\cdot\cdot}{O}}-C=\overset{\cdot\cdot}{\underset{\cdot\cdot}{O}}: \quad \text{or} \quad H-\overset{\cdot\cdot}{\underset{\cdot\cdot}{O}}-C-\overset{\cdot\cdot}{\underset{\cdot\cdot}{O}}:^{-}$$

or
$$H-\overset{+}{\underset{\cdot\cdot}{O}}=C-\overset{\cdot\cdot}{\underset{\cdot\cdot}{O}}:^{-}$$

f. $^{-}CH_2CN$

valence e⁻	
2C x 4 e⁻	= 8
2H x 1 e⁻	= 2
1N x 5 e⁻	= 5
1 for (–) charge	= 1
total e⁻	= 16

$$H-C=C=\overset{\cdot\cdot}{N}:^{-} \quad \text{or} \quad H-\overset{\cdot\cdot}{\underset{}{C}}{}^{-}-C\equiv N:$$
$$\underset{H}{|} \qquad\qquad\qquad \underset{H}{|}$$

1.44 Follow the steps in Answer 1.6 to draw Lewis structures.

a. N_2

[1] N N

[2] Count valence e⁻.

$$\frac{2N \times 5\ e^{-} = 10}{\text{total } e^{-}\ = 10}$$

[3] N—N $\longrightarrow$:N≡N:

2 e⁻ used.　　Complete N octets.

b. $(CH_3OH_2)^{+}$

[1]
$$\begin{array}{ccc} & H & \\ H & C & O & H \\ & H & H \end{array}$$

[2] Count valence e⁻.

1C x 4 e⁻ =	4
5H x 1 e⁻ =	5
1O x 6 e⁻ =	6
total e⁻ =	15

Subtract 1 for
(+) charge = 14

[3]
$$\begin{array}{c} H \\ | \\ H-C-O-H \\ | \quad | \\ H \quad H \end{array}$$

12 e⁻ used.

[4]
$$\begin{array}{c} H \\ | \quad + \\ H-C-\overset{\cdot\cdot}{O}-H \\ | \quad | \\ H \quad H \end{array}$$

Add charge
and lone pair.

c. $(CH_3CH_2)^{-}$

[1]
$$\begin{array}{cccc} & H & & \\ H & C & C & H \\ & H & H \end{array}$$

[2] Count valence e⁻.

2C x 4 e⁻ =	8
5H x 1 e⁻ =	5
total e⁻ =	13

Add 1 for
(–) charge = 14

[3]
$$\begin{array}{c} H \\ | \\ H-C-C-H \\ | \quad | \\ H \quad H \end{array}$$

12 e⁻ used.

[4]
$$\begin{array}{c} H \\ | \quad - \\ H-C-\overset{\cdot\cdot}{C}-H \\ | \quad | \\ H \quad H \end{array}$$

Add charge
and lone pair.

d. HNNH

[1] H N N H

[2] Count valence e⁻.

2H x 1 e⁻ =	2
2N x 5 e⁻ =	10
total e⁻ =	12

[3] H—N—N—H $\longrightarrow$ H—Ñ=Ñ—H

6 e⁻ used.　　Complete N octets.

e. H_6BN

[1]
```
    H H
  H B N H
    H H
```

[2] Count valence e⁻.
1B x 3 e⁻ = 3
6H x 1 e⁻ = 6
1N x 5 e⁻ = 5
––––––––––––––
total e⁻ = 14

[3]
```
    H  H
  H–B–N–H
    H  H
```
14 e⁻ used.

[4]
```
    H  H
       +
  H–B–N⁻H
    H  H
```
Add charges.

1.45 Follow the steps in Answer 1.6 to draw Lewis structures.

a. $(CH_3CH_2)_2O$

[1]
```
  H H   H H
  H C C O C C H
  H H   H H
```

[2] Count valence e⁻.
1O x 6 e⁻ = 6
10H x 1 e⁻ = 10
4C x 4 e⁻ = 16
––––––––––––––
total e⁻ = 32

[3]
```
    H H   H H
  H–C–C–O–C–C–H
    H H   H H
```
28 e⁻ used.

[4]
```
    H H      H H
  H–C–C–Ö–C–C–H
    H H      H H
```
Add lone pairs.

b. CH_2CHCN

[1]
```
  H H
  C C C N
  H
```

[2] Count valence e⁻.
1N x 5 e⁻ = 5
3H x 1 e⁻ = 3
3C x 4 e⁻ = 12
––––––––––––––
total e⁻ = 20

[3]
```
  H H
  C–C–C–N
  H
```
12 e⁻ used.

[4]
```
  H H
  C=C–C≡N̈
  H
```
Add lone pairs and π bonds.

c. $(HOCH_2)_2CO$

[1]
```
      H O H
  H O C C C O H
      H   H
```

[2] Count valence e⁻.
3O x 6 e⁻ = 18
6H x 1 e⁻ = 6
3C x 4 e⁻ = 12
––––––––––––––
total e⁻ = 36

[3]
```
      H O H
  H–O–C–C–C–O–H
      H   H
```
22 e⁻ used.

[4]
```
      H :O: H
      ‖
  H–Ö–C–C–C–Ö–H
      H   H
```
Add lone pairs and π bonds.

d. $(CH_3CO)_2O$

[1]
```
  H O     O H
  H C C O C C H
  H       H
```

[2] Count valence e⁻.
3O x 6 e⁻ = 18
6H x 1 e⁻ = 6
4C x 4 e⁻ = 16
––––––––––––––
total e⁻ = 40

[3]
```
  H O     O H
  H–C–C–O–C–C–H
  H       H
```
24 e⁻ used.

[4]
```
  H :O:   :O: H
    ‖     ‖
  H–C–C–Ö–C–C–H
  H       H
```
Add lone pairs and π bonds.

1.46 Isomers must have a different arrangement of atoms.

a. Two isomers of molecular formula C_3H_7Cl

```
  :C̈l: H H
  H–C–C–C–H
    H H H
```

```
    H :C̈l: H
  H–C–C–C–H
    H  H  H
```

b. Three isomers of molecular formula C_2H_4O

```
  :Ö–H
  H–C=C–H
      H
```

```
    :O: H
    ‖
  H–C–C–H
      H
```

```
      :O:
    H     H
      C–C
    H     H
```

c. Four isomers of molecular formula C_3H_9N

```
    H H H
  H–C–C–C–N̈–H
    H H H H
```

```
    H H   H
  H–C–C–N̈–C–H
    H H   H H
```

```
    H       H
  H–C–N̈–C–H
    H  |  H
    H–C–H
       H
```

```
    H H H
  H–C–C–C–H
    H  |  H
    H–N̈:
       H
```

1.47

Nine isomers of C₃H₆O:

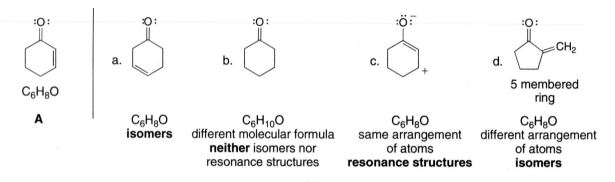

1.48 Use the definition of isomers and resonance structures in Answer 1.11.

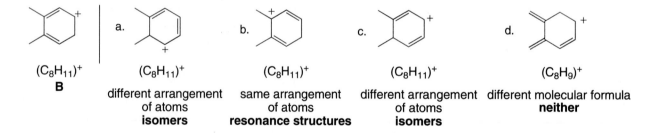

:O:

C₆H₈O

A

a.

C₆H₈O
isomers

b.

C₆H₁₀O
different molecular formula
neither isomers nor
resonance structures

c.

C₆H₈O
same arrangement
of atoms
resonance structures

d.

5 membered
ring

C₆H₈O
different arrangement
of atoms
isomers

1.49 Use the definitions of isomers and resonance structures in Answer 1.11.

(C₈H₁₁)⁺
B

a.

(C₈H₁₁)⁺

different arrangement
of atoms
isomers

b.

(C₈H₁₁)⁺

same arrangement
of atoms
resonance structures

c.

(C₈H₁₁)⁺

different arrangement
of atoms
isomers

d.

(C₈H₉)⁺

different molecular formula
neither

1.50 Use the definitions of isomers and resonance structures in Answer 1.11.

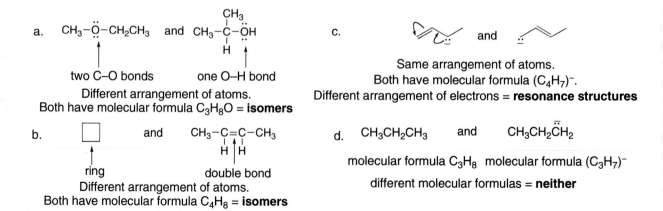

a.   CH₃–Ö–CH₂CH₃ and CH₃–C–ÖH (with CH₃ and H)

two C–O bonds   one O–H bond
Different arrangement of atoms.
Both have molecular formula C₃H₈O = **isomers**

b.   □ and CH₃–C=C–CH₃ (with H)

ring   double bond
Different arrangement of atoms.
Both have molecular formula C₄H₈ = **isomers**

c.   and

Same arrangement of atoms.
Both have molecular formula (C₄H₇)⁻.
Different arrangement of electrons = **resonance structures**

d.   CH₃CH₂CH₃   and   CH₃CH₂C̈H₂

molecular formula C₃H₈ molecular formula (C₃H₇)⁻

different molecular formulas = **neither**

1.51 Compare the resonance structures to see what electrons have "moved." **Use one curved arrow to show the movement of each electron pair.**

a.

$CH_3-\overset{\overset{H}{|}}{\underset{\underset{H}{|}}{C}}-\overset{..}{N}-CH_3 \longleftrightarrow CH_3-\overset{\overset{H}{|}}{C}=\overset{+}{\underset{\underset{H}{|}}{N}}-CH_3$

One electron pair moves = one arrow

b.

$H-\overset{\overset{..\!O\!..}{\|}}{C}-\overset{..}{N}H_2 \longleftrightarrow H-\overset{\overset{..\!\overset{-}{O}\!..}{|}}{C}=\overset{+}{N}H_2$

Two electron pairs move = two arrows

c.

Four electron pairs move = four arrows

1.52 Curved arrow notation shows the movement of an electron pair. The tail begins at an electron pair (a bond or a lone pair) and the head points to where the electron pair moves.

a. $CH_3-\overset{+}{N}\equiv N: \longleftrightarrow CH_3-\overset{..}{N}=\overset{+}{N}:$

b. $CH_3-\overset{\overset{..\!\overset{-}{O}\!..}{|}}{C}=CH-\overset{+}{C}H_2 \longleftrightarrow CH_3-\overset{\overset{:O:}{\|}}{C}-CH=CH_2$

c.

d.

1.53 Use the rules in Answer 1.15.

a. $CH_3-\overset{\overset{:O:}{\|}}{C}-\overset{..\!\overset{-}{O}\!..}{} \longleftrightarrow CH_3-\overset{\overset{:\overset{-}{O}:}{\|}}{C}=\overset{..}{O}:$

Two electron pairs move = two arrows

b. $CH_2=\overset{+}{N}H_2 \longleftrightarrow \overset{+}{C}H_2-\overset{..}{N}H_2$

One electron pair moves = one arrow

c.

Two electron pairs move = two arrows

d. $H-\overset{\overset{+}{:}\!OH}{\underset{|}{C}}-H \longleftrightarrow H-\overset{\overset{:OH}{\|}}{\underset{+}{C}}-H$

One electron pair moves = one arrow

1.54 To draw the **resonance hybrid**, use the rules in Answer 1.16.

Charge is on both O's.

a.

Double bond can be in 2 locations.

b.

partial double bond character

$\overset{\delta^+ \downarrow \delta^+}{CH_2=NH_2}$

Charge is on both atoms.

Double bond can be in 2 locations.

Charge is on both atoms.

c.

d.

Charge is on both atoms.

$H-\overset{\overset{\delta^+ \!:\!OH}{\|}}{\underset{\delta^+}{C}}-H$

C–O bond has partial double bond character.

1.55 For the compounds where the arrangement of atoms is not given, first draw a Lewis structure. Then use the rules in Answer 1.15.

a. O_3

Count valence e⁻.
$3O \times 6\ e^- = 18$
total e⁻ $= 18$

:Ö⁻–Ö⁺=Ö ⟷ Ö=Ö⁺–Ö:⁻

b. NO_3^- (a central N atom)

Count valence e⁻.
$1N \times 5\ e^- = 5$
$3O \times 6\ e^- = 18$
$(-)$ charge $= 1$
total e⁻ $= 24$

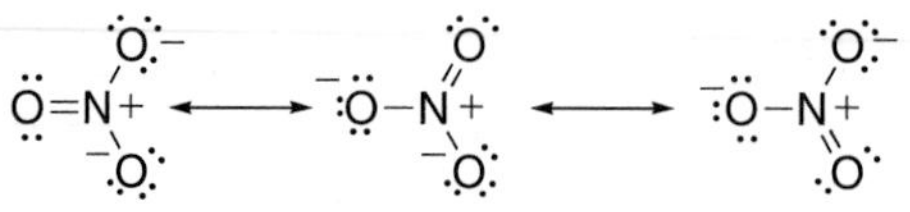

Ö=N⁺–Ö:⁻ ... :Ö⁻–N⁺=Ö ... :Ö⁻–N⁺–Ö:⁻

c. N_3^-

Count valence e⁻.
$3N \times 5\ e^- = 15$
$(-)$ charge $= 1$
total e⁻ $= 16$

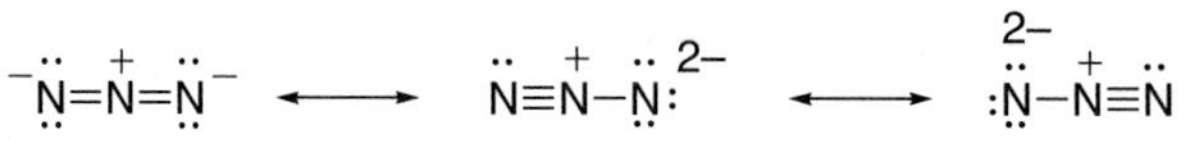

⁻N̈=N⁺=N̈⁻ ⟷ N≡N⁺–N̈:²⁻ ⟷ :N̈²⁻–N⁺≡N

d.

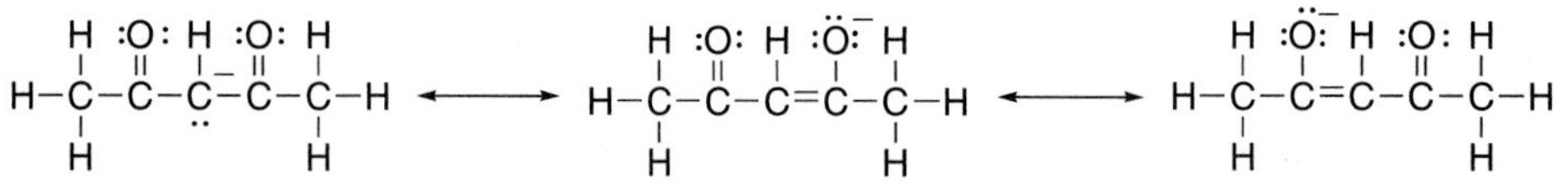

H C(O)(–)...

H–C(H)(H)–C(:O:)–C(:O:⁻)–C–C–H ⟷ H–C–C–C=C–C–H ⟷ H–C–C=C–C–C–H

e.

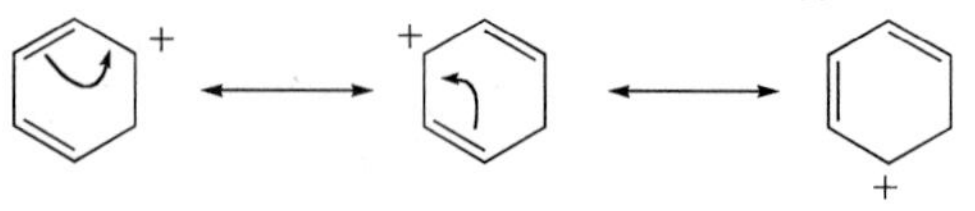

(benzene cation resonance structures)

f. $CH_2=CH–CH^-–CH=CH_2$ ⟷ ⁻$CH_2–CH=CH–CH=CH_2$ ⟷ $CH_2=CH–CH=CH–CH_2^-$

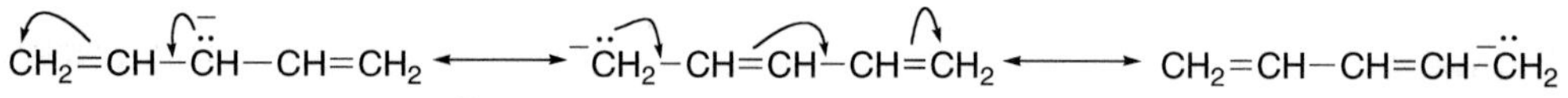

g. (pyran oxygen cation resonance structures)

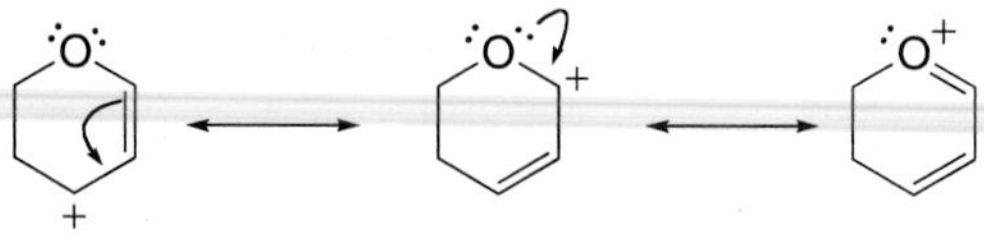

1.56 To draw the **resonance hybrid**, use the rules in Answer 1.16.

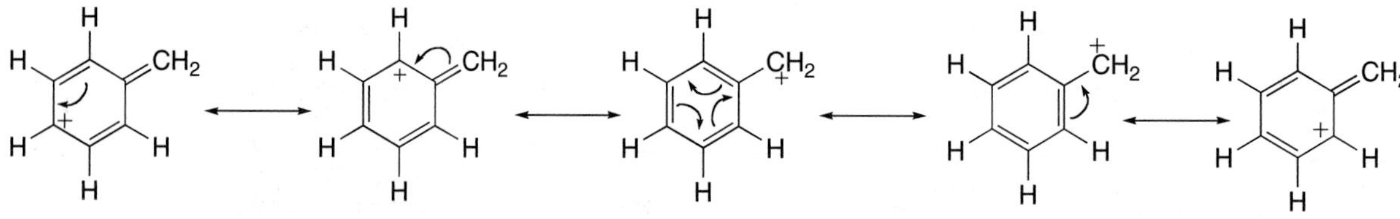

(five resonance structures of styrene cation)

resonance hybrid

(resonance hybrid structure with δ⁺ labels)

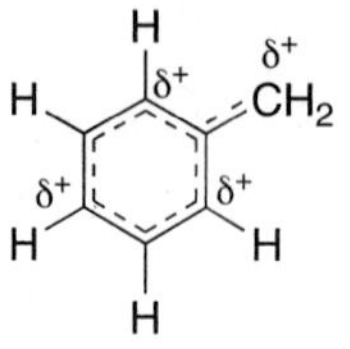

1.57 A "better" resonance structure is one that has more bonds and fewer charges. The better structure is the major contributor and all others are minor contributors.

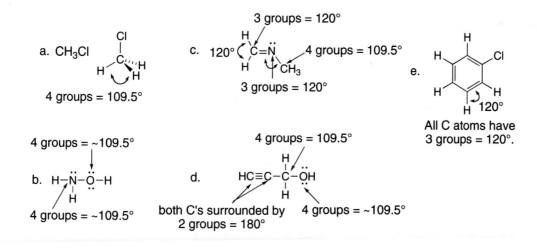

a.   CH₃–C–Ö–CH₃ ⟷ CH₃–C–Ö–CH₃ ⟷ CH₃–C=Ö–CH₃

3 C–O bonds   2 C–O bonds   3 C–O bonds
no charges   2 charges   2 charges
contributes the most = **3**   contributes the least = **1**   **2**

b.   CH₃–C=N–NH₂ ⟷ CH₃–C–N=NH₂ ⟷ CH₃–C–N–NH₂

3 bonds for this N   3 bonds for this N   2 bonds for this N
no charges   2 charges   2 charges
contributes the most = **3**   **2**   contributes the least = **1**

1.58

This C would have 5 bonds.

a. [structure] ✗ +   b. [structure] ⟷ [structure]   c. CH₃CH₂–C≡N: ⟷ CH₃CH₂–C=N: d. [structure] ✗ [structure]

invalid

invalid   This C would have 5 bonds.

[Note: The pentavalent C's in (a) and (d) bear a (–1) formal charge.]

1.59 Use the rules in Answer 1.20.

a. CH₃Cl   [structure with Cl, C, H, H, H]

4 groups = 109.5°

c.   120° [structure: H, C=N, H, CH₃]   4 groups = 109.5°

3 groups = 120°

3 groups = 120°

4 groups = ~109.5°

b. H–N–O–H
     |
     H

4 groups = ~109.5°

d.   HC≡C–C–OH [with H, H]

4 groups = 109.5°

both C's surrounded by 2 groups = 180°   4 groups = ~109.5°

e. [benzene ring structure with Cl, H atoms]

120°

All C atoms have 3 groups = 120°.

1.60 To predict the geometry around an atom, use the rules in Answer 1.19.

a. $CH_3CH_2CH_2CH_3$

4 groups
(4 atoms)
tetrahedral

c. ⬡ ← 3 groups
(3 atoms)
trigonal planar

e. $CH_3-\overset{\overset{O}{\|}}{C}-OH$

3 groups
(3 atoms)
trigonal planar

b. $(CH_3)_2\ddot{N}\overset{..}{\vdots}^-$

4 groups
(2 atoms, 2 lone pairs)
tetrahedral
(bent molecular shape)

d. BF_4^-

4 groups
(4 atoms)
tetrahedral

f. $(CH_3)_3\ddot{N}$

4 groups
(3 atoms, 1 lone pair)
tetrahedral
(trigonal pyramidal molecular shape)

1.61 Each C has two bonds in the plane of the page, one in front of the page (solid wedge) and one behind the page (dashed line).

$CF_3CHClBr$

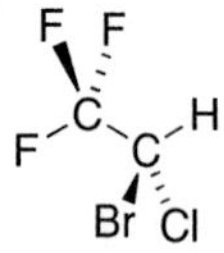

1.62 In shorthand or skeletal drawings, **all line junctions or ends of lines represent carbon atoms**. The C's are all tetravalent. All H's bonded to C's are drawn in the following structures. C's labeled with (*) have no H's bonded to them.

a.

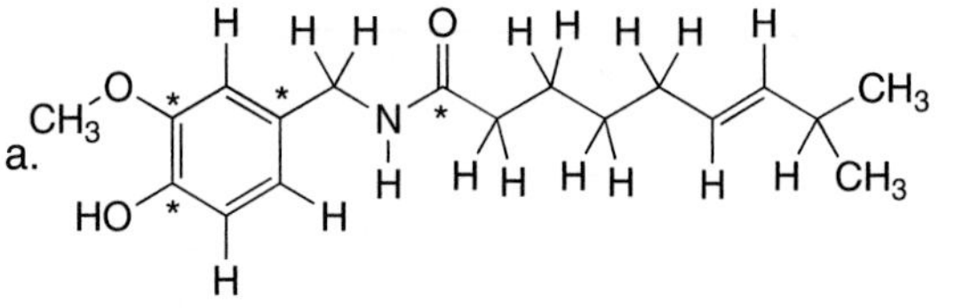

b.

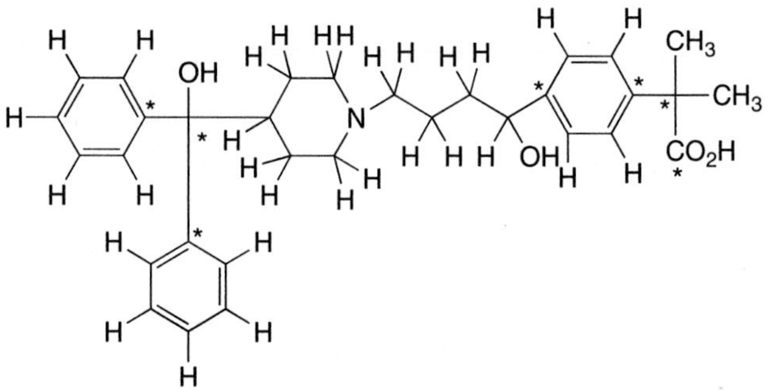

1.63 In shorthand or skeletal drawings, **all line junctions or ends of lines represent carbon atoms**. Convert by writing in all C's, and then adding H's to make the C's tetravalent.

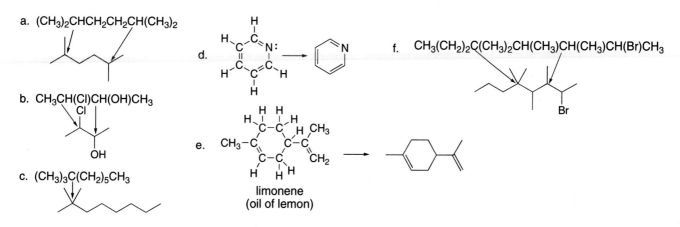

a.

menthol
(isolated from peppermint oil)

b.

myrcene
(isolated from bayberry)

c.

ethambutol
(drug used to treat tuberculosis)

d.

estradiol
(a female sex hormone)

1.64 In skeletal formulas, leave out all C's and H's, except H's bonded to heteroatoms.

a. $(CH_3)_2CHCH_2CH_2CH(CH_3)_2$

b. $CH_3CH(Cl)CH(OH)CH_3$

c. $(CH_3)_3C(CH_2)_5CH_3$

d.

e.

limonene
(oil of lemon)

f. $CH_3(CH_2)_2C(CH_3)_2CH(CH_3)CH(CH_3)CH(Br)CH_3$

1.65 For Lewis structures, all atoms including H's and all lone pairs must be drawn in.

a. CH_3CH_2COOH

b. $CH_3CONHCH_3$

c. CH_3COCH_2Br

d. $(CH_3)_3COH$

e. $(CH_3)_3CCHO$

f. CH_3COCl

g. $CH_3COCH_2CO_2H$

h. $HO_2CCH(OH)CO_2H$

1.66 A charge on a C atom takes the place of one H atom. A negatively charged C has one lone pair, and a positively charged C has none.

a.

b.

c.

d.

1.67

a.

b. $CH_3-C\equiv\overset{+}{N}H$

c.

d.

1.68 Examine each structure to determine the error.

a. $CH_3CH=CH=CHCH_3$

    ↑
This C has 5 bonds.

b. $(CH_3)_3CHCH_2CH_2CH_3$

    ↑
This C has 5 bonds.

c.

This H has 2 bonds.
It should be HO–.

d.

This C has 4 bonds. The negative charge means a lone pair, which gives C 10 electrons.

This C has 3 bonds.

e.

This C has 5 bonds.
It should be CH_3CH_2–.

1.69 To determine the hybridization around the labeled atoms, use the procedure in Answer 1.31.

a. $CH_3\overset{..}{\overset{..}{C}}H_2$

↑
4 groups
(3 atoms, 1 lone pair)
sp^3, tetrahedral

b.

4 groups
(4 atoms)
sp^3, tetrahedral
(Each C has 2 H's.)

c. $(CH_3)_3\overset{..}{\overset{..}{O}}{}^+$

↑
4 groups
(3 atoms, 1 lone pair)
sp^3, tetrahedral

d.

4 groups
(4 atoms)
sp^3, tetrahedral

e. $CH_3-C\equiv C-H$

↑
2 groups
(2 atoms)
sp, linear

f. $CH_2=\overset{....}{N}OCH_3$

↑
3 groups
(2 atoms, 1 lone pair)
sp^2, trigonal planar

g. $CH_3CH=C=CH_2$

3 groups
(3 atoms)
sp^2
trigonal planar

2 groups
(2 atoms)
sp, linear

1.70 To determine what orbitals are involved in bonding, use the procedure in Answer 1.29.

a.

H H
$\leftarrow$ C_{sp^3}–H_{1s}
$\leftarrow$ C_{sp^3}–C_{sp^3}

c.

C_{sp^2}–C_{sp^3} :O:
σ: C_{sp^2}–O_{sp^2}
π: C_p–O_p

b.

C_{sp^2}–H_{1s} →   H
σ: C_{sp^2}–C_{sp^2}
π: C_p–C_p
$\leftarrow$ C_{sp^2}–C_{sp^3}

d.

C_{sp}–C_{sp^2}

H–C≡C–C=$\ddot{N}$–CH_3
         |
         H
σ: C_{sp^3}–N_{sp^2}

C_{sp}–H_{1s} σ: C_{sp}–C_{sp}
π: C_p–C_p
π: C_p–C_p

1.71 To determine what orbitals are involved in bonding, use the procedure in Answer 1.29.

a.

O_{sp^3}–H_{1s} C_{sp^3}–O_{sp^3}

σ: C_{sp^2}–O_{sp^2}
π: C_p–O_p

H CO_2H
:O: :O: :O:

HO   OH

C_{sp^2}–C_{sp^3}

C_{sp^3}–C_{sp}

citric acid
(responsible for the tartness
of citrus fruits)

b.

C_{sp^3}–O_{sp^3} C_{sp^2}–C_{sp^3}

:O:

CH_3 $\ddot{O}$

HO   H

C_{sp^3}–C_{sp^3}

C_{sp^2}–H_{1s}

zingerone
(responsible for the pungent
taste of ginger)

1.72

ketene

σ: C_{sp}–C_{sp^2}
π: C_p–C_p sp sp^2

CH_2=C=O

sp^2 σ: C_{sp}–O_{sp^2}
π: C_p–O_p

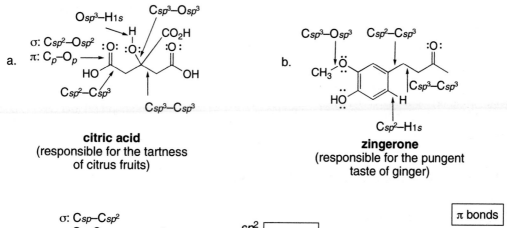

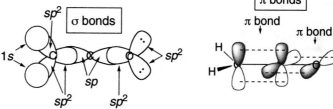

[For clarity, only the large bonding lobes of the hybrid orbitals are drawn.]

1.73

CH_2=$\overset{+}{CH}$ σ: C_{sp^2}–C_{sp}
π: C_p–C_p

sp^2 sp

CH_2=$\overset{..}{\overset{-}{CH}}$ σ: C_{sp^2}–C_{sp^2}
π: C_p–C_p

sp^2 sp^2

1.74 To determine relative bond length, use the rules in Answer 1.34.

a.

triple bond
shortest

double bond
middle

single bond
longest

C_{sp}–H$_{1s}$
highest
% s-character
shortest

b. H–C≡C–CH$_2$–C=C–C–H

C_{sp^2}–H$_{1s}$
middle
% s-character
middle

C_{sp^3}–H$_{1s}$
lowest
% s-character
longest

1.75 To determine relative bond length, use the rules in Answer 1.34.

a. HO⌒⌒⌒NH$_2$
longer
(N is to the left of O in
the second row.)

b. Br⌒⌒⌒Cl
longer
(Br is below Cl
in group 7A.)

c.

single bond
longer

1.76

b. and d.   bond (1)
**longest,
weakest C–C
single bond**

bond (2)

c. shortest, **strongest C–C bond**

a. **shortest C–C
single bond**

e. strongest C–H bond

f. Bond (1) is a C_{sp^3}–C_{sp^3} bond, and bond (2) is a C_{sp^3}–C_{sp^2} bond. Bond
(2) is shorter due to the increased percent s-character in the sp^2
hybridized carbon.

1.77 Remember shorter bonds are stronger bonds. A σ bond formed from two sp^2 hybridized C's is
stronger than a σ bond formed from two sp^3 hybridized C's because the sp^2 hybridized C orbitals
have a higher percent s-character.

1.78 Percent s-character determines the strength of a bond. **The higher percent s-character of an
orbital used to form a bond, the stronger the bond.**

vinyl chloride

CH$_2$=CH—Cl   33% s-character
higher percent s-character
stronger bond

C_{sp^2}

chloroethane (ethyl chloride)

CH$_3$-CH$_2$-Cl
25% s-character

C_{sp^3}

1.79   a. No, a compound with only one polar bond must be polar. The single bond dipole is not
cancelled by another bond dipole, so the molecule as a whole remains polar.

δ$^+$:Br—Cl:δ$^-$

b. Yes, a compound with multiple polar bonds can be nonpolar since the dipoles can cancel each other out, making a nonpolar molecule.

no net dipole

c. No, a compound cannot be polar if it contains only nonpolar bonds. There must be differences in electronegativity to make a compound polar.

H–H   **nonpolar**

1.80 Dipoles result from unequal sharing of electrons in covalent bonds. More electronegative atoms "pull" electron density towards them, making a dipole.

a. δ^+ Br–Cl δ^-

b. δ^+ δ^- NH$_2$–OH

c. δ^+ δ^- CH$_3$–NH$_2$

d. δ^- —Li δ^+

1.81 Use the directions from Answer 1.37.

a. CHBr$_3$ **net dipole**

c. CBr$_4$ **no net dipole**

e.   **net dipole**

b. CH$_3$CH$_2$OCH$_2$CH$_3$ **net dipole**

d.   **net dipole**

f.   **no net dipole**

1.82

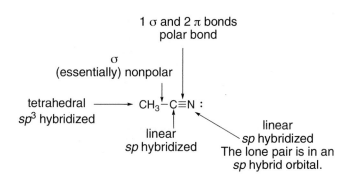

All C–H bonds are nonpolar σ bonds.
All H's use a 1s orbital in bonding.

1.83

    a. *sp²*

    b. Each C is trigonal planar; the ring is flat, drawn as a hexagon.

    c.

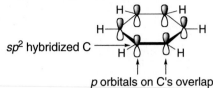

sp² hybridized C

p orbitals on C's overlap

| π bonds |

sp² hybrid orbitals on C

| σ bonds |

[Only the larger bonding lobe of each orbital is drawn.]

    d.

    e. Benzene is stable because of its two resonance structures that contribute equally to the hybrid. [This is only part of the story. We'll learn more about benzene's unusual stability in Chapter 17.]

1.84

    a.

3 groups
sp²
trigonal planar

3 groups
sp²
trigonal planar

4 groups
sp³
tetrahedral

4 groups
sp³
tetrahedral
(trigonal pyramidal molecular shape)

3 groups
sp²
trigonal planar

    b.

All C–O, C–N, C–S, N–H, and O–H bonds are polar and labeled with arrows.
All partial positive charges lie on the C.
All partial negative charges lie on the O, N, or S.
In OH and NH bonds, H bears a δ⁺.

skeletal structure:

    c.

    d.   6 π bonds

    e. **33% *s*-character = *sp²* hybridized**

These C–H bonds have 33% *s*-character.

1.87

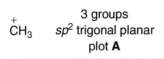

3 groups
sp^2 trigonal planar
plot **A**

The blue region is evidence of
the electron-poor cation.

4 groups
:CH_3 sp^3 tetrahedral
(The molecular shape is trigonal pyramidal.)
plot **B**

The red region is evidence of
the electron-rich anion.

1.88 Polar bonds result from unequal sharing of electrons in covalent bonds. Normally we think of more electronegative atoms "pulling" more of the electron density towards them, making a dipole. In looking at a Csp^2–Csp^3 bond, the atom with a higher percent s-character will "pull" more of the electron density towards it, creating a small dipole.

33% s-character
higher percent s-character
pulls more electron density
more electronegative

δ^- δ^+
=Csp^2—Csp^3—
↑ 25% s-character

1.89

Isomers of C_4H_8:

These two compounds are different
because of restricted rotation around
the C=C (Section 8.2B).

1.90 Carbocation **A** is more stable than carbocation **B** because resonance distributes the positive charge over two carbons. Delocalizing electron density is stabilizing. **B** has no possibility of resonance delocalization.

A   **B**   No resonance structures

1.91

a.   [1]   + :Cl:⁻

b.   [2]   [3]   **X**   phenol   + H—Cl:

1.85

a, b, c.

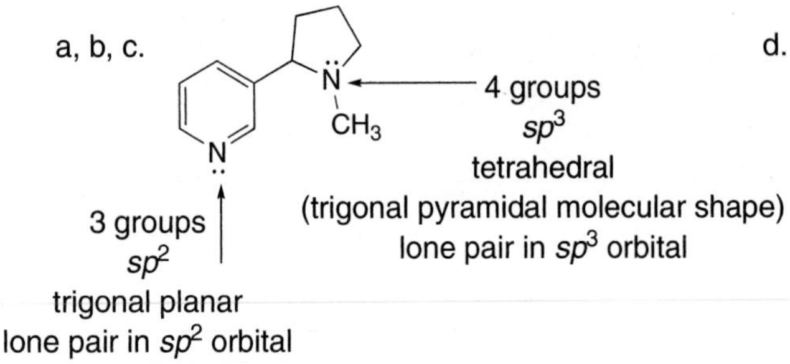

3 groups
sp^2
trigonal planar
lone pair in sp^2 orbital

4 groups
sp^3
tetrahedral
(trigonal pyramidal molecular shape)
lone pair in sp^3 orbital

d.

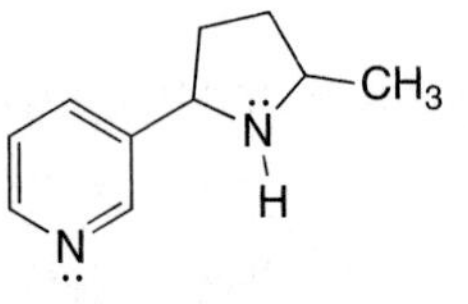

constitutional isomer

e.

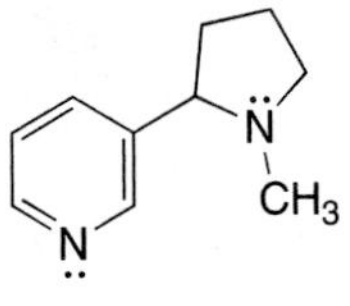

resonance structure

1.86 a.

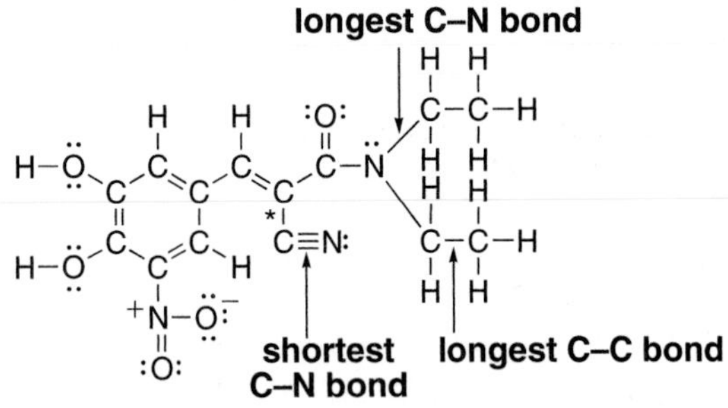

longest C–N bond

shortest C–N bond   longest C–C bond

b. The C–C bonds in the CH_2CH_3 groups are the longest because they are formed from sp^3 hybridized C's.

c. The shortest C–C bond is labeled with a (*) because it is formed from orbitals with the highest percent s-character ($C sp$–$C sp^2$).

d. The longest C–N bond is formed from the sp^3 hybridized C atom bonded to a N atom [labeled in part (a)].

e. The shortest C–N bond is the triple bond (C≡N); increasing the number of electrons between atoms decreases bond length.

f.

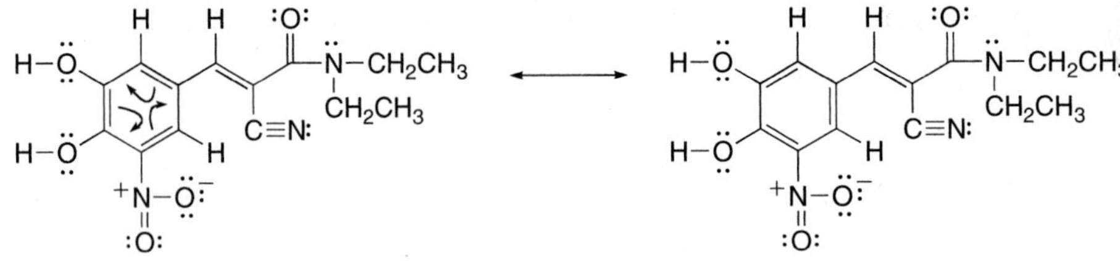

g.

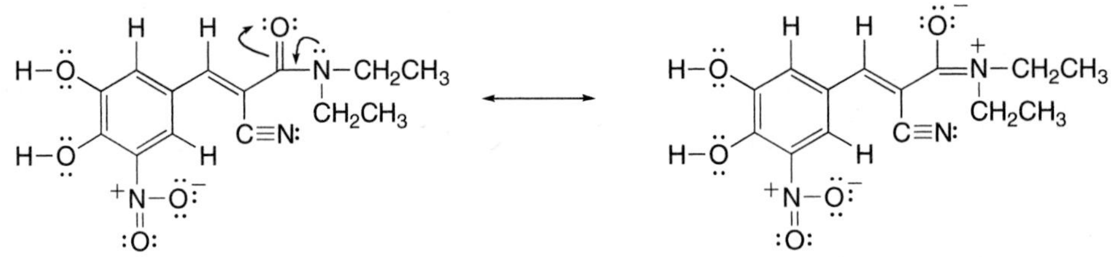

Chapter 2: Acids and Bases

♦ A comparison of Brønsted–Lowry and Lewis acids and bases

Type	Definition	Structural feature	Examples
Brønsted–Lowry acid (2.1)	proton donor	a proton	HCl, H_2SO_4, H_2O, CH_3COOH, TsOH
Brønsted–Lowry base (2.1)	proton acceptor	a lone pair *or* a π bond	^-OH, $^-OCH_3$, H^-, $^-NH_2$, $CH_2=CH_2$
Lewis acid (2.8)	electron pair acceptor	a proton, *or* an unfilled valence shell, *or* a partial (+) charge	BF_3, $AlCl_3$, HCl, CH_3COOH, H_2O
Lewis base (2.8)	electron pair donor	a lone pair *or* a π bond	^-OH, $^-OCH_3$, H^-, $^-NH_2$, $CH_2=CH_2$

♦ Acid–base reactions

[1] A Brønsted–Lowry acid donates a proton to a Brønsted–Lowry base (2.2).

[2] A Lewis base donates an electron pair to a Lewis acid (2.8).

- Electron-rich species react with electron-poor ones.
- Nucleophiles react with electrophiles.

♦ Important facts

- Definition: $pK_a = -\log K_a$. The **lower the** pK_a, the **stronger** the acid (2.3).

$$NH_3 \qquad \text{versus} \qquad H_2O$$
$$pK_a = 38 \qquad\qquad pK_a = 15.7$$
$$\text{lower } pK_a = \text{stronger acid}$$

- The stronger the acid, the weaker the conjugate base (2.3).

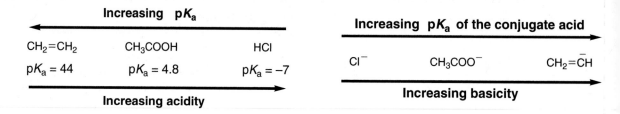

- In proton transfer reactions, equilibrium favors the weaker acid and weaker base (2.4).

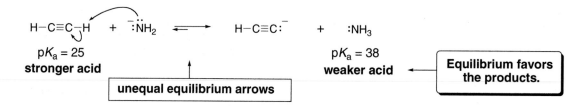

- An acid can be deprotonated by the conjugate base of any acid having a **higher pK$_a$** (2.4).

Acid	pK$_a$	Conjugate base	
CH$_3$COO–H	4.8	CH$_3$COO$^-$	
CH$_3$CH$_2$O–H	16	CH$_3$CH$_2$O$^-$	These bases
HC≡CH	25	HC≡C$^-$	can deprotonate
H–H	35	H$^-$	CH$_3$COO–H.
	higher pK$_a$ than CH$_3$COO–H		

◆ Factors that determine acidity (2.5)

[1] **Element effects** (2.5A)   The acidity of H–A increases both across a row and down a column of the periodic table.

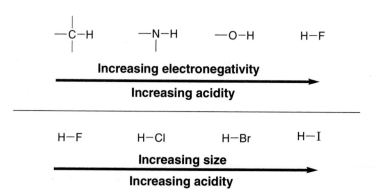

[2] Inductive effects (2.5B)

The acidity of H–A increases with the presence of electron-withdrawing groups in A.

CH_3CH_2OH ⟶ $CH_3CH_2O^-$

weaker acid

No additional electronegative atoms stabilize the conjugate base.

CF_3CH_2OH ⟶

stronger acid

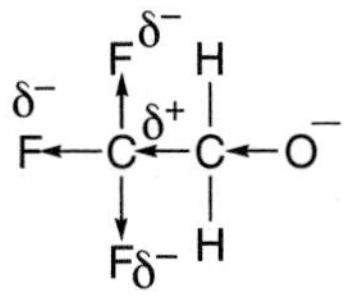

CF₃ withdraws electron density, stabilizing the conjugate base.

[3] Resonance effects (2.5C)

The acidity of H–A increases when the conjugate base A:⁻ is resonance stabilized.

$CH_3CH_2\ddot{O}-H$ ⟶ $CH_3CH_2\ddot{O}\colon^-$

ethanol

ethoxide
conjugate base

only **one** Lewis structure

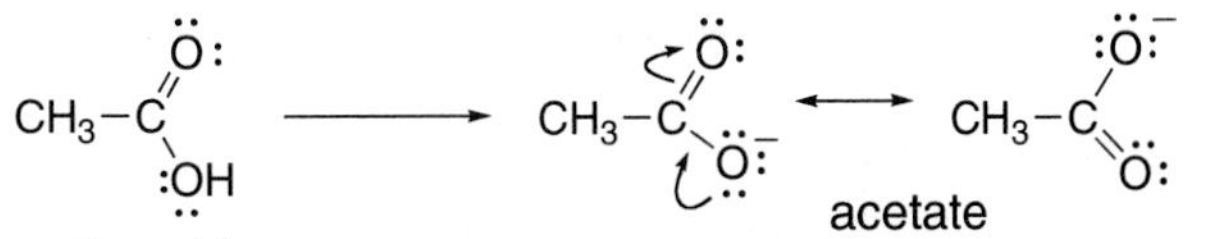

acetic acid
more acidic

acetate
conjugate base

two resonance structures

[4] Hybridization effects (2.5D)

The acidity of H–A increases as the percent *s*-character of the A:⁻ increases.

CH_3CH_3	$CH_2{=}CH_2$	$H{-}C{\equiv}C{-}H$
ethane	ethylene	acetylene
$pK_a = 50$	$pK_a = 44$	$pK_a = 25$

Increasing acidity

Chapter 2: Answers to Problems

2.1 Brønsted–Lowry acids are **proton donors** and must contain a hydrogen atom.
Brønsted–Lowry bases are **proton acceptors** and must have an available electron pair (either a lone pair or a π bond).

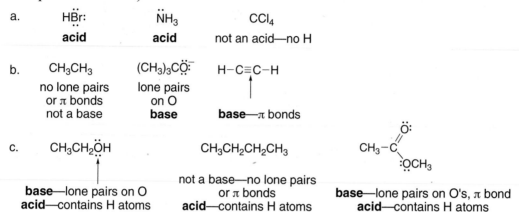

a.
$H\ddot{B}r$: $\ddot{N}H_3$ CCl_4
acid   **acid**   not an acid—no H

b. CH_3CH_3 $(CH_3)_3C\ddot{O}$: $H-C{\equiv}C-H$
no lone pairs   lone pairs
or π bonds   on O
not a base   **base**   **base**—π bonds

c. $CH_3CH_2\ddot{O}H$ $CH_3CH_2CH_2CH_3$ $CH_3-C\overset{\ddot{O}:}{\underset{\ddot{O}CH_3}{}}$

base—lone pairs on O   not a base—no lone pairs   **base**—lone pairs on O's, π bond
acid—contains H atoms   or π bonds   **acid**—contains H atoms
               acid—contains H atoms

2.2 A Brønsted–Lowry base accepts a proton to form the conjugate acid. A Brønsted–Lowry acid loses a proton to form the conjugate base.

a. $NH_3 \longrightarrow NH_4^+$ b. $HBr \longrightarrow Br^-$

      $Cl^- \longrightarrow HCl$ $HSO_4^- \longrightarrow SO_4^{2-}$

      $(CH_3)_2C{=}O \longrightarrow (CH_3)_2C{=}\overset{+}{O}H$ $CH_3OH \longrightarrow CH_3O^-$

2.3 Use the definitions from Answer 2.2.

$CH_2{=}CH_2 \longrightarrow \overset{+}{C}H_2-CH_3$ $CH_2{=}CH_2 \longrightarrow CH_2{=}\ddot{C}H^-$
ethylene   accepts a proton   loses a proton
          conjugate acid   conjugate base

2.4 The Brønsted–Lowry base accepts a proton to form the conjugate acid. The Brønsted–Lowry acid loses a proton to form the conjugate base. Use curved arrows to show the movement of electrons (***NOT protons***). Re-draw the starting materials if necessary to clarify the electron movement.

a. $H-\ddot{C}l$:   + $H_2\ddot{O}$ $\rightleftharpoons$:$\ddot{C}l$:$^-$ + $H_3\ddot{O}^+$
      acid   base   conjugate base   conjugate acid

b. $CH_3-\overset{:O:}{\underset{H}{\overset{\|}{C}}}CH_2$ +   :$\ddot{O}CH_3^-$ $\rightleftharpoons$ $CH_3-\overset{:O:}{\overset{\|}{C}}\overset{-}{C}H_2$ + $CH_3\ddot{O}H$
      acid   base   conjugate base   conjugate acid

2.5 To draw the products:
[1] Find the acid and base.
[2] Transfer a proton from the acid to the base.
[3] Check that the charges on each side of the arrows are balanced.

a. (–)1 charge on each side

b. (–)1 charge on each side

c. net neutral on each side

d. net neutral on each side

2.6 Draw the products in each reaction as in Answer 2.5.

a. CH_3OH $\xrightarrow{HCl}$ $CH_3\overset{+}{O}H_2$ + Cl^-

b. $(CH_3CH_2)_2O$ $\xrightarrow{HCl}$ $(CH_3CH_2)_2\overset{+}{O}H$ + Cl^-

c. $(CH_3)_3N$ $\xrightarrow{HCl}$ $(CH_3)_3\overset{+}{N}H$ + Cl^-

d.

2.7 The smaller the pK_a, the stronger the acid. The larger K_a, the stronger the acid.

a. $CH_3CH_2CH_3$ or CH_3CH_2OH
pK_a = 50   pK_a = 16

↑
smaller pK_a
stronger acid

b. or

K_a = 10^{-10} K_a = 10^{-41}

↓
larger K_a
stronger acid

2.8 To convert from K_a to pK_a, take (–) the log of the K_a; **pK_a = –log K_a.**
To convert pK_a to K_a, take the antilog of (–) the pK_a.

a. $K_a = 10^{-10}$ $K_a = 10^{-21}$ $K_a = 5.2 \times 10^{-5}$ b. p$K_a = 7$ p$K_a = 11$ p$K_a = 3.2$

↓   ↓   ↓   ↓   ↓   ↓

p$K_a = 10$ p$K_a = 21$ p$K_a = 4.3$ $K_a = 10^{-7}$ $K_a = 10^{-11}$ $K_a = 6.3 \times 10^{-4}$

2.9 Since **strong acids form weak conjugate bases**, the basicity of conjugate bases increases with increasing pK_a of their acids. Find the pK_a of each acid from Table 2.1 and then rank the acids in order of increasing pK_a. This will also be the order of increasing basicity of their conjugate bases.

a. Increasing acidity ⟵

H_2O, NH_3, CH_4

pK_a = 15.7   38   50

conjugate bases: $^-OH, ^-NH_2, ^-CH_3$

Increasing basicity ⟶

b. Increasing acidity ⟵

$HC{\equiv}CH, CH_2{=}CH_2, CH_4$

pK_a =   25   44   50

conjugate bases: $^-C{\equiv}CH, ^-CH{=}CH_2, ^-CH_3$

Increasing basicity ⟶

2.10 Use the definitions in Answer 2.9 to compare the acids. The smaller the pK_a, the larger the K_a and the stronger the acid. When a stronger acid dissolves in water, the equilibrium lies farther to the right.

HCO_2H
formic acid
pK_a = 3.8

a. smaller pK_a = larger K_a
b. smaller pK_a = stronger acid
d. stronger acid = equilibrium farther to the right

$(CH_3)_3CCO_2H$
pivalic acid
pK_a = 5.0

c. weaker acid = stronger conjugate base

2.11 To estimate the pK_a of the indicated bond, find a similar bond in the pK_a table (H bonded to the same atom with the same hybridization).

a.

For NH_3, pK_a is 38.
estimated pK_a = 38

b.

For CH_3CH_2OH,
pK_a is 16.
estimated pK_a = 16

c. $BrCH_2COO{-}H$

For CH_3COOH, pK_a is 4.8.
estimated pK_a = 5

2.12 Label the acid and the base and then transfer a proton from the acid to the base. To determine if the reaction will proceed as written, compare the pK_a of the acid on the left with the conjugate acid on the right. **The equilibrium always favors the formation of the weaker acid and the weaker base.**

a. $CH_2{=}CH_2$ + $H{:}^-$ ⇌ $CH_2{=}\ddot{C}H$ + H_2 Equilibrium favors
    acid   base   conjugate base   conjugate acid   the **starting materials.**
    pK_a = 44 pK_a = 35
    weaker acid

b. CH_4 + $:\ddot{O}H$ ⇌ $:CH_3$ + $H_2\ddot{O}:$
    acid   base   conjugate base   conjugate acid   Equilibrium favors
    pK_a = 50 pK_a = 15.7   the **starting materials.**
    weaker acid

c. CH_3COOH + $CH_3CH_2\ddot{O}:^-$ ⇌ CH_3COO^- + $CH_3CH_2\ddot{O}H$ Equilibrium favors
    acid   base   conjugate base   conjugate acid   the **products.**
    pK_a = 4.8 pK_a = 16
                                                                        weaker acid

d. $:\ddot{\underset{..}{Cl}}:^-$ + $CH_3CH_2\ddot{\underset{..}{O}}H$ ⇌ $H\ddot{\underset{..}{Cl}}:$ + $CH_3CH_2\ddot{\underset{..}{O}}:^-$ **Equilibrium favors the starting materials.**

       base   acid   conjugate acid   conjugate base

                     $pK_a = 16$ $pK_a = -7$

                    weaker acid

2.13   An acid can be deprotonated by the conjugate base of any acid with a higher pK_a.

CH_3COOH
$pK_a = 4.8$
Any base having a conjugate acid with a pK_a higher than 4.8 can deprotonate this acid.

Acid	pK_a	Conjugate base	
HCl	−7	Cl⁻	not strong enough
HC≡CH	25	HC≡C⁻	strong enough
H_2	35	H⁻	

$HC≡CH$
$pK_a = 25$
All of these acids have a higher pK_a than HC≡CH, and a conjugate base that can deprotonate HC≡CH.

Acid	pK_a	Conjugate base
H_2	35	H⁻
NH_3	38	⁻NH_2
$CH_2=CH_2$	44	$CH_2=CH^-$
CH_4	50	CH_3^-

2.14   An acid can be deprotonated by the conjugate base of any acid with a higher pK_a.

CH_3CN
$pK_a = 25$
Any base having a conjugate acid with a pK_a higher than 25 can deprotonate this acid.

Base	Conjugate acid	pK_a	
NaH	H_2	35	Only NaH and $NaNH_2$ are strong enough to deprotonate acetonitrile.
Na_2CO_3	HCO_3^-	10.2	
NaOH	H_2O	15.7	
$NaNH_2$	NH_3	38	
$NaHCO_3$	H_2CO_3	6.4	

2.15   The acidity of H–Z **increases across a row and down a column** of the periodic table.

a. NH_3, H_2O

O is farther to the right in the periodic table.
stronger acid

b. HBr, HCl

Br is farther down the periodic table.
stronger acid

c. H_2S, HBr

Br is farther across and down the periodic table.
stronger acid

2.16   Compare the most acidic protons in each compound to determine the stronger acid.

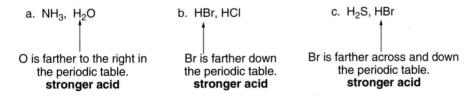

a. $CH_3CH_2CH_2NH_2$ or $(CH_3)_3N$

          N–H bond   C–H bond

N is farther to the right in the periodic table.
stronger acid

b. $CH_3CH_2OCH_3$ or $CH_3CH_2CH_2OH$

       C–H bond   O–H bond

O is farther to the right in the periodic table.
stronger acid

2.17   Look at the element bonded to the acidic H and decide its acidity based on the periodic trends. **Farther right and down the periodic table is more acidic.**

most acidic   most acidic   most acidic

a. $CH_3CH_2CH_2CH_2O\underline{H}$ b. $\underline{H}OCH_2CH_2CH_2NH_2$ c. $(CH_3)_2NCH_2CH_2CH_2N\underline{H}_2$

Molecule contains C–H and O–H bonds. O is farther right; therefore, O–H hydrogen is the most acidic.	Molecule contains C–H, N–H, and O–H bonds. O is farthest right; therefore, O–H hydrogen is the most acidic.	Molecule contains C–H and N–H bonds. N is farther right; therefore, N–H hydrogen is the most acidic.

2.18   The acidity of HA increases across the periodic table. Pseudoephedrine contains C–H, N–H, and O–H bonds. The O–H bond is most acidic.

pseudoephedrine

2.19   **More electronegative atoms stabilize the conjugate base, making the acid stronger.** Compare the electron-withdrawing groups on the acids below to decide which is a stronger acid **(more electronegative groups = more acidic).**

a. $ClCH_2COOH$ or FCH_2COOH

                      more acidic

F is more electronegative than Cl, making the O–H bond in the acid on the right **more acidic.**

c. CH_3COOH or O_2NCH_2COOH

                      more acidic

NO_2 is electron withdrawing, making the O–H bond in the acid on the right **more acidic.**

b. Cl_2CHCH_2OH or $Cl_2CHCH_2CH_2OH$

Cl is closer to the acidic O–H bond. **more acidic**

Cl is farther from the O–H bond.

2.20   **More electronegative groups stabilize the conjugate base, making the acid stronger.**

↓

$HOCH_2CO_2H$ CH_3CO_2H

an α-hydroxy acid   acetic acid

The extra OH group contains an electronegative O, which stabilizes the conjugate base. **stronger acid**

2.21   The acidity of an acid increases when the conjugate base is resonance stabilized. Compare the conjugate bases of acetone and propane to explain why acetone is more acidic.

2 resonance structures more stable conjugate base **Acetone is more acidic.**

acetone
$pK_a = 19.2$

One resonance structure places the (–) charge on the more electronegative O atom. This is especially good.

$$CH_3CH_2CH_3 \xrightarrow{\text{base}} CH_3CH_2\ddot{C}H_2$$

propane
$pK_a = 50$

only one Lewis structure
less stable conjugate base
(Any C–H bond in the starting
material can be removed.)

2.22 The acidity of an acid increases when the conjugate base is resonance stabilized. Acetonitrile has a resonance-stabilized conjugate base, which accounts for its acidity.

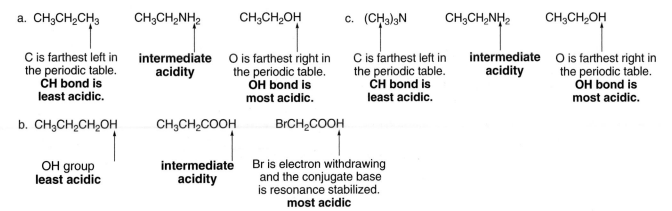

acetonitrile
(one Lewis structure)

The negative charge is stabilized by delocalization on the C and N atoms.

Having the (−) charge on the electronegative N atom adds stability.

2.23 **Increasing percent s-character makes an acid more acidic.** Compare the percent s-character of the carbon atoms in each of the C–H bonds in question. A stronger acid has a weaker conjugate base.

a. $CH_3CH_2-C\equiv C-H$ or $CH_3CH_2CH_2CH_2-H$ b.

sp hybridized C
50% *s*-character
more acidic

sp^3 hybridized C
25% *s*-character

base   base   base   base

$CH_3CH_2-C\equiv C:^-$ or $CH_3CH_2CH_2\ddot{C}H_2$

**stronger
conjugate base**

sp^2 hybridized C
33% *s*-character
more acidic

sp^3 hybridized C
25% *s*-character

or

**stronger
conjugate base**

2.24 To compare the acids, first **look for element effects**. Then identify electron-withdrawing groups, resonance, or hybridization differences.

a. $CH_3CH_2CH_3$ $CH_3CH_2NH_2$ CH_3CH_2OH c. $(CH_3)_3N$ $CH_3CH_2NH_2$ CH_3CH_2OH

C is farthest left in
the periodic table.
**CH bond is
least acidic.**

**intermediate
acidity**

O is farthest right in
the periodic table.
**OH bond is
most acidic.**

C is farthest left in
the periodic table.
**CH bond is
least acidic.**

**intermediate
acidity**

O is farthest right in
the periodic table.
**OH bond is
most acidic.**

b. $CH_3CH_2CH_2OH$ CH_3CH_2COOH $BrCH_2COOH$

OH group
least acidic

**intermediate
acidity**

Br is electron withdrawing
and the conjugate base
is resonance stabilized.
most acidic

2.25 Look at the element bonded to the acidic H and decide its acidity based on the periodic trends. **Farther right and down the periodic table is more acidic.**

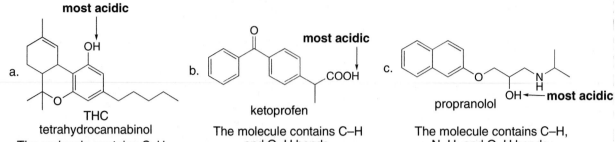

a. THC
tetrahydrocannabinol
The molecule contains C–H and O–H bonds.
O is farther right; therefore, O–H hydrogen is the most acidic.

b. ketoprofen
The molecule contains C–H and O–H bonds.
O is farther right; therefore, O–H hydrogen is the most acidic.

c. propranolol
The molecule contains C–H, N–H, and O–H bonds.
O is farthest right; therefore, O–H hydrogen is the most acidic.

2.26 Draw the products of proton transfer from the acid to the base.

a. $(CH_3)_2CH\ddot{O}–H$ + $Na^+ H:^-$ ⇌ $(CH_3)_2CH\ddot{O}:^- Na^+$ + H_2
   acid   base   conjugate base   conjugate acid

b. $(CH_3)_2CH\ddot{O}–H$ + $H–OSO_3H$ ⇌ $(CH_3)_2CH\overset{+}{\ddot{O}}H_2$ + HSO_4^-
   base   acid   conjugate acid   conjugate base

c. $(CH_3)_2CH\ddot{O}–H$ + $Li^+ {}^-\ddot{N}[CH(CH_3)_2]_2$ ⇌ $(CH_3)_2CH\ddot{O}:^- Li^+$ + $HN[CH(CH_3)_2]_2$
   acid   base   conjugate base   conjugate acid

d. $(CH_3)_2CH\ddot{O}–H$ + $H–O COCH_3$ ⇌ $(CH_3)_2CH\overset{+}{\ddot{O}}H_2$ + $^-OCOCH_3$
   base   acid   conjugate acid   conjugate base

2.27 To cross a cell membrane, amphetamine must be in its neutral (not ionic) form.

amphetamine →protonation by HCl in the stomach→ →deprotonation in the intestines→ **absorption here ↑ in the neutral form**

2.28 **Lewis bases are electron pair donors:** they contain a lone pair or a π bond.

a. $\ddot{N}H_3$ b. $CH_3CH_2CH_3$ c. $H:^-$ d. $H–C≡C–H$

yes - has lone pair

no - no lone pair or π bond

yes - has lone pair

yes - has 2 π bonds

2.29 **Lewis acids are electron pair acceptors.** Most Lewis acids contain a proton or an unfilled valence shell of electrons.

a. BBr_3 b. CH_3CH_2OH c. $(CH_3)_3C^+$ d. Br^-

yes unfilled valence shell on B

yes contains a proton

yes unfilled valence shell on C

no no proton no unfilled valence shell

2.30 Label the Lewis acid and Lewis base and then draw the curved arrows.

a. BF_3 + $CH_3-\overset{..}{\underset{..}{O}}-CH_3$ ⟶

new bond

$F-\overset{\overset{F}{|}}{\underset{\underset{F}{|}}{B}}\overset{+}{-}\overset{..}{O}\overset{CH_3}{\underset{CH_3}{}}$

b. $(CH_3)_2\overset{+}{CH}$ + $\overset{-}{:}\overset{..}{O}H$ ⟶ $(CH_3)_2CH\overset{..}{O}H$

Lewis acid — unfilled valence shell on B

Lewis base — lone pairs on O

Lewis acid — unfilled valence shell on C

Lewis base — lone pairs on O

2.31 A Lewis acid is also called an **electrophile**. When a Lewis base reacts with an electrophile other than a proton, it is called a **nucleophile**. Label the electrophile and nucleophile in the starting materials and then draw the products.

a. $CH_3CH_2-\overset{..}{\underset{..}{O}}-CH_2CH_3$ + BBr_3 ⟶ $CH_3CH_2-\overset{+}{\underset{..}{O}}-CH_2CH_3$

$Br-\overset{\overset{Br}{|}}{\underset{\underset{Br}{|}}{B}}^-$

Lewis base **nucleophile** lone pairs on O

Lewis acid **electrophile** unfilled valence shell on B

b. $\underset{CH_3}{}\overset{\overset{:\overset{..}{O}:}{||}}{C}\underset{CH_3}{}$ + $AlCl_3$ ⟶

$Cl-\overset{\overset{Cl}{|}}{\underset{\underset{:\overset{+}{O}:}{|}}{Al}}^-Cl$

$\underset{CH_3}{}\overset{||}{C}\underset{CH_3}{}$

Lewis base **nucleophile** lone pairs on O

Lewis acid **electrophile** unfilled valence shell on Al

2.32 Draw the product of each reaction by using an electron pair of the Lewis base to form a new bond to the Lewis acid.

a. $CH_3CH_2-\overset{\overset{..}{N}}{\underset{\underset{CH_2CH_3}{|}}{}}-CH_2CH_3$ + $B(CH_3)_3$ ⟶ $CH_3CH_2-\overset{\overset{CH_3-\overset{-}{B}-CH_3}{|}}{\underset{\underset{CH_2CH_3}{|}}{\overset{+}{N}}}-CH_2CH_3$

Lewis base **nucleophile** lone pair on N

Lewis acid **electrophile** unfilled valence shell on B

b. $CH_3CH_2-\overset{\overset{..}{N}}{\underset{\underset{CH_2CH_3}{|}}{}}-CH_2CH_3$ + $^+C(CH_3)_3$ ⟶ $CH_3CH_2-\overset{\overset{CH_3-\overset{}{C}-CH_3}{|}}{\underset{\underset{CH_2CH_3}{|}}{\overset{+}{N}}}-CH_2CH_3$

Lewis base **nucleophile** lone pair on N

Lewis acid **electrophile** unfilled valence shell on C

c. $CH_3CH_2-\overset{\overset{..}{N}}{\underset{\underset{CH_2CH_3}{|}}{}}-CH_2CH_3$ + $AlCl_3$ ⟶ $CH_3CH_2-\overset{\overset{Cl-\overset{-}{Al}-Cl}{|}}{\underset{\underset{CH_2CH_3}{|}}{\overset{+}{N}}}-CH_2CH_3$

Lewis base **nucleophile** lone pair on N

Lewis acid **electrophile** unfilled valence shell on Al

2.33   Curved arrows begin at the Lewis base and point towards the Lewis acid.

Lewis base   **Lewis acid**
contains a π bond   contains a proton

2.34   To draw the conjugate acid of a Brønsted–Lowry base, **add a proton to the base**.

a. $H_2\ddot{O}:$ $\xrightarrow{\ H^+\ }$ $H_3\ddot{O}^+$

d. $CH_3CH_2\ddot{N}HCH_3$ $\xrightarrow{\ H^+\ }$ $CH_3CH_2\overset{+}{N}H_2CH_3$

b. $^-\!:\ddot{N}H_2$ $\xrightarrow{\ H^+\ }$ $\ddot{N}H_3$

e. $CH_3\ddot{O}CH_3$ $\xrightarrow{\ H^+\ }$ $CH_3-\overset{\overset{H}{|}}{\underset{\cdot\cdot}{O}}{}^+\!-CH_3$

c. HCO_3^- $\xrightarrow{\ H^+\ }$ H_2CO_3

f. CH_3COO^- $\xrightarrow{\ H^+\ }$ CH_3COOH

2.35   To draw the conjugate base of a Brønsted–Lowry acid, **remove a proton from the acid.**

a. HCN $\xrightarrow{\ -H^+\ }$ ^-CN

d. $HC\equiv CH$ $\xrightarrow{\ -H^+\ }$ $HC\equiv C^-$

b. HCO_3^- $\xrightarrow{\ -H^+\ }$ CO_3^{2-}

e. CH_3CH_2COOH $\xrightarrow{\ -H^+\ }$ $CH_3CH_2COO^-$

c. $(CH_3)_2\overset{+}{N}H_2$ $\xrightarrow{\ -H^+\ }$ $(CH_3)_2NH$

f. CH_3SO_3H $\xrightarrow{\ -H^+\ }$ $CH_3SO_3^-$

2.36   To draw the products of an acid–base reaction, transfer a proton from the acid (H_2SO_4 in this case) to the base.

a.

b.

c.

d.

2.37   To draw the products of an acid–base reaction, transfer a proton from the acid to the base (^-OH in this case).

a.

b.

$$\text{cyclohexyl}-C(=\overset{\cdot\cdot}{\overset{\cdot\cdot}{O}}:)-\overset{\cdot\cdot}{\underset{\cdot\cdot}{O}}-H \quad + \quad K^+ \;:\overset{\cdot\cdot}{O}H^- \quad \longrightarrow \quad \text{cyclohexyl}-C(=\overset{\cdot\cdot}{\overset{\cdot\cdot}{O}}:)-\overset{\cdot\cdot}{\underset{\cdot\cdot}{O}}:^- \; K^+ \quad + \quad H_2O$$

c.

$$\text{cyclohexyl}-C\equiv C-H \quad + \quad K^+ \;:\overset{\cdot\cdot}{O}H^- \quad \longrightarrow \quad \text{cyclohexyl}-C\equiv C:^- \; K^+ \quad + \quad H_2O$$

d.

$$CH_3-\text{C}_6H_4-\overset{\cdot\cdot}{\underset{\cdot\cdot}{O}}-H \quad + \quad K^+ \;:\overset{\cdot\cdot}{O}H^- \quad \longrightarrow \quad CH_3-\text{C}_6H_4-\overset{\cdot\cdot}{\underset{\cdot\cdot}{O}}:^- K^+ \quad + \quad H_2O$$

2.38 Label the Brønsted–Lowry acid and Brønsted–Lowry base in the starting materials and **transfer a proton from the acid to the base** for the products.

a.
$$CH_3\overset{\cdot\cdot}{\underset{\cdot\cdot}{O}}-H \quad + \quad :NH_2^- \quad \rightleftharpoons \quad CH_3\overset{\cdot\cdot}{\underset{\cdot\cdot}{O}}:^- \quad + \quad \overset{\cdot\cdot}{N}H_3$$
$$\quad \text{acid} \qquad \text{base} \qquad\qquad \text{conjugate base} \quad \text{conjugate acid}$$

b.
$$CH_3CH_2-C(=\overset{\cdot\cdot}{\overset{\cdot\cdot}{O}}:)-\overset{\cdot\cdot}{\underset{\cdot\cdot}{O}}-H \quad + \quad :\overset{\cdot\cdot}{\underset{\cdot\cdot}{O}}-CH_3^- \quad \rightleftharpoons \quad CH_3CH_2-C(=\overset{\cdot\cdot}{\overset{\cdot\cdot}{O}}:)-\overset{\cdot\cdot}{\underset{\cdot\cdot}{O}}:^- \quad + \quad H\overset{\cdot\cdot}{\underset{\cdot\cdot}{O}}-CH_3$$
$$\quad \text{acid} \qquad\qquad\qquad \text{base} \qquad\qquad\qquad \text{conjugate base} \qquad\qquad \text{conjugate acid}$$

c.
$$CH_3CH_2-C\equiv C-H \quad + \quad :H^- \quad \rightleftharpoons \quad CH_3CH_2-C\equiv C:^- \quad + \quad H_2$$
$$\quad \text{acid} \qquad\qquad \text{base} \qquad\qquad \text{conjugate base} \qquad \text{conjugate acid}$$

d.
$$(CH_3CH_2)_3\overset{\cdot\cdot}{N} \quad + \quad H-\overset{\cdot\cdot}{\underset{\cdot\cdot}{Cl}}: \quad \rightleftharpoons \quad (CH_3CH_2)_3\overset{+}{N}H \quad + \quad :\overset{\cdot\cdot}{\underset{\cdot\cdot}{Cl}}:^-$$
$$\quad \text{base} \qquad\qquad \text{acid} \qquad\qquad \text{conjugate acid} \qquad \text{conjugate base}$$

e.
$$CH_3CH_2-\overset{\cdot\cdot}{\underset{\cdot\cdot}{O}}-H \quad + \quad H-\overset{\cdot\cdot}{\underset{\cdot\cdot}{Br}}: \quad \rightleftharpoons \quad CH_3CH_2-\overset{+}{\underset{H}{\overset{\cdot\cdot}{O}}}-H \quad + \quad :\overset{\cdot\cdot}{\underset{\cdot\cdot}{Br}}:^-$$
$$\quad \text{base} \qquad\qquad \text{acid} \qquad\qquad\qquad \text{conjugate acid} \qquad \text{conjugate base}$$

f.
$$CH_3C\equiv C:^- \quad + \quad H-\overset{\cdot\cdot}{O}H \quad \rightleftharpoons \quad CH_3C\equiv CH \quad + \quad H\overset{\cdot\cdot}{O}:^-$$
$$\quad \text{base} \qquad\qquad \text{acid} \qquad\qquad \text{conjugate acid} \quad \text{conjugate base}$$

2.39 Label the acid and base in the starting materials and then draw the products of proton transfer from acid to base.

a.
$$\text{phenyl}-C(=\overset{\cdot\cdot}{\overset{\cdot\cdot}{O}}:)-\overset{\cdot\cdot}{\underset{\cdot\cdot}{O}}-H \quad + \quad \overset{\cdot\cdot}{N}H_3 \quad \rightleftharpoons \quad \text{phenyl}-C(=\overset{\cdot\cdot}{\overset{\cdot\cdot}{O}}:)-\overset{\cdot\cdot}{\underset{\cdot\cdot}{O}}:^- \quad + \quad \overset{+}{N}H_4$$
$$\quad \text{acid} \qquad\qquad\qquad \text{base}$$

b.
$$F_3C-C(=\overset{\cdot\cdot}{\overset{\cdot\cdot}{O}}:)-\overset{\cdot\cdot}{\underset{\cdot\cdot}{O}}-H \quad + \quad Na^+ \; HCO_3^- \quad \rightleftharpoons \quad F_3C-C(=\overset{\cdot\cdot}{\overset{\cdot\cdot}{O}}:)-\overset{\cdot\cdot}{\underset{\cdot\cdot}{O}}:^- \; Na^+ \quad + \quad H_2CO_3$$
$$\quad \text{acid} \qquad\qquad\qquad \text{base}$$

c.
$$CH_3-C\equiv C-H \quad + \quad Na^+ \; :\overset{\cdot\cdot}{N}H_2^- \quad \rightleftharpoons \quad CH_3-C\equiv C:^- \; Na^+ \quad + \quad \overset{\cdot\cdot}{N}H_3$$
$$\quad \text{acid} \qquad\qquad\qquad \text{base}$$

d. $CH_3CH_2-\overset{..}{\underset{..}{O}}-H$ + $K^+ \; \overset{..}{\underset{..}{:O}}H$ ⇌ $CH_3CH_2-\overset{..}{\underset{..}{O}}: \; K^+$ + $H_2\overset{..}{\underset{..}{O}}:$

   acid   **base**

e. $CH_3-\overset{\cdot\cdot}{\underset{\underset{CH_3}{|}}{N}}-H$ + $CH_3-\bigcirc-\overset{O}{\underset{O}{\overset{||}{\underset{||}{S}}}}-O-H$ ⇌ $CH_3-\overset{\overset{H}{|}}{\underset{\underset{CH_3}{|}}{\overset{+}{N}}}-H$ + $CH_3-\bigcirc-SO_3^-$

   base   **acid**

f. $\underset{CH_3\quad\quad CH_3}{\overset{:\overset{..}{O}:}{\overset{||}{C}}}$ + $H-OSO_3H$ ⇌ $\underset{CH_3\quad\quad CH_3}{\overset{:\overset{+}{O}-H}{\overset{||}{C}}}$ + HSO_4^-

     base   **acid**

2.40   Label the acid and base in the starting materials and then draw the products of proton transfer from acid to base.

a. $\bigcirc-CH_2CH(CH_3)\overset{..}{N}H_2$ + $H-Cl$ ⟶ $\bigcirc-CH_2CH(CH_3)\overset{+}{N}H_3$ + $:\overset{..}{\underset{..}{Cl}}:^-$

      base   **acid**

b. $\bigcirc-CH_2CH(CH_3)\overset{..}{\underset{\underset{H}{|}}{N}}-H$ + $Na^+ :H^-$ ⟶ $\bigcirc-CH_2CH(CH_3)\overset{..}{\underset{..}{N}}H$ + H_2

      acid   **base** Na^+

2.41   Draw the products of proton transfer from acid to base.

a. $CH_3O-\bigcirc\bigcirc-\overset{\overset{H\quad CH_3}{\diagdown\,\diagup}}{\underset{COO-H}{C}}$ + $Na^+ :\overset{..}{O}H$ ⟶ $CH_3O-\bigcirc\bigcirc-\overset{\overset{H\quad CH_3}{\diagdown\,\diagup}}{\underset{COO^- \; Na^+}{C}}$ + $H_2\overset{..}{\underset{..}{O}}:$

       acid   **base**

b. $\bigcirc-\overset{\overset{H\quad CH_2CH_2\overset{..}{N}HCH_3}{\diagdown\,\diagup}}{\underset{O}{C}}-\bigcirc-CF_3$ + $H-Cl$ ⟶ $\bigcirc-\overset{\overset{H\quad CH_2CH_2\overset{+}{N}H_2CH_3}{\diagdown\,\diagup}}{\underset{O}{C}}-\bigcirc-CF_3$ + Cl^-

   base   **acid**

2.42   Draw the products of proton transfer from acid to base.

$\overset{CF_3}{\underset{}{\bigcirc}}-CH_2CH(CH_3)\overset{\cdot\cdot}{N}HCH_2CH_3$ + $H-OCOCH_3$ ⟶ $\overset{CF_3}{\underset{}{\bigcirc}}-CH_2CH(CH_3)\overset{+}{\underset{\underset{H}{|}}{N}}HCH_2CH_3$ + $^-OCOCH_3$

         base   **acid**

$\bigcirc-CH_2C(CH_3)_2\overset{..}{N}H_2$ + $H-OCOCH_3$ ⟶ $\bigcirc-CH_2C(CH_3)_2\overset{+}{N}H_3$ + $^-OCOCH_3$

       base   **acid**

2.43   To convert pK_a to K_a, take the antilog of (–) the pK_a.

a. H_2S
$pK_a = 7.0$
$K_a = 10^{-7}$

b. $ClCH_2COOH$
$pK_a = 2.8$
$K_a = 1.6 \times 10^{-3}$

c. HCN
$pK_a = 9.1$
$K_a = 7.9 \times 10^{-10}$

2.44   To convert from K_a to pK_a, take (–) the log of the K_a; $\mathbf{pK_a = -\log K_a}$.

a. ⬡—$CH_2\overset{+}{N}H_3$
$K_a = 4.7 \times 10^{-10}$
$pK_a = 9.3$

b. ⬡—$\overset{+}{N}H_3$
$K_a = 2.3 \times 10^{-5}$
$pK_a = 4.6$

c. CF_3COOH
$K_a = 5.9 \times 10^{-1}$
$pK_a = 0.23$

2.45   An acid can be deprotonated by the conjugate base of any acid with a higher pK_a.

a. H_2O
$pK_a = 15.7$
Any base with a conjugate acid having a pK_a higher than 15.7 can deprotonate it.

Acid	pK_a	Conjugate base	
CH_3CH_2OH	16	$CH_3CH_2O^-$	
$HC{\equiv}CH$	25	$HC{\equiv}C^-$	Strong enough to deprotonate H_2O.
H_2	35	H^-	
NH_3	38	$^-NH_2$	
$CH_2{=}CH_2$	44	$CH_2{=}CH^-$	
CH_4	50	CH_3^-	

b. NH_3
$pK_a = 38$
Any base with a conjugate acid having a pK_a higher than 38 can deprotonate it.

Acid	pK_a	Conjugate base	
$CH_2{=}CH_2$	44	$CH_2{=}CH^-$	Strong enough to deprotonate NH_3.
CH_4	50	CH_3^-	

c. CH_4
$pK_a = 50$
There is no base with a conjugate acid having a pK_a higher than 50 in the table.

2.46   An acid can be deprotonated by the conjugate base of any acid with a higher pK_a.

$CH_3CH_2CH_2C{\equiv}CH$
$pK_a = 25$
Any base having a conjugate acid with a pK_a higher than 25 can deprotonate this acid.

Base	Conjugate acid	pK_a	
H_2O	H_3O^+	–1.7	
NaOH	H_2O	15.7	
$NaNH_2$	NH_3	38	Only $NaNH_2$, NaH, and CH_3Li are strong enough to deprotonate the acid.
NH_3	NH_4^+	9.4	
NaH	H_2	35	
CH_3Li	CH_4	50	

2.47 ^-OH can deprotonate any acid with a $pK_a < 15.7$.

a. HCOOH
$pK_a = 3.8$
stronger acid
deprotonated

b. H_2S
$pK_a = 7.0$
stronger acid
deprotonated

c. ⬡—CH_3
$pK_a = 41$
weaker acid

d. CH_3NH_2
$pK_a = 40$
weaker acid

These acids are too weak to be deprotonated by ^-OH.

2.48 Draw the products and then compare the pK_a of the acid on the left and the conjugate acid on the right. **The equilibrium lies towards the side having the acid with a higher pK_a (weaker acid).**

a.
$$CF_3-C\overset{\ddot{O}:}{\underset{:\overset{..}{O}-H}{}} + \ ^-:\overset{..}{O}CH_2CH_3 \rightleftharpoons CF_3-C\overset{\ddot{O}:}{\underset{:\overset{..}{O}:^-}{}} + \ H\overset{..}{O}CH_2CH_3$$
pK_a = 0.2   pK_a = 16   **products favored**

b.
$$CH_3CH_2-C\overset{\ddot{O}:}{\underset{:\overset{..}{O}-H}{}} + \ Na^+ \ :\overset{..}{\underset{..}{Cl}}:^- \rightleftharpoons CH_3CH_2-C\overset{\ddot{O}:}{\underset{:\overset{..}{O}:^- \ Na^+}{}} + \ H\overset{..}{\underset{..}{Cl}}:$$
pK_a = ~5   pK_a = –7   **starting material favored**

c. $(CH_3)_3C\overset{..}{O}H + H-OSO_3H \rightleftharpoons (CH_3)_3C\overset{+}{\overset{..}{O}}H_2 + HSO_4^-$ **products favored**
pK_a = –9   pK_a = ~ –3

d.
$$\text{(phenol)}\ \overset{\ddot{O}-H}{} + \ Na^+ \ HCO_3^- \rightleftharpoons \text{(phenoxide)}\ \overset{\ddot{O}:^- \ Na^+}{} + \ H_2CO_3$$
pK_a = 10   pK_a = 6.4   **starting material favored**

e. $H-C\equiv C-H + Li^+ \ ^-\overset{..}{C}H_2CH_3 \rightleftharpoons H-C\equiv C:^- \ Li^+ + CH_3CH_3$ **products favored**
pK_a = 25   pK_a = 50

f. $CH_3\overset{..}{N}H_2 + H-OSO_3H \rightleftharpoons CH_3\overset{+}{N}H_3 + HSO_4^-$ **products favored**
pK_a = –9   pK_a = 10.7

2.49 Compare element effects first and then resonance, hybridization, and electron-withdrawing groups to determine the relative strengths of the acids.

a. Acidity increases across a row:
$NH_3 < H_2O < HF$

b. Acidity increases down a column:
$HF < HCl < HBr$

c. increasing acidity: $^-OH < H_2O < H_3O^+$

d. increasing acidity: $NH_3 < H_2O < H_2S$
Compare NH and OH bonds first: acidity increases across a row. OH is more acidic.

Then compare OH and SH bonds: acidity increases down a column. SH is more acidic.

e. Acidity increases across a row:
$CH_3CH_3 < CH_3NH_2 < CH_3OH$

f. increasing acidity: $H_2O < H_2S < HCl$
Compare HCl and SH bonds first: acidity increases across a row. H–Cl is more acidic.

Compare OH and SH bonds: acidity increases down a column. SH is more acidic.

g. $CH_3CH_2CH_3$, $ClCH_2CH_2OH$, CH_3CH_2OH

only C–H bonds   O–H bond and   O–H bond
weakest acid   electron-withdrawing Cl
strongest acid

increasing acidity: $CH_3CH_2CH_3 < CH_3CH_2OH < ClCH_2CH_2OH$

h. $HC\equiv CCH_2CH_3$ $CH_3CH_2CH_2CH_3$ $CH_3C=CCH_3$ (with H H)

sp C–H   all sp^3 C–H sp^2 C–H
strongest acid   **weakest acid**

increasing acidity: $CH_3CH_2CH_2CH_3 < CH_3CH=CHCH_3 < HC\equiv CCH_2CH_3$

2.50 The strongest acid has the weakest conjugate base.

a. Draw the conjugate acid.
Increasing acidity of conjugate acids:
$$CH_3CH_3 < CH_3NH_2 < CH_3OH$$
increasing basicity: $CH_3O^- < CH_3\overset{-}{N}H < CH_3CH_2^-$

b. Draw the conjugate acid.
Increasing acidity of conjugate acids:
$$CH_4 < H_2O < HBr$$
increasing basicity: $Br^- < HO^- < {}^-CH_3$

c. Draw the conjugate acid.
Increasing acidity of conjugate acids:
$$CH_3CH_2OH < CH_3COOH < ClCH_2COOH$$
increasing basicity: $ClCH_2COO^- < CH_3COO^- < CH_3CH_2O^-$

d. Draw the conjugate acid.
Increasing acidity of conjugate acids:

$-CH_2CH_3 <$ $-CH=CH_2 <$ $-C\equiv CH$

increasing basicity:

$-C\equiv C^- <$ $-CH=\overset{-}{C}H <$ $-CH_2\overset{-}{C}H_2$

2.51 **More electronegative atoms stabilize the conjugate base by an electron-withdrawing inductive effect, making the acid stronger.** Thus, an O atom increases the acidity of an acid.

$pK_a = 11.1$ The O atom makes this cation the stronger acid.
$pK_a = 8.33$

2.52 In both molecules the OH proton is the most acidic H. In addition, compare the percent s-character of the carbon atoms in each molecule. Nearby C's with a higher percent s-character can help to stabilize the conjugate base.

$$HC\equiv CCO_2H \qquad\qquad CH_3CH_2CO_2H$$
$$pK_a = 1.8 \qquad\qquad\qquad pK_a = 4.9$$

The sp hybridized C's of the triple bond have a higher percent s-character than an sp^3 hybridized C, so they pull electron density towards them, stabilizing the conjugate base.
stronger acid

2.53

$CH_3CH_2CH_2-H$ $pK_a = 50$

$pK_a = 43$

$pK_a = 19.2$

The negative charge on O is good. This makes this resonance structure especially good.

conjugate base:

$CH_3CH_2\overset{..}{C}H_2$

one Lewis structure
weakest acid

two resonance structures
negative charge delocalized
on two carbons

two resonance structures
negative charge delocalized
on one O and one C
strongest acid

2.54 To draw the conjugate acid, look for the most basic site and protonate it. To draw the conjugate base, look for the most acidic site and remove a proton.

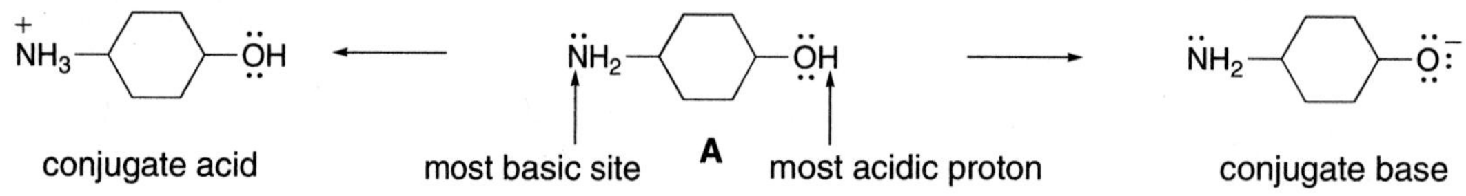

$\overset{+}{N}H_3$—⬡—$\ddot{O}H$ ⟵ $\ddot{N}H_2$—⬡—$\ddot{O}H$ ⟶ $\ddot{N}H_2$—⬡—$\ddot{O}:^-$

conjugate acid   most basic site **A** most acidic proton   conjugate base

2.55 Remove the most acidic proton to form the conjugate base. Protonate the most basic electron pair to form the conjugate acid.

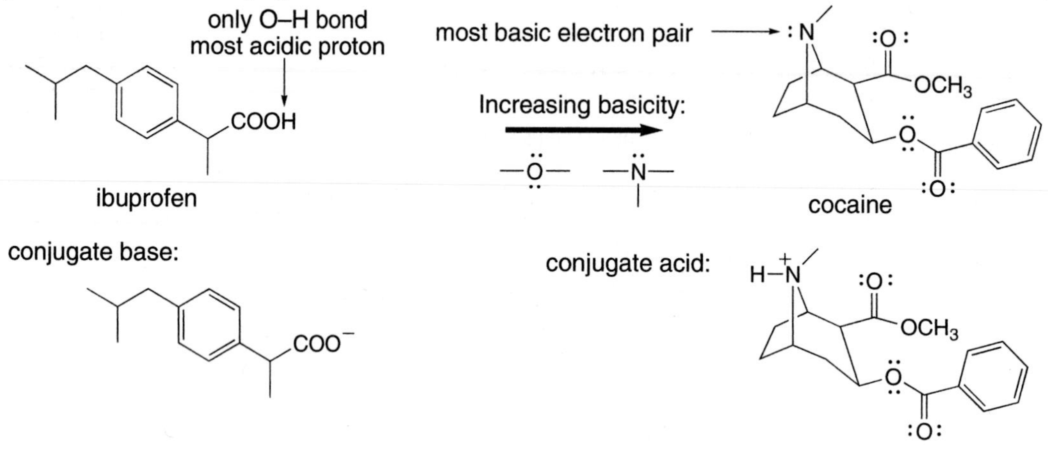

only O–H bond
most acidic proton

most basic electron pair ⟶ :N

Increasing basicity:

—$\ddot{O}$—   —$\ddot{N}$—

ibuprofen   cocaine

conjugate base: conjugate acid:

2.56 **A lower pK_a means a stronger acid.** The pK_a is low for the C–H bond in CH_3NO_2 due to resonance stabilization of the conjugate base.

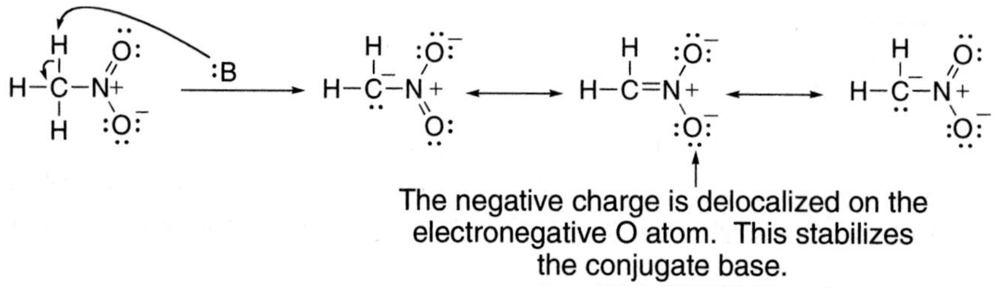

The negative charge is delocalized on the electronegative O atom. This stabilizes the conjugate base.

2.57 Compare the isomers.

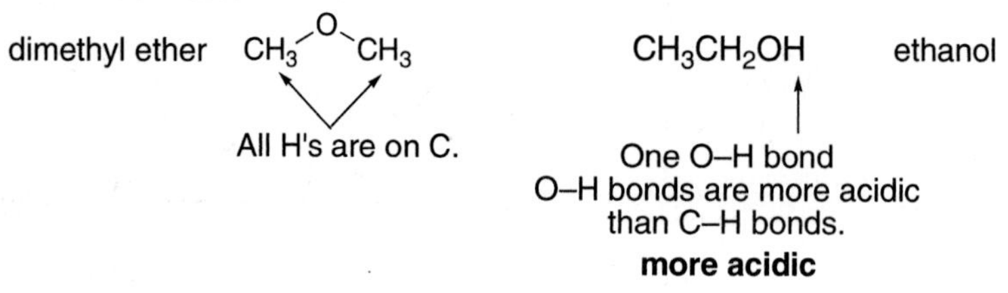

dimethyl ether $CH_3\overset{O}{\diagup}\diagdown CH_3$ CH_3CH_2OH ethanol

All H's are on C.   One O–H bond
O–H bonds are more acidic
than C–H bonds.
more acidic

2.58 Compare the Lewis structures of the conjugate bases when each H is removed. The more stable base makes the proton more acidic.

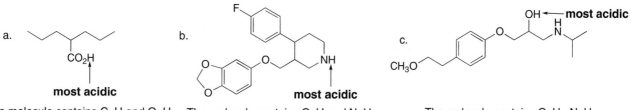

more acidic proton

The negative charge on N is stabilized by resonance. This conjugate base is more stable, so it is formed by removal of the more acidic H.

less acidic proton

no resonance stabilization formed from the weaker acid

2.59 Draw the conjugate base to determine the most acidic hydrogen.

resonance-stabilized conjugate base

Resonance stabilization of the conjugate base makes this the most acidic proton.
In Appendix A, the closest compound for comparison is $CH_3CO_2CH_2CH_3$ with a pK_a of 24.5; therefore, the estimated pK_a of ethyl butanoate is 25.

2.60 Look at the element bonded to the acidic H and decide its acidity based on the periodic trends. **Farther right across a row and down a column of the periodic table is more acidic.**

a.

CO_2H

most acidic

The molecule contains C–H and O–H bonds. O is farther right in the periodic table; therefore, the O–H hydrogen is the most acidic.

b.

most acidic

The molecule contains C–H and N–H bonds. N is farther right in the periodic table; therefore, the N–H hydrogen is the most acidic.

c.

OH ← most acidic

CH_3O

The molecule contains C–H, N–H, and O–H bonds. O is farthest right in the periodic table; therefore, the O–H hydrogen is the most acidic.

2.61 Use element effects, inductive effects, and resonance to determine which protons are the most acidic. The H's of the CH_3 group are least acidic since they are bonded to an sp^3 hybridized C and the conjugate base formed by their removal is not resonance stabilized.

lactic acid

c > b > a
Both O–H protons [(b) and (c)] are more acidic than the C–H proton (a) by the element effect. The most acidic proton has added resonance stabilization when it is removed, making its conjugate base the most stable.

conjugate base by loss of (c):

resonance stabilization
negative charge on O in both resonance structures
This makes (c) most acidic.

conjugate base by loss of (b):

no resonance stabilization, but negative charge on O, an electronegative atom

conjugate base by loss of (a):

This conjugate base has two resonance structures, but one places a negative charge on C.

2.62 *Lewis bases* **are electron pair donors**: they contain a lone pair or a π bond. *Brønsted–Lowry bases* **are proton acceptors**: to accept a proton they need a lone pair or a π bond. This means Lewis bases are also Brønsted–Lowry bases.

a. ← lone pairs on O
    both

b. $CH_3-\ddot{C}l:$ ← lone pairs on Cl
    both

c. **neither** = no lone pairs or π bond

d. π bonds
    both

2.63 A *Lewis acid* **is an electron pair acceptor** and usually contains a proton or an unfilled valence shell of electrons. A *Brønsted–Lowry acid* **is a proton donor** and must contain a hydrogen atom. All Brønsted–Lowry acids are Lewis acids, though the reverse may not be true.

a. H_3O^+

both –
contains a H

b. Cl_3C^+

Lewis acid –
unfilled valence
shell on C

c. BCl_3

Lewis acid –
unfilled valence
shell on B

d. BF_4^-

neither –
no H or unfilled
valence shell

2.64 Label the Lewis acid and Lewis base and then draw the products.

a. :Cl:⁻ + BCl₃ ⟶ Cl–B–Cl
Lewis base   Lewis acid   Cl Cl
                                      new bond

c. CH₃–C(=O)–Cl: + ⁻:OH ⟶ CH₃–C–Cl:
        Lewis acid   **Lewis base**   :OH   new bond

b. (CH₃)(CH₃)C=C(CH₃)(CH₃) + H–OSO₃H ⟶ (CH₃)(CH₃)C⁺–C(CH₃)H + HSO₄⁻
    Lewis base   **Lewis acid**   new bond

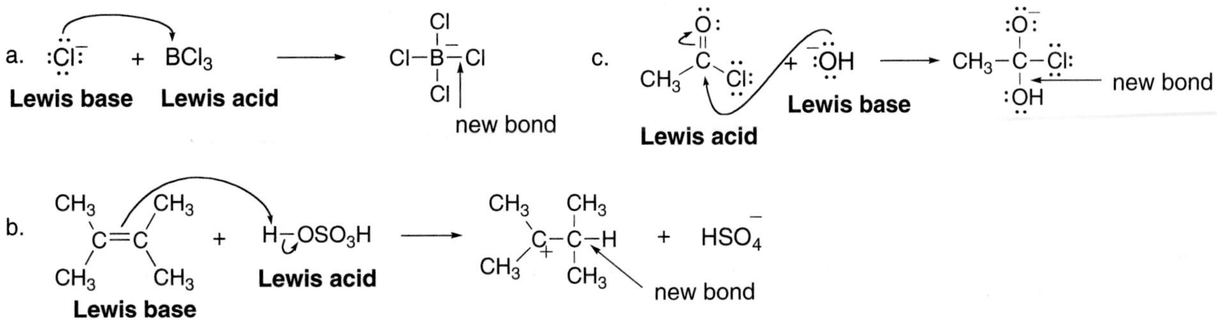

2.65 A Lewis acid is also called an **electrophile**. When a Lewis base reacts with an electrophile other than a proton, it is called a **nucleophile**. Label the electrophile and nucleophile in the starting materials and then draw the products.

a. CH₃CH₂OH + BF₃ ⟶ CH₃CH₂–O⁺–H
    nucleophile   electrophile   BF₃⁻

b. CH₃SCH₃ + AlCl₃ ⟶ CH₃–S⁺–CH₃
    nucleophile   electrophile   AlCl₃⁻

c. (CH₃)₂C=O + BF₃ ⟶ (CH₃)₂C=O⁺–BF₃⁻
    nucleophile   electrophile

d. [cyclohexyl cation] + H₂O ⟶ [cyclohexyl–⁺OH₂]
    electrophile   **nucleophile**

e. :Br–Br: + FeBr₃ ⟶ :Br–Br⁺–Fe–Br (Br, Br⁻)
    nucleophile   electrophile

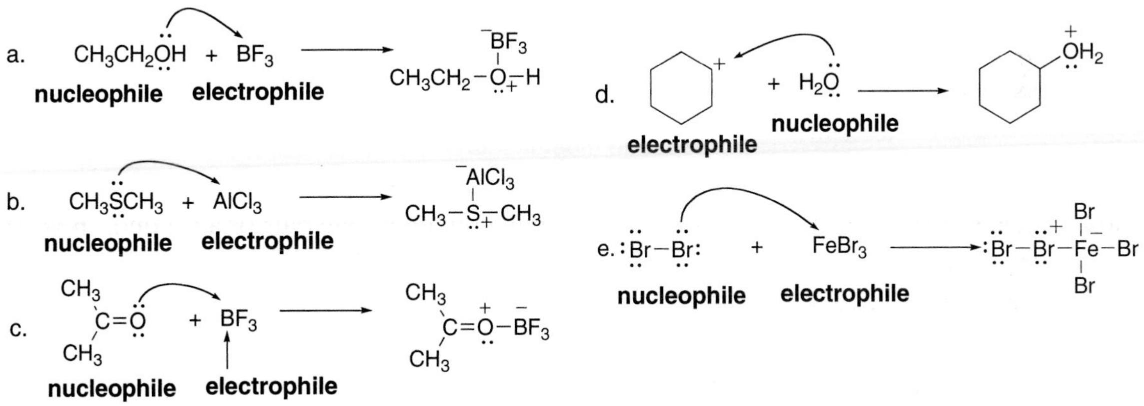

2.66 Draw the product of each reaction.

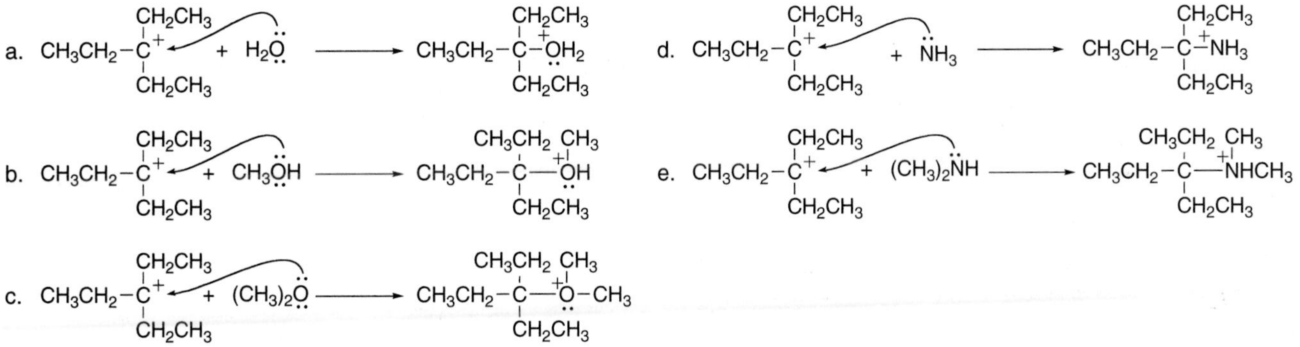

a. CH₃CH₂–C⁺(CH₂CH₃)(CH₂CH₃) + H₂O ⟶ CH₃CH₂–C(CH₂CH₃)(CH₂CH₃)–⁺OH₂

b. CH₃CH₂–C⁺(CH₂CH₃)(CH₂CH₃) + CH₃OH ⟶ CH₃CH₂–C(CH₂CH₃)(CH₃CH₂)–⁺OH

c. CH₃CH₂–C⁺(CH₂CH₃)(CH₂CH₃) + (CH₃)₂O ⟶ CH₃CH₂–C(CH₂CH₃)(CH₃CH₂)–⁺O–CH₃

d. CH₃CH₂–C⁺(CH₂CH₃)(CH₂CH₃) + NH₃ ⟶ CH₃CH₂–C(CH₂CH₃)(CH₂CH₃)–⁺NH₃

e. CH₃CH₂–C⁺(CH₂CH₃)(CH₂CH₃) + (CH₃)₂NH ⟶ CH₃CH₂–C(CH₂CH₃)(CH₃CH₂)–⁺NHCH₃

2.67

nucleophile

a. [(CH₃)₂CH–OH] + H–Br ⟶ **proton transfer** ⟶ [(CH₃)₂CH–⁺OH₂] + :Br:⁻ ⟶ [(CH₃)₂CH–Br] + H₂O:

electrophile

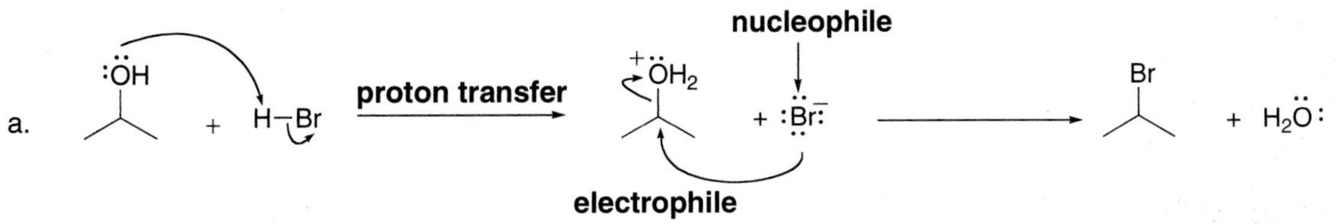

b.

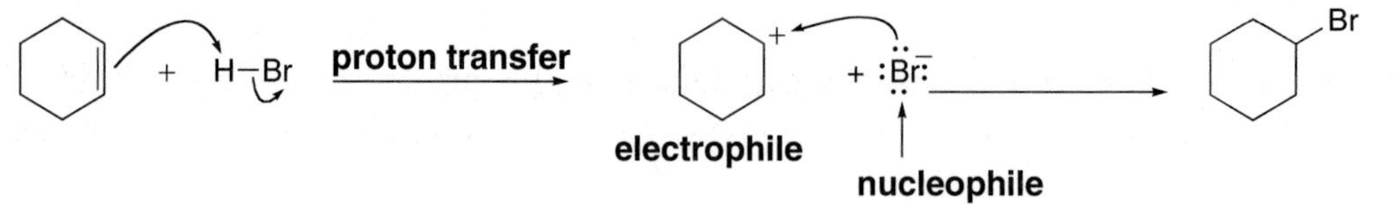

proton transfer

electrophile

nucleophile

c.

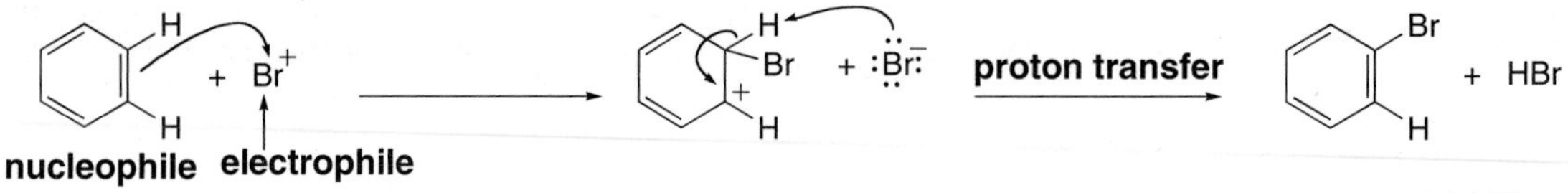

nucleophile   electrophile   **proton transfer**   + HBr

2.68   Draw the products of each reaction. In part (a), ⁻OH pulls off a proton and thus acts as a Brønsted–Lowry base. In part (b), ⁻OH attacks a carbon and thus acts as a Lewis base.

a.

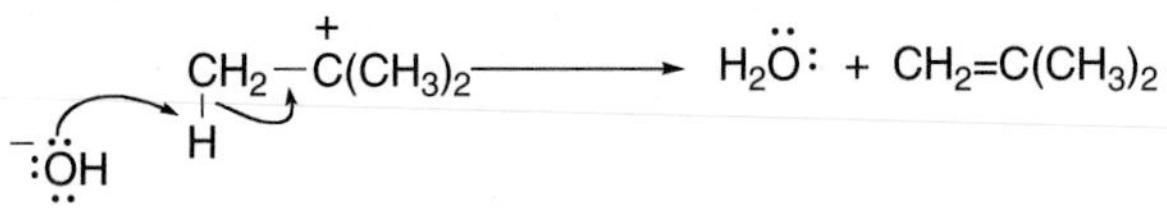

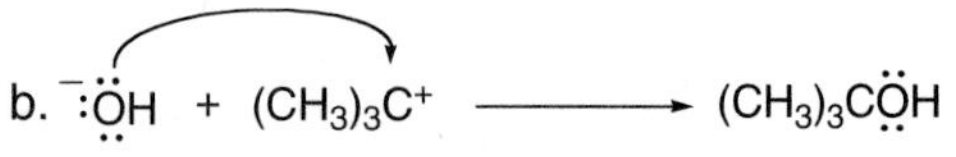

$$CH_2 - \overset{+}{C}(CH_3)_2 \longrightarrow H_2\overset{..}{O}: + CH_2=C(CH_3)_2$$

:ÖH   H

b. :ÖH⁻ + (CH₃)₃C⁺ ⟶ (CH₃)₃CÖH

2.69   Answer each question about esmolol.

a.

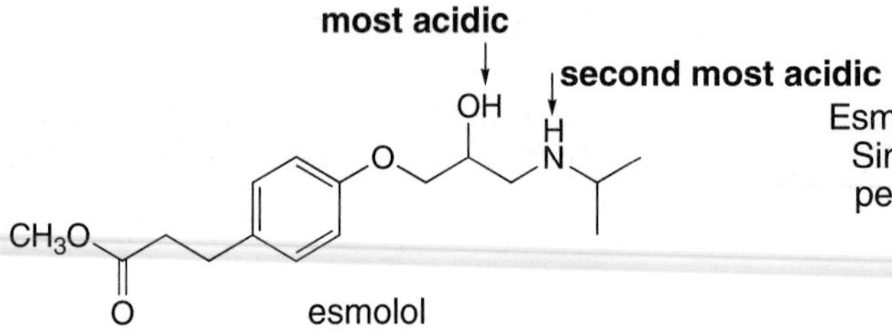

most acidic

second most acidic

esmolol

Esmolol contains C–H, N–H, and O–H bonds. Since acidity increases across a row of the periodic table, the OH bond is most acidic, followed by the NH bond.

b.

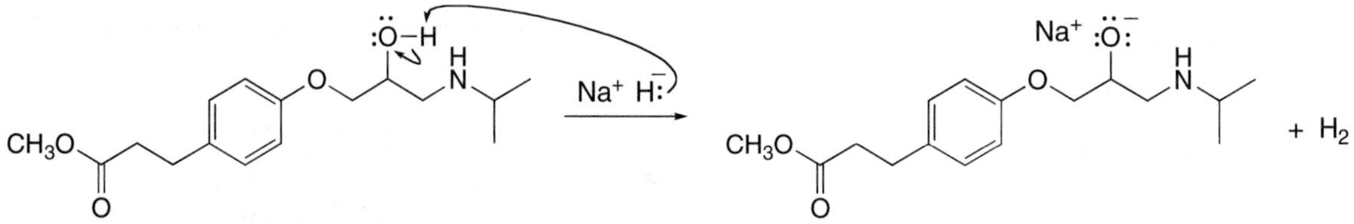

Na⁺ H:⁻   + H₂

c.

H–Cl: +:Cl:⁻

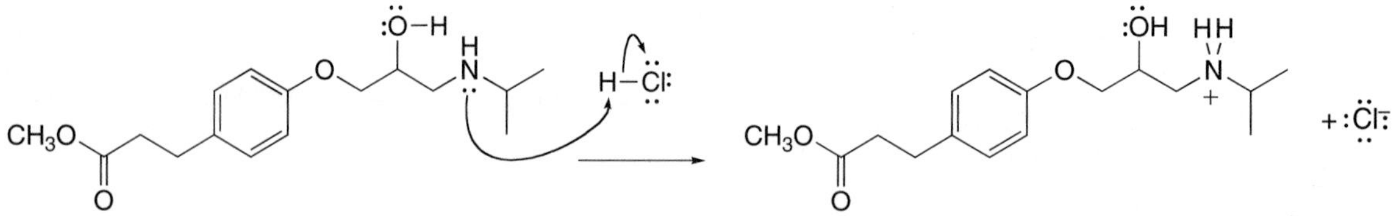

d, e, f.

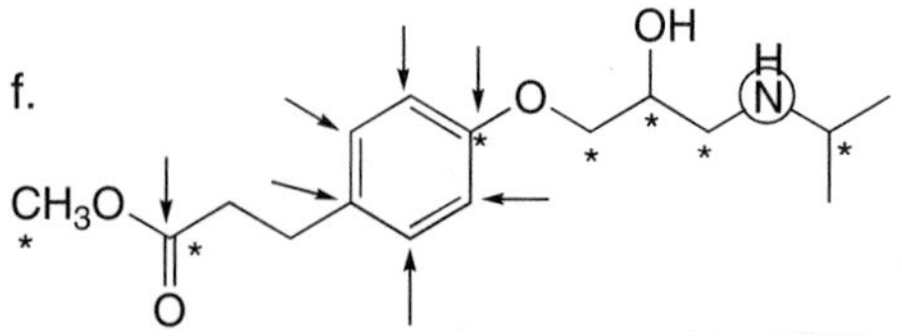

All sp² C are indicated with an arrow.
The N is the only trigonal pyramidal atom.
The δ⁺ C's are indicated with a (*).

2.70 Draw the product of protonation of either O or N and compare the conjugate acids. When acetamide reacts with an acid, the O atom is protonated because it results in a resonance-stabilized conjugate acid.

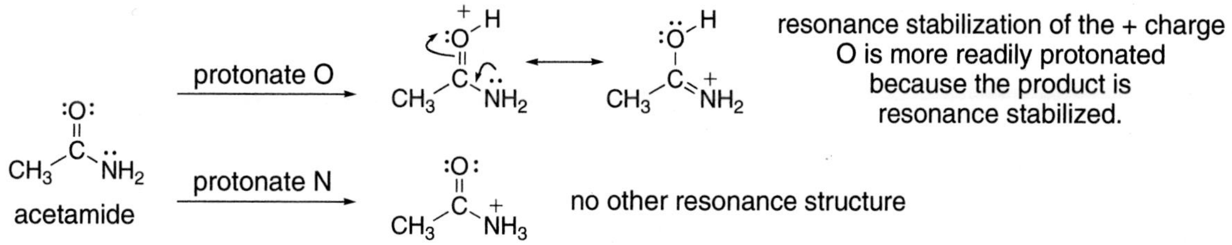
acetamide
protonate O
protonate N
resonance stabilization of the + charge
O is more readily protonated
because the product is
resonance stabilized.
no other resonance structure

2.71

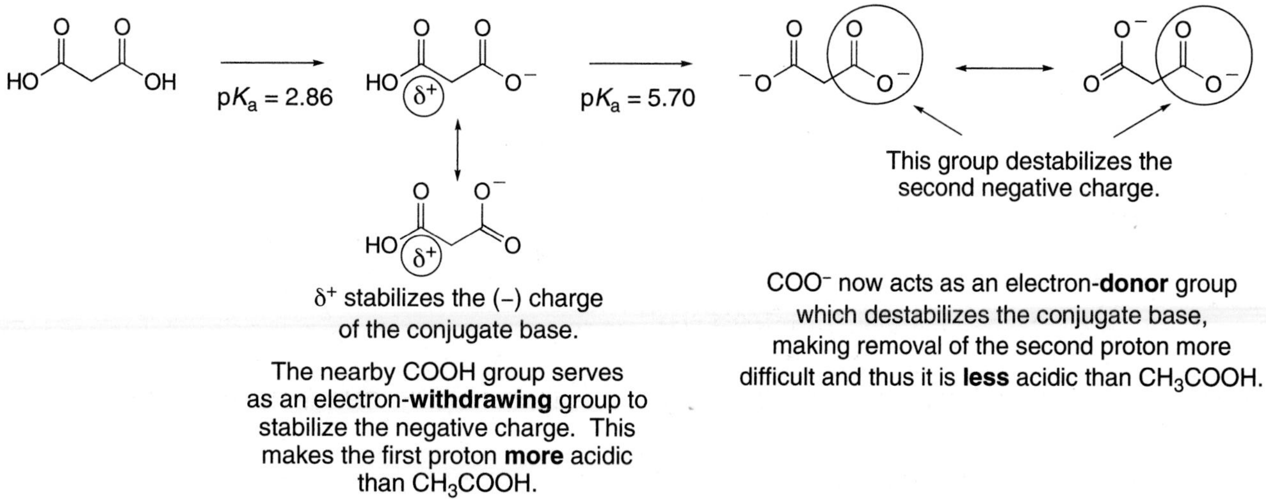
pK_a = 2.86
pK_a = 5.70
This group destabilizes the
second negative charge.
δ+ stabilizes the (–) charge
of the conjugate base.
The nearby COOH group serves
as an electron-**withdrawing** group to
stabilize the negative charge. This
makes the first proton **more** acidic
than CH_3COOH.
COO^- now acts as an electron-**donor** group
which destabilizes the conjugate base,
making removal of the second proton more
difficult and thus it is **less** acidic than CH_3COOH.

2.72 The COOH group of glycine gives up a proton to the basic NH_2 group to form the zwitterion.

a.

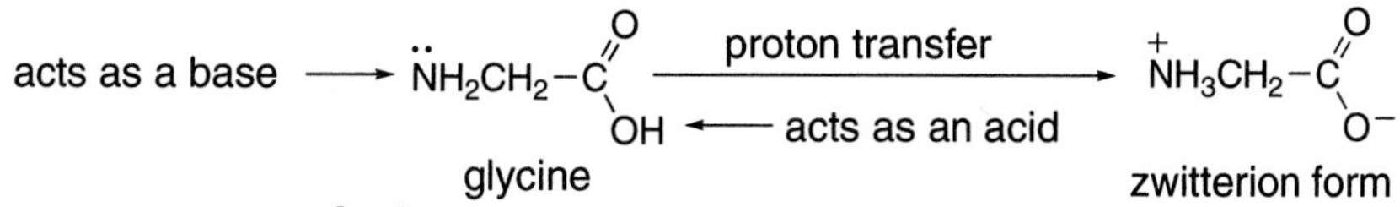
acts as a base
proton transfer
acts as an acid
glycine
zwitterion form

b.

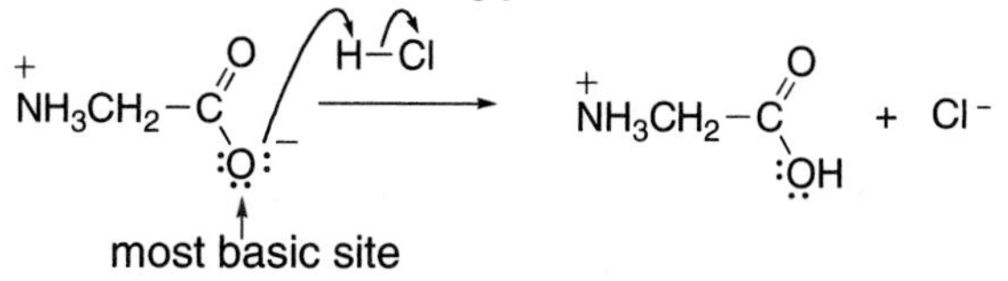
most basic site

c.

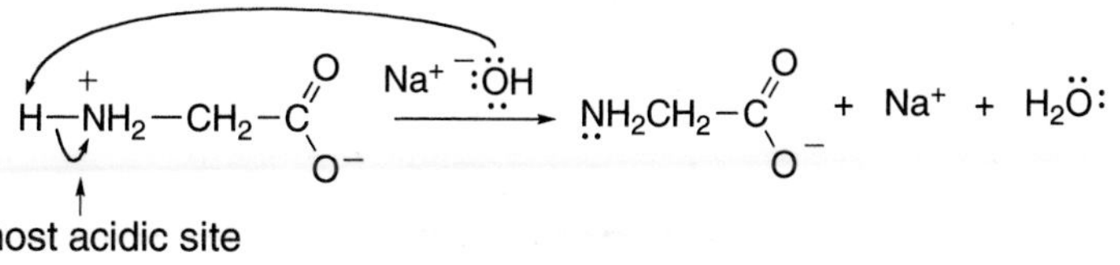
most acidic site

2.73 Use curved arrows to show how the reaction occurs.

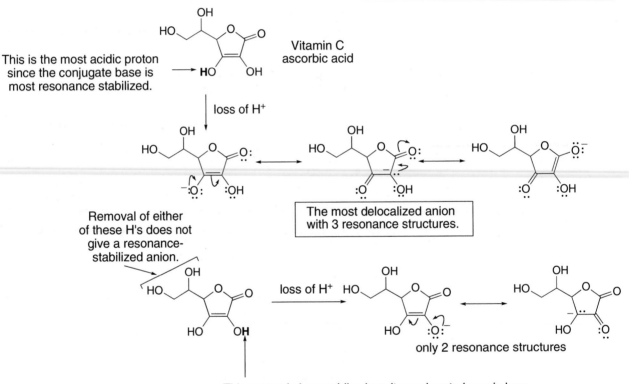

Protonate the negative charge on this carbon to form the product.

2.74 Compare the OH bonds in Vitamin C and decide which one is the most acidic.

Vitamin C
ascorbic acid

This is the most acidic proton since the conjugate base is most resonance stabilized.

loss of H⁺

The most delocalized anion with 3 resonance structures.

Removal of either of these H's does not give a resonance-stabilized anion.

loss of H⁺

only 2 resonance structures

This proton is less acidic since its conjugate base is less resonance stabilized.

Chapter 3: Introduction to Organic Molecules and Functional Groups

♦ Types of intermolecular forces (3.3)

	Type of force	Cause	Examples
Increasing strength ↓	van der Waals (VDW)	Due to the interaction of temporary dipoles • Larger surface area, stronger forces • Larger, more polarizable atoms, stronger forces	All organic compounds
	dipole–dipole (DD)	Due to the interaction of permanent dipoles	$(CH_3)_2C=O$, H_2O
	hydrogen bonding (HB or H-bonding)	Due to the electrostatic interaction of a H atom in an O–H, N–H, or H–F bond with another N, O, or F atom.	H_2O
	ion–ion	Due to the interaction of two ions	NaCl, LiF

♦ Physical properties

Property	Observation
Boiling point (3.4A)	• For compounds of comparable molecular weight, the stronger the forces the higher the bp. $CH_3CH_2CH_2CH_2CH_3$ $CH_3CH_2CH_2CHO$ $CH_3CH_2CH_2CH_2OH$ VDW VDW, DD VDW, DD, HB MW = 72 MW = 72 MW = 74 bp = 36 °C bp = 76 °C bp = 118 °C **Increasing strength of intermolecular forces** **Increasing boiling point** **3.1** For compounds with similar functional groups, the larger the surface area, the higher the bp. $CH_3CH_2CH_2CH_3$ $CH_3CH_2CH_2CH_2CH_3$ bp = 0 °C bp = 36 °C **Increasing surface area** **Increasing boiling point** • For compounds with similar functional groups, the more polarizable the atoms, the higher the bp. CH_3F CH_3I bp = –78 °C bp = 42 °C **Increasing polarizability** **Increasing boiling point**

Melting point (3.4B)	• For compounds of comparable molecular weight, the stronger the forces the higher the mp.

CH$_3$CH$_2$CH$_2$CH$_2$CH$_3$ CH$_3$CH$_2$CH$_2$CHO   CH$_3$CH$_2$CH$_2$CH$_2$OH

VDW   VDW, DD   VDW, DD, HB

MW = 72   MW = 72   MW = 74

mp = –130 °C   mp = –96 °C   mp = –90 °C

→

Increasing strength of intermolecular forces
Increasing melting point

• For compounds with similar functional groups, the more symmetrical the compound, the higher the mp.

CH$_3$CH$_2$CH(CH$_3$)$_2$ (CH$_3$)$_4$C

mp = –160 °C   mp = –17 °C

→

Increasing symmetry
Increasing melting point

Solubility (3.4C)	Types of water-soluble compounds: • Ionic compounds • Organic compounds having ≤ 5 C's, and an O or N atom for hydrogen bonding (for a compound with one functional group). Types of compounds soluble in organic solvents: • Organic compounds regardless of size or functional group. • Examples:

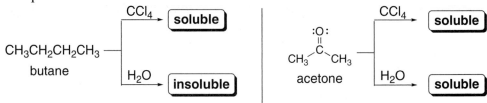

Key: VDW = van der Waals, DD = dipole–dipole, HB = hydrogen bonding
MW = molecular weight

◆ Reactivity (3.8)

- **Nucleophiles react with electrophiles.**
- Electronegative heteroatoms create electrophilic carbon atoms, which tend to react with nucleophiles.
- Lone pairs and π bonds are nucleophilic sites that tend to react with electrophiles.

Chapter 3: Answers to Problems

3.1

3.2 Identify the functional groups based on Tables 3.1, 3.2, and 3.3.

3.3 One possible structure for each functional group:

3.4 One possible structure for each description:

3.5 Summary of forces:
- **All compounds exhibit van der Waals forces (VDW).**
- **Polar molecules have dipole–dipole forces (DD).**
- **Hydrogen bonding (H-bonding) can occur only when a H is bonded to an O, N, or F.**

a.

only nonpolar C–C
and C–H bonds
VDW only

c.　(CH$_3$CH$_2$)$_3$N

- **VDW forces**
- polar C–N bonds - **DD**
- no H on N so
  no H-bonding

e.　CH$_3$CH$_2$CH$_2$COOH

- **VDW forces**
- polar C–O bonds
  and a net dipole - **DD**
- H bonded to O -
  H-bonding

b.

- **VDW forces**
- 2 polar C–O bonds
  and a net dipole - **DD**
- no H on O so
  no H-bonding

d.　CH$_2$=CHCl

- **VDW forces**
- polar C–Cl bond - **DD**

f.　CH$_3$–C≡C–CH$_3$

only nonpolar C–H and
C–C bonds
VDW only

3.6 One principle governs boiling point:
- **Stronger intermolecular forces = higher bp.**
  Increasing intermolecular forces: van der Waals < dipole–dipole < hydrogen bonding

Two factors affect the strength of van der Waals forces, and thus affect bp:
- **Increasing surface area = increasing bp.**
  Longer molecules have a larger surface area. Any branching decreases the surface area of a molecule.
- **Increasing polarizability = increasing bp.**

a.　(CH$_3$)$_2$C=CH$_2$　or　(CH$_3$)$_2$C=O

only VDW　　　　　VDW and DD
　　　　polar, stronger intermolecular forces
　　　　　higher boiling point

c.　CH$_3$(CH$_2$)$_4$CH$_3$ or CH$_3$(CH$_2$)$_5$CH$_3$

longer molecule, more surface area
higher boiling point

b.　CH$_3$CH$_2$COOH or CH$_3$COOCH$_3$

　　　　　　　　　　　no H-bonding
VDW, DD, and H-bonding
stronger intermolecular forces
higher boiling point

d.　CH$_2$=CHCl　or　CH$_2$=CHI

I is more polarizable.
higher boiling point

3.7 Increasing intermolecular forces: van der Waals < dipole–dipole < hydrogen bonding

CH$_3$CH$_2$—C(=O)—NH$_2$

N–H bonds allow for hydrogen bonding.
stronger intermolecular forces
higher boiling point

H—C(=O)—N(CH$_3$)—CH$_3$

no hydrogen bonding
weaker intermolecular forces

3.8

a. [structure: pentane] or [structure with NH₂]   b. [structure] or [structure]

more polar
stronger intermolecular forces
(H-bonding)
higher mp

more spherical
packs better
higher mp

3.9 Compare the intermolecular forces to explain why sodium acetate has a higher melting point than acetic acid.

$$CH_3-C(=O)-OH$$
acetic acid

$$CH_3-C(=O)-O^- \, Na^+$$
sodium acetate

a. VDW, DD, and H-bonding
b. not ionic, lower melting point

a. VDW, DD, ionic bonds
b. Ionic bonds are the strongest: **higher melting point.**

3.10 In the more ordered solid phase, molecules are much closer together than in the less ordered liquid phase. The shape of a molecule determines how closely it can pack in the solid phase so symmetry is important. In the liquid phase, molecules are already farther apart, so symmetry is less important and thus it doesn't affect boiling point.

3.11 A compound is water soluble if it is ionic or if it has an O or N atom and ≤ 5 C's.

a. $CH_3CH_2OCH_2CH_3$

an O atom that
can H-bond with water
≤ 5 C's
water soluble

b. $CH_3CH_2CH_2CH_2CH_3$

nonpolar
not water soluble

c. $(CH_3CH_2CH_2CH_2)_3N$

an N atom that can
H-bond to H_2O, but
> 5 C's
not water soluble

3.12 **Hydrophobic** portions will primarily be hydrocarbon chains. **Hydrophilic** portions will be polar.

Circled regions are **hydrophilic** because they are polar.
All other regions are **hydrophobic** since they have only C and H.

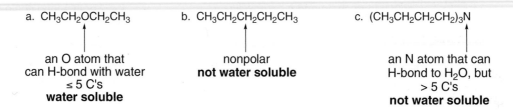

a. norethindrone

b. arachidonic acid

c. benzo[a]pyrene derivative

3.13 Like dissolves like.

- To be **soluble in water**, a molecule must be ionic, or have a polar functional group capable of H-bonding for every 5 C's.
- Organic compounds are generally **soluble in organic solvents** regardless of size or functional group.

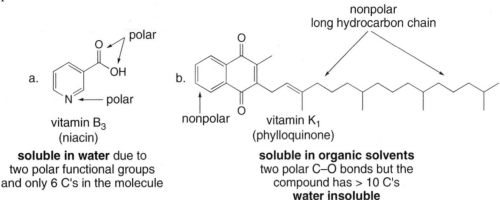

a. vitamin B₃
(niacin)

soluble in water due to
two polar functional groups
and only 6 C's in the molecule

b. vitamin K₁
(phylloquinone)

soluble in organic solvents
two polar C–O bonds but the
compound has > 10 C's
water insoluble

3.14 A soap contains both a long hydrocarbon chain and a carboxylic acid salt.

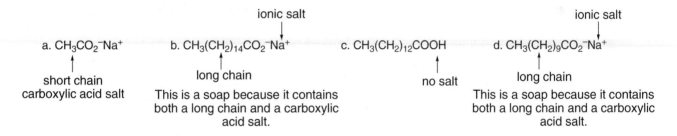

a. $CH_3CO_2^-Na^+$

short chain
carboxylic acid salt

b. $CH_3(CH_2)_{14}CO_2^-Na^+$

ionic salt

long chain
This is a soap because it contains
both a long chain and a carboxylic
acid salt.

c. $CH_3(CH_2)_{12}COOH$

no salt

d. $CH_3(CH_2)_9CO_2^-Na^+$

ionic salt

long chain
This is a soap because it contains
both a long chain and a carboxylic
acid salt.

3.15 Detergents have a polar head consisting of oppositely charged ions, and a nonpolar tail consisting of C–C and C–H bonds, just like soaps do. Detergents clean by having the **hydrophobic ends of molecules surround grease**, while the **hydrophilic portion of the molecule interacts with the polar solvent** (usually water).

a detergent

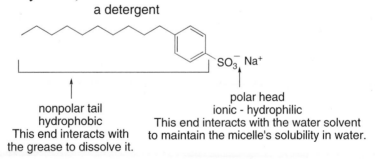

$SO_3^- Na^+$

nonpolar tail
hydrophobic
This end interacts with
the grease to dissolve it.

polar head
ionic - hydrophilic
This end interacts with the water solvent
to maintain the micelle's solubility in water.

3.16

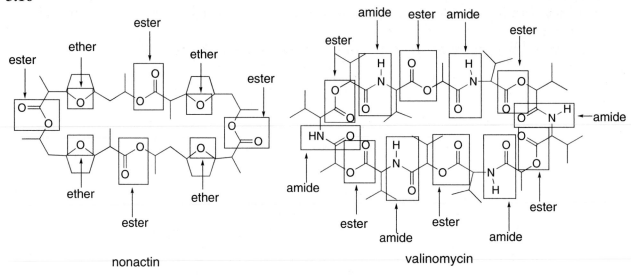

nonactin   valinomycin

3.17 Electronegative heteroatoms like N, O, or X make a carbon atom an *electrophile*.
A lone pair on a heteroatom makes it basic and nucleophilic.
π Bonds create *nucleophilic* sites and are more easily broken than σ bonds.

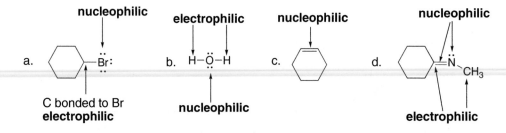

3.18 Electrophiles and nucleophiles react with each other.

a. CH₃CH₂−Br   +   ⁻OH ⟶ **YES**

    electrophile   nucleophile

b. CH₃−C≡C−CH₃ + Br⁻ ⟶ **NO**

       nucleophile   nucleophile

c. CH₃−C(=O)−Cl + ⁻OCH₃ ⟶ **YES**

    electrophile   nucleophile

d. CH₃−C≡C−CH₃ + Br⁺ ⟶ **YES**

       nucleophile   electrophile

3.19 Identify the functional groups based on Tables 3.1, 3.2, and 3.3.

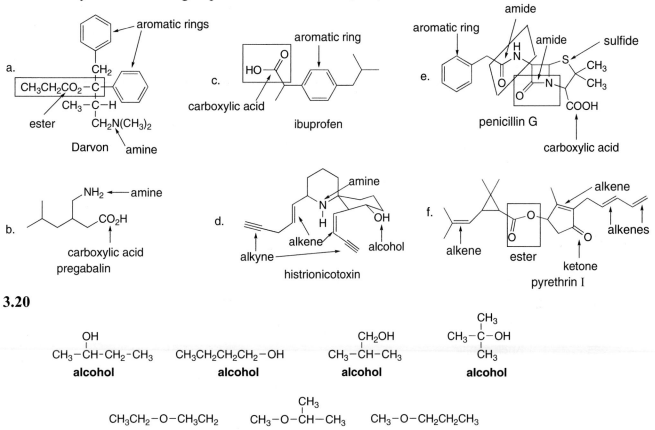

3.20

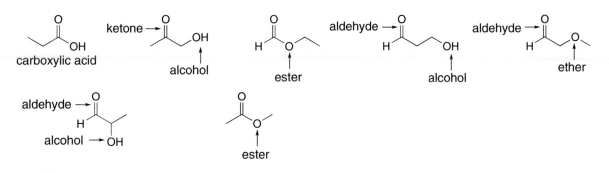

3.21 A cyclic ester is called a lactone. A cyclic amide is called a lactam.

3.22 Draw the constitutional isomers and identify the functional groups.

3.23 Use the rules from Answer 3.5.

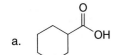

a.

VDW
dipole–dipole
H-bonding (O–H bond)

b.

VDW
dipole–dipole
no H-bonding (no O–H bond)

c.

VDW
dipole–dipole
no H-bonding (no N–H bond)

d.

VDW
no dipole–dipole
(nonpolar C–C, C–H bonds)
no H-bonding (no O, N, F)

3.24 Increasing intermolecular forces: van der Waals < dipole–dipole < H-bonding

a. **increasing intermolecular forces:**

CH_3CH_3 < CH_3Cl < CH_3NH_2

VDW   VDW   VDW
    dipole–dipole   dipole–dipole
             H-bonding

b. **increasing intermolecular forces:**

CH_3Cl < CH_3Br < CH_3I

Increasing polarizability
stronger intermolecular forces

c. **increasing intermolecular forces:**

$(CH_3)_2C=C(CH_3)_2$ < $(CH_3)_2CHCOCH_3$ < $(CH_3)_2CHCOOH$

VDW   VDW   VDW
        dipole–dipole   dipole–dipole
                       H-bonding

d. **increasing intermolecular forces:**

CH_3Cl < CH_3OH < NaCl

VDW   VDW   ionic
dipole–dipole   dipole–dipole
          H-bonding

3.25

CH_3—C ... O- -H—O ... C—CH_3 hydrogen bonding between
       O—H---O   two acetic acid molecules

3.26 **A** = VDW forces; **B** = H-bonding; **C** = ion–ion interactions; **D** = H-bonding; **E** = H-bonding; **F** = VDW forces.

3.27 Use the principles from Answer 3.6.

a. $CH_3(CH_2)_4$—I $CH_3(CH_2)_5$—I $CH_3(CH_2)_6$—I

Increasing size, increasing surface area, increasing boiling point

b. $CH_3CH_2CH_2CH_3$ < $(CH_3)_3N$ < $CH_3CH_2CH_2NH_2$
        VDW   VDW   VDW
             dipole–dipole   dipole–dipole
                       H-bonding

Increasing boiling point

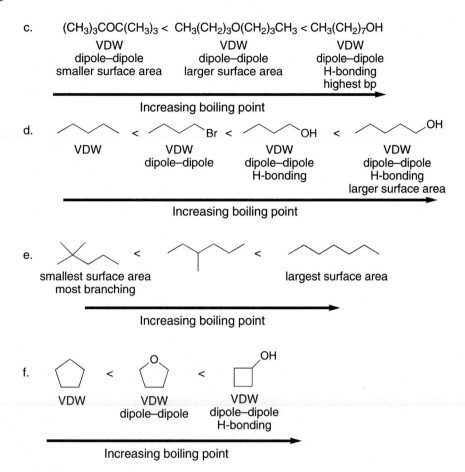

c. $(CH_3)_3COC(CH_3)_3$ < $CH_3(CH_2)_3O(CH_2)_3CH_3$ < $CH_3(CH_2)_7OH$

          VDW   VDW   VDW

  dipole–dipole   dipole–dipole   dipole–dipole

smaller surface area   larger surface area   H-bonding

                                               highest bp

Increasing boiling point

d.

  VDW   VDW   VDW   VDW

           dipole–dipole   dipole–dipole   dipole–dipole

                          H-bonding   H-bonding

                                      larger surface area

Increasing boiling point

e.

smallest surface area   largest surface area

  most branching

Increasing boiling point

f.

  VDW   VDW   VDW

          dipole–dipole   dipole–dipole

                        H-bonding

Increasing boiling point

3.28   In $CH_3CH_2NHCH_3$, there is a N–H bond so the molecules exhibit intermolecular hydrogen bonding, whereas in $(CH_3)_3N$ the N is bonded only to C, so there is no hydrogen bonding. The hydrogen bonding in $CH_3CH_2NHCH_3$ makes it have much **stronger intermolecular forces** than $(CH_3)_3N$. As intermolecular forces increase, the boiling point of a molecule of the same molecular weight increases.

3.29   Stronger forces, higher mp.

            menthone   menthol

              VDW   VDW

       dipole–dipole   dipole–dipole

   lower melting point   H-bonding

                                   stronger forces

                                higher melting point

3.30   Stronger forces, higher mp. More symmetrical compounds, higher mp.

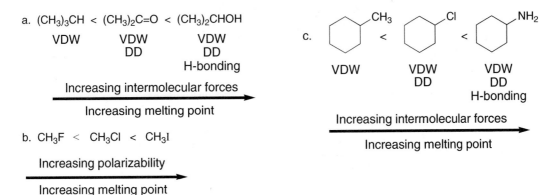

a. $(CH_3)_3CH$ < $(CH_3)_2C=O$ < $(CH_3)_2CHOH$
     VDW   VDW   VDW
                  DD   DD
                              H-bonding

Increasing intermolecular forces →

Increasing melting point

b. CH_3F < CH_3Cl < CH_3I

Increasing polarizability →

Increasing melting point

c.   CH₃   Cl   NH₂
                        <   <

    VDW   VDW   VDW
                 DD   DD
                           H-bonding

Increasing intermolecular forces →

Increasing melting point

3.31

| −119 °C | −118 °C | −91 °C | −25 °C |
| not symmetrical | not symmetrical | symmetrical higher mp | most spherical highest mp |

In both compounds the CH₃ group dangling from the chain makes packing in the solid difficult, so the mp is low.

This molecule can pack somewhat better since it has no CH₃ group dangling from the chain, so the mp is somewhat higher. It also has the most surface area and this increases VDW forces compared to the first two compounds.

This compound packs the best since it is the most spherical in shape, increasing its mp.

3.32   **Boiling point is determined solely by the strength of the intermolecular forces.** Since benzene has a smaller size, it has less surface area and weaker VDW interactions and therefore a lower boiling point than toluene. The increased melting point for benzene can be explained by symmetry: benzene is much more symmetrical than toluene. More symmetrical molecules can pack more tightly together, increasing their melting point. Symmetry has no effect on boiling point.

benzene
bp = 80 °C
mp = 5 °C

very symmetrical
closer packing in solid form
higher mp

and

toluene
bp = 111 °C
mp = −93 °C

less symmetrical
lower mp

3.33 Increasing polarity = increasing water solubility.

Neither compound is very H$_2$O soluble.

a. $\overline{CH_3CH_2CH_2CH_3} \quad < \quad (CH_3)_3CH}$ < CH$_3$OCH$_2$CH$_3$ < CH$_3$CH$_2$CH$_2$OH

VDW	VDW	VDW	VDW
	more spherical	DD	DD
	(This nonpolar, hydrophobic		H-bonding
	molecule is more compact,		
	making it more water soluble than its		
	straight-chain isomer, drawn to the left.)		

b.

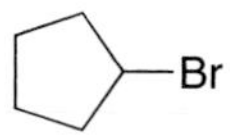

polar	polar	polar and
no H-bonding	H-bonding to H$_2$O,	H-bonding
	not itself	More opportunities
		for H-bonding with its
		O atom and its H on O.

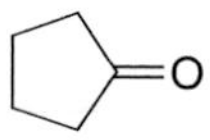

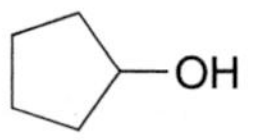

3.34 Look for two things:

- To H-bond to another molecule like itself, the molecule must contain a **H bonded to O, N, or F**.
- To H-bond with water, a molecule need **only contain an O, N, or F**.

Each of these molecules can H-bond to another molecule like itself. Both compounds have N–H bonds.
  b. CH$_3$NH$_2$, e. CH$_3$CH$_2$CH$_2$CONH$_2$

These molecules can H-bond with water. All of these molecules have an O or N atom.
  b. CH$_3$NH$_2$, c. CH$_3$OCH$_3$, d. (CH$_3$CH$_2$)$_3$N, e. CH$_3$CH$_2$CH$_2$CONH$_2$, g. CH$_3$SOCH$_3$, h. CH$_3$CH$_2$COOCH$_3$

3.35 Draw the molecules in question and look at the intermolecular forces involved.

no H bonded to O

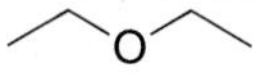

diethyl ether

VDW forces
dipole–dipole forces

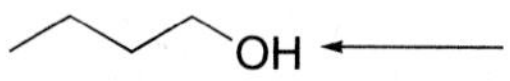

H bonded to O: hydrogen bonding

1-butanol

VDW forces
dipole–dipole forces
H-bonding

- Both have ≤ 5 C's and an electronegative O atom, so they can H-bond to water, making them soluble in water.
- Only 1-butanol can H-bond to another molecule like itself, and this increases its boiling point.

3.36   Use the solubility rule from Answer 3.13.

a.

DDT
no N or O
not water soluble

b.

caffeine
many polar bonds with N and O
atoms
many opportunities for H-bonding
water soluble

c.

mestranol
2 polar functional groups
but > 10 C's
not water soluble

d.

sucrose
many polar bonds with O
11 O's and 12 C's
many opportunities for H-bonding with H_2O
water soluble

e.

aspartame
many polar bonds with N and O atoms
many opportunities for H-bonding
water soluble

f.

carotatoxin
1 polar functional group
but > 10 C's
not water soluble

3.37

$(CH_3)_2CHCH(CH_3)_2$	$CH_3(CH_2)_4CH_3$	$CH_3(CH_2)_5CH_3$	$CH_3(CH_2)_6CH_3$
B	**C**	**D**	**A**
6 C's	6 C's	7 C's	8 C's
Branching makes less surface area, weaker VDW. **lowest bp**	no branching		**highest bp**

C, **D**, and **A** are all long chain hydrocarbons,
but the size increases from **C** to **D** to **A**, increasing
the VDW forces and increasing bp.

3.38   Water solubility is determined by polarity. Polar molecules are soluble in water, while nonpolar
molecules are soluble in organic solvents.

Arrows indicate polar functional groups.

a.

vitamin E
only 2 polar functional groups
many nonpolar C–C and C–H bonds (29 C's)
soluble in organic solvents
insoluble in H_2O

b.

pyridoxine
vitamin B_6
many polar bonds and few nonpolar bonds
soluble in H_2O
It is also **soluble in organic solvents** since it
is organic, but is probably more soluble in H_2O.

3.39 Compare the functional groups in the two components of sunscreen. Dioxybenzone will most likely be washed off in water because it contains two hydroxy groups and is more water soluble.

avobenzone
two ketones
one ether

dioxybenzone
two hydroxy groups
one ketone
one ether
more water soluble

3.40 Because of the O atoms, PEG is capable of hydrogen bonding with water, which makes PEG water soluble and suitable for a product like shampoo. PVC cannot hydrogen bond to water, so PVC is water insoluble, even though it has many polar bonds. Since PVC is water insoluble, it can be used to transport and hold water.

no H-bonding

poly(ethylene glycol)
PEG
water soluble

poly(vinyl chloride)
PVC
water insoluble

3.41 Molecules that dissolve in water are readily excreted from the body in urine whereas less polar molecules that dissolve in organic solvents are soluble in fatty tissue and are retained for longer periods. Compare the solubility properties of THC and ethanol to determine why drug screenings can detect THC and not ethanol weeks after introduction to the body.

tetrahydrocannabinol
THC

THC has relatively few polar
bonds compared to the number
of nonpolar bonds, making it
soluble in organic solvents
and therefore **soluble in fatty tissue.**

CH_3CH_2-OH

ethanol

Ethanol has 1 O atom and
only 2 C's, making it
soluble in water.

Due to their solubilities, **THC is retained much longer in the fatty tissue of the body,** being slowly excreted over many weeks, while ethanol is excreted rapidly in urine after ingestion.

3.42 Compare the intermolecular forces of crack and cocaine hydrochloride. Stronger intermolecular forces increase both the boiling point and the water solubility.

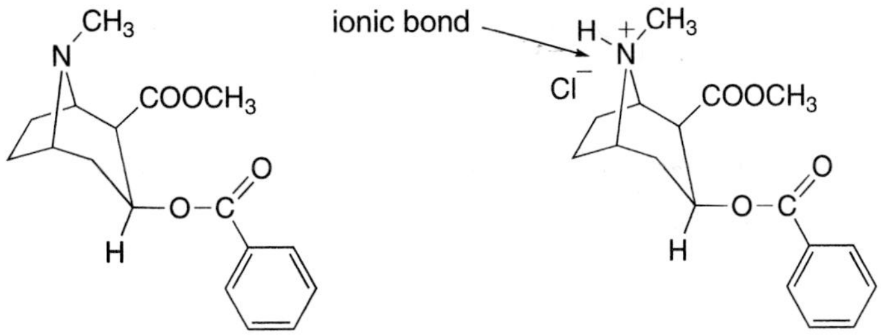

cocaine (crack)
neutral organic molecule

cocaine hydrochloride
a salt

The molecules are identical except for the ionic bond in cocaine hydrochloride. Ionic forces are extremely strong forces, and therefore the cocaine hydrochloride salt has a much **higher boiling point and is more water soluble.** Since the salt is highly water soluble, it can be injected directly into the bloodstream where it dissolves. Crack is smoked because it can dissolve in the organic tissues of the nasal passage and lungs.

3.43 A laundry detergent must have both a highly polar end of the molecule and a nonpolar end of the molecule. The polar end will interact with water, while the nonpolar end surrounds the grease/organic material.

a.

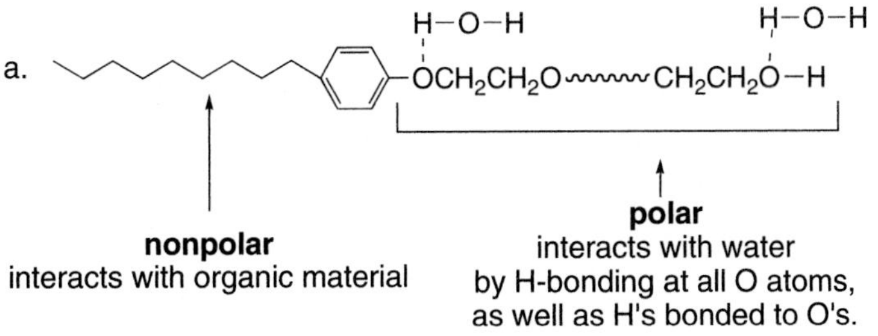

nonpolar
interacts with organic material

polar
interacts with water
by H-bonding at all O atoms,
as well as H's bonded to O's.

b.

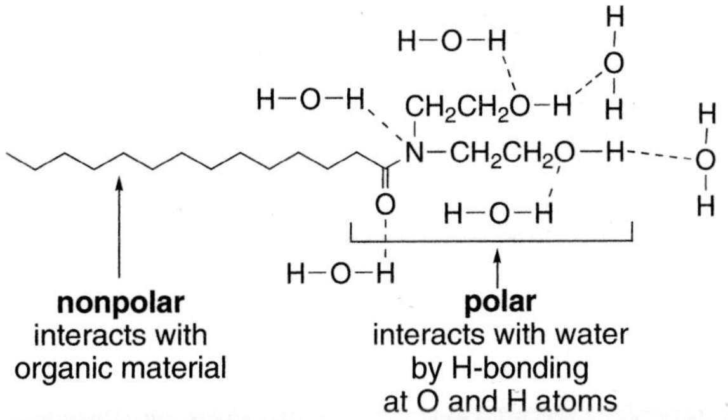

nonpolar
interacts with
organic material

polar
interacts with water
by H-bonding
at O and H atoms

3.44 An emulsifying agent is one that dissolves a compound in a solvent in which it is not normally soluble. In this case the phospholipids can dissolve the oil in its nonpolar tails and bring it into solution in the aqueous vinegar solution. Or, the nonpolar tails dissolve in the oil, and the polar head brings the water-soluble compounds into solution. In any case, the phospholipids make a uniform medium, mayonnaise, from two insoluble layers.

vinegar	oil
aqueous	organic
hydrophilic	hydrophobic

These two ingredients will not mix. The emulsifying agent (egg yolk) has phospholipids that have both hydrophobic and hydrophilic portions, making the mayonnaise uniform.

3.45

a. [structure] → HCl → [structure] + Cl⁻

b. [structure]   [structure]

five functional groups that have
many opportunities for H-bonding
water soluble

ionic salt
more water soluble

c. Since the hydrochloride salt is ionic and therefore more water soluble, it is more readily transported in the bloodstream.

3.46 Use the rules from Answer 3.17.

a. nucleophilic / $\delta^+ \overset{..}{I} : \delta^-$ / electrophilic

b. [cyclohexane =CH₂] nucleophilic

c. nucleophilic / δ^+ δ^+ / electrophilic

d. [styrene structure] All the C=C's are nucleophilic.

e. nucleophilic / δ^+ δ^+ / $CH_3\overset{..}{O}H$ / electrophilic

f. $CH_3 \overset{..}{\underset{\delta^+}{C}} \overset{..}{Cl}:$ with :O: nucleophilic / electrophilic
(All lone pairs on O and Cl are nucleophilic.)

3.47

a.
benzene + Br⁻ ⟶ **NO**
nucleophilic
nucleophilic

b.
cyclohexane–CH₂Cl + ⁻CN ⟶ **YES**
electrophilic   nucleophilic

c.
$$CH_3-\overset{\overset{\textstyle O}{\|}}{C}-CH_3 + {}^-CH_3 \longrightarrow \textbf{YES}$$
electrophilic   nucleophilic

d.
+ ⁻OH ⟶ **NO**
nucleophilic
nucleophilic

e.
+ H₃O⁺ ⟶ **YES**
nucleophilic   electrophilic

3.48 More rigid cell membranes have phospholipids with *fewer* C=C's. Each C=C introduces a bend in the molecule, making the phospholipids pack less tightly. Phospholipids without C=C's can pack very tightly, making the membrane less fluid, and more rigid.

The double bonds introduce kinks in the chain, making packing of the hydrocarbon chains less efficient. This makes the cell membrane formed from them more fluid.

3.49

vancomycin

a. 7 amide groups [regular (unbolded) arrows]
b. OH groups bonded to sp^3 C's are circled. OH groups bonded to sp^2 C's have a square.
c. Despite its size, vancomycin is water soluble because it contains many polar groups and many N and O atoms that can H-bond to H_2O.
d. The most acidic proton is labeled (COOH group).
e. Four functional groups capable of H-bonding are ROH, RCOOH, amides, and amines.

3.50

A

B

The OH and CHO groups are close enough that they can intramolecularly H-bond to each other. Since the two polar functional groups are involved in intramolecular H-bonding, they are less available for H-bonding to H_2O. This makes **A** less H_2O soluble than **B**, whose two functional groups are both available for H-bonding to the H_2O solvent.

The OH and the CHO are too far apart to intramolecularly H-bond to each other, leaving more opportunity to H-bond with solvent.

3.51

a. melting point

fumaric acid

Fumaric acid has its two larger COOH groups on opposite ends of the molecule, and in this way it can pack better in a lattice than maleic acid, giving it a **higher mp**.

b. solubility

maleic acid

Maleic acid is more polar, giving it greater **H_2O solubility.** The bond dipoles in fumaric acid cancel.

c. removal of the first proton (pK_{a1})

loss of 1 proton

loss of 1 proton

In maleic acid, intramolecular H-bonding stabilizes the conjugate base after one H is removed, making maleic acid more acidic than fumaric acid.

Intramolecular H-bonding is not possible here.

d. removal of the second proton (pK_{a2})

Now the dianion is held in close proximity in maleic acid, and this destabilizes the conjugate base. Thus, removing the second H in maleic acid is harder, making it a weaker acid than fumaric acid for removal of the second proton.

The two negative charges are much farther apart. This makes the dianion from fumaric acid more stable and thus pK_{a2} is lower for fumaric acid than maleic acid.

Chapter 4: Alkanes

◆ General facts about alkanes (4.1–4.3)

- Alkanes are composed of **tetrahedral, sp^3** hybridized C's.
- There are two types of alkanes: acyclic alkanes having molecular formula C_nH_{2n+2}, and cycloalkanes having molecular formula C_nH_{2n}.
- Alkanes have only **nonpolar C–C and C–H bonds** and no functional group so they undergo few reactions.
- Alkanes are named with the suffix **-ane.**

◆ Classifying C's and H's (4.1A)

- Carbon atoms are classified by the number of C's bonded to them; **a $1°$ C is bonded to one other C**, and so forth.

- Hydrogen atoms are classified by the type of carbon atom to which they are bonded; **a $1°$ H is bonded to a $1°$ C**, and so forth.

◆ Names of alkyl groups (4.4A)

◆ Conformations in acyclic alkanes (4.9, 4.10)

- Alkane conformations can be classified as **staggered**, **eclipsed**, **anti**, or **gauche** depending on the relative orientation of the groups on adjacent carbons.

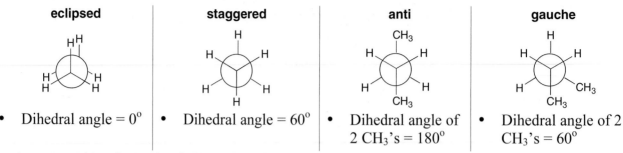

eclipsed	staggered	anti	gauche
• Dihedral angle = 0°	• Dihedral angle = 60°	• Dihedral angle of 2 CH_3's = 180°	• Dihedral angle of 2 CH_3's = 60°

- A staggered conformation is **lower in energy** than an eclipsed conformation.
- An anti conformation is **lower in energy** than a gauche conformation.

◆ Types of strain

- **Torsional strain**—an increase in energy due to eclipsing interactions (4.9).
- **Steric strain**—an increase in energy when atoms are forced too close to each other (4.10).
- **Angle strain**—an increase in energy when tetrahedral bond angles deviate from 109.5° (4.11).

◆ Two types of isomers

[1] **Constitutional isomers**—isomers that differ in the way the atoms are connected to each other (4.1A).

[2] **Stereoisomers**—isomers that differ only in the way atoms are oriented in space (4.13B).

◆ Conformations in cyclohexane (4.12, 4.13)

- Cyclohexane exists as **two chair conformations** in rapid equilibrium at room temperature.
- Each carbon atom on a cyclohexane ring has **one axial** and **one equatorial hydrogen**. Ring-flipping converts axial to equatorial H's, and vice versa.

- In substituted cyclohexanes, groups larger than hydrogen are more stable in the **more roomy equatorial position.**

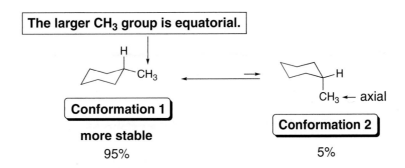

- Disubstituted cyclohexanes with substituents on different atoms exist as two possible stereoisomers.
  - The **cis** isomer has two groups on the **same side** of the ring, either both up or both down.
  - The **trans** isomer has two groups on **opposite sides** of the ring, one up and one down.

trans isomer	**cis** isomer

♦ Oxidation–reduction reactions (4.14)

- **Oxidation** results in an **increase in the number of C–Z bonds** or a **decrease in the number of C–H bonds.**

$$CH_3CH_2-OH \longrightarrow CH_3-\overset{\displaystyle O}{\overset{\|}{C}}-OH$$

ethanol   acetic acid

Increase in C–O bonds = **oxidation**

- **Reduction** results in a **decrease in the number of C–Z bonds** or an **increase in the number of C–H bonds.**

ethylene → ethane

Increase in C–H bonds = **reduction**

Chapter 4: Answers to Problems

4.1 The general molecular formula for an acyclic alkane is C_nH_{2n+2}.

Number of C atoms = n	$2n + 2$	Number of H atoms
23	2(23) + 2 =	48
25	2(25) + 2 =	52
27	2(27) + 2 =	56

4.2 Isopentane has 4 C's in a row with a 1 C branch.

4.3 To classify a carbon atom as 1°, 2°, 3°, or 4° **determine how many carbon atoms it is bonded to** (**1° C** = bonded to **one** other C, **2° C** = bonded to **two** other C's, **3° C** = bonded to **three** other C's, **4° C** = bonded to **four** other C's). Re-draw if necessary to see each carbon clearly.

To classify a hydrogen atom as 1°, 2°, or 3°, **determine if it is bonded to a 1°, 2°, or 3° C (A 1° H is bonded to a 1° C; a 2° H is bonded to a 2° C; a 3° H is bonded to a 3° C).** Re-draw if necessary.

4.4 Use the definition of 1°, 2°, 3°, or 4° carbon atoms from Answer 4.3.

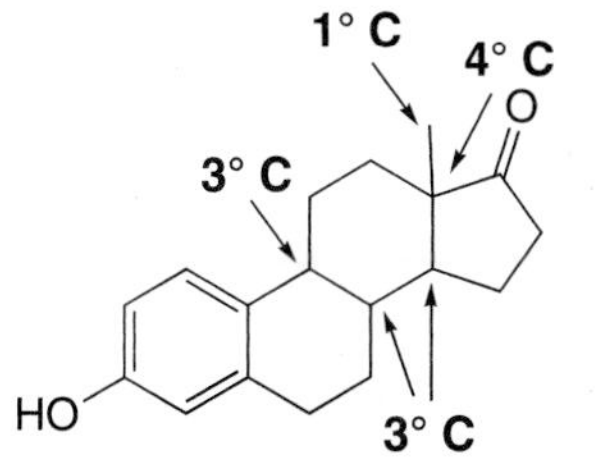
1° C
4° C
O
3° C
HO
3° C

All other tetrahedral C's are **2° C's.**

4.5 Constitutional isomers differ in the way the atoms are connected to each other. To draw all the constitutional isomers:

[1] Draw all of the C's in a long chain.
[2] Take off one C and use it as a substituent. (Don't add it to the end carbon: this re-makes the long chain.)
[3] Take off two C's and use these as substituents, etc.

Five **constitutional isomers** of molecular formula C_6H_{14}:

[1] long chain [2] with one C as a substituent [3] using two C's as substituents

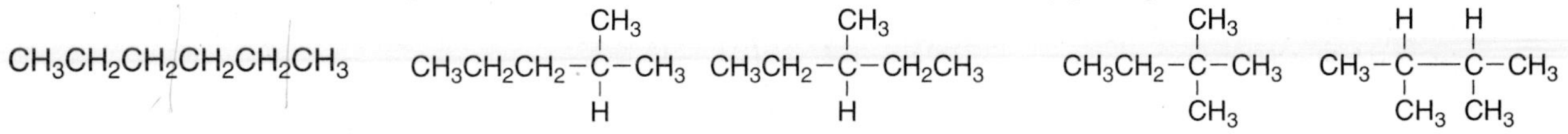
CH3CH2CH2CH2CH2CH3
CH3CH2CH2–C(CH3)(H)–CH3
CH3CH2–C(CH3)(H)–CH2CH3
CH3CH2–C(CH3)(CH3)–CH3
CH3–C(H)(CH3)–C(H)(CH3)–CH3

4.6

Molecular formula C_8H_{18} with one CH_3 substituent:

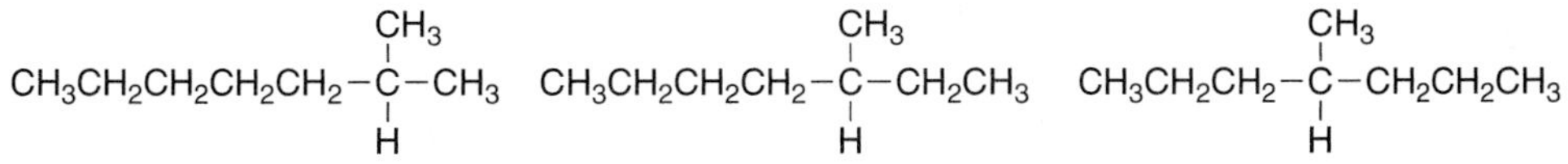
CH3CH2CH2CH2CH2–C(CH3)(H)–CH3
CH3CH2CH2CH2–C(CH3)(H)–CH2CH3
CH3CH2CH2–C(CH3)(H)–CH2CH2CH3

4.7 Draw each alkane to satisfy the requirements.

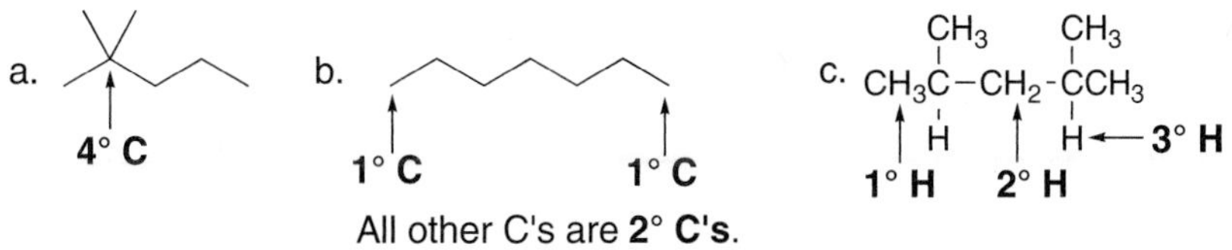
a.
4° C
b.
1° C
1° C
c.
CH3 CH3
CH3C–CH2-CCH3
H H
3° H
1° H
2° H

All other C's are **2° C's**.

4.8 Draw each compound as a skeletal structure to compare the compounds.

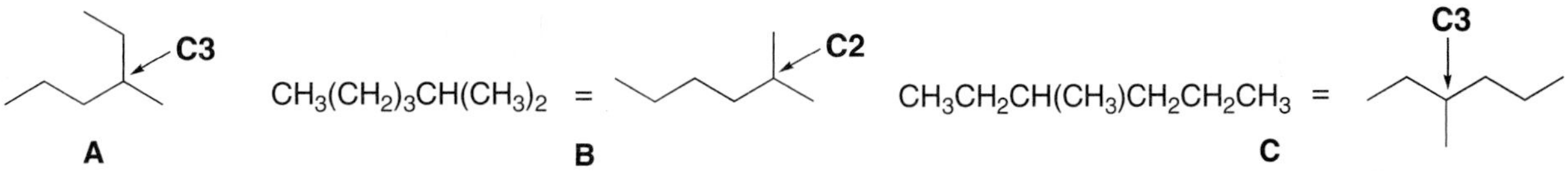
C3
CH3(CH2)3CH(CH3)2 =
C2
CH3CH2CH(CH3)CH2CH2CH3 =
C3

A
CH_3 bonded to C3
identical to compound **C**

B
CH_3 bonded to C2

C
CH_3 bonded to C3
identical to compound **A**

4.9 Use the steps from Answer 4.5 to draw the constitutional isomers.

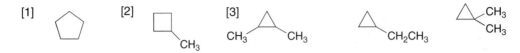

Five **constitutional isomers** of molecular formula C_5H_{10} having one ring:

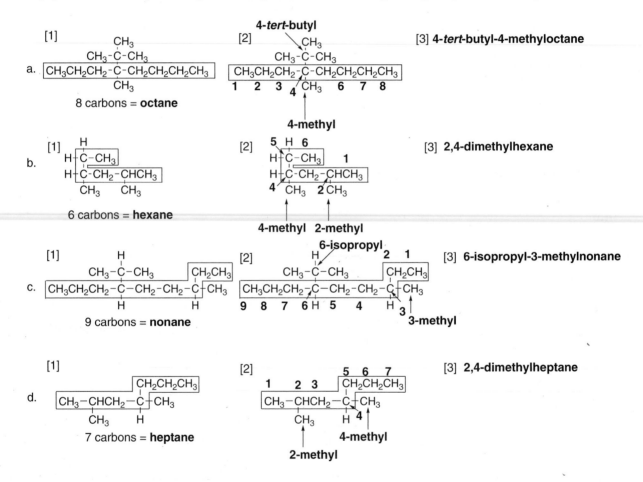

[1]   [2]   [3]

4.10 Follow these steps to name an alkane:
[1] **Name the parent chain** by finding the longest C chain.
[2] **Number the chain** so that the first substituent gets the lower number. Then **name and number all substituents**, giving like substituents a prefix (di, tri, etc.).
[3] **Combine all parts**, alphabetizing the substituents, ignoring all prefixes except *iso*.

4.11 Use the steps in Answer 4.10 to name each alkane.

a. $CH_3CH_2CH(CH_3)CH_2CH_3$

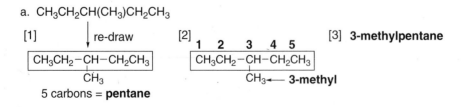

b. (CH₃)₃CCH₂CH(CH₂CH₃)₂

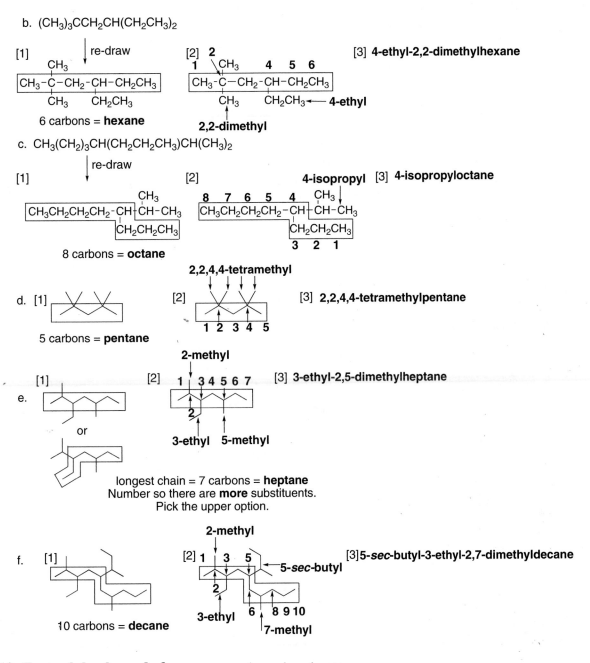

[1]

CH₃

CH₃–C–CH₂–CH–CH₂CH₃

CH₃   CH₂CH₃

6 carbons = **hexane**

[2] **2**

1 ＼CH₃   **4 5 6**

CH₃-C–CH₂-CH–CH₂CH₃

CH₃   CH₂CH₃ ← **4-ethyl**

2,2-dimethyl

[3] **4-ethyl-2,2-dimethylhexane**

c. CH₃(CH₂)₃CH(CH₂CH₂CH₃)CH(CH₃)₂

re-draw

[1]

CH₃

CH₃CH₂CH₂CH₂-CH┼CH–CH₃

CH₂CH₂CH₃

8 carbons = **octane**

[2]

4-isopropyl

8 7 6 5 4 CH₃

CH₃CH₂CH₂CH₂-CH┼CH–CH₃

CH₂CH₂CH₃

3 2 1

2,2,4,4-tetramethyl

[3] **4-isopropyloctane**

d. [1]

5 carbons = **pentane**

[2]

2,2,4,4-tetramethyl

1 2 3 4 5

[3] **2,2,4,4-tetramethylpentane**

e. [1]

or

[2]

2-methyl

1 │ 3 4 5 6 7

2

3-ethyl 5-methyl

[3] **3-ethyl-2,5-dimethylheptane**

longest chain = 7 carbons = **heptane**
Number so there are **more** substituents.
Pick the upper option.

f. [1]

10 carbons = **decane**

[2]

2-methyl

1 3 5

5-sec-butyl

2

3-ethyl **6** **8 9 10**

7-methyl

[3] **5-sec-butyl-3-ethyl-2,7-dimethyldecane**

4.12 To work backwards from a name to a structure:

[1] Find the parent name and draw that number of C's. Use the suffix to identify the functional
group (**-ane = alkane**).

[2] Arbitrarily number the C's in the chain. Add the substituents to the appropriate C's.

[3] Re-draw with H's to make C's have four bonds.

a. 3-methyl**hexane**

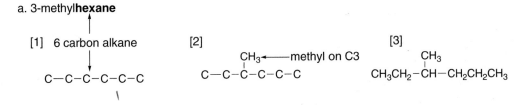

[1] 6 carbon alkane

C—C—C—C—C—C

[2]

CH₃ ←——methyl on C3

C—C—C—C—C—C

[3]

CH₃

CH₃CH₂–CH–CH₂CH₂CH₃

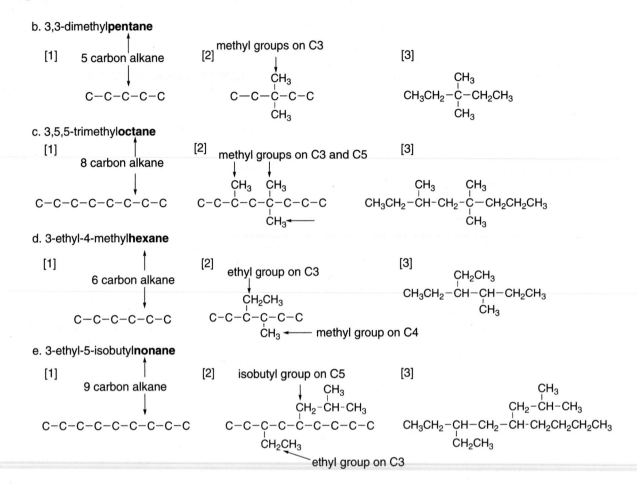

b. 3,3-dimethyl**pentane**

[1] 5 carbon alkane

C—C—C—C—C

[2] methyl groups on C3

C—C—C—C—C with CH₃ above and CH₃ below

[3] CH₃CH₂—C—CH₂CH₃ with CH₃ above and CH₃ below

c. 3,5,5-trimethyl**octane**

[1] 8 carbon alkane

C—C—C—C—C—C—C—C

[2] methyl groups on C3 and C5

C—C—C—C—C—C—C—C

[3] CH₃CH₂—CH—CH₂—C—CH₂CH₂CH₃

d. 3-ethyl-4-methyl**hexane**

[1] 6 carbon alkane

C—C—C—C—C—C

[2] ethyl group on C3

C—C—C—C—C with CH₂CH₃ above and CH₃ below — methyl group on C4

[3] CH₃CH₂—CH—CH—CH₂CH₃ with CH₂CH₃ above and CH₃ below

e. 3-ethyl-5-isobutyl**nonane**

[1] 9 carbon alkane

C—C—C—C—C—C—C—C—C

[2] isobutyl group on C5

C—C—C—C—C—C—C—C—C with CH₂—CH—CH₃ (CH₃) above and CH₂CH₃ below — ethyl group on C3

[3] CH₃CH₂—CH—CH₂—CH-CH₂CH₂CH₂CH₃ with CH₂—CH—CH₃ (CH₃) above and CH₂CH₃ below

4.13 Use the steps in Answer 4.10 to name each alkane.

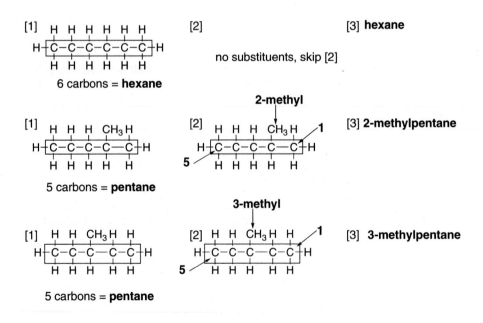

[1] 6 carbons = **hexane**

[2] no substituents, skip [2]

[3] **hexane**

[1] 5 carbons = **pentane**

[2] 2-methyl

[3] **2-methylpentane**

[1] 5 carbons = **pentane**

[2] 3-methyl

[3] **3-methylpentane**

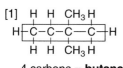

2,2-dimethyl

[1] H H CH₃ H   [2] H H CH₃ H   1   [3] 2,2-dimethylbutane

H—C—C—C—C—H   H—C—C—C—C—H

   H H CH₃ H   4   H H CH₃ H

4 carbons = **butane**

[1] H H H H   [2] H H H H   1   [3] 2,3-dimethylbutane

H—C—C—C—C—H   H—C—C—C—C—H

   H CH₃ CH₃ H   4   H CH₃ CH₃ H

4 carbons = **butane**   2,3-dimethyl

4.14 Follow these steps to name a cycloalkane:

[1] **Name the parent cycloalkane** by counting the C's in the ring and adding cyclo-.

[2] **Numbering:**

   [2a] **Number around the ring** beginning at a substituent and giving the second
        substituent the lower number.

   [2b] **Number to assign the lower number to the substituents alphabetically.**

   [2c] **Name and number all substituents**, giving like substituents a prefix (di, tri, etc.).

[3] **Combine all parts**, alphabetizing the substituents, ignoring all prefixes except *iso*.

   (Remember: If a carbon chain has more C's than the ring, the chain is the parent, and
   the ring is a substituent.)

a.

[1]

6 carbons in ring =
cyclohexane

[2]

1,1-dimethyl

Number so the
substituents are at C1.

[3] 1,1-dimethylcyclohexane

b.

[1]

5 carbons in ring =
cyclopentane

[2]

1,2,3-trimethyl

Number so the first substituent
is at C1, second at C2.

[3] 1,2,3-trimethylcyclopentane

c.

[1]

6 carbons in ring =
cyclohexane

[2]

4-methyl

1-butyl

Number so the earlier alphabetical
substituent is at C1, **b**utyl before **m**ethyl.

[3] 1-butyl-4-methylcyclohexane

d.

[1]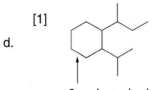

6 carbons in ring =
cyclohexane

[2]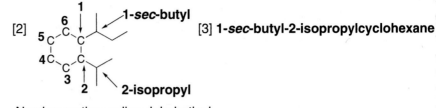

Number so the earlier alphabetical
substituent is at C1, **b**utyl before **i**sopropyl.

[3] **1-*sec*-butyl-2-isopropylcyclohexane**

e.

[1]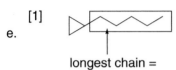

longest chain =
5 carbons =
pentane

[2]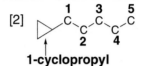

1-cyclopropyl

Number so the
cyclopropyl is at C1.

[3] **1-cyclopropylpentane**

f.

[1]

6 carbons in ring =
cyclohexane

[2]

3-butyl

5 6
4 1
3 2

Number so the two
methyls are at C1.

1,1-dimethyl

[3] **3-butyl-1,1-dimethylcyclohexane**

4.15 To draw the structures, use the steps in Answer 4.12.

a. 1,2-dimethyl**cyclobutane**

[1] 4 carbon cycloalkane

C–C
| |
C–C

[2] methyl groups
on C1 and C2

CH₃ **1**
C–C **4**
| |
C–C **3**
CH₃ **2**

[3]
CH₃

CH₃

b. 1,1,2-trimethyl**cyclopropane**

[1] 3 carbon cycloalkane

C
/ \
C–C

[2]
CH₃
2C
/ \
C–C–CH₃
3 **1** CH₃

3 CH₃'s

[3]
CH₃
/\
/ \––CH₃
CH₃

c. 4-ethyl-1,2-dimethyl**cyclohexane**

[1] 6 carbon cycloalkane

C
C   C
C   C
C

[2]
CH₃CH₂ **4** C **3** C **2** CH₃
\   /
C   C
5 C–C **1** CH₃
6

ethyl
on C4

2 CH₃'s

[3]
CH₃CH₂   CH₃

CH₃

d. 1-*sec*-butyl-3-isopropyl**cyclopentane**

[1] 5 carbon cycloalkane

C
C   C
C–C

[2]
CH₃   CH₃
CH
3C
4C   C **2**
5 C–C **1**
CH–CH₃
CH₃–CH₂

isopropyl

sec-butyl

[3]

e. 1,1,2,3,4-pentamethyl**cycloheptane**

[1] 7 carbon cycloalkane   [2] 5 CH₃'s   [3]

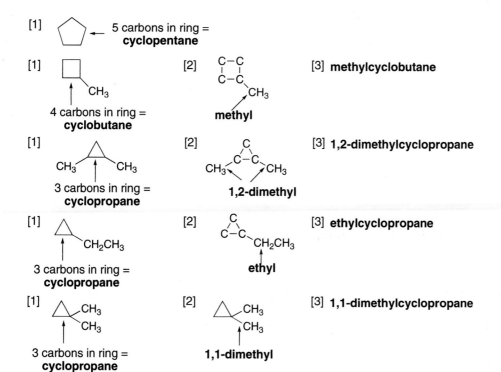

4.16 To name the cycloalkanes, use the steps from Answer 4.14.

[1] 5 carbons in ring = **cyclopentane**

[1] 4 carbons in ring = **cyclobutane**   [2] **methyl**   [3] **methylcyclobutane**

[1] 3 carbons in ring = **cyclopropane**   [2] **1,2-dimethyl**   [3] **1,2-dimethylcyclopropane**

[1] 3 carbons in ring = **cyclopropane**   [2] **ethyl**   [3] **ethylcyclopropane**

[1] 3 carbons in ring = **cyclopropane**   [2] **1,1-dimethyl**   [3] **1,1-dimethylcyclopropane**

4.17 **Compare the number of C's and surface area to determine relative boiling points.** Rules:
[1] Increasing number of C's = increasing boiling point.
[2] Increasing surface area = increasing boiling point (branching decreases surface area).

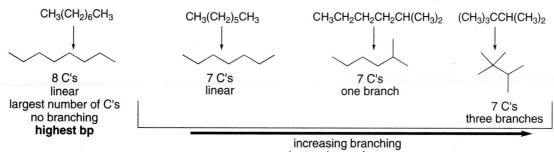

$CH_3(CH_2)_6CH_3$ $CH_3(CH_2)_5CH_3$ $CH_3CH_2CH_2CH_2CH(CH_3)_2$ $(CH_3)_3CCH(CH_3)_2$

8 C's
linear
largest number of C's
no branching
highest bp

7 C's
linear

7 C's
one branch

7 C's
three branches

increasing branching
decreasing surface area
decreasing bp

Increasing boiling point: $(CH_3)_3CCH(CH_3)_2 < CH_3CH_2CH_2CH_2CH(CH_3)_2 < CH_3(CH_2)_5CH_3 < CH_3(CH_2)_6CH_3$

4.18 To draw a Newman projection, visualize the carbons as one in front and one in back of each other. The C–C bond is not drawn. There is only one staggered and one eclipsed conformation.

4.19 Staggered conformations are more stable than eclipsed conformations.

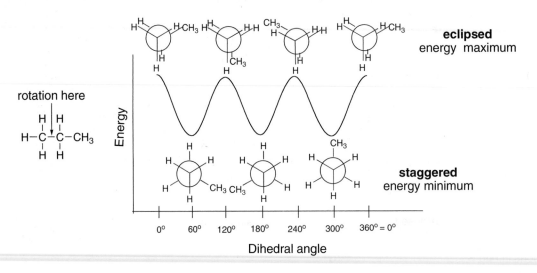

4.20

H,H eclipsing
4.0 kJ/mol of destabilization

4.0 kJ/mol

4.0 kJ/mol

To calculate H,CH₃ destabilization:

14 kJ/mol (total) –
8.0 kJ/mol for 2 H,H eclipsing interactions
= **6 kJ/mol** for one H,CH₃ eclipsing interaction

4.21 To determine the energy of conformations keep two things in mind:
[1] Staggered conformations are more stable than eclipsed conformations.
[2] Minimize steric interactions: keep large groups away from each other.
The highest energy conformation is the eclipsed conformation in which the two largest groups are eclipsed. The lowest energy conformation is the staggered conformation in which the two largest groups are anti.

4.22 To determine the most and least stable conformations, use the rules from Answer 4.21.

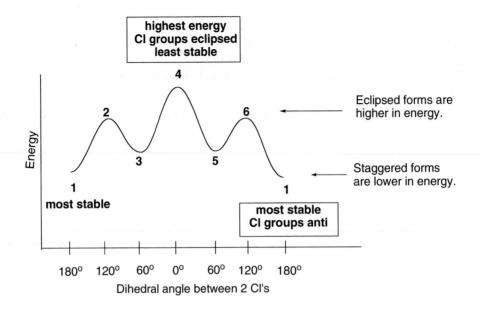

4.23 Add the energy increase for each eclipsing interaction to determine the destabilization.

a.

b.

1 H,H interaction =	4.0 kJ/mol
2 H,CH₃ interactions	
(2 x 6.0 kJ/mol) =	12.0 kJ/mol
Total destabilization =	**16 kJ/mol**

3 H,CH₃ interactions
(3 x 6.0 kJ/mol) = **18 kJ/mol**

Total destabilization

4.24 Two points:
- Axial bonds point up or down, while equatorial bonds point out.
- An *up* carbon has an axial *up* bond, and a *down* carbon has an axial *down* bond.

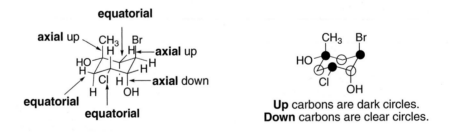

Up carbons are dark circles.
Down carbons are clear circles.

4.25 Draw the second chair conformation by flipping the ring.
- **The *up* carbons become *down* carbons, and the axial bonds become equatorial bonds.**
- **Axial bonds become equatorial, but *up* bonds stay *up*;** i.e., an axial *up* bond becomes an equatorial *up* bond.
- **The conformation with larger groups equatorial is the more stable** conformation and is present in higher concentration at equilibrium.

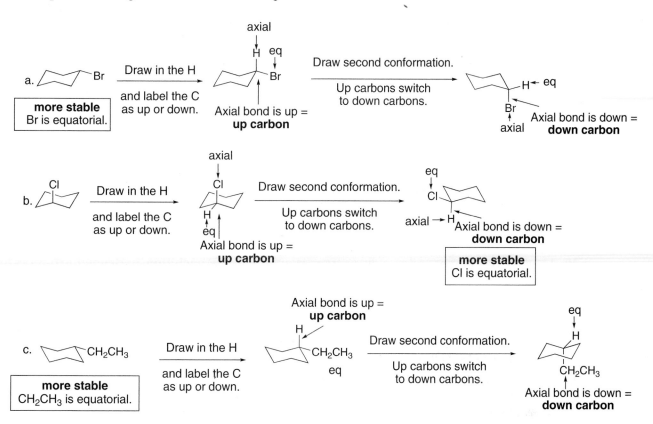

4.26 Larger axial substituents create unfavorable diaxial interactions, whereas equatorial groups have more room and are favored.

H H
  | |
—C—C—H
  | |
  H H

larger substituent
more important to be equatorial

equatorial CH_2CH_3

The H's and CH_3 of the sp^3 hybridized C have severe 1,3-diaxial interactions with the two other axial H's.

—C≡C—H

more compact substituent
less important to be equatorial

equatorial C≡CH

The sp hybridized C's are linear and point down. The 1,3-diaxial interactions with the two other axial H's are less severe.

The axial conformation containing the C≡CH group is not as unstable as the axial conformation containing the CH_2CH_3, so it is present in higher concentration at equilibrium.

4.27 Wedges represent "up" groups in front of the page, and dashes are "down" groups in back of the page. Cis groups are on the same side of the ring, and trans groups are on opposite sides of the ring.

a. **cis-1,2-dimethylcyclopropane**

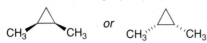

or

cis = **same side of the ring**
both groups on wedges or
both on dashes

b. **trans-1-ethyl-2-methylcyclopentane**

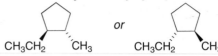

or

trans = **opposite sides of the ring**
one group on a wedge,
one group on a dash

4.28 Cis and trans isomers are stereoisomers.

cis-1,3-diethylcyclobutane

a. **trans-1,3-diethylcyclobutane**

b. **cis-1,2-diethylcyclobutane**

cis = **same side of the ring**
both groups on wedges or
both on dashes

trans = **opposite sides of the ring**
one group on a wedge,
one group on a dash

constitutional isomer
different arrangement of atoms

4.29 To classify a compound as a cis or trans isomer, **classify each non-hydrogen group as up or down. Groups on the same side = cis isomer, groups on opposite sides = trans isomer.**

a.

down bond (up)
(equatorial) H (up)
H

⇌

down bond
(equatorial)

both groups down =
cis isomer

b.

down bond
(up) (equatorial)
H

⇌

up bond
(down) (equatorial)

one group up, one down =
trans isomer

c.

up bond
(axial) (up)
Br H

H
(down)

⇌

down bond
(equatorial)

one group up, one down =
trans isomer

4.30

a.

groups on same side
cis isomer

groups on opposite sides
trans isomer
(one possibility)

c.

trans: CH₃

⇌

both groups equatorial
more stable

two chair conformations for the **trans isomer**

b.

cis:

⇌

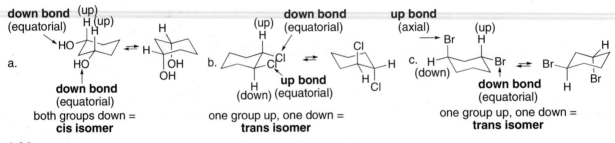

two chair conformations for the **cis isomer**

Same stability since they both have
one equatorial, one axial CH₃ group.

d. The **trans isomer is more stable** because
it can have both methyl groups in the more roomy
equatorial position.

4.31

a.

CH$_2$CH$_3$ ←——axial CH$_2$CH$_3$

CH$_3$

1,1-disubstituted

c.

up ——→ CH$_3$CH$_2$ H

H   CH$_3$ ←—— down (equatorial)

trans-1,3-disubstituted

b.

CH$_3$ ←——up (axial)

H

CH$_2$CH$_3$ ——up

H

cis-1,2-disubstituted

d.

H

up ——→ CH$_3$CH$_2$ CH$_3$ ←—— down
(equatorial)

H

trans-1,4-disubstituted

4.32 *Oxidation* results in an *increase* in the number of C–Z bonds, or a *decrease* in the number of C–H bonds.

Reduction results in a *decrease* in the number of C–Z bonds, or an *increase* in the number of C–H bonds.

a.

$$CH_3-\overset{\overset{\displaystyle O}{\|}}{C}-H \longrightarrow CH_3-\overset{\overset{\displaystyle O}{\|}}{C}-OH$$

Decrease in the number of C–H bonds.
Increase in the number of C–O bonds.
Oxidation

b.

$$CH_3-\overset{\overset{\displaystyle O}{\|}}{C}-CH_3 \longrightarrow CH_3CH_2CH_3$$

Decrease in the number of C–O bonds.
Increase in the number of C–H bonds.
Reduction

c.

$$CH_3-\overset{\overset{\displaystyle O}{\|}}{C}-CH_3 \longrightarrow CH_3-\overset{\overset{\displaystyle HO\quad OH}{\diagdown\diagup}}{C}-CH_3$$

No change in the number of C–O
or C–H bonds. **Neither**

d.

⬡=O ⟶ ⬡—OH

Decrease in the number of C–O bonds.
Increase in the number of C–H bonds.
Reduction

4.33 The products of a combustion reaction of a hydrocarbon are always the same: **CO$_2$ and H$_2$O.**

a.   CH$_3$CH$_2$CH$_3$ +   5 O$_2$ $\xrightarrow{\text{flame}}$ 3 CO$_2$ +   4 H$_2$O   +   heat

b.   ⬡   +   9 O$_2$ $\xrightarrow{\text{flame}}$ 6 CO$_2$ +   6 H$_2$O   +   heat

4.34 Lipids contain many nonpolar C–C and C–H bonds and few polar functional groups.

a. $CH_3(CH_2)_7CH=CH(CH_2)_7COOH$

oleic acid

only one polar functional group
18 carbons
a lipid

b.

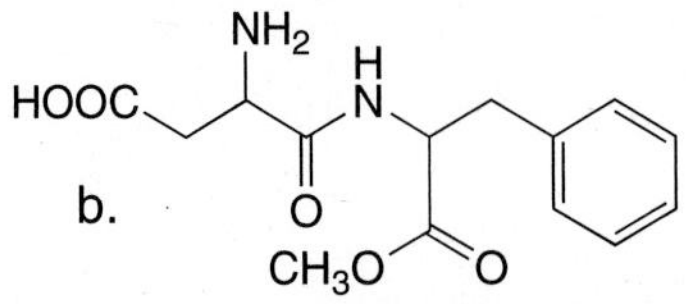

aspartame

many polar functional groups
only 14 carbons
not a lipid

4.35 "Like dissolves like." Beeswax is a lipid, and therefore, it will be more soluble in nonpolar solvents. H_2O is very polar, ethanol is slightly less polar, and chloroform is least polar. Beeswax is most soluble in the least polar solvent.

Increasing polarity

H_2O CH_3CH_2OH $CHCl_3$

Increasing solubility of beeswax

4.36 Use the rules from Answer 4.3.

a. [1]

[2]

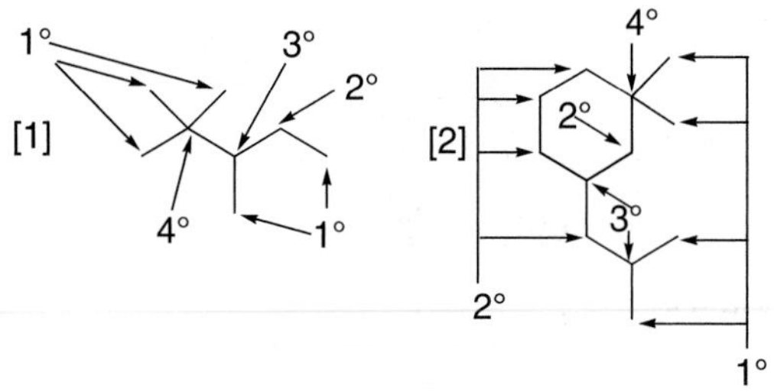

b. [1]

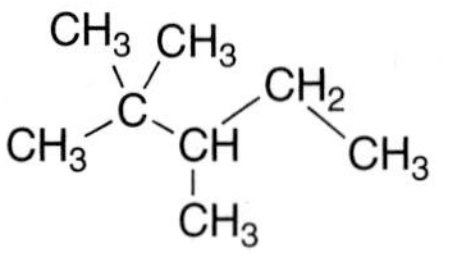

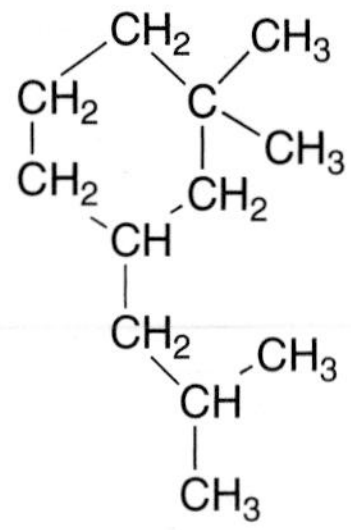

All CH₃'s have 1° H's.
All CH₂'s have 2° H's.
All CH's have 3° H's.

[2]

4.37

One possibility:

a. 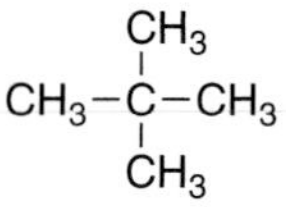 $CH_3-\underset{\underset{CH_3}{|}}{\overset{\overset{CH_3}{|}}{C}}-CH_3$ b. 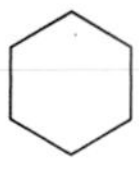   c. $CH_3CH_2CH_3$ d. $(CH_3)_3CH$

4.38 Use the rules from Answer 4.3.

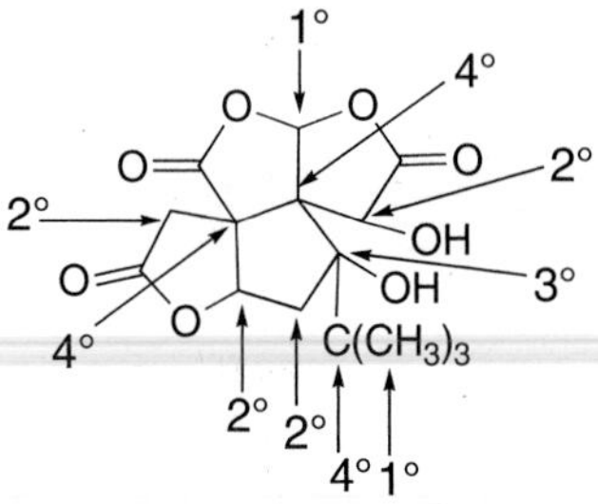

4.39

a. Five constitutional isomers of molecular formula C_4H_8:

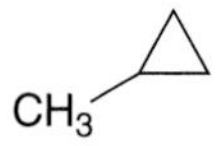

$CH_3CH=CHCH_3$ $CH_2=CHCH_2CH_3$ $\underset{CH_3}{\overset{\overset{CH_2}{||}}{C}}\overset{}{CH_3}$

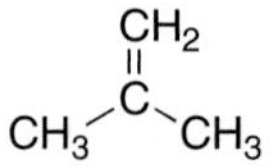

b. Nine constitutional isomers of molecular formula C_7H_{16}:

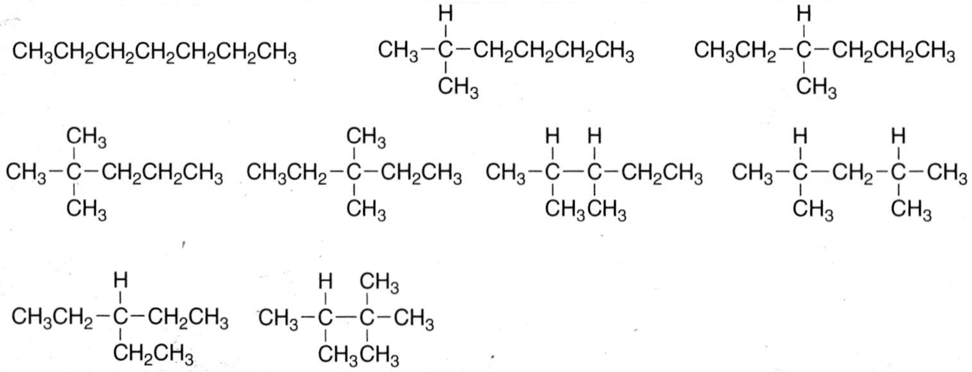

$CH_3CH_2CH_2CH_2CH_2CH_2CH_3$ $CH_3-\underset{\underset{CH_3}{|}}{\overset{\overset{H}{|}}{C}}-CH_2CH_2CH_2CH_3$ $CH_3CH_2-\underset{\underset{CH_3}{|}}{\overset{\overset{H}{|}}{C}}-CH_2CH_2CH_3$

$CH_3-\underset{\underset{CH_3}{|}}{\overset{\overset{CH_3}{|}}{C}}-CH_2CH_2CH_3$ $CH_3CH_2-\underset{\underset{CH_3}{|}}{\overset{\overset{CH_3}{|}}{C}}-CH_2CH_3$ $CH_3-\underset{\underset{CH_3}{|}}{\overset{\overset{H}{|}}{C}}-\underset{\underset{CH_3}{|}}{\overset{\overset{H}{|}}{C}}-CH_2CH_3$ $CH_3-\underset{\underset{CH_3}{|}}{\overset{\overset{H}{|}}{C}}-CH_2-\underset{\underset{CH_3}{|}}{\overset{\overset{H}{|}}{C}}-CH_3$

$CH_3CH_2-\underset{\underset{CH_2CH_3}{|}}{\overset{\overset{H}{|}}{C}}-CH_2CH_3$ $CH_3-\underset{\underset{CH_3CH_3}{|}}{\overset{\overset{H}{|}}{C}}-\underset{\underset{}{|}}{\overset{\overset{CH_3}{|}}{C}}-CH_3$

c. Twelve constitutional isomers of molecular formula C_6H_{12} containing one ring:

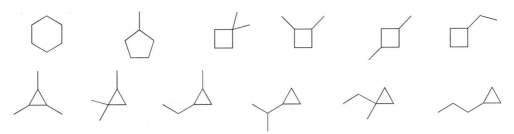

4.40 Use the steps in Answers 4.10 and 4.14 to name the alkanes.

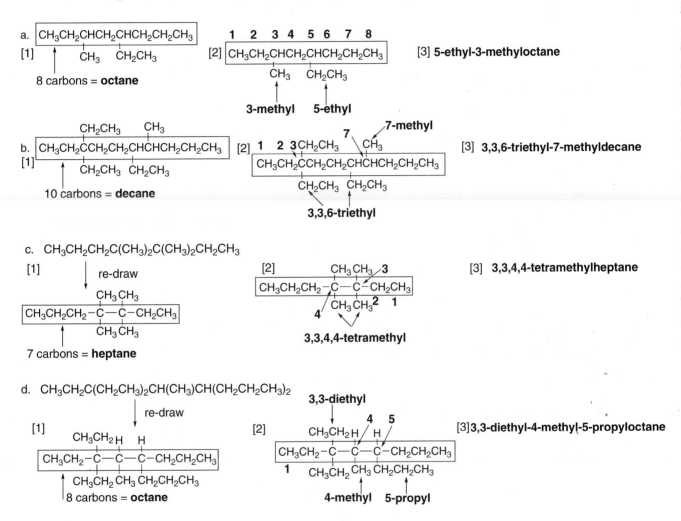

a. | $CH_3CH_2CHCH_2CHCH_2CH_2CH_3$ |
[1] CH_3 CH_2CH_3

8 carbons = **octane**

1 2 3 4 5 6 7 8
[2] | $CH_3CH_2CHCH_2CHCH_2CH_2CH_3$ |
      CH_3 CH_2CH_3

3-methyl   5-ethyl

[3] **5-ethyl-3-methyloctane**

b. CH_2CH_3 CH_3
| $CH_3CH_2CCH_2CH_2CHCHCH_2CH_2CH_3$ |
[1] CH_2CH_3 CH_2CH_3

10 carbons = **decane**

7-methyl
                          7
[2] 1 2 3 CH_2CH_3 CH_3
| $CH_3CH_2CCH_2CH_2CHCHCH_2CH_2CH_3$ |
    CH_2CH_3 CH_2CH_3

3,3,6-triethyl

[3] **3,3,6-triethyl-7-methyldecane**

c. $CH_3CH_2CH_2C(CH_3)_2C(CH_3)_2CH_2CH_3$

[1]   re-draw

        CH_3 CH_3
| $CH_3CH_2CH_2-C-C-CH_2CH_3$ |
        CH_3 CH_3

7 carbons = **heptane**

[2] CH_3 CH_3 3
| $CH_3CH_2CH_2-C-C-CH_2CH_3$ |
        4   CH_3 CH_3 2 1

3,3,4,4-tetramethyl

[3] **3,3,4,4-tetramethylheptane**

d. $CH_3CH_2C(CH_2CH_3)_2CH(CH_3)CH(CH_2CH_2CH_3)_2$

            re-draw

[1] CH_3CH_2 H   H
| $CH_3CH_2-C-C-C-CH_2CH_2CH_3$ |
      CH_3CH_2 CH_3 $CH_2CH_2CH_3$

8 carbons = **octane**

3,3-diethyl
                      4   5
[2] CH_3CH_2 H / H
| $CH_3CH_2-C-C-C-CH_2CH_2CH_3$ |
    1   CH_3CH_2 CH_3 $CH_2CH_2CH_3$

4-methyl   5-propyl

[3] **3,3-diethyl-4-methyl-5-propyloctane**

e. $(CH_3CH_2)_3CCH(CH_3)CH_2CH_2CH_3$

[1]   re-draw

        CH_3CH_2 H
| $CH_3CH_2-C-C-CH_2CH_2CH_3$ |
        CH_3CH_2 CH_3

7 carbons = **heptane**

3,3-diethyl
[2] CH_3CH_2 H 4
| $CH_3CH_2-C-C-CH_2CH_2CH_3$ |
    1   CH_3CH_2 CH_3 6 7

4-methyl

[3] **3,3-diethyl-4-methylheptane**

f. CH₃CH₂CH(CH₃)CH(CH₃)CH(CH₂CH₂CH₃)(CH₂)₃CH₃

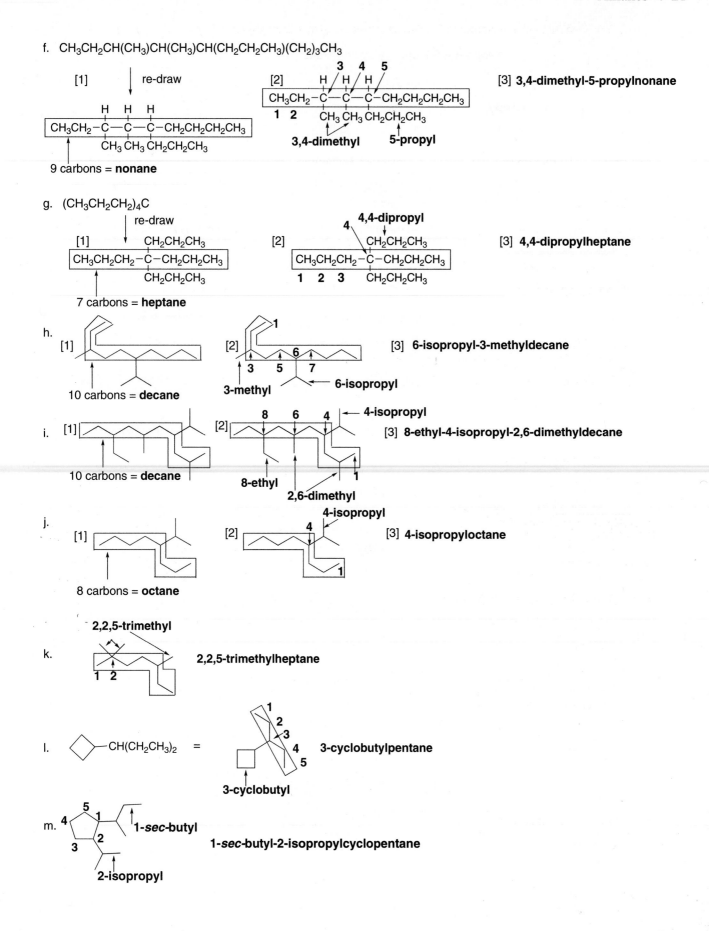

[1] ↓ re-draw

H H H
CH₃CH₂−C−C−C−CH₂CH₂CH₂CH₃
      CH₃ CH₃ CH₂CH₂CH₃

↑ 9 carbons = **nonane**

[2]
        3   4   5
        H / H / H /
CH₃CH₂−C—C—C−CH₂CH₂CH₂CH₃
1   2   CH₃ CH₃ CH₂CH₂CH₃

3,4-dimethyl   **5-propyl**

[3] 3,4-dimethyl-5-propylnonane

g. (CH₃CH₂CH₂)₄C

↓ re-draw

[1]   CH₂CH₂CH₃
CH₃CH₂CH₂−C−CH₂CH₂CH₃
          CH₂CH₂CH₃

↑ 7 carbons = **heptane**

[2]   **4,4-dipropyl**
          4   CH₂CH₂CH₃
CH₃CH₂CH₂−C−CH₂CH₂CH₃
1   2   3   CH₂CH₂CH₃

[3] 4,4-dipropylheptane

h. **[1]**

↑ 10 carbons = **decane**

[2]   1
        3   5   6   7
3-methyl   **6-isopropyl**

[3] 6-isopropyl-3-methyldecane

i. **[1]**

↑ 10 carbons = **decane**

[2]   8   6   4 ← **4-isopropyl**
8-ethyl   **2,6-dimethyl**   1

[3] 8-ethyl-4-isopropyl-2,6-dimethyldecane

j. **[1]**

↑ 8 carbons = **octane**

[2]   **4-isopropyl**
        4
        1

[3] 4-isopropyloctane

k. **2,2,5-trimethyl**

1   2

2,2,5-trimethylheptane

l. ⬦−CH(CH₂CH₃)₂ =

1
2
3
4
5

3-cyclobutyl

3-cyclobutylpentane

m. 5
4   1
3   2

1-sec-butyl

2-isopropyl

1-sec-butyl-2-isopropylcyclopentane

n.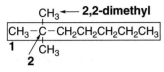

1-isobutyl-3-isopropylcyclohexane

1-isobutyl

3-isopropyl

4.41

CH₃ ← **2,2-dimethyl**

CH₃–C–CH₂CH₂CH₂CH₂CH₃
1 |
     CH₃
2

2,2-dimethylheptane

CH₃ ← **3,3-dimethyl**

CH₃CH₂–C–CH₂CH₂CH₂CH₃
1 |
        CH₃
3

3,3-dimethylheptane

CH₃ ← **4,4-dimethyl**

CH₃CH₂CH₂–C–CH₂CH₂CH₃
1 |
           CH₃
4

4,4-dimethylheptane

H   H

CH₃–C–CH₂–C–CH₂CH₂CH₃
1 | 4 |
2   CH₃   CH₃

2,4-dimethyl

2,4-dimethylheptane

H   H

CH₃–C–CH₂CH₂–C–CH₂CH₃
1 | |
    CH₃   CH₃
2   5

2,5-dimethyl

2,5-dimethylheptane

2   H H   3

CH₃–C–C–CH₂CH₂CH₂CH₃
1 | |
    CH₃CH₃

2,3-dimethyl

2,3-dimethylheptane

H   H

CH₃–C–CH₂CH₂CH₂–C–CH₃
1 | 6 |
2   CH₃   CH₃

2,6-dimethyl

2,6-dimethylheptane

H H

CH₃CH₂–C–C–CH₂CH₂CH₃
1 | |
3 CH₃ CH₃ 4

3,4-dimethyl

3,4-dimethylheptane

H   H

CH₃CH₂–C–CH₂–C–CH₂CH₃
1 | |
3   CH₃   CH₃ 5

3,5-dimethyl

3,5-dimethylheptane

4.42 Use the steps in Answer 4.12 to draw the structures.

a. 3-ethyl-2-methyl**hexane**

[1]   6 C chain ↓

C–C–C–C–C–C

[2]   C–C—C–C–C–C
          | |
         CH₃ CH₂CH₃

      methyl   ethyl on C3
      on C2

[3] CH₃–C–C–CH₂CH₂CH₃
          H H
          | |
         CH₃ CH₂CH₃

b. *sec*-butyl**cyclopentane**

[1]   5 C ring ↓

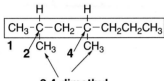

[2]

b. structure of sec-butylcyclopentane

c. 4-isopropyl-2,4,5-trimethyl**heptane**

[1]   7 C chain

C–C–C–C–C–C–C

[2]   isopropyl on C4

      CH₃   CH₃
         \ /
          CH
C–C–C–C—C–C–C
  | | |
  CH₃   CH₃CH₃

methyls on C2, C4, and C5

[3]   CH₃   CH₃
                \ /
                 CH
CH₃–CH–CH₂–C–CHCH₂CH₃
     | |
     CH₃   CH₃ CH₃

d. cyclobutyl**cycloheptane**

[1]   7 C cycloalkane

[2]

d. structure of cyclobutylcycloheptane

e. 3-ethyl-1,1-dimethylcyclohexane

[1] 6 C cycloalkane

[2]

CH$_3$CH$_2$ — C — CH$_3$ ← 2 methyl

ethyl on C3 — C — CH$_3$ ← groups on C1

[3]

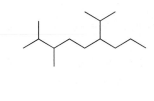

f. 4-butyl-1,1-diethylcyclooctane

[1] 8 C cycloalkane

[2] 2 ethyl groups →

[3]

g. 6-isopropyl-2,3-dimethylnonane

[1] 9 C alkane

[2] methyl   isopropyl

CH$_3$ CH$_3$ – CH – CH$_3$

C – C – C – C – C – C

CH$_3$ ← methyl

[3]

h. 2,2,6,6,7-pentamethyloctane

[1] 8 C alkane

[2] 5 methyl groups

CH$_3$ CH$_3$ CH$_3$ CH$_3$

C – C – C – C – C – C – C

CH$_3$

[3]

i. cis-1-ethyl-3-methylcyclopentane

[1] 5 C ring

[2] CH$_2$CH$_3$ ← ethyl on C1

CH$_3$ ← methyl on C3

or

CH$_2$CH$_3$

CH$_3$

j. trans-1-tert-butyl-4-ethylcyclohexane

[1] 6 C ring

[2] C(CH$_3$)$_3$

CH$_3$CH$_2$

4.43 Draw the compounds.

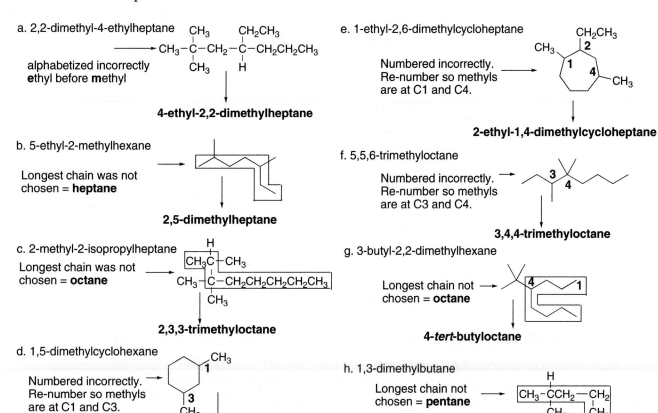

a. 2,2-dimethyl-4-ethylheptane

alphabetized incorrectly
ethyl before **m**ethyl

4-ethyl-2,2-dimethylheptane

b. 5-ethyl-2-methylhexane

Longest chain was not
chosen = **heptane**

2,5-dimethylheptane

c. 2-methyl-2-isopropylheptane
Longest chain was not
chosen = **octane**

2,3,3-trimethyloctane

d. 1,5-dimethylcyclohexane

Numbered incorrectly.
Re-number so methyls
are at C1 and C3.

1,3-dimethylcyclohexane

e. 1-ethyl-2,6-dimethylcycloheptane

Numbered incorrectly.
Re-number so methyls
are at C1 and C4.

2-ethyl-1,4-dimethylcycloheptane

f. 5,5,6-trimethyloctane

Numbered incorrectly.
Re-number so methyls
are at C3 and C4.

3,4,4-trimethyloctane

g. 3-butyl-2,2-dimethylhexane

Longest chain not
chosen = **octane**

4-*tert*-butyloctane

h. 1,3-dimethylbutane

Longest chain not
chosen = **pentane**

2-methylpentane

4.44

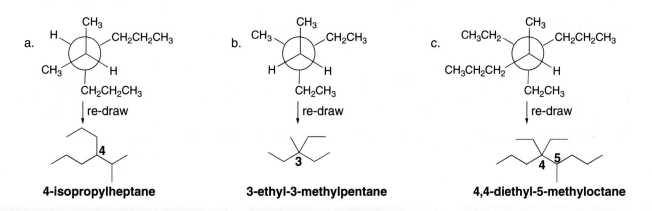

a.

re-draw

4-isopropylheptane

b.

re-draw

3-ethyl-3-methylpentane

c.

re-draw

4,4-diethyl-5-methyloctane

4.45 Use the rules from Answer 4.17.

a. $CH_3CH_2CH_3$ $CH_3CH_2CH_2CH_3$ $CH_3CH_2CH_2CH_2CH_3$
      3C's   4 C's   5 C's
   lowest boiling point   **highest boiling point**

b. $(CH_3)_2CHCH(CH_3)_2$ $CH_3CH_2CH_2CH(CH_3)_2$ $CH_3(CH_2)_4CH_3$
      most branching   least branching
    lowest boiling point   **highest boiling point**

4.46 a.

   $CH_3(CH_2)_6CH_3$
   no branching = higher surface area
   higher boiling point

   $(CH_3)_3CC(CH_3)_3$
   branching = lower surface area
   lower boiling point
   more spherical, better packing =
   higher melting point

b. There is a 159° difference in the melting points, but only a 20° difference in the boiling points because the symmetry in $(CH_3)_3CC(CH_3)_3$ allows it to pack more tightly in the solid, thus requiring more energy to melt. In contrast, once the compounds are in the liquid state, symmetry is no longer a factor, the compounds are isomeric alkanes, and the boiling points are closer together.

4.47

a.

or

1 gauche CH_3,CH_3
= 3.8 kJ/mol
of destabilization

higher energy
2 gauche CH_3,CH_3
3.8 kJ/mol x 2 = 7.6 kJ/mol
of destabilization

b.

or

2 gauche CH_3,CH_3
3.8 kJ/mol x 2 =
7.6 kJ/mol
of destabilization

higher energy
3 eclipsed H,CH_3
6 kJ/mol x 3 = 18 kJ/mol
of destabilization

Energy difference =

7.6 kJ/mol – 3.8 kJ/mol = **3.8 kJ/mol**

Energy difference =

18 kJ/mol – 7.6 kJ/mol = **10.4 kJ/mol**

4.48 Use the rules from Answer 4.21 to determine the most and least stable conformations.

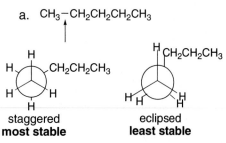

a. $CH_3-CH_2CH_2CH_2CH_3$

staggered
most stable

eclipsed
least stable

All staggered conformations are equal in energy.
All eclipsed conformations are equal in energy.

b. $CH_3CH_2CH_2-CH_2CH_2CH_3$

staggered
ethyl groups anti
most stable

eclipsed
ethyl groups eclipsed
least stable

4.49

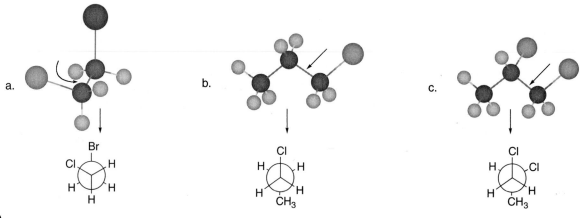

a.

b.

c.

4.50

a.

b.

c.

4.51

[1] $CH_3CH_2-CH_2CH_2CH_3$

1

60°

2

60°

3

60°

60°

6

60°

5

60°

4

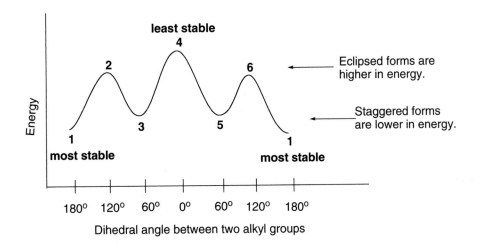

Dihedral angle between two alkyl groups

[2] $CH_3CH_2-\underset{\underset{CH_3}{\uparrow}}{CH}CH_2CH_3$

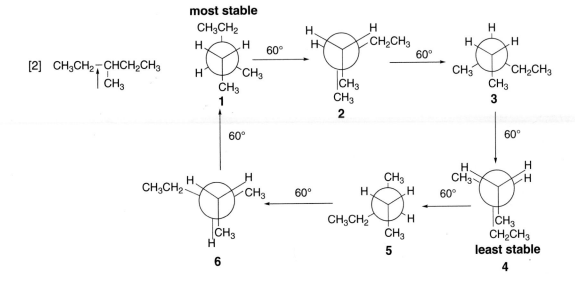

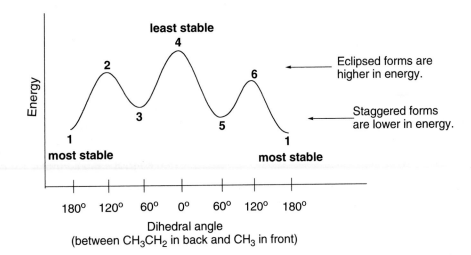

Dihedral angle
(between CH_3CH_2 in back and CH_3 in front)

4.52 Two types of strain:

    4.1 *Torsional strain* is due to eclipsed groups on adjacent carbon atoms.

    4.2 *Steric strain* is due to overlapping electron clouds of large groups (ex: gauche interactions).

a.

two sites
three bulky methyl groups close =
steric strain

b.

eclipsed conformation =
torsional strain

c.

two bulky ethyl groups close =
steric strain
eclipsed conformation =
torsional strain

4.53 The barrier to rotation is equal to the difference in energy between the highest energy eclipsed and lowest energy staggered conformations of the molecule.

    a. $CH_3\!-\!CH(CH_3)_2$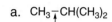

    b. $CH_3\!-\!C(CH_3)_3$

most stable   least stable

most stable   least stable

Destabilization energy =

2 H,CH$_3$ eclipsing interactions
$$2(6.0\ kJ/mol) = 12.0\ kJ/mol$$
1 H,H eclipsing interaction = 4.0 kJ/mol

Total destabilization = **16.0 kJ/mol**

| **16.0 kJ/mol = rotation barrier** |

Destabilization energy =

3 H,CH$_3$ eclipsing interactions
$$3(6.0\ kJ/mol) = 18.0\ kJ/mol$$

Total destabilization = **18.0 kJ/mol**

| **18.0 kJ/mol = rotation barrier** |

4.54

most stable   least stable

2 H,H eclipsing interactions = 2(4.0 kJ/mol) = 8.0 kJ/mol

Since the barrier to rotation is 15 kJ/mol, the difference between this value and the destabilization due to H,H eclipsing is the destabilization due to H,Cl eclipsing.

15.0 kJ/mol – 8.0 kJ/mol = 7.0 kJ/mol

destabilization due to H,Cl eclipsing

4.55 The gauche conformation can intramolecularly hydrogen bond, making it the more stable conformation.

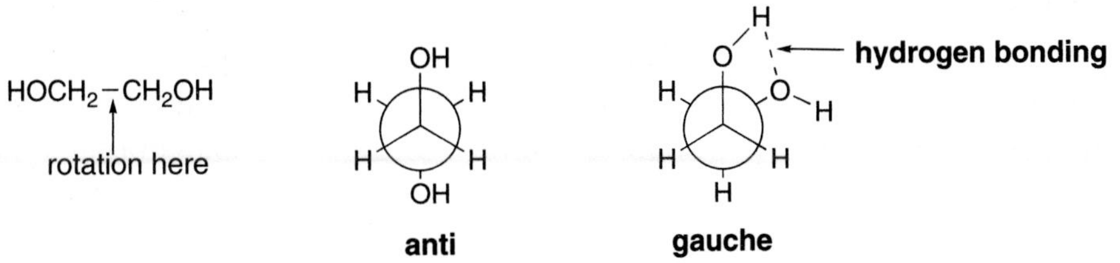

HOCH$_2$–CH$_2$OH

rotation here

anti

hydrogen bonding

gauche

Hydrogen bonding can occur only in the gauche conformation, making it **more stable.**

4.56

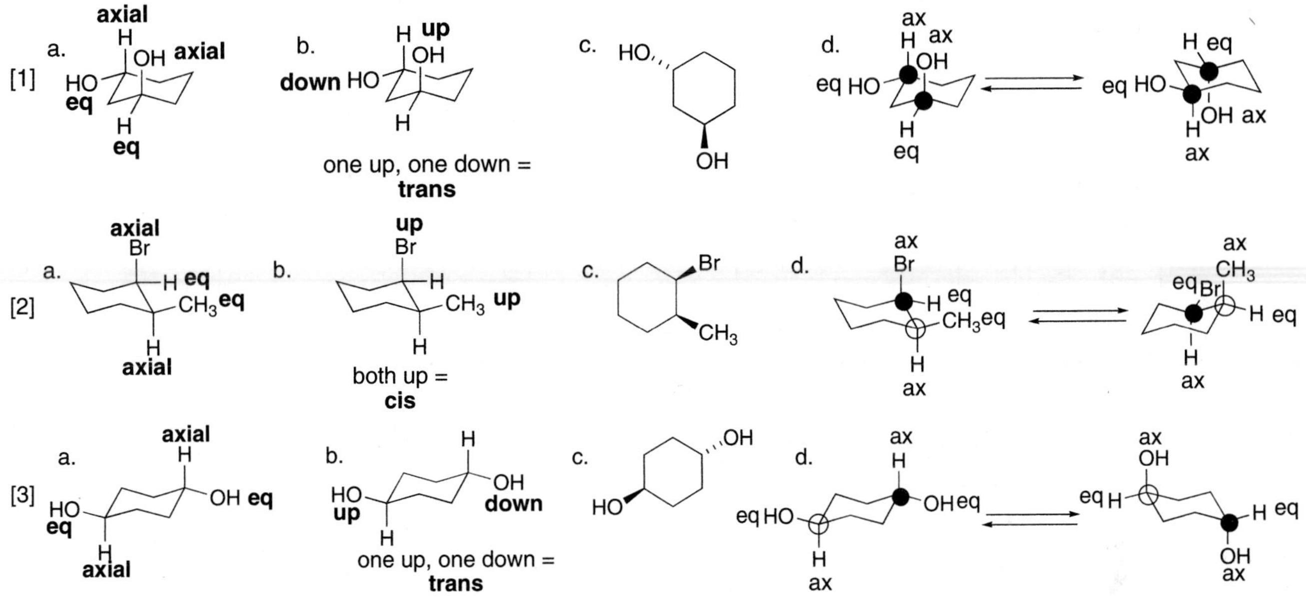

[1]

a.
axial
H
OH axial
HO
eq
H
eq

b.
H up
OH
down HO
H
one up, one down =
trans

c.
HO,,,
OH

d.
ax
H ax
OH
eq HO
H
eq

H eq
eq HO
OH ax
ax

[2]

a.
axial
Br
H eq
CH$_3$ eq
H
axial

b.
up
Br
H
CH$_3$ up
H
both up =
cis

c.
Br
CH$_3$

d.
ax
Br
H eq
CH$_3$ eq
H
ax

ax
CH$_3$
eq Br
H eq
H
ax

[3]

a.
axial
H
OH eq
HO
eq
H
axial

b.
H
OH
HO
up
down
H
one up, one down =
trans

c.
OH
HO

d.
ax
H
OHeq
eq HO
H
ax

ax
OH
eq H
H eq
OH
ax

4.57

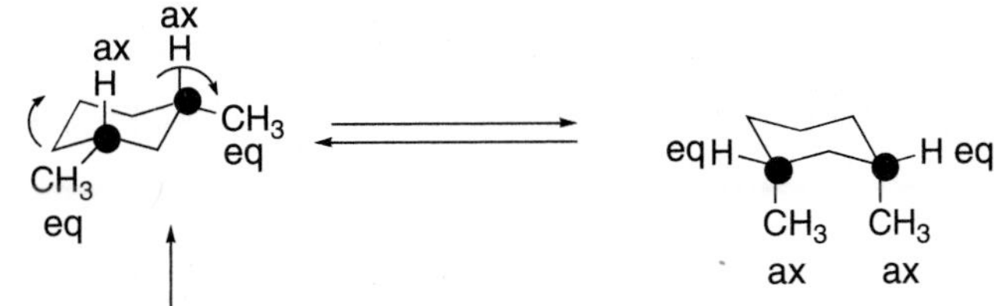

ax
ax H
H
CH$_3$
eq
CH$_3$
eq

eq H
H eq
CH$_3$ CH$_3$
ax ax

both groups equatorial
more stable

4.58

Axial/equatorial substituent location

Disubstituted cyclohexane	Conformation 1	Conformation 2
a. 1,2-cis disubstituted	Axial/equatorial	Equatorial/axial
b. 1,2-trans disubstituted	Axial/axial	Equatorial/equatorial
c. 1,3-cis disubstituted	Axial/axial	Equatorial/equatorial
d. 1,3-trans disubstituted	Axial/equatorial	Equatorial/axial
e. 1,4-cis disubstituted	Axial/equatorial	Equatorial/axial
f. 1,4-trans disubstituted	Axial/axial	Equatorial/equatorial

4.59 A **cis isomer** has two groups on the **same side** of the ring. The two groups can be drawn both up or both down. Only one possibility is drawn. A **trans isomer** has one group on one side of the ring and one group on the other side. Either group can be drawn on either side. Only one possibility is drawn.

[1]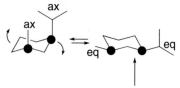

[2]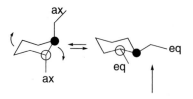

[3]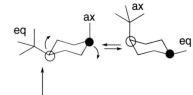

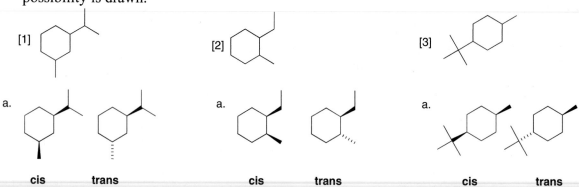

a.

| cis | trans |

a.

| cis | trans |

a.

| cis | trans |

b. cis isomer

both groups equatorial
more stable

b. cis isomer

larger group equatorial
more stable

larger group equatorial
more stable

b. cis isomer

larger group equatorial
more stable

c. trans isomer

larger group equatorial
more stable

c. trans isomer

both groups equatorial
more stable

c. trans isomer

both groups equatorial
more stable

d.

The cis isomer is more
stable than the trans
since one conformation has
both groups equatorial.

d.

The trans isomer is more
stable than the cis
since one conformation has
both groups equatorial.

d.

The trans isomer is more
stable than the cis
since one conformation has
both groups equatorial.

4.60 Compare the isomers by drawing them in chair conformations. Equatorial substituents are more stable. See the definitions in Problem 4.59.

a.

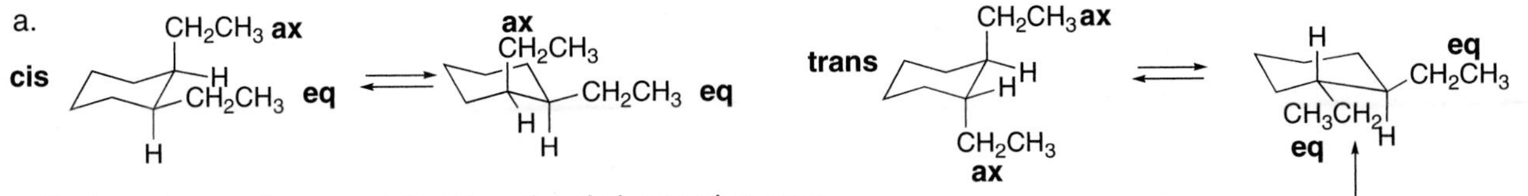

The trans isomer is more stable than the cis isomer because its more stable conformation has two groups equatorial.

both groups equatorial
most stable of all conformations
trans isomer

b. **1-ethyl-3-isopropylcyclohexane**

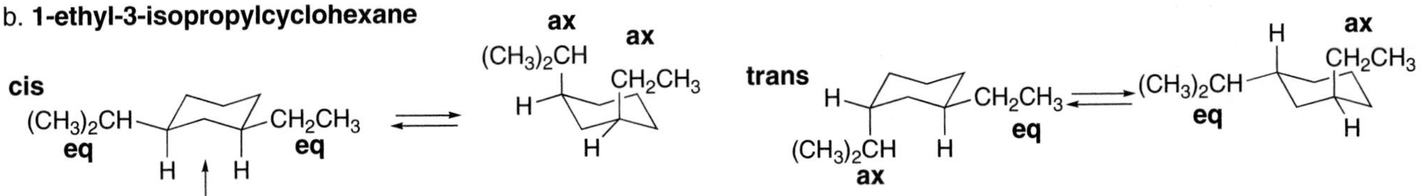

both groups equatorial
most stable of all conformations
cis isomer

The cis isomer is more stable than the trans isomer because its more stable conformation has two groups equatorial.

4.61

Only the more stable conformation is drawn.

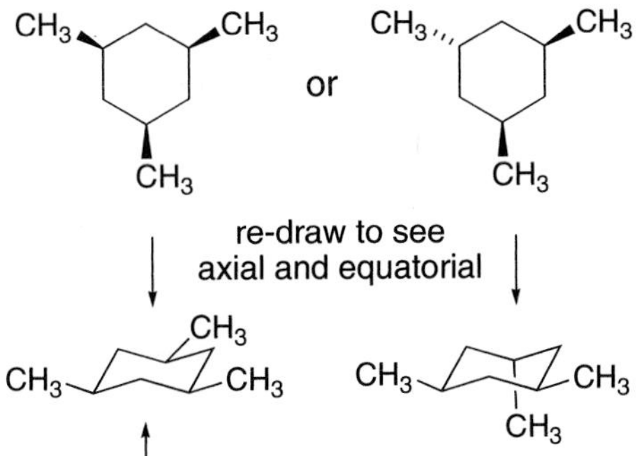

or

re-draw to see
axial and equatorial

more stable
substituents on C1, C3, C5 =
all equatorial

4.62

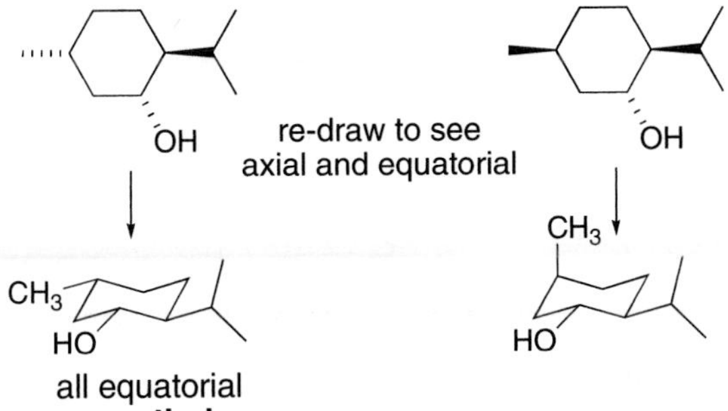

re-draw to see
axial and equatorial

all equatorial
menthol

4.63

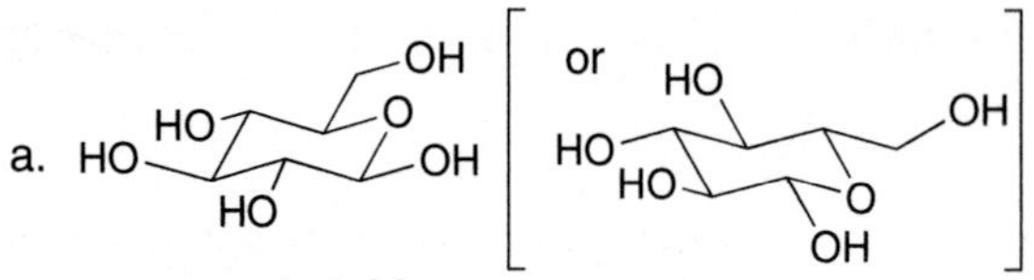

a.

most stable
All groups are equatorial.

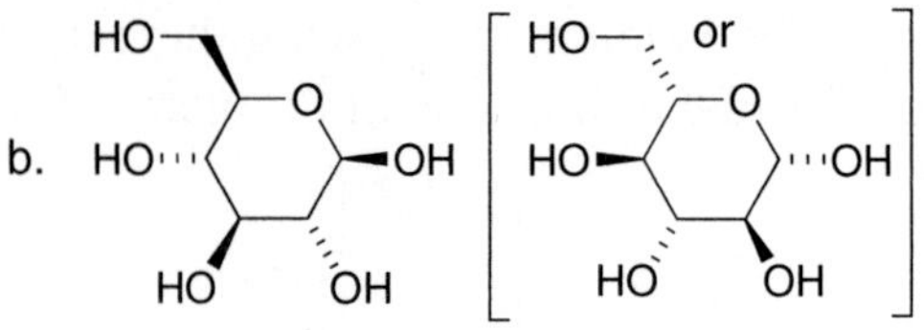

b.

4.64

a.

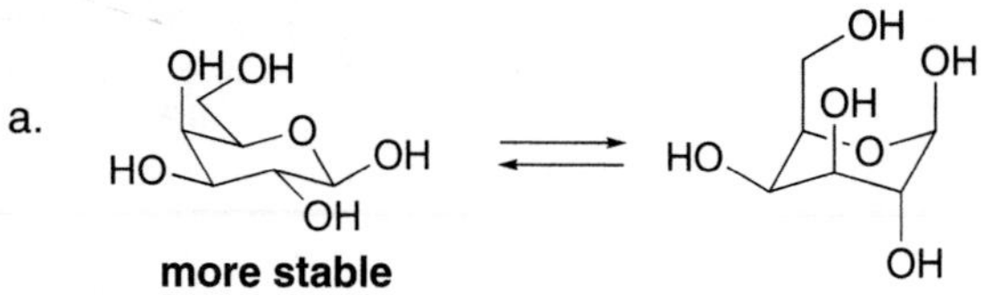

more stable
More groups are equatorial.

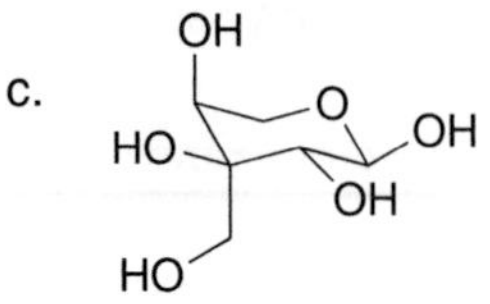

c.

constitutional isomer

b.

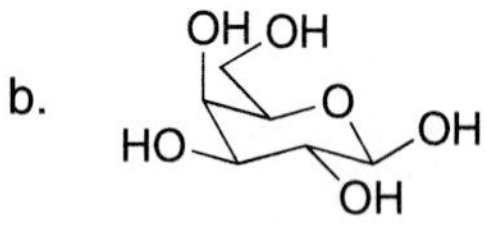

galactose

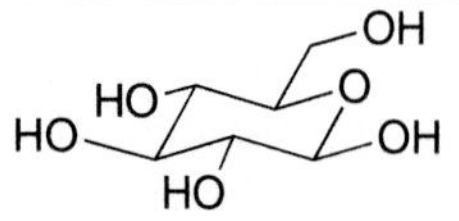

glucose
All groups are equatorial.
more stable

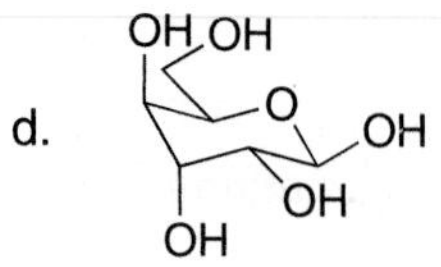

d.

stereoisomer

4.65

a. 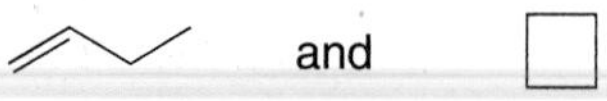 and

same molecular formula C_4H_8
different connectivity
constitutional isomers

d. 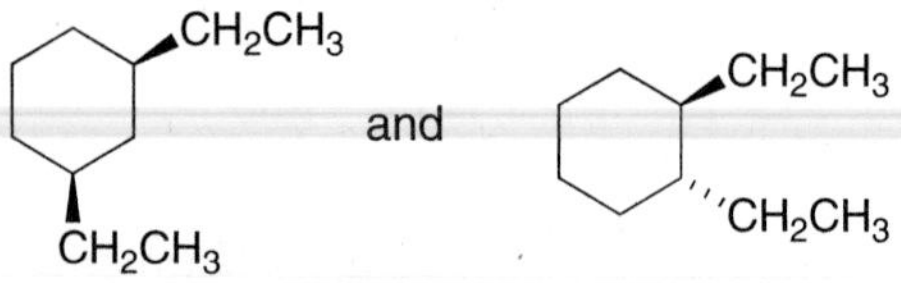 and

same molecular formula $C_{10}H_{20}$
different connectivity
constitutional isomers

b. 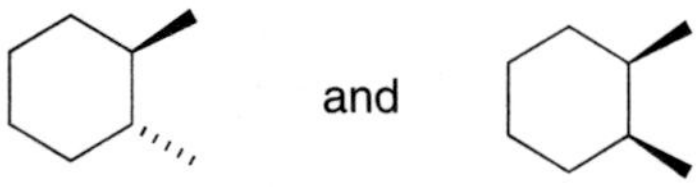 and

different arrangement in three dimensions
stereoisomers

e. 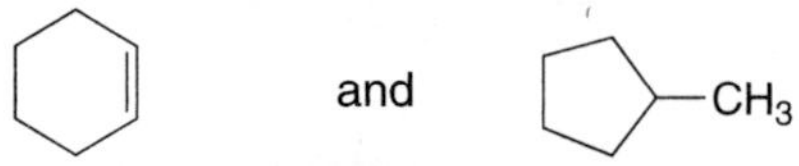 and

molecular formula: C_6H_{10} molecular formula: C_6H_{12}

different molecular formulas
not isomers

c. 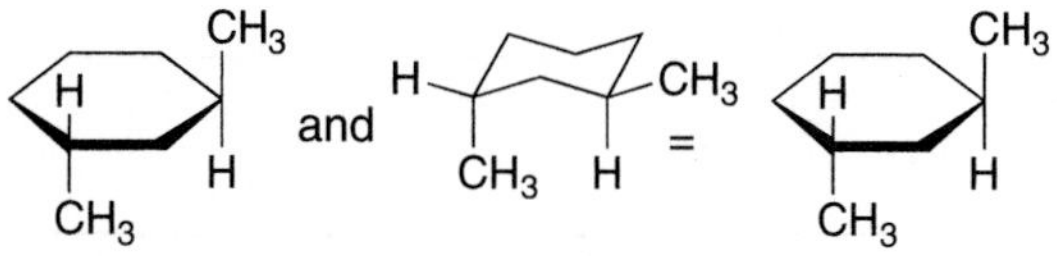 and =

1 down, 1 up =   1 down, 1 up =
trans   **trans**

same arrangement in three dimensions
identical

f. 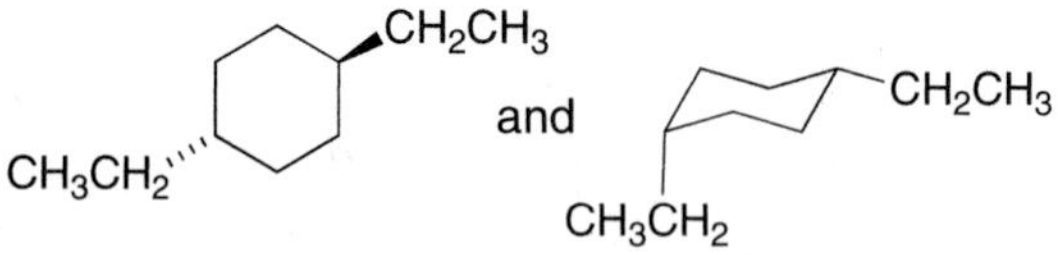 and

1 down, 1 up =   both down =
trans   **cis**

different arrangement in three dimensions
stereoisomers

g.

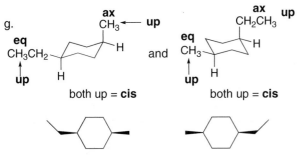

both up = **cis**   both up = **cis**

same arrangement in three dimensions
identical

h.

and

3,4-dimethylhexane   **2,4-dimethylhexane**

same molecular formula C_8H_{18}
different IUPAC names
constitutional isomers

4.66

a.

and

re-draw

CH_3
$CH_3-CH-CH-CH_2-CH_3$
CH_2CH_3

$CH_3 \quad CH_2CH_3$
$CH_3-CH-CH-CH_2CH_3$

3-ethyl-2-methylpentane   3-ethyl-2-methylpentane

same molecular formula
same name
identical molecules

b.

and

re-draw

same molecular formula
different arrangement of atoms
constitutional isomers

4.67

	One possibility:	**constitutional isomer**	**stereoisomer**
a.	**cis**		**trans**
b.	**cis**		**trans**
c.	**cis**		**trans**

4.68

Three constitutional isomers of C_7H_{14}:

1,1-dimethylcyclopentane   1,2-dimethylcyclopentane   1,3-dimethylcyclopentane

4.69 Use the definitions from Answer 4.32 to classify the reactions.

a. CH_3CHO = $\longrightarrow$ CH_3CH_2OH
Decrease in the number of C–O bonds. **Reduction**

d. $CH_2{=}CH_2$ $\longrightarrow$ $H{-}C{\equiv}C{-}H$
Decrease in the number of C–H bonds. **Oxidation**

b.
Increase in the number of C–O bonds. **Oxidation**

e.
Increase in the number of C–Z bonds. **Oxidation**

c. $CH_2{=}CH_2$ $\longrightarrow$ $HOCH_2CH_2OH$
Two new C–O bonds. **Oxidation**

f. CH_3CH_2OH $\longrightarrow$ $CH_2{=}CH_2$
Loss of one C–O bond *and* one C–H bond. **Neither**

4.70 Use the rule from Answer 4.33.

a. $CH_3CH_2CH_2CH_2CH(CH_3)_2$ $\xrightarrow[\text{11 O}_2]{\text{flame}}$ $7\,CO_2 + 8\,H_2O$ + heat

b. $\xrightarrow[\text{(13/2) O}_2]{\text{flame}}$ $4\,CO_2 + 5\,H_2O$ + heat

4.71

a.

benzene   an arene oxide   phenol

[1] increase in C–O bonds
oxidation reaction

[2] loss of 1 C–O bond,
loss of 1 C–H bond
neither

b. Phenol is more water soluble than benzene because it is **polar (contains an O–H group) and can hydrogen bond with water,** whereas benzene is nonpolar and cannot hydrogen bond.

4.72 Lipids contain many nonpolar C–C and C–H bonds and few polar functional groups.

a.

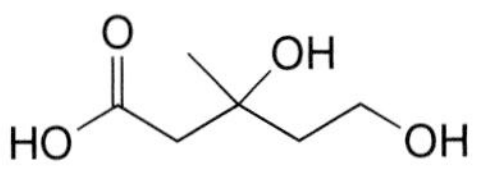

mevalonic acid

many polar functional groups
not a lipid

c.

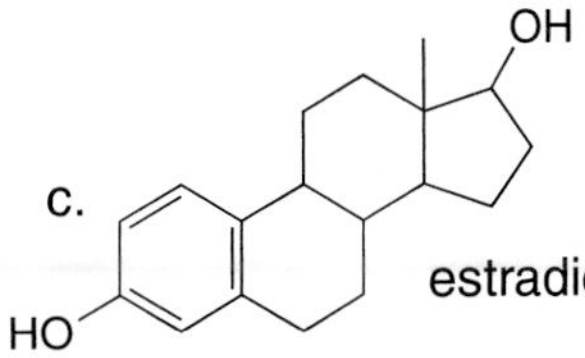

estradiol

few polar functional groups
a lipid

d.

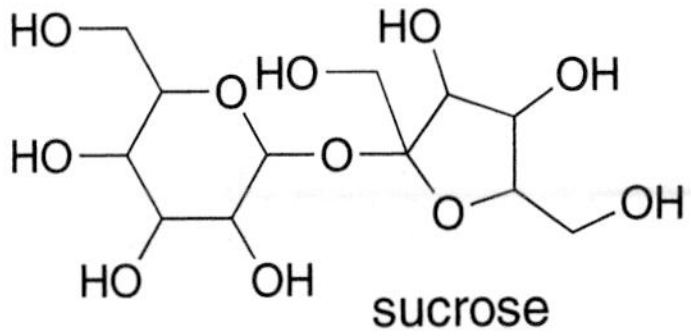

sucrose

many polar functional groups
not a lipid

b.

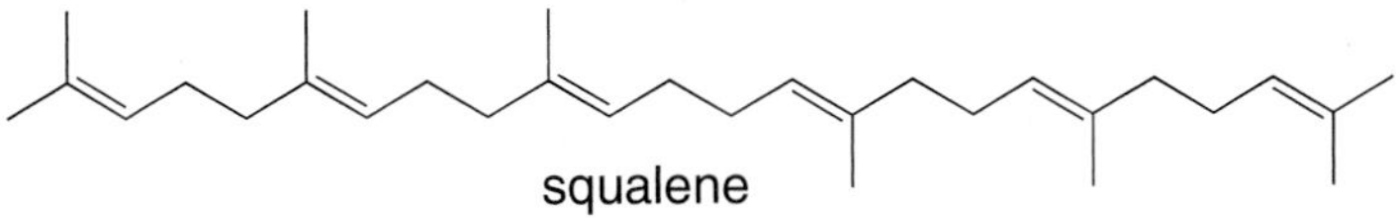

squalene

no polar functional groups
a lipid

4.73

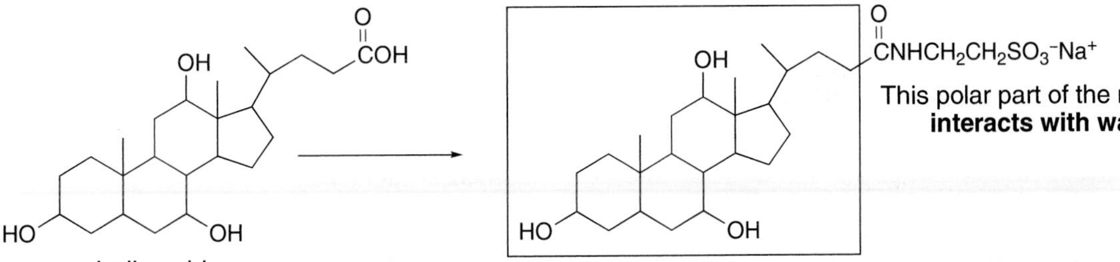

cholic acid
a bile acid

a bile salt

This polar part of the molecule
interacts with water.

This nonpolar part of the molecule
can **interact with lipids** to create
micelles that allow for transport
of lipids through aqueous environments.

4.74 The mineral oil can prevent the body's absorption of important fat-soluble vitamins. The vitamins dissolve in the mineral oil, and are thus not absorbed. Instead, they are expelled with the mineral oil.

4.75 The amide in the four-membered ring has 90° bond angles giving it angle strain, and therefore making it more reactive.

amide

penicillin G

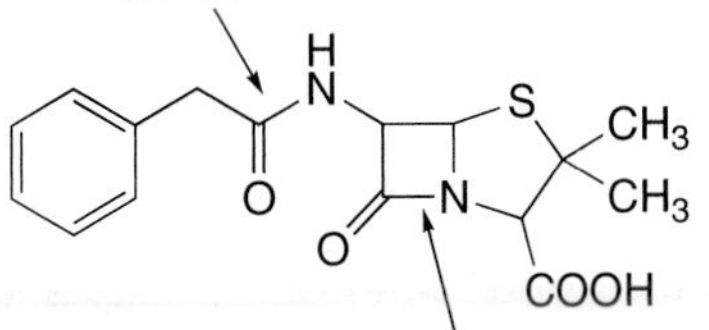

strained amide
more reactive

4.76

Example:

Although I is a much bigger atom than Cl, the C–I bond is also much longer than the C–Cl bond. As a result the eclipsing interaction of the H and I atoms is not very much different in magnitude from the H,Cl eclipsing interaction.

longer bond

4.77

decalin | *trans*-decalin   *cis*-decalin

trans

The trans isomer is more stable since the carbon groups at the ring junction are both in the favorable equatorial position.

1,3-diaxial interaction

cis

This bond is axial, creating unfavorable 1,3-diaxial interactions.

4.78

pentylcyclopentane

(1,1-dimethylpropyl)cyclopentane

(2-methylbutyl)cyclopentane

(2,2-dimethylpropyl)cyclopentane

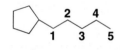

(1-methylbutyl)cyclopentane

(1-ethylpropyl)cyclopentane

(1,2-dimethylpropyl)cyclopentane

(3-methylbutyl)cyclopentane

Chapter 5: Stereochemistry

◆ **Isomers are different compounds with the same molecular formula (5.2, 5.11).**

[1] Constitutional isomers—isomers that differ in the way the atoms are connected to each other. They have:
- different IUPAC names
- the same or different functional groups
- different physical and chemical properties.

[2] Stereoisomers—isomers that differ only in the way atoms are oriented in space. They have the same functional group and the same IUPAC name except for prefixes such as cis, trans, *R*, and *S*.
- **Enantiomers**—stereoisomers that are nonsuperimposable mirror images of each other (5.4).
- **Diastereomers**—stereoisomers that are not mirror images of each other (5.7).

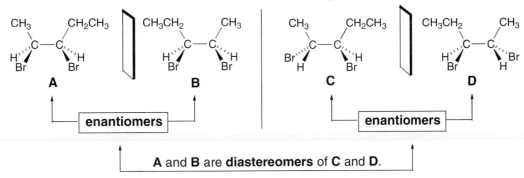

◆ **Assigning priority (5.6)**

- Assign priorities (1, 2, 3, or 4) to the atoms directly bonded to the stereogenic center in order of decreasing atomic number. The atom of *highest* atomic number gets the *highest* priority (1).
- If two atoms on a stereogenic center are the *same*, assign priority based on the atomic number of the atoms bonded to these atoms. *One* atom of higher atomic number determines a higher priority.
- If two isotopes are bonded to the stereogenic center, assign priorities in order of decreasing *mass* number.
- To assign a priority to an atom that is part of a multiple bond, consider a multiply bonded atom as an equivalent number of singly bonded atoms.

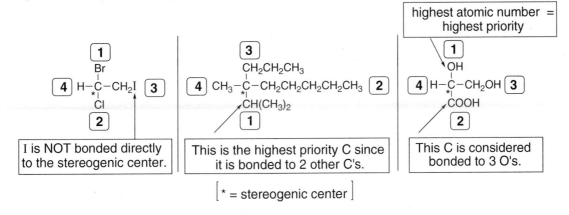

◆ **Some basic principles**

- When a compound and its mirror image are **superimposable**, they are **identical achiral compounds.** A plane of symmetry in one conformation makes a compound achiral (5.3).
- When a compound and its mirror image are **not superimposable**, they are **different chiral compounds** called **enantiomers.** A chiral compound has no plane of symmetry in any conformation (5.3).
- A **tetrahedral stereogenic center** is a carbon atom bonded to four different groups (5.4, 5.5).
- For *n* **stereogenic centers**, the maximum number of stereoisomers is 2^n (5.7).

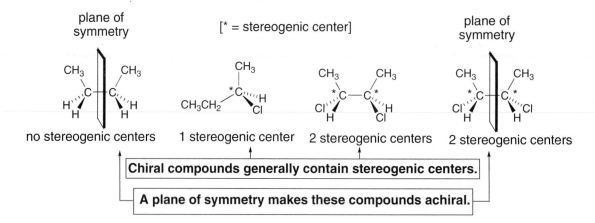

◆ **Optical activity is the ability of a compound to rotate plane-polarized light (5.12).**

- An optically active solution contains a chiral compound.
- An optically inactive solution contains one of the following:
  - an achiral compound with no stereogenic centers.
  - a meso compound—an achiral compound with two or more stereogenic centers.
  - a racemic mixture—an equal amount of two enantiomers.

◆ **The prefixes *R* and *S* compared with *d* and *l***

The prefixes *R* and *S* are labels used in nomenclature. Rules on assigning *R,S* are found in Section 5.6.
- An enantiomer has every stereogenic center opposite in configuration. If a compound with two stereogenic centers has the *R,R* configuration, its enantiomer has the *S,S* configuration.
- A diastereomer of this same compound has either the *R,S* or *S,R* configuration; one stereogenic center has the same configuration and one is opposite.

The prefixes *d* (or +) and *l* (or –) tell the direction a compound rotates plane-polarized light (5.12).
- *d* (or +) stands for dextrorotatory, rotating polarized light clockwise.
- *l* (or –) stands for levorotatory, rotating polarized light counterclockwise.

◆ **The physical properties of isomers compared (5.12)**

Type of isomer	Physical properties
Constitutional isomers	Different
Enantiomers	Identical except the direction of rotation of polarized light
Diastereomers	Different
Racemic mixture	Possibly different from either enantiomer

◆ Equations

- Specific rotation (5.12C):

$$\text{specific rotation} = [\alpha] = \frac{\alpha}{l \times c}$$

α = observed rotation (°)
l = length of sample tube (dm)
c = concentration (g/mL)

$$\begin{bmatrix} \text{dm} = \text{decimeter} \\ 1\ \text{dm} = 10\ \text{cm} \end{bmatrix}$$

- Enantiomeric excess (5.12D):

$$ee = \text{\% of one enantiomer} - \text{\% of other enantiomer}$$

$$= \frac{[\alpha]\ \text{mixture}}{[\alpha]\ \text{pure enantiomer}} \times 100\%$$

Chapter 5: Answers to Problems

5.1 Cellulose consists of long chains held together by intermolecular hydrogen bonds forming sheets that stack in extensive three-dimensional arrays. Most of the OH groups in cellulose are in the interior of this three-dimensional network, unavailable for hydrogen bonding to water. Thus, even though cellulose has many OH groups, its three-dimensional structure prevents many of the OH groups from hydrogen bonding with the solvent and this makes it water insoluble.

5.2 **Constitutional isomers** have atoms bonded to different atoms.
Stereoisomers differ only in the three-dimensional arrangement of atoms.

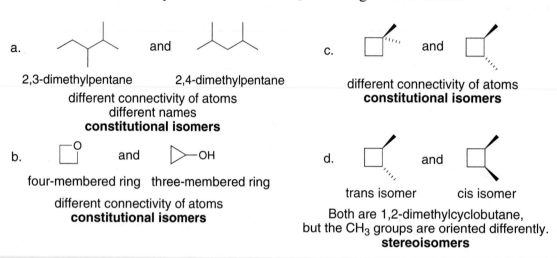

a.   and

2,3-dimethylpentane   2,4-dimethylpentane
different connectivity of atoms
different names
constitutional isomers

c.   and

different connectivity of atoms
constitutional isomers

b.   and   —OH

four-membered ring   three-membered ring

different connectivity of atoms
constitutional isomers

d.   and

trans isomer   cis isomer

Both are 1,2-dimethylcyclobutane,
but the CH_3 groups are oriented differently.
stereoisomers

5.3 Draw the mirror image of each molecule by drawing a mirror plane and then drawing the molecule's reflection. **A chiral molecule is one that is not superimposable on its mirror image.** A molecule with one stereogenic center is always chiral. A molecule with zero stereogenic centers is not chiral (in general).

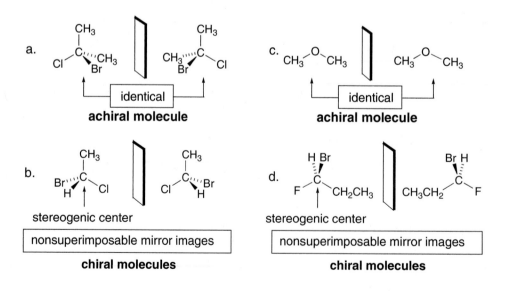

a.   identical
    achiral molecule

c.   identical
    achiral molecule

b.   stereogenic center
    nonsuperimposable mirror images
    chiral molecules

d.   stereogenic center
    nonsuperimposable mirror images
    chiral molecules

5.4 A plane of symmetry cuts the molecule into **two identical halves**.

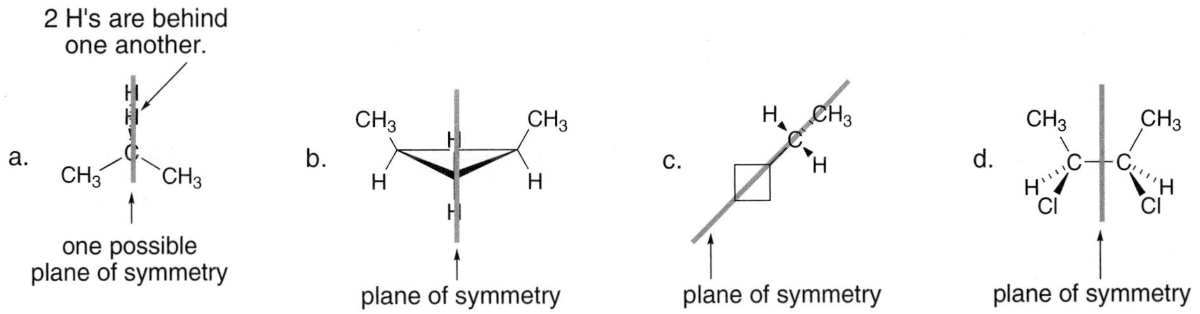

a.

2 H's are behind one another.

one possible plane of symmetry

b.

plane of symmetry

c.

plane of symmetry

d.

plane of symmetry

5.5 Rotate around the middle C–C bond so that the Br atoms are eclipsed.

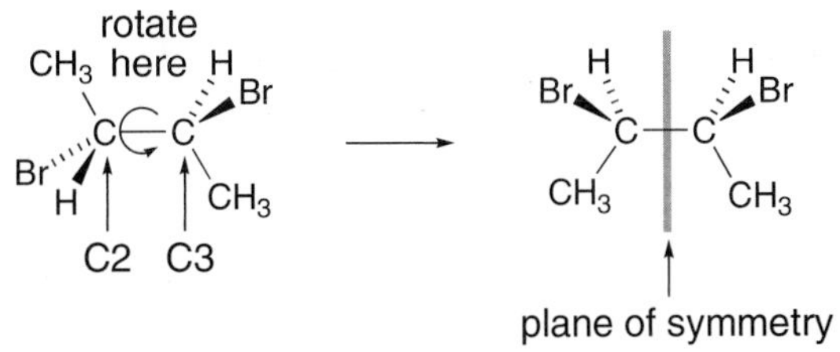

plane of symmetry

5.6 To locate a stereogenic center, omit:

All C's with 2 or more H's, all sp and sp^2 hybridized atoms, and all heteroatoms. (In Chapter 25, we will learn that the N atoms of ammonium salts [$R_4N^+X^-$] can sometimes be stereogenic centers.) Then evaluate any remaining atoms: a tetrahedral stereogenic center has a carbon bonded to **four different groups**.

a. 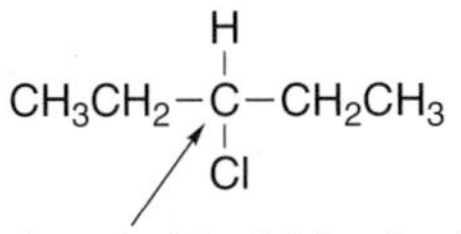
$$CH_3CH_2-\overset{\overset{H}{|}}{\underset{|}{C}}-CH_2CH_3$$
Cl

bonded to 2 identical
ethyl groups
0 stereogenic centers

b. $(CH_3)_3CH$

0 stereogenic centers

c. 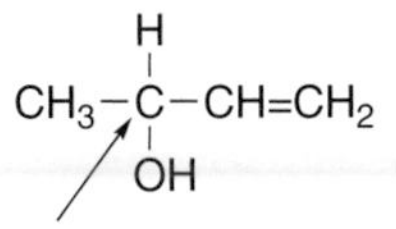
$$CH_3-\overset{\overset{H}{|}}{\underset{|}{C}}-CH=CH_2$$
OH

This C is bonded to
4 different groups.
1 stereogenic center

d. $CH_3CH_2CH_2OH$

0 stereogenic centers

e. 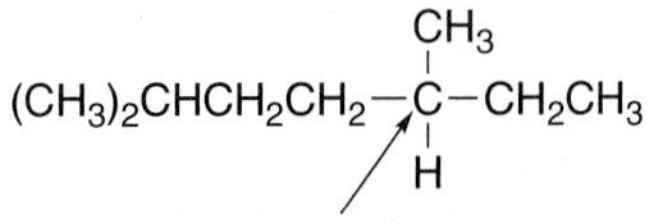 $(CH_3)_2CHCH_2CH_2-\overset{\overset{CH_3}{|}}{\underset{|}{C}}-CH_2CH_3$
H

This C is bonded to
4 different groups.
1 stereogenic center

f. 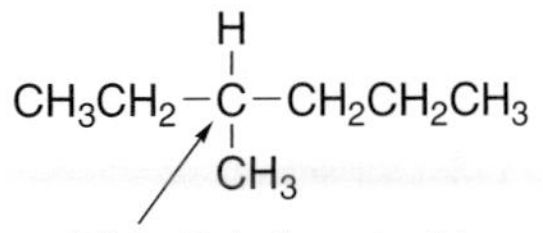
$$CH_3CH_2-\overset{\overset{H}{|}}{\underset{|}{C}}-CH_2CH_2CH_3$$
CH_3

This C is bonded to
4 different groups.
1 stereogenic center

5.7 Use the directions from Answer 5.6 to locate the stereogenic centers.

a. CH₃CH₂CH₂—C—CH₃ with H and OH
stereogenic center

b. (CH₃)₂CHCH₂—C—COOH with H and NH₂
stereogenic center

c.
both C's bonded to 4
different groups
2 stereogenic centers

d.
3 C's bonded to 4
different groups
3 stereogenic centers

5.8 Use the directions from Answer 5.6 to locate the stereogenic centers.

aliskiren

4 C's bonded to 4
different groups
4 stereogenic centers

5.9 Find the C bonded to four different groups in each molecule. At the stereogenic center, draw two bonds in the plane of the page, one in front (on a wedge), and one behind (on a dash). Then draw the mirror image (enantiomer).

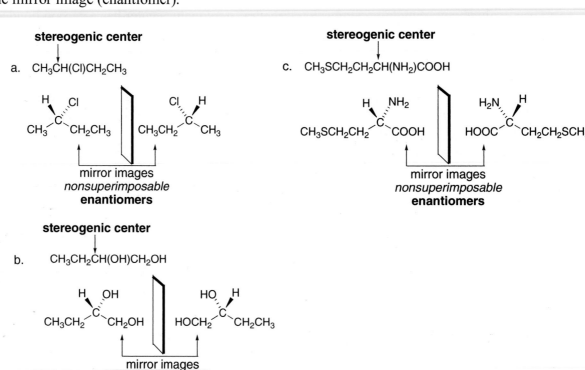

stereogenic center

a. CH₃CH(Cl)CH₂CH₃

mirror images
nonsuperimposable
enantiomers

stereogenic center

b. CH₃CH₂CH(OH)CH₂OH

mirror images
nonsuperimposable
enantiomers

stereogenic center

c. CH₃SCH₂CH₂CH(NH₂)COOH

mirror images
nonsuperimposable
enantiomers

5.10 Draw the chiral molecule with only C and H atoms.

$$CH_3-\overset{\overset{\displaystyle CH_2CH_3}{|}}{\underset{\underset{\displaystyle H}{|}}{C}}-CH_2CH_2CH_3 \quad or \quad CH_3-\overset{\overset{\displaystyle CH_2CH_3}{|}}{\underset{\underset{\displaystyle H}{|}}{C}}-CH(CH_3)_2$$

5.11 Use the directions from Answer 5.6 to locate the stereogenic centers.

a.
C bonded to
H and 3 different C's:
1 stereogenic center

c.
4 C's bonded to 4
different groups:
4 stereogenic centers

e.
C bonded to
H, 2 different O's and 1 C:
1 stereogenic center

b.
Each labeled C
is bonded to:
H, Cl, CH₂, CHCl:
2 stereogenic centers

d.
—NH₂
—CO₂H   **no stereogenic centers**

5.12

a.

cholesterol

b.

simvastatin

All stereogenic C's are circled. Each C is sp^3
hybridized and bonded to 4 different groups.

5.13 Assign priority based on atomic number: atoms with a higher atomic number get a higher priority.
If two atoms are the same, look at what they are bonded to and assign priority based on the atomic
number of these atoms.

a. –CH₃, –CH₂CH₃
higher priority

c. –H, –D
higher mass
higher priority

e. –CH₂CH₂Cl, –CH₂CH(CH₃)₂
higher priority

b. –I, –Br
higher priority

d. –CH₂Br, –CH₂CH₂Br
higher priority

f. –CH₂OH, –CHO = $-\overset{\overset{\displaystyle H}{|}}{\underset{\underset{\displaystyle O}{|}}{C}}=O$ = $-\overset{\overset{\displaystyle H}{|}}{\underset{\underset{\displaystyle C}{|}}{C}}\overset{O}{\underset{}{}}$

2 H's, 1 O   **2 O's, 1 H**   2 C–O bonds

C bonded to 2 O's has
higher priority.

5.14 Rank by decreasing priority. Lower atomic number = lower priority.

Highest priority = 1, Lowest priority = 4

a. –COOH C = second lowest atomic number — priority **3**

 –H H = lowest atomic number **4**

 –NH₂ N = second highest atomic number **2**

 –OH O = highest atomic number **1**

decreasing priority: –OH, –NH₂, –COOH, –H

c. –CH₂CH₃ C bonded to 2 H's + **1 C** priority **2**

 –CH₃ C bonded to 3 H's **3**

 –H H = lowest atomic number **4**

 –CH(CH₃)₂ C bonded to 1 H + **2 C's** **1**

decreasing priority: –CH(CH₃)₂, –CH₂CH₃, –CH₃, –H

b. –H H = lowest atomic number priority **4**

 –CH₃ C bonded to 3 H's **3**

 –Cl Cl = highest atomic number **1**

 –CH₂Cl C bonded to 2 H's + **1 Cl** **2**

decreasing priority: –Cl, –CH₂Cl, –CH₃, –H

d. –CH=CH₂ C bonded to 1 H + **2 C's** priority **2**

 –CH₃ C bonded to 3 H's **3**

 –C≡CH C bonded to **3 C's** **1**

 –H H = lowest atomic number **4**

decreasing priority: –C≡CH, –CH=CH₂, –CH₃, –H

5.15 To assign *R* or *S* to the molecule, first rank the groups. The lowest priority group must be oriented behind the page. If tracing a circle from (1) → (2) → (3) proceeds in the clockwise direction, the stereogenic center is labeled *R*; if the circle is counterclockwise, it is labeled *S*.

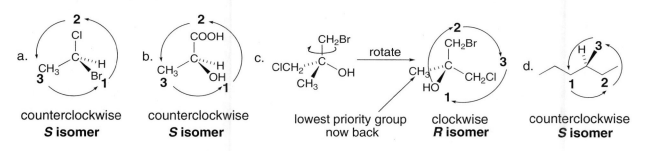

a. counterclockwise **S isomer**   b. counterclockwise **S isomer**   c. lowest priority group now back / clockwise **R isomer**   d. counterclockwise **S isomer**

5.16

clopidogrel

clockwise **R isomer**

counterclockwise **S isomer** **Plavix**

5.17 The maximum number of stereoisomers = 2^n where *n* = the number of stereogenic centers.

a. 3 stereogenic centers
 $2^3 = 8$ stereoisomers

b. 8 stereogenic centers
 $2^8 = 256$ stereoisomers

5.18

a. $CH_3CH_2CH(Cl)CH(OH)CH_2CH_3$

2 stereogenic centers = 4 possible stereoisomers

CH₃CH₂   CH₂CH₃
C—C
H···· Cl   OH ·H
A

CH₃CH₂   CH₂CH₃
C—C
H···· HO   Cl ·H
B

CH₃CH₂   CH₂CH₃
C—C
H···· Cl   H ·OH
C

CH₃CH₂   CH₂CH₃
C—C
HO···· H   Cl ·H
D

b. $CH_3CH(Br)CH_2CH(Cl)CH_3$

2 stereogenic centers = 4 possible stereoisomers

H   Cl H   Br
A

Br   H   Cl   H
B

H   Cl Br   H
C

H   Br   Cl   H
D

5.19

a. $CH_3CH(OH)CH(OH)CH_3$

2 stereogenic centers = 4 possible stereoisomers

CH₃   CH₃
C—C
HO···· H   OH ·H
A

CH₃   CH₃
C—C
H···· HO   H ·OH
B

CH₃   CH₃
C—C
H···· HO   H ·OH
C

CH₃   CH₃
C—C
H···· HO   H ·OH
C

↑ identical ↑

C is a meso compound.

A and **B** are enantiomers.
Pairs of diastereomers: **A** and **C**, **B** and **C**.

b. $CH_3CH(OH)CH(Cl)CH_3$

2 stereogenic centers = 4 possible stereoisomers

CH₃   CH₃
C—C
HO···· H   Cl ·H
A

CH₃   CH₃
C—C
H···· Cl   OH ·H
B

CH₃   CH₃
C—C
H···· HO   Cl ·H
C

CH₃   CH₃
C—C
H···· Cl   OH ·H
D

Pairs of enantiomers: **A** and **B**, **C** and **D**.
Pairs of diastereomers: **A** and **C**, **A** and **D**,
B and **C**, **B** and **D**.

5.20 **A meso compound must have at least two stereogenic centers. Usually a meso compound has a plane of symmetry.** You may have to rotate around a C–C bond to see the plane of symmetry clearly.

a.
CH₃CH₂ | CH₂CH₃
C—C
HO···· H   H ·OH

2 stereogenic centers
plane of symmetry
meso compound

b.
CH₃   OH ·H
C—C
HO···· H   CH₃

2 stereogenic centers
no plane of symmetry
not a meso compound

c.
H   Br

Br   H

→ rotate →

H   Br Br   H

2 stereogenic centers
plane of symmetry
meso compound

5.21 Use the definition in Answer 5.20 to draw the meso compounds.

a. BrCH$_2$CH$_2$CH(Cl)CH(Cl)CH$_2$CH$_2$Br   b. HO ⌐⌐⌐ OH   c. H$_2$N ⌐⌐⌐ NH$_2$

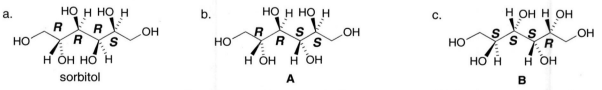

plane of symmetry   plane of symmetry   plane of symmetry

5.22 The enantiomer must have the exact opposite *R,S* designations. Diastereomers with two stereogenic centers have one center the same and one different.

If a compound is **R,S:**

Its enantiomer is: **S,R** ◄————— Exact opposite: *R* and *S* interchanged.

Its diastereomers are: **R,R and S,S** ◄————— One designation remains the same, the other changes.

5.23 The enantiomer must have the exact opposite *R,S* designations. For diastereomers, at least one of the *R,S* designations is the same, but not all of them.

a. (2*R*,3*S*)-2,3-hexanediol and (2*R*,3*R*)-2,3-hexanediol

One changes; one remains the same:
diastereomers

b. (2*R*,3*R*)-2,3-hexanediol and (2*S*,3*S*)-2,3-hexanediol

Both *R*'s change to *S*'s:
enantiomers

c. (2*R*,3*S*,4*R*)-2,3,4-hexanetriol and (2*S*,3*R*,4*R*)-2,3,4-hexanetriol

Two change; one remains the same:
diastereomers

5.24 The enantiomer must have the exact opposite *R,S* designations. For diastereomers, at least one of the *R,S* designations is the same, but not all of them.

a.

HO H HO H
R   *R*
HO   OH
    R   *S*
H OH HO H
sorbitol

b.

HO H HO H
R   *S*
HO   OH
    R   *S*
H OH   H OH
A

One changes; three remain the same.
diastereomer

c.

H OH H OH
S   *S*
HO   OH
    S   *R*
HO H   H OH
B

All stereogenic centers change.
enantiomers

5.25 Meso compounds generally have a plane of symmetry. They cannot have just one stereogenic center.

a.

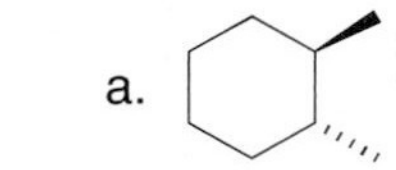

no plane of symmetry
not a meso compound

b.

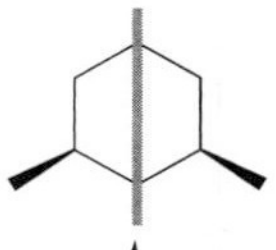

plane of symmetry
meso compound

c.

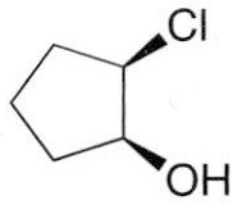

no plane of symmetry
not a meso compound

5.26

a. 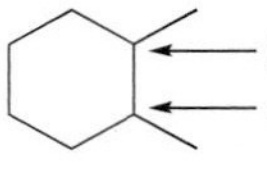 — 2 stereogenic centers =
— 4 stereoisomers maximum

c.

Draw the cis and trans isomers:

cis

CH₃

CH₃

A

identical

CH₃

CH₃

trans

CH₃

''CH₃

B

CH₃

CH₃''

C

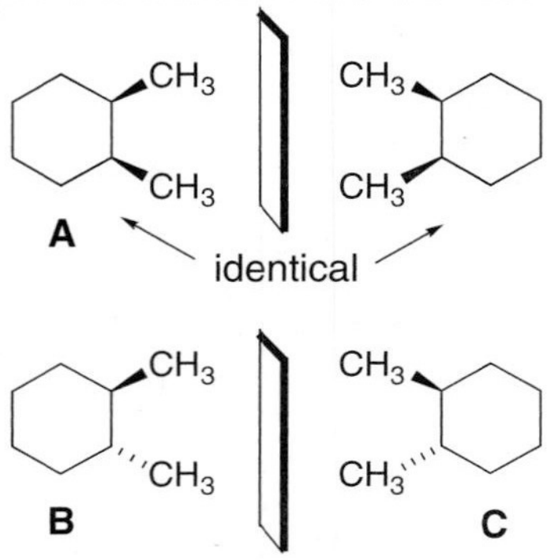

Pair of enantiomers: **B** and **C**.
Pairs of diastereomers: **A** and **B**, **A** and **C**.

Only 3 stereoisomers exist.

c.

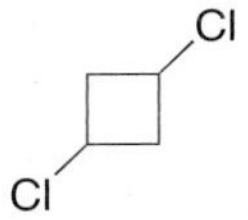

Draw the cis and trans isomers:

Cl

Cl

A

identical

Cl

Cl

Cl

Cl

B

identical

Cl

Cl

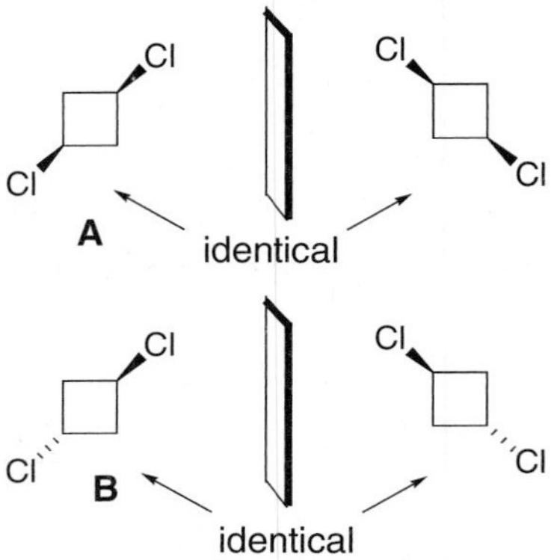

Pair of diastereomers: **A** and **B**.

Only 2 stereoisomers exist.

b. 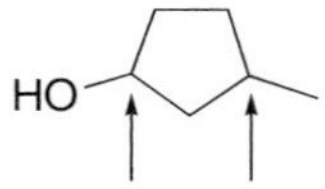 HO — 2 stereogenic centers =
4 stereoisomers maximum

Draw the cis and trans isomers:

cis

CH₃ OH

A

HO CH₃

B

trans

CH₃ ''OH

C

HO'' CH₃

D

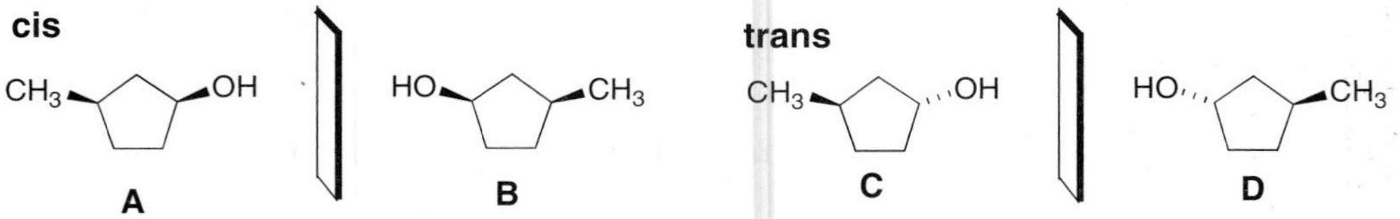

Pairs of enantiomers: **A** and **B**, **C** and **D**.
Pairs of diastereomers: **A** and **C**, **A** and **D**,
B and **C**, **B** and **D**.

All 4 stereoisomers exist.

5.27 Four facts:
- **Enantiomers** are mirror image isomers.
- **Diastereomers** are stereoisomers that are not mirror images.
- **Constitutional isomers** have the same molecular formula but the atoms are bonded to different atoms.
- **Cis and trans isomers** are always diastereomers.

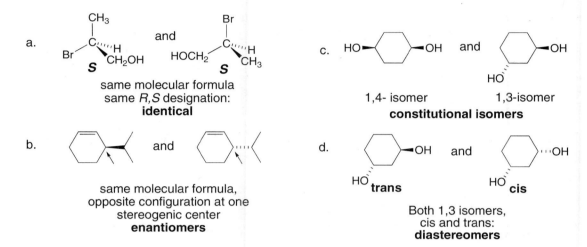

a.

same molecular formula
same *R,S* designation:
identical

b.

same molecular formula,
opposite configuration at one
stereogenic center
enantiomers

c.

1,4- isomer   1,3-isomer
constitutional isomers

d.

trans   **cis**

Both 1,3 isomers,
cis and trans:
diastereomers

5.28

COOH
│
CH₃—C⁗H
      NH₂

(*S*)-alanine
[α] = +8.5
mp = 297 °C

a. Mp = same as the *S* isomer.
b. The mp of a racemic mixture is often different from the melting point of the enantiomers.
c. –8.5, same as *S* but opposite sign
d. Zero. A racemic mixture is optically inactive.
e. Solution of pure (*S*)-alanine: **optically active**
   Equal mixture of (*R*)- and (*S*)-alanine: **optically inactive**
   75% (*S*)- and 25% (*R*)-alanine: **optically active**

5.29

$$[\alpha] = \frac{\alpha}{l \times c}$$

α = observed rotation
l = length of tube (dm)
c = concentration (g/mL)

$$[\alpha] = \frac{10°}{1 \text{ dm} \times (1 \text{ g}/10 \text{ mL})} = +100 = \textbf{specific rotation}$$

5.30 Enantiomeric excess = *ee* = % of one enantiomer – % of other enantiomer.

a. 95% – 5% = **90% *ee***   b. 85% – 15% = **70% *ee***

5.31

a. 90% *ee* means 90% excess of **A**, and 10% racemic mixture of **A** and **B** (5% each); therefore, **95% A and 5% B.**

b. 99% *ee* means 99% excess of **A**, and 1% racemic mixture of **A** and **B** (0.5% each); therefore, **99.5% A and 0.5% B.**

c. 60% *ee* means 60% excess of **A**, and 40% racemic mixture of **A** and **B** (20% each); therefore, **80% A and 20% B.**

5.32

$$ee = \frac{[\alpha] \text{ mixture}}{[\alpha] \text{ pure enantiomer}} \times 100\%$$

a. $\frac{+10}{+24} \times 100\% = 42\%$ *ee*

b. $\frac{[\alpha] \text{ solution}}{+24} \times 100\% = 80\%$ *ee*

   $[\alpha]$ solution = +19.2

5.33

a. $\frac{[\alpha] \text{ mixture}}{+3.8} \times 100\% = 60\%$ *ee*

   $[\alpha]$ mixture = +2.3

b. % one enantiomer – % other enantiomer = *ee*
   80% – 20% = 60% *ee*

   80% dextrorotatory (+) enantiomer
   20% levorotatory (–) enantiomer

5.34• Enantiomers have the same physical properties (mp, bp, solubility), and rotate the plane of polarized light to an equal but opposite extent.
- **Diastereomers have different physical properties.**
- **A racemic mixture is optically inactive.**

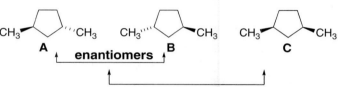

trans isomers   cis isomer

A and **B** are diastereomers of **C**.

a. The bp's of **A** and **B** are the same. The bp's of **A** and **C** are different.
b. Pure **A**: optically active
   Pure **B**: optically active
   Pure **C**: optically inactive
   Equal mixture of **A** and **B**: optically inactive
   Equal mixture of **A** and **C**: optically active
c. There would be two fractions: one containing **A** and **B** (optically inactive), and one containing **C** (optically inactive).

5.35 Use the definitions from Answer 5.2.

a.  and

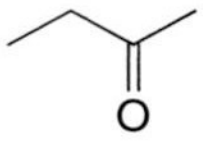

same molecular formula C_4H_8O
different connectivity
constitutional isomers

c. 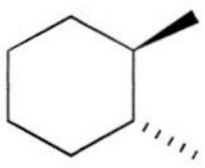 and

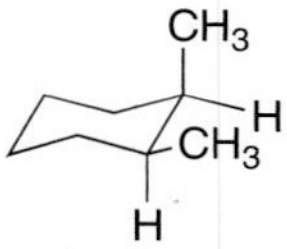

one up, one down   both up
trans   **cis**

Both compounds are
1,2-dimethylcyclohexane.
one cis, one trans = **stereoisomers**

b. 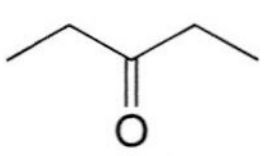 and

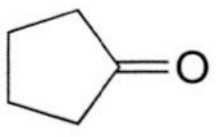

$C_5H_{10}O$ C_5H_8O
different molecular formulas
not isomers

d. 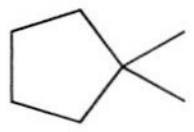 and

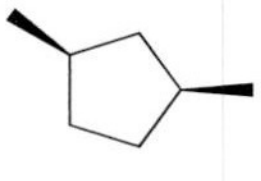

same molecular formula C_7H_{14}
different connectivity
constitutional isomers

5.36 Use the definitions from Answer 5.3.

a.

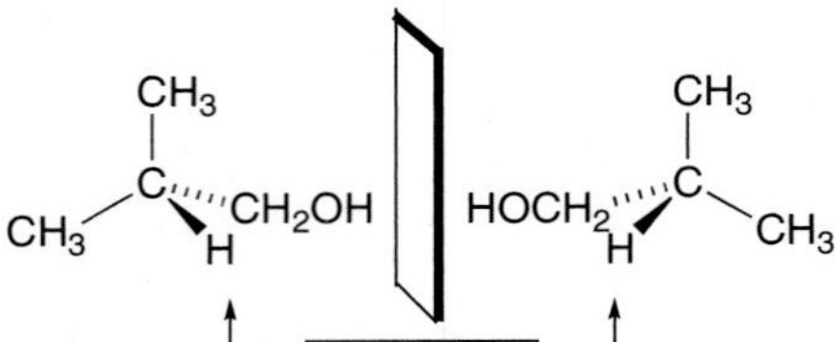

identical
achiral

c.

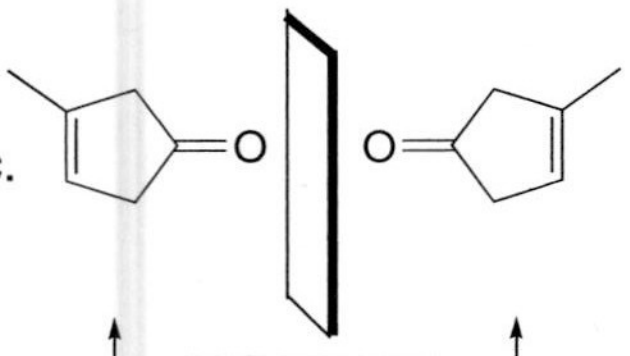

identical
achiral

e.

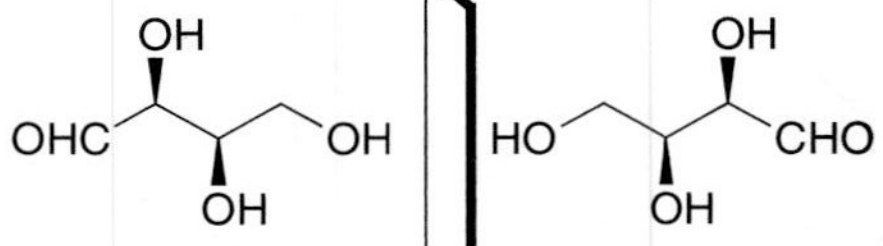

threose

chiral

b.

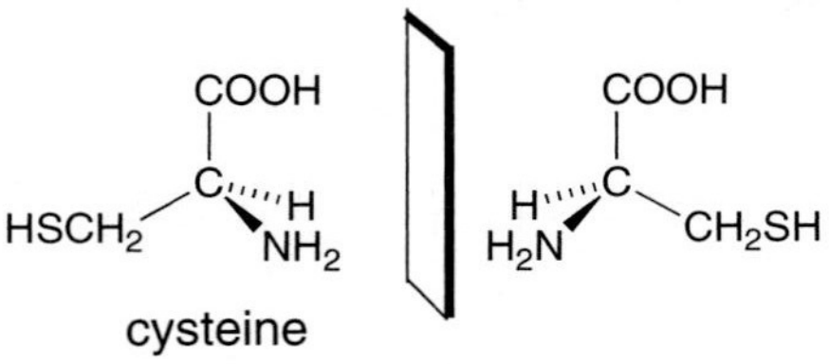

cysteine

chiral

d.

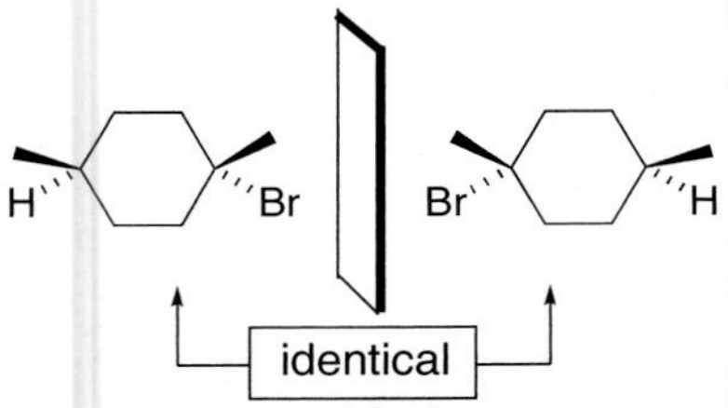

identical
achiral

5.37

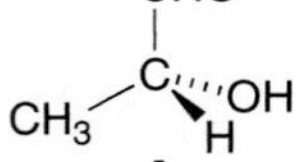

A
R isomer

a. 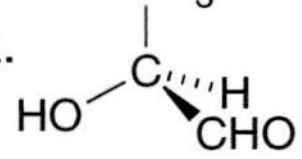
S
enantiomer

b. 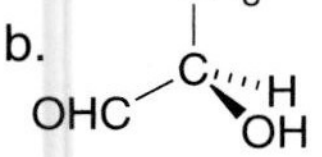
R
identical

c. 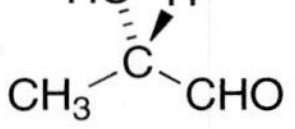
S
enantiomer

5.38 A plane of symmetry cuts the molecule into **two identical halves**.

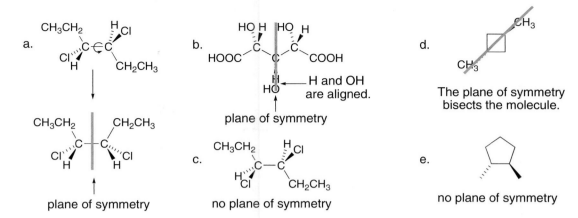

a.

b.

H and OH
are aligned.

plane of symmetry

d.

The plane of symmetry
bisects the molecule.

c.

no plane of symmetry

e.

no plane of symmetry

plane of symmetry

5.39 Use the directions from Answer 5.6 to locate the stereogenic centers.

a. $CH_3CH_2CH_2CH_2CH_2CH_3$
   All C's have 2 or more H's.
   0 stereogenic centers

b.
   $CH_3CH_2O{-}\overset{\overset{\displaystyle H}{|}}{\underset{\underset{\displaystyle CH_3}{|}}{C}}{-}CH_2CH_3$
   1 stereogenic center

c. $(CH_3)_2CHCH(OH)CH(CH_3)_2$
   0 stereogenic centers

d. $(CH_3)_2CHCH_2{-}\overset{\overset{\displaystyle H}{|}}{\underset{\underset{\displaystyle CH_3}{|}}{C}}{-}CH_2{-}\overset{\overset{\displaystyle H}{|}}{\underset{\underset{\displaystyle CH_3}{|}}{C}}{-}\overset{\overset{\displaystyle H}{|}}{\underset{\underset{\displaystyle CH_3}{|}}{C}}{-}CH_2CH_3$
   3 stereogenic centers

e. $CH_3{-}\overset{\overset{\displaystyle H}{|}}{\underset{\underset{\displaystyle D}{|}}{C}}{-}CH_2CH_3$
   bonded to 4 different groups
   1 stereogenic center

f.
   Each indicated C bonded to 4 different groups =
   6 stereogenic centers

g.
   bonded to 4 different groups
   1 stereogenic center

h.
   All C's have 2 or more H's, or
   are sp^2 hybridized.
   0 stereogenic centers

i.
   Each indicated C bonded to
   4 different groups =
   2 stereogenic centers

j.
   Each indicated C bonded to 4 different groups =
   5 stereogenic centers

5.40 Stereogenic centers are circled.

Eight constitutional isomers:

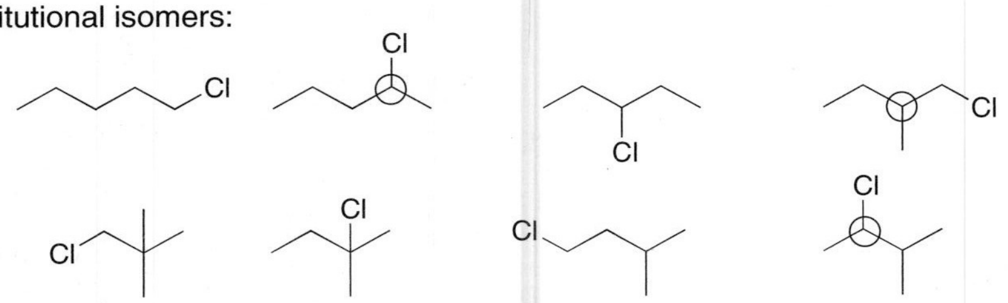

5.41

a.

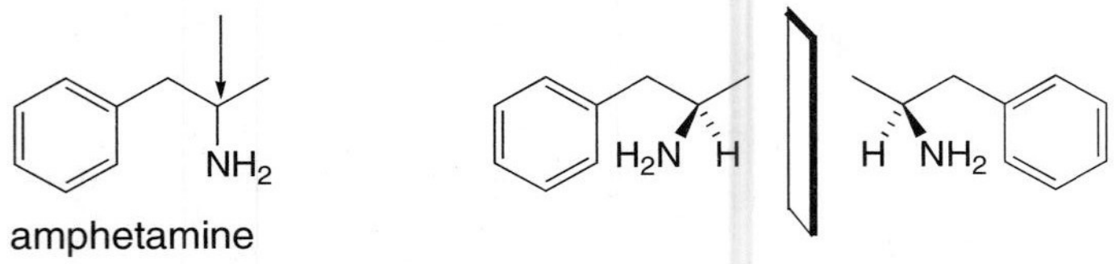

amphetamine

b.

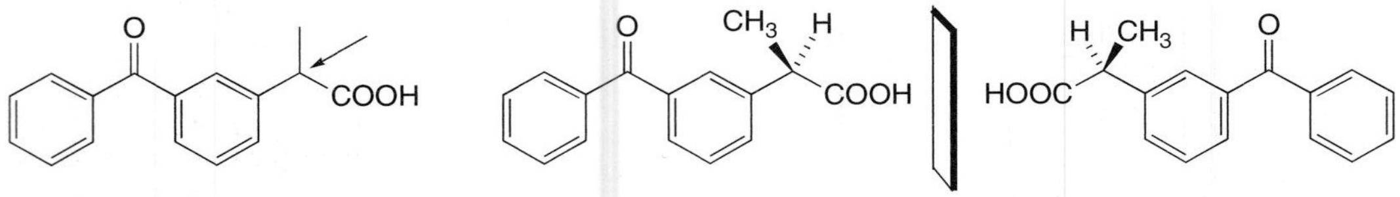

ketoprofen

5.42 Draw a molecule to fit each description.

a.

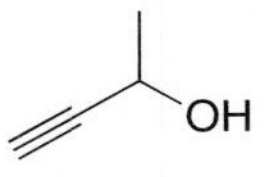

alcohol

b.

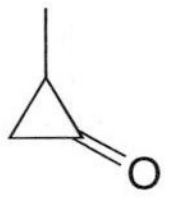

ketone

c.

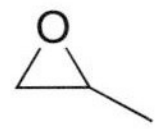

cyclic ether

5.43 Assign priority based on the rules in Answer 5.13.

a. –OH, –NH$_2$

higher atomic number
higher priority

b. –CD$_3$, –CH$_3$

D higher mass than H
higher priority

c. –CH(CH$_3$)$_2$, –CH$_2$OH

C bonded to O
higher priority

d. –CH$_2$Cl, –CH$_2$CH$_2$CH$_2$Br

C bonded to Cl
higher priority

e. –CHO, –COOH

C has 3 bonds to O.
higher priority

f. –CH$_2$NH$_2$, –NHCH$_3$

higher atomic number
higher priority

5.44 Assign priority based on the rules in Answer 5.13.

a. –F > –OH > –NH$_2$ > –CH$_3$

b. –(CH$_2$)$_3$CH$_3$ > –CH$_2$CH$_2$CH$_3$ > –CH$_2$CH$_3$ > –CH$_3$

c. –NH$_2$ > –CH$_2$NHCH$_3$ > –CH$_2$NH$_2$ > –CH$_3$

d. –COOH > –CHO > –CH$_2$OH > –H

e. –Cl > –SH > –OH > –CH$_3$

f. –C≡CH > –CH=CH$_2$ > –CH(CH$_3$)$_2$ > –CH$_2$CH$_3$

5.45 Use the rules in Answer 5.15 to assign *R* or *S* to each stereogenic center.

a.

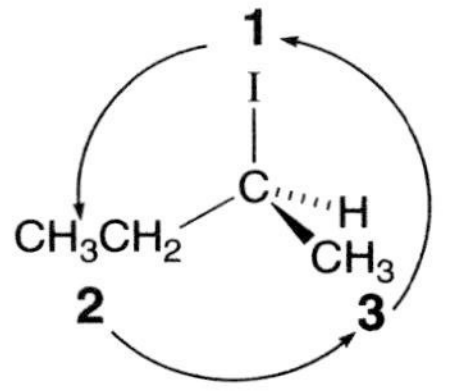

counterclockwise
S isomer

c.

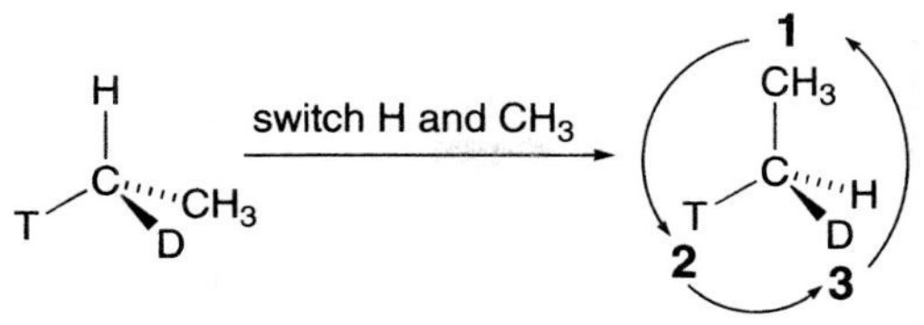

switch H and CH$_3$

counterclockwise
It looks like an *S* isomer, but we
must reverse the answer, *S* to *R*.
R isomer

b.

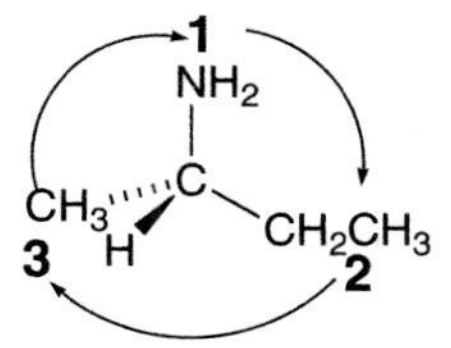

clockwise, but H in front
S isomer

d.

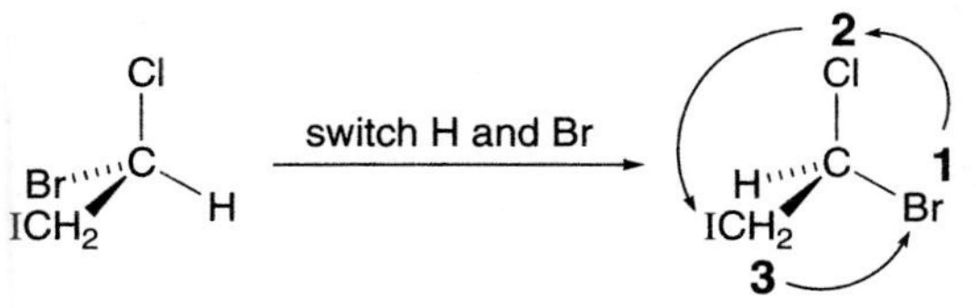

switch H and Br

counterclockwise
It looks like an *S* isomer, but we
must reverse the answer, *S* to *R*.
R isomer

e.

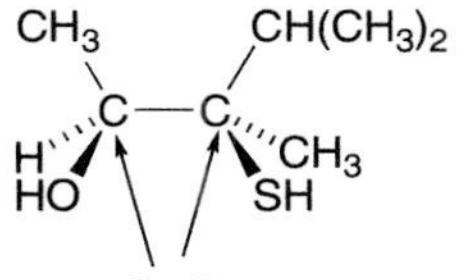

S, R

f.

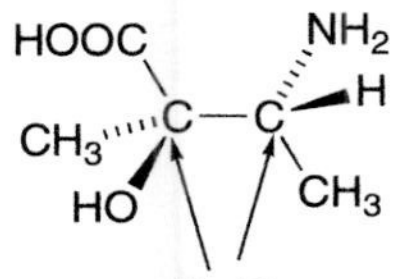

R, R

g.

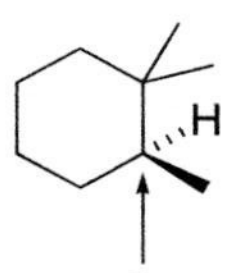

S

h.

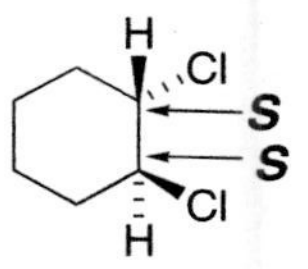

S
S

5.46

a. (3*R*)-3-methylhexane

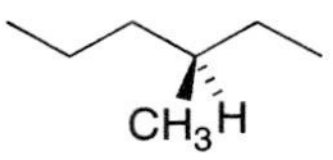

b. (4*R*,5*S*)-4,5-diethyloctane

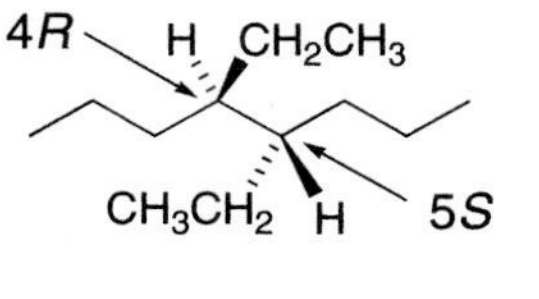

c. (3*R*,5*S*,6*R*)-5-ethyl-3,6-dimethylnonane

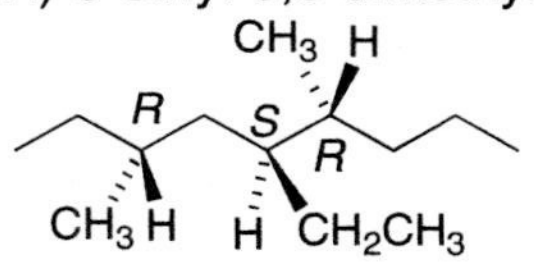

d. (3*S*,6*S*)-6-isopropyl-3-methyldecane

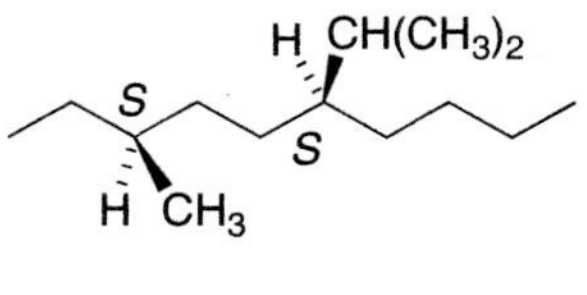

5.47

a.

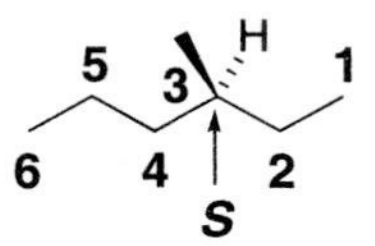

(3S)-3-methylhexane

b.

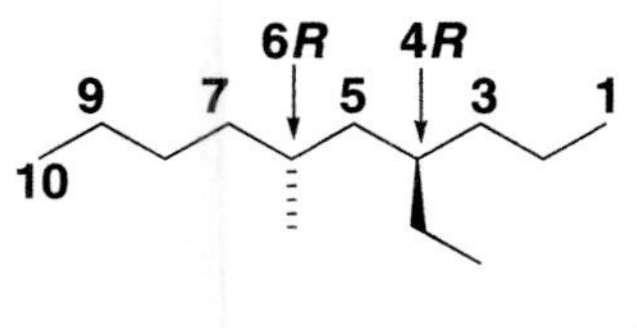

(4R,6R)-4-ethyl-6-methyldecane

c.

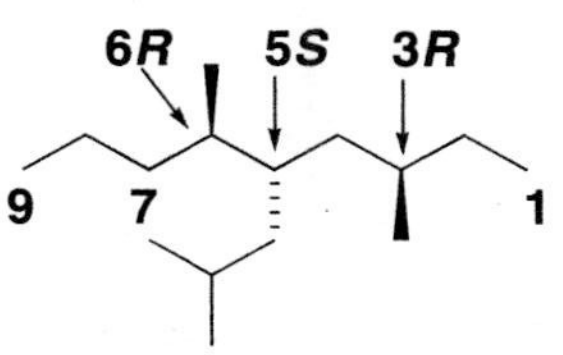

(3R,5S,6R)-5-isobutyl-3,6-dimethylnonane

5.48 Two enantiomers of the amino acid leucine.

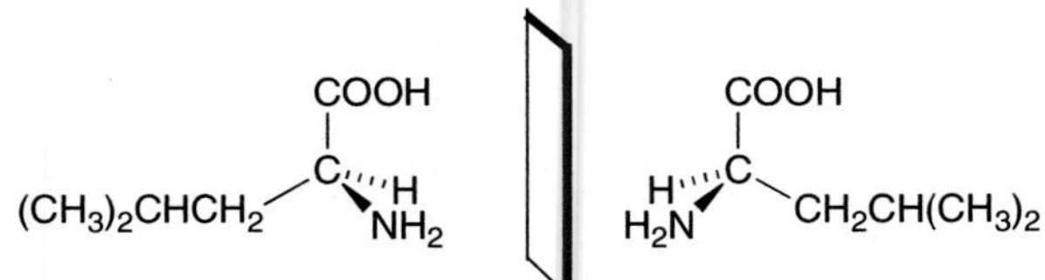

S isomer
naturally occurring

R isomer

5.49

a. 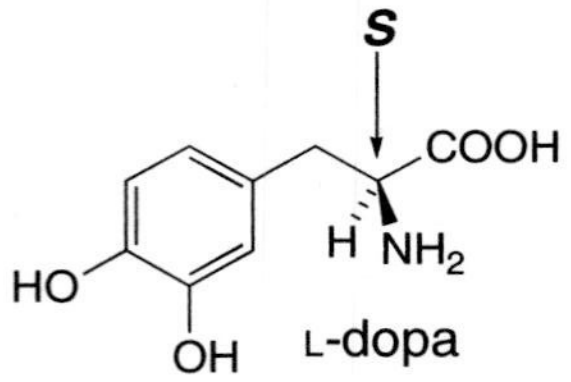
L-dopa

b. 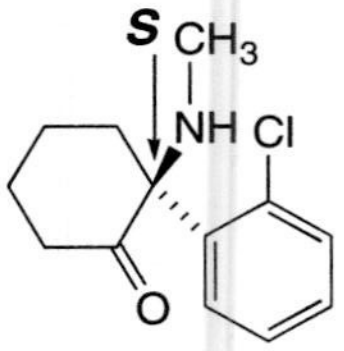
ketamine

c. 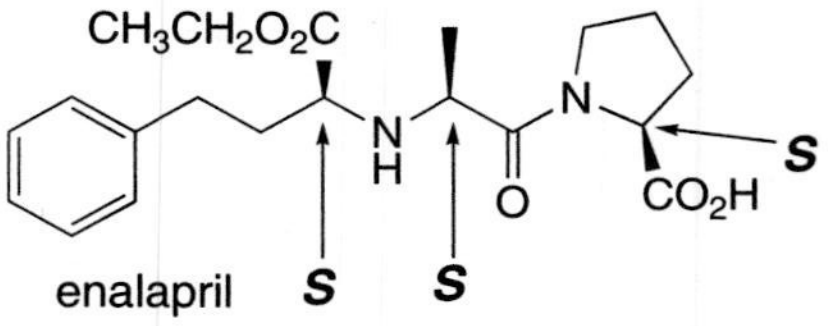
enalapril **S S**

5.50

methylphenidate

R,R

S,S

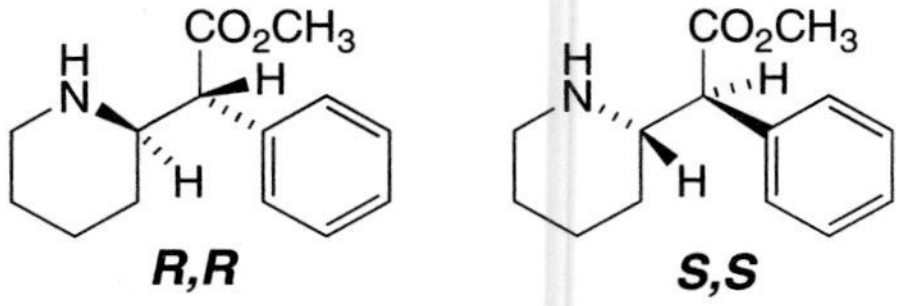

5.51

a. 1R,2S
ephedrine

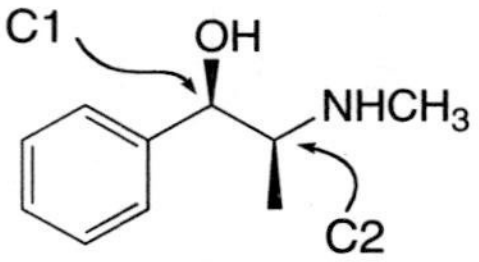

b. 1S,2S
pseudoephedrine

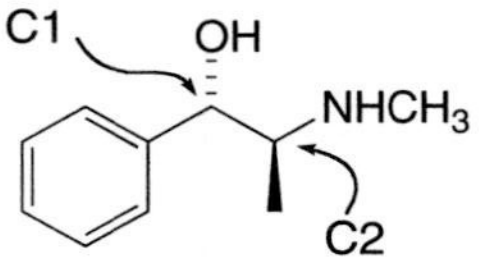

c. Ephedrine and pseudoephedrine
are diastereomers (one stereogenic
center is the same; one is different).

d. 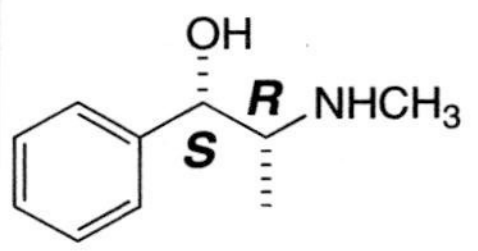
e. ⟵ enantiomer of ephedrine

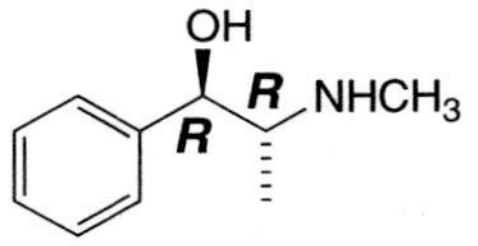

⟵ diastereomer of ephedrine

5.52

a. 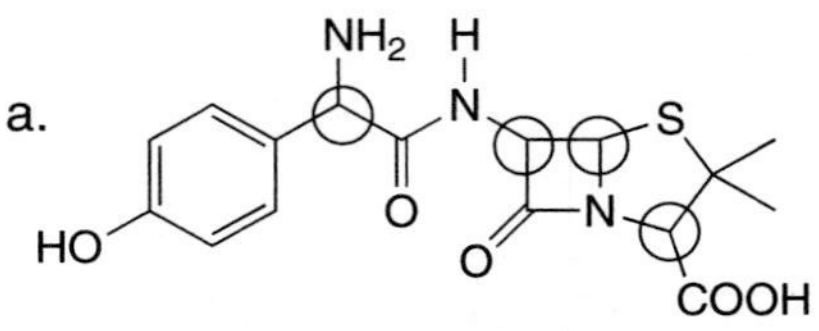
amoxicillin

b. 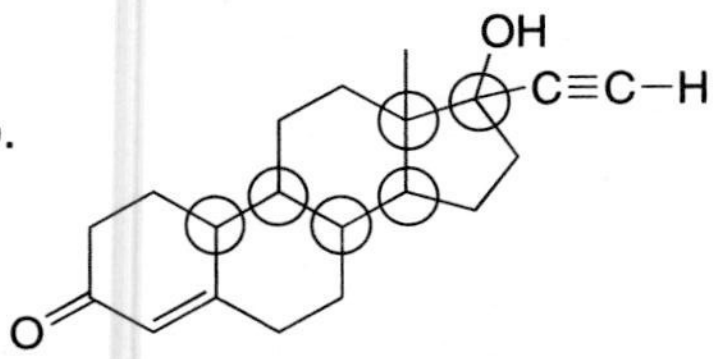
norethindrone

c. 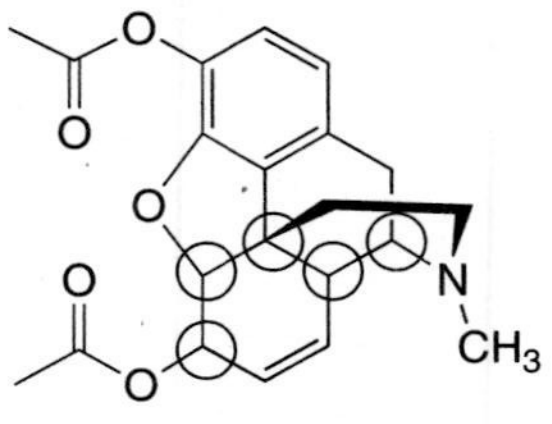
heroin

5.53

a. $CH_3CH(OH)CH(OH)CH_2CH_3$

2 stereogenic centers
$2^2 = 4$ possible stereoisomers

b. $CH_3CH_2CH_2CH(CH_3)_2$

0 stereogenic centers

c.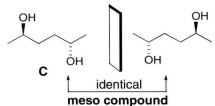

4 stereogenic centers
$2^4 = 16$ possible stereoisomers

5.54

a. $CH_3CH(OH)CH(OH)CH_2CH_3$

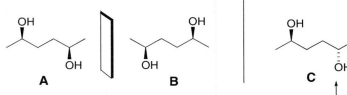

Pairs of enantiomers: **A** and **B**, **C** and **D**.
Pairs of diastereomers: **A** and **C**, **A** and **D**, **B** and **C**, **B** and **D**.

b. $CH_3CH(OH)CH_2CH_2CH(OH)CH_3$

Pair of enantiomers: **A** and **B**.
Pairs of diastereomers: **A** and **C**, **B** and **C**.

c. $CH_3CH(Cl)CH_2CH(Br)CH_3$

Pairs of enantiomers: **A** and **B**, **C** and **D**.
Pairs of diastereomers: **A** and **C**, **A** and **D**, **B** and **C**, **B** and **D**.

d. $CH_3CH(Br)CH(Br)CH(Br)CH_3$

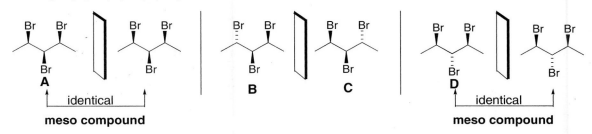

Pair of enantiomers: **B** and **C**.
Pairs of diastereomers: **A** and **B**, **A** and **C**, **A** and **D**, **B** and **D**, **C** and **D**.

5.55

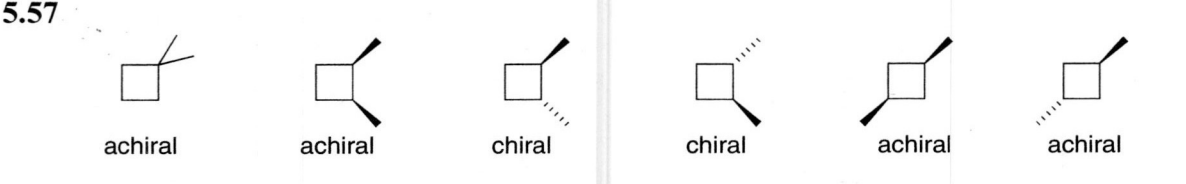

a.

HOCH₂   CH₃   CH₃   CH₂OH   CH₃   CH₂OH   or   CH₃   CH₂OH

enantiomer   **diastereomer**   **diastereomer**

b.

enantiomer   **diastereomer**

c.

enantiomer   **diastereomer**   or   **diastereomer**

d.

enantiomer   **diastereomer**   or   **diastereomer**

5.56

a.

A   identical   **B**   **C**

meso compound

Pair of enantiomers: **B** and **C**.
Pairs of diastereomers: **A** and **B**, **A** and **C**.

b.

A   identical   **B**   identical

Pair of diastereomers: **A** and **B**.
Meso compounds: **A** and **B**.

c.

A   **B**   **C**   **D**

Pairs of enantiomers: **A** and **B**, **C** and **D**.
Pairs of diastereomers: **A** and **C**, **A** and **D**, **B** and **C**, **B** and **D**.

5.57

achiral   achiral   chiral   chiral   achiral   achiral

5.58 Explain each statement.

a. All molecules have a mirror image, but only chiral molecules have enantiomers. **A** is not chiral, and therefore, does not have an enantiomer.

b. **B** has one stereogenic center, and therefore, has an enantiomer. Only compounds with two or more stereogenic centers have diastereomers.

c. **C** is chiral and has two stereogenic centers, and therefore, has both an enantiomer and a diastereomer.

d. **D** has two stereogenic centers, but is a meso compound. Therefore, it has a diastereomer, but no enantiomer since it is achiral.

plane of symmetry

e. **E** has two stereogenic centers, but is a meso compound. Therefore, it has a diastereomer, but no enantiomer since it is achiral.

plane of symmetry

5.59

C2 C3

D-erythrose
2R,3R

a. 2S,3R diastereomer

b. 2R,3R identical

c. 2S,3S enantiomer

d. 2R,3S diastereomer

5.60 Re-draw each Newman projection and determine the *R,S* configuration. Then determine how the molecules are related.

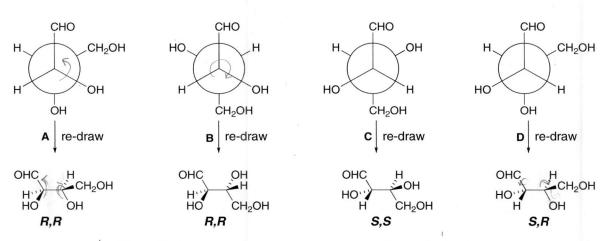

A | re-draw

B | re-draw

C | re-draw

D | re-draw

R,R

R,R

S,S

S,R

a. **A** and **B** are identical.
b. **A** and **C** are enantiomers.

c. **A** and **D** are diastereomers.
d. **C** and **D** are diastereomers.

5.61

A (trans, *R*) and **B** (cis, *R*) are diastereomers.
A (trans, *R*) and **C** (trans, *R*) are identical molecules.
A (trans, *R*) and **D** (trans, *S*) are enantiomers.
A (trans, *R*) and **E** are constitutional isomers.

5.62

a.   and

enantiomers

b.   CH₃ / CH₃   and

same molecular formula
different connectivity
constitutional isomers

c.   CH₃ — CHO   and   OHC — CH₃

2*R*,3*S*   2*R*,3*R*

one different configuration
diastereomers

d.   and

different molecular formulas
not isomers

e.   Cl   and   Cl

mirror images
not superimposable
enantiomers

f.   Cl / I—C—Br / H   and   Cl / H—C—Br / I

enantiomers

g.   CH₃   CH₃ / H—C—C—Br / Br   H   and   CH₃   H / H—C—C—Br / Br   CH₃

2*S*,3*S*   2*S*,3*S*
identical

h.   H / HO—⬡—OH / H   and   OH / HO—⬡—H / H

1,4-trans   1,4-cis

diastereomers

i.   and

6 H's   12 H's
different molecular formulas
not isomers

j.   H / BrCH₂—C—CH₂OH / CH₃   and   CH₃ / HOCH₂—C—H / BrCH₂

enantiomers

k.   and   H—⬡—CH₃ / CH₃   H

1,3-cis   1,3-trans

diastereomers

l.   HO CH₃ / H—C—CH₂Br   Br on end   and   CH₃ — C — CH₂OH / Br   H   Br in middle

different connectivity
constitutional isomers

5.63

a. **A** and **B** are constitutional isomers.
A and **C** are constitutional isomers.
B and **C** are diastereomers (cis and trans).
C and **D** are enantiomers.

b.

plane of symmetry

A
A has two
planes of symmetry.
achiral

B
achiral

C
chiral

D
chiral

mirror images and not
superimposable
enantiomers

c. Alone, **C** and **D** would be optically active.
d. **A** and **B** have a plane of symmetry.
e. **A** and **B** have different boiling points.
B and **C** have different boiling points.
C and **D** have the same boiling point.
f. **B** is a meso compound.
g. An equal mixture of **C** and **D** is optically inactive because it is a racemic mixture.
An equal mixture of **B** and **C** would be optically active.

5.64

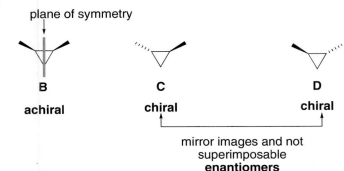

quinine

$$ee = \frac{[\alpha] \text{ mixture}}{[\alpha] \text{ pure enantiomer}} \times 100\%$$

quinine = **A**
quinine's enantiomer = **B**

a.

$$\frac{-50}{-165} \times 100\% = 30\% \, ee$$

b. 30% *ee* = 30% excess one compound (**A**)
remaining 70% = mixture of 2 compounds (35% each **A** and **B**)
Amount of **A** = 30 + 35 = **65%**
Amount of **B** = **35%**

$$\frac{-83}{-165} \times 100\% = 50\% \, ee$$

50% *ee* = 50% excess one compound (**A**)
remaining 50% = mixture of 2 compounds (25% each **A** and **B**)
Amount of **A** = 50 + 25 = **75%**
Amount of **B** = **25%**

$$\frac{-120}{-165} \times 100\% = 73\% \, ee$$

73% *ee* = 73% excess of one compound (**A**)
remaining 27% = mixture of 2 compounds (13.5% each **A** and **B**)
Amount of **A** = 73 + 13.5 = **86.5%**
Amount of **B** = **13.5%**

c. $[\alpha] = +165$
d. 80% − 20% = 60% *ee*

e. $60\% = \dfrac{[\alpha] \text{ mixture}}{-165} \times 100\%$

$[\alpha]$ mixture = −99

5.65

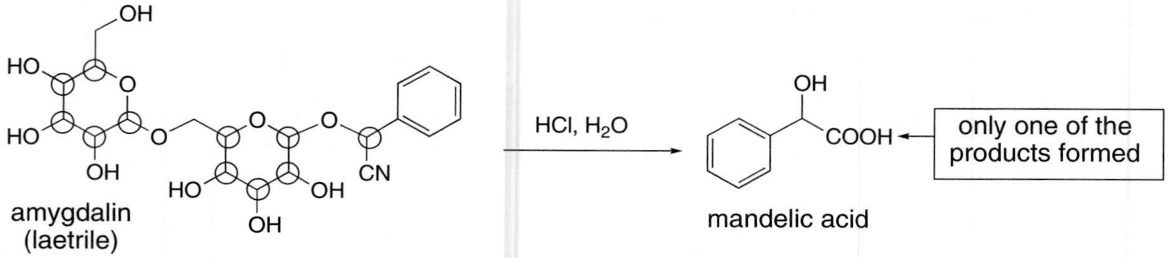

amygdalin
(laetrile)

$\xrightarrow{\text{HCl, H}_2\text{O}}$

mandelic acid

only one of the products formed

a. The 11 stereogenic centers are circled. Maximum number of stereoisomers = 2^{11} = 2048
b. Enantiomers of mandelic acid:

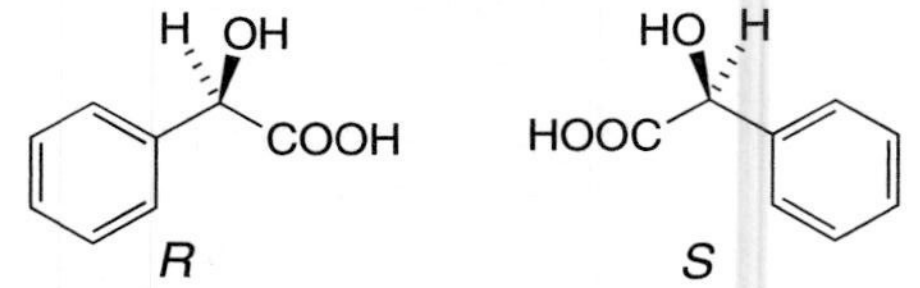

R   S

c. 60% − 40% = 20% *ee*
   20% = [α] mixture/−154 x 100%
   [α] mixture = −31

d. *ee* = $\dfrac{+50}{+154}$ x 100% = 32% *ee*

   [α] for (*S*)-mandelic acid = +154

32% excess of the *S* enantiomer
68% of racemic *R* and *S* = 34% *S* and 34% *R*

S enantiomer: 32% + 34% = 66%
R enantiomer = 34%

5.66

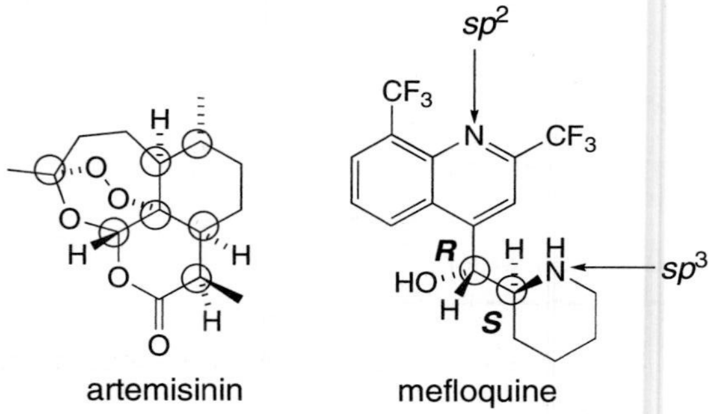

artemisinin   mefloquine

a. Each stereogenic center is circled.
b. The stereogenic centers in mefloquine are labeled.
c. Artemisinin has 7 stereogenic centers.
   $2^n = 2^7 = 128$ possible stereoisomers
d. One N atom in mefloquine is sp^2 and one is sp^3.
e. Two molecules of artemisinin cannot intermolecularly H-bond because there are no O–H or N–H bonds.

f.

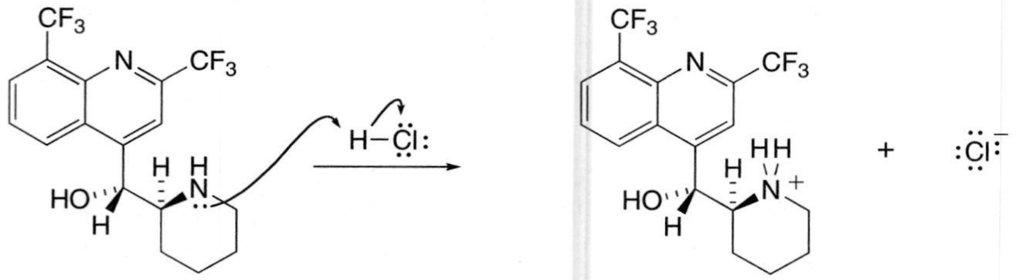

5.67

a. Each stereogenic center is circled.

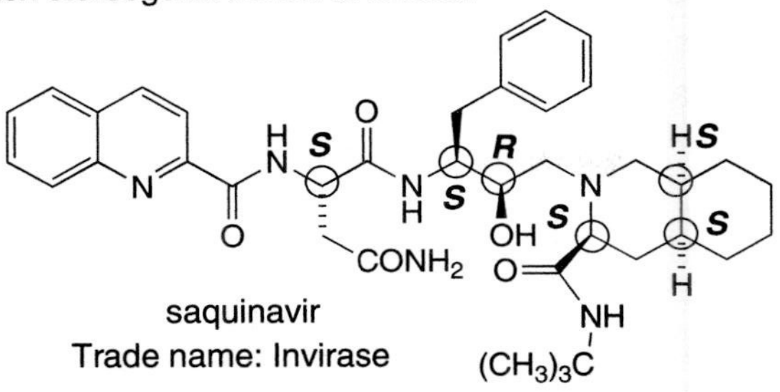

saquinavir
Trade name: Invirase

c. diastereomer

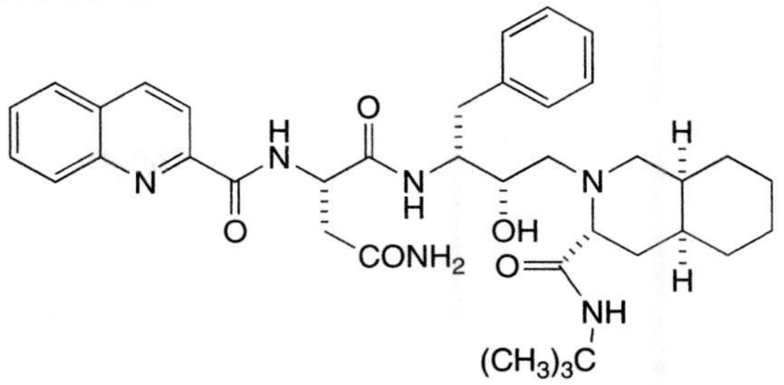

b. enantiomer

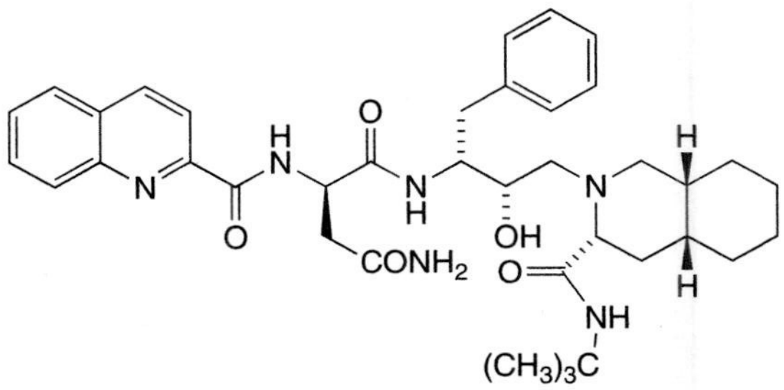

d. constitutional isomer

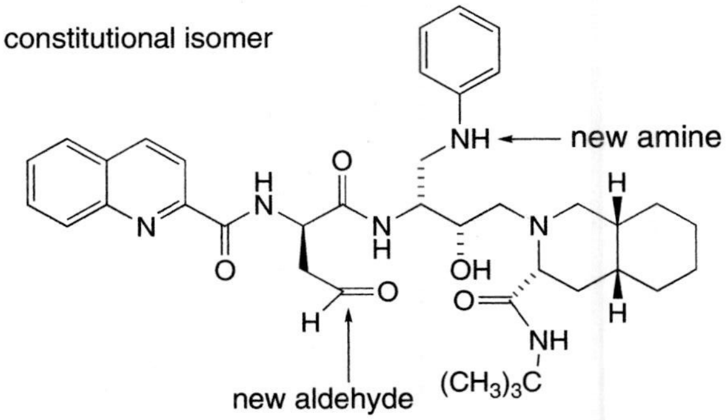

new amine

new aldehyde

5.68

a.

re-draw

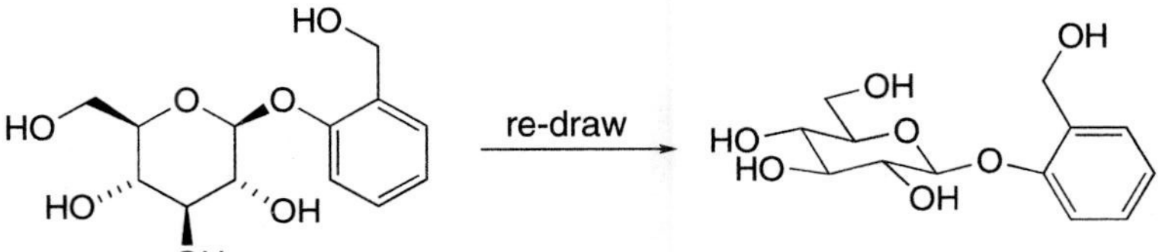

salicin

more stable ring
all groups equatorial

b. diastereomer

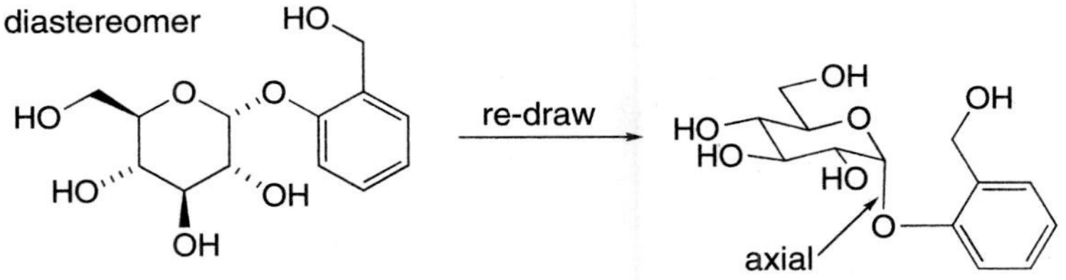

re-draw

axial

All other groups on the ring are equatorial.

c. enantiomer

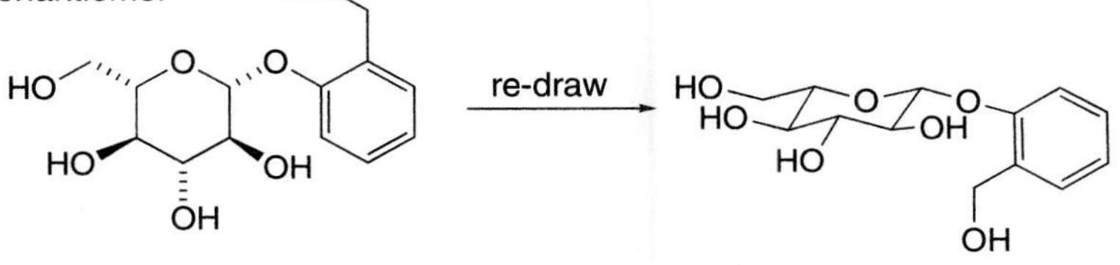

re-draw

In the enantiomer, all groups are still equatorial, but all down bonds are up bonds and all up bonds are down bonds.

5.69 Allenes contain an *sp* hybridized carbon atom doubly bonded to two other carbons. This makes the double bonds of an allene perpendicular to each other. When each end of the allene has two like substituents, the allene contains two planes of symmetry and it is achiral. When each end of the allene has two different groups, the allene has no plane of symmetry and it becomes chiral.

These two substituents are at 90° to these two substituents.
Allene **A** contains two planes of symmetry,
making it **achiral.**

$HC{\equiv}C-C{\equiv}C\!+\!CH{=}C{=}CH\!+\!CH{=}CH-CH{=}CHCH_2CO_2H$

allene → **mycomycin**
re-draw

The substituents on each end of the allene in mycomycin
are different. Therefore, mycomycin is **chiral.**

5.70

discodermolide

a. The 13 tetrahedral stereogenic centers are circled.

b. Because there is restricted rotation around a C–C double bond, groups on the end of the double bond cannot interconvert. Whenever the substituents on each end of the double bond are different from each other, the double bond is a stereogenic site. Thus, the following two double bonds are isomers:

These compounds are isomers.

There are three stereogenic double bonds in discodermolide, labeled with arrows.

c. The maximum number of stereoisomers for discodermolide must include the 13 tetrahedral stereogenic centers and the three double bonds. Maximum number of stereoisomers = 2^{16} = 65,536.

5.71

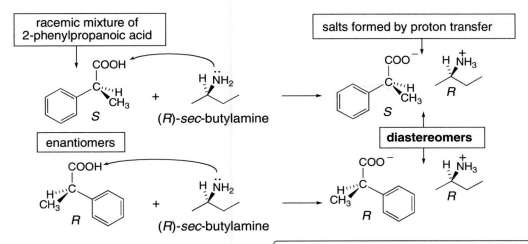

These salts are **diastereomers,** and they are now separable by physical methods since they have different physical properties.

Chapter 6: Understanding Organic Reactions

◆ Writing organic reactions (6.1)

- Use curved arrows to show the movement of electrons. Full-headed arrows are used for electron pairs and half-headed arrows are used for single electrons.

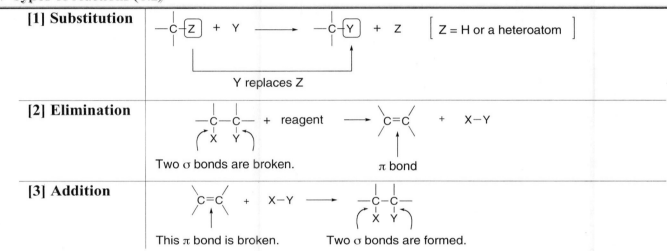

half-headed arrows   full-headed arrow

- Reagents can be drawn either on the left side of an equation or over an arrow. Catalysts are drawn over or under an arrow.

◆ Types of reactions (6.2)

[1] Substitution	![substitution reaction] Z = H or a heteroatom Y replaces Z
[2] Elimination	Two σ bonds are broken. π bond
[3] Addition	This π bond is broken. Two σ bonds are formed.

◆ Important trends

Values compared	Trend
Bond dissociation energy and **bond strength**	The **higher** the bond dissociation energy, the **stronger** the bond (6.4). Increasing size of the halogen → CH_3-F CH_3-Cl CH_3-Br CH_3-I $\Delta H^o = $ 456 kJ/mol 351 kJ/mol 293 kJ/mol 234 kJ/mol ← Increasing bond strength

E_a and **reaction rate**	The *larger* the energy of activation, the *slower* the reaction (6.9A).

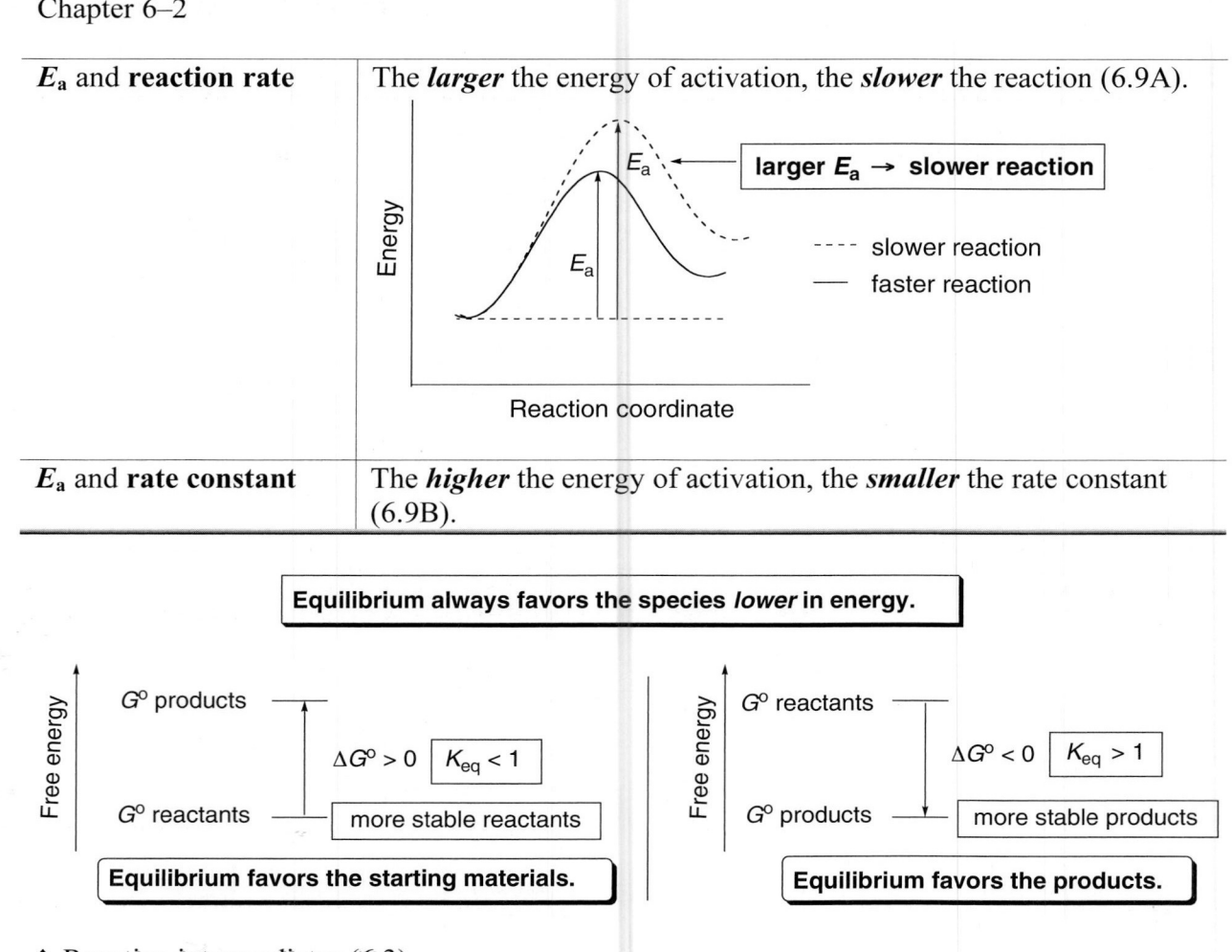

E_a and **rate constant**	The *higher* the energy of activation, the *smaller* the rate constant (6.9B).

◆ Reactive intermediates (6.3)

- Breaking bonds generates reactive intermediates.
- Homolysis generates radicals with unpaired electrons.
- Heterolysis generates ions.

Reactive intermediate	General structure	Reactive feature	Reactivity
radical	—C·	unpaired electron	electrophilic
carbocation	—C+	positive charge; only six electrons around C	electrophilic
carbanion	—C:⁻	net negative charge; lone electron pair on C	nucleophilic

◆ Energy diagrams (6.7, 6.8)

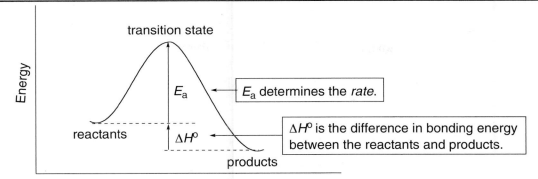

◆ Conditions favoring product formation (6.5, 6.6)

Variable	Value	Meaning
K_{eq}	$K_{eq} > 1$	More product than starting material is present at equilibrium.
ΔG^{o}	$\Delta G^{o} < 0$	The energy of the products is **lower** than the energy of the reactants.
ΔH^{o}	$\Delta H^{o} < 0$	Bonds in the products are **stronger** than bonds in the reactants.
ΔS^{o}	$\Delta S^{o} > 0$	The product is **more disordered** than the reactant.

◆ Equations (6.5, 6.6)

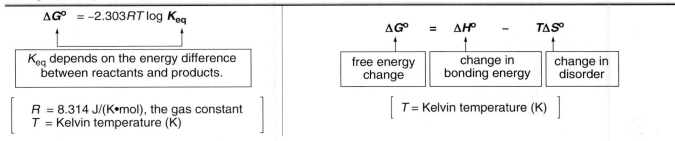

◆ Factors affecting reaction rate (6.9)

Factor	Effect
energy of activation	higher E_a → slower reaction
concentration	higher concentration → faster reaction
temperature	higher temperature → faster reaction

Chapter 6: Answers to Problems

6.1 [1] In a **substitution reaction**, one group replaces another.
[2] In an **elimination reaction**, elements of the starting material are lost and a π bond is formed.
[3] In an **addition reaction**, elements are added to the starting material.

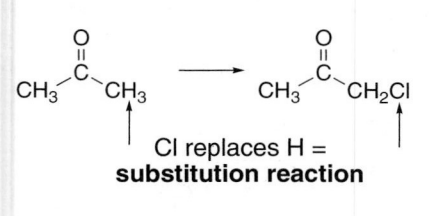

a. Br replaces OH =
substitution reaction

b. addition of 2 H's
addition reaction

c. Cl replaces H =
substitution reaction

d. $CH_3CH_2CH(OH)CH_3$ ⟶ $CH_3CH=CHCH_3$
elements lost
(H + OH)   π bond formed
elimination reaction

6.2 Alkenes undergo addition reactions.

$CH_2=CH_2$ $\xrightarrow{HCl}$ CH_3CH_2Cl

addition of 1 H and 1 Cl
addition reaction

$CH_2=CH_2$ $\xrightarrow{Br_2}$ $BrCH_2CH_2Br$

addition of 2 Br's
addition reaction

6.3 **Heterolysis** means one atom gets both of the electrons when a bond is broken. A carbocation is a C with a positive charge, and a carbanion is a C with a negative charge.

a. heterolysis
$CH_3{-}C{-}\ddot{O}H$ with CH_3 groups

Electrons go to the more
electronegative atom, O.

$CH_3{-}C{+}$:ÖH
carbocation

b. heterolysis

Electrons go to the more
electronegative atom, Br.

+ :Br:⁻
carbocation

c. heterolysis
$CH_3CH_2{-}Li$

Electrons go to the more
electronegative atom, C.

$CH_3\ddot{C}H_2$ Li⁺
carbanion

6.4 Use **full-headed arrows** to show the movement of electron pairs, and **half-headed arrows** to show the movement of single electrons.

a. $(CH_3)_3C{-}\overset{+}{N}{\equiv}N:$ ⟶ $(CH_3)_3C^+$ + :N≡N:

b. ·CH₃ + ·CH₃ ⟶ $CH_3{-}CH_3$

c. $CH_3{-}\overset{+}{C}{-}CH_3$ (with CH_3) + :Br:⁻ ⟶ $CH_3{-}C{-}Br:$ (with CH_3)

d. HÖ–ÖH ⟶ 2 HÖ·

6.5 Increasing number of electrons between atoms = increasing bond strength = increasing bond dissociation energy = decreasing bond length.

Increasing size of an atom = increasing bond length = decreasing bond strength.

a. H–Cl   or   H–Br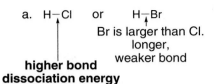
Br is larger than Cl.
longer,
weaker bond
higher bond dissociation energy

b. CH_3–OH   or CH_3–SH
S is larger than O.
longer,
weaker bond
higher bond dissociation energy

c. $(CH_3)_2C{=}O$ or CH_3–OCH_3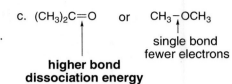
single bond
fewer electrons
higher bond dissociation energy

6.6 **To determine ΔH^o for a reaction**:

[1] Add the bond dissociation energies for all bonds *broken* in the equation (+ values).

[2] Add the bond dissociation energies for all of the bonds *formed* in the equation (– values).

[3] *Add the energies together* to get the ΔH^o for the reaction.

A positive ΔH^o means the reaction is **endothermic**. **A negative ΔH^o** means the reaction is **exothermic**.

a. CH_3CH_2–Br + H_2O $\longrightarrow$ CH_3CH_2–OH + HBr

[1] Bonds broken

	ΔH^o (kJ/mol)
CH_3CH_2–Br	+ 285
H–OH	+ 498
Total	+ 783 kJ/mol

[2] Bonds formed

	ΔH^o (kJ/mol)
CH_3CH_2–OH	– 389
H–Br	– 368
Total	– 757 kJ/mol

[3] Overall ΔH^o =

| sum in Step [1] |
| + |
| sum in Step [2] |

+ 783   kJ/mol
– 757   kJ/mol

ANSWER: + 26   kJ/mol
endothermic

b. CH_4 + Cl_2 $\longrightarrow$ CH_3Cl + HCl

[1] Bonds broken

	ΔH^o (kJ/mol)
CH_3–H	+ 435
Cl–Cl	+ 242
Total	+ 677 kJ/mol

[2] Bonds formed

	ΔH^o (kJ/mol)
CH_3–Cl	– 351
H–Cl	– 431
Total	– 782 kJ/mol

[3] Overall ΔH^o =

| sum in Step [1] |
| + |
| sum in Step [2] |

+ 677   kJ/mol
– 782   kJ/mol

ANSWER: – 105   kJ/mol
exothermic

6.7 Use the directions from Answer 6.6. In determining the number of bonds broken or formed, you must take into account the coefficients needed to balance an equation.

a. $CH_4 + 2 O_2 \longrightarrow CO_2 + 2 H_2O$

[1] **Bonds broken**		
		ΔH^o (kJ/mol)
CH_3-H	$+ 435 \times 4 =$	$+ 1740$
$O-O$	$+ 497 \times 2 =$	$+ 994$
Total		$+ 2734$ kJ/mol

[2] **Bonds formed**		
		ΔH^o (kJ/mol)
$OC-O$	$- 535 \times 2 =$	$- 1070$
$HO-H$	$- 498 \times 4 =$	$- 1992$
Total		$- 3062$ kJ/mol

[3] Overall $\Delta H^o =$

sum in Step [1]
+
sum in Step [2]

$+ 2734$ kJ/mol
$- 3062$ kJ/mol
_____

ANSWER: $- 328$ kJ/mol

b. $2 CH_3CH_3 + 7 O_2 \longrightarrow 4 CO_2 + 6 H_2O$

[1] **Bonds broken**		
		ΔH^o (kJ/mol)
CH_3CH_2-H	$+ 410 \times 12 =$	$+ 4920$
$O-O$	$+ 497 \times 7 =$	$+ 3479$
$C-C$	$+ 368 \times 2 =$	$+ 736$
Total		$+ 9135$ kJ/mol

[2] **Bonds formed**		
		ΔH^o (kJ/mol)
$OC-O$	$- 535 \times 8 =$	$- 4280$
$HO-H$	$- 498 \times 12 =$	$- 5976$
Total		$- 10256$ kJ/mol

[3] Overall $\Delta H^o =$

sum in Step [1]
+
sum in Step [2]

$+ 9135$ kJ/mol
$- 10256$ kJ/mol
_____

ANSWER: $- 1121$ kJ/mol

6.8 Use the following relationships to answer the questions:
$K_{eq} = 1$ then $\Delta G^o = 0$; $K_{eq} > 1$ then $\Delta G^o < 0$; $K_{eq} < 1$ then $\Delta G^o > 0$

a. A negative value of ΔG^o means the equilibrium favors the product and K_{eq} is > 1. Therefore $K_{eq} = 1000$ is the answer.

b. A lower value of ΔG^o means a larger value of K_{eq}, and the products are more favored. $K_{eq} = 10^{-2}$ is larger than $K_{eq} = 10^{-5}$, so ΔG^o is lower.

6.9 Use the relationships from Answer 6.8.

a. $K_{eq} = 5.5$. $K_{eq} > 1$ means that the equilibrium favors the **product**.

b. $\Delta G^o = 40$ kJ/mol. A positive ΔG^o means the equilibrium favors the **starting material**.

6.10 When the product is lower in energy than the starting material, the equilibrium favors the product. When the starting material is lower in energy than the product, the equilibrium favors the starting material.

    a. $\Delta G°$ **is positive** so the equilibrium favors the starting material. Therefore the *starting material is lower in energy than the product.*

    b. K_{eq} **is > 1** so the equilibrium favors the product. Therefore the *product is lower in energy than the starting material.*

    c. $\Delta G°$ **is negative** so the equilibrium favors the product. Therefore the *product is lower in energy than the starting material.*

    d. K_{eq} **is < 1** so the equilibrium favors the starting material. Therefore *the starting material is lower in energy than the product.*

6.11

    a. The K_{eq} is > 1 and therefore the **product** (the conformation on the right) is favored at equilibrium.

    b. The $\Delta G°$ for this process must be **negative** since the product is favored.

    c. $\Delta G°$ is somewhere between 0 and −6 kJ/mol.

6.12 A positive $\Delta H°$ favors the starting material. A negative $\Delta H°$ favors the product.

    a. $\Delta H°$ is positive (80 kJ/mol). The starting material is favored.

    b. $\Delta H°$ is negative (−40 kJ/mol). The product is favored.

6.13

    a. **False.** The reaction is endothermic.

    b. **True.** This assumes that $\Delta G°$ is approximately equal to $\Delta H°$.

    c. **False.** $K_{eq} < 1$

    d. **True.**

    e. **False.** The starting material is favored at equilibrium.

6.14

    a. **True.**

    b. **False.** $\Delta G°$ for the reaction is negative.

    c. **True.**

    d. **False.** The bonds in the product are stronger than the bonds in the starting material.

    e. **True.**

6.15

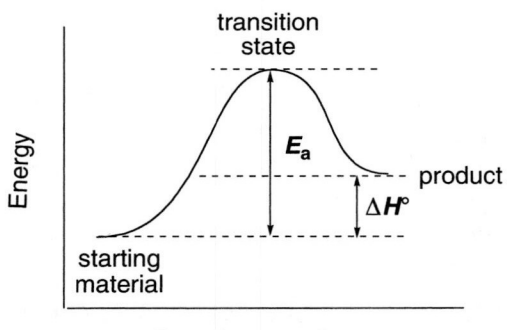

6.16 A transition state is drawn with dashed lines to indicate the partially broken and partially formed bonds. Any atom that gains or loses a charge contains a partial charge in the transition state.

a. $CH_3-\overset{\overset{\displaystyle CH_3}{|}}{\underset{\underset{\displaystyle CH_3}{|}}{\overset{+}{C}}}-OH_2 \longrightarrow CH_3-\overset{\overset{\displaystyle CH_3}{|}}{\underset{\underset{\displaystyle CH_3}{|}}{C}}+ \; + \; H_2O$

transition state: $\left[CH_3 \overset{\overset{\displaystyle CH_3}{|}}{\underset{\underset{\displaystyle CH_3}{|}}{\overset{\delta+}{C}}}---\overset{\delta+}{\ddot{O}}H_2 \right]^{\ddagger}$

b. $CH_3O-H \; + \; ^-OH \longrightarrow CH_3O^- \; + \; H_2O$

transition state: $\left[CH_3\overset{\delta-}{\ddot{O}}---H---\overset{\delta-}{\ddot{O}}H \right]^{\ddagger}$

6.17

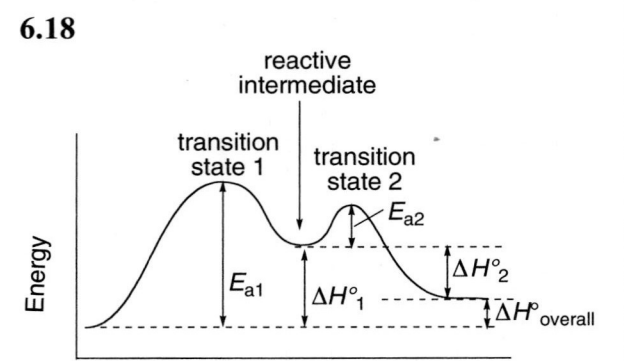

a. Reaction **A–C** is exothermic. Reaction **A–B** is endothermic.
b. Reaction **A–C** is faster.
c. Reaction **A–C** generates a lower-energy product.
d. See labels.
e. See labels.
f. See labels.

6.18

a. Two steps since there are two energy barriers.
b. See labels.
c. See labels.
d. One reactive intermediate is formed (see label).
e. The first step is rate determining since its transition state is at higher energy.
f. The overall reaction is endothermic since the energy of the products is higher than the energy of the reactants.

6.19

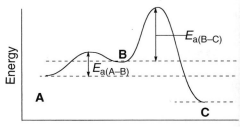

relative energies: **C < A < B**
B → C is rate-determining.

6.20 E_a**, concentration, and temperature affect reaction rate.** $\Delta H°$, $\Delta G°$, and K_{eq} do not affect reaction rate.

a. $E_a = $ **4 kJ/mol** corresponds to a faster reaction rate.
b. A temperature of **25 °C** will have a faster reaction rate since a higher temperature corresponds to a faster reaction.
c. **No change**: K_{eq} does not affect reaction rate.
d. **No change**: $\Delta H°$ does not affect reaction rate.

6.21 The E_a of an endothermic reaction is at least as large as its $\Delta H°$ because the E_a essentially "includes" the $\Delta H°$ in its total. The E_a measures the difference between the energy of the starting material and the energy of the transition state, and in an endothermic reaction, the energy of the products is somewhere in between these two values.

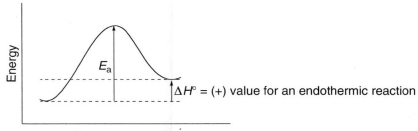

$\Delta H° = (+)$ value for an endothermic reaction

6.22

a. **False.** The reaction occurs at the same rate as a reaction with $K_{eq} = 8$ and $E_a = 80$ kJ/mol.
b. **False.** The reaction is slower than a reaction with $K_{eq} = 0.8$ and $E_a = 40$ kJ/mol.
c. **True.**
d. **True.**
e. **False**. The reaction is endothermic.

6.23 All reactants in the rate equation determine the rate of the reaction.

[1] rate = $k[CH_3CH_2Br][^-OH]$

a. Tripling the concentration of CH_3CH_2Br only → **The rate is tripled.**
b. Tripling the concentration of ^-OH only → **The rate is tripled.**
c. Tripling the concentration of both $CH_3CH_2CH_2Br$ and ^-OH → **The rate increases by a factor of 9 (3 × 3 = 9).**

[2] rate = $k[(CH_3)_3COH]$

a. Doubling the concentration of $(CH_3)_3COH$ → **The rate is doubled.**
b. Increasing the concentration of $(CH_3)_3COH$ by a factor of 10 → **The rate increases by a factor of 10.**

6.24 The rate equation is determined by the rate-determining step.

a. CH_3CH_2-Br + ^-OH ⟶ $CH_2=CH_2$ + H_2O + Br^- one step
rate = $k[CH_3CH_2Br][^-OH]$

b. $(CH_3)_3C-Br$ $\xrightarrow{\text{slow}}$ $(CH_3)_3C^+$ + Br^- $\xrightarrow[\text{fast}]{^-OH}$ $(CH_3)_2C=CH_2$ + H_2O
two steps
The slow step determines the rate equation.
rate = $k[(CH_3)_3CBr]$

6.25 A catalyst is not used up or changed in the reaction. It only speeds up the reaction rate.

a.   OH and H are added to the starting material.
$CH_2=CH_2$ $\xrightarrow[H_2SO_4]{H_2O}$ CH_3CH_2OH
H_2SO_4 is not used up = **catalyst.**

b. I^- not used up = **catalyst.**
CH_3Cl $\xrightarrow[^-OH]{I^-}$ CH_3OH
^-OH substitutes for Cl^-.

c. H_2 adds to the starting material.
$\xrightarrow[Pt]{H_2}$
Pt not used up = **catalyst.**

6.26 Use the directions from Answer 6.1.

a.   elements lost (H + OH)   π bond formed
elimination reaction

b.   Cl replaces H = **substitution reaction**

c.   addition of 2 H's **addition reaction**

d.   H replaces Cl = **substitution reaction**

6.27

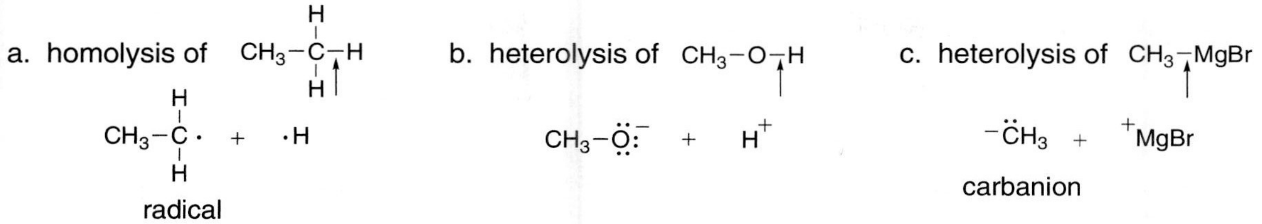

a. homolysis of   CH₃–C–H (with H's)

CH₃–C· + ·H
radical

b. heterolysis of   CH₃–O–H

CH₃–Ö:⁻ + H⁺

c. heterolysis of   CH₃–MgBr

⁻:CH₃ + ⁺MgBr
carbanion

6.28 Use the rules in Answer 6.4 to draw the arrows.

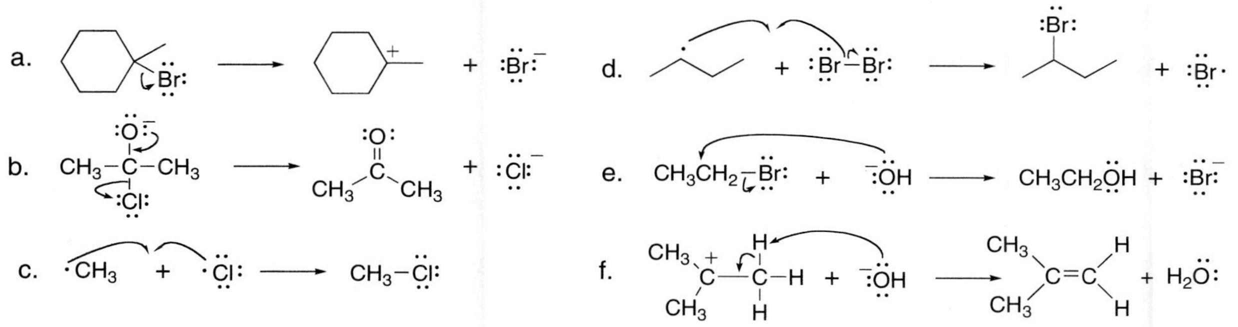

a. [cyclohexane with methyl and Br] ⟶ [cyclohexyl cation] + :Br:⁻

b. CH₃–C(–CH₃)(:O:)(:Cl:) ⟶ CH₃–C(=O)–CH₃ + :Cl:⁻

c. ·CH₃ + ·Cl: ⟶ CH₃–Cl:

d. [alkene structure] + :Br–Br: ⟶ [product with :Br:] + :Br·

e. CH₃CH₂–Br: + ⁻:OH ⟶ CH₃CH₂ÖH + :Br:⁻

f. [(CH₃)₂C–C(H)(CH₃)–H] + ⁻:OH ⟶ [(CH₃)₂C=C(H)(CH₃)] + H₂Ö:

6.29

a. [cyclohexyl–I:] + ⁻:OH ⟶ [cyclohexyl–ÖH] + :I:⁻

b. CH₃–C(:O:⁻)(CH₂CH₂CH₃)(:OCH₂CH₃) ⟶ CH₃–C(=O)–CH₂CH₂CH₃ + ⁻:OCH₂CH₃

c. ⁻:OH + H–C(H)(H)–C(H)(:Br:)–H ⟶ [H₂C=CH₂] + H₂Ö: + :Br:⁻

d. [benzyl CHO with H] + ·Cl: ⟶ [benzyl C·H] + HCl:

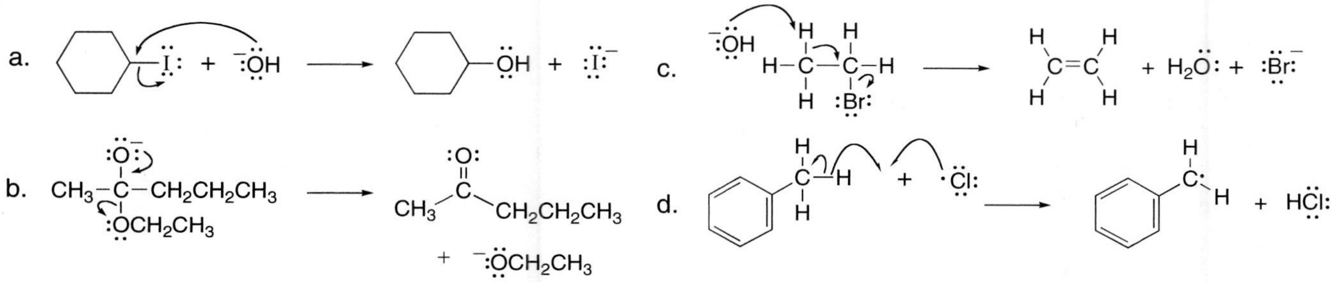

6.30 Draw the curved arrows to identify the product **X**.

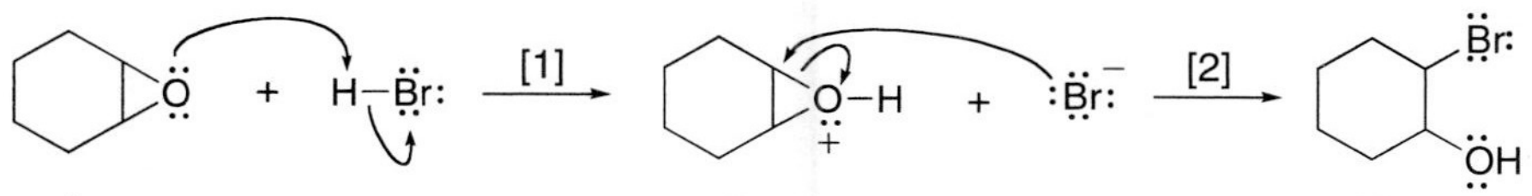

[epoxide A] + H–Br: —[1]→ [protonated epoxide B] O–H + :Br:⁻ —[2]→ [X: cyclohexane with Br and ÖH]

A   B   X

6.31 Follow the curved arrows to identify the intermediate **Y**.

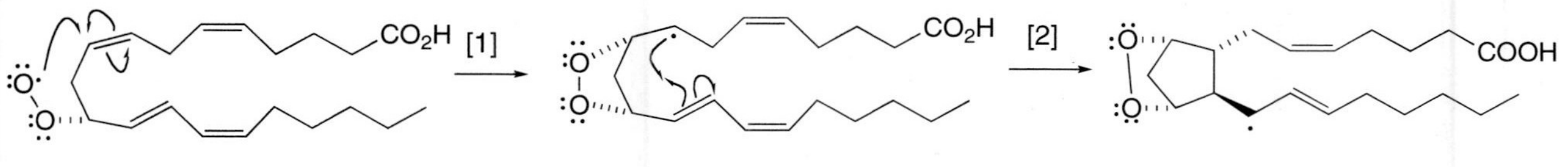

[structure C with CO₂H] —[1]→ [structure D with CO₂H] —[2]→ [structure Y with COOH]

C   D   Y

6.32 Use the rules from Answer 6.5.

a.   I—CCl$_3$ Br—CCl$_3$ Cl—CCl$_3$ b. H$_2$N—NH$_2$ HN=NH   N≡N

largest halogen   **intermediate**   smallest halogen   single bond   double bond   triple bond

weakest bond **bond strength** **strongest bond**   **weakest bond**   **intermediate** **strongest bond**

                                                                            bond strength

6.33 Use the directions from Answer 6.6.

a. CH_3CH_2—H + Br_2 ⟶ CH_3CH_2—Br + HBr

[1] Bonds broken

	ΔH^o (kJ/mol)
CH_3CH_2—H	+ 410
Br—Br	+ 192
Total	+ 602 kJ/mol

[2] Bonds formed

	ΔH^o (kJ/mol)
CH_3CH_2—Br	– 285
H—Br	– 368
Total	– 653 kJ/mol

[3] Overall ΔH^o =

+ 602 kJ/mol
– 653 kJ/mol
ANSWER: – 51 kJ/mol

b. ·OH + CH_4 ⟶ ·CH_3 + H_2O

[1] Bonds broken

	ΔH^o (kJ/mol)
CH_3—H	+ 435 kJ/mol

[2] Bonds formed

	ΔH^o (kJ/mol)
H—OH	– 498 kJ/mol

[3] Overall ΔH^o =

+ 435 kJ/mol
– 498 kJ/mol
ANSWER: – 63 kJ/mol

c. CH_3—OH + HBr ⟶ CH_3—Br + H_2O

[1] Bonds broken

	ΔH^o (kJ/mol)
CH_3—OH	+ 389
H—Br	+ 368
Total	+ 757 kJ/mol

[2] Bonds formed

	ΔH^o (kJ/mol)
CH_3—Br	– 293
H—OH	– 498
Total	– 791 kJ/mol

[3] Overall ΔH^o =

+ 757 kJ/mol
– 791 kJ/mol
ANSWER: – 34 kJ/mol

d. ·Br + CH_4 ⟶ ·H + CH_3Br

[1] Bonds broken

	ΔH^o (kJ/mol)
CH_3—H	+ 435 kJ/mol

[2] Bonds formed

	ΔH^o (kJ/mol)
CH_3—Br	–293 kJ/mol

[3] Overall ΔH^o =

+ 435 kJ/mol
– 293 kJ/mol
ANSWER: + 142 kJ/mol

6.34

propane

$CH_3{-}CH_2CH_3 \longrightarrow \cdot CH_3 \; + \; \cdot CH_2CH_3$

$\Delta H^\circ = 356$ kJ/mol

This bond is formed from two sp^3 hybridized C's.

propene

$CH_3{-}CH{=}CH_2 \longrightarrow \cdot CH_3 \; + \; \cdot CH{=}CH_2$

$\Delta H^\circ = 385$ kJ/mol

This bond is formed from one sp^2 and one sp^3 hybridized C. The higher percent s-character in one C makes a stronger bond; thus the bond dissociation energy is higher.

6.35

$CH_2{=}CH{-}\overset{\displaystyle H}{\underset{\displaystyle H}{C}}{-}H \longrightarrow CH_2{=}CH{-}\overset{\displaystyle H}{C}{-}H \longleftrightarrow \cdot CH_2{-}CH{=}\overset{\displaystyle H}{C}{-}H$

hybrid: $\quad CH_2{=}CH{=}\overset{\displaystyle H}{\underset{\displaystyle \delta^\cdot}{C}}{-}H$ $\quad \delta^\cdot$

6.36 The more stable radical is formed by a reaction with a smaller ΔH°.

$CH_3{-}CH_2{-}\overset{\displaystyle H}{\underset{\displaystyle H}{C}}{-}H \longrightarrow CH_3{-}CH_2{-}\overset{\displaystyle \cdot}{\underset{\displaystyle H}{C}}{-}H \quad \Delta H^\circ = 410$ kJ/mol = less stable radical

This C–H bond is stronger. **A**

$CH_3{-}\overset{\displaystyle H}{\underset{\displaystyle H}{C}}{-}CH_3 \longrightarrow CH_3{-}\overset{\displaystyle \cdot}{\underset{\displaystyle H}{C}}{-}CH_3 \quad \Delta H^\circ = 397$ kJ/mol = more stable radical

This C–H bond is weaker. **B**

Since the bond dissociation for cleavage of the C–H bond to form radical **A** is higher, more energy must be added to form it. This makes **A** higher in energy and therefore less stable than **B**.

6.37 Use the bond dissociation energy for the C–C σ bond in ethane as an estimate of the σ bond strength in ethylene. Then you can estimate the π bond strength as well.

$CH_3{-}CH_3$

$\Delta H^\circ = 368$ kJ/mol

$CH_2{=}CH_2$

$\Delta H^\circ = 635$ kJ/mol

$635 - 368 = 267$ kJ/mol = π bond

6.38 Use the rules from Answer 6.10.
   a. $K_{eq} = 0.5$. K_{eq} is less than one so the **starting material** is favored.
   b. $\Delta G^\circ = -100$ kJ/mol. ΔG° is less than 0 so the **product** is favored.
   c. $\Delta H^\circ = 8.0$ kJ/mol. ΔH° is positive, so the **starting material** is favored.
   d. $K_{eq} = 16$. K_{eq} is greater than one so the **product** is favored.
   e. $\Delta G^\circ = 2.0$ kJ/mol. ΔG° is greater than zero so the **starting material** is favored.
   f. $\Delta H^\circ = 200$ kJ/mol. ΔH° is positive so the **starting material** is favored.
   g. $\Delta S^\circ = 8$ J/(K•mol). ΔS° is greater than zero so the **product** is more disordered and favored.
   h. $\Delta S^\circ = -8$ J/(K•mol). ΔS° is less than zero so the **starting material** is more disordered and favored.

6.39

    a. A negative $\Delta G°$ must have $K_{eq} > 1$. $K_{eq} = 10^2$.

    b. $K_{eq} = [\text{products}]/[\text{reactants}] = [1]/[5] = 0.2 = K_{eq}$. $\Delta G°$ is positive.

    c. A negative $\Delta G°$ has $K_{eq} > 1$, and a positive $\Delta G°$ has $K_{eq} < 1$. $\Delta G° = -8$ kJ/mol will have a larger K_{eq}.

6.40

R	K_{eq}
$-CH_3$	18
$-CH_2CH_3$	23
$-CH(CH_3)_2$	38
$-C(CH_3)_3$	4000

    a. The equatorial conformation is always present in the larger amount at equilibrium since the K_{eq} for all R groups is greater than 1.

    b. The cyclohexane with the $-C(CH_3)_3$ group will have the greatest amount of equatorial conformation at equilibrium since this group has the highest K_{eq}.

    c. The cyclohexane with the $-CH_3$ group will have the greatest amount of axial conformation at equilibrium since this group has the lowest K_{eq}.

    d. The cyclohexane with the $-C(CH_3)_3$ group will have the most negative $\Delta G°$ since it has the largest K_{eq}.

    e. The larger the R group, the more favored the equatorial conformation.

    f. The K_{eq} for *tert*-butylcyclohexane is much higher because the *tert*-butyl group is bulkier than the other groups. With a *tert*-butyl group, a CH_3 group is always oriented over the ring when the group is axial, creating severe 1,3-diaxial interactions. With all other substituents, the larger CH_3 groups can be oriented away from the ring, placing a H over the ring, making the 1,3-diaxial interactions less severe. Compare:

tert-butylcyclohexane

severe 1,3-diaxial interactions with the CH_3 group and the axial H's

isopropylcyclohexane

less severe 1,3-diaxial interactions

6.41 Calculate K_{eq}, and then find the percentage of axial and equatorial conformations present at equilibrium.

fluorocyclohexane
1 part

1.5 parts

    a. $G° = -5.9\log K_{eq}$
            $G° = -1.0$ kJ/mol
            -1.0 kJ/mol $= -5.9\log K_{eq}$
            $K_{eq} = 1.5$

    b. $K_{eq} = [\text{products}]/[\text{reactants}]$
            $1.5 = [\text{products}]/[\text{reactants}]$
            $1.5[\text{reactants}] = [\text{products}]$
            $[\text{reactants}] = 0.4 = 40\%$ axial
            $[\text{products}] = 0.6 = 60\%$ equatorial

6.42 Reactions resulting in an increase in entropy are favored. When a single molecule forms two molecules, there is an increase in entropy.

a. [structure] ⟶ [structure] + [structure]   increased number of molecules
$\Delta S°$ is positive.
products favored

b. $CH_3\cdot$ + $CH_3\cdot$ ⟶ CH_3CH_3 decreased number of molecules
$\Delta S°$ is negative.
starting material favored

c. $(CH_3)_2C(OH)_2$ ⟶ $(CH_3)_2C=O$ + H_2O increased number of molecules
$\Delta S°$ is positive.
products favored

d. CH_3COOCH_3 + H_2O ⟶ CH_3COOH + CH_3OH no change in the number of molecules
neither favored

6.43 Use the directions in Answer 6.16 to draw the transition state. Nonbonded electron pairs are drawn in at reacting sites.

a. [reaction with cyclohexyl–Br: ⟶ cyclohexyl cation + :Br:⁻]

transition state: $\left[\begin{array}{c}\delta^+ \ \delta^-\text{Br:}\end{array}\right]^{\ddagger}$ [cyclohexane ring structure]

b. BF_3 + $:\overset{..}{\underset{..}{Cl}}:^-$ ⟶   F–B–Cl: (with F substituents)

transition state: $\left[\begin{array}{c}F\\ |\\ F–B\text{-}\text{-}\overset{..}{\underset{..}{Cl}}:\\ |\delta^- \ \delta^-\\ F\end{array}\right]^{\ddagger}$

c. [cyclohexyl–OH + :NH₂⁻ ⟶ cyclohexyl–O:⁻ + NH₃]

transition state: $\left[\begin{array}{c}\delta^- \quad \delta^-\\ \overset{..}{\underset{..}{O}}\text{-}\text{-}H\text{-}\text{-}NH_2\end{array}\right]^{\ddagger}$ [cyclohexane ring]

d. [reaction: (CH₃)₂C–CH(H)–H + H₂Ö: ⟶ (CH₃)₂C=CH(CH₃)(H) + H₃Ö⁺]

transition state: $\left[\begin{array}{c}CH_3 \ \delta^+ \quad H \quad \delta^+\\ C\text{=}\text{=}C\text{-}\text{-}H\text{-}\text{-}\overset{..}{\underset{..}{O}}H_2\\ CH_3 \qquad H\end{array}\right]^{\ddagger}$

6.44

a.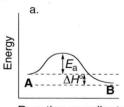

Reaction coordinate
• one step **A → B**
• exothermic since **B** lower than **A**
• low E_a (small energy barrier)

b.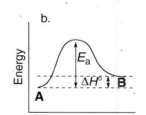

Reaction coordinate
• one step **A → B**
• endothermic since **B** higher than **A**
• high E_a (large energy barrier)

c.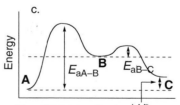

Reaction coordinate $\Delta H°_{overall}$
• two steps
• **A** lowest energy
• **B** highest energy
• $E_{a(A-B)}$ **is rate-determining**, since the transition state for Step [1] is higher in energy.

d.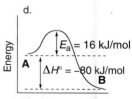

$E_a = 16$ kJ/mol
$\Delta H° = -80$ kJ/mol

Reaction coordinate
• one step **A → B**
• exothermic since **B** lower than **A**

6.45

a. 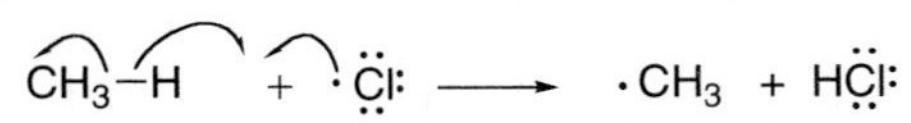 CH_3-H + $\cdot\ddot{Cl}:$ $\longrightarrow$ $\cdot CH_3$ + $H\ddot{Cl}:$

b. $\cdot Cl + CH_4$ $\longrightarrow$ $\cdot CH_3 + HCl$

[1] Bonds broken		**[2] Bonds formed**		**[3] Overall ΔH° =**
	ΔH° (kJ/mol)		ΔH° (kJ/mol)	+ 435 kJ/mol
CH_3-H	+ 435 kJ/mol	$H-Cl$	– 431 kJ/mol	– 431 kJ/mol
				ANSWER: + 4 kJ/mol

c.

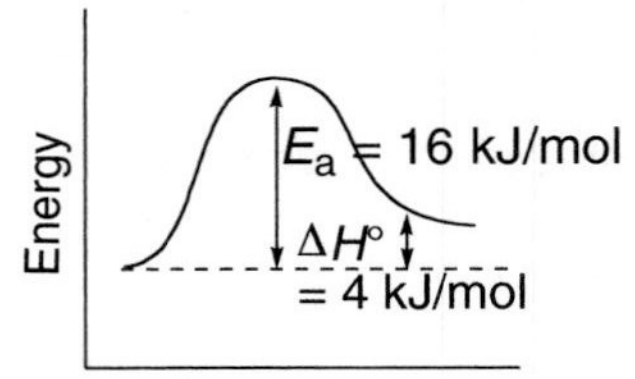

d. The E_a for the reverse reaction is the difference in energy between the products and the transition state, 12 kJ/mol.

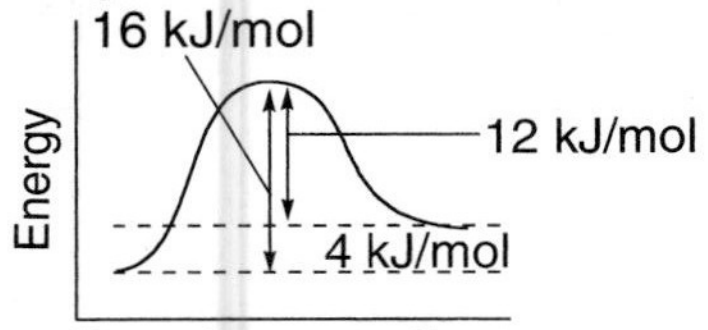

6.46

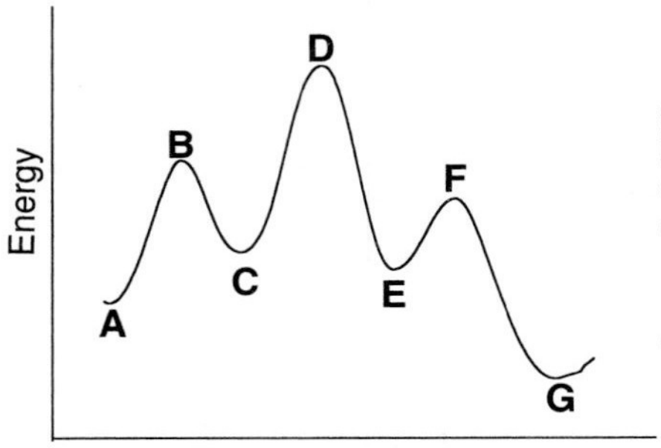

a. **B, D,** and **F** are transition states.
b. **C** and **E** are reactive intermediates.
c. The overall reaction has **three steps.**
d. **A–C** is endothermic.
   C–E is exothermic.
   E–G is exothermic.
e. The overall reaction is exothermic.

6.47

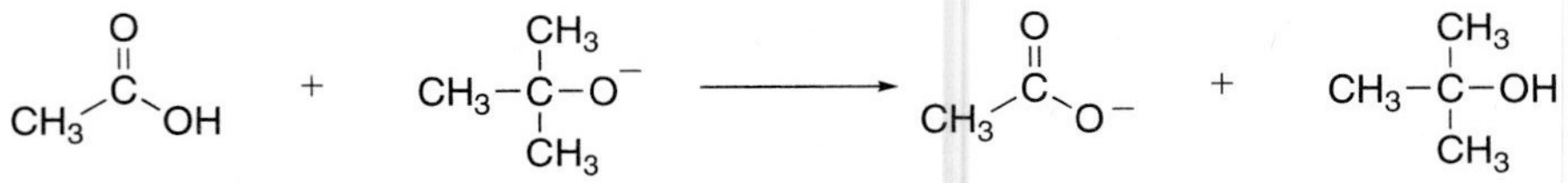

Since pK_a (CH_3CO_2H) = 4.8 and pK_a [$(CH_3)_3COH$] = 18, the weaker acid is formed as product, and equilibrium favors the products. Thus, ΔH° is negative, and the products are lower in energy than the starting materials.

transition state:

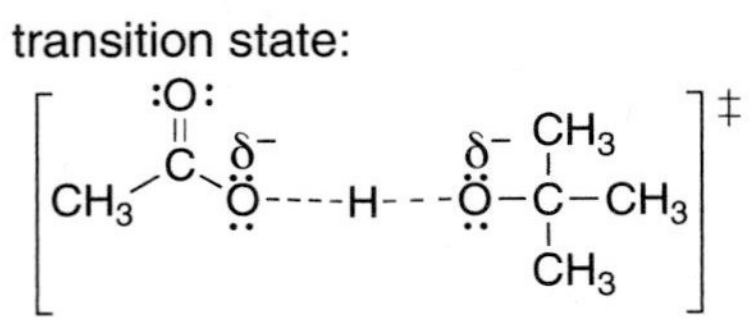

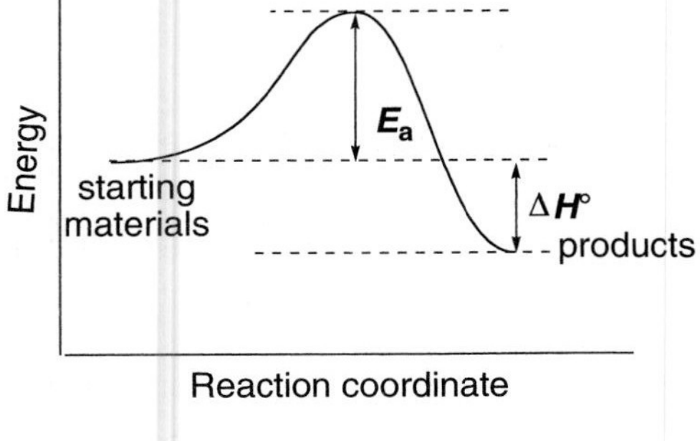

6.48

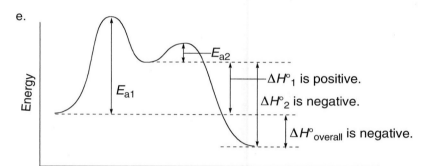

a. Step [1] breaks one π bond and the H–Cl bond, and one C–H bond is formed. The $\Delta H°$ for this step should be positive since more bonds are broken than formed.

b. Step [2] forms one bond. The $\Delta H°$ for this step should be negative since one bond is formed and none is broken.

c. Step [1] is rate-determining since it is more difficult.

d. Transition state for Step [1]:   Transition state for Step [2]:

e.

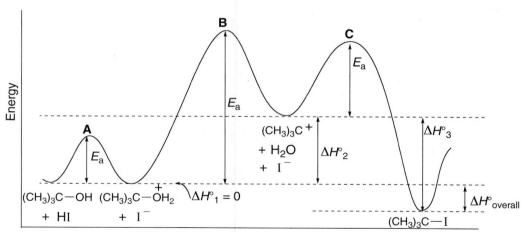

6.49

a. The reaction has three steps, since there are three energy barriers.

b. See above.

c. Transition state **A** (see graph for location):

Transition state **B:**

Transition state **C:**

$$\left[\begin{array}{c} (CH_3)_3C-\overset{\delta+}{\overset{..}{O}}H \\[4pt] \underset{H----\overset{..}{\underset{..}{I}}:}{\overset{\delta-}{}} \end{array} \right]^{\ddagger}$$

$$\left[(CH_3)_3\underset{\delta+}{C}----\underset{\delta+}{\overset{..}{O}}H_2 \right]^{\ddagger}$$

$$\left[(CH_3)_3\overset{\delta+}{C}---\overset{\delta-}{\underset{..}{\overset{..}{I}}}: \right]^{\ddagger}$$

d. Step [2] is rate-determining since this step has the highest energy transition state.

6.50 E_a, concentration, catalysts, rate constant, and temperature affect reaction rate so (c), (d), (e), (g), and (h) affect rate.

6.51

a. **rate = $k[CH_3Br][NaCN]$**
b. Double $[CH_3Br]$ = **rate doubles.**
c. Halve $[NaCN]$ = **rate halved.**
d. Increase both $[CH_3Br]$ and $[NaCN]$ by factor of 5 = [5][5] = **rate increases by a factor of 25.**

6.52

acetyl chloride

[1] slow

[2] fast

methyl acetate

a. Only the slow step is included in the rate equation: **Rate = $k[CH_3O^-][CH_3COCl]$**
b. CH_3O^- is in the rate equation. Increasing its concentration by 10 times would increase the rate by **10 times.**
c. When both reactant concentrations are increased by 10 times, the rate increases by **100 times** **(10 × 10 = 100).**
d. This is a **substitution reaction** (OCH_3 substitutes for Cl).

6.53

a. **True**: Increasing temperature increases reaction rate.
b. **True**: If a reaction is fast, it has a large rate constant.
c. **False: Corrected** - There is no relationship between $\Delta G°$ and reaction rate.
d. **False: Corrected** - When the E_a is large, *the rate constant is small.*
e. **False: Corrected** - There is no relationship between K_{eq} and reaction rate.
f. **False: Corrected** - Increasing the concentration of a reactant increases the rate of a reaction *only if the reactant appears in the rate equation.*

6.54

a. The first mechanism has one step: **Rate = $k[(CH_3)_3CI][^-OH]$**
b. The second mechanism has two steps, but only the first step would be in the rate equation since it is slow and therefore rate-determining: **Rate = $k[(CH_3)_3CI]$**
c. Possibility [1] is second order; possibility [2] is first order.

d. These rate equations can be used to show which mechanism is plausible by changing the concentration of ⁻OH. If this affects the rate, possibility [1] is reasonable. If it does not affect the rate, possibility [2] is reasonable.

e.

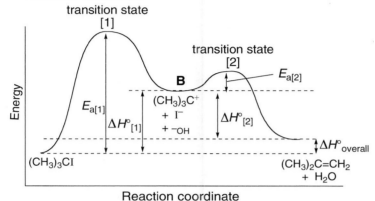

A = (CH₃)₃CI + ⁻OH
B = (CH₃)₂C=CH₂ + I⁻ + H₂O

f.

6.55 The difference in both the acidity and the bond dissociation energy of CH_3CH_3 versus $HC{\equiv}CH$ is due to the same factor: percent *s*-character. The difference results because one process is based on homolysis and one is based on heterolysis.
Bond dissociation energy:

$CH_3CH_2{-}H$

sp³ hybridized
25% *s*-character

$HC{\equiv}C{-}H$

sp hybridized
50% *s*-character
Higher percent *s*-character makes
this bond shorter and stronger.

Acidity: To compare acidity, we must compare the stability of the conjugate bases:

$CH_3\bar{C}H_2$

sp³ hybridized
25% *s*-character

$HC{\equiv}C^{\,\bar{}}$

sp hybridized
50% *s*-character
Now a higher percent *s*-character
stabilizes the conjugate base making
the starting acid more acidic.

6.56

a. [chemical structures showing resonance-stabilized benzylic radical formation]

b. [chemical structures showing radical formation]

c. C–H$_a$ is weaker the C–H$_b$ since the carbon radical formed when the C–H$_a$ bond is broken is highly resonance stabilized. This means the bond dissociation energy for C–H$_a$ is lower.

6.57 In Reaction [1], the number of molecules of reactants and products stays the same, so entropy is not a factor. In Reaction [2], a single molecule of starting material forms two molecules of products, so entropy increases. This makes $\Delta G°$ more favorable, thus increasing K_{eq}.

6.58

[chemical reaction: acetic acid + CH$_3$CH$_2$OH ⇌ ethyl acetate + H$_2$O] $K_{eq} = 4$

ethyl acetate

To increase the yield of ethyl acetate, H$_2$O can be removed from the reaction mixture, or there can be a large excess of one of the starting materials.

6.59

a. [chemical structures: phenol homolysis with resonance structures]

phenol

resonance stabilized
less energy for homolysis

$CH_3CH_2{-}O{-}H \longrightarrow CH_3CH_2{-}\overset{..}{\underset{..}{O}}\cdot$

ethanol   no resonance stabilization

Less energy is required for cleavage of C$_6$H$_5$O–H because homolysis forms the more stable radical.

b. [structure: phenol C–O bond]   [structure: CH$_3$CH$_2$–O–H]

Csp^2–O
higher % s-character
shorter bond

Csp^3–O
lower % s-character
longer bond

Chapter 7: Alkyl Halides and Nucleophilic Substitution

◆ General facts about alkyl halides

- Alkyl halides contain a halogen atom X bonded to an sp^3 hybridized carbon (7.1).
- Alkyl halides are named as halo alkanes, with the halogen as a substituent (7.2).
- Alkyl halides have a polar C–X bond, so they exhibit dipole–dipole interactions but are incapable of intermolecular hydrogen bonding (7.3).
- The polar C–X bond containing an electrophilic carbon makes alkyl halides reactive towards nucleophiles and bases (7.5).

◆ The central theme (7.6)

- Nucleophilic substitution is one of the two main reactions of alkyl halides. A nucleophile replaces a leaving group on an sp^3 hybridized carbon.

$$R-X \quad + \quad :Nu^- \longrightarrow \quad R-Nu \quad + \quad X:^-$$

nucleophile   leaving group

| The electron pair in the C–Nu bond comes from the nucleophile. |

- One σ bond is broken and one σ bond is formed.
- There are two possible mechanisms: S_N1 and S_N2.

◆ S_N1 and S_N2 mechanisms compared

	S_N2 mechanism	S_N1 mechanism
[1] Mechanism	• One step (7.11B)	• Two steps (7.13B)
[2] Alkyl halide	• Order of reactivity: CH_3X > RCH_2X > R_2CHX > R_3CX (7.11D)	• Order of reactivity: R_3CX > R_2CHX > RCH_2X > CH_3X (7.13D)
[3] Rate equation	• rate = $k[RX][:Nu^-]$ • second-order kinetics (7.11A)	• rate = $k[RX]$ • first-order kinetics (7.13A)
[4] Stereochemistry	• backside attack of the nucleophile (7.11C) • inversion of configuration at a stereogenic center	• trigonal planar carbocation intermediate (7.13C) • racemization at a stereogenic center
[5] Nucleophile	• favored by stronger nucleophiles (7.17B)	• favored by weaker nucleophiles (7.17B)
[6] Leaving group	• better leaving group → faster reaction (7.17C)	• better leaving group → faster reaction (7.17C)
[7] Solvent	• favored by polar aprotic solvents (7.17D)	• favored by polar protic solvents (7.17D)

Increasing rate of the S_N1 reaction →

$$H-\underset{\underset{H}{|}}{\overset{\overset{H}{|}}{C}}-Br \qquad R-\underset{\underset{H}{|}}{\overset{\overset{H}{|}}{C}}-Br \qquad R-\underset{\underset{R}{|}}{\overset{\overset{H}{|}}{C}}-Br \qquad R-\underset{\underset{R}{|}}{\overset{\overset{R}{|}}{C}}-Br$$

| methyl | 1° | 2° | 3° |

S_N2 (methyl, 1°)

2° : both S_N1 and S_N2

3° : S_N1

← **Increasing rate of an S_N2 reaction**

◆ Important trends

- The best leaving group is the weakest base. Leaving group ability increases across a row and down a column of the periodic table (7.7).

← **Increasing basicity**

:NH₃   H₂Ö:

Increasing leaving group ability →

← **Increasing basicity**

F⁻   Cl⁻   Br⁻   I⁻

Increasing leaving group ability →

- Nucleophilicity decreases across a row of the periodic table (7.8A).

For 2nd row elements with the same charge:

CH_3^- $^-NH_2$ ^-OH F^-

← **Increasing basicity**
Increasing nucleophilicity

- Nucleophilicity decreases down a column of the periodic table in polar aprotic solvents (7.8C).

Down a column of the periodic table

F^- Cl^- Br^- I^-

← **Increasing nucleophilicity in polar aprotic solvents**

- Nucleophilicity increases down a column of the periodic table in polar protic solvents (7.8C).

Down a column of the periodic table

F^- Cl^- Br^- I^-

Increasing nucleophilicity in polar protic solvents →

- The stability of a carbocation increases as the number of R groups bonded to the positively charged carbon increases (7.14).

$\overset{+}{C}H_3$ $R\overset{+}{C}H_2$ $R_2\overset{+}{C}H$ $R_3\overset{+}{C}$

| methyl | 1° | 2° | 3° |

Increasing carbocation stability →

◆ Important principles

Principle	Example
• Electron-donating groups (such as R groups) stabilize a positive charge (7.14A).	• $3°$ Carbocations (R_3C^+) are more stable than $2°$ carbocations (R_2CH^+), which are more stable than $1°$ carbocations (RCH_2^+).
• Steric hindrance decreases nucleophilicity but not basicity (7.8B).	• $(CH_3)_3CO^-$ is a stronger base but a weaker nucleophile than $CH_3CH_2O^-$.
• Hammond postulate: In an endothermic reaction, the more stable product is formed faster. In an exothermic reaction, this fact is not necessarily true (7.15).	• S_N1 reactions are faster when more stable (more substituted) carbocations are formed, because the rate-determining step is endothermic.
• Planar, sp^2 hybridized atoms react with reagents from both sides of the plane (7.13C).	• A trigonal planar carbocation reacts with nucleophiles from both sides of the plane.

Chapter 7: Answers to Problems

7.1 Classify the alkyl halide as 1°, 2°, or 3° **by counting the number of carbons bonded directly to the carbon bonded to the halogen.**

C bonded to 1 C
1° alkyl halide

C bonded to 2 C's
2° alkyl halide

a. $CH_3CH_2CH_2CH_2CH_2-Br$

b.

c. $CH_3-\overset{\underset{\textstyle CH_3}{|}}{\underset{\underset{\textstyle CH_3}{|}}{C}}-\overset{\underset{\textstyle Cl}{|}}{C}HCH_3$

d.

C bonded to 3 C's
3° alkyl halide

C bonded to 3 C's
3° alkyl halide

7.2 Use the directions from Answer 7.1.

This F is bonded to a C which is ⟶ not bonded to any other C's. Therefore, it cannot be classified as 1°, 2°, or 3°.

bonded to C bonded to 3 C's
3° alkyl halide

bonded to C bonded to 2 C's
2° alkyl halide

7.3 Draw a compound of molecular formula $C_6H_{13}Br$ to fit each description.

a. Br

b. Br

c. Br

1° alkyl halide
one stereogenic center

2° alkyl halide
two stereogenic centers

3° alkyl halide
no stereogenic centers

7.4 To name a compound with the IUPAC system:
[1] **Name the parent** chain by finding the longest carbon chain.
[2] **Number the chain** so the first substituent gets the lower number. Then **name and number all substituents**, giving like substituents a prefix (di, tri, etc.). **To name the halogen substituent, change the -ine ending to -o.**
[3] **Combine all parts**, alphabetizing substituents, and ignoring all prefixes except iso.

a. $(CH_3)_2CHCH(Cl)CH_2CH_3$

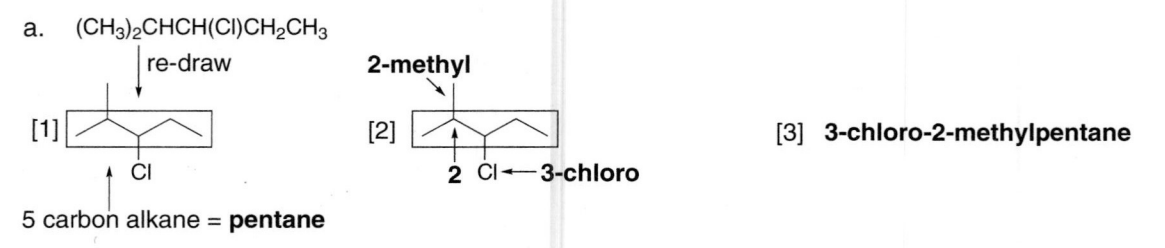

re-draw

2-methyl

[1]   Cl

[2]   2 Cl ⟵ **3-chloro**

[3] **3-chloro-2-methylpentane**

5 carbon alkane = **pentane**

b.

[1] 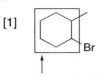 7 carbon alkane = **heptane**

[2] **2-bromo** **5,5-dimethyl**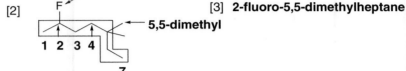

[3] **2-bromo-5,5-dimethylheptane**

c.

[1] 6 carbon cycloalkane = **cyclohexane**

[2] **2-methyl** **1-bromo**

[3] **1-bromo-2-methylcyclohexane**

d.

[1] 7 carbon alkane = **heptane**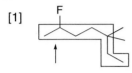

[2] **2-fluoro** **5,5-dimethyl** 1 2 3 4 7

[3] **2-fluoro-5,5-dimethylheptane**

7.5 To work backwards from a name to a structure:

[1] Find the parent name and draw that number of carbons. Use the suffix to identify the functional group (**-ane = alkane**).

[2] Arbitrarily number the carbons in the chain. Add the substituents to the appropriate carbon.

a. 3-chloro-2-methyl**hexane**

[1] 6 carbon alkane

1 2 3 4 5 6

[2] **methyl at C2**
chloro at C3

b. 4-ethyl-5-iodo-2,2-dimethyl**octane**

[1] 8 carbon alkane

1 2 3 4 5 6 7 8

[2] **ethyl at C4**
2 methyls at C2 — iodo at C5

c. *cis*-1,3-dichloro**cyclopentane**

[1] 5 carbon cycloalkane

[2] chloro groups at C1 and C3, both on the same side

Cl Cl
C1 C3

d. 1,1,3-tribromo**cyclohexane**

[1] 6 carbon cycloalkane

[2] 3 Br groups
Br
C3 Br
Br
C1

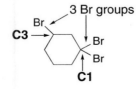

e. **propyl** chloride

    [1] 3 carbon alkyl group   [2] **chloride on end**

        $CH_3CH_2CH_2-$ $CH_3CH_2CH_2-Cl$

f. **sec-butyl** bromide

    [1] 4 carbon alkyl group   [2] **bromide**

        $CH_3-CHCH_2CH_3$ $CH_3-CHCH_2CH_3$
                                                    Br

7.6 Boiling points of alkyl halides increase as the size (and polarizability) of X increases. Remember: **stronger intermolecular forces = higher boiling point.**

  a. $CH_3CH_2CH_2F$ $CH_3CH_3CH_2Cl$ $CH_3CH_2CH_2I$

      smallest halogen   middle size halogen   largest halogen
      least polarizable   **intermediate**   most polarizable
    lowest boiling point   **boiling point**   **highest boiling point**

  b. $CH_3(CH_2)_4CH_3$ $CH_3(CH_2)_5Br$ $CH_3(CH_2)_5OH$

      weakest forces   VDW, DD forces   OH is capable of hydrogen bonding.
       nonpolar   **intermediate**   strongest forces
    lowest boiling point   **boiling point**   **highest boiling point**

7.7 a. Because an sp^2 hybridized C has a higher percent s-character than an sp^3 hybridized C, it holds electron density closer to C. This pulls a little more electron density towards C, away from Cl, and thus a C_{sp^2}–Cl bond is less polar than a C_{sp^3}–Cl bond.

  b.

    lowest boiling point sp^3 C–Cl bond   larger halogen, sp^3 C–Br bond
                    intermediate   **highest boiling point**
                    boiling point

7.8 Since more polar molecules are more water soluble, look for polarity differences between methoxychlor and DDT.

methoxychlor   DDT

2 methoxy groups   2 chloro groups
more polar   less polar
The O atoms can hydrogen bond to H_2O.   readily soluble in an organic medium
more biodegradable

7.9   To draw the products of a nucleophilic substitution reaction:
   [1] **Find the *sp*³ hybridized electrophilic carbon** with a leaving group.
   [2] **Find the nucleophile** with lone pairs or electrons in π bonds.
   [3] **Substitute the nucleophile for the leaving group** on the electrophilic carbon.

a.

b.

c.

d.

7.10 Use the steps from Answer 7.9 and then draw the proton transfer reaction.

a.

b. (CH₃)₃C—Cl   +   H₂Ö: —substitution→   (CH₃)₃C—⁺Ö—H   +   :Ċl:⁻   —proton transfer→   (CH₃)₃C—Ö—H   +   HCl

7.11 Draw the structure of CPC using the steps from Answer 7.9.

7.12 Compare the leaving groups based on these trends:
- Better leaving groups are weaker bases.
- A neutral leaving group is always better than its conjugate base.

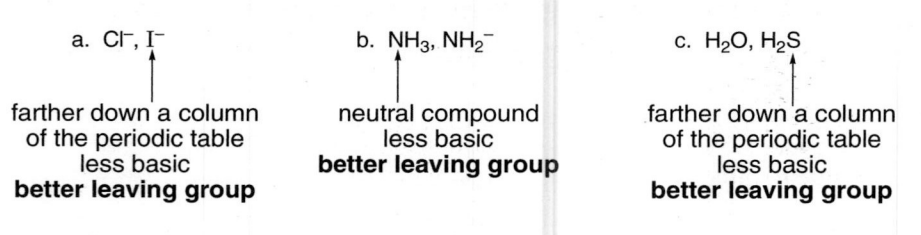

a. Cl⁻, I⁻

farther down a column
of the periodic table
less basic
better leaving group

b. NH₃, NH₂⁻

neutral compound
less basic
better leaving group

c. H₂O, H₂S

farther down a column
of the periodic table
less basic
better leaving group

7.13 Good leaving groups include Cl⁻, Br⁻, I⁻, H₂O.

a. CH₃CH₂CH₂$-$Br

Br⁻ is a **good
leaving group.**

b. CH₃CH₂CH₂OH

No good leaving group.
⁻OH is too strong a base.

c. CH₃CH₂CH₂$-$O̟H₂⁺

H₂O is a **good
leaving group.**

d. CH₃CH₃

No good leaving group.
H⁻ is too strong a base.

7.14 To decide whether the equilibrium favors the starting material or the products, **compare the nucleophile and the leaving group.** The reaction proceeds towards the weaker base.

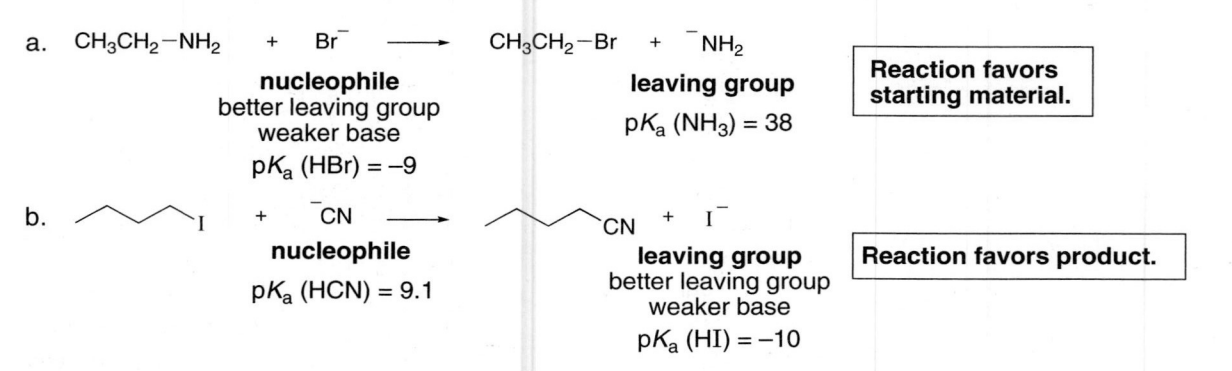

a. CH₃CH₂$-$NH₂ + Br⁻ ⟶ CH₃CH₂$-$Br + ⁻NH₂

nucleophile
better leaving group
weaker base
pKₐ (HBr) = –9

leaving group
pKₐ (NH₃) = 38

**Reaction favors
starting material.**

b. ⟍⟋⟍I + ⁻CN ⟶ ⟍⟋⟍CN + I⁻

nucleophile
pKₐ (HCN) = 9.1

leaving group
better leaving group
weaker base
pKₐ (HI) = –10

Reaction favors product.

7.15 It is not possible to convert CH₃CH₂CH₂OH to CH₃CH₂CH₂Cl by nucleophilic substitution with NaCl because ⁻OH is a stronger base and poorer leaving group than Cl⁻. The equilibrium favors the reactants, not the products.

CH₃CH₂CH₂OH + Na⁺Cl⁻ ⟶✕ CH₃CH₂CH₂Cl + Na⁺ ⁻OH

weaker base

stronger base

7.16 Use these three rules to find the stronger nucleophile in each pair:

[1] Comparing two nucleophiles having the *same attacking atom*, **the stronger base is a stronger nucleophile**.

[2] **Negatively charged nucleophiles** are always **stronger than their conjugate acids**.

[3] **Across a row of the periodic table, nucleophilicity decreases** when comparing species of similar charge.

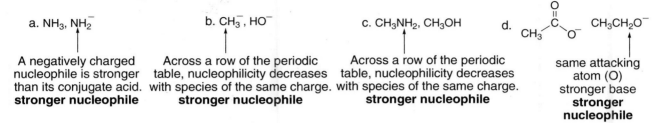

a. NH_3, NH_2^-

A negatively charged
nucleophile is stronger
than its conjugate acid.
stronger nucleophile

b. CH_3^-, HO^-

Across a row of the periodic
table, nucleophilicity decreases
with species of the same charge.
stronger nucleophile

c. CH_3NH_2, CH_3OH

Across a row of the periodic
table, nucleophilicity decreases
with species of the same charge.
stronger nucleophile

d.

same attacking
atom (O)
stronger base
**stronger
nucleophile**

7.17 *Polar protic solvents* are capable of H-bonding, and therefore must contain a **H bonded to an electronegative O or N**. *Polar aprotic solvents* are incapable of H-bonding, and therefore do not contain any O–H or N–H bonds.

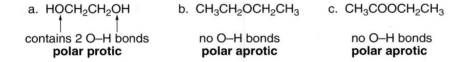

a. $HOCH_2CH_2OH$

contains 2 O–H bonds
polar protic

b. $CH_3CH_2OCH_2CH_3$

no O–H bonds
polar aprotic

c. $CH_3COOCH_2CH_3$

no O–H bonds
polar aprotic

7.18 • In *polar protic solvents*, **the trend in nucleophilicity is opposite to the trend in basicity** down a column of the periodic table so that nucleophilicity increases.

• In *polar aprotic solvents*, **the trend is identical to basicity** so that nucleophilicity decreases down a column.

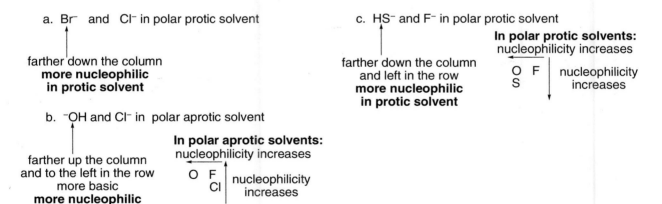

a. Br^- and Cl^- in polar protic solvent

farther down the column
**more nucleophilic
in protic solvent**

c. HS^- and F^- in polar protic solvent

farther down the column
and left in the row
**more nucleophilic
in protic solvent**

In polar protic solvents:
nucleophilicity increases

O F
S   nucleophilicity
increases

b. ^-OH and Cl^- in polar aprotic solvent

farther up the column
and to the left in the row
more basic
more nucleophilic

In polar aprotic solvents:
nucleophilicity increases

O F
Cl   nucleophilicity
increases

7.19 The stronger base is the stronger nucleophile except in polar protic solvents when nucleophilicity increases down a column. For other rules, see Answers 7.16 and 7.18.

a.

H_2O	^-OH	$^-NH_2$
no charge	negatively charged	negatively charged
weakest nucleophile	**intermediate nucleophile**	farther left in periodic table
		strongest nucleophile

b.

Br^-	F^-	^-OH
Basicity decreases down a column in polar aprotic solvents.	Basicity decreases across a row.	**strongest nucleophile**
weakest nucleophile	**intermediate nucleophile**	

c.

H_2O	CH_3COO^-	^-OH
weakest nucleophile	weaker base than ^-OH	**strongest nucleophile**
	intermediate nucleophile	

7.20 To determine what nucleophile is needed to carry out each reaction, look at the product to see what has replaced the leaving group.

a. $(CH_3)_2CHCH_2CH_2-Br \longrightarrow (CH_3)_2CHCH_2CH_2-SH$
SH replaces Br.
HS^- is needed.

b. $(CH_3)_2CHCH_2CH_2-Br \longrightarrow (CH_3)_2CHCH_2CH_2-OCH_2CH_3$
OCH_2CH_3 replaces Br.
$CH_3CH_2O^-$ is needed.

c. $(CH_3)_2CHCH_2CH_2-Br \longrightarrow (CH_3)_2CHCH_2CH_2-OCOCH_3$
$OCOCH_3$ replaces Br.
CH_3COO^- is needed.

d. $(CH_3)_2CHCH_2CH_2-Br \longrightarrow (CH_3)_2CHCH_2CH_2-C\equiv CH$
$C\equiv CH$ replaces Br.
$HC\equiv C^-$ is needed.

7.21 The general rate equation for an S_N2 reaction is rate = $k[RX][:Nu^-]$.

a. [RX] is tripled, and $[:Nu^-]$ stays the same: **rate triples.**
b. Both [RX] and $[:Nu^-]$ are tripled: **rate increases by a factor of 9 (3 × 3 = 9).**
c. [RX] is halved, and $[:Nu^-]$ stays the same: **rate halved.**
d. [RX] is halved, and $[:Nu^-]$ is doubled: **rate stays the same (1/2 × 2 = 1).**

7.22 The transition state in an S_N2 reaction has **dashed bonds to both the leaving group and the nucleophile**, and must contain partial charges.

a. $CH_3CH_2CH_2-Cl$ + $^-OCH_3$ $\longrightarrow$ $CH_3CH_2CH_2-OCH_3$ + Cl^- $\left[\begin{array}{c} CH_3CH_2CH_2 \text{---} \ddot{\underset{..}{Cl}}:\delta^- \\ | \\ CH_3\underset{..}{\ddot{O}}:\delta^- \end{array} \right]^{\ddagger}$

b. ⌇⌇Br + ^-SH $\longrightarrow$ ⌇⌇SH + Br^- $\left[\begin{array}{c} \delta^-\ :\ddot{S}H \\ | \\ \ddot{\underset{..}{Br}}:\delta^- \end{array} \right]^{\ddagger}$

7.23 All S_N2 reactions have one step.

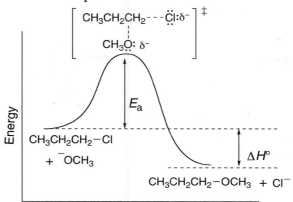

7.24 To draw the products of S_N2 reactions, **replace the leaving group by the nucleophile, and then draw the stereochemistry with *inversion* at the stereogenic center**.

a. CH₃CH₂ ... C—Br + ⁻OCH₂CH₃ ⟶ CH₃CH₂O—C ... (with D and CH₂CH₃ and H)

b. [cyclopentane ring]—I + ⁻CN ⟶ [cyclopentane ring] ... CN

7.25 *Increasing* the number of R groups *increases* crowding of the transition state and *decreases* the rate of an S_N2 reaction.

a. CH₃CH₂—Cl or CH₃—Cl

   1° alkyl halide   methyl halide
               faster reaction

b. [structure with Cl] or [structure with Cl]

   2° alkyl halide   1° alkyl halide
               faster reaction

c. [cyclohexane]—Br or [cyclohexane with methyl]—Br

   2° alkyl halide   3° alkyl halide
   faster reaction

7.26

> These three methyl groups make the alkyl halide sterically hindered. This slows the rate of an S_N2 reaction even though it is a 1° alkyl halide.

          CH₃
       CH₃—C—CH₂Br
          CH₃

7.27

A   CH₃—SR₂⁺   loss of a proton   nicotine

+ SR₂

7.28 In a first-order reaction, **the rate changes with any change in [RX]**. The rate is independent of any change in [nucleophile].

a. [RX] is tripled, and [:Nu⁻] stays the same: **rate triples**.

b. Both [RX] and [:Nu⁻] are tripled: **rate triples**.

c. [RX] is halved, and [:Nu⁻] stays the same: **rate halved**.

d. [RX] is halved, and [:Nu⁻] is doubled: **rate halved**.

7.29 In S$_N$1 reactions, racemization always occurs at a stereogenic center. Draw two products, with the two possible configurations at the stereogenic center.

a.

leaving group   nucleophile

$$(CH_3)_2CH-\underset{CH_2CH_3}{\overset{CH_3}{C}}-Br \xrightarrow{\ H_2O\ } (CH_3)_2CH-\underset{CH_2CH_3}{\overset{CH_3}{C}}\cdots OH \ + \ (CH_3)_2CH-\underset{OH}{\overset{CH_3}{C}}\cdots CH_2CH_3 \ + \ HBr$$

enantiomers

b.

nucleophile

$$\text{CH}_3\text{COO}^-$$

diastereomers

leaving group

7.30 Carbocations are classified by the number of R groups bonded to the carbon: 0 R groups = methyl, 1 R group = 1°, 2 R groups = 2°, and 3 R groups = 3°.

a.

2 R groups
2° carbocation

b. $(CH_3)_3C\overset{+}{C}H_2$

1 R group
1° carbocation

c.

3 R groups
3° carbocation

d.

2 R groups
2° carbocation

7.31 For carbocations: Increasing number of R groups = Increasing stability.

$$CH_3CH_2CH_2\overset{+}{C}H_2$$

1° carbocation
least stable

$$CH_3\overset{+}{C}HCH_2CH_3$$

2° carbocation
intermediate stability

$$CH_3-\underset{CH_3}{\overset{+}{C}}-CH_3$$

3° carbocation
most stable

7.32 For carbocations: Increasing number of R groups = Increasing stability.

a. $(CH_3)_2CHCH_2\overset{+}{C}H_2$

1° carbocation
least stable

$(CH_3)_2CH\overset{+}{C}HCH_3$

2° carbocation
intermediate stability

$(CH_3)_2\overset{+}{C}CH_2CH_3$

3° carbocation
most stable

b.

1° carbocation
least stable

2° carbocation
intermediate stability

3° carbocation
most stable

7.33 The rate of an S_N1 reaction increases with increasing alkyl substitution.

a.　$(CH_3)_3CBr$　or　$(CH_3)_3CCH_2Br$

　　3° alkyl halide　　　　1° alkyl halide
　faster S_N1 reaction　**slower S_N1 reaction**

b.

　2° alkyl halide　　　　　3° alkyl halide
　slower S_N1 reaction　**faster S_N1 reaction**

c.

　3° alkyl halide　　　　2° alkyl halide
　faster S_N1 reaction　**slower S_N1 reaction**

7.34 • For **methyl and 1° alkyl halides**, only S_N2 will occur.
　　　• For **2° alkyl halides**, S_N1 and S_N2 will occur.
　　　• For **3° alkyl halides**, only S_N1 will occur.

a.

2° alkyl halide
S_N1 and S_N2

b.

1° alkyl halide
S_N2

c.

2° alkyl halide
S_N1 and S_N2

d.

3° alkyl halide
S_N1

7.35 • Draw the product of nucleophilic substitution for each reaction.
　　　• For **methyl and 1° alkyl halides**, only S_N2 will occur.
　　　• For **2° alkyl halides**, S_N1 and S_N2 will occur and other factors determine which mechanism operates.
　　　• For **3° alkyl halides**, only S_N1 will occur.

a.

3° alkyl halide
only S_N1

b.

1° alkyl halide
only S_N2

c.

Strong nucleophile
favors S_N2.

2° alkyl halide
**Both S_N1 and S_N2
are possible.**

d.

Weak nucleophile
favors S_N1.

2° alkyl halide
**Both S_N1 and S_N2
are possible.**

7.36 First decide whether the reaction will proceed via an S_N1 or S_N2 mechanism. Then draw the products with stereochemistry.

a.

2° alkyl halide
S_N1 and S_N2

Weak nucleophile
favors S_N1.

enantiomers

S_N1 = racemization at the stereogenic C

b. + ⁻:C≡C–H ⟶   + Cl⁻   **S_N2 = inversion at the stereogenic C**

1° alkyl halide
S_N2 only

7.37 Compounds with better leaving groups react faster. Weaker bases are better leaving groups.

    a. $CH_3CH_2CH_2Cl$ or $CH_3CH_2CH_2I$
                           ↑
                     weaker base
              better leaving group

c. $(CH_3)_3C-OH$ or $(CH_3)_3C-\overset{+}{O}H_2$
                               ↑
                        weaker base
                 better leaving group

    b. $(CH_3)_3CBr$ or $(CH_3)_3CI$
                       ↑
            weaker base
       better leaving group

d. $CH_3CH_2CH_2OH$ or $CH_3CH_2CH_2-OCOCH_3$
                                      ↑
                            weaker base
                      better leaving group

7.38   • **Polar protic solvents** favor the S_N1 mechanism by solvating the intermediate carbocation and halide.
          • **Polar aprotic solvents** favor the S_N2 mechanism by making the nucleophile stronger.

a. CH_3CH_2OH	b. CH_3CN	c. CH_3COOH	d. $CH_3CH_2OCH_2CH_3$
polar protic solvent	polar aprotic solvent	polar protic solvent	polar aprotic solvent
contains an O–H bond	no O–H or N–H bond	contains an O–H bond	no O–H or N–H bond
favors S_N1	**favors S_N2**	**favors S_N1**	**favors S_N2**

7.39 Compare the solvents in the reactions below. **For the solvent to increase the reaction rate of an S_N1 reaction, the solvent must be *polar protic*.**

a. $(CH_3)_3CBr$ + H_2O $\xrightarrow[\text{or}\ (CH_3)_2C=O]{H_2O}$ $(CH_3)_3COH$ + HBr

       3° RX – S_N1 reaction

              H_2O
          Polar protic solvent increases the rate of an S_N1 reaction.

b.   + CH_3OH $\xrightarrow[\text{DMSO}]{CH_3OH\ \text{or}}$ + HCl

   3° RX – S_N1 reaction

            CH_3OH
        Polar protic solvent increases the rate of an S_N1 reaction.

c.   + ⁻OH $\xrightarrow[\text{DMF}]{H_2O\ \text{or}}$ + Br⁻

   1° RX – S_N2 reaction

          DMF [$HCON(CH_3)_2$]
        Polar aprotic solvent increases the rate of an S_N2 reaction.

d.   + CH_3O^- $\xrightarrow[\text{HMPA}]{CH_3OH\ \text{or}}$ + Cl⁻

   2° RX   strong nucleophile
       S_N2 reaction

         HMPA [$(CH_3)_2N]_3P=O$]
        Polar aprotic solvent increases the rate of an S_N2 reaction.

7.40 To predict whether the reaction follows an S_N1 or S_N2 mechanism:
   [1] **Classify RX as a methyl, 1°, 2°, or 3° halide.** (Methyl, 1° = S_N2; 3° = S_N1; 2° = either.)
   [2] **Classify the nucleophile as strong or weak.** (Strong favors S_N2; weak favors S_N1.)
   [3] **Classify the solvent as polar protic or polar aprotic**. (Polar protic favors S_N1; polar aprotic favors S_N2.)

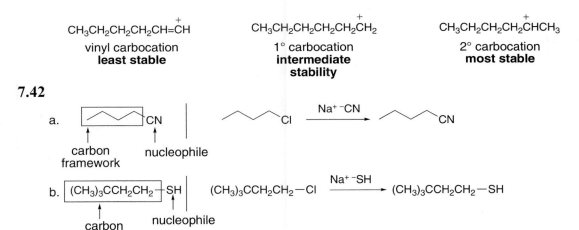

a. —CH$_2$Br + CH$_3$CH$_2$O$^-$ ⟶ —CH$_2$OCH$_2$CH$_3$ + Br$^-$ S_N2 reaction

1° alkyl halide
S_N2

b.   ▬Br + N$_3^-$ ⟶   ''N$_3$ + Br$^-$ S_N2 reaction = *inversion* at the stereogenic center
The leaving group was "up."
The nucleophile attacks from below.

2° alkyl halide   Strong nucleophile
S_N1 or S_N2   **favors S_N2.**

c.   I + CH$_3$OH ⟶   OCH$_3$ + HI S_N1 reaction

3° alkyl halide   Weak nucleophile
S_N1   **favors S_N1.**

d.   Cl + H$_2$O ⟶   OH +   HO + HCl S_N1 reaction
    Weak nucleophile   forms **two enantiomers.**

3° alkyl halide   **favors S_N1.**
S_N1

7.41 Vinyl carbocations are even less stable than 1° carbocations.

CH$_3$CH$_2$CH$_2$CH$_2$CH=$\overset{+}{\text{C}}$H   CH$_3$CH$_2$CH$_2$CH$_2$CH$_2$$\overset{+}{\text{C}}H_2$ CH$_3$CH$_2$CH$_2$CH$_2$$\overset{+}{\text{C}}HCH_3$

vinyl carbocation   1° carbocation   2° carbocation
least stable   **intermediate**   **most stable**
                                stability

7.42

a.   ☐CN   Cl $\xrightarrow{\text{Na}^+ {}^-\text{CN}}$ CN

     ↑   ↑
  carbon   nucleophile
  framework

b. ☐(CH$_3$)$_3$CCH$_2$CH$_2$┤SH☐   (CH$_3$)$_3$CCH$_2$CH$_2$—Cl $\xrightarrow{\text{Na}^+ {}^-\text{SH}}$ (CH$_3$)$_3$CCH$_2$CH$_2$—SH

     ↑   ↑
  carbon   nucleophile
  framework

c.

carbon framework / nucleophile

$$CH_3CH_2{-}C{\equiv}C{-}H \quad | \quad CH_3CH_2{-}Cl \xrightarrow{\ Na^+\ ^-C{\equiv}CH\ } CH_3CH_2{-}C{\equiv}C{-}H$$

d.

carbon framework / nucleophile

7.43

$$CH_3\ddot{O}{:}^- + Cl{-}CH_2CH_3 \longrightarrow CH_3\ddot{O}CH_2CH_3 \qquad CH_3CH_2\ddot{O}{:}^- + Cl{-}CH_3 \longrightarrow CH_3\ddot{O}CH_2CH_3$$

7.44 Use the directions from Answer 7.4 to name the compounds.

a.
[1]
$$CH_3{-}\underset{CH_3}{\overset{CH_3}{C}}{-}CH_2CH_2{-}F$$
4 carbon alkane = **butane**

[2]
$$CH_3{-}\underset{\overset{|}{3\ CH_3}}{\overset{CH_3}{C}}{-}CH_2CH_2{-}F$$
3,3-dimethyl
1-fluoro

[3] **1-fluoro-3,3-dimethylbutane**

b.
[1] I
6 carbon alkane = **hexane**

[2] 3-ethyl→ 2-methyl → 1-iodo
3 2 1

[3] **3-ethyl-1-iodo-2-methylhexane**

c.
[1] $(CH_3)_3CCH_2Br$
$$CH_3{-}\underset{CH_3}{\overset{CH_3}{C}}{-}CH_2{-}Br$$
3 carbon alkane = **propane**

[2]
$$CH_3{-}\underset{\overset{|}{2\ CH_3}}{\overset{CH_3}{C}}{-}CH_2{-}Br$$
2,2-dimethyl 1-bromo

[3] **1-bromo-2,2-dimethylpropane**

d.
[1] Br Cl
8 carbon alkane = **octane**

[2]
6-methyl 6-bromo 2-chloro

[3] **6-bromo-2-chloro-6-methyloctane**

e.
[1] Br / I
5 carbon cycloalkane = **cyclopentane**

[2] 1-bromo → 3-iodo
2

[3] *cis*-**1-bromo-3-iodocyclopentane**

f.
[1] Cl / Cl
6 carbon cycloalkane = **cyclohexane**

[2] *trans*-1,2-dichloro
1 2

[3] *trans*-**1,2-dichlorocyclohexane**

g.[1] (CH₃)₃CCH₂CH(Cl)CH₂Cl

5 carbon alkane = **pentane**

[2]

$$CH_3-\overset{\overset{\displaystyle CH_3}{|}}{\underset{\underset{\displaystyle CH_3}{|}}{C}}-CH_2-\overset{\overset{\displaystyle H}{|}}{\underset{\underset{\displaystyle Cl}{|}}{C}}-CH_2Cl$$

4,4-dimethyl   **1,2-dichloro**

[3] **1,2-dichloro-4,4-dimethylpentane**

h.[1]

6 carbon alkane = **hexane**
(Indicate the *R/S*
designation also)

[2]

4,4-dimethyl   **(2R)-2-iodo**

[3] **(2R)-2-iodo-4,4-dimethylhexane**

Clockwise
R

7.45 To work backwards to a structure, use the directions in Answer 7.5.

a. **isopropyl** bromide

Br ◄—— Bromine on middle C
| makes it an isopropyl group.
CH₃–CHCH₃

b. 3-bromo-4-ethyl**heptane**

4-ethyl
3-bromo

c. 1,1-dichloro-2-methyl**cyclohexane**

1,1-dichloro
2-methyl

d. *trans*-1-chloro-3-iodo**cyclobutane**

1-chloro
3-iodo

e. 1-bromo-4-ethyl-3-fluoro**octane**

4-ethyl
1-bromo   F ◄—3-fluoro

f. (3S)-3-iodo-2-methyl**nonane**

2-methyl
3S   3-iodo

g. (1R,2R)-*trans*-1-bromo-2-chloro**cyclohexane**

1R   Br ◄— 1-bromo
Cl ◄— 2-chloro
2R

h. (5R)-4,4,5-trichloro-3,3-dimethyl**decane**

5R   3,3-dimethyl
4,4,5-trichloro

7.46

a. CH₃–C(CH₃)₂–CH₂CH₂F **1° halide**

b. **1° halide**

c. (CH₃)₃CCH₂Br   **1° halide**

d. **3° halide** **2° halide**

e. Br← **2° halide**   I← **2° halide**

f. Cl← Cl← Both are **2° halides.**

g. (CH₃)₂CCH₂–C(H₂)–C(H)(Cl)–Cl ← **1° halide** **2° halide**

h. **2° halide**

7.47

1-chloro
1
Cl
1-chloropentane

1-chloro
2 **1**
Cl
1-chloro-2,2-dimethylpropane

3-chloro → Cl
3
3-chloropentane

2-methyl → 2-chloro ↓ Cl
2
2-chloro-2-methylbutane

3-methyl →
Cl **1** **3**
1-chloro
1-chloro-3-methylbutane

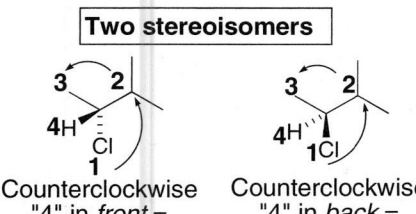

Two stereoisomers

2
2-chloro → Cl *
2-chloropentane
[* denotes stereogenic center]

2   3
1 Cl H 4
Clockwise
"4" in *back* =
R

2   3
1 Cl H 4
Clockwise
"4" in *front* =
S

Two stereoisomers

3-methyl →
2 **3**
2-chloro → Cl *
2-chloro-3-methylbutane
[* denotes stereogenic center]

3   2
4H
1
Counterclockwise
"4" in *front* =
R

3   2
4H
1Cl
Counterclockwise
"4" in *back* =
S

Two stereoisomers

2-methyl →
1-chloro ↓
Cl *
Cl
1-chloro-2-methylbutane
[* denotes stereogenic center]

3   4
H
Cl
2   1
Clockwise
"4" in *front* =
S

3   4
H
Cl
2   1
Clockwise
"4" in *back* =
R

7.48 Use the directions from Answer 7.6.

a. $(CH_3)_3CBr$ or $CH_3CH_2CH_2CH_2Br$ b. ⤴ or ⤴   c. ⬡ or ⬡

larger surface area =
stronger intermolecular forces =
higher boiling point

larger halide = more polarizable =
higher boiling point

nonpolar
only VDW forces

more polar =
higher boiling point

7.49

a. $CH_3CH_2CH_2CH_2-Br$ + ^-OH ⟶ $CH_3CH_2CH_2CH_2OH$ + Br^-

b. $CH_3CH_2CH_2CH_2-Br$ + ^-SH ⟶ $CH_3CH_2CH_2CH_2SH$ + Br^-

c. $CH_3CH_2CH_2CH_2-Br$ + ^-CN ⟶ $CH_3CH_2CH_2CH_2CN$ + Br^-

d. $CH_3CH_2CH_2CH_2-Br$ + $^-OCH(CH_3)_2$ ⟶ $CH_3CH_2CH_2CH_2OCH(CH_3)_2$ + Br^-

e. $CH_3CH_2CH_2CH_2-Br$ + $^-C\equiv CH$ ⟶ $CH_3CH_2CH_2CH_2C\equiv CH$ + Br^-

f. $CH_3CH_2CH_2CH_2-Br$ + $H_2\ddot{O}:$ ⟶ $CH_3CH_2CH_2CH_2\overset{+}{O}H_2$ + Br^- ⟶ $CH_3CH_2CH_2CH_2OH$ + HBr

g. $CH_3CH_2CH_2CH_2-Br$ + $\ddot{N}H_3$ ⟶ $CH_3CH_2CH_2CH_2\overset{+}{N}H_3$ + Br^- ⟶ $CH_3CH_2CH_2CH_2NH_2$ + HBr

h. $CH_3CH_2CH_2CH_2-Br$ + $Na^+ I^-$ ⟶ $CH_3CH_2CH_2CH_2I$ + $Na^+ Br^-$

i. $CH_3CH_2CH_2CH_2-Br$ + $Na^+ N_3^-$ ⟶ $CH_3CH_2CH_2CH_2N_3$ + $Na^+ Br^-$

7.50 Use the steps from Answer 7.9 and then draw the proton transfer reaction, when necessary.

a.

leaving group   nucleophile   + Cl^-

b.

leaving group   + $Na^+ {}^-CN$ nucleophile   ⟶ CN + NaI

c.

leaving group   + $H_2\ddot{O}:$ nucleophile   ⟶ OH + HI

d.

leaving group   + $CH_3CH_2\ddot{O}H$ nucleophile   ⟶ OCH_2CH_3 + HCl

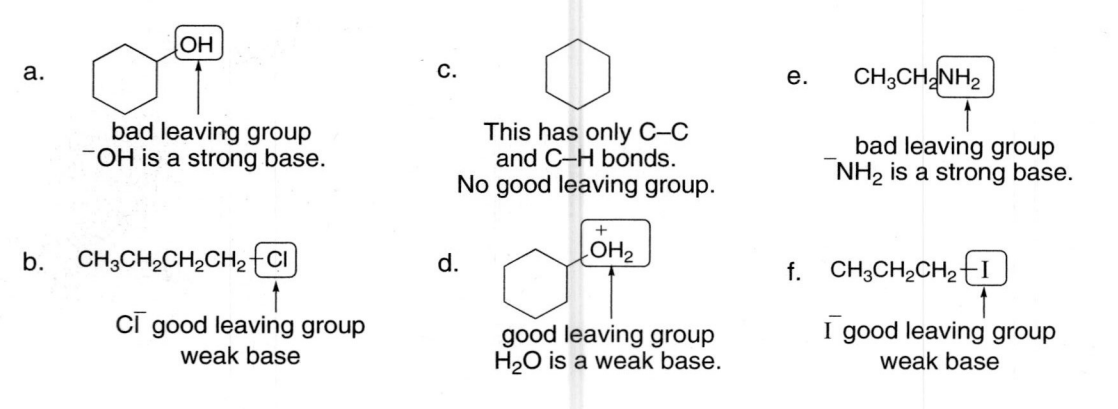

e. leaving group   nucleophile   + NaBr

f. leaving group   nucleophile

7.51 A good leaving group is a weak base.

a.

OH

bad leaving group
^-OH is a strong base.

c.

This has only C–C
and C–H bonds.
No good leaving group.

e. $CH_3CH_2NH_2$

bad leaving group
$^-NH_2$ is a strong base.

b. $CH_3CH_2CH_2CH_2{-}Cl$

Cl^- good leaving group
weak base

d.

$^+OH_2$

good leaving group
H_2O is a weak base.

f. $CH_3CH_2CH_2{-}I$

I^- good leaving group
weak base

7.52 Use the rules from Answer 7.12.

a. increasing leaving group ability: $^-NH_2 < {}^-OH < F^-$

most basic — least basic
worst leaving group — **best leaving group**

c. increasing leaving group ability: $Cl^- < Br^- < I^-$

most basic — least basic
worst leaving group — **best leaving group**

b. increasing leaving group ability: $^-NH_2 < {}^-OH < H_2O$

most basic — least basic
worst leaving group — **best leaving group**

d. increasing leaving group ability: $NH_3 < H_2O < H_2S$

most basic — least basic
worst leaving group — **best leaving group**

7.53 Compare the nucleophile and the leaving group in each reaction. The reaction will occur if it proceeds towards the weaker base. Remember that the stronger the acid (lower pK_a), the weaker the conjugate base.

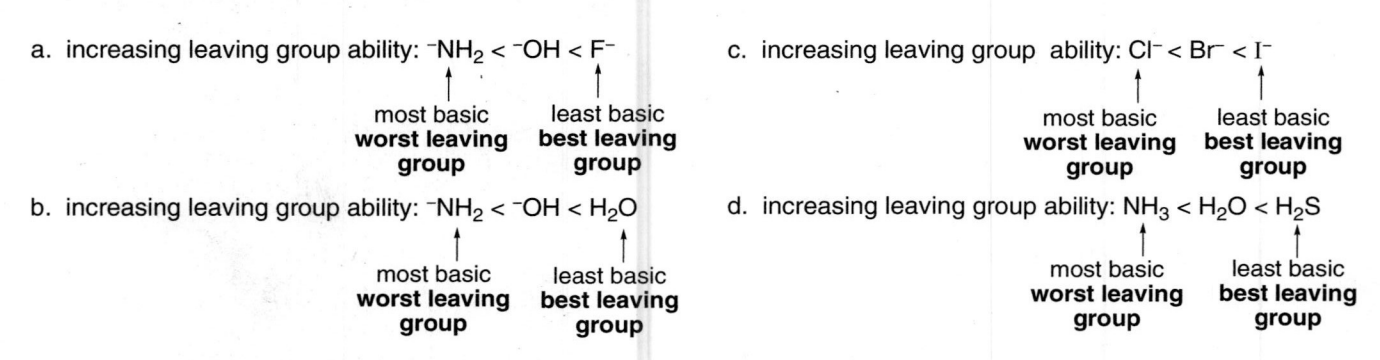

a.

NH_2 + I^- ✗

weaker base
pK_a (HI) = –10

I + $^-NH_2$

stronger base
pK_a (NH$_3$) = 38

Reaction will not occur.

b. CH_3CH_2I + CH_3O^- ⟶ $CH_3CH_2OCH_3$ + I^-

stronger base
pK_a (CH$_3$OH) = 15.5

weaker base
pK_a (HI) = –10

Reaction will occur.

c.

weaker base
pK_a (HF) = 3.2

stronger base
pK_a (H_2O) = 15.7

| Reaction will not occur. |

d.

weaker base
pK_a (HI) = –10

stronger base
pK_a (HCN) = 9.1

| Reaction will not occur. |

7.54

A

a.

SCH_3 replaces Br.
$^-SCH_3$ is needed.

c.

$C≡CCH_3$ replaces Br.
$^-C≡CCH_3$ is needed.

b.

$OCH(CH_3)_2$ replaces Br.
$^-OCH(CH_3)_2$ needed.

d.

$N(CH_3)_3$ replaces Br.
$N(CH_3)_3$ is needed.

7.55 Use the directions in Answer 7.16.

a. Across a row of the periodic table nucleophilicity decreases.

$^-OH < {}^-NH_2 < CH_3^-$

b. • In a **polar protic solvent** (CH_3OH), nucleophilicity *increases down a column* of the periodic table, so: ^-SH is more nucleophilic than ^-OH.
• *Negatively charged species* are more nucleophilic than neutral species so ^-OH is more nucleophilic than H_2O.

$H_2O < {}^-OH < {}^-SH$

c. • In a **polar protic solvent** (CH_3OH), nucleophilicity *increases down a column* of the periodic table, so: $CH_3CH_2S^-$ is more nucleophilic than $CH_3CH_2O^-$.
• For two species with the same attacking atom, the more basic is the more nucleophilic so $CH_3CH_2O^-$ is more nucleophilic than CH_3COO^-.

$CH_3COO^- < CH_3CH_2O^- < CH_3CH_2S^-$

d. Compare the nucleophilicity of N, S, and O. In a polar aprotic solvent (acetone), nucleophilicity parallels basicity.

$CH_3SH < CH_3OH < CH_3NH_2$

e. In a **polar aprotic solvent** (acetone), nucleophilicity parallels basicity. Across a row and down a column of the periodic table nucleophilicity decreases.

$Cl^- < F^- < {}^-OH$

f. Nucleophilicity decreases across a row so ^-SH is more nucleophilic than Cl^-.
In a **polar protic solvent** (CH_3OH), nucleophilicity increases down a column so Cl^- is more nucleophilic than F^-.

$F^- < Cl^- < {}^-SH$

7.56 *Polar protic solvents* **are capable of hydrogen bonding,** and therefore must contain a H bonded to an electronegative O or N. *Polar aprotic solvents* **are incapable of hydrogen bonding,** and therefore do not contain any O–H or N–H bonds.

a. $(CH_3)_2CHOH$

contains O–H bond
protic

c. CH_2Cl_2

no O–H or N–H bond
aprotic

e. $N(CH_3)_3$

no O–H or N–H bond
aprotic

b. CH_3NO_2

no O–H or N–H bond
aprotic

d. NH_3

contains N–H bond
protic

f. $HCONH_2$

contains an N–H bond
protic

7.57

The amine N is more nucleophilic since the electron pair is localized on the N.

The amide N is less nucleophilic since the electron pair is delocalized by resonance.

7.58

a. Mechanism:

1° alkyl halide
S$_N$2 reaction

b. Energy diagram:

Reaction coordinate

c. Transition state:

d. Rate equation: one step reaction with both nucleophile and alkyl halide in the only step:
   rate = k[R–Br][$^-$CN]

e. [1] The leaving group is changed from Br$^-$ to I$^-$:
   Leaving group becomes less basic → a better leaving group → faster reaction.
   [2] The solvent is changed from acetone to CH_3CH_2OH:
   Solvent changed to polar protic → decreases reaction rate.
   [3] The alkyl halide is changed from $CH_3(CH_2)_4Br$ to $CH_3CH_2CH_2CH(Br)CH_3$:
   Changed from 1° to 2° alkyl halide → the alkyl halide gets more crowded and the reaction rate decreases.
   [4] The concentration of $^-$CN is increased by a factor of 5.
   Reaction rate will increase by a factor of 5.
   [5] The concentration of both the alkyl halide and $^-$CN are increased by a factor of 5:
   Reaction rate will increase by a factor of 25 (5 x 5 = 25).

7.65

a. Mechanism:
   S_N1 only

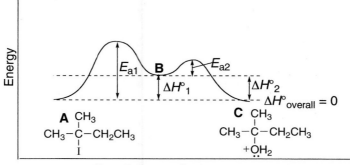

b. Energy diagram:

c. Transition states:

d. rate equation: **rate = $k[(CH_3)_2CICH_2CH_3]$**

e. [1] Leaving group changed from I⁻ to Cl⁻: **rate decreases** since I⁻ is a better leaving group.
   [2] Solvent changed from H_2O (polar protic) to DMF (polar aprotic):
       rate decreases since polar protic solvent favors S_N1.
   [3] Alkyl halide changed from 3° to 2°: **rate decreases** since 2° carbocations are less stable.
   [4] [H_2O] increased by factor of five: **no change in rate** since H_2O is not in rate equation.
   [5] [R–X] and [H_2O] increased by factor of five: **rate increases** by a factor of five. (Only the concentration of R–X affects the rate.)

7.66 The rate of an S_N1 reaction increases with increasing alkyl substitution.

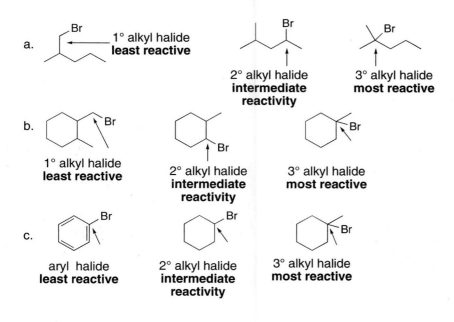

7.67 The rate of an S$_N$1 reaction increases with increasing alkyl substitution, polar protic solvents, and better leaving groups.

a. $(CH_3)_3CCl$ + H_2O $\longrightarrow$

$(CH_3)_3CI$ + H_2O $\longrightarrow$ **better leaving group** **faster** reaction

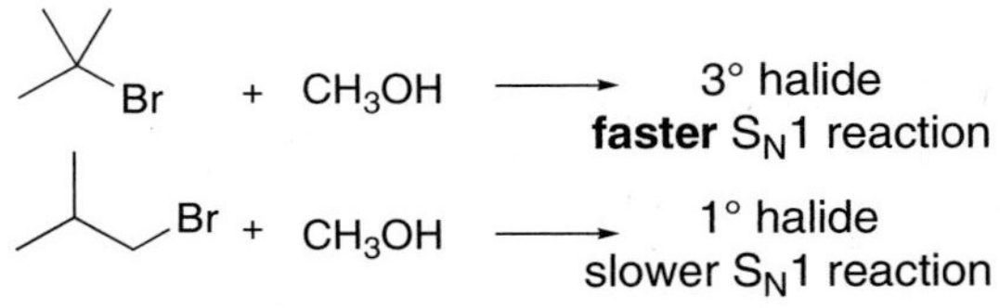

c. [structure] Cl + H_2O $\longrightarrow$ aryl halide slower reaction

[structure] Cl + H_2O $\longrightarrow$ 2° halide **faster** S$_N$1 reaction

b. [structure] Br + CH_3OH $\longrightarrow$ 3° halide **faster** S$_N$1 reaction

[structure] Br + CH_3OH $\longrightarrow$ 1° halide slower S$_N$1 reaction

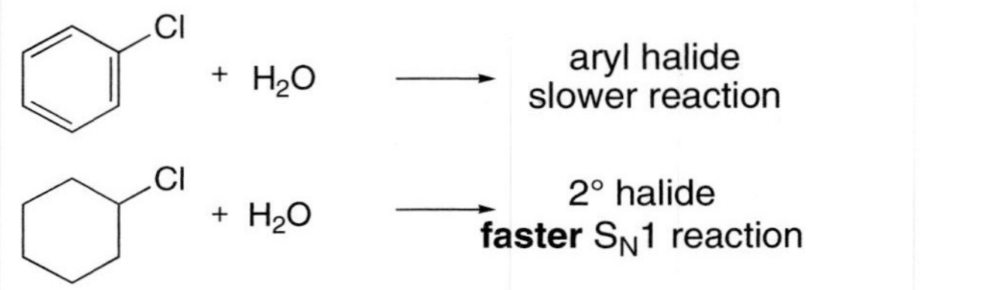

d. [structure] I + CH_3CH_2OH $\xrightarrow{CH_3CH_2OH}$ polar protic solvent **faster** reaction

[structure] I + CH_3CH_2OH $\xrightarrow{DMSO}$ polar aprotic solvent slower reaction

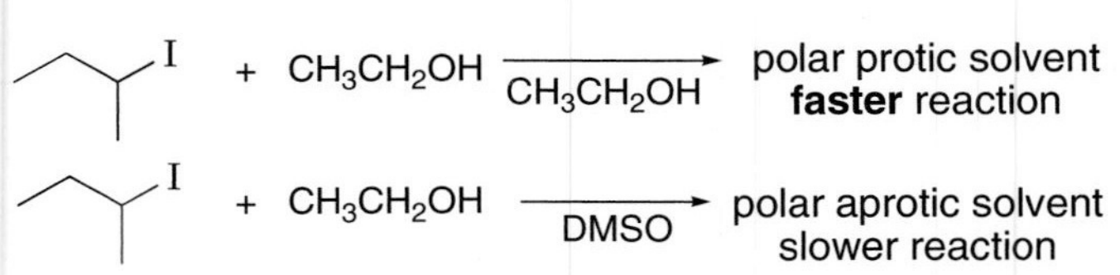

7.68

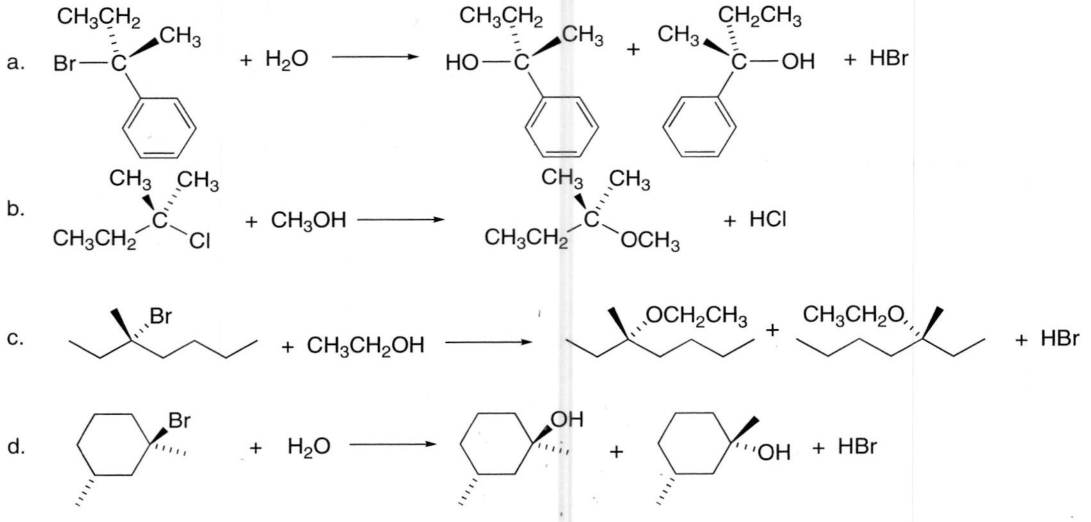

a. [structure] Br—C— with CH$_3$CH$_2$, CH$_3$, phenyl + H_2O $\longrightarrow$ HO—C— + CH$_3$—C—OH + HBr

b. [structure] CH$_3$CH$_2$—C—Cl with CH$_3$, CH$_3$ + CH_3OH $\longrightarrow$ CH$_3$CH$_2$—C—OCH$_3$ + HCl

c. [structure] Br + CH_3CH_2OH $\longrightarrow$ OCH$_2$CH$_3$ + CH$_3$CH$_2$O + HBr

d. [structure] Br + H_2O $\longrightarrow$ OH + OH + HBr

7.69 The 1° alkyl halide is also allylic, so it forms a resonance-stabilized carbocation. Increasing the stability of the carbocation by resonance, increases the rate of the S$_N$1 reaction.

$CH_3CH_2CH_2CH=CHCH_2Br \xrightarrow{CH_3OH} CH_3CH_2CH_2CH=CHCH_2OCH_3$ + $CH_3CH_2CH_2CHCH=CH_2$ + HBr
　　$\overset{|}{O}CH_3$

$CH_3CH_2CH_2CH=CHCH_2 \longrightarrow CH_3CH_2CH_2CH=CH-CH_2 \longleftrightarrow CH_3CH_2CH_2CH-CH=CH_2$ + Br$^-$
　　　　　　$\overset{|}{Br}$　　　　　　　　　　　　　　　$\overset{+}{}$　　　　　　　　　　　　　　　$\overset{+}{}$
　　　　　　　　　　　　　　　　　resonance-stabilized carbocation

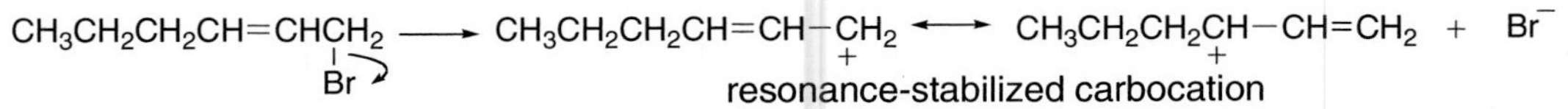

Use each resonance structure individually to continue the mechanism:

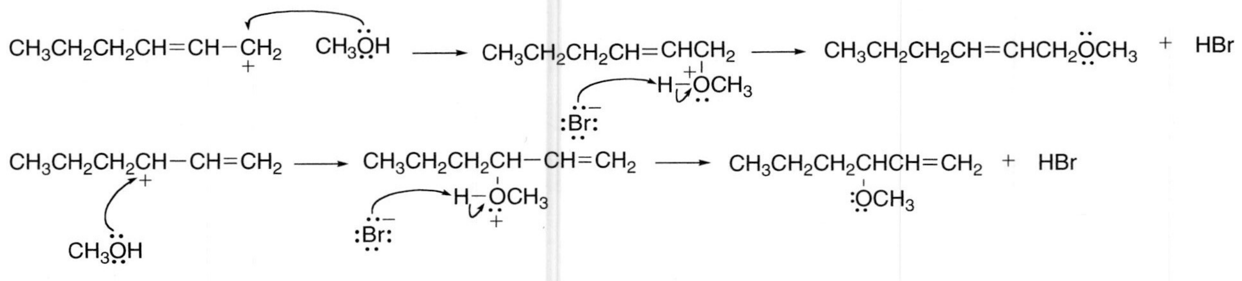

$CH_3CH_2CH_2CH=CH-CH_2$ $CH_3\overset{..}{O}H$ $\longrightarrow$ $CH_3CH_2CH_2CH=CHCH_2$ $\longrightarrow$ $CH_3CH_2CH_2CH=CHCH_2\overset{..}{O}CH_3$ + HBr
　　　　　　　　　$\overset{+}{}$　　　　　　　　　　　　　　　　　　　H$\overset{+|}{-}\overset{..}{O}CH_3$
　　　　　　　　　　　　　　　　　　　　　　　　:Br:

$CH_3CH_2CH_2CH-CH=CH_2$ $\longrightarrow$ $CH_3CH_2CH_2CH-CH=CH_2$ $\longrightarrow$ $CH_3CH_2CH_2CHCH=CH_2$ + HBr
　　　　$\overset{+}{}$　　　　　　　　　　　　　　　　　H$-\overset{|}{O}CH_3$　　　　　　　　　　$\overset{|}{:}\overset{..}{O}CH_3$
　　$CH_3\overset{..}{O}H$　　　　　　　　　　:Br:

7.70

a.

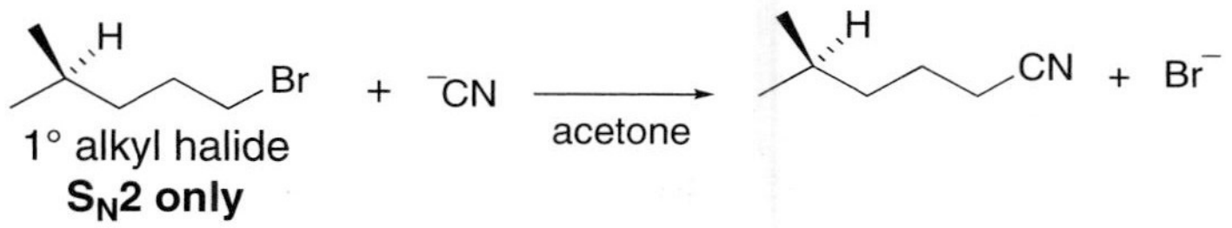
H
Br
+ ⁻CN
acetone
H
CN + Br⁻
1° alkyl halide
SN2 only

b.

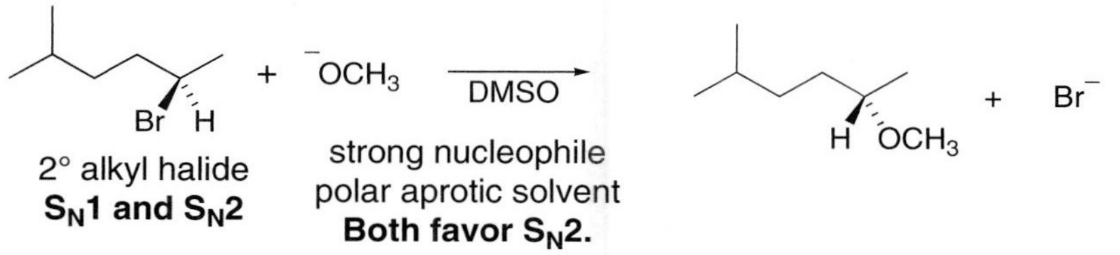
+ ⁻OCH3
DMSO
Br H
2° alkyl halide
SN1 and SN2
strong nucleophile
polar aprotic solvent
Both favor SN2.
H OCH3
+ Br⁻

reaction at a stereogenic center
inversion of configuration

c.

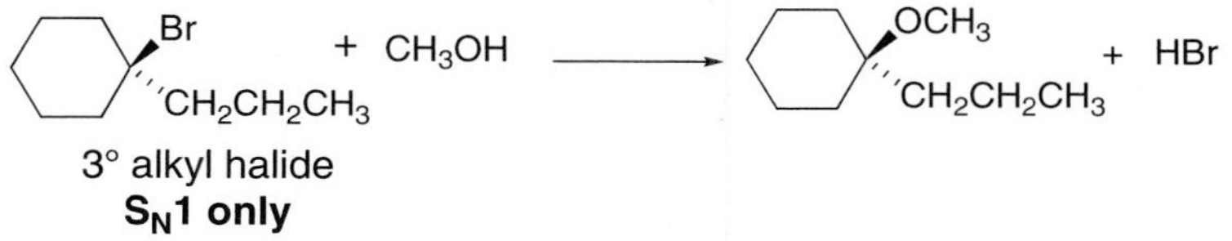
Br
+ CH3OH
CH2CH2CH3
3° alkyl halide
SN1 only
OCH3
+ HBr
CH2CH2CH3

d.

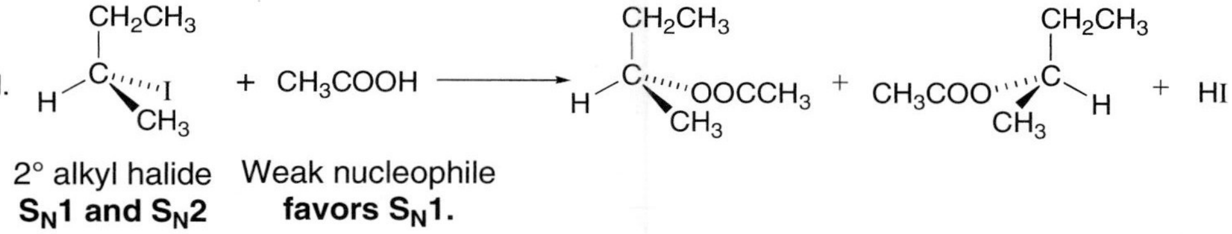
CH2CH3
H C I
CH3
+ CH3COOH
2° alkyl halide
SN1 and SN2
Weak nucleophile
favors SN1.
CH2CH3
H C OOCCH3
CH3
+
CH2CH3
CH3COO C H
CH3
+ HI

reaction at a stereogenic center
racemization of product

e.

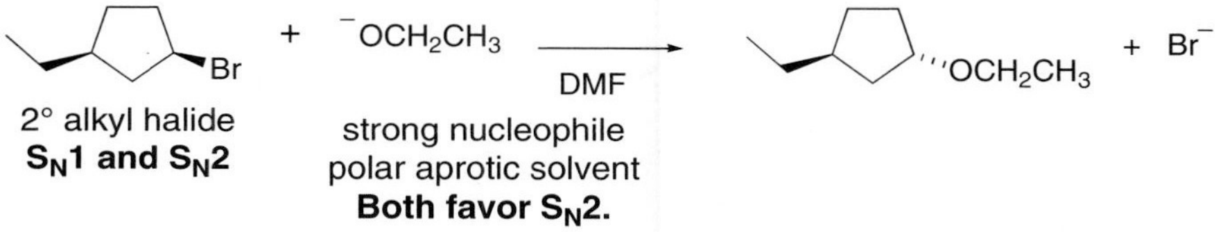
Br
+ ⁻OCH2CH3
DMF
2° alkyl halide
SN1 and SN2
strong nucleophile
polar aprotic solvent
Both favor SN2.
OCH2CH3
+ Br⁻

reaction at a stereogenic center
inversion of configuration

f.

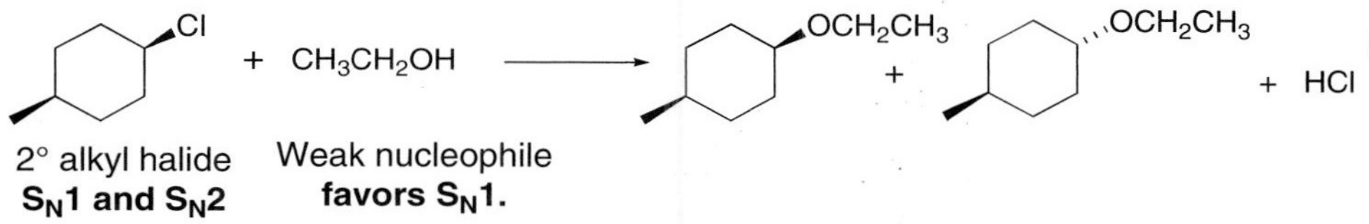
Cl
+ CH3CH2OH
2° alkyl halide
SN1 and SN2
Weak nucleophile
favors SN1.
OCH2CH3
+
OCH2CH3
+ HCl

two products – **diastereomers**
Nucleophile attacks
from above and below.

7.71

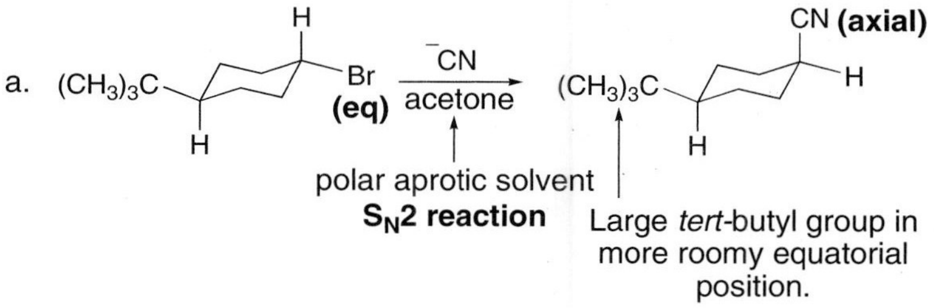
a. (CH3)3C
H
H
Br
(eq)
⁻CN
acetone
polar aprotic solvent
SN2 reaction
(CH3)3C
CN (axial)
H
H
Large tert-butyl group in
more roomy equatorial
position.

inversion (equatorial to axial)

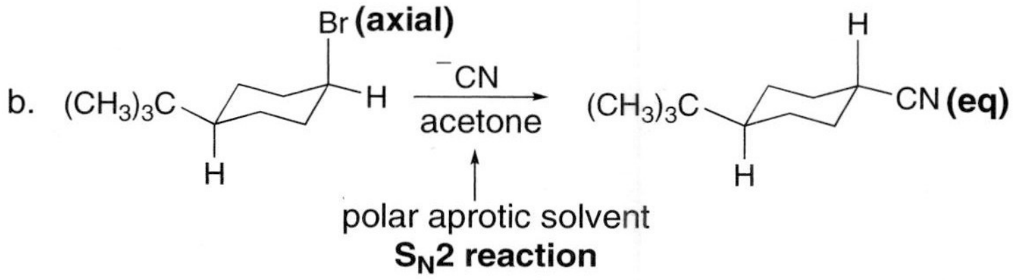
b. (CH3)3C
Br (axial)
H
H
⁻CN
acetone
polar aprotic solvent
SN2 reaction
(CH3)3C
H
CN (eq)
H

inversion (axial to equatorial)

7.72

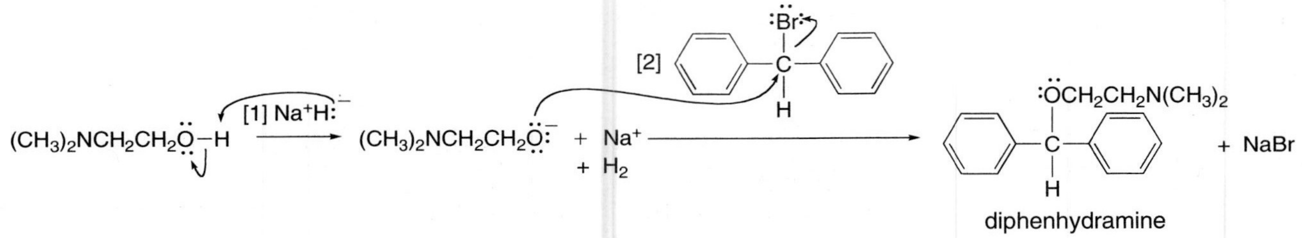

[1] Na+H:−
[2]
(CH3)2NCH2CH2O–H
(CH3)2NCH2CH2O:− + Na+ + H2
:Br:
H
:OCH2CH2N(CH3)2
C
H
+ NaBr
diphenhydramine

7.73 First decide whether the reaction will proceed via an S_N1 or S_N2 mechanism (Answer 7.40), and then draw the mechanism.

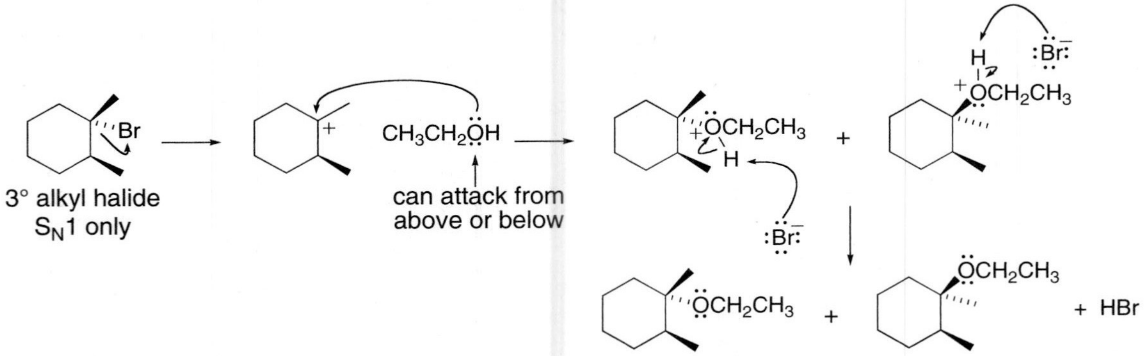

Br
3° alkyl halide
SN1 only
CH3CH2OH
can attack from
above or below
OCH2CH3
H
:Br:
+
+ HBr

7.74

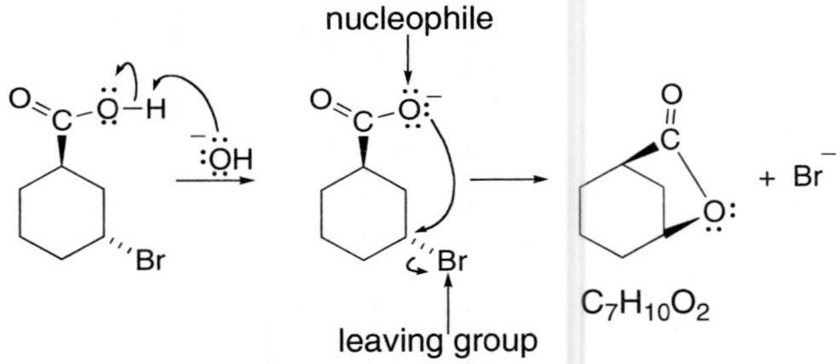

nucleophile
:OH
Br
leaving group
C7H10O2
+ Br−

7.75

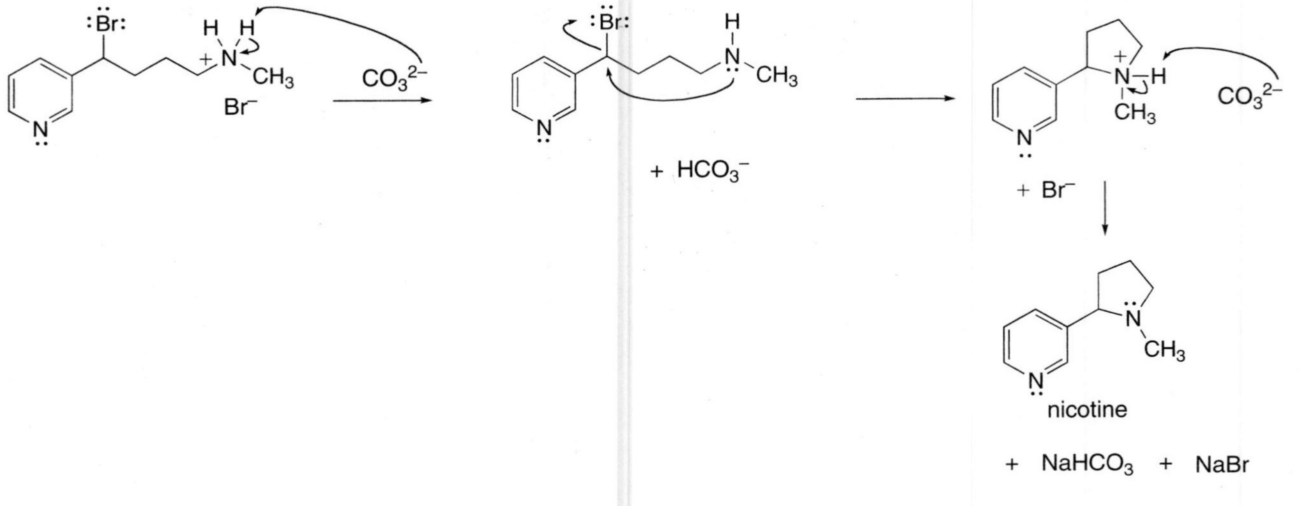

:Br:
H H
N
CH3
Br−
CO32−
+ HCO3−
N–H
CH3
+ Br−
nicotine
+ NaHCO3 + NaBr

7.76

a. Hexane is nonpolar and therefore few nucleophiles will dissolve in it.

b. $(CH_3)_3CO^-$ is a stronger base than $CH_3CH_2O^-$:

The three electron-donating CH_3 groups add electron density to the negative charge of the conjugate base, destabilizing it and making it a stronger base.

c. By the Hammond postulate, the S_N1 reaction is faster with RX that form more stable carbocations.

$(CH_3)_3C +$

3° Carbocation is stabilized by three electron-donor CH_3 groups.

$(CH_3)_2C +$
CF_3 Although this carbocation is also 3°, the three electron-withdrawing F atoms destabilize the positive charge. Since the carbocation is less stable, the reaction to form it is slower.

d. The identity of the nucleophile does not affect the rate of S_N1 reactions since the nucleophile does not appear in the rate-determining step.

e.

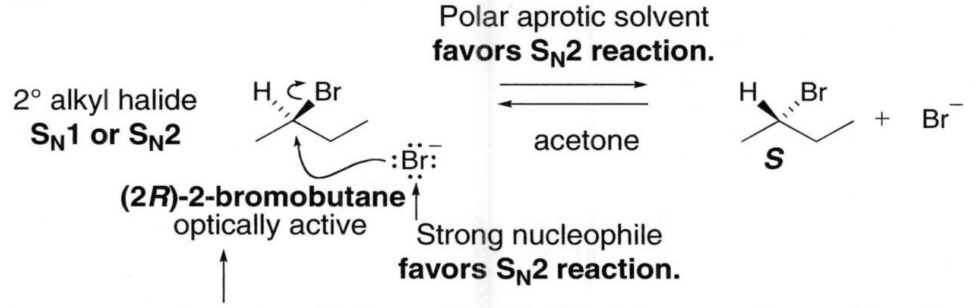

Polar aprotic solvent **favors S_N2 reaction.**

2° alkyl halide **S_N1 or S_N2**

H C Br

:Br:⁻

(2R)-2-bromobutane optically active

acetone

H Br

S + Br⁻

Strong nucleophile **favors S_N2 reaction.**

This compound reacts with Br⁻ until a 50:50 mixture results, making the mixture optically inactive. Then either compound can react with Br⁻ and the mixture remains optically inactive.

7.77

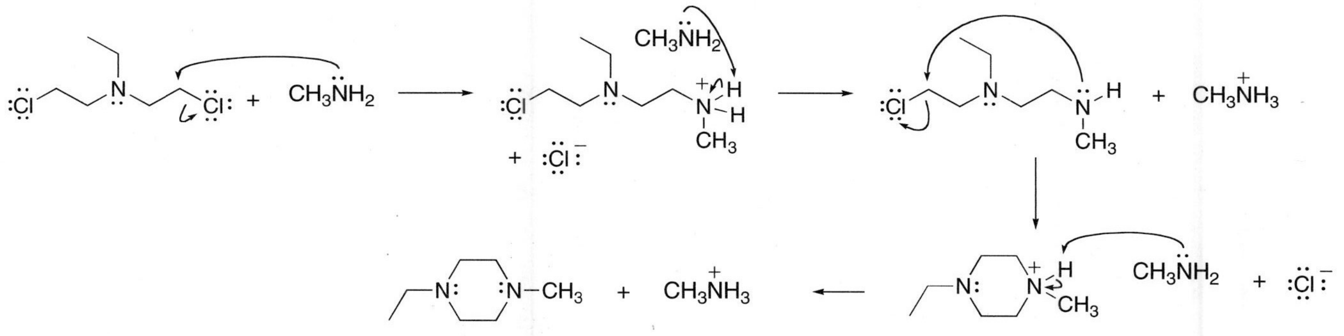

7.78 In the first reaction, substitution occurs at the stereogenic center. Since an achiral, planar carbocation is formed, the nucleophile can attack from either side, thus generating a racemic mixture.

3° alkyl halide

CH_3OH

S_N1

two steps

(6R)-6-bromo-2,6-dimethylnonane

+ Br⁻

achiral,
planar carbocation

OCH_3

+

OCH_3

racemic mixture
optically inactive

In the second reaction, the starting material contains a stereogenic center, but the nucleophile does not attack at that carbon. Since a bond to the stereogenic center is not broken, the configuration is retained and a chiral product is formed.

3° alkyl halide

Br

CH_3OH

S_N1

(5R)-2-bromo-2,5-dimethylnonane

+ Br⁻

Reaction does not occur at
the stereogenic center.

two steps

OCH_3 / configuration retained

optically active

7.79

a.

The nucleophile has replaced the leaving group.
Missing reagent:

⁻O

b.

Cl

C≡CH

The nucleophile has replaced the leaving group.
Missing reagent: ⁻C≡CH

c.

N₃⁻

N₃

The nucleophile has replaced the halide.
Starting material: Cl

d.

⁻SH

SH

The nucleophile has replaced the halide.
Starting material:

Cl

The leaving group must have the opposite orientation to the position of the nucleophile in the product.

7.80 To devise a synthesis, look for the carbon framework and the functional group in the product. **The carbon framework is from the alkyl halide and the functional group is from the nucleophile.**

a.

SH

carbon
framework

functional
group

Cl Na⁺ ⁻SH

SH

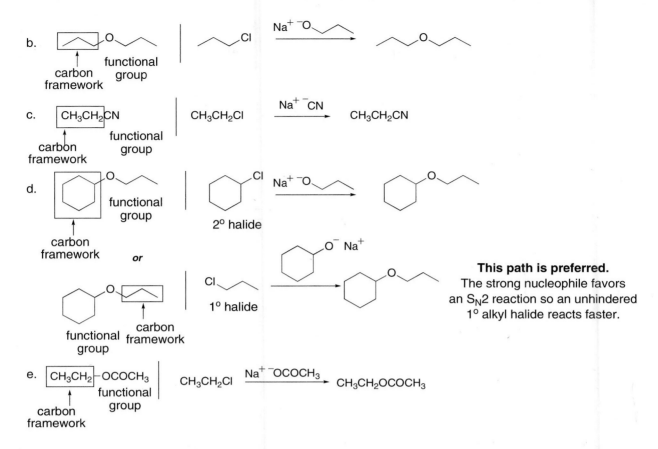

b.

functional group

carbon framework

c.

CH₃CH₂CN

functional group

carbon framework

CH₃CH₂Cl $\xrightarrow{Na^+ \ ^-CN}$ CH₃CH₂CN

d.

functional group

carbon framework

2° halide

or

functional group carbon framework

1° halide

This path is preferred.
The strong nucleophile favors an S_N2 reaction so an unhindered 1° alkyl halide reacts faster.

e. CH₃CH₂—OCOCH₃

functional group

carbon framework

CH₃CH₂Cl $\xrightarrow{Na^+ \ ^-OCOCH_3}$ CH₃CH₂OCOCH₃

7.81 Work backwards to determine the alkyl chloride needed to prepare benzalkonium chloride **A**.

a. —CH₂N(CH₃)₂ + CH₃(CH₂)₁₇Cl ⟶

B

—CH₂—N⁺(CH₃)—(CH₂)₁₇CH₃   Cl⁻

A

b. CH₃(CH₂)₁₇N(CH₃)₂ + —CH₂Cl ⟶

C

—CH₂—N⁺(CH₃)—(CH₂)₁₇CH₃   Cl⁻

A

7.82

$\xrightarrow{Na^+ \ :\ddot{O}CH_3}$

C

B
very crowded 3° halide

E

$\ddot{O}\text{:}^- Na^+$ CH₃I ⟶ OCH₃

D   **A**

unhindered methyl halide

E

preferred method
The strong nucleophile favors S_N2 reaction so the alkyl halide should be unhindered for a faster reaction.

7.83

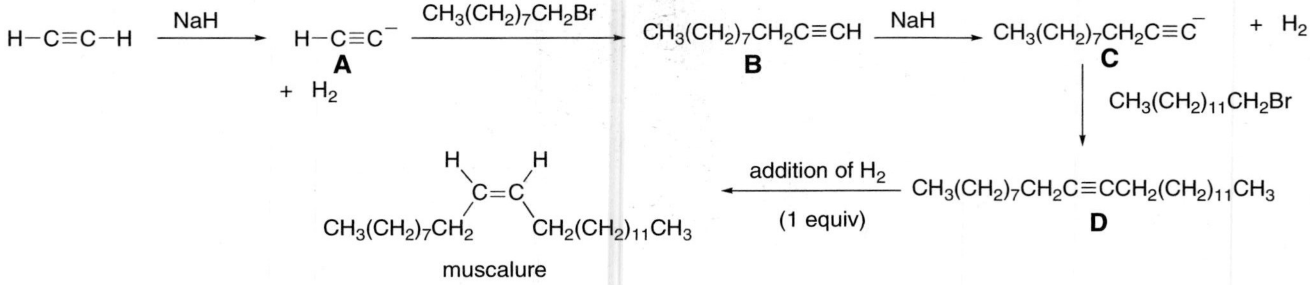
NaH
H−C≡C−H
H−C≡C⁻
A
+ H2
CH3(CH2)7CH2Br
CH3(CH2)7CH2C≡CH
B
NaH
CH3(CH2)7CH2C≡C⁻
C
+ H2
CH3(CH2)11CH2Br
CH3(CH2)7CH2C≡CCH2(CH2)11CH3
D
addition of H2
(1 equiv)
CH3(CH2)7CH2
CH2(CH2)11CH3
muscalure

7.84

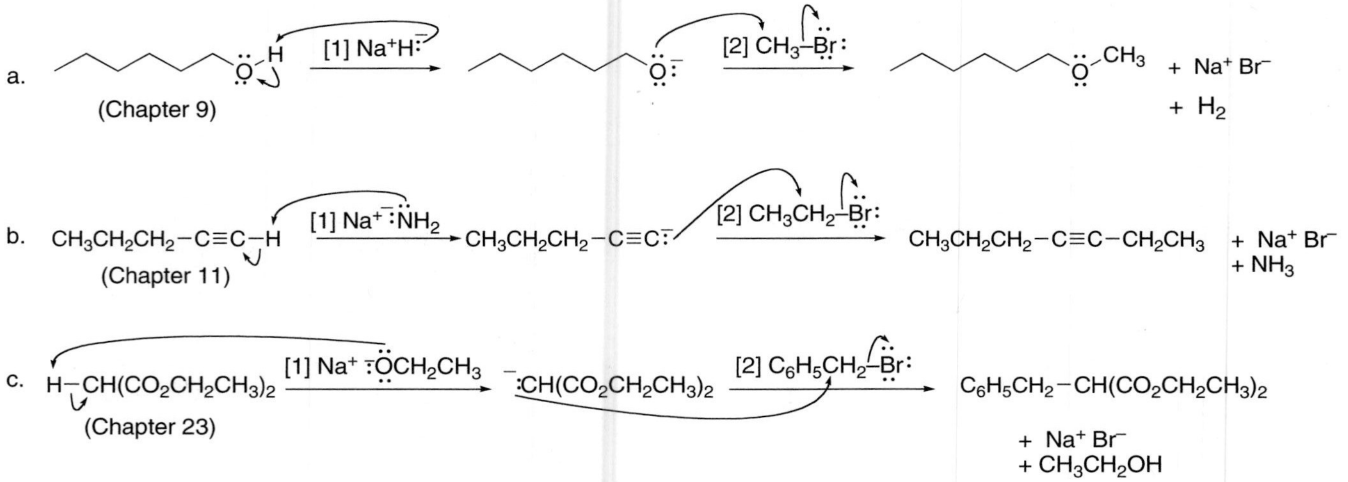
a.
[1] Na+H:⁻
[2] CH3–Br:
+ Na+ Br⁻
+ H2
(Chapter 9)
b. CH3CH2CH2−C≡C−H
[1] Na+ :NH2⁻
CH3CH2CH2−C≡C:⁻
[2] CH3CH2–Br:
CH3CH2CH2−C≡C−CH2CH3
+ Na+ Br⁻
+ NH3
(Chapter 11)
c. H−CH(CO2CH2CH3)2
[1] Na+ ⁻:OCH2CH3
⁻:CH(CO2CH2CH3)2
[2] C6H5CH2–Br:
C6H5CH2−CH(CO2CH2CH3)2
(Chapter 23)
+ Na+ Br⁻
+ CH3CH2OH

7.85

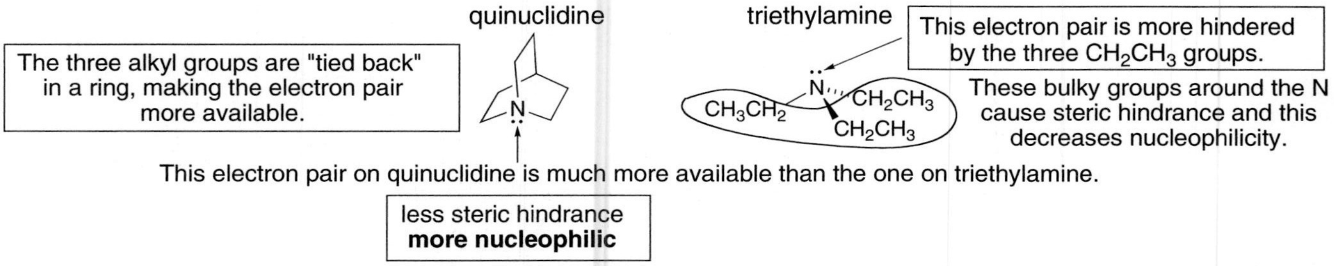
quinuclidine
triethylamine
The three alkyl groups are "tied back" in a ring, making the electron pair more available.
This electron pair is more hindered by the three CH2CH3 groups.
CH3CH2
CH2CH3
CH2CH3
These bulky groups around the N cause steric hindrance and this decreases nucleophilicity.
This electron pair on quinuclidine is much more available than the one on triethylamine.
less steric hindrance
more nucleophilic

7.86

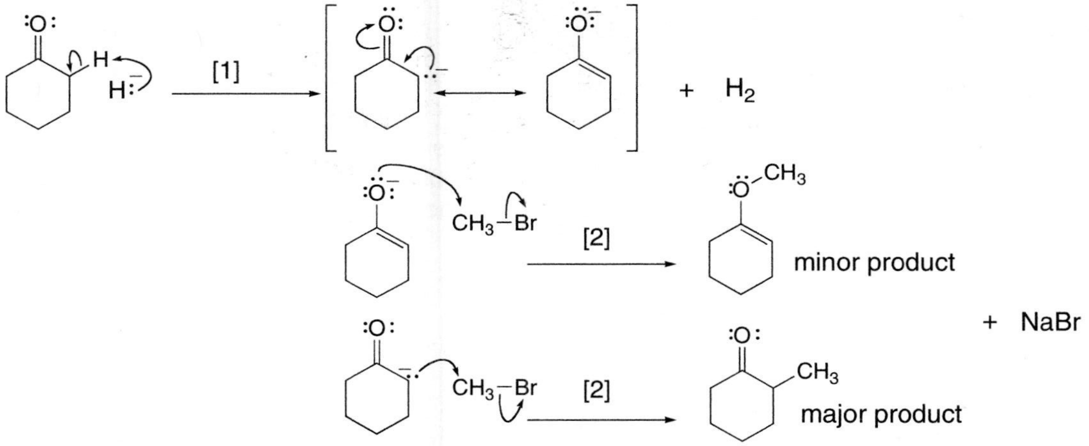

7.87

a.

base → (CH₃)₃C [intermediate with Br and O^-] → intramolecular S_N2 → (CH₃)₃C [epoxide product]

b.

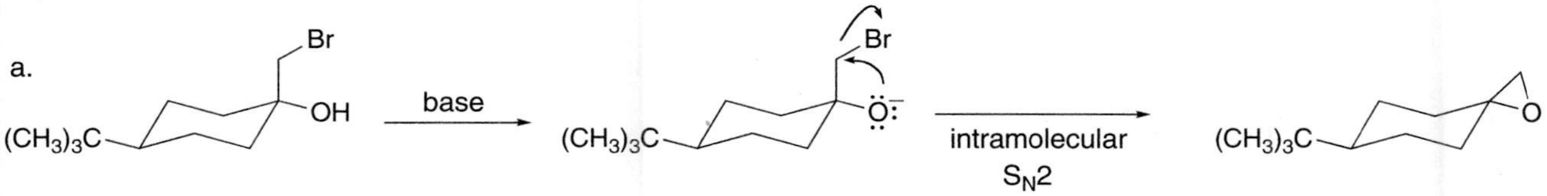

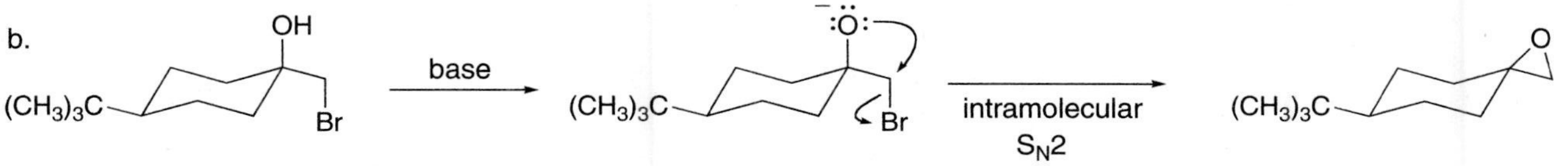

base → intramolecular S_N2

c.

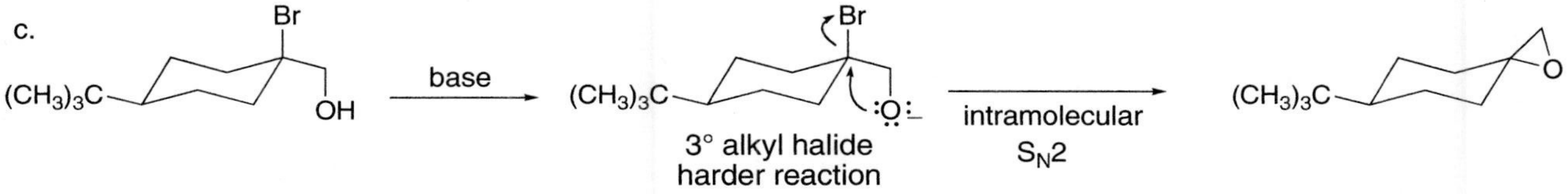

base → 3° alkyl halide harder reaction → intramolecular S_N2

d.

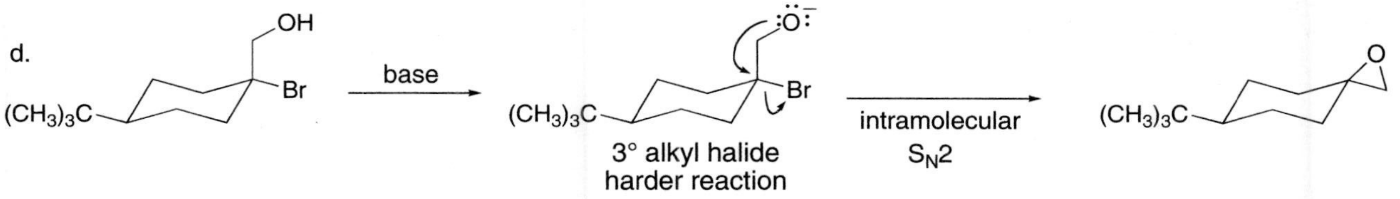

base → 3° alkyl halide harder reaction → intramolecular S_N2

7.88

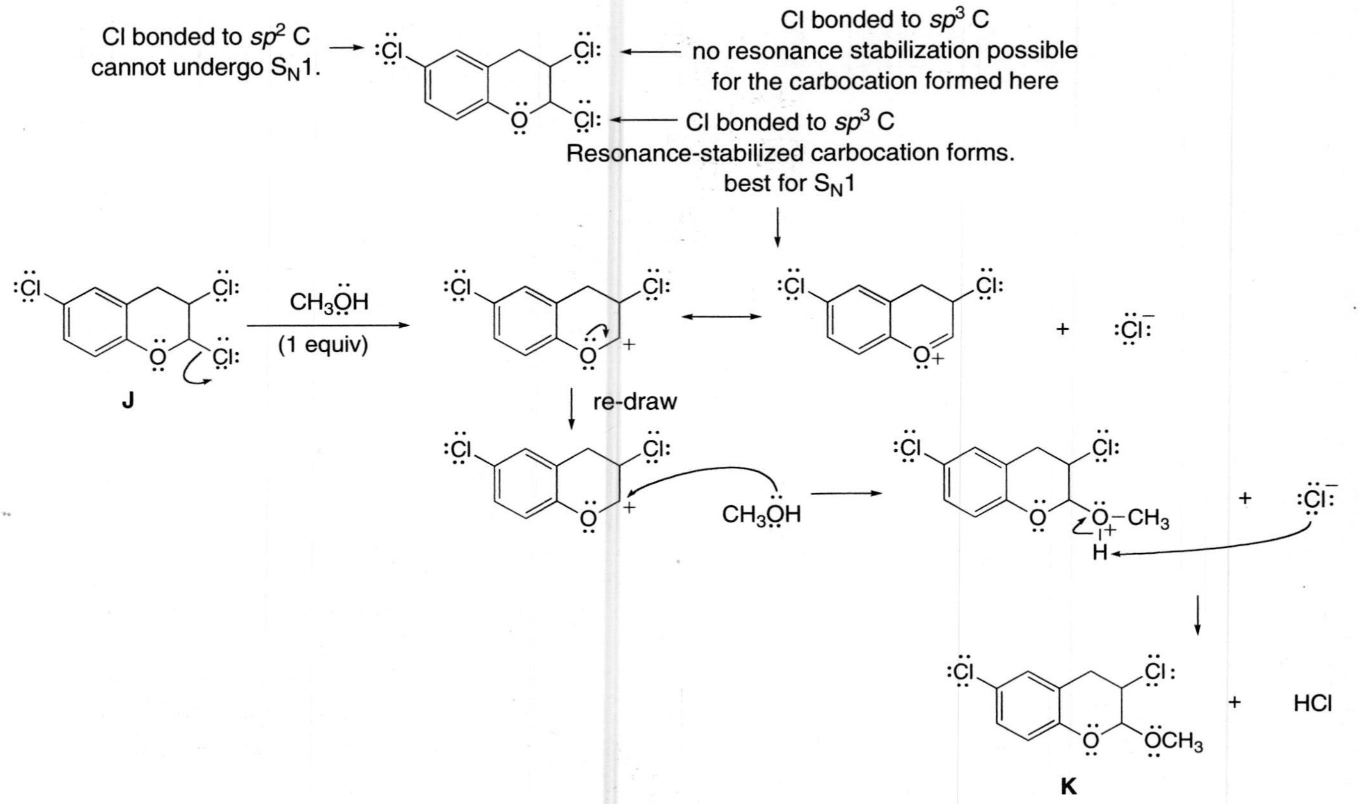

Cl bonded to sp^2 C
cannot undergo S_N1.

Cl bonded to sp^3 C
no resonance stabilization possible
for the carbocation formed here

Cl bonded to sp^3 C
Resonance-stabilized carbocation forms.
best for S_N1

$CH_3\ddot{O}H$
(1 equiv)

J

re-draw

$CH_3\ddot{O}H$

$+$ $:\ddot{Cl}:^-$

$+$ $:\ddot{Cl}:^-$

$+$ HCl

K

Chapter 8: Alkyl Halides and Elimination Reactions

◆ A comparison between nucleophilic substitution and β-elimination

Nucleophilic substitution—A nucleophile attacks a carbon atom (7.6).

substitution product

good leaving group

β-Elimination—A base attacks a proton (8.1).

elimination product

good leaving group

Similarities	Differences
• In both reactions RX acts as an electrophile, reacting with an electron-rich reagent. • Both reactions require a **good leaving group X:⁻** willing to accept the electron density in the C–X bond.	• In substitution, a nucleophile attacks a single carbon atom. • In elimination, a Brønsted–Lowry base removes a proton to form a π bond, and two carbons are involved in the reaction.

◆ The importance of the base in E2 and E1 reactions (8.9)

The strength of the base determines the mechanism of elimination.
* Strong bases favor E2 reactions.
* Weak bases favor E1 reactions.

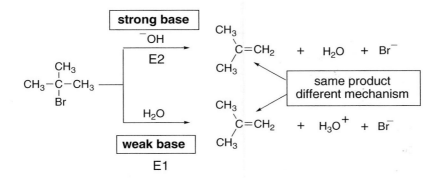

♦ **E1 and E2 mechanisms compared**

	E2 mechanism	E1 mechanism
[1] Mechanism	• one step (8.4B)	• two steps (8.6B)
[2] Alkyl halide	• rate: $R_3CX > R_2CHX >$ RCH_2X (8.4C)	• rate: $R_3CX > R_2CHX >$ RCH_2X (8.6C)
[3] Rate equation	• rate $= k[RX][B:]$ • second-order kinetics (8.4A)	• rate $= k[RX]$ • first-order kinetics (8.6A)
[4] Stereochemistry	• anti periplanar arrangement of H and X (8.8)	• trigonal planar carbocation intermediate (8.6B)
[5] Base	• favored by strong bases (8.4B)	• favored by weak bases (8.6C)
[6] Leaving group	• better leaving group $\rightarrow$ faster reaction (8.4B)	• better leaving group $\rightarrow$ faster reaction (Table 8.4)
[7] Solvent	• favored by polar aprotic solvents (8.4B)	• favored by polar protic solvents (Table 8.4)
[8] Product	• more substituted alkene favored (Zaitsev rule, 8.5)	• more substituted alkene favored (Zaitsev rule, 8.6C)

♦ **Summary chart on the four mechanisms: S_N1, S_N2, E1, or E2**

Alkyl halide type	Conditions	Mechanism
$1°$ RCH_2X	strong nucleophile strong bulky base	S_N2 E2
$2°$ R_2CHX	strong base and nucleophile strong bulky base weak base and nucleophile	$S_N2 + E2$ E2 $S_N1 + E1$
$3°$ R_3CX	weak base and nucleophile strong base	$S_N1 + E1$ E2

♦ **Zaitsev rule**

• β-Elimination affords the more stable product having the more substituted double bond.
• Zaitsev products predominate in E2 reactions except when a cyclohexane ring prevents trans diaxial arrangement.

8.1 • The carbon bonded to the leaving group is the **α carbon**. Any carbon bonded to it is a **β carbon**.

 • **To draw the products of an elimination reaction:** Remove the leaving group from the α carbon and a H from the β carbon and form a π bond.

a. $\overset{\beta}{C}H_3CH_2CH_2\overset{\alpha}{C}H_2CH_2-Cl$ $\xrightarrow{K^+ \ ^-OC(CH_3)_3}$ $CH_3CH_2CH_2CH=CH_2$

b. $\xrightarrow{K^+ \ ^-OC(CH_3)_3}$ $(CH_3CH_2)_2C=CH_2$ + $CH_3CH=C(CH_3)CH_2CH_3$

c. $\xrightarrow{K^+ \ ^-OC(CH_3)_3}$

8.2 **Alkenes are classified by the number of carbon atoms bonded to the double bond.** A monosubstituted alkene has one carbon atom bonded to the double bond, a disubstituted alkene has two carbon atoms bonded to the double bond, etc.

a.
4 C's bonded to C=C
tetrasubstituted
2 C's bonded to each C=C
disubstituted
3 C's bonded to each C=C
trisubstituted
vitamin A

b.
vitamin D$_3$
3 C's bonded to each C=C
trisubstituted
2 C's bonded to the C=C
disubstituted

8.3 To have stereoisomers at a C=C, the two groups on each end of the double bond must be different from each other.

two different groups
(CH$_3$CH$_2$ and H)
two different groups
(H and CH$_3$)
two different groups
(cyclohexyl and H)
two different groups
(cyclohexyl and H)

a.

b. $CH_3CH_2CH=CHCH_3$

c.

two CH$_3$ groups
no stereoisomers
possible

stereoisomers possible

stereoisomers possible

8.4 Two definitions:
- **Constitutional isomers** differ in the connectivity of the atoms.
- **Stereoisomers** differ only in the 3-D arrangement of atoms in space.

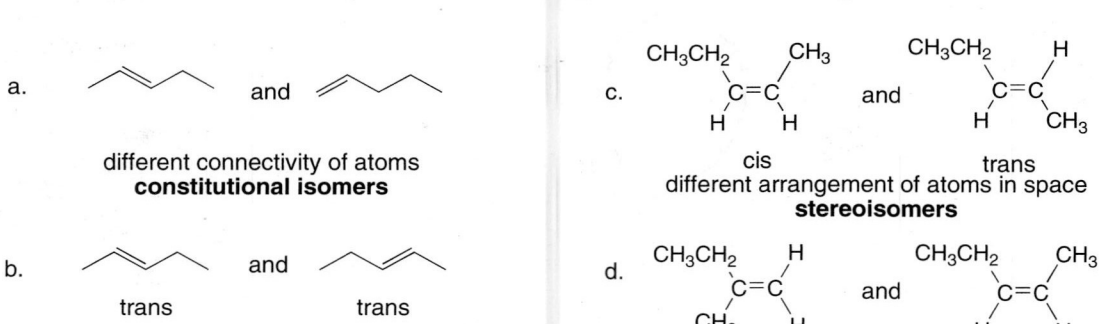

a.

and

different connectivity of atoms
constitutional isomers

b.

trans   and   trans

identical

c.

cis   trans
different arrangement of atoms in space
stereoisomers

d.

and

different connectivity of atoms
constitutional isomers

8.5 Two rules to predict the relative stability of alkenes:
[1] Trans alkenes are generally more stable than cis alkenes.
[2] The stability of an alkene increases as the number of R groups on the C=C increases.

a.

monosubstituted   or   disubstituted
more stable

b.

cis   or   trans
more stable

c.

trisubstituted   or   disubstituted
more stable

8.6 Use the rules from Answer 8.5 to explain the energy differences.

cis-2-butene   *trans*-2-butene   *cis*-2,2,5,5-tetramethyl-3-hexene   *trans*-2,2,5,5-tetramethyl-3-hexene

less steric interaction between
smaller CH_3 groups
smaller energy difference

more steric interaction between larger
tert-butyl groups in the cis isomer
larger difference in stability

8.7

A   B

Alkene **A** is more stable than alkene **B** because the double bond in **A** is in a six-membered ring. The double bond in **B** is in a four-membered ring, which has considerable angle strain due to the small ring size.

8.8 In an E2 mechanism, four bonds are involved in the single step. Use curved arrows to show these simultaneous actions:
[1] The base attacks a hydrogen on a β carbon.
[2] A π bond forms.
[3] The leaving group comes off.

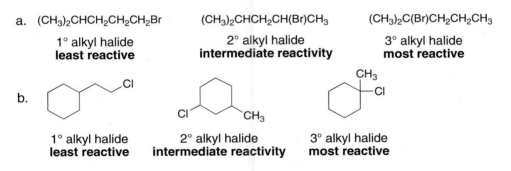

8.9 For E2 elimination to occur there must be at least one hydrogen on a β carbon.

β carbon ⟶ CH₃–C–C–H no H's on β carbon inert to E2 elimination

8.10 In both cases, the rate of elimination decreases.

stronger base
faster reaction

a. CH_3CH_2-Br + $^-OC(CH_3)_3$ ⟶

CH_3CH_2-Br + ^-OH ⟶

better leaving group
faster reaction

b. CH_3CH_2-Br + $^-OC(CH_3)_3$ ⟶

CH_3CH_2-Cl + $^-OC(CH_3)_3$ ⟶

8.11 As the number of R groups on the carbon with the leaving group increases, the rate of an E2 reaction increases.

a. $(CH_3)_2CHCH_2CH_2CH_2Br$ $(CH_3)_2CHCH_2CH(Br)CH_3$ $(CH_3)_2C(Br)CH_2CH_2CH_3$

1° alkyl halide
least reactive

2° alkyl halide
intermediate reactivity

3° alkyl halide
most reactive

b.

1° alkyl halide
least reactive

2° alkyl halide
intermediate reactivity

3° alkyl halide
most reactive

8.12 Use the following characteristics of an E2 reaction to answer the questions:
[1] E2 reactions are second order and one step.
[2] More substituted halides react faster.
[3] Reactions with strong bases or better leaving groups are faster.
[4] Reactions with polar aprotic solvents are faster.

Rate equation: rate = k[RX][Base]

    a. tripling the concentration of the alkyl halide = **rate triples**

    b. halving the concentration of the base = **rate halved**

    c. changing the solvent from CH_3OH to DMSO = **rate increases** (Polar aprotic solvent is better for E2.)

    d. changing the leaving group from I^- to Br^- = **rate decreases** (I^- is a better leaving group.)

    e. changing the base from ^-OH to H_2O = **rate decreases** (weaker base)

    f. changing the alkyl halide from CH_3CH_2Br to $(CH_3)_2CHBr$ = **rate increases** (More substituted halide reacts faster.)

8.13 The Zaitsev rule states: In a β-elimination reaction, the major product has the more substituted double bond.

a.

$CH_3-\underset{\underset{H}{|}}{\overset{\overset{CH_3}{|}}{C}}-\underset{\underset{Br}{|}}{\overset{\overset{H}{|}}{C}}^{\alpha}-CH_2CH_3$ $\xrightarrow{\text{loss of H and Br}}$ $(CH_3)_2C=CHCH_2CH_3$ + $(CH_3)_2CHCH=CHCH_3$

                                                                 trisubstituted   disubstituted

                                                            major product   **minor product**

b.

$\xrightarrow{\text{loss of H and Br}}$

                             trisubstituted   tetrasubstituted   disubstituted

                             minor product   **major product**   **minor product**

c.

$\xrightarrow{\text{loss of H and Cl}}$ + $CH_3CH_2CH_2CH_2CH=CHCH_3$

                                 monosubstituted   disubstituted

                                 minor product   **major product**

d.

$\xrightarrow{\text{loss of H and Cl}}$ trisubstituted **ONLY product**

8.14 An E1 mechanism has two steps:

    [1] The leaving group comes off, creating a carbocation.

    [2] A base pulls off a proton from a β carbon, and a π bond forms.

$CH_3-\underset{\underset{Cl}{|}}{\overset{\overset{CH_3}{|}}{C}}-CH_2CH_3$ + $CH_3\ddot{O}H$ $\xrightarrow{[1]}$ $CH_3-\overset{+}{\underset{|}{C}}-CH-CH_3 \quad CH_3\ddot{O}H$ + Cl^- $\xrightarrow{[2]}$ $(CH_3)_2C=CHCH_3$ + $CH_3\overset{+}{O}H_2$ + Cl^-

transition state [1]:

$$\left[\begin{array}{c} CH_3 \\ | \\ CH_3-\underset{\delta+}{C}-CH_2CH_3 \\ | \\ :\ddot{C}l:\delta- \end{array} \right]^{\ddagger}$$

transition state [2]:

$$\left[\begin{array}{c} CH_3 \\ | \\ CH_3-\underset{\delta+}{C}{=\!=\!=}CH-CH_3 \\ | \\ H-\!-\!-\underset{\delta+}{\ddot{O}}CH_3 \\ | \\ H \end{array} \right]^{\ddagger}$$

8.15 The Zaitsev rule states: In a β-elimination reaction, the major product has the more substituted double bond.

a.

CH$_3$CH$_2$—C—CH$_2$CH$_3$ + H$_2$O ⟶

trisubstituted
major product

+

disubstituted

b.

+ CH$_3$OH ⟶

tetrasubstituted
major product

+

disubstituted

+

trisubstituted

8.16 Use the following characteristics of an **E1 reaction** to answer the questions:
[1] E1 reactions are first order and two steps.
[2] More substituted halides react faster.
[3] Weaker bases are preferred.
[4] Reactions with better leaving groups are faster.
[5] Reactions in polar protic solvents are faster.

Rate equation: rate = k[RX]. The base doesn't affect rate.
 a. doubling the concentration of the alkyl halide = **rate doubles**
 b. doubling the concentration of the base = **no change** (Base is not in the rate equation.)
 c. changing the alkyl halide from $(CH_3)_3CBr$ to $CH_3CH_2CH_2Br$ = **rate decreases** (More substituted halides react faster.)
 d. changing the leaving group from Cl^- to Br^- = **rate increases** (better leaving group)
 e. changing the solvent from DMSO to CH_3OH = **rate increases** (Polar protic solvent favors E1.)

8.17 Both S$_N$1 and E1 reactions occur by forming a carbocation. To draw the products:
[1] **For the S$_N$1 reaction,** substitute the nucleophile for the leaving group.

[2] **For the E1 reaction**, remove a proton from a β carbon and create a new π bond.

a.

b.

8.18 E2 reactions occur with anti periplanar geometry. **The anti periplanar arrangement uses a** *staggered* **conformation and has the H and X on** *opposite sides* **of the C–C bond.**

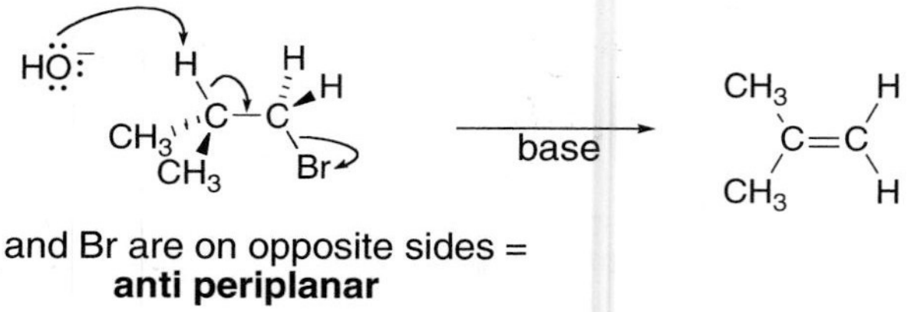

H and Br are on opposite sides =
anti periplanar

8.19 The E2 elimination reactions will occur in the anti periplanar orientation as drawn. To draw the product of elimination, maintain the orientation of the remaining groups around the C=C.

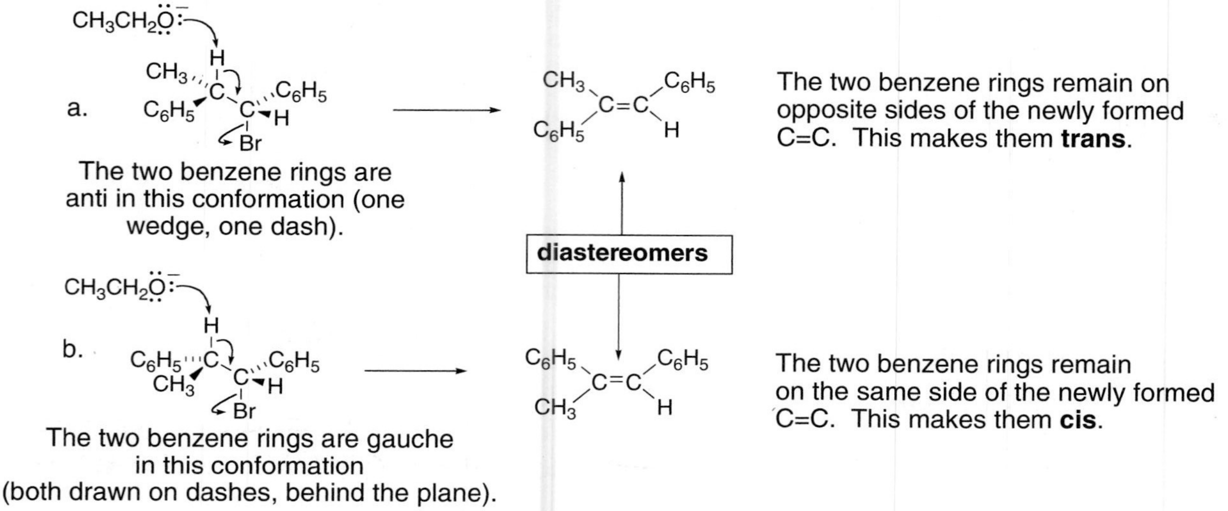

a. The two benzene rings are anti in this conformation (one wedge, one dash).

The two benzene rings remain on opposite sides of the newly formed C=C. This makes them **trans**.

diastereomers

b. The two benzene rings are gauche in this conformation (both drawn on dashes, behind the plane).

The two benzene rings remain on the same side of the newly formed C=C. This makes them **cis**.

8.20 Note: The Zaitsev products predominate in E2 elimination *except* when substituents on a cyclohexane ring prevent a **trans diaxial** arrangement of H and X.

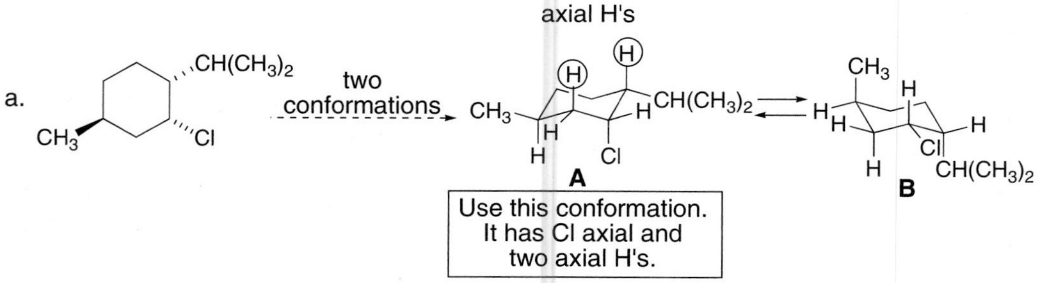

axial H's

a.

two conformations

A

Use this conformation. It has Cl axial and two axial H's.

B

[loss of H(β_2) + Cl]   [loss of H(β_1) + Cl]

two different axial H's

re-draw   re-draw

disubstituted   trisubstituted
major product

b.

two
conformations

A   B

Use this conformation.
It has Cl axial and
one axial H.

B   =

only one axial H
on a β carbon

[loss of H(β_1) + Cl]   disubstituted
only product

8.21 Draw the chair conformations of *cis*-1-chloro-2-methylcyclohexane and its trans isomer. For E2
elimination reactions to occur, **there must be a H and X trans diaxial to each other.**

Two conformations of the cis isomer:	Two conformations of the trans isomer:

A
reacting conformation (axial Cl)

This reacting conformation has only one
group axial, making it more stable and
present in a higher concentration than **B**.
This makes a **faster elimination reaction
with the cis isomer.**

B
reacting conformation (axial Cl)

This conformation is less stable than **A**,
since both CH_3 and Cl are axial.
**This slows the rate of elimination
from the trans isomer.**

8.22 **E2 reactions are favored by strong negatively charged bases** and occur with 1°, 2°, and 3° halides, with 3° being the most reactive.

   E1 reactions are favored by weaker neutral bases and do not occur with 1° halides since they would have to form highly unstable carbocations.

a. $CH_3-\underset{\underset{Cl}{|}}{\overset{\overset{CH_3}{|}}{C}}-CH_3$ + $^-OCH_3$ $\longrightarrow$

   strong negatively
   charged base
   E2

b.   (cyclohexyl)—I   + H_2O $\longrightarrow$

   weak neutral
   base
   E1

c.   (cyclohexyl with $C(CH_3)_3$ and Cl)   + CH_3OH $\longrightarrow$

   weak neutral
   base
   E1

d. CH_3CH_2Br + $^-OC(CH_3)_3$ $\longrightarrow$

   strong negatively
   charged base
   E2

8.23 Draw the alkynes that result from removal of two equivalents of HX.

a.   (cyclohexyl)$-\underset{\underset{H}{|}}{\overset{\overset{Cl}{|}}{C}}-\underset{\underset{H}{|}}{\overset{\overset{Cl}{|}}{C}}-CH_2CH_3$ $\xrightarrow{^-NH_2}$ (cyclohexyl)$-C\equiv C-CH_2CH_3$

c. $CH_3-\underset{\underset{Br}{|}}{\overset{\overset{Br}{|}}{C}}-CH_2CH_3$ $\xrightarrow{^-NH_2}$ $CH_3C\equiv CCH_3$

   $+$ $HC\equiv CCH_2CH_3$

b. $CH_3CH_2CH_2CHCl_2$ $\xrightarrow[\text{DMSO}]{KOC(CH_3)_3}$ $CH_3CH_2C\equiv CH$

d.   (Ph)$-\underset{\underset{Br}{|}}{\overset{\overset{Br}{|}}{CH}}-CH-$(Ph) $\xrightarrow{^-NH_2}$ (Ph)$-C\equiv C-$(Ph)

8.24

a.   (chain)—Cl $\xrightarrow{K^+ {}^-OC(CH_3)_3}$ (alkene chain)

   1° halide   strong bulky base
   S_N2 or E2   **E2**

b. $CH_3-\underset{\underset{Cl}{|}}{\overset{\overset{H}{|}}{C}}-CH_2CH_3$ $\xrightarrow{^-OH}$ $CH_3-\underset{\underset{OH}{|}}{\overset{\overset{H}{|}}{C}}-CH_2CH_3$ + $CH_3-CH=CHCH_3$ + $CH_2=CH-CH_2CH_3$

   2° halide   strong base S_N2 product   disubstituted   monosubstituted
   any mechanism S_N2 and E2   **major E2 product**   **minor E2 product**

c.   (cyclohexyl with CH_2CH_3 and I) $\xrightarrow[\text{weak base}]{CH_3CH_2OH}$ (cyclohexyl with CH_2CH_3 and OCH_2CH_3) + (cyclohexene with $CHCH_3$) + (cyclohexene with CH_2CH_3)

   3° halide S_N1 and E1 S_N1 product   E1 product   E1 product
   no S_N2

d.   (chain)$-\underset{\underset{Cl}{|}}{C}-$(CH_3) $\xrightarrow[\text{CH}_3\text{CH}_2\text{OH}]{\text{strong base}\atop CH_3CH_2O^-}$ (alkene) + (alkene)

   3° halide   **E2**   **major E2 product**   **minor E2 product**
   no S_N2

8.25

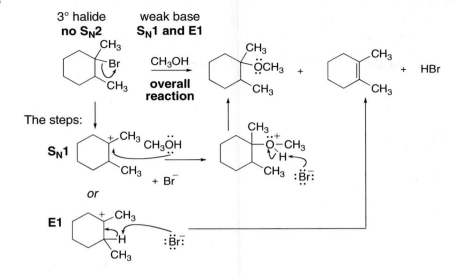

8.26

a. $CH_3CH_2CH_2CH_2CH_2CH_2Br \longrightarrow CH_3CH_2CH_2CH_2CH=CH_2$

b.

Br

$\longrightarrow CH_3CH_2CH_2CH_2CH=CHCH_2CH_3$ + $CH_3CH_2CH_2CH_2CH_2CH=CHCH_3$

CH₃

c. $CH_3CH_2CHCHCH_3 \longrightarrow CH_3CH_2CH=C(CH_3)_2$ + $CH_3CH=CHCH(CH_3)_2$

Cl

d.

CH₂—I $\longrightarrow$ =CH₂

8.27 To give only one product in an elimination reaction, **the starting alkyl halide must have only one type of β carbon with H's.**

a. $CH_2=CHCH_2CH_2CH_3 \longleftarrow$ $\overset{\alpha}{CH_2}-\overset{\beta}{CH_2}CH_2CH_2CH_3$

Cl

d. (structure with CH₃) $\longleftarrow$ (structure with β α Cl, CH₃) Two β carbons are identical.

b. $(CH_3)_2CHCH=CH_2 \longleftarrow$ $(CH_3)_2\overset{\beta}{CH}\overset{\alpha}{CH_2}-CH_2Cl$

c. (=CH₂ cyclohexane) $\longleftarrow$ (CH₂Cl cyclohexane, β)

e. (cyclopentene–C(CH₃)₃) $\longleftarrow$ (β Cl, C(CH₃)₃, α, β) Two β carbons are identical.

8.28 To have stereoisomers, the two groups on each end of the double bond must be different from each other.

farnesene

a.

two methyl groups no stereoisomers | two different groups at each end **can have stereoisomers** | 2 H's — no stereoisomers

geranial

b. CHO

two methyl groups no stereoisomers | two different groups at each end **can have stereoisomers**

8.29 Use the definitions in Answer 8.4.

a. (cyclohexene CH₃) and (cyclohexane =CH₂)

different connectivity **constitutional isomers**

c. (trans trans) and (trans trans)

identical

b. CH_3CH_2 \ /CH₃ C=C / \ CH₃ CH₂CH₃ and CH_3CH_2 \ /CH₂CH₃ C=C / \ CH₃ CH₃

stereoisomers

d. (structures with H, CH₃, CH₃) and (structures with CH₃, H, CH₃)

stereoisomers

8.30 There are three different isomers. Cis and trans isomers are diastereomers.

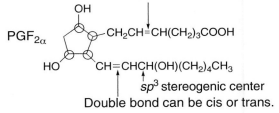

A
constitutional isomer
of B and C

B   **C**

↑   ↑
diastereomers

8.31

Double bond can be cis or trans.

PGF$_{2\alpha}$ —CH$_2$CH=CH(CH$_2$)$_3$COOH

a. five sp^3 stereogenic centers (four circled, one labeled)
b. Two double bonds can both be cis or trans.
c. 2^7 = 128 stereoisomers possible

HO   CH=CHCH(OH)(CH$_2$)$_4$CH$_3$

sp^3 stereogenic center

Double bond can be cis or trans.

8.32 Use the rules from Answer 8.5 to rank the alkenes.

a. CH$_2$=CHCH$_2$CH$_2$CH$_3$

CH$_3$CH$_2$, CH$_3$ / C=C / H , H

CH$_3$CH$_2$, H / C=C / H , CH$_3$

monosubstituted

disubstituted
cis

disubstituted
trans

least stable

**intermediate
stability**

most stable

b.   CH$_2$=CHCH(CH$_3$)$_2$ CH$_2$=C(CH$_3$)CH$_2$CH$_3$ (CH$_3$)$_2$C=CHCH$_3$

monosubstituted

disubstituted

trisubstituted

least stable

**intermediate
stability**

most stable

8.33 A larger negative value for $\Delta H°$ means the reaction is more exothermic. Since both 1-butene and *cis*-2-butene form the same product (butane), these data show that 1-butene was higher in energy to begin with, **since more energy is released in the hydrogenation reaction.**

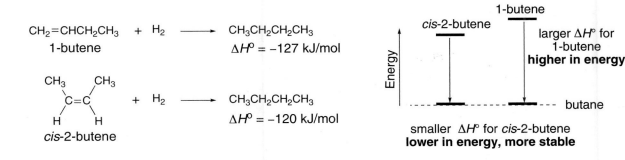

CH$_2$=CHCH$_2$CH$_3$ + H$_2$ ⟶   CH$_3$CH$_2$CH$_2$CH$_3$
1-butene $\Delta H° = -127$ kJ/mol

CH$_3$, CH$_3$ / C=C / H , H   + H$_2$ ⟶   CH$_3$CH$_2$CH$_2$CH$_3$
cis-2-butene $\Delta H° = -120$ kJ/mol

1-butene
cis-2-butene

larger $\Delta H°$ for
1-butene
higher in energy

Energy

butane

smaller $\Delta H°$ for *cis*-2-butene
lower in energy, more stable

8.34

a.

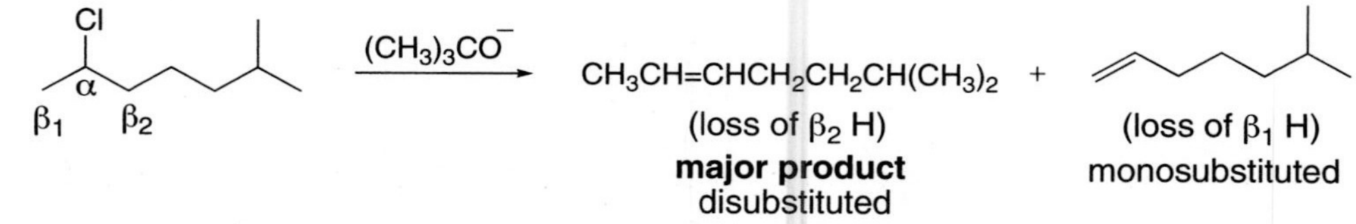

$(CH_3)_3CO^-$

$CH_3CH=CHCH_2CH_2CH(CH_3)_2$ +

(loss of β_2 H)
major product
disubstituted

(loss of β_1 H)
monosubstituted

b.

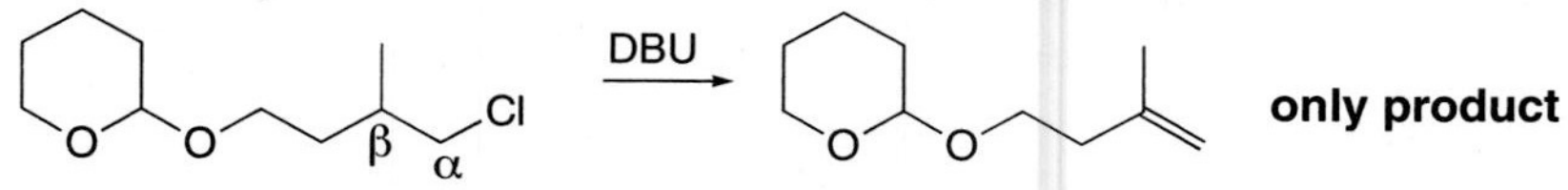

DBU

only product

c.

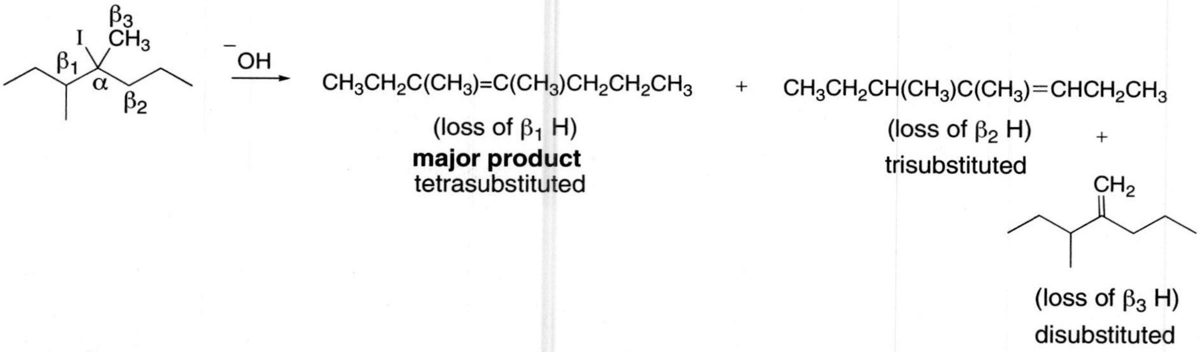

^-OH

$CH_3CH_2C(CH_3)=C(CH_3)CH_2CH_2CH_3$

(loss of β_1 H)
major product
tetrasubstituted

+

$CH_3CH_2CH(CH_3)C(CH_3)=CHCH_2CH_3$

(loss of β_2 H)
trisubstituted

+

(loss of β_3 H)
disubstituted

d.

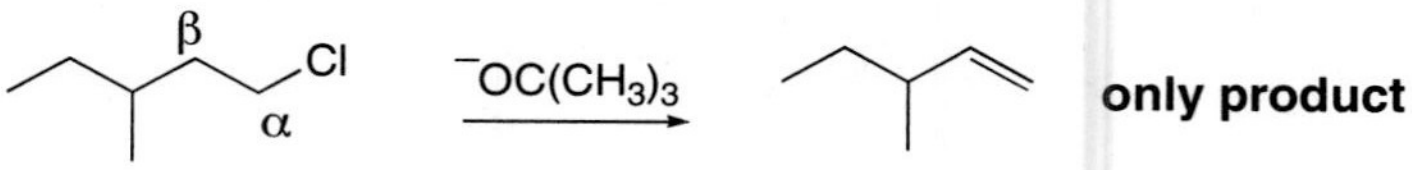

$^-OC(CH_3)_3$

only product

e.

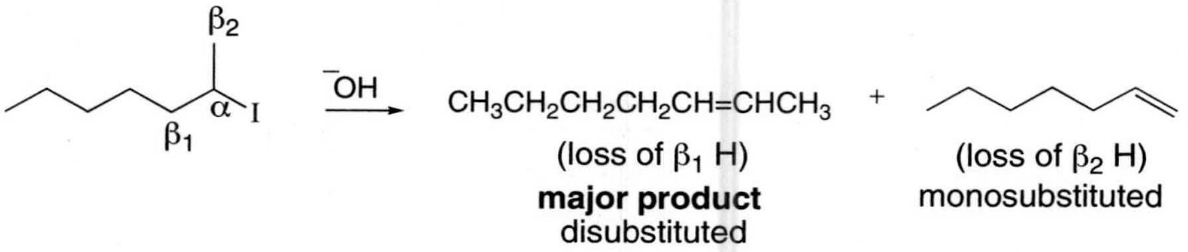

^-OH

$CH_3CH_2CH_2CH_2CH=CHCH_3$

(loss of β_1 H)
major product
disubstituted

+

(loss of β_2 H)
monosubstituted

f.

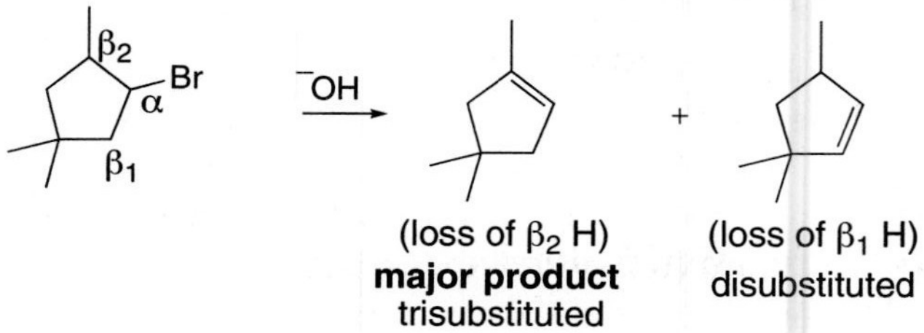

^-OH

(loss of β_2 H)
major product
trisubstituted

+

(loss of β_1 H)
disubstituted

8.35 To give only one alkene as the product of elimination, the alkyl halide must have either:
• only one β carbon with a hydrogen atom
• all identical β carbons so the resulting elimination products are identical

a. 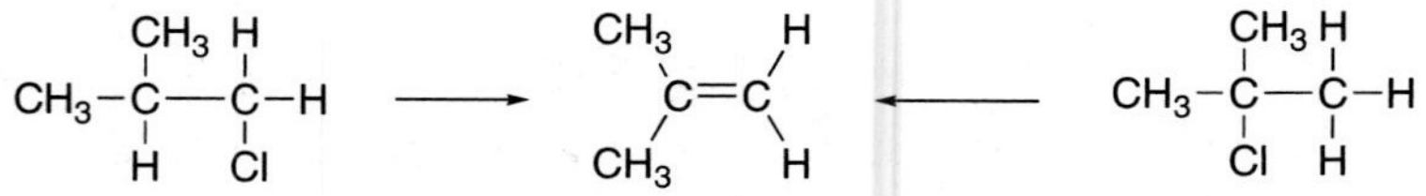 $CH_3-\overset{CH_3}{\underset{H}{C}}-\overset{H}{\underset{Cl}{C}}-H$ → $\underset{CH_3}{\overset{CH_3}{C}}=\underset{H}{\overset{H}{C}}$ ← $CH_3-\overset{CH_3}{\underset{Cl}{C}}-\overset{H}{\underset{H}{C}}-H$

b.

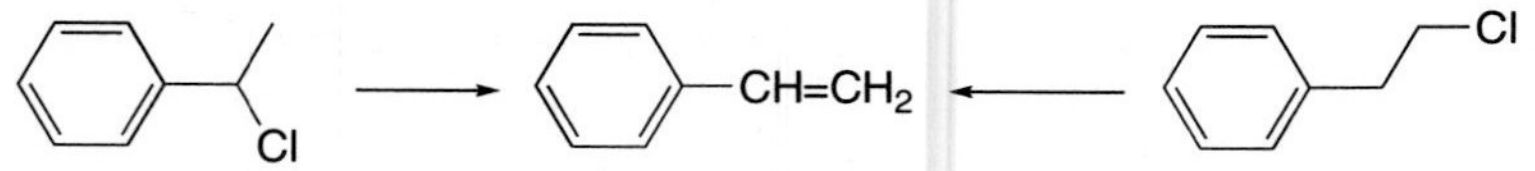

c.

8.36 Draw the products of the E2 reaction and compare the number of C's bonded to the C=C.

A yields a trisubstituted alkene as the major product and a disubstituted alkene as minor product. **B** yields a disubstituted alkene as the major product and a monosubstituted alkene as minor product. Since the major and minor products formed from **A** have more alkyl groups on the C=C (making them more stable) than those formed from **B**, **A** reacts faster in an elimination reaction.

8.37

a. Mechanism:

by-products

b. Rate = k[R–Br][$^-$OC(CH$_3$)$_3$]

    [1] Solvent changed to DMF (polar aprotic) = **rate increases**

    [2] [$^-$OC(CH$_3$)$_3$] decreased = **rate decreases**

    [3] Base changed to $^-$OH = **rate decreases** (weaker base)

    [4] Halide changed to 2° = **rate increases** (More substituted RX reacts faster.)

    [5] Leaving group changed to I$^-$ = **rate increases** (better leaving group)

8.38

1-chloro-1-methyl-
cyclopropane

The dehydrohalogenation of an alkyl halide usually forms the more stable alkene. In this case **A** is more stable than **B** even though **A** contains a disubstituted C=C whereas **B** contains a trisubstituted C=C. The double bond in **B** is part of a three-membered ring, and is less stable than **A** because of severe angle strain around both C's of the double bond.

8.39

a. $CH_3CH_2CH_2-\overset{\text{H}}{\underset{\text{Br}}{\text{C}}}-\text{C}_6H_5$ — KOH →

trans isomer more stable
major product

b. (structure) — NaOCH$_2$CH$_3$ →

trans isomer more stable
major product

8.40

a. (structure with CH$_3$, Cl, CH$_3$) →
(tetrasubstituted) + (trisubstituted) + (disubstituted)

tetrasubstituted   trisubstituted   disubstituted
major product

b. (structure with Br) →
(trisubstituted) + (trisubstituted) + (disubstituted)

trisubstituted   disubstituted
This isomer is more stable —
large groups farther away.   trisubstituted
major product

c. (structure with Cl) →
(disubstituted) + (trisubstituted)

disubstituted   trisubstituted
major product

8.41 Use the rules from Answer 8.22.

a. (structure with Br) — $^-OCH_3$ / strong base **E2** →
$CH_3CH=CHCH_3$ + $CH_3CH_2CH=CH_2$
2° halide   (cis and trans)

b. (structure with Br) — CH_3OH / weak base **E1** →
$CH_3CH=CHCH_3$ + $CH_3CH_2CH=CH_2$
2° halide   (cis and trans)

c. (structure with I) — $^-OC(CH_3)_3$ / strong base **E2** →
(structure)
1° halide

d. (structure with CH$_2$CH$_2$CH$_3$, Cl, CH$_3$) — H_2O / weak base **E1** →
(=CHCH$_2$CH$_3$, CH$_3$) + (CH$_2$CH$_2$CH$_3$, CH$_3$) + (CH$_2$CH$_2$CH$_3$, CH$_3$)
3° halide

e. (structure with Cl) — ^-OH / strong base **E2** →
(structure)
2° halide

f. (structure with Cl) — ^-OH / strong base **E2** →
(structure) + (structure)
2° halide

8.42 The order of reactivity is the same for both E2 and E1: $1° < 2° < 3°$

a.

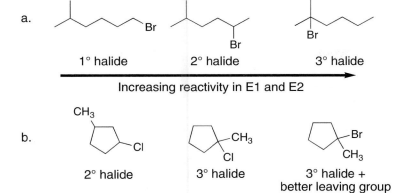

1° halide   2° halide   3° halide

Increasing reactivity in E1 and E2

b.

2° halide   3° halide   3° halide + better leaving group

Increasing reactivity in E1 and E2

8.43

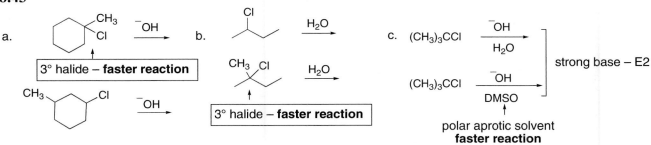

a.

3° halide – **faster reaction**

b.

3° halide – **faster reaction**

c. $(CH_3)_3CCl$ $\overline{}OH$ / H_2O

    $(CH_3)_3CCl$ $\overline{}OH$ / DMSO

strong base – E2

polar aprotic solvent **faster reaction**

8.44

bromocyclodecane   *cis*-cyclodecene

In a ten-membered ring, the cis isomer is more stable and, therefore, the preferred elimination product. The trans isomer is less stable because strain is introduced when two ends of the double bond are connected in a trans arrangement in this medium-sized ring.

8.45 With the strong base $\overline{}OCH_2CH_3$, the mechanism is E2, whereas with dilute base, the mechanism is E1. E2 elimination proceeds with anti periplanar arrangement of H and X. In the E1 mechanism there is no requirement for elimination to proceed with anti periplanar geometry. In this case the major product is always the most stable, more substituted alkene. Thus, **C** is the major product under E1 conditions. (In Chapter 9, we will learn that additional elimination products may form in the E1 reaction due to carbocation rearrangement.)

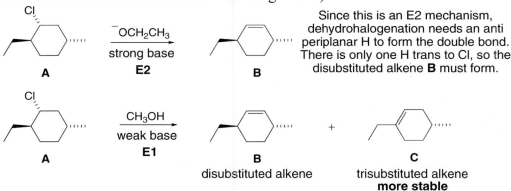

A $\overline{}OCH_2CH_3$ strong base **E2**   **B**

Since this is an E2 mechanism, dehydrohalogenation needs an anti periplanar H to form the double bond. There is only one H trans to Cl, so the disubstituted alkene **B** must form.

A CH_3OH weak base **E1**   **B** disubstituted alkene   +   **C** trisubstituted alkene **more stable**

8.46 H and Br must be anti during the E2 elimination. Rotate if necessary to make them anti; then eliminate.

a.

b.

c.

8.47

a.

b.

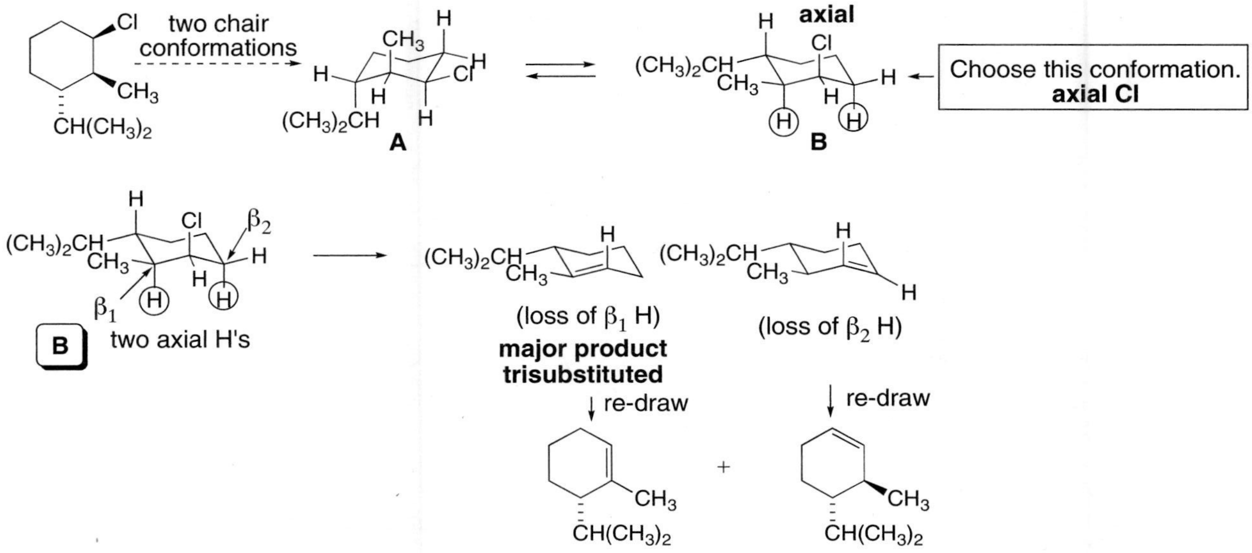

two chair conformations

A

axial

Choose this conformation.
axial Cl

B

B two axial H's

(CH₃)₂CH β_1

(loss of β_1 H)
major product
trisubstituted

(loss of β_2 H)

↓ re-draw

↓ re-draw

CH₃
CH(CH₃)₂

+

CH₃
CH(CH₃)₂

c.

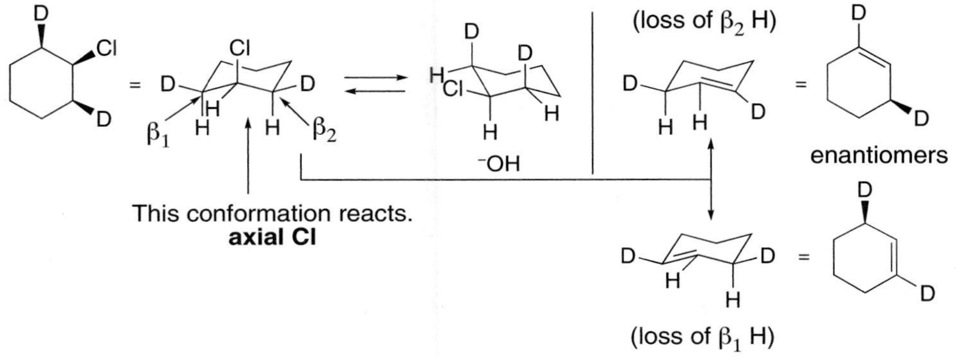

=

β_1 β_2

This conformation reacts.
axial Cl

⁻OH

(loss of β_2 H)

=

enantiomers

(loss of β_1 H)

d.

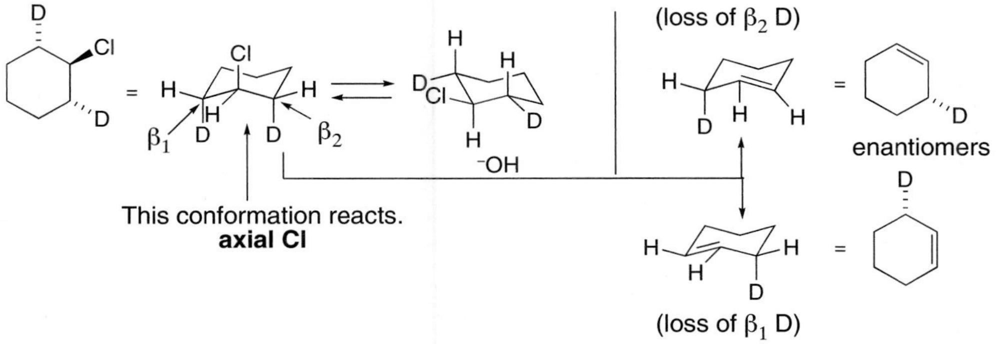

=

β_1 β_2

This conformation reacts.
axial Cl

⁻OH

(loss of β_2 D)

=

enantiomers

(loss of β_1 D)

8.48

a.

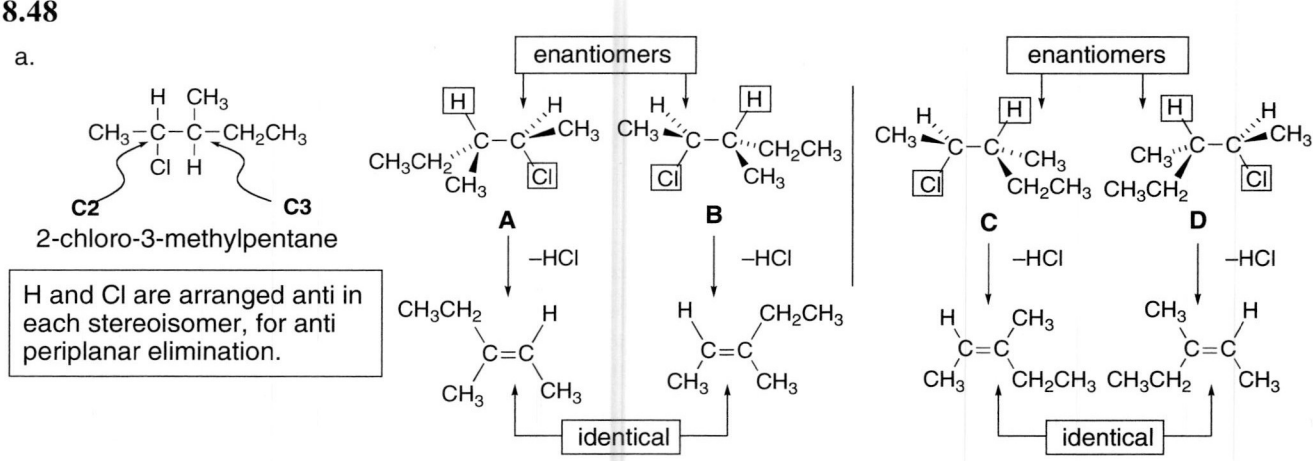

2-chloro-3-methylpentane

> H and Cl are arranged anti in each stereoisomer, for anti periplanar elimination.

b. Two different alkenes are formed as products.
c. The products are diastereomers: Two enantiomers (**A** and **B**) give identical products. **A** and **B** are diastereomers of **C** and **D**. Each pair of enantiomers gives a single alkene. Thus diastereomers give diastereomeric products.

8.49 The trans isomer reacts faster. During elimination, Br must be axial to give trans diaxial elimination. In the trans isomer, the more stable conformation has the bulky *tert*-butyl group in the more roomy equatorial position. In the cis isomer, elimination can occur only when both the *tert*-butyl and Br groups are axial, a conformation that is not energetically favorable.

cis

This conformation must react, but it contains two axial groups.

trans

preferred conformation

8.50

a.

CH_2CHCl_2 $\xrightarrow[\text{(2 equiv)}]{\text{NaNH}_2}$ $C{\equiv}CH$

b. $CH_3CH_2{-}\overset{\overset{\displaystyle CH_3}{|}}{\underset{\underset{\displaystyle CH_3}{|}}{C}}{-}CHCH_2Br \xrightarrow[\text{(2 equiv)}]{\text{NaNH}_2} CH_3CH_2{-}\overset{\overset{\displaystyle CH_3}{|}}{\underset{\underset{\displaystyle CH_3}{|}}{C}}{-}C{\equiv}CH$

c. $CH_3{-}\overset{\overset{\displaystyle Cl}{|}}{\underset{\underset{\displaystyle Cl}{|}}{C}}{-}CH_2CH_3 \xrightarrow[\text{(excess)}]{\text{NaNH}_2} HC{\equiv}C{-}CH_2CH_3 \quad + \quad CH_3{-}C{\equiv}C{-}CH_3$

d.

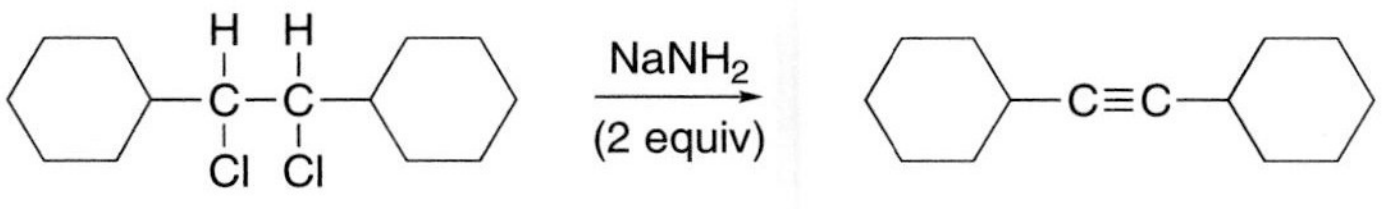

8.51

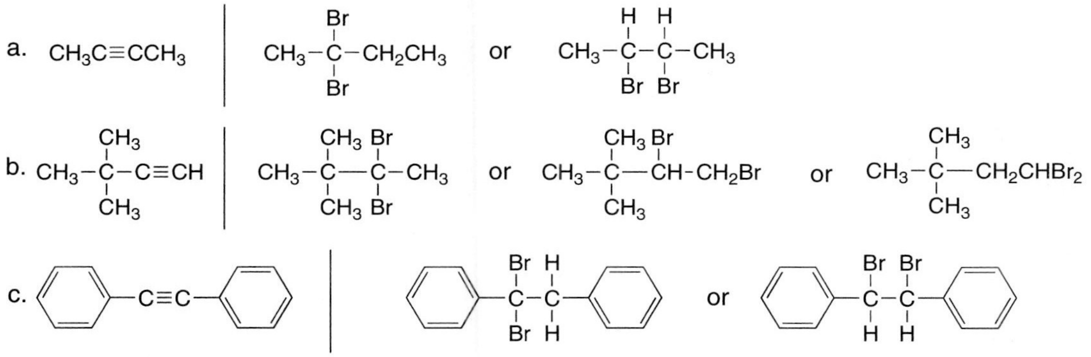

a. $CH_3C{\equiv}CCH_3$ |

b.

c.

8.52

2,3-dibromobutane → $CH_3-C{\equiv}C-CH_3$ (**A**, sp sp) $CH_3-CH{=}C{=}CH_2$ (**B**, sp) $CH_2{=}CH-CH{=}CH_2$ (**C**)

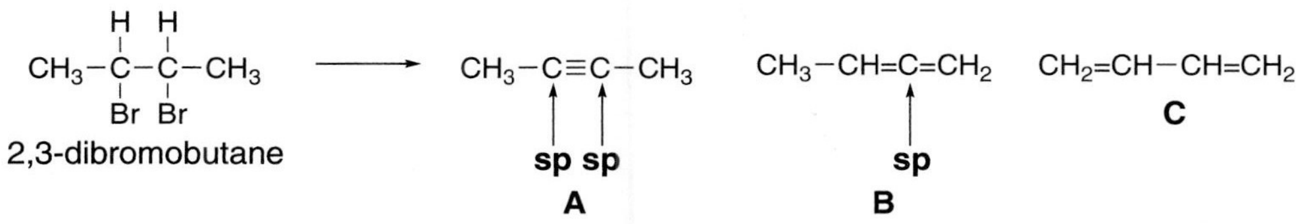

8.53 Use the "Summary chart on the four mechanisms: S_N1, S_N2, E1, or E2" on p. 8–2 to answer the questions.

a. Both S_N1 and E1 involve carbocation intermediates.
b. Both S_N1 and E1 have two steps.
c. S_N1, S_N2, E1, and E2 have increased reaction rates with better leaving groups.
d. Both S_N2 and E2 have increased rates when changing from CH_3OH (protic solvent) to $(CH_3)_2SO$ (aprotic solvent).
e. In S_N1 and E1 reactions, the rate depends on only the alkyl halide concentration.
f. Both S_N2 and E2 are concerted reactions.
g. CH_3CH_2Br and NaOH react by an S_N2 mechanism.
h. Racemization occurs in S_N1 reactions.
i. In S_N1, E1, and E2 mechanisms, 3° alkyl halides react faster than 1° or 2° halides.
j. E2 and S_N2 reactions follow second-order rate equations.

8.54

a.

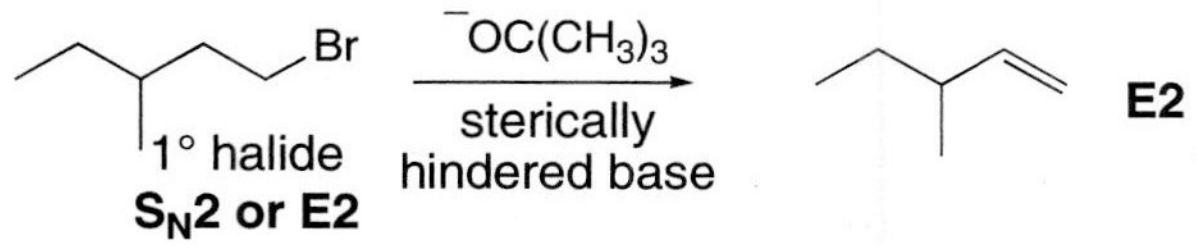

b.

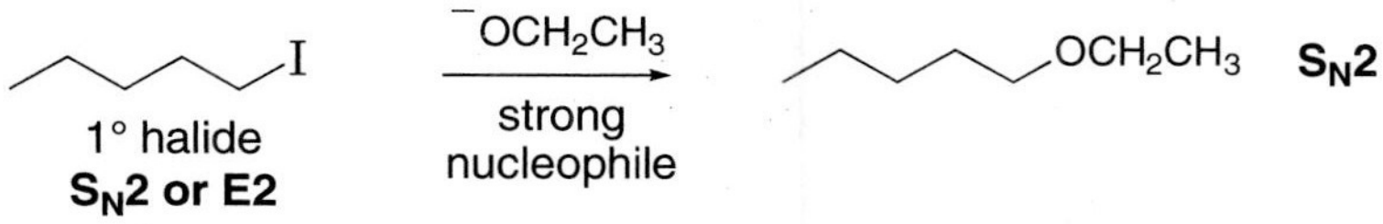

c.

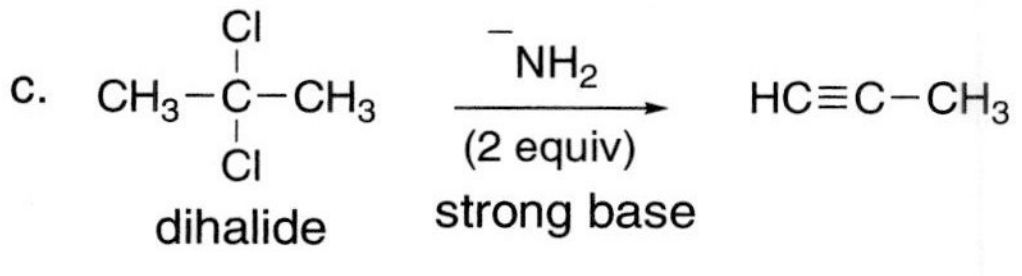

d.
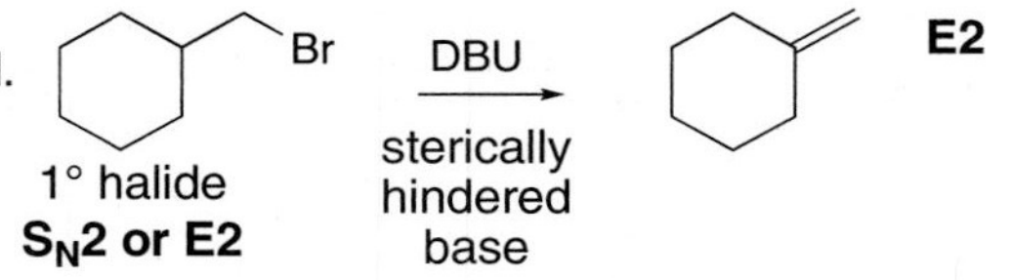
Br
DBU
sterically
hindered
base
E2
1° halide
SN2 or E2

e.
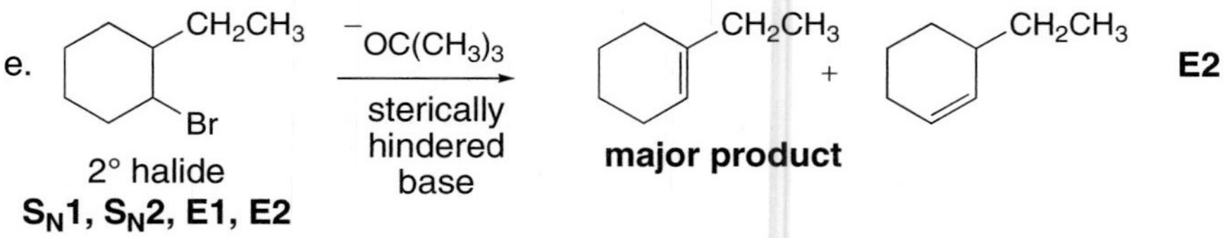
CH2CH3
Br
−OC(CH3)3
sterically
hindered
base
CH2CH3
CH2CH3
E2
major product
2° halide
SN1, SN2, E1, E2

f.
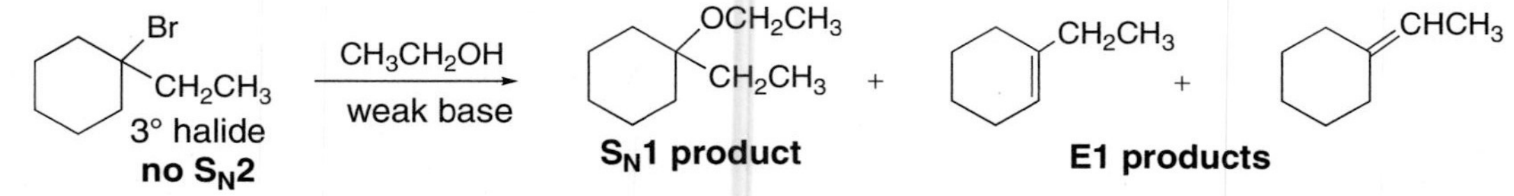
Br
CH2CH3
3° halide
no SN2
CH3CH2OH
weak base
OCH2CH3
CH2CH3
SN1 product
CH2CH3
CHCH3
E1 products

g. $(CH_3)_2CH-CH(Br)CH_2Br \xrightarrow{\text{2 NaNH}_2} (CH_3)_2CH-C\equiv CH$

dihalide

h.
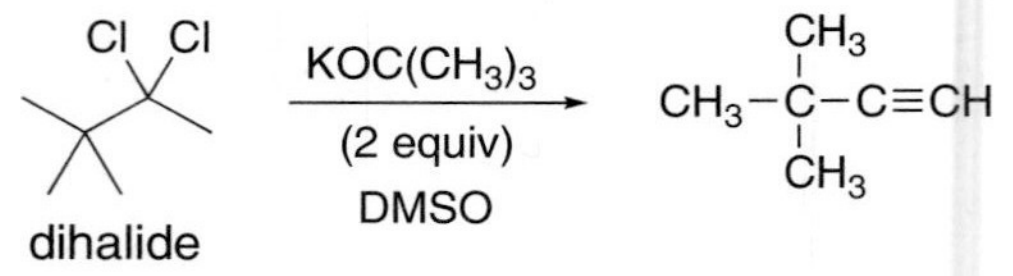
Cl Cl
dihalide
KOC(CH3)3
(2 equiv)
DMSO
CH3
CH3–C–C≡CH
CH3

i.
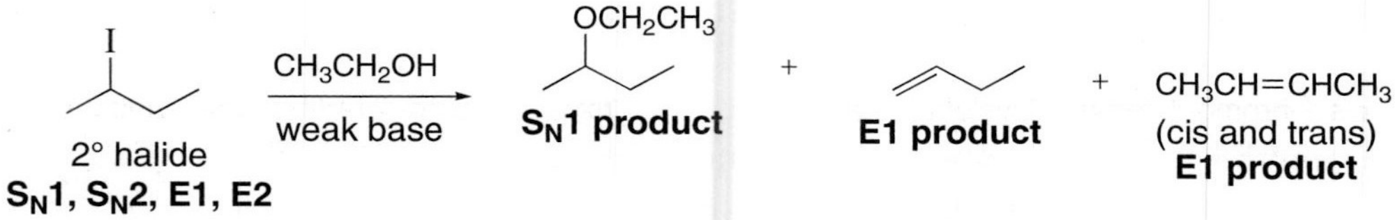
I
2° halide
SN1, SN2, E1, E2
CH3CH2OH
weak base
OCH2CH3
SN1 product
E1 product
CH3CH=CHCH3
(cis and trans)
E1 product

j.
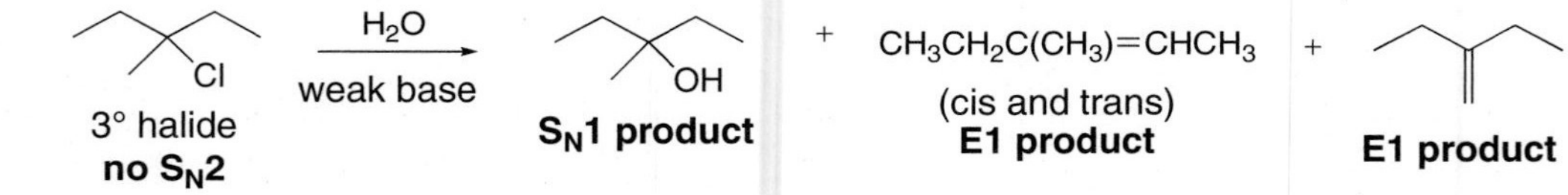
Cl
3° halide
no SN2
H2O
weak base
OH
SN1 product
CH3CH2C(CH3)=CHCH3
(cis and trans)
E1 product
E1 product

8.55 [1] $NaOCOCH_3$ is a good nucleophile and weak base, and substitution is favored. [3] $KOC(CH_3)_3$ is a strong, bulky base that reacts by E2 elimination when there is a β hydrogen in the alkyl halide.

a. $CH_3Cl \xrightarrow{\text{[1] NaOCOCH}_3} CH_3OCOCH_3$

$\xrightarrow{\text{[2] NaOCH}_3} CH_3OCH_3$

$\xrightarrow{\text{[3] KOC(CH}_3)_3} CH_3OC(CH_3)_3$

b.
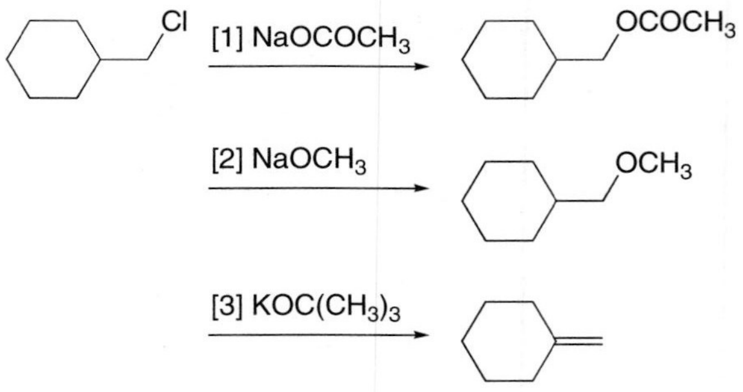
Cl
[1] NaOCOCH3
OCOCH3
[2] NaOCH3
OCH3
[3] KOC(CH3)3

c. ⬠—Cl $\xrightarrow{\text{[1] NaOCOCH}_3}$ ⬠—OCOCH₃

$\xrightarrow{\text{[2] NaOCH}_3}$ ⬠—OCH₃ + ⬠

SN2   E2

$\xrightarrow{\text{[3] KOC(CH}_3)_3}$ ⬠

d. ⬠—Cl $\xrightarrow{\text{[1] NaOCOCH}_3}$ ⬠—OCOCH₃

$\xrightarrow{\text{[2] NaOCH}_3}$ ⬠

$\xrightarrow{\text{[3] KOC(CH}_3)_3}$ ⬠

8.56

a. two enantiomers: (CH₃)₃C—⬡   (CH₃)₃C⋯⬡

                                      A   **B**

b. The bulky *tert*-butyl group anchors the cyclohexane ring and occupies the more roomy
equatorial position. The cis isomer has the Br atom axial, while the trans isomer has the Br
atom equatorial. For dehydrohalogenation to occur on a halo cyclohexane, the halogen must
be axial to afford trans diaxial elimination of H and X. The cis isomer readily reacts since the
Br atom is axial. The only way for the trans isomer to react is for the six-membered ring to flip
into a highly unstable conformation having both (CH₃)₃C and Br axial. Thus, the trans isomer
reacts much more slowly.

(CH₃)₃C— [Br, trans diaxial, H H]   (CH₃)₃C— [Br]

    cis-1-bromo-4-*tert*-butylcyclohexane   *trans*-1-bromo-4-*tert*-butylcyclohexane

c. two products: **C** = (CH₃)₃C—⟨⟩—OCH₃   **D** = (CH₃)₃C—⟨⟩—OCH₃

d. *cis*-1-Bromo-4-*tert*-butylcyclohexane reacts faster. With the strong nucleophile ⁻OCH₃,
backside attack occurs by an SN2 reaction, and with the cis isomer, the nucleophile can
approach from the equatorial direction, avoiding 1,3-diaxial interactions.

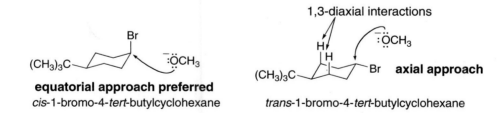

(CH₃)₃C— [Br ⁻:ÖCH₃]

equatorial approach preferred
cis-1-bromo-4-*tert*-butylcyclohexane

1,3-diaxial interactions
H :ÖCH₃
H
(CH₃)₃C— Br **axial approach**

trans-1-bromo-4-*tert*-butylcyclohexane

e. The bulky base ⁻OC(CH₃)₃ favors elimination by an E2 mechanism, affording a mixture of
two enantiomers **A** and **B**. The strong nucleophile ⁻OCH₃ favors nucleophilic substitution by
an SN2 mechanism. Inversion of configuration results from backside attack of the
nucleophile.

8.57

a.

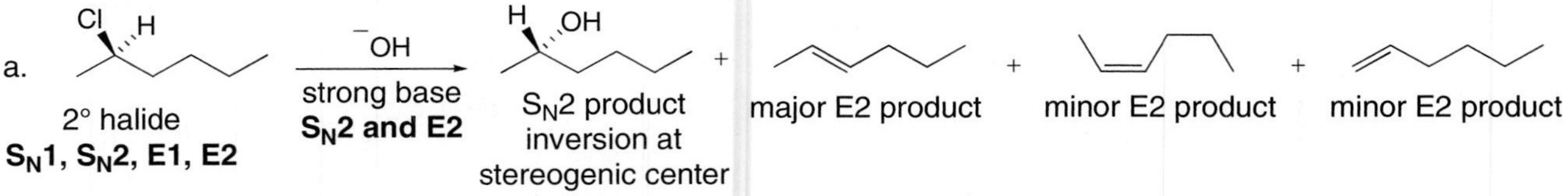

2° halide
S$_N$1, S$_N$2, E1, E2

$^-$OH
strong base
S$_N$2 and E2

S$_N$2 product
inversion at
stereogenic center

+ major E2 product + minor E2 product + minor E2 product

b.

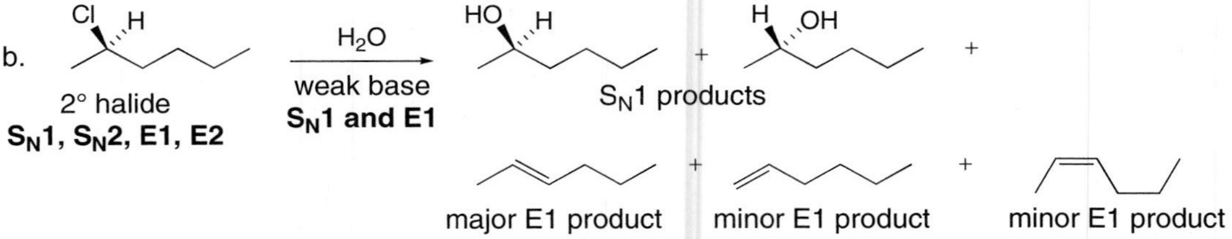

2° halide
S$_N$1, S$_N$2, E1, E2

H$_2$O
weak base
S$_N$1 and E1

S$_N$1 products

+

major E1 product + minor E1 product + minor E1 product

c.

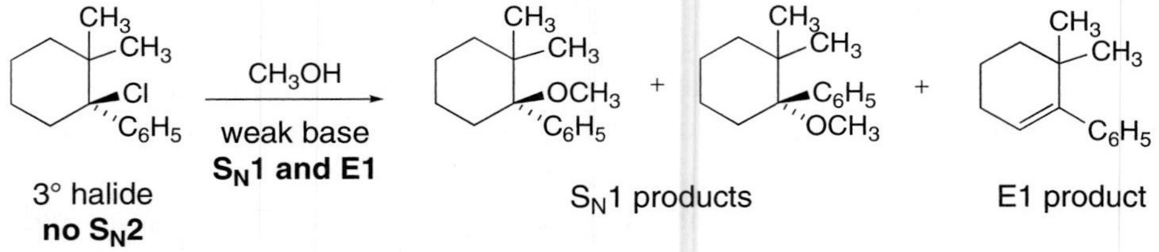

3° halide
no S$_N$2

CH$_3$OH
weak base
S$_N$1 and E1

S$_N$1 products

E1 product

d.

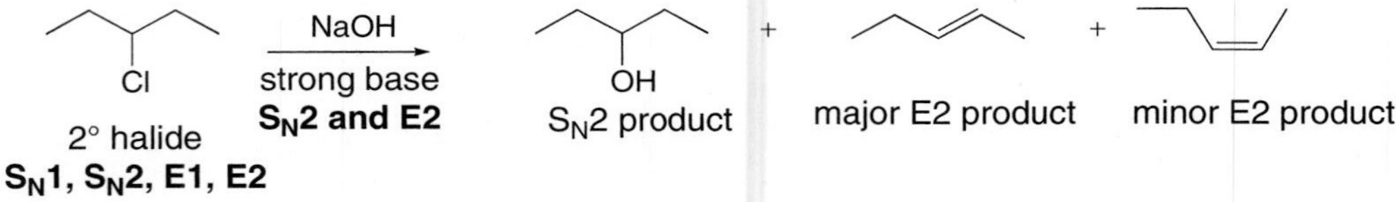

2° halide
S$_N$1, S$_N$2, E1, E2

NaOH
strong base
S$_N$2 and E2

S$_N$2 product

+ major E2 product + minor E2 product

e.

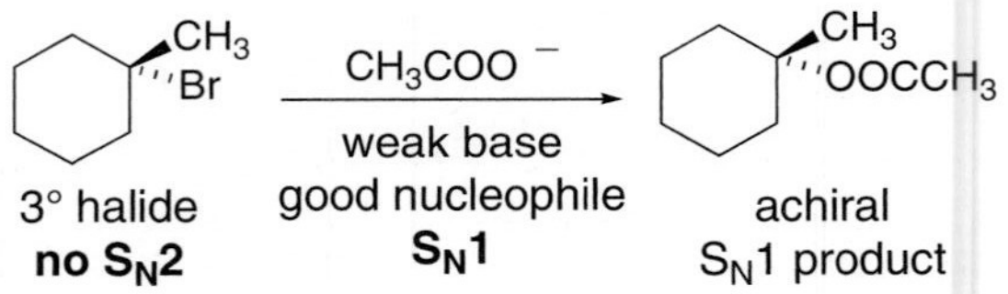

3° halide
no S$_N$2

CH$_3$COO$^-$
weak base
good nucleophile
S$_N$1

achiral
S$_N$1 product

f.

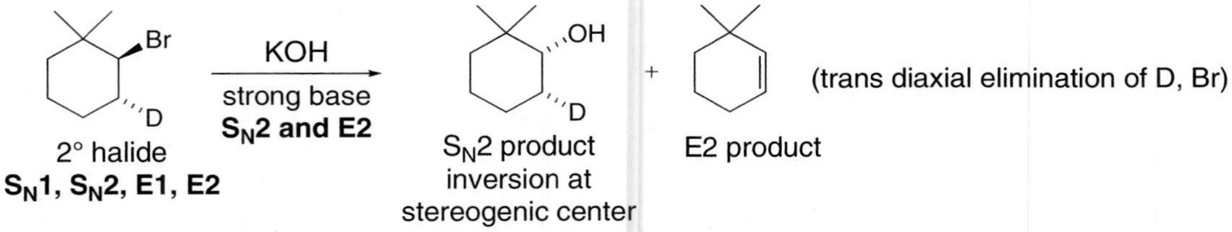

2° halide
S$_N$1, S$_N$2, E1, E2

KOH
strong base
S$_N$2 and E2

S$_N$2 product
inversion at
stereogenic center

E2 product

(trans diaxial elimination of D, Br)

8.58

a.

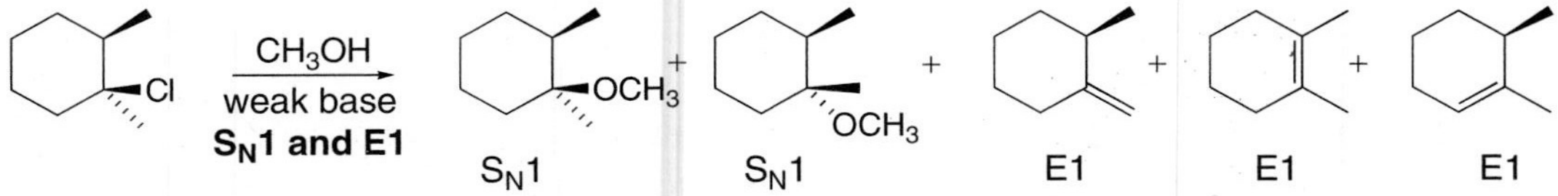

CH$_3$OH
weak base
S$_N$1 and E1

S$_N$1 + S$_N$1 + E1 + E1 + E1

b.

KOH
strong base
E2

+ +

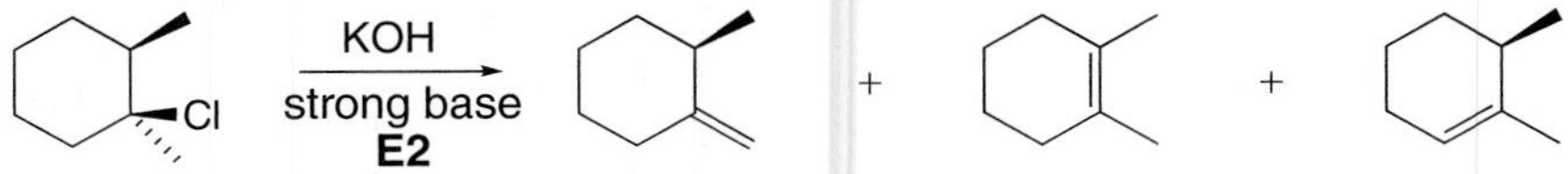

8.59

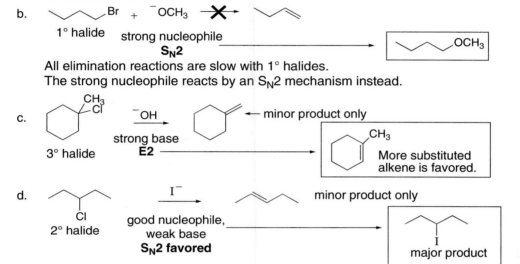

a.

3° halide

strong bulky base
E2

major product
more substituted alkene

No substitution occurs with a strong bulky base and a 3° RX. The C with the leaving group is too crowded for an S$_N$2 substitution to occur. Elimination occurs instead by an E2 mechanism.

b.

1° halide   strong nucleophile
S$_N$2

All elimination reactions are slow with 1° halides.
The strong nucleophile reacts by an S$_N$2 mechanism instead.

c.

←— minor product only

strong base
E2

3° halide

More substituted
alkene is favored.

d.

minor product only

good nucleophile,
weak base
S$_N$2 favored

2° halide

major product

The 2° halide can react by an E2 or S$_N$2 reaction with a negatively charged nucleophile or base. Since I$^-$ is a weak base, substitution by an S$_N$2 mechanism is favored.

8.60

3° halide, weak base:
S_N1 and E1

a.

Any base (such as CH_3CH_2OH or Cl^-) can be used to remove a proton to form an alkene. If Cl^- is used, HCl is formed as a reaction by-product. If CH_3CH_2OH is used, $(CH_3CH_2OH_2)^+$ is formed instead.

b.

8.61 Draw the products of each reaction with the 1° alkyl halide.

8.62

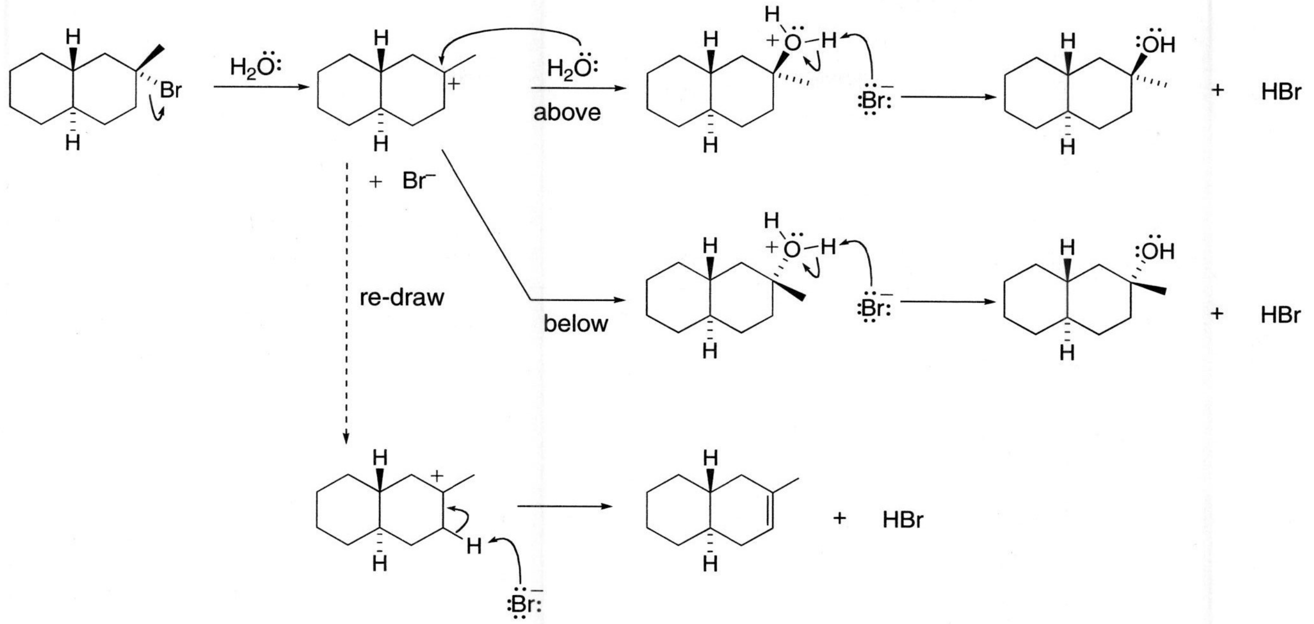

8.63

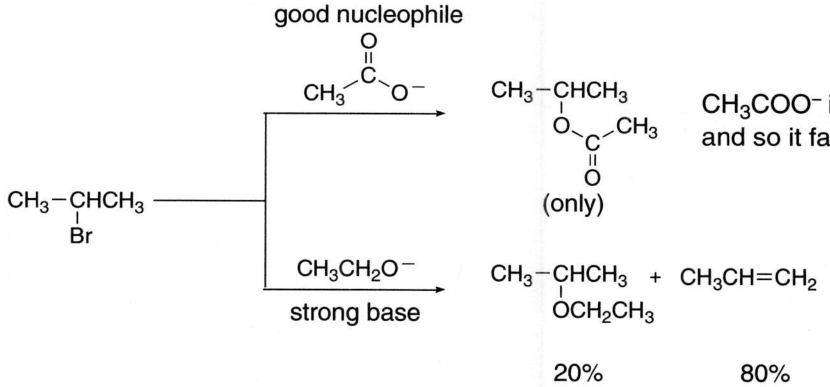

good nucleophile

CH_3COO^- is a good nucleophile and a weak base and so it favors substitution by S_N2.

The strong base gives both S_N2 and E2 products, but since the 2° RX is somewhat hindered to substitution, the E2 product is favored.

8.64

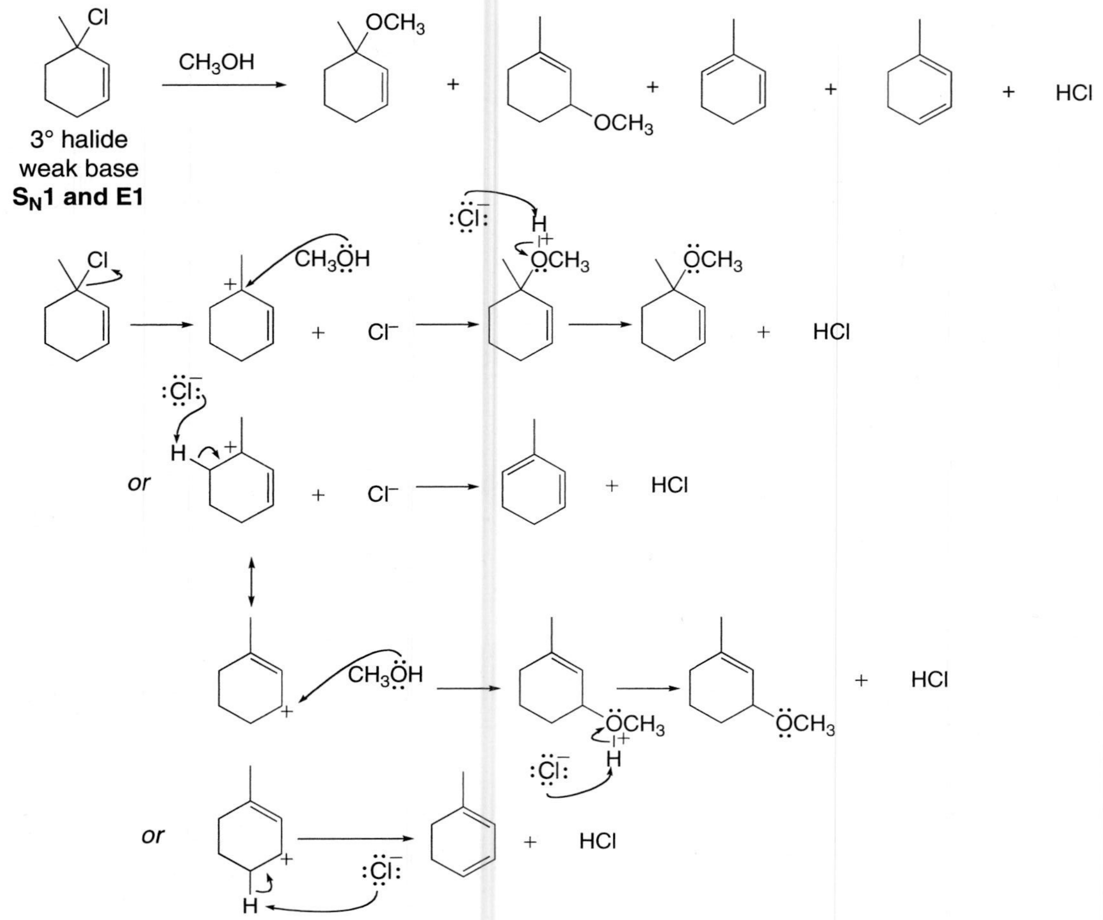

3° halide
weak base
S_N1 and E1

8.65 E2 elimination needs a leaving group and a hydrogen in the **trans diaxial** position.

Two different conformations:

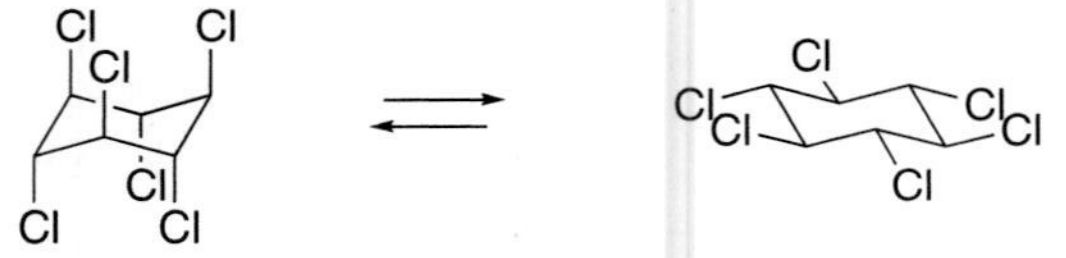

This conformation has Cl's axial, but no H's axial.   This conformation has no Cl's axial.

For elimination to occur, a cyclohexane must have a H and Cl in the trans diaxial arrangement. Neither conformation of this isomer has both atoms—H and Cl—axial; thus, this isomer only slowly loses HCl by elimination.

8.66

H and Br are **anti periplanar.**
Elimination can occur.

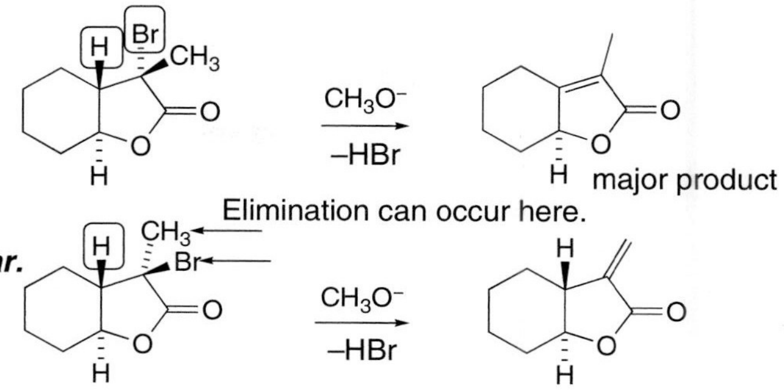

CH$_3$O$^-$
—HBr

major product

H (in the ring) and Br are **NOT anti periplanar.**
Elimination can**not** occur using this H.
Instead elimination must occur with the
H on the CH$_3$ group.

Elimination can occur here.

CH$_3$O$^-$
—HBr

Elimination cannot occur in the ring
because the required anti periplanar geometry is not present.

8.67

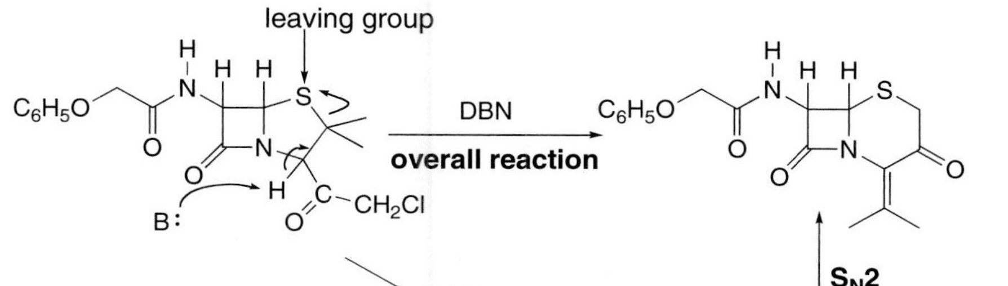

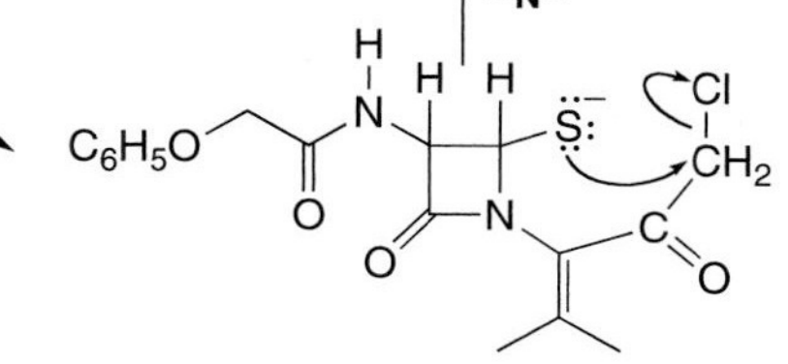

leaving group

DBN

overall reaction

E2

A sequence of two reactions forms the
final product: E2 elimination opens the
five-membered ring. Then the sulfur
nucleophile displaces the Cl$^-$ leaving group
to form the six-membered ring.

S$_N$2

8.68

a.

Δ

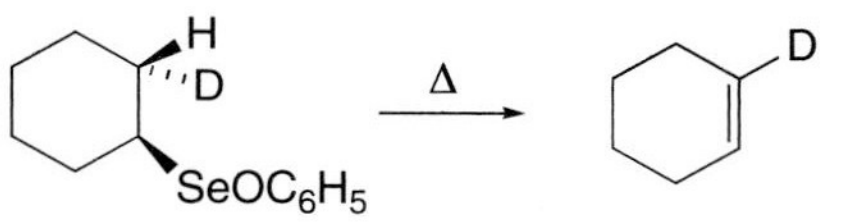

SeOC$_6$H$_5$ and H are on the
same side of the ring.
syn elimination

b.

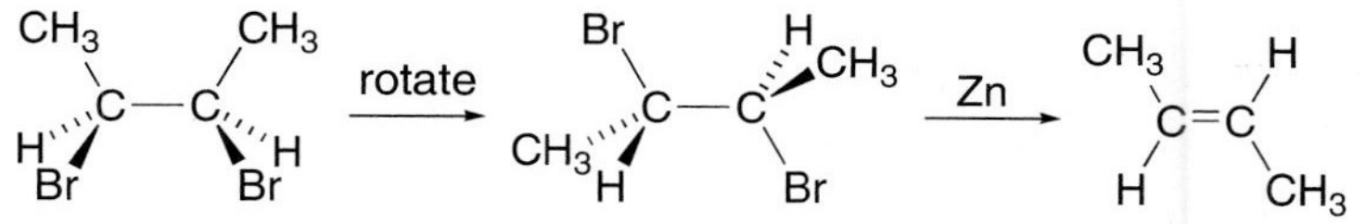

rotate

Zn

Both Br atoms are on the opposite
sides of the C–C bond.
anti elimination

Chapter 9: Alcohols, Ethers, and Epoxides

◆ General facts about ROH, ROR, and epoxides

- All three compounds contain an O atom that is sp^3 hybridized and tetrahedral (9.2).

- All three compounds have polar C–O bonds, but only alcohols have an O–H bond for intermolecular hydrogen bonding (9.4).

- Alcohols and ethers do not contain a good leaving group. Nucleophilic substitution can occur only after the OH (or OR) group is converted to a better leaving group (9.7A).

- Epoxides have a leaving group located in a strained three-membered ring, making them reactive to strong nucleophiles and acids HZ that contain a nucleophilic atom Z (9.15).

◆ A new reaction of carbocations (9.9)

- Less stable carbocations rearrange to more stable carbocations by shift of a hydrogen atom or an alkyl group. Besides rearrangement, carbocations also react with nucleophiles (7.13) and bases (8.6).

◆ Preparation of alcohols, ethers, and epoxides (9.6)

[1] Preparation of alcohols

- The mechanism is S_N2.
- The reaction works best for CH_3X and 1° RX.

[2] Preparation of alkoxides (a Brønsted–Lowry acid–base reaction)

$$R-O-H + Na^+H^- \longrightarrow \boxed{R-O^-}Na^+ + H_2$$

alkoxide

[3] Preparation of ethers (Williamson ether synthesis)

$$R-X + \boxed{^-OR'} \longrightarrow R\boxed{OR'} + X^-$$

- The mechanism is S_N2.
- The reaction works best for CH_3X and 1° RX.

[4] Preparation of epoxides (Intramolecular S_N2 reaction)

halohydrin

- A two-step reaction sequence:
  [1] Removal of a proton with base forms an alkoxide.
  [2] Intramolecular S_N2 reaction forms the epoxide.

♦ Reactions of alcohols

[1] Dehydration to form alkenes

[a] Using strong acid (9.8, 9.9)

- Order of reactivity: $R_3COH > R_2CHOH > RCH_2OH$.
- The mechanism for 2° and 3° ROH is E1; carbocations are intermediates and rearrangements occur.
- The mechanism for 1° ROH is E2.
- The Zaitsev rule is followed.

[b] Using $POCl_3$ and pyridine (9.10)

- The mechanism is E2.
- No carbocation rearrangements occur.

[2] Reaction with HX to form RX (9.11)

$$R-OH + H-X \longrightarrow \boxed{R-X} + H_2O$$

- Order of reactivity: $R_3COH > R_2CHOH > RCH_2OH$.
- The mechanism for 2° and 3° ROH is S_N1; carbocations are intermediates and rearrangements occur.
- The mechanism for CH_3OH and 1° ROH is S_N2.

[3] Reaction with other reagents to form RX (9.12)

R—OH + SOCl₂ $\xrightarrow[\text{pyridine}]{}$ R—Cl

R—OH + PBr₃ $\longrightarrow$ R—Br

- Reactions occur with CH_3OH and $1°$ and $2°$ ROH.
- The reactions follow an S_N2 mechanism.

[4] Reaction with tosyl chloride to form alkyl tosylates (9.13A)

R—OH + Cl—S(=O)(=O)—⟨ ⟩—CH₃ $\xrightarrow{\text{pyridine}}$ R—O—S(=O)(=O)—⟨ ⟩—CH₃

R—OTs

- The C–O bond is not broken so the configuration at a stereogenic center is retained.

◆ Reactions of alkyl tosylates

Alkyl tosylates undergo either substitution or elimination depending on the reagent (9.13B).

- Substitution is carried out with strong :Nu⁻ so the mechanism is S_N2.

- Elimination is carried out with strong bases so the mechanism is E2.

◆ Reactions of ethers

Only one reaction is useful: Cleavage with strong acids (9.14)

R—O—R' + H—X $\longrightarrow$ R—X + R'—X + H₂O
(2 equiv)
(X = Br or I)

- With $2°$ and $3°$ R groups, the mechanism is S_N1.
- With CH_3 and $1°$ R groups the mechanism is S_N2.

◆ Reactions of epoxides

Epoxide rings are opened with nucleophiles :Nu⁻ and acids HZ (9.15).

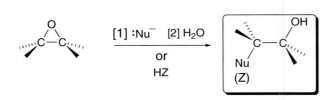

[1] :Nu⁻ [2] H₂O
or
HZ

- The reaction occurs with backside attack, resulting in trans or anti products.
- With :Nu⁻, the mechanism is S_N2, and nucleophilic attack occurs at the less substituted C.
- With HZ, the mechanism is between S_N1 and S_N2, and attack of Z⁻ occurs at the more substituted C.

Chapter 9: Answers to Problems

9.1 • **Alcohols** are classified as 1°, 2°, or 3°, depending on the number of carbon atoms bonded to the carbon with the OH group.
  • **Symmetrical ethers** have two identical R groups, and **unsymmetrical ethers** have R groups that are different.

1° alcohol   **2° alcohol**   **symmetrical ether**   **unsymmetrical ether**

3° alcohol   **1° alcohol**   **unsymmetrical ether**

9.2 Use the definition in Answer 9.1 to classify each OH group in cortisol.

2° alcohol ⟶ HO   OH ⟵ 1° alcohol

'''OH

3° alcohol

cortisol

9.3 To name an alcohol:
  [1] **Find the longest chain that has the OH group as a substituent**. Name the molecule as a derivative of that number of carbons by changing the *-e* ending of the alkane to the suffix *-ol*.
  [2] **Number the carbon chain to give the OH group the lower number**. When the OH group is bonded to a ring, the ring is numbered beginning with the OH group, and the "1" is usually omitted.
  [3] Apply the other rules of nomenclature to complete the name.

a. [1] OH   [2] OH   [3] **3,3-dimethyl-1-pentanol**

5 carbons = **pentanol**

b. [1] CH₃   [2] CH₃ ⟵ **2-methyl**   [3] *cis*-**2-methylcyclohexanol**

OH   OH

6 carbon ring = **cyclohexanol**   1

c. [1] OH   [2] OH   [3] **5-ethyl-6-methyl-3-nonanol**

9 carbons = **nonanol**   ⟵ **6-methyl**   **3**

5-ethyl

9.4 To work backwards from a name to a structure:
[1] Find the parent name and draw its structure.
[2] Add the substituents to the long chain.

a. 7,7-dimethyl-4-**octanol**

c. 2-*tert*-butyl-3-methyl**cyclohexanol**

b. 5-methyl-4-propyl-3-**heptanol**

d. *trans*-1,2-**cyclohexanediol**

9.5 **To name simple ethers:**
[1] Name both alkyl groups bonded to the oxygen.
[2] Arrange these names alphabetically and add the word ***ether***. For symmetrical ethers, name the alkyl group and add the prefix ***di***.

To name ethers using the IUPAC system:
[1] Find the two alkyl groups bonded to the ether oxygen. The smaller chain becomes the substituent, named as an alkoxy group.
[2] Number the chain to give the lower number to the first substituent.

a. **common name:**

$CH_3-O-CH_2CH_2CH_2CH_3$

methyl   butyl

butyl methyl ether

IUPAC name:

$CH_3-O\!\mid\!CH_2CH_2CH_2CH_3$ ← larger group – 4 C's
butane

substituent:
methoxy

1-methoxybutane

b. **common name:**

OCH_3
methyl

cyclohexyl

cyclohexyl methyl ether

IUPAC name:

OCH_3 ← substituent –
methoxy

larger group – 6 C's
cyclohexane

methoxycyclohexane

c. **common name:**

$CH_3CH_2CH_2-O-CH_2CH_2CH_3$

propyl   propyl

dipropyl ether

IUPAC name:

$CH_3CH_2CH_2-O\!\mid\!CH_2CH_2CH_3$

propoxy   propane

1-propoxypropane

9.6 Name each ether using the rules from Answer 9.5.

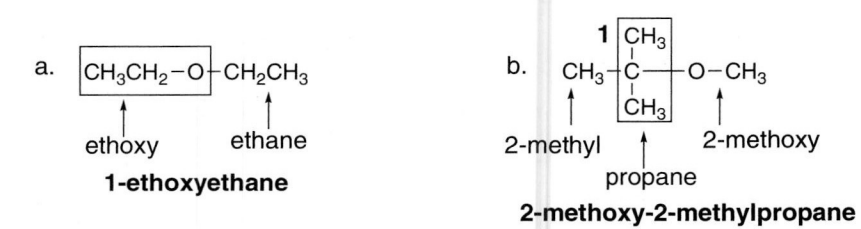

a.
CH₃CH₂−O−CH₂CH₃

ethoxy   ethane

1-ethoxyethane

b.
1 CH₃
CH₃−C−O−CH₃
CH₃

2-methyl   2-methoxy
propane

2-methoxy-2-methylpropane

9.7 **Three ways to name epoxides**:
[1] Epoxides are named as derivatives of oxirane, the simplest epoxide.
[2] Epoxides can be named by considering the oxygen as a substituent called an **epoxy** group, bonded to a hydrocarbon chain or ring. Use two numbers to designate which two atoms the oxygen is bonded to.
[3] Epoxides can be named as **alkene oxides** by mentally replacing the epoxide oxygen by a double bond. Name the alkene (Chapter 10) and add the word *oxide*.

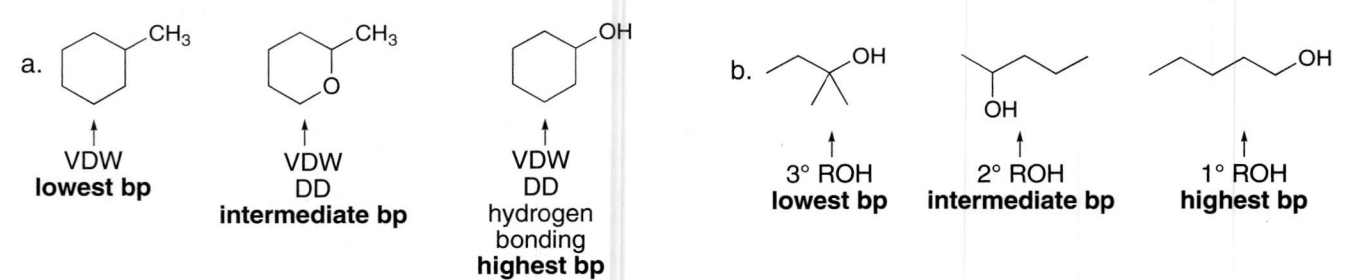

a.

Three possibilities:
[1] **methyloxirane**
[2] **1,2-epoxypropane**
[3] **propene oxide**

c.

Three possibilities:
[1] *cis*-**2-methyl-3-propyloxirane**
[2] *cis*-**2,3-epoxyhexane**
[3] *cis*-**2-hexene oxide**

b.

1
CH₃ ←—— 1-methyl
O ←—— epoxy group
2

Two possibilities:
[1] 6 carbons = cyclohexane
     1,2-epoxy-1-methylcyclohexane
[2] **1-methylcyclohexene oxide**

9.8 Two rules for boiling point:
[1] **The stronger the forces the higher the bp.**
[2] **Bp increases as the extent of the hydrogen bonding increases.** For alcohols with the same number of carbon atoms: hydrogen bonding and bp's increase: 3° ROH < 2° ROH < 1° ROH.

a.

CH₃

VDW
lowest bp

CH₃
O

VDW
DD
intermediate bp

OH

VDW
DD
hydrogen
bonding
highest bp

b.

OH

3° ROH
lowest bp

OH
OH

2° ROH
intermediate bp

OH

1° ROH
highest bp

9.9 Draw dimethyl ether and ethanol and analyze their intermolecular forces to explain the observed trend.

dimethyl ether	ethanol	
CH$_3$–O–CH$_3$	CH$_3$CH$_2$OH	Both molecules contain an O atom and can hydrogen bond with water. They have fewer than 5 C's and are therefore **water soluble**.
VDW	VDW	
DD	DD	
no HB	**HB**	
much lower bp	Two molecules of CH$_3$CH$_2$OH can hydrogen bond to each other. **stronger forces = much higher bp**	

9.10 Strong nucleophiles (like ⁻CN) favor S$_N$2 reactions. The use of crown ethers in nonpolar solvents increases the nucleophilicity of the anion, and this increases the rate of the S$_N$2 reaction. The nucleophile does not appear in the rate equation for the S$_N$1 reaction. Nonpolar solvents cannot solvate carbocations so this disfavors S$_N$1 reactions as well.

9.11 Compare the number of carbons and functional groups to determine if each compound is soluble in water, an organic solvent, or both.

eplerenone
24 C's and 6 O's (from four functional groups)
soluble in organic solvents
probably borderline water solubility

tiotropium bromide
19 C's, 4 O's
a salt – water soluble
Because it has many C's, it is also probably soluble in organic solvent.

9.12 Draw the products of substitution in the following reactions by substituting OH or OR for X in the starting material.

a. CH$_3$CH$_2$CH$_2$CH$_2$–Br + ⁻OH ⟶ CH$_3$CH$_2$CH$_2$CH$_2$–OH + Br⁻ **alcohol**

b. ~~~~Cl + ⁻OCH$_3$ ⟶ ~~~~OCH$_3$ + Cl⁻ **unsymmetrical ether**

c. ⬡–CH$_2$CH$_2$–I + ⁻OCH(CH$_3$)$_2$ ⟶ ⬡–CH$_2$CH$_2$–OCH(CH$_3$)$_2$ + I⁻ **unsymmetrical ether**

d. ⌲Br + ⁻OCH$_2$CH$_3$ ⟶ ⌲OCH$_2$CH$_3$ + Br⁻ **unsymmetrical ether**

9.13 To synthesize an ether using a Williamson ether synthesis:
 [1] First find the two possible alkoxides and alkyl halides needed for nucleophilic substitution.
 [2] Classify the alkyl halides as 1°, 2°, or 3°. The favored path has the less hindered halide.

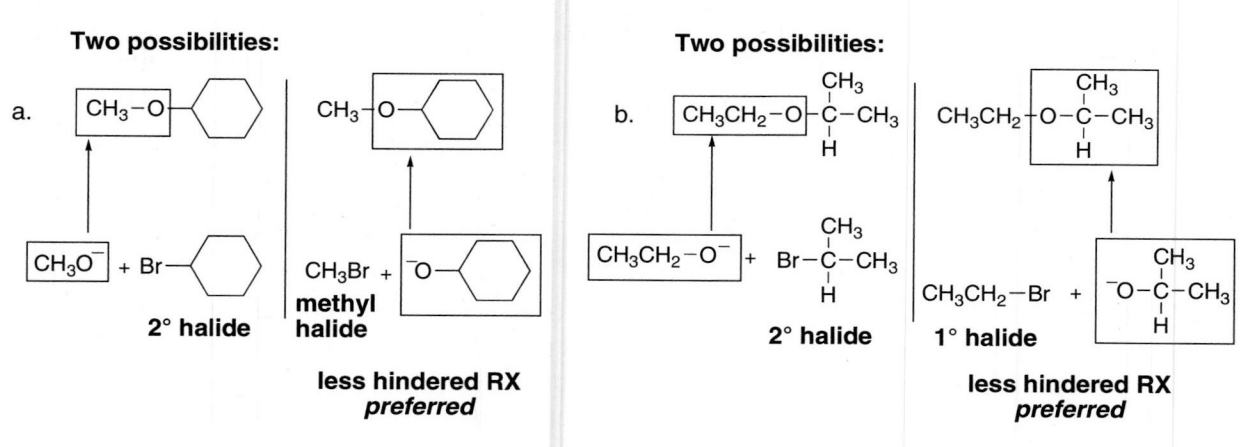

9.14 NaH and NaNH₂ are strong bases that will remove a proton from an alcohol, creating a nucleophile.

a. $CH_3CH_2CH_2-\ddot{O}-H$ + $Na^+ H\colon^-$ ⟶ $CH_3CH_2CH_2-\ddot{O}\colon^-$ Na^+ + H_2

b. [cyclopentyl]–C(CH₃)(H)–Ö–H + Na⁺ :NH₂ ⟶ [cyclopentyl]–C(CH₃)(H)–Ö:⁻ + Na⁺ + NH₃

c. [structure] –Ö–H   Na⁺ H:⁻ ⟶ [structure] –Ö:⁻   CH₃CH₂CH₂–Br + Na⁺ + H₂

⟶ [structure] –Ö–CH₂CH₂CH₃ + Br⁻

d. [cyclohexyl with Ö–H and Br] Na⁺ H:⁻ ⟶ [cyclohexyl with Ö:⁻ and Br] + Na⁺ + H₂ ⟶ [epoxide] Ö + Br⁻

$C_6H_{10}O$

9.15 Dehydration follows the Zaitsev rule, so the more stable, more substituted alkene is the major product.

a. $CH_3-\overset{H}{\underset{OH}{C}}-CH_3$ $\xrightarrow{TsOH}$ $CH_2=CH-CH_3$ + H_2O

b. [structure] –OH $\xrightarrow{TsOH}$ $\underset{CH_3}{\overset{CH_3CH_2}{C}}=CHCH_3$ + $\underset{CH_2}{[structure]}$ + H_2O

trisubstituted   disubstituted
major product   **minor product**

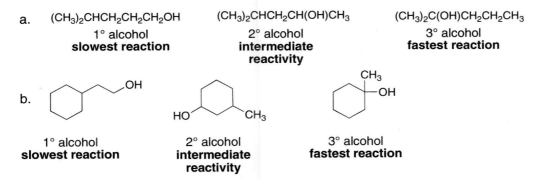

c.

trisubstituted   disubstituted
major product   minor product

9.16 The rate of dehydration increases as the number of R groups increases.

a.   (CH₃)₂CHCH₂CH₂CH₂OH   (CH₃)₂CHCH₂CH(OH)CH₃   (CH₃)₂C(OH)CH₂CH₂CH₃
        1° alcohol   2° alcohol   3° alcohol
        slowest reaction   **intermediate**   **fastest reaction**
                                              reactivity

b.

        1° alcohol   2° alcohol   3° alcohol
        slowest reaction   **intermediate**   **fastest reaction**
                                              reactivity

9.17 There are three steps in the E1 mechanism for dehydration of alcohols and three transition states.

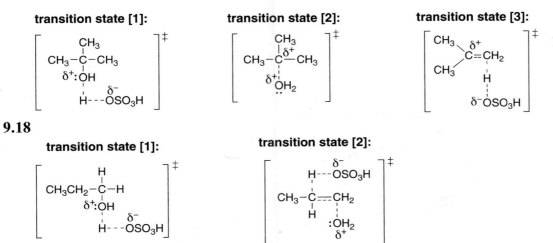

transition state [1]:   **transition state [2]:**   **transition state [3]:**

9.18

transition state [1]:   **transition state [2]:**

9.19

rearranged 3° carbocation

This alkene is also formed in addition to **Y** from the rearranged carbocation.

+ H₂SO₄

The initially formed 2° carbocation gives two alkenes:

or

HSO₄⁻

9.20

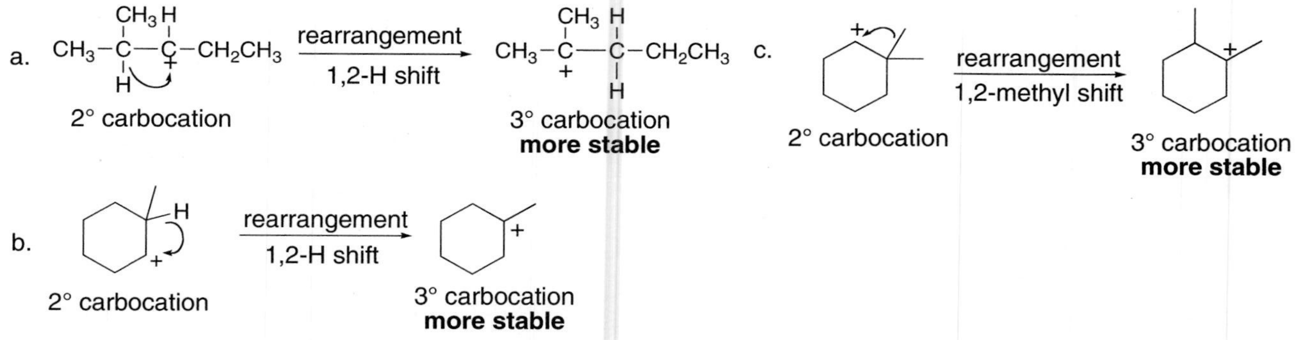

a.

2° carbocation → 3° carbocation **more stable**

via rearrangement, 1,2-H shift

c.

2° carbocation → 3° carbocation **more stable**

via rearrangement, 1,2-methyl shift

b.

2° carbocation → 3° carbocation **more stable**

via rearrangement, 1,2-H shift

9.21

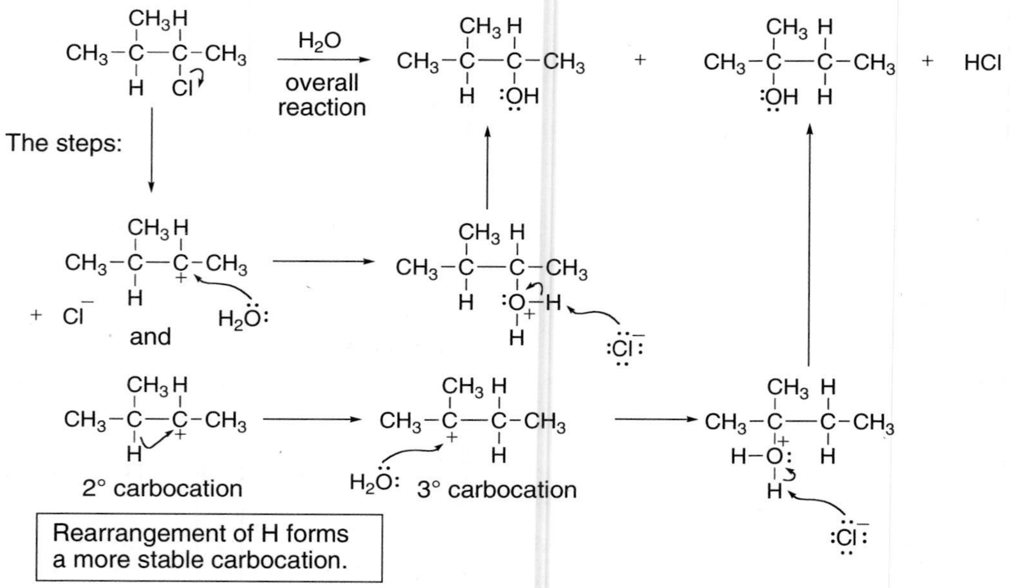

The steps:

$+ \; Cl^-$ and $H_2\ddot{O}:$

2° carbocation → 3° carbocation

Rearrangement of H forms a more stable carbocation.

9.22

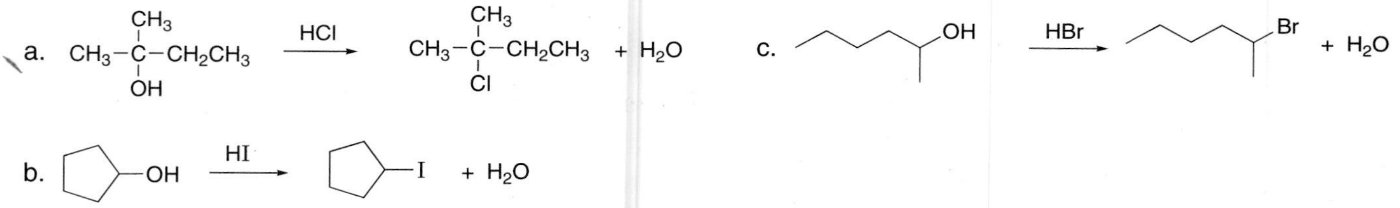

a.

$CH_3-\underset{OH}{\underset{|}{\overset{CH_3}{\overset{|}{C}}}}-CH_2CH_3 \xrightarrow{HCl} CH_3-\underset{Cl}{\underset{|}{\overset{CH_3}{\overset{|}{C}}}}-CH_2CH_3 + H_2O$

b.

$\xrightarrow{HI}$ + H_2O

c.

$\xrightarrow{HBr}$ + H_2O

9.23 • **CH₃OH and 1° alcohols** follow an S_N2 mechanism, which results in inversion of configuration.
• **Secondary (2°) and 3° alcohols** follow an S_N1 mechanism, which results in racemization at a stereogenic center.

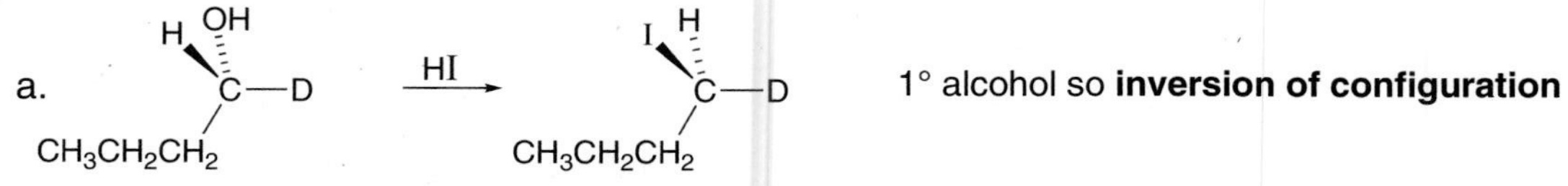

a. $\xrightarrow{HI}$ 1° alcohol so **inversion of configuration**

b. $\xrightarrow{\text{HBr}}$ 3° alcohol, so Br⁻ attacks from above and below. The product is achiral.

achiral starting material   achiral product

c. $\xrightarrow{\text{HCl}}$ + 3° alcohol = **racemization**

9.24

a. $\xrightarrow{\text{HCl}}$

c. $\xrightarrow{\text{HCl}}$

(product formed after a 1,2-H shift)

b. $\xrightarrow{\text{HCl}}$

(product formed after a 1,2-CH₃ shift)

9.25 Substitution reactions of alcohols using $SOCl_2$ proceed by an S_N2 mechanism. Therefore, there is **inversion of configuration** at a stereogenic center.

$\xrightarrow[\text{pyridine}]{SOCl_2}$ Reactions using $SOCl_2$ proceed by an S_N2 mechanism = **inversion of configuration.**

R   *S*

9.26 Substitution reactions of alcohols using PBr_3 proceed by an S_N2 mechanism. Therefore, there is inversion of configuration at a stereogenic center.

$\xrightarrow{PBr_3}$ Reactions using PBr_3 proceed by an S_N2 mechanism = **inversion of configuration.**

R   *S*

9.27 Stereochemistry for conversion of ROH to RX by reagent:
[1] **HX**—with 1°, S_N2, so inversion of configuration; with 2° and 3°, S_N1, so racemization.
[2] **SOCl₂**—S_N2, so inversion of configuration.
[3] **PBr₃**—S_N2, so inversion of configuration.

a. $\xrightarrow[\text{pyridine}]{SOCl_2}$

c. $\xrightarrow{PBr_3}$

S_N2 = **inversion**

b. $\xrightarrow{HI}$ +

3° alcohol, S_N1 = **racemization**

9.28 To do a two-step synthesis with this starting material:
  [1] Convert the OH group into a good leaving group (by using either PBr₃ or SOCl₂).
  [2] Add the nucleophile for the S_N2 reaction.

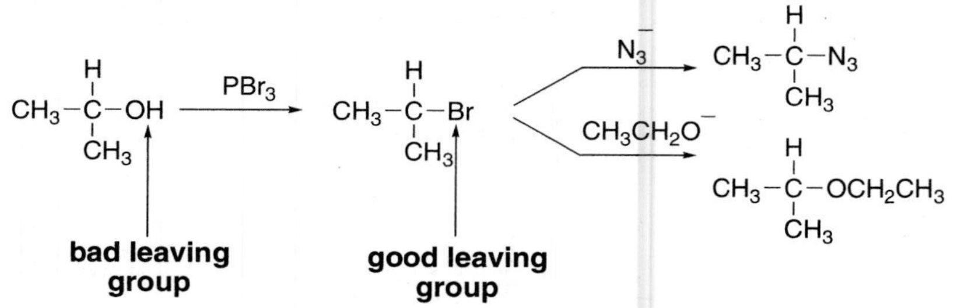

9.29

a. 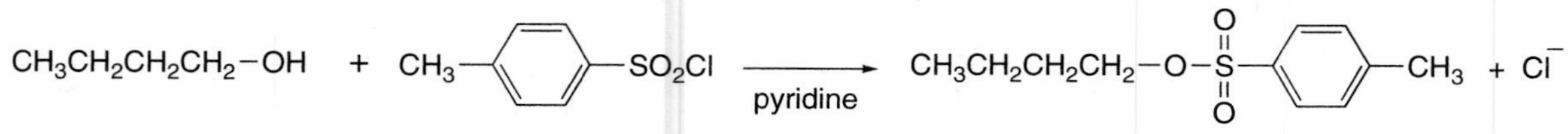 $CH_3CH_2CH_2CH_2-OH$ + $CH_3-\!\!\bigcirc\!\!-SO_2Cl$ $\xrightarrow{\text{pyridine}}$ $CH_3CH_2CH_2CH_2-O-\overset{\overset{O}{\|}}{\underset{\underset{O}{\|}}{S}}-\!\!\bigcirc\!\!-CH_3$ + Cl^-

b.

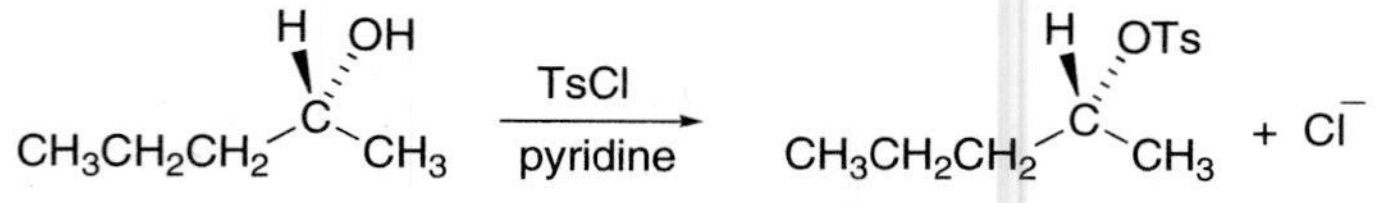

9.30

a.

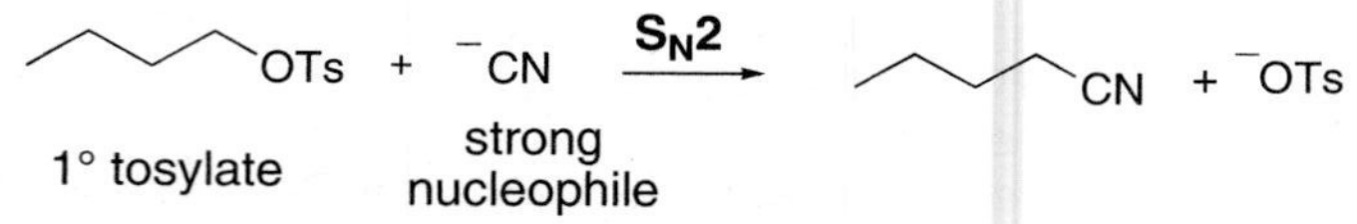

b. $CH_3CH_2CH_2-OTs$ + $K^+\ {}^-OC(CH_3)_3$ $\xrightarrow{\text{E2}}$ $CH_3CH=CH_2$ + $K^+\ {}^-OTs$ + $HOC(CH_3)_3$

1° tosylate   strong bulky base

c.

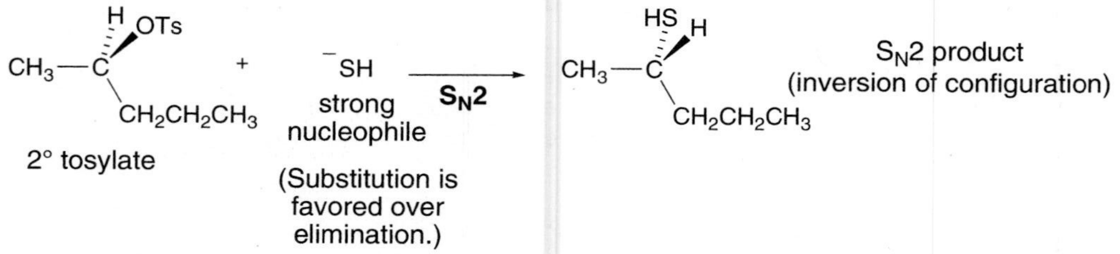

9.31

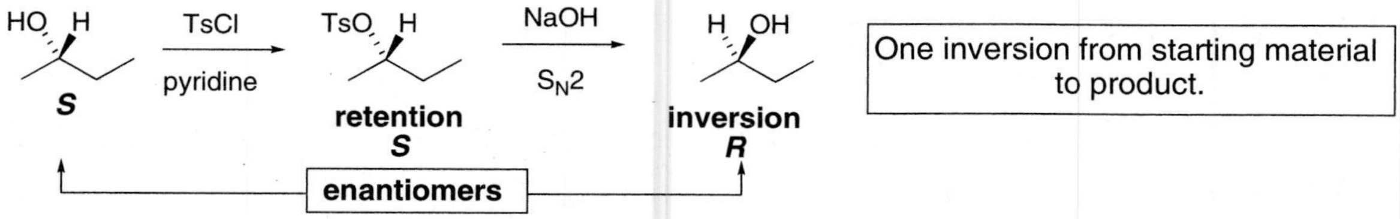

One inversion from starting material to product.

9.32 These reagents can be classified as:

[1] $SOCl_2$, PBr_3, HCl, and HBr replace OH with X by a substitution reaction.

[2] Tosyl chloride (TsCl) makes OH a better leaving group by converting it to OTs.

[3] Strong acids (H_2SO_4) and $POCl_3$ (pyridine) result in elimination by dehydration.

a. $CH_3-\overset{\overset{H}{|}}{\underset{\underset{CH_3}{|}}{C}}-OH \xrightarrow[\text{pyridine}]{SOCl_2} CH_3-\overset{\overset{H}{|}}{\underset{\underset{CH_3}{|}}{C}}-Cl$

d. $CH_3-\overset{\overset{H}{|}}{\underset{\underset{CH_3}{|}}{C}}-OH \xrightarrow{HBr} CH_3-\overset{\overset{H}{|}}{\underset{\underset{CH_3}{|}}{C}}-Br$

b. $CH_3-\overset{\overset{H}{|}}{\underset{\underset{CH_3}{|}}{C}}-OH \xrightarrow[\text{pyridine}]{TsCl} CH_3-\overset{\overset{H}{|}}{\underset{\underset{CH_3}{|}}{C}}-OTs$

e. $CH_3-\overset{\overset{H}{|}}{\underset{\underset{CH_3}{|}}{C}}-OH \xrightarrow[\text{[2] NaCN}]{\text{[1] } PBr_3} CH_3-\overset{\overset{H}{|}}{\underset{\underset{CH_3}{|}}{C}}-CN$

c. $CH_3-\overset{\overset{H}{|}}{\underset{\underset{CH_3}{|}}{C}}-OH \xrightarrow{H_2SO_4} CH_2=CHCH_3$

f. $CH_3-\overset{\overset{H}{|}}{\underset{\underset{CH_3}{|}}{C}}-OH \xrightarrow[\text{pyridine}]{POCl_3} CH_2=CHCH_3$

9.33

a. $CH_3CH_2-O-CH_2CH_3 \xrightarrow{HBr} 2\ CH_3CH_2-Br + H_2O$

c. ⬡$-O-CH_3 \xrightarrow{HBr}$ ⬡$-Br$
$+ CH_3Br + H_2O$

b. $CH_3-\overset{\overset{CH_3}{|}}{\underset{\underset{H}{|}}{C}}-O-CH_2CH_3 \xrightarrow{HBr} CH_3-\overset{\overset{CH_3}{|}}{\underset{\underset{H}{|}}{C}}-Br + CH_3CH_2Br + H_2O$

9.34 Ether cleavage can occur by either an S_N1 or S_N2 mechanism, but neither mechanism can occur when the ether O atom is bonded to an aromatic ring. An S_N1 reaction would require formation of a highly unstable carbocation on a benzene ring, a process that does not occur. An S_N2 reaction would require backside attack through the plane of the aromatic ring, which is also not possible. Thus cleavage of the Ph–OCH_3 bond does not occur.

⬡$-OCH_3 \xrightarrow{HBr}$ ⬡$-OH + CH_3Br$ $\left[⬡-Br\right]$

anisole   phenol   bromobenzene
NOT formed

S_N1: S_N2:

highly unstable carbocation $+ CH_3OH$

:Br:⁻

9.35 Compare epoxides and cyclopropane. For a compound to be reactive towards nucleophiles, it must be electrophilic.

$\overset{\delta-}{:\ddot{O}:}$
$\delta+ \triangle \delta+$

epoxide

cyclopropane

O is electronegative and pulls electron density away from C's. This makes them electrophilic and reactive with nucleophiles.

Cyclopropane has all C's and H's, so all nonpolar bonds. There are no electrophilic C's so it will not react with nucleophiles.

9.36 Two rules for reaction of an epoxide:

[1] Nucleophiles attack from the **back side** of the epoxide.

[2] Negatively charged nucleophiles attack at the **less substituted carbon**.

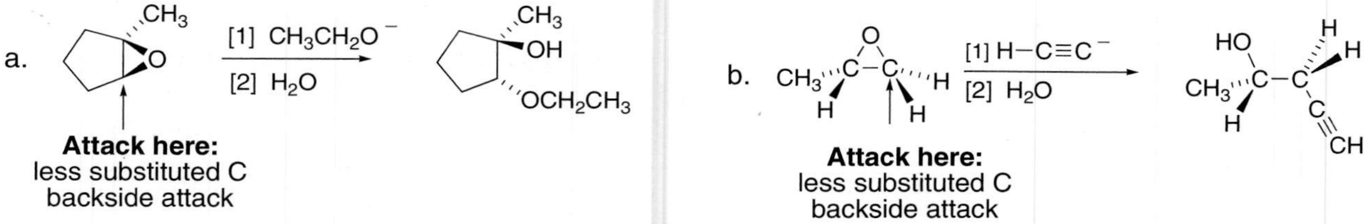
a.
[1] $CH_3CH_2O^-$
[2] H_2O
Attack here:
less substituted C
backside attack
b.
[1] $H{-}C{\equiv}C^-$
[2] H_2O
Attack here:
less substituted C
backside attack

9.37 In both isomers, ⁻OH attacks from the back side at either C–O bond.

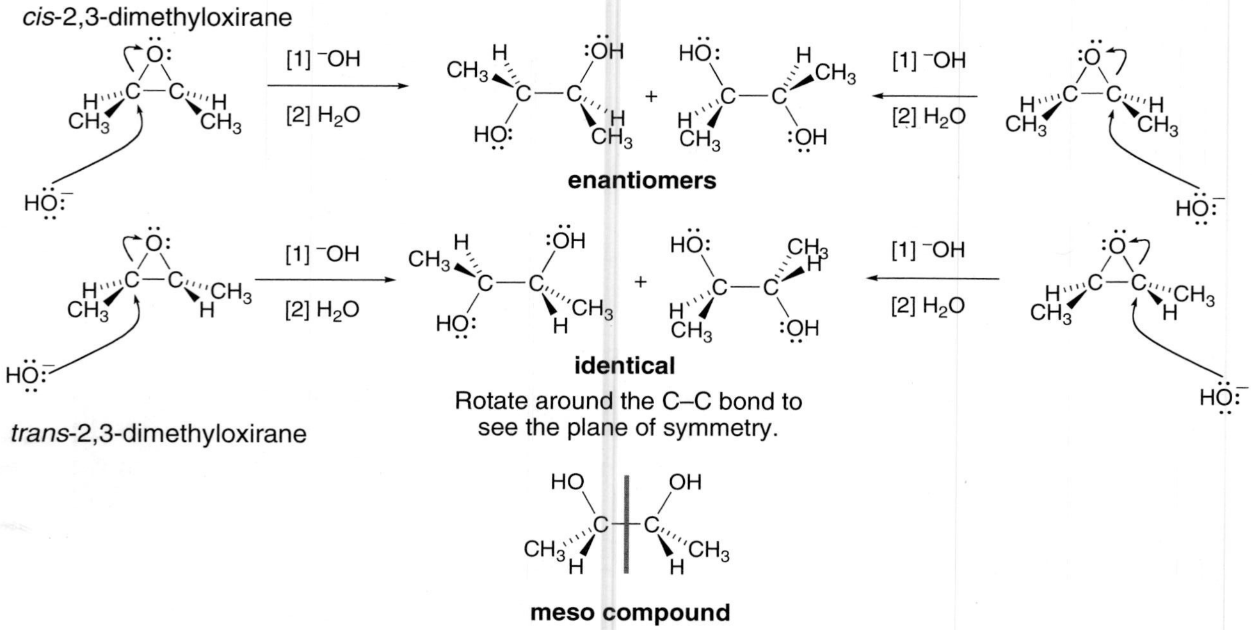
cis-2,3-dimethyloxirane
[1] ⁻OH
[2] H_2O
enantiomers
identical
Rotate around the C–C bond to
see the plane of symmetry.
trans-2,3-dimethyloxirane
meso compound

9.38 Remember the difference between negatively charged nucleophiles and neutral nucleophiles:

- **Negatively charged nucleophiles attack first**, followed by protonation, and the nucleophile attacks at the **less substituted carbon**.
- **Neutral nucleophiles have protonation first**, followed by nucleophilic attack at the **more substituted carbon**.

BUT – trans or anti products are always formed regardless of the nucleophile.

a.

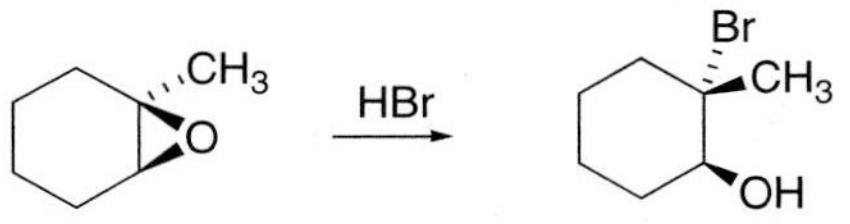

neutral nucleophile:
attack at **more**
substituted C

b.

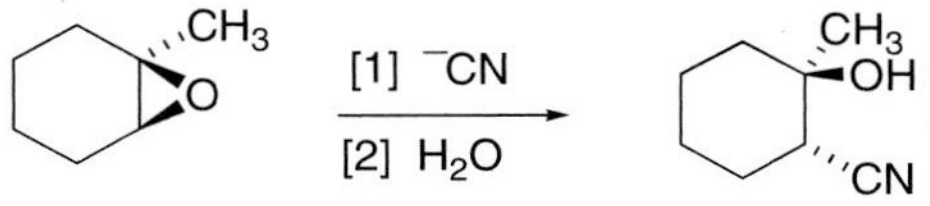

negatively charged
nucleophile:
attack at **less**
substituted C

c.

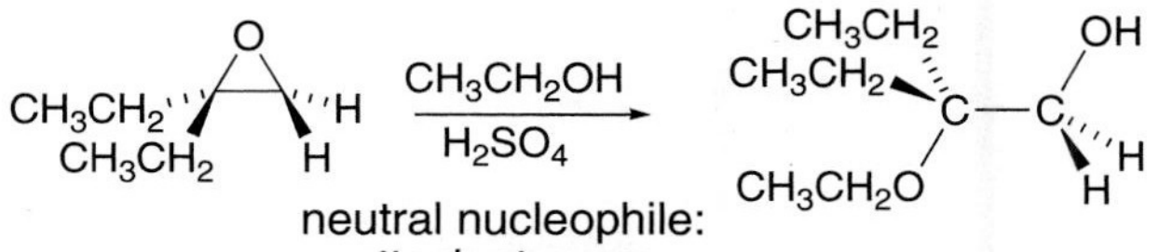

neutral nucleophile:
attack at **more**
substituted C

d.

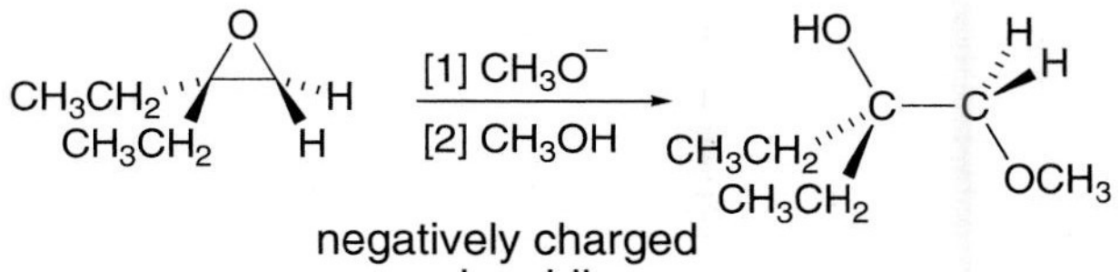

negatively charged
nucleophile:
attack at **less**
substituted C

9.39 Draw the structure of each alcohol, using the definitions in Answer 9.1.

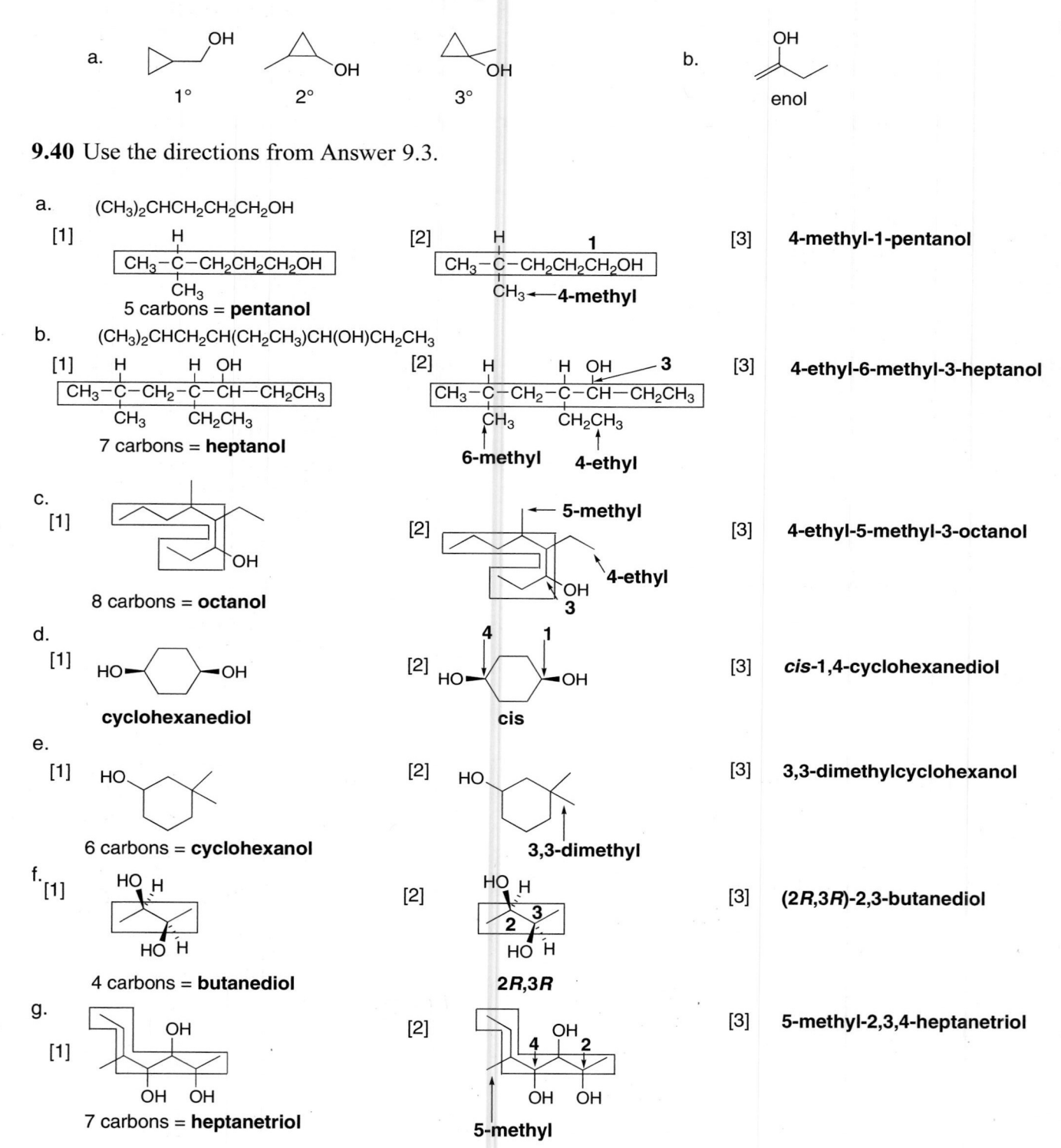

a.

1°   2°   3°

b.

OH

enol

9.40 Use the directions from Answer 9.3.

a.   (CH₃)₂CHCH₂CH₂CH₂OH

[1]   CH₃—C—CH₂CH₂CH₂OH
                  |
                  CH₃
         5 carbons = **pentanol**

[2]   CH₃—C—CH₂CH₂CH₂OH
                  |
                  CH₃ ←—**4-methyl**

[3]   **4-methyl-1-pentanol**

b.   (CH₃)₂CHCH₂CH(CH₂CH₃)CH(OH)CH₂CH₃

[1]   CH₃—C—CH₂—C—CH—CH₂CH₃
                  | |
                  CH₃   CH₂CH₃
         7 carbons = **heptanol**

[2]   CH₃—C—CH₂—C—CH—CH₂CH₃
                  | |
                  CH₃   CH₂CH₃
         6-methyl   **4-ethyl**

[3]   **4-ethyl-6-methyl-3-heptanol**

c.
[1]   8 carbons = **octanol**

[2]   5-methyl   4-ethyl   3

[3]   **4-ethyl-5-methyl-3-octanol**

d.
[1]   HO—⬡—OH
         cyclohexanediol

[2]   HO—⬡—OH   **cis**

[3]   ***cis*-1,4-cyclohexanediol**

e.
[1]   HO
         6 carbons = **cyclohexanol**

[2]   HO   **3,3-dimethyl**

[3]   **3,3-dimethylcyclohexanol**

f.
[1]   HO H
         HO H
         4 carbons = **butanediol**

[2]   HO H   2 3
         HO H   **2R,3R**

[3]   **(2R,3R)-2,3-butanediol**

g.
[1]   OH
         OH OH
         7 carbons = **heptanetriol**

[2]   OH   4   2
         OH OH   **5-methyl**

[3]   **5-methyl-2,3,4-heptanetriol**

h.

[1]

HO''''⟨cyclopentane ring⟩'''CH(CH₃)₂

5 carbons = **cyclopentanol**

[2]

HO''''⟨cyclopentane ring⟩'''CH(CH₃)₂

3-isopropyl

[3] ***trans*-3-isopropylcyclopentanol**

9.41 Use the rules from Answers 9.5 and 9.7.

a.

⟨cyclohexyl⟩–O–⟨cyclohexyl⟩

dicyclohexyl ether

b.

← **4,4-dimethyl**

OCH₂CH₂CH₃

longest chain = **heptane**

substituent = **3-propoxy**

4,4-dimethyl-3-propoxyheptane

c.

⟨structure⟩–O–

ethyl isobutyl ether
or **1-ethoxy-2-methylpropane**

d.

O

1,2-epoxy-2-methylhexane
or **2-butyl-2-methyloxirane**
or **2-methylhexene oxide**

e.

CH₂CH₃

O ← **epoxy**

2

5 carbons = **cyclopentane**

1,2-epoxy-1-ethylcyclopentane
or **1-ethylcyclopentene oxide**

f.

CH₃ CH₃
CH₃–C–O–C–CH₃
CH₃ CH₃

***tert*-butyl**   ***tert*-butyl**
di-*tert*-butyl ether

9.42 Use the directions from Answer 9.4.

a. 4-ethyl-3-**heptanol**

3   4
OH

b. *trans*-2-methyl**cyclohexanol**

1
OH
or
''CH₃

OH
''CH₃

c. 2,3,3-trimethyl-2-**butanol**

HO   3
2

d. 6-*sec*-butyl-7,7-diethyl-4-**decanol**

OH
6
4
7

e. 3-chloro-1,2-**propanediol**

1   3
HO   2   CI
OH

f. diisobutyl ether

⟨structure⟩–O–⟨structure⟩

g. 1,2-epoxy-1,3,3-trimethyl**cyclohexane**

i. (2R,3S)-3-isopropyl-2-**hexanol**

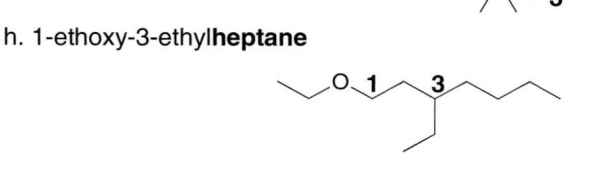

h. 1-ethoxy-3-ethyl**heptane**

j. (2S)-2-ethoxy-1,1-dimethyl**cyclopentane**

9.43

Eight constitutional isomers of molecular formula $C_5H_{12}O$ containing an OH group:

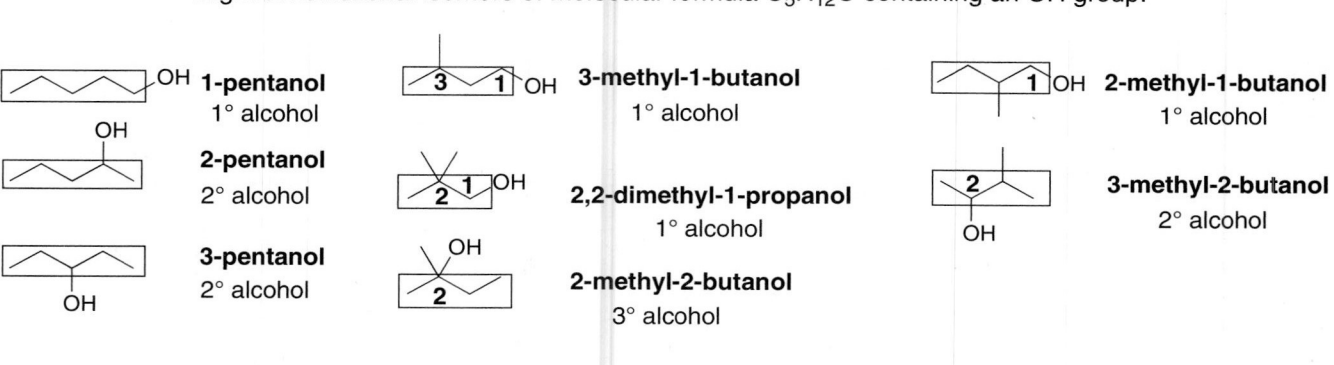

9.44 Use the boiling point rules from Answer 9.8.

a. $CH_3CH_2OCH_3$ $(CH_3)_2CHOH$ $CH_3CH_2CH_2OH$

ether
no hydrogen bonding
lowest bp

2° alcohol
hydrogen bonding
intermediate bp

1° alcohol
hydrogen bonding
highest bp

b. $CH_3CH_2CH_2CH_2CH_2CH_3$ $CH_3CH_2CH_2CH_2CH_2CH_2OH$ $HOCH_2CH_2CH_2CH_2CH_2CH_2OH$

no OH group
lowest water solubility

one OH group
intermediate water solubility

two OH groups
highest water solubility

9.45 Melting points depend on intermolecular forces and symmetry. $(CH_3)_2CHCH_2OH$ has a lower melting point than $CH_3CH_2CH_2CH_2OH$ because branching decreases surface area and makes $(CH_3)_2CHCH_2OH$ less symmetrical so it packs less well. Although $(CH_3)_3COH$ has the most branching and least surface area, it is the most symmetrical so it packs best in a crystalline lattice, giving it the highest melting point.

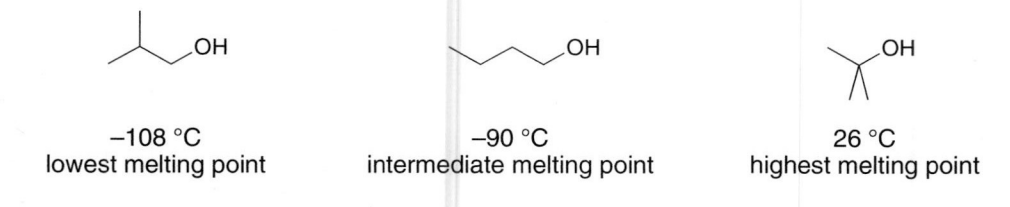

−108 °C
lowest melting point

−90 °C
intermediate melting point

26 °C
highest melting point

9.46 Stronger intermolecular forces increase boiling point. All of the compounds can hydrogen bond, but both diols have more opportunity for hydrogen bonding since they have two OH groups, making their bp's higher than the bp of 1-butanol. 1,2-Propanediol can also intramolecularly hydrogen bond. Intramolecular hydrogen bonding decreases the amount of intermolecular hydrogen bonding, so the bp of 1,2-propanediol is somewhat lower.

Increasing boiling point →

1-butanol	1,2-propanediol	1,3-propanediol
118 °C	187 °C	215 °C

9.47

a. $CH_3CH_2CH_2OH \xrightarrow{H_2SO_4} CH_3CH=CH_2 + H_2O$

b. $CH_3CH_2CH_2OH \xrightarrow{NaH} CH_3CH_2CH_2O^- \ Na^+ + H_2$

c. $CH_3CH_2CH_2OH \xrightarrow[ZnCl_2]{HCl} CH_3CH_2CH_2Cl + H_2O$

d. $CH_3CH_2CH_2OH \xrightarrow{HBr} CH_3CH_2CH_2Br + H_2O$

e. $CH_3CH_2CH_2OH \xrightarrow[pyridine]{SOCl_2} CH_3CH_2CH_2Cl$

f. $CH_3CH_2CH_2OH \xrightarrow{PBr_3} CH_3CH_2CH_2Br$

g. $CH_3CH_2CH_2OH \xrightarrow[pyridine]{TsCl} CH_3CH_2CH_2OTs$

h. $CH_3CH_2CH_2OH \xrightarrow{[1]\ NaH} CH_3CH_2CH_2O^- \ Na^+ \xrightarrow{[2]\ CH_3CH_2Br} CH_3CH_2CH_2OCH_2CH_3$

i. $CH_3CH_2CH_2OH \xrightarrow{[1]\ TsCl} CH_3CH_2CH_2OTs \xrightarrow{[2]\ NaSH} CH_3CH_2CH_2SH$

9.48

a. (cyclohexane with OH) $\xrightarrow{NaH}$ (cyclohexane with O$^-$ Na$^+$) + H$_2$

b. (cyclohexane with OH) $\xrightarrow{NaCl}$ N.R.

c. (cyclohexane with OH) $\xrightarrow{HBr}$ (cyclohexane with Br)

d. (cyclohexane with OH) $\xrightarrow{HCl}$ (cyclohexane with Cl)

e. (cyclohexane with OH) $\xrightarrow{H_2SO_4}$ (cyclohexene) + (methylenecyclohexane) + H$_2$O

f. (cyclohexane with OH) $\xrightarrow{NaHCO_3}$ N.R.

g. (cyclohexane with OH) $\xrightarrow{[1]\ NaH}$ (cyclohexane with O$^-$) $\xrightarrow{[2]\ CH_3CH_2Br}$ (cyclohexane with O–CH$_2$CH$_3$)

h. (cyclohexane with OH) $\xrightarrow[pyridine]{POCl_3}$ (cyclohexene) + (methylenecyclohexane)

9.49 Dehydration follows the Zaitsev rule, so the more stable, more substituted alkene is the major product.

a.

TsOH

tetrasubstituted
major product

disubstituted

b.

CH₂CH₃
OH

TsOH

CH₂CH₃

+

CHCH₃

c.

OH

TsOH

trisubstituted
major product

disubstituted

d. CH₃CH₂CH₂CH₂OH $\xrightarrow{\text{TsOH}}$ CH₃CH₂CH=CH₂

e.

OH

TsOH

tetrasubstituted
major product

+

disubstituted

> Two products formed
> by carbocation rearrangement

9.50 The more stable alkene is the major product.

OH

$\xrightarrow{\text{H}_2\text{SO}_4}$

trans and disubstituted
major product

+

monosubstituted

+

cis and disubstituted

9.51 OTs is a good leaving group and will easily be replaced by a nucleophile. Draw the products by substituting the nucleophile in the reagent for OTs in the starting material.

a. CH₃CH₂CH₂CH₂–OTs $\xrightarrow[\mathbf{S_N2}]{\text{CH}_3\text{SH}}$ CH₃CH₂CH₂CH₂–SCH₃ + HOTs

b. CH₃CH₂CH₂CH₂–OTs $\xrightarrow[\mathbf{S_N2}]{\text{NaOCH}_2\text{CH}_3}$ CH₃CH₂CH₂CH₂–OCH₂CH₃ + Na⁺ ⁻OTs

c. CH₃CH₂CH₂CH₂–OTs $\xrightarrow[\mathbf{S_N2}]{\text{NaOH}}$ CH₃CH₂CH₂CH₂–OH + Na⁺ ⁻OTs

d. CH₃CH₂CH₂CH₂–OTs $\xrightarrow[\mathbf{E2}]{\text{K}^+ \ ^-\text{OC(CH}_3)_3}$ CH₃CH₂CH=CH₂ + (CH₃)₃COH + K⁺ ⁻OTs

9.52

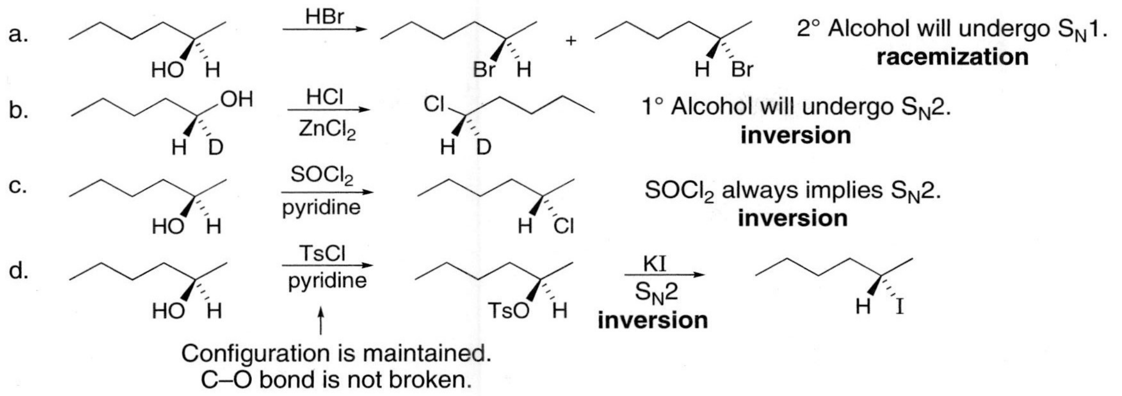
a.
HBr
HO H
Br H
+
H Br
2° Alcohol will undergo S_N1.
racemization
b.
OH
HCl
ZnCl2
H D
Cl
H D
1° Alcohol will undergo S_N2.
inversion
c.
SOCl2
pyridine
HO H
H Cl
SOCl2 always implies S_N2.
inversion
d.
TsCl
pyridine
HO H
TsO H
KI
S_N2
inversion
H I
Configuration is maintained.
C–O bond is not broken.

9.53

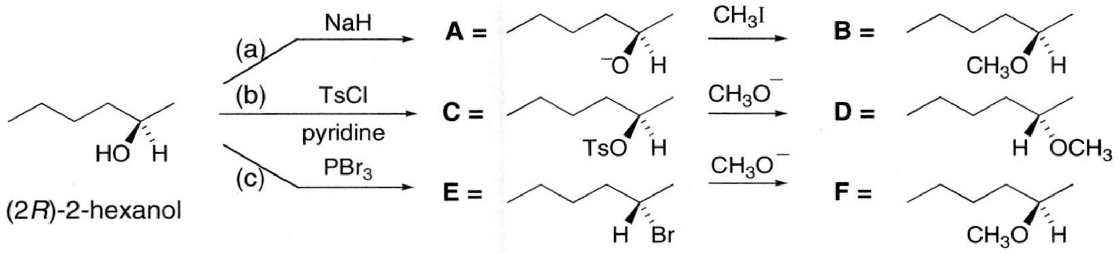
(a)
NaH
A =
−O H
CH3I
B =
CH3O H
(b)
TsCl
pyridine
C =
TsO H
CH3O−
D =
H OCH3
(c)
PBr3
E =
H Br
CH3O−
F =
CH3O H
HO H
(2R)-2-hexanol

Routes (a) and (c) given identical products, labeled **B** and **F**.

9.54 Acid-catalyzed dehydration follows an E1 mechanism for 2° and 3° ROH with an added step to make a good leaving group. The three steps are:
[1] Protonate the oxygen to make a good leaving group.
[2] Break the C–O bond to form a carbocation.
[3] Remove a β hydrogen to form the π bond.

a.

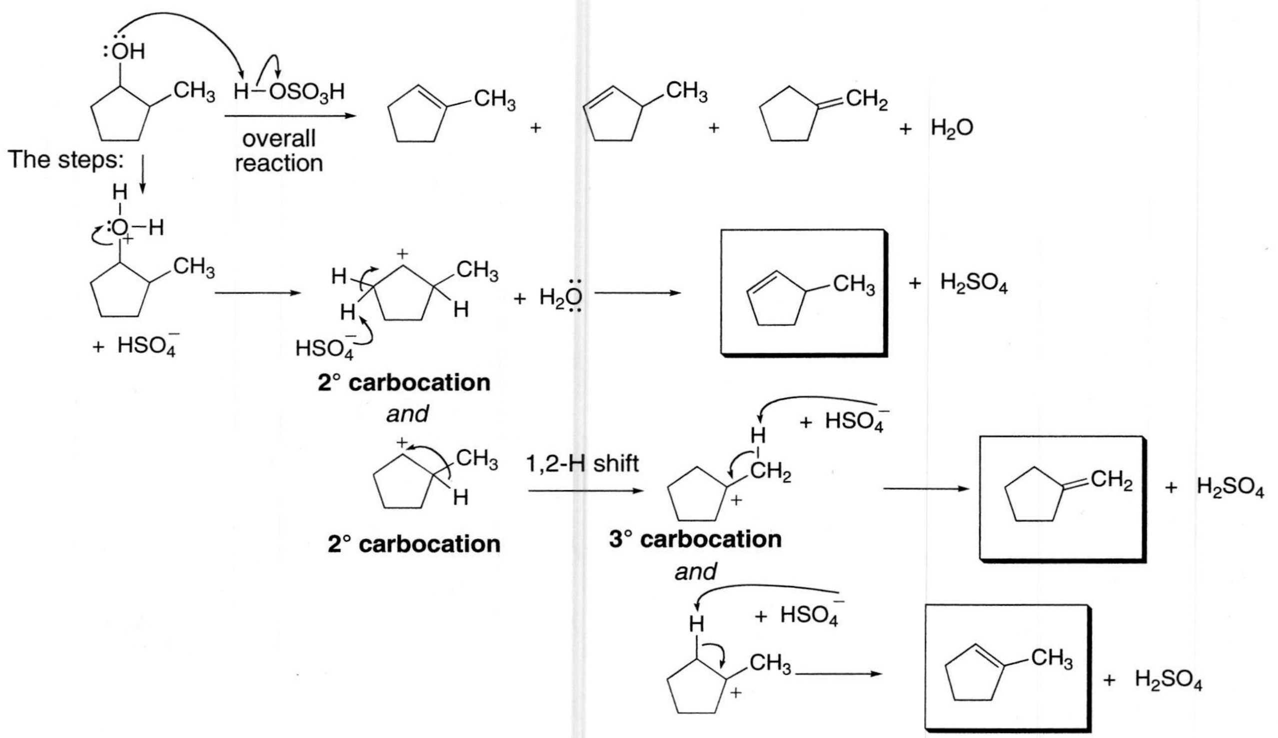
:ÖH
CH3
H–OSO3H
overall
reaction
CH3
+
CH3
+
CH2
+ H2O
The steps:
H
:O–H
CH3
+ HSO4−
H
H
CH3
H
HSO4−
+ H2Ö
CH3
+ H2SO4
2° carbocation
and
CH3
1,2-H shift
H
H
CH2
+ HSO4−
CH2
+ H2SO4
2° carbocation
3° carbocation
and
H
+ HSO4−
CH3
CH3
+ H2SO4

b.

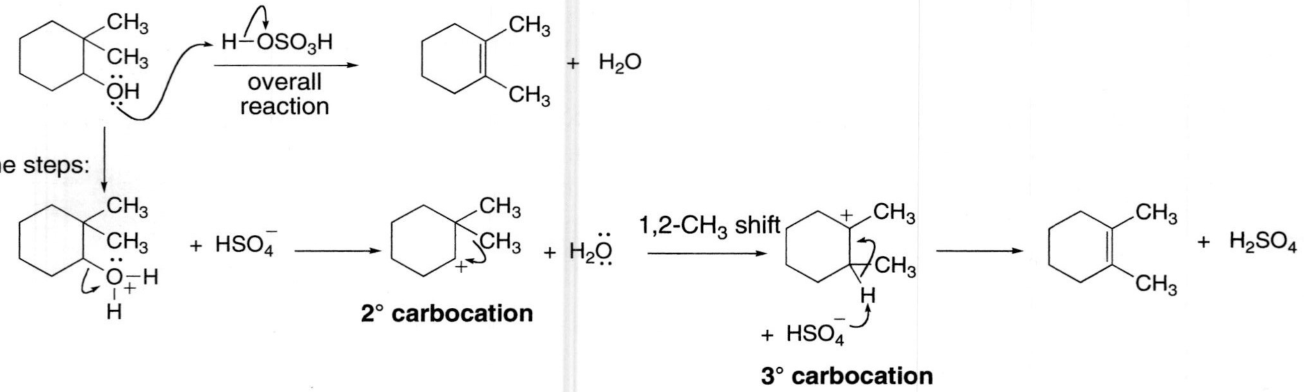
CH3
CH3
ÖH
H–OSO3H
overall
reaction
CH3
CH3
+ H2O
The steps:
CH3
CH3
O–H
H
+ HSO4−
CH3
CH3
+ H2Ö
2° carbocation
1,2-CH3 shift
CH3
CH3
H
+ HSO4−
3° carbocation
CH3
CH3
+ H2SO4

9.55

a.

A	B	C	D
(E1 with acid)	(E2 with base)	(E2 with base)	(E1 with acid)

3,3-dimethyl-cyclopentene

major product

b. The best starting material to form 3,3-dimethylcyclopentene would be **C** since the alkene can be formed as a single product by an E2 elimination with base.

9.56 With POCl₃ (pyridine), elimination occurs by an E2 mechanism. Since only one carbon has a β hydrogen, only one product is formed. With H₂SO₄, the mechanism of elimination is E1. A 2° carbocation rearranges to a 3° carbocation, which has three pathways for elimination.

V

+ Cl⁻

W

+ ⁻OPOCl₂

V

+ HSO₄⁻

2° carbocation

+ H₂O

1,2–CH₃ shift

3° carbocation

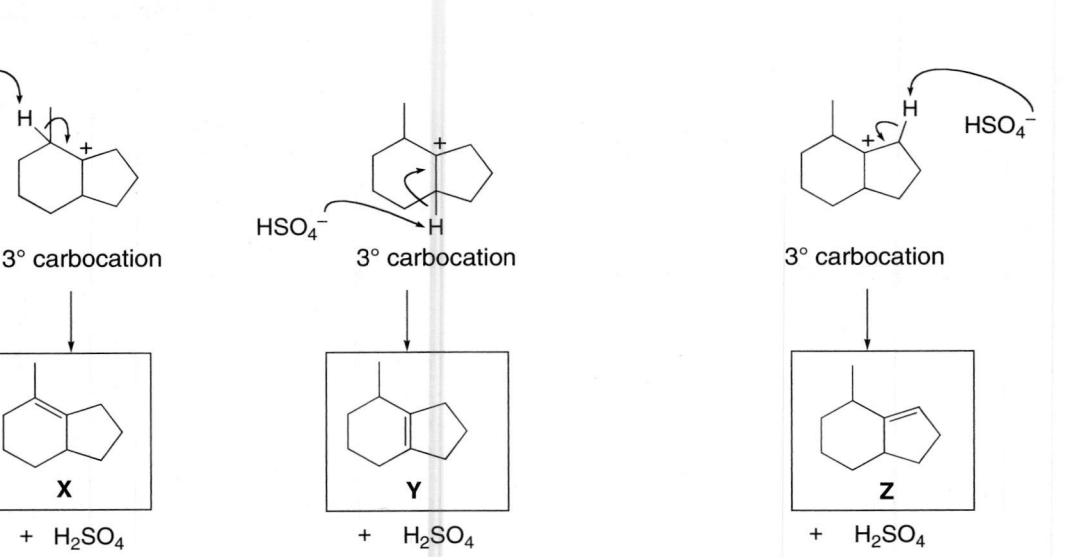

3° carbocation   3° carbocation   3° carbocation

X + H_2SO_4 **Y** + H_2SO_4 **Z** + H_2SO_4

9.57 To draw the mechanism:

[1] Protonate the oxygen to make a good leaving group.
[2] Break the C–O bond to form a carbocation.
[3] Look for possible rearrangements to make a more stable carbocation.
[4] Remove a β hydrogen to form the π bond.

Dark and light circles are meant to show where the carbons in the starting material appear in the product.

2° carbocation   3° carbocation

9.58

a.
$$\text{HO, H} \xrightarrow{\text{HBr}} \text{Br, H (S)} + \text{H, Br (R)} \quad \begin{array}{l} 2° \text{ alcohol} \\ S_N1 = \text{racemization} \end{array}$$

b.
$$\text{HO, H} \xrightarrow{\text{PBr}_3} \text{H, Br (R)} \quad \begin{array}{l} \text{PBr}_3 \text{ follows} \\ S_N2 = \text{inversion.} \end{array}$$

c.
$$\text{HO, H} \xrightarrow{\text{HCl}} \text{Cl, H (S)} + \text{H, Cl (R)} \quad \begin{array}{l} 2° \text{ alcohol} \\ S_N1 = \text{racemization} \end{array}$$

d.
$$\text{HO, H} \xrightarrow[\text{pyridine}]{\text{SOCl}_2} \text{H, Cl (R)} \quad \begin{array}{l} \text{SOCl}_2 \text{ follows} \\ S_N2 = \text{inversion.} \end{array}$$

9.59

3-methyl-2-butanol → [1] HBr / [2] –H$_2$O → 2° carbocation → 1,2-H shift → 3° carbocation → :Br:$^-$ → The 2° alcohol reacts by an S$_N$1 mechanism to form a carbocation that rearranges.

2-methyl-1-propanol HÖ— → HBr → H$_2$Ö$^+$:Br:$^-$ S$_N$2 no carbocation → Br + H$_2$Ö

The 1° alcohol reacts with HBr by an S$_N$2 mechanism. **no carbocation intermediate = no rearrangement possible**

9.60 Conversion of a 1° alcohol into a 1° alkyl chloride occurs by an S$_N$2 mechanism. S$_N$2 mechanisms occur more readily in polar aprotic solvents by making the nucleophile stronger. No added ZnCl$_2$ is necessary.

$$R-OH \xrightarrow[\text{HMPA}]{\text{HCl}} R-Cl$$

↑
polar aprotic solvent
This makes Cl$^-$ a better nucleophile.

9.61

A → H–Cl → ÖH$_2$$^+$ + :Cl:$^-$ → two resonance structures for the carbocation + H$_2$Ö

:Cl:$^-$ → B (Cl)

:Cl:$^-$ → C (Cl)

9.62

CH$_3$CH$_2$CH$_2$CH$_2$OH $\xrightarrow[\text{overall reaction}]{\text{H}_2\text{SO}_4,\ \text{NaBr}}$ CH$_3$CH$_2$CH$_2$CH$_2$Br + CH$_3$CH$_2$CH=CH$_2$

+ CH$_3$CH$_2$CH$_2$CH$_2$OCH$_2$CH$_2$CH$_2$CH$_3$

Step [1] for all products: Formation of a good leaving group

CH$_3$CH$_2$CH$_2$CH$_2$—ÖH + H—OSO$_3$H → CH$_3$CH$_2$CH$_2$CH$_2$—Ö$^+$—H + $^-$HSO$_4$
                                                                          |
                                                                          H

Formation of CH$_3$CH$_2$CH$_2$CH$_2$Br:

CH$_3$CH$_2$CH$_2$CH$_2$—Ö—H → CH$_3$CH$_2$CH$_2$CH$_2$Br + H$_2$Ö
              |$^+$
              H
:Br:$^-$

Formation of CH₃CH₂CH=CH₂:

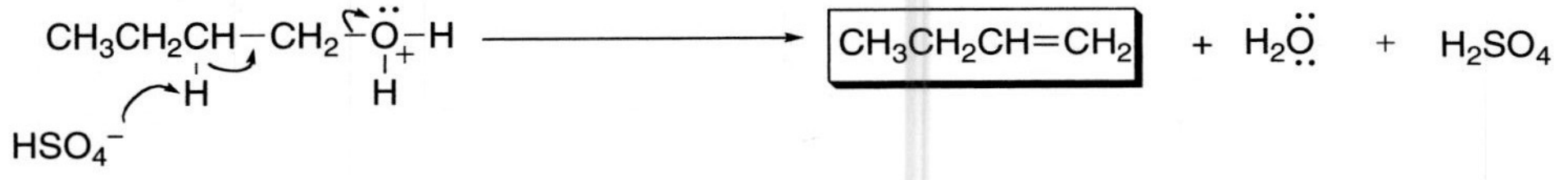

$CH_3CH_2CH-CH_2-O-H$ $\longrightarrow$ $\boxed{CH_3CH_2CH=CH_2}$ + H_2O + H_2SO_4

HSO_4^-

Ether formation (from the protonated alcohol):

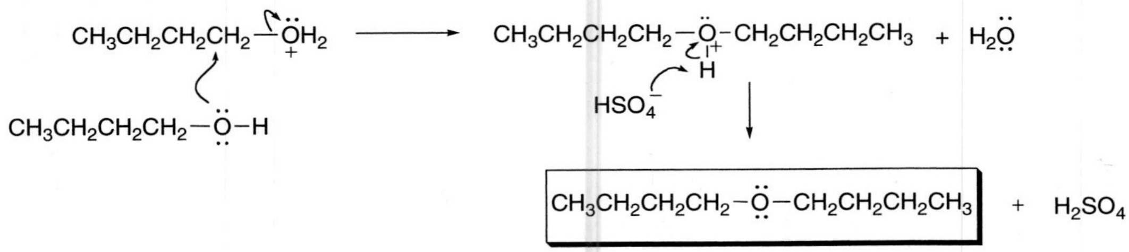

$CH_3CH_2CH_2CH_2-OH_2$ $\longrightarrow$ $CH_3CH_2CH_2CH_2-O-CH_2CH_2CH_2CH_3$ + H_2O

$CH_3CH_2CH_2CH_2-O-H$

HSO_4^-

$\boxed{CH_3CH_2CH_2CH_2-O-CH_2CH_2CH_2CH_3}$ + H_2SO_4

9.63

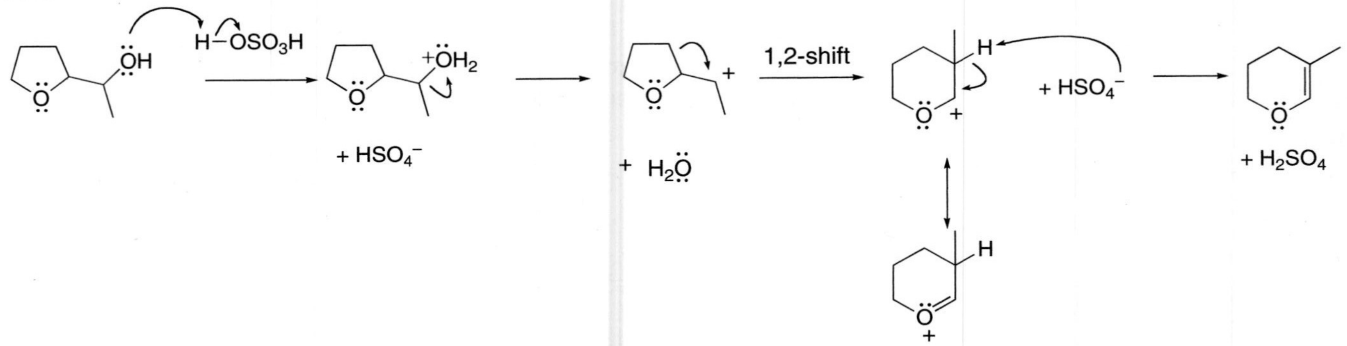

+ HSO₄⁻   + H₂O   + HSO₄⁻   + H₂SO₄

1,2-shift

9.64

+ HSO₄⁻   + H₂O   + H₂SO₄

HSO₄⁻

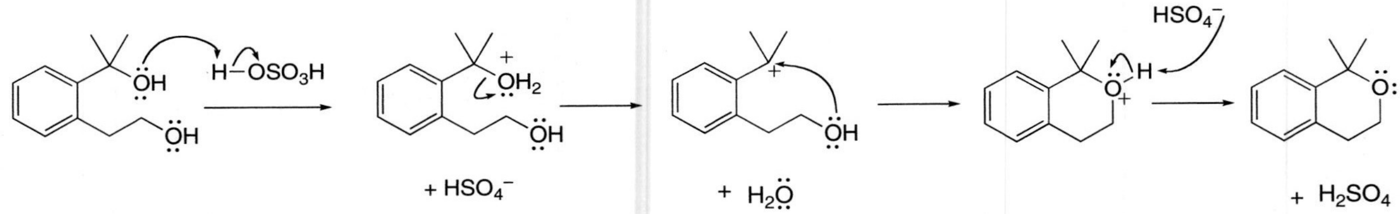

9.65

a.

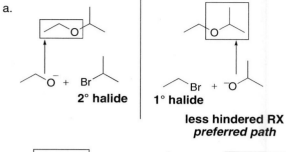

2° halide | **1° halide**

less hindered RX
preferred path

c.

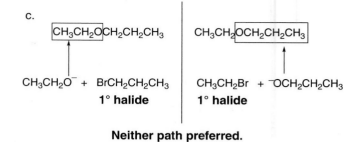

CH₃CH₂O⁻ + BrCH₂CH₂CH₃ | CH₃CH₂Br + ⁻OCH₂CH₂CH₃

1° halide | **1° halide**

Neither path preferred.

b.

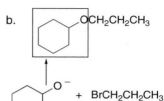

+ BrCH₂CH₂CH₃

1° halide | **2° halide**

less hindered RX
preferred path

9.66 A tertiary halide is too hindered and an aryl halide too unreactive to undergo a Williamson ether synthesis.

Two possible sets of starting materials:

aryl halide
unreactive in S_N2

3° alkyl halide
too sterically hindered for S_N2

9.67

a. (CH₃)₃COCH₂CH₂CH₃ $\xrightarrow[\text{(2 equiv)}]{\text{HBr}}$ (CH₃)₃CBr + BrCH₂CH₂CH₃ + H₂O

c. ⬠—OCH₃ $\xrightarrow[\text{(2 equiv)}]{\text{HBr}}$ ⬠—Br + CH₃Br + H₂O

b. ⬡—O—⬡ $\xrightarrow[\text{(2 equiv)}]{\text{HBr}}$ 2 ⬡—Br + H₂O

9.68

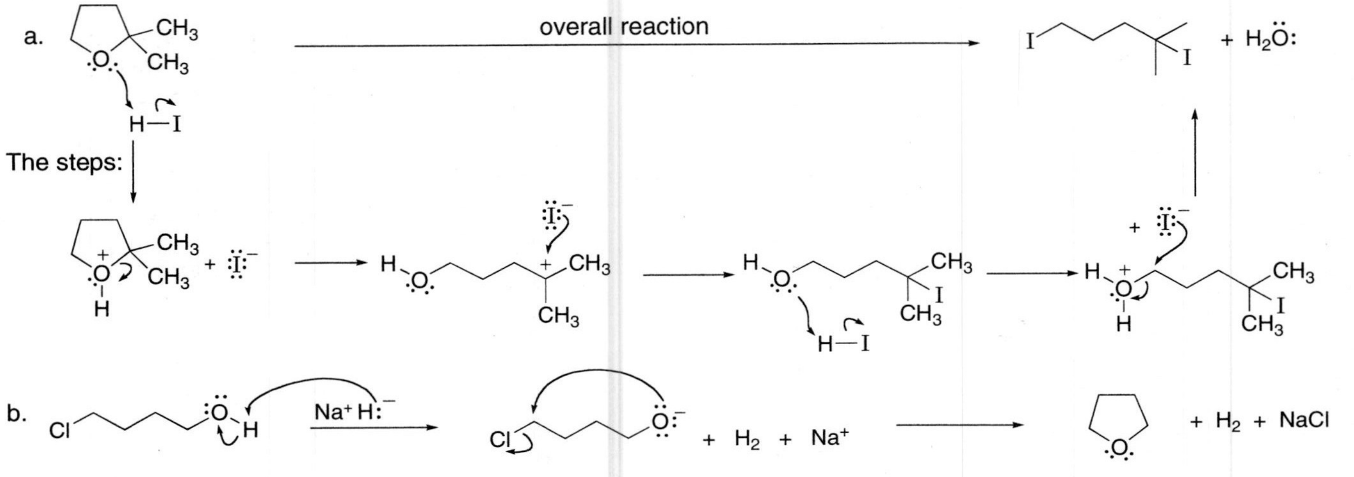

a.

The steps:

b.

9.69

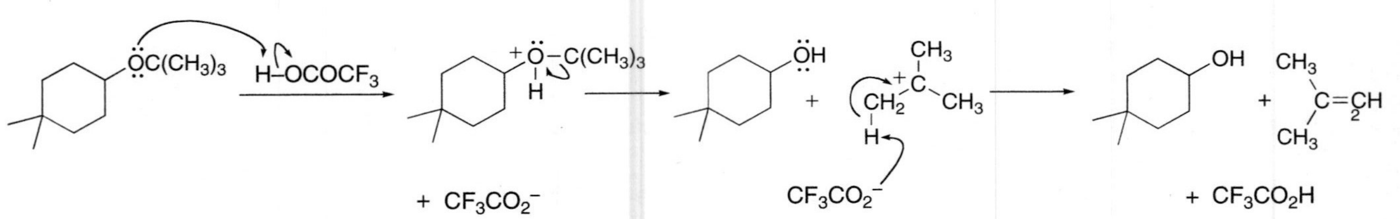

9.70

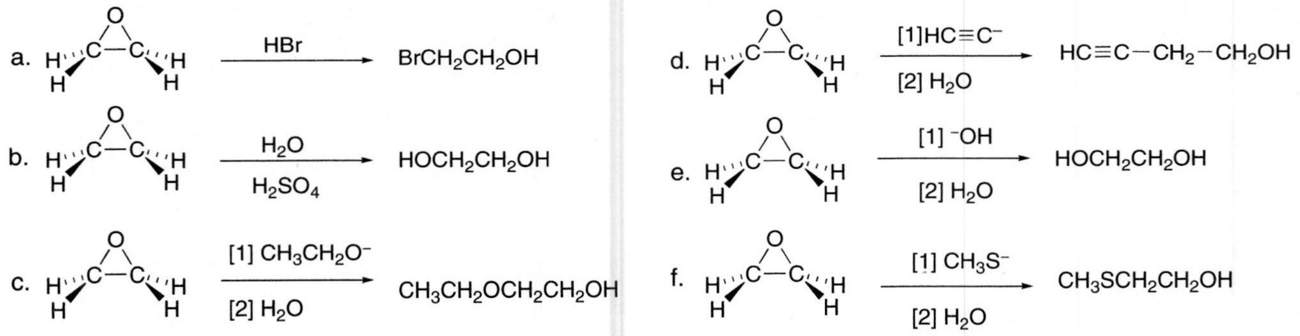

a.

$\xrightarrow{\text{HBr}}$ $BrCH_2CH_2OH$

b.

$\xrightarrow[\text{H}_2\text{SO}_4]{\text{H}_2\text{O}}$ $HOCH_2CH_2OH$

c.

$\xrightarrow[\text{[2] H}_2\text{O}]{\text{[1] CH}_3\text{CH}_2\text{O}^-}$ $CH_3CH_2OCH_2CH_2OH$

d.

$\xrightarrow[\text{[2] H}_2\text{O}]{\text{[1]HC}\equiv\text{C}^-}$ $HC\equiv C-CH_2-CH_2OH$

e.

$\xrightarrow[\text{[2] H}_2\text{O}]{\text{[1] }^-\text{OH}}$ $HOCH_2CH_2OH$

f.

$\xrightarrow[\text{[2] H}_2\text{O}]{\text{[1] CH}_3\text{S}^-}$ $CH_3SCH_2CH_2OH$

9.71

a.

$\xrightarrow[\text{H}_2\text{SO}_4]{\text{CH}_3\text{CH}_2\text{OH}}$

b.

$\xrightarrow[\text{[2] H}_2\text{O}]{\text{[1] CH}_3\text{CH}_2\text{O}^-\text{ Na}^+}$

c.

$\xrightarrow{\text{HBr}}$

d.

$\xrightarrow[\text{[2] H}_2\text{O}]{\text{[1] NaCN}}$

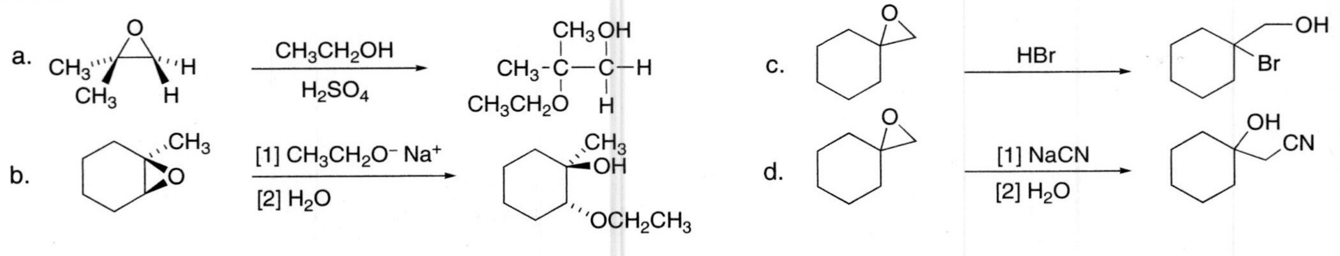

9.72

a. 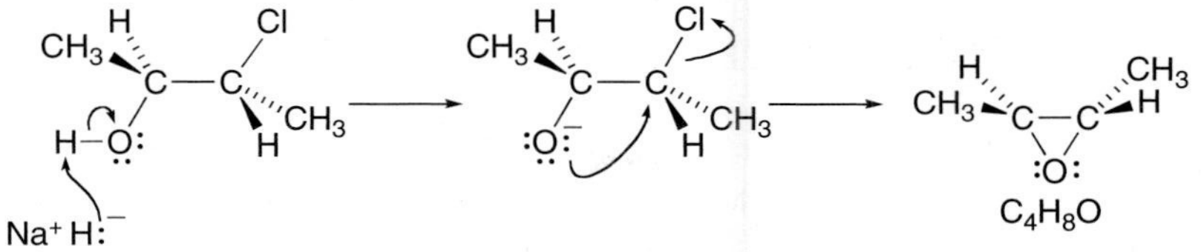
The 2 CH_3 groups are anti in the starting material, making them trans in the product.

C_4H_8O

b. 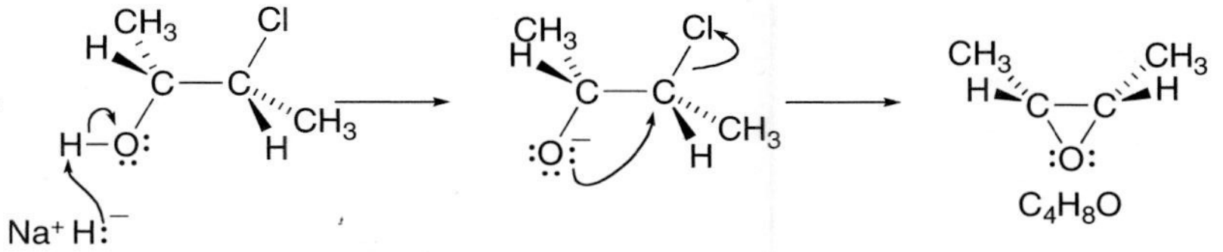
The 2 CH_3 groups are gauche in the starting material, making them cis in the product.

C_4H_8O

c.
backside attack

C_4H_8O

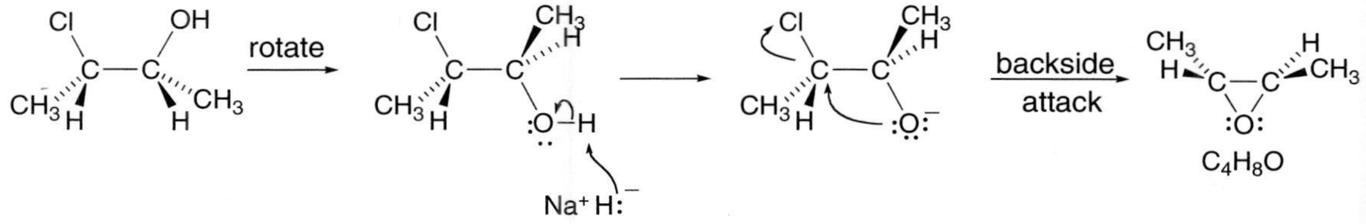

9.73

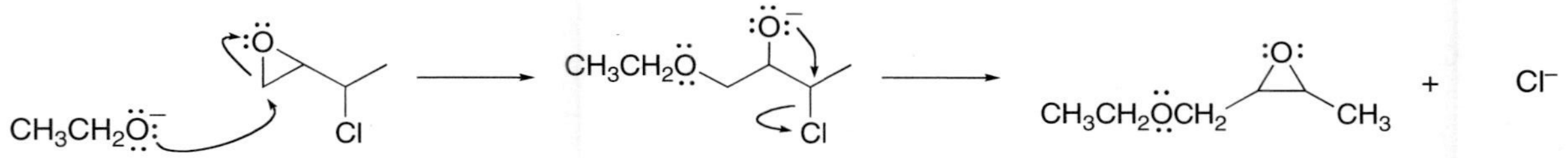

9.74

a. (1R,2R)-2-isobutylcyclopentanol

b. 2° alcohol

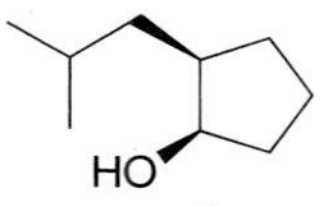

A

c. stereoisomer

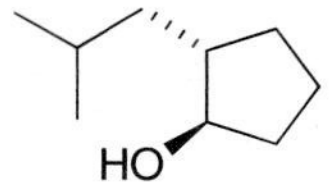

(1R,2S)-2-isobutylcyclopentanol

d. constitutional isomer

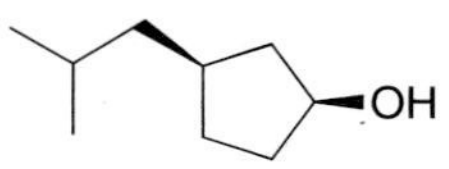

(1S,3S)-3-isobutylcyclopentanol

e. constitutional isomer with an ether

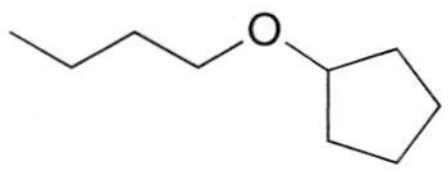

butoxycyclopentane

f.

[1]
NaH

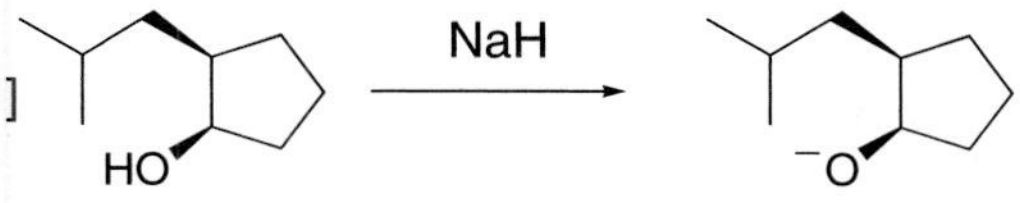

A

[2]
H_2SO_4 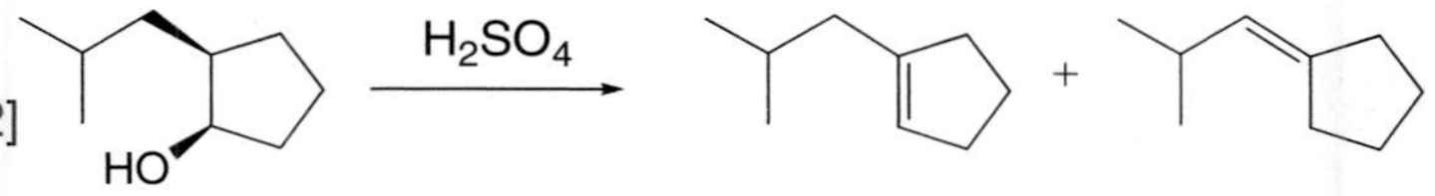

[3]
$POCl_3$ 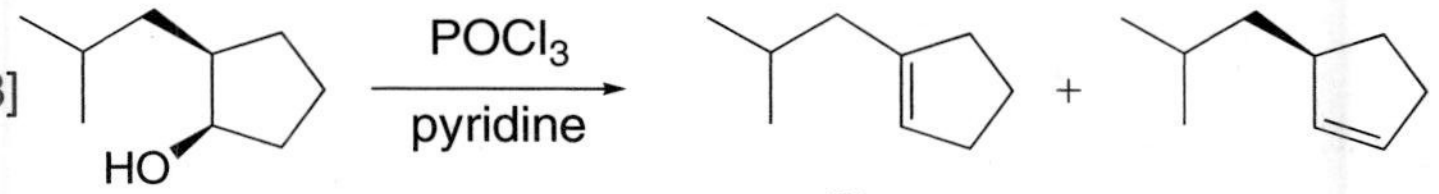
pyridine

[4]
HCl

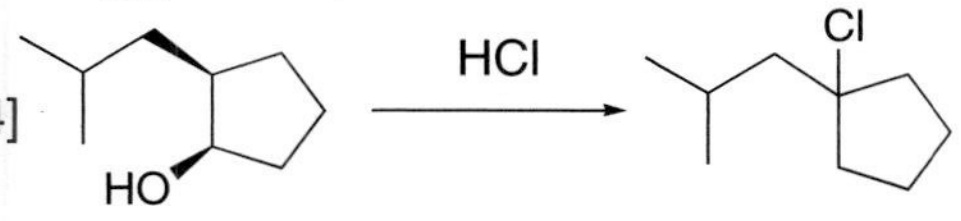

[5]
$SOCl_2$
pyridine

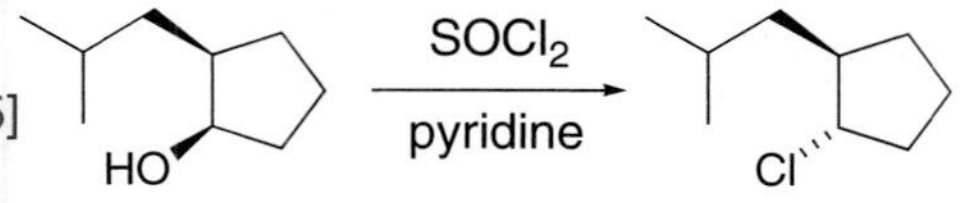

[6]
TsCl
pyridine

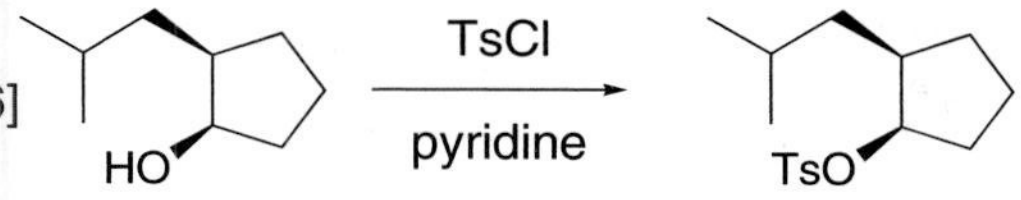

9.75

a. Bulky base favors E2.

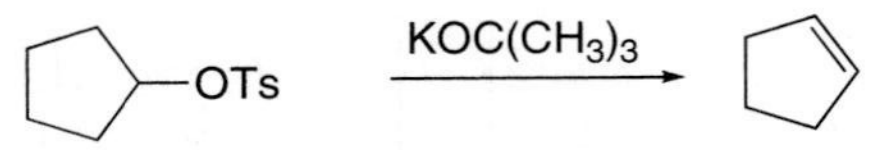

KOC(CH₃)₃

b. Keep the stereochemistry at the stereogenic center [*] the same here since no bond is broken to it.

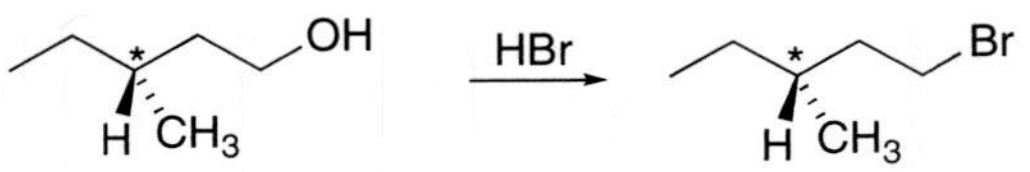

HBr

c. **identical**

[1] ⁻CN
[2] H₂O

d.

KCN
S_N2
inversion

e.

PBr₃

S_N2
inversion

f.

TsCl
pyridine

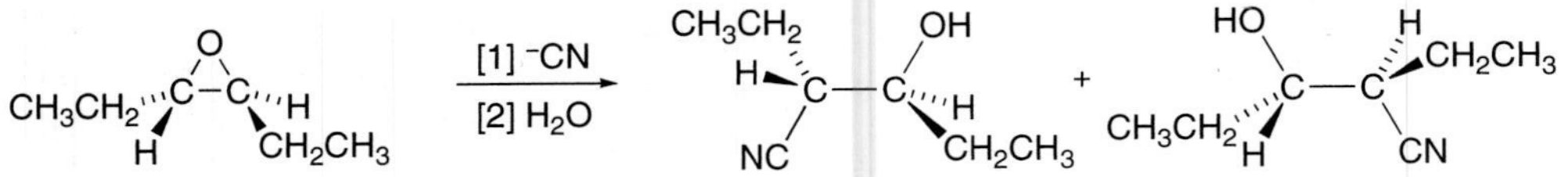

CH₃CO₂⁻

g.

HBr

h.

[1] NaOCH₃
[2] H₂O

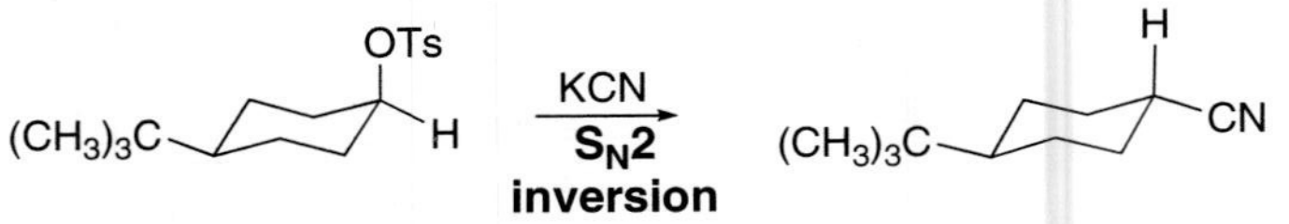

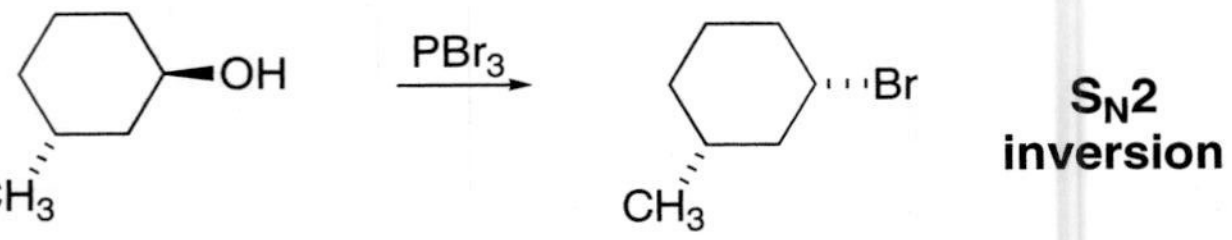

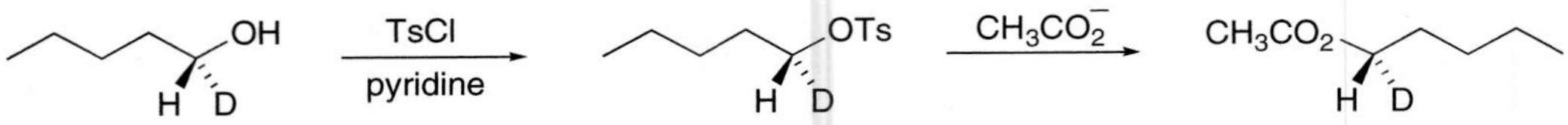

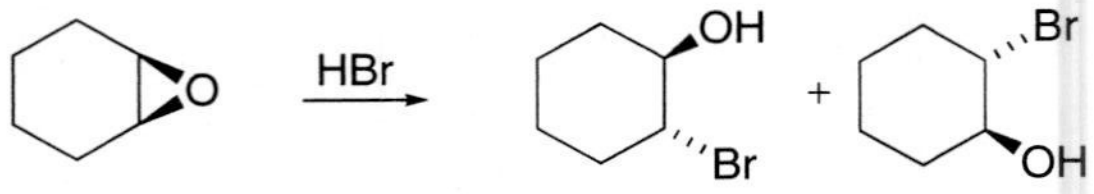

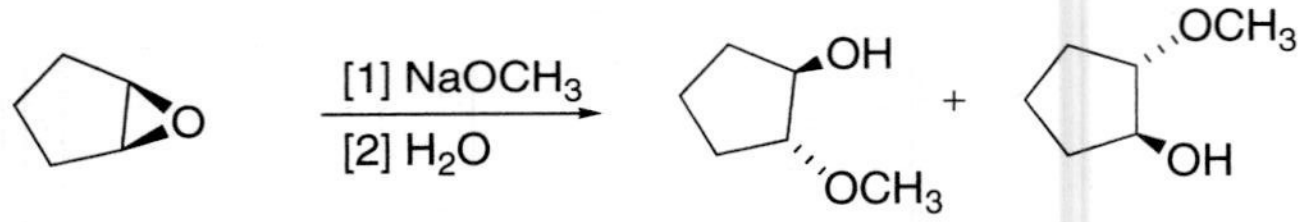

i.

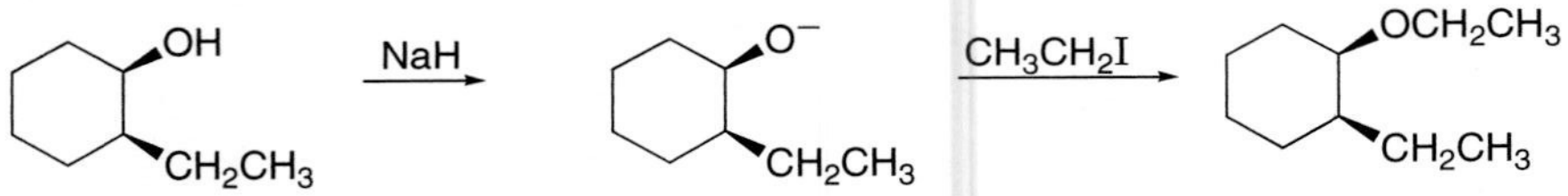

NaH

CH₃CH₂I

j.

HI
(2 equiv)

CH₃CH₂–C(CH₃)₂–I + I–CH₃ + H₂O

9.76

a.

HBr
or
PBr₃

b.

HCl, ZnCl₂
or
SOCl₂, pyridine

c.

[1] Na⁺ H⁻

[2] CH₃CH₂Cl

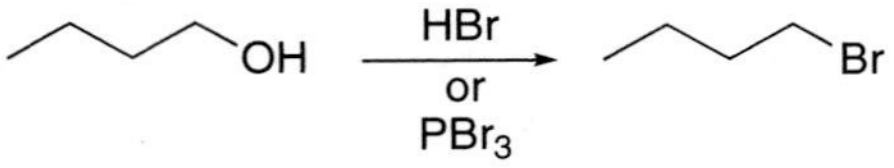

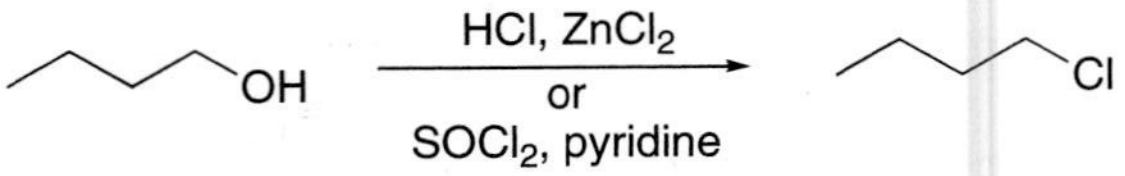

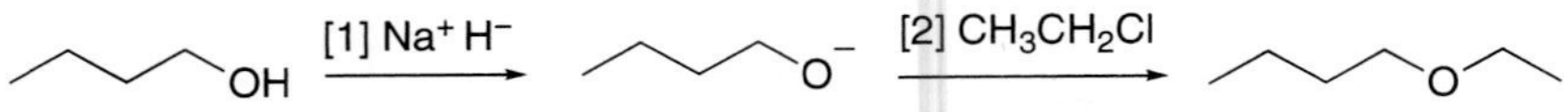

d.

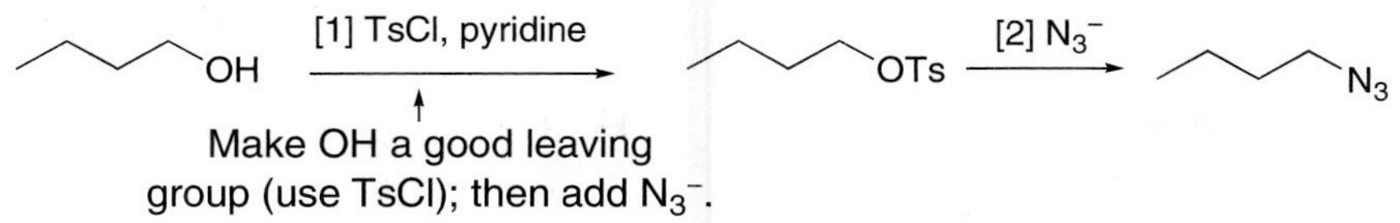

[1] TsCl, pyridine

↑
Make OH a good leaving
group (use TsCl); then add N_3^-.

[2] N_3^-

9.77

a.

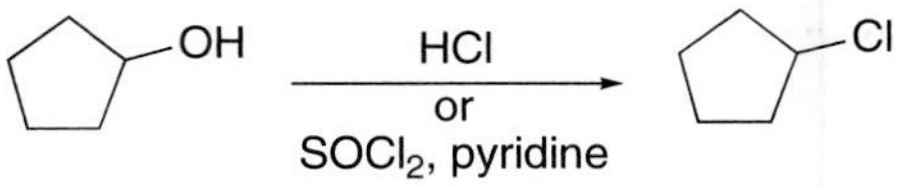

HCl
or
$SOCl_2$, pyridine

b.

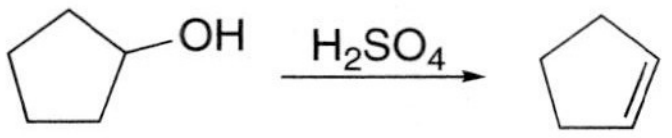

H_2SO_4

c.

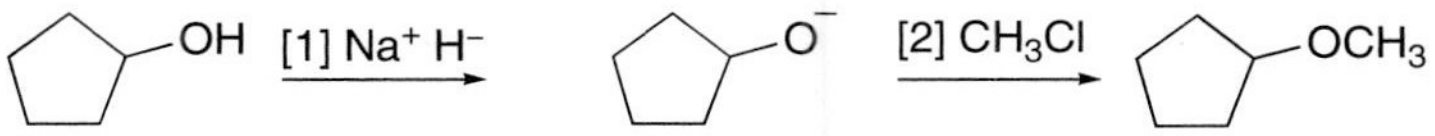

[1] $Na^+ H^-$

[2] CH_3Cl

d.

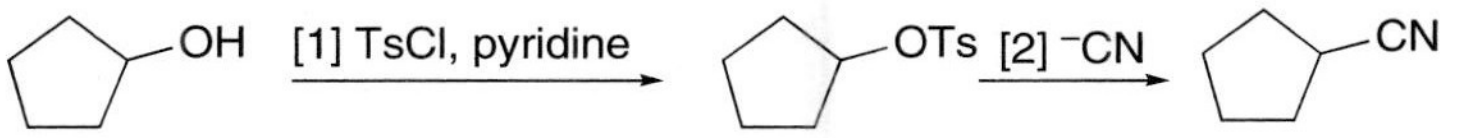

[1] TsCl, pyridine

[2] ^-CN

Make OH a good
leaving group (use TsCl);
then add ^-CN.

9.78

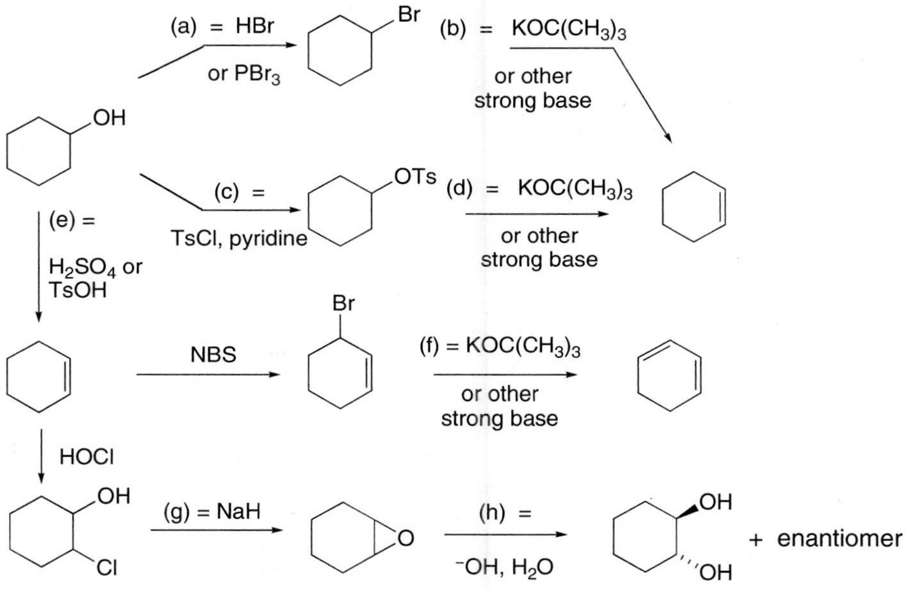

(a) = HBr
or PBr_3

(b) = $KOC(CH_3)_3$
or other
strong base

(c) =
TsCl, pyridine

(d) = $KOC(CH_3)_3$
or other
strong base

(e) =
H_2SO_4 or
TsOH

NBS

(f) = $KOC(CH_3)_3$
or other
strong base

HOCl

(g) = NaH

(h) =
^-OH, H_2O

+ enantiomer

9.79

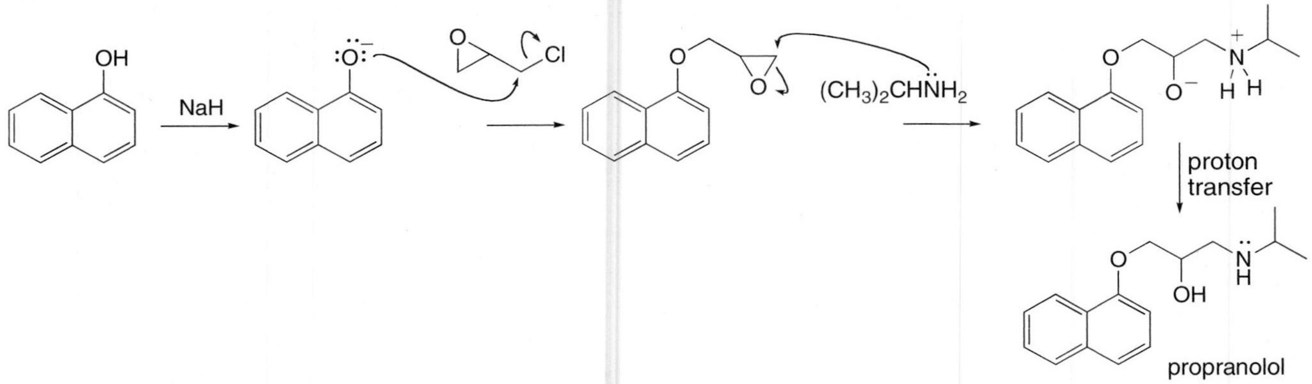

9.80

a. All 2° OH groups on stereogenic centers are circled (40 stereogenic centers).

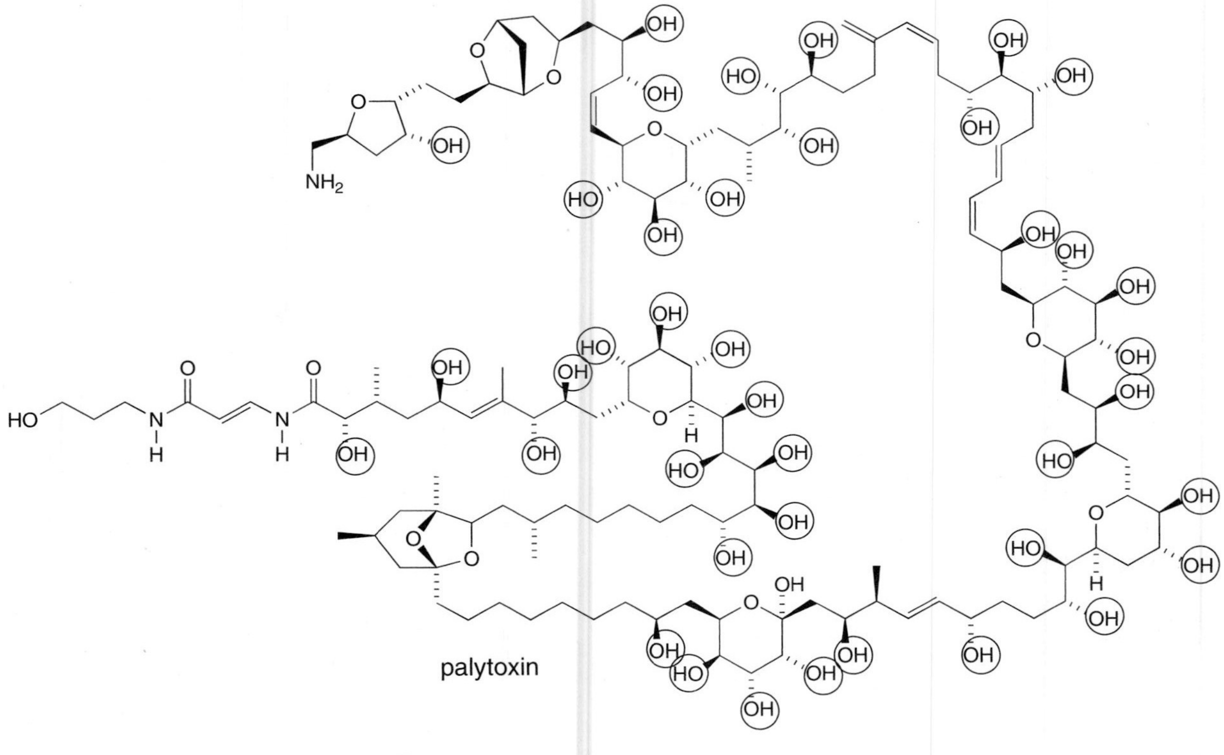

palytoxin

This carbocation is resonance stabilized, so loss of H_2O to form it is easier than loss of H_2O from a 2° alcohol, where the carbocation is not resonance stabilized.

b.

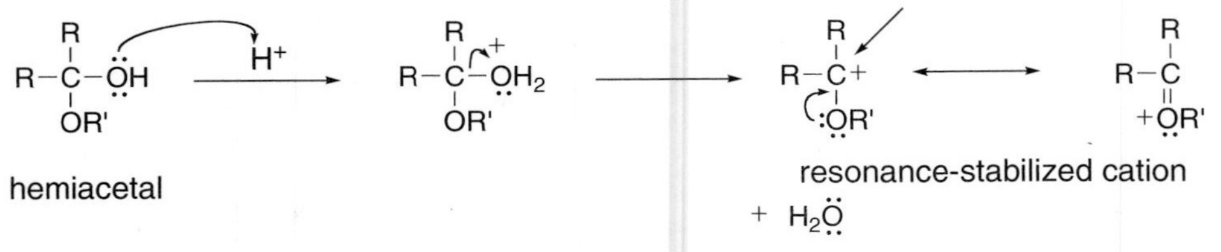

hemiacetal

resonance-stabilized cation

$+ H_2O$

9.81 If the base is not bulky, it can react as a nucleophile and open the epoxide ring. The bulky base cannot act as a nucleophile, and will only remove the proton.

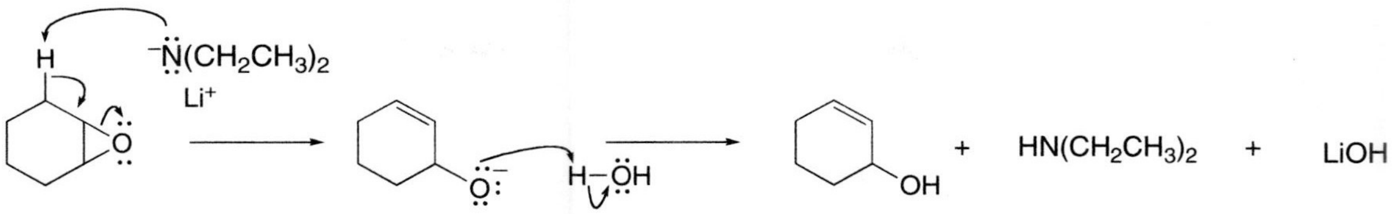

9.82 First form the 2° carbocation. Then lose a proton to form each product.

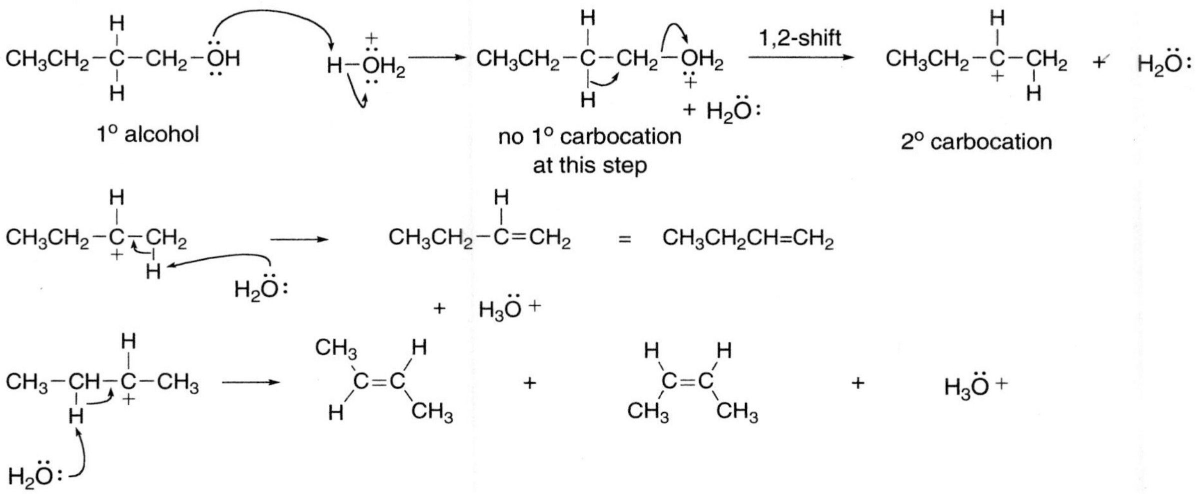
1° alcohol
no 1° carbocation
at this step
1,2-shift
2° carbocation

9.83

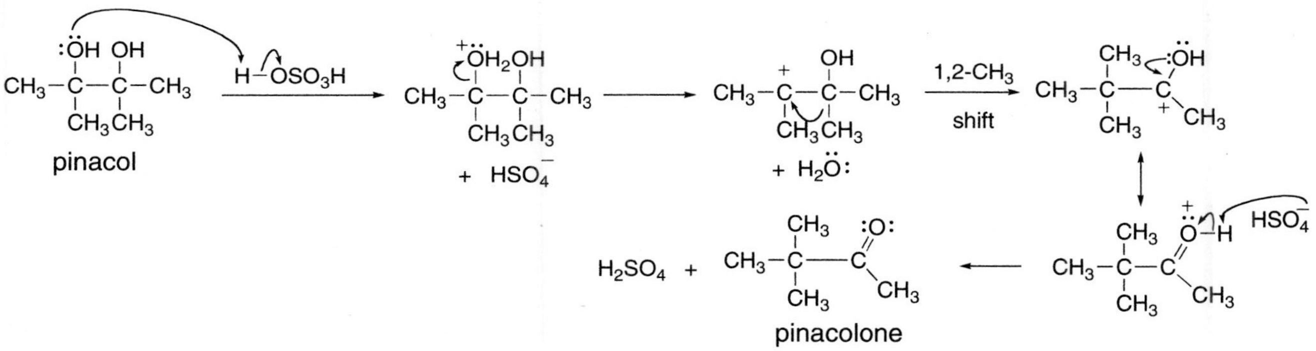
pinacol
1,2-CH3
shift
pinacolone

9.84

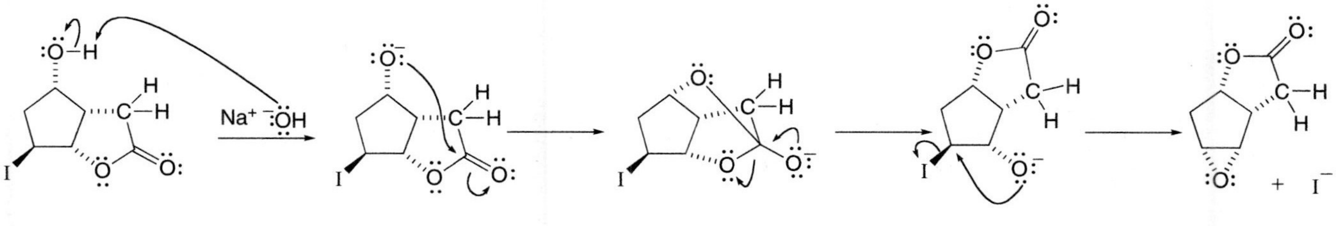

Chapter 10: Alkenes

◆ General facts about alkenes

- Alkenes contain a carbon–carbon double bond consisting of a stronger σ bond and a weaker π bond. Each carbon is sp^2 hybridized and trigonal planar (10.1).
- Alkenes are named using the suffix *-ene* (10.3).
- Alkenes with different groups on each end of the double bond exist as a pair of diastereomers, identified by the prefixes *E* and *Z* (10.3B).

Two higher priority groups on
opposite sides

Two higher priority groups on
the **same** side

E isomer
(2*E*)-3-methyl-2-pentene

Z isomer
(2*Z*)-3-methyl-2-pentene

- Alkenes have weak intermolecular forces, giving them low mp's and bp's, and making them water insoluble. A cis alkene is more polar than a trans alkene, giving it a slightly higher boiling point (10.4).

cis-2-butene

trans-2-butene

more polar isomer ⟶

⟵ **less polar isomer**

a small net dipole
higher bp

no net dipole
lower bp

- Since a π bond is electron rich and much weaker than a σ bond, alkenes undergo addition reactions with electrophiles (10.8).

◆ Stereochemistry of alkene addition reactions (10.8)

A reagent XY adds to a double bond in one of three different ways:

- **Syn addition**—X and Y add from the same side.

  - Syn addition occurs in **hydroboration.**

- **Anti addition**—X and Y add from opposite sides.

  - Anti addition occurs in **halogenation** and **halohydrin formation.**

- **Both syn and anti addition** occur when carbocations are intermediates.

  and

  - Syn and anti addition occur in **hydrohalogenation** and **hydration.**

◆ **Addition reactions of alkenes**

[1] **Hydrohalogenation**—Addition of HX (X = Cl, Br, I) (10.9–10.11)

$RCH=CH_2$ + H—X ⟶ | R—CH—CH$_2$ | X H | **alkyl halide**

- The mechanism has two steps.
- Carbocations are formed as intermediates.
- Carbocation rearrangements are possible.
- Markovnikov's rule is followed. H bonds to the less substituted C to form the more stable carbocation.
- Syn and anti addition occur.

[2] **Hydration** and related reactions—Addition of H_2O or ROH (10.12)

$RCH=CH_2$ + H—OH $\xrightarrow{H_2SO_4}$ | R—CH—CH$_2$ | OH H | **alcohol**

$RCH=CH_2$ + H—OR $\xrightarrow{H_2SO_4}$ | R—CH—CH$_2$ | OR H | **ether**

For both reactions:
- The mechanism has three steps.
- Carbocations are formed as intermediates.
- Carbocation rearrangements are possible.
- Markovnikov's rule is followed. H bonds to the less substituted C to form the more stable carbocation.
- Syn and anti addition occur.

[3] **Halogenation**—Addition of X_2 (X = Cl or Br) (10.13–10.14)

$RCH=CH_2$ + X—X ⟶ | R—CH—CH$_2$ | X X | **vicinal dihalide**

- The mechanism has two steps.
- Bridged halonium ions are formed as intermediates.
- No rearrangements occur.
- Anti addition occurs.

[4] **Halohydrin formation**—Addition of OH and X (X = Cl, Br) (10.15)

$RCH=CH_2$ + X—X $\xrightarrow{H_2O}$ | R—CH—CH$_2$ | OH X | **halohydrin**

- The mechanism has three steps.
- Bridged halonium ions are formed as intermediates.
- No rearrangements occur.
- X bonds to the less substituted C.
- Anti addition occurs.
- NBS in DMSO and H_2O adds Br and OH in the same fashion.

[5] **Hydroboration–oxidation**—Addition of H_2O (10.16)

$RCH=CH_2$ $\xrightarrow[\text{[2] } H_2O_2, \text{ HO}^-]{\text{[1] BH}_3 \text{ or 9-BBN}}$ | R—CH—CH$_2$ | H OH | **alcohol**

- Hydroboration has a one-step mechanism.
- No rearrangements occur.
- OH bonds to the less substituted C.
- Syn addition of H_2O results.

Chapter 10: Answers to Problems

10.1

Six alkenes of molecular formula C_5H_{10}:

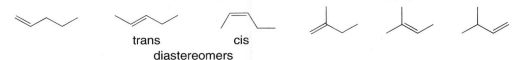

trans   cis

diastereomers

10.2 To determine the number of degrees of unsaturation:
[1] Calculate the maximum number of H's ($2n + 2$).
[2] Subtract the actual number of H's from the maximum number.
[3] Divide by two.

a. C_2H_2
[1] maximum number of H's = $2n + 2 = 2(2) + 2 = 6$
[2] subtract actual from maximum = $6 - 2 = 4$
[3] divide by two = $4/2 = $ **2 degrees of unsaturation**

b. C_6H_6
[1] maximum number of H's = $2n + 2 = 2(6) + 2 = 14$
[2] subtract actual from maximum = $14 - 6 = 8$
[3] divide by two = $8/2 = $ **4 degrees of unsaturation**

c. C_8H_{18}
[1] maximum number of H's = $2n + 2 = 2(8) + 2 = 18$
[2] subtract actual from maximum = $18 - 18 = 0$
[3] divide by two = $0/2 = $ **0 degrees of unsaturation**

d. C_7H_8O
Ignore the O.
[1] maximum number of H's = $2n + 2 = 2(7) + 2 = 16$
[2] subtract actual from maximum = $16 - 8 = 8$
[3] divide by two = $8/2 = $ **4 degrees of unsaturation**

e. $C_7H_{11}Br$
Because of Br, add one more H (11 + 1 H = 12 H's).
[1] maximum number of H's = $2n + 2 = 2(7) + 2 = 16$
[2] subtract actual from maximum = $16 - 12 = 4$
[3] divide by two = $4/2 = $ **2 degrees of unsaturation**

f. C_5H_9N
Because of N, subtract one H (9 − 1 H = 8 H's).
[1] maximum number of H's = $2n + 2 = 2(5) + 2 = 12$
[2] subtract actual from maximum = $12 - 8 = 4$
[3] divide by two = $4/2 = $ **2 degrees of unsaturation**

10.3

One possibility for C_6H_{10}:

a. a compound that has 2 π bonds

b. a compound that has 1 ring and 1 π bond

c. a compound with 2 rings

d. a compound with 1 triple bond

10.4 To name an alkene:
[1] Find the longest chain that contains the double bond. Change the ending from *-ane* to *-ene*.
[2] Number the chain to give the double bond the lower number. The alkene is named by the first number.
[3] Apply all other rules of nomenclature.

To name a cycloalkene:
[1] When a double bond is located in a ring, it is always located between C1 and C2. Omit the "1" in the name. Change the ending from *-ane* to *-ene*.
[2] Number the ring clockwise or counterclockwise to give the first substituent the lower number.
[3] Apply all other rules of nomenclature.

a. [1] $CH_2=CHCH(CH_3)CH_2CH_3$

     5 C chain with double bond
     pentene

     3-methyl

[2] $CH_2=CHCHCH_2CH_3$ (with CH_3 branch)

     1-pentene

[3] **3-methyl-1-pentene**

b. [1] $(CH_3CH_2)_2C=CHCH_2CH_2CH_3$

     7 C chain with double bond
     heptene

[2] CH_3CH_2 **3** $C=CHCH_2CH_2CH_3$ CH_3CH_2

     3-ethyl   **3-heptene**

[3] **3-ethyl-3-heptene**

c. [1]

     5 C chain with double bond
     pentene

[2] **2-ethyl** **4-methyl** **1-** ... **1-pentene**

[3] **2-ethyl-4-methyl-1-pentene**

d. [1]

     5 C ring with a double bond
     cyclopentene

[2] **4** **1** **3** **3,4-dimethyl**

[3] **3,4-dimethylcyclopentene**

e. [1]

     6 C ring with a double bond
     cyclohexene

[2] **1-methyl** **5-tert-butyl** **1**

[3] **5-*tert*-butyl-1-methylcyclohexene**

10.5 Use the rules from Answer 10.4 to name the compounds. Enols are named to give the OH the lower number. Compounds with two C=C's are named with the suffix *-adiene*.

a. [1] OH

     6 C chain with double bond
     hexene

[2] **3** **1** OH **4-ethyl**

[3] **4-ethyl-3-hexen-1-ol**

b. [1] OH

     8 C chain with double bond
     octene

[2] **1** **4** OH **5-ethyl** **6-methyl**

[3] **5-ethyl-6-methyl-7-octen-4-ol**

c. [1]

     7 C chain with two double bonds
     heptadiene

[2] **5** **2** **6-methyl** **2-methyl**

[3] **2,6-dimethyl-2,5-heptadiene**

10.6 To label an alkene as *E* or *Z*:

    [1] **Assign priorities** to the two substituents *on each end* using the rules for *R,S* nomenclature.

    [2] **Assign *E* or *Z*** depending on the location of the two higher priority groups.

        • The *E* prefix is used when the two higher priority groups are on **opposite sides**.

        • The *Z* prefix is used when the two higher priority groups are on the **same side** of the double bond.

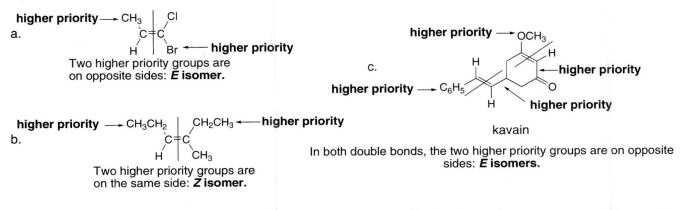

10.7 To name an alkene: First follow the rules from Answer 10.4. Then, when necessary, assign an *E* or *Z* prefix based on priority, as in 10.6.

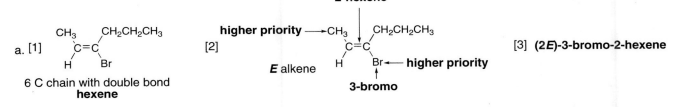

10.8 To work backwards from a name to a structure:
    [1] Find the parent name and functional group and draw, remembering that the double bond is between C1 and C2 for cycloalkenes.
    [2] Add the substituents to the appropriate carbons.

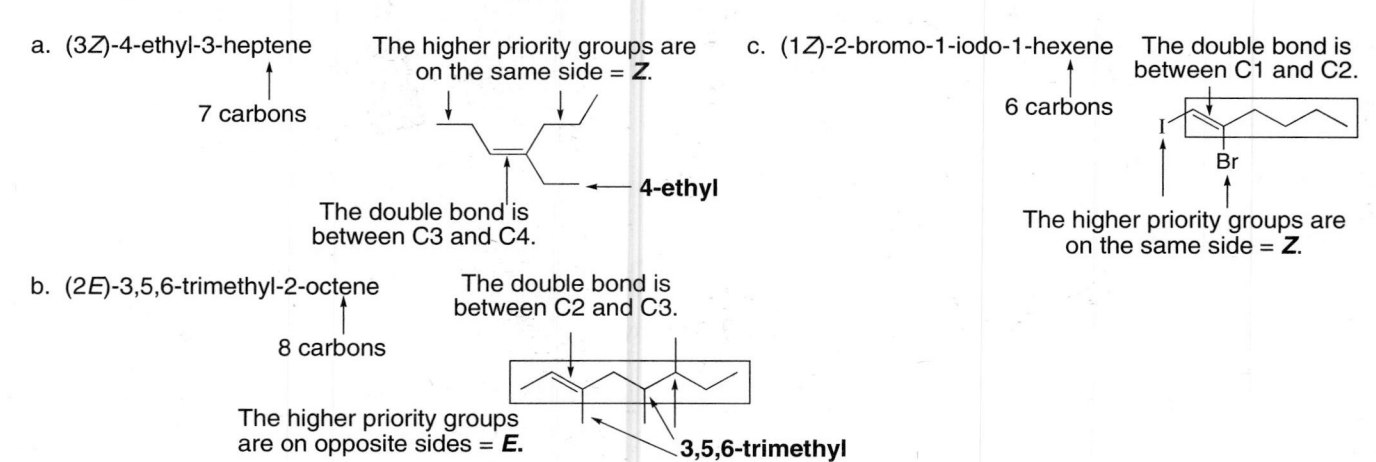

a. (3*Z*)-4-ethyl-3-heptene   The higher priority groups are on the same side = **Z**.
    7 carbons
    4-ethyl
    The double bond is between C3 and C4.

c. (1*Z*)-2-bromo-1-iodo-1-hexene   The double bond is between C1 and C2.
    6 carbons
    The higher priority groups are on the same side = **Z**.

b. (2*E*)-3,5,6-trimethyl-2-octene   The double bond is between C2 and C3.
    8 carbons
    The higher priority groups are on opposite sides = **E**.
    3,5,6-trimethyl

10.9   Draw all of the stereoisomers and then use the rules from Answer 10.6 to name each diene.

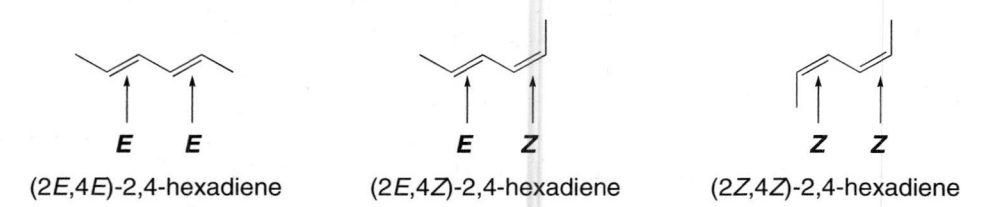

    E   **E**   **E**   **Z**   **Z**   **Z**
(2*E*,4*E*)-2,4-hexadiene   (2*E*,4*Z*)-2,4-hexadiene   (2*Z*,4*Z*)-2,4-hexadiene

10.10   To rank the isomers by increasing boiling point:
    Look for polarity differences: *small net dipoles* make an alkene more polar, giving it a higher boiling point than an alkene with *no net dipole*. Cis isomers have a higher boiling point than their trans isomers.

All dipoles cancel.	Two dipoles cancel.	Two dipoles reinforce.
smallest surface area	no net dipole	net dipole
no net dipole	trans isomer	cis isomer
lowest bp	**intermediate bp**	**highest bp**

10.11   Recall from Section 5.13B that the odor of a molecule is determined more by shape than by functional groups. That is why the *R* and *S* isomers of limonene smell so differently.

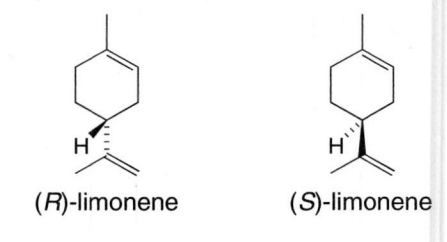

    (*R*)-limonene   (*S*)-limonene

10.12   Increasing number of double bonds = decreasing melting point.

stearidonic acid
4 double bonds
lowest melting point

stearic acid
no double bonds
highest melting point

linolenic acid
3 double bonds
intermediate melting point

10.13

a.   [structure]OH $\xrightarrow{H_2SO_4}$ [cyclopentene] + [methylenecyclopentane]

b.   [structure with Br] $\xrightarrow{NaOCH_2CH_3}$ $CH_2=CHCH_2CH_2CH_2CH_3$
+
$CH_3CH=CHCH_2CH_2CH_3$

10.14   To draw the products of an addition reaction:
[1] Locate the two bonds that will be broken in the reaction. Always break the π bond.
[2] Draw the product by forming two new σ bonds.

a.   [cyclopentene] $\xrightarrow{HCl}$ [product] two new σ bonds

c.   [structure with CH$_3$ groups] $\xrightarrow{HCl}$ [product] two new σ bonds

b. $CH_3CH_2CH_2CH=CHCH_2CH_2CH_3$ $\xrightarrow{HCl}$ $CH_3CH_2CH_2-\overset{H}{\underset{H}{C}}-\overset{H}{\underset{Cl}{C}}-CH_2CH_2CH_3$
two new σ bonds

10.15   Addition reactions of HX occur in two steps:
[1] The double bond attacks the H atom of HX to form a carbocation.
[2] X⁻ attacks the carbocation to form a C–X bond.

transition state
step [1]:

transition state
step [2]:

10.16 Addition to alkenes follows Markovnikov's rule: When HX adds to an unsymmetrical alkene, the H bonds to the C that has more H's to begin with.

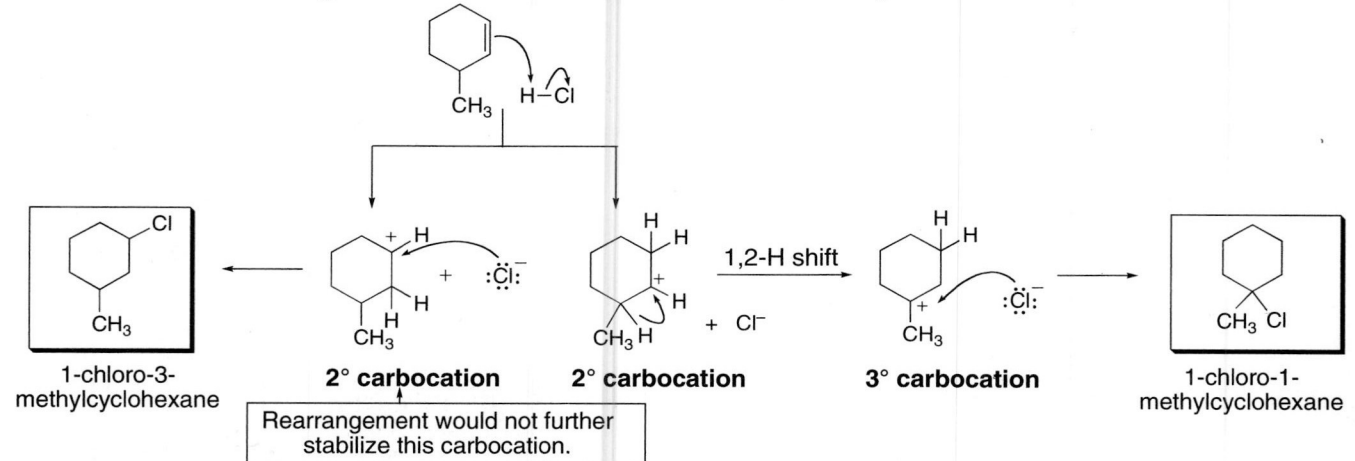

a.

no H's
Cl adds here.

one H
H adds here.

CH₃

H

HCl →

CH₃
Cl
H
H

c.

2 H's
H adds here.

no H's
Cl adds here.

HCl →

CH₂—H
Cl
CH₃

b.

no H's
Cl adds here

CH₃

CH₃
C=CH₂

2 H's
H adds here.

HCl →

CH₃
Cl—C—CH₂
CH₃ H

10.17 To determine which alkene will react faster, draw the carbocation that forms in the rate-determining step. The more stable, more substituted carbocation, the lower the E_a to form it and the faster the reaction.

CH₃   H
  C=C
CH₃   H
H–X

→

CH₃
+C—CH₂
CH₃ H
3° carbocation
faster reaction

CH₃   H
  C=C
H   H
H–X

→

+
CH₃CHCH₂
H
2° carbocation
slower reaction

10.18 Look for rearrangements of a carbocation intermediate to explain these results.

CH₃   H–Cl

1-chloro-3-methylcyclohexane

2° carbocation

+   :C̈l:⁻

2° carbocation   + Cl⁻

1,2-H shift →

3° carbocation

:C̈l:⁻

1-chloro-1-methylcyclohexane

Rearrangement would not further stabilize this carbocation.

10.19 Addition of HX to alkenes involves formation of carbocation intermediates. Rearrangement of the carbocation will occur if it forms a more stable carbocation.

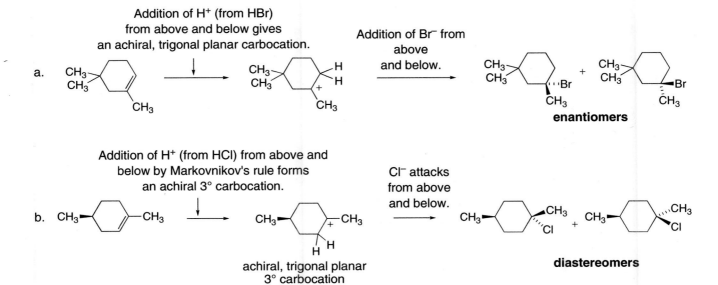

a.

3° carbocation
no rearrangement

b.

(cis or trans)   2° carbocation   2° carbocation

Rearrangement would not further
stabilize either carbocation.
no rearrangement

c.

(cis or trans)   2° carbocation   2° carbocation   **1,2-H shift**   3° carbocation

Rearrangement would not further
stabilize the carbocation.   **rearrangement**   **more stable**

10.20 To draw the products, remember that addition of HX proceeds via a carbocation intermediate.

a.

Addition of H⁺ (from HBr)
from above and below gives
an achiral, trigonal planar carbocation.

Addition of Br⁻ from
above
and below.

enantiomers

b.

Addition of H⁺ (from HCl) from above and
below by Markovnikov's rule forms
an achiral 3° carbocation.

Cl⁻ attacks
from above
and below.

achiral, trigonal planar
3° carbocation

diastereomers

10.21 The product of syn addition will have H and Cl both up or down (both on wedges or both dashes), while the product of anti addition will have one up and one down (one wedge, one dash).

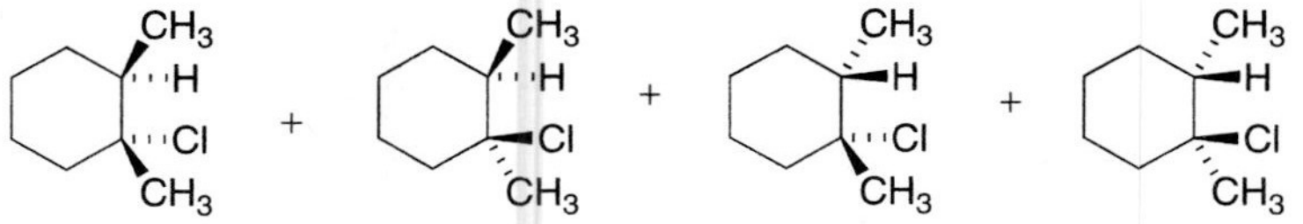

A	**B**	**C**	**D**
syn addition	anti addition	anti addition	syn addition

10.22

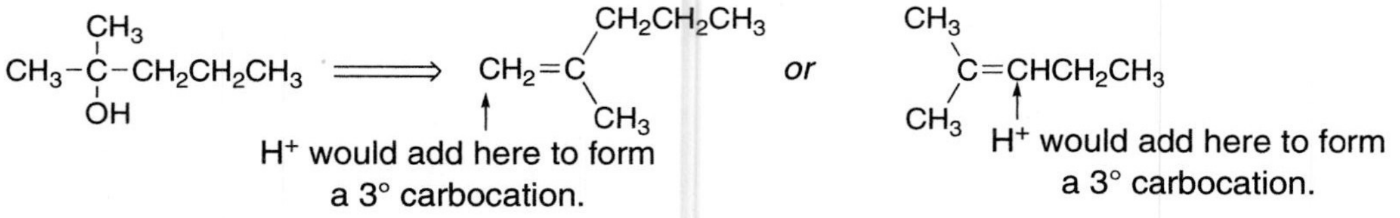

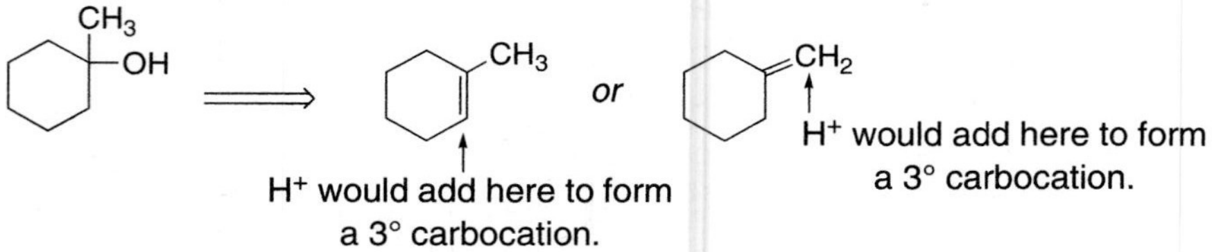

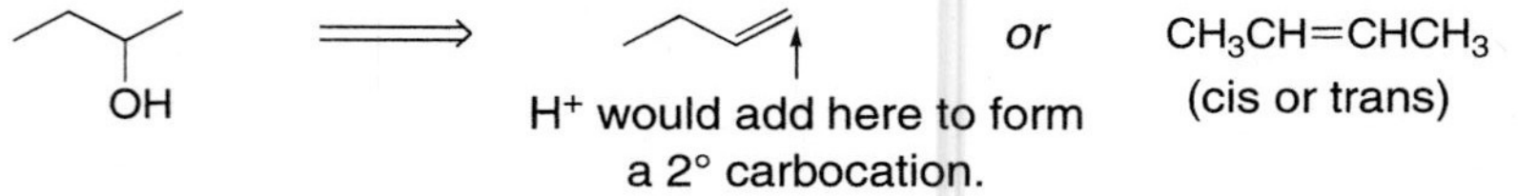

a.

H⁺ would add here to form a 3° carbocation.   *or*   H⁺ would add here to form a 3° carbocation.

b.

H⁺ would add here to form a 3° carbocation.   *or*   H⁺ would add here to form a 3° carbocation.

c.

H⁺ would add here to form a 2° carbocation.   *or* $CH_3CH=CHCH_3$ (cis or trans)

10.23

1-pentene H_2O / H_2SO_4 →   **enantiomers**

10.24 **The two steps in the mechanism for the halogenation of an alkene are:**
[1] Addition of X⁺ to the alkene to form a bridged halonium ion
[2] Nucleophilic attack by X⁻

transition state [1]:   transition state [2]:

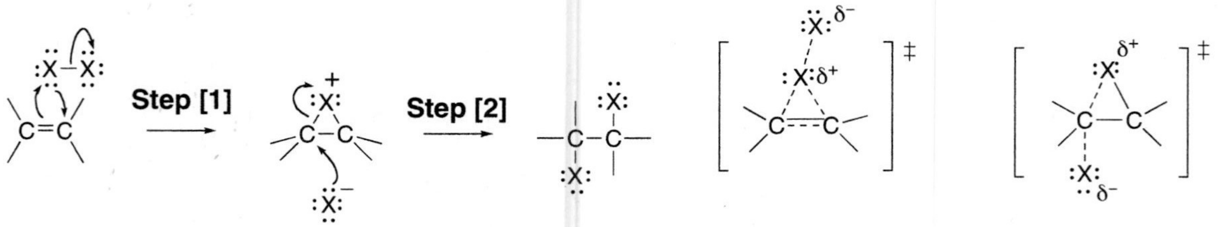

10.25 Halogenation of an alkene adds two elements of X in an anti fashion.

a.

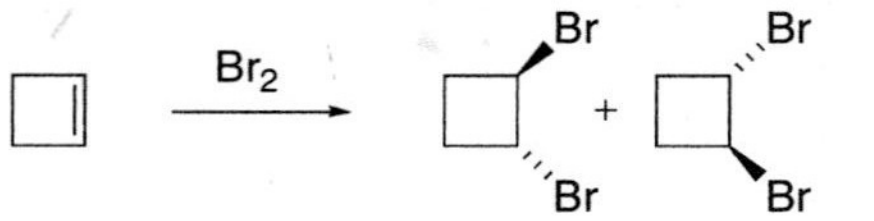

b.

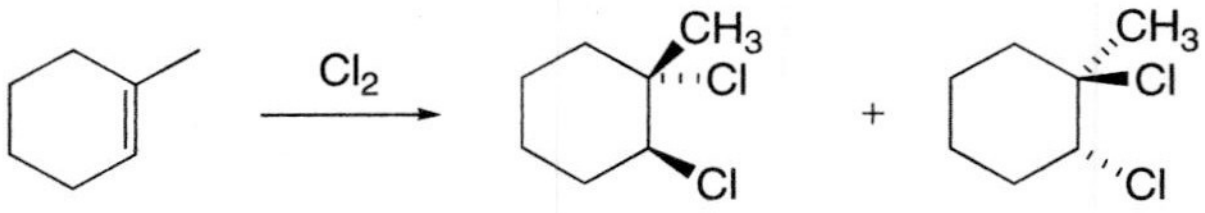

10.26 To draw the products of halogenation of an alkene, remember that the halogen adds to both ends of the double bond but only anti addition occurs.

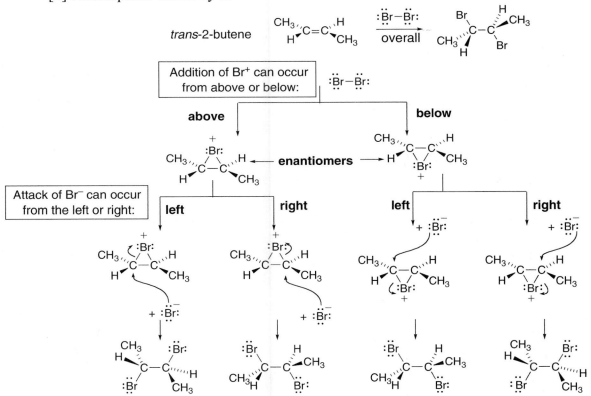

10.27 **The two steps in the mechanism for the halogenation of an alkene are:**
[1] Addition of X^+ to the alkene to form a bridged halonium ion
[2] Nucleophilic attack by X^-

10.28 Halohydrin formation adds the elements of X and OH across the double bond in an anti fashion. The reaction is regioselective so X ends up on the carbon that had more H's to begin with.

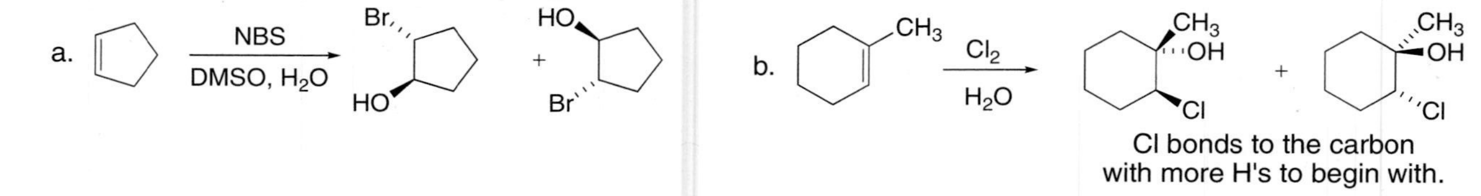

a.

b.

Cl bonds to the carbon with more H's to begin with.

10.29

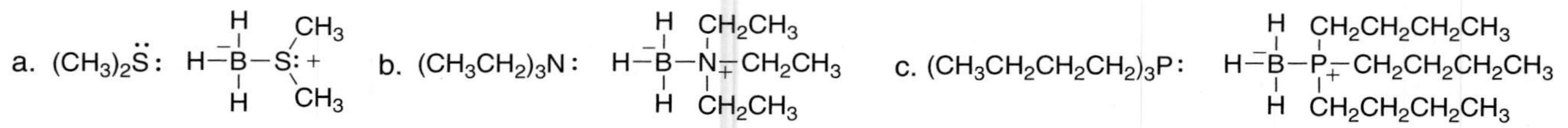

a. $(CH_3)_2\ddot{S}:$

b. $(CH_3CH_2)_3N:$

c. $(CH_3CH_2CH_2CH_2)_3P:$

10.30 In hydroboration the boron atom is the electrophile and becomes bonded to the carbon atom that had more H's to begin with.

a.

C with more H's.
B will add here.

c.

C with more H's.
B will add here.

b.

C with more H's.
B will add here.

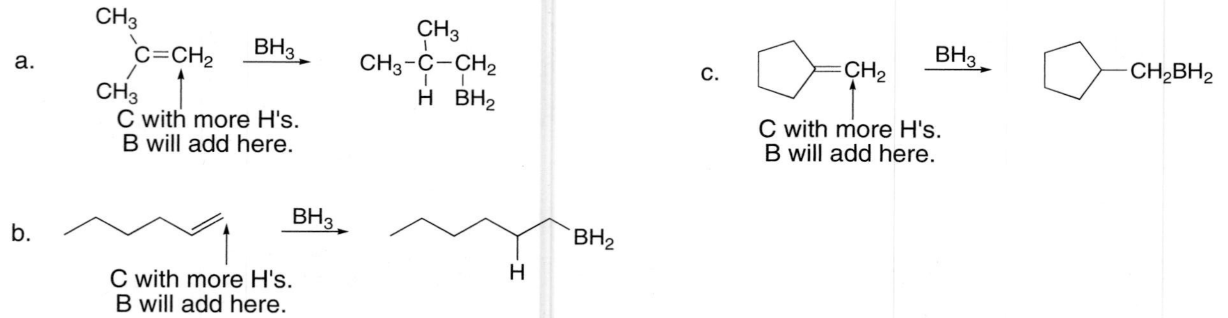

10.31 The hydroboration–oxidation reaction occurs in two steps:
[1] Syn addition of BH_3, with the boron on the less substituted carbon atom
[2] OH replaces the BH_2 with retention of configuration.

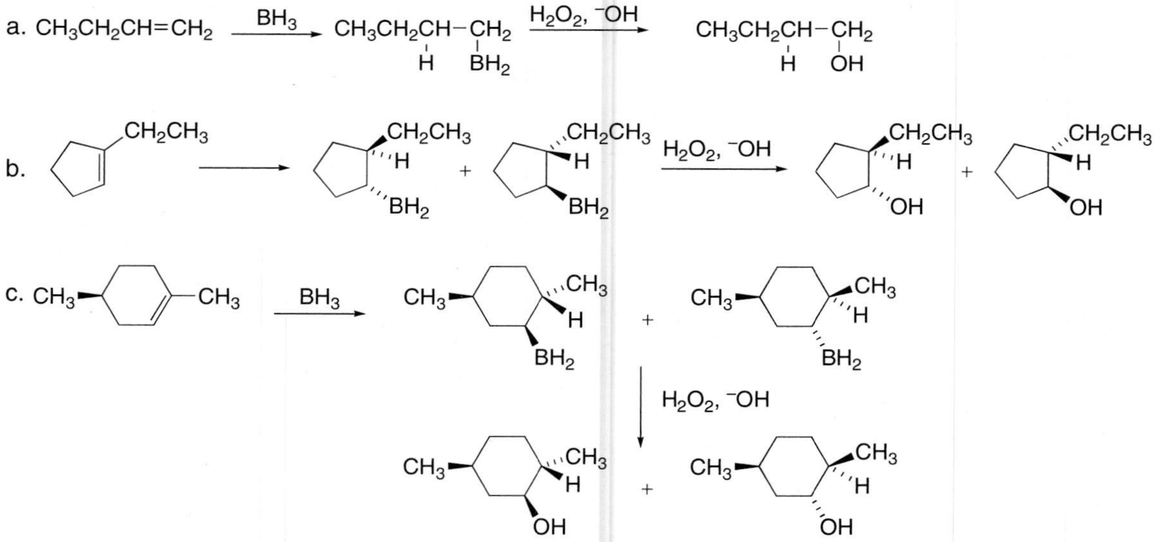

a.

b.

c.

10.32 Remember that hydroboration results in addition of OH on the less substituted C.

a.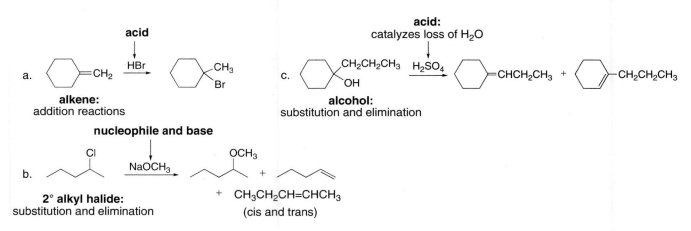

c.

b. (*E* or *Z* isomer can be used.)

10.33

a. $\overset{}{\bigcirc}=CH_2 \quad \xrightarrow[H_2SO_4]{H_2O}$ (OH, CH₃ product)   Hydration places the OH on the more substituted carbon.

$\overset{}{\bigcirc}=CH_2 \quad \xrightarrow[\text{[2] } H_2O_2, \text{ HO}^-]{\text{[1] } BH_3}$ (—CH₂OH product)   Hydroboration–oxidation places the OH on the less substituted carbon.

b. $\xrightarrow[H_2SO_4]{H_2O}$ (OH product)   Hydration places the OH on the more substituted carbon.

   $\xrightarrow[\text{[2] } H_2O_2, \text{ HO}^-]{\text{[1] } BH_3}$ (OH product)   Hydroboration–oxidation places the OH on the less substituted carbon.

c. $\xrightarrow[H_2SO_4]{H_2O}$ (OH product)   Hydration places the OH on the more substituted carbon.

   $\xrightarrow[\text{[2] } H_2O_2, \text{ HO}^-]{\text{[1] } BH_3}$ (HO product)   Hydroboration–oxidation places the OH on the less substituted carbon.

10.34 There are always two steps in this kind of question:
[1] **Identify the functional group and decide what types of reactions it undergoes** (for example, substitution, elimination, or addition).
[2] **Look at the reagent and determine if it is an electrophile, nucleophile, acid, or base.**

a.   **acid**   HBr   (CH₃, Br product)

alkene:
addition reactions

c.   **acid:** catalyzes loss of H_2O $CH_2CH_2CH_3$, OH $\xrightarrow{H_2SO_4}$ (=CHCH₂CH₃ + —CH₂CH₂CH₃ products)

alcohol:
substitution and elimination

b.   **nucleophile and base**   Cl $\xrightarrow{NaOCH_3}$ OCH₃ +   + $CH_3CH_2CH=CHCH_3$ (cis and trans)

2° alkyl halide:
substitution and elimination

10.35 To devise a synthesis:
[1] Look at the starting material and decide what reactions it can undergo.
[2] Look at the product and decide what reactions could make it.

a.

alkyl halide
Can undergo
substitution and elimination.

halohydrin:
Can form from an
alkene with Cl_2 and H_2O.

OH is added to
the more substituted C.

b.

alcohol
Can undergo
substitution and elimination.

alcohol
Can be formed by
substitution and addition.

(major product)

OH is added to
the less substituted C.

10.36 Use the directions from Answer 10.2 to calculate degrees of unsaturation.

a. C_3H_4
[1] maximum number of H's = $2n + 2 = 2(3) + 2 = 8$
[2] subtract actual from maximum = $8 - 4 = 4$
[3] divide by 2 = 4/2 = **2 degrees of unsaturation**

b. C_6H_8
[1] maximum number of H's = $2n + 2 = 2(6) + 2 = 14$
[2] subtract actual from maximum = $14 - 8 = 6$
[3] divide by 2 = 6/2 = **3 degrees of unsaturation**

c. $C_{40}H_{56}$
[1] maximum number of H's = $2n + 2 = 2(40) + 2 = 82$
[2] subtract actual from maximum = $82 - 56 = 26$
[3] divide by 2 = 26/2 = **13 degrees of unsaturation**

d. C_8H_8O
Ignore the O.
[1] maximum number of H's = $2n + 2 = 2(8) + 2 = 18$
[2] subtract actual from maximum = $18 - 8 = 10$
[3] divide by 2 = 10/2 = **5 degrees of unsaturation**

e. $C_{10}H_{16}O_2$
Ignore both O's.
[1] maximum number of H's = $2n + 2 = 2(10) + 2 = 22$
[2] subtract actual from maximum = $22 - 16 = 6$
[3] divide by 2 = 6/2 = **3 degrees of unsaturation**

f. C_8H_9Br
Because of Br, add one H (9 + 1 = 10 H's).
[1] maximum number of H's = $2n + 2 = 2(8) + 2 = 18$
[2] subtract actual from maximum = $18 - 10 = 8$
[3] divide by 2 = 8/2 = **4 degrees of unsaturation**

g. C_8H_9ClO
Ignore the O; count Cl as one more H (9 + 1 = 10 H's).
[1] maximum number of H's = $2n + 2 = 2(8) + 2 = 18$
[2] subtract actual from maximum = $18 - 10 = 8$
[3] divide by 2 = 8/2 = **4 degrees of unsaturation**

h. C_7H_9Br
Because of Br, add one H (9 + 1 = 10 H's).
[1] maximum number of H's = $2n + 2 = 2(7) + 2 = 16$
[2] subtract actual from maximum = $16 - 10 = 6$
[3] divide by 2 = 6/2 = **3 degrees of unsaturation**

i. $C_7H_{11}N$
Because of N, subtract one H (11 − 1 = 10 H's).
[1] maximum number of H's = $2n + 2 = 2(7) + 2 = 16$
[2] subtract actual from maximum = $16 - 10 = 6$
[3] divide by 2 = 6/2 = **3 degrees of unsaturation**

j. C_4H_8BrN
Because of Br, add one H, but subtract one for N
(8 + 1 − 1 = 8 H's).
[1] maximum number of H's = $2n + 2 = 2(4) + 2 = 10$
[2] subtract actual from maximum = $10 - 8 = 2$
[3] divide by 2 = 2/2 = **1 degree of unsaturation**

10.37 First determine the number of degrees of unsaturation in the compound. Then decide which combinations of rings and π bonds could exist.

$C_{10}H_{14}$
[1] maximum number of H's = $2n + 2 = 2(10) + 2 = 22$
[2] subtract actual from maximum = $22 - 14 = 8$
[3] divide by two = 8/2 = **4 degrees of unsaturation**

possibilities:
  4 π bonds
  3 π bonds + 1 ring
  2 π bonds + 2 rings
  1 π bond + 3 rings
  4 rings

10.38 The statement is incorrect because in naming isomers with more than two groups on a double bond, one must use an *E/Z* label, rather than a cis/trans label.

higher priority → Cl

higher priority

N(CH₂CH₃)₂

enclomiphene
E isomer

higher priority

higher priority

N(CH₂CH₃)₂

zuclomiphene
Z isomer

10.39 Name the alkenes using the rules in Answers 10.4 and 10.6.

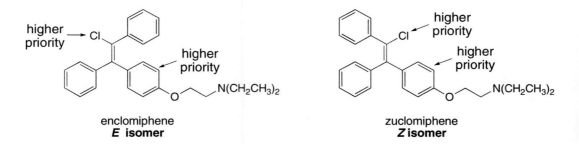

a.   CH₂=CHCH₂CH(CH₃)CH₂CH₃

6 C chain with a double bond =
hexene

CH₂=CHCH₂CHCH₂CH₃ with CH₃
1-hexene 4-methyl

4-methyl-1-hexene

b.

8 C chain with a double bond =
octene

2-methyl →
5-ethyl
2-octene

5-ethyl-2-methyl-2-octene

c.

5 C chain with a double bond =
pentene

2-isopropyl
1-pentene
4-methyl

2-isopropyl-4-methyl-1-pentene

d.
CH₃   CH₃
   C=C
H   CH₂CH(CH₃)₂

6 C chain with a double bond =
hexene

CH₃   CH₃ ← **3-methyl**
  C=C   CH₃ ← **5-methyl**
H   CH₂CHCH₃
2-hexene

higher priority → CH₃   CH₃ **(2E)-3,5-dimethyl-2-hexene**
  C=C
H   CH₂CH(CH₃)₂ ← higher priority
Higher priority groups
are on opposite sides =
***E* alkene.**

e.

6 C ring with a double bond =
cyclohexene

1-ethyl
5-isopropyl

1-ethyl-5-isopropylcyclohexene

Chapter 10–16

f.

5 C ring with a double bond =
cyclopentene

2-methyl ⟶

1-*sec*-butyl

1-*sec*-butyl-2-methylcyclopentene

g.

OH

7 C chain with a double bond =
heptene

4-isopropyl ⟶

E **double bond** (higher priority groups on
opposite sides with bold bonds)

3-ol ⟶ OH

4-heptene

(4*E*)-4-isopropyl-4-hepten-3-ol

h.

OH

6 C ring with a double bond =
cyclohexene

2-cyclohexene

1-ol

5-*sec*-butyl

5-*sec*-butyl-2-cyclohexenol

10.40 Use the directions from Answer 10.8.

a. (3*E*)-4-ethyl-3-**heptene**

7 carbons

4-ethyl ⟶

3

Higher priority groups on
opposite sides = *E*.

b. 3,3-dimethyl**cyclopentene**

5 carbon ring

3,3-dimethyl

1

c. *cis*-4-**octene**

8 carbons

4

Higher priority groups on
the same side = cis.

d. 4-vinyl**cyclopentene**

5 carbon ring

2 1

4 ⟶

e. (2*Z*)-3-isopropyl-2-**heptene**

7 carbons

2

3-isopropyl

Higher priority groups on
the same side = *Z*.

f. *cis*-3,4-dimethyl**cyclopentene**

5 carbon ring

1

4 3

3,4-dimethyl

or

g. *trans*-2-**heptene**

7 carbons

2

Higher priority groups on
opposite sides = trans.

4-propyl

1-isopropyl

h. 1-isopropyl-4-propyl**cyclohexene**

6 carbon ring

10.41

a.

(2*E*,4*S*)-4-methyl-2-nonene
A

(2*E*,4*R*)-4-methyl-2-nonene
B

(2*Z*,4*S*)-4-methyl-2-nonene
C

(2*Z*,4*R*)-4-methyl-2-nonene
D

b. **A** and **B** are enantiomers. **C** and **D** are enantiomers.
c. Pairs of diastereomers: **A** and **C**, **A** and **D**, **B** and **C**, **B** and **D**.

10.42

a.

(1*E*,4*R*)-1,4-dimethylcyclodecene

b.

(1*E*,4*S*)-1,4-dimethylcyclodecene
enantiomer

c.

(1*Z*,4*S*)-1,4-dimethylcyclodecene
diastereomer

(1*Z*,4*R*)-1,4-dimethylcyclodecene
diastereomer

10.43 Name the alkene from which the epoxide can be derived and add the word *oxide*.

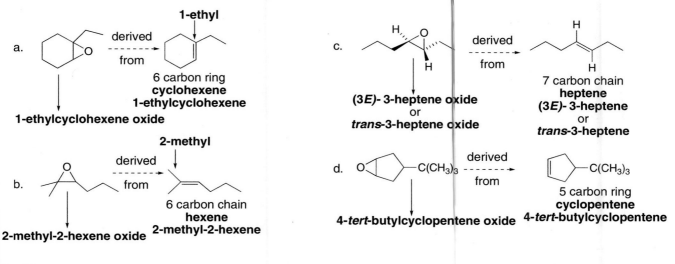

a.
1-ethyl
derived
from
6 carbon ring
cyclohexene
1-ethylcyclohexene

1-ethylcyclohexene oxide

b.
2-methyl
derived
from
6 carbon chain
hexene
2-methyl-2-hexene

2-methyl-2-hexene oxide

c.
derived
from
7 carbon chain
heptene
(3*E*)- 3-heptene
or
***trans*-3-heptene**

(3*E*)- 3-heptene oxide
or
***trans*-3-heptene oxide**

d. O—C(CH₃)₃
derived
from
—C(CH₃)₃
5 carbon ring
cyclopentene
4-*tert*-butylcyclopentene

4-*tert*-butylcyclopentene oxide

10.44

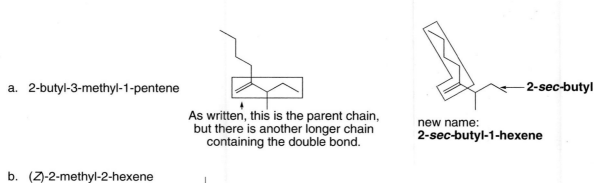

a. 2-butyl-3-methyl-1-pentene

As written, this is the parent chain,
but there is another longer chain
containing the double bond.

2-***sec*-butyl**

new name:
2-*sec*-butyl-1-hexene

b. (*Z*)-2-methyl-2-hexene

Two groups on one end of the C=C
are the same (2 CH₃'s), so no *E* and *Z* isomers are possible.

new name:
2-methyl-2-hexene

c. (*E*)-1-isopropyl-1-butene

As written, this is the parent chain, but there is another longer chain containing the double bond.

new name:
(3*E*)-2-methyl-3-hexene

d. 5-methylcyclohexene

As written the methyl is at C5. Re-number to put it at C4.

new name:
4-methylcyclohexene

e. 4-isobutyl-2-methylcylohexene

As written this methyl is at C2. Re-number to put it at C1.

new name:
5-isobutyl-1-methylcyclohexene

f. 1-*sec*-butyl-2-cyclopentene

This has the double bond between C2 and C3. Cycloalkenes must have the double bond between C1 and C2. Re-number.

new name:
3-*sec*-butylcyclopentene

g. 1-cyclohexen-4-ol

The numbering is incorrect. When a compound contains both a double bond and an OH group, number the C skeleton to give the OH group the lower number.

3-cyclohexenol (The "1" can be omitted.)

h. 3-ethyl-3-octen-5-ol

The numbering is incorrect. When a compound contains both a double bond and an OH group, number the C skeleton to give the OH group the lower number.

6-ethyl-5-octen-4-ol

10.45

a. and b.

bongkrekic acid

c. Since there are 7 double bonds and 2 tetrahedral stereogenic centers, $2^9 = 512$ possible stereoisomers.

10.46

2-methyl-1-pentene   (2*E*)-4-methyl-2-pentene   4-methyl-1-pentene   (2*E*)-3-methyl-2-pentene   (3*R*)-3-methyl-1-pentene

2-methyl-2-pentene   (2*Z*)-4-methyl-2-pentene   2-ethyl-1-butene   (2*Z*)-3-methyl-2-pentene   (3*S*)-3-methyl-1-pentene

10.47

stearic acid

hIghest melting point
no double bonds

elaidic acid

intermediate melting point
one *E* double bond

oleic acid

lowest melting point
one *Z* double bond

10.48

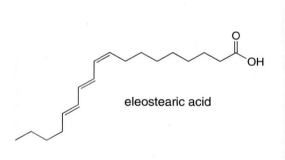

eleostearic acid

a.

all trans double bonds
higher melting point

b.

all cis double bonds
lower melting point

10.49

a.

A

A has one tetrahedral stereogenic center, labeled with an asterisk [*].

B

b.

R

H

S

H

10.50

transition state [1]:

$$\left[\begin{array}{c} :\overset{..}{Br}: \; \delta^- \\ :\overset{..}{Br}: \; \delta^+ \\ H \quad \quad H \\ C = C \\ H \quad \quad H \end{array} \right]^{\ddagger}$$

transition state [2]:

$$\left[\begin{array}{c} \delta^+ \\ :Br: \\ H \quad \quad H \\ C - C \\ H \quad \quad H \\ :\overset{..}{Br}: \\ \delta^- \end{array} \right]^{\ddagger}$$

Energy — Reaction coordinate

[1]

[2]

A

CH₂=CH₂ + Br₂

BrCH₂CH₂Br

10.51 The more negative the $\Delta H°$, the larger the K_{eq} assuming entropy changes are comparable. Calculate the $\Delta H°$ for each reaction and compare.

$$CH_2=CH_2 \; + \; HI \longrightarrow CH_3CH_2-I$$

[1] Bonds broken

	$\Delta H°$ (kJ/mol)
C–C π bond	+ 267
H–I	+ 297
Total	+ 564 kJ/mol

[2] Bonds formed

	$\Delta H°$ (kJ/mol)
CH₂ICH₂–H	– 410
C–I	– 222
Total	– 632 kJ/mol

[3] Overall $\Delta H°$ =

sum in Step [1]
+
sum in Step [2]

+ 564 kJ/mol

– 632 kJ/mol

– 68 kJ/mol

$$CH_2{=}CH_2 \ + \ HCl \ \longrightarrow \ CH_3CH_2{-}Cl$$

[1] Bonds broken		**[2] Bonds formed**		**[3] Overall $\Delta H^\circ =$**

	ΔH° (kJ/mol)		ΔH° (kJ/mol)
C–C π bond	+ 267	$CH_2ClCH_2{-}H$	– 410
H–Cl	+ 431	C–Cl	– 339
Total	+ 698 kJ/mol	Total	– 749 kJ/mol

[3] Overall $\Delta H^\circ =$

sum in Step [1]
+
sum in Step [2]

+ 698	kJ/mol
– 749	kJ/mol

– 51 kJ/mol

Compare the ΔH°:
Addition of HI: **–68 kJ/mol** more negative ΔH°, larger K_{eq}.
Addition of HCl: **–51 kJ/mol**

10.52

a. cyclohexene + HBr → 1-bromocyclohexane (H, Br)

f. cyclohexene + Br₂, H₂O → trans-2-bromocyclohexanol (Br, OH) + enantiomer (Br, OH)

b. cyclohexene + HI → 1-iodocyclohexane (H, I)

g. cyclohexene + NBS, aqueous DMSO → trans-2-bromocyclohexanol (Br, OH) + enantiomer (Br, OH)

c. cyclohexene + H₂O / H₂SO₄ → cyclohexanol (H, OH)

h. cyclohexene + [1] BH₃ [2] H₂O₂, HO⁻ → cyclohexanol (H, OH)

d. cyclohexene + CH₃CH₂OH / H₂SO₄ → ethoxycyclohexane (H, OCH₂CH₃)

i. cyclohexene + [1] 9-BBN [2] H₂O₂, HO⁻ → cyclohexanol (H, OH)

e. cyclohexene + Cl₂ → trans-1,2-dichlorocyclohexane (Cl, Cl) + enantiomer (Cl, Cl)

10.53

a. $(CH_3)_2C{=}CH_2$ + HBr → $CH_3{-}\underset{\underset{Br}{|}}{\overset{\overset{CH_3}{|}}{C}}{-}CH_3$

f. $(CH_3)_2C{=}CH_2$ + Br₂, H₂O → $CH_3{-}\underset{\underset{OH}{|}}{\overset{\overset{CH_3}{|}}{C}}{-}CH_2Br$

b. $(CH_3)_2C{=}CH_2$ + HI → $CH_3{-}\underset{\underset{I}{|}}{\overset{\overset{CH_3}{|}}{C}}{-}CH_3$

g. $(CH_3)_2C{=}CH_2$ + NBS, aqueous DMSO → $CH_3{-}\underset{\underset{OH}{|}}{\overset{\overset{CH_3}{|}}{C}}{-}CH_2Br$

c. $(CH_3)_2C{=}CH_2$ + H₂O / H₂SO₄ → $CH_3{-}\underset{\underset{OH}{|}}{\overset{\overset{CH_3}{|}}{C}}{-}CH_3$

h. $(CH_3)_2C{=}CH_2$ + [1] BH₃ [2] H₂O₂, HO⁻ → $CH_3{-}\underset{\underset{H}{|}}{\overset{\overset{CH_3}{|}}{C}}{-}CH_2OH$

d. $(CH_3)_2C{=}CH_2$ + CH₃CH₂OH / H₂SO₄ → $CH_3{-}\underset{\underset{OCH_2CH_3}{|}}{\overset{\overset{CH_3}{|}}{C}}{-}CH_3$

i. $(CH_3)_2C{=}CH_2$ + [1] 9-BBN [2] H₂O₂, HO⁻ → $CH_3{-}\underset{\underset{H}{|}}{\overset{\overset{CH_3}{|}}{C}}{-}CH_2OH$

e. $(CH_3)_2C{=}CH_2$ + Cl₂ → $CH_3{-}\underset{\underset{Cl}{|}}{\overset{\overset{CH_3}{|}}{C}}{-}CH_2Cl$

10.54

a. 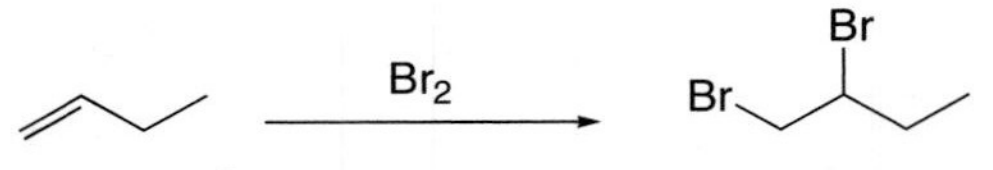 Br₂ → **Halogenation**

b. 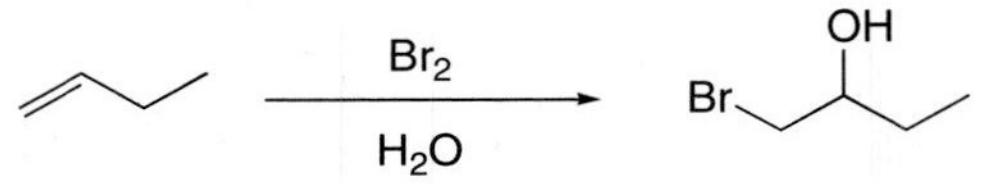 Br₂ / H₂O → **Halohydrin formation**: Br adds to the C that had more H's to begin with.

c. 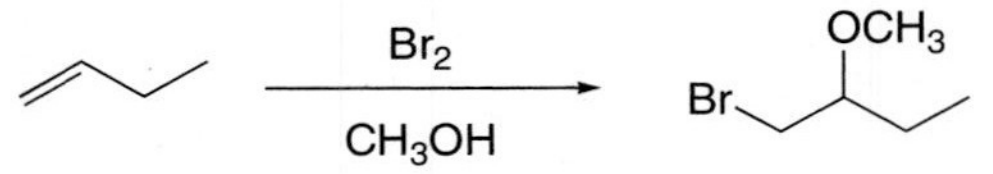 Br₂ / CH₃OH → Same as halohydrin formation, except CH₃OH in place of H₂O.

10.55

a.

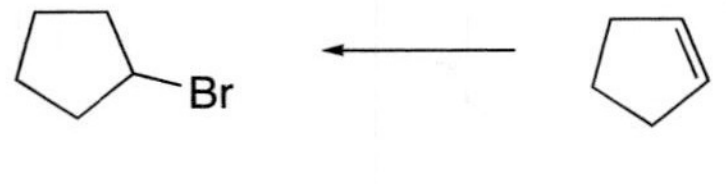

b.

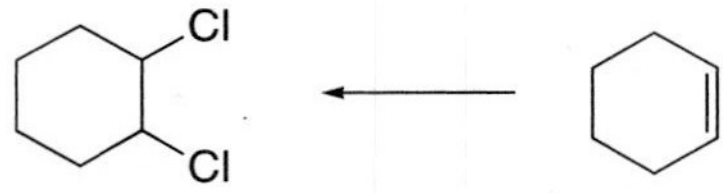

c.

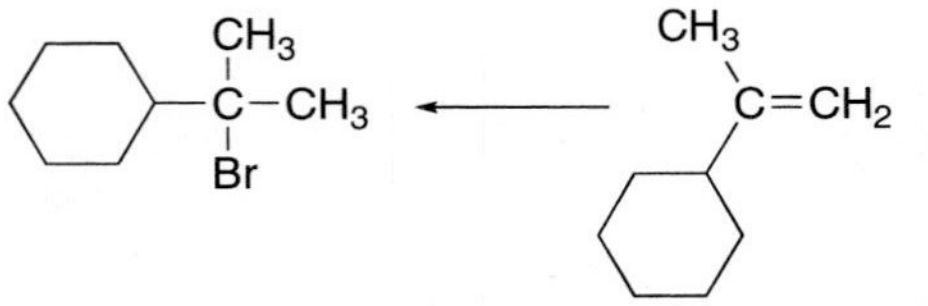

d.

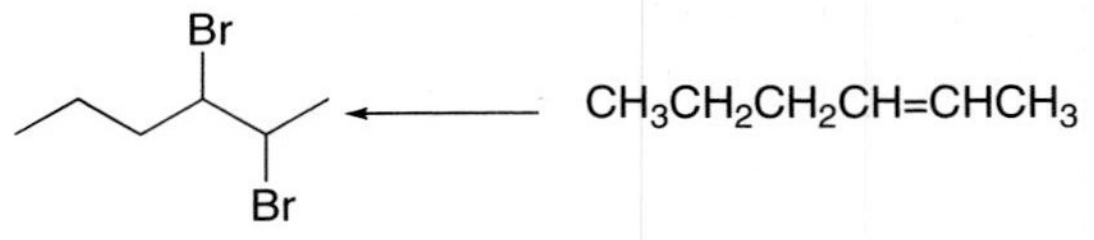

e.

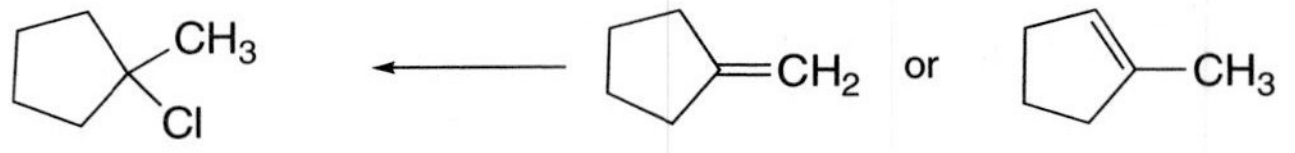

f. 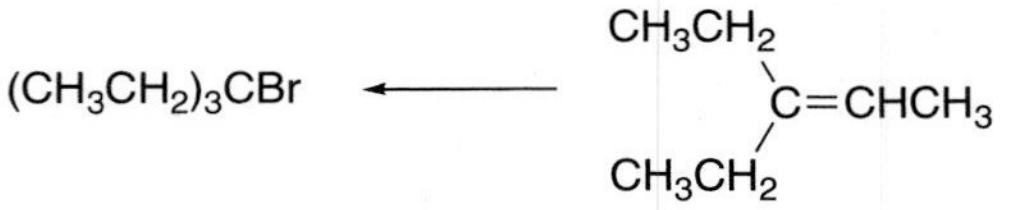 $(CH_3CH_2)_3CBr$ ⟵

10.56 Hydroboration–oxidation results in addition of an OH group on the less substituted carbon, whereas acid-catalyzed addition of H₂O results in the addition of an OH group on the more substituted carbon.

a.

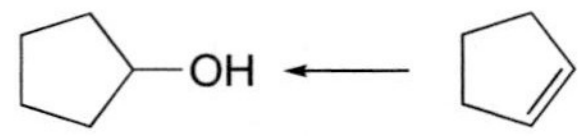

hydroboration–oxidation
and
acid-catalyzed addition

b.

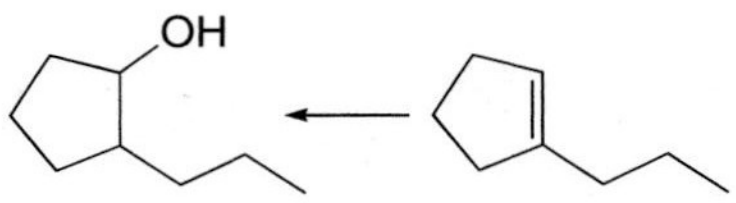

hydroboration–oxidation

c.

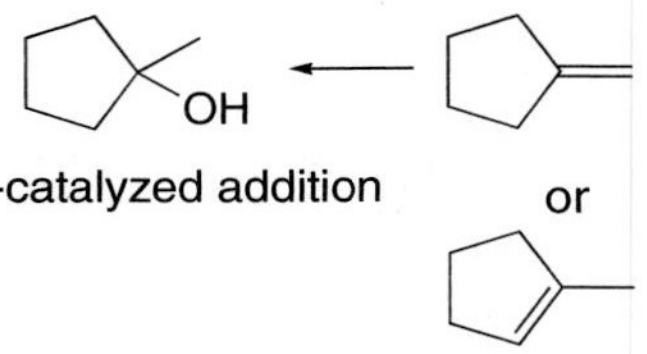

acid-catalyzed addition   or

d.

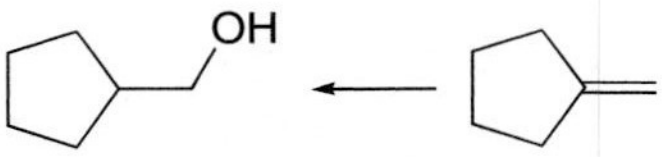

hydroboration–oxidation

e.

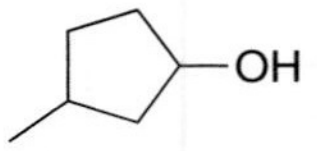

Both methods would give
product mixtures.

10.57

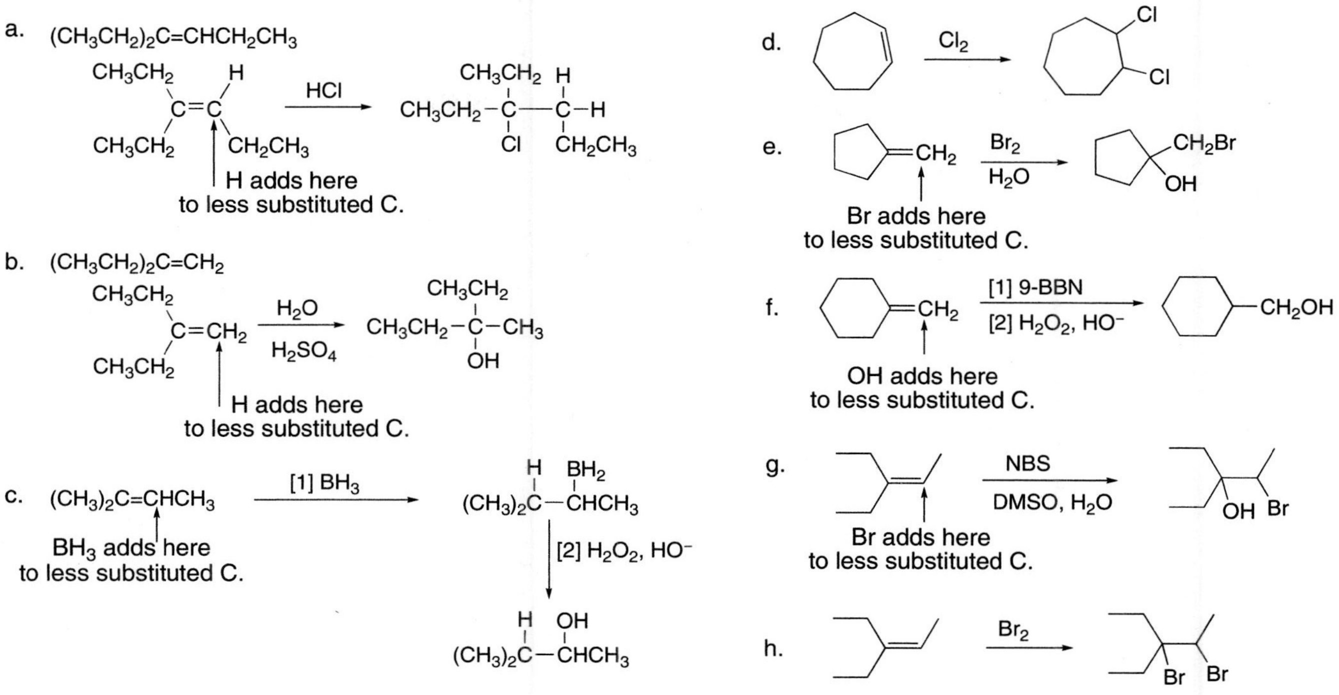

a. $(CH_3CH_2)_2C=CHCH_2CH_3$

HCl

H adds here
to less substituted C.

b. $(CH_3CH_2)_2C=CH_2$

H_2O
H_2SO_4

H adds here
to less substituted C.

c. $(CH_3)_2C=CHCH_3$

BH$_3$ adds here
to less substituted C.

[1] BH$_3$

[2] H_2O_2, HO$^-$

d.

Cl$_2$

e.

Br_2
H_2O

Br adds here
to less substituted C.

f.

[1] 9-BBN
[2] H_2O_2, HO$^-$

OH adds here
to less substituted C.

g.

NBS
DMSO, H_2O

Br adds here
to less substituted C.

h.

Br_2

10.58

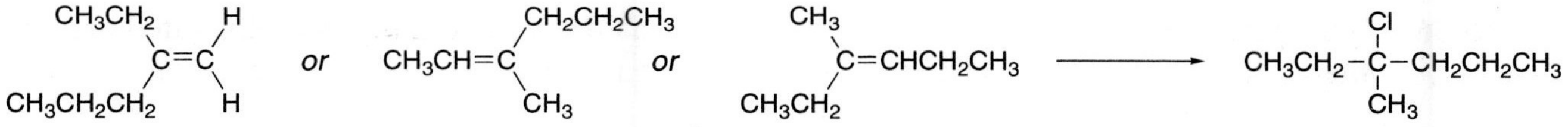

or or → $CH_3CH_2-\overset{Cl}{\underset{CH_3}{C}}-CH_2CH_2CH_3$

10.59

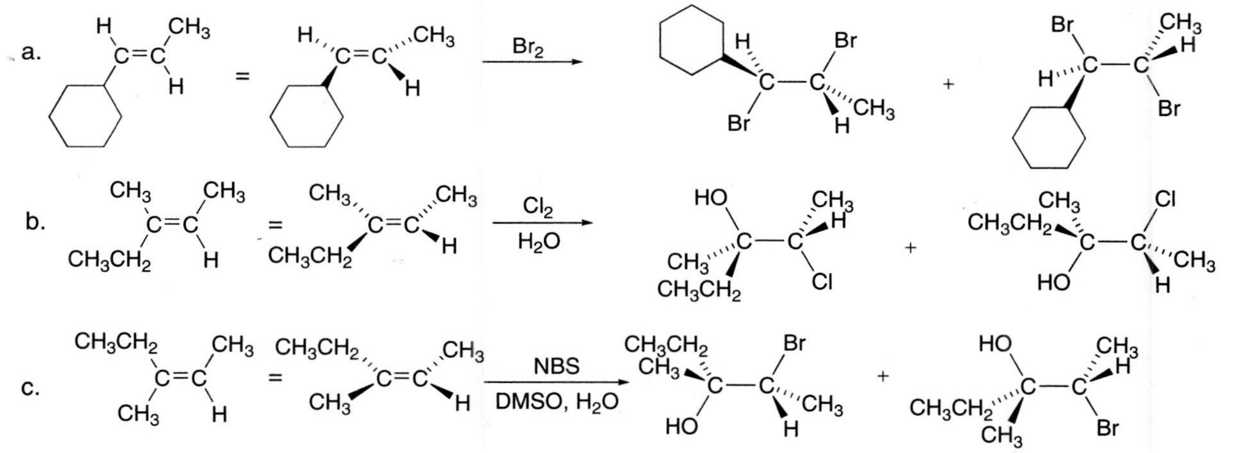

a.

=

Br_2

+

b.

=

Cl_2
H_2O

+

c.

=

NBS
DMSO, H_2O

+

10.60

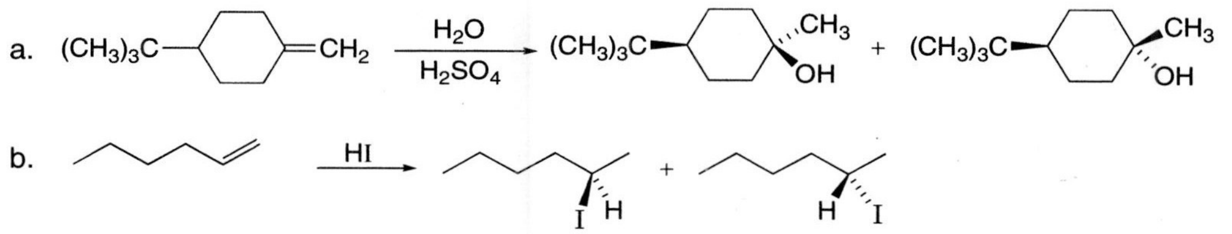

a. $(CH_3)_3C$—

H_2O
H_2SO_4

+

b.

HI

+

c. 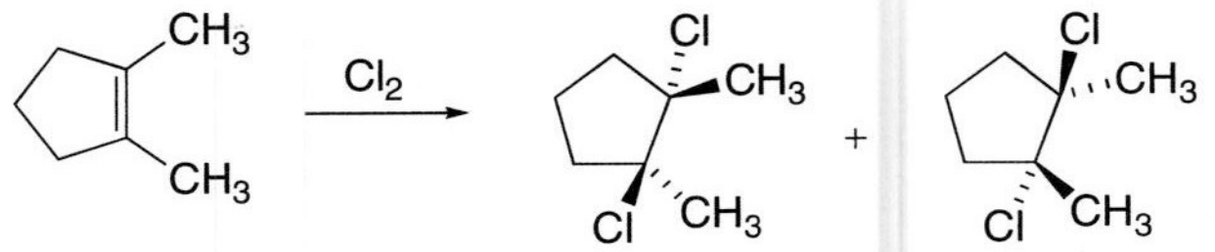

CH_3
CH_3
Cl_2
Cl
CH_3
Cl
CH_3
+
Cl
CH_3
Cl
CH_3

only anti addition

d. 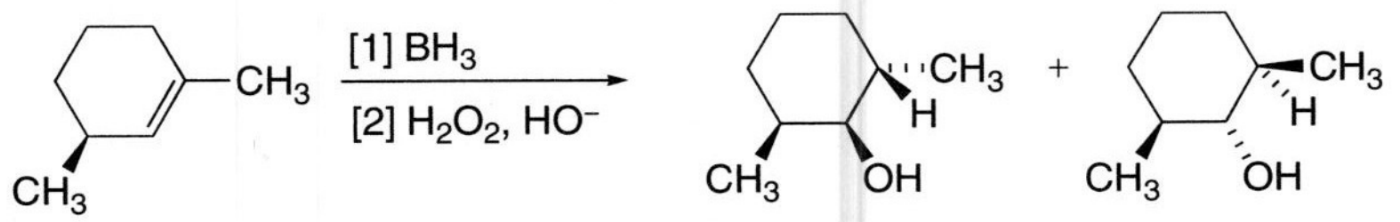

CH_3
CH_3
[1] BH_3
[2] H_2O_2, HO^-
CH_3
H
CH_3
OH
+
CH_3
H
CH_3
OH

only syn addition

e. 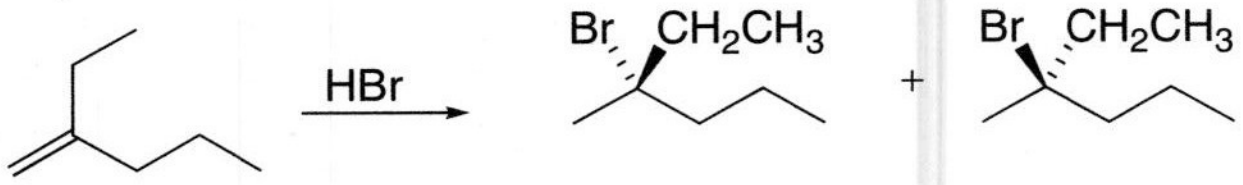

HBr
Br
CH_2CH_3
+
Br
CH_2CH_3

f. 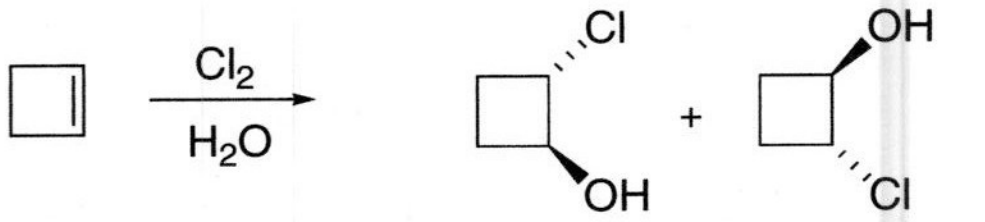

Cl_2
H_2O
Cl
OH
+
OH
Cl

only anti addition

g. 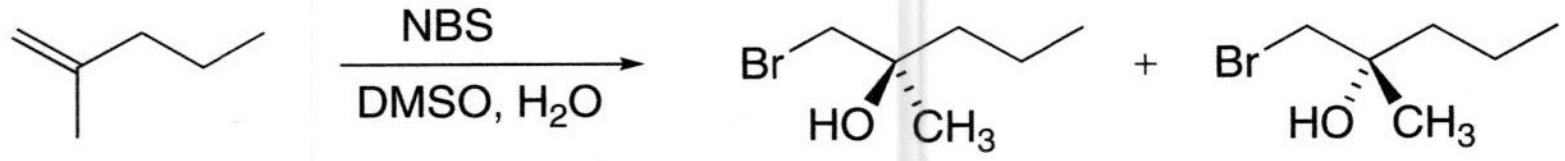

NBS
DMSO, H_2O
Br
HO
CH_3
+
Br
HO
CH_3

h. 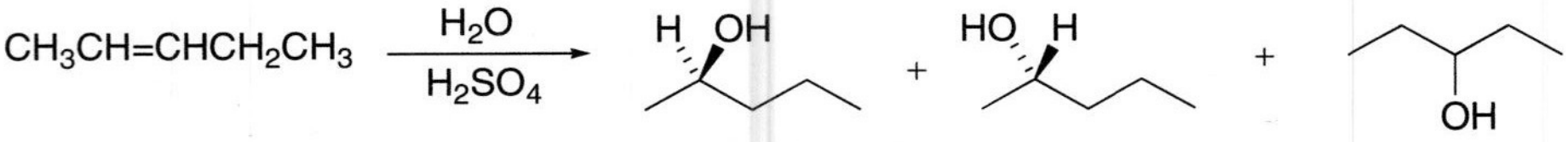

$CH_3CH{=}CHCH_2CH_3$
H_2O
H_2SO_4
H
OH
+
HO
H
+
OH

10.61

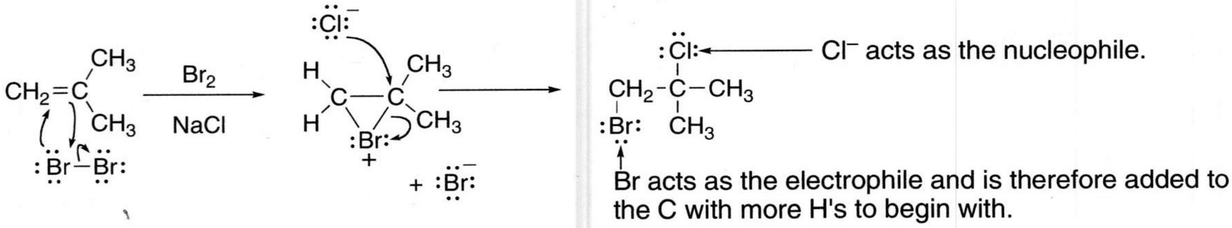

$CH_2{=}C$
CH_3
CH_3
:Br–Br:
Br_2
NaCl
:Cl:⁻
H
H
CH_3
CH_3
:Br:
+
+ :Br:⁻
:Cl:
CH_2–C–CH_3
:Br:
CH_3
Cl⁻ acts as the nucleophile.
Br acts as the electrophile and is therefore added to the C with more H's to begin with.

10.62 Draw each reaction. (a) The cis isomer of 4-octene gives two enantiomers on addition of Br_2. (b) The trans isomer gives a meso compound.

a. 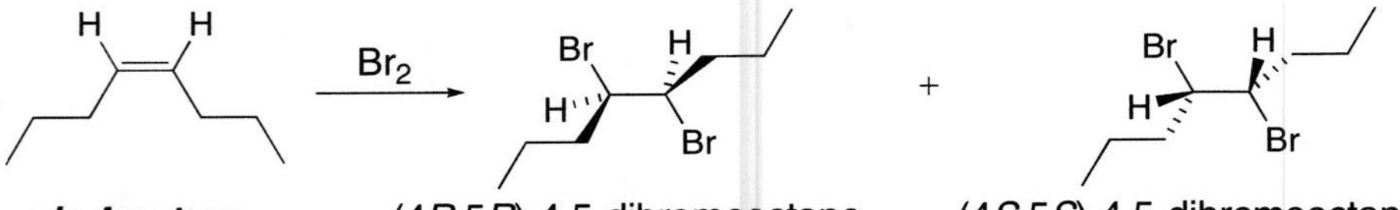

H
H
Br_2
Br
H
H
Br
+
Br
H
H
Br

***cis*-4-octene** (4*R*,5*R*)-4,5-dibromooctane (4*S*,5*S*)-4,5-dibromooctane

enantiomers

b. 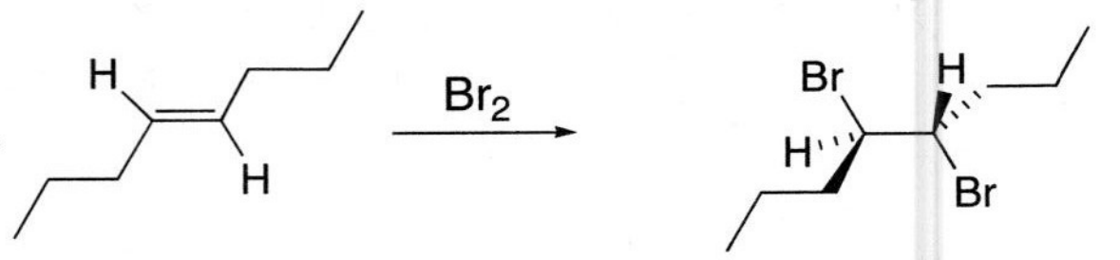

H
H
Br_2
Br
H
H
Br

***trans*-4-octene** (4*R*,5*S*)-4,5-dibromooctane

meso compound

10.63

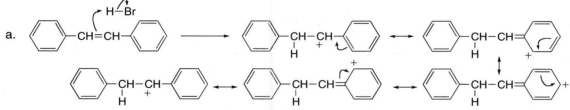

By protonation of the alkene, the cis and trans isomers produce **identical** carbocation intermediates.

Both *cis*- and *trans*-3-hexene give the same racemic mixture of products, so the reaction is not stereospecific.

10.64 The alkene that forms the more stable carbocation reacts faster, according to the Hammond postulate.

a.

This 2° carbocation is resonance stabilized, making it more stable, so the starting alkene reacts **faster** with HBr.

$$CH_3-CH{=}CH-CH_3 \longrightarrow CH_3-\overset{}{C}H-\overset{+}{C}H-CH_3$$

**2° carbocation
no resonance stabilization**

b.

$$CH_2{=}C\overset{CH_3}{\underset{CH_3}{}} \quad \xrightarrow[\text{addition}]{\text{Markovnikov}} \quad CH_2-\overset{+}{C}(CH_3)_2 \quad \textbf{faster}$$

3° carbocation

$$CH_2{=}C\overset{CH_3}{\underset{CH_2OCH_3}{}} \quad \xrightarrow[\text{addition}]{\text{Markovnikov}} \quad CH_2-\overset{CH_3}{\underset{CH_2-OCH_3}{C}}+$$

3° carbocation

This carbocation is still 3°, but the nearby electronegative O atom withdraws electron density from the carbocation, destabilizing it. Thus, the reaction to form this carbocation occurs more slowly.

10.65

a.

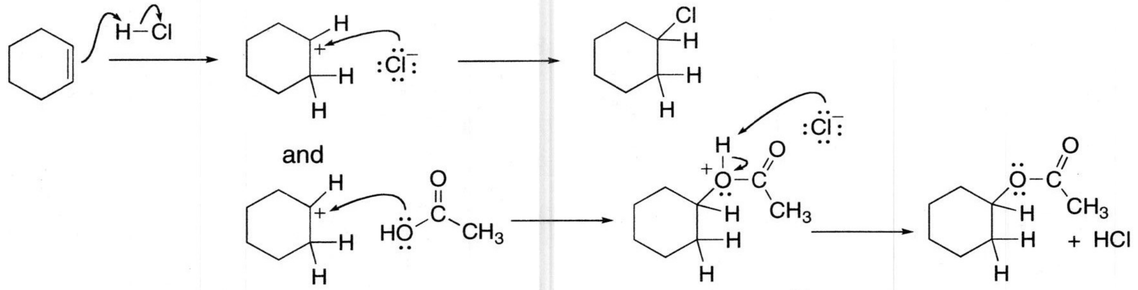

and

b.

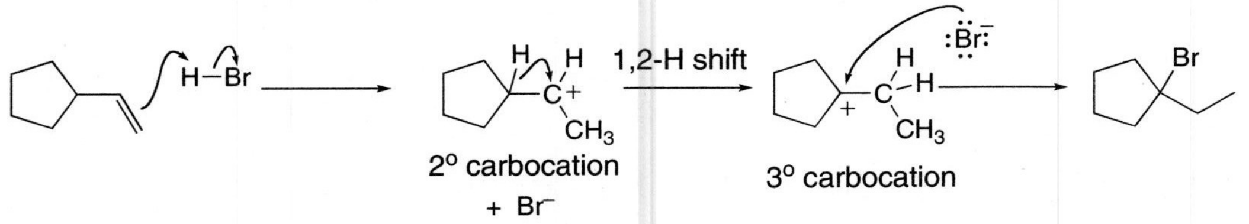

2° carbocation
+ Br⁻

3° carbocation

10.66

a.

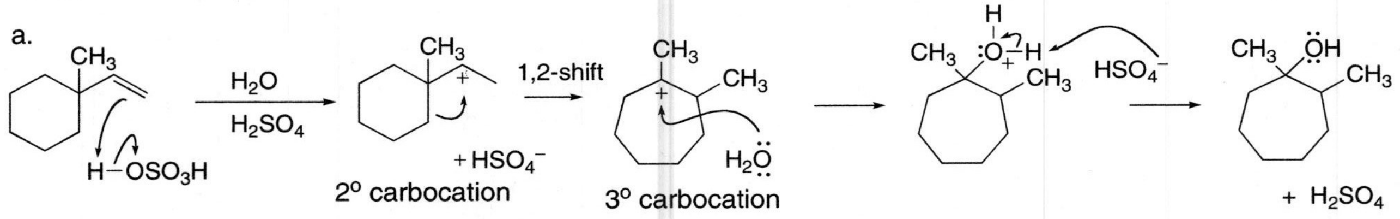

2° carbocation
+ HSO₄⁻

3° carbocation

+ H₂SO₄

b.

+ HSO₄⁻

+ H₂SO₄

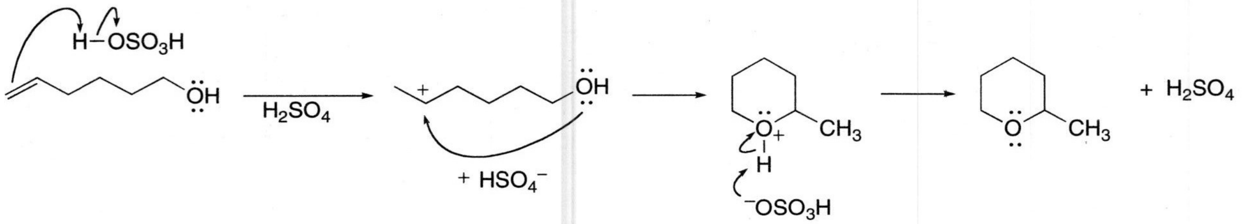

10.67

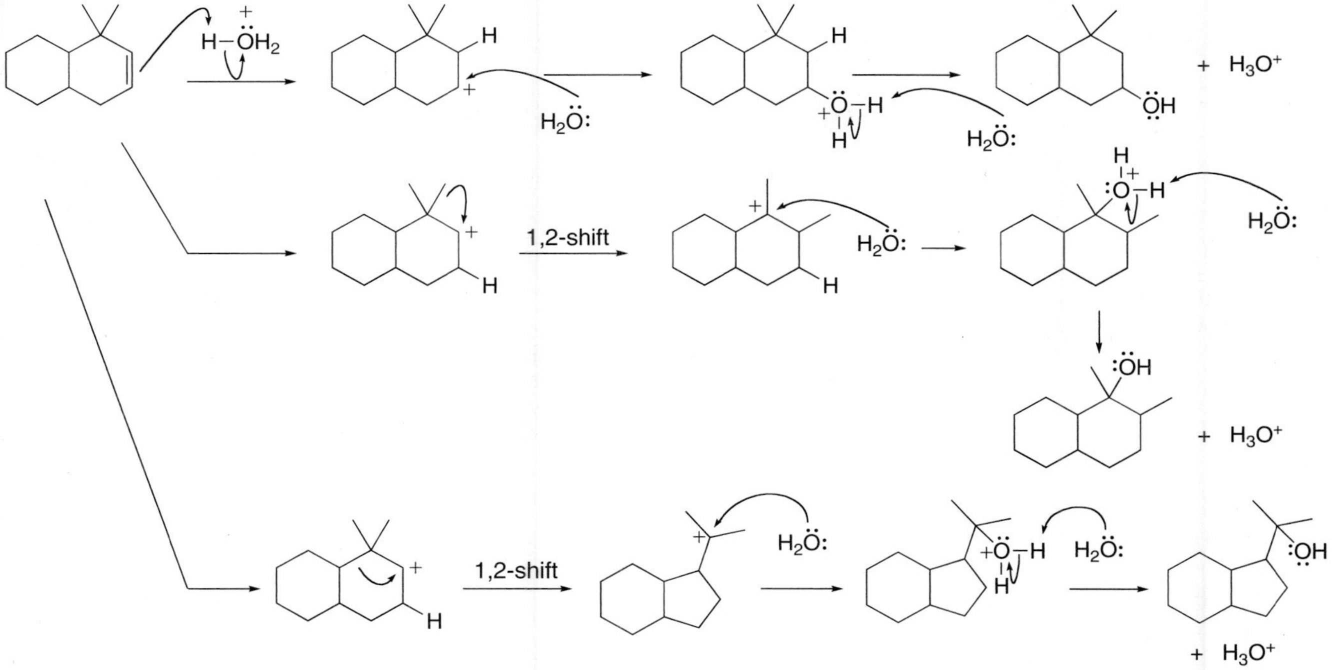

10.68 The isomerization reaction occurs by protonation and deprotonation.

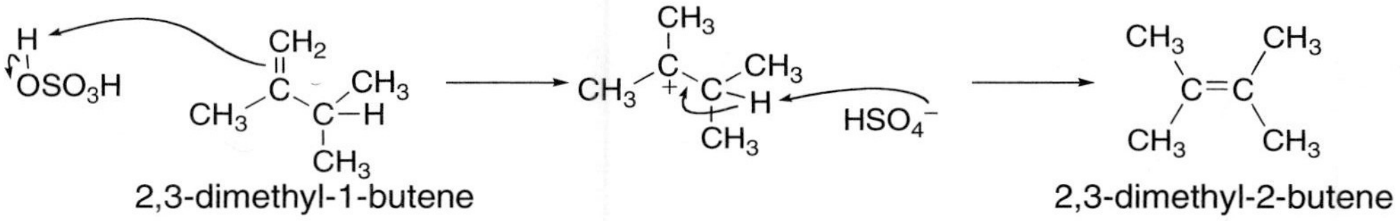

2,3-dimethyl-1-butene   2,3-dimethyl-2-butene

10.69

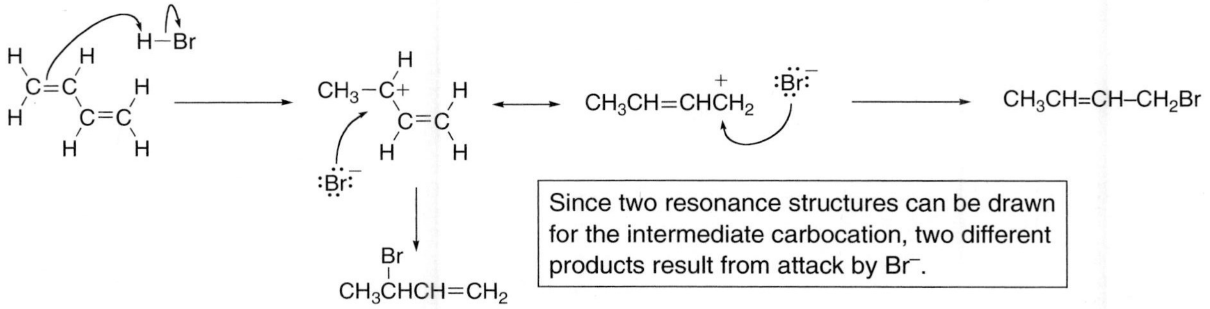

Since two resonance structures can be drawn for the intermediate carbocation, two different products result from attack by Br⁻.

10.70

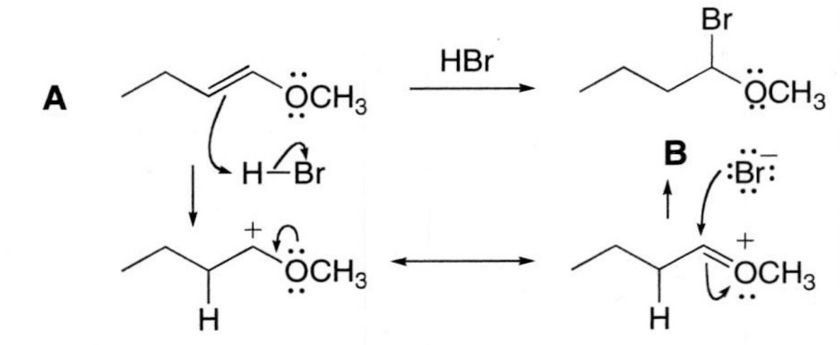

This carbocation is resonance stabilized by the O atom, and therefore preferentially forms and results in **B**.

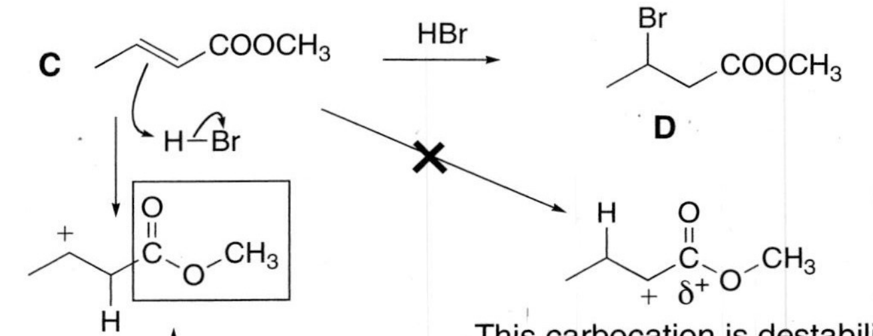

This carbocation is formed preferentially and results in product **D**. It is not destabilized by an adjacent electron-withdrawing $COOCH_3$ group.

This carbocation is destabilized by the δ^+ on the adjacent C, so it does not form.

10.71

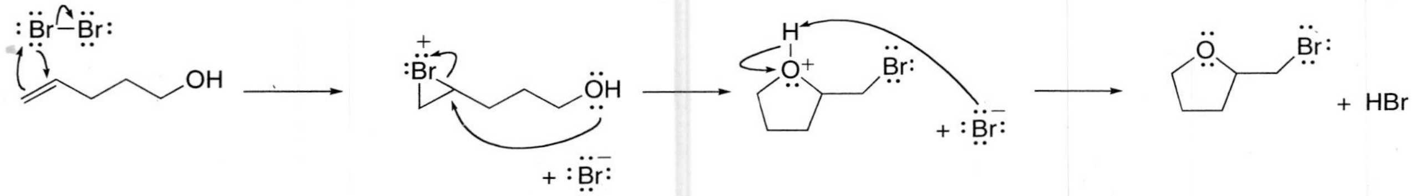

10.72

a.

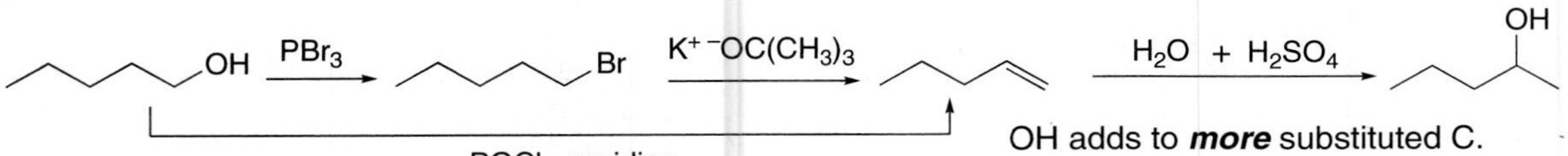

OH adds to ***more*** substituted C.

b. $CH_3-CBr(H)-CH_3 \xrightarrow{K^+\ {}^-OC(CH_3)_3} CH_3-CH{=}CH_2 \xrightarrow{Br_2} CH_3-CBr(H)-CH_2Br$

c.

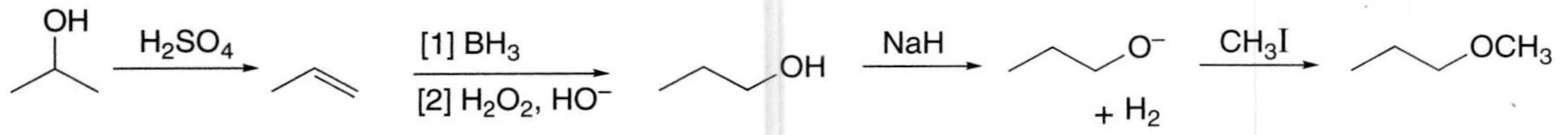

d. $CH_3-CH(CH_3)-CH_2I \xrightarrow{K^+\ {}^-OC(CH_3)_3} (CH_3)_2C{=}CH_2 \xrightarrow{HCl} CH_3-C(CH_3)_2-Cl$

e.

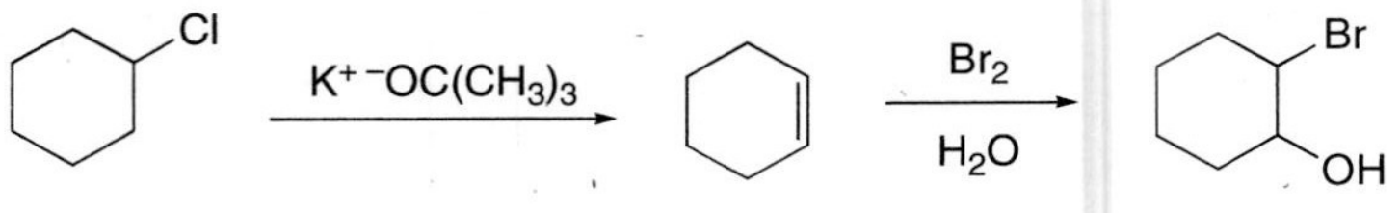

f. $CH_3CH{=}CH_2 \xrightarrow{Br_2} CH_3CHBr-CH_2Br \xrightarrow[(2\ equiv)]{NaNH_2} CH_3C{\equiv}CH$

10.73

a.

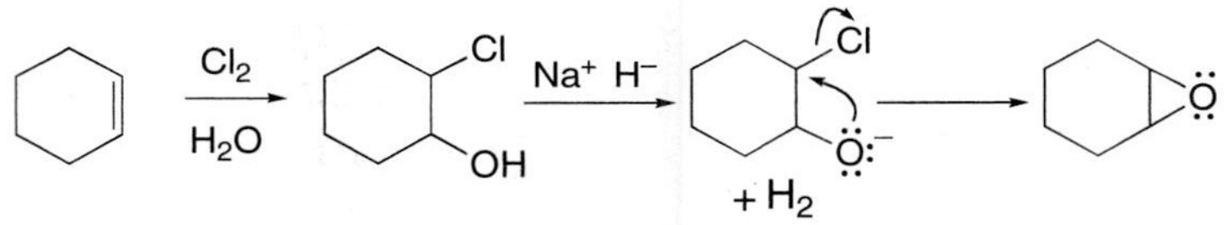

b.

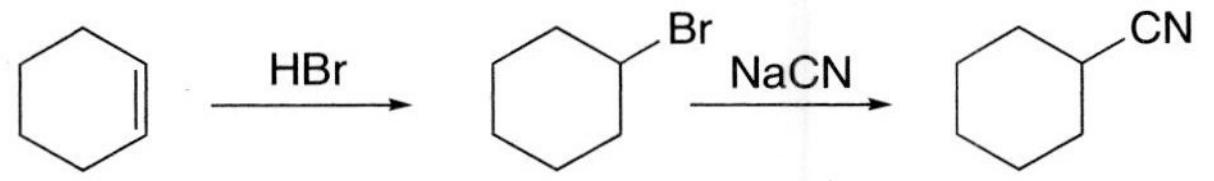

c. 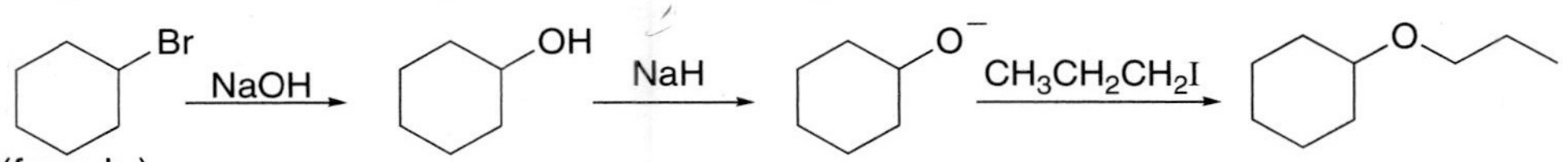
(from b.)

d. 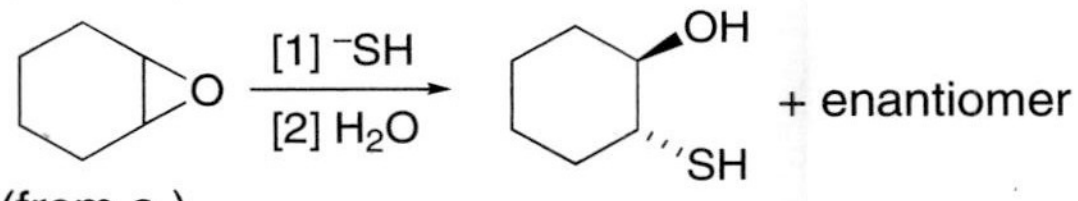
(from a.)

e. 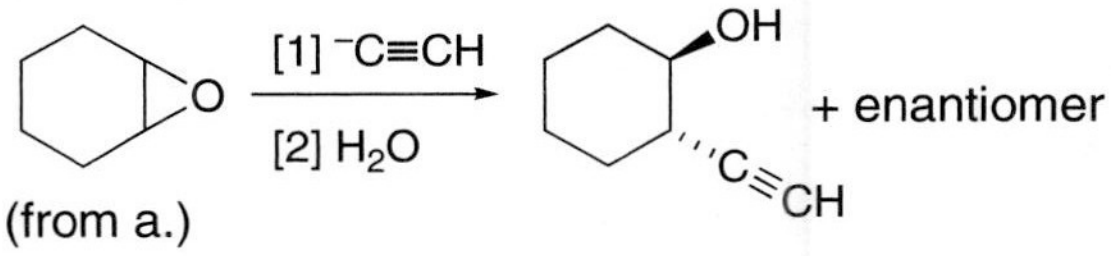
(from a.)

10.74

a.

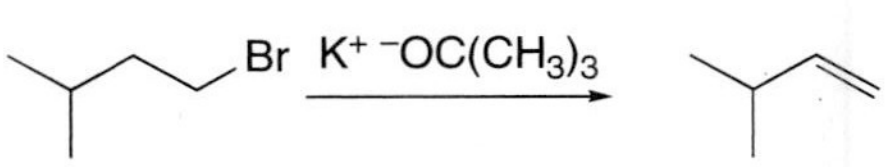

b. 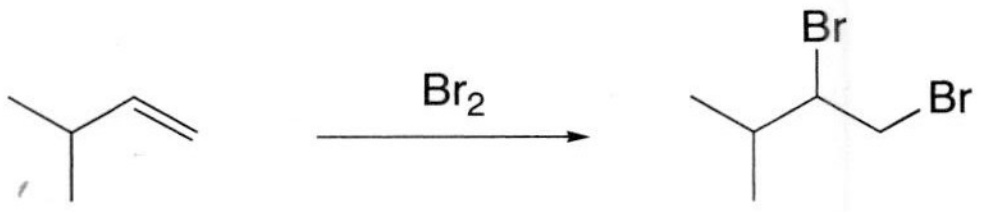
(from a.)

c. 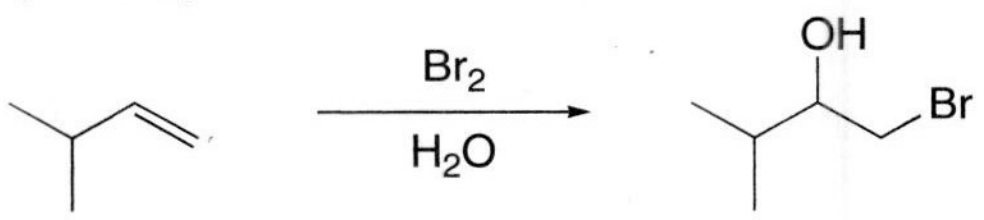
(from a.)

d. 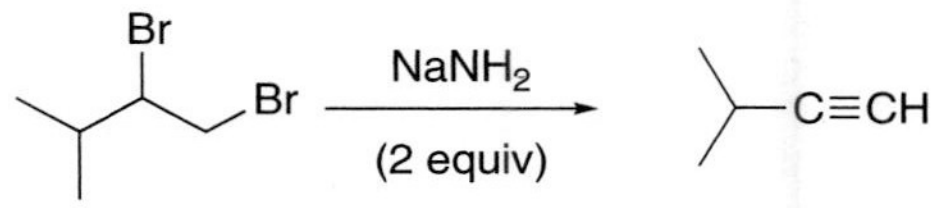
(from b.)

e. 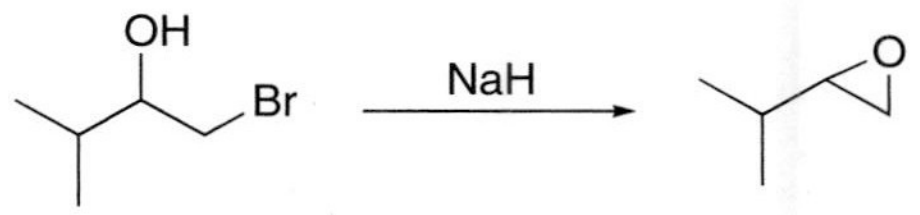
(from c.)

10.75

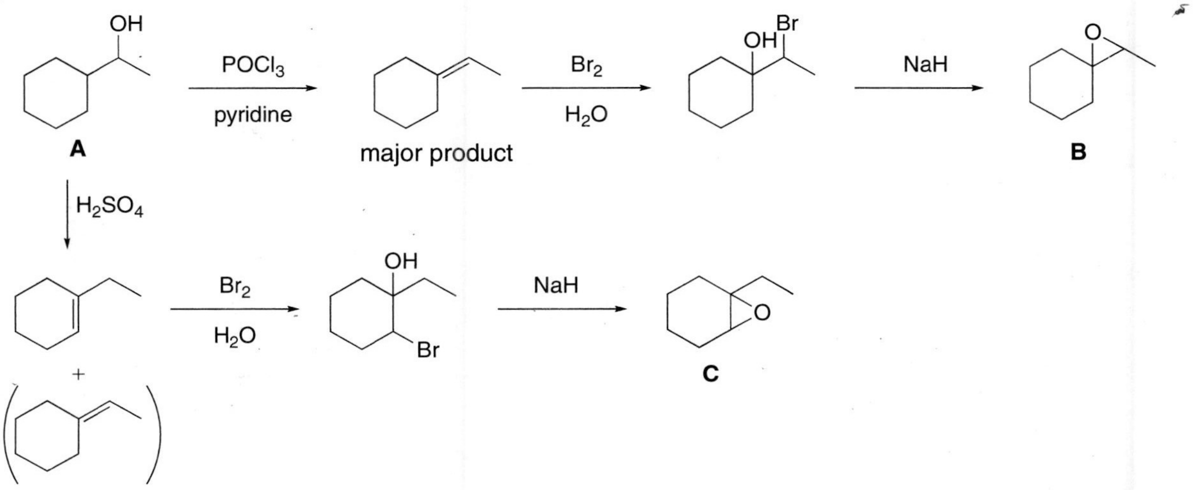

10.76

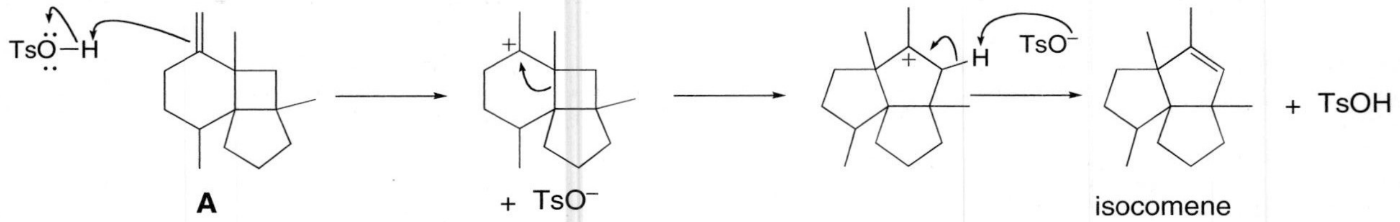

10.77

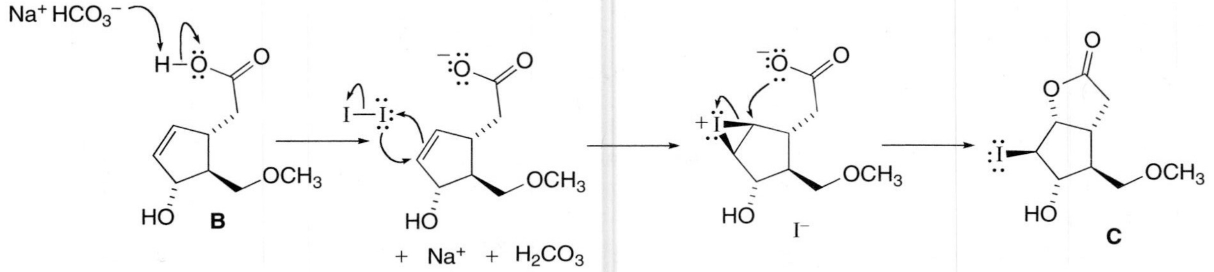

10.78

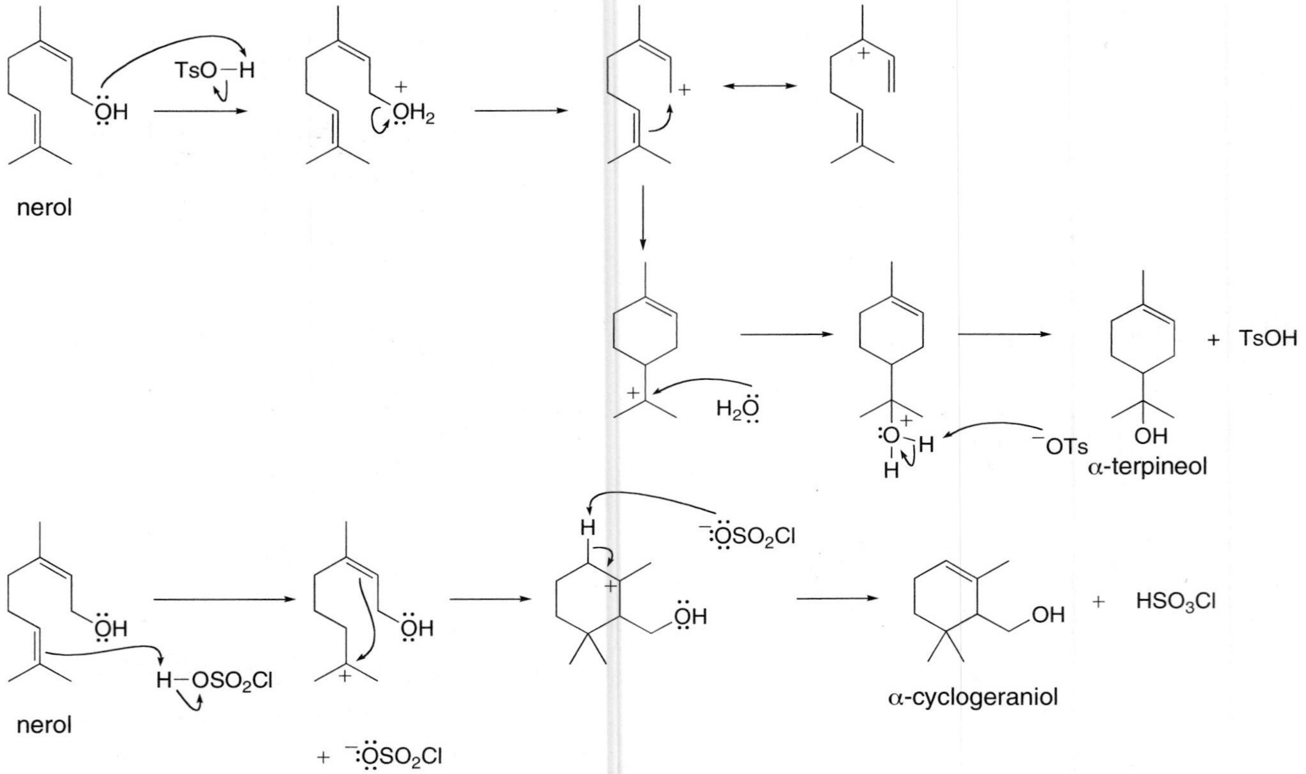

10.79

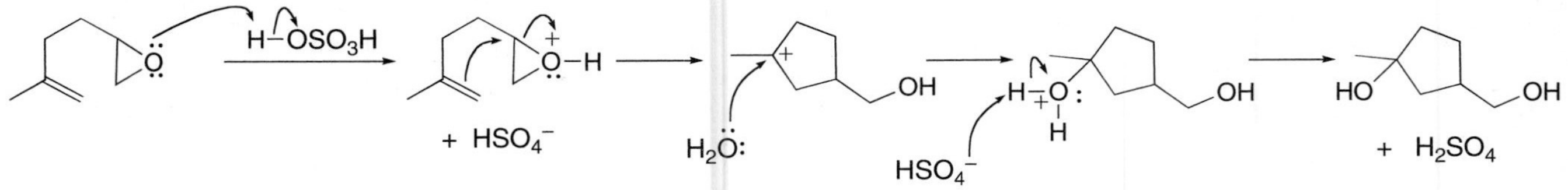

Chapter 11: Alkynes

◆ General facts about alkynes

- Alkynes contain a carbon–carbon triple bond consisting of a strong σ bond and two weak π bonds. Each carbon is *sp* hybridized and linear (11.1).

$$H-C\equiv C-H$$
acetylene

$$= \qquad 180°$$
sp hybridized

- Alkynes are named using the suffix **-yne** (11.2).
- Alkynes have weak intermolecular forces, giving them low mp's and low bp's, and making them water insoluble (11.3).
- Since its weaker π bonds make an alkyne electron rich, alkynes undergo addition reactions with electrophiles (11.6).

◆ Addition reactions of alkynes

[1] Hydrohalogenation—Addition of HX (X = Cl, Br, I) (11.7)

$$R-C\equiv C-H \xrightarrow[\text{(2 equiv)}]{H-X}$$

X H
| |
R−C−C−H
| |
X H
geminal dihalide

- Markovnikov's rule is followed. H bonds to the less substituted C in order to form the more stable carbocation.

[2] Halogenation—Addition of X_2 (X = Cl or Br) (11.8)

$$R-C\equiv C-H \xrightarrow[\text{(2 equiv)}]{X-X}$$

X X
| |
R−C−C−H
| |
X X
tetrahalide

- Bridged halonium ions are formed as intermediates.
- Anti addition of X_2 occurs.

[3] Hydration—Addition of H_2O (11.9)

$$R-C\equiv C-H \xrightarrow[\substack{H_2SO_4 \\ HgSO_4}]{H_2O}$$

$$\left[\begin{array}{c} R \quad\quad H \\ C=C \\ HO \quad\quad H \end{array} \right]$$
enol

$$\rightleftharpoons$$

O
||
R−C−CH₃
ketone

- Markovnikov's rule is followed. H bonds to the less substituted C in order to form the more stable carbocation.
- The unstable enol that is first formed rearranges to a carbonyl group.

[4] Hydroboration–oxidation—Addition of H_2O (11.10)

$$R-C\equiv C-H \xrightarrow[\text{[2] } H_2O_2,\ HO^-]{\text{[1] } BH_3} \left[\underset{\text{enol}}{\overset{R}{\underset{H}{C}}}=\overset{H}{\underset{OH}{C}} \right] \rightleftharpoons \underset{\text{aldehyde}}{R-\overset{H}{\underset{H}{C}}-\overset{O}{C}-H}$$

- The unstable enol, first formed after oxidation, rearranges to a carbonyl group.

◆ Reactions involving acetylide anions

[1] Formation of acetylide anions from terminal alkynes (11.6B)

$$R-C\equiv C-H \quad + \quad :B \rightleftharpoons \boxed{R-C\equiv C:^-} + HB^+$$

- Typical bases used for the reaction are $NaNH_2$ and NaH.

[2] Reaction of acetylide anions with alkyl halides (11.11A)

$$H-C\equiv C:^- \quad + \quad R-X \longrightarrow \boxed{H-C\equiv C-R} + X^-$$

- The reaction follows an S_N2 mechanism.
- The reaction works best with CH_3X and RCH_2X.

[3] Reaction of acetylide anions with epoxides (11.11B)

$$H-C\equiv C:^- \xrightarrow[\text{[2] } H_2O]{\text{[1] } \triangle O} \boxed{H-C\equiv C-CH_2CH_2OH}$$

- The reaction follows an S_N2 mechanism.
- Ring opening occurs from the back side at the less substituted end of the epoxide.

Chapter 11: Answers to Problems

11.1 • An *internal alkyne* has the triple bond somewhere in the *middle* of the carbon chain.
   • A *terminal alkyne* has the triple bond at the *end* of the carbon chain.

$$HC{\equiv}C-CH_2CH_2CH_3 \qquad CH_3-C{\equiv}C-CH_2CH_3 \qquad HC{\equiv}C-\underset{\underset{CH_3}{|}}{CH}-CH_3$$

   terminal alkyne   internal alkyne   terminal alkyne

11.2

santalbic acid

Increasing bond strength: a < c < b

11.3 Like alkenes, the larger the number of alkyl groups bonded to the *sp* hybridized C, the more stable the alkyne. This makes internal alkynes more stable than terminal alkynes.

11.4 To name an alkyne:
   [1] Find the longest chain that contains both atoms of the triple bond, change the *-ane* ending of the parent name to *-yne*, and number the chain to give the first carbon of the triple bond the lower number.
   [2] Name all substituents following the other rules of nomenclature.

a. $H-C{\equiv}C-CH_2\underset{\underset{CH_2CH_2CH_3}{|}}{\overset{\overset{CH_2CH_2CH_3}{|}}{C}}CH_2CH_2CH_3$

4,4-dipropyl-1-heptyne

c. $CH_2{=}CHCH_2\underset{\underset{CH_2CH_3}{|}}{\overset{\overset{CH_2CH_3}{|}}{CH}}C{\equiv}C\underset{\underset{CH_3}{|}}{\overset{\overset{CH_3}{|}}{C}}CH_2CH_2CH_3$

4-ethyl-7,7-dimethyl-1-decen-5-yne

(Number to give the lower number to the first site of unsaturation.)

b. $CH_3C{\equiv}CC\underset{\underset{CH_3}{|}}{Cl}CH_2CH_3$

4-chloro-4-methyl-2-hexyne

d.

(The longest chain must contain both functional groups.)

3-isopropyl-1,5-octadiyne

11.5 To work backwards from a name to a structure:
   [1] Find the parent name and the functional group.
   [2] Add the substituents to the appropriate carbon.

a. *trans*-2-ethynyl**cyclopentanol**
      5 C ring with OH at C1

OH ←——OH on C1

C≡CH ←—— ethynyl at C2   or

b. 4-*tert*-butyl-5-**decyne**

  10 C chain with
  a triple bond

tert-butyl at C4

triple bond at C5

c. 3-methyl**cyclononyne**

  9 C ring with a
  triple bond at C1

3-methyl   triple bond at C1

11.6 Two factors cause the boiling point increase. The linear *sp* hybridized C's of the alkyne allow for more van der Waals attraction between alkyne molecules. Also, since a triple bond is more polarizable than a double bond, this increases the van der Waals forces between two molecules as well.

11.7 To convert an alkene to an alkyne:
  [1] Make a vicinal dihalide from the alkene by addition of X_2.
  [2] Add base to remove two equivalents of HX and form the alkyne.

a. $Br_2CH(CH_2)_4CH_3$ $\xrightarrow{Na^+ \, {}^-NH_2}$ $\left[BrCH{=}CHCH_2CH_2CH_2CH_3 \right.$ not isolated $\left. \right]$ $\xrightarrow{Na^+ \, {}^-NH_2}$ $HC{\equiv}CCH_2CH_2CH_2CH_3$

b. $CH_2{=}CCl(CH_2)_3CH_3$ $\xrightarrow{Na^+ \, {}^-NH_2}$ $HC{\equiv}CCH_2CH_2CH_2CH_3$

c. $CH_2{=}CH(CH_2)_3CH_3$ $\xrightarrow{Cl_2}$ $\underset{\underset{Cl \quad Cl}{|\quad\;\;|}}{CH_2CHCH_2CH_2CH_2CH_3}$ $\xrightarrow[\text{(2 equiv)}]{Na^+ \, {}^-NH_2}$ $HC{\equiv}CCH_2CH_2CH_2CH_3$

11.8 Acetylene has a pK_a of 25, so **bases having a conjugate acid with a pK_a *above* 25 will be able to deprotonate it.**

a. CH_3NH^- [pK_a $(CH_3NH_2) = 40$]
  $pK_a > 25 =$ **Can deprotonate acetylene.**
b. CO_3^{2-} [pK_a $(HCO_3^-) = 10.2$]
  $pK_a < 25 =$ **Cannot deprotonate acetylene.**

c. $CH_2{=}CH^-$ [pK_a $(CH_2{=}CH_2) = 44$]
  $pK_a > 25 =$ **Can deprotonate acetylene.**
d. $(CH_3)_3CO^-$ {pK_a [$(CH_3)_3COH$] = 18}
  $pK_a < 25 =$ **Cannot deprotonate acetylene.**

11.9 To draw the products of reactions with HX:
- Add two moles of HX to the triple bond, following Markovnikov's rule.
- Both X's end up on the more substituted C.

a. $CH_3CH_2CH_2CH_2-C{\equiv}C-H$ $\xrightarrow{\text{2 HBr}}$ $CH_3CH_2CH_2CH_2-\overset{\displaystyle Br}{\underset{\displaystyle Br}{C}}-CH_3$

b. $CH_3-C{\equiv}C-CH_2CH_3$ $\xrightarrow{\text{2 HBr}}$ $CH_3-CH_2-\overset{\displaystyle Br}{\underset{\displaystyle Br}{C}}-CH_2CH_3$ + $CH_3-\overset{\displaystyle Br}{\underset{\displaystyle Br}{C}}-CH_2-CH_2CH_3$

c. ⬠$-C{\equiv}CH$ $\xrightarrow{\text{2 HBr}}$ ⬠$-\overset{\displaystyle Br}{\underset{\displaystyle Br}{C}}-CH_3$

11.10

a. ⬡$\overset{+}{-}\ddot{C}l:$ ⟷ ⬡$=\overset{+}{\ddot{C}l}:$ c. ⬡$\ddot{C}:NH$ ⟷ ⬡$\overset{+}{N}H$

b. $CH_3-\overset{..}{\underset{..}{O}}-\overset{+}{C}H_2$ ⟷ $CH_3-\overset{+}{\underset{..}{O}}=CH_2$

11.11 Addition of one equivalent of X_2 to alkynes forms trans dihalides.
Addition of two equivalents of X_2 to alkynes forms tetrahalides.

$CH_3CH_2-C{\equiv}C-CH_2CH_3$ $\xrightarrow{\text{2 Br}_2}$ $CH_3CH_2-\overset{\displaystyle Br\ Br}{\underset{\displaystyle Br\ Br}{C-C}}-CH_2CH_3$

$CH_3CH_2-C{\equiv}C-CH_2CH_3$ $\xrightarrow{\text{Cl}_2}$ $\overset{\displaystyle Cl\qquad CH_2CH_3}{\underset{\displaystyle CH_3CH_2\qquad Cl}{C=C}}$ **trans dihalide**

11.12

$CH_3-C{\equiv}C-CH_3$ $\xrightarrow{\text{Cl}_2}$ $\overset{\displaystyle Cl\qquad CH_3}{\underset{\displaystyle CH_3\qquad Cl}{C=C}}$ The two Cl atoms are electron withdrawing, making the π bond less electron rich and therefore less reactive with an electrophile.

11.13 **To draw the keto form of each enol:**
[1] Change the C–OH to a C=O at one end of the double bond.
[2] At the other end of the double bond, add a proton.

a. (structure with CH_2 and OH) ⇌ (keto structure with C, $H\ H$, H, O)
new C–H bond

b. (structure with OH and H) → (keto structure with O, $H\ H$)
new C–H bond

c. (cyclohexene with $-OH$ and H) ⇌ (cyclohexanone with $=O$, H, H)
new C–H bond

11.14 Treatment of alkynes with H_2O, H_2SO_4, and $HgSO_4$ yields ketones.

$$CH_3-C\equiv C-CH_2CH_3 \xrightarrow[\text{H}_2\text{SO}_4,\ \text{HgSO}_4]{\text{H}_2\text{O}}$$

$$CH_3CH=C(OH)CH_2CH_3$$
+
$$CH_3C(OH)=CHCH_2CH_3$$
Two enols form.

$\rightleftharpoons$ +

two ketones after tautomerization

11.15

a.

enol tautomers

b.

constitutional isomers,
but not tautomers

11.16 Reaction with H_2O, H_2SO_4, and $HgSO_4$ adds the oxygen to the *more* substituted carbon. Reaction with [1] BH_3, [2] H_2O_2, ^-OH adds the oxygen to the *less* substituted carbon.

a. $(CH_3)_2CHCH_2-C\equiv C-H \xrightarrow[\text{H}_2\text{SO}_4,\ \text{HgSO}_4]{\text{H}_2\text{O}}$ Forms a **ketone.** H_2O is added with the O atom on the *more* substituted carbon.

$(CH_3)_2CHCH_2-C\equiv C-H \xrightarrow[\text{[2] H}_2\text{O}_2,\ \text{HO}^-]{\text{[1] BH}_3}$ Forms an **aldehyde.** H_2O is added with the O atom on the *less* substituted carbon.

b. $\xrightarrow[\text{H}_2\text{SO}_4,\ \text{HgSO}_4]{\text{H}_2\text{O}}$ Forms a **ketone.** H_2O is added with the O atom on the *more* substituted carbon.

$\xrightarrow[\text{[2] H}_2\text{O}_2,\ \text{HO}^-]{\text{[1] BH}_3}$ Forms an **aldehyde.** H_2O is added with the O atom on the *less* substituted carbon.

11.17

a. $H-C\equiv C-H \xrightarrow{\text{[1] NaH}} H-C\equiv C:^- + H_2 \xrightarrow{\text{[2] (CH}_3)_2\text{CHCH}_2-\text{Cl}} (CH_3)_2CHCH_2-C\equiv C-H + NaCl$

b.

11.18

a. $(CH_3)_2CHCH_2C\equiv CH$

$CH_3-\overset{\overset{\displaystyle CH_3}{|}}{\underset{\underset{\displaystyle H}{|}}{C}}-CH_2-\text{\Large\}}-C\equiv C-H \implies CH_3-\overset{\overset{\displaystyle CH_3}{|}}{\underset{\underset{\displaystyle H}{|}}{C}}-CH_2Cl + {}^-C\equiv C-H$

1° RX

> **terminal alkyne**
> only one possibility

b. $CH_3C\equiv CCH_2CH_2CH_2CH_3$

$CH_3\text{-\}-C\equiv C-\text{\}-CH_2CH_2CH_2CH_3 \implies$
[1]   [2]

[1] $CH_3Cl + {}^-C\equiv C-CH_2CH_2CH_2CH_3$

[2] $CH_3-C\equiv C^- + Cl-CH_2CH_2CH_2CH_3$
1° RX

> **internal alkyne**
> two possibilities

c. $(CH_3)_3CC\equiv CCH_2CH_3$

$CH_3-\overset{\overset{\displaystyle CH_3}{|}}{\underset{\underset{\displaystyle CH_3}{|}}{C}}-C\equiv C-\text{\}-CH_2CH_3 \implies CH_3-\overset{\overset{\displaystyle CH_3}{|}}{\underset{\underset{\displaystyle CH_3}{|}}{C}}-C\equiv C^- + Cl-CH_2CH_3$
1° RX

> **internal alkyne**
> only one possibility

$\overset{\displaystyle \bf X}{\implies} CH_3-\overset{\overset{\displaystyle CH_3}{|}}{\underset{\underset{\displaystyle CH_3}{|}}{C}}-Cl + {}^-C\equiv CCH_2CH_3$

3° RX
too crowded for S_N2 reaction

> The 3° alkyl halide
> would undergo elimination.

11.19

$HC\equiv C-H \xrightarrow{Na^+ H:^-} HC\equiv C:^- \xrightarrow{CH_3CH_2-Br} H-C\equiv C-CH_2CH_3 \xrightarrow{Na^+ H:^-} {}^-:C\equiv C-CH_2CH_3 + H_2$

$+ H_2$ $+ Na^+Br^-$

$+ Na^+$ $(CH_3)_2CHCH_2\overset{\overset{\displaystyle H}{|}}{\underset{\underset{\displaystyle H}{|}}{C}}-Br$

$(CH_3)_2CHCH_2CH_2-C\equiv C-CH_2CH_3 + Br^-$

11.20

$CH_3-\overset{\overset{\displaystyle CH_3}{|}}{\underset{\underset{\displaystyle CH_3}{|}}{C}}-C\equiv C-\overset{\overset{\displaystyle CH_3}{|}}{\underset{\underset{\displaystyle CH_3}{|}}{C}}-CH_3 \implies CH_3-\overset{\overset{\displaystyle CH_3}{|}}{\underset{\underset{\displaystyle CH_3}{|}}{C}}-C\equiv C^- + X-\overset{\overset{\displaystyle CH_3}{|}}{\underset{\underset{\displaystyle CH_3}{|}}{C}}-CH_3$

> The 3° alkyl halide is too crowded for nucleophilic substitution. Instead, it would undergo elimination with the acetylide anion.

2,2,5,5-tetramethyl-3-hexyne

11.21

a. $\xrightarrow[\text{[2] } H_2O]{\text{[1] } {}^-:C\equiv C-H}$

> Epoxide is drawn up, so the **acetylide anion attacks from below at less substituted C.**

b. $\xrightarrow[\text{[2] } H_2O]{\text{[1] } {}^-:C\equiv C-H}$ +

enantiomers

> Backside attack of the nucleophile ($^-C\equiv CH$) at either C since both ends are equally substituted.

11.22

a. $CH_3CH_2CH_2Br$ $\xrightarrow{CH_3CH_2C \equiv C^- Na^+}$ $CH_3CH_2CH_2C \equiv CCH_2CH_3$

b. $(CH_3)_2CHCH_2CH_2Cl$ $\xrightarrow{CH_3CH_2C \equiv C^- Na^+}$ $(CH_3)_2CHCH_2CH_2C \equiv CCH_2CH_3$

c. $(CH_3CH_2)_3CCl$ $\xrightarrow{CH_3CH_2C \equiv C^- Na^+}$ $(CH_3CH_2)_2C = CHCH_3$ + $CH_3CH_2C \equiv CH$

d. $BrCH_2CH_2CH_2CH_2OH$ $\xrightarrow{CH_3CH_2C \equiv C^- Na^+}$ $CH_3CH_2C \equiv CH$ + $BrCH_2CH_2CH_2CH_2O^- Na^+$

e. $\xrightarrow{CH_3CH_2C \equiv C^- Na^+}$ $\xrightarrow{H_2O}$ $CH_3CH_2C \equiv CCH_2CH_2OH$

f. $\xrightarrow{CH_3CH_2C \equiv C^- Na^+}$ $\xrightarrow{H_2O}$ $CH_3CH_2C \equiv CCH_2\underset{OH}{CH}CH_3$

11.23 To use a retrosynthetic analysis:
[1] **Count the number of carbon atoms** in the starting material and product.
[2] **Look at the functional groups** in the starting material and product.
   Determine what types of reactions can form the product.
   Determine what types of reactions the starting material can undergo.
[3] **Work backwards** from the product to make the starting material.
[4] Write out the synthesis in the synthetic direction.

$$CH_3CH_2C \equiv CCH_2CH_3 \overset{?}{\Longrightarrow} HC \equiv CH$$
$$\text{6 C's} \qquad\qquad \text{2 C's}$$

$$CH_3CH_2C \equiv CCH_2CH_3 \Longrightarrow CH_3CH_2C \equiv C^- + CH_3CH_2Br \Longrightarrow HC \equiv C^- + CH_3CH_2Br$$

$HC \equiv C-H$ $\xrightarrow{Na^+ H:^-}$ $HC \equiv C:^-$ CH_3CH_2-Br $CH_3CH_2C \equiv C-H$ $\xrightarrow{Na^+ H:^-}$ $CH_3CH_2C \equiv C:^-$ CH_3CH_2-Br $CH_3CH_2C \equiv CCH_2CH_3$

11.24

$$CH_3CH_2CH_2-\overset{O}{\underset{H}{C}} \Longrightarrow HC \equiv CH$$

product:
4 carbons, aldehyde functional group
(can be made by hydroboration–oxidation of
a terminal alkyne)

starting material:
2 carbons, C≡C functional group
(can form an acetylide
anion by reaction with NaH)

Retrosynthetic
analysis: $CH_3CH_2CH_2-\overset{O}{\underset{H}{C}} \Longrightarrow CH_3CH_2C \equiv C-H \Longrightarrow H-C \equiv C-H$

Forward direction: $H-C \equiv C-H$ $\xrightarrow{Na^+ H:^-}$ $:C \equiv C-H$ $\xrightarrow{CH_3CH_2-Br}$ $CH_3CH_2C \equiv C-H$ $\xrightarrow[\text{[2] } H_2O_2, HO^-]{\text{[1] } BH_3}$ $CH_3CH_2CH_2-\overset{O}{\underset{H}{C}}$

11.25

a., b. **erlotinib**

H
most acidic C–H proton

HN

shortest C–C single bond
$C_{sp}–C_{sp^2}$

c.
d. * = *sp* hybridized C
e. 3 < 1 < 2

(1)   (2)

* * * *

O

HO

(3)

phomallenic acid C

most acidic proton

11.26 Use the rules from Answer 11.4 to name the alkynes.

1-hexyne   3-hexyne   3-methyl-1-pentyne   3,3-dimethyl-1-butyne

2-hexyne   4-methyl-1-pentyne   4-methyl-2-pentyne

11.27 Use the rules from Answer 11.4 to name the alkynes.

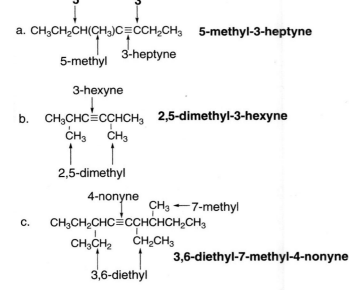

5   3
a. CH₃CH₂CH(CH₃)C≡CCH₂CH₃   **5-methyl-3-heptyne**

5-methyl   3-heptyne

b.   3-hexyne
    CH₃CHC≡CCHCH₃   **2,5-dimethyl-3-hexyne**
      CH₃   CH₃

    2,5-dimethyl

c.   4-nonyne   CH₃ ← 7-methyl
    CH₃CH₂CHC≡CCHCHCH₂CH₃
        CH₃CH₂   CH₂CH₃

        3,6-diethyl

    3,6-diethyl-7-methyl-4-nonyne

d.   HC≡C–CH(CH₂CH₃)CH₂CH₂CH₃   **3-ethyl-1-hexyne**
    1-hexyne   3-ethyl

e.   CH₃ ← 3-methyl
    CH₃CH₂–C–C≡CH   **3-ethyl-3-methyl-1-hexyne**
        CH₂CH₂CH₃
    3-ethyl

f.   CH₃CH₂C≡CCH₂C≡CCH₃   **2,5-octadiyne**
            5   2

g.   2   5-ethyl
                          6   **(2E)-4,5-diethyl-2-decen-6-yne**
         4-ethyl

h.   6-methyl   1
                      ≡ ← ethynyl
    1-ethynyl-6-methylcyclohexene

11.28 Use the directions from Answer 11.5 to draw each structure.

a. 5,6-dimethyl-2-**heptyne**

2-heptyne →

b. 5-*tert*-butyl-6,6-dimethyl-3-**nonyne**

←—**6,6-dimethyl**

3-nonyne

5-*tert*-butyl

c. (4*S*)-4-chloro-2-**pentyne**

Cl ←— **4-chloro**
⋯H   *S* configuration

2-pentyne

d. *cis*-1-ethynyl-2-methyl**cyclopentane**

1-ethynyl
↓
C≡CH   or   ⋯C≡CH

←— **2-methyl**

e. 3,4-dimethyl-1,5-**octadiyne**

3,4-dimethyl

diyne   diyne

f. (6*Z*)-6-methyl-6-**octen-1-yne**

←— **6-methyl**

Z

1

11.29 Keto–enol tautomers are constitutional isomers in equilibrium that differ in the location of a double bond and a hydrogen. The OH in an enol must be bonded to a C=C.

a.
$$CH_3-\overset{O}{\underset{}{C}}-CH_3 \quad and \quad CH_2=\overset{OH}{\underset{}{C}}-CH_3$$

• C=O   • C=C
• one more CH bond   • OH on C=C

keto–enol tautomers

b.

and

OH is not bonded
to the C=C.

NOT keto–enol tautomers

c.

⁀⁀OH   and   ⁀⁀H
                          ‖
                          O

• C=C   • C=O
• OH on C=C   • one more CH bond

keto–enol tautomers

d.

and

OH ←——— OH is not bonded
to the C=C.

NOT keto–enol tautomers

11.30 To draw the enol form of each keto form:
   [1] Change the C=O to a C–OH.
      [2] Change one single C–C bond to a double bond, making sure the OH group is bonded to the C=C.

a. 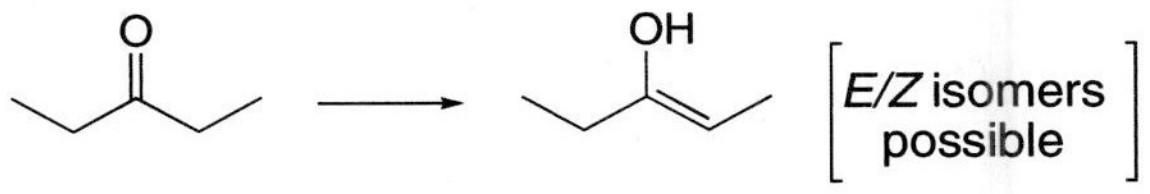 [structure] ⟶ [structure] [E/Z isomers possible]   c. 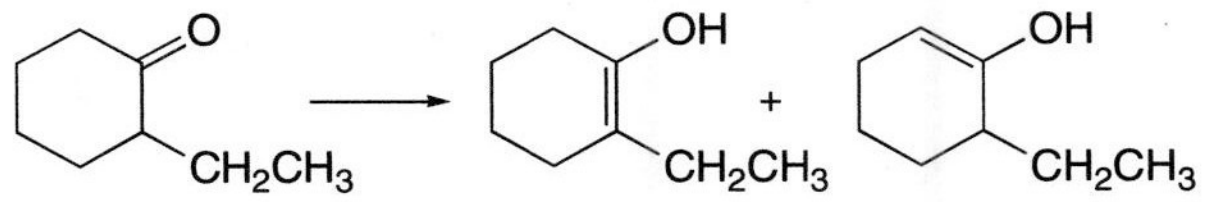 [structure] ⟶ [structure] + [structure]

b. CH₃CH₂CHO = [structure] ⟶ [structure] [E/Z isomers possible]

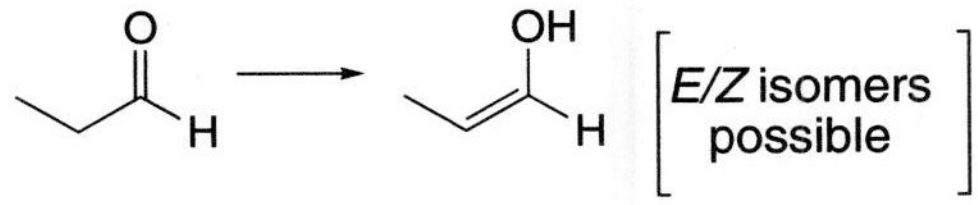

11.31 Use the directions from Answer 11.13 to draw each keto form.

a. 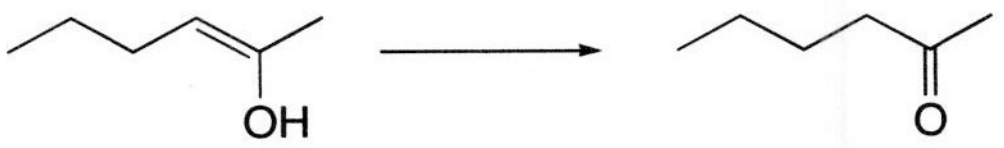 [structure] ⟶ [structure]   c. 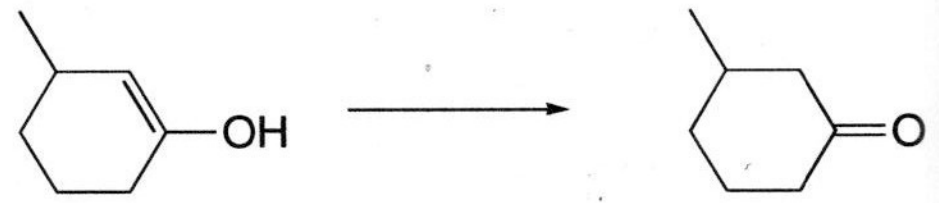 [structure] ⟶ [structure]

b. 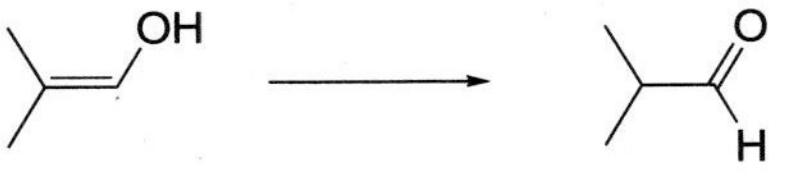 [structure] ⟶ [structure]

11.32 Tautomers are constitutional isomers that are in equilibrium and differ in the location of a double bond and a hydrogen atom.

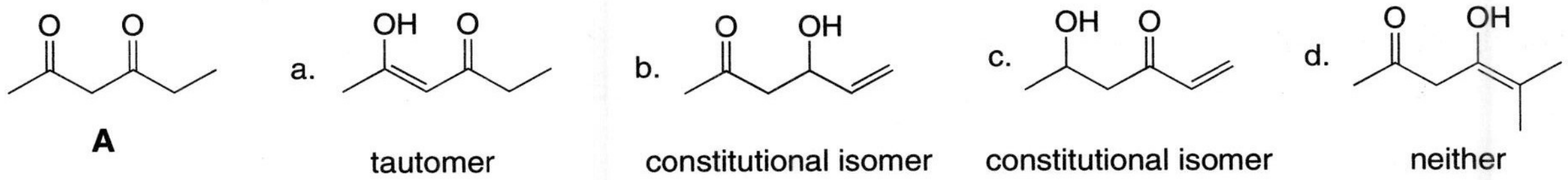

[structure] **A**   a. [structure] tautomer   b. [structure] constitutional isomer   c. [structure] constitutional isomer   d. [structure] neither

11.33

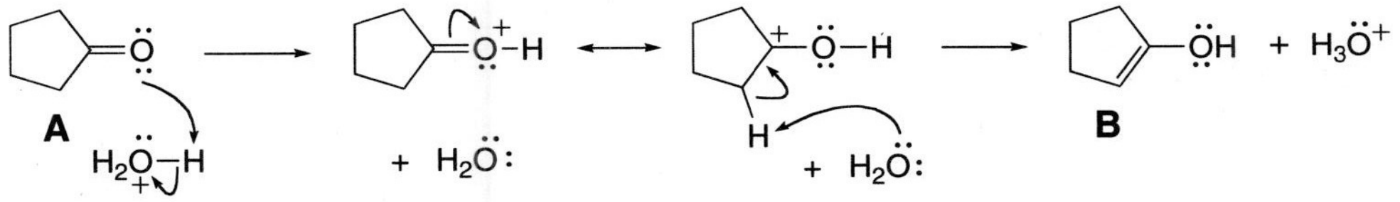

[mechanism] A ... B

11.34

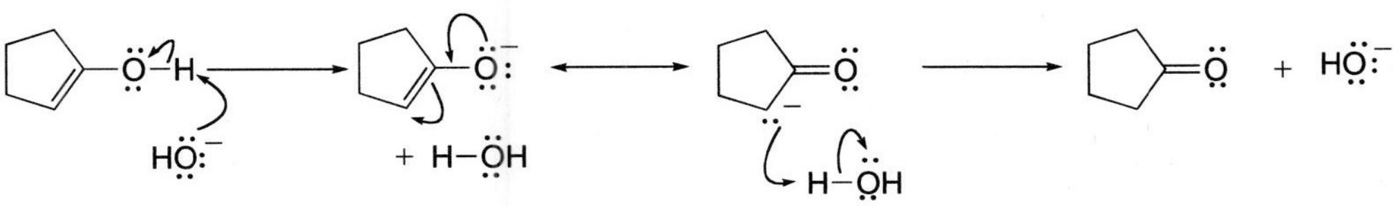

[mechanism]

11.35

[mechanism] **X** enamine ... **Y** imine

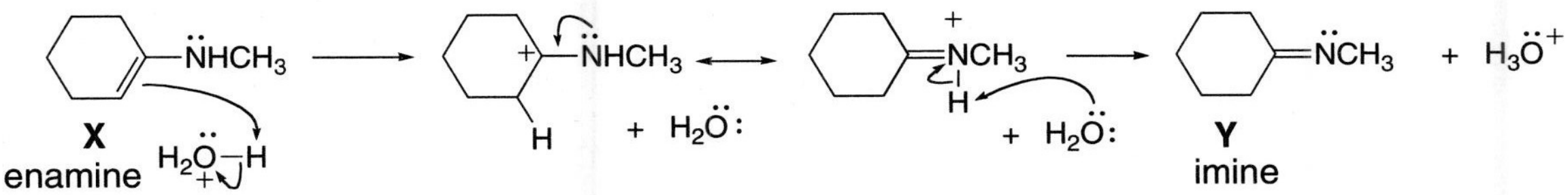

11.36 The equilibrium always favors the formation of the weaker acid and the weaker base.

a. $HC\equiv C^- + CH_3OH \rightleftharpoons HC\equiv CH + CH_3O^-$

      $pK_a = 15.5$ $pK_a = 25$
                          weaker acid
      Equilibrium favors products.

c. $HC\equiv CH + Na^+Br^- \longleftarrow HC\equiv C^- Na^+ + HBr$

      $pK_a = 25$ $pK_a = -9$
      weaker acid
   Equilibrium favors
   starting materials.

b. $CH_3C\equiv CH + CH_3^- \rightleftharpoons CH_3C\equiv C^- + CH_4$

   $pK_a = 25$ $pK_a = 50$
                        weaker acid
  Equilibrium favors products.

d. $CH_3CH_2C\equiv C^- + CH_3COOH \rightleftharpoons CH_3CH_2C\equiv CH + CH_3COO^-$

           $pK_a = 4.8$ $pK_a = 25$
                      weaker acid
  Equilibrium favors products.

11.37

$HC\equiv CCH_2CH_2CH_2CH_3$

a. $\xrightarrow[\text{(2 equiv)}]{HCl}$ $CH_3-\overset{\overset{Cl}{|}}{\underset{\underset{Cl}{|}}{C}}-CH_2CH_2CH_2CH_3$

b. $\xrightarrow[\text{(2 equiv)}]{HBr}$ $CH_3-\overset{\overset{Br}{|}}{\underset{\underset{Br}{|}}{C}}CH_2CH_2CH_2CH_3$

c. $\xrightarrow[\text{(2 equiv)}]{Cl_2}$ $\overset{\overset{Cl}{|}}{\underset{\underset{Cl}{|}}{H}}C-\overset{\overset{Cl}{|}}{\underset{\underset{Cl}{|}}{C}}CH_2CH_2CH_2CH_3$

d. $\xrightarrow[\text{H}_2\text{SO}_4,\ \text{HgSO}_4]{H_2O}$ $\overset{H}{\underset{H}{C}}=\overset{OH}{\underset{CH_2CH_2CH_2CH_3}{C}}$ $\longrightarrow$ $CH_3-\overset{O}{\overset{||}{C}}-CH_2CH_2CH_2CH_3$

e. $\xrightarrow[\text{[2] H}_2\text{O}_2,\ \text{HO}^-]{\text{[1] BH}_3}$ $H-\overset{O}{\overset{||}{C}}-CH_2CH_2CH_2CH_2CH_3$

f. $\xrightarrow{NaH}$ $Na^+{}^-C\equiv CCH_2CH_2CH_2CH_3$

g. $\xrightarrow[\text{[2] CH}_3\text{CH}_2\text{Br}]{\text{[1] }^-\text{NH}_2}$ $CH_3CH_2C\equiv CCH_2CH_2CH_2CH_3$

h. $\xrightarrow[\text{[3] H}_2\text{O}]{\substack{\text{[1] }^-\text{NH}_2 \\ \text{[2] } \triangle\text{O}}}$ $HOCH_2CH_2-C\equiv CCH_2CH_2CH_2CH_3$

11.38

a. $\xrightarrow[\text{(2 equiv)}]{HBr}$ (product)

b. $\xrightarrow[\text{(2 equiv)}]{Br_2}$ (product)

c. $\xrightarrow[\text{H}_2\text{SO}_4]{H_2O}$ (product)

d. $\xrightarrow[\text{[2] H}_2\text{O}_2,\ \text{HO}^-]{\text{[1] BH}_3}$ (product)

11.39

a. $(CH_3CH_2)_3C-C\equiv CH \xrightarrow[\text{H}_2\text{SO}_4,\ \text{HgSO}_4]{H_2O} (CH_3CH_2)_3C-\overset{O}{\overset{||}{C}}-CH_3$

b. $(CH_3CH_2)_3C-C\equiv CH \xrightarrow[\text{[2] H}_2\text{O}_2,\ \text{HO}^-]{\text{[1] BH}_3} (CH_3CH_2)_3C-CH_2CHO$

c. $(CH_3CH_2)_3C-C\equiv CH \xrightarrow[\text{(2 equiv)}]{HCl} (CH_3CH_2)_3CCCl_2CH_3$

d. $(CH_3CH_2)_3C-C\equiv CH \xrightarrow[\text{[2] CH}_3\text{CH}_2\text{Br}]{\text{[1] NaH}} (CH_3CH_2)_3C-C\equiv CCH_2CH_3$

11.40 Reaction rate (which is determined by E_a) and enthalpy ($\Delta H°$) are not related. More exothermic reactions are not necessarily faster. Since the addition of HX to an alkene forms a more stable carbocation in an endothermic, rate-determining step, this carbocation is formed faster by the Hammond postulate.

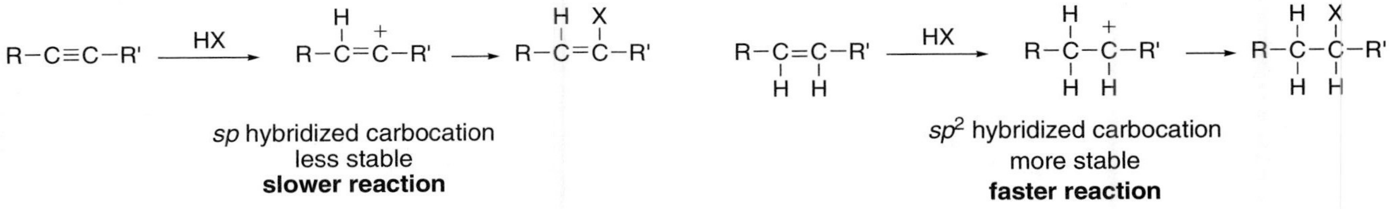

sp hybridized carbocation
less stable
slower reaction

*sp*² hybridized carbocation
more stable
faster reaction

11.41

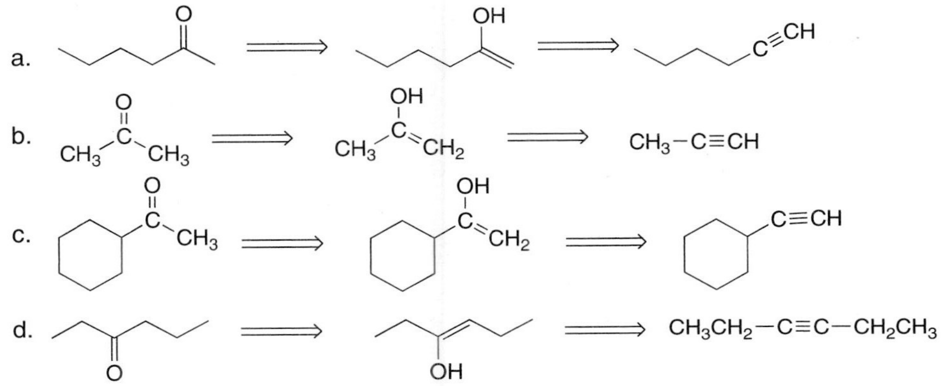

a.

b.

c.

d.

11.42 To determine what two alkynes could yield the given ketone, work backwards by drawing the enols and then the alkynes.

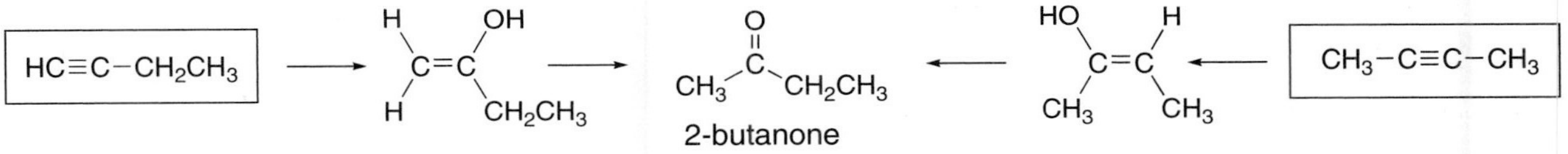

2-butanone

11.43

a.

b.

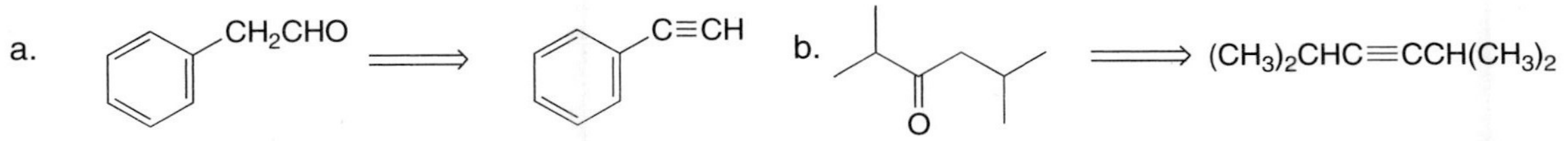

11.44 Equilibrium favors the weaker acid. $CH_3CH_2CH_2CH_2^-Li^+$ is a strong enough base to remove the proton of an alkyne because its conjugate acid, $CH_3CH_2CH_2CH_3$, is weaker than a terminal alkyne.

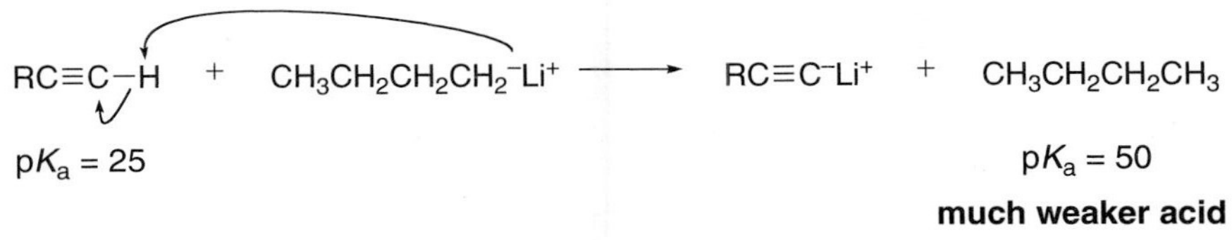

$RC{\equiv}C{-}H$ + $CH_3CH_2CH_2CH_2^-Li^+$ $\longrightarrow$ $RC{\equiv}C^-Li^+$ + $CH_3CH_2CH_2CH_3$

$pK_a = 25$ $pK_a = 50$

much weaker acid

11.45

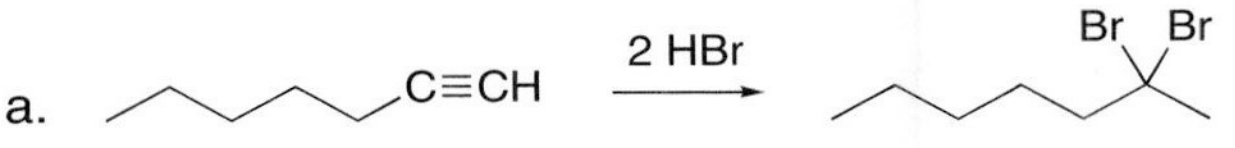

a.

b. $(CH_3)_3CC\equiv CH$ $\xrightarrow{2\ Cl_2}$ $(CH_3)_3\overset{\overset{\displaystyle Cl}{|}}{\underset{\underset{\displaystyle Cl}{|}}{C}}-CHCl_2$

c. Ph–CH=CH–Ph $\xrightarrow{[1]\ Cl_2}$ Ph–CHCl–CHCl–Ph $\xrightarrow[\text{(2 equiv)}]{[2]\ NaNH_2}$ Ph–C≡C–Ph

d. $\sim\!\!\!\sim\!\!C\equiv CH$ $\xrightarrow[\text{[2] }H_2O_2,\ HO^-]{\text{[1] }BH_3}$ aldehyde product

e. cyclopentyl–C≡CH $\xrightarrow[\text{(1 equiv)}]{HCl}$ cyclopentyl–C(Cl)=CH₂

f. $HC\equiv C^- + D_2O \longrightarrow HC\equiv CD + DO^-$

g. cyclopentyl–C≡C–CH₃ $\xrightarrow[H_2SO_4]{H_2O}$ cyclopentyl–CH₂–C(=O)–CH₃ + cyclopentyl–C(=O)–CH₂CH₃

h. $CH_3CH_2C\equiv C^- + CH_3CH_2CH_2{-}OTs \longrightarrow CH_3CH_2C\equiv CCH_2CH_2CH_3 + {}^-OTs$

i. epoxide $\xrightarrow{[1]\ HC\equiv C^-}$ alkoxide intermediate $\xrightarrow{[2]\ HO{-}H}$ hydroxy product

j. $CH_3C\equiv C{-}H$ $\xrightarrow{[1]\ Na^+\ H^-}$ $CH_3C\equiv C^-$ $\xrightarrow[\substack{2°\ \text{halide}\\ E2}]{[2]}$ cyclopentene + $CH_3C\equiv C{-}H$

k. $CH_3CH_2C\equiv C{-}H$ $\xrightarrow{[1]\ Na^+\ {}^-NH_2}$ $CH_3CH_2C\equiv C^-$ $\xrightarrow{[2]\ cyclohexyl-CH_2{-}I}$ $CH_3CH_2C\equiv C{-}CH_2$–cyclohexyl

l. Ph–C≡C–H $\xrightarrow{[1]\ Na^+\ H^-}$ Ph–C≡C$^-$ $\xrightarrow{[2]\ \text{epoxide}}$ Ph–C≡CCH₂CH₂O$^-$ $\xrightarrow{[3]\ HO{-}H}$ Ph–C≡C–CH₂CH₂OH

11.46

cyclohexyl–CH₂CH₂Br $\xrightarrow{KOC(CH_3)_3}$ cyclohexyl–CH=CH₂ $\xrightarrow{Br_2}$ cyclohexyl–CHBr–CHBr–H $\xrightarrow[\substack{DMSO\\ (2\ equiv)}]{KOC(CH_3)_3}$ cyclohexyl–C≡CH

A   **B**   **C**   **D** $\Big\downarrow NaNH_2$

cyclohexyl–C≡CCH₃ $\xleftarrow{CH_3I}$ cyclohexyl–C≡C$^-$

    E

11.47

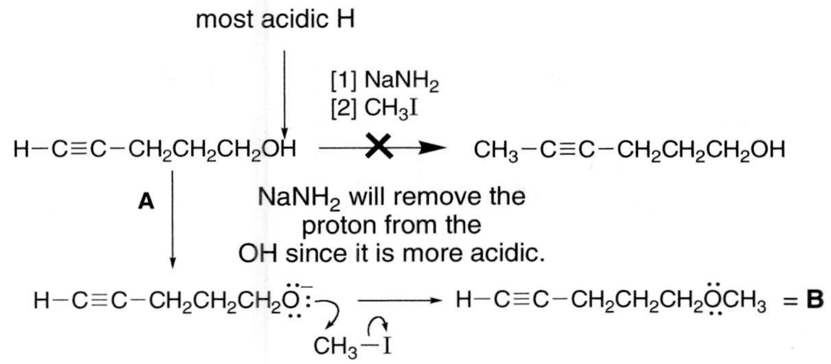

most acidic H

$$H-C\equiv C-CH_2CH_2CH_2\overset{\downarrow}{O}H \xrightarrow[]{\begin{array}{l}[1]\ NaNH_2\\ [2]\ CH_3I\end{array}} \times \quad CH_3-C\equiv C-CH_2CH_2CH_2OH$$

A

NaNH₂ will remove the
proton from the
OH since it is more acidic.

$$H-C\equiv C-CH_2CH_2CH_2\overset{\cdot\cdot}{\underset{\cdot\cdot}{O}}{:}^- \longrightarrow H-C\equiv C-CH_2CH_2CH_2\overset{\cdot\cdot}{\underset{\cdot\cdot}{O}}CH_3 = \mathbf{B}$$

$$CH_3-I$$

11.48

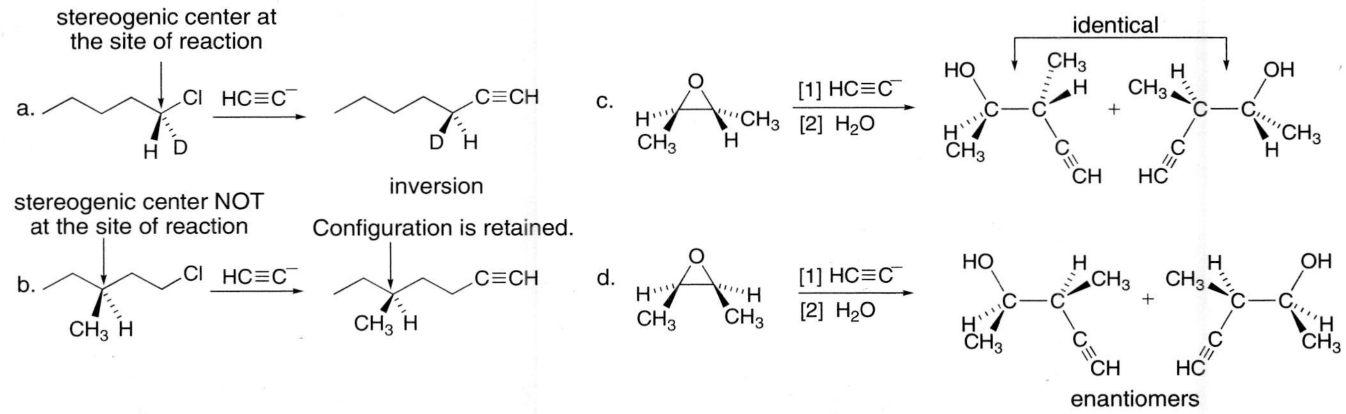

stereogenic center at
the site of reaction

a.

inversion
Configuration is retained.

stereogenic center NOT
at the site of reaction

b.

c. identical

d. enantiomers

11.49

A →[TsCl / pyridine] →[HC≡C⁻ / S_N2] B

retention   inversion

A →[PBr₃ / S_N2] →[HC≡C⁻ / S_N2] C

inversion   inversion

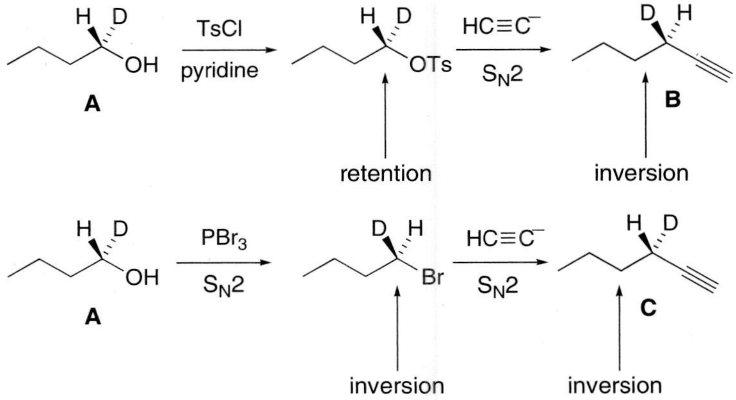

11.50

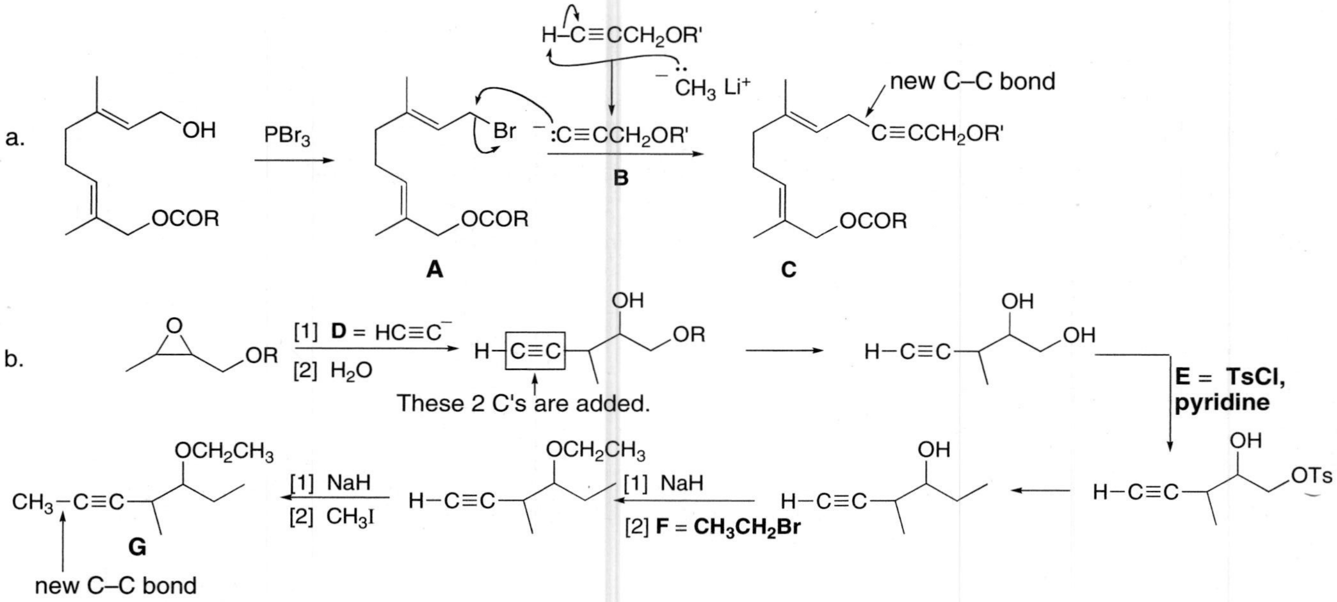

a.

b.

These 2 C's are added.

E = TsCl, pyridine

G

new C–C bond

11.51

CH$_3$CH$_2$–C≡C–H

1-butyne

CH$_3$–C≡C–CH$_3$

2-butyne

major product
more substituted alkyne

(*E* and *Z* isomers possible)

Reaction by-products:

2 $\overset{..}{N}H_3$ + 2 Br⁻

CH$_3$CH=C=CH$_2$

1,2-butadiene

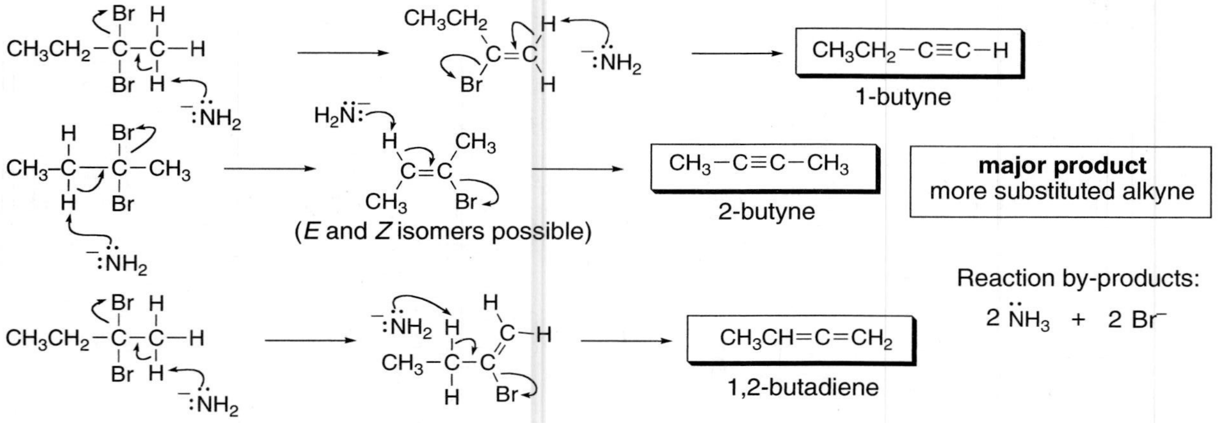

11.52 Draw two diagrams to show σ and π bonds.

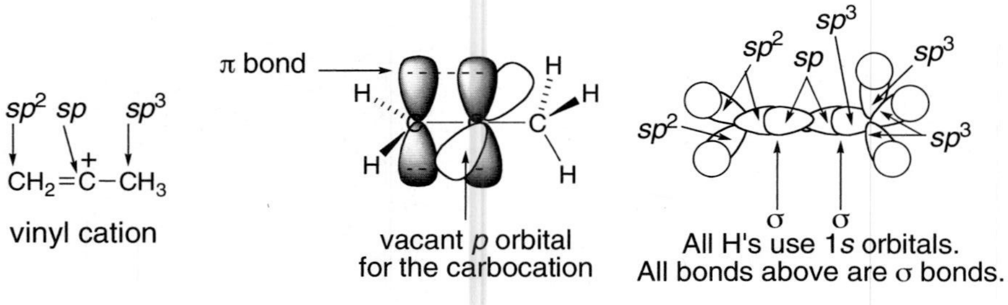

sp^2 sp sp^3

CH$_2$=C–CH$_3$

vinyl cation

π bond

vacant *p* orbital
for the carbocation

All H's use 1*s* orbitals.
All bonds above are σ bonds.

The positive charge in a vinyl carbocation resides on a carbon that is *sp* hybridized, while in (CH$_3$)$_2$CH⁺, the positive charge is located on an *sp²* hybridized carbon. The higher percent *s*-character on carbon destabilizes the positive charge in the vinyl cation. Moreover, the positively charged carbocation is now bonded to an *sp²* hybridized carbon, which donates electrons less readily than an *sp³* hybridized carbon.

11.53 A carbanion is more stable when its lone pair is in an orbital with a higher percentage of the smaller *s* orbital. A carbocation is more stable when its positive charge is due to a vacant orbital with a lower percentage of the smaller *s* orbital. In $HC\equiv C^+$, the positively charged C uses two *p* orbitals to form two π bonds. If the σ bond is formed using an *sp* hybrid orbital, the second hybrid orbital would have to remain vacant, a highly unstable situation. See also Problem 11.52.

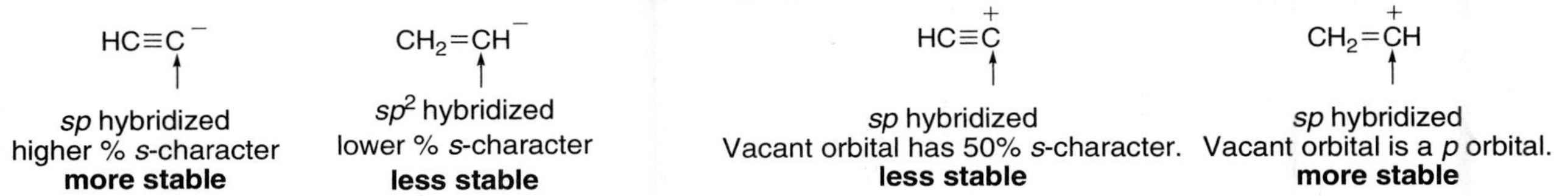
HC≡C⁻
sp hybridized
higher % s-character
more stable
CH₂=CH⁻
sp² hybridized
lower % s-character
less stable
HC≡C⁺
sp hybridized
Vacant orbital has 50% s-character.
less stable
CH₂=CH⁺
sp hybridized
Vacant orbital is a p orbital.
more stable

11.54

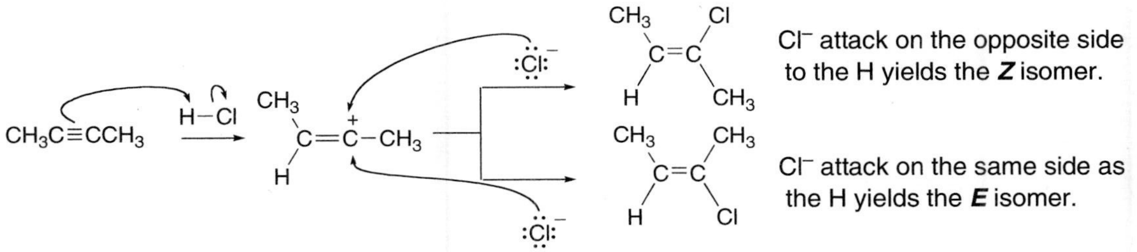
CH₃C≡CCH₃
H–Cl
Cl⁻ attack on the opposite side to the H yields the Z isomer.
Cl⁻ attack on the same side as the H yields the E isomer.

11.55

a.

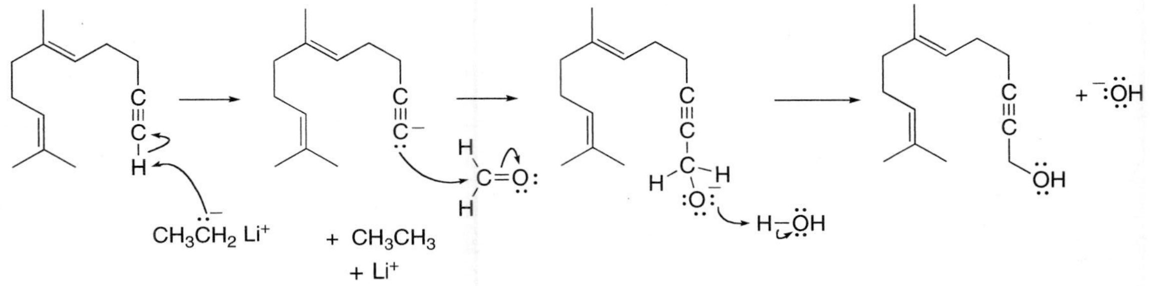
CH₃CH₂ Li⁺
+ CH₃CH₃
+ Li⁺
H–OH
+ ⁻:OH

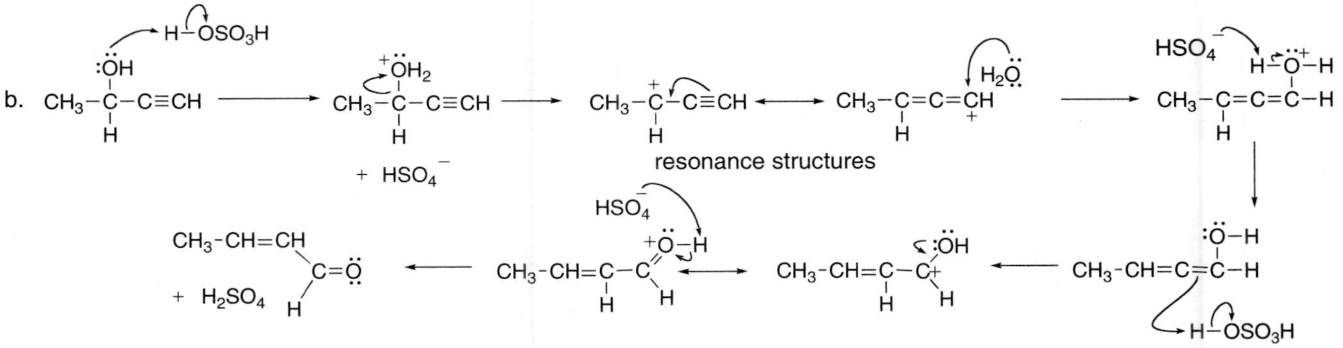
b.
H–OSO₃H
+ HSO₄⁻
resonance structures
HSO₄⁻
H₂O
+ H₂SO₄

11.56

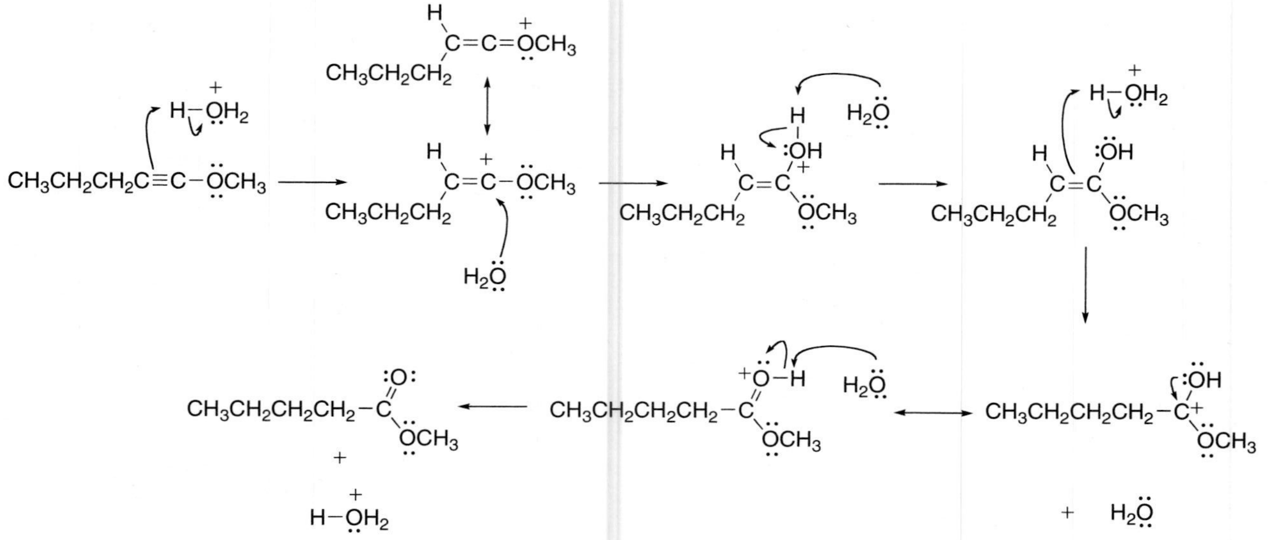

11.57

a.

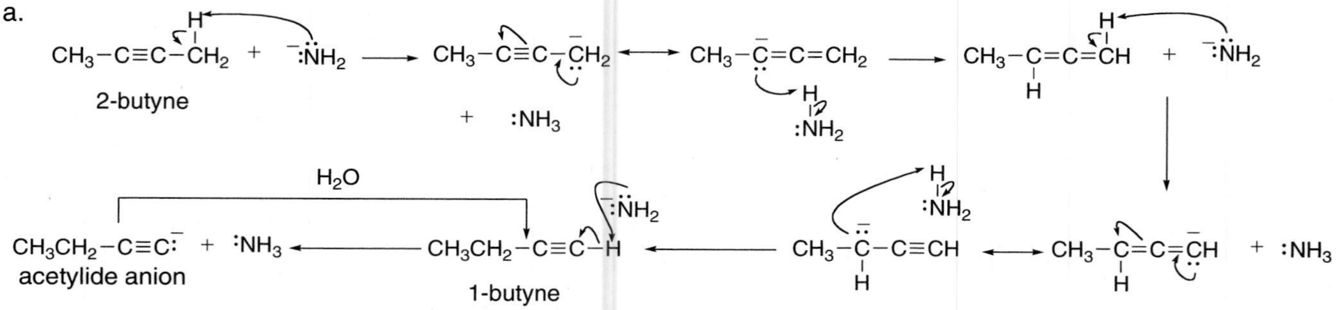

b. A more stable internal alkyne can be isomerized to a less stable terminal alkyne under these reaction conditions because when $CH_3CH_2C{\equiv}CH$ is first formed, it contains an *sp* hybridized C–H bond, which is more acidic than any proton in CH_3–$C{\equiv}C$–CH_3. Under the reaction conditions, this proton is removed with base. Formation of the resulting acetylide anion drives the equilibrium to favor its formation. Protonation of this acetylide anion gives the less stable terminal alkyne.

11.58

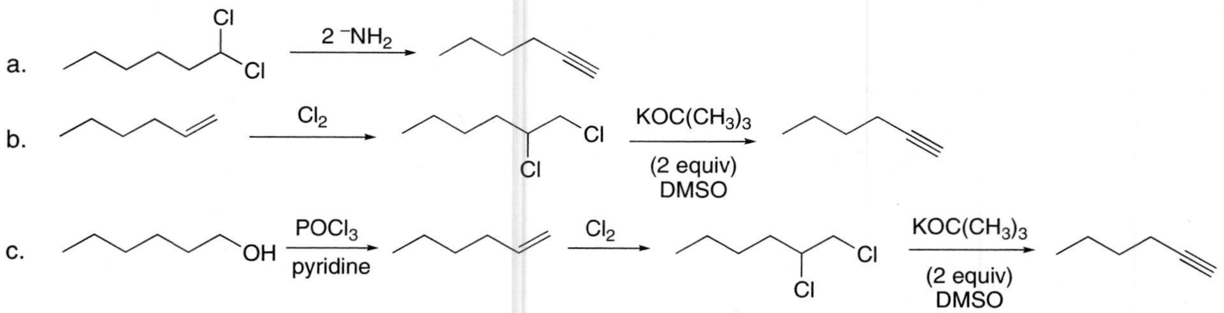

11.59

a. $C_6H_5CH_2CH_2Br$ $\xrightarrow{KOC(CH_3)_3}$ $C_6H_5CH{=}CH_2$ $\xrightarrow{Br_2}$ $C_6H_5CHBrCH_2Br$ $\xrightarrow[\text{excess}]{NaNH_2}$ $C_6H_5C{\equiv}CH$

b. $C_6H_5CHBrCH_3$ $\xrightarrow{KOC(CH_3)_3}$ $C_6H_5CH=CH_2$ $\xrightarrow{Br_2}$ $C_6H_5CHBrCH_2Br$ $\xrightarrow[\text{excess}]{NaNH_2}$ $C_6H_5C{\equiv}CH$

c. $C_6H_5CH_2CH_2OH$ $\xrightarrow{H_2SO_4}$ $C_6H_5CH=CH_2$ $\xrightarrow{Br_2}$ $C_6H_5CHBrCH_2Br$ $\xrightarrow[\text{excess}]{NaNH_2}$ $C_6H_5C{\equiv}CH$

11.60 The alkyl halides must be methyl or 1°.

a. $HC{\equiv}C{-}CH_2CH_2CH(CH_3)_2$ $\Longrightarrow$ $HC{\equiv}C{:}^-$ + $Cl{-}CH_2CH_2CH(CH_3)_2$ $\boxed{\text{1° RX}}$

b. $CH_3{-}C{\equiv}C{-}\underset{CH_3}{\overset{CH_3}{C}}{-}CH_2CH_3$ $\Longrightarrow$ $CH_3{-}Cl$ + $^-{:}C{\equiv}C{-}\underset{CH_3}{\overset{CH_3}{C}}{-}CH_2CH_3$

c. $\langle\text{cyclohexyl}\rangle{-}C{\equiv}C{-}CH_2CH_2CH_3$ $\Longrightarrow$ $\langle\text{cyclohexyl}\rangle{-}C{\equiv}C{:}^-$ + $Cl{-}CH_2CH_2CH_3$ $\boxed{\text{1° RX}}$

11.61

a. $HC{\equiv}C{-}H$ $\xrightarrow{Na^+ H^-}$ $HC{\equiv}C^-$ $\xrightarrow{(CH_3)_2CHCH_2{-}Cl}$ $(CH_3)_2CHCH_2C{\equiv}CH$

b. $HC{\equiv}C{-}H$ $\xrightarrow{Na^+ H^-}$ $HC{\equiv}C^-$ $\xrightarrow{CH_3CH_2CH_2{-}Cl}$ $\boxed{CH_3CH_2CH_2C{\equiv}CH}$ $\xrightarrow{NaH}$ $CH_3CH_2CH_2C{\equiv}C^-$

$\downarrow CH_3CH_2CH_2{-}Cl$

$CH_3CH_2CH_2C{\equiv}CCH_2CH_2CH_3$

c. $CH_3CH_2CH_2C{\equiv}CH$ (from b.) $\xrightarrow[\text{[2] } H_2O_2,\ HO^-]{\text{[1] } BH_3}$ $CH_3CH_2CH_2CH_2CHO$

d. $CH_3CH_2CH_2C{\equiv}CH$ (from b.) $\xrightarrow[\substack{H_2SO_4 \\ HgSO_4}]{H_2O}$ $CH_3CH_2CH_2\overset{\overset{O}{\|}}{C}CH_3$

e. $CH_3CH_2CH_2C{\equiv}CH$ (from b.) $\xrightarrow{2\ HCl}$ $CH_3CH_2CH_2CCl_2CH_3$

f. $CH_3CH_2CH_2C{\equiv}CCH_2CH_2CH_3$ (from b.) $\xrightarrow[H_2SO_4,\ HgSO_4]{H_2O}$ $CH_3CH_2CH_2\overset{\overset{O}{\|}}{C}CH_2CH_2CH_3$

11.62

a. $CH_3CH_2CH=CH_2$ $\xrightarrow{Cl_2}$ $CH_3CH_2\underset{}{\overset{Cl}{C}}H{-}\underset{}{\overset{Cl}{C}}H_2$ $\xrightarrow{2\ ^-NH_2}$ $CH_3CH_2C{\equiv}CH$

b. $CH_3CH_2C{\equiv}CH$ (from a.) $\xrightarrow[\text{(2 equiv)}]{HBr}$ $CH_3CH_2CBr_2CH_3$

c. $CH_3CH_2C{\equiv}CH$ (from a.) $\xrightarrow[\text{(2 equiv)}]{Cl_2}$ $CH_3CH_2CCl_2CHCl_2$

d. $CH_3CH_2CH=CH_2$ $\xrightarrow{Br_2}$ $CH_3CH_2CHBrCH_2Br$

e. $CH_3CH_2C\equiv CH$ $\xrightarrow{NaH}$ $CH_3CH_2C\equiv C^-$ $\xrightarrow{CH_3CH_2CH_2-Br}$ $CH_3CH_2C\equiv C-CH_2CH_2CH_3$
(from a.)

f. $CH_3CH_2C\equiv C^-$ $\xrightarrow[\text{[2] } H_2O]{\text{[1]} \triangle O}$ $CH_3CH_2C\equiv C-CH_2CH_2OH$
(from e.)

g. $CH_3CH_2C\equiv C^-$ $\xrightarrow[\text{[2] } H_2O]{\text{[1]}}$
(from e.)

(+ enantiomer)

11.63

a. $HC\equiv CH$ $\xrightarrow{NaH}$ $HC\equiv C^-$ $\xrightarrow{CH_3CH_2CH_2CH_2CH_2CH_2Br}$ $CH_3CH_2CH_2CH_2CH_2CH_2C\equiv CH$

b. $CH_3CH_2CH_2CH_2CH_2CH_2C\equiv CH$ $\xrightarrow{NaH}$ $CH_3CH_2CH_2CH_2CH_2CH_2C\equiv C^-$ $\xrightarrow{CH_3CH_2Br}$ $CH_3CH_2CH_2CH_2CH_2CH_2C\equiv CCH_2CH_3$
(from a.)

c. $CH_3CH_2CH_2CH_2CH_2CH_2C\equiv C^-$ $\xrightarrow[\text{[2] } H_2O]{\text{[1]} \triangle O}$ $CH_3CH_2CH_2CH_2CH_2CH_2C\equiv CCH_2CH_2OH$
(from b.)

d. $CH_3CH_2CH_2CH_2CH_2CH_2C\equiv CCH_2CH_2OH$ $\xrightarrow[\text{[2] } CH_3CH_2Br]{\text{[1] NaH}}$ $CH_3CH_2CH_2CH_2CH_2CH_2C\equiv CCH_2CH_2OCH_2CH_3$
(from c.)

11.64

11.65

11.66

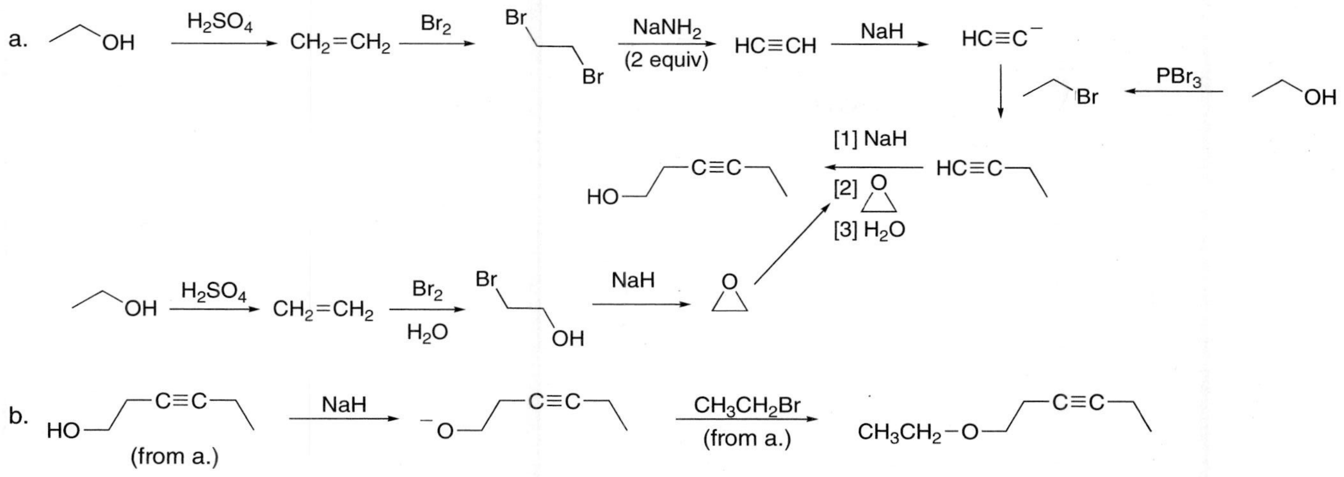
a.
OH
H_2SO_4
$CH_2=CH_2$
Br_2
Br
Br
$NaNH_2$
(2 equiv)
HC≡CH
NaH
HC≡C⁻
PBr_3
Br
OH
[1] NaH
[2]
[3] H_2O
HC≡C
HO
C≡C
H_2O
OH
b.
(from a.)
CH_3CH_2Br
CH_3CH_2–O

11.67

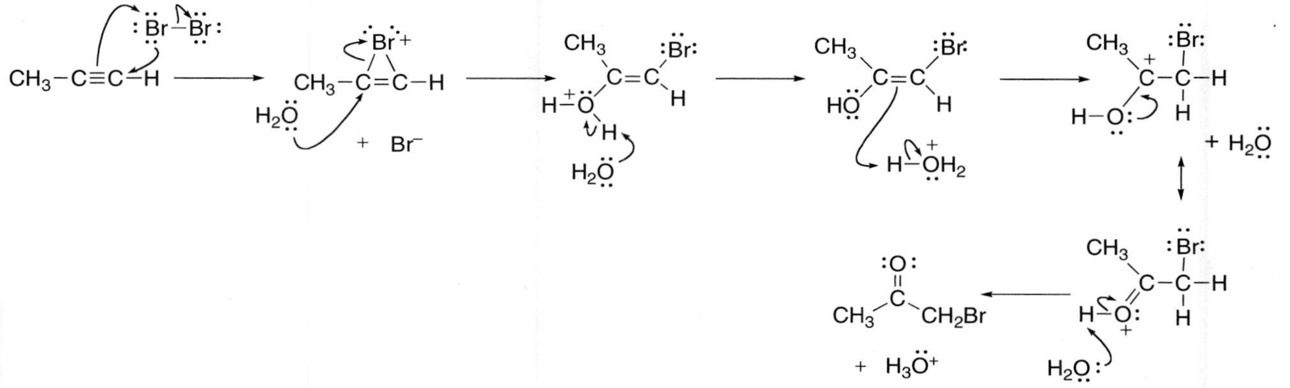
CH_3–C≡C–H
CH_3
:Br:
+ Br⁻
H_2O
+ H_2O
+ H_3O^+
CH_2Br

11.68

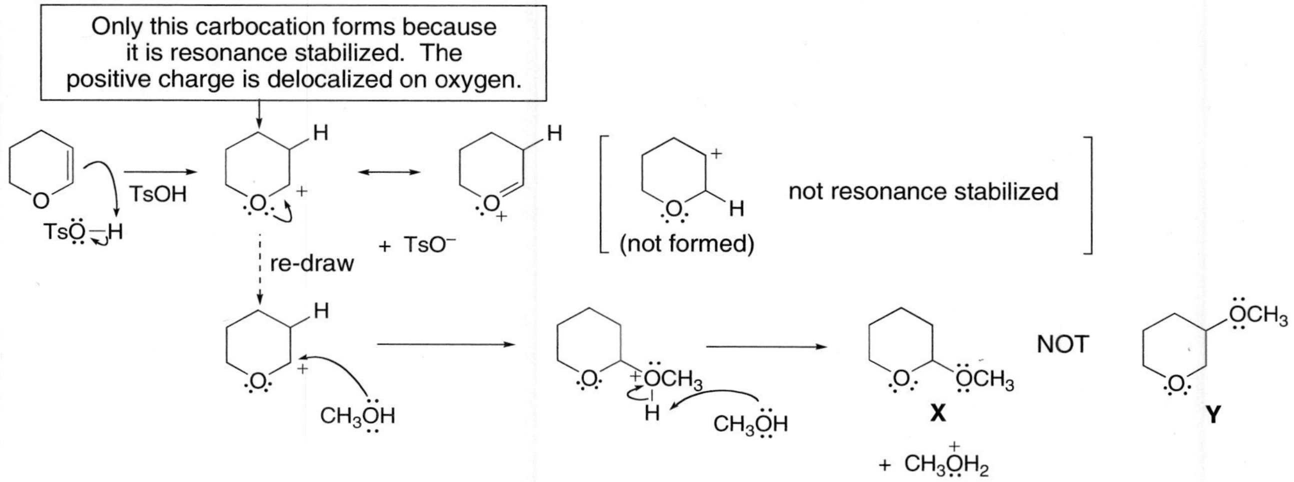
Only this carbocation forms because it is resonance stabilized. The positive charge is delocalized on oxygen.
TsOH
+ TsO⁻
re-draw
not resonance stabilized
(not formed)
CH_3OH
OCH_3
X
NOT
Y
+ CH_3OH_2

11.69

+ HCOOH   resonance structures   enol

Chapter 12: Oxidation and Reduction

◆ **Summary: Terms that describe reaction selectivity**

- A **regioselective reaction** forms predominately or exclusively one constitutional isomer (Section 8.5).

major product
trisubstituted alkene

minor product
disubstituted alkene

- A **stereoselective reaction** forms predominately or exclusively one stereoisomer (Section 8.5).

trans alkene
major product

cis alkene
minor product

- An **enantioselective reaction** forms predominately or exclusively one enantiomer (Section 12.15).

allylic alcohol

Sharpless reagent

or

One enantiomer is favored.

◆ **Definitions of oxidation and reduction**

Oxidation reactions result in:
- an increase in the number of C–Z bonds, *or*
- a decrease in the number of C–H bonds.

Reduction reactions result in:
- a decrease in the number of C–Z bonds, *or*
- an increase in the number of C–H bonds.

[Z = an element more electronegative than C]

◆ **Reduction reactions**

[1] Reduction of alkenes—Catalytic hydrogenation (12.3)

R—CH=CH-R $\xrightarrow[\text{Pd, Pt, or Ni}]{\text{H}_2}$

alkane

- **Syn addition** of H$_2$ occurs.
- Increasing alkyl substitution on the C=C decreases the rate of reaction.

[2] Reduction of alkynes

[a] $R-C\equiv C-R$ $\xrightarrow{\text{2 H}_2}{\text{Pd-C}}$

H H
| |
R–C–C–R
| |
H H
alkane

- Two equivalents of H_2 are added and four new C–H bonds are formed (12.5A).

[b] $R-C\equiv C-R$ $\xrightarrow{\text{H}_2}{\text{Lindlar catalyst}}$

R   R
  \\   /
   C=C
  /   \\
 H   H
cis alkene

- **Syn addition** of H_2 occurs, forming a **cis** alkene (12.5B).
- The Lindlar catalyst is deactivated so that reaction stops after one equivalent of H_2 has been added.

[c] $R-C\equiv C-R$ $\xrightarrow{\text{Na}}{\text{NH}_3}$

R   H
  \\   /
   C=C
  /   \\
 H   R
trans alkene

- **Anti addition** of H_2 occurs, forming a **trans** alkene (12.5C).

[3] Reduction of alkyl halides (12.6)

$R-X$ $\xrightarrow{\text{[1] LiAlH}_4}{\text{[2] H}_2\text{O}}$ R–H **alkane**

- The reaction follows an S_N2 mechanism.
- CH_3X and RCH_2X react faster than more substituted RX.

[4] Reduction of epoxides (12.6)

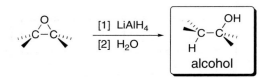

$\xrightarrow{\text{[1] LiAlH}_4}{\text{[2] H}_2\text{O}}$

OH
  |
C–C
|
H
alcohol

- The reaction follows an S_N2 mechanism.
- In unsymmetrical epoxides, H^- (from $LiAlH_4$) attacks at the less substituted carbon.

◆ Oxidation reactions

[1] Oxidation of alkenes

[a] Epoxidation (12.8)

$C=C$ + RCO_3H $\longrightarrow$

O
/ \\
C–C
epoxide

- The mechanism has **one step**.
- **Syn addition** of an O atom occurs.
- The reaction is stereospecific.

[b] Anti dihydroxylation (12.9A)

$C=C$ $\xrightarrow{\text{[1] RCO}_3\text{H}}{\text{[2] H}_2\text{O (H}^+ \text{ or HO}^-)}$

HO
   \\
    C–C
   /   \\
        OH
1,2-diol

- Ring opening of an epoxide intermediate with ^-OH or H_2O forms a 1,2-diol with two OH groups added in an **anti** fashion.

[c] Syn dihydroxylation (12.9B)

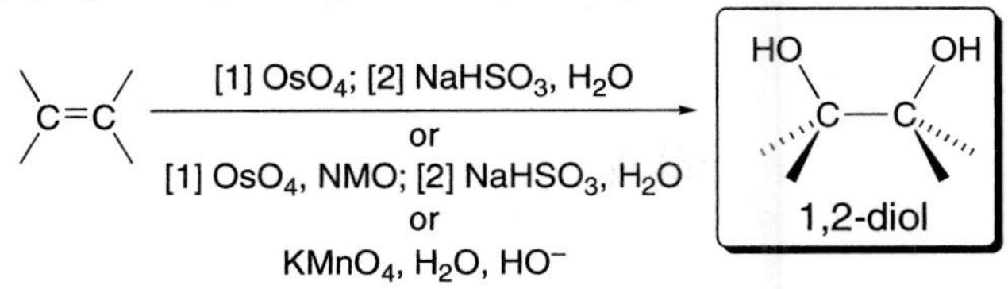

- Each reagent adds two new C–O bonds to the C=C in a **syn** fashion.

[d] Oxidative cleavage (12.10)

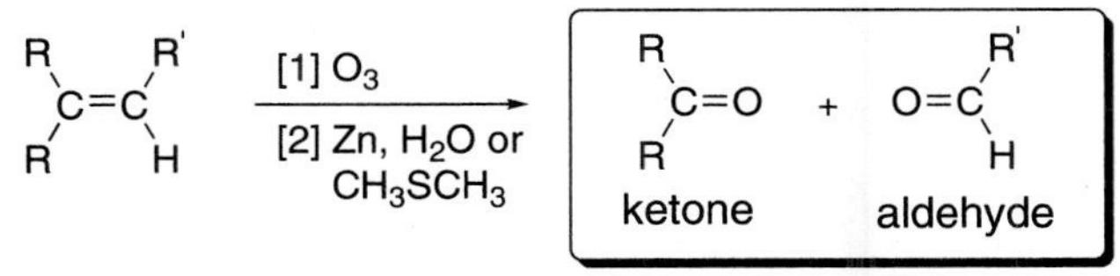

- Both the σ and π bonds of the alkene are cleaved to form two carbonyl groups.

[2] Oxidative cleavage of alkynes (12.11)

[a] R−C≡C−R' internal alkyne

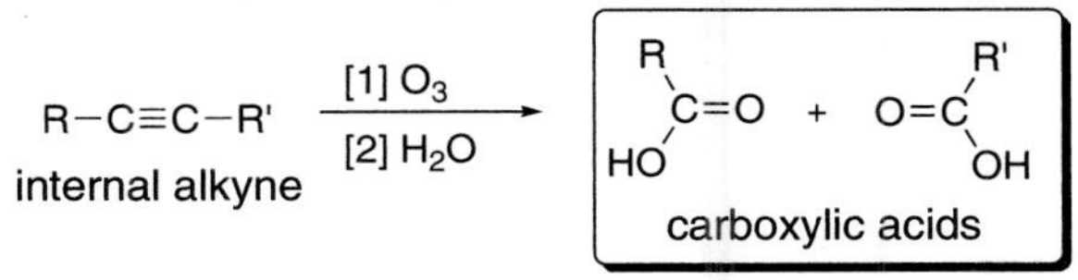

- The σ bond and both π bonds of the alkyne are cleaved.

[b] R−C≡C−H terminal alkyne

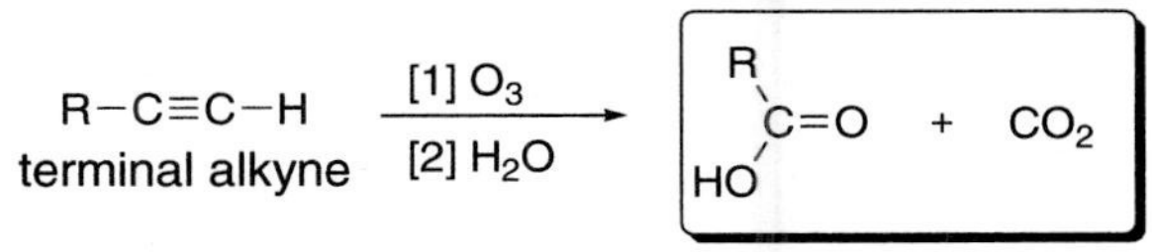

[3] Oxidation of alcohols (12.12, 12.13)

[a]

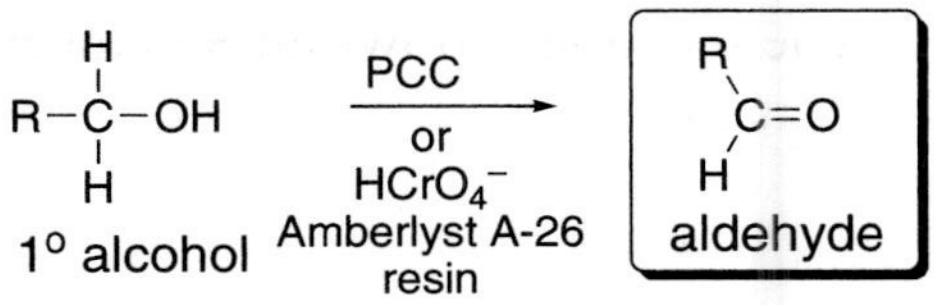

1° alcohol

- Oxidation of a 1° alcohol with PCC or HCrO$_4^-$ (Amberlyst A-26 resin) stops at the aldehyde stage. Only one C–H bond is replaced by a C–O bond.

[b]

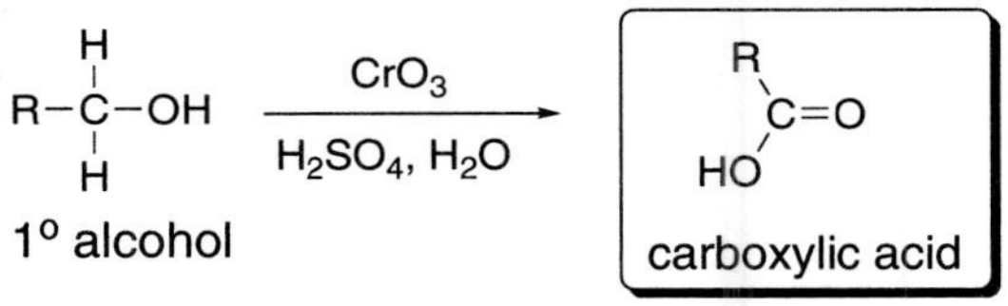

1° alcohol

- Oxidation of a 1° alcohol under harsher reaction conditions—CrO$_3$ (or Na$_2$Cr$_2$O$_7$ or K$_2$Cr$_2$O$_7$) + H$_2$O + H$_2$SO$_4$—affords a RCOOH. Two C–H bonds are replaced by two C–O bonds.

[c]

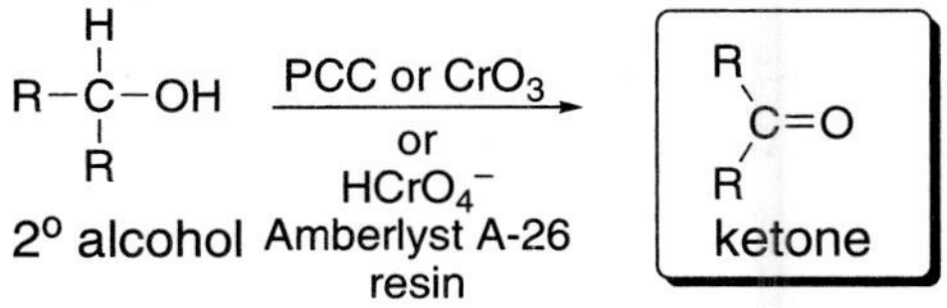

2° alcohol

- Since a 2° alcohol has only one C–H bond on the carbon bearing the OH group, all Cr^{6+} reagents—PCC, CrO$_3$, Na$_2$Cr$_2$O$_7$, K$_2$Cr$_2$O$_7$, or HCrO$_4^-$ (Amberlyst A-26 resin)—oxidize a 2° alcohol to a ketone.

[4] Asymmetric epoxidation of allylic alcohols (12.15)

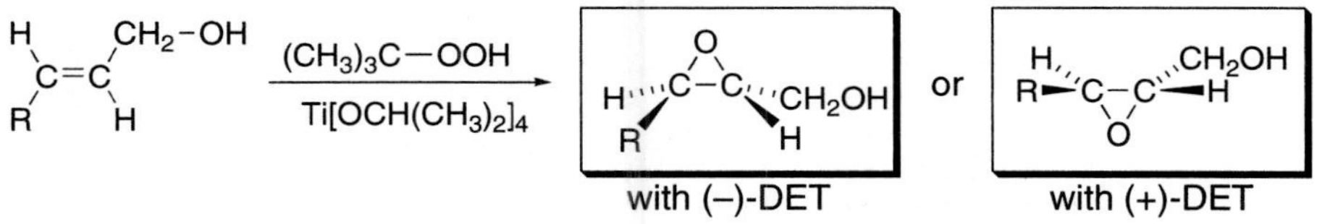

Chapter 12: Answers to Problems

12.1 *Oxidation* results in an *increase* in the number of C–Z bonds (usually C–O bonds) *or* a *decrease* in the number of C–H bonds.

Reduction results in a *decrease* in the number of C–Z bonds (usually C–O bonds) *or* an *increase* in the number of C–H bonds.

a. ⟶ **oxidation**

b. ⟶ **reduction**

c. CH_3—C(=O)—CH_3 ⟶ CH_3—C(=O)—OCH_3 **oxidation**

d. $CH_2{=}CH_2$ ⟶ CH_3CH_2Cl **neither**
1 new C–H bond
and 1 new C–Cl bond

12.2 **Hydrogenation** is the addition of hydrogen. When alkenes are hydrogenated, they are *reduced* by the addition of H_2 to the π bond. To draw the alkane product, add a H to each C of the double bond.

a. CH_3, $CH_2CH(CH_3)_2$ $C{=}C$ CH_3, H ⟶ CH_3—C(CH_3)($CH_2CH(CH_3)_2$)—C(H)(H)—H

b. ⟶

c. ⟶

12.3 Draw the alkenes that form each alkane when hydrogenated.

a. [structure] or [structure] or [structure] $\xrightarrow{H_2, \text{Pd-C}}$ [structure]

b. [structure] or [structure] $\xrightarrow{H_2, \text{Pd-C}}$ [structure]

c. [structures] or ... or ... or ... $\xrightarrow{H_2, \text{Pd-C}}$ [structure]

12.4 Cis alkenes are less stable than trans alkenes, so they have larger heats of hydrogenation. Increasing alkyl substitution increases the stability of a C=C, decreasing the heat of hydrogenation.

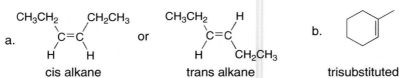

a. CH_3CH_2, CH_2CH_3 $C{=}C$ H, H **or** CH_3CH_2, H $C{=}C$ H, CH_2CH_3

cis alkane
less stable
***larger* heat of hydrogenation**

trans alkane

b. [structure] **or** [structure]

trisubstituted

disubstituted
less stable
***larger* heat of hydrogenation**

12.5 Hydrogenation products must be identical to use hydrogenation data to evaluate the relative stability of the starting materials.

2-methyl-2-pentene

3-methyl-1-pentene

Different products are formed.
Hydrogenation data can't be used to determine the relative stability of the starting materials.

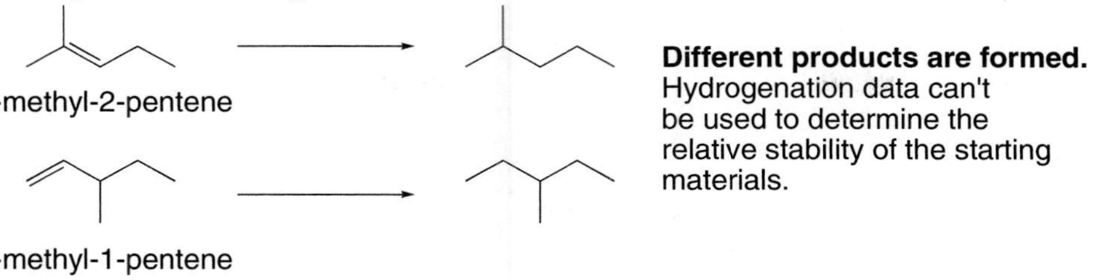

12.6 Increasing alkyl substitution on the C=C decreases the rate of hydrogenation.

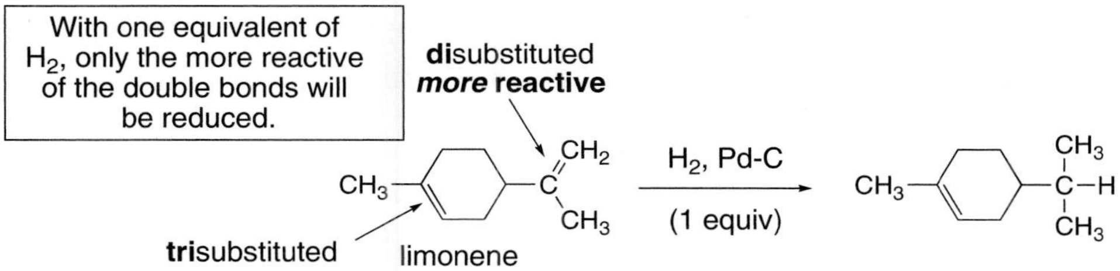

With one equivalent of H_2, only the more reactive of the double bonds will be reduced.

disubstituted *more* **reactive**

trisubstituted   limonene

H_2, Pd-C
(1 equiv)

12.7

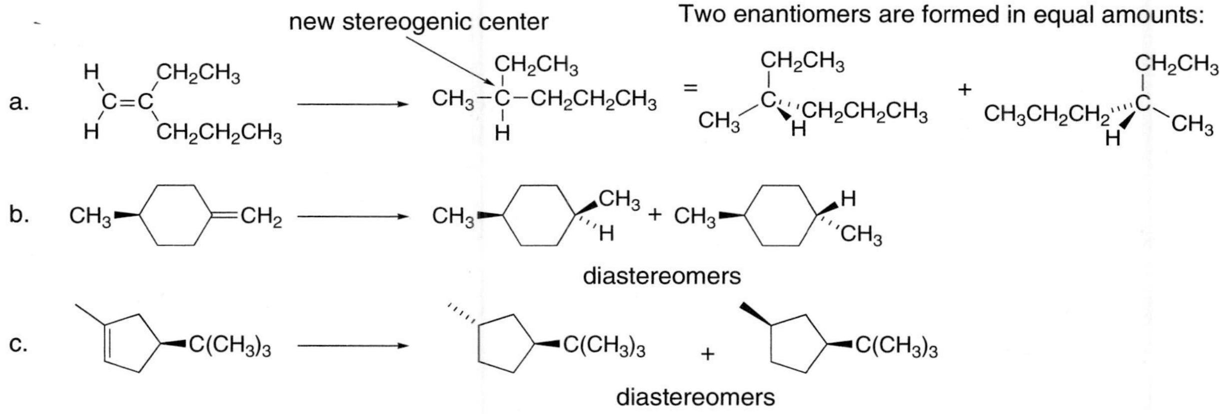

new stereogenic center

Two enantiomers are formed in equal amounts:

a.

b.

diastereomers

c.

diastereomers

12.8

Compound	Molecular formula before hydrogenation	Molecular formula after hydrogenation	Number of rings	Number of π bonds
A	$C_{10}H_{12}$	$C_{10}H_{16}$	3	2
B	C_4H_8	C_4H_{10}	0	1
C	C_6H_8	C_6H_{12}	1	2

12.9

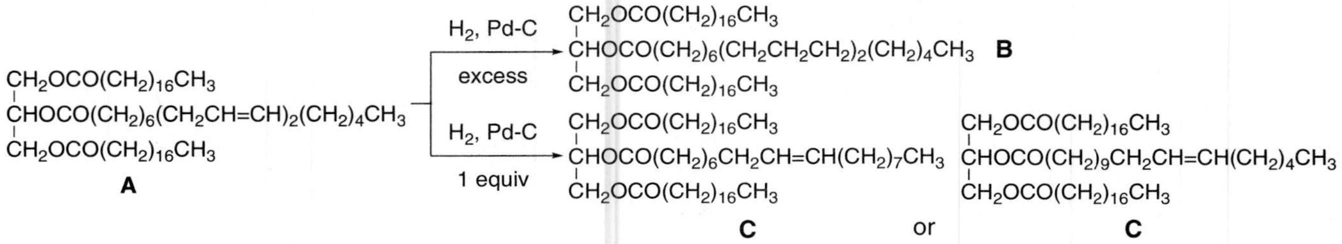

A has 2 double bonds.
lowest melting point

C has 1 double bond.
intermediate melting point

B has 0 double bonds.
highest melting point

12.10 Hydrogenation of $HC \equiv CCH_2CH_2CH_3$ and $CH_3C \equiv CCH_2CH_3$ yields the same compound. The heat of hydrogenation is larger for $HC \equiv CCH_2CH_2CH_3$ than for $CH_3C \equiv CCH_2CH_3$ because internal alkynes are more stable (lower in energy) than terminal alkynes.

12.11

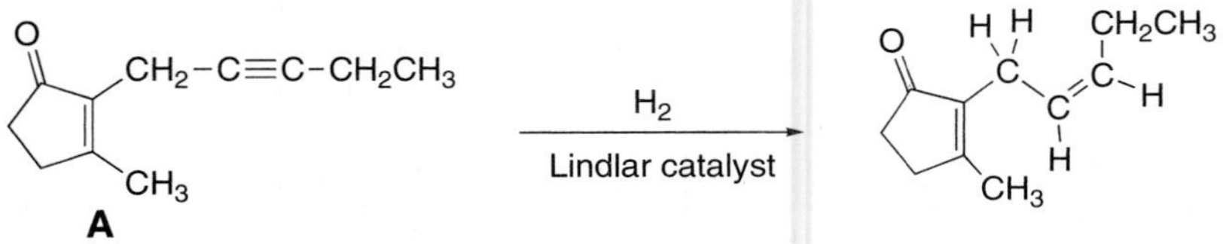

cis-jasmone
(perfume component
isolated from jasmine flowers)

12.12 To draw the products of catalytic hydrogenation remember:
- H_2 (excess), Pd will reduce **alkenes and alkynes to alkanes**.
- H_2 (excess), Lindlar catalyst will reduce *only* **alkynes to cis alkenes**.

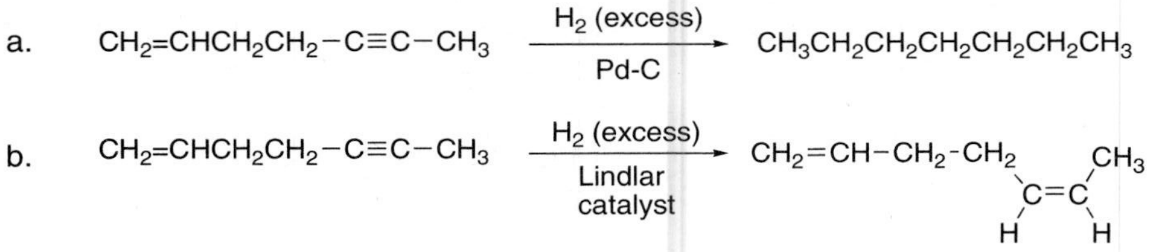

a.

b.

12.13 Use the directions from Answer 12.12.

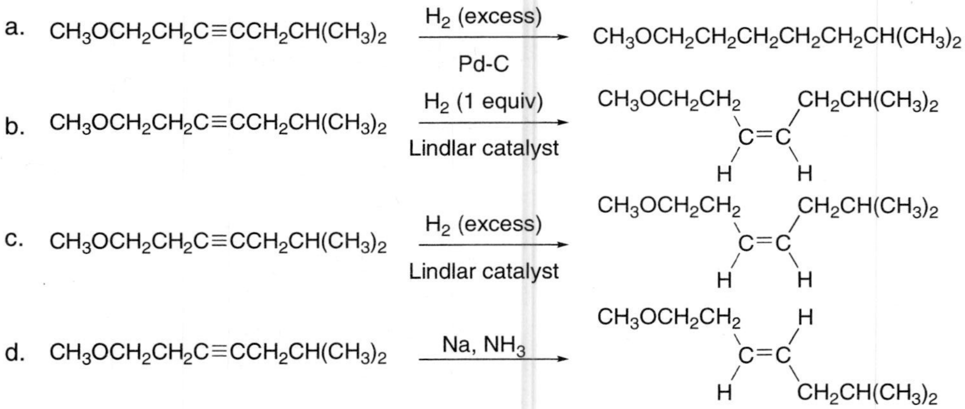

a.

b.

c.

d.

12.14

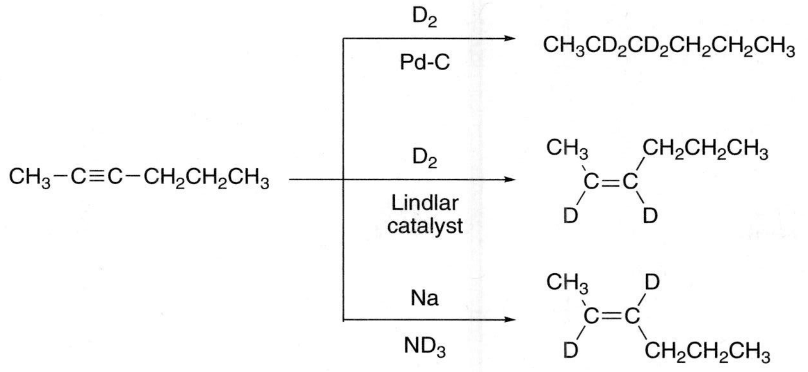

$$CH_3-C\equiv C-CH_2CH_2CH_3$$

D₂, Pd-C → $CH_3CD_2CD_2CH_2CH_2CH_3$

D₂, Lindlar catalyst → (cis product with CH₃ and CH₂CH₂CH₃ on same side, D and D)

Na, ND₃ → (trans product with CH₃ and D, D and CH₂CH₂CH₃)

12.15

A →(H₂, Lindlar catalyst)→ R B

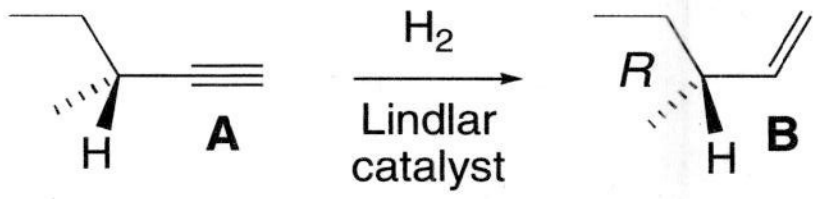

12.16 LiAlH₄ reduces alkyl halides to alkanes and epoxides to alcohols.

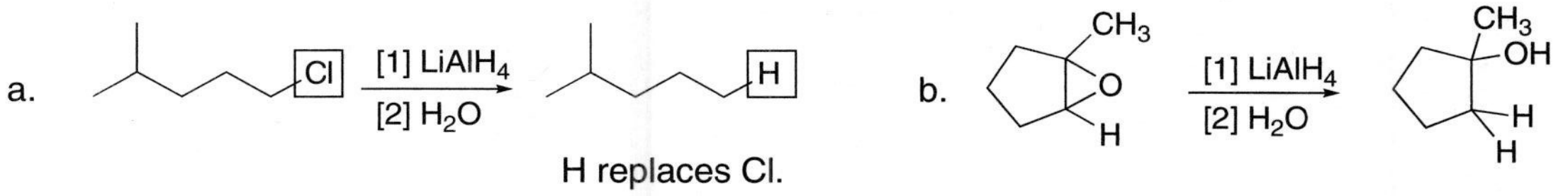

a.
[1] LiAlH₄, [2] H₂O

H replaces Cl.

b.
[1] LiAlH₄, [2] H₂O

12.17 To draw the product, add an O atom across the π bond of the C=C.

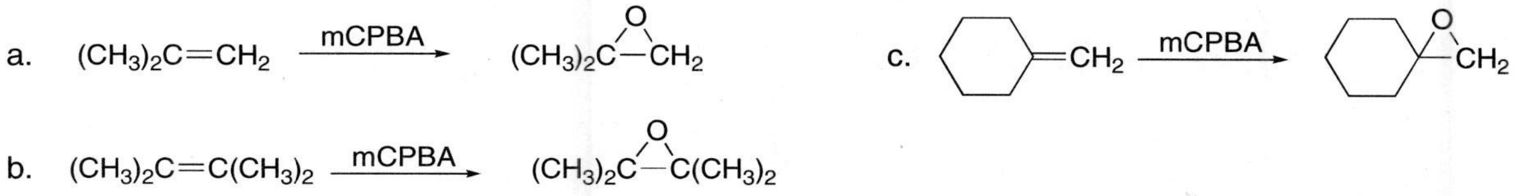

a. $(CH_3)_2C=CH_2$ —(mCPBA)→ $(CH_3)_2C\overset{O}{-}CH_2$

b. $(CH_3)_2C=C(CH_3)_2$ —(mCPBA)→ $(CH_3)_2C\overset{O}{-}C(CH_3)_2$

c. =CH₂ —(mCPBA)→

12.18 For epoxidation reactions:
- There are two possible products: O adds from above and below the double bond.
- Substituents on the C=C retain their original configuration in the products.

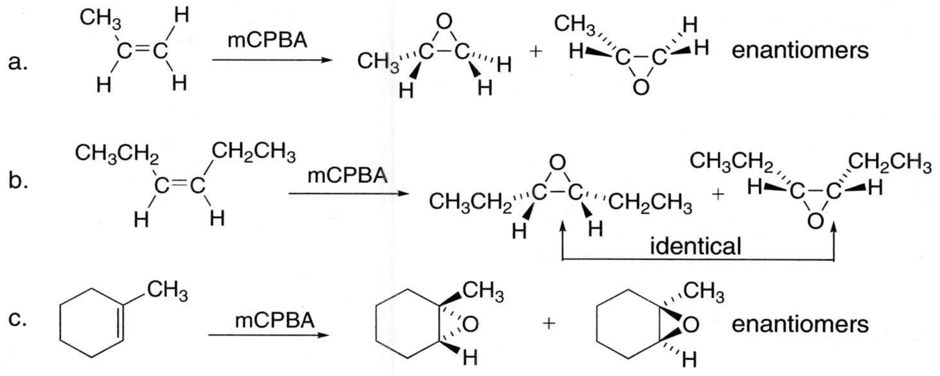

a.
—(mCPBA)→ (two epoxides) enantiomers

b.
—(mCPBA)→ (two epoxides) identical

c.
—(mCPBA)→ (two epoxides) enantiomers

12.19 Treatment of an alkene with a peroxyacid followed by H_2O, HO^- adds two hydroxy groups in an **anti** fashion. *cis*-2-Butene and *trans*-2-butene yield different products of dihydroxylation. *cis*-2-Butene gives a mixture of two enantiomers and *trans*-2-butene gives a meso compound. The reaction is stereospecific because two stereoisomeric starting materials give different products that are also stereoisomers of each other.

12.20 Treatment of an alkene with OsO_4 adds two hydroxy groups in a **syn** fashion. *cis*-2-Butene and *trans*-2-butene yield different stereoisomers in this dihydroxylation, so the reaction is stereospecific.

12.21 To draw the oxidative cleavage products:
- **Locate all the π bonds** in the molecule.
- **Replace all C=C's with *two* C=O's.**

a. $(CH_3)_2C \overset{\swarrow}{=} CHCH_2CH_2CH_2CH_3$
  Replace this π bond with two C=O's.
  $\xrightarrow[\text{[2] Zn, H}_2O]{\text{[1] O}_3}$ $(CH_3)_2C=O$ + $O=CHCH_2CH_2CH_2CH_3$
  ketone aldehyde
  One **ketone** and one **aldehyde** are formed.

b. $\xrightarrow[\text{[2] Zn, H}_2O]{\text{[1] O}_3}$ Two **aldehydes** are formed.

c. $\xrightarrow[\text{[2] Zn, H}_2O]{\text{[1] O}_3}$ A **dicarbonyl** compound is formed.

12.22 To find the alkene that yields the oxidative cleavage products:
- **Find the two carbonyl groups** in the products.
- **Join the two carbonyl carbons** together with a double bond. This is the double bond that was broken during ozonolysis.

a. $(CH_3)_2C=O$ + $(CH_3CH_2)_2C=O$ $\Longrightarrow$ $(CH_3)_2C=C(CH_2CH_3)_2$

Join these two C's.

b. (cyclohexanone) + CH_3CHO $\Longrightarrow$ (cyclohexylidene=CHCH_3)

Join these two C's.

c.
$$\begin{array}{c} CH_3 \\ C=O \\ CH_3 \end{array} \text{ only} \Longrightarrow \begin{array}{c} CH_3 \quad CH_3 \\ C=C \\ CH_3 \quad CH_3 \end{array}$$

With only one product, the alkene must be symmetrical around the double bond. Join this C to the same C in another identical molecule.

12.23

a. (1,4-cyclohexadiene) $\xrightarrow[\text{[2] CH}_3\text{SCH}_3]{\text{[1] O}_3}$ (OHC–CHO) + (OHC–CH_2–CHO)

c. (methylenecyclohexene type) $\xrightarrow[\text{[2] CH}_3\text{SCH}_3]{\text{[1] O}_3}$ (H–CO–CH_2CH_2CH_2–CO–CH_2–CHO) + $H_2C=O$ (H–CHO–H)

b. (cyclohexadiene) $\xrightarrow[\text{[2] CH}_3\text{SCH}_3]{\text{[1] O}_3}$ (H–CO–CH_2–CHO) + (CH_3–CO–CH_2–CHO)

d. (cyclohexadiene) $\xrightarrow[\text{[2] CH}_3\text{SCH}_3]{\text{[1] O}_3}$ (H–CO–CH_2–CHO) + (OHC–CH(CH_3)–CHO)

12.24 To draw the products of oxidative cleavage of alkynes:
- **Locate the triple bond**.
- For internal alkynes, **convert the *sp* hybridized C to COOH.**
- For terminal alkynes, the *sp* hybridized **C–H becomes CO₂.**

a. $CH_3CH_2–C\equiv C–CH_2CH_2CH_3$ $\xrightarrow[\text{[2] H}_2\text{O}]{\text{[1] O}_3}$ $CH_3CH_2–CO–OH$ + $HO–CO–CH_2CH_2CH_3$

internal alkyne

b. (Ph–C≡C–Ph) $\xrightarrow[\text{[2] H}_2\text{O}]{\text{[1] O}_3}$ (Ph–CO–OH) + (HO–CO–Ph)

internal alkyne

identical compounds

c. $H–C\equiv C–CH_2–CH_2–C\equiv C–CH_3$ $\xrightarrow[\text{[2] H}_2\text{O}]{\text{[1] O}_3}$ CO_2 + $HO–CO–CH_2CH_2–CO–OH$ + $HO–CO–CH_3$

terminal alkyne   internal alkyne

12.25

a. $CO_2 + CH_3(CH_2)_8CO_2H$

$\Downarrow$

$CH_3(CH_2)_8-C\equiv CH$

b. $CH_3CH_2CH(CH_3)CO_2H$

$\Downarrow$

$CH_3CH_2CH-C\equiv C-CHCH_2CH_3$
    | |
    CH_3 CH_3

c. $CH_3CH_2CO_2H, HO_2CCH_2CO_2H, CH_3CO_2H$

$\Downarrow$

$CH_3CH_2-C\equiv C-CH_2-C\equiv C-CH_3$

d. $HO_2C(CH_2)_{14}CO_2H \Longrightarrow$

12.26 For **oxidation of alcohols**, remember:
- **1° Alcohols** are oxidized to aldehydes with PCC.
- **1° Alcohols** are oxidized to carboxylic acids with oxidizing agents like CrO_3 or $Na_2Cr_2O_7$.
- **2° Alcohols** are oxidized to ketones with all Cr^{6+} reagents.

a.

c.

b.

d.

12.27 Upon treatment with $HCrO_4^-$ Amberlyst A-26 resin:
- **1° Alcohols** are oxidized to aldehydes.
- **2° Alcohols** are oxidized to ketones.

a.

c.

b. $HO-\!\!\!\bigcirc\!\!\!-OH$

12.28

a.

The by-products of the reaction with sodium hypochlorite are water and table salt (NaCl), as opposed to the by-products with $HCrO_4^-$ Amberlyst A-26 resin, which contain carcinogenic Cr^{3+} metal.

b. Oxidation with NaOCl has at least two advantages over oxidation with CrO_3, H_2SO_4 and H_2O. Since no Cr^{6+} is used as oxidant, there are no Cr by-products that must be disposed of. Also, CrO_3 oxidation is carried out in corrosive inorganic acids (H_2SO_4) and oxidation with NaOCl avoids this.

12.29

ethylene glycol

12.30 To draw the products of a **Sharpless epoxidation**:
- With the C=C horizontal, draw the allylic alcohol with the OH on the **top right** of the alkene.
- Add the new oxygen **above** the plane if (–)-DET is used and **below** the plane if (+)-DET is used.

a.

(+)-DET adds O
below the plane.

b.

(–)-DET adds O
above the plane.

12.31 Sharpless epoxidation needs an ***allylic alcohol*** as the starting material. Alkenes with no allylic OH group will not undergo reaction with the Sharpless reagent.

This alkene is part of an **allylic alcohol** and will be epoxidized.

geraniol

This alkene is **not** part of an allylic alcohol and will not be epoxidized.

12.32 Use the rules from Answer 12.1.

a. [structure: cyclohexyl–C≡CH] ⟶ [structure: cyclohexyl–CH$_2$CH$_3$] **reduction**

b. [structure: CH$_3$CH$_2$CH$_2$CH$_2$OH] ⟶ [structure: carboxylic acid] **oxidation**

c. CH_3CH_2Br ⟶ $CH_2=CH_2$ **neither**
   1 C–H and 1 C–Br
   bond are removed.

d. [epoxide structure] ⟶ [cyclohexanol with OH] **reduction**

e. $CH_2=CH_2$ ⟶ $ClCH_2CH_2Cl$ **oxidation**
   (2 new C–Cl bonds)

f. HO–[benzene ring]–OH ⟶ O=[quinone ring]=O **oxidation**

12.33 Use the principles from Answer 12.2 and draw the products of syn addition of H$_2$ from above and below the C=C.

a. [pent-1-ene structure] $\xrightarrow[\text{Pd-C}]{\text{H}_2}$ [pentane structure]

b. [structure: cyclohexene with CH$_3$ and CH$_3$] $\xrightarrow[\text{Pd-C}]{\text{H}_2}$ [cis dimethylcyclohexane with CH$_3$] + [trans dimethylcyclohexane with CH$_3$]

c. [structure: cyclohexene with CH$_2$CH$_3$ and CH$_3$] $\xrightarrow[\text{Pd-C}]{\text{H}_2}$ [structure with CH$_2$CH$_3$ and CH$_3$] + [structure with CH$_2$CH$_3$ and CH$_3$]

d. [structure: (CH$_3$)$_2$CH–C(CH$_3$CH$_2$)=CH$_2$] $\xrightarrow[\text{Pd-C}]{\text{H}_2}$ [structure: (CH$_3$)$_2$CH, H, C, CH$_2$CH$_3$, CH$_3$] + [structure: (CH$_3$)$_2$CH, H, C, CH$_2$CH$_3$, CH$_3$]

12.34 Increasing alkyl substitution increases alkene stability, decreasing the heat of hydrogenation.

2-methyl-2-butene
trisubstituted
smallest ΔH° = –112 kJ/mol

2-methyl-1-butene
disubstituted
intermediate ΔH° = –119 kJ/mol

3-methyl-1-butene
monosubstituted
largest ΔH° = –127 kJ/mol

12.35

A possible structure:

a. Compound **A**: molecular formula C$_5$H$_8$: hydrogenated to C$_5$H$_{10}$.
   2 degrees of unsaturation, 1 is hydrogenated.
   1 ring and 1 π bond ⟶ [methylenecyclobutane structure]

b. Compound **B**: molecular formula C$_{10}$H$_{16}$: hydrogenated to C$_{10}$H$_{18}$.
   3 degrees of unsaturation, 1 is hydrogenated.
   2 rings and 1 π bond ⟶ [octahydronaphthalene structure]

c. Compound **C**: molecular formula C$_8$H$_8$: hydrogenated to C$_8$H$_{16}$.
   5 degrees of unsaturation, 4 are hydrogenated.
   1 ring and 4 π bonds ⟶ [cyclooctatetraene structure]

12.36

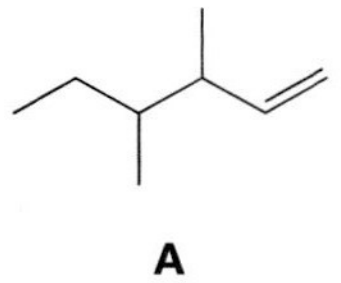

a. **mono**substituted
   largest heat of hydrogenation
b. **fastest** reaction rate

A

c.
$\xrightarrow[\text{[2] Zn, H}_2\text{O}]{\text{[1] O}_3}$

a. **tetra**substituted
   smallest heat of hydrogenation
b. **slowest** reaction rate

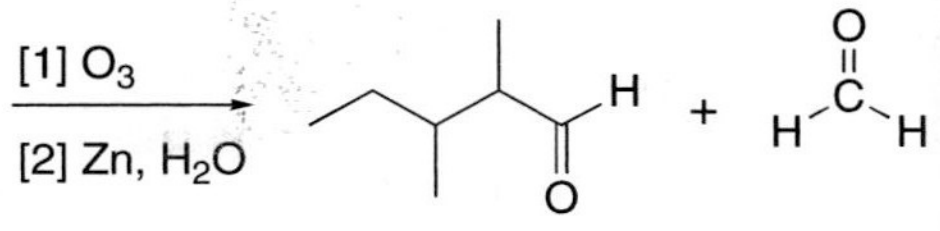

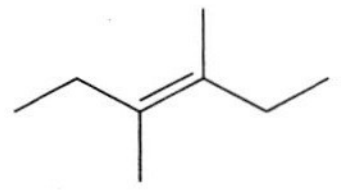

B

c.
$\xrightarrow[\text{[2] Zn, H}_2\text{O}]{\text{[1] O}_3}$

identical

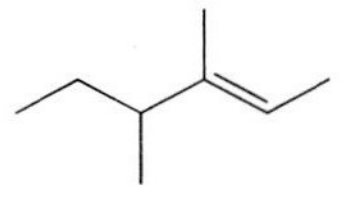

a. **tri**substituted
   intermediate heat of hydrogenation
b. **intermediate** reaction rate

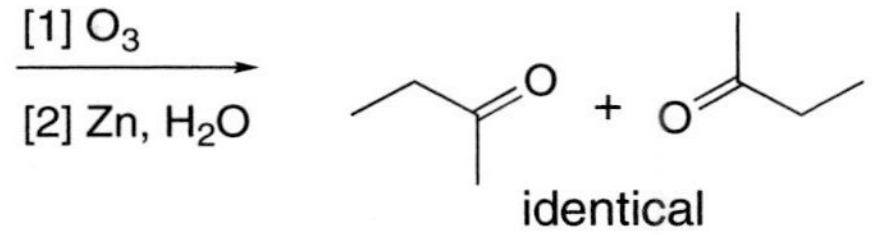

C

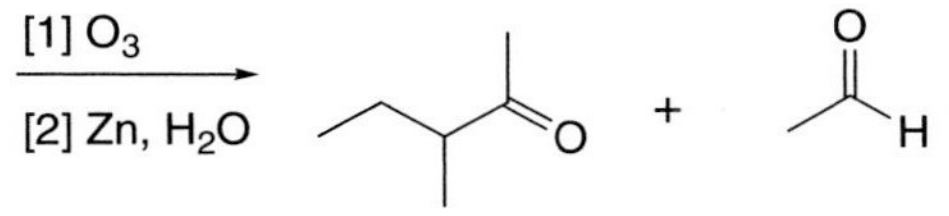

c.
$\xrightarrow[\text{[2] Zn, H}_2\text{O}]{\text{[1] O}_3}$

12.37 Work backwards to find the alkene that will be hydrogenated to form 3-methylpentane.

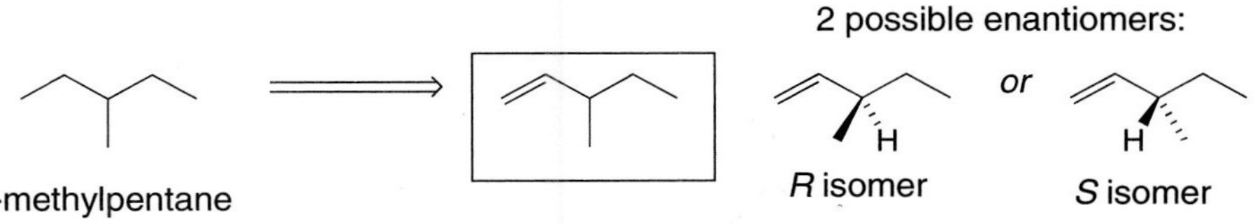

3-methylpentane

2 possible enantiomers:

or

R isomer   *S* isomer

12.38

a. stearidonic acid $\xrightarrow[\text{Pd-C}]{\text{H}_2 \text{ (excess)}}$ stearic acid

b. $\xrightarrow[\text{Pd-C}]{\text{H}_2 \text{ (1 equiv)}}$ + + + +,

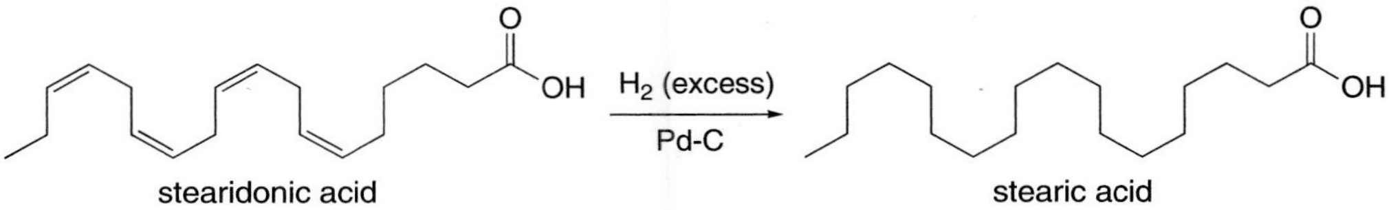

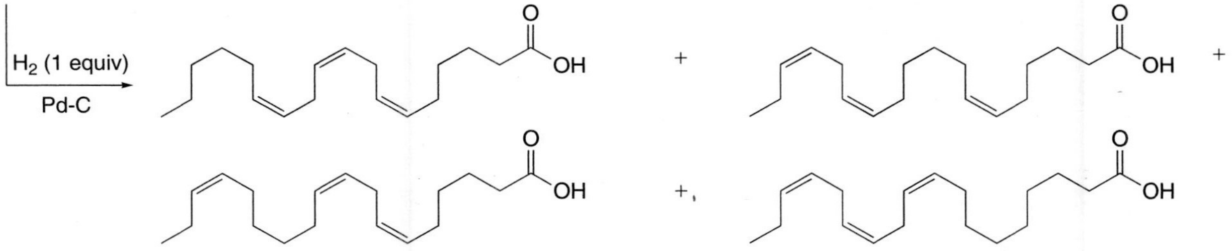

c.

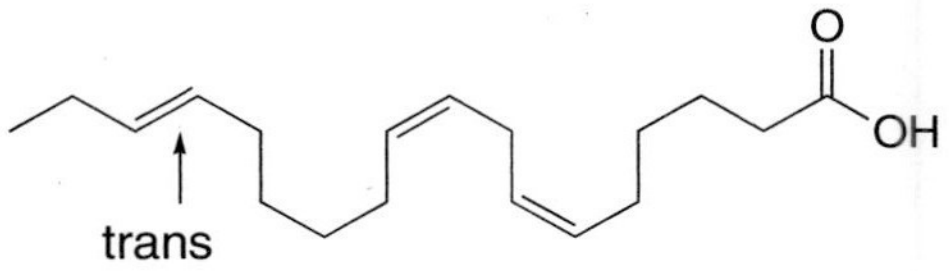

trans
one possibility

d.

stearidonic acid
4 cis C=C's

<

product in (b)
3 cis C=C's

<

product in (c)
2 cis and one trans C=C's

12.39

a. $\xrightarrow[\text{Pd-C}]{\text{H}_2}$

b. $\xrightarrow[\text{Lindlar catalyst}]{\text{H}_2}$ no reaction

c. $\xrightarrow[\text{NH}_3]{\text{Na}}$ no reaction

d. $\xrightarrow{\text{CH}_3\text{CO}_3\text{H}}$

e. $\xrightarrow[\text{[2] H}_2\text{O, HO}^-]{\text{[1] CH}_3\text{CO}_3\text{H}}$ + anti addition

f. $\xrightarrow[\text{[2] NaHSO}_3, \text{H}_2\text{O}]{\text{[1] OsO}_4 + \text{NMO}}$ syn addition

g. $\xrightarrow[\text{H}_2\text{O, HO}^-]{\text{KMnO}_4}$ syn addition

h. $\xrightarrow[\text{[2] H}_2\text{O}]{\text{[1] LiAlH}_4}$ no reaction

i. $\xrightarrow[\text{[2] CH}_3\text{SCH}_3]{\text{[1] O}_3}$

j. $\xrightarrow[\substack{\text{Ti[OCH(CH}_3)_2]_4 \\ (-)\text{-DET}}]{(\text{CH}_3)_3\text{COOH}}$ no reaction

k. $\xrightarrow{\text{mCPBA}}$

l. $\xrightarrow[\text{[2] H}_2\text{O}]{\text{[1] LiAlH}_4}$

12.40

a. $\text{CH}_3\text{CH}_2\text{CH}_2-\text{C}\equiv\text{C}-\text{CH}_2\text{CH}_2\text{CH}_3$ $\xrightarrow[\text{Pd-C}]{\text{H}_2 \text{ (excess)}}$

b. $\text{CH}_3\text{CH}_2\text{CH}_2-\text{C}\equiv\text{C}-\text{CH}_2\text{CH}_2\text{CH}_3$ $\xrightarrow[\text{Lindlar catalyst}]{\text{H}_2}$ cis alkene

c. $\text{CH}_3\text{CH}_2\text{CH}_2-\text{C}\equiv\text{C}-\text{CH}_2\text{CH}_2\text{CH}_3$ $\xrightarrow[\text{NH}_3]{\text{Na}}$ trans alkene

d. $\text{CH}_3\text{CH}_2\text{CH}_2-\text{C}\equiv\text{C}-\text{CH}_2\text{CH}_2\text{CH}_3$ $\xrightarrow[\text{[2] H}_2\text{O}]{\text{[1] O}_3}$

identical

12.41

a. $\xrightarrow[\text{Pd-C}]{\text{H}_2}$ [* = new stereogenic center]

+

Two enantiomers are formed.

12.42

12.43

a. PCC →

b. $CH_3CH_2CH_2CH_2OH$ —PCC→ $CH_3CH_2CH_2\overset{O}{\underset{}{C}}H$

c. —OH $\xrightarrow[\text{H}_2\text{SO}_4,\ \text{H}_2\text{O}]{\text{CrO}_3}$ —OH

12.44

a. —OH $\xrightarrow[\text{pyridine}]{[1]\ \text{SOCl}_2}$ —Cl $\xrightarrow[\text{[3] H}_2\text{O}]{\text{[2] LiAlH}_4}$

b. =CH$_2$ $\xrightarrow[\text{[2] NaHSO}_3,\ \text{H}_2\text{O}]{\text{[1] OsO}_4}$ —OH / CH$_2$OH

c. =CH$_2$ $\xrightarrow{\text{[1] mCPBA}}$ (epoxide) $\xrightarrow[\text{[3] H}_2\text{O}]{\text{[2] LiAlH}_4}$ —OH / CH$_3$

d. $\xrightarrow[\substack{\text{Lindlar}\\\text{catalyst}}]{\text{H}_2}$

12.45

a. H$_2$, Pd-C

b. mCPBA

d. [1] LiAlH$_4$ / [2] H$_2$O

f. PBr$_3$ or HBr

h. [1] LiAlH$_4$ / [2] H$_2$O

c. KMnO$_4$ H$_2$O, HO$^-$ or [1] OsO$_4$ [2] NaHSO$_3$, H$_2$O

e. H$_2$O ($^-$OH)

g. CrO$_3$, H$_2$SO$_4$, H$_2$O or PCC or HCrO$_4^-$, Amberlyst A-26 resin

(+ enantiomer)

12.46 Alkenes treated with [1] OsO$_4$ followed by NaHSO$_3$ in H$_2$O will undergo syn addition, whereas alkenes treated with [2] CH$_3$CO$_3$H followed by $^-$OH in H$_2$O will undergo anti addition.

a.

[1] $\xrightarrow[\text{[2] NaHSO}_3,\ \text{H}_2\text{O}]{\text{[1] OsO}_4}$

[2] $\xrightarrow[\substack{\text{[2] }^-\text{OH, H}_2\text{O}\\\text{anti addition}}]{\text{[1] CH}_3\text{CO}_3\text{H}}$ → rotate →

b. [1]

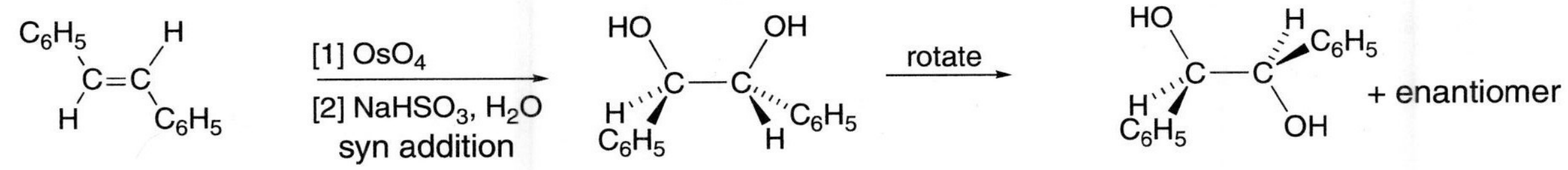

C6H5 / H C=C / H C6H5

[1] OsO4
[2] NaHSO3, H2O
syn addition

HO OH
H—C—C—C6H5
C6H5 H

rotate

HO H C6H5
H—C—C—OH
C6H5

+ enantiomer

[2]

C6H5 C6H5 C=C / H H

[1] CH3CO3H
[2] ⁻OH, H2O

HO H C6H5
C—C
H OH
C6H5

+ enantiomer

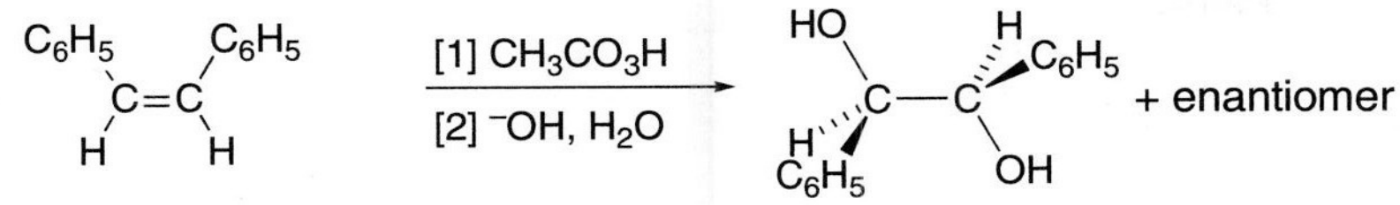

c. [1]

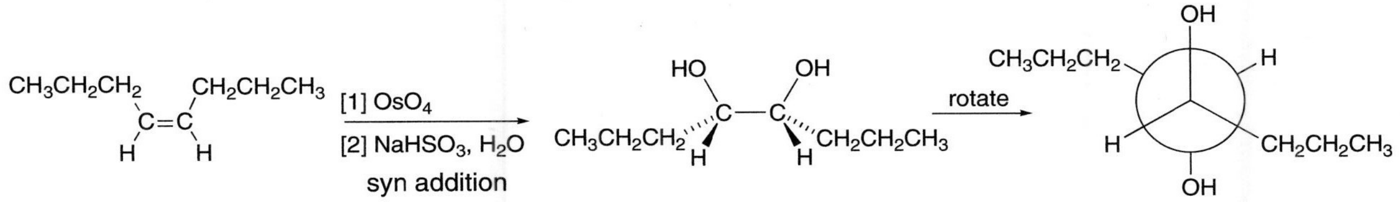

CH3CH2CH2 CH2CH2CH3 C=C / H H

[1] OsO4
[2] NaHSO3, H2O
syn addition

HO OH
CH3CH2CH2—C—C—CH2CH2CH3
H H

rotate

CH3CH2CH2 OH H
H CH2CH2CH3
OH

[2]

CH3CH2CH2 H C=C / H CH2CH2CH3

[1] CH3CO3H
[2] ⁻OH, H2O

CH3CH2CH2 OH H
H CH2CH2CH3
OH

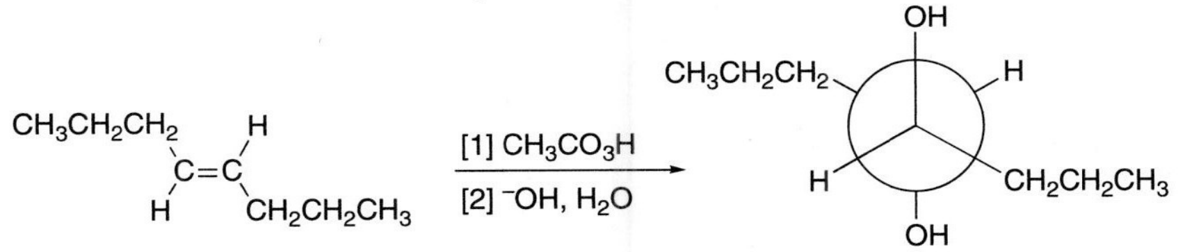

12.47

A

H3Al⁻H
Li⁺

H :O:⁻
H—OH

+ AlH3 + Li⁺

:OH

+ ⁻:OH

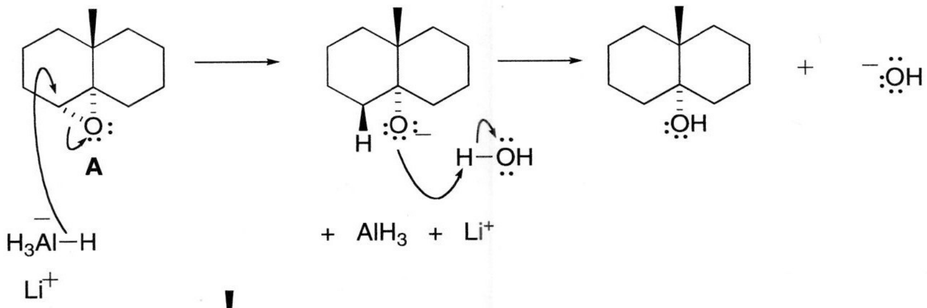

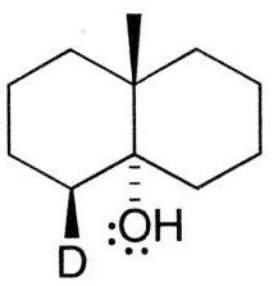

D⁻ (from LiAlD4) opens the epoxide ring from the back side, so it is oriented on a wedge in the final product.

D :OH

12.48

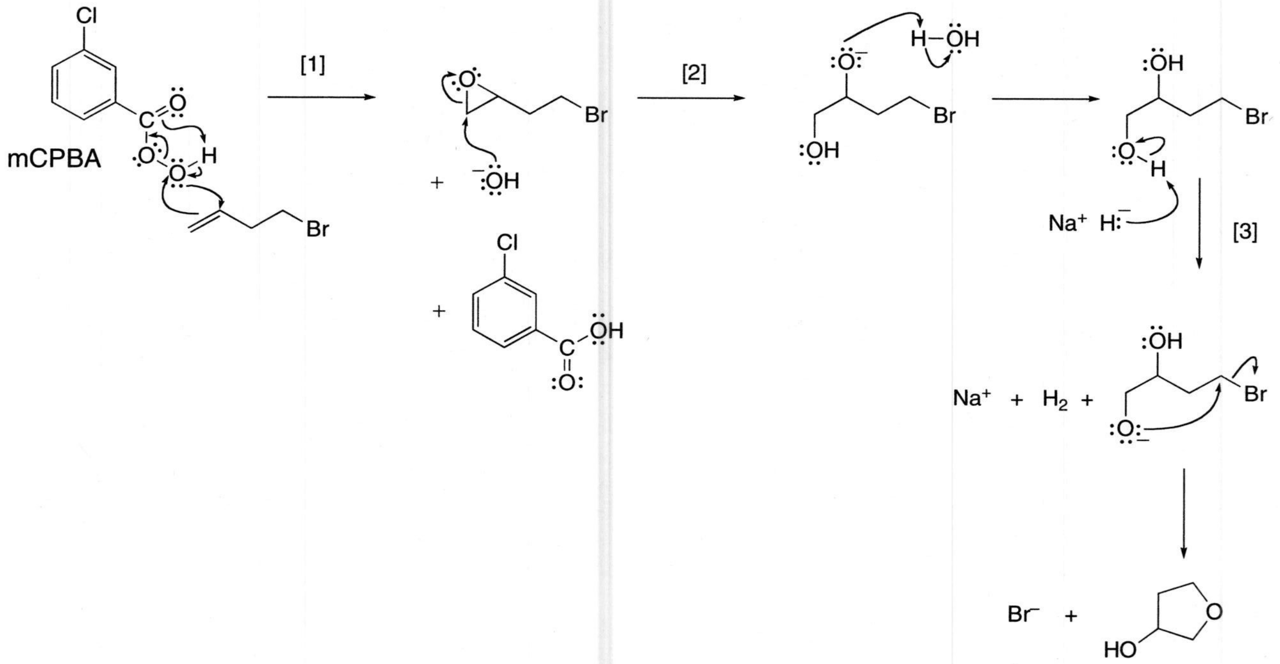

12.49 Use the directions from Answer 12.21.

a. $(CH_3CH_2)_2C=CHCH_2CH_3$ $\xrightarrow[\text{[2] } CH_3SCH_3]{\text{[1] } O_3}$ $(CH_3CH_2)_2C=O$ + $O=CHCH_2CH_3$

b.

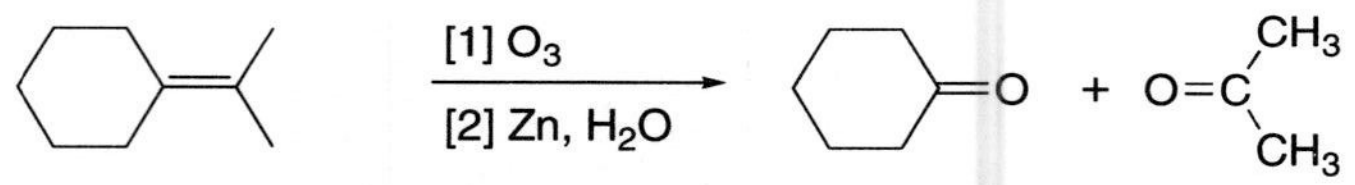

c.

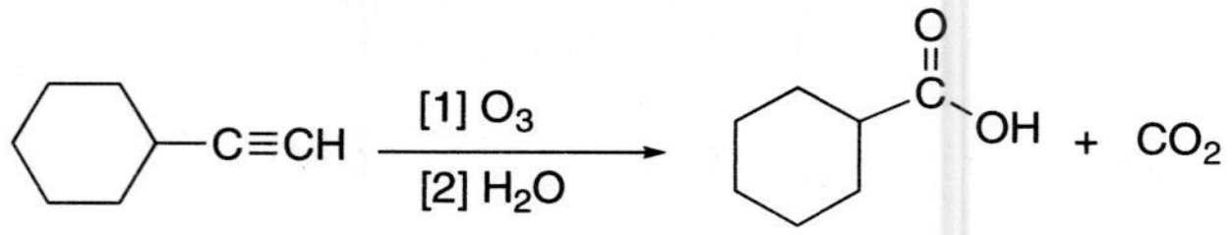

d.

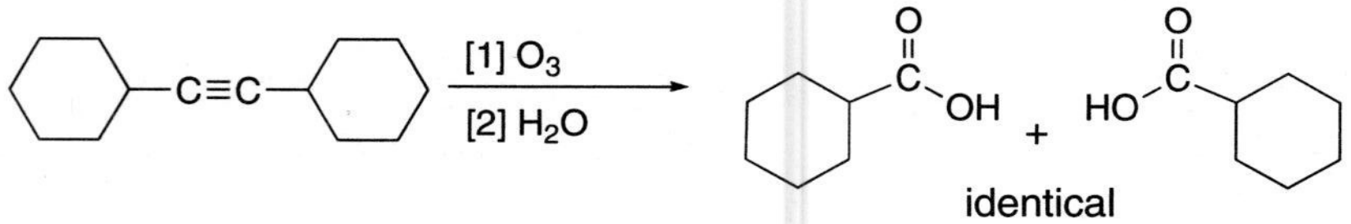

identical

12.50

a. $(CH_3)_2C=O$ and $CH_2=O$ $\Longrightarrow$ $(CH_3)_2C=CH_2$

b.

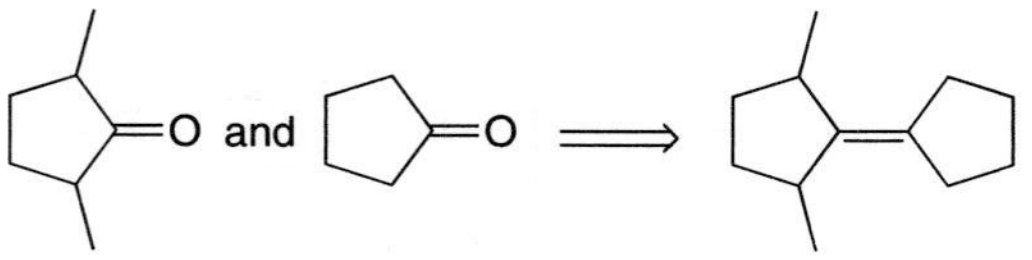

c. $CH_3CH_2CH_2CHO$ only $\Longrightarrow$ $CH_3CH_2CH_2CH=CHCH_2CH_2CH_3$

Join this C to the same C
in another identical molecule.

d.

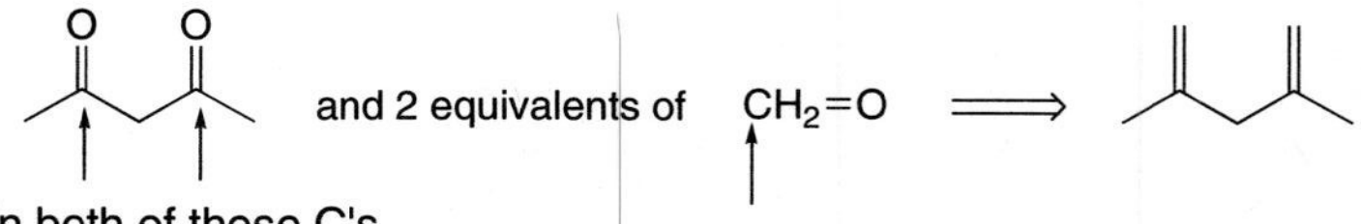

and 2 equivalents of $CH_2=O$ $\Longrightarrow$

Join both of these C's
to a C from formaldehyde.

formaldehyde C

12.51 Use the directions from Answer 12.22.

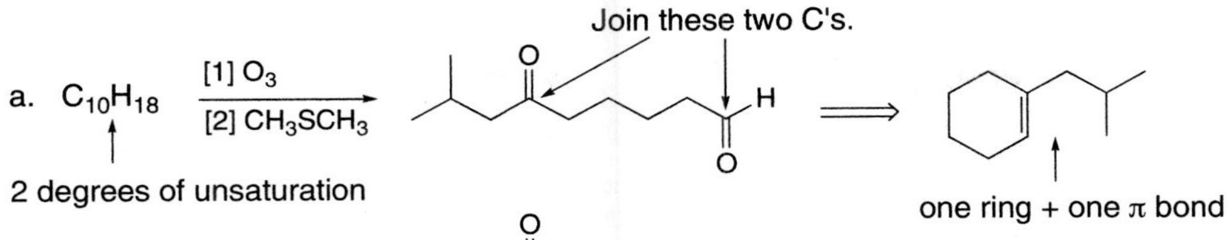

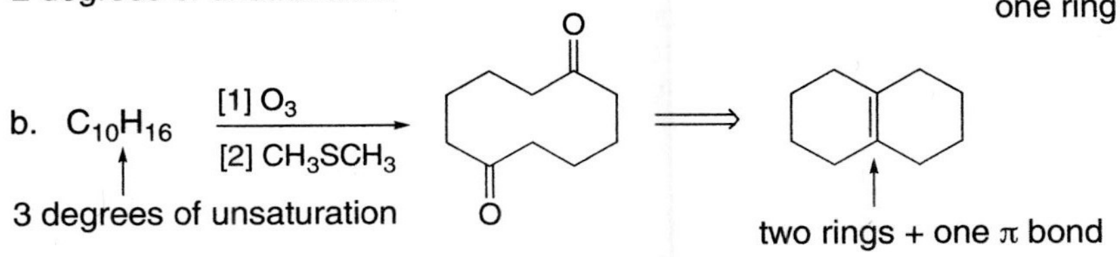

a. $C_{10}H_{18}$ $\xrightarrow[\text{[2] CH}_3\text{SCH}_3]{\text{[1] O}_3}$

2 degrees of unsaturation

Join these two C's.

one ring + one π bond

b. $C_{10}H_{16}$ $\xrightarrow[\text{[2] CH}_3\text{SCH}_3]{\text{[1] O}_3}$

3 degrees of unsaturation

two rings + one π bond

12.52

Join these two C's.

a. $CH_3CH_2CH_2CH_2COOH$ and CO_2 $\Longrightarrow$ $CH_3CH_2CH_2CH_2C\equiv CH$

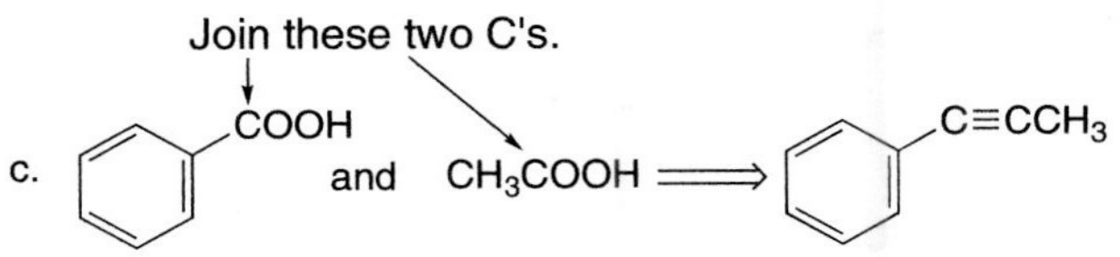

Join these two C's.

c.   COOH   and CH_3COOH $\Longrightarrow$ $C\equiv CCH_3$

Join these two C's.

b. CH_3CH_2COOH and $CH_3CH_2CH_2COOH$ $\Longrightarrow$ $CH_3CH_2C\equiv CCH_2CH_2CH_3$

12.53

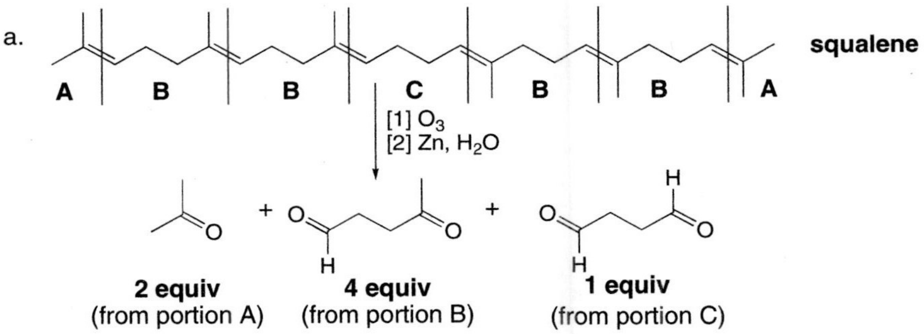

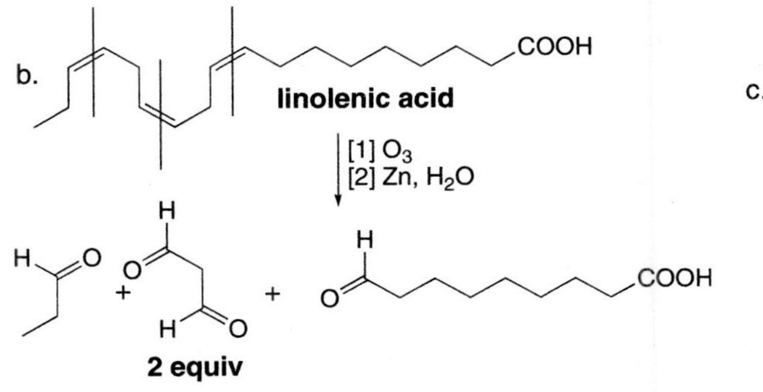

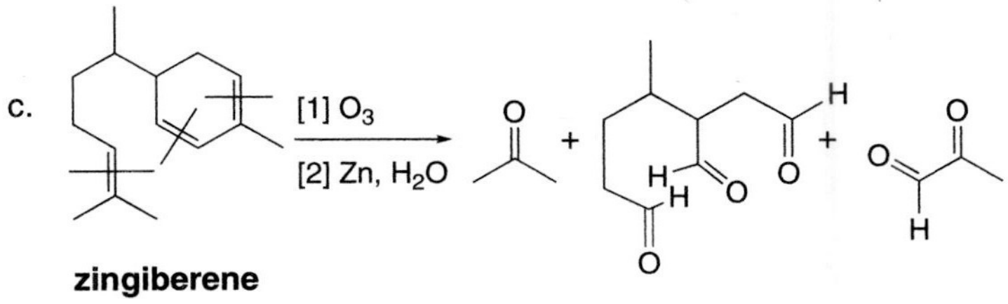

a.   **squalene**

A   B   B   C   B   B   A

$\xrightarrow[\text{[2] Zn, H}_2\text{O}]{\text{[1] O}_3}$

2 equiv   **4 equiv**   **1 equiv**
(from portion A)   (from portion B)   (from portion C)

b.   **linolenic acid**   COOH

$\xrightarrow[\text{[2] Zn, H}_2\text{O}]{\text{[1] O}_3}$

2 equiv   COOH

c.   **zingiberene** $\xrightarrow[\text{[2] Zn, H}_2\text{O}]{\text{[1] O}_3}$

12.54

a. C_8H_{12} **A** $\xrightarrow[\text{[2] CH}_3\text{SCH}_3]{\text{[1] O}_3}$

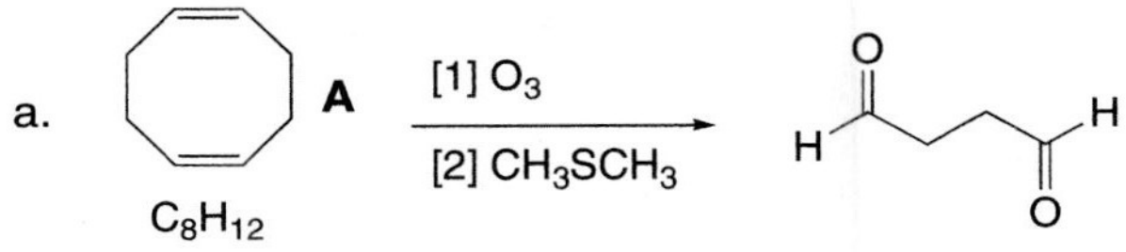

b.

C_6H_{10} **B** $\xrightarrow[\text{Pd-C}]{\text{H}_2\text{ (excess)}}$ (isohexane)   (4-methyl-1-pentyne) $\xrightarrow[\text{[2] CH}_3\text{I}]{\text{[1] NaNH}_2}$ C_7H_{12} **C**

12.55

$C_{10}H_{16}$ $\xrightarrow[\text{Pd-C}]{\text{H}_2}$ 2,6-dimethyloctane

3 degrees of unsaturation

The hydrogenation reaction tells you that both oximene and myrcene have 3 π bonds (and no rings). Use this carbon backbone and add in the double bonds based on the oxidative cleavage products.

Oximene: $(CH_3)_2C{=}O$ $CH_2{=}O$ $CH_2(CHO)_2$ $CH_3{-}\overset{\overset{\text{O}}{\|}}{C}{-}CHO$ $\Longrightarrow$

Myrcene: $(CH_3)_2C{=}O$ $CH_2{=}O$ (2 equiv) $H{-}\overset{\overset{\text{O}}{\|}}{C}{-}CH_2CH_2{-}\overset{\overset{\text{O}}{\|}}{C}{-}CHO$ $\Longrightarrow$

12.56

a.

C_7H_{12} **A** $\xrightarrow[\text{Pd-C}]{\text{H}_2\text{ (2 equiv)}}$

oxidative cleavage ↓

$CH_3{-}\overset{\overset{\text{O}}{\|}}{C}{-}OH$ + other product(s)

A $\xrightarrow[\text{Lindlar catalyst}]{\text{H}_2}$ **B**

↓ Na, NH₃

C

b. **A** does not react with NaH because it is not a terminal alkyne.

12.57

$C_{10}H_{18}O$ **A** $\xrightarrow{\text{H}_2\text{SO}_4}$ $C_{10}H_{16}$ **B**   + $C_{10}H_{16}$ **C** $\xrightarrow[\text{Pd-C}]{\text{H}_2}$ decalin

↓ ozonolysis ↓

D   **E** $C_{10}H_{16}O_2$

12.58 Since hydrogenation of DHA forms $CH_3(CH_2)_{20}COOH$, DHA is a 22-carbon fatty acid. The ozonolysis products show where the double bonds are located.

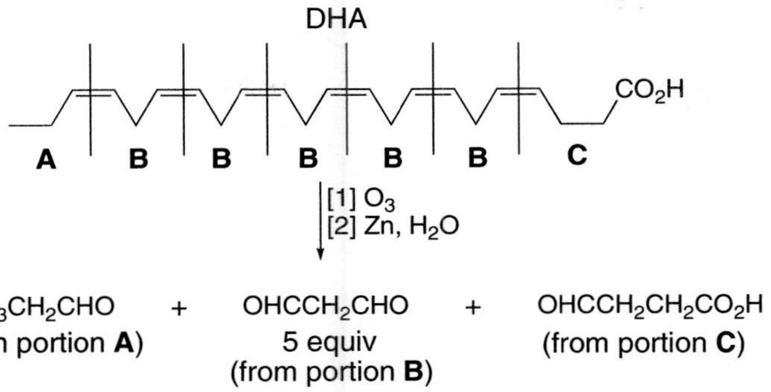
DHA
CO_2H
A
B
B
B
B
B
C
[1] O_3
[2] Zn, H_2O
CH_3CH_2CHO (from portion **A**)
+
$OHCCH_2CHO$ 5 equiv (from portion **B**)
+
$OHCCH_2CH_2CO_2H$ (from portion **C**)

12.59 The stereogenic center (labeled with *) in both structures can be *R* or *S*.

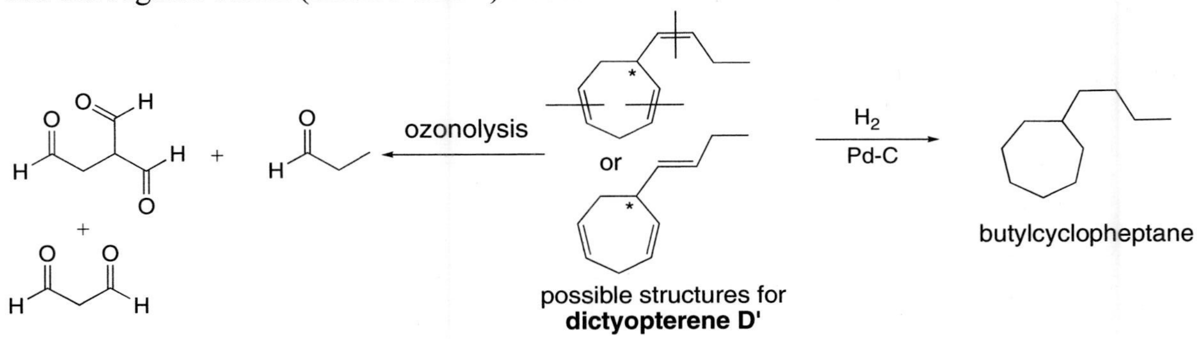
ozonolysis
or
H_2
Pd-C
butylcycloheptane
possible structures for
dictyopterene D'

12.60

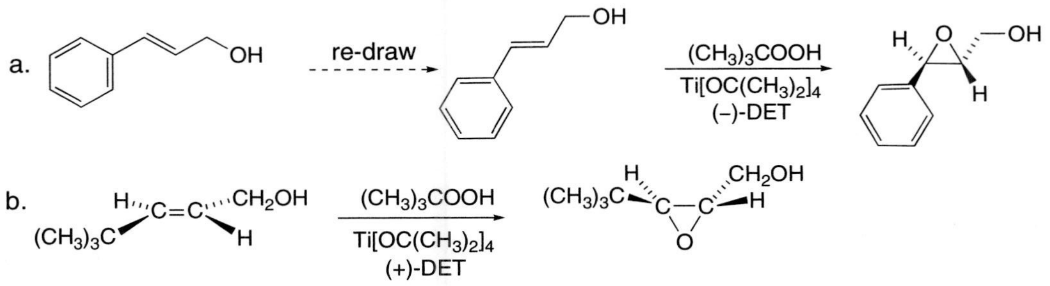
a.
OH
re-draw
$(CH_3)_3COOH$
$Ti[OC(CH_3)_2]_4$
(–)-DET
b.
$(CH_3)_3C$
CH_2OH
$(CH_3)_3COOH$
$Ti[OC(CH_3)_2]_4$
(+)-DET

12.61

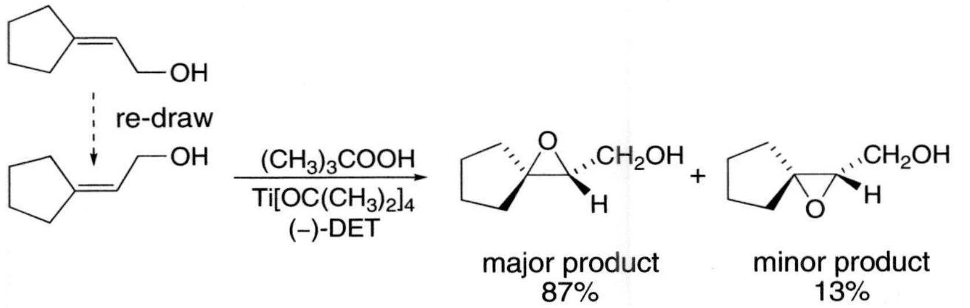
OH
re-draw
$(CH_3)_3COOH$
$Ti[OC(CH_3)_2]_4$
(–)-DET
CH_2OH
major product
87%
minor product
13%

enantiomeric excess =
% one enantiomer – % second enantiomer

ee = 87% – 13% = **74%**

12.62

a.

Replace this O to make an alkene.

$(-)$-DET

c.

Replace this O to make an alkene.

$(+)$-DET

b.

$(-)$-DET

Replace this O to make an alkene.

12.63 Use retrosynthetic analysis to devise a synthesis of each hydrocarbon from acetylene.

a. $CH_3CH_2CH{=}CH_2 \Longrightarrow CH_3CH_2CH{\equiv}CH \Longrightarrow {}^-C{\equiv}CH \Longrightarrow HC{\equiv}CH$

$HC{\equiv}CH \xrightarrow{NaH} {}^-C{\equiv}CH \xrightarrow{CH_3CH_2Cl} CH_3CH_2C{\equiv}CH \xrightarrow[\text{Lindlar catalyst}]{H_2} CH_3CH_2CH{=}CH_2$

b.

$\Longrightarrow CH_3{-}C{\equiv}C{-}CH_3 \Longrightarrow CH_3{-}C{\equiv}CH \Longrightarrow {}^-C{\equiv}CH \Longrightarrow HC{\equiv}CH$

$HC{\equiv}CH \xrightarrow{NaH} {}^-C{\equiv}CH \xrightarrow{CH_3Cl} CH_3{-}C{\equiv}CH \xrightarrow{NaH} CH_3{-}C{\equiv}C^- \xrightarrow{CH_3Cl} CH_3{-}C{\equiv}C{-}CH_3 \xrightarrow[\text{Lindlar catalyst}]{H_2}$

c.

$\Longrightarrow CH_3{-}C{\equiv}C{-}CH_3 \Longrightarrow CH_3{-}C{\equiv}CH \Longrightarrow {}^-C{\equiv}CH \Longrightarrow HC{\equiv}CH$

$HC{\equiv}CH \xrightarrow{NaH} {}^-C{\equiv}CH \xrightarrow{CH_3Cl} CH_3{-}C{\equiv}CH \xrightarrow{NaH} CH_3{-}C{\equiv}C^- \xrightarrow{CH_3Cl} CH_3{-}C{\equiv}C{-}CH_3 \xrightarrow[NH_3]{Na}$

d. $(CH_3)_2CHCH_2CH_2CH_2CH_2CH(CH_3)_2 \Longrightarrow HC{\equiv}CCH_2CH(CH_3)_2 \Longrightarrow {}^-C{\equiv}CH \Longrightarrow HC{\equiv}CH$

$HC{\equiv}CH \xrightarrow{NaH} {}^-C{\equiv}CH \xrightarrow{Cl\diagdown} HC{\equiv}C\diagdown \xrightarrow{NaH} {}^-C{\equiv}C\diagdown \xrightarrow{Cl\diagdown} \diagdown C{\equiv}C\diagdown \xrightarrow[\text{Pd-C}]{H_2 \text{ (2 equiv)}}$

12.64

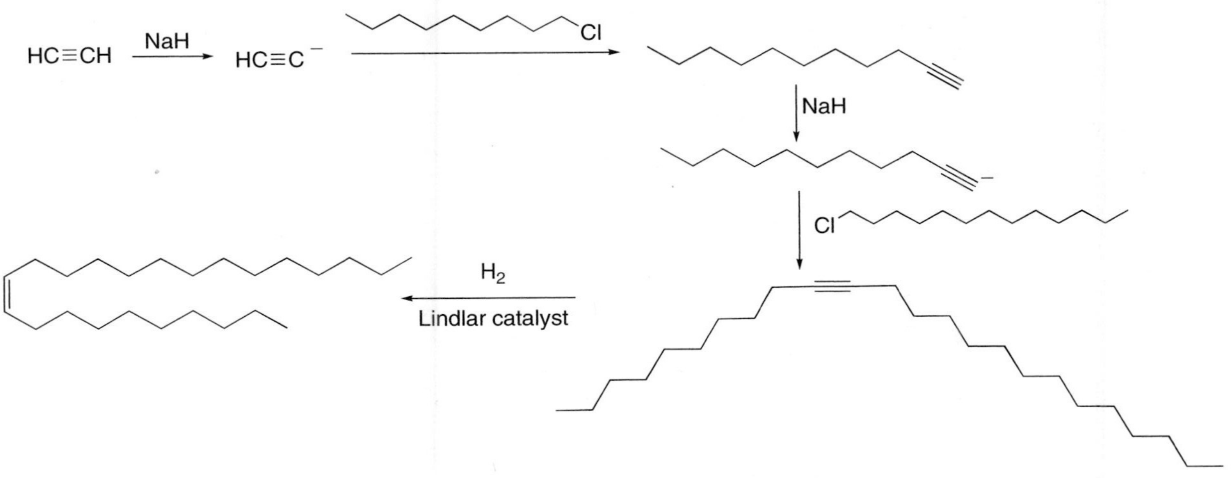

12.65

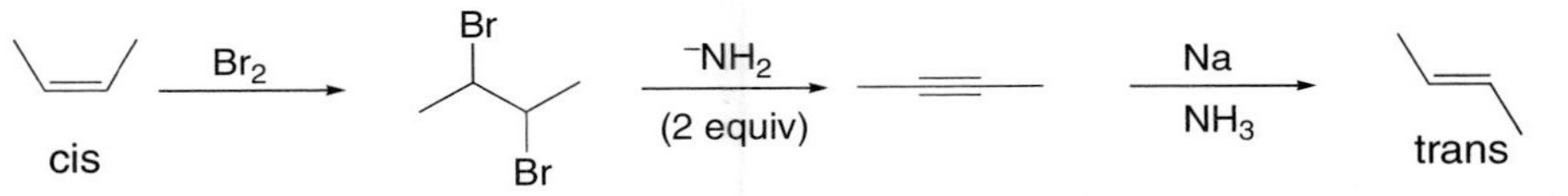

12.66

a.

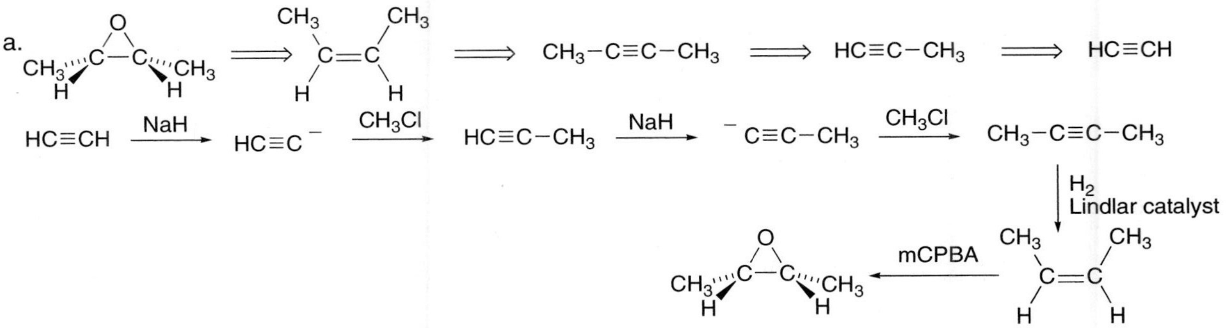

b.

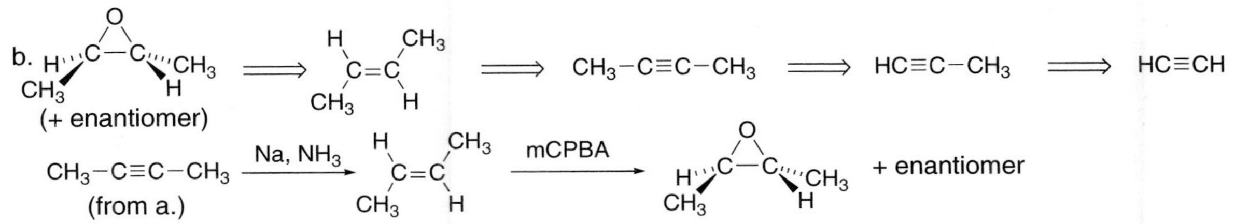

c.

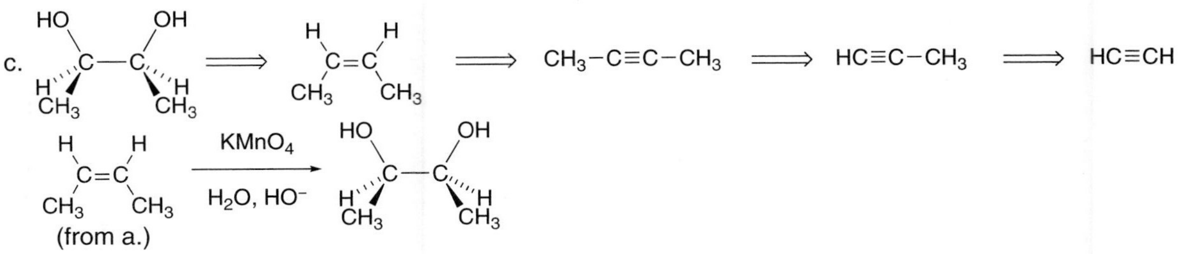

d.

$$HO-\underset{\underset{CH_3}{|}}{\overset{\overset{H}{|}}{C}}-\underset{\underset{H}{|}}{\overset{\overset{OH}{|}}{C}}-CH_3 \quad \Longrightarrow \quad \underset{CH_3}{\overset{H}{C}}=\underset{H}{\overset{CH_3}{C}} \quad \Longrightarrow \quad CH_3-C\equiv C-CH_3 \quad \Longrightarrow \quad HC\equiv C-CH_3 \quad \Longrightarrow \quad HC\equiv CH$$

(+ enantiomer)

$$\underset{CH_3}{\overset{H}{C}}=\underset{H}{\overset{CH_3}{C}} \quad \xrightarrow[\text{H}_2\text{O, HO}^-]{\text{KMnO}_4} \quad HO-\underset{\underset{CH_3}{|}}{\overset{\overset{H}{|}}{C}}-\underset{\underset{H}{|}}{\overset{\overset{OH}{|}}{C}}-CH_3 \quad \text{(+ enantiomer)}$$

(from b.)

12.67

a. $C_6H_5CH=CH_2 \xrightarrow[\text{[2] H}_2\text{O, }^-\text{OH}]{\text{[1] BH}_3} C_6H_5CH_2-CH_2OH \xrightarrow{\text{PCC}} C_6H_5CH_2-CHO$

b. $C_6H_5CH=CH_2 \xrightarrow[\text{H}_2\text{SO}_4]{\text{H}_2\text{O}} C_6H_5\overset{\overset{HO}{|}}{C}H-CH_3 \xrightarrow{\text{PCC}} C_5H_5\overset{\overset{O}{||}}{C}CH_3$

c. $C_6H_5CH=CH_2 \xrightarrow[\text{[2] H}_2\text{O, }^-\text{OH}]{\text{[1] BH}_3} C_6H_5CH_2-CH_2OH \xrightarrow[\text{H}_2\text{SO}_4, \text{H}_2\text{O}]{\text{CrO}_3} C_6H_5CH_2-COOH$

d. $C_6H_5CH=CH_2 \xrightarrow{\text{mCPBA}} C_6H_5\underset{H}{\overset{\triangle}{C}}\underset{H}{\overset{O}{C}}H \xrightarrow[\text{[2] H}_2\text{O}]{\begin{array}{c}HC\equiv CH\\ \downarrow \text{NaH}\\ \text{[1]} ^-C\equiv CH\end{array}} C_6H_5\overset{\overset{HO}{|}}{C}H-CH_2C\equiv CH$

12.68

1-pentene

(2E)-2-hexene

12.69

a. $CH_3CH_2CH=CH_2 \xrightarrow[\text{[2] H}_2\text{O}_2, \text{HO}^-]{\text{[1] 9-BBN or BH}_3} CH_3CH_2CH_2CH_2OH \xrightarrow[\text{H}_2\text{SO}_4, \text{H}_2\text{O}]{\text{CrO}_3} CH_3CH_2CH_2COOH$

b. $CH_3CH_2CH_2CH_2OH \xrightarrow[\text{pyridine}]{\text{POCl}_3} CH_3CH_2CH=CH_2 \xrightarrow{\text{mCPBA}} CH_3CH_2\overset{\triangle}{C}H-CH_2 \xrightarrow[\text{[2] H}_2\text{O}]{\text{[1] LiAlH}_4} CH_3CH_2\overset{\overset{HO}{|}}{C}H-CH_3$

$\xrightarrow{\text{H}_2\text{O, H}_2\text{SO}_4}$

$\xrightarrow[\text{H}_2\text{SO}_4, \text{H}_2\text{O}]{\text{CrO}_3} CH_3CH_2\overset{\overset{O}{||}}{C}CH_3$

c.

d.

12.70

a. $HC\equiv CH$ $\xrightarrow{NaH}$ $HC\equiv C^-$ $\xrightarrow{CH_3CH_2Br}$ $HC\equiv C-CH_2CH_3$ $\xrightarrow{NaH}$ $^-C\equiv C-CH_2CH_3$

$\xleftarrow[\text{Lindlar catalyst}]{H_2}$ $HO-CH_2CH_2-C\equiv C-CH_2CH_3$ $\xleftarrow{H_2O}$ $^-O-CH_2CH_2-C\equiv C-CH_2CH_3$

b. (from a.) $\xrightarrow[NH_3]{Na}$

c. (from b.) $\xrightarrow{PBr_3}$

d. (from b.) $\xrightarrow{PCC}$

12.71

a. $\xrightarrow[\text{Lindlar catalyst}]{H_2}$ $\xrightarrow[\text{[2] } H_2O_2, HO^-]{\text{[1] 9-BBN or BH}_3}$

b. (from a.) $\xrightarrow{mCPBA}$ $\xrightarrow[\text{[2] } H_2O]{\text{[1] LiAlH}_4}$

c. $\xrightarrow{NaH}$ $\xrightarrow{CH_3Cl}$ $\xrightarrow[NH_3]{Na}$ $\xrightarrow[H_2O, HO^-]{KMnO_4}$

d. (from c.) $\xrightarrow{mCPBA}$ (+ enantiomer)

12.72

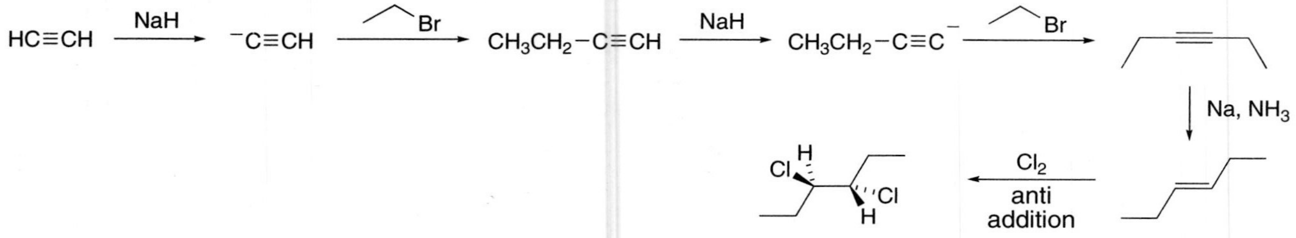

(3R,4S)-3,4-dichlorohexane

12.73

a.

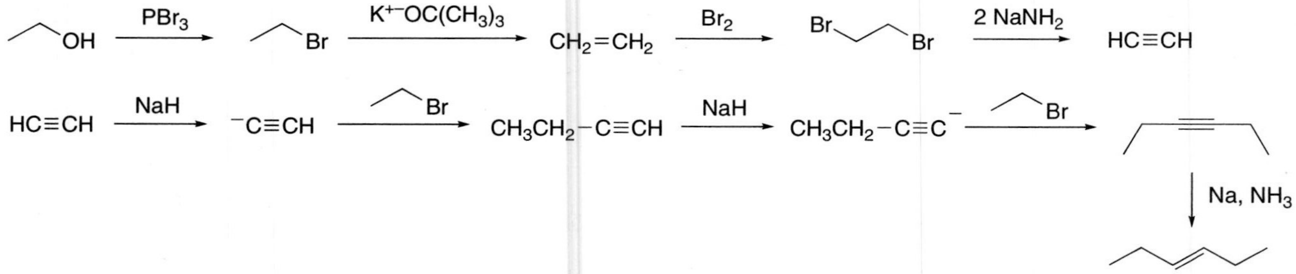

b.

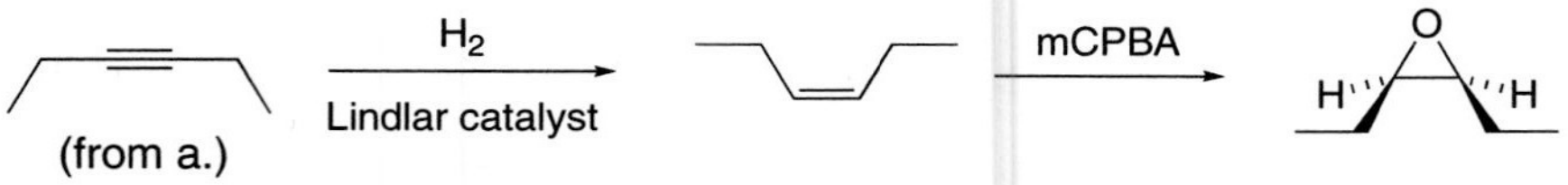

c.

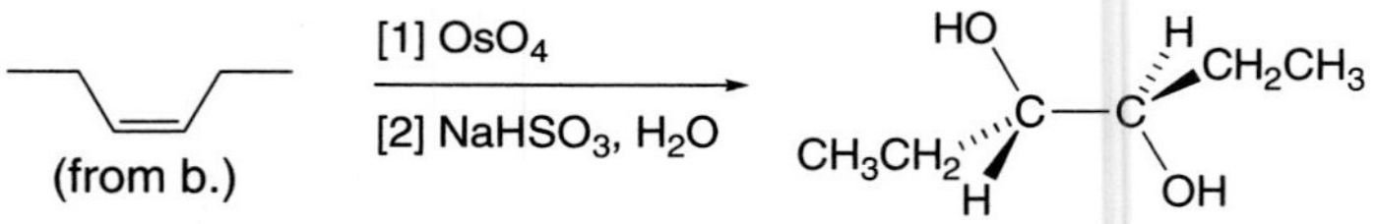

12.74

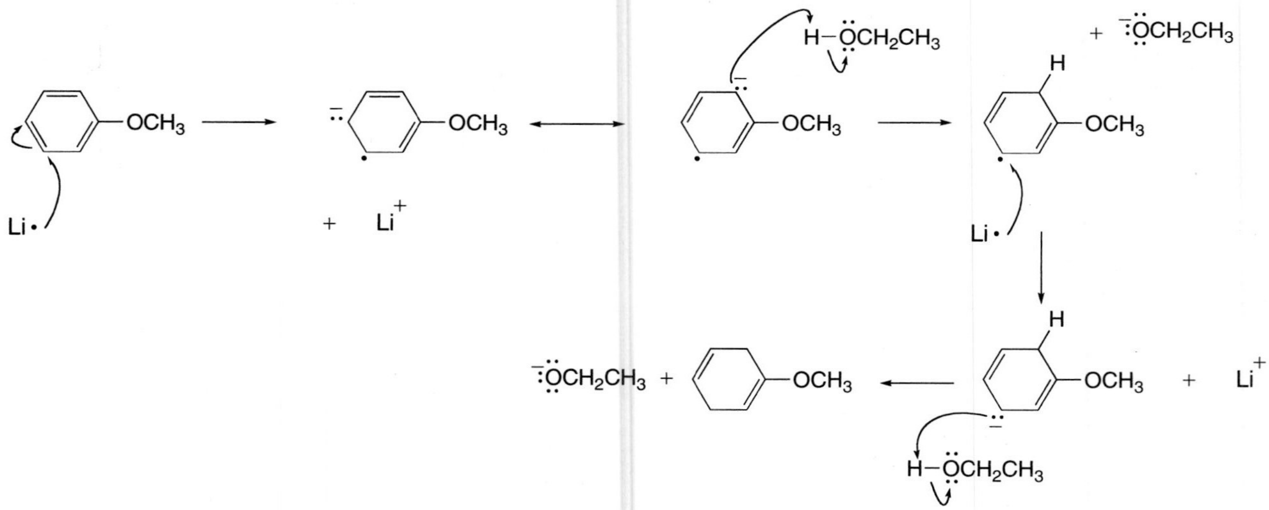

12.75

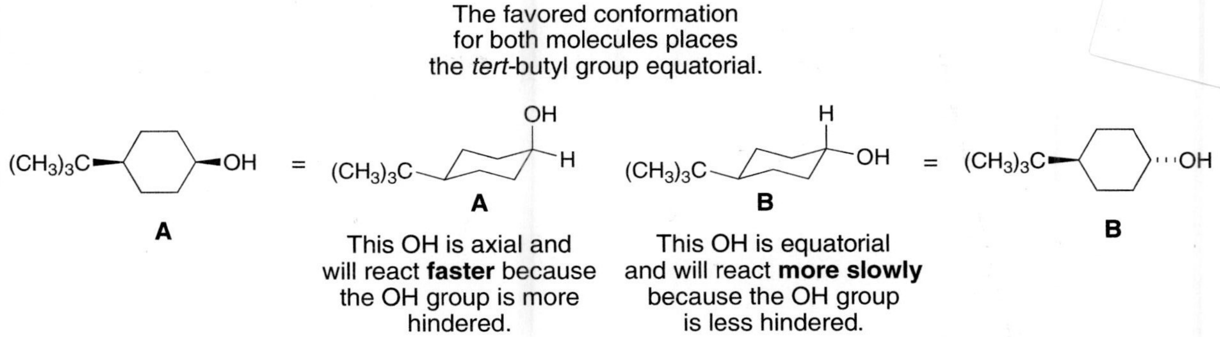

The favored conformation
for both molecules places
the *tert*-butyl group equatorial.

A

This OH is axial and
will react **faster** because
the OH group is more
hindered.

This OH is equatorial
and will react **more slowly**
because the OH group
is less hindered.

B

12.76

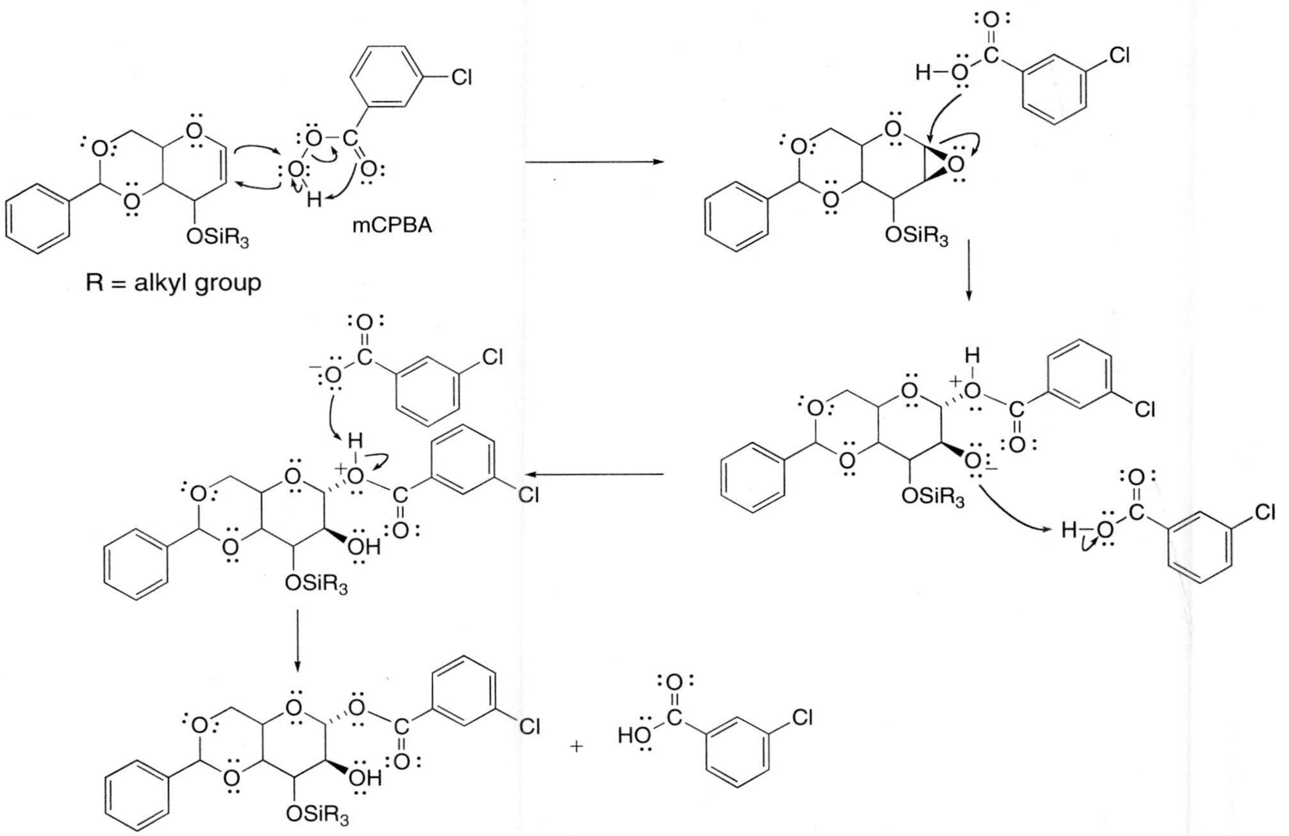

R = alkyl group

mCPBA

12.77

a.

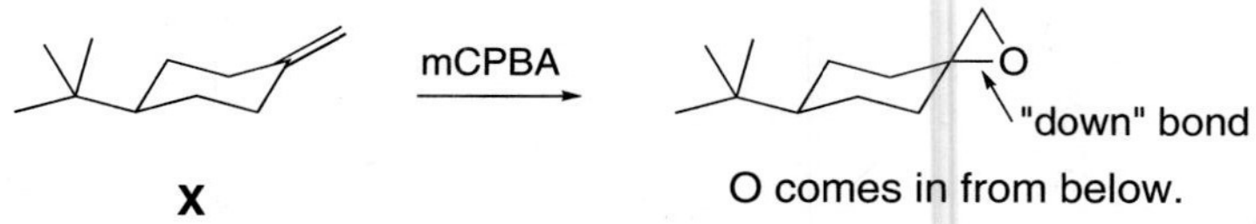
mCPBA
"down" bond
X
O comes in from below.

b.

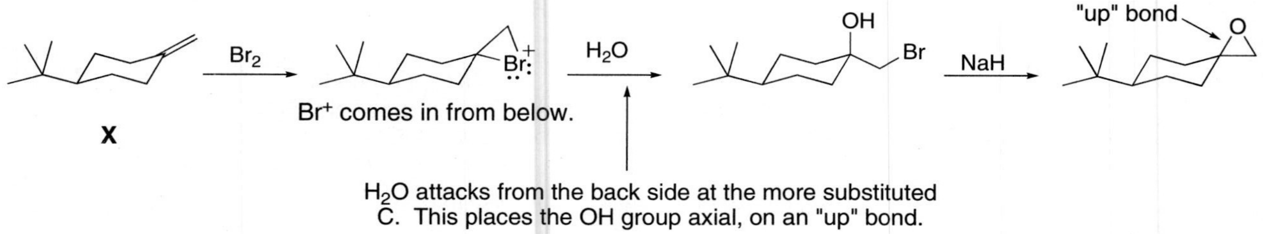
Br2
Br
H2O
OH
Br
NaH
"up" bond
O
X
Br+ comes in from below.
H2O attacks from the back side at the more substituted
C. This places the OH group axial, on an "up" bond.

13.9

benzene
C_6H_6 $m/z = 78$

toluene
C_7H_8 $m/z = 92$

p-xylene
C_8H_{10} $m/z = 106$

GC–MS analysis:
Three peaks in the gas chromatogram.
Order of peaks: benzene, toluene, p-xylene,
in order of increasing bp.
Molecular ions observed in the three mass spectra:
78, 92, 106.

13.10 Wavelength and frequency are inversely proportional. The higher frequency light will have a shorter wavelength.
   a. Light having a λ of 10^2 nm has a higher ν than light with a λ of 10^4 nm.
   b. Light having a λ of 100 nm has a higher ν than light with a λ of 100 μm.
   c. Blue light has a higher ν than red light.

13.11 The **energy of a photon** is *proportional* to its **frequency**, and inversely proportional to its wavelength.
   a. Light having a ν of 10^8 Hz is of higher energy than light having a ν of 10^4 Hz.
   b. Light having a λ of 10 nm is of higher energy than light having a λ of 1000 nm.
   c. Blue light is of higher energy than red light.

13.12 The larger the energy difference between two states, the higher the frequency of radiation needed for absorption. The 400 kJ/mol transition requires a higher ν of radiation than a 20 kJ/mol transition.

13.13 Higher wavenumbers are proportional to higher frequencies and higher energies.
   a. IR light with a wavenumber of 3000 cm^{-1} is higher in energy than IR light with a wavenumber of 1500 cm^{-1}.
   b. IR light having a λ of 10 μm is higher in energy than IR light having a λ of 20 μm.

13.14 Stronger bonds absorb at a higher wavenumber. Bonds to lighter atoms (H versus D) absorb at higher wavenumber.

   a. $CH_3-C\equiv C-CH_2CH_3$ or $CH_2=C(CH_3)_2$
            stronger bond
         higher wavenumber

   b. CH_3-H or CH_3-D
            lighter atom H
         higher wavenumber

13.15 Cyclopentane and 1-pentene are both composed of C–C and C–H bonds, but 1-pentene also has a C=C bond. This difference will give the IR of 1-pentene an additional peak at 1650 cm^{-1} (for the C=C). 1-Pentene will also show C–H absorptions for sp^2 hybridized C–H bonds at 3150–3000 cm^{-1}.

13.16 Look at the functional groups in each compound below to explain how each IR is different.

A   **B**   **C**

C=O peak at ~1700 cm^{-1} C=C peak at 1650 cm^{-1} O–H peak at 3200–3600 cm^{-1}
                              C_{sp^2}–H at 3150–3000 cm^{-1}

13.5

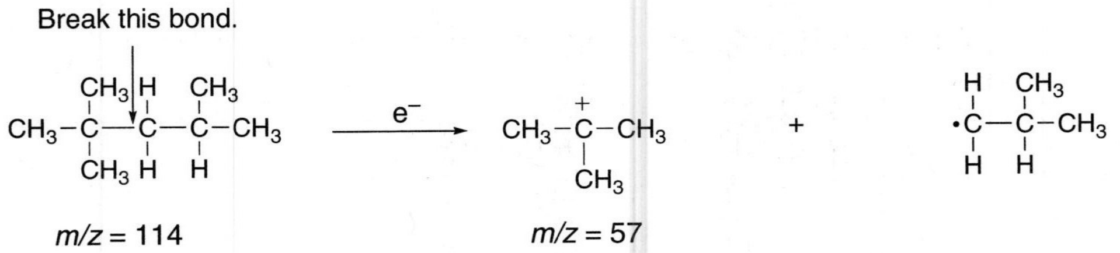

Break this bond.

$m/z = 114$ → e^- → $m/z = 57$ +

This 3° carbocation is more stable than others that can form, and is therefore the most abundant fragment.

13.6

a.

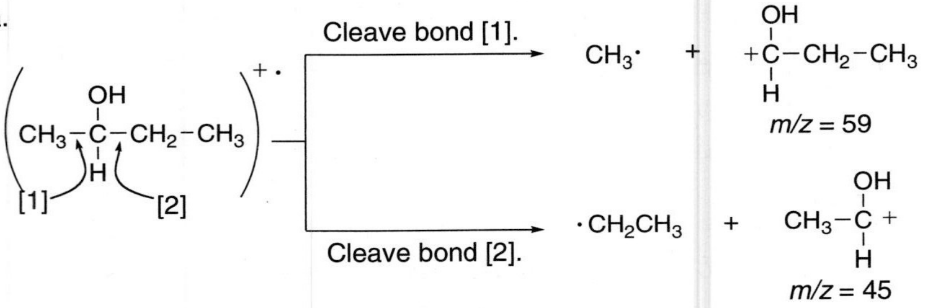

Cleave bond [1].

$CH_3\cdot$ + $^+C(OH)H-CH_2-CH_3$ $m/z = 59$

Cleave bond [2].

$\cdot CH_2CH_3$ + $CH_3-C(OH)H^+$ $m/z = 45$

b.

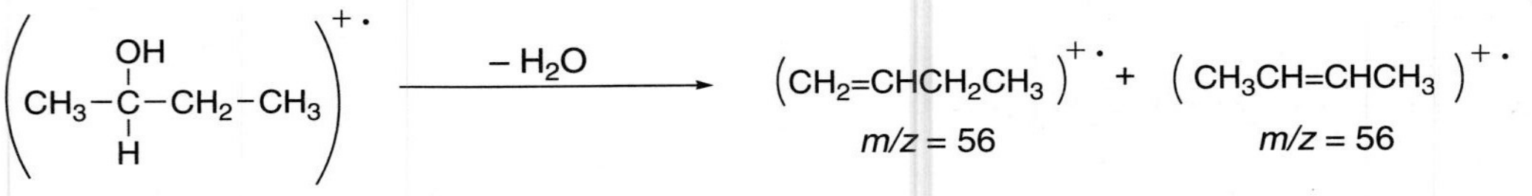

$- H_2O$ →

$(CH_2=CHCH_2CH_3)^{+\cdot}$ + $(CH_3CH=CHCH_3)^{+\cdot}$
$m/z = 56$ $m/z = 56$

13.7

a.

→ $CH_3CH_2-\overset{+}{C}=O$ + $O=\overset{+}{C}-$cyclohexyl

(from cleavage of bond [2])   (from cleavage of bond [1])

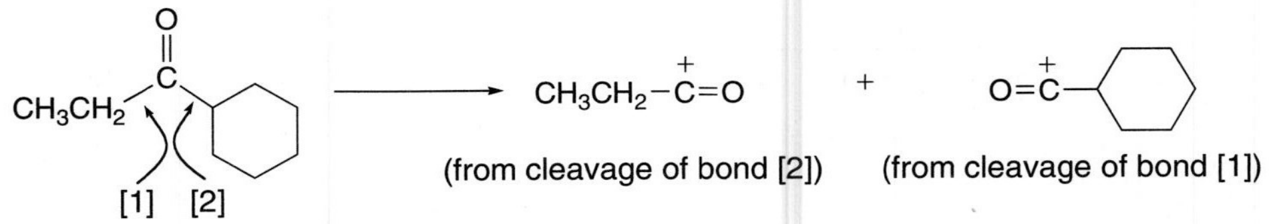

b. $CH_3CH_2CH_2CH_2CH_2CH_2OH$ → $^+CH_2OH$

c. $CH_3CH_2CH_2CHO$ → $CH_3CH_2CH_2\overset{+}{C}=O$ + $H\overset{+}{C}=O$

13.8   Use the exact mass values given in Table 13.1 to calculate the exact mass of each compound.

$C_7H_5NO_3$
mass: 151.0270

$C_8H_9NO_2$
mass: 151.0634
compound **X**

$C_{10}H_{17}N$
mass: 151.1362

Chapter 13: Answers to Problems

13.1 The molecular ion formed from each compound is equal to its molecular weight.

  a. C_3H_6O
  molecular weight = **58**
  molecular ion (m/z) = **58**

  b. $C_{10}H_{20}$
  molecular weight = **140**
  molecular ion (m/z) = **140**

  c. $C_8H_8O_2$
  molecular weight = **136**
  molecular ion (m/z) = **136**

  d. $C_{10}H_{15}N$
  molecular weight = **149**
  molecular ion (m/z) = **149**

13.2 Some possible formulas for each molecular ion:
  a. Molecular ion at 72: C_5H_{12}, C_4H_8O, $C_3H_4O_2$
  b. Molecular ion at 100: C_8H_4, C_7H_{16}, $C_6H_{12}O$, $C_5H_8O_2$
  c. Molecular ion at 73: $C_4H_{11}N$, $C_2H_7N_3$

13.3 To calculate the molecular ions you would expect for compounds with Cl, calculate the molecular weight using each of the two most common isotopes of Cl (^{35}Cl and ^{37}Cl). Do the same for Br, using ^{79}Br and ^{81}Br.

  a. $C_4H_9{}^{35}Cl$ = **92**
  $C_4H_9{}^{37}Cl$ = **94**
  Two peaks in 3:1 ratio at m/z 92 and 94

  d. $C_4H_{11}N$ = **73**
  One peak at m/z 73

  b. C_3H_7F = **62**
  One peak at m/z 62

  e. $C_4H_4N_2$ = **80**
  One peak at m/z 80

  c. $C_6H_{11}{}^{79}Br$ = **162**
  $C_6H_{11}{}^{81}Br$ = **164**
  Two peaks in a 1:1 ratio at m/z 162 and 164

13.4 After calculating the mass of the molecular ion, draw the structure and determine which C–C bond is broken to form fragments of the appropriate mass-to-charge ratio.

Chapter 13: Mass Spectrometry and Infrared Spectroscopy

◆ Mass spectrometry (MS)

- Mass spectrometry measures the molecular weight of a compound (13.1A).
- The mass of the molecular ion (**M**) = the molecular weight of a compound. Except for isotope peaks at M + 1 and M + 2, the molecular ion has the highest mass in a mass spectrum (13.1A).
- The base peak is the tallest peak in a mass spectrum (13.1A).
- A compound with an odd number of N atoms gives an odd molecular ion. A compound with an even number of N atoms (including zero) gives an even molecular ion (13.1B).
- Organic chlorides show two peaks for the molecular ion (M and M + 2) in a 3:1 ratio (13.2).
- Organic bromides show two peaks for the molecular ion (M and M + 2) in a 1:1 ratio (13.2).
- The fragmentation of radical cations formed in a mass spectrometer gives lower molecular weight fragments, often characteristic of a functional group (13.3).
- High-resolution mass spectrometry gives the molecular formula of a compound (13.4A).

◆ Electromagnetic radiation

- The wavelength and frequency of electromagnetic radiation are *inversely* related by the following equations: $\lambda = c/\nu$ or $\nu = c/\lambda$ (13.5).
- The energy of a photon is proportional to its frequency; the higher the frequency the higher the energy: $E = h\nu$ (13.5).

◆ Infrared spectroscopy (IR, 13.6 and 13.7)

- Infrared spectroscopy identifies functional groups.
- IR absorptions are reported in wavenumbers:

$$\text{wavenumber} = \tilde{\nu} = 1/\lambda$$

- The functional group region from **4000–1500 cm^{-1}** is the most useful region of an IR spectrum.
- C–H, O–H, and N–H bonds absorb at high frequency, $\geq$ 2500 cm^{-1}.
- As bond strength increases, the wavenumber of an absorption increases; thus triple bonds absorb at higher wavenumber than double bonds.

$$
\begin{array}{cc}
\text{C=C} & \text{C}\equiv\text{C} \\
\sim 1650 \text{ cm}^{-1} & \sim 2250 \text{ cm}^{-1}
\end{array}
$$

Increasing bond strength
Increasing $\tilde{\nu}$

- The higher the percent *s*-character, the stronger the bond, and the higher the wavenumber of an IR absorption.

$$
\begin{array}{ccc}
-\overset{|}{\underset{|}{\text{C}}}-\text{H} & =\text{C}\overset{/}{\underset{\text{H}}{\diagdown}} & \equiv\text{C}-\text{H} \\
\text{C}sp^3-\text{H} & \text{C}sp^2-\text{H} & \text{C}sp-\text{H} \\
25\% \text{ } s\text{-character} & 33\% \text{ } s\text{-character} & 50\% \text{ } s\text{-character} \\
3000\text{–}2850 \text{ cm}^{-1} & 3150\text{–}3000 \text{ cm}^{-1} & 3300 \text{ cm}^{-1}
\end{array}
$$

Increasing percent *s*-character
Increasing $\tilde{\nu}$

13.17

a. Compound **A** has peaks at ~3150 (sp^2 hybridized C–H), 3000–2850 (sp^3 hybridized C–H), and 1650 (C=C) cm^{-1}.

b. Compound **B** has a peak at 3000–2850 (sp^3 hybridized C–H) cm^{-1}.

13.18 All compounds show an absorption at 3000–2850 cm^{-1} due to the sp^3 hybridized C–H bonds. Additional peaks in the functional group region for each compound are shown.

a.

no additional peaks

b.

OH

O–H bond at 3600–3200 cm^{-1}

c.

Csp^2–H at 3150–3000 cm^{-1}
C=C bond at 1650 cm^{-1}

d.

=O

C=O bond at ~1700 cm^{-1}

e.

Csp^2–H at 3150–3000 cm^{-1}
C=O at ~1700 cm^{-1}
C=C at 1650 cm^{-1}
O–H above 3000 cm^{-1}

[The OH of a COOH is much broader than the OH of an alcohol and occurs at 3500–2500 cm^{-1} (see Chapter 19).]

f.

CH$_3$O
HO

O–H at 3600–3200 cm^{-1}
N–H at 3500–3200 cm^{-1}
Csp^2–H at 3150–3000 cm^{-1}
C=O at ~1700 cm^{-1}
C=C at 1650 cm^{-1}
aromatic ring at 1600, 1500 cm^{-1}

13.19 Possible structures are (a) $CH_3COOCH_2CH_3$ and (c) $CH_3CH_2COOCH_3$. Compounds (b) and (d) also have an OH group that would give a strong absorption at ~3600–3200 cm^{-1}, which is absent in the IR spectrum of **X**, thus excluding them as possibilities.

13.20

a. Hydrocarbon with a molecular ion at $m/z = 68$
IR absorptions at 3310 cm^{-1} = Csp–H bond
3000–2850 cm^{-1} = Csp^3–H bonds
2120 cm^{-1} = C≡C bond
Molecular formula: C_5H_8

H−C≡C−CH$_2$CH$_2$CH$_3$ or H−C≡C−CHCH$_3$
|
CH$_3$

b. Compound with C, H, and O with a molecular ion at $m/z = 60$
IR absorptions at 3600–3200 cm^{-1} = O–H bond
3000–2850 cm^{-1} = Csp^3–H bonds
Molecular formula: C_3H_8O

CH$_3$CH$_2$CH$_2$−O−H or CH$_3$CH−O−H
|
CH$_3$

13.21

a.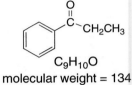

molecular formula: C_6H_6
molecular ion (m/z): **78**

c.

molecular formula: $C_5H_{10}O$
molecular ion (m/z): **86**

e. $(CH_3)_3CCH(Br)CH(CH_3)_2$

molecular formula: $C_8H_{17}Br$
molecular ions (m/z): **192, 194**

b.

molecular formula: $C_{10}H_{16}$
molecular ion (m/z): **136**

d. Cl

molecular formula: $C_5H_{11}Cl$
molecular ions (m/z): **106, 108**

13.22

$CH_2CH_2CH_3$

C_9H_{12}
molecular weight = 120

CH_2CH_3
(O)

$C_9H_{10}O$
molecular weight = 134

OCH_2CH_3

$C_8H_{10}O$
molecular weight = 122

13.23 Examples are given for each molecular ion.
  a. molecular ion 102: C_8H_6, $C_6H_{14}O$, $C_5H_{10}O_2$, $C_5H_{14}N_2$
  b. molecular ion 98: C_8H_2, C_7H_{14}, $C_6H_{10}O$, $C_5H_6O_2$
  c. molecular ion 119: C_8H_9N, $C_6H_5N_3$
  d. molecular ion 74: C_6H_2, $C_4H_{10}O$, $C_3H_6O_2$

13.24 Likely molecular formula, C_8H_{16} (one degree of unsaturation—one ring or one π bond).

Four structures with m/z = 112

13.25

$CH_3-CH-CO_2CH_3$
$\quad\quad |$
$\quad\quad Cl$
B

$C_4H_7O_2Cl$
molecular weight: **122, 124**
should show 2 peaks for the
molecular ion with a **3:1 ratio**

Mass spectrum [1]

CH_3
OCH_3
C

$C_8H_{10}O$
molecular weight: **122**

Mass spectrum [2]

$CH_3CH_2CH_2Br$
A

C_3H_7Br
molecular weight: **122, 124**
should show 2 peaks for the
molecular ion with a **1:1 ratio**

Mass spectrum [3]

13.26

Possible structures
C_7H_{12}
(exact mass 96.0940)

13.27

a.

(from cleavage of bond [1])   (from cleavage of bond [2])

b.

(from cleavage of bond [1])   (from cleavage of bond [2])

c.

(from cleavage of bond [1])   (from cleavage of bond [2])

13.28

a.

$\left(C_6H_5-CH=CH_2\right)^{+\bullet}$ $\xleftarrow{-H_2O}$ $C_6H_5-\overset{H}{\underset{H}{C}}-\overset{H}{\underset{H}{C}}-OH$ $\xrightarrow[\text{Cleave bond [1].}]{e^-}$ $C_6H_5-\overset{H}{\underset{H}{C}}+$ (resonance-stabilized carbocation)

$m/z = 104$ $m/z = 122$ $m/z = 91$

b.

$\left(CH_2=\overset{CH_3}{\underset{}{C}}-\overset{H}{\underset{H}{C}}=\overset{H}{\underset{}{C}}\right)^{+\bullet}$ $\xleftarrow{-H_2O}$ $CH_2=\overset{CH_3}{\underset{}{C}}-\overset{H}{\underset{H}{C}}-\overset{H}{\underset{H}{C}}-OH$

$m/z = 68$ $m/z = 86$

$\xrightarrow[\text{Cleave bond [1].}]{e^-}$ $CH_2=\overset{+}{C}-\overset{H}{\underset{H}{C}}-\overset{H}{\underset{H}{C}}-OH$

$m/z = 71$

$\xrightarrow[\text{Cleave bond [2].}]{e^-}$ $CH_2=\overset{+}{C}-CH_3$

$m/z = 41$

$\xrightarrow[\text{Cleave bond [3].}]{e^-}$ $^+CH_2-OH$

$m/z = 31$

13.29

Cleave bond [1].

$\cdot CH_3$ + $CH_3CH_2-\overset{+}{\underset{CH_2CH_3}{C}}-CH_2CH_2CH_2CH_3$ m/z = 127

Cleave bond [2].

$\cdot CH_2CH_2CH_2CH_3$ + $CH_3CH_2-\overset{CH_3}{\underset{CH_2CH_3}{C}}+$ m/z = 85

Cleave bond [3].

$\cdot CH_2CH_3$ + $+\overset{CH_3}{\underset{CH_2CH_3}{C}}-CH_2CH_2CH_2CH_3$ m/z = 113

13.30

Ketone **A**

m/z = 128

↓ α cleavage

$\cdot CH_2CH_3$ +

m/z = 99

This is ketone **A** since α cleavage gives a fragment with m/z of 99.

Ketone **B**

m/z = 128

↓ α cleavage

$\cdot CH_3$ +

m/z = 113

This is ketone **B** since α cleavage gives a fragment with m/z of 113.

13.31 One possible structure is drawn for each set of data:

a. A compound that contains a benzene ring and has a molecular ion at m/z = 107

C_7H_9N

b. A hydrocarbon that contains only sp^3 hybridized carbons and a molecular ion at m/z = 84

C_6H_{12}

c. A compound that contains a carbonyl group and gives a molecular ion at m/z = 114

CH_3 $CH_2CH_2CH_2CH_2CH_3$
$C_7H_{14}O$

d. A compound that contains C, H, N, and O and has an exact mass for the molecular ion at 101.0841

CH_3 $NHCH_2CH_2CH_3$
$C_5H_{11}NO$

13.32 Use the values given in Table 13.1 to calculate the exact mass of each compound. $C_8H_{11}NO_2$ (exact mass 153.0790) is the correct molecular formula.

13.33 Molecules with an odd number of N's have an odd number of H's, making the molecular ion odd as well.

13.34 Two isomers such as $CH_2=CHCH_2CH_2CH_2CH_3$ and $(CH_3)_2C=CHCH_2CH_3$ have the same molecular formulas and therefore give the same exact mass, so they are not distinguishable by their exact mass spectra.

13.35 α Cleavage of a 1° alcohol (RCH_2OH) forms an alkyl radical R• and a resonance-stabilized carbocation with $m/z = 31$.

$$+CH_2OH \longleftrightarrow CH_2=\overset{+}{O}H \qquad m/z = 31$$

resonance-stabilized carbocation

13.36 An ether fragments by α cleavage because the resulting carbocation is resonance stabilized.

13.37

a. $(CH_3)_2C=O$ or $(CH_3)_2CH-OH$

stronger bond
higher $\tilde{v}$ absorption

b. $(CH_3)_2C=NCH_3$ or $(CH_3)_2CH-NCH_3$

stronger bond
higher $\tilde{v}$ absorption

c.

stronger bond
higher $\tilde{v}$ absorption

13.38 Locate the functional groups in each compound. Use Table 13.2 to determine what IR absorptions each would have.

a. Csp^3–H at 2850–3000 cm^{-1}

d. Csp^3–H at 2850–3000 cm^{-1}
C=O at 1700 cm^{-1}

b. —C≡CH Csp–H at 3300 cm^{-1}
Csp^3–H at 2850–3000 cm^{-1}
C–C triple bond at 2250 cm^{-1}

e. —OH O–H at 3200–3600 cm^{-1}
Csp^2–H at 3000–3150 cm^{-1}
Csp^3–H at 2850–3000 cm^{-1}
C=C at 1650 cm^{-1}

c. O–H at 3200–3600 cm^{-1}
Csp^3–H at 2850–3000 cm^{-1}

f. O–H at > 3000 cm^{-1}
Csp^2–H at 3000–3150 cm^{-1}
C=O at ~1700 cm^{-1}
phenyl group at 1600, 1500 cm^{-1}

The OH of the RCOOH is even broader than the OH of an alcohol (3500–2500 cm^{-1}), as we will learn in Chapter 19.

13.39

a. 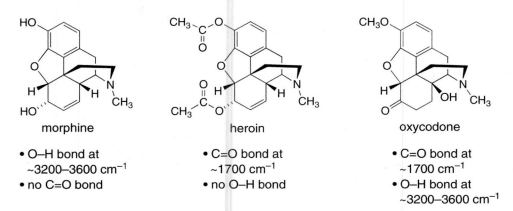 and HC≡CCH$_2$CH$_2$CH$_3$

C=C bond
1650 cm^{-1}
C$_{sp^2}$–H at 3150–3000 cm^{-1}

C≡C bond
2250 cm^{-1}
C$_{sp}$–H at 3300 cm^{-1}

b. CH$_3$CH$_2$C(=O)OH and CH$_3$C(=O)OCH$_3$

O–H bond
> 3000 cm^{-1}
[See note on OH in Answer 13.38f.]

no O–H bond

c. CH$_3$CH$_2$C(=O)CH$_3$ and CH$_3$CH=CHCH$_2$OH

C=O bond
1700 cm^{-1}

O–H bond
3200–3600 cm^{-1}
C$_{sp^2}$–H at 3150–3000 cm^{-1}
C=C bond at 1650 cm^{-1}

d. cyclohexane with OCH$_3$, OCH$_3$ and CH$_3$(CH$_2$)$_5$C(=O)OCH$_3$

no C=O bond

C=O bond
~1700 cm^{-1}

e. CH$_3$C≡CCH$_3$ and CH$_3$CH$_2$C≡CH

no C≡C absorption
due to symmetry

C$_{sp}$–H bond
3300 cm^{-1}
C≡C bond at ~2250 cm^{-1}

f. HC≡CCH$_2$N(CH$_2$CH$_3$)$_2$ and CH$_3$(CH$_2$)$_5$C≡N

C$_{sp}$–H bond
3300 cm^{-1}

13.40 The IR absorptions above 1500 cm^{-1} are different for each of the narcotics.

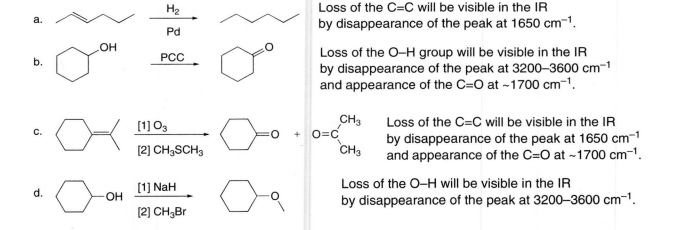

morphine

• O–H bond at
  ~3200–3600 cm^{-1}
• no C=O bond

heroin

• C=O bond at
  ~1700 cm^{-1}
• no O–H bond

oxycodone

• C=O bond at
  ~1700 cm^{-1}
• O–H bond at
  ~3200–3600 cm^{-1}

13.41 Look for a **change in functional groups** from starting material to product to see how IR could be used to determine when the reaction is complete.

a. (alkene) →[H$_2$ / Pd] (alkane)

Loss of the C=C will be visible in the IR by disappearance of the peak at 1650 cm^{-1}.

b. (cyclohexanol with OH) →[PCC] (cyclohexanone with O)

Loss of the O–H group will be visible in the IR by disappearance of the peak at 3200–3600 cm^{-1} and appearance of the C=O at ~1700 cm^{-1}.

c. (alkene) →[[1] O$_3$ / [2] CH$_3$SCH$_3$] (cyclohexanone) + O=C(CH$_3$)(CH$_3$)

Loss of the C=C will be visible in the IR by disappearance of the peak at 1650 cm^{-1} and appearance of the C=O at ~1700 cm^{-1}.

d. (cyclohexanol –OH) →[[1] NaH / [2] CH$_3$Br] (cyclohexyl –O–CH$_3$)

Loss of the O–H will be visible in the IR by disappearance of the peak at 3200–3600 cm^{-1}.

13.42 In addition to Csp^3–H at ~3000–2850 cm^{-1}:

Spectrum [1]:

$CH_2=C(CH_3)CH_2CH_2CH_2CH_3$ **(B)**
    C=C peak at 1650 cm^{-1}
    Csp^2–H at ~3150 cm^{-1}

Spectrum [2]:

$(CH_3CH_2)_3COH$ **(F)**
    OH at 3600–3200 cm^{-1}

Spectrum [3]:

$(CH_3)_2CHOCH(CH_3)_2$ **(D)**
    No other peaks above 1500 cm^{-1}

Spectrum [4]:

—CH(CH$_3$)$_2$ **(C)**
    Csp^2–H at ~3150 cm^{-1}
    Phenyl peaks at 1600 and 1500 cm^{-1}

Spectrum [5]:

$CH_3CH_2CH_2CH_2COOH$ **(A)**
    OH at ~3500–2500 cm^{-1}
    C=O at ~1700 cm^{-1}

Spectrum [6]:

$CH_3COOC(CH_3)_3$ **(E)**
    C=O at ~1700 cm^{-1}

13.43 In addition to Csp^3–H at ~3000–2850 cm^{-1}:

~1700 cm^{-1} ~1700 cm^{-1} (OH) 3200–3600 cm^{-1} (OH) 3200–3600 cm^{-1} (C=C)1650 cm^{-1}
    (C=C)1650 cm^{-1} (Csp^2–H) 3150–3000 cm^{-1}
    (Csp^2–H) 3150–3000 cm^{-1}

No enols (such as CH$_3$CH=CHOH) are drawn since these compounds are not stable.

No additional peaks above 1500 cm^{-1}

13.44

a. Compound with a molecular ion at $m/z = 72$
IR absorption at 1725 cm^{-1} = C=O bond
Molecular formula: C_4H_8O

b. Compound with a molecular ion at $m/z = 55$
The odd molecular ion means an odd number of N's present. Molecular formula: C_3H_5N
IR absorption at 2250 cm^{-1} = C≡N bond
    CH$_3$CH$_2$C≡N

c. Compound with a molecular ion at $m/z = 74$
IR absorption at 3600–3200 cm^{-1} = O–H bond
Molecular formula: $C_4H_{10}O$

13.45

Chiral hydrocarbon with a molecular ion at $m/z = 82$
Molecular formula: C_6H_{10}

IR absorptions at 3300 cm^{-1} = Csp–H bond
3000–2850 cm^{-1} = Csp^3–H bonds
2250 cm^{-1} = C≡C bond

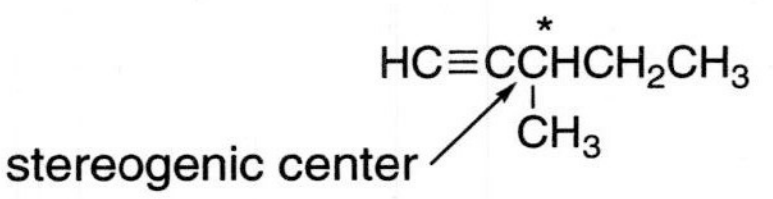

HC≡CCHCH$_2$CH$_3$

stereogenic center ← * CH$_3$

Two possible enantiomers:

or

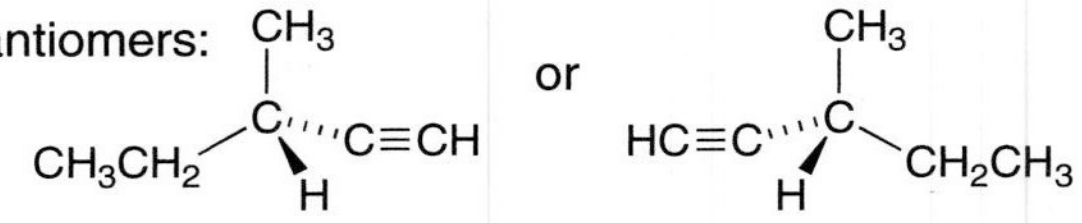

13.46 The chiral compound **Y** has a strong absorption at 2970–2840 cm^{-1} in its IR spectrum due to sp^3 hybridized C–H bonds. The two peaks of equal intensity at 136 and 138 indicate the presence of a Br atom. The molecular formula is C_4H_9Br. Only one constitutional isomer of this molecular formula has a stereogenic center:

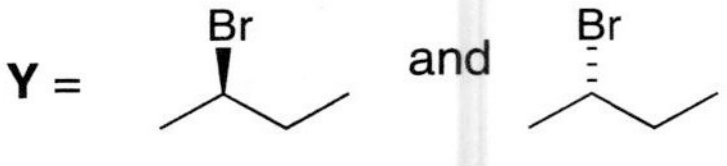

Y =

and

two possible enantiomers

13.47

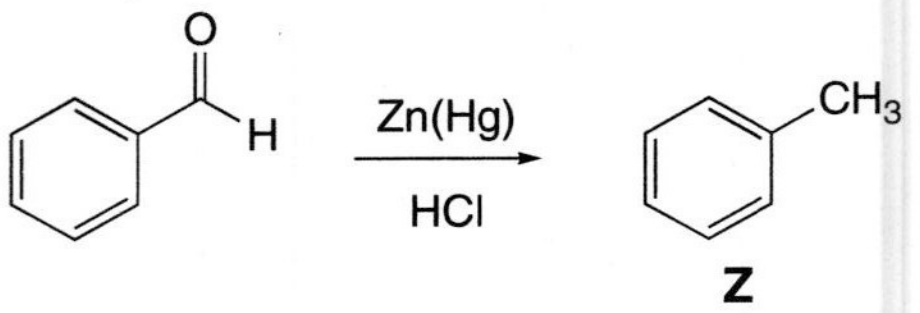

Zn(Hg)
—————
HCl

Z

$m/z = 92$; molecular formula C_7H_8
IR absorptions at:
3150–2950 cm^{-1} = Csp^3–H and Csp^2–H bonds
1605 cm^{-1} and 1496 cm^{-1} due to phenyl group

13.48

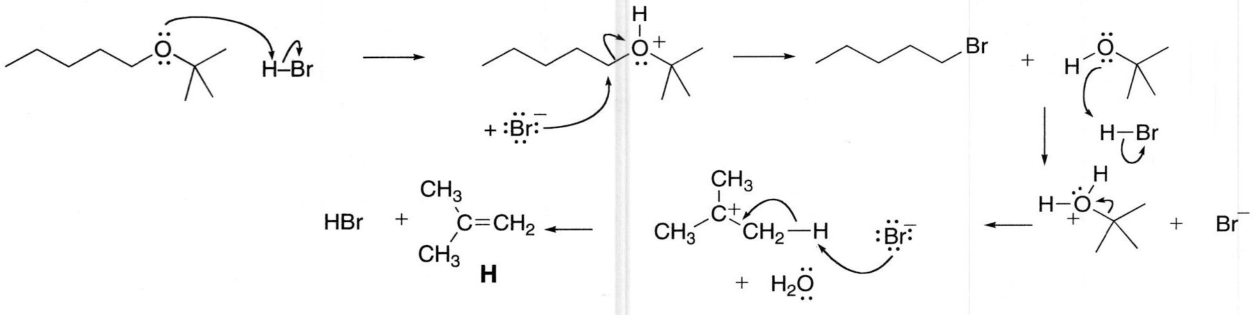

13.49

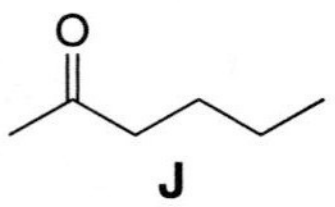

J
$C_6H_{12}O$
$m/z = 100$
IR absorption at 2962 cm^{-1} = Csp^3–H bonds
1718 cm^{-1} = C=O bond

fragments:

α cleavage product
$m/z = 43$

α cleavage product
$m/z = 85$

The fragment at $m/z = 57$ could be due to $(C_4H_9)^+$ or $(C_3H_5O)^+$.

13.50

K

C_7H_9N
$m/z = 107$
IR absorptions at 3373 and 3290 cm^{-1} = N–H
3062 cm^{-1} = C_{sp^2}–H bonds
2920 cm^{-1} = C_{sp^3}–H bonds
1600 cm^{-1} = benzene ring

The odd molecular ion indicates
the presence of a N atom.

L
C_7H_6O
$m/z = 106$

$m/z = 105$ $m/z = 77$

IR absorption at 3068 cm^{-1} = C_{sp^2}–H bonds on ring
2850 cm^{-1} = C_{sp^3}–H bond
2820 cm^{-1} and 2736 cm^{-1} = C–H of RCHO (Appendix E)
1703 cm^{-1} = C=O bond
1600 cm^{-1} = aromatic ring

13.51

Possible structures of **P**:

C_7H_7ClO

$m/z = 142, 144$
IR absorption at 3096–2837 cm^{-1} = C_{sp^3}–H bonds and C_{sp^2}–H bonds
1582 cm^{-1} and 1494 cm^{-1} = benzene ring

The peak at M + 2 shows the presence of Cl or Br. Since
Cl$_2$ is a reactant, the compound presumably contains Cl.

13.52 The mass spectrum has a molecular ion at 71. The odd mass suggests the presence of an odd number of N atoms; likely formula, C_4H_9N. The IR absorption at ~3300 cm^{-1} is due to N–H and the 3000–2850 cm^{-1} is due to sp^3 hybridized C–H bonds.

W

13.53 The α,β-unsaturated carbonyl compound has three resonance structures, two of which place a single bond between the C and O atoms. This means that the C–O bond has partial single bond character, making it weaker than a regular C=O bond, and moving the absorption to lower wavenumber.

three resonance structures for 2-cyclohexenone

13.54

a. and b.

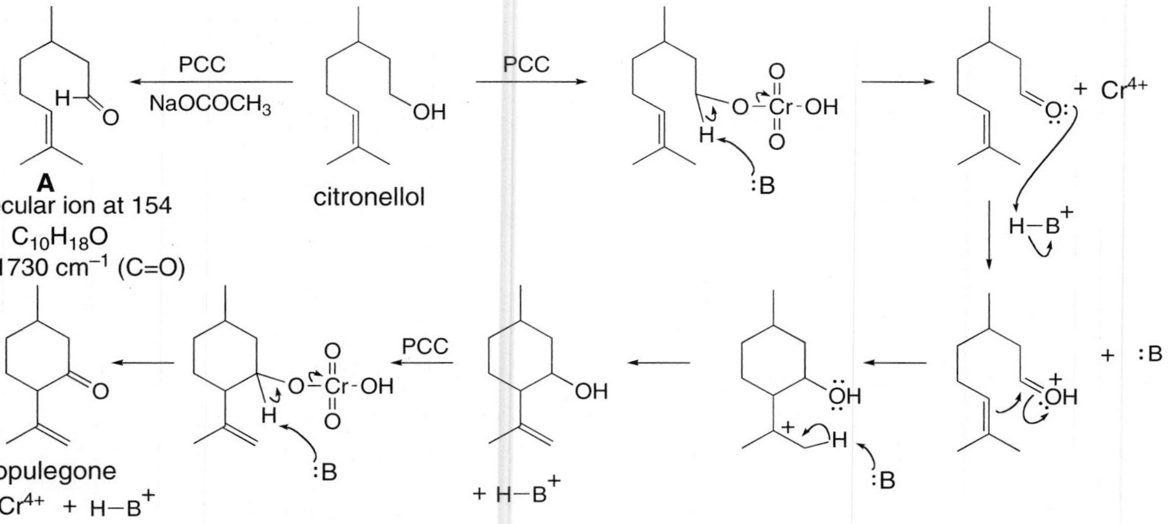

A
molecular ion at 154
$C_{10}H_{18}O$
IR at 1730 cm^{-1} (C=O)

citronellol

PCC
NaOCOCH$_3$

PCC

+ Cr^{4+}

H–B$^+$

+ :B

isopulegone
+ Cr^{4+} + H–B$^+$

PCC

+ H–B$^+$

a. and c.

isopulegone
+ H–B$^+$

H$_2$Ö:

H–OH

B

+ $^-$:ÖH

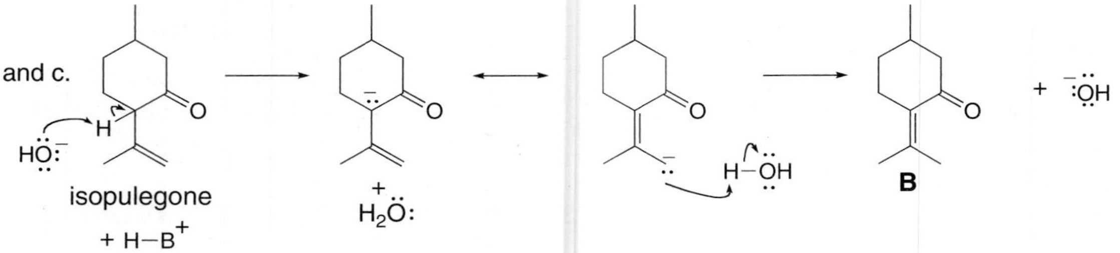

Chapter 14: Nuclear Magnetic Resonance Spectroscopy

♦ ^{1}H NMR spectroscopy

[1] The **number of signals** equals the number of different types of protons (14.2).

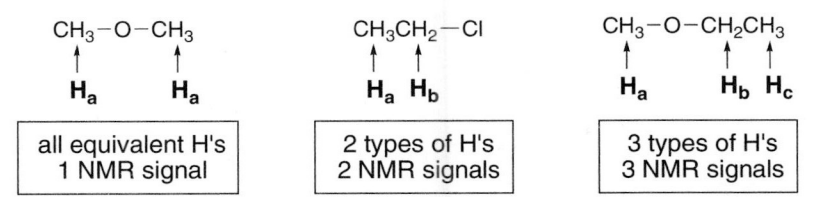

| all equivalent H's
1 NMR signal | 2 types of H's
2 NMR signals | 3 types of H's
3 NMR signals |

[2] The **position of a signal** (its chemical shift) is determined by shielding and deshielding effects.
- Shielding shifts an absorption upfield; deshielding shifts an absorption downfield.
- Electronegative atoms withdraw electron density, deshield a nucleus, and shift an absorption downfield (14.3).

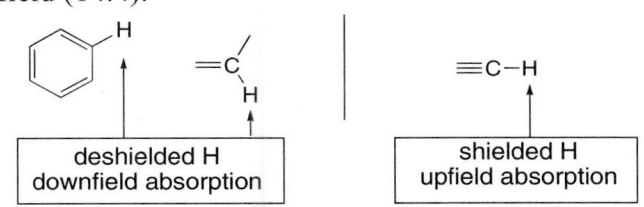

- Loosely held π electrons can either shield or deshield a nucleus. Protons on benzene rings and double bonds are deshielded and absorb downfield, whereas protons on triple bonds are shielded and absorb upfield (14.4).

| deshielded H
downfield absorption | shielded H
upfield absorption |

[3] The **area under an NMR signal** is proportional to the number of absorbing protons (14.5).

[4] **Spin-spin splitting** tells about nearby nonequivalent protons (14.6–14.8).
- Equivalent protons do not split each other's signals.
- A set of n nonequivalent protons on the same carbon or adjacent carbons split an NMR signal into $n + 1$ peaks.
- OH and NH protons do not cause splitting (14.9).
- When an absorbing proton has two sets of nearby nonequivalent protons that are equivalent to each other, use the $n + 1$ rule to determine splitting.
- When an absorbing proton has two sets of nearby nonequivalent protons that are not equivalent to each other, the number of peaks in the NMR signal = $(n + 1)(m + 1)$. In flexible alkyl chains, peak overlap often occurs, resulting in $n + m + 1$ peaks in an NMR signal.

♦ ^{13}C NMR spectroscopy (14.11)

[1] The number of signals equals the number of different types of carbon atoms. All signals are single lines.

[2] The relative position of ^{13}C signals is determined by shielding and deshielding effects.
- Carbons that are sp^3 hybridized are shielded and absorb upfield.
- Electronegative elements (N, O, and X) shift absorptions downfield.
- The carbons of alkenes and benzene rings absorb downfield.
- Carbonyl carbons are highly deshielded, and absorb farther downfield than other carbon types.

Chapter 14: Answers to Problems

14.1 Use the formula δ = [observed chemical shift (Hz)/ν of the NMR (MHz)] to calculate the chemical shifts.

a. **CH$_3$ protons:**
δ = [1715 Hz] / [500 MHz]
= **3.43 ppm**

OH proton:
δ = [1830 Hz] / [500 MHz]
= **3.66 ppm**

b. The positive direction of the δ scale is downfield from TMS. The CH$_3$ protons absorb upfield from the OH proton.

14.2 Calculate the chemical shifts as in Answer 14.1.

a. one signal:
δ = [1017 Hz] / [300 MHz]
= **3.39 ppm**

second signal:
δ = [1065 Hz] / [300 MHz]
= **3.55 ppm**

b. one signal:
3.39 = [x Hz] / [500 MHz]
x = **1695 Hz**

second signal:
3.55 = [x Hz] / [500 MHz]
x = **1775 Hz**

14.3 To determine if two H's are equivalent replace each by an atom X. If this yields the same compound or mirror images, the two H's are equivalent. Each kind of H will give one NMR signal.

a. CH$_3$CH$_3$

1 kind of H
1 NMR signal

b. CH$_3$CH$_2$CH$_3$

2 kinds of H's
2 NMR signals

c. CH$_3$CH$_2$CH$_2$CH$_3$

2 kinds of H's
2 NMR signals

d. (CH$_3$)$_2$CHCH(CH$_3$)$_2$

2 kinds of H's
2 NMR signals

e. CH$_3$CH$_2$CO$_2$CH$_2$CH$_3$

4 kinds of H's
4 NMR signals

f. CH$_3$OCH$_2$CH(CH$_3$)$_2$

4 kinds of H's
4 NMR signals

g. CH$_3$CH$_2$OCH$_2$CH$_3$

2 kinds of H's
2 NMR signals

h. CH$_3$CH$_2$CH$_2$OH

4 kinds of H's
4 NMR signals

14.4

CH$_3$CH$_2$CH$_2$CH$_2$CH$_2$CH$_2$CH$_2$CH$_2$Cl

Each C is a different distance from the Cl. This makes each C different, and each set of H's different. There are 8 different kinds of protons.

14.5 Draw in all of the H's and compare them. If two H's are cis and trans to the same group, they are equivalent.

a. 4 identical H's

2 NMR signals

b.

4 NMR signals

c.

3 NMR signals

14.6 If replacement of H with X yields enantiomers, the protons are **enantiotopic**.
If replacement of H with X yields diastereomers, the protons are **diastereotopic**. In general, if the compound has **one stereogenic center**, the protons in a CH$_2$ group are **diastereotopic**.

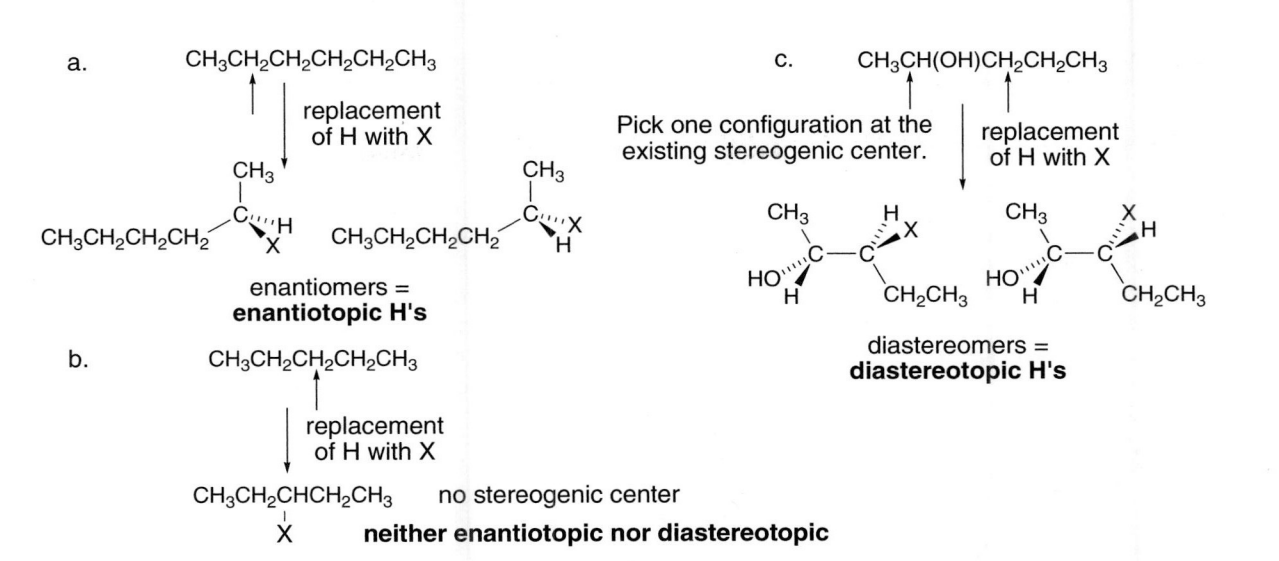

a. CH₃CH₂CH₂CH₂CH₂CH₃
replacement of H with X
enantiomers = **enantiotopic H's**

b. CH₃CH₂CH₂CH₂CH₃
replacement of H with X
CH₃CH₂CHCH₂CH₃ no stereogenic center
X **neither enantiotopic nor diastereotopic**

c. CH₃CH(OH)CH₂CH₂CH₃
Pick one configuration at the existing stereogenic center.
replacement of H with X
diastereomers = **diastereotopic H's**

14.7   The two protons of a CH_2 group are different from each other if the compound has one stereogenic center. Replace one proton with X and compare the products.

a. The stereogenic center makes the H's in the CH_2 group diastereotopic and therefore different from each other.

stereogenic center

$H_d \rightarrow$ CH₃–C–C–CH₃ $\leftarrow H_a$

Cl H $\leftarrow H_b$
$H_c \rightarrow$ H   H $\leftarrow H_e$

5 NMR signals

b.

H_c stereogenic center

$H_d \rightarrow$ H H
Cl–C–C–O–CH₃ $\leftarrow H_a$
$H_e \rightarrow$ H CH₃ $\leftarrow H_b$

5 NMR signals

c.

stereogenic center

H_b H_c

$H_a \rightarrow$ CH₃–C–C–C–CH₃ $\leftarrow H_e$

H H H $\leftarrow H_d$
Br H H
H_f H_g

7 NMR signals

14.8   Decreased electron density deshields a nucleus and the absorption goes downfield. Absorption also shifts downfield with increasing alkyl substitution.

a. FC**H₂**CH₂C**H₂**Cl
F is more electronegative than Cl. The CH_2 group adjacent to the F is more deshielded and the H's will absorb farther downfield.

b. CH₃C**H₂**CH₂C**H₂**OCH₃
The CH_2 group adjacent to the O will absorb farther downfield because it is closer to the electronegative O atom.

c. C**H₃**OC(C**H₃**)₃
The CH_3 group bonded to the O atom will absorb farther downfield.

14.9

a. ClC**H₂**C**H₂**C**H₂**Br
H_a H_b H_c
3 types of protons:
$H_b < H_c < H_a$

b. CH₃OC**H₂**OC(C**H₃**)₃
H_a H_b H_c
3 types of protons:
$H_c < H_a < H_b$

c. CH₃–C(=O)–C**H₂**CH₃
H_a H_b H_c
3 types of protons:
$H_c < H_a < H_b$

14.10

a. $CH_3-C\equiv C-H$ $CH_3CH=CH_2$ $CH_3CH_2CH_3$
              H_a H_b H_c

H_c protons are shielded because they are bonded to an sp^3 C.
H_a is shielded because it is bonded to an sp C.
H_b protons are deshielded because they are bonded to an sp^2 C.

$H_c < H_a < H_b$

b. $CH_3\overset{\overset{O}{\|}}{C}OCH_2CH_3$
         H_a H_b H_c

H_c protons are shielded because they are bonded to an sp^3 C.
H_a protons are deshielded slightly because the CH_3 group is bonded to a C=O.
H_b protons are deshielded because the CH_2 group is bonded to an O atom.

$H_c < H_a < H_b$

14.11 An integration ratio of 2:3 means that there are two types of hydrogens in the compound, and that the ratio of one type to another type is 2:3.

a. CH_3CH_2Cl
2 types of H's
3:2 - YES

b. $CH_3CH_2CH_3$
2 types of H's
6:2 or 3:1 - no

c. $CH_3CH_2OCH_2CH_3$
2 types of H's
6:4 or 3:2 - YES

d. $CH_3OCH_2CH_2OCH_3$
2 types of H's
6:4 or 3:2 - YES

14.12 To determine how many protons give rise to each signal:
- Divide the total number of integration units by the total number of protons to find the number of units per H.
- Divide each integration value by this value and round to the nearest whole number.

$C_8H_{14}O_2$
total number of integration units = 14 + 12 + 44 = 70 units
total number of protons = 14 H's
70 units/14 H's = **5 units per H**

Signal [A] = 14/5 = **3 H**
Signal [B] = 12/5 = **2 H**
Signal [C] = 44/5 = **9 H**

14.13

downfield absorption
closer to O

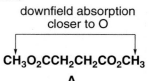

$CH_3O_2CCH_2CH_2CO_2CH_3$
A
ratio of absorbing signals 2:3
Signal [1] = **4 H = 2.64**
Signal [2] = **6 H = 3.69** ←——— 6 H's with downfield absorption

downfield absorption
closer to O

$CH_3CO_2CH_2CH_2O_2CCH_3$
B
ratio of absorbing signals 3:2
Signal [1] = **6 H = 2.09**
Signal [2] = **4 H = 4.27** ←——— 4 H's with downfield absorption

14.14 To determine the **splitting pattern** for a molecule:
- Determine the number of different kinds of protons.
- Nonequivalent protons on the same C or adjacent C's split each other.
- Apply the $n + 1$ rule.

a.

$$CH_3CH_2 \overset{\overset{\displaystyle O}{\|}}{C} Cl$$

H_a H_b

H_a: 3 peaks - triplet
H_b: 4 peaks - quartet

c.

$$CH_3 \overset{\overset{\displaystyle O}{\|}}{C} CH_2CH_2Br$$

H_a H_b H_c

H_a: 1 peak - singlet
H_b: 3 peaks - triplet
H_c: 3 peaks - triplet

e.

$$CH_3CH_2 \quad H \leftarrow H_a$$
$$C=C$$
$$CH_3 \quad H \leftarrow H_b$$

H_a: 2 peaks - doublet
H_b: 2 peaks - doublet

b.

$$H \leftarrow H_b$$
$$CH_3 - C - Br$$
$$Br$$
$$H_a$$

H_a: 2 peaks - doublet
H_b: 4 peaks - quartet

d.

$$H_a \rightarrow H \quad Cl$$
$$C=C$$
$$Br \quad H \leftarrow H_b$$

H_a: 2 peaks - doublet
H_b: 2 peaks - doublet

f. $ClCH_2CH(OCH_3)_2$

H_a H_b

H_a: 2 peaks - doublet
H_b: 3 peaks - triplet

14.15 Identical protons do not split each other.

$$H_a \rightarrow H \quad Cl$$
$$Cl - C - C - Cl$$
$$Cl \quad H \leftarrow H_a$$

H_a: 1 singlet
All protons are equivalent.

$$H_a \rightarrow H \quad Cl$$
$$Br - C - C - Cl$$
$$Br \quad H \leftarrow H_b$$

H_a: 2 peaks - doublet
H_b: 2 peaks - doublet

14.16 Use the directions from Answer 14.14.

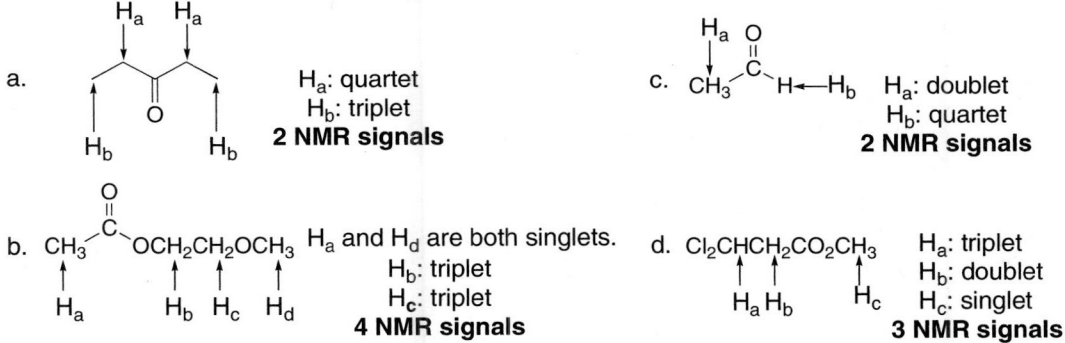

a.

H_a H_a

H_b H_b

H_a: quartet
H_b: triplet
2 NMR signals

c.

$$CH_3 \overset{\overset{\displaystyle H_a}{\downarrow}}{\overset{\overset{\displaystyle O}{\|}}{C}} H \leftarrow H_b$$

H_a: doublet
H_b: quartet
2 NMR signals

b. $CH_3 \overset{\overset{\displaystyle O}{\|}}{C} OCH_2CH_2OCH_3$

H_a H_b H_c H_d

H_a and H_d are both singlets.
H_b: triplet
H_c: triplet
4 NMR signals

d. $Cl_2CHCH_2CO_2CH_3$

$H_a H_b$ H_c

H_a: triplet
H_b: doublet
H_c: singlet
3 NMR signals

14.17 CH_3CH_2Cl

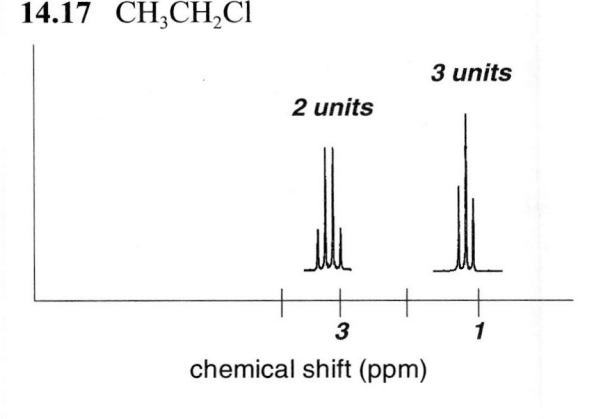

3 units

2 units

3   1

chemical shift (ppm)

There are two kinds of protons, and they can split each other. The CH_3 signal will be split by the CH_2 protons into $2 + 1 = 3$ peaks. It will be upfield from the CH_2 protons since it is farther from the Cl. The CH_2 signal will be split by the CH_3 protons into $3 + 1 = 4$ peaks. It will be downfield from the CH_3 protons since the CH_2 protons are closer to the Cl. The ratio of integration units will be 3:2.

14.18

a. $(CH_3)_2CHCO_2CH_3$

split by 6 equivalent H's
6 + 1 = **7 peaks**

b. $CH_3CH_2CH_2CH_2CH_3$
  $\quad\quad\;\; H_a \;\; H_b \;\; H_c$

H_a: split by 2 H's
3 peaks
H_c: split by 4 equivalent H's
5 peaks
H_b: split by 2 sets of H's
(3 + 1)(2 + 1) = **12 peaks (maximum)**
Since this is a flexible alkyl chain, the signal
due to H_b will have peak overlap, and
3 + 2 + 1 = **6 peaks** will likely be visible.

c.
Cl, CH₂Br on C=C; $H_a \to H$, $H \leftarrow H_b$

H_a: split by 1 H
2 peaks
H_b: split by 2 sets of H's
(1 + 1)(2 + 1) = **6 peaks**

d.
$H_a \to H$, $H \leftarrow H_b$ on C=C; Br, $H \leftarrow H_c$ (all H's)

H_a: split by 2 different H's
(1+1)(1+1) = **4 peaks**
H_b: split by 2 different H's
(1+1)(1+1) = **4 peaks**
H_c: split by 2 different H's
(1+1)(1+1) = **4 peaks**

14.19

a. $CH_3OCH_2CH_3$
  $\quad\;\; H_a \;\; H_b \;\; H_c$

H_a: singlet at ~3 ppm
H_b: quartet at ~3.5 ppm
H_c: triplet at ~1 ppm

b.
$CH_3CH_2\overset{\overset{\displaystyle O}{\|}}{C}OCH(CH_3)_2$
  $\quad H_a \; H_b \quad\quad H_c \; H_d$

H_a: triplet at ~1 ppm
H_b: quartet at ~2 ppm
H_c: septet at ~3.5 ppm
H_d: doublet at ~1 ppm

c. $CH_3OCH_2CH_2CH_2OCH_3$
  $\quad\;\; H_a \;\; H_b \;\; H_c \;\; H_b \;\; H_a$

H_a: singlet at ~3 ppm
H_b: triplet at ~3.5 ppm
H_c: quintet at ~1.5 ppm

d.
CH_3CH_2 and CH_2CH_3 with H_b, $C=C$, H_a, H, H, H_c

H_a: triplet at ~1 ppm
H_b: multiplet (8 peaks) at ~2.5 ppm
H_c: triplet at ~5 ppm

14.20

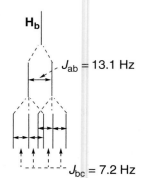

Hb

J_{ab} = 13.1 Hz

J_{bc} = 7.2 Hz

Cl, Cl on C=C; H_a, H_c
2 H_c protons

trans-1,3-dichloropropene

Splitting diagram for H_b

1 **trans** H_a proton splits H_b into
1 + 1 = 2 peaks
a doublet

2 H_c protons split H_b into
2 + 1 = 3 peaks
Now it's a doublet of triplets.

14.21

Cl, Cl on C=C; $H \leftarrow H_b$, $CH_3 \leftarrow H_a$
A

H_a: 1.75 ppm, doublet, 3 H, J = 6.9 Hz
H_b: 5.89 ppm, quartet, 1 H, J = 6.9 Hz

$C_3H_4Cl_2$

$ClCH_2$, $H \leftarrow$ doublet on C=C; singlet Cl, $H \leftarrow$ doublet
B

signal at 4.16 ppm, singlet, 2 H
signal at 5.42 ppm, doublet, 1 H, J = 1.9 Hz
signal at 5.59 ppm, doublet, 1 H, J = 1.9 Hz

14.22 Remember that OH (or NH) protons do not split other signals, and are not split by adjacent protons.

a. (CH₃)₃CCH₂OH
singlet → (CH₃)
singlet → (CH₂)
singlet → (OH)
3 NMR signals

b. CH₃CH₂CH₂OH
triplet → (CH₃)
triplet → (CH₂)
singlet → (OH)
12 peaks (maximum)
6 peaks (more likely, resulting from peak overlap)
4 NMR signals

c. (CH₃)₂CHNH₂
doublet → (CH₃)
singlet → (NH₂)
7 peaks
3 NMR signals

14.23

H_d
5 H's on benzene ring

A

H_a: doublet at ~1.4 due to the CH_3 group, split into two peaks by one adjacent nonequivalent H (H_c).
H_b: singlet at ~2.7 due to the OH group. OH protons are not split by nor do they split adjacent protons.
H_c: quartet at ~4.7 due to the CH group, split into four peaks by the adjacent CH_3 group.
H_d: Five protons on the benzene ring.

14.24 Use these steps to propose a structure consistent with the molecular formula, IR, and NMR data.
- Calculate the **degrees of unsaturation**.
- Use the IR data to determine what types of **functional groups** are present.
- Determine the number of different **types of protons**.
- Calculate the **number of H's** giving rise to each signal.
- Analyze the **splitting pattern** and put the molecule together.
- Use the **chemical shift** information to check the structure.

- Molecular formula $C_7H_{14}O_2$
  $2n + 2 = 2(7) + 2 = 16$
  $16 - 14 = 2/2 = $ **1 degree of unsaturation**
  1 π bond or 1 ring

- IR peak at 1740 cm^{-1}
  C=O absorption is around 1700 cm^{-1} (causes the degree of unsaturation).
  No signal at 3200–3600 cm^{-1} means there is no O–H bond.

- NMR data:

absorption	ppm	integration	
singlet	1.2	26	--------→ 26 units/3 units per H = **9 H's**
triplet	1.3	10	-------→ 10 units/3 units per H = **3 H's** (probably a CH_3 group)
quartet	4.1	6	-------→ 6 units/3 units per H = **2 H's** (probably a CH_2 group)

- 3 kinds of H's
- number of H's per signal
  total integration units: 26 + 10 + 6 = 42 units
  42 units / 14 H's = 3 units per H
- look at the splitting pattern

The singlet (9 H) is likely from a *tert*-butyl group:

$$-\overset{\overset{\displaystyle CH_3}{|}}{\underset{\underset{\displaystyle CH_3}{|}}{C}}-CH_3$$

The CH_3 and CH_2 groups split each other: CH_3-CH_2-

- Join the pieces together.

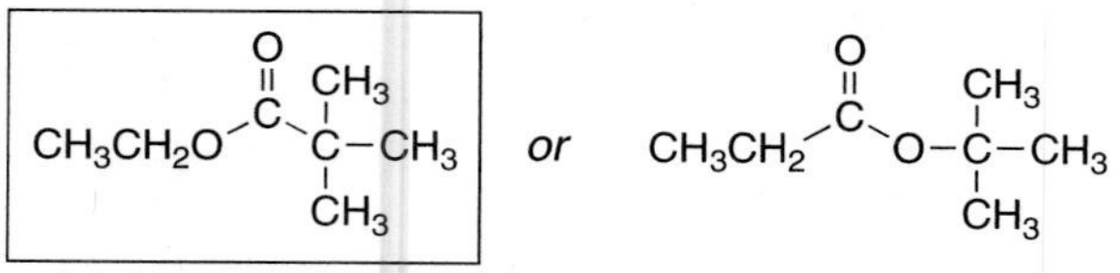

or

Pick this structure due to the chemical shift data.
The CH_2 group is shifted downfield (4 ppm), so it
is close to the electron-withdrawing O.

14.25

- Molecular formula: C_3H_8O

- Calculate degrees of unsaturation
$$2n + 2 = 2(3) + 2 = 8$$
$$8 - 8 = \textbf{0 degrees of unsaturation}$$

- IR peak at 3200–3600 cm^{-1}
- NMR data:
  - doublet at ~1.2 (6 H)
  - singlet at ~2.2 (1 H)
  - septet at ~4 (1 H)

➤ Peak at 3200–3600 cm^{-1} is due to an **O–H bond**.

3 types of H's
septet from 1 H ⟵———— split by 6 H's
singlet from 1 H
doublet from 6 H's ⟵———— split by 1 H
from the O–H proton

➤ Put information together:

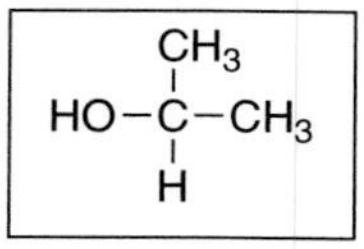

14.26

a.

Absorption [**A**]:	singlet at ~3.8 ppm	CH_3O-
Absorption [**B**]:	multiplet at ~3.6 ppm	CH_2N
Absorption [**C**]:	triplet at ~2.9 ppm	CH_2 adjacent to five-membered ring
Absorption [**D**]:	singlet at ~1.9 ppm	$CH_3C=O$

b.

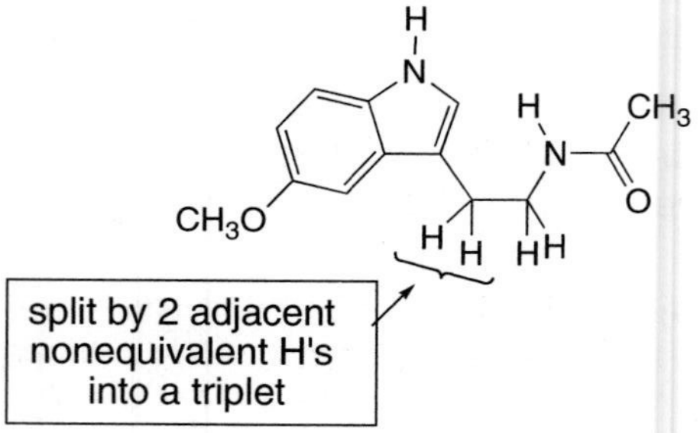

split by 2 adjacent
nonequivalent H's
into a triplet

14.27 Identify each compound from the ^{1}H NMR data.

a.

$CH_2=CHCOCH_3$ $\xrightarrow{\text{HCl}}$

triplet at 3.6

CH_3 — singlet

A triplet at 3.05

b.

$(CH_3)_2C=O$ $\xrightarrow[\text{H}_2\text{O}]{\text{base}}$

singlet at 2.5

singlet at 1.3

singlet at 3.8

singlet at 2.2

B

14.28 Each different kind of carbon atom will give a different ^{13}C NMR signal.

a. $CH_3CH_2CH_2CH_3$
C_a C_b C_b C_a

2 kinds of C's

2 ^{13}C NMR signals

b.

CH_3CH_2 OCH_3

Each C is different.
4 kinds of C's

4 ^{13}C NMR signals

c. $CH_3CH_2CH_2-O-CH_2CH_2CH_3$
C_a C_b C_c C_c C_b C_a

same groups on both sides of O
3 kinds of C's

3 ^{13}C NMR signals

d.

CH_3CH_2 H

Each C is different.
4 kinds of C's

4 ^{13}C NMR signals

14.29

a.

H_b H_a H_d
H_c H H H
H—C—C—C—H
H_c H Cl Cl

4 ^{1}H NMR signals

H_a H_b H_a
H_a H H H H_a
H—C—C—C—H
Cl   Cl
H_b

2 ^{1}H NMR signals

H_b
H_c H H Cl H_a
H—C—C—C—H
H_c H H Cl
H_b

3 ^{1}H NMR signals

H Cl H
H—C—C—C—H
H Cl H

all H's identical

1 ^{1}H NMR signal

b.

H H H
H—C—C—C—H
H Cl Cl

Each C is different.
3 kinds of C's

3 ^{13}C NMR signals

H H H
H—C—C—C—H
C_a Cl H Cl C_a
C_b

2 kinds of C's

2 ^{13}C NMR signals

H H Cl
H—C—C—C—H
H H Cl

Each C is different.
3 kinds of C's

3 ^{13}C NMR signals

H Cl H
H—C—C—C—H
C_a H Cl H C_a
C_b

2 kinds of C's

2 ^{13}C NMR signals

c. Although the number of ^{13}C signals cannot be used to distinguish these isomers, each isomer exhibits a different number of signals in its ^{1}H NMR spectrum. As a result, the isomers are distinguishable by ^{1}H NMR spectroscopy.

14.30 Electronegative elements shift absorptions downfield. The carbons of alkenes and benzene rings, and carbonyl carbons are also shifted downfield.

a. $CH_3CH_2OCH_2CH_3$

The CH_2 group is closer to the electronegative O and will be farther downfield.

b. $BrCH_2CHBr_2$

The C of the $CHBr_2$ group has two bonds to electronegative Br atoms and will be farther downfield.

c.

The carbonyl carbon is highly deshielded and will be farther downfield.

d. $CH_3CH=CH_2$

The CH_2 group is part of a double bond and will be farther downfield.

14.31

a. In order of lowest to highest chemical shift:

$$\overset{C_a \; C_b \; C_c \; C_d}{\underset{\underset{OH}{|}}{CH_3CHCH_2CH_3}}$$

$$C_d < \; C_a < C_c < C_b$$

b. In order of lowest to highest chemical shift:

$$(CH_3CH_2)_2C=O$$
$$C_a \; C_b \quad C_c$$

$$C_a < C_b < C_c$$

14.32

- molecular formula $C_4H_8O_2$
  $2n + 2 = 2(4) + 2 = 10$
  $10 - 8 = 2/2 = $ **1 degree of unsaturation**

- no IR peaks at 3200–3600 or 1700 cm^{-1}
  no O–H or C=O

- ^{1}H NMR spectrum at 3.69 ppm
  only one kind of proton

- ^{13}C NMR spectrum at 67 ppm
  only one kind of carbon

This structure satisifies all the data.
One ring is one degree of
unsaturation. All carbons and
protons are identical.

14.33

- molecular formula C_4H_8O
  $2n + 2 = 2(4) + 2 = 10$
  $10 - 8 = 2/2 = $ **1 degree of unsaturation**

- ^{13}C NMR signal at > 160 ppm due to
  C=O

- molecular formula C_4H_8O
  $2n + 2 = 2(4) + 2 = 10$
  $10 - 8 = 2/2 = $ **1 degree of unsaturation**

- all ^{13}C NMR signals at < 160 ppm
  NO C=O

14.34 Use the directions from Answer 14.3.

a. $(CH_3)_3CH$
2 kinds of H's

b. $(CH_3)_3CC(CH_3)_3$
1 kind of H

c. $CH_3CH_2OCH_2CH_2CH_2CH_2CH_3$
7 kinds of H's

d. $H_a \rightarrow CH_3$ $H \leftarrow H_c$
$\quad\quad C=C$
$H_b \rightarrow H \quad\quad H \leftarrow H_d$
4 kinds of H's

e.
5 kinds of H's

f. $CH_3CH_2 \quad H$
$\quad\quad C=C$ **3 kinds** of H's
$\quad H \quad CH_2CH_3$

g. $CH_3CH_2CH_2OCH_2CH_2CH_3$
3 kinds of H's

h. $H_a \rightarrow CH_3 \quad CH_2CH_3 \leftarrow H_d$
$\quad\quad C=C \quad H_c$
$H_b \rightarrow CH_3 \quad Br$
4 kinds of H's

i. $H_b \rightarrow H \quad H \leftarrow H_d$
$CH_3 - C - C - CH_3 \leftarrow H_f$
$\quad H_a \uparrow HO \quad H \leftarrow H_e$
$\quad H_c$
6 kinds of H's

j. $H_a \rightarrow H \quad\quad CH_3 \leftarrow H_c$
$H_b \rightarrow H \quad O \quad H \leftarrow H_d$
4 kinds of H's

14.35

a. $H_b \{ \begin{array}{c} H \\ H \end{array} \quad \begin{array}{c} CH_3 \\ CH_3 \end{array} \} H_a$
$H_c \quad H \quad H \quad H_b$
3 kinds of protons

b. $H_e \quad H_d \quad H_b$
$H \quad H$
$H_d \rightarrow H \quad CH_2CH_3 \leftarrow H_a$
$H \quad H$
$H_f \rightarrow H \quad H_g \quad H_f \quad H \leftarrow H_c$
7 kinds of protons

c. $H_c \rightarrow H \quad H \leftarrow H_c$
$H_a \rightarrow CH_3 \quad CH_3 \leftarrow H_a$
$\quad H \quad H$
$H_b \rightarrow H \quad H_d \quad H_d \quad H \leftarrow H_b$
4 kinds of protons

d. $H_d \rightarrow H \quad H \leftarrow H_c$
$H_a \rightarrow CH_3 \quad H \leftarrow H_b$
$\quad H \quad H$
$H_b \rightarrow H \quad H_c \quad H_d \quad CH_3 \leftarrow H_a$
4 kinds of protons

14.36

a. caffeine
4 NMR signals

b. vanillin
6 NMR signals

c. thymol
7 NMR signals
equivalent

d. capsaicin
15 NMR signals
equivalent

14.37

δ (in ppm) = [observed chemical shift (Hz)] / ν of the NMR (MHz)]

a. 2.5 = x Hz/300 MHz
    x = **750 Hz**
b. ppm = 1200 Hz/300 MHz
    = **4 ppm**
c. 2.0 = x Hz/300 MHz
    x = **600 Hz**

14.38

2.16 = x Hz/500 MHz
x = 1080 Hz (chemical shift of acetone in Hz)
1080 Hz + 1570 Hz = **2650 Hz**
2650 Hz/500 MHz = **5.3 ppm**, chemical shift of the CH$_2$Cl$_2$ signal

14.39 Use the directions from Answer 14.8.

a. CH$_3$CH$_2$CH$_2$CH$_2$CH$_3$ or CH$_3$CH$_2$CH$_2$OCH$_3$

Adjacent O deshields the H's.
farther downfield

c. CH$_3$OCH$_2$CH$_3$
    or
Increasing alkyl substitution
farther downfield

b. CH$_3$CH$_2$CH$_2$I or CH$_3$CH$_2$CH$_2$F

More electronegative F
deshields the H's.
farther downfield

d. CH$_3$CH$_2$CHBr$_2$ or CH$_3$CH$_2$CH$_2$Br

Two electronegative
Br's deshield the H.
farther downfield

14.40 Use the directions from Answer 14.12.

[total number of integration units] / [total number of protons]
[13 + 33 + 73] / 10 = ~12 units per proton

Signal of 13 units is from **1 H.**
Signal of 33 units is from **3 H's.**
Signal of 73 units is from **6 H's.**

14.41

a. CH$_3$CO$_2$C(CH$_3$)$_3$ and CH$_3$CO$_2$CH$_3$
    H$_a$ H$_b$ H$_a$ H$_b$

    H$_a$: H$_b$ = 1:3   H$_a$: H$_b$ = 1:1
    different ratio of peak areas
    H$_b$ in CH$_3$CO$_2$CH$_3$ is farther downfield
    than all H's in CH$_3$CO$_2$C(CH$_3$)$_3$.

b. CH$_3$OCH$_2$CH$_2$OCH$_3$ and CH$_3$OCH$_2$OCH$_3$
    H$_a$ H$_b$ H$_b$ H$_a$ H$_a$ H$_b$ H$_a$

    H$_a$: H$_b$ = 3:2   H$_a$: H$_b$ = 3:1
    different ratio of peak areas

c. CH$_3$—[ring]—CH$_3$ and CH$_3$—[ring]—CH$_3$
    (H$_b$ H$_b$ / H$_a$ / H$_b$ H$_b$)   (H$_b$ / CH$_3$←H$_a$ / ←H$_b$ / CH$_3$←H$_a$)

    H$_a$: H$_b$ = 3:2   H$_a$: H$_b$ = 3:1
    different ratio of peak areas

14.42 The following compounds give one singlet in a ^{1}H NMR spectrum:

CH$_3$CH$_3$ CH$_3$−C≡C−CH$_3$ Cl—CH$_2$—C(Br)(Br)—CH$_2$—Cl   (CH$_3$)(CH$_3$)C=C(CH$_3$)(CH$_3$)   (CH$_3$)$_3$C—C(=O)—C(CH$_3$)$_3$

14.43

$$CH_3CH_2CH_2CH_2CH_2CH_3$$
$$H_a \quad H_b \quad H_c \quad H_c \quad H_b \quad H_a$$

3 signals:
H_a: split by 2 H_b protons - triplet
H_c: split by 2 H_b protons - triplet
H_b: split by 3 H_a + 2 H_c protons - 12 peaks (maximum)
Since H_b is located in a flexible alkyl chain, peak overlap occurs, so that only 3 + 2 + 1 = 6 peaks will likely be observed.

$$CH_3CH_2CHCH_2CH_3$$
$$H_a \quad H_b \quad CH_3 \quad H_b \quad H_a$$
$$H_d$$
(with H_c above)

4 signals:
H_a: split by 2 H_b protons - triplet
H_b: split by 3 H_a + 1 H_c protons - 8 peaks (maximum)
H_c: split by 4 H_b + 3 H_d protons - 20 peaks (maximum)
H_d: split by 1 H_c proton - doublet
Since H_b and H_c are located in a flexible alkyl chain, it is likely that peak overlap occurs, so that the following is observed: H_b (3 + 1 + 1 = 5 peaks) and H_c (4 + 3 + 1 = 8 peaks).

$$CH_3 \leftarrow H_a$$
$$CH_3CHCH_2CH_2CH_3$$
$$H_a \quad H_b \quad H_c \quad H_d \quad H_e$$

5 signals:
H_a: split by 1 H_b proton - doublet
H_b: split by 6 H_a + 2 H_c protons - 21 peaks (maximum)
H_c: split by 1 H_b + 2 H_d protons - 6 peaks (maximum)
H_d: split by 2 H_c + 3 H_e protons - 12 peaks (maximum)
H_e: split by 2 H_d protons - triplet
Since H_b, H_c, and H_d are located in a flexible alkyl chain, it is likely that peak overlap occurs, so that the following is observed:
H_b (6 + 2 + 1 = 9 peaks), H_c (1 + 2 + 1 = 4 peaks), and H_d (2 + 3 + 1 = 6 peaks).

$$H_a \quad H_a$$
$$CH_3 \quad CH_3$$
$$CH_3CH-CHCH_3$$
$$H_a \quad H_b \quad H_b \quad H_a$$

2 signals:
H_a: split by 1 H_b proton - doublet
H_b: split by 6 H_a protons - septet

$$CH_3 \leftarrow H_c$$
$$CH_3CH_2-C-CH_3 \leftarrow H_c$$
$$CH_3 \leftarrow H_c$$
$$H_a \quad H_b$$

3 signals:
H_a: split by 2 H_b protons - triplet
H_b: split by 3 H_a protons - quartet
H_c: no splitting - singlet

14.44

a. $CH_3CH(OCH_3)_2$

CH_3 protons split by 1 H = **doublet**
CH proton split by 3 H's = **quartet**

b.
$$O$$
$$CH_3OCH_2CH_2-C-OCH_3$$
both CH_2 groups split each other = **triplets**

c.
[benzene ring]—CH_2CH_3

CH_3 protons split by 2 H's = **triplet**
CH_2 protons split by 3 H's = **quartet**

d. $CH_3OCH_2CHCl_2$

CH_2 protons split by 1 H = **doublet**
CH proton split by 2 H's = **triplet**

e.
$$O$$
$$(CH_3)_2CH-C-OCH_2CH_3$$
$$H_a \quad H_b \quad H_c \quad H_d$$

H_a protons split by 1 H = **doublet**
H_b proton split by 6 H's = **septet**
H_c protons split by 3 H's = **quartet**
H_d protons split by 2 H's = **triplet**

f. $HOCH_2CH_2CH_2OH$
$$H_a \quad H_b$$

H_a protons split by 2 CH_2 groups = **quintet**
H_b protons split by 2 H's = **triplet**

g. $CH_3CH_2CH_2CH_2OH$
$$H_a \quad H_b \quad H_c \quad H_d$$

H_a protons split by 2 H's = **triplet**
H_b protons split by CH_3 + CH_2 protons = **12 peaks** (maximum)
H_c protons split by 2 different CH_2 groups = **9 peaks** (maximum)
H_d protons split by 2 H's = **triplet**
Since H_b and H_c are located in a flexible alkyl chain, it is likely that peak overlap occurs, so that the following is observed: H_b (3 + 2 + 1 = 6 peaks), H_c (2 + 2 + 1 = 5 peaks).

h.
$$O$$
$$CH_3CH_2CH_2-C-OH$$
$$H_a \quad H_b \quad H_c$$

H_a protons split by 2 H's = **triplet**
H_c protons split by 2 H's = **triplet**
H_b protons split by CH_3 + CH_2 protons = **12 peaks** (maximum)
Since H_b is located in a flexible alkyl chain, it is likely that peak overlap occurs, so that only 3 + 2 + 1 = 6 peaks will be observed.

i.
$$O$$
$$CH_3CH_2-C-H$$
$$H_a \quad H_b$$

H_a: split by CH_3 group + H_b = **8 peaks** (maximum)
H_b: split by 2 H's = **triplet**

j.

CH$_3$ H←H$_a$
 \   /
  C=C
 /   \
CH$_3$CH$_2$ H←H$_b$

H$_a$: split by 1 H = **doublet**
H$_b$: split by 1 H = **doublet**

k.

CH$_3$ H←H$_a$
 \   /
  C=C
 /   \
Br   H← H$_b$

H$_a$: split by 1 H = **doublet**
H$_b$: split by 1 H = **doublet**

l.

CH$_3$ H←H$_a$
 \   /
  C=C
 /   \
H$_c$→ H   H←H$_b$

H$_a$: split by H$_b$ + H$_c$ -
doublet of doublets (4 peaks)
H$_b$: split by H$_a$ + H$_c$ -
doublet of doublets (4 peaks)
H$_c$: split by CH$_3$, H$_a$ + H$_b$ - **16 peaks**

14.45

H$_a$→ H   Br
   \   /
    C=C
   /   \
H$_b$→ H   CO$_2$CH$_3$

H$_a$: split by 1 H = **doublet**
H$_b$: split by 1 H = **doublet**
H$_a$ and H$_b$ are geminal.

Br   H←H$_a$
 \   /
  C=C
 /   \
H$_b$→H   CO$_2$CH$_3$

H$_a$: split by 1 H = **doublet**
H$_b$: split by 1 H = **doublet**
H$_a$ and H$_b$ are trans.

Both compounds exhibit two doublets for the H's on the C=C, but the
coupling constants ($J_{geminal}$ and J_{trans}) are different. $J_{geminal}$ is much smaller
than J_{trans} (0–3 Hz versus 11–18 Hz).

14.46

H$_b$ H$_a$
  \   /
   C=C
  /   \
H$_c$ CN

J_{ab} = 11.8 Hz
J_{bc} = 0.9 Hz
J_{ac} = 18 Hz

H$_a$: doublet of doublets at 5.7 ppm. Two large J values are seen for the H's cis
(J_{ab} = 11.8 Hz) and trans (J_{ac} = 18 Hz) to H$_a$.
H$_b$: doublet of doublets at ~6.2 ppm. One large J value is seen for the cis H
(J_{ab} = 11.8 Hz). The geminal coupling (J_{bc} = 0.9 Hz) is hard to see.
H$_c$: doublet of doublets at ~6.6 ppm. One large J value is seen for the trans H
(J_{ac} = 18 Hz). The geminal coupling (J_{bc} = 0.9 Hz) is hard to see.

| **Splitting diagram for H$_a$** |

1 **trans** H$_c$ proton splits H$_a$ into
1 + 1 = 2 peaks
a doublet

1 **cis** H$_b$ proton splits H$_a$ into
1 + 1 = 2 peaks
Now it's a doublet of doublets.

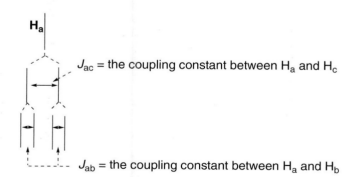

H$_a$

J_{ac} = the coupling constant between H$_a$ and H$_c$

J_{ab} = the coupling constant between H$_a$ and H$_b$

14.47

Four constitutional isomers of **C$_4$H$_9$Br**:

4 different C's   4 different C's   2 different C's   3 different C's

14.48 Only two compounds in Problem 14.42 give one signal in their ^{13}C NMR spectrum:

$$CH_3CH_3$$

14.49

The O atom of an ester donates electron density so the carbonyl carbon has less δ^+, making it less deshielded than the carbonyl carbon of an aldyhyde or ketone. Therefore, the carbonyl carbon of an aldehyde or ketone is more deshielded and absorbs farther downfield.

14.50

a. $HC(CH_3)_3$

2 signals

d.

7 signals

g.

5 signals

b.

5 signals

e. CH_3CH_2 $\quad$ CH_2CH_3

$C=C$

$H \quad\quad H$

3 signals

h.

4 signals

c. $CH_3OCH(CH_3)_2$

3 signals

f. OH

7 signals

i.

3 signals

14.51

a. $CH_3CH_2 \overset{\overset{\displaystyle O}{\|}}{C} OH$

$\uparrow \quad \uparrow \quad\quad \uparrow$

$C_a \quad C_b \quad\quad C_c$

$C_a < C_b < C_c$

b. $CH_3CH_2\overset{\overset{\displaystyle OH}{|}}{C}HCH_2CH_3$

$\uparrow \quad\quad \uparrow \; \uparrow$

$C_a \quad\quad C_b \; C_c$

$C_a < C_b < C_c$

c. CH_2CH_3

$C_a \quad \uparrow \quad \uparrow$

$\quad\quad C_b \; C_c$

$C_c < C_b < C_a$

d. $CH_2=CHCH_2CH_2CH_2Br$

$\uparrow \quad\quad\quad\quad \uparrow \quad \uparrow$

$C_a \quad\quad\quad\quad C_b \; C_c$

$C_b < C_c < C_a$

14.52

$\quad$ 19 ppm $\quad$ 62 ppm

$\quad\quad \downarrow \quad\quad \downarrow$

a. $CH_3CH_2CH_2CH_2OH$

$\quad \uparrow \quad\quad\quad \uparrow$

14 ppm $\quad$ 35 ppm

$\quad$ 16 ppm $\quad$ 205 ppm

$\quad\quad \downarrow \quad\quad \downarrow$

b. $(CH_3)_2CHCHO$

$\quad\quad\quad\quad \uparrow$

$\quad\quad\quad\quad$ 41 ppm

$\quad$ 143 ppm $\quad$ 23 ppm

$\quad\quad \downarrow \quad\quad \downarrow$

c. $CH_2=CHCH(OH)CH_3$

$\quad \uparrow \quad\quad\quad \uparrow$

113 ppm $\quad$ 69 ppm

14.53

a.

3 different C's
3 signals

b.

3 signals

4 signals

5 signals

14.54 Use the directions from Answer 14.24.

a. **C$_4$H$_8$Br$_2$**: 0 degrees of unsaturation
IR peak at 3000–2850 cm^{-1}: **C$_{sp^3}$–H bonds**
NMR: singlet at 1.87 ppm (6 H) (2 CH$_3$ groups)
  singlet at 3.86 ppm (2 H) (CH$_2$ group)

$$CH_3-\underset{\underset{Br}{|}}{\overset{\overset{CH_3}{|}}{C}}-CH_2Br$$

b. **C$_3$H$_6$Br$_2$**: 0 degrees of unsaturation
IR peak at 3000–2850 cm^{-1}: **C$_{sp^3}$–H bonds**
NMR: quintet at 2.4 ppm (split by 2 CH$_2$ groups)
  triplet at 3.5 ppm (split by 2 H's)

Br⌒⌒Br

c. **C$_5$H$_{10}$O$_2$**: **1 degree of unsaturation**
IR peak at 1740 cm^{-1}: **C=O**
NMR: triplet at 1.15 ppm (3 H) (CH$_3$ split by 2 H's)
  triplet at 1.25 ppm (3 H) (CH$_3$ split by 2 H's)
  quartet at 2.30 ppm (2 H) (CH$_2$ split by 3 H's)
  quartet at 4.72 ppm (2 H) (CH$_2$ split by 3 H's)

$$CH_3CH_2\overset{\overset{O}{\|}}{C}O-CH_2CH_3$$

d. **C$_6$H$_{14}$O**: 0 degrees of unsaturation
IR peak at 3600–3200 cm^{-1}: **O–H**
NMR: triplet at 0.8 ppm (6 H) (2 CH$_3$ groups
  split by CH$_2$ groups)
  singlet at 1.0 ppm (3 H) (CH$_3$)
  quartet at 1.5 ppm (4 H) (2 CH$_2$ groups split
  by CH$_3$ groups)
  singlet at 1.6 ppm (1 H) (O–H proton)

$$CH_3CH_2-\underset{\underset{OH}{|}}{\overset{\overset{CH_3}{|}}{C}}-CH_2CH_3$$

e. **C$_6$H$_{14}$O**: 0 degrees of unsaturation
IR peak at 3000–2850 cm^{-1}: **C$_{sp^3}$–H bonds**
NMR: doublet at 1.10 ppm (integration = 30 units)
  (from 12 H's)
  septet at 3.60 ppm (integration = 5 units)
  (from 2 H's)

$$CH_3-\underset{\underset{CH_3}{|}}{\overset{\overset{H}{|}}{C}}-O-\underset{\underset{CH_3}{|}}{\overset{\overset{H}{|}}{C}}-CH_3$$

f. **C$_3$H$_6$O**: **1 degree of unsaturation**
IR peak at 1730 cm^{-1}: **C=O**
NMR: triplet at 1.11 ppm
  multiplet at 2.46 ppm
  triplet at 9.79 ppm

$$CH_3CH_2\overset{\overset{O}{\|}}{C}H$$

14.55

Two isomers of $C_9H_{10}O$: 5 degrees of unsaturation (benzene ring likely)

Compound A:
  IR absorption at 1742 cm^{-1}: **C=O**
  NMR data:
    Absorptions:
    singlet at 2.15 (3 H) (CH$_3$ group)
    singlet at 3.70 (2 H) (CH$_2$ group)
    broad singlet at 7.20 (5 H)
        (likely a monosubstituted benzene ring)

Compound B:
  IR absorption at 1688 cm^{-1}: **C=O**
  NMR data:
    Absorptions:
    triplet at 1.22 (3 H) (CH$_3$ group split by 2 H's)
    quartet at 2.98 (2 H) (CH$_2$ group split by 3 H's)
    multiplet at 7.28–7.95 (5 H)
        (likely a monosubstituted benzene ring)

14.56

Compound C:
  molecular ion 146 (molecular formula $C_6H_{10}O_4$)
  IR absorption at 1762 cm^{-1}: **C=O**
  ^{1}H NMR data:
    Absorptions:
    H$_a$: doublet at 1.47 (3 H) (CH$_3$ group adjacent to CH)
    H$_b$: singlet at 2.07 (6 H) (2 CH$_3$ groups)
    H$_c$: quartet at 6.84 (1 H adjacent to CH$_3$)

14.57

Compound D:
  molecular ion 84 (molecular formula C_5H_8O)
  IR absorptions at 3600–3200 cm^{-1}: OH
                  3303 cm^{-1}: C$_{sp}$–H
                  2938 cm^{-1}: C$_{sp^3}$–H
                  2120 cm^{-1}: C≡C
  ^{1}H NMR data:
    Absorptions:
    H$_a$: singlet at 1.53 (6 H) (2 CH$_3$ groups)
    H$_b$: singlet at 2.37 (1 H)
    H$_c$: singlet at 2.43 (1 H) ⎬ alkynyl CH and OH

14.58

Compound E:
$C_4H_8O_2$:
   1 degree of unsaturation
IR absorption at 1743 cm^{-1}: **C=O**
NMR data:

     total integration units/# H's
     (23 + 29 + 30)/8 = ~10 units per H

     H_a: quartet at 4.1 (23 units - **2 H**)
     H_b: singlet at 2.0 (29 units - **3 H**)
     H_c: triplet at 1.4 (30 units - **3 H**)

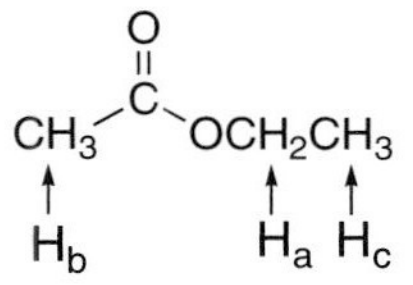

Compound F:
$C_4H_8O_2$:
   1 degree of unsaturation
IR absorption at 1730 cm^{-1}: **C=O**
NMR data:

     total integration units/# H's
     (18 + 30 + 31)/8 = ~10 units per H

     H_a: singlet at 4.1 (18 units - **2 H**)
     H_b: singlet at 3.4 (30 units - **3 H**)
     H_c: singlet at 2.1 (31 units - **3 H**)

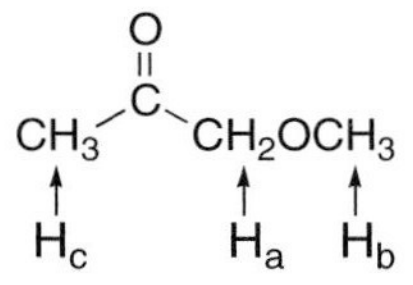

14.59

Compound H:
$C_8H_{11}N$:
   4 degrees of unsaturation
IR absorptions at 3365 cm^{-1}: N–H
                 3284 cm^{-1}: N–H
                 3026 cm^{-1}: C_{sp^2}–H
                 2932 cm^{-1}: C_{sp^3}–H
                 1603 cm^{-1}: due to benzene
                 1497 cm^{-1}: due to benzene
NMR data:

  multiplet at 7.2–7.4 ppm, **5 H** on a benzene ring
  H_a: triplet at 2.9 ppm, **2 H**, split by 2 H's
  H_b: triplet at 2.8 ppm, **2 H**, split by 2 H's
  H_c: singlet at 1.1 ppm, **2 H**, no splitting (on NH$_2$)

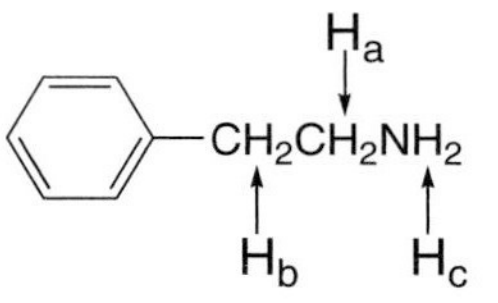

Compound I:
$C_8H_{11}N$:
   4 degrees of unsaturation
IR absorptions at 3367 cm^{-1}: N–H
                 3286 cm^{-1}: N–H
                 3027 cm^{-1}: C_{sp^2}–H
                 2962 cm^{-1}: C_{sp^3}–H
                 1604 cm^{-1}: due to benzene
                 1492 cm^{-1}: due to benzene
NMR data:

  multiplet at 7.2–7.4 ppm, **5 H** on a benzene ring
  H_a: quartet at 4.1 ppm, **1 H**, split by 3H's
  H_b: singlet at 1.45 ppm, **2 H**, no splitting (NH$_2$)
  H_c: doublet at 1.4 ppm, **3 H**, split by 1 H

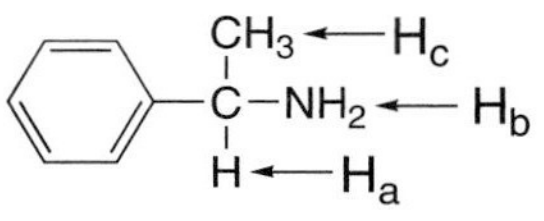

14.60

a. **$C_9H_{10}O_2$:**
   5 degrees of unsaturation
   IR absorption at 1718 cm^{-1}: **C=O**
   NMR data:

     multiplet at 7.4–8.1 ppm, **5 H** on a benzene ring
     quartet at 4.4 ppm, **2 H**, split by 3 H's
     triplet at 1.3 ppm, **3 H**, split by 2 H's

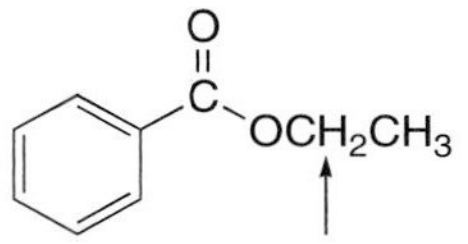

     downfield due to the O atom

b. **C_9H_{12}:**
   4 degrees of unsaturation
   IR absorption at 2850–3150 cm^{-1}:
   C–H bonds
   NMR data:

     singlet at 7.1–7.4 ppm, **5 H**, benzene
     septet at 2.8 ppm, **1 H**, split by 6 H's
     doublet at 1.3 ppm, **6 H**, split by 1 H

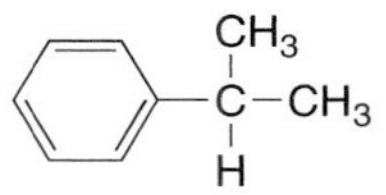

14.61

a. Compound **J** has a molecular ion at 72: molecular formula **C₄H₈O**
   1 degree of unsaturation
   IR spectrum at 1710 cm⁻¹: **C=O**
   ¹H NMR data (ppm):
   1.0 (triplet, 3 H), split by 2 H's
   2.1 (singlet, 3 H)
   2.4 (quartet, 2 H), split by 3 H's

b. Compound **K** has a molecular ion at 88: molecular formula **C₅H₁₂O**
   0 degrees of unsaturation
   IR spectrum at 3600–3200 cm⁻¹: **O–H bond**
   ¹H NMR data (ppm):
   0.9 (triplet, 3 H), split by 2 H's
   1.2 (singlet, 6 H), due to 2 CH₃ groups
   1.5 (quartet, 2 H), split by 3 H's
   1.6 (singlet, 1 H), due to the OH proton

14.62

Compound **L** has a molecular ion at 90: molecular formula **C₄H₁₀O₂**
   0 degrees of unsaturation
   IR absorptions at 2992 and 2941 cm⁻¹: C_{sp^3}–H
   ¹H NMR data (ppm):
   Hₐ: 1.2 (doublet, 3 H), split by 1 H
   H_b: 3.3 (singlet, 6 H), due to 2 CH₃ groups
   H_c: 4.8 (quartet, 1 H), split by 3 adjacent H's

total integration units/# H's
(25 + 46 + 7)/10 = ~8 units per H

14.63

14.64

Compound **O** has a molecular formula **$C_{10}H_{12}O$.**

5 degrees of unsaturation

IR absorption at 1687 cm^{-1}

^{1}H NMR data (ppm):

H_a: 1.0 (triplet, 3 H), due to CH_3 group, split by 2 adjacent H's

H_b: 1.7 (sextet, 2 H), split by CH_3 and CH_2 groups

H_c: 2.9 (triplet, 2 H), split by 2 H's

7.4–8.0 (multiplet, 5 H), benzene ring

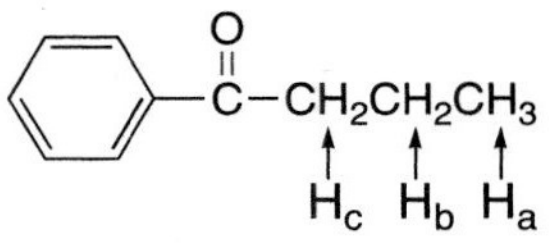

14.65

Compound **P** has a molecular formula **$C_5H_9ClO_2$.**

1 degree of unsaturation

^{13}C NMR shows 5 different C's, including a C=O.

^{1}H NMR data (ppm):

H_a: 1.3 (triplet, 3 H), split by 2 H's

H_b: 2.8 (triplet, 2 H), split by 2 H's

H_c: 3.7 (triplet, 2 H), split by 2 H's

H_d: 4.2 (quartet, 2 H), split by CH_3 group

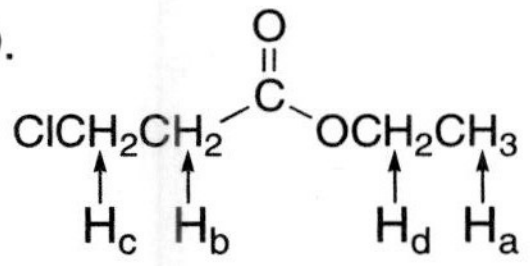

14.66

Compound Q: Molecular ion at 86.

Molecular formula: **$C_5H_{10}O$:**

1 degree of unsaturation

IR absorption at ~1700 cm^{-1}: **C=O**

NMR data:

H_a: doublet at 1.1 ppm, 2 CH_3 groups split by 1 H

H_b: singlet at 2.1 ppm, CH_3 group

H_c: septet at 2.6 ppm, 1 H split by 6 H's

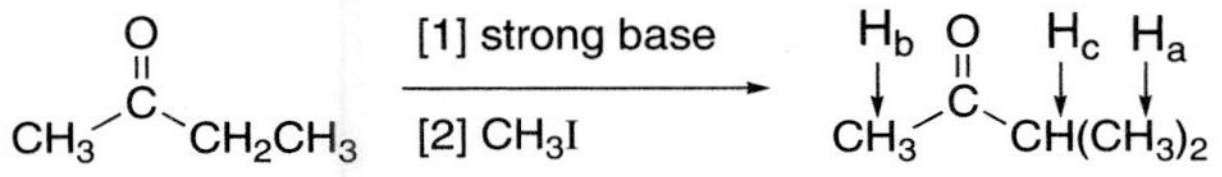

14.67

a. Compound **R**, the odor of banana: $C_7H_{14}O_2$

1 degree of unsaturation

^{1}H NMR (ppm):

H_a: 0.93 (doublet, 6 H)

H_b: 1.52 (multiplet, 2 H)

H_c: 1.69 (multiplet, 1 H)

H_d: 2.04 (singlet, 3 H)

H_e: 4.10 (triplet, 2 H)

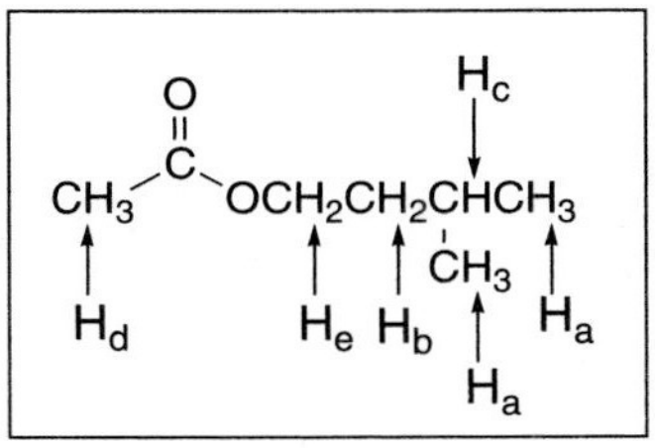

b. Compound **S**, the odor of rum: $C_7H_{14}O_2$

1 degree of unsaturation

^{1}H NMR (ppm):

H_a: 0.94 (doublet, 6 H)

H_b: 1.15 (triplet, 3 H)

H_c: 1.91 (multiplet, 1 H)

H_d: 2.33 (quartet, 2 H)

H_e: 3.86 (doublet, 2 H)

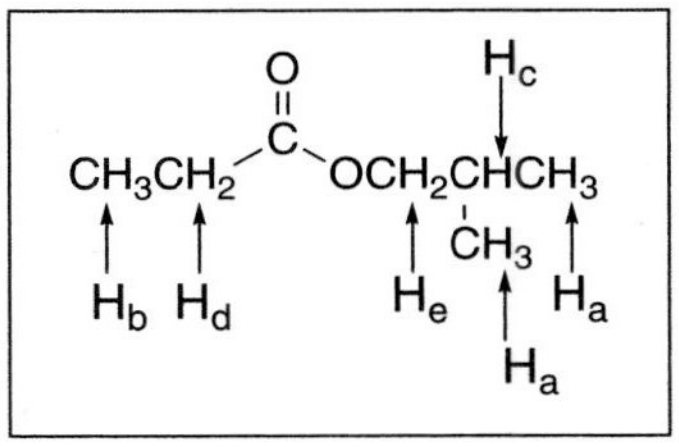

14.68

C₆H₁₂:

1 degree of unsaturation

K⁺ ⁻OC(CH₃)₃

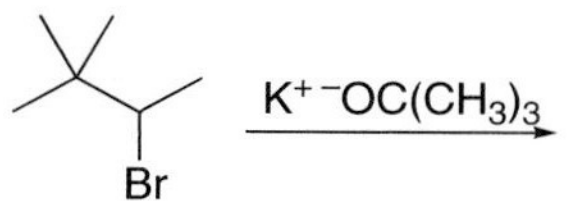

T   ¹H NMR of **T** (ppm):

Hₐ: 1.01 (singlet, 9 H)

H_b: 4.82 (doublet of doublets, 1 H, J = 10, 1.7 Hz)

H_c: 4.93 (doublet of doublets, 1 H, J = 18, 1.7 Hz)

H_d: 5.83 (doublet of doublets, 1 H, J = 18, 10 Hz)

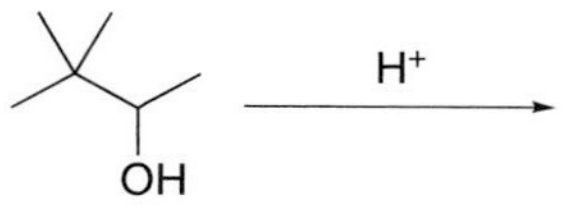

H⁺

U   ¹H NMR of **U**: 1.60 (singlet) ppm.

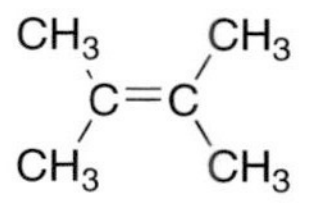

All H's are identical, so there is only one singlet in the NMR.

14.69 Both **A** and **B** have the same molecular ion—since they are isomers—and show a C=O peak in their IR spectra. ¹H NMR spectroscopy is the best way to distinguish the two compounds.

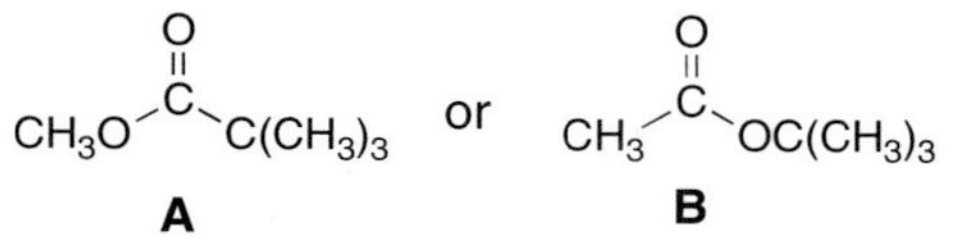

Both **A** and **B** have two singlets in a 3:1 ratio in their ¹H NMR spectra.   But **A** has a peak at ~3 ppm due to the deshielded CH₃ group bonded to the O atom. **B** has no proton that is so deshielded. Both of its singlets are in the 1–2.5 ppm region.

14.70

a.

C₆H₁₂O₂:

1 degree of unsaturation

IR peak at 1740 cm⁻¹: **C=O**

¹H NMR 2 signals: 2 types of H's

¹³C NMR: 4 signals: 4 kinds of C's, including one at ~170 ppm due a C=O

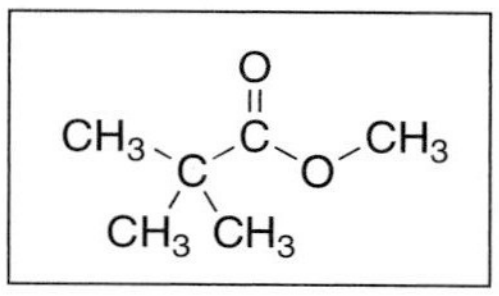

b.

C₆H₁₀:

2 degrees of unsaturation

IR peak at 3000 cm⁻¹: **C_{sp³}–H bonds**

peak at 3300 cm⁻¹: **C_{sp}–H bond**

peak at ~2150 cm⁻¹: **C≡C bond**

¹³C NMR: 4 signals: 4 kinds of C's

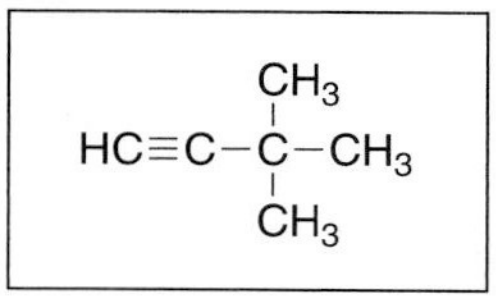

14.71

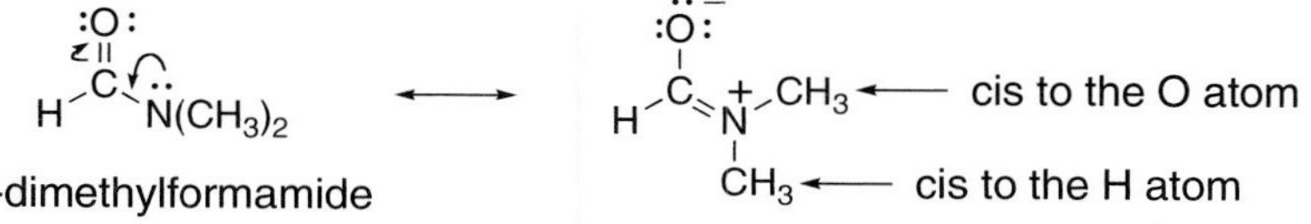

N,N-dimethylformamide

cis to the O atom

cis to the H atom

A second resonance structure for *N,N*-dimethylformamide places the two CH₃ groups in different environments. One CH₃ group is cis to the O atom, and one is cis to the H atom.   This gives rise to two different absorptions for the CH₃ groups.

14.72

18-Annulene has 18 π electrons that create an **induced magnetic field** similar to the 6 π electrons of benzene. 18-Annulene has 12 protons that are oriented on the outside of the ring (labeled H_o), and 6 protons that are oriented inside the ring (labeled H_i). The induced magnetic field reinforces the external field in the vicinity of the protons on the outside of the ring. These protons are deshielded and so they absorb downfield (8.9 ppm). In contrast, the induced magnetic field is opposite in direction to the applied magnetic field in the vicinity of the protons on the inside of the ring. This shields the protons and the absorption is therefore very far upfield, even higher than TMS (−1.8 ppm).

14.73

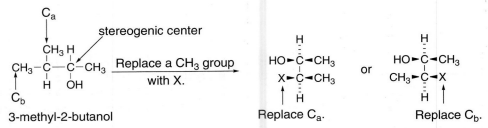

3-methyl-2-butanol

stereogenic center

Replace a CH_3 group with X.

Replace C_a.　Replace C_b.

The CH_3 groups are not equivalent to each other, since replacement of each by X forms two diastereomers.

Thus, every C in this compound is different and there are five ^{13}C signals.

14.74

One P atom splits each nearby CH_3 into a doublet by the $n + 1$ rule, making two doublets.

All 6 H_a protons are equivalent.

Chapter 15: Radical Reactions

◆ General features of radicals

- A radical is a reactive intermediate with an unpaired electron (15.1).
- A carbon radical is sp^2 hybridized and trigonal planar (15.1).
- The stability of a radical increases as the number of C's bonded to the radical carbon increases (15.1).

least stable			most stable
$\dot{C}H_3$	$R\dot{C}H_2$	$R_2\dot{C}H$	$R_3\dot{C}$
	1°	2°	3°

→ **Increasing alkyl substitution**
Increasing radical stability

- Allylic radicals are stabilized by resonance, making them more stable than 3° radicals (15.10).

$CH_2{=}CH{-}\dot{C}H_2 \quad\longleftrightarrow\quad \dot{C}H_2{-}CH{=}CH_2$

two resonance structures for the allyl radical

◆ Radical reactions

[1] Halogenation of alkanes (15.4)

$$R{-}H \xrightarrow[\text{hv or } \Delta]{X_2} \boxed{\begin{array}{c} R{-}X \\ \text{alkyl halide} \end{array}}$$

X = Cl or Br

- The reaction follows a radical chain mechanism.
- The weaker the C–H bond, the more readily the H is replaced by X.
- Chlorination is faster and less selective than bromination (15.6).
- Radical substitution results in racemization at a stereogenic center (15.8).

[2] Allylic halogenation (15.10)

$$CH_2{=}CH{-}CH_3 \xrightarrow[\text{hv or ROOR}]{NBS} \boxed{\begin{array}{c} CH_2{=}CHCH_2Br \\ \text{allylic halide} \end{array}}$$

- The reaction follows a radical chain mechanism.

[3] Radical addition of HBr to an alkene (15.13)

$$RCH{=}CH_2 \xrightarrow[\substack{\text{hv, } \Delta\text{, or} \\ \text{ROOR}}]{HBr} \boxed{\begin{array}{c} \underset{\substack{| \\ H}}{\overset{\substack{H \\ |}}{R{-}C}}{-}\underset{\substack{| \\ Br}}{\overset{\substack{H \\ |}}{C}}{-}H \\ \text{alkyl bromide} \end{array}}$$

- A radical addition mechanism is followed.
- Br bonds to the less substituted carbon atom to form the more substituted, more stable radical.

[4] Radical polymerization of alkenes (15.14)

$$CH_2{=}CHZ \xrightarrow{ROOR} \boxed{\begin{array}{c} \text{polymer} \end{array}}$$

- A radical addition mechanism is followed.

Chapter 15: Answers to Problems

15.1 1° Radicals are on C's bonded to one other C; 2° radicals are on C's bonded to two other C's; 3° radicals are on C's bonded to three other C's.

a. $CH_3CH_2-\overset{\bullet}{C}HCH_2CH_3$ b.   c.   d.

      2° radical   **3° radical**   **2° radical**   **1° radical**

15.2 The stability of a radical increases as the number of alkyl groups bonded to the radical carbon increases. Draw the most stable radical.

a. $(CH_3)_2\overset{\bullet}{C}CH_2CH_3$ b. $(CH_3)_3C\overset{\bullet}{C}HCH_3$ c. $(CH_3)_3C\overset{\bullet}{C}H_2$ d.

15.3 Reaction of a radical with:
- an alkane abstracts a hydrogen atom and creates a new carbon radical.
- an alkene generates a new bond to one carbon and a new carbon radical.
- another radical forms a bond.

a. $CH_3-CH_3 \xrightarrow{\;:\overset{\bullet\bullet}{\underset{\bullet\bullet}{Cl}}\cdot\;} CH_3-\overset{\bullet}{C}H_2 + H-\overset{\bullet\bullet}{\underset{\bullet\bullet}{Cl}}:$

c. $:\overset{\bullet\bullet}{\underset{\bullet\bullet}{Cl}}\cdot \xrightarrow{\;:\overset{\bullet\bullet}{\underset{\bullet\bullet}{Cl}}\cdot\;} :\overset{\bullet\bullet}{\underset{\bullet\bullet}{Cl}}-\overset{\bullet\bullet}{\underset{\bullet\bullet}{Cl}}:$

b. $CH_2=CH_2 \xrightarrow{\;:\overset{\bullet\bullet}{\underset{\bullet\bullet}{Cl}}\cdot\;} \overset{\bullet}{C}H_2-\overset{:\overset{\bullet\bullet}{\underset{\bullet\bullet}{Cl}}:}{\underset{}{C}}H_2$

d. $:\overset{\bullet\bullet}{\underset{\bullet\bullet}{Cl}}\cdot + \cdot\overset{\bullet\bullet}{\underset{\bullet\bullet}{O}}-\overset{\bullet\bullet}{\underset{\bullet\bullet}{O}}\cdot \longrightarrow :\overset{\bullet\bullet}{\underset{\bullet\bullet}{Cl}}-\overset{\bullet\bullet}{\underset{\bullet\bullet}{O}}-\overset{\bullet\bullet}{\underset{\bullet\bullet}{O}}\cdot$

15.4 **Monochlorination** is a radical substitution reaction in which a Cl replaces a H generating an alkyl halide.

a.

b. $CH_3CH_2CH_2CH_2CH_2CH_3 \xrightarrow[\Delta]{Cl_2} ClCH_2CH_2CH_2CH_2CH_2CH_3 + CH_3-\overset{H}{\underset{Cl}{C}}-CH_2CH_2CH_2CH_3 + CH_3CH_2-\overset{H}{\underset{Cl}{C}}-CH_2CH_2CH_3$

c. $(CH_3)_3CH \xrightarrow[\Delta]{Cl_2} CH_3-\overset{Cl}{\underset{CH_3}{C}}-CH_3 + CH_3-\overset{H}{\underset{CH_3}{C}}-CH_2Cl$

15.5

A $\xrightarrow[\Delta]{Cl_2}$

B $\xrightarrow[\Delta]{Cl_2}$ + +

15.6

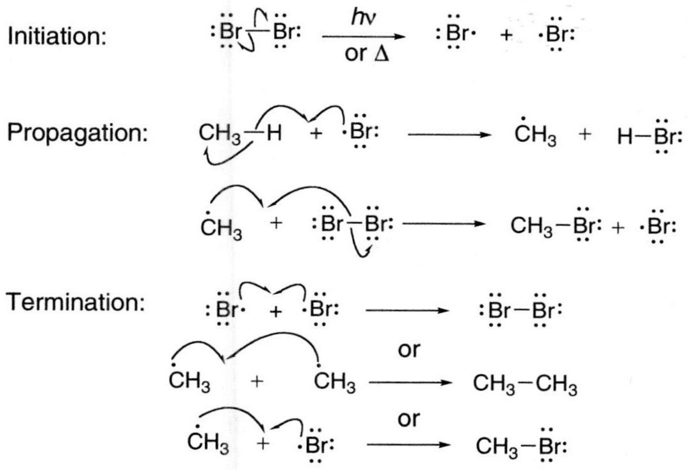

Initiation:

$:\ddot{Br}-\ddot{Br}: \xrightarrow[\text{or } \Delta]{h\nu} :\ddot{Br}\cdot + \cdot\ddot{Br}:$

Propagation:

$CH_3-H + \cdot\ddot{Br}: \longrightarrow \dot{C}H_3 + H-\ddot{Br}:$

$\dot{C}H_3 + :\ddot{Br}-\ddot{Br}: \longrightarrow CH_3-\ddot{Br}: + \cdot\ddot{Br}:$

Termination:

$:\ddot{Br}\cdot + \cdot\ddot{Br}: \longrightarrow :\ddot{Br}-\ddot{Br}:$

or

$\dot{C}H_3 + \dot{C}H_3 \longrightarrow CH_3-CH_3$

or

$\dot{C}H_3 + \cdot\ddot{Br}: \longrightarrow CH_3-\ddot{Br}:$

15.7

Step 1:

$CH_3-H + \cdot\ddot{Br}: \longrightarrow \dot{C}H_3 + H-\ddot{Br}: \qquad \Delta H^\circ = +67 \text{ kJ/mol}$

1 bond broken
+435 kJ/mol

1 bond formed
−368 kJ/mol

Step 2:

$\dot{C}H_3 + :\ddot{Br}-\ddot{Br}: \longrightarrow CH_3-Br + \cdot\ddot{Br}: \qquad \Delta H^\circ = -101 \text{ kJ/mol}$

1 bond broken
+192 kJ/mol

1 bond formed
−293 kJ/mol

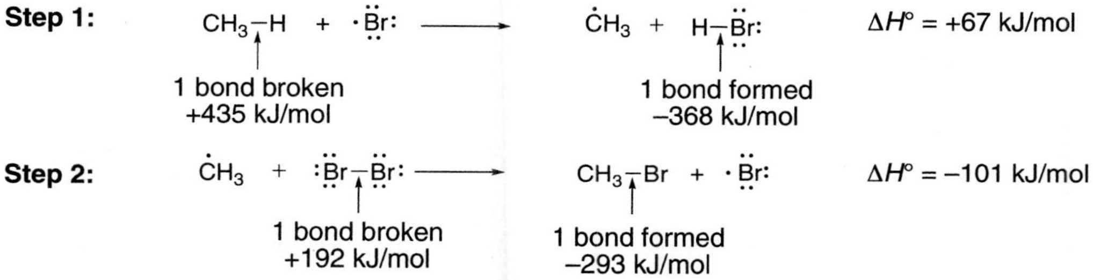

15.8 The rate-determining step for halogenation reactions is formation of $CH_3\cdot + HX$.

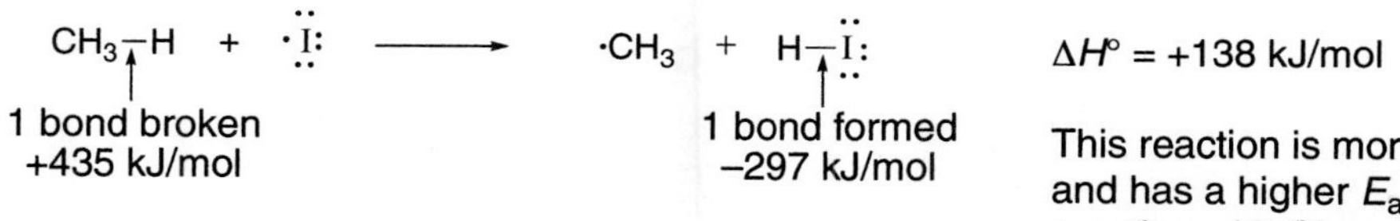

$CH_3-H + \cdot\ddot{I}: \longrightarrow \cdot CH_3 + H-\ddot{I}: \qquad \Delta H^\circ = +138 \text{ kJ/mol}$

1 bond broken
+435 kJ/mol

1 bond formed
−297 kJ/mol

This reaction is more endothermic and has a higher E_a than a similar reaction with Cl_2 or Br_2.

15.9 The **weakest C–H bond** in each alkane is the **most readily cleaved** during radical halogenation.

a.

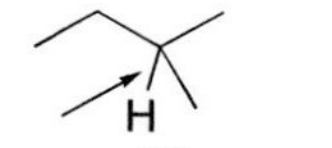

H
3°
most reactive

b.

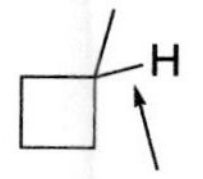

H
3°
most reactive

c. 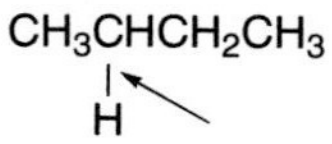 $CH_3CHCH_2CH_3$

H
2°
most reactive

15.10 To draw the product of bromination:

- Draw out the starting material and find the most reactive C–H bond (on the most substituted C).
- The major product is formed by **cleavage of the *weakest* C–H bond.**

a. 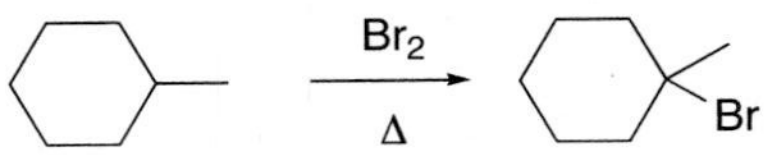

c. 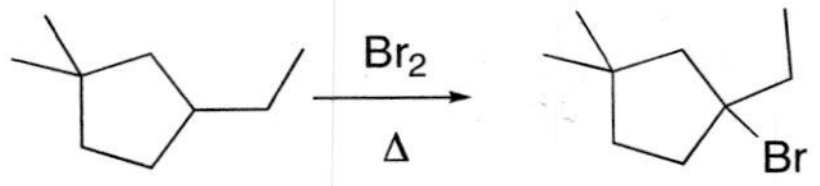

b. 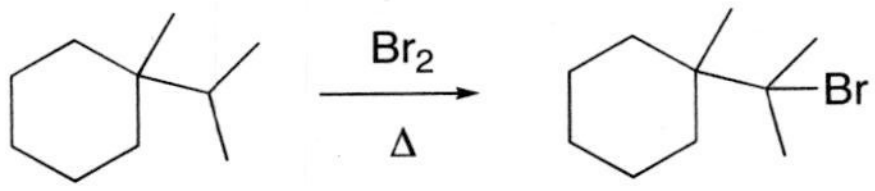

d. 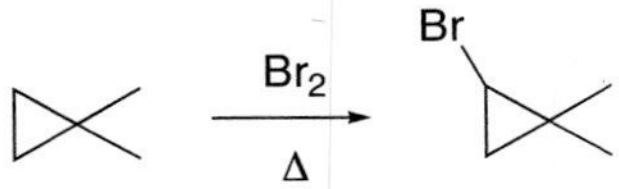

15.11

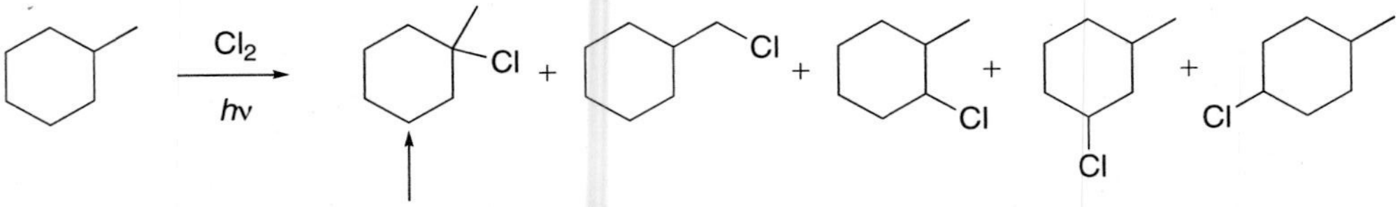

This is the desired product,
1-chloro-1-methylcyclohexane,
but many other products are formed.

15.12 If 1° C–H and 3° C–H bonds were equally reactive there would be nine times as much $(CH_3)_2CHCH_2Cl$ as $(CH_3)_3CCl$ since the ratio of 1° to 3° H's is 9:1. The fact that the ratio is only 63:37 shows that the 1° C–H bond is less reactive than the 3° C–H bond. $(CH_3)_2CHCH_2Cl$ is still the major product, though, because there are nine 1° C–H bonds and only one 3° C–H bond.

15.13

a. 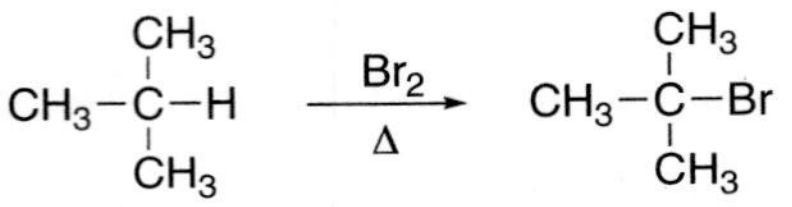

c. 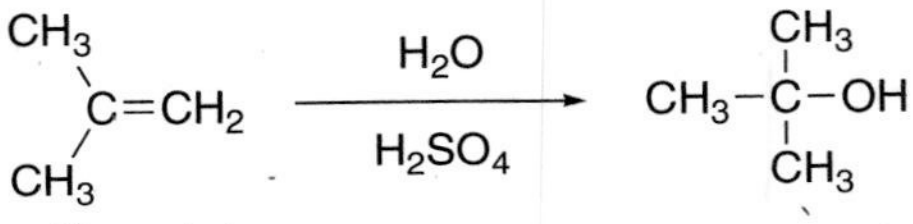

(from b.)

b. 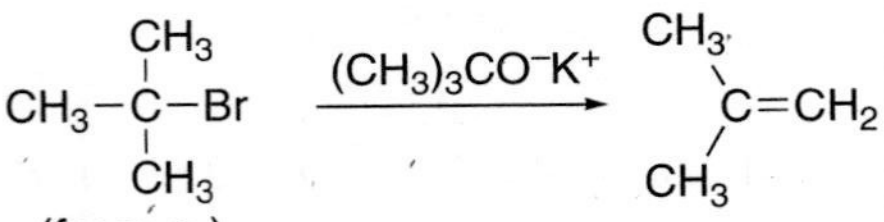

(from a.)

d. 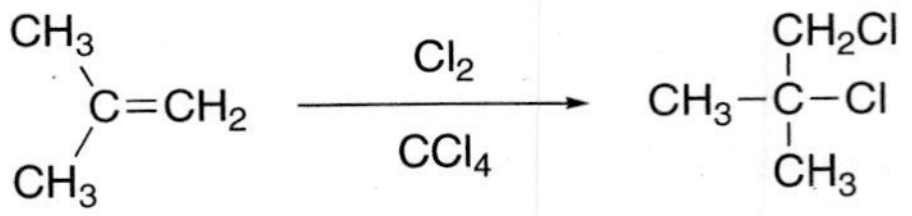

(from b.)

15.14

a. 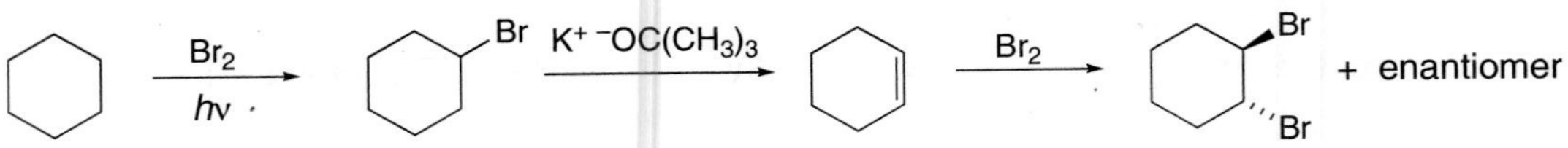

b. 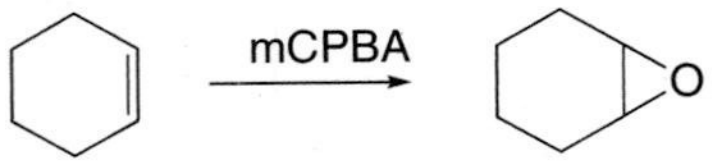

(from a.)

15.15 Since the reaction does not occur at the stereogenic center, leave it as is.

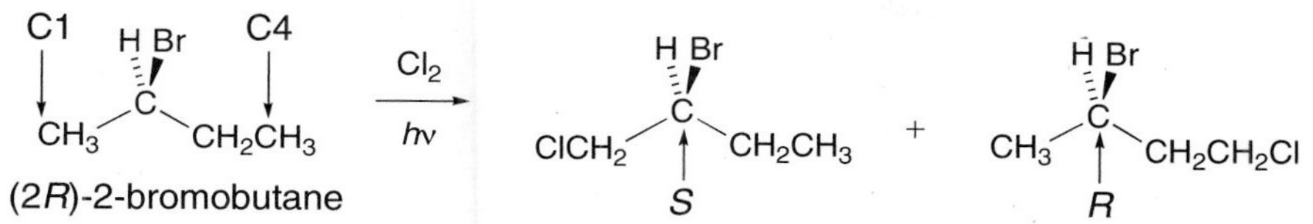

15.16

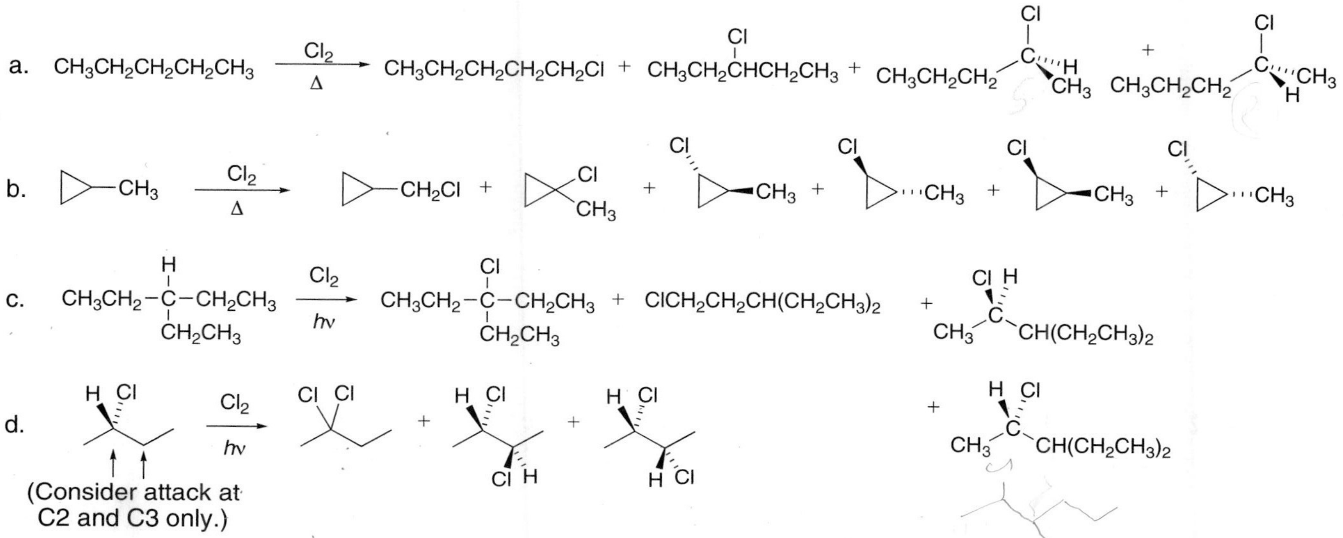

a. $CH_3CH_2CH_2CH_2CH_3$ $\xrightarrow[\Delta]{Cl_2}$ $CH_3CH_2CH_2CH_2CH_2Cl$ + $CH_3CH_2CHClCH_2CH_3$ + ...

b.

c.

d.

(Consider attack at C2 and C3 only.)

15.17

Chain propagation:

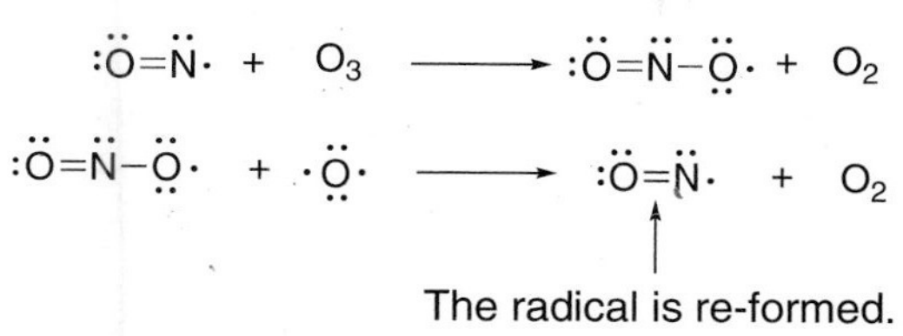

$$:\ddot{O}=\dot{N}\cdot\ +\ O_3 \longrightarrow :\ddot{O}=\ddot{N}-\ddot{O}\cdot\ +\ O_2$$

$$:\ddot{O}=\ddot{N}-\ddot{O}\cdot\ +\ \cdot\ddot{O}\cdot \longrightarrow :\ddot{O}=\dot{N}\cdot\ +\ O_2$$

The radical is re-formed.

15.18 Draw the resonance structure by moving the π bond and the unpaired electron. The hybrid is drawn with dashed lines for bonds that are in one resonance structure but not another. The symbol δ· is used on any atom that has an unpaired electron in any resonance structure.

a. 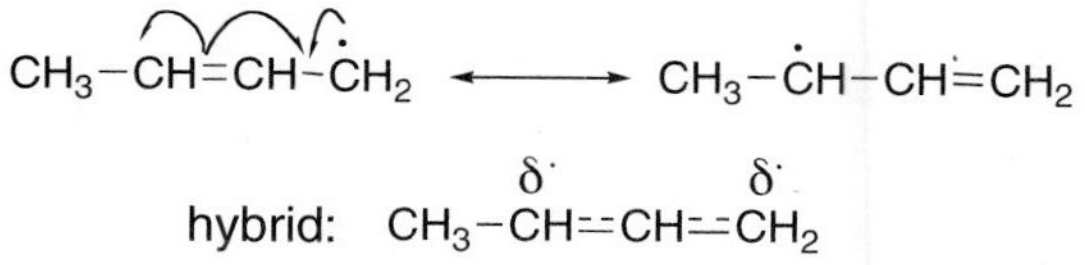 $CH_3-CH=CH-CH_2\cdot \longleftrightarrow CH_3-\dot{C}H-CH=CH_2$

hybrid: $CH_3-\overset{\delta\cdot}{CH}=\!=CH=\!=\overset{\delta\cdot}{CH_2}$

c.

hybrid:

b.

hybrid:

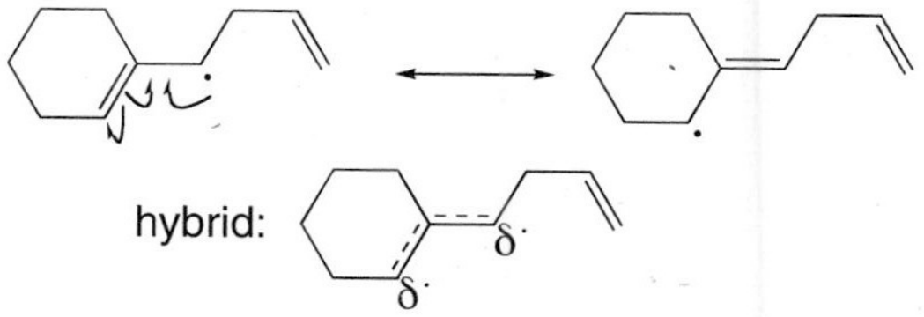

d.

hybrid:

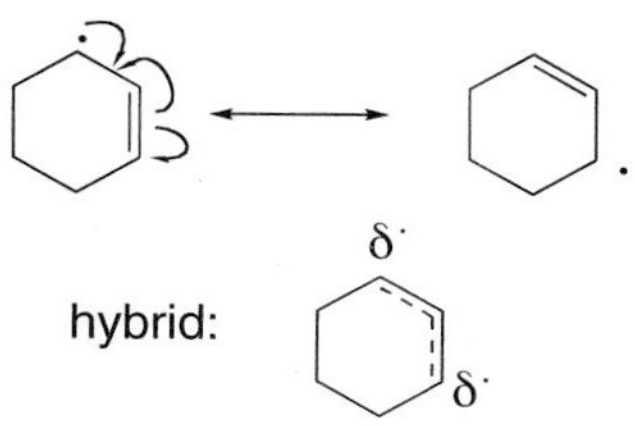

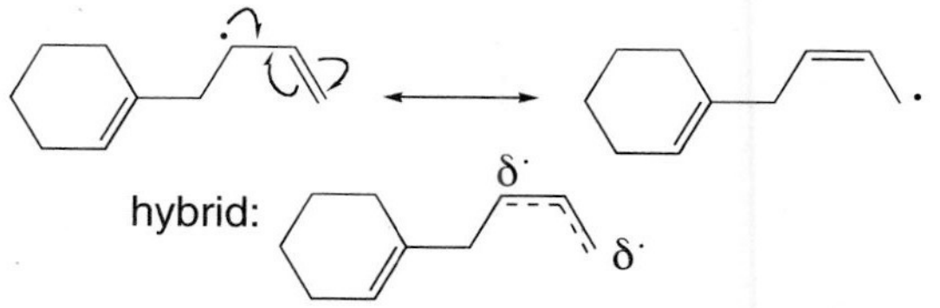

15.19 Reaction of an alkene with NBS or Br$_2$ + *h*v yields allylic substitution products.

a. [cyclopentene] $\xrightarrow[\text{hv}]{\text{NBS}}$ [3-bromocyclopentene, with Br]

b. $CH_2=CH-CH_3 \xrightarrow[\text{hv}]{\text{NBS}} CH_2=CH-CH_2Br$

c. $CH_2=CH-CH_3 \xrightarrow{\text{Br}_2} CH_2CHCH_3$ with Br Br

15.20

a. $CH_3-CH=CH-CH_3 \xrightarrow[\text{hv}]{\text{NBS}} CH_3-CH=CH-CH_2Br$
+
$CH_3CH(Br)CH=CH_2$

b. [cyclohexene with two CH$_3$] $\xrightarrow[\text{hv}]{\text{NBS}}$ Br—[ring with CH$_3$ CH$_3$] + [ring with CH$_3$ CH$_3$ and Br]

c. $CH_2=C(CH_2CH_3)_2 \xrightarrow[\text{hv}]{\text{NBS}} CH_2=C$ with CHBrCH$_3$ and CH$_2$CH$_3$
+
$BrCH_2C(CH_2CH_3)=CHCH_3$

15.21

[cyclohexene with CH$_3$] $\xrightarrow[\text{hv}]{\text{NBS}}$ [ring with CH$_3$, Br] + [ring with CH$_3$, —Br] + [ring with CH$_3$, Br] + [ring with CH$_3$, Br]

15.22 Reaction of an alkene with NBS + *h*v yields allylic substitution products.

a. [cycloheptene with Br] $\Longrightarrow$ [cycloheptene] one possible product: **high yield**

b. $CH_3CH_2CH=CHCH_2Br$ c. [cyclohexene with CH$_2$Br]

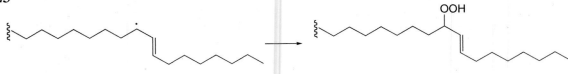

Cannot be made in high yield by allylic halogenation.
Any alkene starting material would yield a mixture of allylic halides.

15.23

[long chain structure] $\longrightarrow$ [long chain structure with OOH]

second resonance structure

15.24 The weakest C–H bond is most readily cleaved. To draw the hydroperoxide products, add OOH to each carbon that bears a radical in one of the resonance structures.

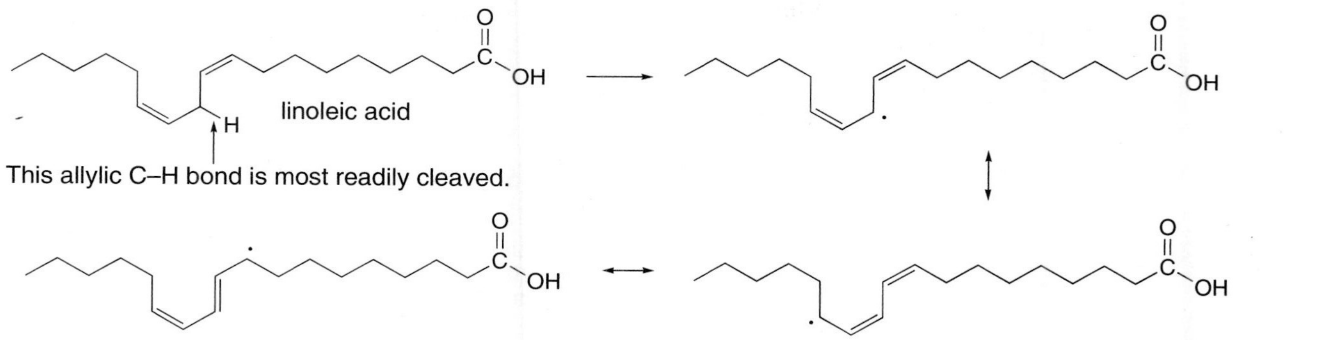

linoleic acid

This allylic C–H bond is most readily cleaved.

hydroperoxide products:

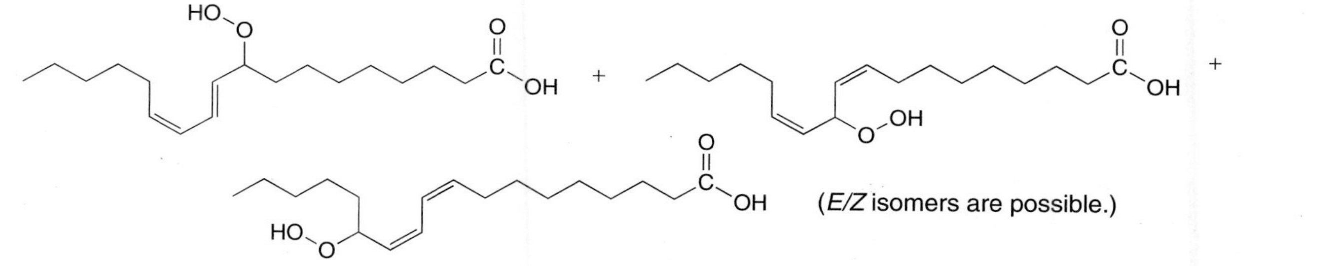

+ (*E/Z* isomers are possible.)

15.25

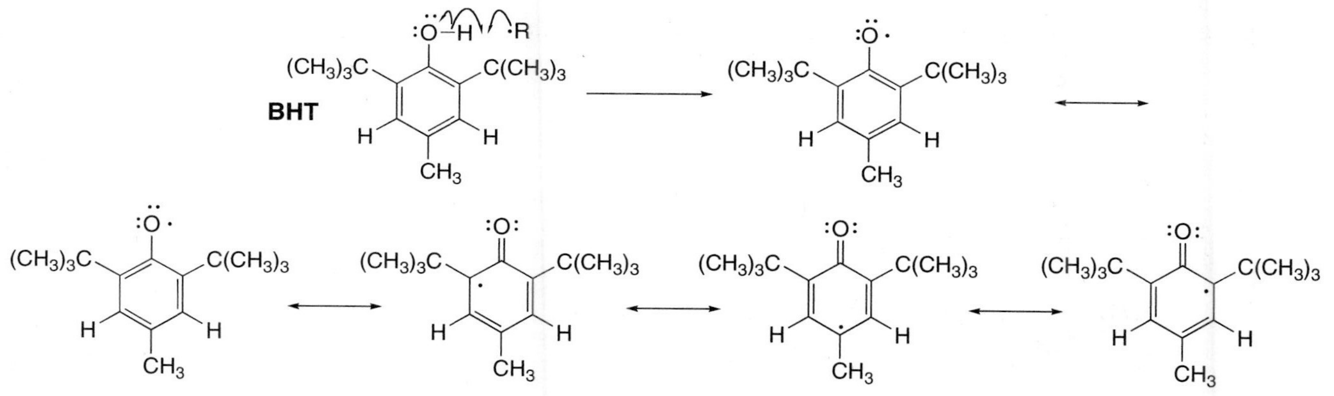

BHT

15.26

a. $CH_2{=}CHCH_2CH_2CH_2CH_3 \xrightarrow{HBr} CH_3CHCH_2CH_2CH_2CH_3$
$\qquad\qquad\qquad\qquad\qquad\qquad\qquad\qquad\quad |$
$\qquad\qquad\qquad\qquad\qquad\qquad\qquad\qquad\ Br$

$CH_2{=}CHCH_2CH_2CH_2CH_3 \xrightarrow[ROOR]{HBr} CH_2BrCH_2CH_2CH_2CH_2CH_3$

b.

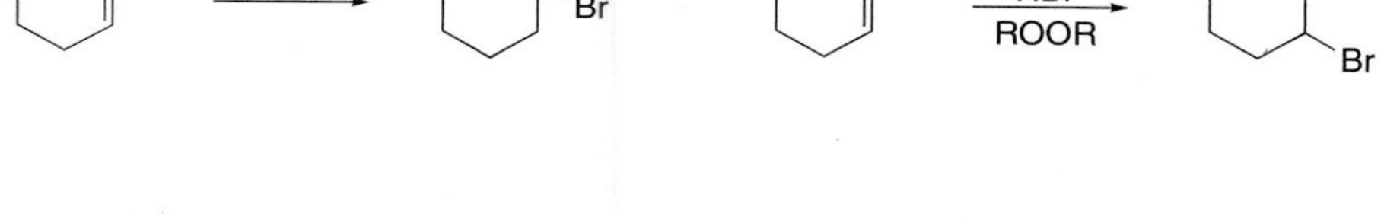

c.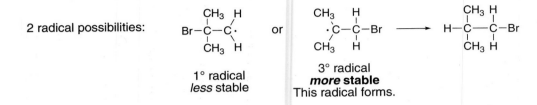

15.27 In addition of HBr under radical conditions:
- Br· adds first to form the more stable radical.
- Then H· is added to the carbon radical.

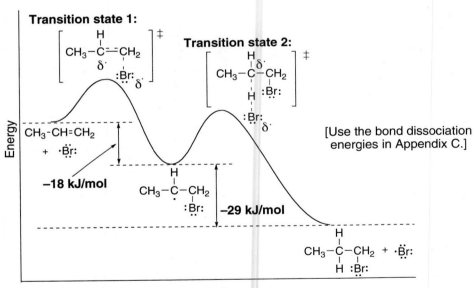

15.28

a.

b.

c.

15.29

Transition state 1:

Transition state 2:

[Use the bond dissociation energies in Appendix C.]

−18 kJ/mol

−29 kJ/mol

Reaction coordinate

15.30

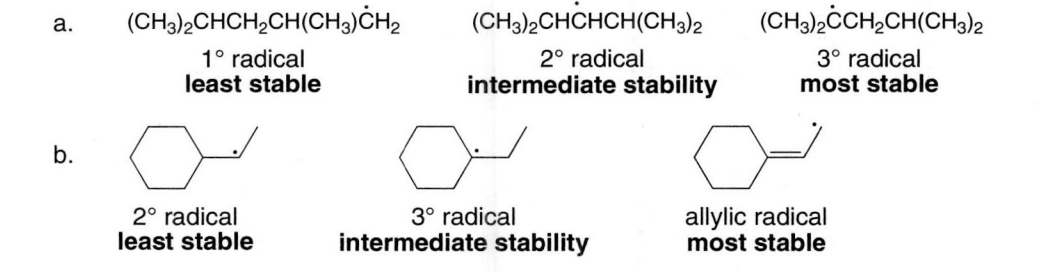

a. CH₂=C(H)(C₆H₅) ⟶ polystyrene

b. poly(vinyl acetate) ⟹ CH₂=CH−OCOCH₃

15.31

Initiation:

Propagation:

Repeat Step [3] over and over. new C–C bond

Termination: [4] [one possibility]

15.32

a. increasing bond strength: 2 < 3 < 1

b. and c.

·CH₂−C(CH₃)(H)−CHCH₃(H)
1° radical
least stable

H−CH₂−C(CH₃)(H)−·CHCH₃
2° radical
intermediate stability

H−CH₂−C(CH₃)−CHCH₃(H)
3° radical
most stable

d. increasing ease of H abstraction: 1 < 3 < 2

15.33 Use the directions from Answer 15.2 to rank the radicals.

a. (CH₃)₂CHCH₂CH(CH₃)ĊH₂
1° radical
least stable

(CH₃)₂CHĊHCH(CH₃)₂
2° radical
intermediate stability

(CH₃)₂ĊCH₂CH(CH₃)₂
3° radical
most stable

b.

2° radical
least stable

3° radical
intermediate stability

allylic radical
most stable

15.34 Draw the radical formed by cleavage of the benzylic C–H bond. Then draw all of the resonance structures. Having more resonance structures (five in this case) makes the radical more stable, and the benzylic C–H bond weaker.

benzylic C–H bond
bond dissociation energy = 356 kJ/mol

15.35

H_a = bonded to an sp^3 3° carbon
H_b = bonded to an allylic carbon
H_c = bonded to an sp^3 1° carbon
H_d = bonded to an sp^3 2° carbon

Increasing ease of abstraction:
$H_c < H_d < H_a < H_b$

15.36

a.

b. $(CH_3)_3CCH_2CH_2CH_2CH_3$ $\longrightarrow$ $(CH_3)_3CCH_2CH_2CH_2CH_2Cl$ + $(CH_3)_3CCH_2CH_2\overset{Cl}{\underset{}{C}}HCH_3$ + $(CH_3)_3CCH_2\overset{Cl}{\underset{}{C}}HCH_2CH_3$

$+$ $(CH_3)_2\overset{CH_2Cl}{\underset{}{C}}CH_2CH_2CH_2CH_3$ + $(CH_3)_3\overset{Cl}{\underset{}{C}}CHCH_2CH_2CH_3$

c.

d.

15.37 To draw the product of bromination:
 • Draw out the starting material and find the most reactive C–H bond (on the most substituted C).
 • The major product is formed by **cleavage of the *weakest* C–H bond**.

a.

b. $(CH_3)_3CCH_2CH(CH_3)_2$ $\longrightarrow$ $(CH_3)_3CCH_2\overset{}{\underset{Br}{C}}(CH_3)_2$

c.

d. $(CH_3)_3CCH_2CH_3$ $\longrightarrow$ $(CH_3)_3C\overset{Br}{\underset{}{C}}HCH_3$

15.38 Draw all of the alkane isomers of C_6H_{14} and their products on chlorination. Then determine which letter corresponds to which alkane.

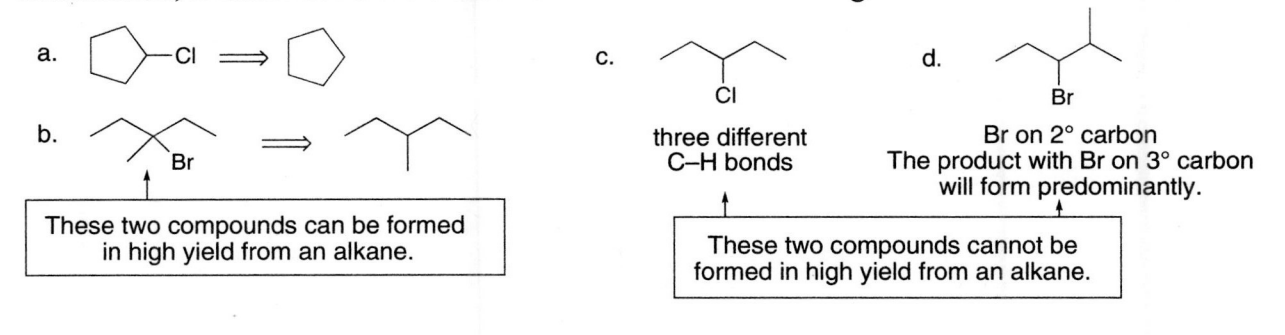

15.39 Halogenation replaces a C–H bond with a C–X bond. To find the alkane needed to make each of the alkyl halides, replace the X with a H.

a.

b.

c.

d. $(CH_3)_3CCH_2Cl \Longrightarrow (CH_3)_3CCH_3$

15.40 For an alkane to yield one major product on monohalogenation with Cl_2, all of the hydrogens must be identical in the starting material. For an alkane to yield one major product on bromination, it must have a more substituted carbon in the starting material.

a.

b.

These two compounds can be formed in high yield from an alkane.

c.

three different C–H bonds

d.

Br on 2° carbon
The product with Br on 3° carbon will form predominantly.

These two compounds cannot be formed in high yield from an alkane.

15.41 Chlorination with two equivalents of Cl_2 yields a variety of products.

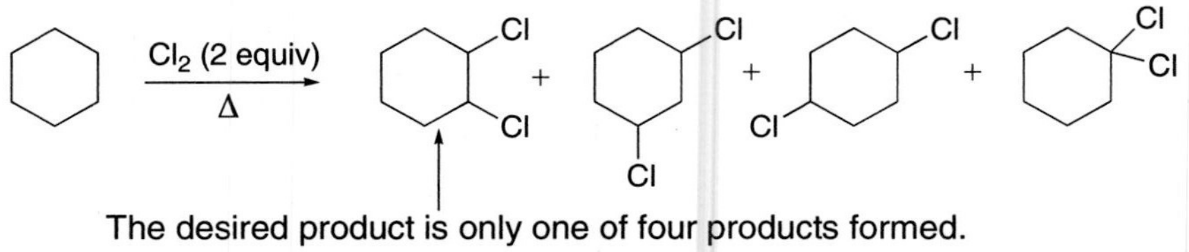

The desired product is only one of four products formed.

15.42 In bromination, the predominant (or exclusive) product is formed by cleavage of the weaker C–H bond to form the more stable radical intermediate.

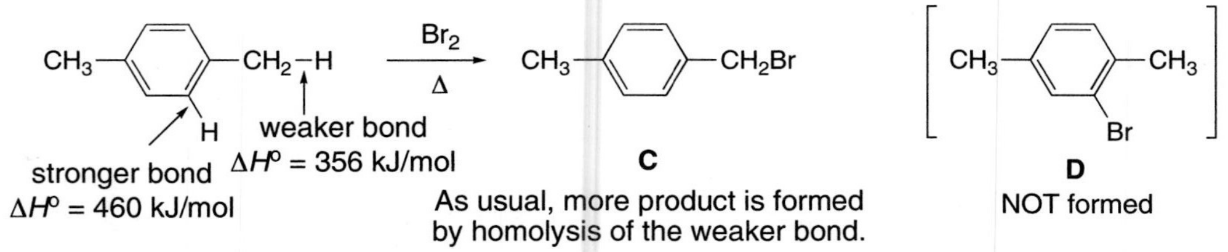

stronger bond $\Delta H^o = 460$ kJ/mol

weaker bond $\Delta H^o = 356$ kJ/mol

C

As usual, more product is formed by homolysis of the weaker bond.

D

NOT formed

15.43 Chlorination is not selective so a mixture of products results. Bromination is selective, and the major product is formed by cleavage of the weakest C–H bond.

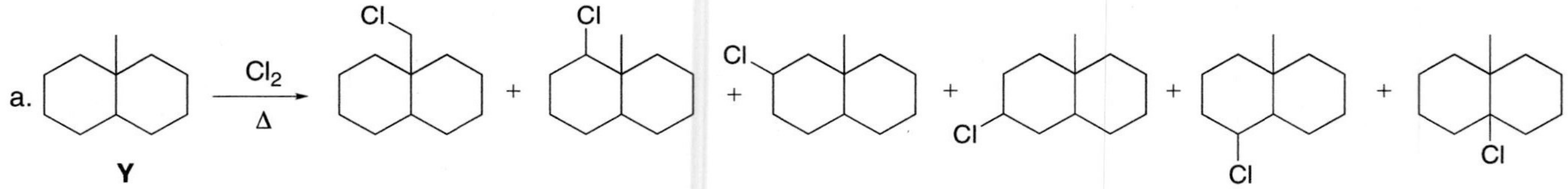

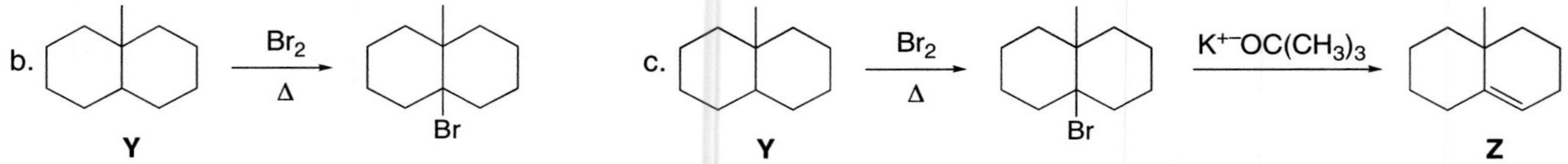

a.

b.

c.

15.44 Draw the resonance structures by moving the π bonds and the radical.

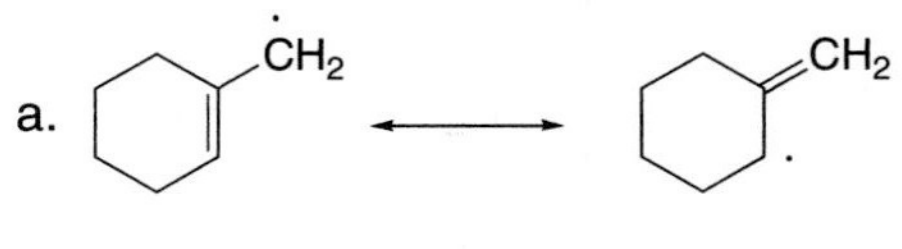

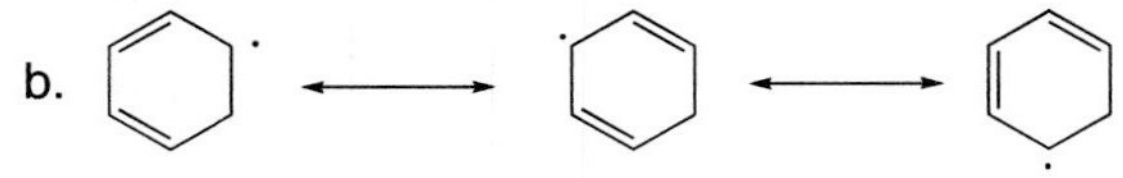

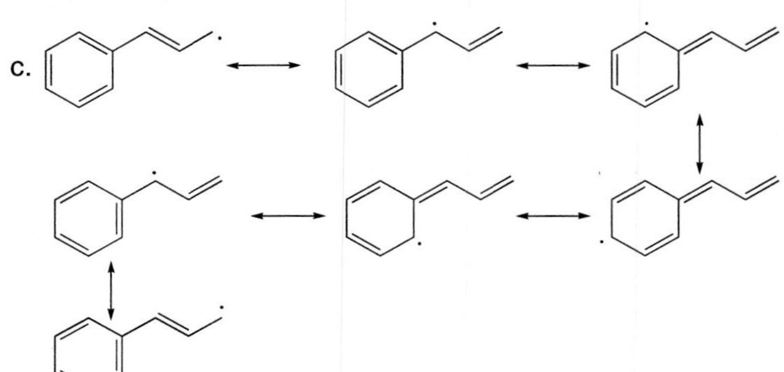

a.

b.

c.

15.45 Reaction of an alkene with NBS + *h*v yields allylic substitution products.

a.

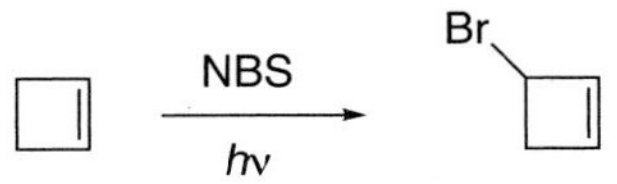

b. $CH_3CH_2CH{=}CHCH_2CH_3 \xrightarrow[h\nu]{NBS} CH_3CH_2CH{=}CHCHCH_3 \ + \ CH_3CH_2CHCH{=}CHCH_3$ (each with Br)

c. $(CH_3)_2C{=}CHCH_3 \xrightarrow[h\nu]{NBS} (CH_3)_2C{=}CHCH_2Br \ + \ BrCH_2C(CH_3){=}CHCH_3 \ + \ CH_2{=}C(CH_3)CH(Br)CH_3$
$+$
$(CH_3)_2C(Br)CH{=}CH_2$

d.

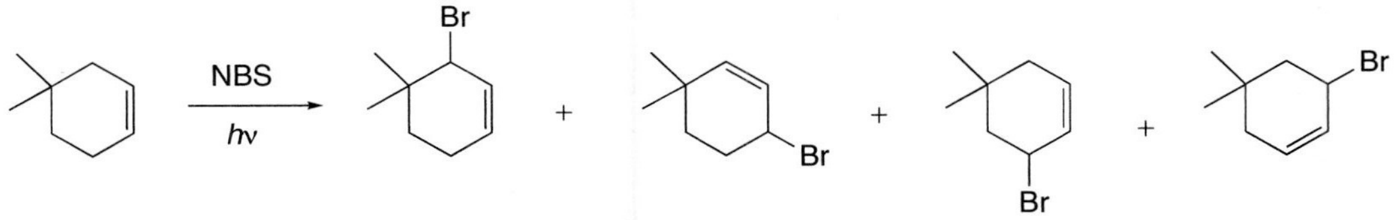

15.46 It is not possible to form 5-bromo-1-methylcyclopentene in good yield by allylic bromination because several other products are formed. 1-Methylcyclopentene has three different types of allylic hydrogens (labeled with *), all of which can be removed during radical bromination.

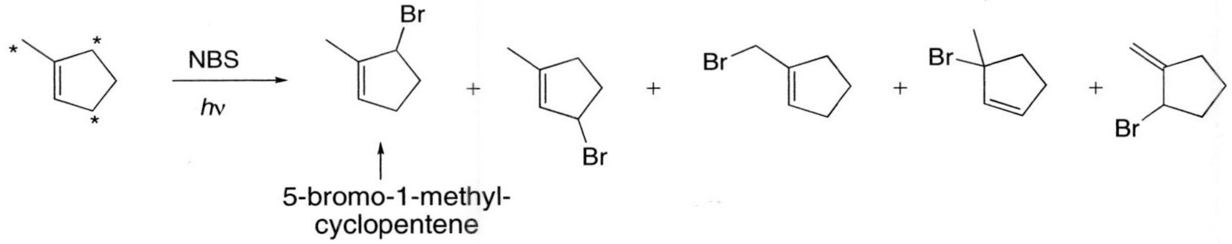

5-bromo-1-methyl-
cyclopentene

15.47

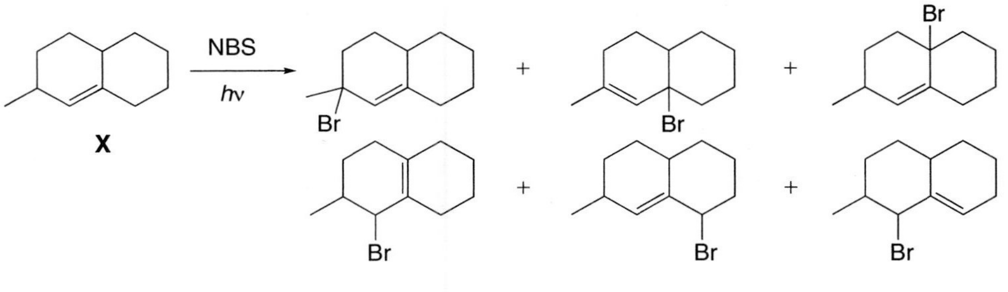

15.48

a.

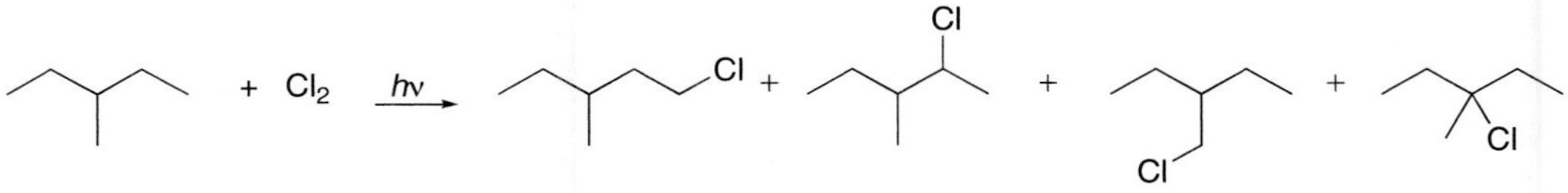

b.

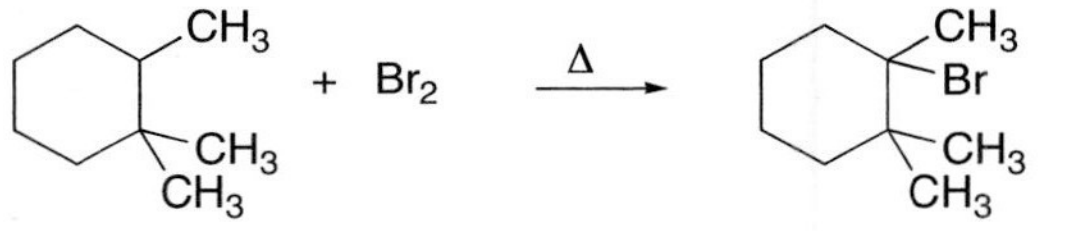

(major product)

c. 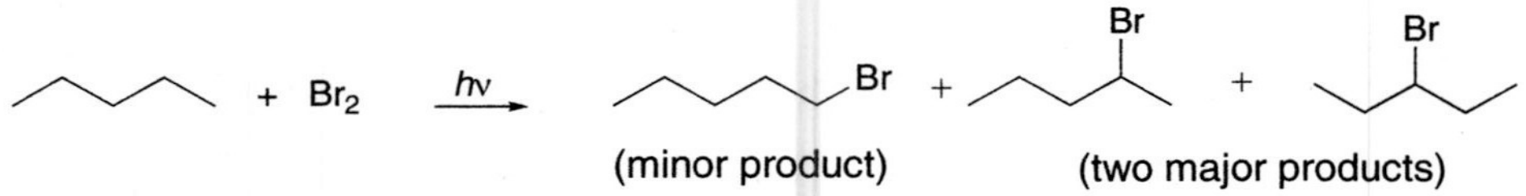 + Br$_2$ $\xrightarrow{h\nu}$ (minor product) + + (two major products)

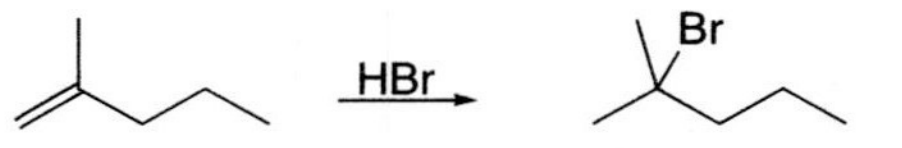

d. + NBS $\xrightarrow{h\nu}$ +

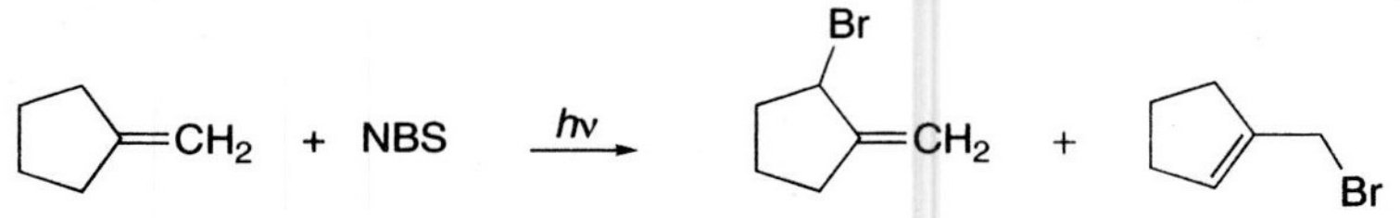

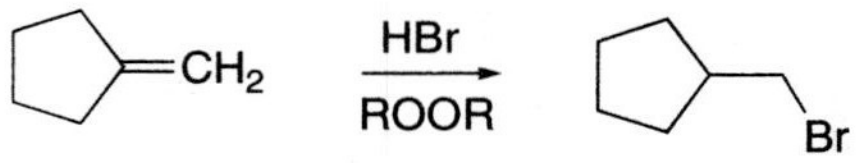

e. $\xrightarrow{HBr}$

f. $\xrightarrow[ROOR]{HBr}$

g. $\xrightarrow[ROOR]{HBr}$

h. $\xrightarrow{Br_2}$

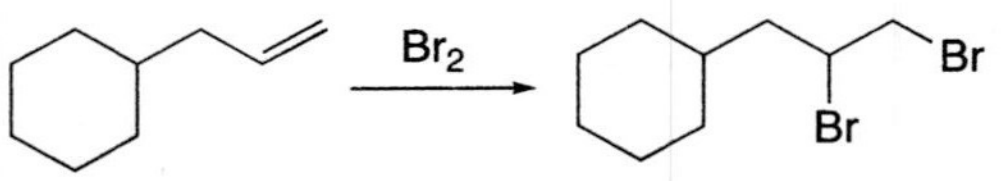

i. 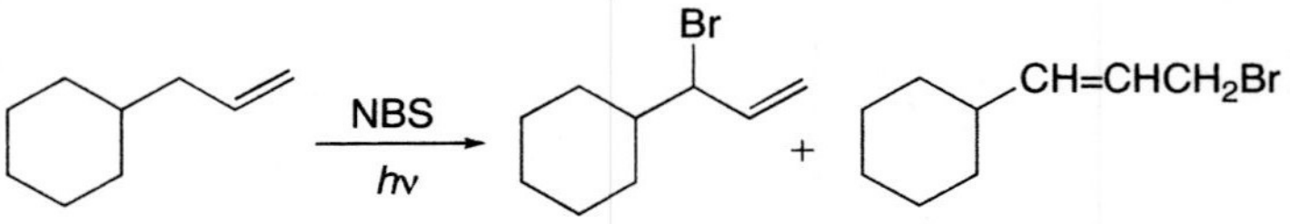 $\xrightarrow[h\nu]{NBS}$ + CH=CHCH$_2$Br

15.49

a. 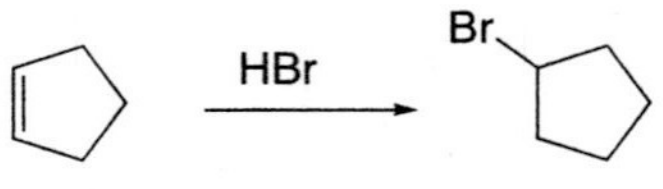 $\xrightarrow{HBr}$

b. 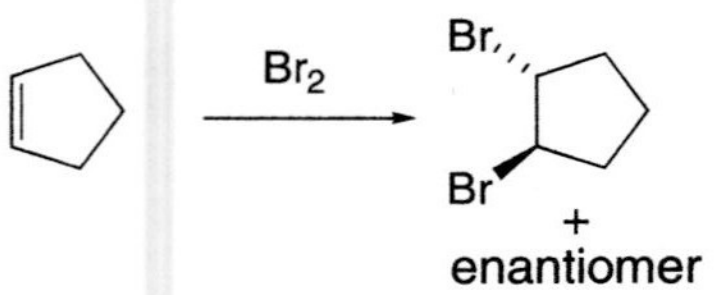 $\xrightarrow{Br_2}$ + enantiomer

c. 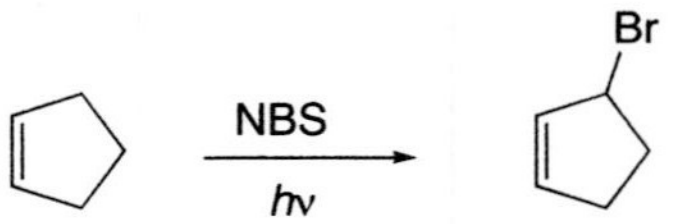 $\xrightarrow[h\nu]{NBS}$

15.50

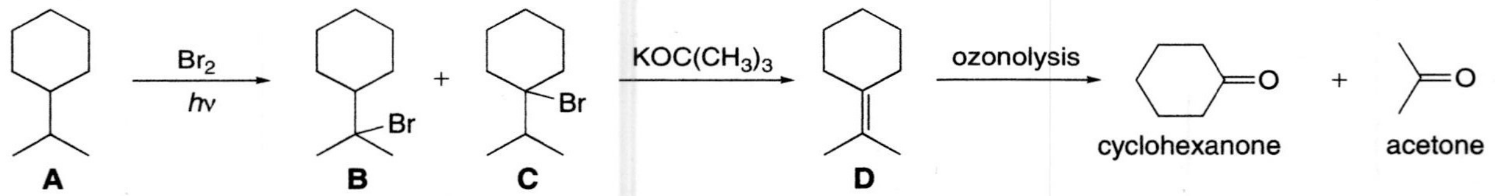

$\xrightarrow[h\nu]{Br_2}$ + $\xrightarrow{KOC(CH_3)_3}$ $\xrightarrow{ozonolysis}$ cyclohexanone + acetone

A   B   C   D

15.51

a. 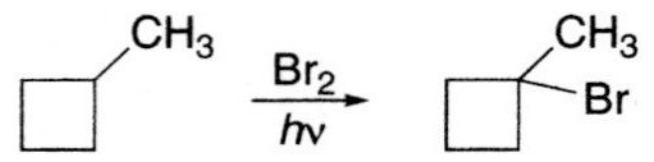 $\xrightarrow[h\nu]{Br_2}$

b. 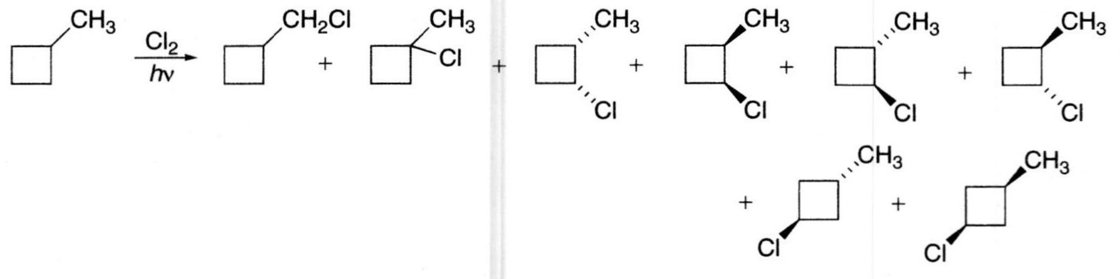 $\xrightarrow[h\nu]{Cl_2}$ + + + + + + +

c.

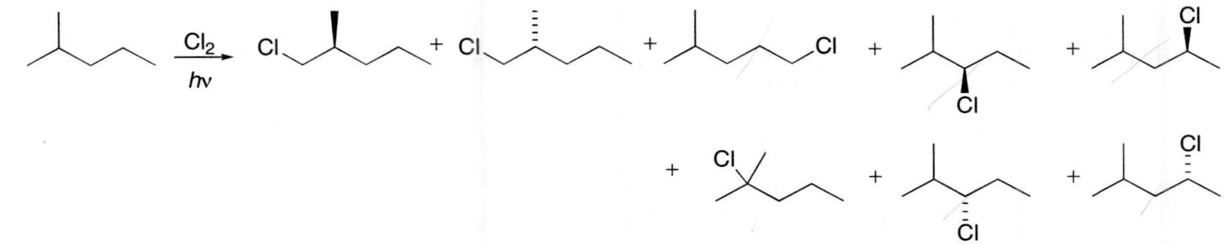

d.

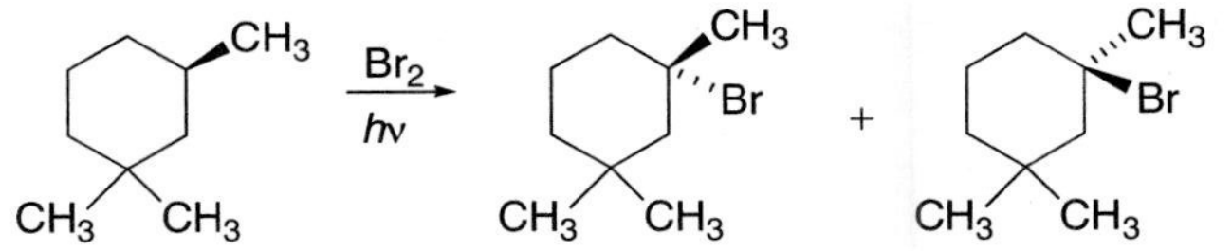

e.

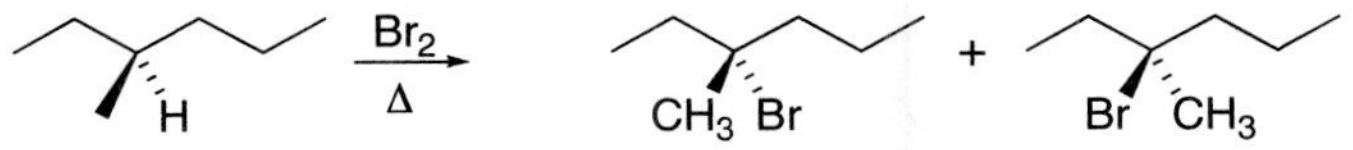

f.

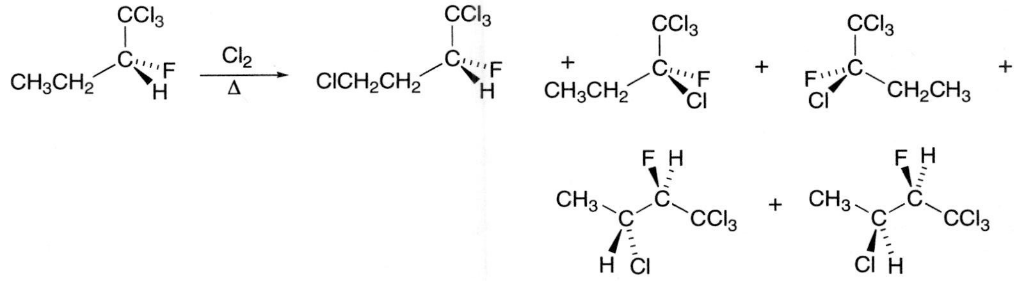

15.52

a.

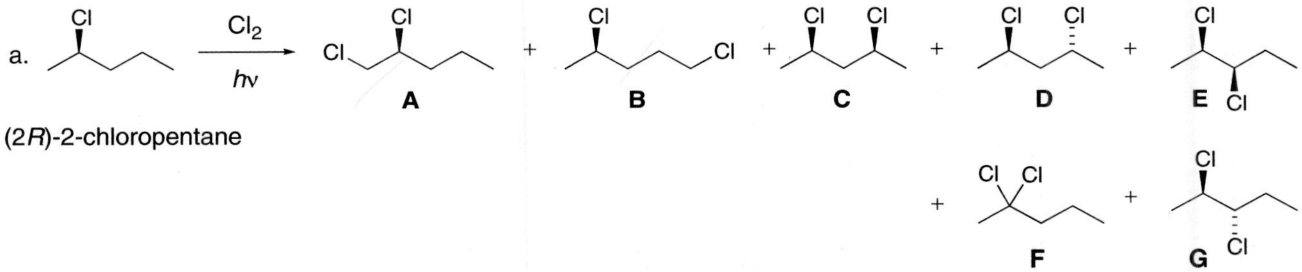

(2R)-2-chloropentane

A   B   C   D   E

F   G

b. There would be seven fractions, since each molecule drawn has different physical properties.
c. Fractions **A**, **B**, **D**, **E**, and **G** would show optical activity.

15.53

a.

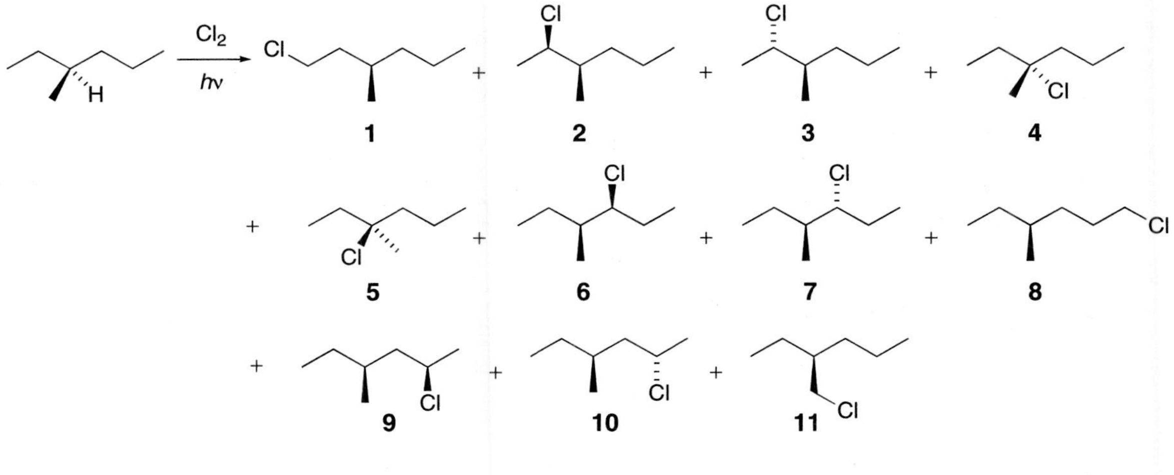

1   2   3   4

5   6   7   8

9   10   11

a. There would be 10 fractions, since **4** and **5** (two enantiomers) would be in one fraction.
b. All fractions except the one that contains **4** and **5** would be optically active.

15.54

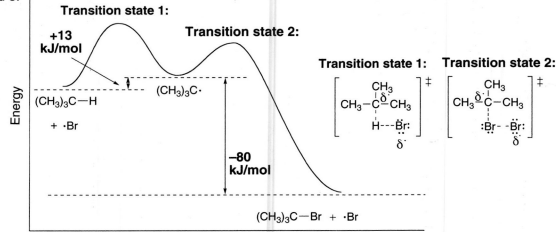

15.55

a.

$$CH_3-\underset{\underset{H}{|}}{\overset{\overset{CH_3}{|}}{C}}-CH_3 \ + \ Br_2 \ \xrightarrow{\ h\nu\ } \ CH_3-\underset{\underset{Br}{|}}{\overset{\overset{CH_3}{|}}{C}}-CH_3 \ + \ HBr$$

C–H bond broken	Br–Br bond broken	C–Br bond formed	H–Br bond formed
+381 kJ/mol	+192 kJ/mol	–272 kJ/mol	–368 kJ/mol

total bonds broken = +573 kJ/mol   total bonds formed = –640 kJ/mol $\Delta H° = -67$ kJ/mol

b. Initiation: $:\!\ddot{Br}\!-\!\ddot{Br}\!: \ \xrightarrow[\text{or } \Delta]{h\nu} \ :\!\ddot{Br}\!\cdot \ + \ \cdot\ddot{Br}\!:$

Propagation: $(CH_3)_3C\!-\!H \ + \ \cdot\ddot{Br}\!: \ \longrightarrow \ (CH_3)_3C\cdot \ + \ H\!-\!\ddot{Br}\!:$

c. $\Delta H° =$ (bonds broken) – (bonds formed)
= (+381 kJ/mol) + (–368 kJ/mol)
= +13 kJ/mol

$(CH_3)_3C\cdot \ + \ :\!\ddot{Br}\!-\!\ddot{Br}\!: \ \longrightarrow \ (CH_3)_3C\!-\!Br \ + \ \cdot\ddot{Br}\!:$

$\Delta H° =$ (bonds broken) – (bonds formed)
= (+192 kJ/mol) + (–272 kJ/mol)
= –80 kJ/mol

Termination: $:\!\ddot{Br}\!\cdot \ + \ \cdot\ddot{Br}\!: \ \longrightarrow \ :\!\ddot{Br}\!-\!\ddot{Br}\!:$
(one possibility)

d. and e.

Transition state 1:

+13 kJ/mol

Transition state 2:

Energy

$(CH_3)_3C\!-\!H$
+ $\cdot Br$

$(CH_3)_3C\cdot$

–80 kJ/mol

$(CH_3)_3C\!-\!Br \ + \ \cdot Br$

Reaction coordinate

Transition state 1:

$$\left[\ CH_3-\underset{\underset{\overset{|}{H}\text{---}\ddot{Br}:}{\overset{|}{\overset{\delta\cdot}{C}}}}{\overset{\overset{CH_3}{|}}{}}-CH_3 \ \right]^{\ddagger}$$

Transition state 2:

$$\left[\ CH_3\overset{\delta\cdot}{-}\underset{\underset{:\ddot{Br}\text{- -}\ddot{Br}:}{\overset{|}{\overset{\delta\cdot}{C}}}}{\overset{\overset{CH_3}{|}}{}}-CH_3 \ \right]^{\ddagger}$$

15.56

Initiation:

NBS

Propagation:

(from NBS)

(from NBS)

+ H—Br:

Termination: :Br· + ·Br: ⟶ :Br—Br:
(one possibility)

15.57 Calculate the $\Delta H°$ for the propagation steps of the reaction of CH_4 with I_2 to show why it does not occur at an appreciable rate.

$$CH_3—H + ·I: \longrightarrow ·CH_3 + H—I: \quad \Delta H° = +138 \text{ kJ/mol}$$

+435 kJ/mol   −297 kJ/mol

> This step is highly endothermic, making it difficult for chain propagation to occur over and over again.

$$·CH_3 + :I—I: \longrightarrow CH_3—I + ·I: \quad \Delta H° = −83 \text{ kJ/mol}$$

+151 kJ/mol   −234 kJ/mol

15.58 Calculate $\Delta H°$ for each of these steps, and use these values to explain why this alternate mechanism is unlikely.

[1] $CH_4 + :Cl· \longrightarrow CH_3Cl + H·$
        C–H bond broken   C–Cl bond formed $\Delta H° = +84$ kJ/mol
        +435 kJ/mol   −351 kJ/mol

> The $\Delta H°$ for Step [1] is very endothermic, making this mechanism unlikely.

[2] $H· + Cl_2 \longrightarrow HCl + :Cl·$
        Cl–Cl bond broken   H–Cl bond formed $\Delta H° = −189$ kJ/mol
        +242 kJ/mol   −431 kJ/mol

15.59

3,3-dimethyl-1-butene   2° carbocation   3° carbocation   2-bromo-2,3-dimethylbutane

+ Br⁻

1,2-CH_3 shift

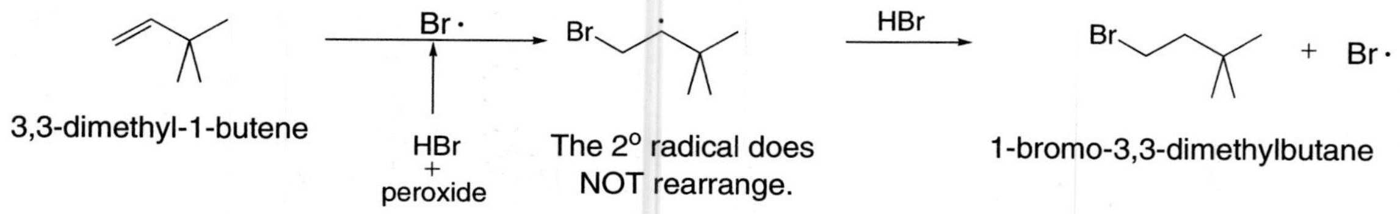
Br·
Br
HBr
Br
+ Br·
3,3-dimethyl-1-butene
HBr
+
peroxide
The 2° radical does
NOT rearrange.
1-bromo-3,3-dimethylbutane

Addition of HBr without added peroxide occurs by an ionic mechanism and forms a 2° carbocation, which rearranges to a more stable 3° carbocation. The addition of H^+ occurs first, followed by Br^-. Addition of HBr with added peroxide occurs by a radical mechanism and forms a 2° radical that does not rearrange. In the radical mechanism Br· adds first, followed by H·.

15.60

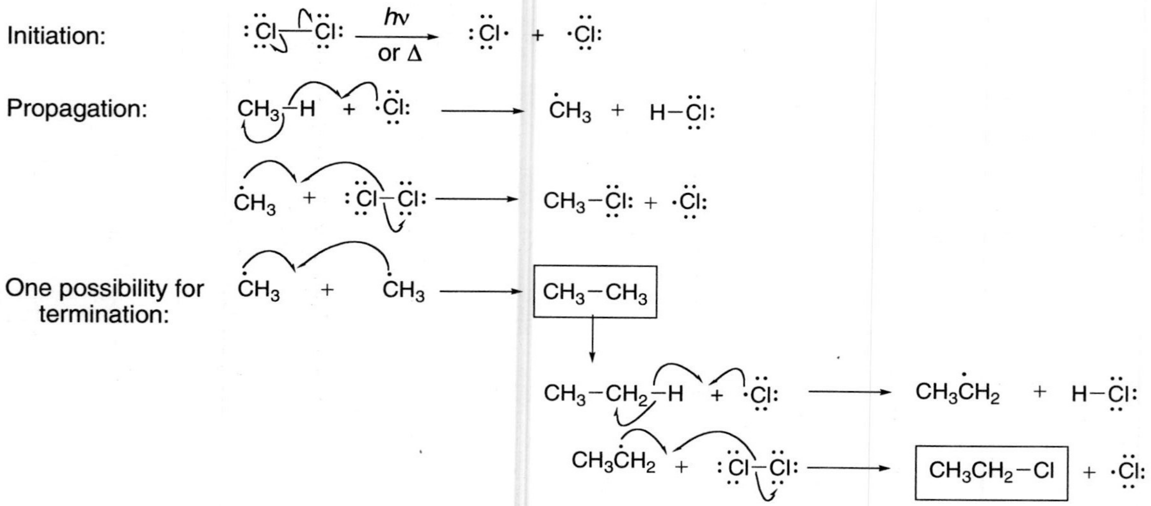
Initiation:
hν
or Δ
Propagation:
One possibility for
termination:
CH3–CH3
CH3CH2–Cl

15.61

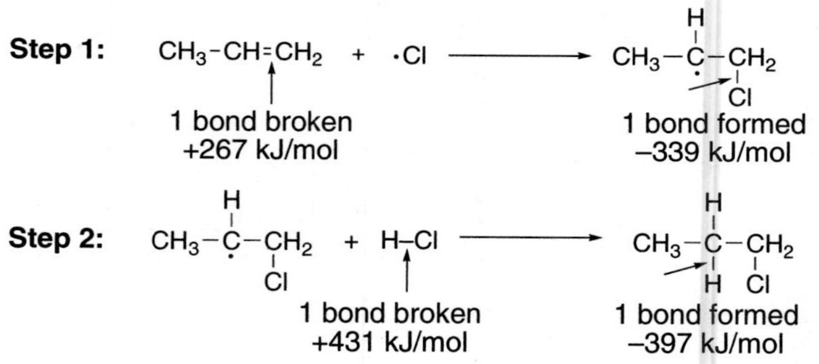
Step 1:
CH3-CH=CH2 + ·Cl
1 bond broken
+267 kJ/mol
1 bond formed
–339 kJ/mol
Step 2:
+ H–Cl
1 bond broken
+431 kJ/mol
1 bond formed
–397 kJ/mol

$\Delta H_1° = -72$ kJ/mol

$\Delta H_2° = +34$ kJ/mol
This step of propagation is endothermic. It prohibits chain propagation from occurring over and over.

15.62

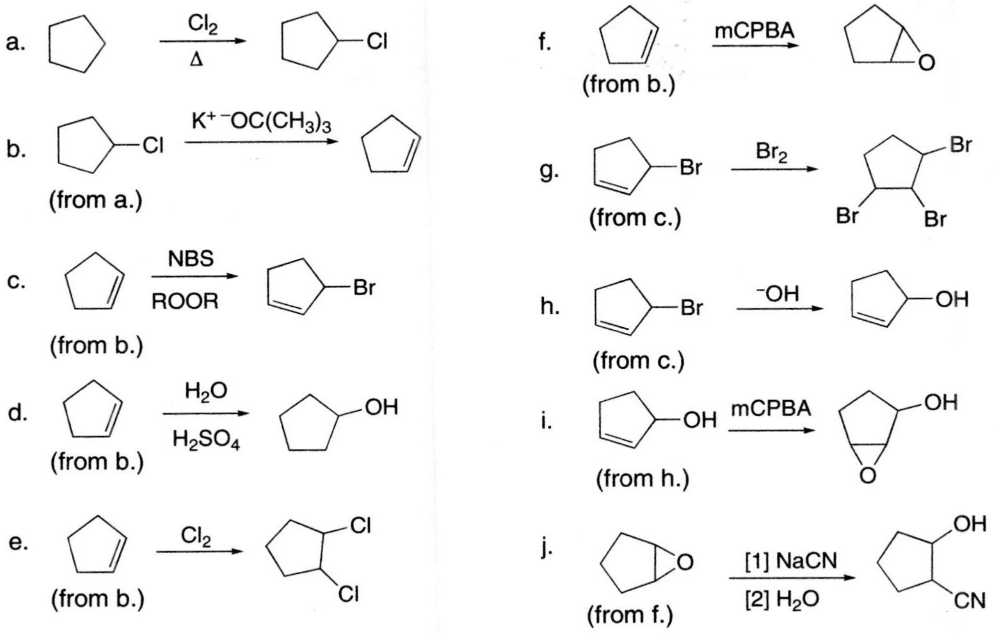

a.

Cl₂, Δ

b. (from a.)

K⁺ ⁻OC(CH₃)₃

c. (from b.)

NBS / ROOR

d. (from b.)

H₂O / H₂SO₄

e. (from b.)

Cl₂

f. (from b.)

mCPBA

g. (from c.)

Br₂

h. (from c.)

⁻OH

i. (from h.)

mCPBA

j. (from f.)

[1] NaCN
[2] H₂O

15.63

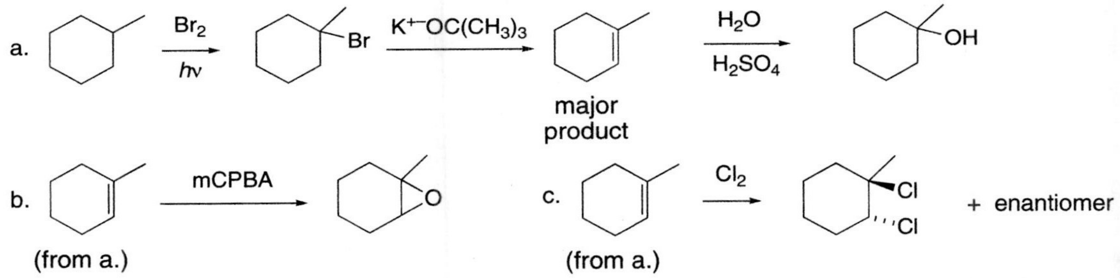

a.

Br₂ / hv

K⁺ ⁻OC(CH₃)₃

major product

H₂O / H₂SO₄

b. (from a.)

mCPBA

c. (from a.)

Cl₂

+ enantiomer

15.64

a. $CH_3CH_2CH_3$

Br₂ / hv

CH_3CHCH_3 (Br)

K⁺ ⁻OC(CH₃)₃

$CH_3CH=CH_2$

Br₂

$CH_3-C-C-H$ (Br Br / H H)

K⁺ ⁻OC(CH₃)₃ (2 equiv) DMSO

$CH_3C≡CH$

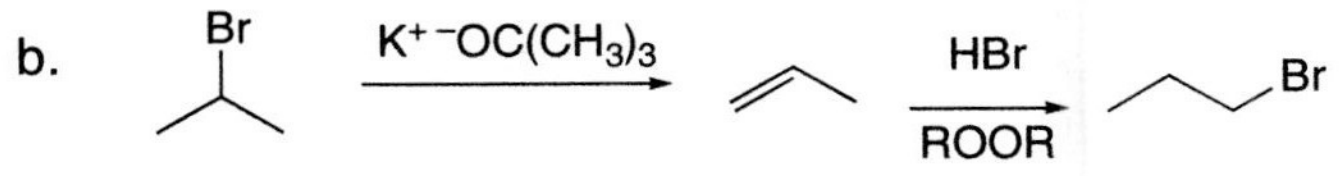

b.

K⁺ ⁻OC(CH₃)₃

HBr / ROOR

c. $HC≡CH$

NaH

$HC≡C^-$

CH_3CH_2Br

H₂ / Lindlar catalyst

HBr / ROOR

15.65

a. CH_3-CH_3

Br₂ / hv

CH_3-CH_2Br

K⁺ ⁻OC(CH₃)₃

$CH_2=CH_2$

Br₂

$BrCH_2-CH_2Br$

2 NaNH₂

$HC≡CH$

b. $HC≡CH$ (from a.)

NaH

$HC≡C^-$

CH_3-CH_2Br (from a.)

$HC≡CCH_2CH_3$

c. 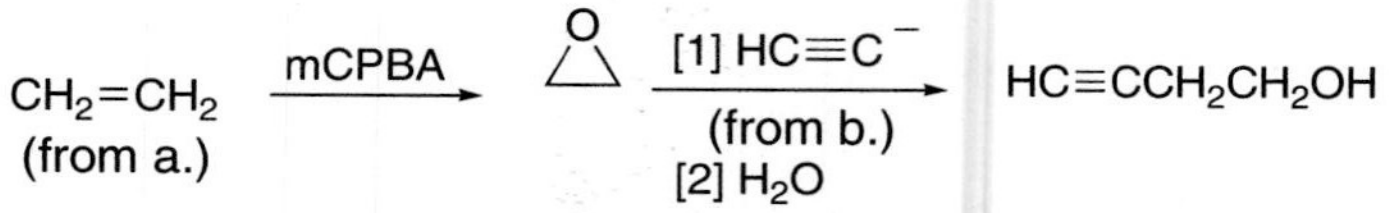 $CH_2=CH_2$ $\xrightarrow{mCPBA}$ [epoxide] $\xrightarrow[\text{[2] } H_2O]{\text{[1] } HC\equiv C^- \text{ (from b.)}}$ $HC\equiv CCH_2CH_2OH$
(from a.)

d. 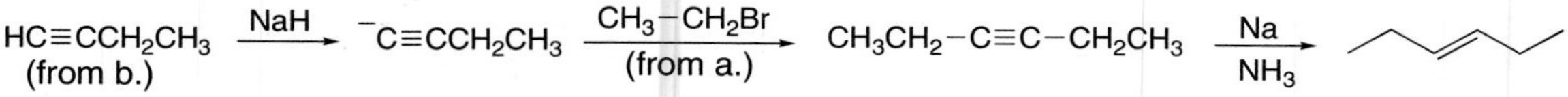 $HC\equiv CCH_2CH_3$ $\xrightarrow{NaH}$ $^-C\equiv CCH_2CH_3$ $\xrightarrow[\text{(from a.)}]{CH_3-CH_2Br}$ $CH_3CH_2-C\equiv C-CH_2CH_3$ $\xrightarrow[NH_3]{Na}$ [trans-alkene]
(from b.)

e. 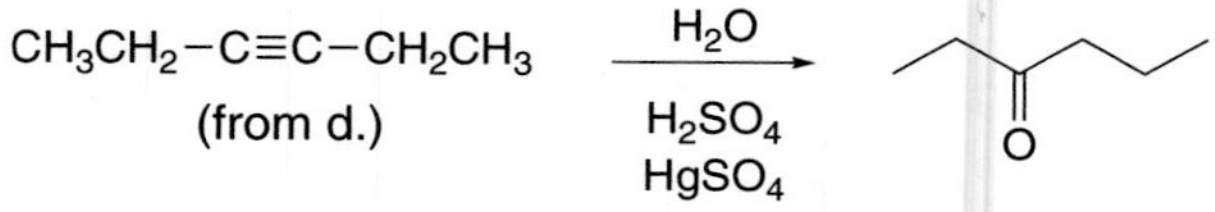 $CH_3CH_2-C\equiv C-CH_2CH_3$ $\xrightarrow[\substack{H_2SO_4 \\ HgSO_4}]{H_2O}$ [ketone]
(from d.)

15.66

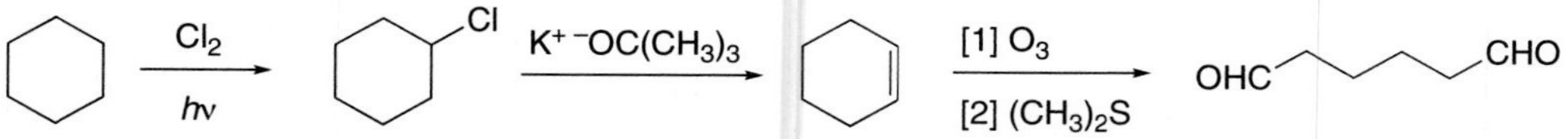

15.67

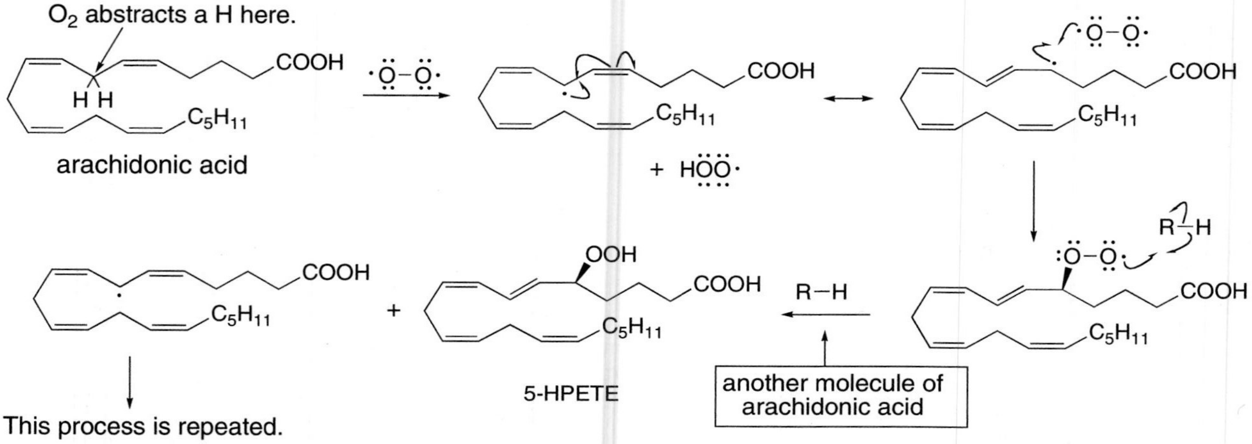

O_2 abstracts a H here.

arachidonic acid

$+ \; HO\ddot{O}\cdot$

5-HPETE

another molecule of arachidonic acid

This process is repeated.

15.68

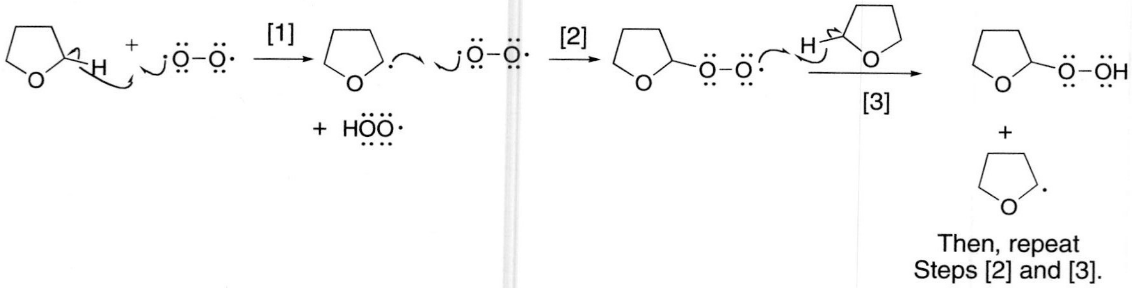

$+ \; H\ddot{O}\ddot{O}\cdot$

Then, repeat Steps [2] and [3].

15.69

a.

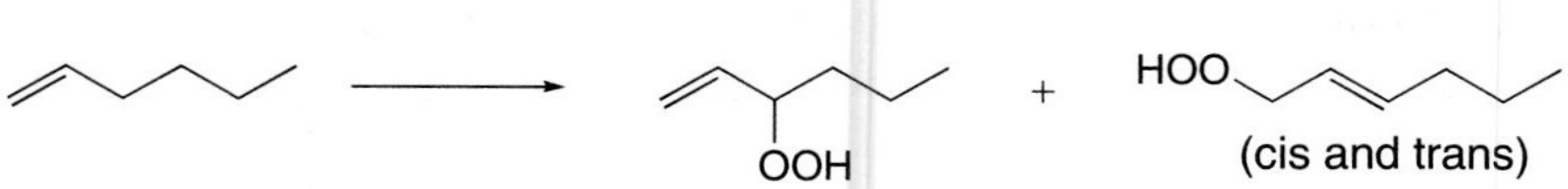

(cis and trans)

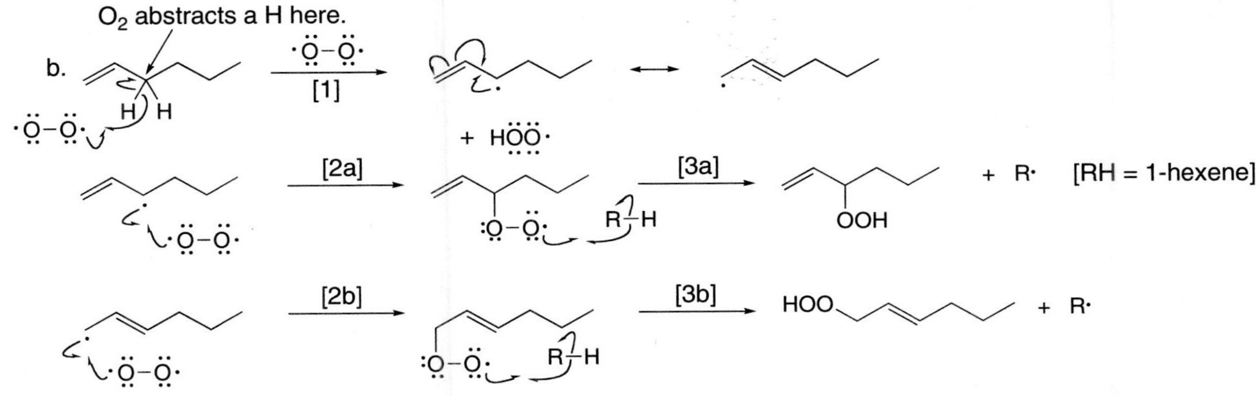
O2 abstracts a H here.
b.
[1]
+ HOO·
[2a]
[3a]
OOH
+ R·
[RH = 1-hexene]
[2b]
[3b]
HOO
+ R·

Repeat Steps [2] and [3] again and again.

15.70

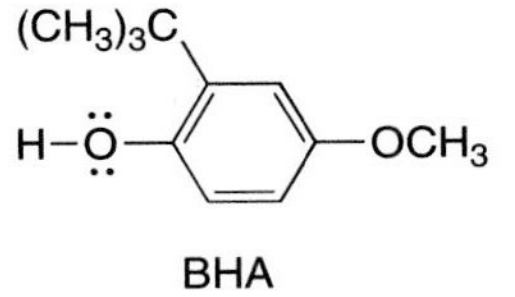
(CH3)3C
OCH3
BHA

Abstraction of the phenol H produces a resonance-stabilized radical.

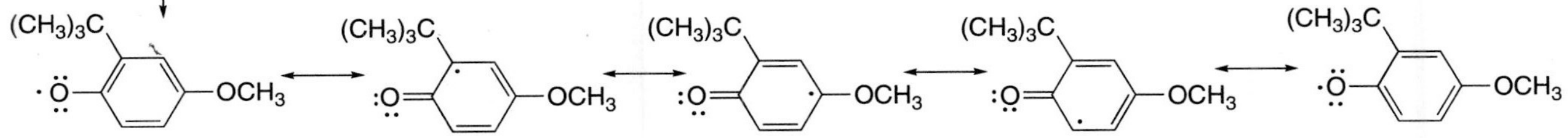
(CH3)3C
OCH3

15.71 Abstraction of the labeled H forms a highly resonance-stabilized radical. Four of the possible resonance structures are drawn.

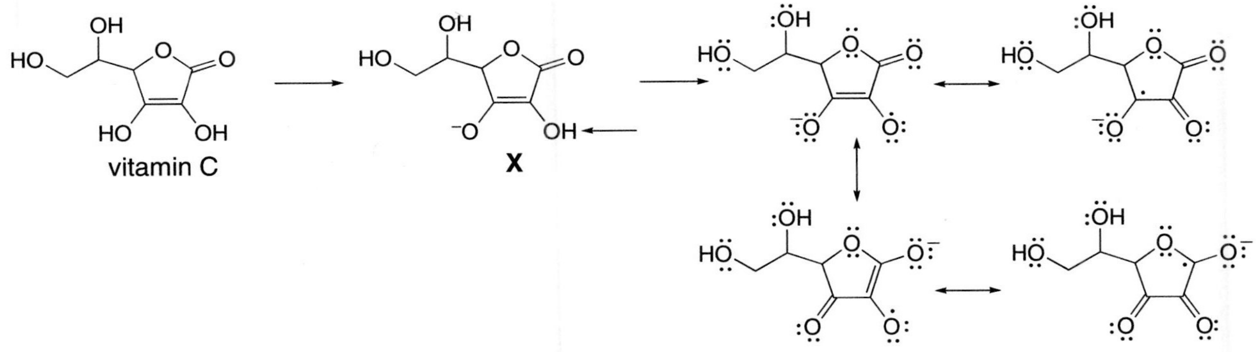
OH
HO
HO
OH
vitamin C
X

15.72 The monomers used in radical polymerization always contain double bonds.

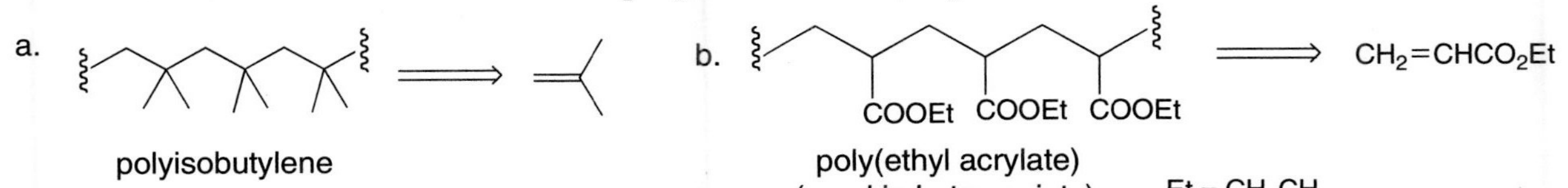
a.
polyisobutylene
b.
COOEt COOEt COOEt
CH2=CHCO2Et
poly(ethyl acrylate)
(used in Latex paints)
Et = CH2CH3

15.73

a.

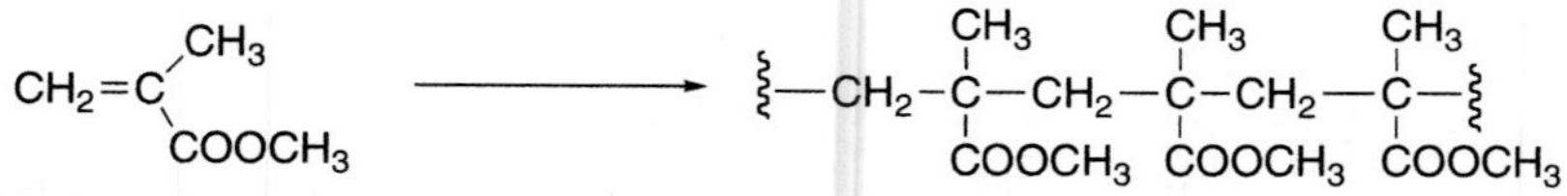

methyl methacrylate   PMMA

b.

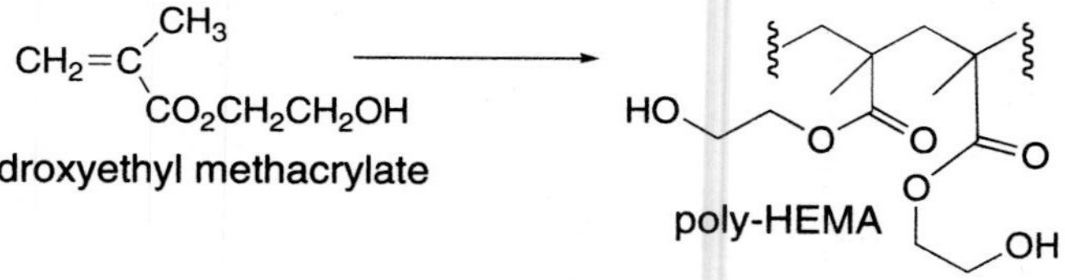

hydroxyethyl methacrylate

poly-HEMA

15.74

Overall reaction: $CH_2=CHCN$ $\xrightarrow{ROOR}$

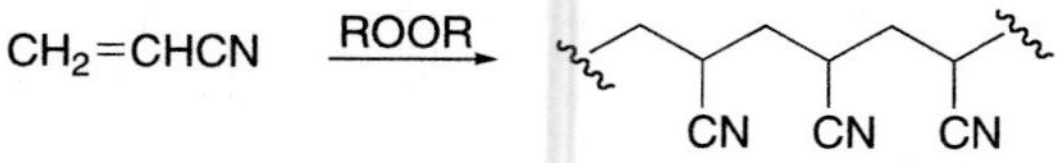

Initiation:

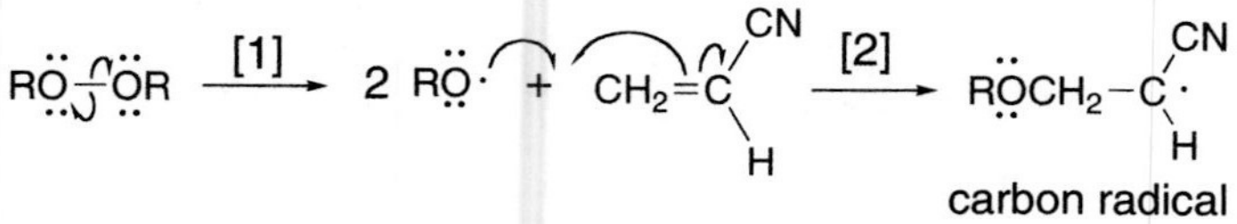

carbon radical

Propagation:

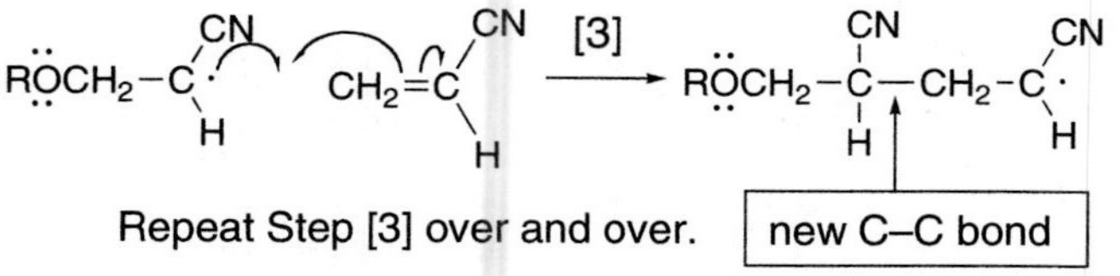

Repeat Step [3] over and over.   new C–C bond

Termination:

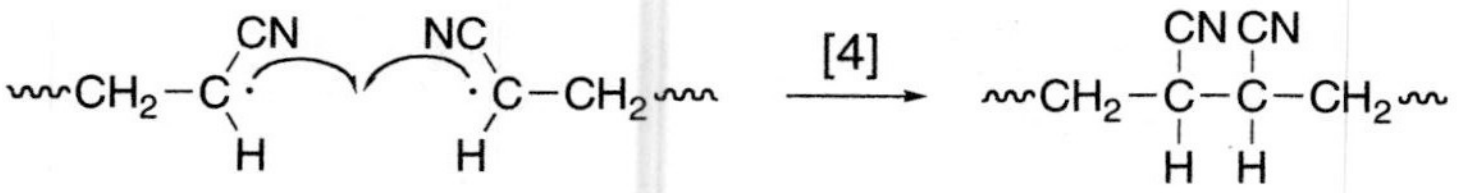

[one possibility]

15.75

a. CH_3O—⬡—$CH=CH_2$

A

b. The OCH_3 group stabilizes an intermediate carbocation by resonance. This makes **A** react faster than styrene in cationic polymerization.

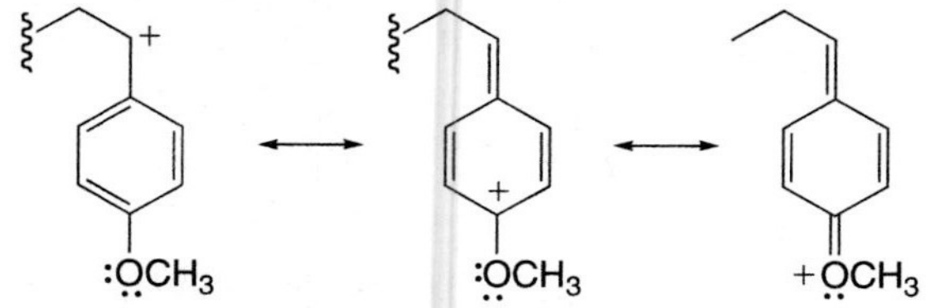

three of the possible resonance structures

15.76

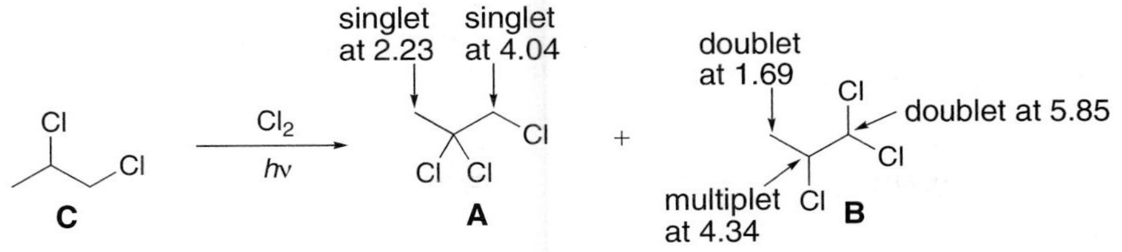

15.77

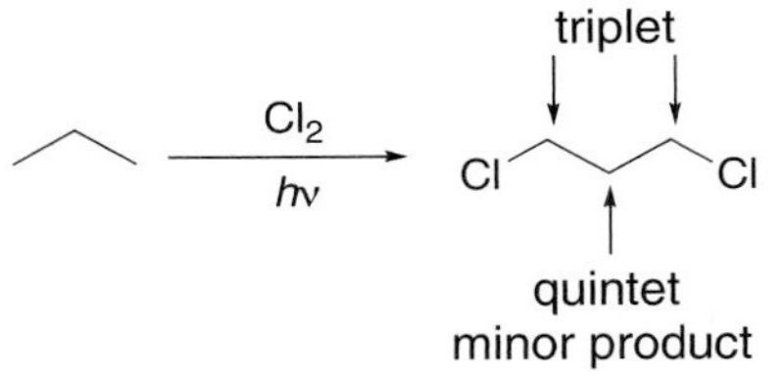

molecular formula $C_3H_6Cl_2$
Integration:
(57 units + 29 units)/6 H's = 14 units per H
one signal is 57 units/14 units per H = 4 H's
second signal is 29 units/14 units per H = 2 H's
^{1}H NMR data:
quintet at 2.2 (2 H's) split by 4 H's
triplet at 3.7 (4 H's) split by 2 H's

15.78

a. $CH_3CH_3 \xrightarrow[h\nu]{Cl_2} ClCH_2CH_2Cl$ (**X**; equivalent H's, singlet) + CH_3CHCl_2 (**Y**; doublet, quartet)

b.

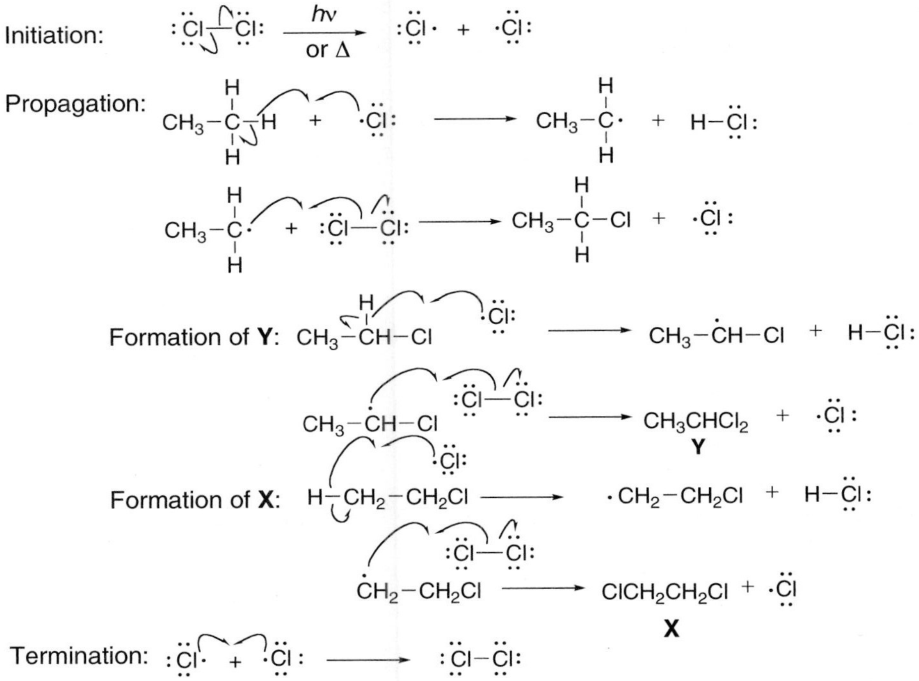

15.79

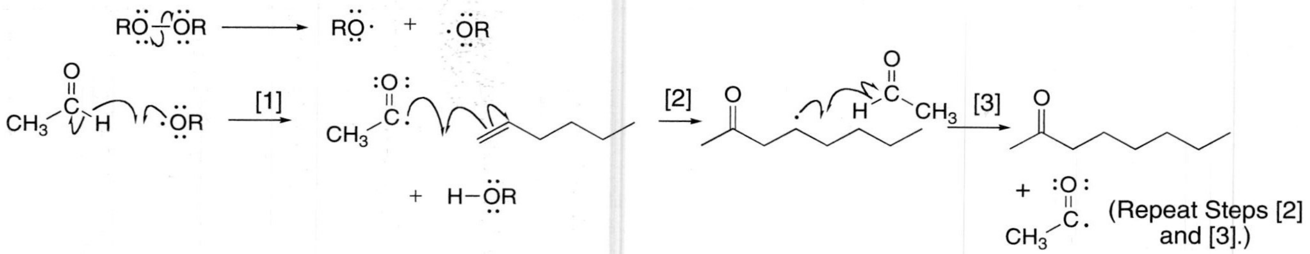

15.80

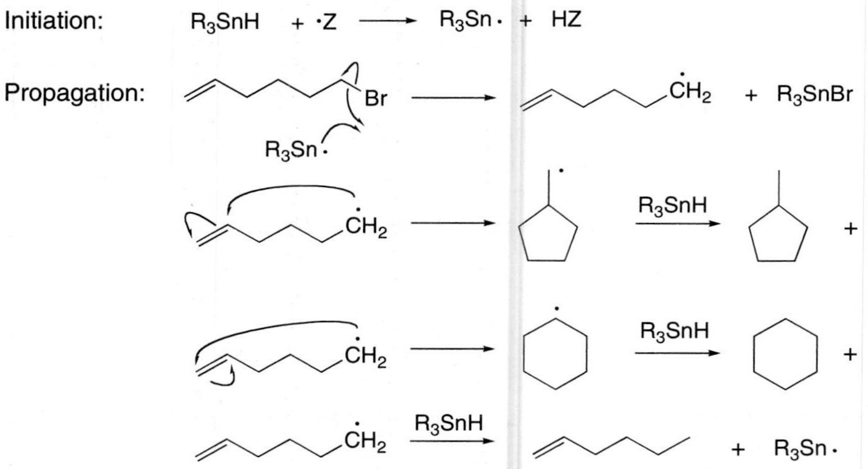

Initiation: $R_3SnH + \cdot Z \longrightarrow R_3Sn\cdot + HZ$

Propagation:

15.81

A

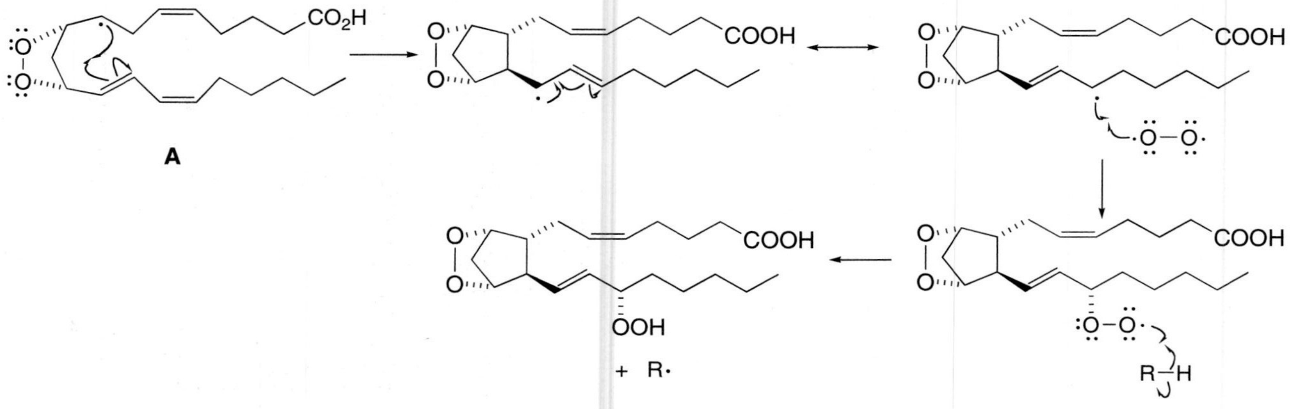

Chapter 16: Conjugation, Resonance, and Dienes

◆ Conjugation and delocalization of electron density

- The overlap of p orbitals on three or more adjacent atoms allows electron density to delocalize, thus adding stability (16.1).
- An allyl carbocation ($CH_2=CHCH_2^+$) is more stable than a $1°$ carbocation because of p orbital overlap (16.2).
- In any system X=Y–Z:, Z is sp^2 hybridized to allow the lone pair to occupy a p orbital, making the system conjugated (16.5).

◆ Four common examples of resonance (16.3)

[1] The three atom "allyl" system:

$$X=Y-\underset{*}{Z} \longleftrightarrow \underset{*}{X}-Y=Z \qquad * = +, -, •, \text{ or } ••$$

[2] Conjugated double bonds:

[3] Cations having a positive charge adjacent to a lone pair:

$$\ddot{X}-\overset{+}{Y} \longleftrightarrow \overset{+}{X}=Y$$

[4] Double bonds having one atom more electronegative than the other:

$$X=Y \longleftrightarrow \overset{+}{X}-\overset{-}{Y}: \quad \left[\text{Electronegativity of Y > X}\right]$$

◆ Rules on evaluating the relative "stability" of resonance structures (16.4)

[1] Structures with more bonds and fewer charges are more stable.

more stable resonance structure →

all neutral atoms
one more bond

charge separation

[2] Structures in which every atom has an octet are more stable.

$$CH_3-\ddot{O}-\overset{+}{C}H_2 \longleftrightarrow CH_3-\overset{+}{O}=CH_2$$

more stable resonance structure ←

All 2nd row elements have an octet.

[3] Structures that place a negative charge on a more electronegative element are more stable.

The (–) charge is on the more electronegative O atom.

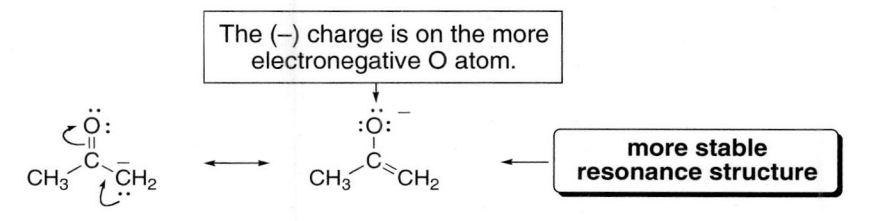

more stable resonance structure ←

◆ The unusual properties of conjugated dienes

[1] The C–C σ bond joining the two double bonds is unusually short (16.8).

[2] Conjugated dienes are more stable than similar isolated dienes. $\Delta H°$ of hydrogenation is smaller for a conjugated diene than for an isolated diene converted to the same product (16.9).

[3] The reactions are unusual:
- Electrophilic addition affords products of 1,2-addition and 1,4-addition (16.10, 16.11).
- Conjugated dienes undergo the Diels–Alder reaction, a reaction that does not occur with isolated dienes (16.12–16.14).

[4] Conjugated dienes absorb UV light in the 200–400 nm region. As the number of conjugated π bonds increases, the absorption shifts to longer wavelength (16.15).

◆ Reactions of conjugated dienes

[1] Electrophilic addition of HX (X = halogen) (16.10–16.11)

$$CH_2{=}CH{-}CH{=}CH_2 \xrightarrow[\text{(1 equiv)}]{HX}$$

$\underset{\text{	}}{CH_2}{-}\underset{\text{	}}{CH}{-}CH{=}CH_2$	+	$\underset{\text{	}}{CH_2}{-}CH{=}CH{-}\underset{\text{	}}{CH_2}$
H X		H X
1,2-product		**1,4-product**
kinetic product		thermodynamic product

- The mechanism has two steps.
- Markovnikov's rule is followed. Addition of H^+ forms the more stable allylic carbocation.
- The 1,2-product is the kinetic product. When H^+ adds to the double bond, X^- adds to the end of the allylic carbocation to which it is closer (C2 not C4). The kinetic product is formed faster at low temperature.
- The thermodynamic product has the more substituted, more stable double bond. The thermodynamic product predominates at equilibrium. With 1,3-butadiene, the thermodynamic product is the 1,4-product.

[2] Diels–Alder reaction (16.12–16.14)

1,3-diene   dienophile

> The three new bonds
> are labeled in **bold**.

- The reaction forms two σ and one π bond in a six-membered ring.
- The reaction is initiated by heat.
- The mechanism is concerted: all bonds are broken and formed in a single step.
- The diene must react in the *s*-cis conformation (16.13A).
- Electron-withdrawing groups in the dienophile increase the reaction rate (16.13B).
- The stereochemistry of the dienophile is retained in the product (16.13C).
- Endo products are preferred (16.13D).

16.1 **Isolated dienes** have two double bonds separated by two or more σ bonds.
Conjugated dienes have two double bonds separated by only one σ bond.

a.

One σ bond separates
two double bonds =
conjugated diene

b.

Two σ bonds separate
two double bonds =
isolated diene

c.

One σ bond separates
two double bonds =
conjugated diene

d.

Four σ bonds separate
two double bonds =
isolated diene

16.2

isolated isolated

OOH

O

isolated

C

OH

5-HPETE

conjugated

16.3 **Conjugation** occurs when there are overlapping *p* orbitals on three or more adjacent atoms.
Double bonds separated by 2 σ bonds are not conjugated.

a. 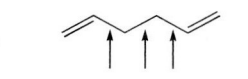 $CH_2=CH-CH=CH-CH=CH_2$

All of the carbon atoms are *sp²*
hybridized. Each π bond is
separated by only one σ bond.
conjugated

c.

O

The two π bonds are
separated by only one σ bond.
conjugated

e.

+

This carbon is not *sp²* hybridized.
NOT conjugated

b.

The two π bonds are
separated by three σ bonds.
NOT conjugated

d.

+

Three adjacent carbon atoms are *sp²* hybridized
and have an unhybridized *p* orbital.
conjugated

16.4 Two resonance structures differ only in the placement of electrons. All σ bonds stay in the same
place. Nonbonded electrons and π bonds can be moved. To draw the hybrid:
  • Use a dashed line between atoms that have a π bond in one resonance structure and not the
    other.
  • Use a δ symbol for atoms with a charge or radical in one structure but not the other.

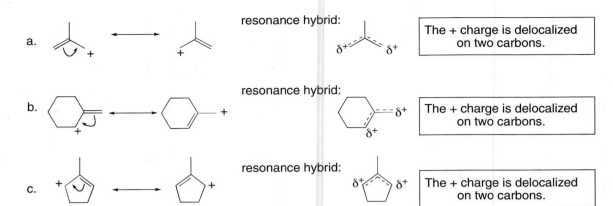

a.

resonance hybrid:

The + charge is delocalized on two carbons.

b.

resonance hybrid:

The + charge is delocalized on two carbons.

c.

resonance hybrid:

The + charge is delocalized on two carbons.

16.5 Each different kind of carbon atom will give a different ^{13}C signal. When a carbocation is delocalized as in structure **B**, carbons become equivalent.

A
5 different kinds of C
5 ^{13}C NMR signals

B
4 different kinds of C
4 ^{13}C NMR signals

The two δ^+ carbons are identical.

16.6 S_N1 reactions proceed via a carbocation intermediate. Draw the carbocation formed on loss of Cl and compare. The more stable the carbocation, the faster the S_N1 reaction.

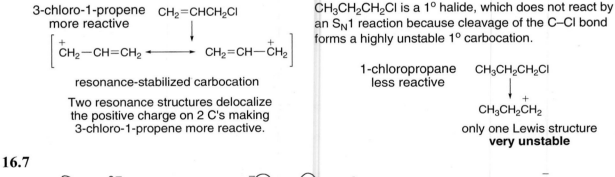

3-chloro-1-propene $CH_2=CHCH_2Cl$
more reactive

$[\overset{+}{C}H_2-CH=CH_2 \longleftrightarrow CH_2=CH-\overset{+}{C}H_2]$

resonance-stabilized carbocation

Two resonance structures delocalize the positive charge on 2 C's making 3-chloro-1-propene more reactive.

$CH_3CH_2CH_2Cl$ is a 1° halide, which does not react by an S_N1 reaction because cleavage of the C–Cl bond forms a highly unstable 1° carbocation.

1-chloropropane $CH_3CH_2CH_2Cl$
less reactive

$CH_3CH_2\overset{+}{C}H_2$

only one Lewis structure
very unstable

16.7

a. $CH_2=CH-\overset{..}{\underset{..}{C}}H-CH=CH_2 \longleftrightarrow \overset{..}{\underset{..}{C}}H_2-CH=CH-CH=CH_2 \longleftrightarrow CH_2=CH-CH=CH-\overset{..}{\underset{..}{C}}H_2$

Move the charge and the double bond.

b. CH_3CH_2—C(=O:)—C=CH_3, H $\longleftrightarrow$ CH_3CH_2—C(—:O:⁻)=C—CH_3, H

Move the charge and the double bond.

c. $CH_3-\overset{+}{C}H-\overset{..}{\underset{..}{C}}l: \longleftrightarrow CH_3-CH=\overset{+}{\underset{..}{C}}l:$

Move the lone pair.

d.

Move the charge and the double bond.

16.8 To compare the resonance structures remember:

- Resonance structures with **more bonds** are better.
- Resonance structures in which **every atom has an octet** are better.
- Resonance structures with **neutral atoms** are better than those with charge separation.
- Resonance structures that place a **negative charge on a more electronegative atom** are better.

a.

least stable

one more bond
All atoms have an octet.
better resonance structure
intermediate stability

most stable

c.

least stable

one more bond
better resonance structure
intermediate stability

most stable

b.

least stable

negative charge on the
more electronegative atom
better resonance structure
intermediate stability

hybrid
most stable

d.

least stable

one more bond
better resonance structure
intermediate stability

most stable

16.9

largest contribution
more bonds and all
atoms have octets

16.10

a.

A
no charges
All atoms have an octet.
4 bonds (C–C + C–O)
most stable
lowest energy
resonance structure

B
2 charges
C does not have an octet.
3 bonds (C–C + C–O)
least stable
highest energy

C
2 charges
All atoms have an octet.
4 bonds (C–C + C–O)
intermediate stability
intermediate energy

b.

D
resonance hybrid
lower in energy
than any single
resonance form

c. In order of increasing energy: **D < A < C < B**

16.11 Remember that in any allyl system, there must be *p* orbitals to delocalize the lone pair.

a.

sp² hybridized
trigonal planar geometry

b.

sp² hybridized
trigonal planar geometry

c.

sp² hybridized
trigonal planar geometry

16.12 The **s-cis** conformation has two double bonds on the **same side** of the single bond.
The **s-trans** conformation has two double bonds on **opposite sides** of the single bond.

a. (2*E*,4*E*)-2,4-**octadiene** in the *s*-trans conformation

double bonds on opposite sides
s-trans

b. (3*E*,5*Z*)- 3,5-**nonadiene** in the *s*-cis conformation

Z
E

double bonds on the same side
s-cis

c. (3*Z*,5*Z*)-4,5-dimethyl-3,5-**decadiene** in
both the *s*-cis and *s*-trans conformations

Z

Z

s-cis

Z

Z

s-trans

16.13

s-cis

conjugated

conjugated

s-trans

conjugated

E E

OH

isolated

isolated

Z

Z

O

OH

HO

Z

isolated

isolated

isolated

16.14 Bond length depends on hybridization and percent *s*-character. Bonds with a higher percent
s-character have smaller orbitals and are shorter.

$HC \equiv C - C \equiv CH$

$CH_2 = CH - CH = CH_2$

$CH_3 - CH_3$

sp hybridized carbons
50% *s*-character
shortest bond

sp² hybridized carbons
33% *s*-character
intermediate length

sp³ hybridized carbons
25% *s*-character
longest bond

16.15 Two equivalent resonance structures delocalize the π bond and the negative charge.

$CH_3 - C$

$CH_3 - C$

hybrid:

$CH_3 - C$

These bond lengths are equal
because they are identical.

16.16 The **less stable** (higher energy) **diene** has the **larger heat of hydrogenation.** Isolated dienes are higher in energy than conjugated dienes, so they will have a larger heat of hydrogenation.

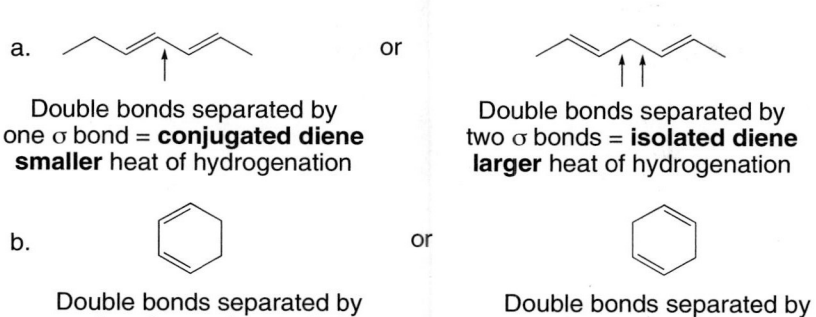

a.

    Double bonds separated by
    one σ bond = **conjugated diene**
    smaller heat of hydrogenation

or

    Double bonds separated by
    two σ bonds = **isolated diene**
    larger heat of hydrogenation

b.

    Double bonds separated by
    one σ bond = **conjugated diene**
    smaller heat of hydrogenation

or

    Double bonds separated by
    two σ bonds = **isolated diene**
    larger heat of hydrogenation

16.17 Isolated dienes are higher in energy than conjugated dienes. Compare the location of the double bonds in the compounds below.

0 conjugated double bonds
least stable

2 conjugated double bonds
intermediate stability

3 conjugated double bonds
most stable

16.18 Conjugated dienes react with HX to form 1,2- and 1,4-products.

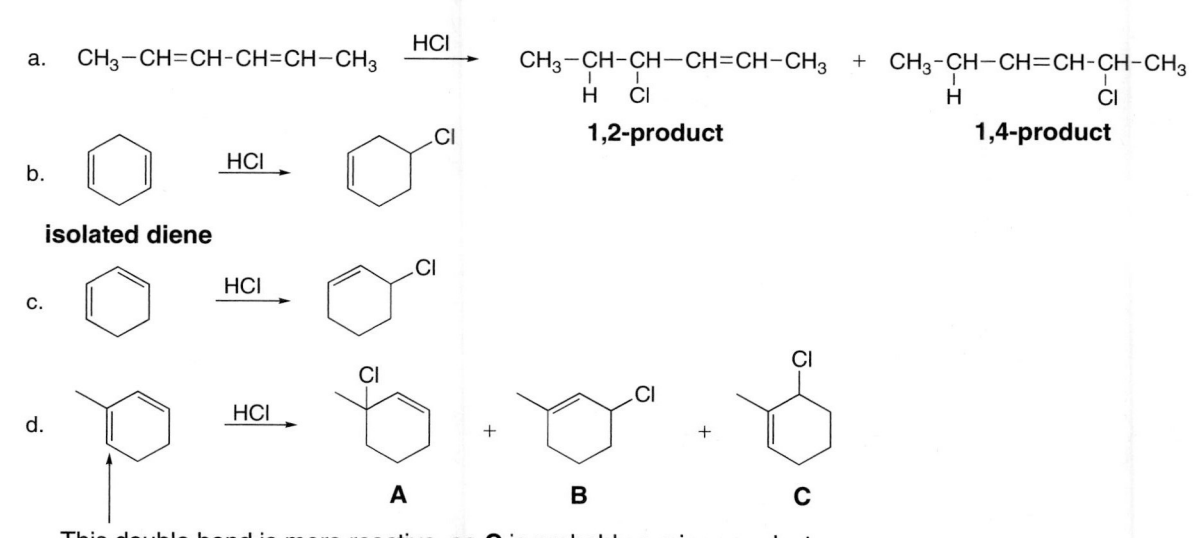

a. $CH_3-CH=CH-CH=CH-CH_3$ →(HCl) $CH_3-CH-CH-CH=CH-CH_3$ + $CH_3-CH-CH=CH-CH-CH_3$

          1,2-product   **1,4-product**

b.   **isolated diene** →(HCl)

c.   →(HCl)

d.   →(HCl) **A** + **B** + **C**

This double bond is more reactive, so **C** is probably a minor product because it results from HCl addition to the less reactive double bond.

16.19 The mechanism for addition of DCl has two steps:

[1] **Addition of D⁺** forms a resonance-stabilized carbocation.

[2] **Nucleophilic attack of Cl⁻** forms 1,2- and 1,4-products.

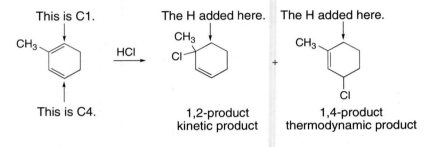

16.20 Label the products as 1,2- or 1,4-products. The 1,2-product is the kinetic product, and the 1,4-product, which has the more substituted double bond, is the thermodynamic product.

This is C1.

This is C4.

The H added here.   The H added here.

1,2-product
kinetic product

1,4-product
thermodynamic product

16.21 To draw the products of a Diels–Alder reaction:

[1] Find the 1,3-diene and the dienophile.

[2] Arrange them so the diene is on the left and the dienophile is on the right.

[3] Cleave three bonds and use arrows to show where the new bonds will be formed.

a.

diene   dienophile

re-draw   Δ

b.

diene   dienophile
Rotate to
make it s-cis.

re-draw   Δ

c.

dienophile   diene
Rotate to
make it s-cis.

re-draw   Δ

16.22 For a diene to be reactive in a Diels–Alder reaction, **a diene must be able to adopt an *s*-cis conformation.**

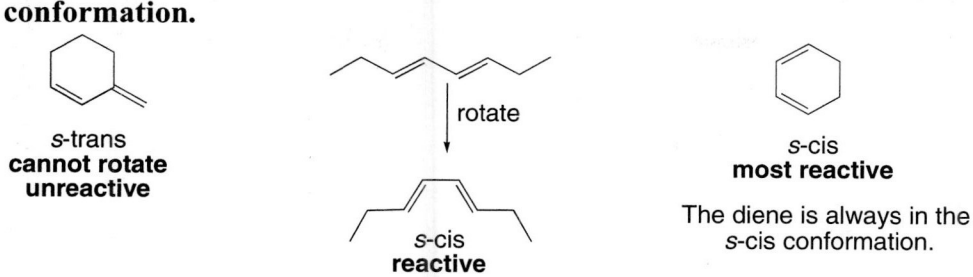

s-trans
cannot rotate
unreactive

rotate

s-cis
reactive

s-cis
most reactive

The diene is always in the
s-cis conformation.

16.23

(2*Z*,4*Z*)-2,4-hexadiene
sterically hindered

(2*E*,4*E*)-2,4-hexadiene
sterically unhindered
more reactive

Steric interactions between the two CH_3 groups make it difficult
for the diene to adopt the needed *s*-cis conformation.

16.24 Electron-withdrawing substituents in the dienophile increase the reaction rate.

$CH_2=CH_2$

$CH_2=C$ H, COOH

H, H, HOOC, COOH (C=C)

no electron-withdrawing groups
least reactive

one electron-withdrawing group
intermediate reactivity

two electron-withdrawing groups
most reactive

16.25 A cis dienophile forms a cis-substituted cyclohexene.
A trans dienophile forms a trans-substituted cyclohexene.

a.

CH_3OOC $COOCH_3$ (C=C) H H
cis dienophile

Δ

COOCH_3 / COOCH_3 + COOCH_3 / COOCH_3
cis-substituted products

b.

CH_3OOC H (C=C) H $COOCH_3$
trans dienophile

Δ

COOCH_3 / COOCH_3 + COOCH_3 / COOCH_3
trans-substituted products

c.

cis dienophile

Δ

H O ... H O + H O ... H O
identical
cis-substituted product

16.26 The **endo product** (with the substituents under the plane of the new six-membered ring) is the preferred product.

a.

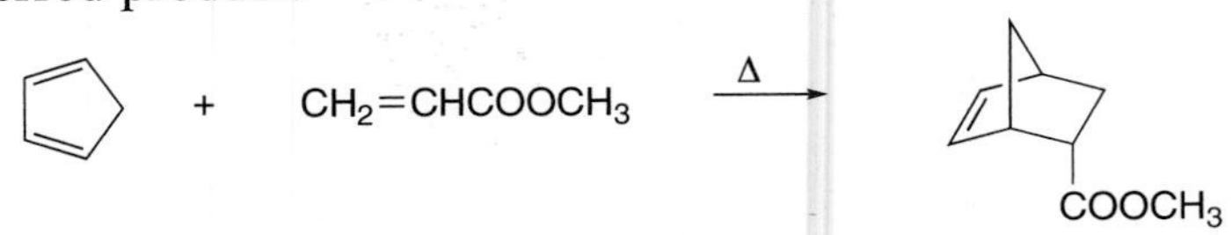

COOCH₃   **endo** substituent

b.

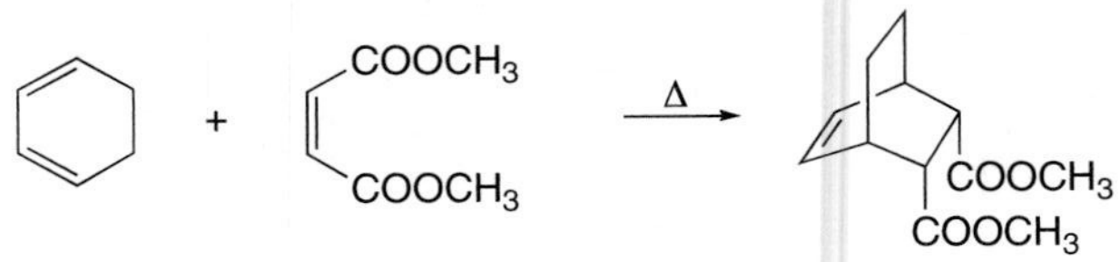

both groups **endo**

16.27 To find the diene and dienophile needed to make each of the products:
[1] Find the six-membered ring with a C–C double bond.
[2] Draw three arrows to work backwards.
[3] Follow the arrows to show the diene and dienophile.

a.

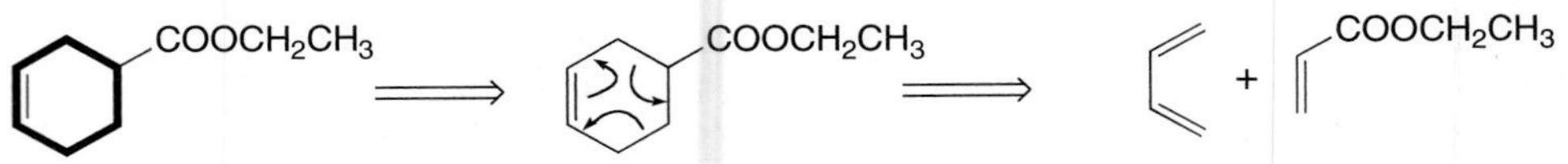

b.

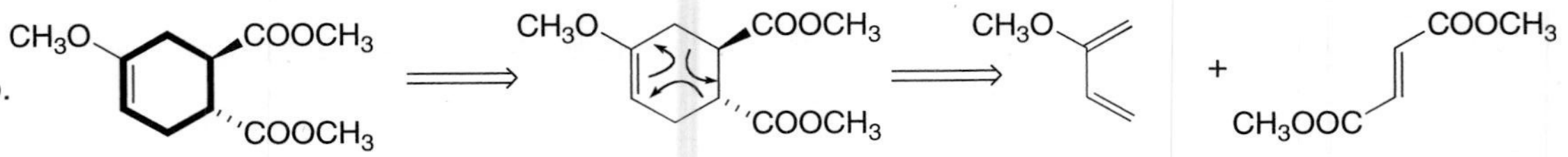

c.

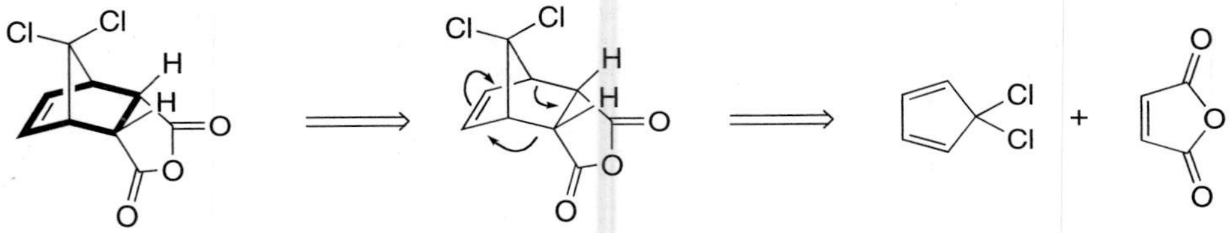

16.28

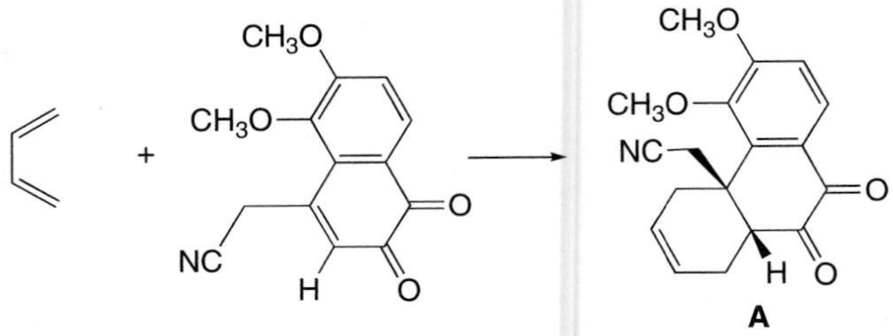

(+ enantiomer)

16.29 Conjugated molecules absorb light at a longer wavelength than molecules that are not conjugated.

a.

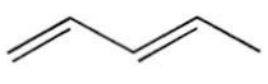

conjugated
longer wavelength

not conjugated

b.

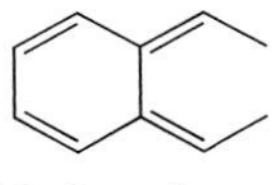

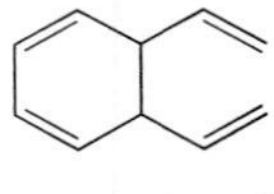

all double bonds conjugated
longer wavelength

one set of
conjugated dienes

16.30 Sunscreens contain conjugated systems to absorb UV radiation from sunlight. Look for conjugated systems in the compounds below.

a.

conjugated system
could be a sunscreen

b.

not a conjugated system

c.

conjugated system
could be a sunscreen

16.31 Use the definition from Answer 16.1.

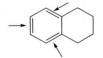

**3 π bonds with only
1 σ bond between
conjugated**

CH₂=CHC≡N

**2 multiple bonds with only
1 σ bond between
conjugated**

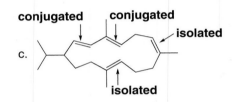

**1 π bond with
no adjacent *sp²* hybridized atoms
NOT conjugated**

**3 π bonds with 2 or
more σ bonds between
NOT conjugated**

CH₂

This C is *sp²*.

**1 π bond with
an adjacent *sp²* hybridized atom**

The lone pair occupies a *p* orbital,
so there are *p* orbitals on
three adjacent atoms.

conjugated

CH₂OCH₃

**1 π bond with
no adjacent *sp²* hybridized atoms
NOT conjugated**

16.32 Use the definition from Answer 16.1.

a.

conjugated

isolated

conjugated

conjugated

b.

conjugated

isolated   isolated   conjugated

c.

conjugated   conjugated

isolated

isolated

16.33 Although 2,3-di-*tert*-butyl-1,3-butadiene has four adjacent *p* orbitals, the bulky *tert*-butyl groups prevent the diene from adopting the *s*-cis conformation needed for the Diels–Alder reaction. Thus, this diene does not undergo a characteristic reaction of conjugated dienes.

16.34

a. (CH₃)₂C⁺—CH=CH₂ ⟷ (CH₃)₂C=CH—CH₂⁺

b.

c.   CH₂⁺ ⟷ =CH₂

d.

e. CH₃Ö—CH=CH—CH₂⁺ ⟷ CH₃Ö⁺—CH—CH=CH₂

CH₃Ö⁺=CH—CH=CH₂

f.

g.   N(CH₃)₂ ⟷ N⁺(CH₃)₂

h.

16.35

a.

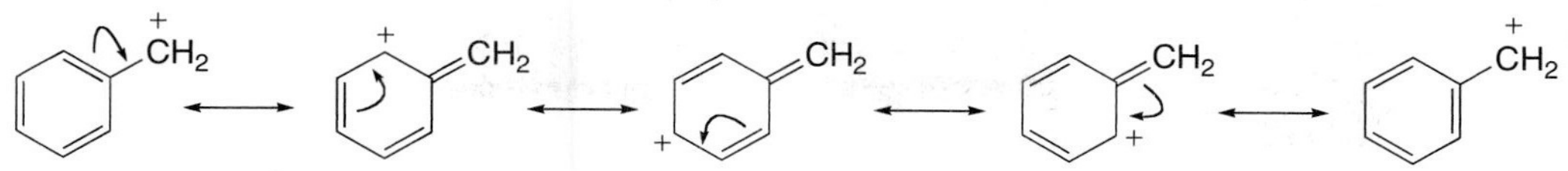

b.

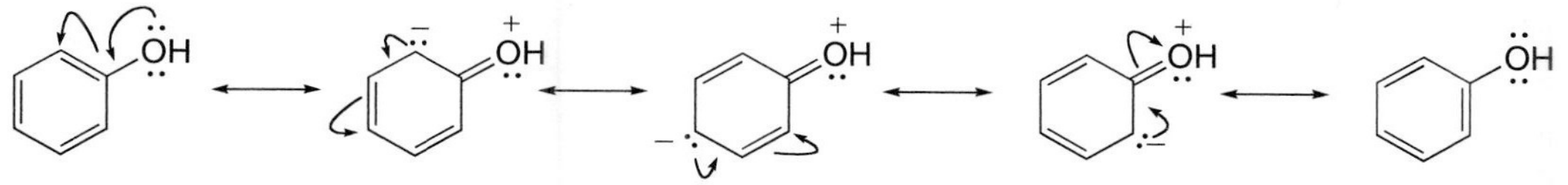

c.

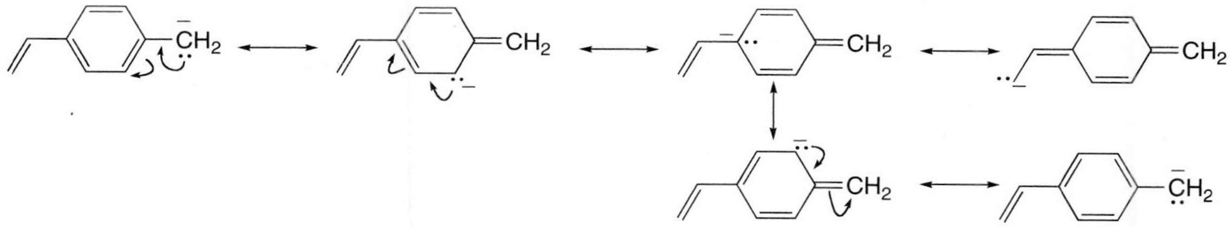

d.

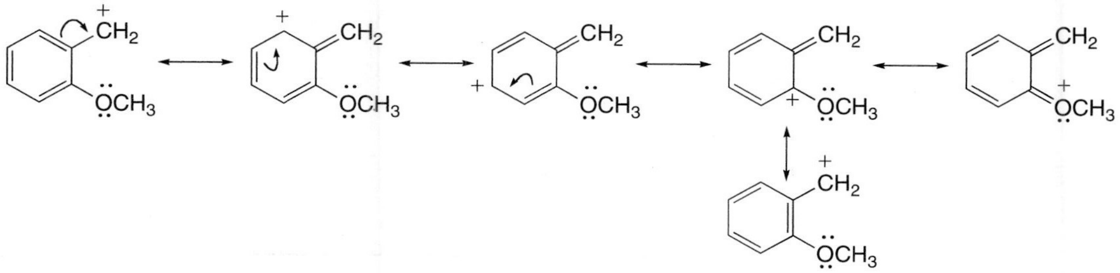

16.36

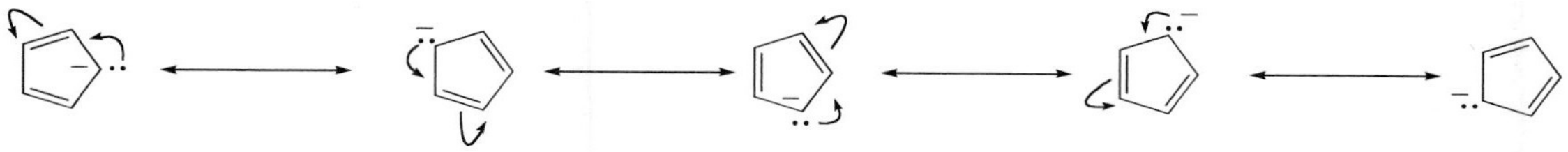

resonance hybrid:

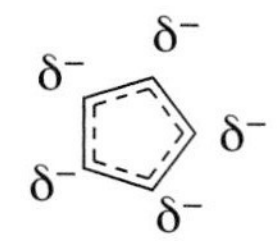

Five resonance structures delocalize the negative charge
on five C's making them all equivalent.

All of the carbons are identical in the anion.

16.37 The N atom of $C_6H_5CH_2NH_2$ is surrounded by four groups and is sp^3 hybridized. Although the N atom of $C_6H_5NH_2$ is surrounded by four groups, it is also bonded to a benzene ring. To be conjugated with the benzene ring, the N atom must be sp^2 hybridized and its lone pairs must occupy a p orbital. In this way the lone pair can be delocalized, as shown in one resonance structure.

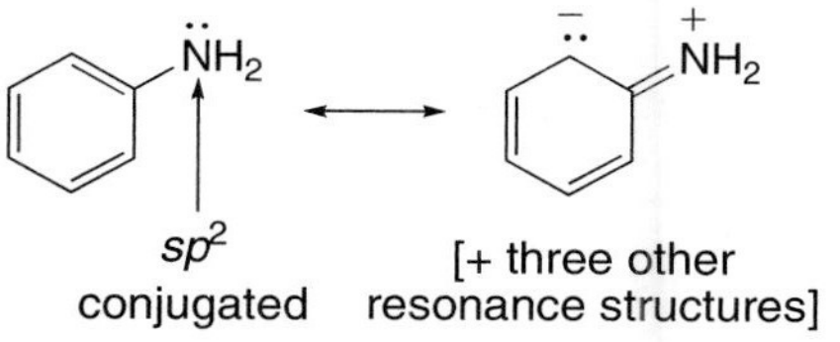

sp^2
conjugated

[+ three other resonance structures]

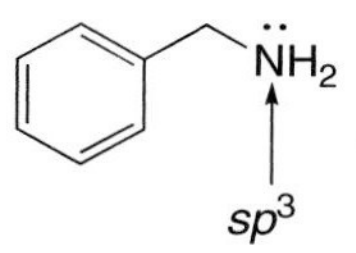

sp^3

The N atom is *not* conjugated with the benzene ring.

16.38

a.

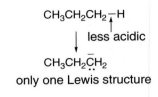

CH₂=CH—CH₂┬H
           ↑ more acidic
↓

C̄H₂—CH=CH₂ ⟷ CH₂=CH—C̄H₂

Resonance stabilization delocalizes the
negative charge on 2 C's after loss of a proton.
This makes propene more acidic than propane.

CH₃CH₂CH₂┬H
      ↑ less acidic
↓

CH₃CH₂C̄H₂
only one Lewis structure

b. Draw the products of cleavage of the bond.

ethane CH₃—CH₃

↓

·CH₃ + ·CH₃

Two unstable radicals form.

1-butene CH₃┬CH₂CH=CH₂
         ↑
         ↓

·CH₃ + ·CH₂—CH=CH₂ ⟷ CH₂=CH—ĊH₂

One resonance-stabilized radical forms.
This makes the bond dissociation
energy lower because a more stable radical is formed.

16.39 Use the directions from Answer 16.12.

a. (3*Z*)-1,3-pentadiene in the *s*-trans conformation

double bonds on opposite sides
***s*-trans**

b. (2*E*,4*Z*)-1-bromo-3-methyl-2,4-hexadiene
Br

c. (2*E*,4*E*,6*E*)-2,4,6-octatriene

d. (2*E*,4*E*)-3-methyl-2,4-hexadiene in the *s*-cis conformation

***s*-cis**

16.40

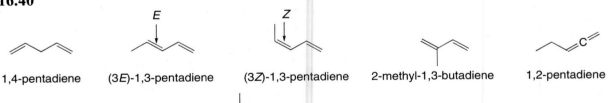

                E   *Z*

1,4-pentadiene   (3*E*)-1,3-pentadiene   (3*Z*)-1,3-pentadiene   2-methyl-1,3-butadiene   1,2-pentadiene

3-methyl-1,2-butadiene   2,3-pentadiene

16.41

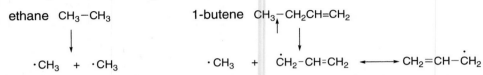

    2*E*,4*E*   2*E*,4*Z*   2*Z*,4*E*   2*Z*,4*Z*

16.42

a. 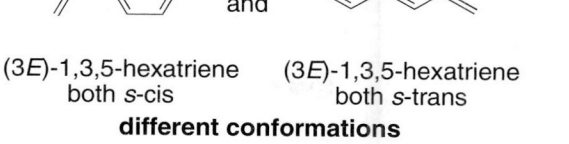 and

(3*E*)-1,3,5-hexatriene
both *s*-cis

(3*E*)-1,3,5-hexatriene
both *s*-trans

different conformations

c. 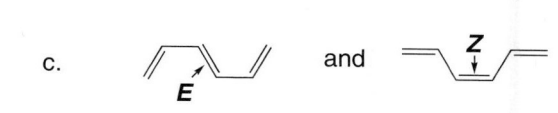 and

(3*E*)-1,3,5-hexatriene   (3*Z*)-1,3,5-hexatriene

different stereoisomers

b. 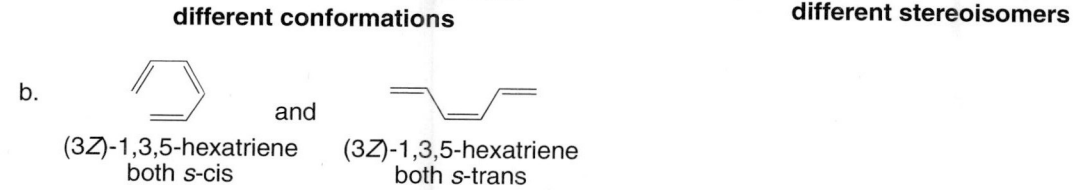 and

(3*Z*)-1,3,5-hexatriene
both *s*-cis

(3*Z*)-1,3,5-hexatriene
both *s*-trans

different conformations

16.43 Use the directions from Answer 16.16 and recall that more substituted double bonds are more stable.

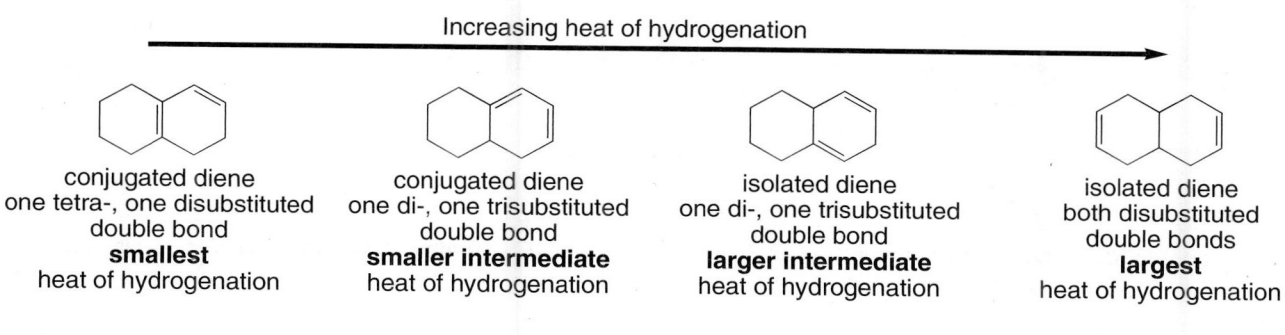

Increasing heat of hydrogenation

conjugated diene
one tetra-, one disubstituted
double bond
smallest
heat of hydrogenation

conjugated diene
one di-, one trisubstituted
double bond
smaller intermediate
heat of hydrogenation

isolated diene
one di-, one trisubstituted
double bond
larger intermediate
heat of hydrogenation

isolated diene
both disubstituted
double bonds
largest
heat of hydrogenation

16.44 Conjugated dienes react with HX to form 1,2- and 1,4-products.

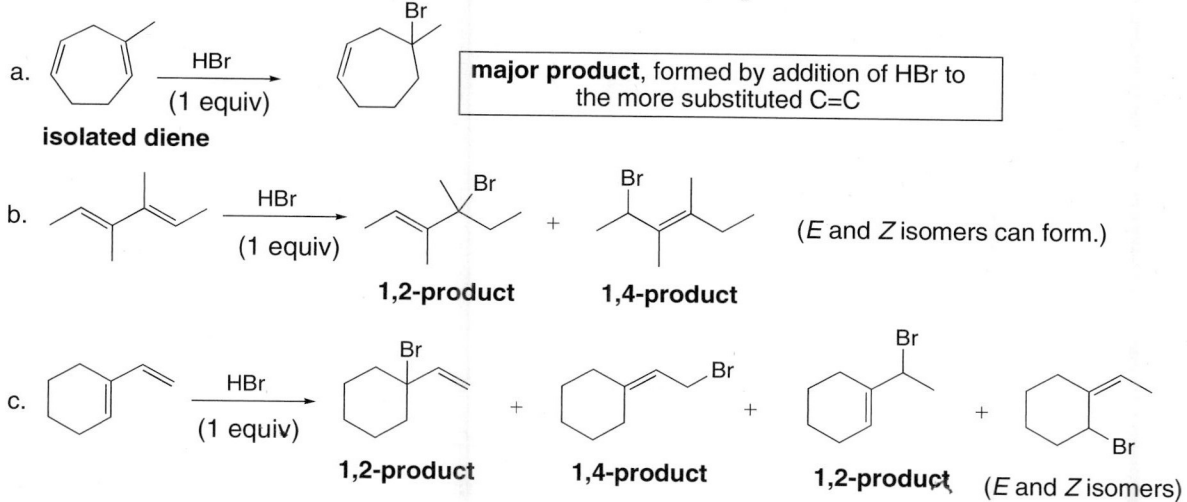

a. $\xrightarrow[\text{(1 equiv)}]{\text{HBr}}$

isolated diene

major product, formed by addition of HBr to
the more substituted C=C

b. $\xrightarrow[\text{(1 equiv)}]{\text{HBr}}$ +

(*E* and *Z* isomers can form.)

1,2-product   **1,4-product**

c. $\xrightarrow[\text{(1 equiv)}]{\text{HBr}}$ +   +   +

1,2-product   **1,4-product**   **1,2-product** (*E* and *Z* isomers)
1,4-product

16.45

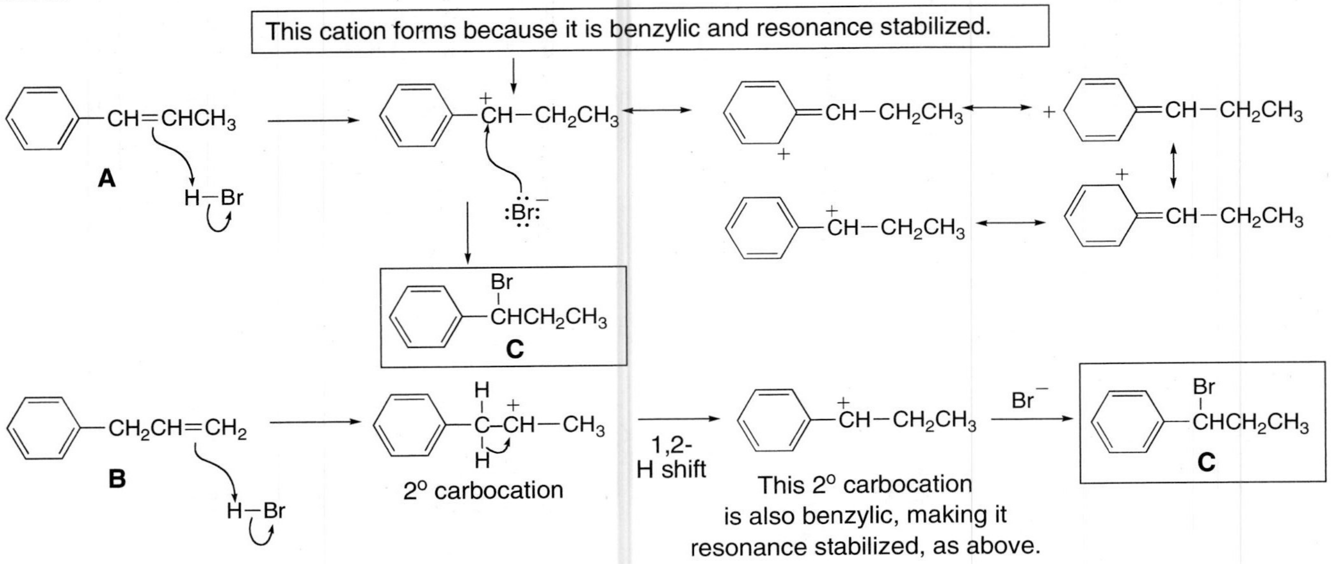
This cation forms because it is benzylic and resonance stabilized.
A
B
C
2° carbocation
1,2-
H shift
This 2° carbocation
is also benzylic, making it
resonance stabilized, as above.

16.46 To draw the mechanism for reaction of a diene with HBr and ROOR, recall from Chapter 15 that when an alkene is treated with HBr under these radical conditions, the Br ends up on the carbon with more H's to begin with.

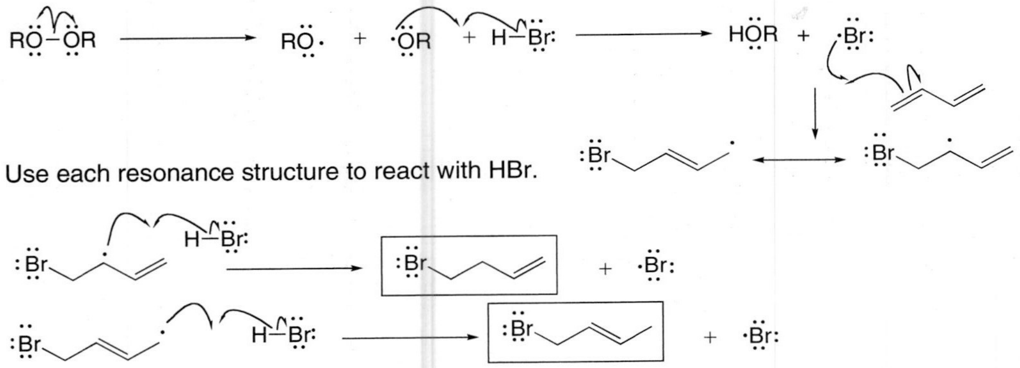
Use each resonance structure to react with HBr.

16.47

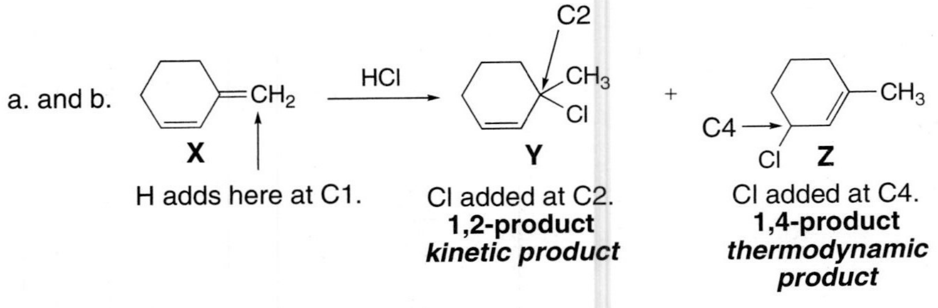
a. and b.
HCl
X
Y
Z
C2
C4
H adds here at C1.
Cl added at C2.
1,2-product
kinetic product
Cl added at C4.
1,4-product
thermodynamic
product

Y is the kinetic product because of the proximity effect. H and Cl add across two adjacent atoms.
Z is the thermodynamic product because it has a more stable trisubstituted double bond.

Addition occurs at the labeled double bond due to the stability of the carbocation intermediate.

If addition occurred at the other C=C, the following allylic carbocation would form:

c.

The two resonance structures for this allylic cation are 3° and 2° carbocations.

more stable intermediate Addition occurs here.

The two resonance structures for this allylic cation are 1° and 2° carbocations.

less stable

16.48 Addition of HCl at the terminal double bond forms a carbocation that is highly resonance stabilized since it is both allylic and benzylic. Such stabilization does not occur when HCl is added to the other double bond. This gives rise to two products of electrophilic addition.

1,2-product

1,4-product

(+ three more resonance structures that delocalize the positive charge onto the benzene ring)

16.49 There are two possible products:

disubstituted C=C

$(CH_3)_2C-CH-CH=C(CH_3)_2$ + $(CH_3)_2C-CH=CH-C(CH_3)_2$
  | | | |
  H   Br   H   Br

trisubstituted C=C

1,2-product   **1,4-product**

The 1,2-product is always the kinetic product because of the proximity effect. In this case, it is also the thermodynamic (more stable) product because it contains a more highly substituted C=C (trisubstituted) than the 1,4-product (disubstituted). Thus, the 1,2-product is the major product at high and low temperature.

16.50 The electron pairs on O can be donated to the double bond through resonance. This increases the electron density of the double bond, making it less electrophilic and therefore less reactive in a Diels–Alder reaction.

$CH_2=CH-OCH_3$ ⟷ $\overline{C}H_2-CH=\overset{+}{O}CH_3$

methyl vinyl ether

This C now bears a net negative charge.

Chapter 16–18

16.51 Use the directions from Answer 16.21.

a.

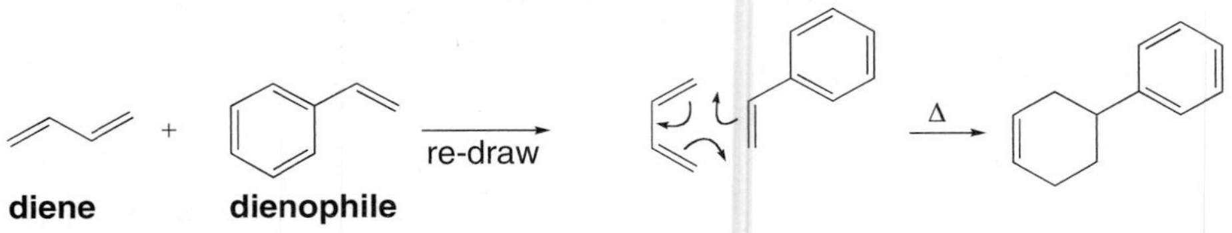

diene   dienophile

b.

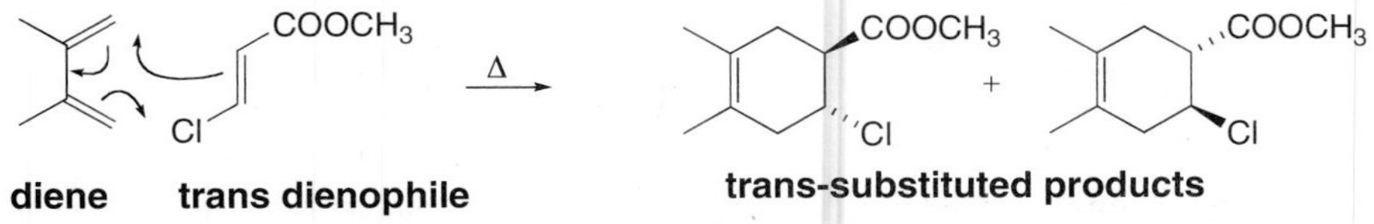

diene   trans dienophile   trans-substituted products

c.

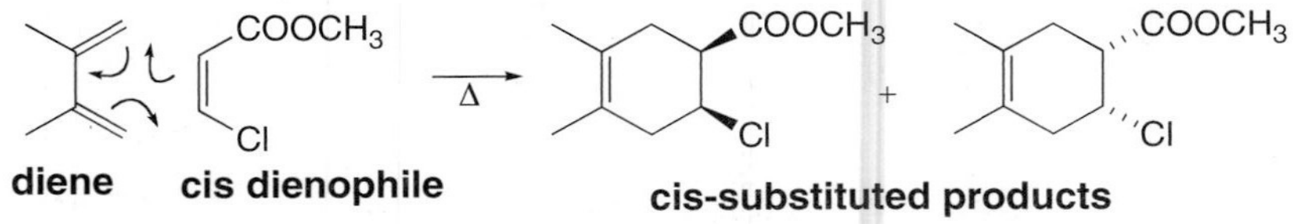

diene   cis dienophile   cis-substituted products

d.

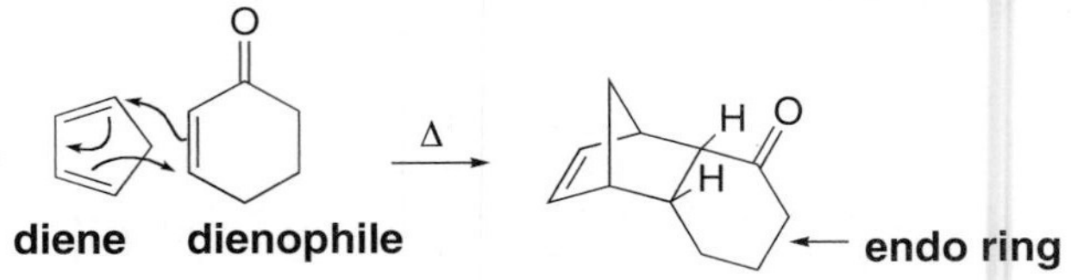

diene   dienophile   ← endo ring

e.

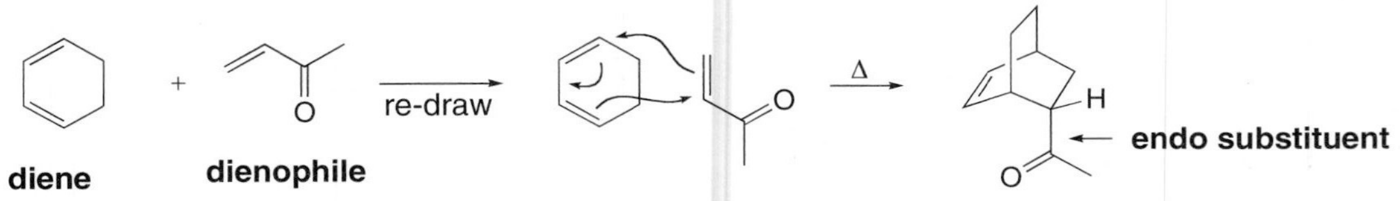

diene   dienophile   endo substituent →

f.

excess

diene   dienophile

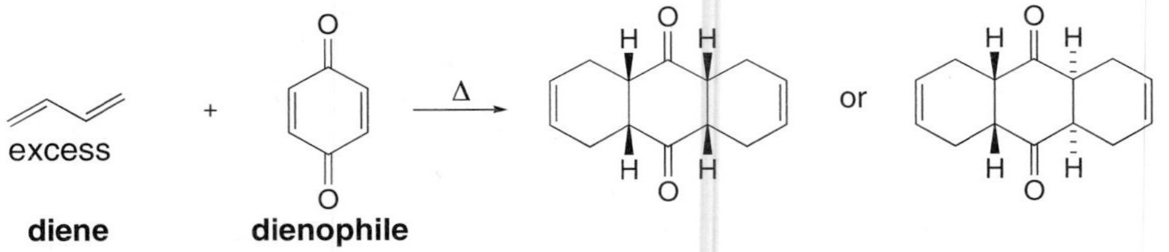

Both C=C's of the dienophile react so two Diels–Alder reactions take place.

16.52 Use the directions from Answer 16.27.

a.

b.

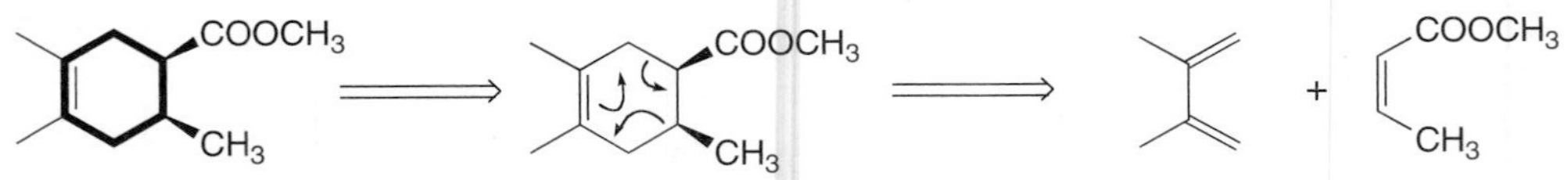

c.

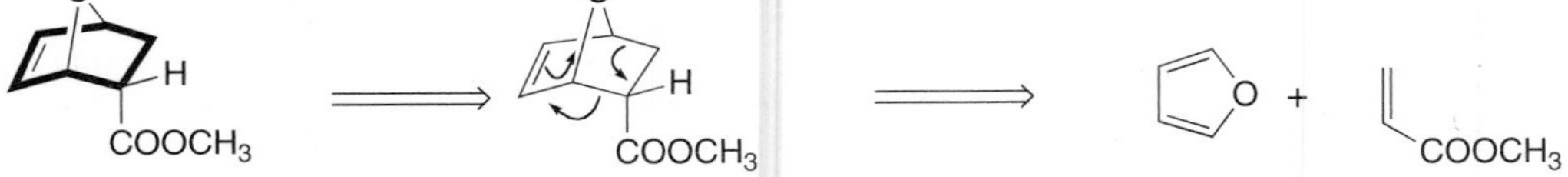

d.

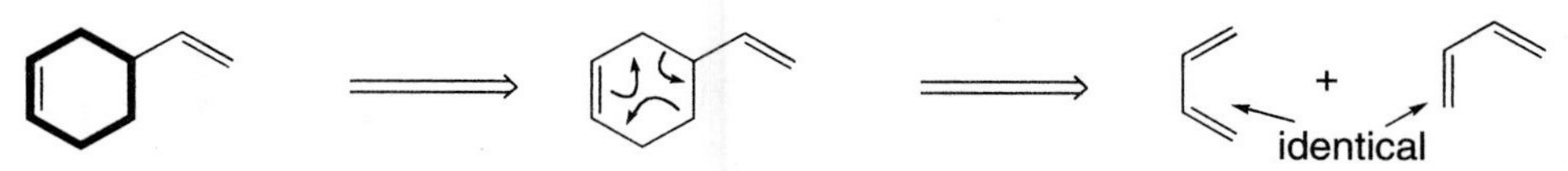

e.

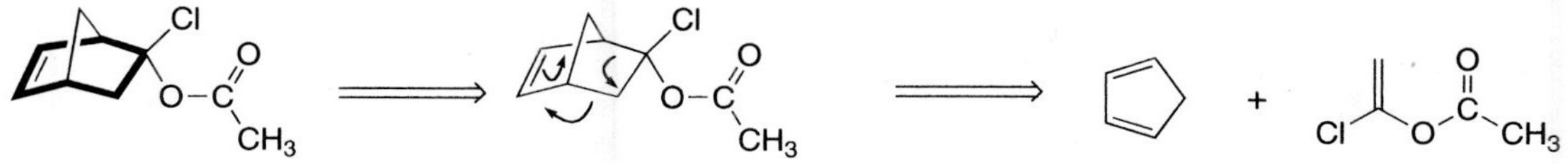

f.

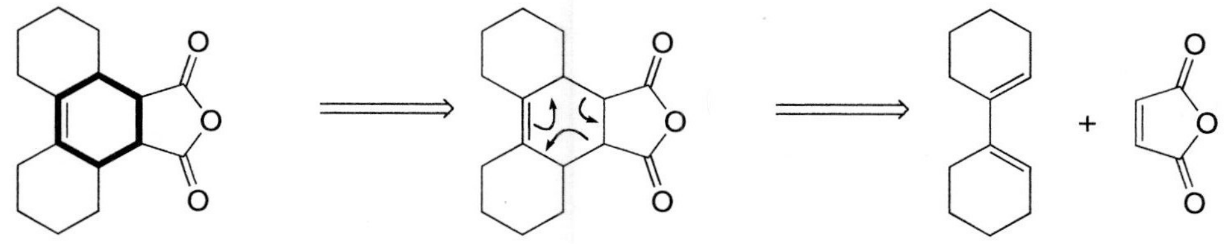

16.53

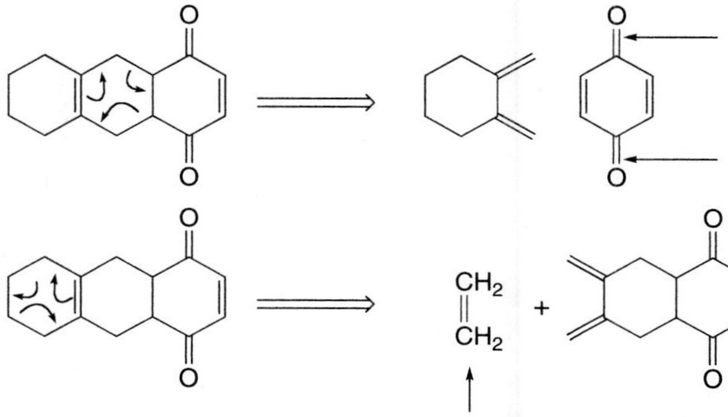

This pathway is **preferred** because the dienophile has electron-withdrawing C=O groups that make it more reactive.

no electron-withdrawing groups
less reactive

16.54

a.

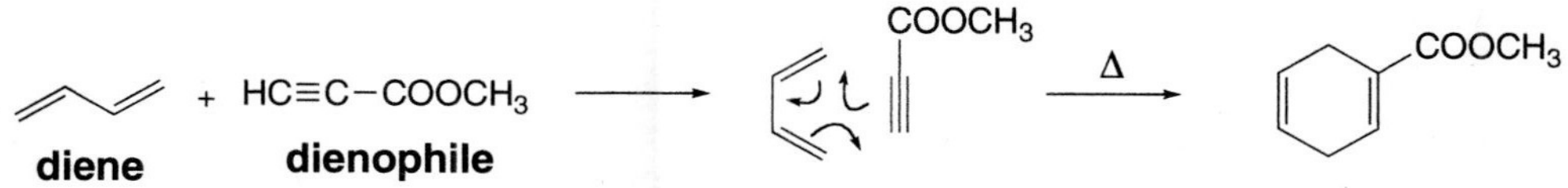

  diene   **dienophile**

b.

  diene   **dienophile**

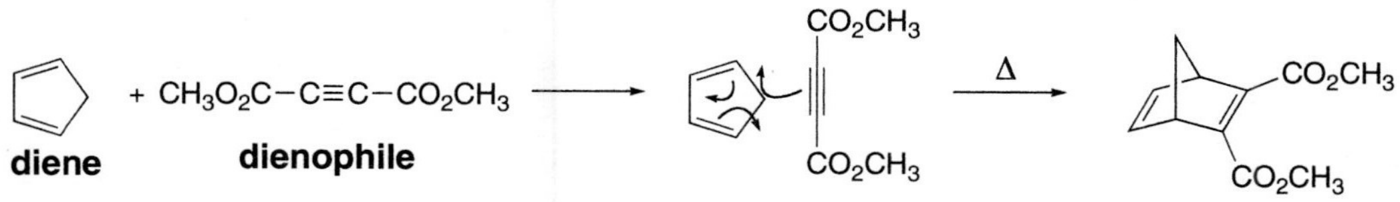

16.55

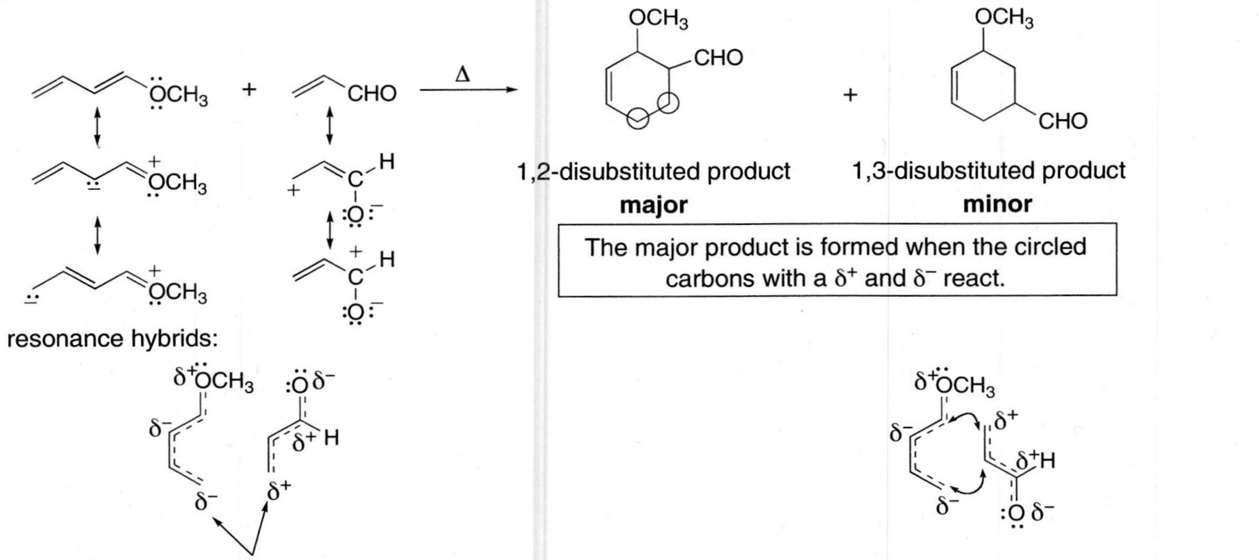

1,2-disubstituted product
major

+

1,3-disubstituted product
minor

The major product is formed when the circled carbons with a δ+ and δ− react.

resonance hybrids:

For the 1,2-product, carbons with unlike charges would react. This is favored because the electron-rich and the electron-poor C's can bond.

For the 1,3-product, there are no partial charges of opposite sign on reacting carbons. This arrangement is less attractive.

16.56

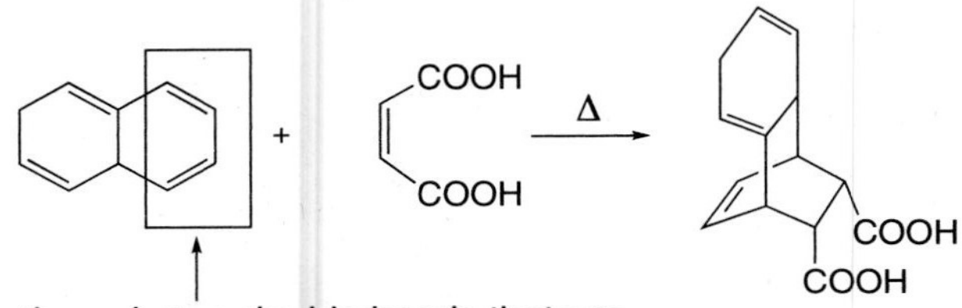

+

.COOH

.COOH

Δ

COOH
COOH

These are the only two double bonds that are conjugated and have the *s*-cis conformation needed for a Diels–Alder reaction.

16.57

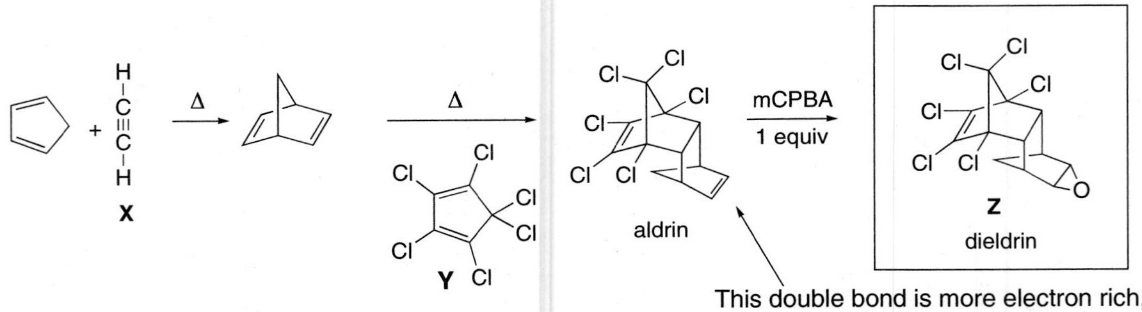

X

Δ

Δ

Y

aldrin

mCPBA
1 equiv

Z

dieldrin

This double bond is more electron rich, so it is epoxidized more readily.

16.58

In each problem, the synthesis must begin with the preparation of cyclopentadiene from dicyclopentadiene.

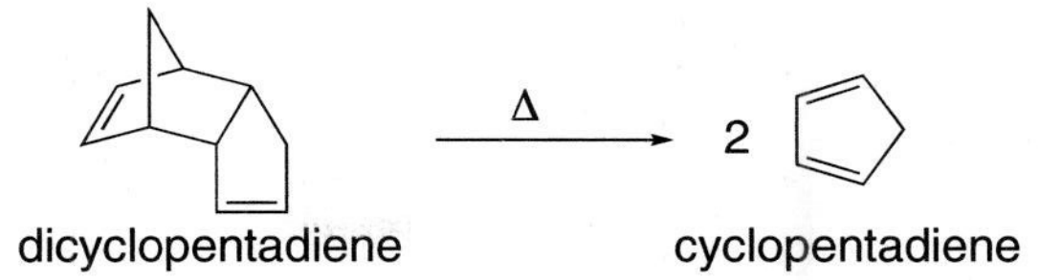

dicyclopentadiene $\xrightarrow{\Delta}$ 2 cyclopentadiene

a.

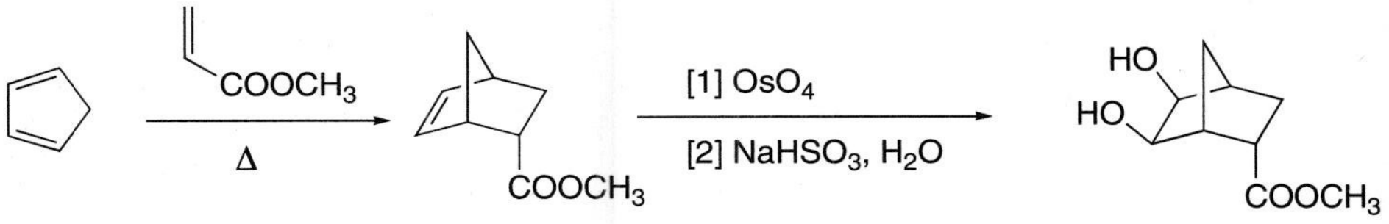

b.

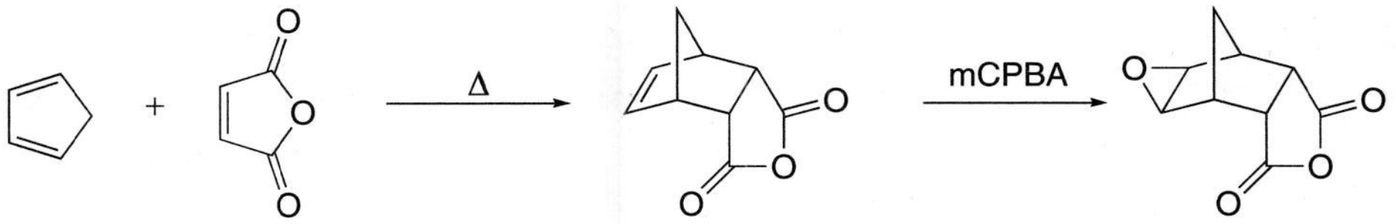

c.

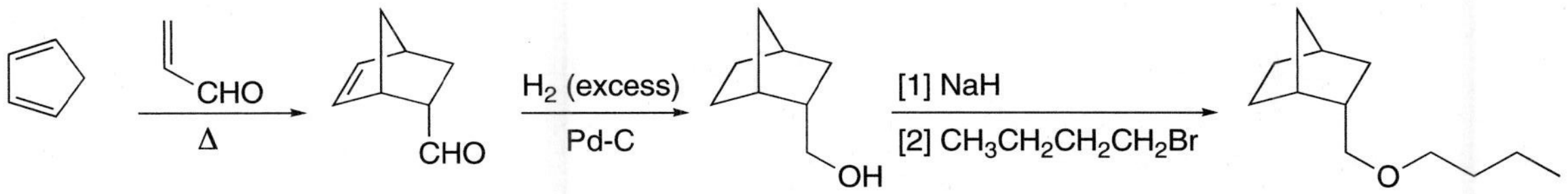

16.59

a.

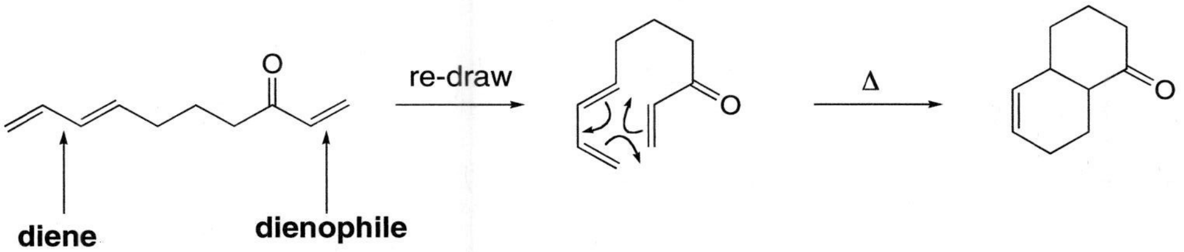

diene   dienophile

b.

diene   dienophile

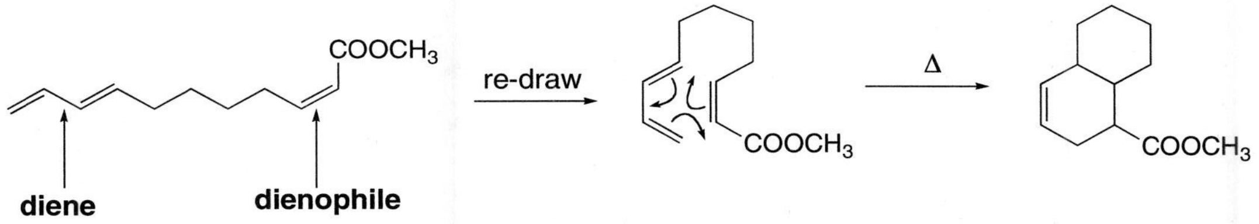

16.60 A transannular Diels–Alder reaction forms a tricyclic product from a monocyclic starting material.

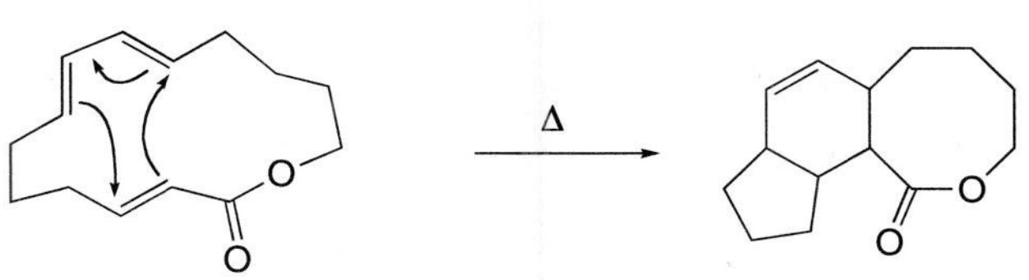

16.61

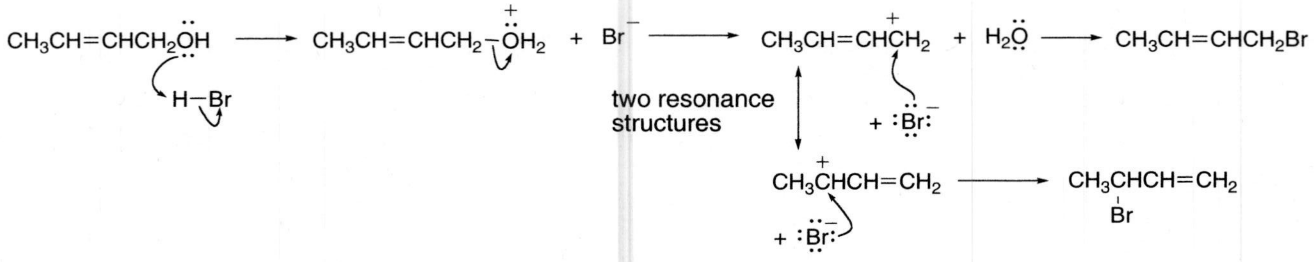

16.62

a. $(CH_3)_2C=CHCH_2CH_2CH=CH_2$ $\xrightarrow{HCl}$ $(CH_3)_2C(Cl)CH_2CH_2CH_2CH=CH_2$ + $(CH_3)_2C=CHCH_2CH_2CHClCH_3$

    isolated diene   **major product**   **minor product**

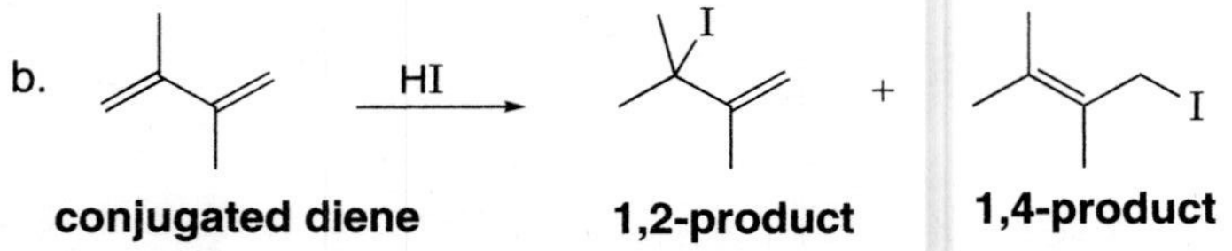

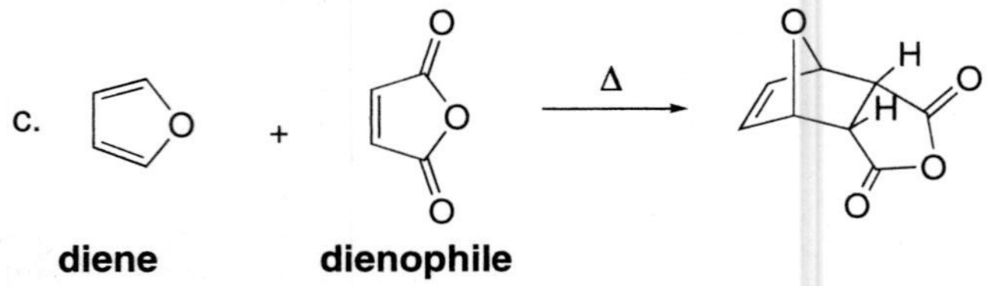

b.

    conjugated diene   **1,2-product**   **1,4-product**

c.

    diene   **dienophile**

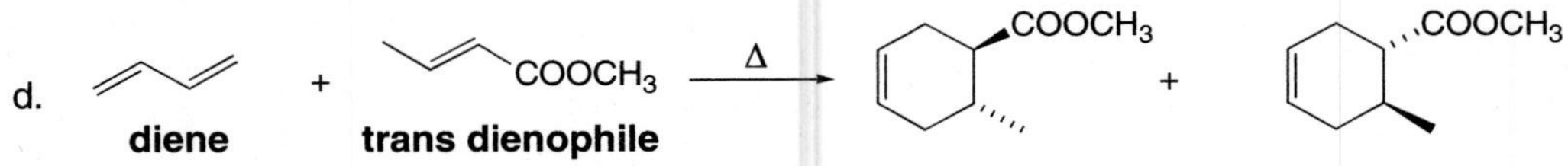

d.

    diene   **trans dienophile**

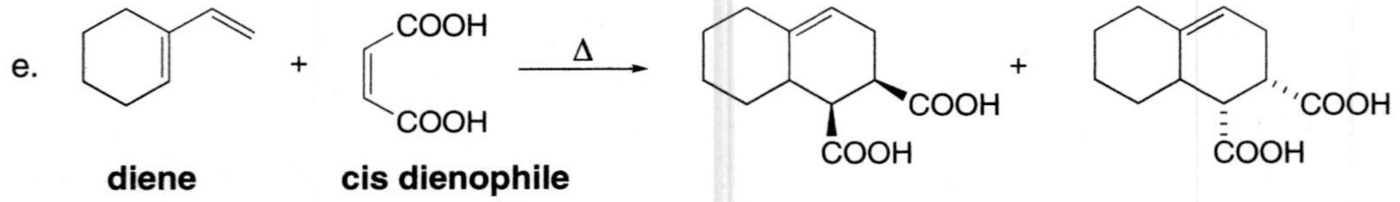

e.

    diene   **cis dienophile**

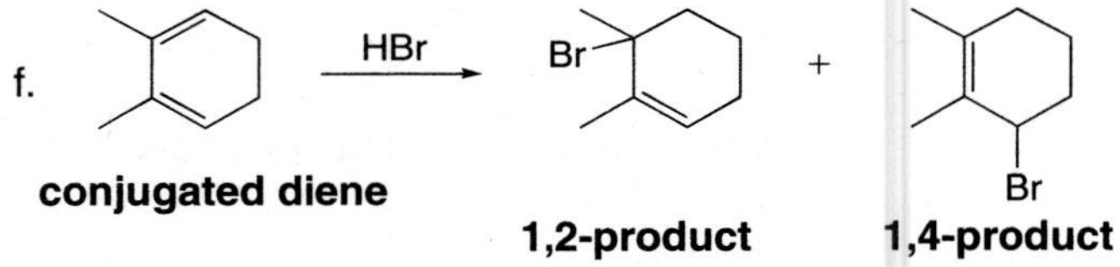

f.

    conjugated diene   **1,2-product**   **1,4-product**

16.63 The mechanism is E1, with formation of a resonance-stabilized carbocation.

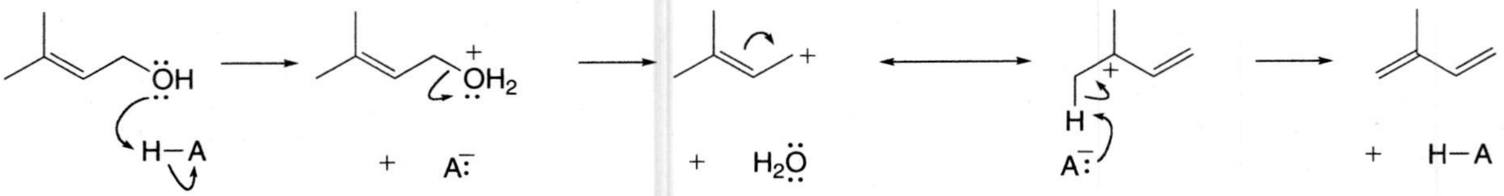

16.64

a.

loss of H (from β_1 C) + Cl
conjugated
more stable
major product

loss of H (from β_2 C) + Cl
more substituted

b. Dehydrohalogentaion generally forms the more stable product. In this reaction, loss of H from the β_1 carbon forms a more stable conjugated diene, so this product is preferred even though it does not contain the more substituted C=C.

16.65

singlet at 1.42 ppm

isoprene

mCPBA

Each H is a doublet of doublets in the 5.2–5.4 ppm region.

doublet of doublets at 5.5–5.7 ppm

two doublets at 2.7–2.9 ppm

16.66

1 H: doublet of doublets at 6.0 ppm

The IR shows an OH absorption at 3200–3600 cm^{-1}.

B

OH peak at 1.5 ppm

6 H: singlet at 1.3 ppm

Each H is a doublet of doublets in the 4.9–5.2 ppm region.

16.67

isolated diene
shortest wavelength
1

2 conjugated bonds
intermediate
wavelength
2

3 conjugated bonds
intermediate
wavelength
3

4 conjugated bonds
longest wavelength
4

16.68

The phenol makes ferulic acid an antioxidant. Loss of H forms a highly stabilized phenoxy radical that inhibits radical formation during oxidation.

ferulic acid

The highly conjugated π system makes it a sunscreen.

16.69

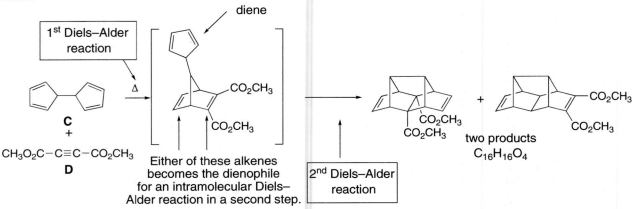

16.70

16.71

16.72 Retro Diels–Alder reaction forms a conjugated diene. Intramolecular Diels–Alder reaction then forms **N**.

Chapter 17: Benzene and Aromatic Compounds

◆ Comparing aromatic, antiaromatic, and nonaromatic compounds (17.7)

- **Aromatic compound**
  - A cyclic, planar, completely conjugated compound that contains $4n + 2$ π electrons ($n = 0, 1, 2, 3$, and so forth).
  - An aromatic compound is more stable than a similar acyclic compound having the same number of π electrons.

- **Antiaromatic compound**
  - A cyclic, planar, completely conjugated compound that contains $4n$ π electrons ($n = 0, 1, 2, 3$, and so forth).
  - An antiaromatic compound is less stable than a similar acyclic compound having the same number of π electrons.

- **A compound that is not aromatic**
  - A compound that lacks one (or more) of the requirements to be aromatic or antiaromatic.

◆ Properties of aromatic compounds

- Every carbon has a p orbital to delocalize electron density (17.2).
- They are unusually stable. ΔH° for hydrogenation is much less than expected, given the number of degrees of unsaturation (17.6).
- They do not undergo the usual addition reactions of alkenes (17.6).
- ^{1}H NMR spectra show highly deshielded protons because of ring currents (17.4).

◆ Examples of aromatic compounds with 6 π electrons (17.8)

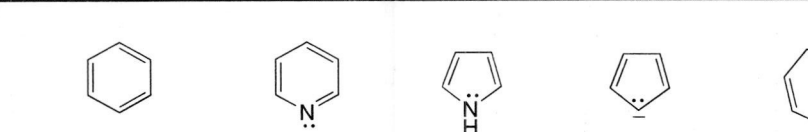

| benzene | pyridine | pyrrole | cyclopentadienyl anion | tropylium cation |

◆ Examples of compounds that are not aromatic (17.8)

 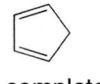

not cyclic   not planar   not completely
                                        conjugated

17.1   Move the electrons in the π bonds to draw all major resonance structures.

diphenhydramine

17.2   Look at the hybridization of the atoms involved in each bond. Carbons in a benzene ring are surrounded by three groups and are sp^2 hybridized.

a.

C_{sp^2}–C_{sp^3}

C_{sp^2}–H_{1s}

C_{sp^2}–C_{sp^2}
C_p–C_p
shortest of all
the indicated bonds
in (a) and (b)

b.

C_{sp^2}–C_{sp^2}

C_{sp^2}–C_{sp^2}
C_p–C_p

17.3

- To name a benzene ring with **one substituent**, name the substituent and add the word *benzene*.
- To name a **disubstituted ring**, select the correct prefix (ortho = 1,2; meta = 1,3; para = 1,4) and alphabetize the substituents. Use a common name if it is a derivative of that monosubstituted benzene.
- To name a **polysubstituted ring**, number the ring to give the lowest possible numbers and then follow other rules of nomenclature.

a.

isopropyl group

PhCH(CH₃)₂

isopropylbenzene

b.

1 CH_2CH_3 ← ethyl

I 4
3   2

iodo Two groups are 1,4 = para.
***p*-ethyliodobenzene**

c.

Two groups are 1,3 = meta.

butyl group

phenol

***m*-butylphenol**

d.

CH_3

Br → 2-bromo

Cl 5

→toluene   (CH₃ group must be at the "1" position,
if the molecule is named as a toluene derivative.)

5-chloro

2-bromo-5-chlorotoluene

17.4 Work backwards to draw the structures from the names.

a. isobutylbenzene

c. *cis*-1,2-diphenylcyclohexane

e. 4-chloro-1,2-diethylbenzene

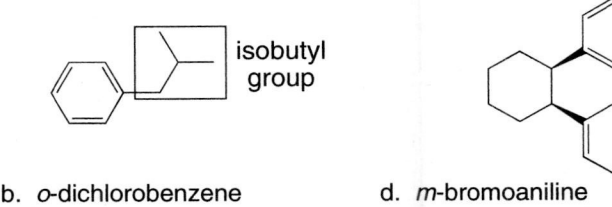

isobutyl group

b. *o*-dichlorobenzene

d. *m*-bromoaniline

f. 3-*tert*-butyl-2-ethyltoluene

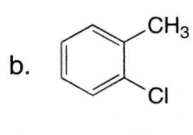

aniline

17.5

1,2,3-trichlorobenzene   1,2,4-trichlorobenzene   1,3,5-trichlorobenzene

17.6

Molecular formula $C_{10}H_{14}O_2$: 4 degrees of unsaturation
IR absorption at 3150–2850 cm^{-1}: sp^2 and sp^3 hybridized C–H bonds
NMR absorptions (ppm):
    1.4 (triplet, 6 H)
    4.0 (quartet, 4 H)
    6.8 (singlet, 4 H)

17.7 Count the different types of carbons to determine the number of ^{13}C NMR signals.

a.

4 types of C's in the benzene ring
6 signals

b.

All C's are different.
7 signals

c.

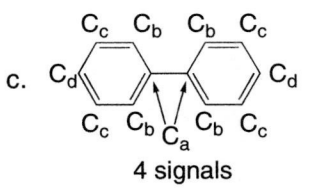

4 signals

17.8 Each of the three isomeric trichlorobenzenes exhibits a different number of ^{13}C NMR signals.

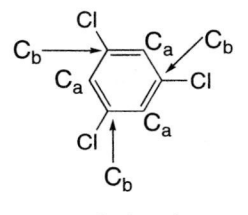

4 signals   6 signals   2 signals

17.9   The less stable compound has a larger heat of hydrogenation.

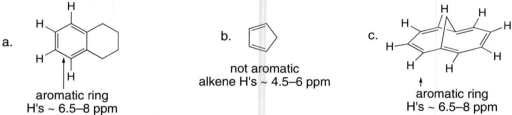

    A   B

benzene ring, more stable   no benzene ring, less stable
     smaller ΔH°   larger ΔH°

17.10   The protons on sp^2 hybridized carbons in aromatic hydrocarbons are highly deshielded and absorb at 6.5–8 ppm whereas hydrocarbons that are not aromatic show an absorption at 4.5–6 ppm, typical of protons bonded to the C=C of an alkene.

a.

aromatic ring
H's ~ 6.5–8 ppm

b.

not aromatic
alkene H's ~ 4.5–6 ppm

c.

aromatic ring
H's ~ 6.5–8 ppm

17.11   To be aromatic, a ring must have $4n + 2$ π electrons.

16 π e⁻	20 π e⁻	22 π e⁻
$4n$	$4n$	$4n + 2$
$4(4) = 16$	$4(5) = 20$	$4(5) + 2 = 22$
antiaromatic	**antiaromatic**	**aromatic**

17.12   Annulenes have alternating double and single bonds. An odd number of carbon atoms in the ring would mean there would be two adjacent single bonds. Therefore an annulene having an odd number of carbon atoms cannot exist.

17.13

17.14   In determining if a heterocycle is aromatic, count a nonbonded electron pair if it makes the ring aromatic in calculating $4n + 2$. Lone pairs on atoms already part of a multiple bond cannot be delocalized in a ring, and so they are never counted in determining aromaticity.

a.

Count one lone pair from O.
$4n + 2 = 4(1) + 2 = 6$
aromatic

b.

no lone pair from O
$4n + 2 = 4(1) + 2 = 6$
aromatic

c.

With one lone
pair from each O
there would be 8 electrons.
If O's are sp^3 hybridized, the
ring is not completely
conjugated.
not aromatic

d.

Both N atoms are part of
a double bond, so the lone
pairs cannot be counted:
there are 6 electrons.
$4n + 2 = 4(1) + 2 = 6$
aromatic

17.30

a. *p*-dichlorobenzene

d. *o*-bromonitrobenzene

g. 2-phenyl-2-propen-1-ol

b. *m*-chlorophenol

e. 2,6-dimethoxytoluene

h. *trans*-1-benzyl-3-phenylcyclopentane

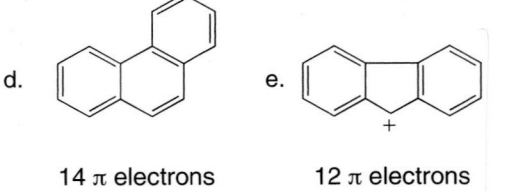

or

c. *p*-iodoaniline

f. 2-phenyl-1-butene

17.31

a. constitutional isomers of molecular formula **C₈H₉Cl**, and b. names of the trisubstituted benzenes

stereoisomers
for this isomer

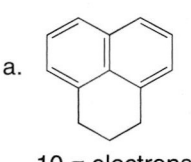 2-chloro-1,3-dimethylbenzene   1-chloro-2,3-dimethylbenzene   4-chloro-1,2-dimethylbenzene

1-chloro-2,4-dimethylbenzene   1-chloro-3,5-dimethylbenzene   2-chloro-1,4-dimethylbenzene

c. stereoisomers

17.32 Count the electrons in the π bonds. Each π bond holds two electrons.

a.

b.

c.

d.

e.

10 π electrons   7 π electrons   10 π electrons   14 π electrons   12 π electrons

17.33 To be aromatic, the compounds must be cyclic, planar, completely conjugated, and have $4n + 2$ π electrons.

a. Circled C's are not sp^2.
not completely conjugated
not aromatic

d. Circled C is not sp^2.
not completely conjugated
not aromatic

b. 14 π electrons in outer ring
aromatic

e. 12 π electrons
does **not** have $4n + 2$
π electrons
not aromatic

c. 4 benzene rings
joined together
aromatic

f. 12 π electrons
does **not** have $4n + 2$
π electrons
not aromatic

17.34 In determining if a heterocycle is aromatic, count a nonbonded electron pair if it makes the ring aromatic in calculating $4n +2$. Lone pairs on atoms already part of a multiple bond cannot be delocalized in a ring, and so they are never counted in determining aromaticity.

a. 6 π electrons
counting a lone pair from S
$4(1) + 2 = 6$
aromatic

c. **not aromatic**

e. **not aromatic**

g. 6 π electrons
counting the lone
pair from N
$4(1) + 2 = 6$
aromatic

b. 6 π electrons
counting a lone pair from O
$4(1) + 2 = 6$
aromatic

d. 10 π electrons
$4(2) + 2 = 10$
aromatic

f. 6 π electrons,
counting a lone pair from O
$4(1) + 2 = 6$
aromatic

h. 10 π electrons
$4(2) + 2 = 10$
aromatic

Count
these 2e⁻.

These lone
pairs are on
doubly
bonded N
atoms, so they
can't be
counted.

17.35

a. Circled C's are
not sp^2.
not aromatic

c. Circled C is
not sp^2.
not aromatic

b. 10 π electrons
in 10-membered ring
$4(2) + 2 = 10$
aromatic

d. 4 π electrons
$4(1) = 4$
antiaromatic

17.36

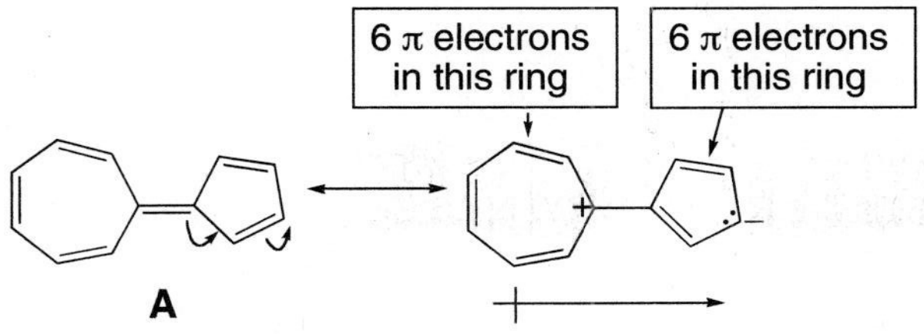

A resonance structure can be drawn for **A** that places a negative charge in the five-membered ring and a positive charge in the seven-membered ring. This resonance structure shows that each ring has 6 π electrons, making it aromatic. The molecule possesses a dipole such that the seven-membered ring is electron deficient and the five-membered ring is electron rich.

17.37 Each compound is completely conjugated. A compound with $4n + 2$ π electrons is especially stable, while a compound with $4n$ π electrons is especially unstable.

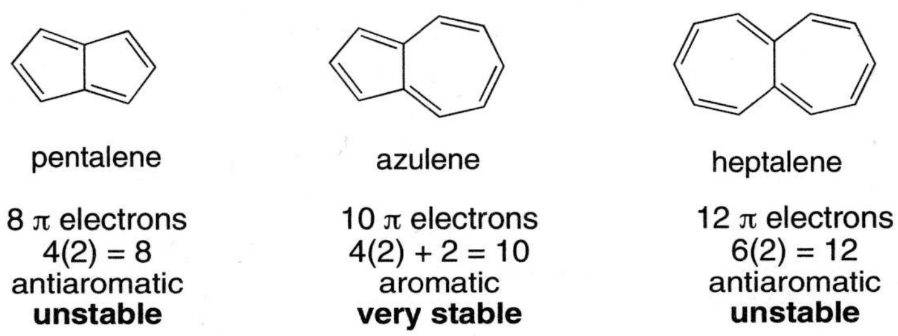

pentalene	azulene	heptalene
8 π electrons	10 π electrons	12 π electrons
4(2) = 8	4(2) + 2 = 10	6(2) = 12
antiaromatic	aromatic	antiaromatic
unstable	**very stable**	**unstable**

17.38 Benzene has C–C bonds of equal length, intermediate between a C–C double and single bond. Cyclooctatetraene is not planar and not aromatic so its double bonds are localized.

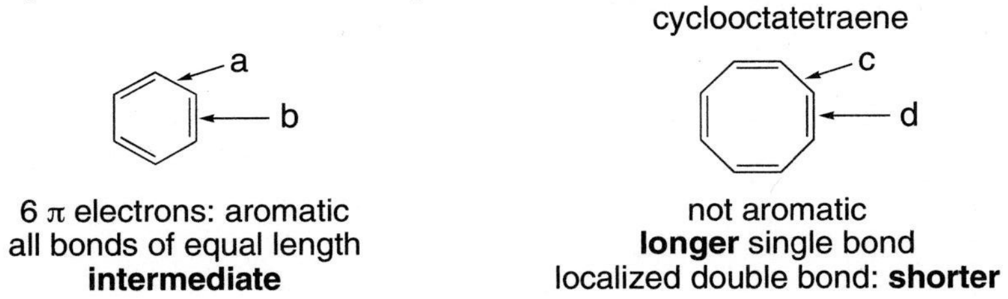

6 π electrons: aromatic
all bonds of equal length
intermediate

not aromatic
longer single bond
localized double bond: **shorter**

d < a = b < c

17.39

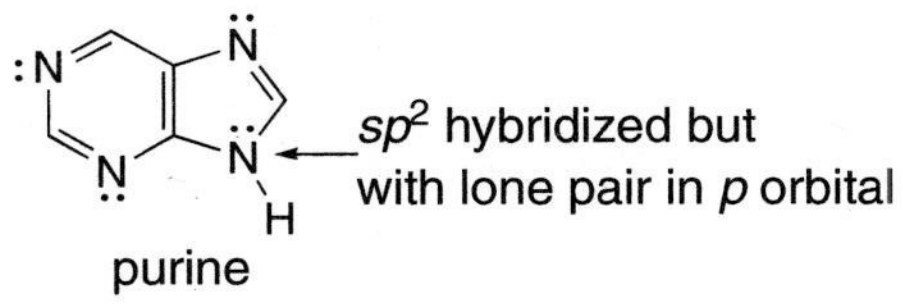

purine

sp^2 hybridized but with lone pair in p orbital

a. Each N atom is sp^2 hybridized.
b. The three unlabeled N atoms are sp^2 hybridized with lone pairs in one of the sp^2 hybrid orbitals. The labeled N has its lone pair in a p orbital.
c. 10 π electrons
d. Purine is cyclic, planar, completely conjugated, and has 10 π electrons [4(2) + 2] so it is aromatic.

17.40

a. 16 total π electrons

b. 14 π electrons delocalized in the ring. [Note: Two of the electrons in the triple bond are localized between two C's, perpendicular to the π electrons delocalized in the ring.]

c. By having two of the *p* orbitals of the C–C triple bond co-planar with the *p* orbitals of all the C=C's, the total number of π electrons delocalized in the ring is 14. 4(3) + 2 = 14, so the ring is **aromatic.**

17.41 The rate of an S$_N$1 reaction increases with increasing stability of the intermediate carbocation.

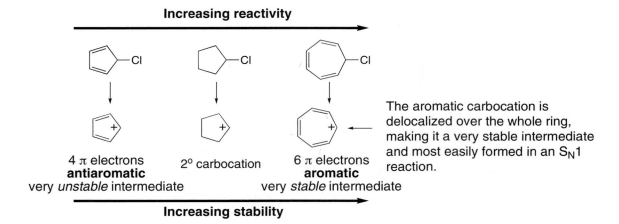

17.42

17.43 α-Pyrone reacts like benzene because it is aromatic. A second resonance structure can be drawn showing how the ring has six π electrons. Thus, α-pyrone undergoes reactions characteristic of aromatic compounds—that is, substitution rather than addition.

α-pyrone   6 π electrons

17.44

a.

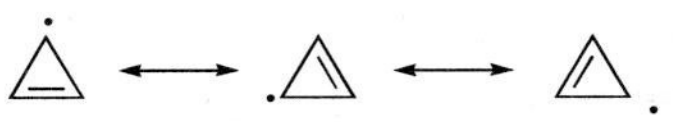

cyclopropenyl radical

b.

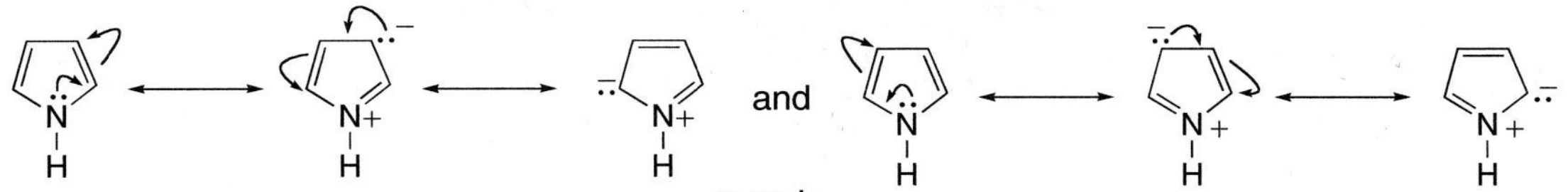

and

pyrrole

c.

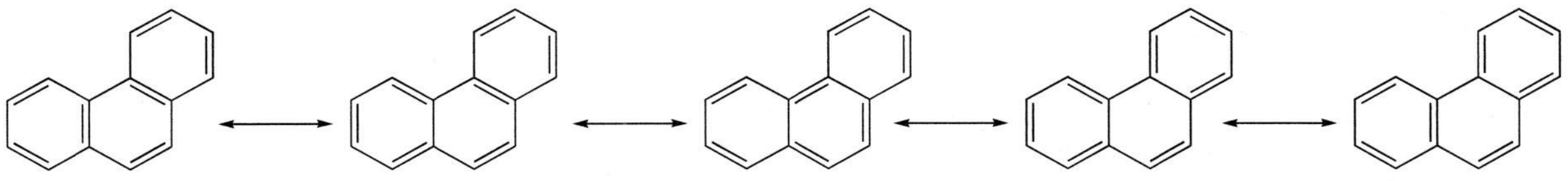

phenanthrene

17.45

Naphthalene can be drawn as three resonance structures:

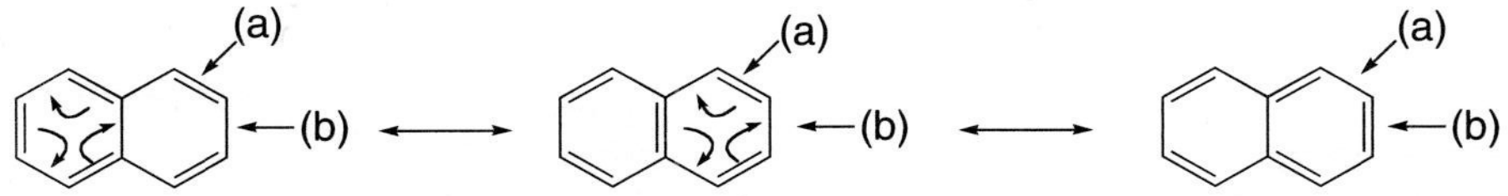

In two of the resonance structures bond (a) is a double bond, and
bond (b) is a single bond. Therefore, bond (b) has more single
bond character, making it longer.

17.46

a.

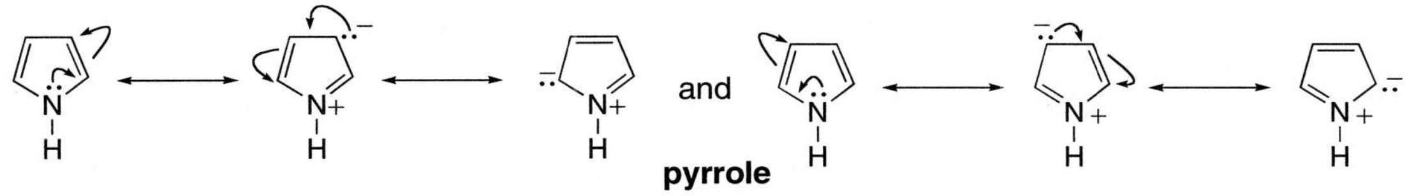

and

pyrrole

Pyrrole is less resonance stabilized than benzene
because four of the resonance structures have
charges, making them less good.

b.

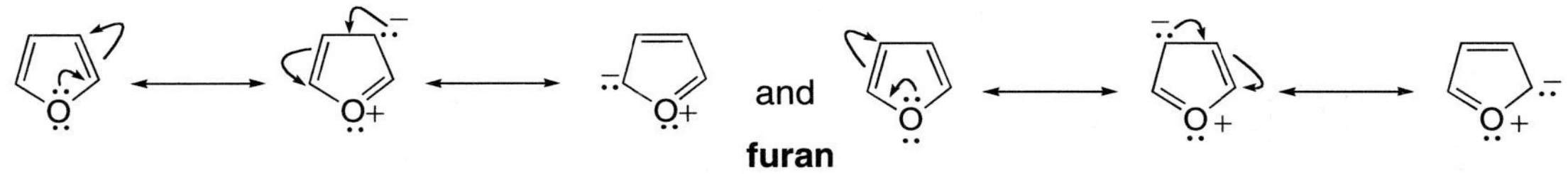

and

furan

Furan is less resonance stabilized than pyrrole because its O atom is
less basic, so it donates electron density less "willingly." Thus,
charge-separated resonance forms are more minor contributors to
the hybrid than the charge-separated resonance forms of pyrrole.

17.47 The compound with the more stable conjugate base is the stronger acid. Draw and compare the conjugate bases of each pair of compounds.

conjugate bases

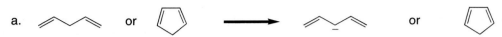

a. ⟨⟩ or ⟨⟩ ⟶ ⟨⟩ or ⟨⟩

more acidic

resonance-stabilized but not aromatic

6 π electrons, aromatic
more stable conjugate base
Its acid is more acidic.

b. ⟨⟩ or △ ⟶ ⟨⟩ or △

more acidic

6 π electrons, aromatic
more stable conjugate base
Its acid is more acidic.

antiaromatic
highly destabilized
conjugate base

17.48

indene + NaNH₂ ⟶ [indenyl anion] + NH₃
Na⁺

The conjugate base of indene has 10 π electrons making it aromatic
and very stable. Therefore, indene is more acidic than many hydrocarbons.

17.49

A

H$_b$ is most acidic because its
conjugate base is aromatic (6 π electrons).

B

H$_c$ is least acidic because its
conjugate base has 8 π electrons,
making it antiaromatic.

17.50

pyrrole ⟶ conjugate base

Both pyrrole and the conjugate base of pyrrole have 6 π
electrons in the ring, making them both aromatic. Thus,
deprotonation of pyrrole does not result in a gain of aromaticity
since the starting material is aromatic to begin with.

cyclopentadiene ⟶ conjugate base
more acidic

Cyclopentadiene is not aromatic, but the conjugate base has 6 π
electrons and is therefore aromatic. This makes the C–H bond in
cyclopentadiene more acidic than the N–H bond in pyrrole, since
deprotonation of cyclopentadiene forms an aromatic conjugate base.

17.51

a.

pyrrole + H⁺ → **A** (Protonation at C2)

or

+ H⁺ → **B**

Protonation at C2 forms conjugate acid **A** because the positive charge can be delocalized by resonance. There is no resonance stabilization of the positive charge in **B**.

b.

A
$pK_a = 0.4$

Loss of a proton from **A** (which is not aromatic) gives two electrons to N, and forms pyrrole, which has six π electrons that can then delocalize in the five-membered ring, making it aromatic. This makes deprotonation a highly favorable process, and **A** more acidic.

C
$pK_a = 5.2$

Both **C** and its conjugate base pyridine are aromatic. Since **C** has six π electrons, it is already aromatic to begin with, so there is less to be gained by deprotonation, and **C** is thus less acidic than **A**.

17.52

a.

3 antibonding MOs
2 nonbonding MOs
3 bonding MOs

cyclooctatetraene and its 8 π electrons

cyclooctatetraene → (2 K) → dianion of cyclooctatetraene

b. Even if cyclooctatetraene were flat, it has two unpaired electrons in its HOMOs (nonbonding MOs) so it cannot be aromatic.

c. The dianion has 10 π electrons.

d. The two additional electrons fill the nonbonding MOs; that is, all the bonding and nonbonding MOs are filled with electrons in the dianion.

e. The dianion is aromatic since its HOMOs are completely filled, and it has no electrons in antibonding MOs.

17.53

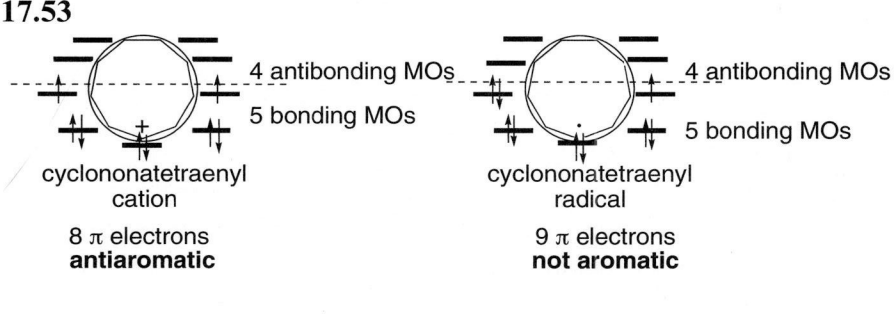

4 antibonding MOs
5 bonding MOs

cyclononatetraenyl cation

8 π electrons
antiaromatic

4 antibonding MOs
5 bonding MOs

cyclononatetraenyl radical

9 π electrons
not aromatic

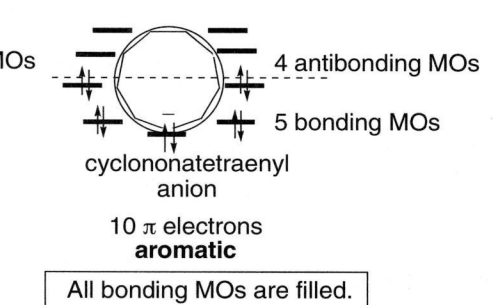

4 antibonding MOs
5 bonding MOs

cyclononatetraenyl anion

10 π electrons
aromatic

All bonding MOs are filled.

17.54 The number of different types of C's = the number of signals.

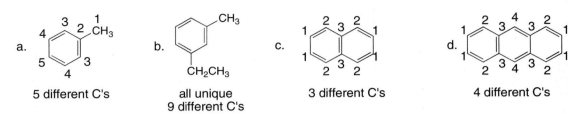

a. 5 different C's

b. all unique
9 different C's

c. 3 different C's

d. 4 different C's

17.55 Draw the three isomers and count the different types of carbons in each. Then match the structures with the data.

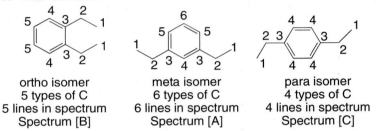

ortho isomer
5 types of C
5 lines in spectrum
Spectrum [B]

meta isomer
6 types of C
6 lines in spectrum
Spectrum [A]

para isomer
4 types of C
4 lines in spectrum
Spectrum [C]

17.56

a. $C_{10}H_{14}$: IR absorptions at 3150–2850 (sp^2 and sp^3 hybridized C–H), 1600, and 1500 (due to a benzene ring) cm^{-1}

^{1}H NMR data:

Absorption	ppm	# of H's	Explanation
doublet	1.2	6	6 H's adjacent to 1 H
singlet	2.3	3	CH_3
septet	3.1	1	1 H adjacent to 6 H's
multiplet	7–7.4	4	a disubstituted benzene ring

$(CH_3)_2CH$ group

You can't tell from these data where the two groups are on the benzene ring. They are not para, since the para arrangement usually gives two sets of distinct peaks (resembling two doublets) so there are two possible structures—ortho and meta isomers.

or

b. C_9H_{12}: ^{13}C NMR signals at 21, 127, and 138 ppm → means three different types of C's.
^{1}H NMR shows 2 types of H's: 9 H's probably means 3 CH_3 groups; the other 3 H's are very deshielded so they are bonded to a benzene ring.
Only one possible structure fits:

CH_3

CH_3 CH_3

c. C_8H_{10}: IR absorptions at 3108–2875 (sp^2 and sp^3 hybridized C–H), 1606, and 1496 (due to a benzene ring) cm^{-1}

^{1}H NMR data:

Absorption	ppm	# of H's	Explanation	Structure:
triplet	1.3	3	3 H's adjacent to 2 H's	
quartet	2.7	2	2 H's adjacent to 3 H's	
multiplet	7.3	5	a monosubstituted benzene ring	

17.57

a. Compound **A**: Molecular formula $C_8H_{10}O$

IR absorption at 3150–2850 (sp^2 and sp^3 hybridized C–H) cm^{-1}

^{1}H NMR data:

Absorption	ppm	# of H's	Explanation	Structure:
triplet	1.4	3	3 H's adjacent to 2 H's	
quartet	3.95	2	2 H's adjacent to 3 H's	
multiplet	6.8–7.3	5	a monosubstituted benzene ring	

b. Compound **B**: Molecular formula $C_9H_{10}O_2$

IR absorption at 1669 (C=O) cm^{-1}

^{1}H NMR data:

Absorption	ppm	# of H's	Explanation	Structure:
singlet	2.5	3	CH_3 group	
singlet	3.8	3	CH_3 group	
doublet	6.9	2	2 H's on a benzene ring	
doublet	7.9	2	2 H's on a benzene ring	

It would be hard to distinguish these two compounds with the given data.

17.58

3 equivalent H's
singlet at ~2.3 ppm

* 3 other H's on benzene ring
(arrows with *)
at ~6.9 ppm

thymol

1 H
septet at ~3.2 ppm

6 equivalent H's
doublet at ~1.2 ppm

IR absorptions:
3500–3200 cm^{-1} (O–H)
3150–2850 cm^{-1} (C–H bonds)
1621 and 1585 cm^{-1} (benzene ring)

basic structure of
thymol

Thymol must have this basic structure given the NMR and IR data since it is a trisubstituted benzene ring with one singlet and two doublets in the NMR at ~6.9 ppm. However, which group [OH, CH₃, or CH(CH₃)₂] corresponds to X, Y, and Z is not readily distinguished with the given data. The correct structure for thymol is given.

17.59

The induced magnetic field by the circulating π electrons opposes the applied field in this vicinity, shifting the absorption upfield to a lower chemical shift than other sp^3 C–H protons.
In this case the protons absorb upfield from TMS, an unusual phenomenon for C–H protons.

| Induced magnetic field

2.84 ppm ⟶ H

H H ◀
–0.6 ppm
(upfield from TMS)

H

B_0

| Induced magnetic field

The induced magnetic field by the circulating π electrons reinforces the applied field in this vicinity, shifting the absorption downfield to a somewhat higher chemical shift.

17.60

C₁ C₂ C₃ C₄

CH₃
N
CH₃   OCH₂CH₃

C₂ C₃

¹³C NMR has four lines that are located in the aromatic region (~110–155 ppm), corresponding to the four different types of carbons in the aromatic ring of the para isomer. The ortho and meta isomers have six different C's, and so six lines would be expected for each of them.

17.61

a.

HO   OH
OCH₃   curcumin   OCH₃

⟶

HO sp^3 OH
OCH₃   keto form   OCH₃

The enol form is more stable because the enol double bond makes a highly conjugated system. The enol OH can also intramolecularly hydrogen bond to the nearby carbonyl O atom.

b.

The enol O–H proton is more acidic than an alcohol O–H proton because the conjugate base is resonance stabilized.

c. Curcumin is colored because it has many conjugated π electrons, which shift absorption of light from the UV to the visible region.

d. Curcumin is an antioxidant because it contains a phenol. Homolytic cleavage affords a resonance-stabilized phenoxy radical, which can inhibit oxidation from occurring, much like vitamin E and BHT in Chapter 15.

(+ other resonance structures)

phenoxy radical
Resonance delocalizes the radical on the ring and C chain of curcumin.

17.62

a. Pyrazole rings are aromatic because they have 6 π electrons—two from the lone pair on the N atom that is not part of the double bond, and four from the double bonds.

b.

c.

d. The N atom in the NH bond in the pyrazole ring is sp^2 hybridized with 33% s-character, increasing the acidity of the N–H bond. The N–H bond of CH_3NH_2 contains an sp^3 hybridized N atom.

17.63 A second resonance structure for **A** shows that the ring is completely conjugated and has 6 π electrons, making it aromatic and especially stable. A similar charge-separated resonance structure for **B** makes the ring completely conjugated, but gives the ring 4 π electrons, making it antiaromatic and especially unstable.

A 6 π electrons aromatic stable

B 4 π electrons antiaromatic not stable

17.64 The conversion of carvone to carvacrol involves acid-catalyzed isomerization of two double bonds and tautomerization of a ketone to an enol tautomer. In this case the enol form is part of an aromatic phenol. Each isomerization of a C=C involves Markovnikov addition of a proton, followed by deprotonation.

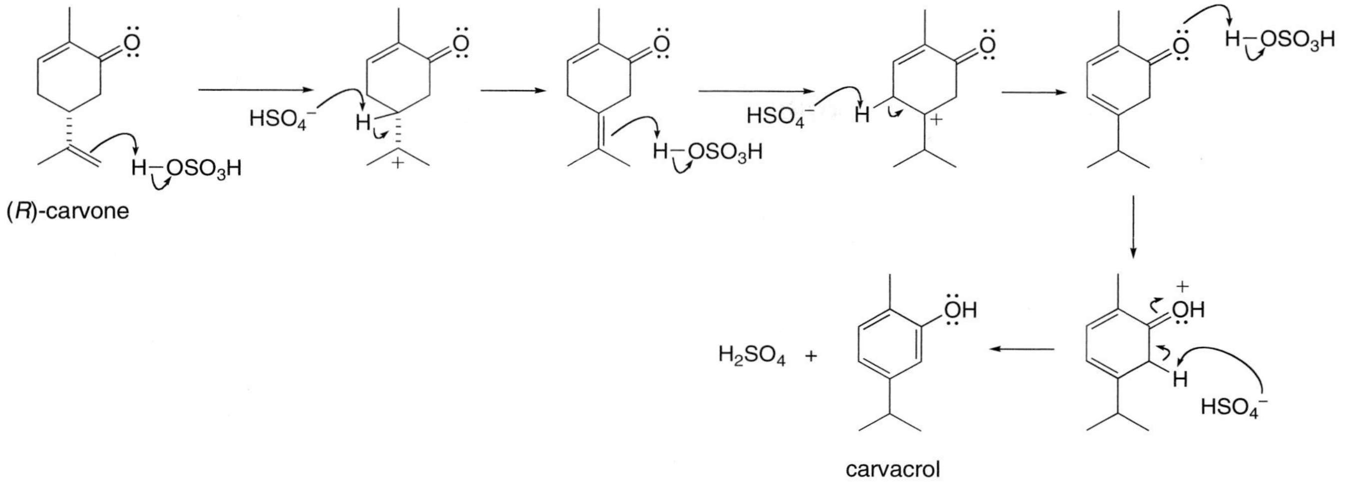
H–OSO3H
(R)-carvone
HSO4−
H
H–OSO3H
HSO4−
H
H–OSO3H
OH
H
HSO4−
H2SO4 +
OH
carvacrol

17.65 Resonance structures for triphenylene:

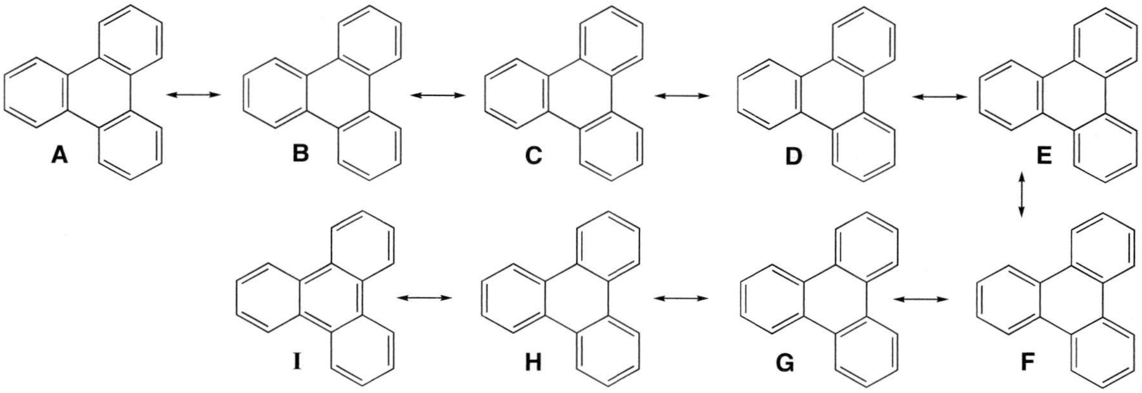
A
B
C
D
E
F
G
H
I

Resonance structures **A–H** all keep three double and three single bonds in the three six-membered rings on the periphery of the molecule. This means that each ring behaves like an isolated benzene ring undergoing substitution rather than addition because the π electron density is delocalized within each six-membered ring. Only resonance structure **I** does not have this form. Each C–C bond of triphenylene has four (or five) resonance structures in which it is a single bond and four (or five) resonance structures in which it is a double bond.

Resonance structures for phenanthrene:

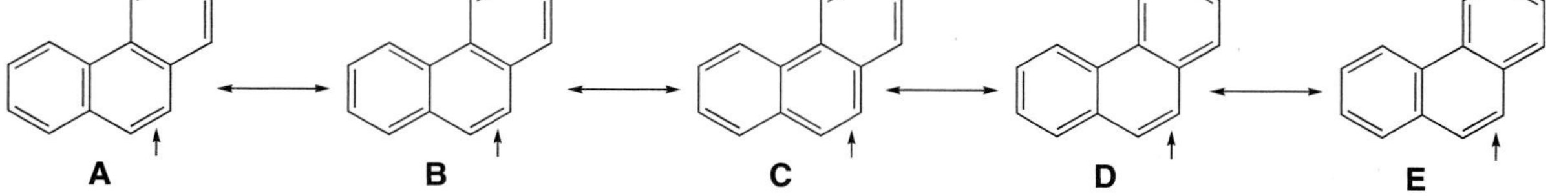
A
B
C
D
E

With phenanthrene, however, four of the five resonance structures keep a double bond at the labeled C's. (Only **C** does not.) This means that these two C's have more double bond character than other C–C bonds in phenanthrene, making them more susceptible to addition rather than substitution.

17.66

X

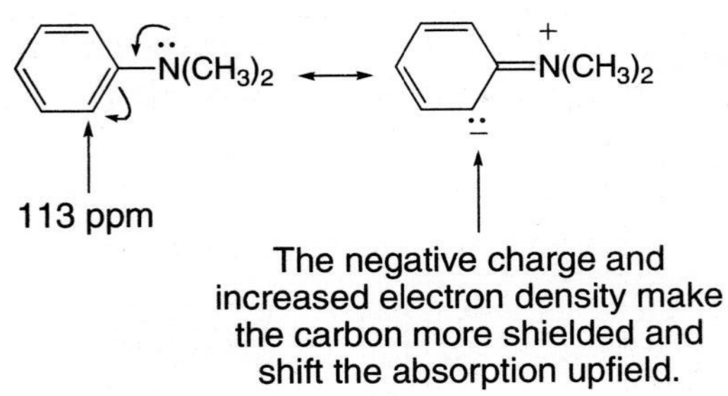
N(CH3)2
+
N(CH3)2
113 ppm
The negative charge and
increased electron density make
the carbon more shielded and
shift the absorption upfield.

Y

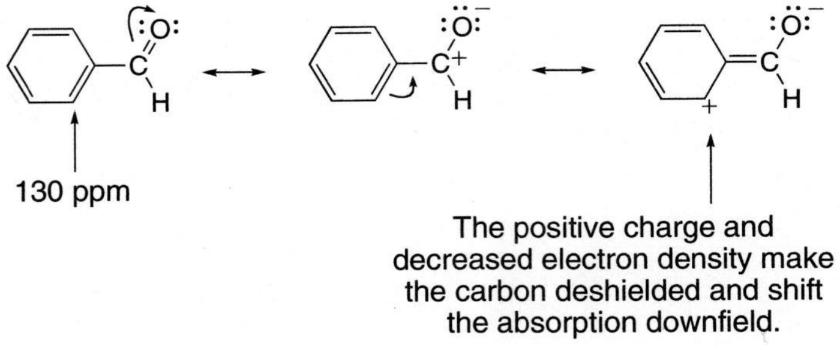
:O:
C
H
130 ppm
The positive charge and
decreased electron density make
the carbon deshielded and shift
the absorption downfield.

Chapter 18: Electrophilic Aromatic Substitution

◆ Mechanism of electrophilic aromatic substitution (18.2)

- Electrophilic aromatic substitution follows a two-step mechanism. Reaction of the aromatic ring with an electrophile forms a carbocation, and loss of a proton regenerates the aromatic ring.
- The first step is rate-determining.
- The intermediate carbocation is stabilized by resonance; a minimum of three resonance structures can be drawn. The positive charge is always located ortho or para to the new C–E bond.

| (+) ortho to E | (+) para to E | (+) ortho to E |

◆ Three rules describing the reactivity and directing effects of common substituents (18.7–18.9)

[1] All ortho, para directors except the halogens activate the benzene ring.
[2] All meta directors deactivate the benzene ring.
[3] The halogens deactivate the benzene ring.

◆ Summary of substituent effects in electrophilic aromatic substitution (18.6–18.9)

	Substituent	Inductive effect	Resonance effect	Reactivity	Directing effect
[1]	R = alkyl	donating	none	activating	ortho, para
[2]	Z = N or O	withdrawing	donating	activating	ortho, para
[3]	X = halogen	withdrawing	donating	deactivating	ortho, para
[4]	Y (δ+ or +)	withdrawing	withdrawing	deactivating	meta

◆ **Five examples of electrophilic aromatic substitution**

[1] Halogenation–Replacement of H by Cl or Br (18.3)

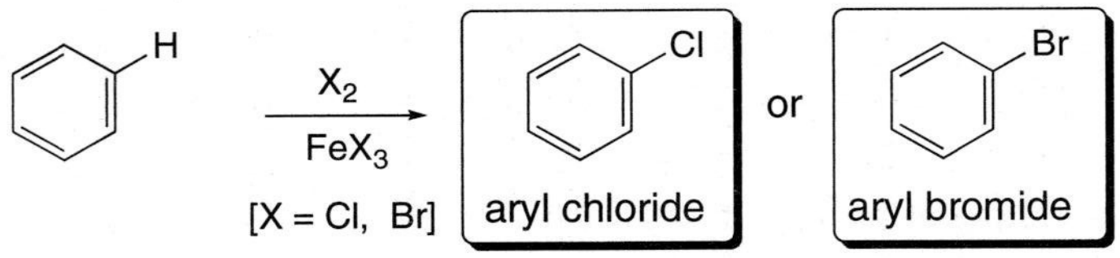

- Polyhalogenation occurs on benzene rings substituted by OH and NH_2 (and related substituents) (18.10A).

[2] Nitration–Replacement of H by NO_2 (18.4)

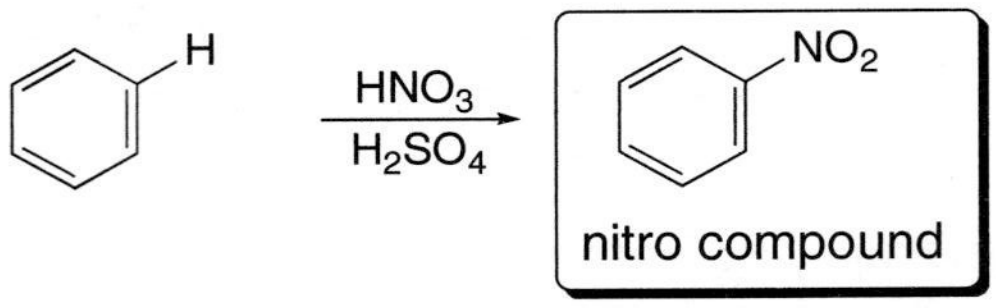

nitro compound

[3] Sulfonation–Replacement of H by SO_3H (18.4)

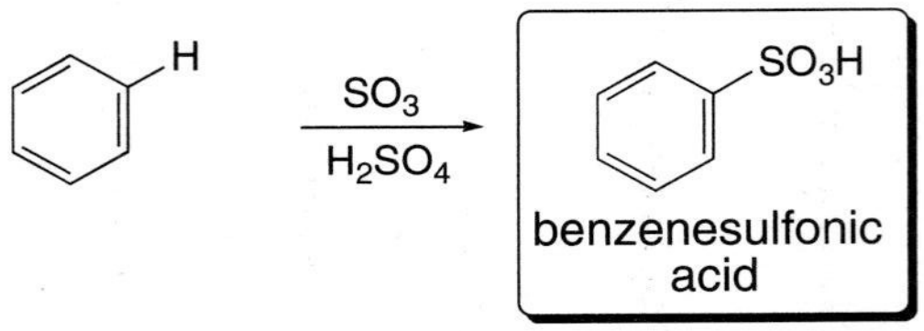

benzenesulfonic acid

[4] Friedel–Crafts alkylation–Replacement of H by R (18.5)

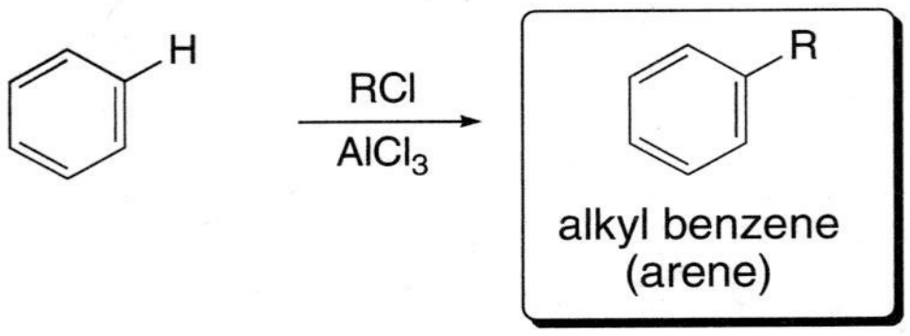

alkyl benzene
(arene)

- Rearrangements can occur.
- Vinyl halides and aryl halides are unreactive.
- The reaction does not occur on benzene rings substituted by meta deactivating groups or NH_2 groups (18.10B).
- Polyalkylation can occur.

Variations:

[1] with alcohols

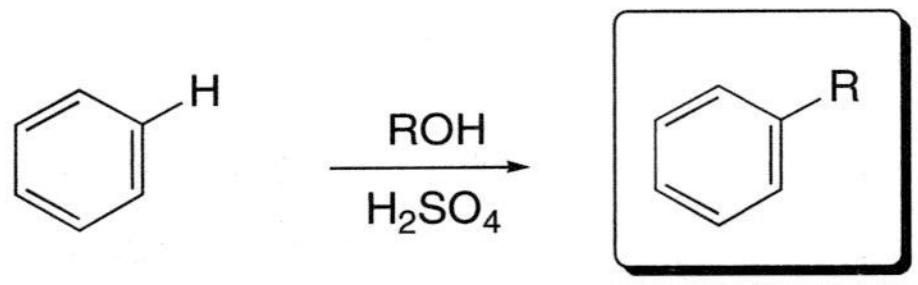

[2] with alkenes

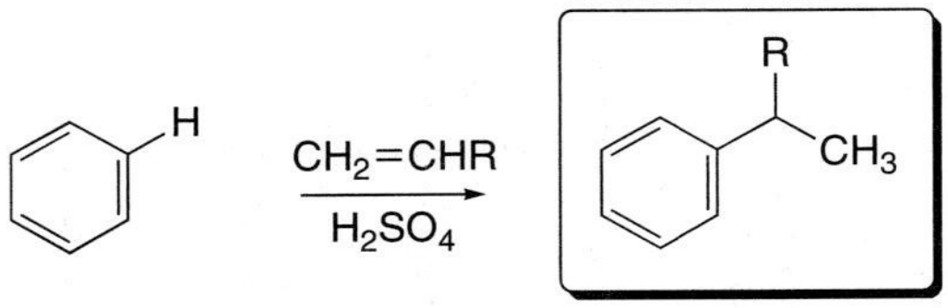

[5] Friedel–Crafts acylation–Replacement of H by RCO (18.5)

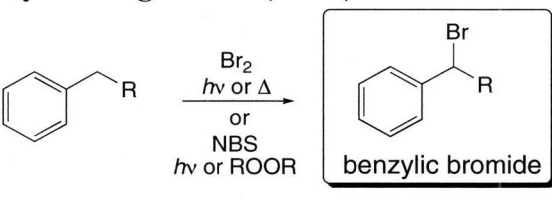

- The reaction does not occur on benzene rings substituted by meta deactivating groups or NH_2 groups (18.10B).

◆ **Other reactions of benzene derivatives**

[1] Benzylic halogenation (18.13)

[2] Oxidation of alkyl benzenes (18.14A)

- A benzylic C–H bond is needed for reaction.

[3] Reduction of ketones to alkyl benzenes (18.14B)

[4] Reduction of nitro groups to amino groups (18.14C)

Chapter 18: Answers to Problems

18.1   The π electrons of benzene are delocalized over the six atoms of the ring, increasing benzene's stability and making them less available for electron donation. With an alkene, the two π electrons are localized between the two C's making them more nucleophilic and thus more reactive with an electrophile than the delocalized electrons in benzene.

18.2

18.3   Reaction with Cl_2 and $FeCl_3$ as the catalyst occurs in two parts. First is the formation of an electrophile, followed by a two-step substitution reaction.

18.4   There are two parts in the mechanism. The first part is formation of an electrophile. The second part is a two-step substitution reaction.

18.5 Friedel–Crafts alkylation results in the transfer of an alkyl group from a halogen to a benzene ring. In Friedel–Crafts acylation an acyl group is transferred from a halogen to a benzene ring.

a.

b.

c.

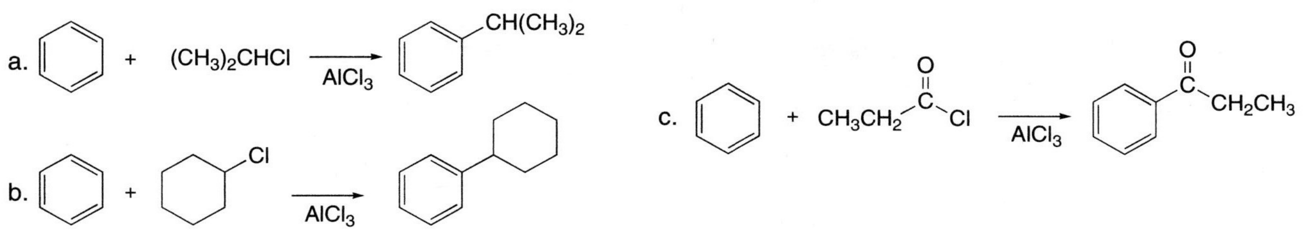

18.6 Remember that an acyl group is transferred from a Cl atom to a benzene ring. To draw the acid chloride, substitute a Cl for the benzene ring.

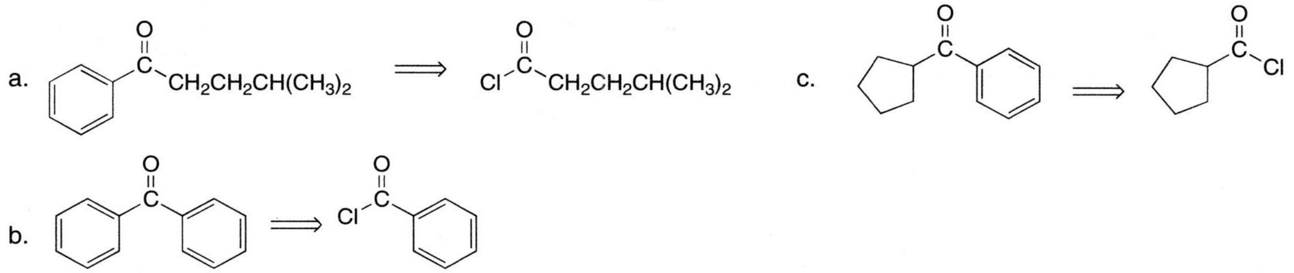

a.

b.

c.

18.7

[1] $CH_3CH_2-\ddot{C}l: + AlCl_3 \longrightarrow CH_3CH_2-\overset{+}{\ddot{C}}l-\overset{-}{AlCl_3}$
electrophile

[2]

resonance-stabilized carbocation $+ AlCl_4^-$

[3] $+ HCl + AlCl_3$

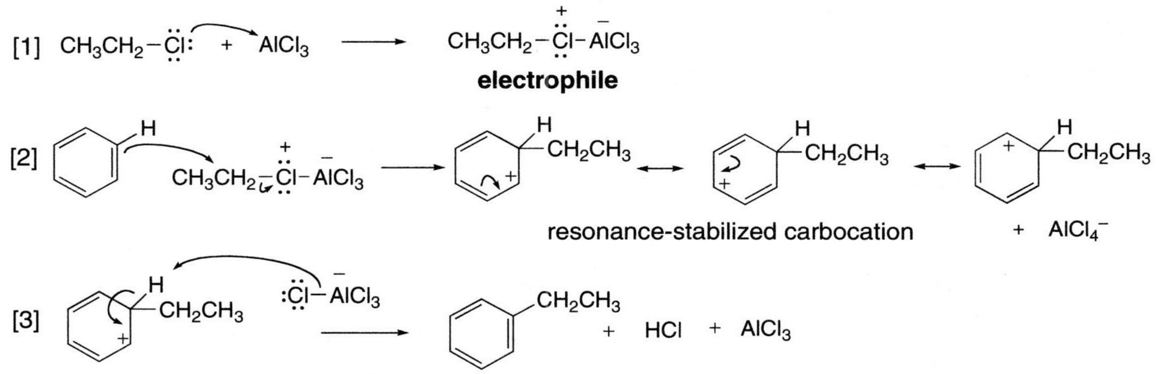

18.8 To be reactive in a Friedel–Crafts alkylation reaction, the X must be bonded to an sp^3 hybridized carbon atom.

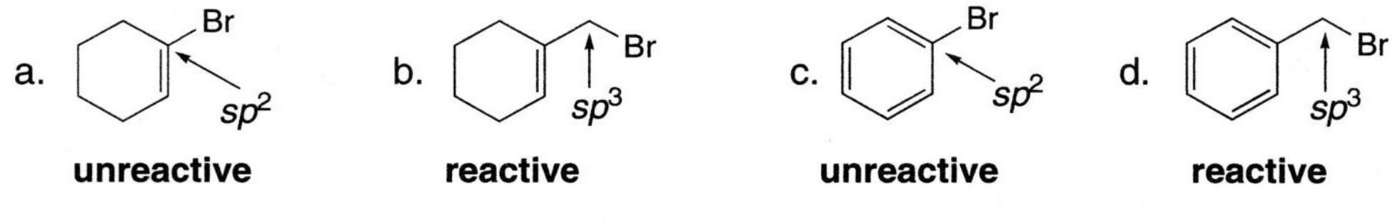

a. sp^2
unreactive

b. sp^3
reactive

c. sp^2
unreactive

d. sp^3
reactive

18.9 The product has an "unexpected" carbon skeleton, so rearrangement must have occurred.

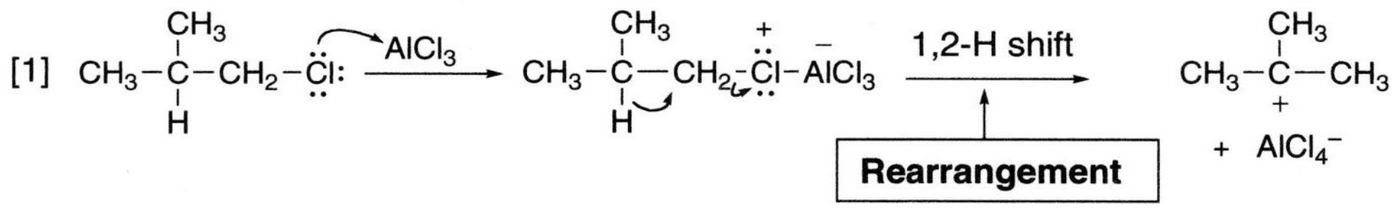

[1]

Rearrangement $+ AlCl_4^-$

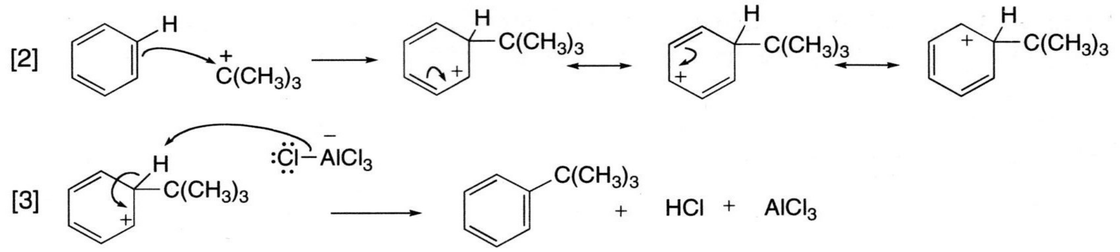

[2]

[3]

18.10 Rearrangements do not occur with acylium ions formed in a Friedel–Craft acylation because the acylium ion is resonance stabilized.

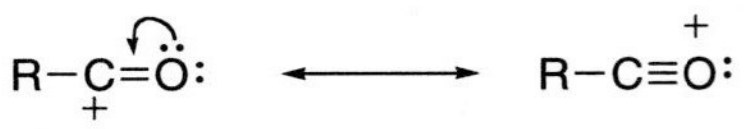

18.11 Both alkenes and alcohols can form carbocations for Friedel–Crafts alkylation reactions.

a.

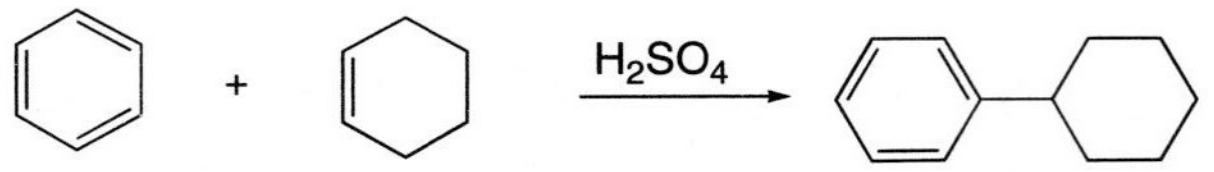

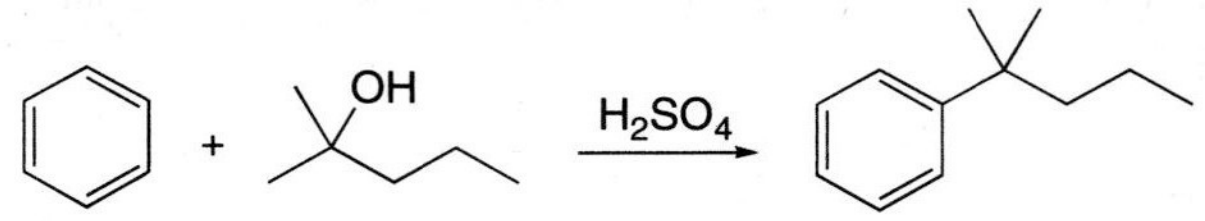

b.

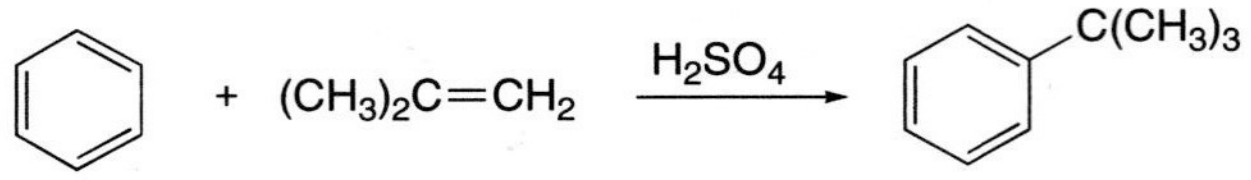

c.

d.

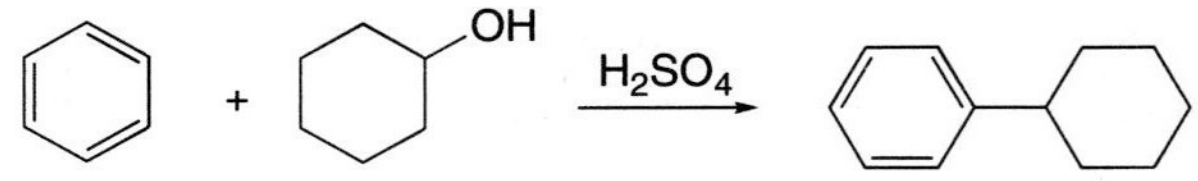

18.12

[1]

[2]

[3]

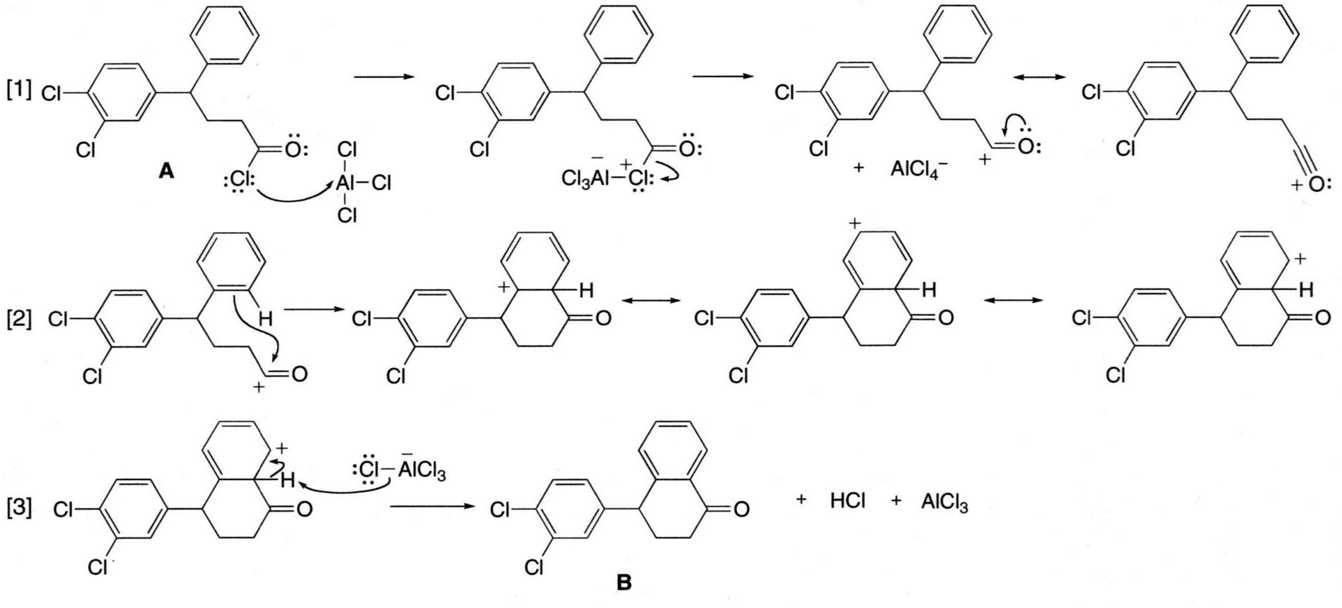

18.13 In parts (b) and (c), a 1,2-shift occurs to afford a rearrangement product.

a.

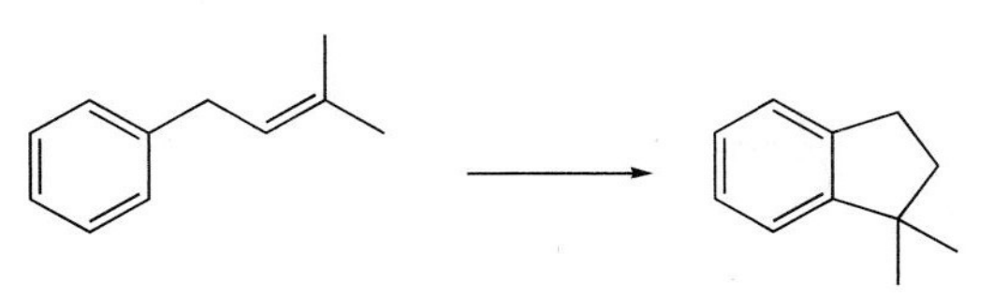

c.

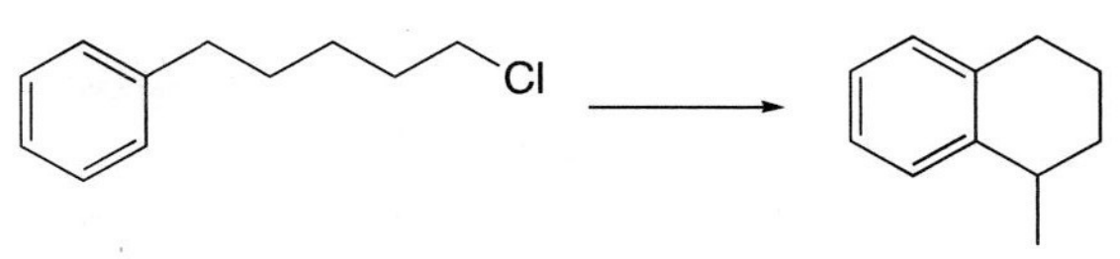

b.

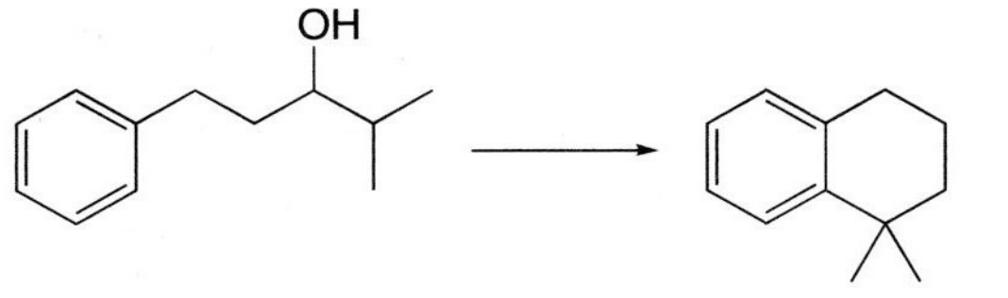

18.14

a. $-CH_2CH_2CH_2CH_3$	b. $-Br$	c. $-OCH_2CH_3$
alkyl group	halide	electronegative O
electron donating	**electron withdrawing**	**electron withdrawing**

18.15 Electron-donating groups place a negative charge in the benzene ring. Draw the resonance structures to show how $-OCH_3$ puts a negative charge in the ring. Electron-withdrawing groups place a positive charge in the benzene ring. Draw the resonance structures to show how $-COCH_3$ puts a positive charge in the ring.

a.

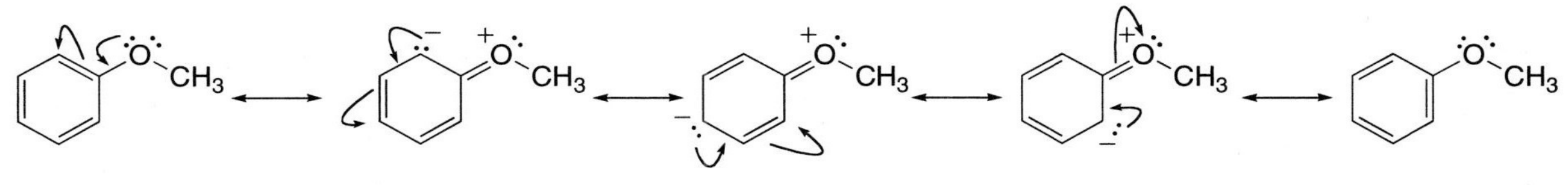

b.

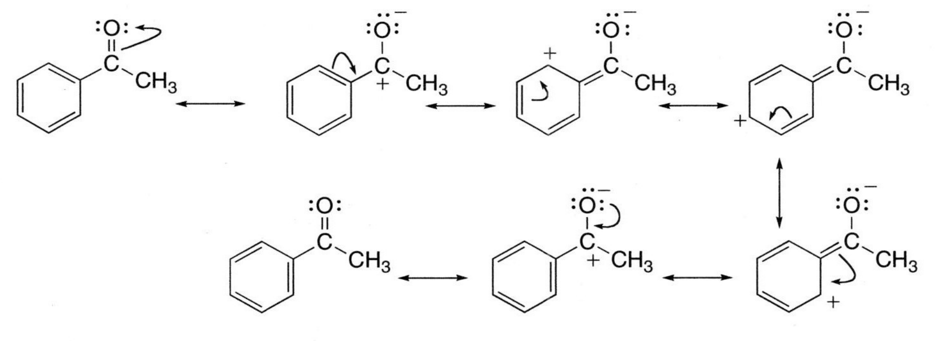

18.16 To classify each substituent, look at the atom bonded directly to the benzene ring. All R groups and Z groups (except halogens) are electron donating. All groups with a positive charge, δ^+, or halogens are electron withdrawing.

a.

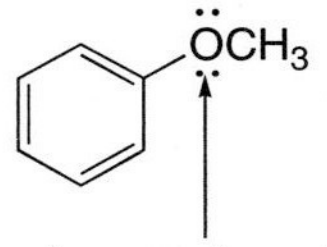

lone pair on O
electron donating

b.

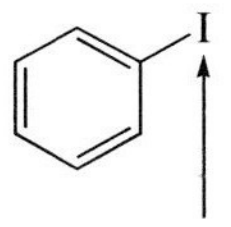

halogen
electron withdrawing

c.

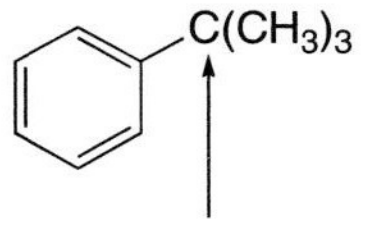

R group
electron donating

18.17 First classify the substituents in the starting material as: ortho, para activating, ortho, para deactivating, or meta deactivating. Then draw the products.

a.

lone pair on O
o,p activating

ortho product   **para product**

b.

halogen
o,p deactivating

ortho product   **para product**

c.

meta deactivating   **meta product**

18.18 **Electron-donating groups** make the compound *react faster* than benzene in electrophilic aromatic substitution. **Electron-withdrawing groups** make the compound *react more slowly* than benzene in electrophilic aromatic substitution.

a.

electron withdrawing
reacts slower

b.

electron withdrawing
reacts slower

c.

lone pairs on O
electron donating
reacts faster

d.

halogen
electron withdrawing
reacts slower

e.

R group
electron donating
reacts faster

18.19 **Electron-donating groups** make the compound *more reactive* than benzene in electrophilic aromatic substitution. **Electron-withdrawing groups** make the compound *less reactive* than benzene in electrophilic aromatic substitution.

a.

R group
electron donating
more reactive

b.

two OH's
electron donating
more reactive

c.

C with 2 electronegative O's
electron withdrawing
less reactive

d.

electron withdrawing
less reactive

18.20

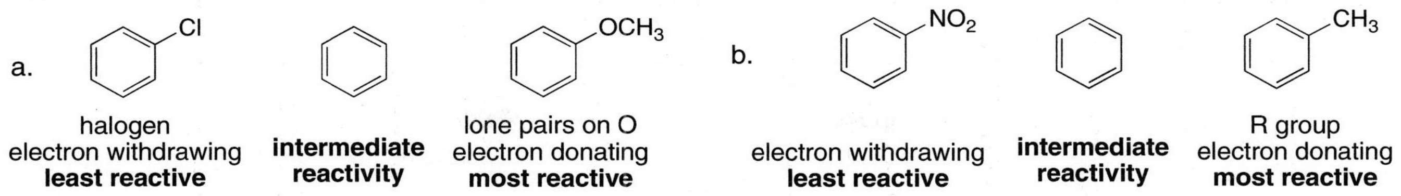
a.
Cl
halogen
electron withdrawing
least reactive
intermediate reactivity
OCH3
lone pairs on O
electron donating
most reactive
b.
NO2
electron withdrawing
least reactive
intermediate reactivity
CH3
R group
electron donating
most reactive

18.21 Especially stable resonance structures have all atoms with an octet. Carbocations with additional electron donor R groups are also more stable structures. Especially unstable resonance structures have adjacent like charges.

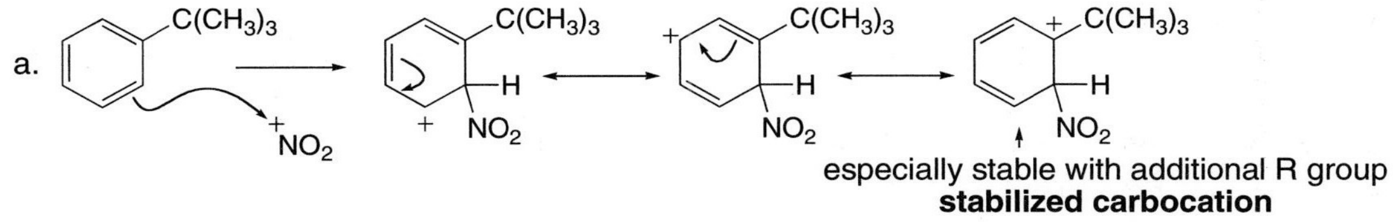
a.
C(CH3)3
+NO2
H
NO2
especially stable with additional R group
stabilized carbocation

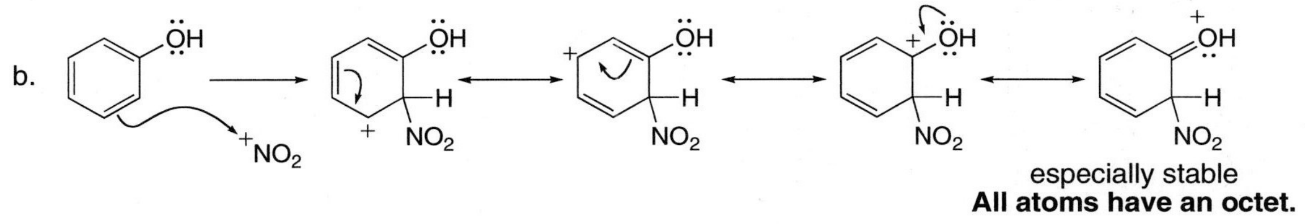
b.
OH
+NO2
H
NO2
especially stable
All atoms have an octet.

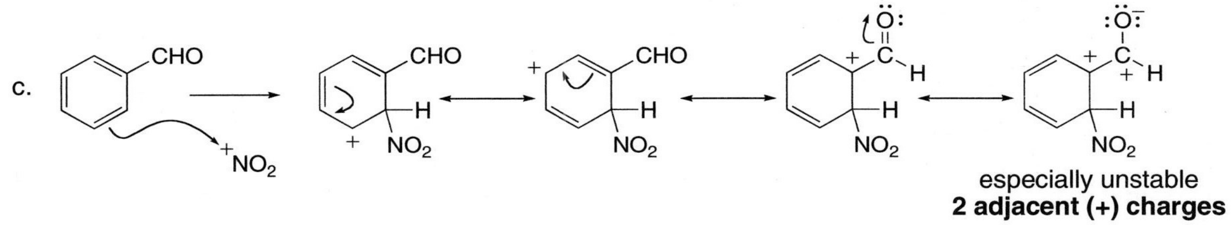
c.
CHO
+NO2
H
NO2
especially unstable
2 adjacent (+) charges

18.22

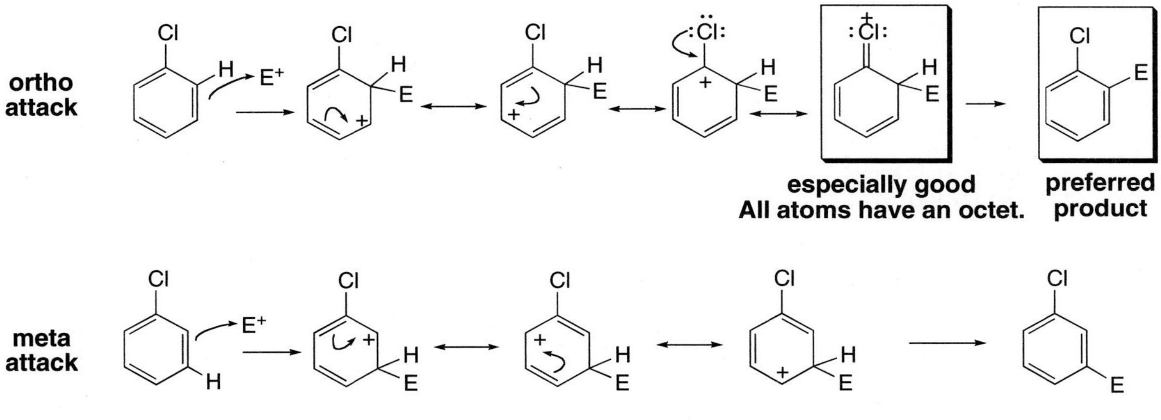
ortho attack
Cl
H
E+
E
especially good
All atoms have an octet.
preferred product
meta attack

**especially good
All atoms have an octet.**

**preferred
product**

18.23 Polyhalogenation occurs with highly activated benzene rings containing OH, NH$_2$, and related groups with a catalyst.

18.24 Friedel–Crafts reactions do not occur with strongly deactivating substituents including NO$_2$, or with NH$_2$, NR$_2$, or NHR groups.

18.25 To draw the product of reaction with these disubstituted benzene derivatives and HNO$_3$, H$_2$SO$_4$ remember:

- If the two directing effects reinforce each other, the new substituent will be on the position reinforced by both.
- If the directing effects oppose each other, the stronger activator wins.
- No substitution occurs between two meta substituents.

a.

o,p (arrow pointing to OCH₃)

4-methoxy benzoate reacting with HNO_3 / H_2SO_4 to give nitro product

meta (pointing to COOCH₃)

b.

o,p (strong) (pointing to OCH₃)

HNO_3 / H_2SO_4 giving two products

o,p
oppose
products due to
OCH₃ directing effects

c.

o,p (pointing to CH₃)

meta (pointing to NO₂)

HNO_3 / H_2SO_4 giving two products

d.

o,p (pointing to Cl and Br)

HNO_3 / H_2SO_4 giving two products

18.26

a.

benzene → SO_3, H_2SO_4 → benzenesulfonic acid → $Cl_2, FeCl_3$ → 3-chlorobenzenesulfonic acid

Put meta director on first.

b.

benzene → $ClCOCH_3$ / $AlCl_3$ → acetophenone → HNO_3, H_2SO_4 → 3-nitroacetophenone

c.

phenol → $CH_3Cl, AlCl_3$ → p-cresol (+ ortho isomer) → Br_2 → 2-bromo-4-methylphenol

Br goes ortho to
the stronger activator.

18.27

This reaction proceeds via a radical bromination mechanism and two radicals are possible: **A** (2°
and benzylic) and **B** (1°). Since **B** (which leads to $C_6H_5CH_2CH_2Br$) is much less stable, this
radical is not formed so only $C_6H_5CH(Br)CH_3$ is formed as product.

ethylbenzene → **A** (·CHCH₃) **2° and benzylic** or **B** (CH₂CH₂·) **1°**

A → (Br)CHCH₃ only product

B → CH₂CH₂Br not formed

18.28 **Radical substitution** occurs at the carbon adjacent to the benzene ring (at the **benzylic position**).

a.

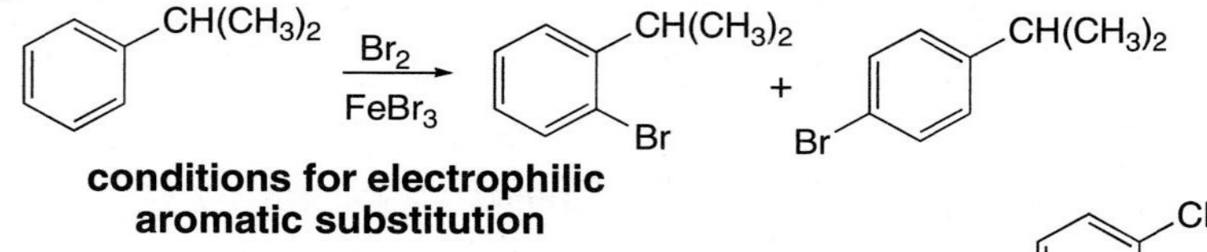

conditions for electrophilic aromatic substitution

b.

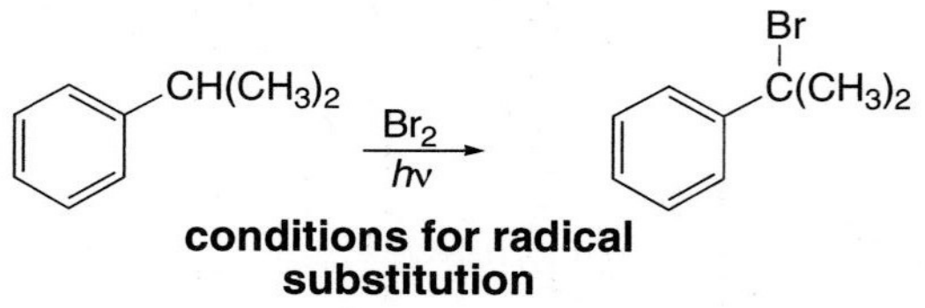

conditions for radical substitution

c.

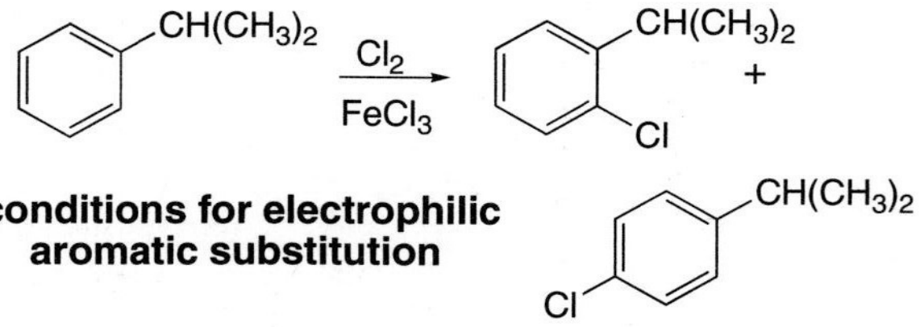

conditions for electrophilic aromatic substitution

18.29

a.

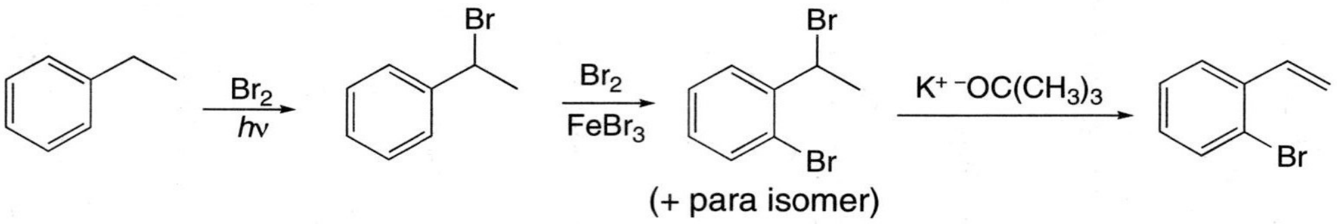

(+ para isomer)

b.

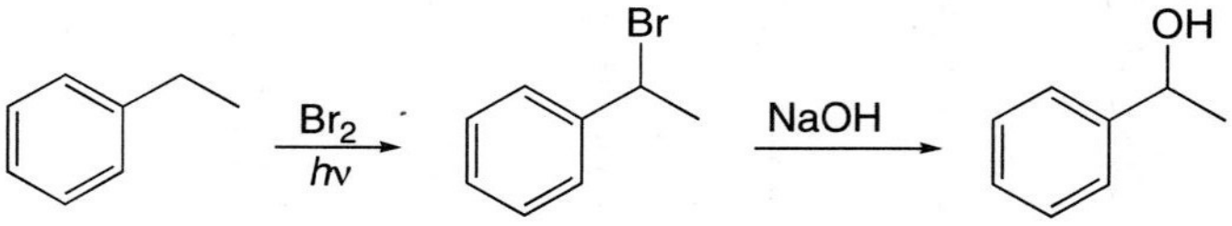

c.

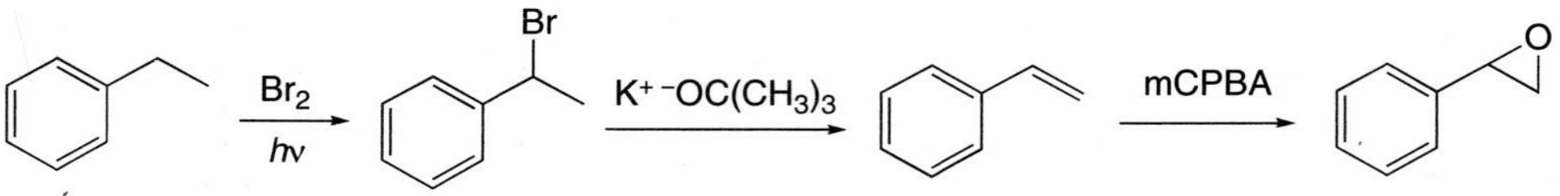

d.

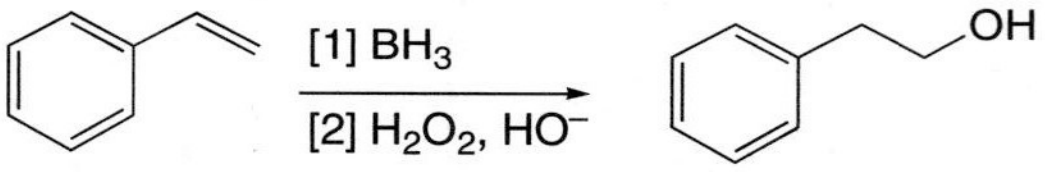

(from c.)

18.30 First use an acylation reaction, and then reduce the carbonyl group to form the alkyl benzenes.

a.

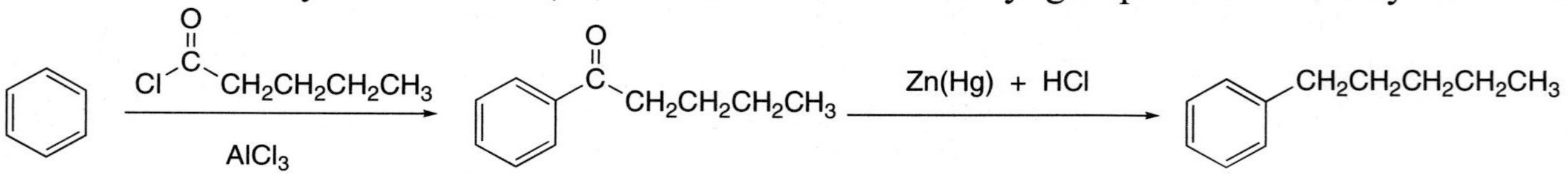

b.

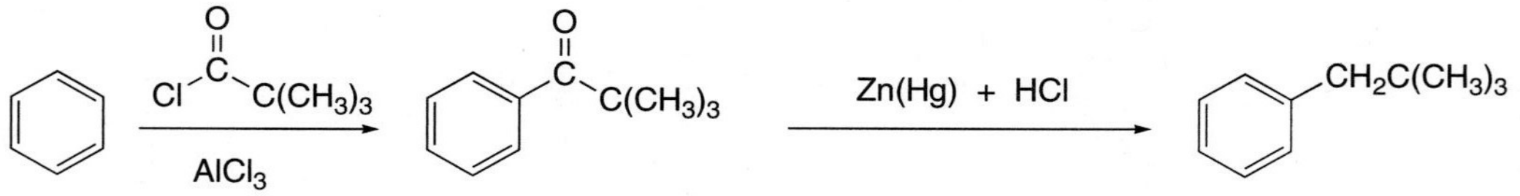

18.31

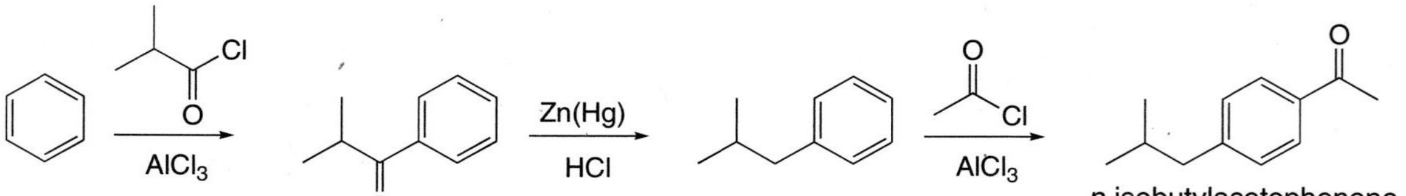

p-isobutylacetophenone
(+ ortho isomer)

18.32 When R = CH$_3$ in C$_6$H$_5$CH$_2$R, the product can be made by two different Friedel–Crafts reactions.

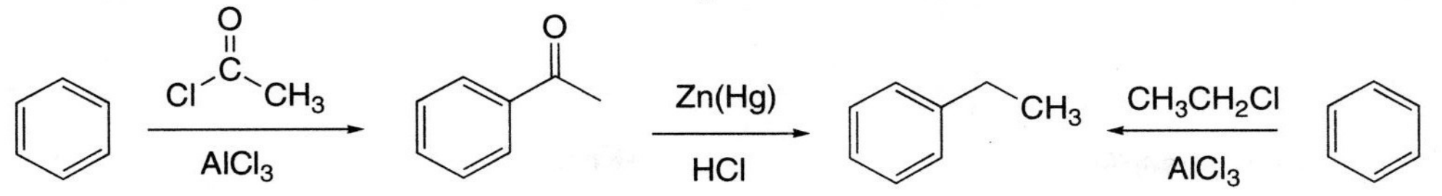

18.33

a.

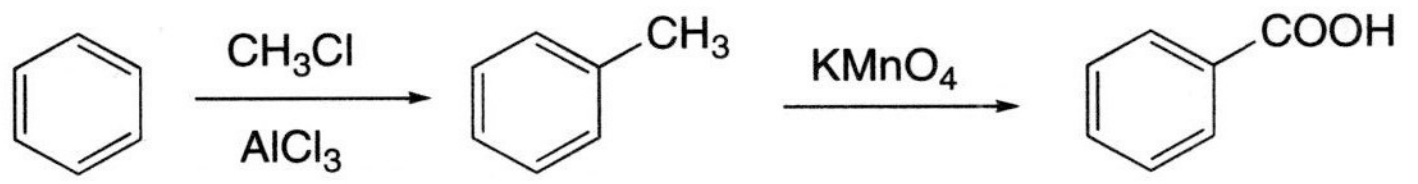

b.

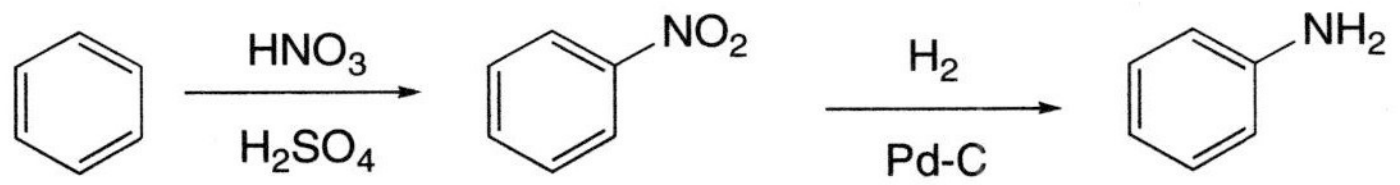

c.

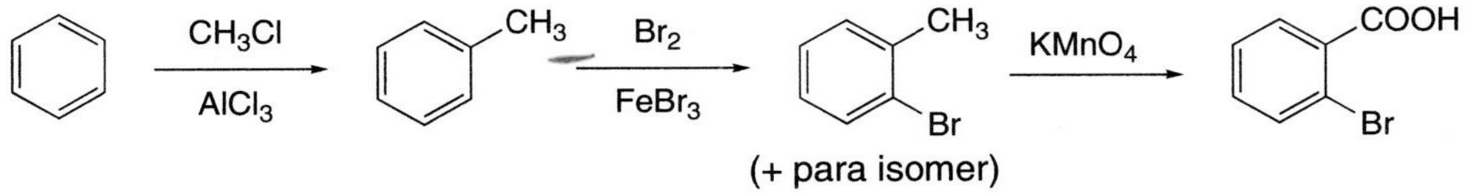

(+ para isomer)

18.34

a.

(+ ortho isomer)

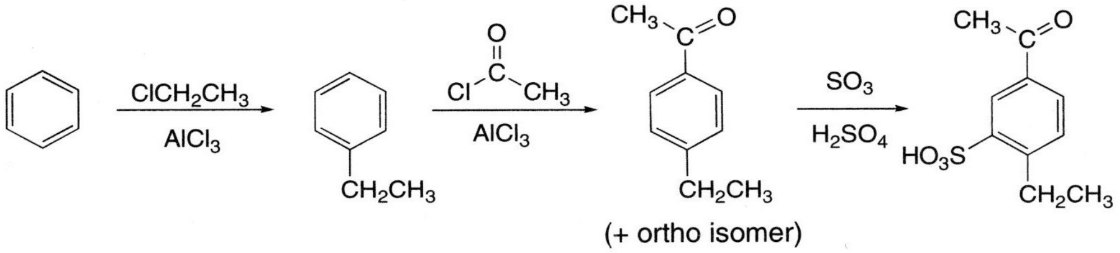

b.

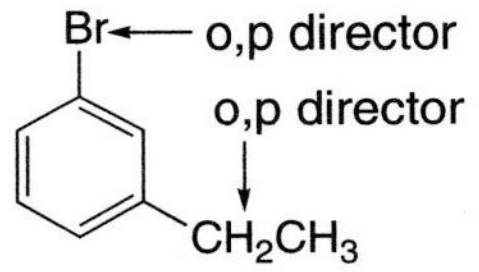

Both are o,p directors, but they are meta to each other. The alkyl group must be obtained by reduction of a carbonyl.

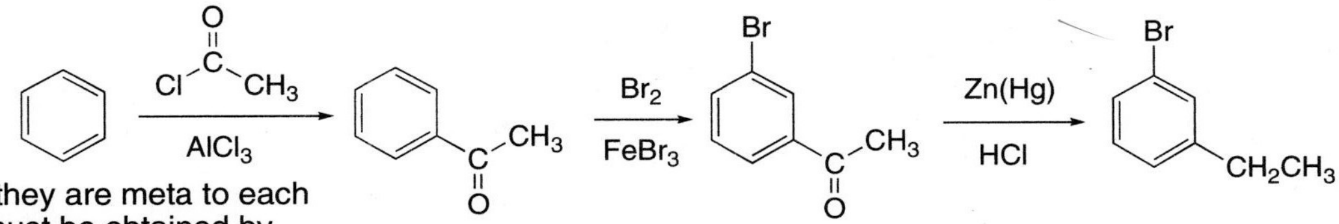

c.

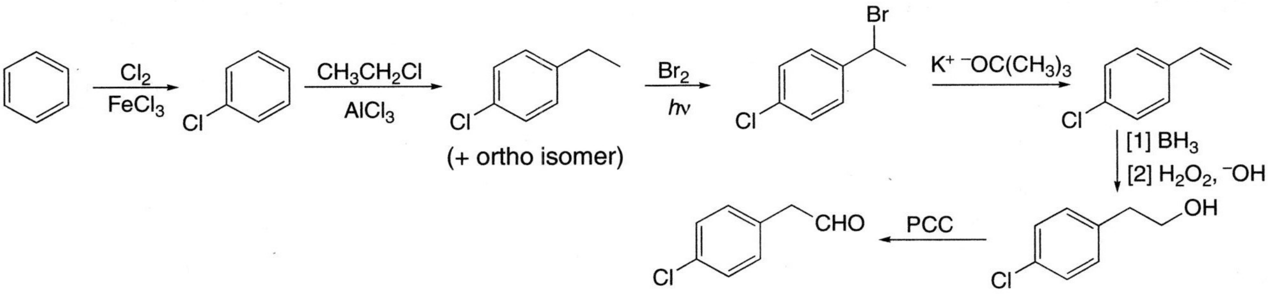

(+ ortho isomer)

18.35 OH is an ortho, para director.

a.

b.

c.

d.

e.

f.

g.

h. (+ ortho isomer) $\xrightarrow[\text{HCl}]{\text{Sn}}$ (+ ortho isomer)

i. (+ ortho isomer) $\xrightarrow[\text{HCl}]{\text{Zn(Hg)}}$ (+ ortho isomer)

j. (+ ortho isomer) $\xrightarrow[\text{}^-\text{OH}]{\text{NH}_2\text{NH}_2}$ (+ ortho isomer)

k. (+ ortho isomer) $\xrightarrow[h\nu]{\text{Br}_2}$ (+ ortho isomer)

l. (+ ortho isomer) $\xrightarrow{\text{KMnO}_4}$ (+ ortho isomer)

18.36 CN is a meta director that deactivates the benzene ring.

a.

b.

c.

d. $\xrightarrow[\text{AlCl}_3]{\text{CH}_3\text{CH}_2\text{CH}_2\text{Cl}}$ No reaction

e. $\xrightarrow[\text{AlCl}_3]{\text{CH}_3\text{COCl}}$ No reaction

18.37

a.

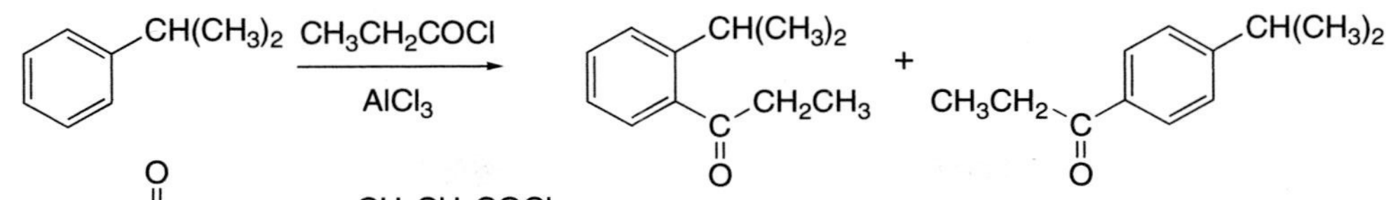

b.

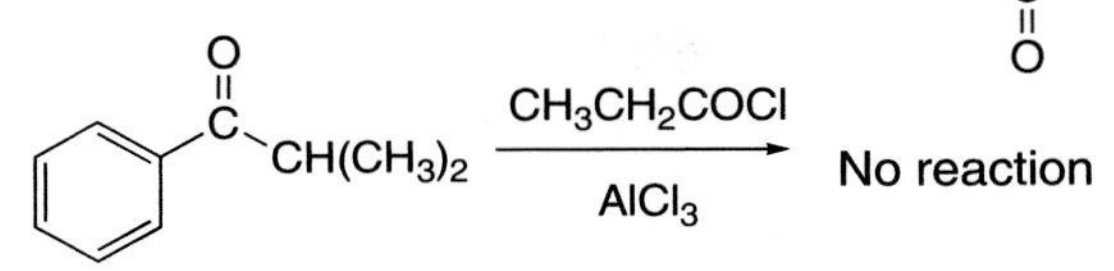

No reaction

c.

No Friedel–Crafts reaction

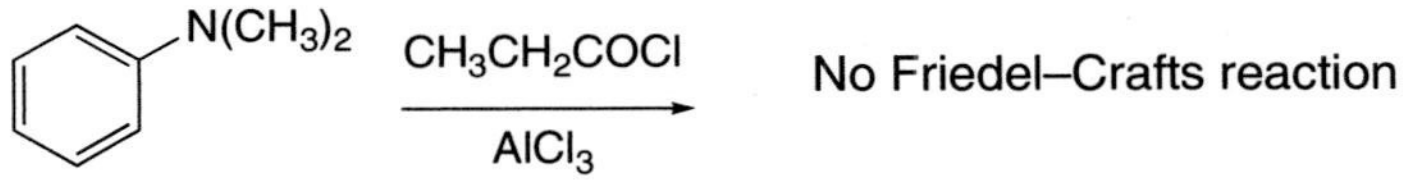

d.

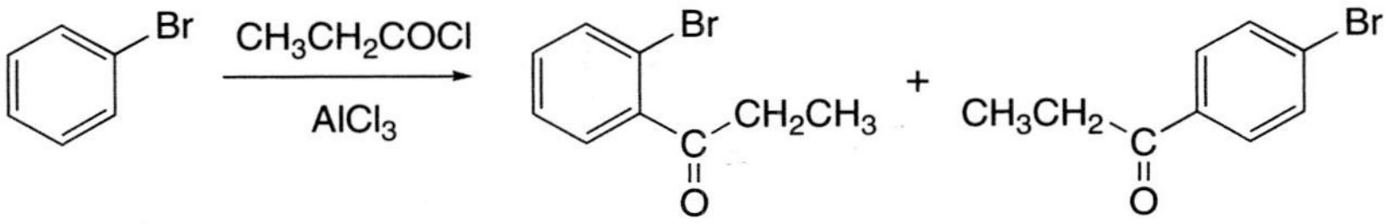

e.

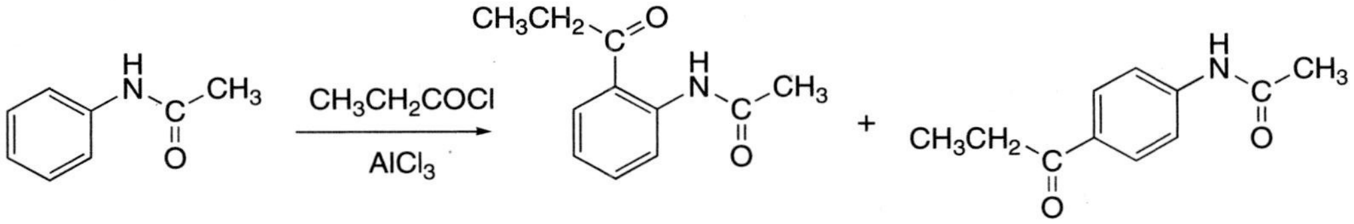

18.38

a.

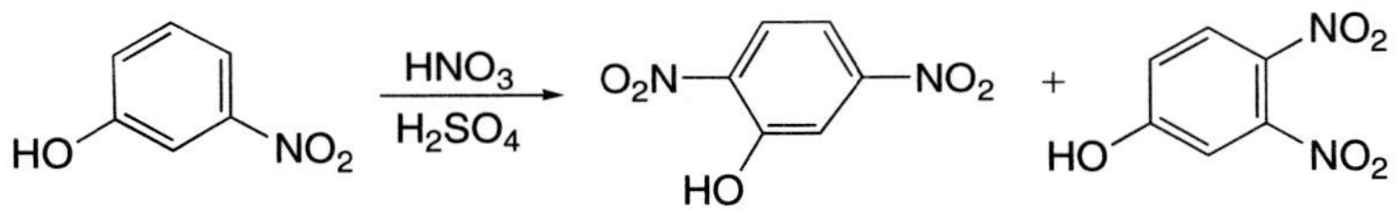

b.

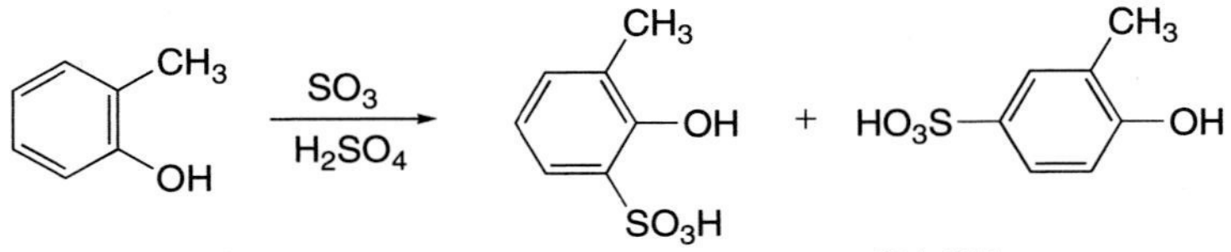

c.

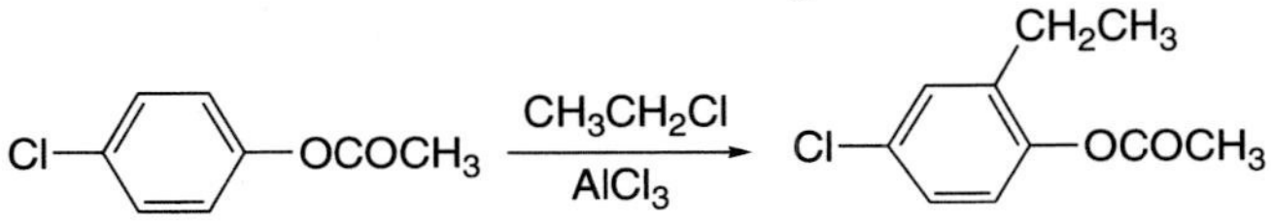

d.

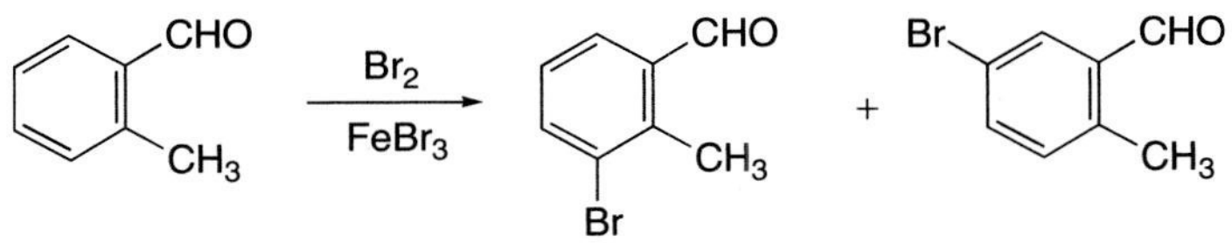

e.

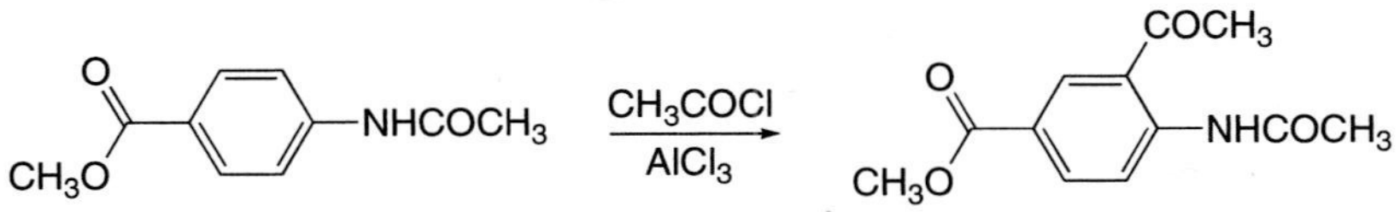

f.

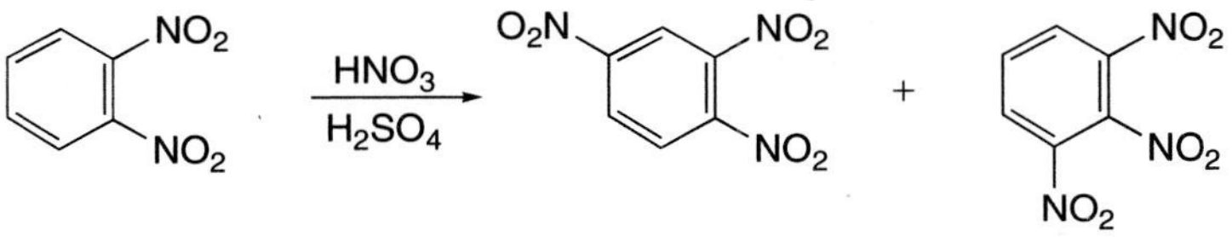

g.

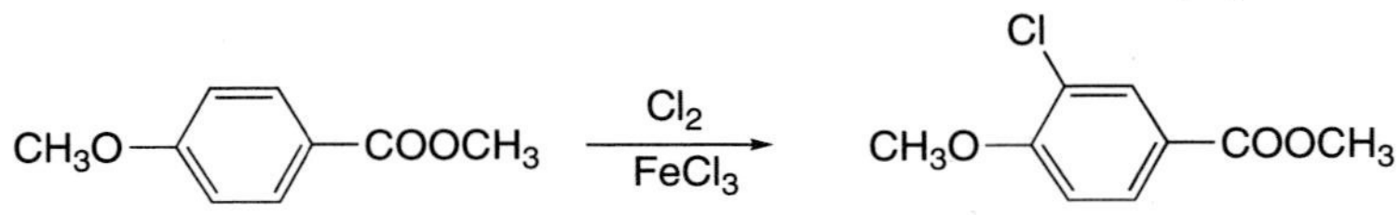

h.

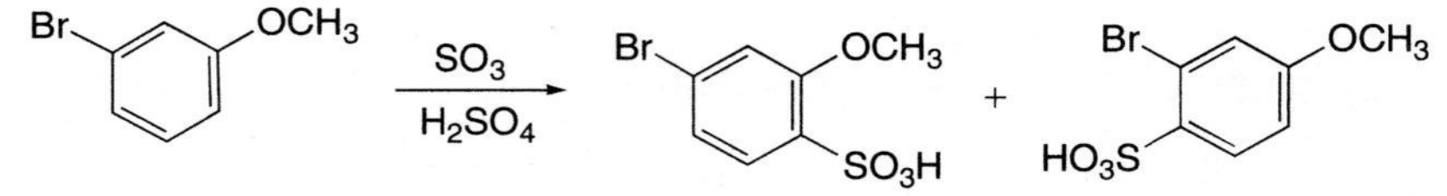

18.39 Watch out for rearrangements.

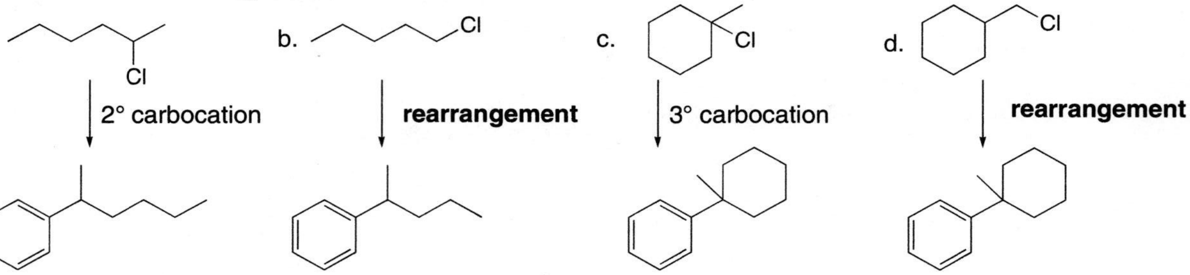

a.

2° carbocation

b.

rearrangement

c.

3° carbocation

d.

rearrangement

18.40

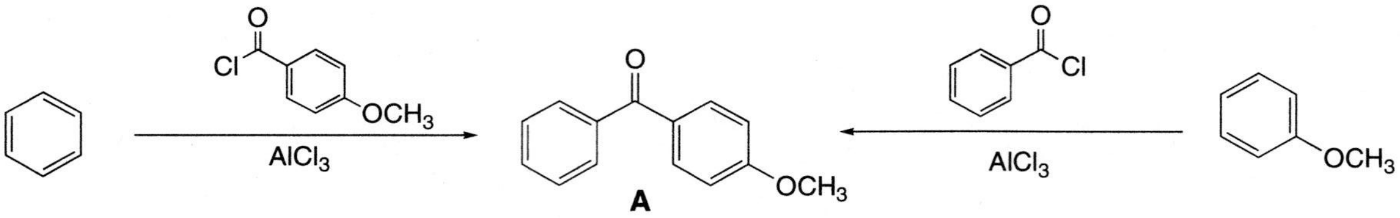

18.41

a.

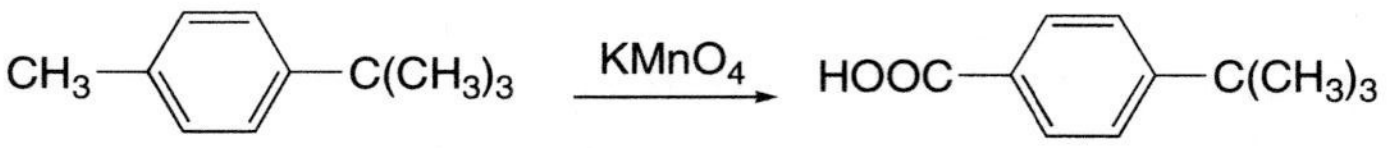

b.

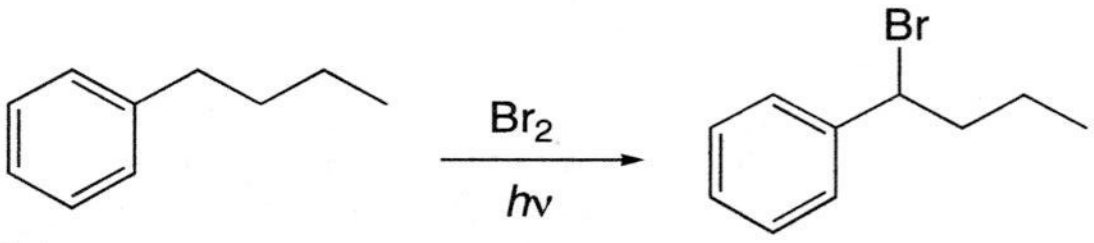

c.

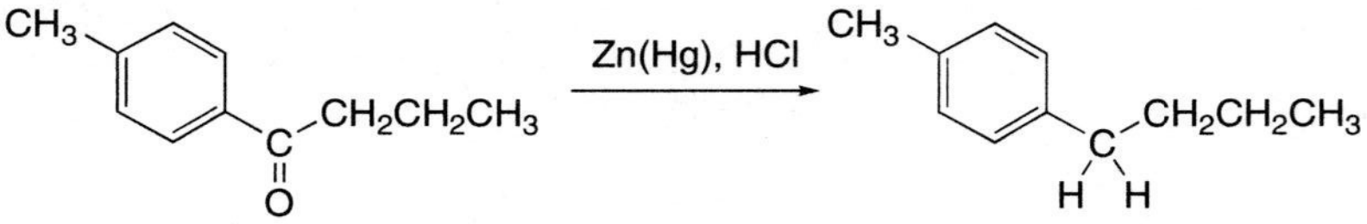

d.

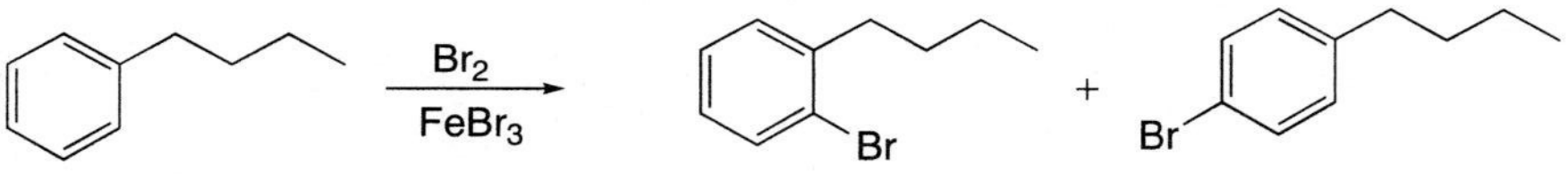

e.

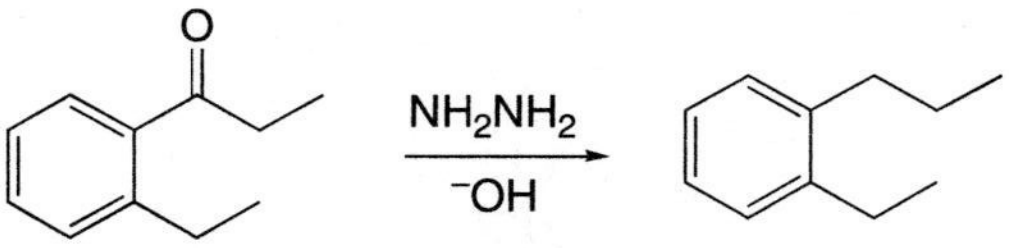

f.

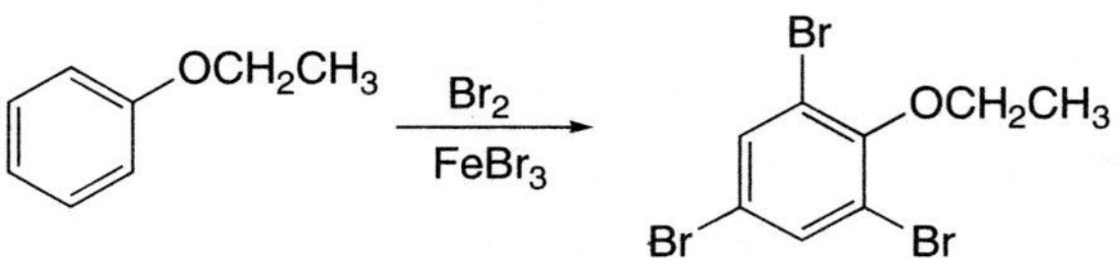

18.42

C bonded to 2 H's must use acylation followed by reduction.

a.

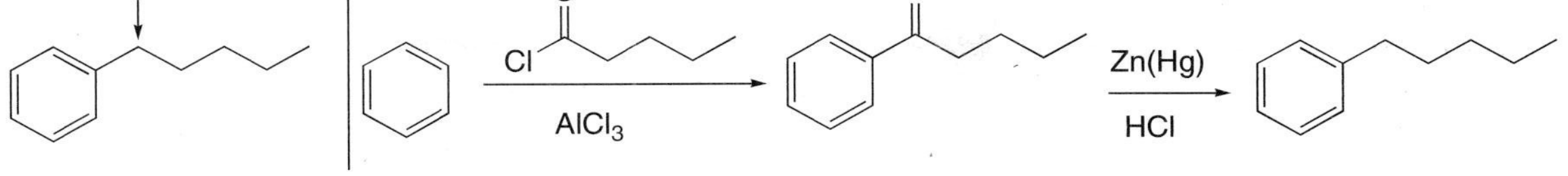
O
Cl
AlCl$_3$
O
Zn(Hg)
HCl

C bonded to 1 H can be added directly by alkylation.

b.

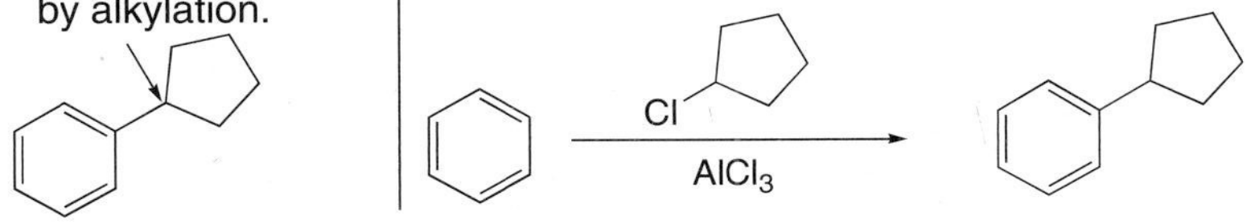
Cl
AlCl$_3$

c.

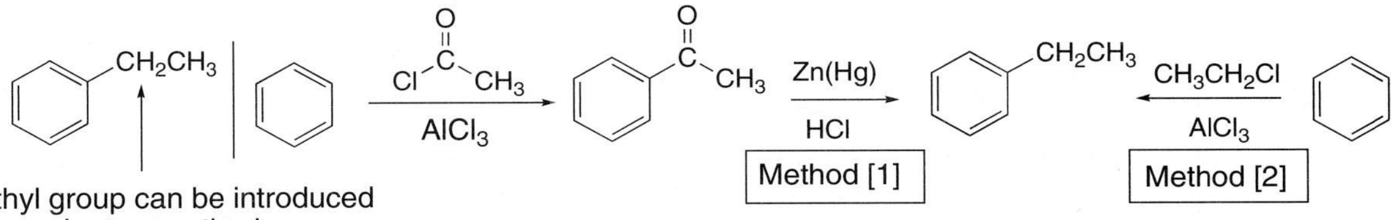
CH$_2$CH$_3$
O
Cl C CH$_3$
AlCl$_3$
O
C CH$_3$
Zn(Hg)
HCl
Method [1]
CH$_2$CH$_3$
CH$_3$CH$_2$Cl
AlCl$_3$
Method [2]

Ethyl group can be introduced by two methods.

d.

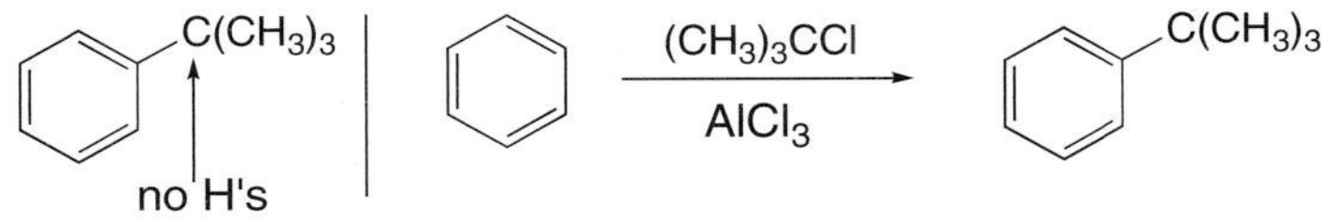
C(CH$_3$)$_3$
no H's
(CH$_3$)$_3$CCl
AlCl$_3$
C(CH$_3$)$_3$

18.43

a.

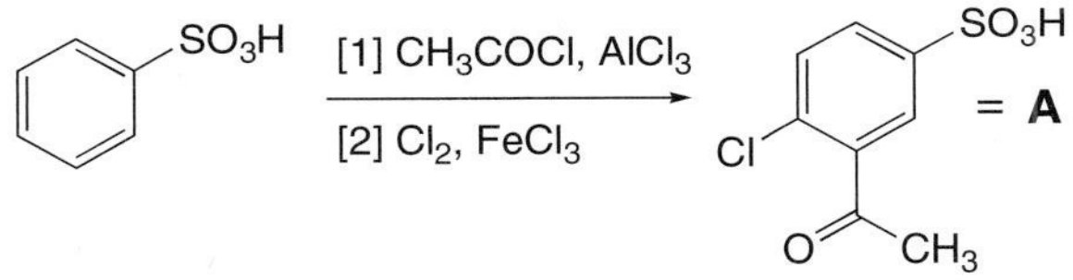
SO$_3$H
[1] CH$_3$COCl, AlCl$_3$
[2] Cl$_2$, FeCl$_3$
SO$_3$H
Cl
O CH$_3$
= **A**

Step [1] won't work because a Friedel–Crafts reaction can't be done on a deactivated benzene ring, as is the case with the SO_3H substituent. Even if Step [1] did work, the second step would introduce Cl meta to SO_3H, not para as drawn.

Alternate synthesis:

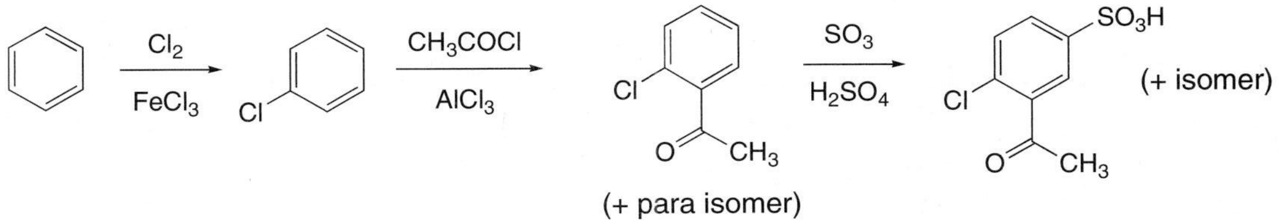
Cl$_2$
FeCl$_3$
Cl
CH$_3$COCl
AlCl$_3$
Cl
O CH$_3$
(+ para isomer)
SO$_3$
H$_2$SO$_4$
SO$_3$H
Cl
O CH$_3$
(+ isomer)

b.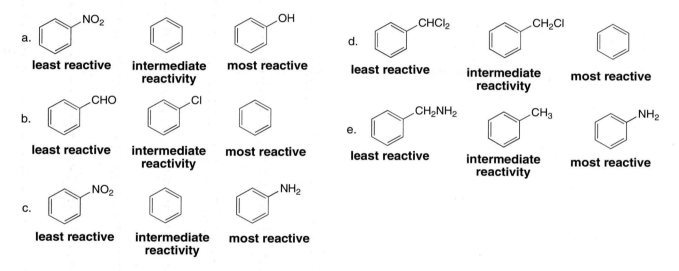

Benzene ring with OCH₃   [1] CH₃CH₂CH₂CH₂Cl, AlCl₃
                          ————————————————————————→   Product ring with OCH₃, CH₃CH₂CH₂CH₂, NO₂   = **B**
                          [2] HNO₃, H₂SO₄

Step [1] involves a Friedel–Crafts alkylation using a 1° alkyl halide that will undergo rearrangement, so that a butyl group will not be introduced as a side chain.

Alternate synthesis:

Phenol OH   [1] NaH
            ——————→   anisole OCH₃   CH₃(CH₂)₂COCl
            [2] CH₃Cl   ——————————→   CH₃CH₂CH₂–C(=O)– ring –OCH₃   (+ ortho isomer)
                                     AlCl₃

Zn(Hg), HCl

CH₃CH₂CH₂CH₂– ring –OCH₃

HNO₃
—————
H₂SO₄

CH₃CH₂CH₂CH₂– ring with OCH₃, NO₂

B

18.44 Use the directions from Answer 18.19 to rank the compounds.

a.   NO₂   benzene   OH
     least reactive   intermediate reactivity   most reactive

d.   CHCl₂   CH₂Cl   benzene
     least reactive   intermediate reactivity   most reactive

b.   CHO   Cl   benzene
     least reactive   intermediate reactivity   most reactive

e.   CH₂NH₂   CH₃   NH₂
     least reactive   intermediate reactivity   most reactive

c.   NO₂   benzene   NH₂
     least reactive   intermediate reactivity   most reactive

18.45 Electron-withdrawing groups place a positive charge in the benzene ring. Draw the resonance structures to show how NO₂ puts a positive charge in the ring, giving it an electron-withdrawing resonance effect. Electron-donating groups place a negative charge in the benzene ring. Draw the resonance structures to show how F puts a negative charge in the ring, giving it an electron-donating resonance effect.

a.

b.

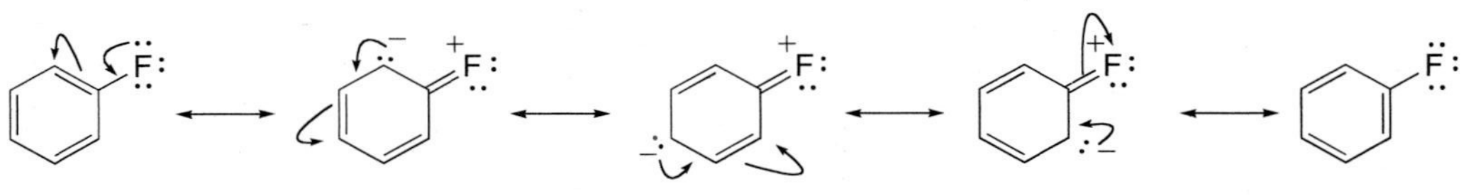

18.46

[1]

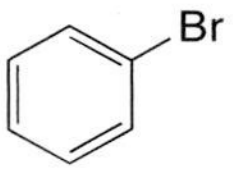

a. withdraw
b. donate
c. less
d. deactivate

[2]

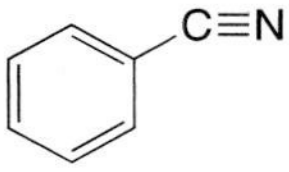

a. withdraw
b. withdraw
c. less
d. deactivate

[3]

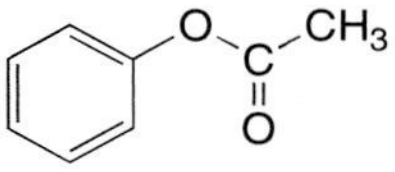

a. withdraw
b. donate
c. more
d. activate

18.47

a.

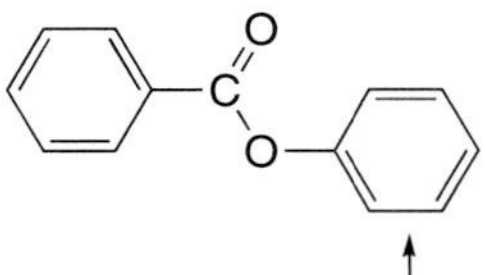

more electron rich
due to O atom
more reactive

b.

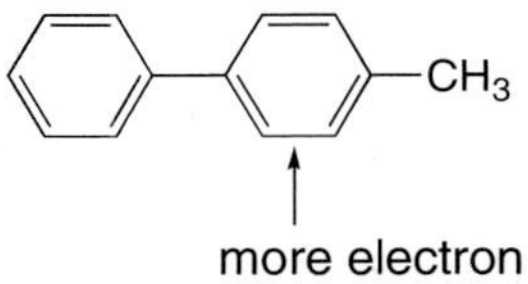

more electron rich
due to CH₃ group
more reactive

c.

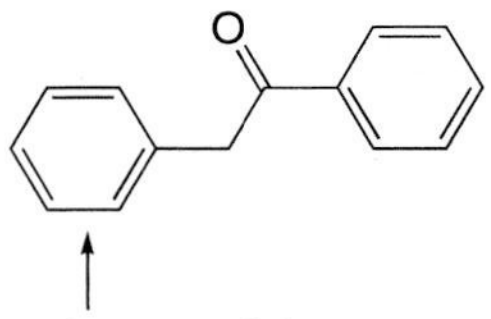

more electron rich
due to C atom
more reactive

18.48

a.

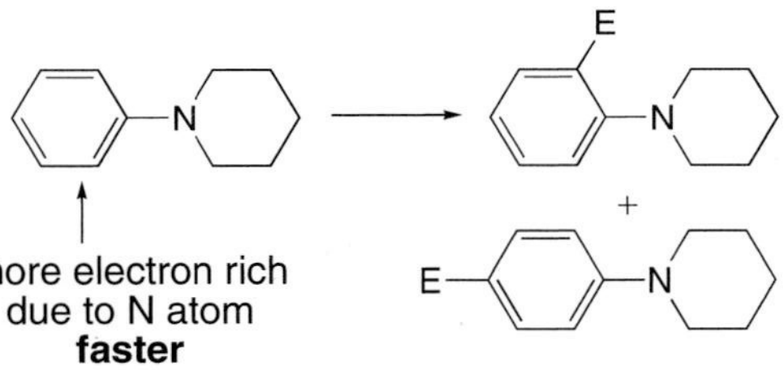

more electron rich
due to N atom
faster

b.

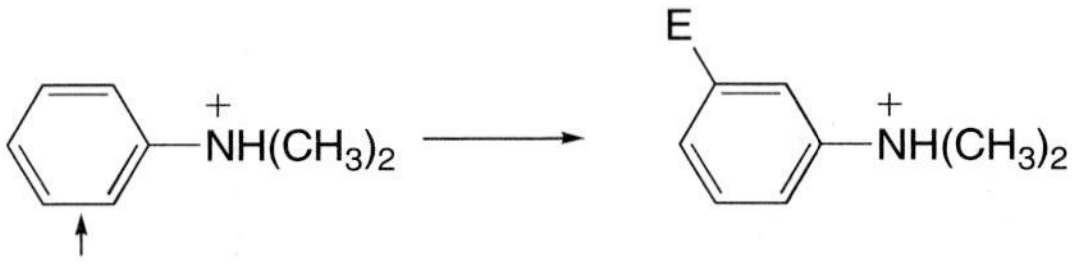

less electron rich
due to (+) charge on N
slower

c.

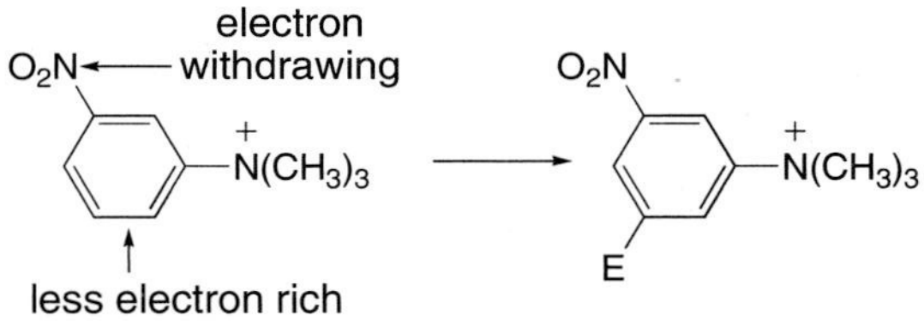

less electron rich
due to (+) charge on N and
electron-withdrawing NO₂ group
slower

d.

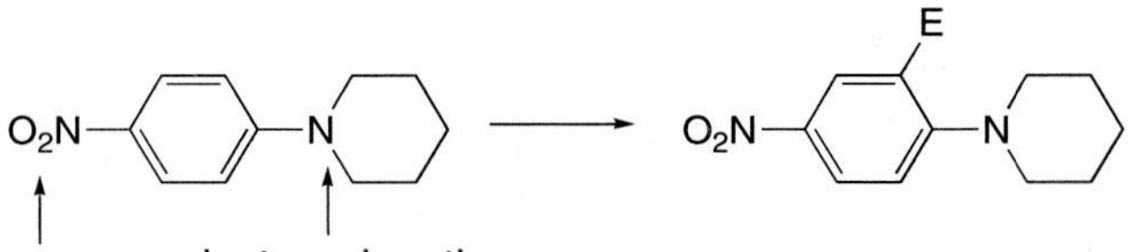

electron   electron donating
withdrawing

Effects cancel out.
similar in reactivity to benzene

18.49

a.

A benzene ring is an electron-rich substituent that stabilizes an intermediate positive charge by an electron-donating resonance effect. As a result, it activates a benzene ring toward reaction with electrophiles.

ortho, para director

Ortho and para products are isolated.

With ortho and para attack there is additional resonance stabilization that delocalizes the positive charge onto the second benzene ring. Such additional stabilization is not possible with meta attack.

Ortho attack:

Meta attack:

Para attack:

b.

ortho, para director

Ortho and para products are isolated.

With ortho and para attack there is additional resonance stabilization that delocalizes the positive charge onto the nitroso group. Such additional stabilization is not possible with meta attack. This makes –NO an ortho, para director. Since the N atom bears a partial (+) charge (because it is bonded to a more electronegative O atom), the –NO group inductively withdraws electron density, thus deactivating the benzene ring towards electrophilic attack. In this way, the –NO group resembles the halogens. Thus, the electron-donating resonance effect makes –NO an o,p director, but the electron-withdrawing inductive effect makes it a deactivator.

Ortho attack:

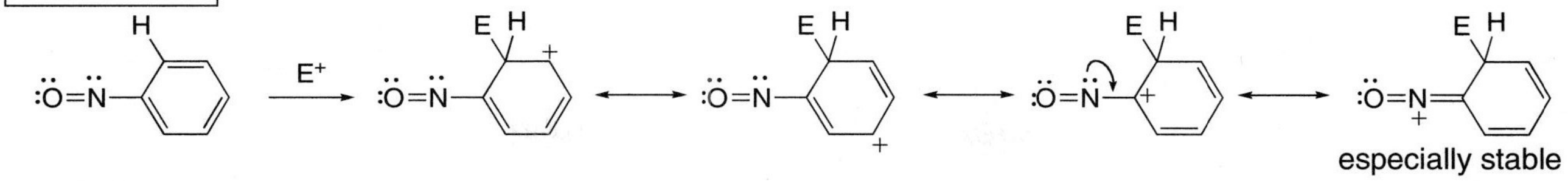

especially stable

Meta attack:

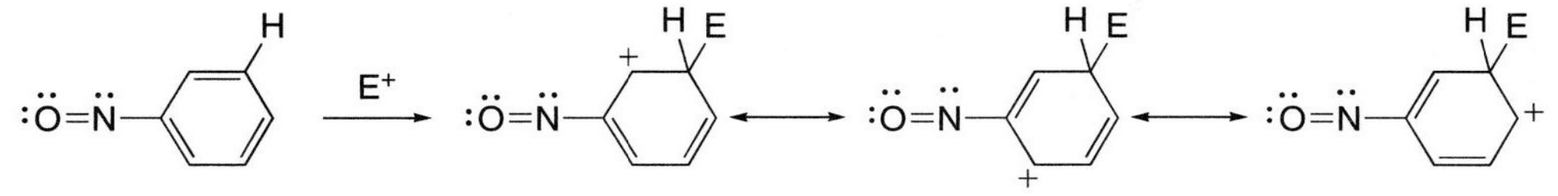

Para attack:

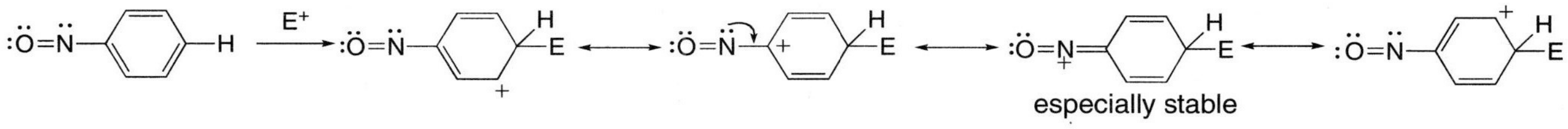

especially stable

18.50

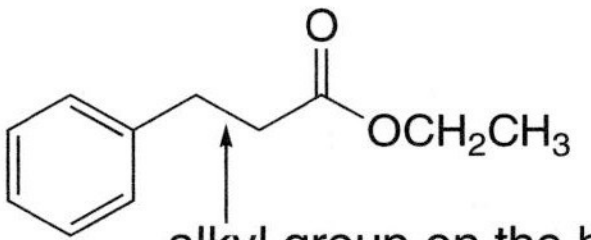

alkyl group on the benzene ring
R stabilizes (+) charges on the o,p positions by an
electron-donating inductive effect. This group behaves
like any other R group so that ortho and para products
are formed in electrophilic aromatic substitution.

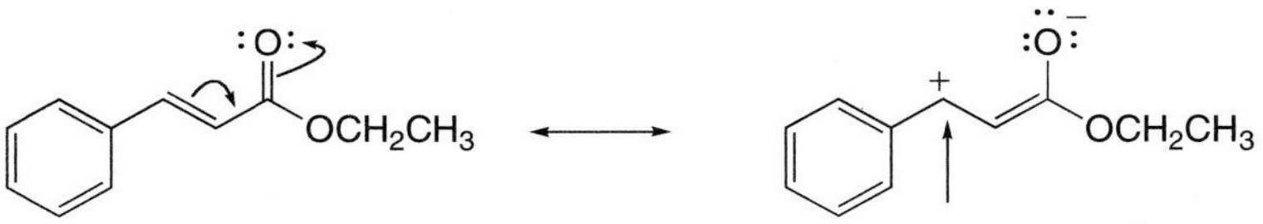

(+) charge on atom bonded to the benzene ring
Drawing resonance structures in electrophilic aromatic substitution
results in especially unstable structures for attack at the o,p positions—
two (+) charges on adjacent atoms. This doesn't happen with meta
attack, so meta attack is preferred. This is identical to the situation
observed with all meta directors.

18.51 Under the acidic conditions of nitration, the N atom of the starting material gets protonated, so the atom directly bonded to the benzene ring bears a (+) charge. This makes it a meta director, so the new NO_2 group is introduced meta to it.

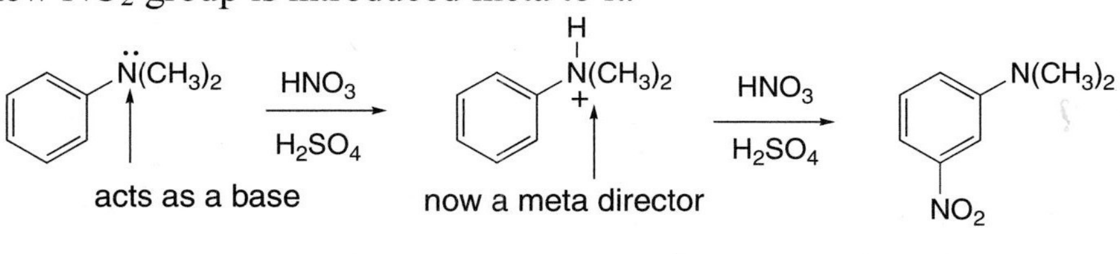

acts as a base   now a meta director

18.52

a.

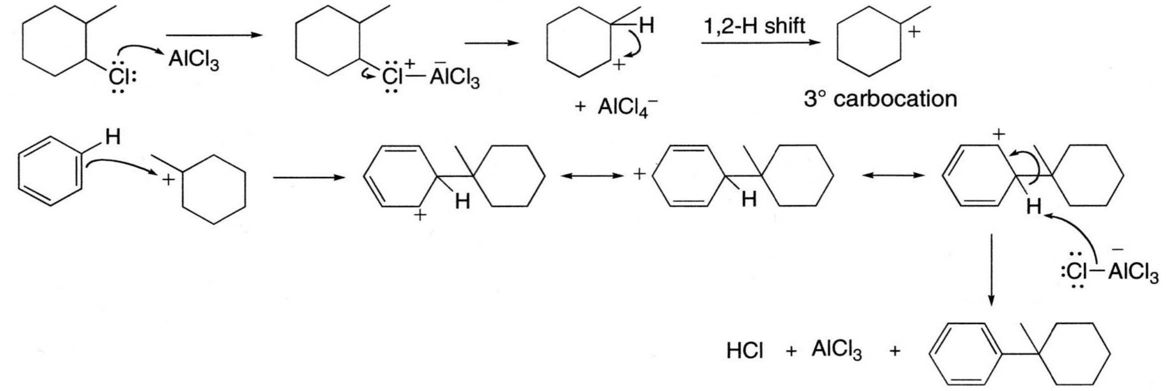

b.

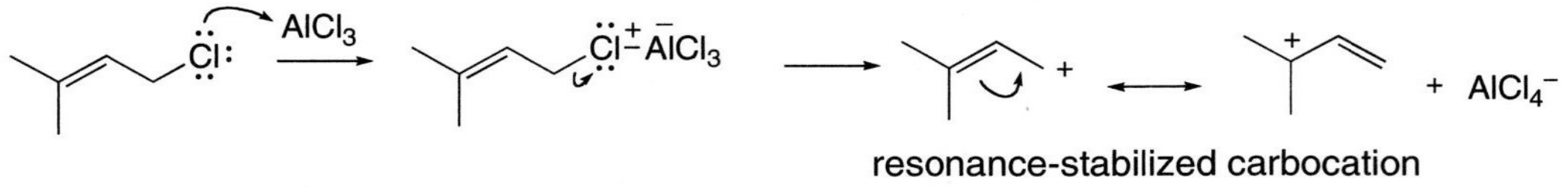

resonance-stabilized carbocation

Use both resonance forms to show how two products are formed.

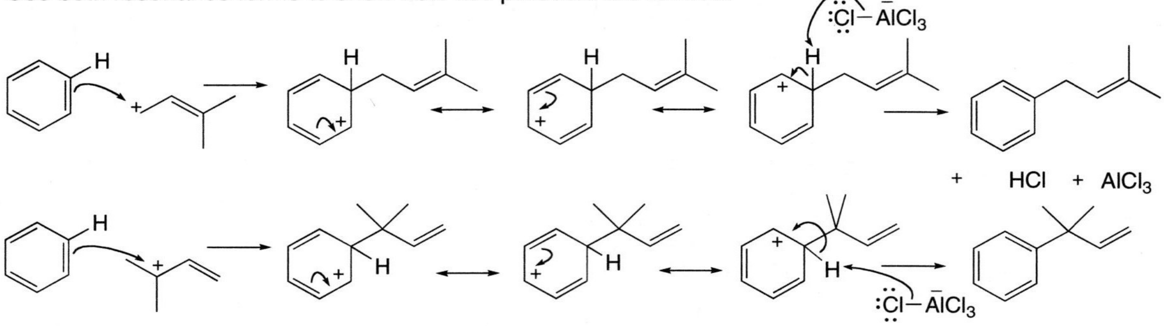

18.53

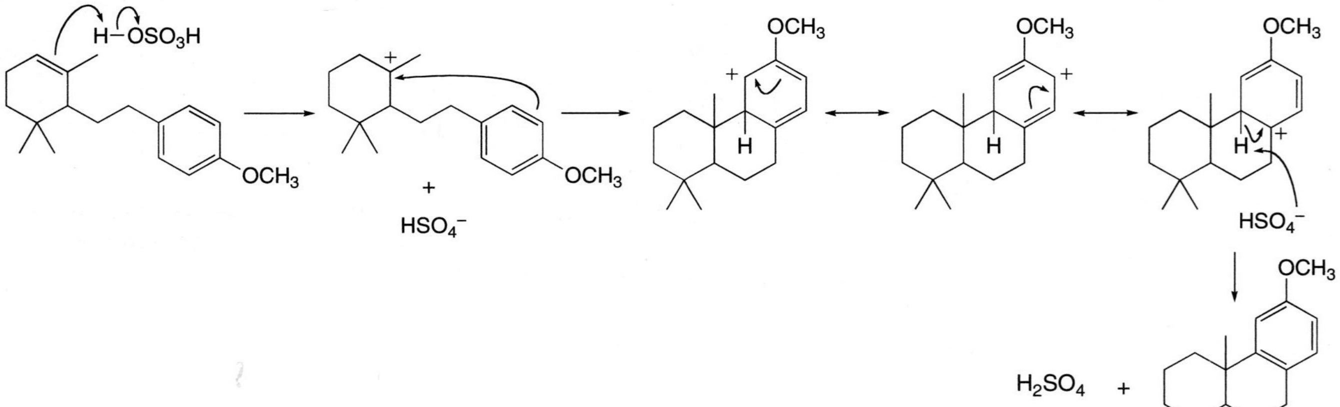

18.54

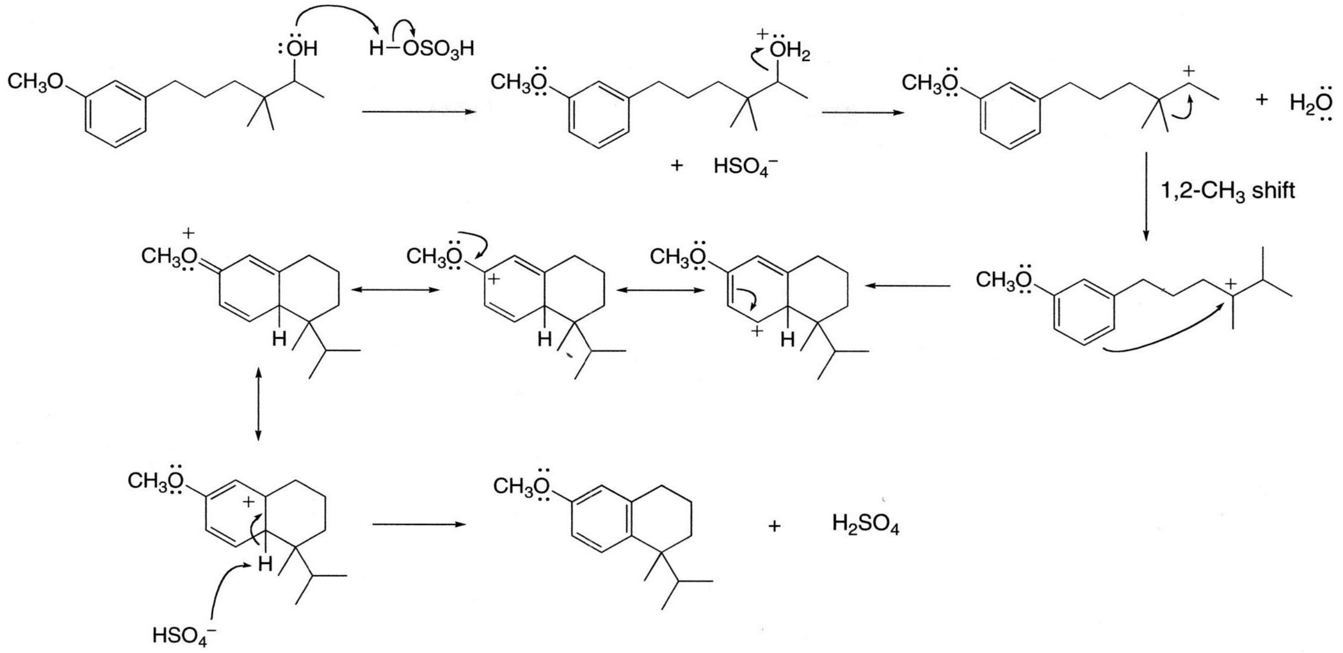

18.55

a. The product has one stereogenic center.

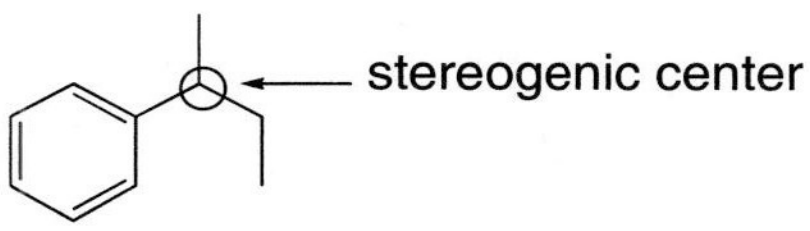

stereogenic center

b. The mechanism for Friedel–Crafts alkylation with this 2° halide involves formation of a trigonal planar carbocation. Since the carbocation is achiral, it reacts with benzene with equal probability from two possible directions (above and below) to afford an optically inactive, racemic mixture of two products.

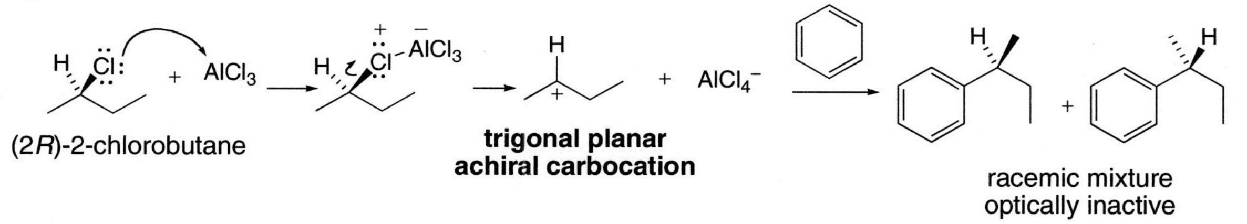

(2R)-2-chlorobutane

trigonal planar achiral carbocation

racemic mixture optically inactive

18.56

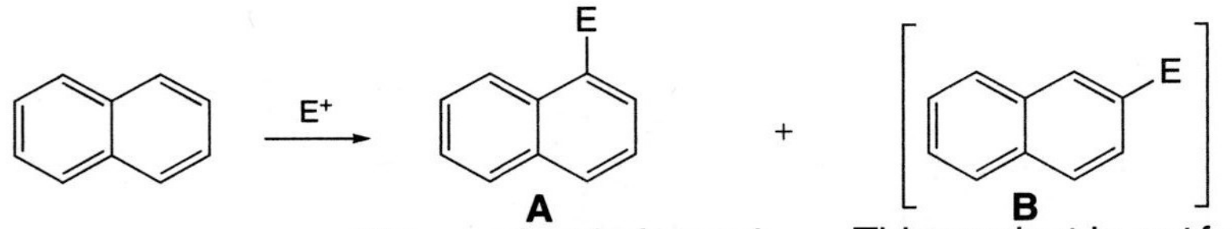
E
E⁺
E
+
E

A This product is formed. **B** This product is *not* formed.

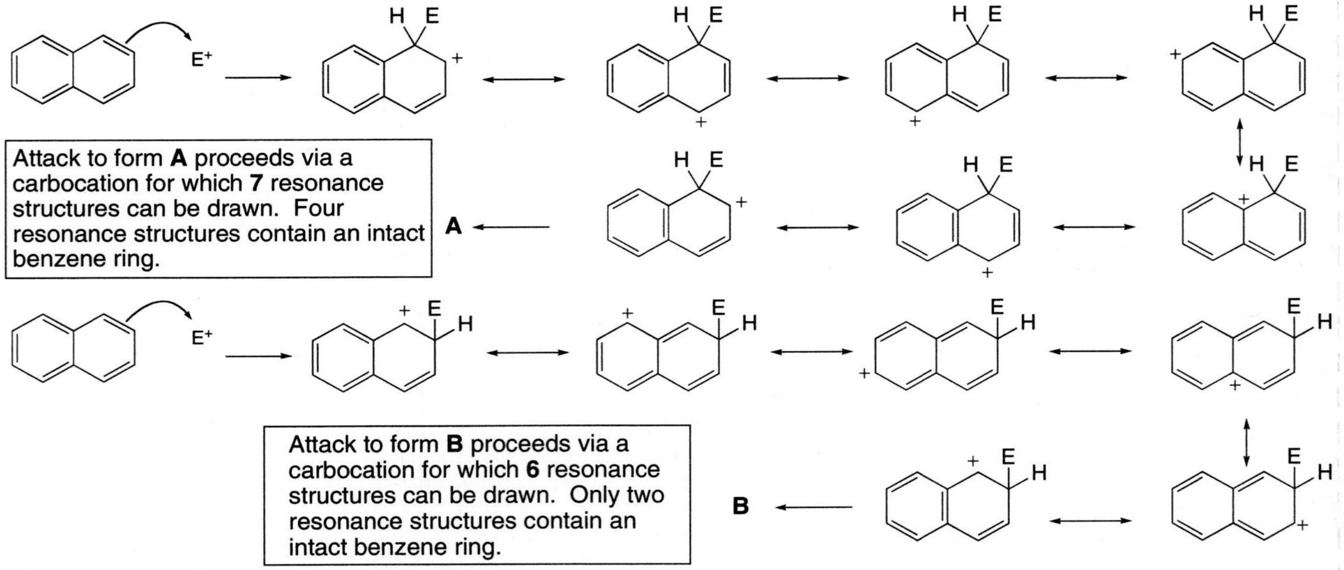
Attack to form **A** proceeds via a carbocation for which **7** resonance structures can be drawn. Four resonance structures contain an intact benzene ring.
A
Attack to form **B** proceeds via a carbocation for which **6** resonance structures can be drawn. Only two resonance structures contain an intact benzene ring.
B

A reaction that occurs by way of the more stable carbocation is preferred so product **A** is formed.

18.57

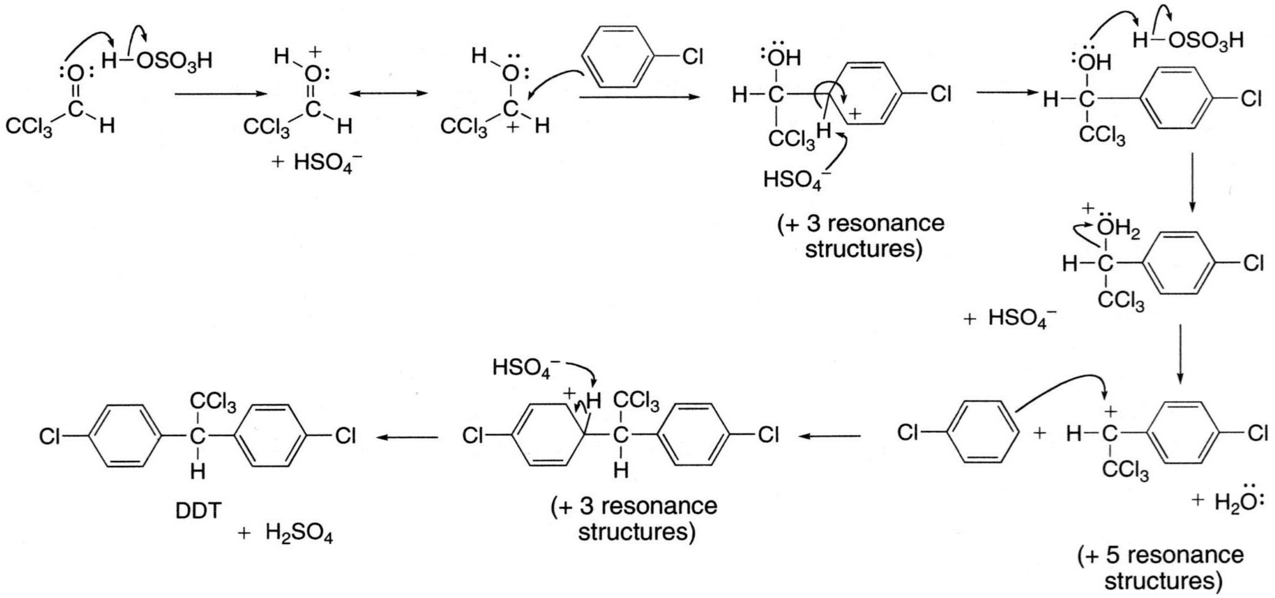
+ HSO4⁻
(+ 3 resonance structures)
+ HSO4⁻
(+ 5 resonance structures)
(+ 3 resonance structures)
DDT
+ H2SO4

18.58

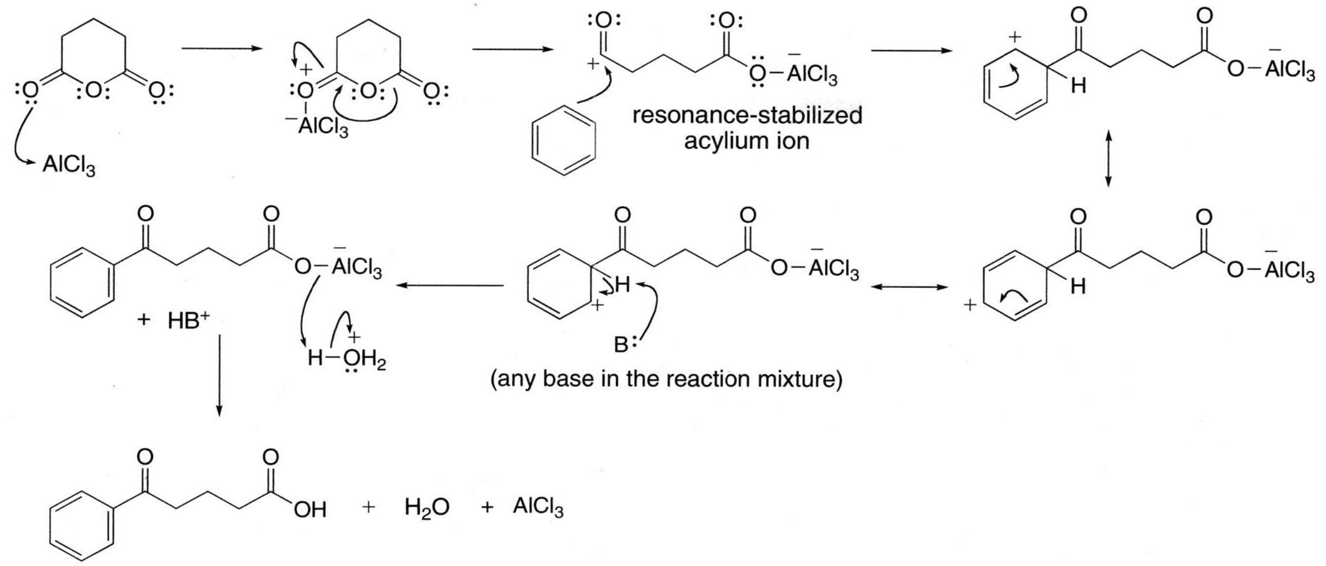

18.59 Benzyl bromide forms a resonance-stabilized intermediate that allows it to react rapidly under S_N1 conditions.

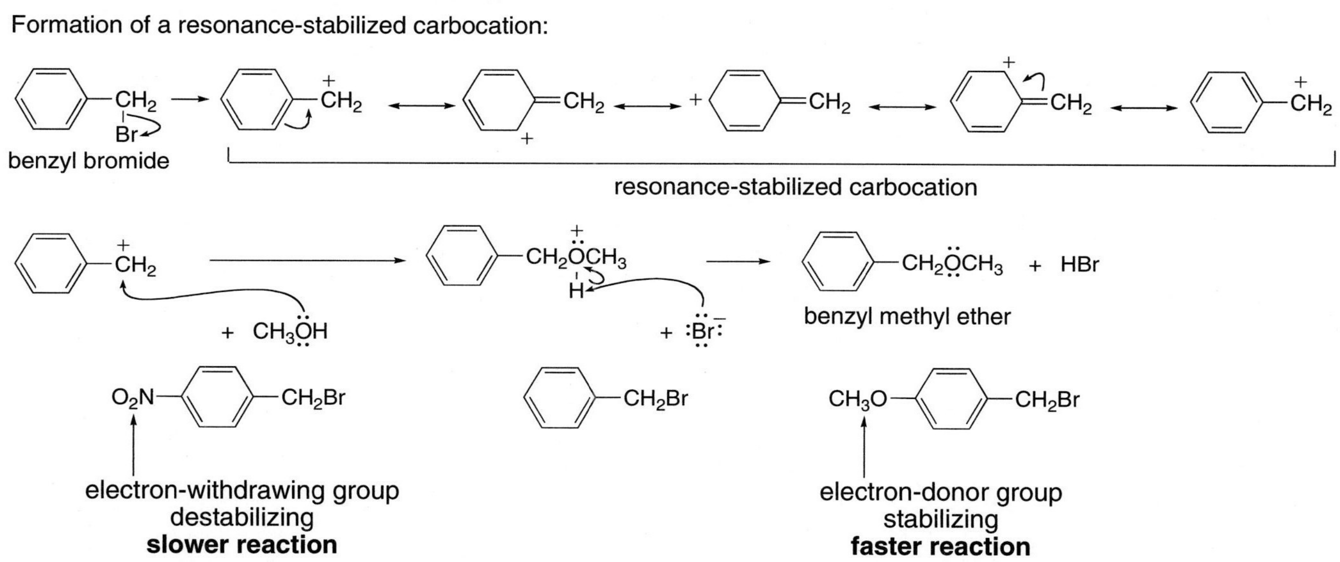

Formation of a resonance-stabilized carbocation:

The electron-withdrawing NO_2 group will destabilize the carbocation so the benzylic halide will be less reactive, while the electron-donating OCH_3 group will stabilize the carbocation, so the benzylic halide will be more reactive.

18.60 Addition of HBr will afford only one alkyl bromide because the intermediate carbocation leading to its formation is resonance stabilized.

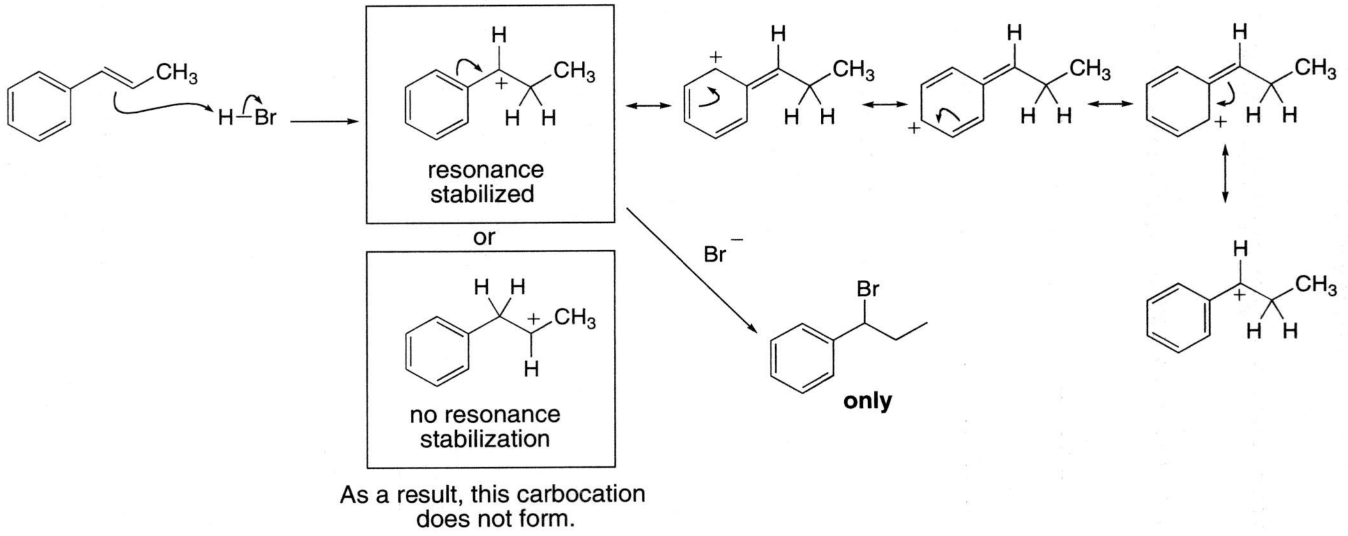

18.61

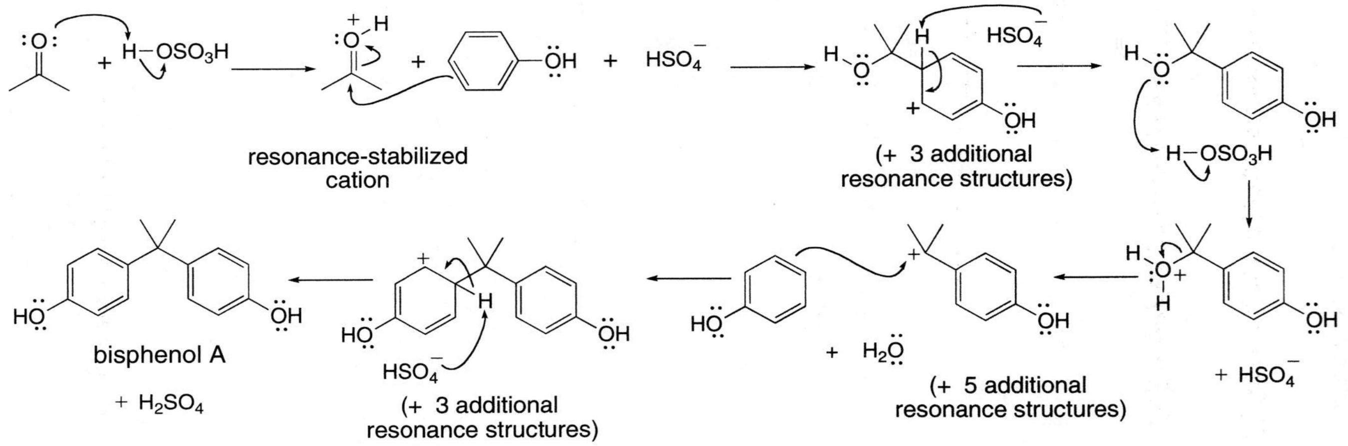

18.62

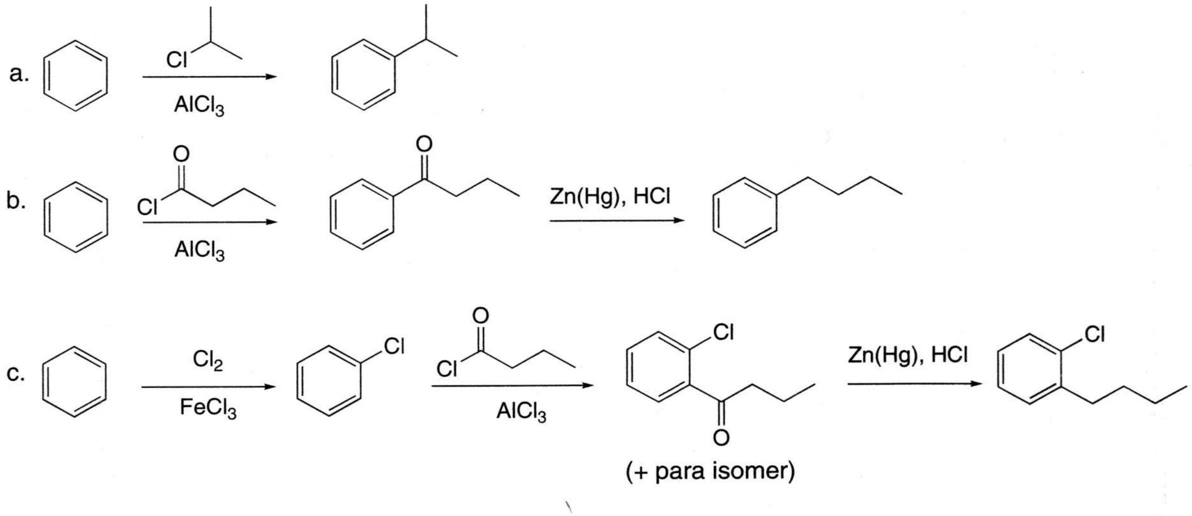

d.

$$\text{benzene} \xrightarrow[\text{H}_2\text{SO}_4]{\text{HNO}_3} \text{nitrobenzene (NO}_2\text{)} \xrightarrow[\text{FeBr}_3]{\text{Br}_2} \text{m-bromonitrobenzene (NO}_2, \text{Br)}$$

e.

$$\text{benzene} \xrightarrow[\text{FeBr}_3]{\text{Br}_2} \text{bromobenzene (Br)} \xrightarrow[\text{H}_2\text{SO}_4]{\text{HNO}_3} \text{o-bromonitrobenzene (NO}_2, \text{Br)} \quad (+ \text{ para isomer})$$

f.

$$\text{benzene} \xrightarrow[\text{AlCl}_3]{\text{CH}_3\text{Cl}} \text{toluene (CH}_3\text{)} \xrightarrow[\text{H}_2\text{SO}_4]{\text{SO}_3} \text{(CH}_3, \text{SO}_3\text{H)} \xrightarrow{\text{KMnO}_4} \text{(COOH, SO}_3\text{H)}$$

(+ para isomer)

g.

$$\text{benzene} \xrightarrow[\text{AlCl}_3]{\text{CH}_3\text{Cl}} \text{toluene (CH}_3\text{)} \xrightarrow{\text{KMnO}_4} \text{benzoic acid (COOH)} \xrightarrow[\text{H}_2\text{SO}_4]{\text{HNO}_3} \text{(COOH, NO}_2\text{)}$$

h.

$$\text{benzene} \xrightarrow[\text{AlCl}_3]{\text{CH}_3\text{Cl}} \text{toluene (CH}_3\text{)} \xrightarrow[\text{FeBr}_3]{\text{Br}_2} \text{Br–(CH}_3\text{)} \xrightarrow{\text{KMnO}_4} \text{Br–(COOH)}$$

(+ ortho isomer)

i.

$$\text{benzene} \xrightarrow[\text{AlCl}_3]{\text{Cl–C(=O)–CH}_2\text{CH}_3} \text{(C(=O)CH}_2\text{CH}_3\text{)} \xrightarrow[\text{FeBr}_3]{\text{Br}_2} \text{(C(=O)CH}_2\text{CH}_3, \text{Br)} \xrightarrow{\text{Zn(Hg), HCl}} \text{(CH}_2\text{CH}_2\text{CH}_3, \text{Br)} \xrightarrow[\text{H}_2\text{SO}_4]{\text{HNO}_3} \text{(NO}_2, \text{CH}_2\text{CH}_2\text{CH}_3, \text{Br)}$$

(+ isomer)

j.

$$\text{benzene} \xrightarrow[\text{AlCl}_3]{\text{CH}_3\text{Cl}} \text{toluene (CH}_3\text{)} \xrightarrow[\text{H}_2\text{SO}_4]{\text{SO}_3} \text{(CH}_3, \text{SO}_3\text{H)} \xrightarrow[\text{FeCl}_3]{\text{Cl}_2 \text{ (excess)}} \text{(CH}_3, \text{Cl, Cl, SO}_3\text{H)} \xrightarrow{\text{KMnO}_4} \text{(COOH, Cl, Cl, SO}_3\text{H)}$$

(+ ortho isomer)

k.

$$\text{benzene} \xrightarrow[\text{AlCl}_3]{\text{(CH}_3\text{)}_2\text{CHCl}} \text{cumene} \xrightarrow[\text{H}_2\text{SO}_4]{\text{HNO}_3} \text{(CH(CH}_3\text{)}_2, \text{NO}_2\text{)} \xrightarrow[\text{FeCl}_3]{\text{Cl}_2} \text{(Cl, CH(CH}_3\text{)}_2, \text{NO}_2\text{)} \xrightarrow[\text{h}\nu]{\text{Br}_2} \text{(Cl, C(CH}_3\text{)}_2\text{Br, NO}_2\text{)} \xrightarrow{\text{K}^+ \text{ }^-\text{OC(CH}_3\text{)}_3} \text{(Cl, C(CH}_3\text{)=CH}_2, \text{NO}_2\text{)}$$

(+ para isomer)   (+ isomer)

18.63

a.

b.

c.

d.

e.

f.

g.

18.64

a.

b. product in (a) $\xrightarrow{\text{NaOH}}$

c. product in (a) $\xrightarrow{\ ^-\text{OC(CH}_3)_3\ }$

d. product in (b) $\xrightarrow{\text{PCC}}$

e.

f.

g.

h.

i.

j.

k.

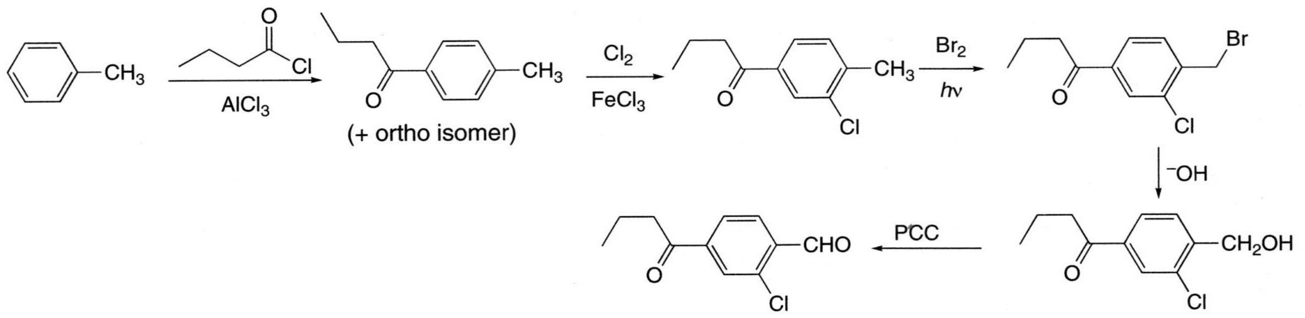

18.65

a.

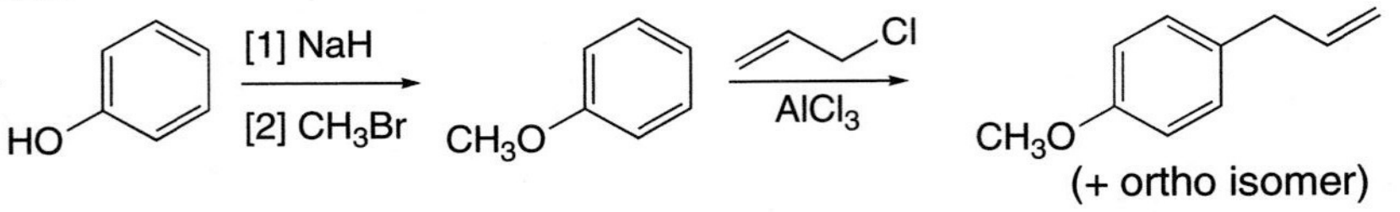

(+ ortho isomer)

b.

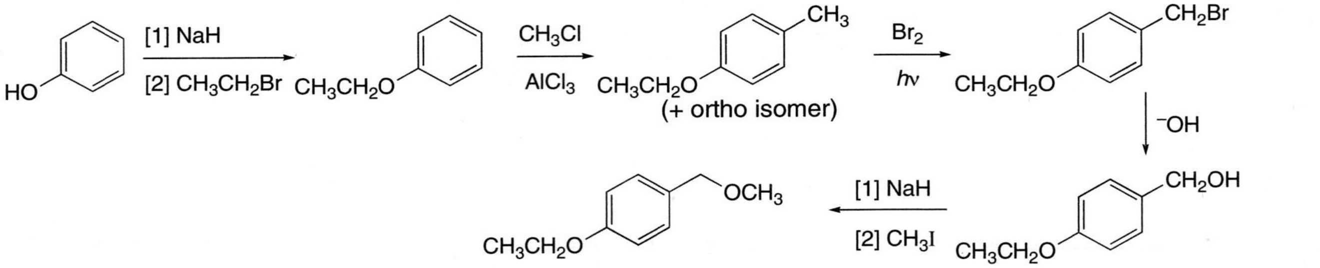

c.

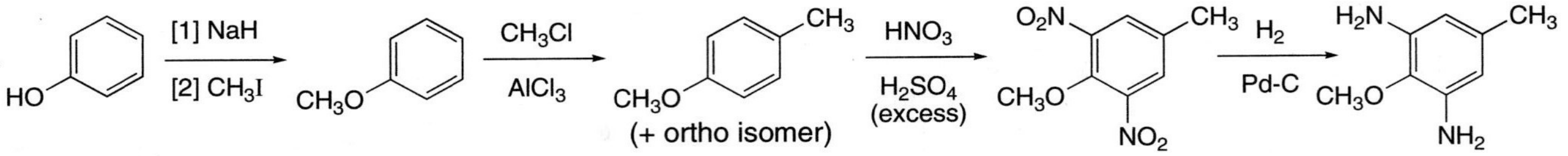

18.66

a.

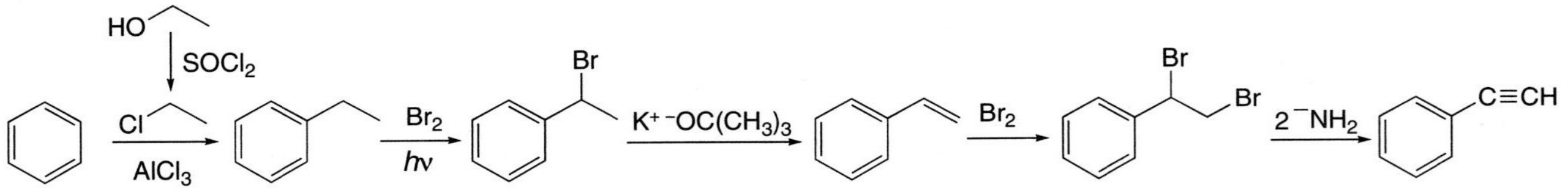

b.

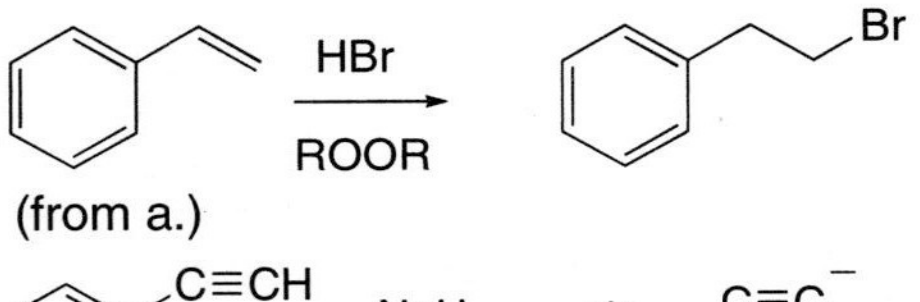

(from a.)

c.

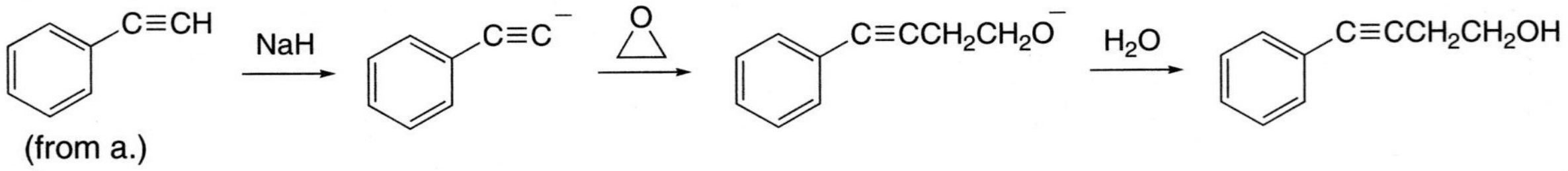

(from a.)

d.

(from a.)

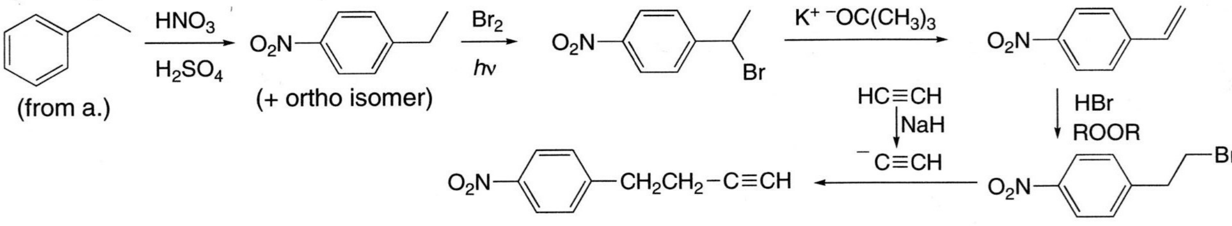

e.

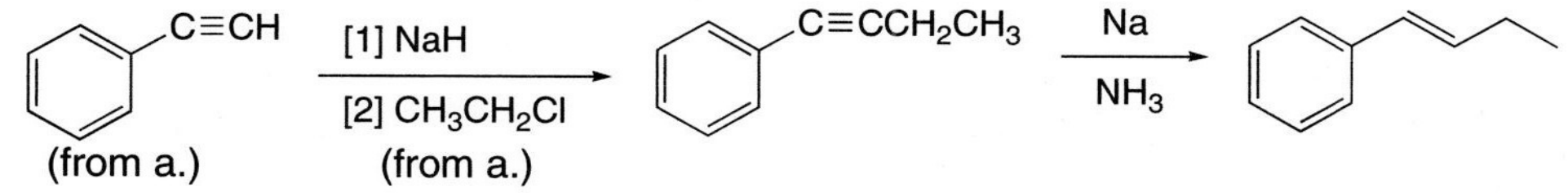

f.

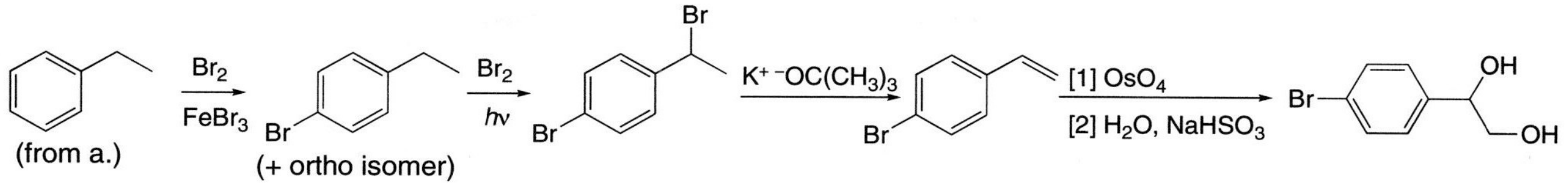

g.

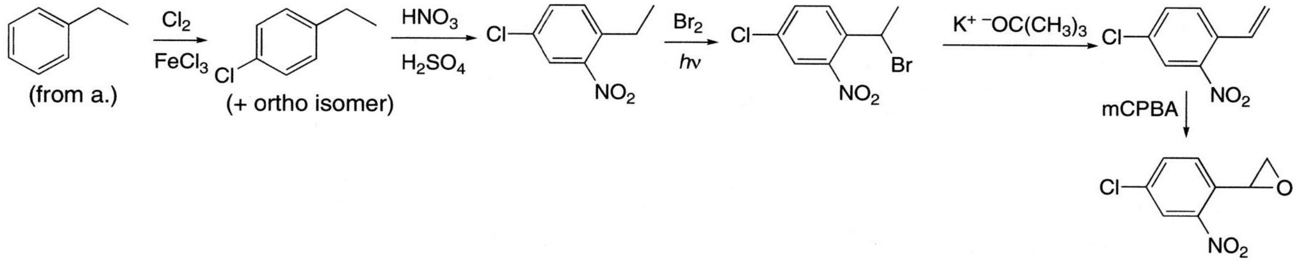

18.67

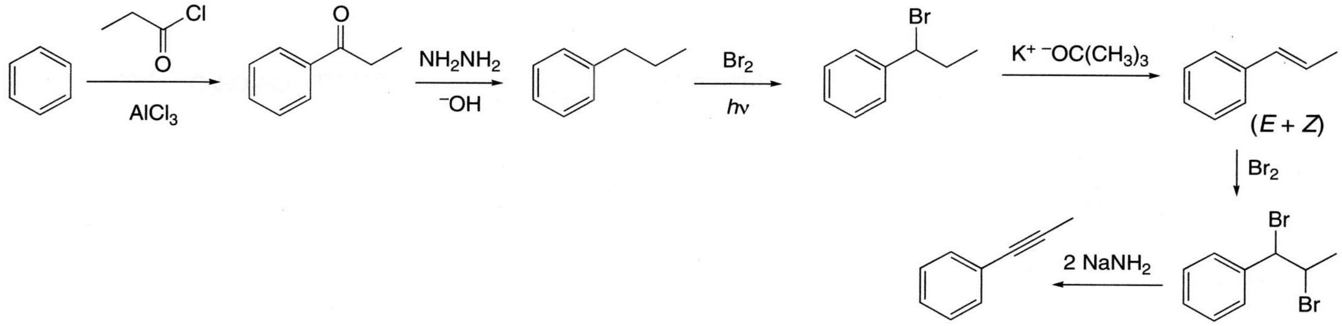

18.68 One possibility:

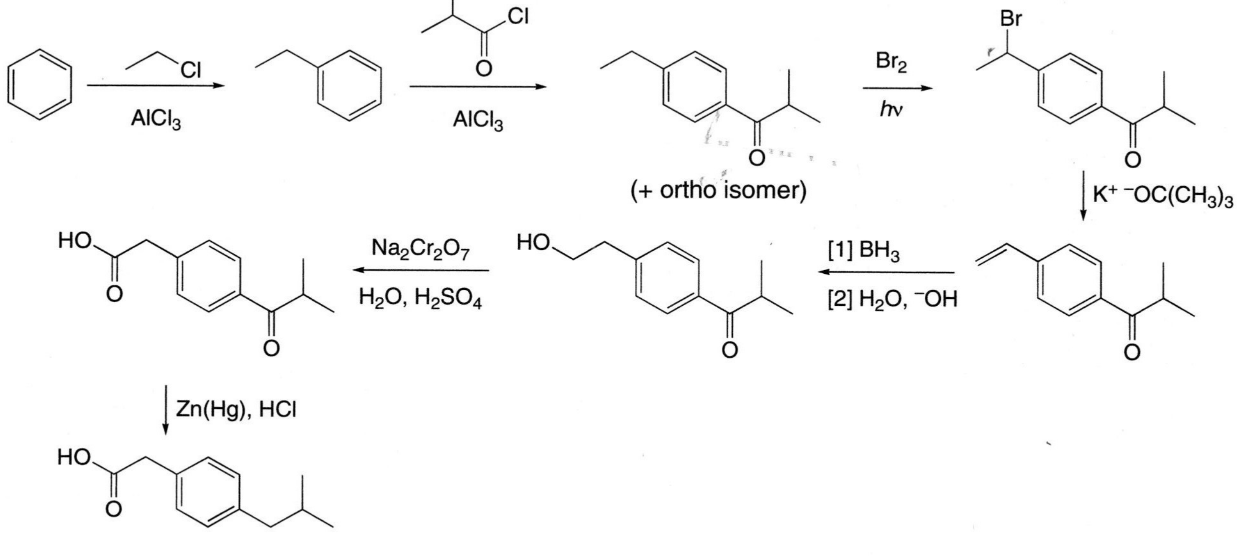

ibufenac

18.69

18.70 Use integration data and the molecular formula to determine the number of H's that give rise to each signal (Section 14.5, *How To*).

^{1}H NMR data of compound **A** (C_8H_9Br):

Absorption	ppm	# of H's	Explanation	Structure:
triplet	1.2	3	3 H's adjacent to 2 H's	
quartet	2.6	2	2 H's adjacent to 3 H's	
two signals	7.1 and 7.4	2 + 2	para disubstituted benzene	

^{1}H NMR data of compound **B** (C_8H_9Br):

Absorption	ppm	# of H's	Explanation	Structure:
triplet	3.1	2	2 H's adjacent to 2 H's	
triplet	3.5	2	2 H's adjacent to 2 H's	
multiplet	7.1–7.4	5	monosubstituted benzene	

18.71 IR absorption at 1717 cm^{-1} means compound **C** has a C=O.

^{1}H NMR data of compound **C** ($C_{10}H_{12}O$):

Absorption	ppm	# of H's	Explanation	Structure:
singlet	2.1	3	3 H's	
triplet	2.8	2	2 H's adjacent to 2 H's	
triplet	2.9	2	2 H's adjacent to 2 H's	
multiplet	7.1–7.4	5	monosubstituted benzene	

18.72

^{1}H NMR data of compound **X** ($C_{10}H_{12}O$):

Absorption	ppm	# of H's	Explanation	Structure:
doublet	1.3	6	6 H's adjacent to 1 H	
septet	3.5	1	1 H adjacent to 6 H's	
multiplet	7.4–8.1	5	monosubstituted benzene	

^{1}H NMR data of compound **Y** ($C_{10}H_{14}$):

Absorption	ppm	# of H's	Explanation
doublet	0.9	6	6 H's adjacent to 1 H
multiplet	1.8	1	1 H adjacent to many H's
doublet	2.5	2	2 H's adjacent to 1 H
multiplet	7.1–7.3	5	monosubstituted benzene

Structure:

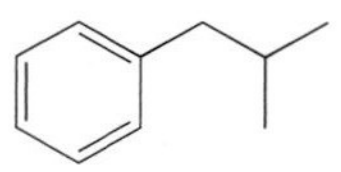

18.73

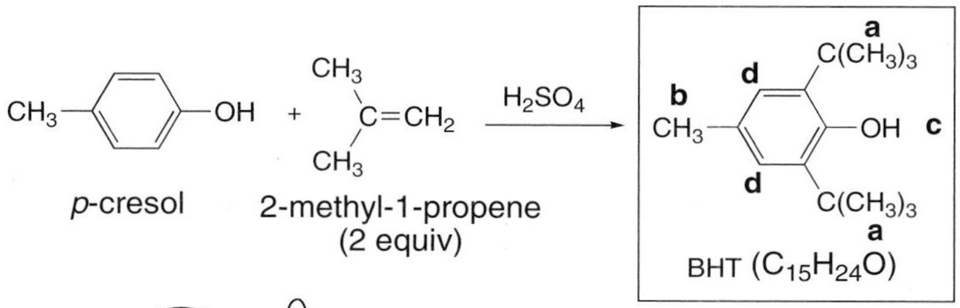

^{1}H NMR spectral data:

1.4 (singlet, 18 H) (a)
2.27 (singlet, 3 H) (b)
5.0 (singlet, 1 H) (c)
7.0 (singlet, 2 H) (d) ppm

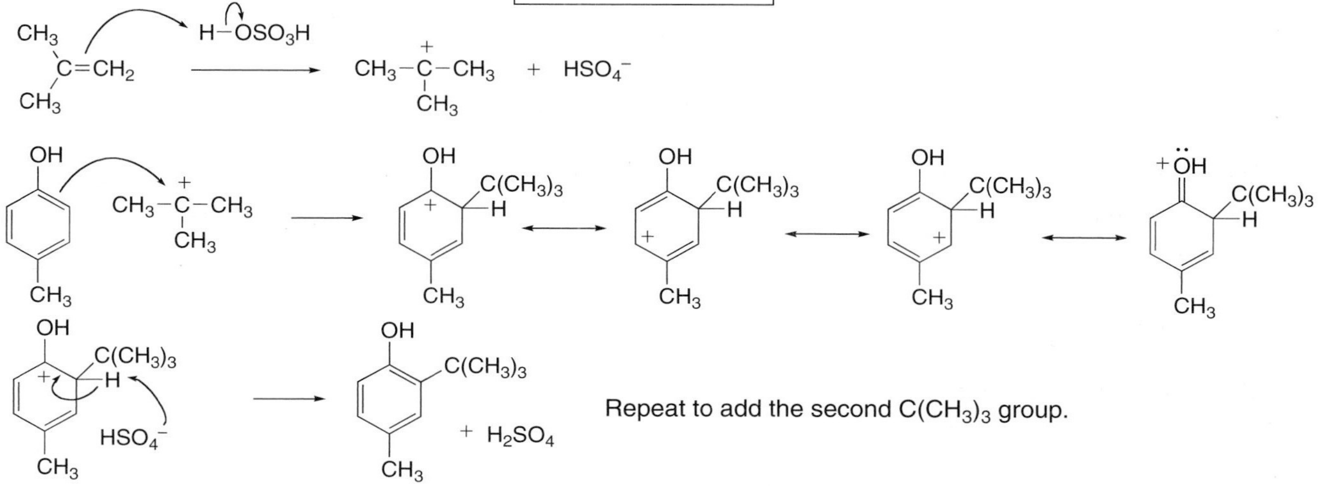

Repeat to add the second $C(CH_3)_3$ group.

18.74

Molecular formula (**Z**): C_9H_9ClO
IR absorption at 1683 cm^{-1}: C=O
^{1}H NMR spectral data:

Absorption	ppm	# of H's	Explanation
triplet	1.2	3	3 H's adjacent to 2 H's
quartet	2.9	2	2 H's adjacent to 3 H's
multiplet	7.2–8.0	4	disubstituted benzene

Structure:

18.75 Five resonance structures can be drawn for phenol, three of which place a negative charge on the ortho and para carbons. These illustrate that the electron density at these positions is increased, thus shielding the protons at these positions, and shifting the absorptions to lower chemical shift. Similar resonance structures cannot be drawn with a negative charge at the meta position, so it is more deshielded and absorbs farther downfield, at higher chemical shift.

(–) charges on the ortho and para positions

18.76 a. Pyridine: The electron-withdrawing inductive effect of N makes the ring electron poor. Also, electrophiles E$^+$ can react with N, putting a positive charge on the ring. This makes the ring less reactive with another positively charged species.

To understand why substitution occurs at C3, compare the stability of the carbocation formed by attack at C2 and C3.

Electrophilic attack on N:

Electrophilic attack at C2:

Electrophilic attack at C3:

less reactive than benzene

N does not have an octet.
(+) charge on an electronegative N atom
poor resonance structure
attack at C2 does not occur

better resonance structures

Since attack at C3 forms a more stable carbocation, attack at C3 occurs. Attack at C4 generates a carbocation of similar stability to attack at C2, so attack at C4 does not occur.

b. Pyrrole is more reactive than benzene because the C's are more electron rich. The lone pair on N has an electron-donating resonance effect.

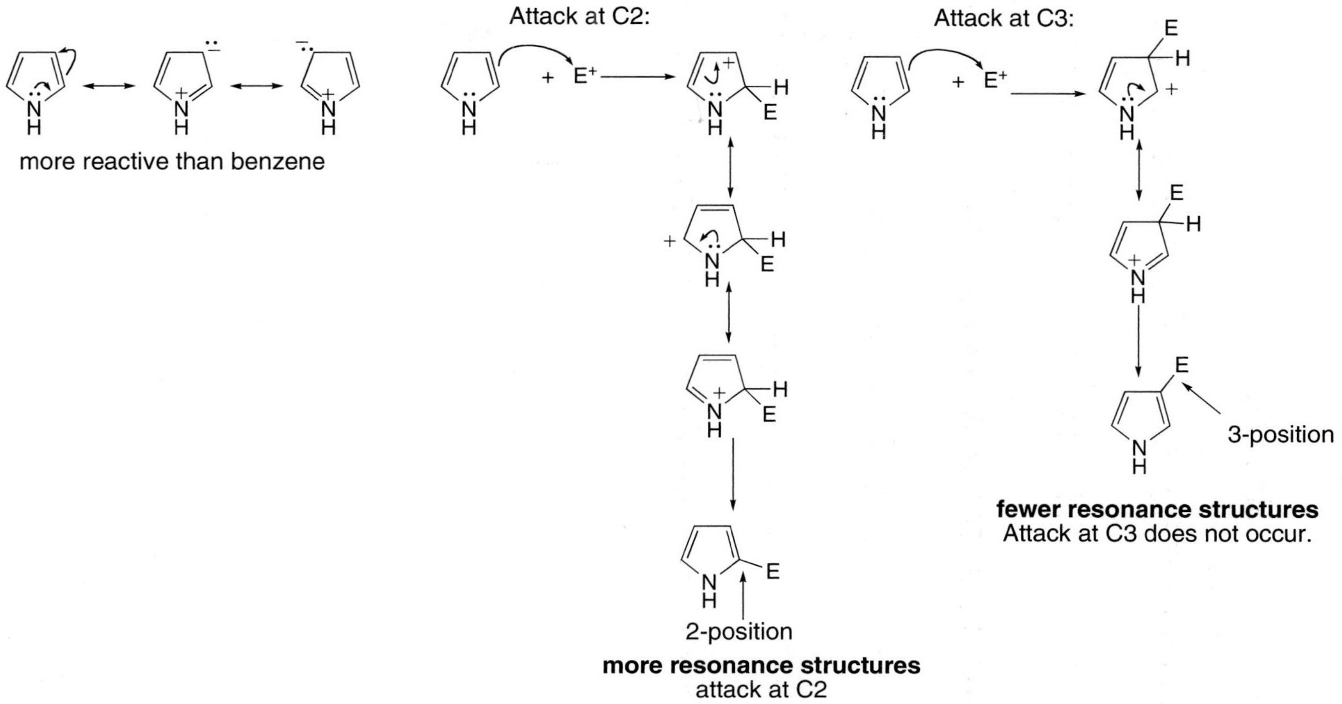

more reactive than benzene

Attack at C2:

Attack at C3:

2-position
more resonance structures
attack at C2

3-position

fewer resonance structures
Attack at C3 does not occur.

Since attack at C2 forms a more stable carbocation, electrophilic substitution occurs at C2.

18.77

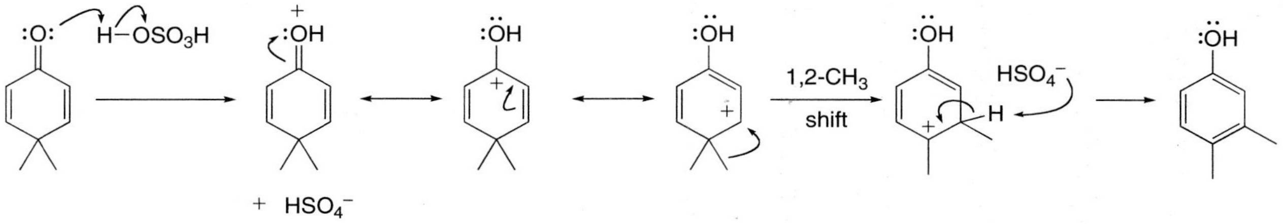

18.78 Draw a stepwise mechanism for the following intramolecular reaction, which was used in the synthesis of the female sex hormone estrone.

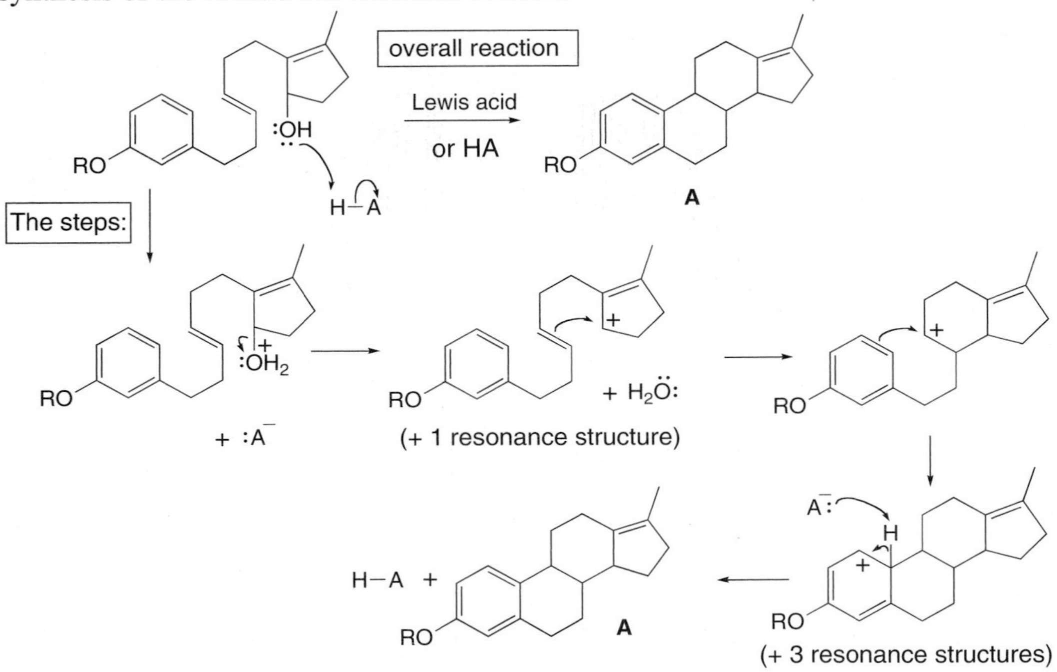

18.79

a. The reaction could follow a two-step mechanism: [1] addition of the nucleophile to form a carbanion, followed by [2] elimination of the leaving group.

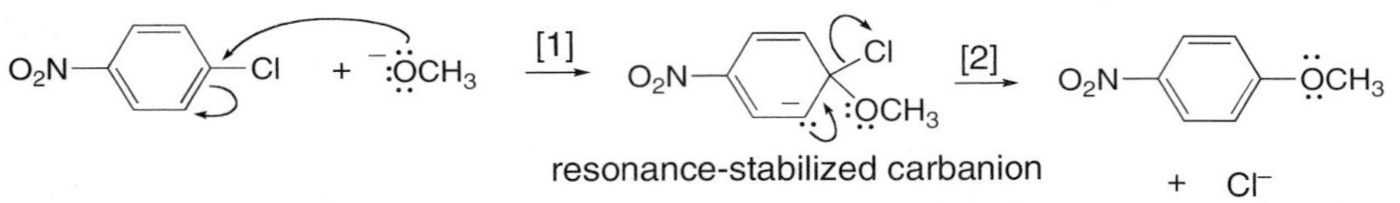

b. The NO_2 group stabilizes the negatively charged intermediate by an electron-withdrawing inductive effect and by resonance.

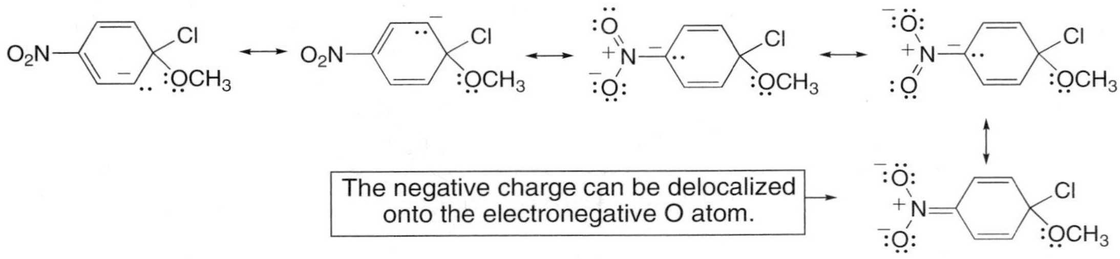

The negative charge can be delocalized onto the electronegative O atom.

c. *m*-Chloronitrobenzene does not undergo this reaction because no resonance structure can be drawn that delocalizes the negative charge of the reactive intermediate onto the O atom of the NO_2 group.

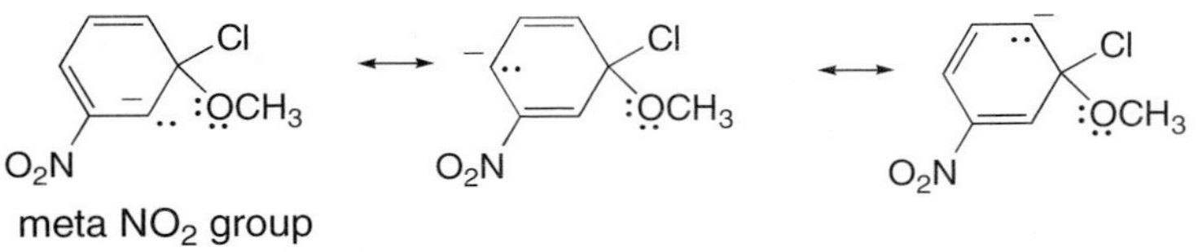

meta NO_2 group

Chapter 19: Carboxylic Acids and the Acidity of the O–H Bond

◆ General facts

- Carboxylic acids contain a carboxy group (COOH). The central carbon is sp^2 hybridized and trigonal planar (19.1).
- Carboxylic acids are identified by the suffixes -*oic acid*, *carboxylic acid*, or -*ic acid* (19.2).
- Carboxylic acids are polar compounds that exhibit hydrogen bonding interactions (19.3).

◆ Summary of spectroscopic absorptions (19.4)

IR absorptions	C=O	~1710 cm^{-1}
	O–H	3500–2500 cm^{-1} (very broad and strong)
^{1}H NMR absorptions	O–H	10–12 ppm (highly deshielded proton)
	C–H α to COOH	2–2.5 ppm (somewhat deshielded Csp^3–H)
^{13}C NMR absorption	C=O	170–210 ppm (highly deshielded carbon)

◆ General acid–base reaction of carboxylic acids (19.9)

$pK_a \approx 5$ carboxylate anion

- Carboxylic acids are especially acidic because carboxylate anions are resonance stabilized.
- For equilibrium to favor the products, the base must have a conjugate acid with a $pK_a > 5$. Common bases are listed in Table 19.3.

◆ Factors that affect acidity

Resonance effects. A carboxylic acid is more acidic than an alcohol or phenol because its conjugate base is more effectively stabilized by resonance (19.9).

ROH   R–C

$pK_a = 16\text{–}18$ $pK_a = 10$ $pK_a \approx 5$

Increasing acidity

Inductive effects. Acidity increases with the presence of electron-withdrawing groups (like the electronegative halogens) and decreases with the presence of electron-donating groups (like polarizable alkyl groups) (19.10).

Substituted benzoic acids.

- Electron-donor groups (D) make a substituted benzoic acid less acidic than benzoic acid.
- Electron-withdrawing groups (W) make a substituted benzoic acid more acidic than benzoic acid.

less acidic	$pK_a = 4.2$	**more acidic**
higher pK_a		**lower** pK_a
$pK_a > 4.2$		$pK_a < 4.2$

Increasing acidity

◆ Other facts

- Extraction is a useful technique for separating compounds having different solubility properties. Carboxylic acids can be separated from other organic compounds by extraction, because aqueous base converts a carboxylic acid into a water-soluble carboxylate anion (19.12).
- A sulfonic acid (RSO_3H) is a strong acid because it forms a weak, resonance-stabilized conjugate base on deprotonation (19.13).
- Amino acids have an amino group on the α carbon to the carboxy group [$RCH(NH_2)COOH$]. Amino acids exist as zwitterions at pH $\approx$ 6. Adding acid forms a species with a net (+1) charge [$RCH(NH_3)COOH$]$^+$. Adding base forms a species with a net (−1) charge [$RCH(NH_2)COO$]$^-$ (19.14).

19.1 To name a carboxylic acid:
[1] Find the longest chain containing the COOH group and change the -e ending to -oic acid.
[2] Number the chain to put the COOH carbon at C1, but omit the number from the name.
[3] Follow all other rules of nomenclature.

a.
$$\overset{3}{\underset{\underset{\underset{1}{CH_3}}{|}}{\overset{CH_3}{|}}}{CH_3CH_2CH_2-\overset{|}{\underset{|}{C}}-\underset{1}{CH_2COOH}}$$
Number the chain to put COOH at C1.
6 carbon chain = **hexanoic acid**
3,3-dimethylhexanoic acid

c.
$$CH_3CH_2-\overset{\overset{4}{H}}{\underset{CH_3CH_2}{\underset{|}{C}}}-CH_2-\overset{\overset{2}{H}}{\underset{CH_2CH_3}{\underset{|}{C}}}\overset{1}{-COOH}$$
Number the chain to put COOH at C1.
6 carbon chain = **hexanoic acid**
2,4-diethylhexanoic acid

b.
$$CH_3-\overset{\overset{4}{H}}{\underset{\underset{}{Cl}}{\underset{|}{C}}}-CH_2CH_2\underset{1}{COOH}$$
Number the chain to put COOH at C1.
5 carbon chain = **pentanoic acid**
4-chloropentanoic acid

d.
Number the chain to put COOH at C1.
9 carbon chain = **nonanoic acid**
4-isopropyl-6,8-dimethylnonanoic acid

19.2
a. 2-bromo**butanoic acid**

c. 3,3,4-trimethyl**heptanoic acid**

e. 3,4-diethyl**cyclohexanecarboxylic acid**

b. 2,3-dimethyl**pentanoic acid**

d. 2-*sec*-butyl-4,4-diethyl**nonanoic acid**

f. 1-isopropyl**cyclobutane-carboxylic acid**

19.3

a. α-methoxy**valeric acid**

c. α,β-dimethyl**caproic acid**

b. β-phenyl**propionic acid**

d. α-chloro-β-methyl**butyric acid**

19.4

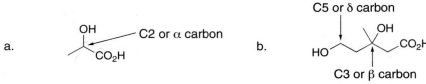

a.

IUPAC: 2-hydroxy**propanoic acid**
common: α-hydroxy**propionic acid**

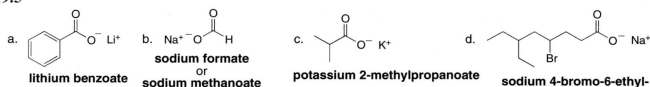

b.

IUPAC: 3,5-dihydroxy-3-methyl**pentanoic acid**
common: β,δ-dihydroxy-β-methyl**valeric acid**

19.5

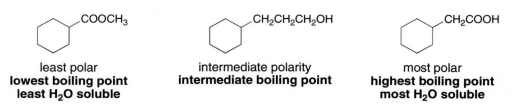

a.   lithium benzoate

b.   **sodium formate**
or
sodium methanoate

c.   **potassium 2-methylpropanoate**

d.   **sodium 4-bromo-6-ethyl-
octanoate**

19.6

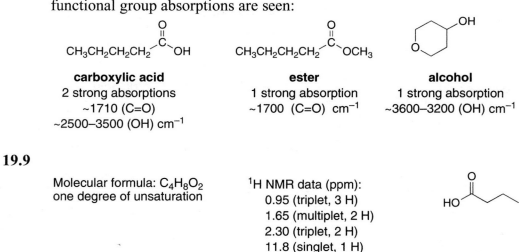

2-propyl**pentanoic acid**   sodium 2-propyl**pentanoate**

19.7   More polar molecules have a higher boiling point and are more water soluble.

      COOCH₃ → $COOCH_3$

least polar
**lowest boiling point
least H₂O soluble**

CH₂CH₂CH₂OH → $CH_2CH_2CH_2OH$

intermediate polarity
intermediate boiling point

CH₂COOH → CH_2COOH

most polar
**highest boiling point
most H₂O soluble**

19.8   Look for functional group differences to distinguish the compounds by IR. Besides sp^3 hybridized C–H bonds at 3000–2850 cm^{-1} (which all three compounds have), the following functional group absorptions are seen:

$CH_3CH_2CH_2CH_2$—$\overset{O}{\overset{\|}{C}}$—$OH$

carboxylic acid
2 strong absorptions
~1710 (C=O)
~2500–3500 (OH) cm^{-1}

$CH_3CH_2CH_2CH_2$—$\overset{O}{\overset{\|}{C}}$—$OCH_3$

ester
1 strong absorption
~1700 (C=O) cm^{-1}

alcohol
1 strong absorption
~3600–3200 (OH) cm^{-1}

19.9

Molecular formula: $C_4H_8O_2$
one degree of unsaturation

^{1}H NMR data (ppm):
  0.95 (triplet, 3 H)
  1.65 (multiplet, 2 H)
  2.30 (triplet, 2 H)
  11.8 (singlet, 1 H)

19.10

H–C(=O)–OH with Ha (on H–C) and Hb (on OH)

HO–C(=O)–CH2–C(=O)–OH with Hc (on OH's), Hd Hd (on CH2)

2 singlets
1:1 ratio

2 singlets
1:1 ratio

Although both compounds have an absorption at 10–12 ppm in their ^{1}H NMR spectra (due to H$_b$ and H$_c$), H$_a$, which is bonded directly to the carbonyl carbon, is much farther downfield than H$_d$ because it is more deshielded.

19.11

PGF$_{2\alpha}$
a prostaglandin

enantiomer

There are five tetrahedral stereogenic centers. Both double bonds can exhibit cis–trans isomerism. Therefore, there are $2^7 = 128$ stereoisomers.

19.12   1° Alcohols are converted to carboxylic acids by oxidation reactions.

a. (pentanoic acid) $\Longrightarrow$ (pentanol, CH$_2$OH)

b. $(CH_3)_2CH$–C(=O)–OH $\Longrightarrow$ $(CH_3)_2CH$–CH$_2$–OH

c. (cyclohexyl–COOH) $\Longrightarrow$ (cyclohexyl–CH$_2$OH)

19.13

a. (benzyl–CH$_2$OH) **A** $\xrightarrow[\text{H}_2\text{SO}_4,\ \text{H}_2\text{O}]{\text{Na}_2\text{Cr}_2\text{O}_7}$ (benzoic acid, COOH)

b. $CH_3C{\equiv}CCH_3$ **B** $\xrightarrow[\text{[2] H}_2\text{O}]{\text{[1] O}_3}$ CH_3COOH (2 equiv)

c. O_2N–(benzene)–CH_3 **C** $\xrightarrow{\text{KMnO}_4}$ O_2N–(benzene)–COOH

(Any R group with benzylic H's can be present para to NO$_2$.)

d. OH ← 2° OH ... OH ↑ 1° OH **D** $\xrightarrow[\text{H}_2\text{SO}_4,\ \text{H}_2\text{O}]{\text{CrO}_3}$ (keto acid, CO$_2$H)

19.14

a. (cyclohexyl–COOH) $\xrightarrow{\text{NaOH}}$ (cyclohexyl–COO$^-$ Na$^+$) + H$_2$O

b. CH_3–(benzene)–OH $\xrightarrow{\text{NaOCH}_3}$ CH_3–(benzene)–O$^-$ Na$^+$ + HOCH$_3$

c. CH_3–C(CH$_3$)(CH$_3$)–OH $\xrightarrow{\text{NaH}}$ CH_3–C(CH$_3$)(CH$_3$)–O$^-$ Na$^+$ + H$_2$

d. (benzene–COOH) $\xrightarrow{\text{NaHCO}_3}$ (benzene–COO$^-$ Na$^+$) + H$_2$CO$_3$

19.15 CH$_3$COOH has a pK_a of 4.8. Any base with a conjugate acid with a pK_a higher than 4.8 can deprotonate it.

a. F$^-$ pK_a (HF) = 3.2 **not strong enough**

b. (CH$_3$)$_3$CO$^-$ pK_a [(CH$_3$)$_3$COH]= 18 **strong enough**

c. CH$_3^-$ pK_a (CH$_4$) = 50 **strong enough**

d. $^-$NH$_2$ pK_a (NH$_3$) = 38 **strong enough**

e. Cl$^-$ pK_a (HCl) = –7.0 **not strong enough**

19.16

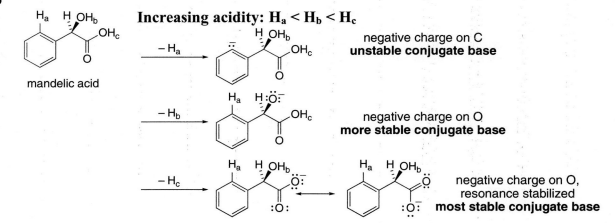

Increasing acidity: **H$_a$ < H$_b$ < H$_c$**

−H$_a$ → negative charge on C **unstable conjugate base**

−H$_b$ → negative charge on O **more stable conjugate base**

−H$_c$ → negative charge on O, resonance stabilized **most stable conjugate base**

mandelic acid

19.17 Electron-withdrawing groups make an acid more acidic, lowering its pK_a.

CH$_3$CH$_2$—COOH
least acidic
pK_a = 4.9

ICH$_2$—COOH
one electron-withdrawing group
intermediate acidity
pK_a = 3.2

CF$_3$—COOH
three electron-withdrawing F's
most acidic
pK_a = 0.2

19.18 Acetic acid has an electron-donating methyl group bonded to the carboxy group. The CH$_3$ group both stabilizes the acid and destabilizes the nearby negative charge on the conjugate base, making CH$_3$COOH less acidic (with a higher pK_a) than HCOOH.

electron-donating CH$_3$ group ⟶

acetic acid
CH$_3$ stabilizes the
partial positive charge.

conjugate base
CH$_3$ destabilizes the
negative charge.
a less stable conjugate base

19.19

a. CH$_3$COOH
least acidic

HSCH$_2$COOH
intermediate acidity

HOCH$_2$COOH
most acidic

b. ICH$_2$CH$_2$COOH
least acidic

ICH$_2$COOH
intermediate acidity

I$_2$CHCOOH
most acidic

19.20

a.

CH_3—[benzene ring]—COOH
least acidic

[benzene ring]—COOH
intermediate acidity

Cl—[benzene ring]—COOH
most acidic

b. CH_3O—[benzene ring]—COOH
least acidic

CH_3—[benzene ring]—COOH
intermediate acidity

[structure with C=O and CH_3]—[benzene ring]—COOH
most acidic

19.21

[structure A: phenol ring with OH, HO, and long alkyl chain]

A

Phenol **A** has a higher pK_a than phenol because of its substituents. Both the OH and CH_3 are electron-donating groups, which make the conjugate base less stable. Therefore, the acid is **less acidic.**

19.22 To separate compounds by an extraction procedure, they must have different solubility properties.

a. $CH_3(CH_2)_6COOH$ and $CH_3CH_2CH_2CH_2CH=CH_2$: **YES.** The acid can be extracted into aqueous base, while the alkene will remain in an organic layer.

b. $CH_3CH_2CH_2CH_2CH=CH_2$ and $(CH_3CH_2CH_2)_2O$: **NO.** Both compounds are soluble in organic solvents and insoluble in water. Neither is acidic enough to be extracted into aqueous base.

c. $CH_3(CH_2)_6COOH$ and NaCl: one carboxylic acid, one salt: **YES.** The carboxylic acid is soluble in an organic solvent while the salt is soluble in water.

d. NaCl and KCl: two salts: **NO.**

19.23 To separate compounds by an aqueous extraction technique, compounds must have different solubility properties. CH_3CH_2COOH and $CH_3CH_2CH_2OH$ are low molecular weight organic compounds that can hydrogen bond to water, so they are water soluble. They also both dissolve in organic solvents. As a result, they are inseparable because of their similar solubility properties.

19.24

weaker conjugate base
better leaving group
↓
CF_3SO_3H ⟶ $CF_3SO_3^-$

CF_3 is electron *withdrawing.*
stronger acid
lower pK_a

stronger conjugate base
worse leaving group
↓
CH_3SO_3H ⟶ $CH_3SO_3^-$

CH_3 is electron *donating.*
weaker acid
higher pK_a

19.25

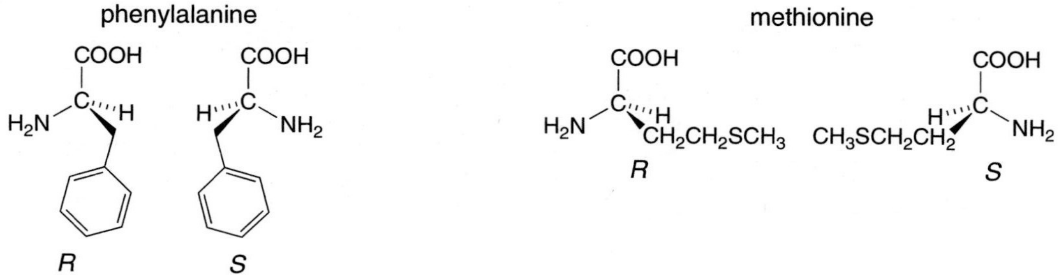

phenylalanine

R S

methionine

R S

19.26 Since amino acids exist as zwitterions (i.e., salts), they are too polar to be soluble in organic solvents like diethyl ether. Thus, they are soluble in water.

19.27

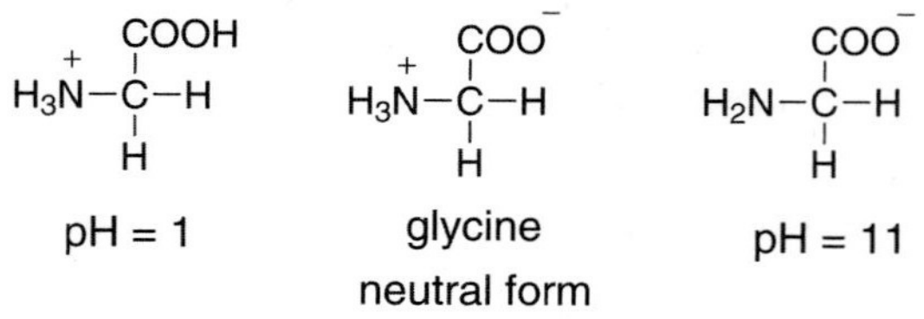

pH = 1   glycine neutral form   pH = 11

19.28

$$pI = \frac{pK_a(COOH) + pK_a(NH_3{}^+)}{2} = \frac{(2.58) + (9.24)}{2} = 5.91$$

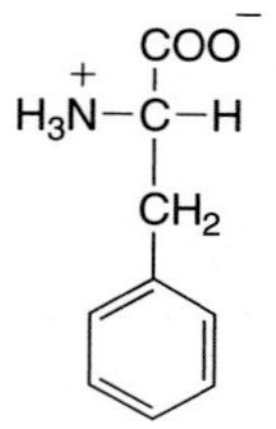

19.29

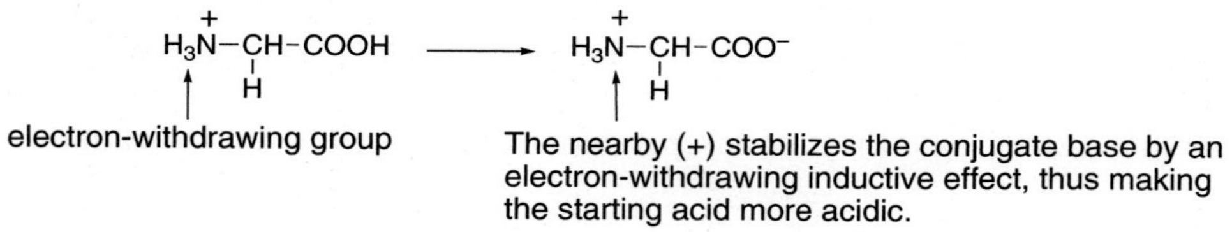

electron-withdrawing group

The nearby (+) stabilizes the conjugate base by an electron-withdrawing inductive effect, thus making the starting acid more acidic.

19.30 Use the directions from Answer 19.1 to name the compounds.

a. $(CH_3)_2CHCH_2CH_2CO_2H$ 4-methylpentanoic acid

b. $BrCH_2COOH$ 2-bromoacetic acid
        or 2-bromoethanoic acid

c. 
    4,4,5,5-tetramethyloctanoic acid

d. $CH_3CH_2CH_2COO^-Li^+$ lithium butanoate

e. 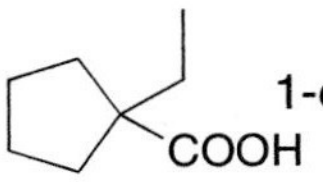   1-ethylcyclopentanecarboxylic acid

f. 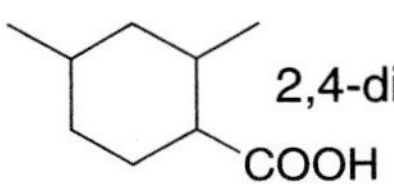   2,4-dimethylcyclohexanecarboxylic acid

g. 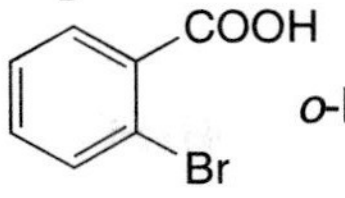   o-bromobenzoic acid

h. 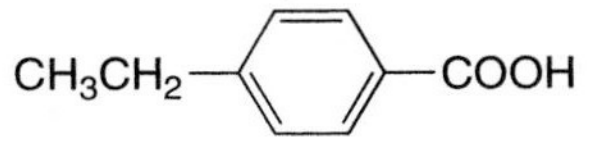 CH_3CH_2——COOH   p-ethylbenzoic acid

i. 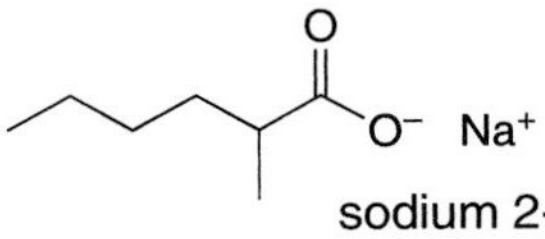
        sodium 2-methylhexanoate

j. 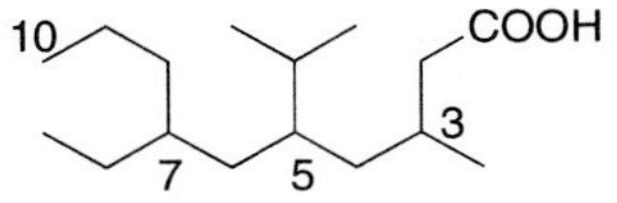
        7-ethyl-5-isopropyl-3-methyldecanoic acid

19.31

a. 3,3-dimethylpentanoic acid

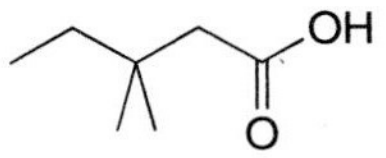

b. 4-chloro-3-phenylheptanoic acid

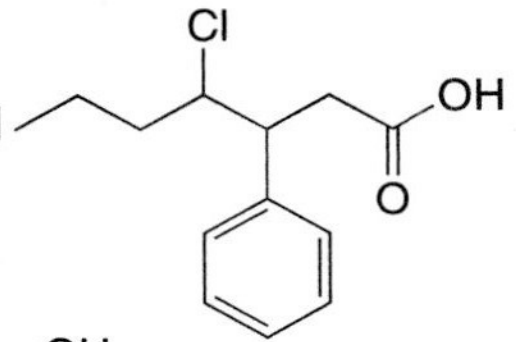

c. (2R)-2-chloropropanoic acid

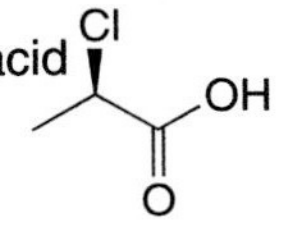

d. β,β-dichloropropionic acid

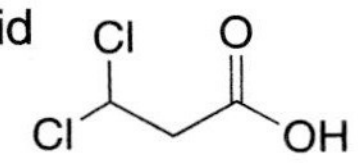

e. m-hydroxybenzoic acid

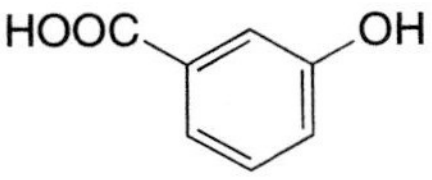

f. o-chlorobenzoic acid

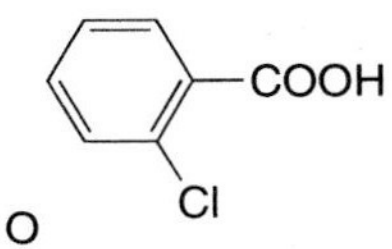

g. potassium acetate

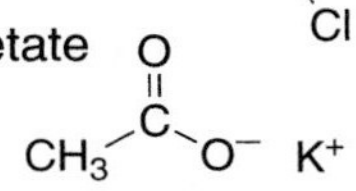

h. sodium α-bromobutyrate

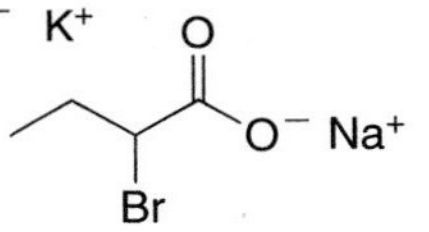

i. 2,2-dichloropentanedioic acid

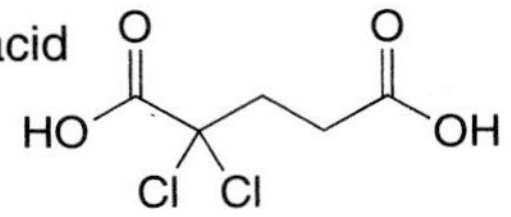

j. 4-isopropyl-2-methyloctanedioic acid

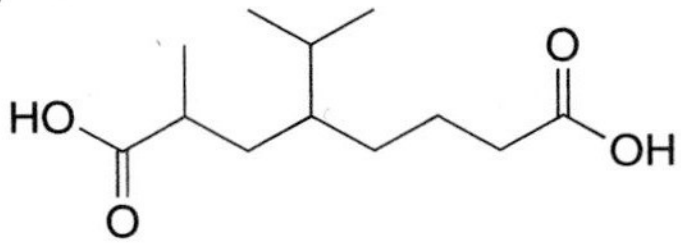

19.32

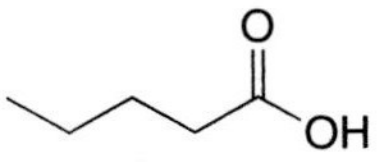

pentanoic acid

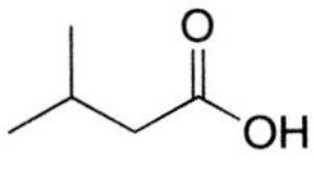

3-methylbutanoic acid

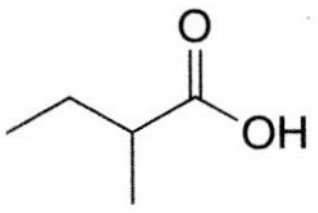

2-methylbutanoic acid

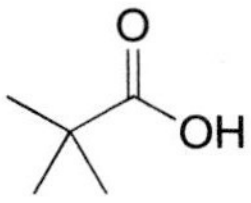

2,2-dimethylpropanoic acid

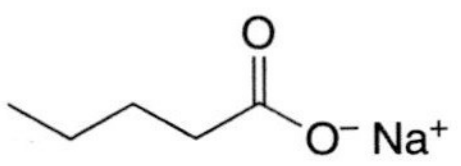

sodium pentanoate

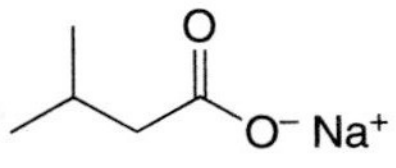

sodium 3-methylbutanoate

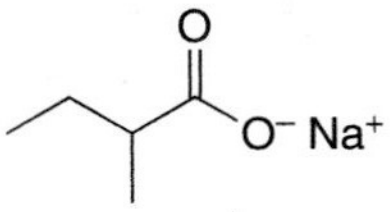

sodium 2-methylbutanoate

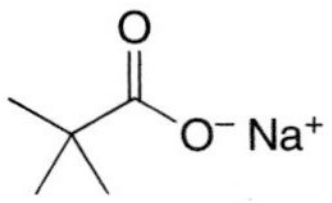

sodium
2,2-dimethylpropanoate

19.33

a.

lowest boiling point   **intermediate boiling point**   **highest boiling point**

b.

lowest boiling point   **intermediate boiling point**   **highest boiling point**

19.34

a.

c.

b. $(CH_3)_2CH$—

—CH_3 $\xrightarrow{KMnO_4}$ HOOC—

—COOH   d. $CH_3(CH_2)_6CH_2OH$ $\xrightarrow[H_2SO_4, H_2O]{Na_2Cr_2O_7}$ $CH_3(CH_2)_6COOH$

19.35

a.

b. $HC\equiv CH$ $\xrightarrow[{[2]\ CH_3I}]{[1]\ NaNH_2}$ $HC\equiv CCH_3$ $\xrightarrow[{[2]\ CH_3CH_2I}]{[1]\ NaNH_2}$ $CH_3CH_2C\equiv CCH_3$ $\xrightarrow[{[2]\ H_2O}]{[1]\ O_3}$ CH_3CH_2COOH + CH_3COOH
$\qquad\qquad\qquad\qquad$ **C** $\qquad\qquad\qquad\qquad$ **D** $\qquad\qquad\qquad\qquad$ **E** + **F**

c.

19.36

Bases: [1] ^-OH pK_a (H_2O) = 15.7; [2] $CH_3CH_2^-$ pK_a (CH_3CH_3) = 50; [3] $^-NH_2$ pK_a (NH_3) = 38;
[4] NH_3 pK_a (NH_4^+) = 9.4; [5] $HC\equiv C^-$ pK_a ($HC\equiv CH$) = 25.

a. CH_3—

—COOH

pK_a = 4.3
All of the bases
can deprotonate this.

b. Cl—

—OH

pK_a = 9.4
^-OH, $CH_3CH_2^-$, $^-NH_2$, and $HC\equiv C^-$
can deprotonate this.

c. $(CH_3)_3COH$

pK_a = 18
$CH_3CH_2^-$, $^-NH_2$, and $HC\equiv C^-$
can deprotonate this.

19.37

a.

—COOH + K^+ $^-OC(CH_3)_3$ $\rightleftharpoons$

—COO^- K^+ + $HOC(CH_3)_3$
pK_a = 18

Reaction favors products.

pK_a = 4.2

b.

OH + NH_3 $\rightleftharpoons$

O^- + NH_4^+
pK_a = 9.4

Reaction favors reactants.

$pK_a \approx 16$

c. + Na$^+$ NH$_2^-$ ⇌ + NH$_3$ + Na$^+$ | Reaction favors products. |

pK_a = 10   pK_a = 38

d. + CH$_3^-$Li$^+$ ⇌ + CH$_4$ | Reaction favors products. |

pK_a ≈ 4   pK_a = 50

e. + Na$^+$H$^-$ ⇌ + H$_2$ | Reaction favors products. |

pK_a ≈ 16   pK_a = 35

f. CH$_3$——OH + Na$_2$CO$_3$ ⇌ CH$_3$——O$^-$ Na$^+$ + Na$^+$ HCO$_3^-$ | With the same pK_a for the starting acid and the conjugate acid, an equal amount of starting materials and products is present. |

pK_a = 10.2   pK_a = 10.2

19.38 The stronger acid has a lower pK_a and a weaker conjugate base.

a. or

carboxylic acid   alcohol
stronger acid   weaker acid
lower pK_a   higher pK_a
weaker conjugate base   **stronger conjugate base**

c. CH$_3$——COOH or Cl——COOH

CH$_3$ is electron donating.   Cl is electron withdrawing.
weaker acid   stronger acid
higher pK_a **lower pK_a**
stronger conjugate base   weaker conjugate base

b. ClCH$_2$COOH or FCH$_2$COOH

weaker acid   F is more electronegative.
higher pK_a stronger acid
stronger conjugate base   **lower pK_a**
                    weaker conjugate base

d. NCCH$_2$COOH or CH$_3$COOH

CN is electron withdrawing.   weaker acid
stronger acid   higher pK_a
lower pK_a   **stronger conjugate base**
weaker conjugate base

19.39

a.

least acidic   Br is electronegative   Cl more electronegative
               intermediate acidity   **most acidic**

b.

least acidic   **intermediate acidity**   **most acidic**

c.

least acidic   **intermediate acidity**   **most acidic**

d.

least acidic   **intermediate acidity**   **most acidic**

19.40

a. BrCH$_2$COO$^-$ BrCH$_2$CH$_2$COO$^-$ (CH$_3$)$_3$CCOO$^-$

weakest base   intermediate basicity   strongest base

c.

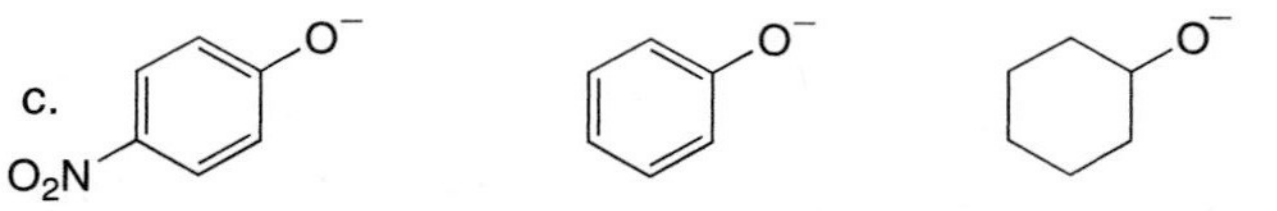

weakest base   intermediate basicity   strongest base

b.

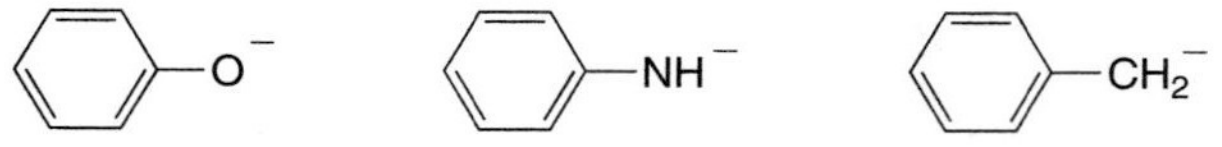

weakest base   intermediate basicity   strongest base

19.41

Increasing acidity

pK$_a$ values	ICH$_2$COOH least acidic 3.12	BrCH$_2$COOH 2.86	FCH$_2$COOH 2.66	F$_2$CHCOOH 1.24	F$_3$CCOOH most acidic 0.28

19.42 The OH of the phenol group in morphine is more acidic than the OH of the alcohol (pK$_a$ ≈ 10 versus pK$_a$ ≈ 16). KOH is basic enough to remove the phenolic OH, the most acidic proton.

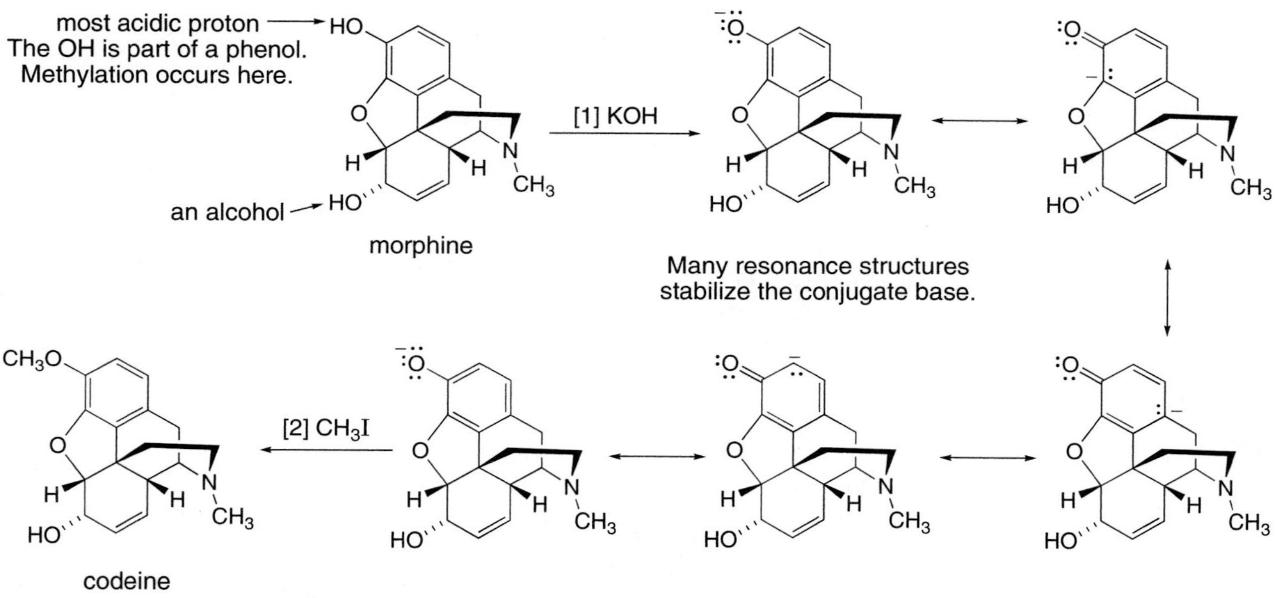

most acidic proton ⟶
The OH is part of a phenol.
Methylation occurs here.

an alcohol ⟶

morphine

[1] KOH

Many resonance structures
stabilize the conjugate base.

[2] CH$_3$I

codeine

19.43

a. The negative charge on the conjugate base of *p*-nitrophenol is delocalized on the NO$_2$ group, stabilizing the conjugate base, and making *p*-nitrophenol more acidic than phenol (where the negative charge is delocalized only around the benzene ring).

p-nitrophenol
pK$_a$ = 7.2

two of the possible resonance structures for the conjugate base (See part b. for all the possible resonance structures.)

phenol
pK$_a$ = 10

b. In the para isomer, the negative charge of the conjugate base is delocalized over both the benzene ring and onto the NO$_2$ group, whereas in the meta isomer it cannot be delocalized onto the NO$_2$ group. This makes the conjugate base from the para isomer more highly resonance stabilized, and the para substituted phenol more acidic than its meta isomer.

pK$_a$ = 7.2
p-nitrophenol

negative charge on
two O atoms
very good resonance structure
more stable conjugate base
stronger acid

pK$_a$ = 8.3
m-nitrophenol

19.44 A CH$_3$O group has an electron-withdrawing inductive effect and an electron-donating resonance effect. In 2-methoxyacetic acid, the OCH$_3$ group is bonded to an *sp*^{3} hybridized C, so there is no way to donate electron density by resonance. The CH$_3$O group withdraws electron density because of the electronegative O atom, stabilizing the conjugate base, and making CH$_3$OCH$_2$COOH a stronger acid than CH$_3$COOH.

more acidic acid   more stable conjugate base

In *p*-methoxybenzoic acid, the CH$_3$O group is bonded to an *sp*^{2} hybridized C, so it can donate electron density by a resonance effect. This destabilizes the conjugate base, making the starting material less acidic than C$_6$H$_5$COOH.

less acidic acid

like charges nearby
less stable conjugate base

19.45

The O in **A** is more electronegative than the N in **C** so there is a stronger electron-withdrawing inductive effect. This stabilizes the conjugate base of **A**, making **A** more acidic than **C**.

A
pK$_a$ = 3.2

B
pK$_a$ = 3.9

C
pK$_a$ = 4.4

Since the O in **A** is closer to the COOH group than the O atom in **B**, there is a stronger electron-withdrawing inductive effect. This makes **A** more acidic than **B**.

19.46

D

E

C

– H$^+$

– H$^+$

– H$^+$

Since the benzene ring is bonded to the α carbon (not the carbonyl carbon), this compound is not much different than any alkyl-substituted carboxylic acid.
least acidic

The electron withdrawing-inductive effect of the NO$_2$ group helps stabilize the COO$^-$ group.
intermediate acidity

Since the NO$_2$ group is bonded to a benzene ring that is bonded directly to the carbonyl group, inductive effects and resonance effects stabilize the conjugate base. For example, a resonance structure can be drawn that places a (+) charge close to the COO$^-$ group.
most acidic

Two of the resonance structures for the conjugate base of **C**:

unlike charges nearby
stabilizing

19.47

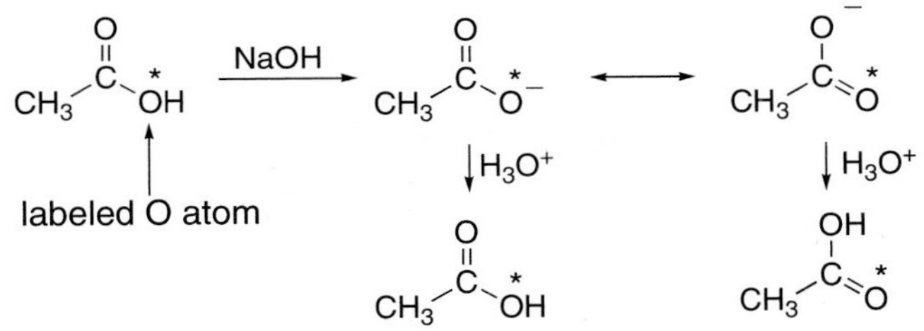

The resonance-stabilized carboxylate anion can now be protonated on either O atom, the one with the label and the one without the label.

The label is now in two different locations.

19.48

a.

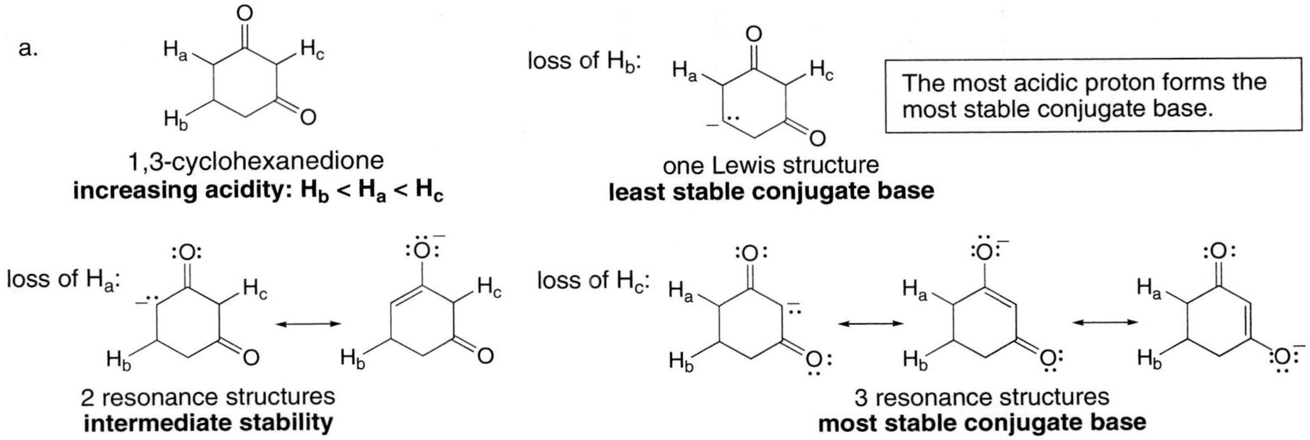

1,3-cyclohexanedione
increasing acidity: H$_b$ < H$_a$ < H$_c$

loss of H$_b$:

one Lewis structure
least stable conjugate base

The most acidic proton forms the most stable conjugate base.

loss of H$_a$:

2 resonance structures
intermediate stability

loss of H$_c$:

3 resonance structures
most stable conjugate base

b.

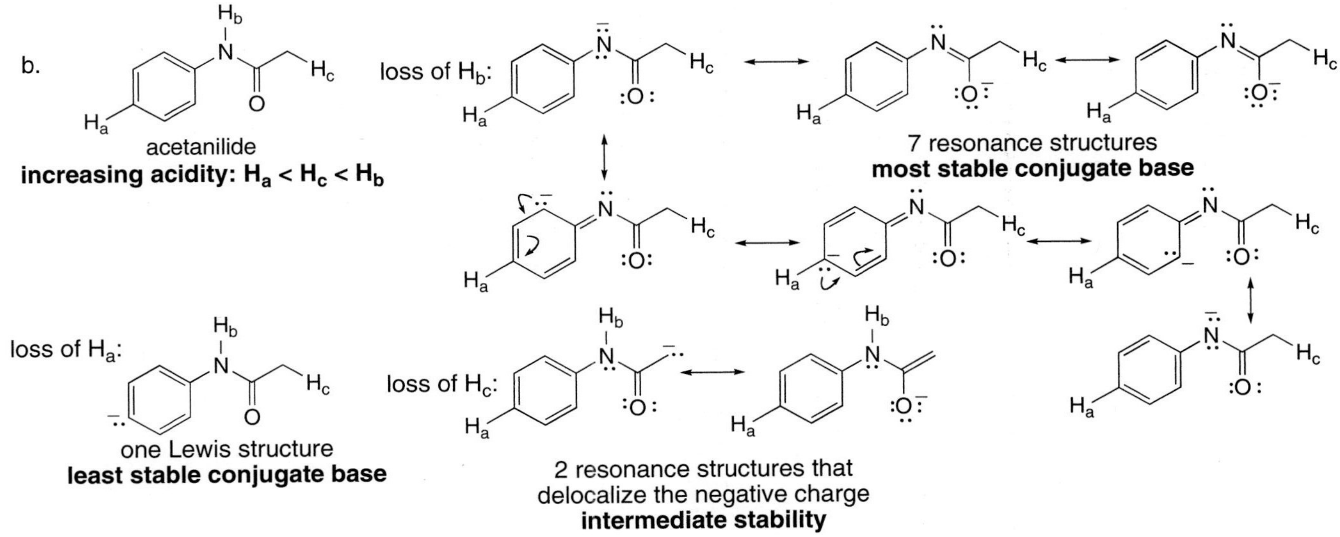

acetanilide
increasing acidity: H$_a$ < H$_c$ < H$_b$

loss of H$_b$:

7 resonance structures
most stable conjugate base

loss of H$_a$:

one Lewis structure
least stable conjugate base

loss of H$_c$:

2 resonance structures that delocalize the negative charge
intermediate stability

19.49

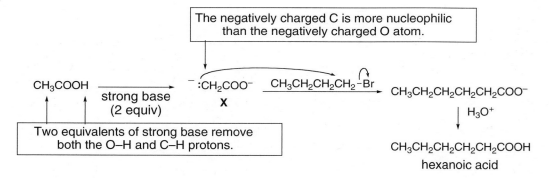

O O are double-bonded carbonyl structure

The conjugate base is resonance stabilized. Two of the structures place a negative charge on an O atom.
weaker conjugate base
stronger acid

HO⌒⌒OH ⟶ HO⌒⌒O⁻   The conjugate base has only one Lewis structure.
stronger conjugate base
weaker acid

19.50 As usual, compare the stability of the conjugate bases. With RSO_3H, loss of a proton forms a conjugate base that has three resonance structures, all of which are equivalent and place a negative charge on a more electronegative O atom. With the conjugate base of RCOOH, there are only two of these resonance structures. Thus, the conjugate base RSO_3^- is more highly resonance stabilized than $RCOO^-$, so RSO_3H is a stronger acid than RCOOH.

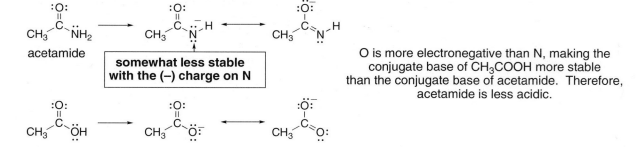

19.51

The negatively charged C is more nucleophilic than the negatively charged O atom.

CH_3COOH →(strong base (2 equiv))→ $^-:CH_2COO^-$ **X** →($CH_3CH_2CH_2CH_2$–Br)→ $CH_3CH_2CH_2CH_2CH_2COO^-$
↓ H_3O^+
$CH_3CH_2CH_2CH_2CH_2COOH$
hexanoic acid

Two equivalents of strong base remove both the O–H and C–H protons.

19.52

:O:
‖
CH_3–C–$\ddot{N}H_2$ ⟶   :O:
acetamide

:O:
‖
CH_3–C–$\ddot{N}$–H   ⟷   :Ö⁻:
CH_3–C=$\ddot{N}$–H

somewhat less stable with the (–) charge on N

O is more electronegative than N, making the conjugate base of CH_3COOH more stable than the conjugate base of acetamide. Therefore, acetamide is less acidic.

:O:
‖
CH_3–C–$\ddot{O}H$ ⟶   :O:
‖
CH_3–C–$\ddot{O}^-$ ⟷   :Ö⁻:
CH_3–C–$\ddot{O}$:

19.53

A

B

- Dissolve both compounds in CH_2Cl_2.
- Add 10% $NaHCO_3$ solution. This makes a carboxylate anion ($C_{10}H_7COO^-$) from **B**, which dissolves in the aqueous layer. The other compound (**A**) remains in the CH_2Cl_2.
- Separate the layers.

19.54

OH   and   OH

- Dissolve both compounds in CH_2Cl_2.
- Add 10% NaOH solution. This converts C_6H_5OH into a phenoxide anion, $C_6H_5O^-$, which dissolves in the aqueous solution. The alcohol remains in the organic layer (neutral) since it is not acidic enough to be deprotonated to any significant extent by NaOH.
- Separate the layers.

19.55 To separate two compounds in an aqueous extraction, one must be water soluble (or be able to be converted into a water-soluble ionic compound by an acid–base reaction), and the other insoluble. 1-Octanol has greater than 5 C's, making it insoluble in water. Octane is an alkane, also insoluble in water. Neither compound is acidic enough to be deprotonated by a base in aqueous solution. Since their solubility properties are similar, they cannot be separated by an extraction procedure.

19.56

a. Molecular formula: $C_3H_5ClO_2$ ——→ one double bond or ring
   IR: 3500–2500 cm^{-1}, 1714 cm^{-1} ——→ C=O and O–H
   NMR data: 2.87 (triplet, 2 H), 3.76 (triplet, 2 H), and 11.8 (singlet, 1 H) ppm

   $ClCH_2CH_2$—C(=O)—OH

b. Molecular formula: $C_8H_8O_3$ ——→ 5 double bonds or rings
   IR: 3500–2500 cm^{-1}, 1688 cm^{-1} ——→ C=O and O–H
   NMR data: 3.8 (singlet, 3 H), 7.0 (doublet, 2 H), 7.9 (doublet, 2 H), and 12.7 (singlet, 1 H) ppm

   CH_3O—⟨benzene⟩—COOH

   ↑ para disubstituted benzene ring ↑

c. Molecular formula: $C_8H_8O_3$ ——→ 5 double bonds or rings
   IR: 3500–2500 cm^{-1}, 1710 cm^{-1} ——→ C=O and O–H
   NMR data: 4.7 (singlet, 2 H), 6.9–7.3 (multiplet, 5 H), and 11.3 (singlet, 1 H) ppm

   ⟨benzene⟩—OCH_2COOH

   ↑ monosubstituted benzene ring

19.57

Compound **A**: Molecular formula $C_4H_8O_2$ (one degree of unsaturation)
IR absorptions at 3600–3200 (O–H), 3000–2800 (C–H), and 1700 (C=O) cm^{-1}
^1H NMR data:

Absorption	ppm	# of H's	Explanation	Structure:
singlet	2.2	3	a CH_3 group	
singlet	2.55	1	1 H adjacent to none or OH	
triplet	2.7	2	2 H's adjacent to 2 H's	
triplet	3.9	2	2 H's adjacent to 2 H's	

$$CH_3\overset{\overset{O}{\parallel}}{\underset{\mathbf{A}}{C}}CH_2CH_2OH$$

Compound **B**: Molecular formula $C_4H_8O_2$ (one degree of unsaturation)
IR absorptions at 3500–2500 (O–H) and 1700 (C=O) cm^{-1}
^1H NMR data:

Absorption	ppm	# of H's	Explanation	Structure:
doublet	1.6	6	6 H's adjacent to 1 H	
septet	2.3	1	1 H adjacent to 6 H's	
singlet (very broad)	10.7	1	OH of RCOOH	

$$CH_3-\overset{\overset{CH_3}{|}}{\underset{\underset{\mathbf{B}}{H}}{C}}-COOH$$

19.58

Compound **C**: Molecular formula $C_4H_8O_3$ (one degree of unsaturation)
IR absorptions at 3600–2500 (O–H) and 1734 (C=O) cm^{-1}
^1H NMR data:

Absorption	ppm	# of H's	Explanation	Structure:
triplet	1.2	3	a CH_3 group adjacent to 2 H's	
quartet	3.6	2	2 H's adjacent to 3 H's	
singlet	4.1	2	2 H's	
singlet	11.3	1	OH of COOH	

19.59

Compound **D**: Molecular formula $C_9H_9ClO_2$ (five degrees of unsaturation)
^{13}C NMR data: 30, 36, 128, 130, 133, 139, 179 = 7 different types of C's
^1H NMR data:

Absorption	ppm	# of H's	Explanation	Structure:
triplet	2.7	2	2 H's adjacent to 2 H's	
triplet	2.9	2	2 H's adjacent to 2 H's	
two signals	7.2	4	on benzene ring	
singlet	11.7	1	OH of COOH	

19.60

Molecular formula $C_6H_{12}O_2$ (1 double bond due to COOH)
^1H NMR: 1.1 (singlet), 2.2 (singlet), and 11.9 (singlet) ppm

$$CH_3-\overset{\overset{CH_3}{|}}{\underset{\underset{CH_3}{|}}{C}}-CH_2COOH$$

19.61

Molecular formula: $C_8H_6O_4$: 6 degrees of unsaturation
IR 1692 cm^{-1} (C=O)
^{1}H NMR 8.2 and 10.0 ppm (singlets)

↑ ↑
aromatic H   COOH

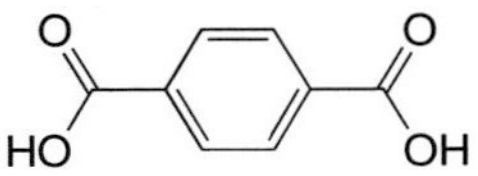

19.62

A 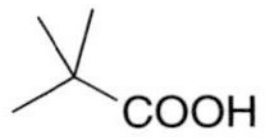   3 different C's
Spectrum [2]: peaks at 27, 39, 186 ppm

B    5 different C's
Spectrum [1]: peaks at 14, 22, 27, 34, 181 ppm

C 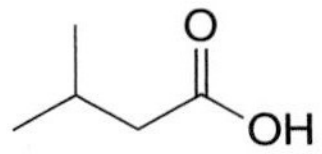   4 different C's
Spectrum [3]: peaks at 22, 26, 43, 180 ppm

19.63

GBL: Molecular formula $C_4H_6O_2$ (two degrees of unsaturation)
IR absorption at 1770 (C=O) cm^{-1}
^{1}H NMR data:

Absorption	ppm	# of H's	Explanation	Structure:
multiplet	2.28	2	2 H's adjacent to several H's	
triplet	2.48	2	2 H's adjacent to 2 H's	
triplet	4.35	2	2 H's adjacent to 2 H's	**GBL**

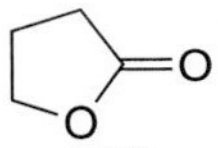

19.64

threonine

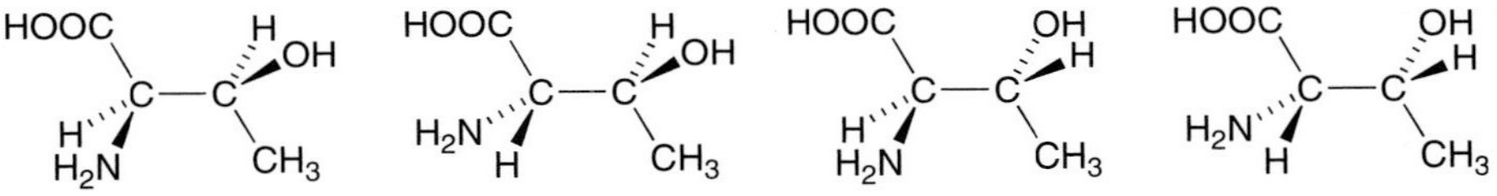

2R,3S   2S,3S   2R,3R   2S,3R
naturally
occurring

19.65

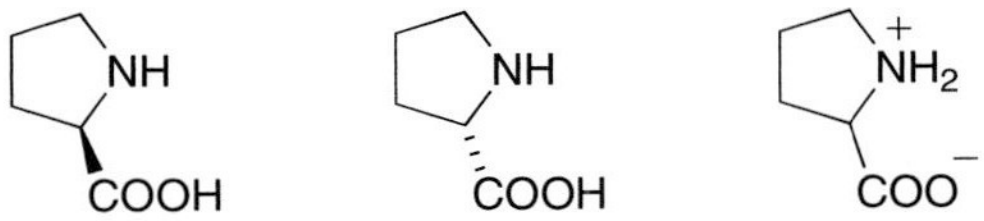

proline   enantiomer   zwitterion

19.66

a. methionine   b. serine

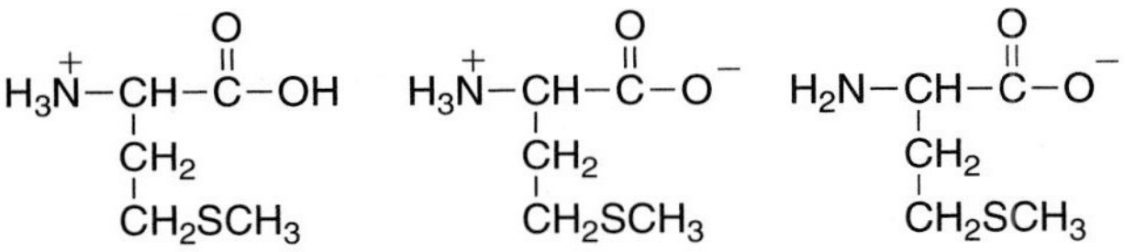

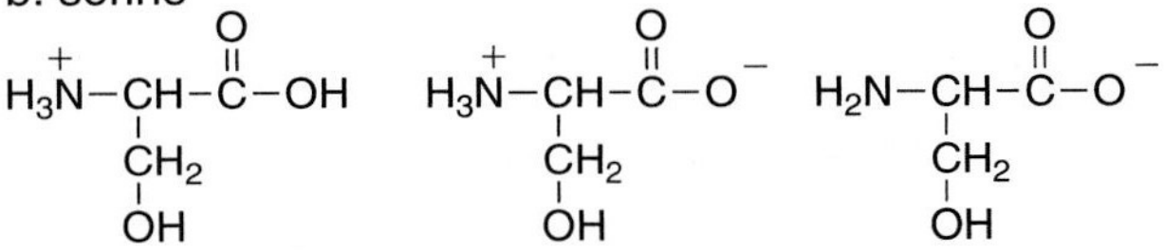

pH = 1   pH = 6   pH = 11   pH = 1   pH = 6   pH = 11
form at isoelectric point   form at isoelectric point

19.67

a. cysteine $pI = \dfrac{pK_a(\text{COOH}) + pK_a(\text{NH}_3{}^+)}{2} = (2.05) + (10.25) / 2 = \textbf{6.15}$

b. methionine $pI = \dfrac{pK_a(\text{COOH}) + pK_a(\text{NH}_3{}^+)}{2} = (2.28) + (9.21) / 2 = \textbf{5.75}$

19.68

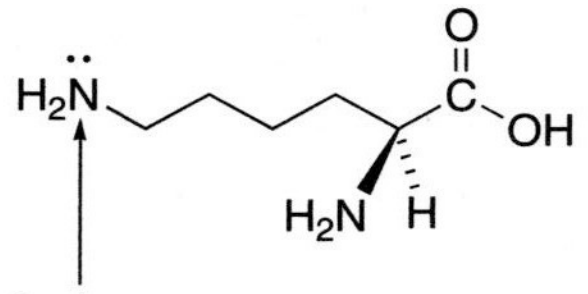

lysine
This lone pair is localized
on the N atom, making it a
base.

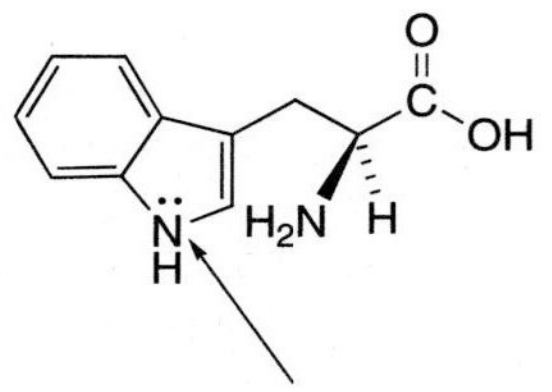

tryptophan
This lone pair is delocalized in the π system to give 10 π
electrons, making it aromatic. This is similar to pyrrole
(Chapter 17). Since these electrons are delocalized in the
aromatic system, this N atom in tryptophan is not basic.

19.69   The first equivalent of NH_3 acts as a base to remove a proton from the carboxylic acid. A second
equivalent then acts as a nucleophile to displace X to form the ammonium salt of the amino acid.

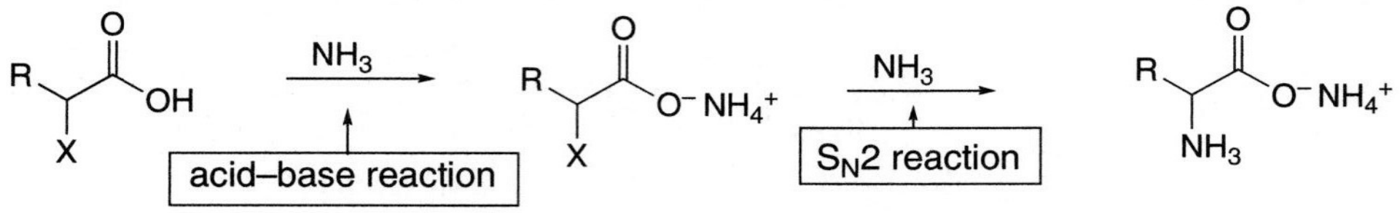

19.70

a. At pH = 1, the net
charge is (+1).

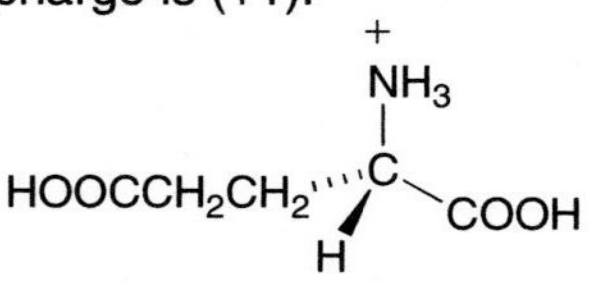

b. increasing pH: As base is added, the
most acidic proton is removed first,
then the next most acidic proton, and
so forth.

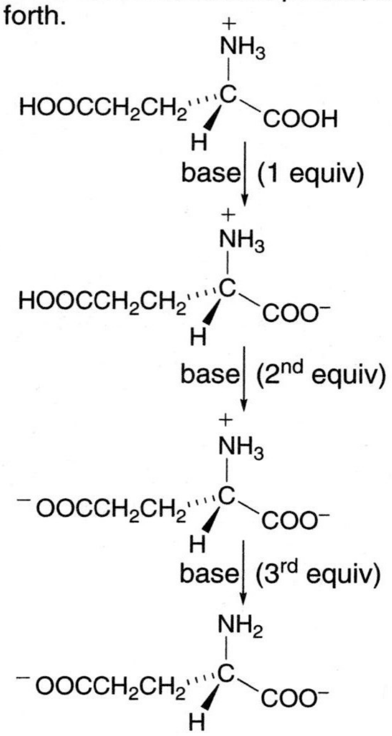

c. monosodium glutamate

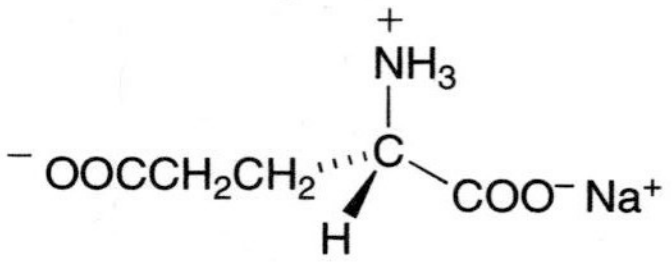

19.71 The first equivalent of NaH removes the most acidic proton; that is, the OH proton on the phenol. The resulting phenoxide can then act as a nucleophile to displace I to form a substitution product. With two equivalents, both OH protons are removed. In this case the more nucleophilic O atom is the stronger base; that is, the alkoxide derived from the alcohol (not the phenoxide), so this negatively charged O atom reacts first in a nucleophilic substitution reaction.

19.72

p-hydroxybenzoic acid
less acidic than benzoic acid

like charges on nearby atoms
destabilizing

The OH group donates electron density by its resonance effect and this destabilizes the conjugate base, making the acid less acidic than benzoic acid.

o-hydroxybenzoic acid
more acidic than benzoic acid

Intramolecular hydrogen bonding stabilizes the conjugate base, making the acid more acidic than benzoic acid.

19.73

2-hydroxybutanedioic acid
increasing acidity:
$H_d < H_c < H_b < H_e < H_a$

H_a and H_e must be the two most acidic protons since they are part of carboxylic acids. Loss of a proton forms a resonance-stabilized carboxylate anion that has the negative charge delocalized on two O atoms. H_a is more acidic than H_e because the nearby OH group on the α carbon increases acidity by an electron-withdrawing inductive effect. H_b is the next most acidic proton because the conjugate base places a negative charge on the electronegative O atom, but it is not resonance stabilized.

The least acidic H's are H_c and H_d since these H's are bonded to C atoms. The electronegative O atom further acidifies H_c by an electron-withdrawing inductive effect.

Chapter 20: Introduction to Carbonyl Chemistry

◆ Reduction reactions

---

[1] Reduction of aldehydes and ketones to 1° and 2° alcohols (20.4)

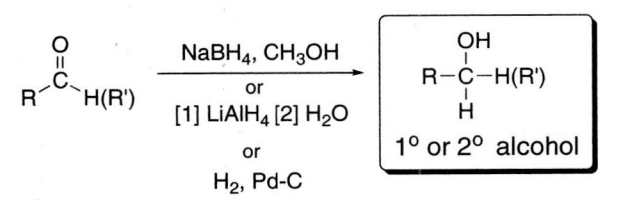

[2] Reduction of α,β-unsaturated aldehydes and ketones (20.4C)

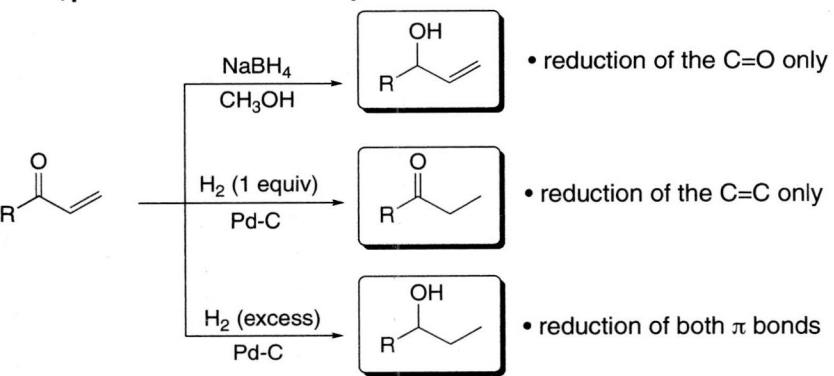

[3] Enantioselective ketone reduction (20.6)

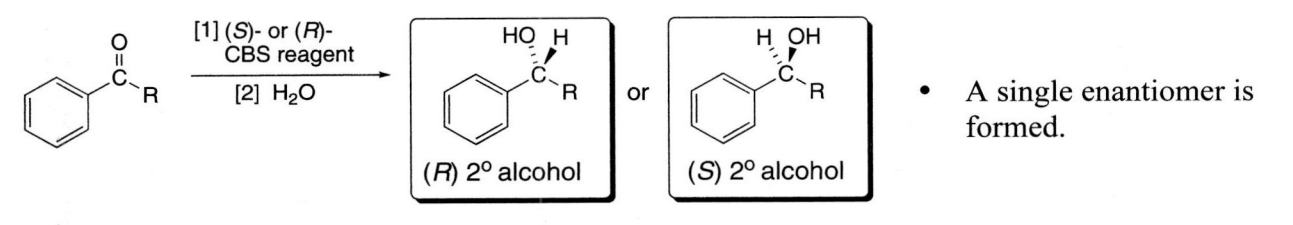

[4] Reduction of acid chlorides (20.7A)

- LiAlH$_4$, a strong reducing agent, reduces an acid chloride to a 1° alcohol.

- With LiAlH[OC(CH$_3$)$_3$]$_3$, a milder reducing agent, reduction stops at the aldehyde stage.

[5] Reduction of esters (20.7A)

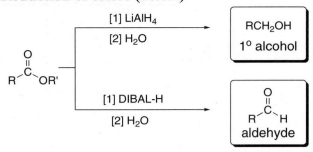

- LiAlH$_4$, a strong reducing agent, reduces an ester to a 1° alcohol.

- With DIBAL-H, a milder reducing agent, reduction stops at the aldehyde stage.

[6] Reduction of carboxylic acids to 1° alcohols (20.7B)

$$\underset{R}{\overset{O}{\underset{\parallel}{C}}}\text{OH} \xrightarrow[\text{[2] H}_2\text{O}]{\text{[1] LiAlH}_4} \boxed{\begin{array}{c}\text{RCH}_2\text{OH}\\ \text{1° alcohol}\end{array}}$$

[7] Reduction of amides to amines (20.7B)

$$\underset{R}{\overset{O}{\underset{\parallel}{C}}}\underset{|}{N} \xrightarrow[\text{[2] H}_2\text{O}]{\text{[1] LiAlH}_4} \boxed{\begin{array}{c}\text{RCH}_2-\underset{|}{N}-\\ \text{amine}\end{array}}$$

◆ Oxidation reactions

Oxidation of aldehydes to carboxylic acids (20.8)

$$\underset{R}{\overset{O}{\underset{\parallel}{C}}}\text{H} \xrightarrow[\substack{\text{or}\\ \text{Ag}_2\text{O, NH}_4\text{OH}}]{\text{CrO}_3,\ \text{Na}_2\text{Cr}_2\text{O}_7,\ \text{K}_2\text{Cr}_2\text{O}_7,\ \text{KMnO}_4} \boxed{\begin{array}{c}\underset{R}{\overset{O}{\underset{\parallel}{C}}}\text{OH}\\ \text{carboxylic acid}\end{array}}$$

- All Cr^{6+} reagents except PCC oxidize RCHO to RCOOH.
- Tollens reagent (Ag$_2$O + NH$_4$OH) oxidizes RCHO only. Primary (1°) and secondary (2°) alcohols do not react with Tollens reagent.

◆ Preparation of organometallic reagents (20.9)

[1] Organolithium reagents:

$$R-X\ +\ 2\,\text{Li} \longrightarrow \boxed{R-\text{Li}}\ +\ \text{LiX}$$

[2] Grignard reagents:

$$R-X\ +\ \text{Mg} \xrightarrow[(\text{CH}_3\text{CH}_2)_2\text{O}]{} \boxed{R-\text{Mg-X}}$$

[3] Organocuprate reagents:

$$R-X\ +\ 2\,\text{Li} \longrightarrow R-\text{Li}\ +\ \text{LiX}$$

$$2\,R-\text{Li}\ +\ \text{CuI} \longrightarrow \boxed{R_2\text{Cu}^-\ \text{Li}^+}\ +\ \text{LiI}$$

[4] Lithium and sodium acetylides:

$$R-C\equiv C-H \xrightarrow{Na^+ \ ^-NH_2} \boxed{R-C\equiv C^- \ Na^+} + NH_3$$
a sodium acetylide

$$R-C\equiv C-H \xrightarrow{R-Li} \boxed{R-C\equiv C-Li} + R-H$$
a lithium acetylide

◆ Reactions with organometallic reagents

[1] Reaction as a base (20.9C)

$$R-M + H-\overset{..}{\underset{..}{O}}-R \longrightarrow \boxed{R-H} + M^+ \ ^-\overset{..}{\underset{..}{O}}-R$$

- RM = RLi, RMgX, R_2CuLi
- This acid–base reaction occurs with H_2O, ROH, RNH_2, R_2NH, RSH, RCOOH, $RCONH_2$, and RCONHR.

[2] Reaction with aldehydes and ketones to form 1°, 2°, and 3° alcohols (20.10)

$$\underset{R}{\overset{O}{\underset{\|}{C}}}H(R') \xrightarrow[\text{[2] } H_2O]{\text{[1] } R''MgX \text{ or } R''Li} \boxed{\underset{R''}{\overset{OH}{\underset{|}{R-C-H}}}(R')}$$
1°, 2°, or 3° alcohol

[3] Reaction with esters to form 3° alcohols (20.13A)

$$\underset{R}{\overset{O}{\underset{\|}{C}}}OR' \xrightarrow[\text{[2] } H_2O]{\substack{\text{[1] } R''Li \text{ or } R''MgX \\ \text{(2 equiv)}}} \boxed{\underset{R''}{\overset{OH}{\underset{|}{R-C-R''}}}}$$
3° alcohol

[4] Reaction with acid chlorides (20.13)

$$\underset{R}{\overset{O}{\underset{\|}{C}}}Cl$$

$$\xrightarrow[\text{[2] } H_2O]{\substack{\text{[1] } R''Li \text{ or } R''MgX \\ \text{(2 equiv)}}} \boxed{\underset{R''}{\overset{OH}{\underset{|}{R-C-R''}}}}$$
3° alcohol

$$\xrightarrow[\text{[2] } H_2O]{\text{[1] } R'_2CuLi} \boxed{\underset{R}{\overset{O}{\underset{\|}{C}}}R'}$$
ketone

- More reactive organometallic reagents—R"Li and R"MgX—add two equivalents of R" to an acid chloride to form a 3° alcohol with two identical R" groups.

- Less reactive organometallic reagents—R'_2CuLi—add only one equivalent of R' to an acid chloride to form a ketone.

[5] Reaction with carbon dioxide—Carboxylation (20.14A)

$$R-MgX \xrightarrow[\text{[2] } H_3O^+]{\text{[1] } CO_2}$$

R—C with =O and OH

carboxylic acid

[6] Reaction with epoxides (20.14B)

$$\xrightarrow[\text{[2] } H_2O]{\text{[1] RLi, RMgX, or } R_2CuLi}$$

alcohol

[7] Reaction with α,β-unsaturated aldehydes and ketones (20.15B)

[1] R'Li or R'MgX
[2] H_2O

OH
R—C—C=C
R'
allylic alcohol

[1] R'$_2$CuLi
[2] H_2O

O H
R—C—C—C—
R'
ketone

- More reactive organometallic reagents— R'Li and R'MgX—react with α,β-unsaturated carbonyls by 1,2-addition.

- Less reactive organometallic reagents— R'$_2$CuLi— react with α,β-unsaturated carbonyls by 1,4-addition.

◆ **Protecting groups (20.12)**

[1] Protecting an alcohol as a *tert*-butyldimethylsilyl ether

$$R-O-H \; + \; Cl-\underset{\underset{CH_3}{|}}{\overset{\overset{CH_3}{|}}{Si}}-C(CH_3)_3$$

[Cl—TBDMS]

$$R-O-\underset{\underset{CH_3}{|}}{\overset{\overset{CH_3}{|}}{Si}}-C(CH_3)_3$$

[R—O—TBDMS]
tert-butyldimethylsilyl ether

[2] Deprotecting a *tert*-butyldimethylsilyl ether to re-form an alcohol

$$R-O-\underset{\underset{CH_3}{|}}{\overset{\overset{CH_3}{|}}{Si}}-C(CH_3)_3 \xrightarrow{(CH_3CH_2CH_2CH_2)_4N^+\,F^-}$$

[R—O—TBDMS]

R—O—H + $F-\underset{\underset{CH_3}{|}}{\overset{\overset{CH_3}{|}}{Si}}-C(CH_3)_3$

[F—TBDMS]

Chapter 20: Answers to Problems

20.1

a.

α-sinensal

[1] $C_{sp^2} - C_{sp^2}$
[2] σ: $C_{sp^2} - O_{sp^2}$ π: $C_p - O_p$
[3] $C_{sp^2} - C_{sp^2}$

b. The O is sp^2 hybridized.
   Both lone pairs occupy sp^2 hybrid orbitals.

20.2 A carbonyl compound with a reasonable leaving group undergoes substitution reactions. Those without good leaving groups undergo addition.

a.
$CH_3\overset{O}{\underset{}{C}}CH_3$

no good leaving group
addition reactions

b.
$CH_3CH_2CH_2\overset{O}{\underset{}{C}}Cl$

Cl–good leaving group
substitution reactions

c.
$CH_3\overset{O}{\underset{}{C}}OCH_3$

OCH_3–reasonable leaving group
substitution reactions

d.

no good leaving group
addition reactions

20.3 A carbonyl compound with a reasonable leaving group (NR_2 or OR bonded to the C=O) undergoes substitution reactions. Those without good leaving groups undergo addition.

no good leaving group
addition reactions

All other C=O's have leaving groups.
substitution reactions

20.4 Aldehydes are more reactive than ketones. In carbonyl compounds with leaving groups, the better the leaving group, the more reactive the carbonyl compound.

a. $CH_3CH_2CH_2\overset{O}{\underset{}{C}}H$ or $CH_3CH_2CH_2\overset{O}{\underset{}{C}}CH_3$

less hindered carbonyl
more reactive

b. $CH_3CH_2\overset{O}{\underset{}{C}}CH_3$ or $CH_3CH(CH_3)\overset{O}{\underset{}{C}}CH_2CH_3$

less hindered carbonyl
more reactive

c. $CH_3CH_2\overset{O}{\underset{}{C}}Cl$ or $CH_3\overset{O}{\underset{}{C}}OCH_3$

better leaving group
more reactive

d. $CH_3\overset{O}{\underset{}{C}}OCH_3$ or $CH_3\overset{O}{\underset{}{C}}NHCH_3$

better leaving group
more reactive

20.5 NaBH₄ reduces aldehydes to 1° alcohols, and ketones to 2° alcohols.

a. $CH_3CH_2CH_2$—CHO $\xrightarrow[\text{CH}_3\text{OH}]{\text{NaBH}_4}$ $CH_3CH_2CH_2$—CH(OH)H

c. (isobutyl ketone) $\xrightarrow[\text{CH}_3\text{OH}]{\text{NaBH}_4}$ (alcohol product)

b. (cyclohexanone) $\xrightarrow[\text{CH}_3\text{OH}]{\text{NaBH}_4}$ (cyclohexanol)

20.6 1° Alcohols are prepared from aldehydes and 2° alcohols are from ketones.

a. (2-pentanol) $\Rightarrow$ (2-pentanone)

b. (2-methylcyclohexanol) $\Rightarrow$ (2-methylcyclohexanone)

c. (benzyl alcohol) $\Rightarrow$ (benzaldehyde)

20.7

(1-methylcyclohexanol)   3° Alcohols cannot be made by reduction of a carbonyl group, because they do not contain a H on the C with the OH.

1-methylcyclohexanol

20.8

a. (ketone with alkene) $\xrightarrow[\text{[2] H}_2\text{O}]{\text{[1] LiAlH}_4}$ (alcohol with alkene)

d. (ketone with alkene) $\xrightarrow[\text{Pd-C}]{\text{H}_2\text{ (excess)}}$ (saturated alcohol)

b. (ketone with alkene) $\xrightarrow[\text{CH}_3\text{OH}]{\text{NaBH}_4}$ (alcohol with alkene)

e. (ketone with alkene) $\xrightarrow[\text{CH}_3\text{OH}]{\text{NaBH}_4 \text{ (excess)}}$ (alcohol with alkene)

c. (ketone with alkene) $\xrightarrow[\text{Pd-C}]{\text{H}_2 \text{ (1 equiv)}}$ (saturated ketone)

f. (ketone with alkene) $\xrightarrow[\text{CH}_3\text{OH}]{\text{NaBD}_4}$ (D-labeled alcohol with alkene)

20.9

a. (3-hexanone) $\xrightarrow[\text{CH}_3\text{OH}]{\text{NaBH}_4}$ (R-3-hexanol) + (S-3-hexanol)

b. (butanal, CHO) $\xrightarrow[\text{CH}_3\text{OH}]{\text{NaBH}_4}$ (butanol, OH)

c. $(CH_3)_3C$—(cyclohexanone) $\xrightarrow[\text{CH}_3\text{OH}]{\text{NaBH}_4}$ $(CH_3)_3C$—(cyclohexanol, ···OH) + $(CH_3)_3C$—(cyclohexanol, OH)

20.10

(aryl ketone with Cl chain) **A** $\xrightarrow[\text{[2] H}_2\text{O}]{\text{[1] (S)-CBS reagent}}$ (chiral alcohol HO H with Cl chain) **B**

20.11

Part [1]: Nucleophilic substitution of H for Cl

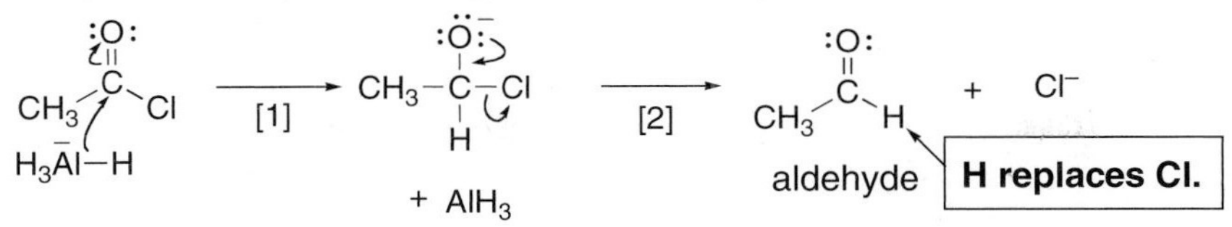

Part [2]: Nucleophilic addition of H⁻ to form an alcohol

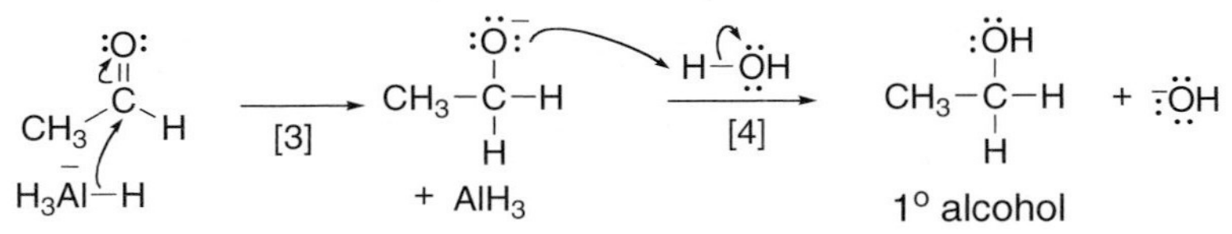

20.12 Acid chlorides and esters can be reduced to 1° alcohols. Keep the carbon skeleton the same in drawing an ester and acid chloride precursor.

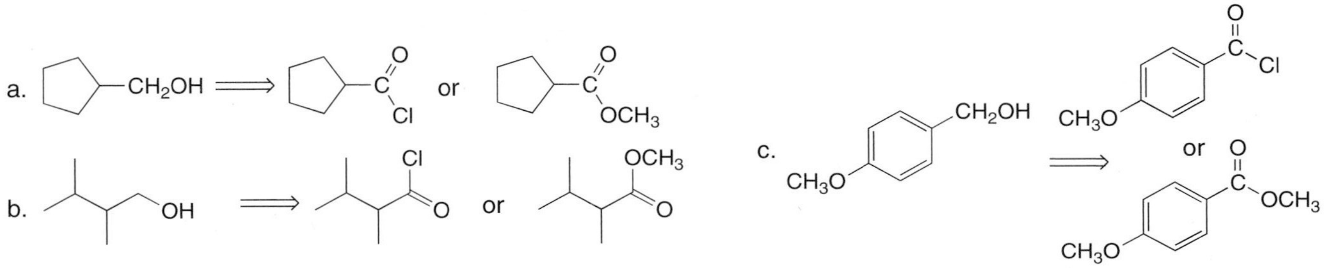

20.13

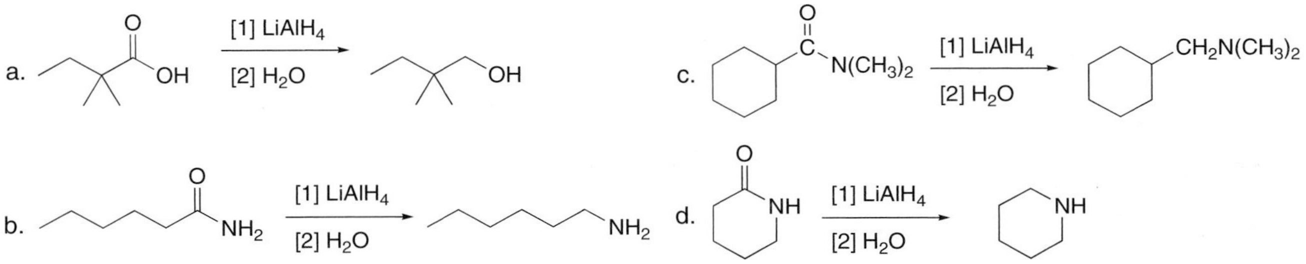

20.14

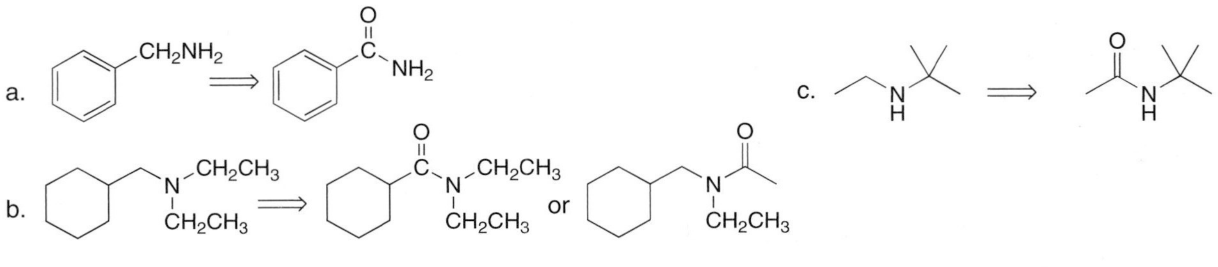

20.15

a. [structures] $COOCH_3$
[1] LiAlH$_4$ / [2] H$_2$O → OH ... OH + HOCH$_3$
NaBH$_4$ / CH$_3$OH → OH ... $COOCH_3$

c. CH_3O ... O
[1] LiAlH$_4$ / [2] H$_2$O → CH_3O ... OH
NaBH$_4$ / CH$_3$OH → CH_3O ... OH

b. CH_3O ... OH
[1] LiAlH$_4$ / [2] H$_2$O → HO ... OH + HOCH$_3$
NaBH$_4$ / CH$_3$OH → **Neither functional group reduced**

20.16

a. [structure] $CO_2CH_2CH_3$
$\xrightarrow[\text{(1 equiv)}\ \text{Pd-C}]{H_2}$ [structure] $CO_2CH_2CH_3$

b. [structure] $CO_2CH_2CH_3$
$\xrightarrow[\text{(2 equiv)}\ \text{Pd-C}]{H_2}$ [structure] $CO_2CH_2CH_3$ OH

c. [structure] $CO_2CH_2CH_3$
$\xrightarrow[\text{[2] H}_2\text{O}]{\text{[1] LiAlH}_4}$ [structure] OH ... OH + CH_3CH_2OH

d. [structure] $CO_2CH_2CH_3$
$\xrightarrow[\text{CH}_3\text{OH}]{\text{NaBH}_4}$ [structure] $CO_2CH_2CH_3$ OH

20.17 Tollens reagent reacts only with aldehydes.

a. [benzene]—CH_2OH
$\xrightarrow{\text{Ag}_2\text{O, NH}_4\text{OH}}$ **No reaction**
$\xrightarrow{\text{Na}_2\text{Cr}_2\text{O}_7\ \text{H}_2\text{SO}_4,\ \text{H}_2\text{O}}$ [benzene]—COOH

b. [structure] OH ... CHO
$\xrightarrow{\text{Ag}_2\text{O, NH}_4\text{OH}}$ OH ... $\overset{O}{C}$—OH
$\xrightarrow{\text{Na}_2\text{Cr}_2\text{O}_7\ \text{H}_2\text{SO}_4,\ \text{H}_2\text{O}}$ O ... $\overset{O}{C}$—OH

20.18

a. **B** $\xrightarrow[\text{CH}_3\text{OH}]{\text{NaBH}_4}$

b. **B** $\xrightarrow[\text{[2] H}_2\text{O}]{\text{[1] LiAlH}_4}$

c. **B** $\xrightarrow{\text{PCC}}$

d. **B** $\xrightarrow{\text{Ag}_2\text{O, NH}_4\text{OH}}$

e. **B** $\xrightarrow[\text{H}_2\text{SO}_4,\ \text{H}_2\text{O}]{\text{CrO}_3}$

20.19

a. $\text{CH}_3\text{CH}_2\text{Br} + 2\,\text{Li} \longrightarrow \text{CH}_3\text{CH}_2\text{Li} + \text{LiBr}$

b. $\text{CH}_3\text{CH}_2\text{Br} + \text{Mg} \longrightarrow \text{CH}_3\text{CH}_2\text{MgBr}$

c. $\text{CH}_3\text{CH}_2\text{Br} + 2\,\text{Li} \longrightarrow \text{CH}_3\text{CH}_2\text{Li} + \text{LiBr}$

$2\,\text{CH}_3\text{CH}_2\text{Li} + \text{CuI} \longrightarrow \text{LiCu(CH}_2\text{CH}_3)_2 + \text{LiI}$

20.20

$\text{HC}\equiv\text{CCH}_2\text{CH}_2\text{CH}_2\text{CH}_2\text{CH}_2\text{CH}_3 + \text{NaH} \longrightarrow \text{Na}^{+-}\text{C}\equiv\text{CCH}_2\text{CH}_2\text{CH}_2\text{CH}_2\text{CH}_2\text{CH}_3 + \text{H}_2 \longleftarrow \text{hydrogen gas}$

$\text{HC}\equiv\text{CCH}_2\text{CH}_2\text{CH}_2\text{CH}_2\text{CH}_2\text{CH}_3 + \text{CH}_3\text{MgBr} \longrightarrow \text{BrMgC}\equiv\text{CCH}_2\text{CH}_2\text{CH}_2\text{CH}_2\text{CH}_2\text{CH}_3 + \text{CH}_4 \longleftarrow \text{methane gas}$

20.21

a. + $\text{H}_2\text{O} \longrightarrow$ + LiOH

b. $\text{CH}_3-\overset{\text{CH}_3}{\underset{\text{CH}_3}{\text{C}}}-\text{MgBr} + \text{H}_2\text{O} \longrightarrow \text{CH}_3-\overset{\text{CH}_3}{\underset{\text{CH}_3}{\text{C}}}-\text{H} + \text{HOMgBr}$

c. + $\text{H}_2\text{O} \longrightarrow$ + HOMgBr

d. $\text{CH}_3\text{CH}_2\text{C}\equiv\text{C}-\text{Li} + \text{H}_2\text{O} \longrightarrow \text{CH}_3\text{CH}_2\text{C}\equiv\text{CH} + \text{LiOH}$

20.22 To draw the product, add the benzene ring to the carbonyl carbon and protonate the oxygen.

20.23 To draw the products, add the alkyl or phenyl group to the carbonyl carbon and protonate the oxygen.

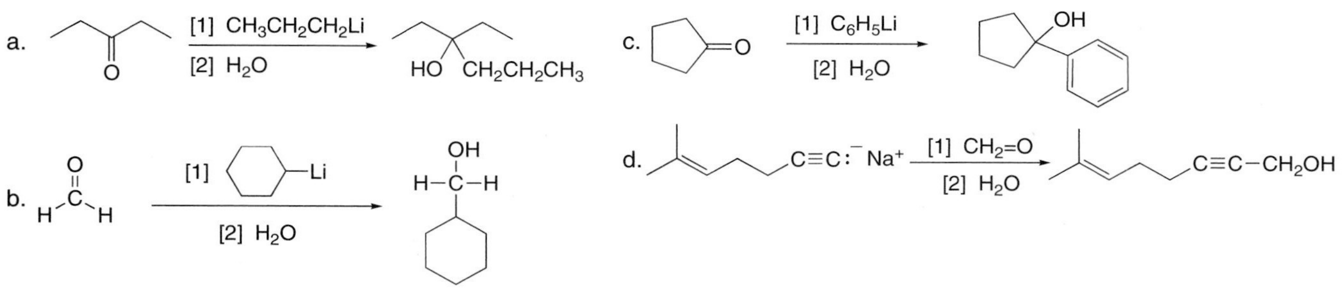

a. [1] $CH_3CH_2CH_2Li$ [2] H_2O

c. [1] C_6H_5Li [2] H_2O

b. [1] cyclohexyl–Li [2] H_2O

d. [1] $CH_2=O$ [2] H_2O

20.24 Addition of RM always occurs from above and below the plane of the molecule.

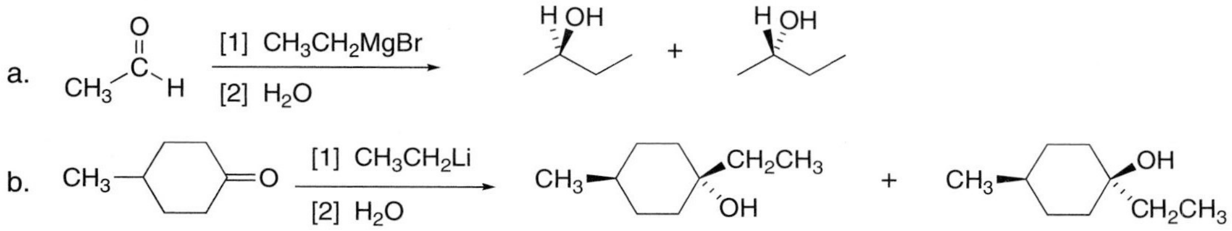

a. [1] CH_3CH_2MgBr [2] H_2O

b. [1] CH_3CH_2Li [2] H_2O

20.25

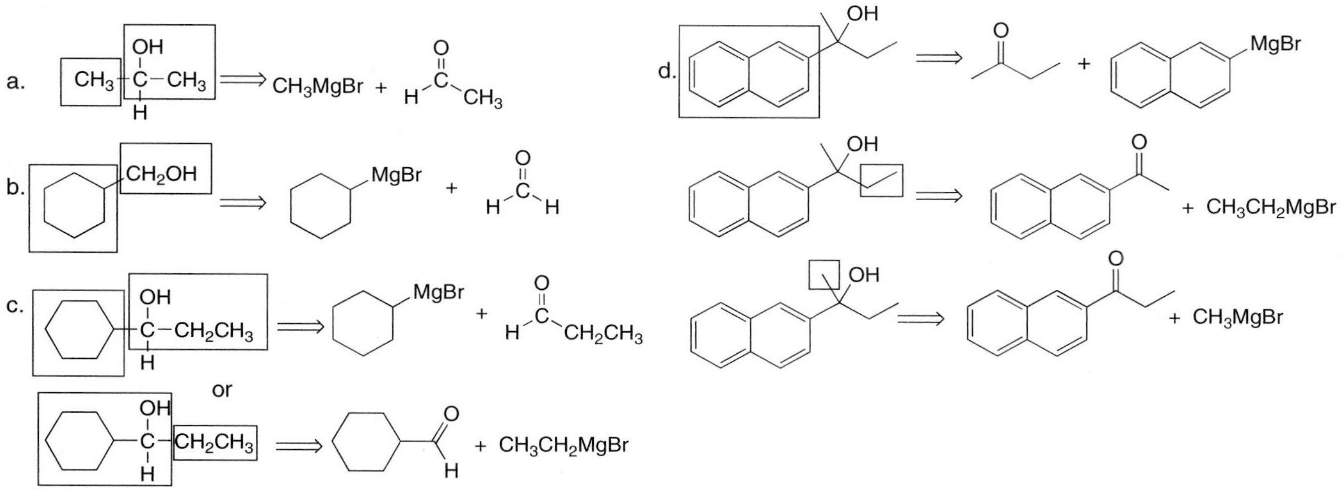

a.

b.

c.

or

d.

20.26

a.

linalool
(three methods)

$+ CH_3Li$

$Li +$

$+$ Li

lavandulol

$+ CH_2=O$

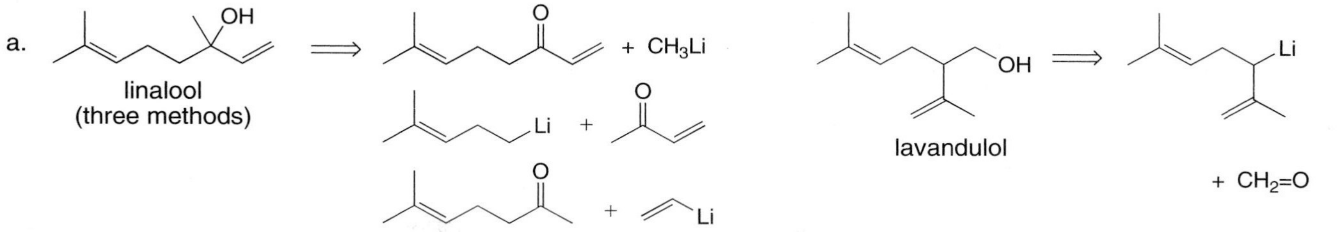

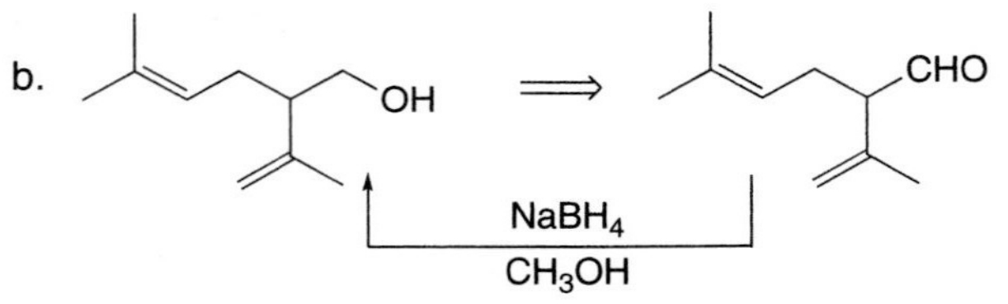

b.

NaBH$_4$
CH$_3$OH

c. Linalool is a 3° ROH. Therefore, it has no H on the carbon with the OH group, and cannot be prepared by reduction of a carbonyl compound.

20.27

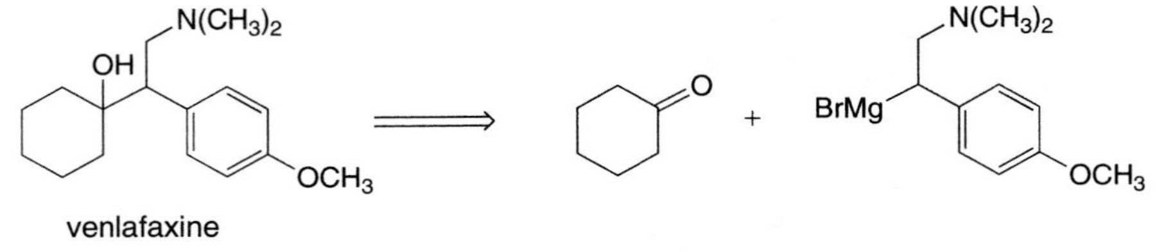

venlafaxine

20.28

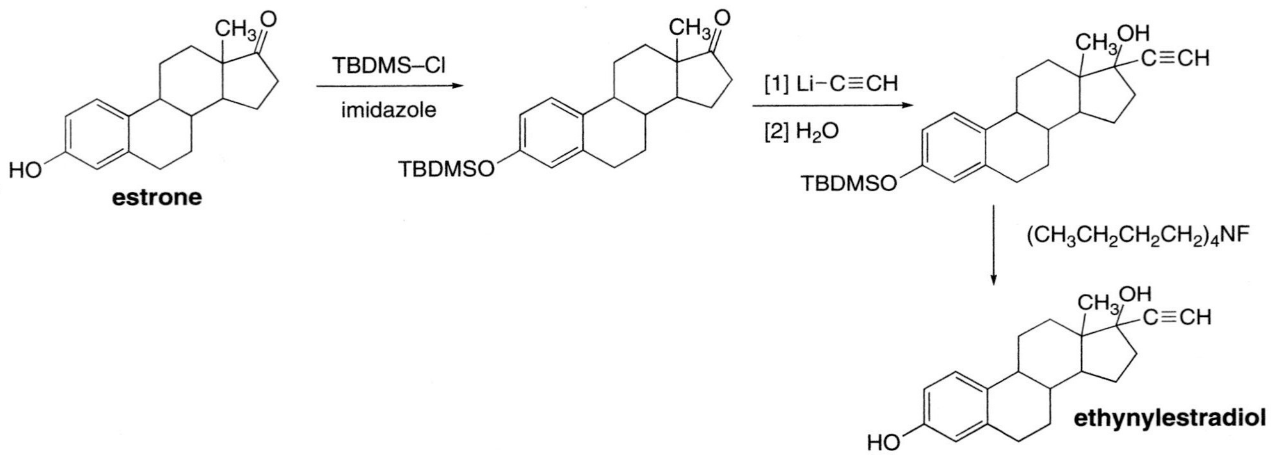

estrone

TBDMS–Cl
imidazole

[1] Li–C≡CH
[2] H$_2$O

(CH$_3$CH$_2$CH$_2$CH$_2$)$_4$NF

ethynylestradiol

20.29

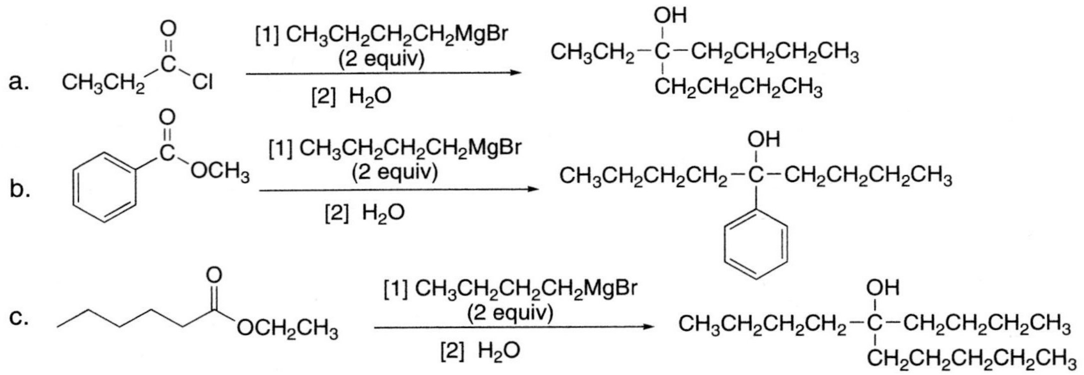

a.
[1] CH$_3$CH$_2$CH$_2$CH$_2$MgBr (2 equiv)
[2] H$_2$O

CH$_3$CH$_2$–C–CH$_2$CH$_2$CH$_2$CH$_3$ (with OH above and CH$_2$CH$_2$CH$_2$CH$_3$ below)

b.
[1] CH$_3$CH$_2$CH$_2$CH$_2$MgBr (2 equiv)
[2] H$_2$O

CH$_3$CH$_2$CH$_2$CH$_2$–C–CH$_2$CH$_2$CH$_2$CH$_3$ (with OH above and phenyl below)

c.
[1] CH$_3$CH$_2$CH$_2$CH$_2$MgBr (2 equiv)
[2] H$_2$O

CH$_3$CH$_2$CH$_2$CH$_2$–C–CH$_2$CH$_2$CH$_2$CH$_3$ (with OH above and CH$_2$CH$_2$CH$_2$CH$_2$CH$_3$ below)

20.30

a.

CH$_3$O–C–CH$_3$ + cyclohexyl–MgBr (2 equiv)

b. (CH$_3$CH$_2$CH$_2$)$_3$COH

CH$_3$O–C–CH$_2$CH$_2$CH$_3$ + CH$_3$CH$_2$CH$_2$MgBr (2 equiv)

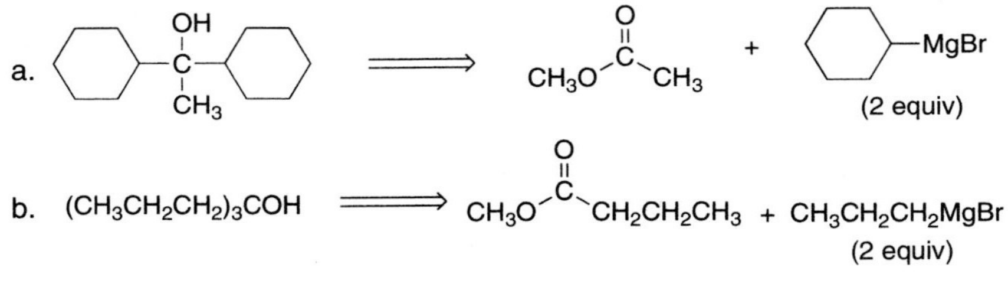

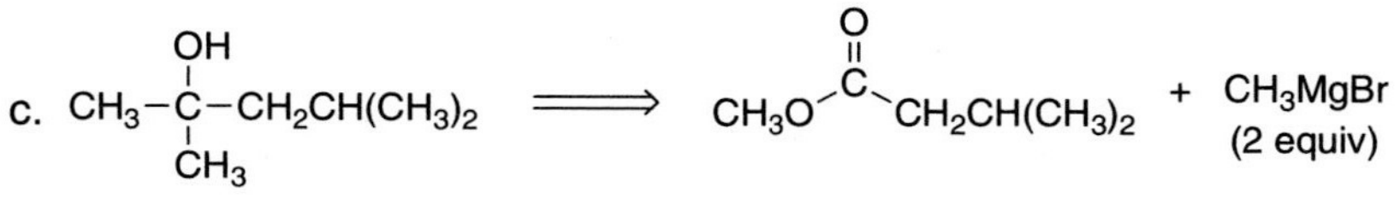

c.

20.31 The R group of the organocuprate has replaced the Cl on the acid chloride.

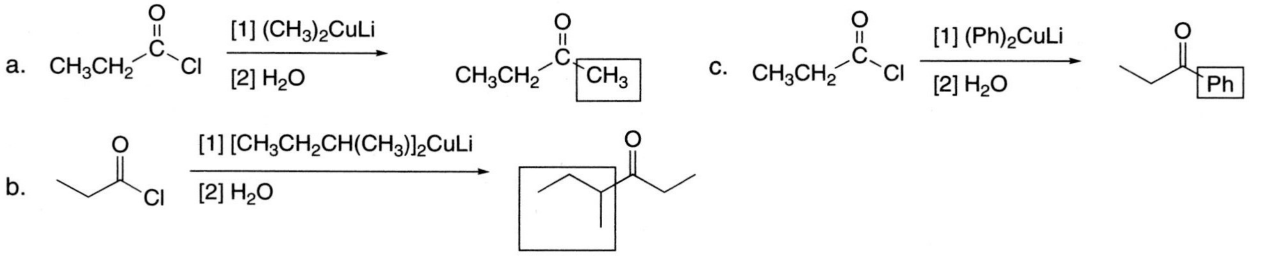

a.

c.

b.

20.32

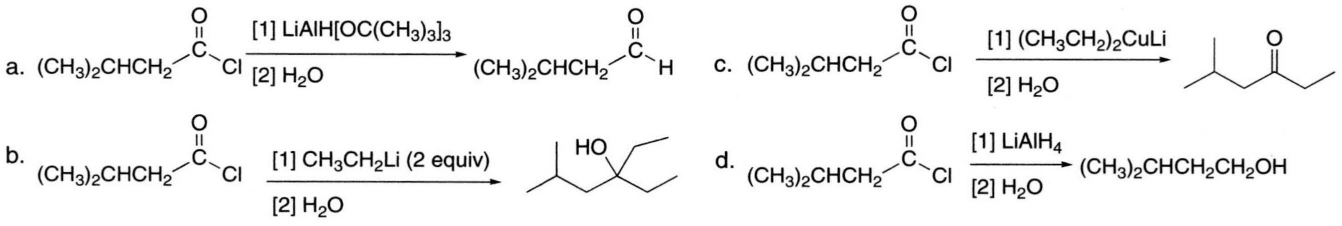

a.

c.

b.

d.

20.33

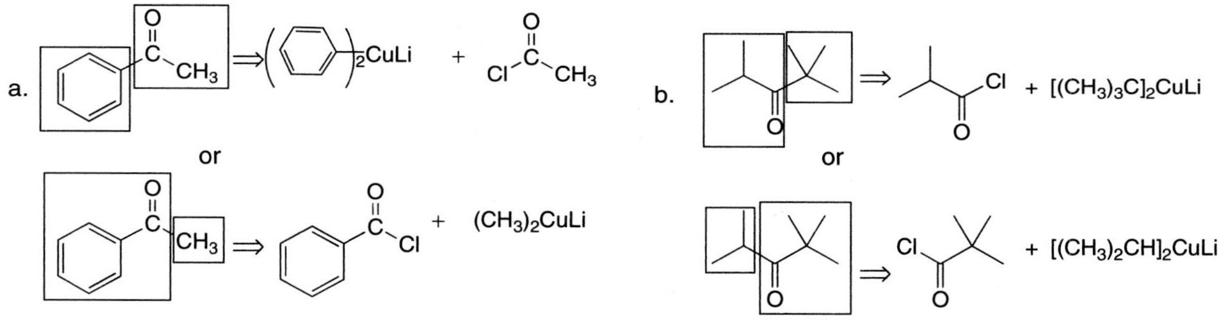

a.

b.

or

or

20.34

a.

b.

c.

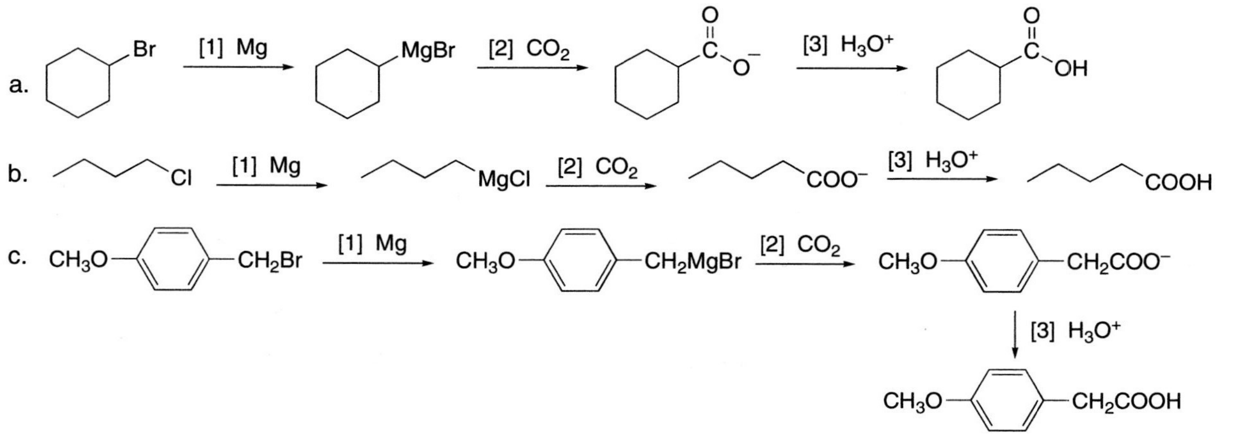

20.35

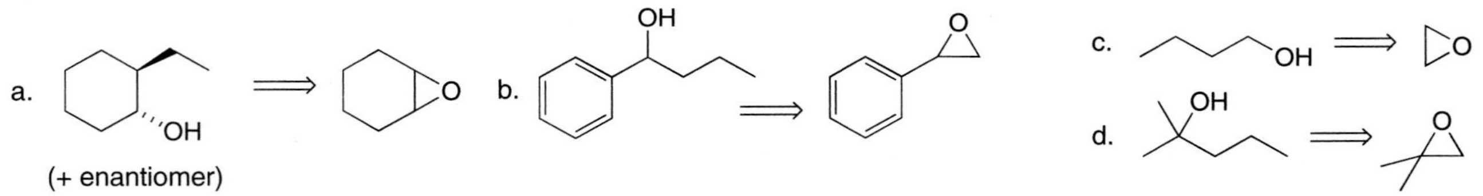

a.

(+ enantiomer)

b.

c.

d.

20.36 The characteristic reaction of α,β-unsaturated carbonyl compounds is nucleophilic addition. Grignard and organolithium reagents react by 1,2-addition and organocuprate reagents react by 1,4-addition.

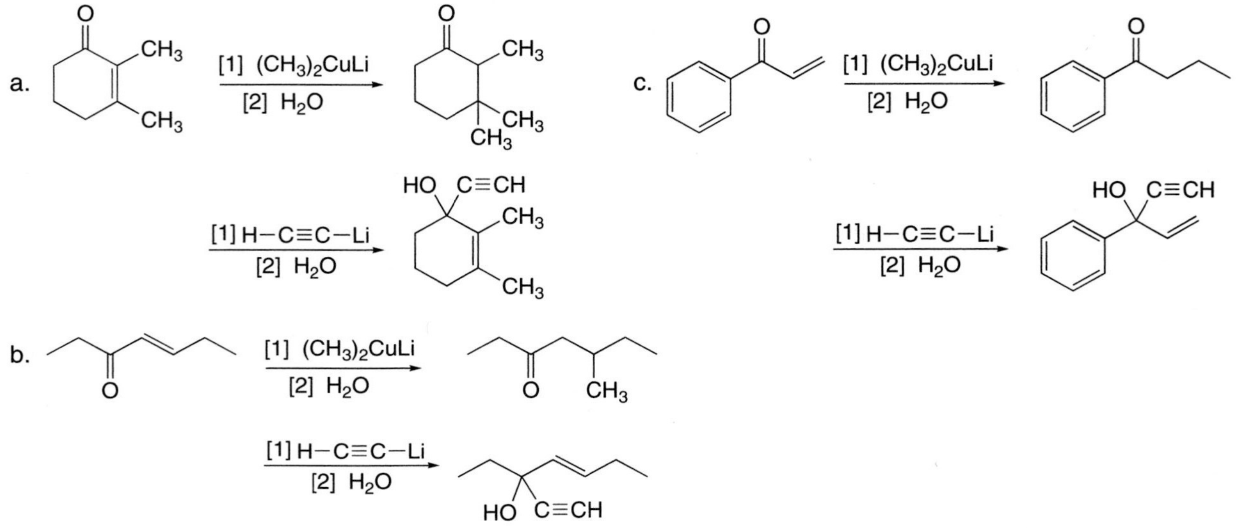

a.

b.

c.

20.37

a.

b.

c.

(from a.)

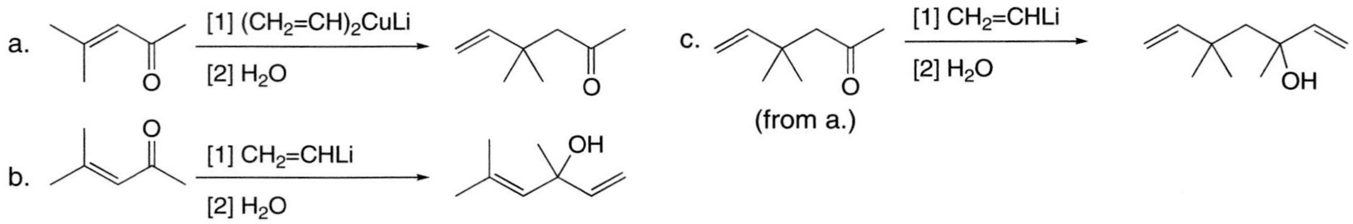

20.38

a. CH_3CH_2OH $\xrightarrow[PBr_3]{HBr \ or}$ CH_3CH_2Br $\xrightarrow{Mg}$ CH_3CH_2MgBr

[cyclohexanol]—OH $\xrightarrow{PCC}$ [cyclohexanone]=O $\xrightarrow[{[2] \ H_2O}]{[1] \ CH_3CH_2MgBr}$ [1-ethylcyclohexanol]OH

b. [1-ethylcyclohexanol]OH $\xrightarrow{HBr}$ [1-bromo-1-ethylcyclohexane]Br

(from a.)

c. [1-ethylcyclohexanol]OH $\xrightarrow{H_2SO_4}$ [ethylidenecyclohexane] + [ethylcyclohexene] $\xrightarrow[Pd-C]{H_2}$ [ethylcyclohexane]

(from a.)

d. [cyclohexanol]—OH $\xrightarrow[PBr_3]{HBr \ or}$ [bromocyclohexane]—Br $\xrightarrow{Mg}$ [cyclohexylMgBr]—MgBr $\Big\}$ $\xrightarrow{H_2O}$ [1-cyclohexylethanol]OH $\xrightarrow{PCC}$ [cyclohexyl methyl ketone]O

CH_3CH_2OH $\xrightarrow{PCC}$ CH_3CHO

e. [cyclohexyl]—MgBr $\xrightarrow{\overset{O}{\triangle}}$ $\xrightarrow{H_2O}$ [2-cyclohexylethanol]—OH

(from d.)

CH_3CH_2OH $\xrightarrow{H_2SO_4}$ $CH_2{=}CH_2$ $\xrightarrow{mCPBA}$ [epoxide]

20.39

a. [pentanal]H $\xrightarrow[CH_3OH]{NaBH_4}$ [pentanol]OH

b. [pentanal]H $\xrightarrow[{[2] \ H_2O}]{[1] \ LiAlH_4}$ [pentanol]OH

c. [pentanal]H $\xrightarrow[Pd-C]{H_2}$ [pentanol]OH

d. [pentanal]H $\xrightarrow{PCC}$ **No reaction**

e. [pentanal]H $\xrightarrow[H_2SO_4, \ H_2O]{Na_2Cr_2O_7}$ [pentanoic acid]OH

f. [pentanal]H $\xrightarrow[NH_4OH]{Ag_2O}$ [pentanoic acid]OH

g. [pentanal]H $\xrightarrow[{[2] \ H_2O}]{[1] \ CH_3MgBr}$ [2-hexanol]OH

h. [pentanal]H $\xrightarrow[{[2] \ H_2O}]{[1] \ C_6H_5Li}$ [1-phenyl-1-pentanol]OH

i. [pentanal]H $\xrightarrow[{[2] \ H_2O}]{[1] \ (CH_3)_2CuLi}$ **No reaction**

j. [pentanal]H $\xrightarrow[{[2] \ H_2O}]{[1] \ HC{\equiv}CNa}$ [oct-1-yn-4-ol]OH

k. [pentanal]H $\xrightarrow[{[2] \ H_2O}]{[1] \ CH_3C{\equiv}CLi}$ [oct-2-yn-4-ol]OH

l. [1-pentanol]OH $\xrightarrow[imidazole]{TBDMSCl}$ [silyl ether]O–TBDMS

20.40

a. $\xrightarrow[\text{CH}_3\text{OH}]{\text{NaBH}_4}$

g. $\xrightarrow[\text{[2] H}_2\text{O}]{\text{[1] CH}_3\text{MgBr}}$

b. $\xrightarrow[\text{[2] H}_2\text{O}]{\text{[1] LiAlH}_4}$

h. $\xrightarrow[\text{[2] H}_2\text{O}]{\text{[1] C}_6\text{H}_5\text{Li}}$ C₆H₅ OH → C_6H_5

c. $\xrightarrow[\text{Pd-C}]{\text{H}_2}$

i. $\xrightarrow[\text{[2] H}_2\text{O}]{\text{[1] (CH}_3)_2\text{CuLi}}$ **No reaction**

d. $\xrightarrow{\text{PCC}}$ **No reaction**

j. $\xrightarrow[\text{[2] H}_2\text{O}]{\text{[1] HC}\equiv\text{CNa}}$

e. $\xrightarrow[\text{H}_2\text{SO}_4,\ \text{H}_2\text{O}]{\text{Na}_2\text{Cr}_2\text{O}_7}$ **No reaction**

k. $\xrightarrow[\text{[2] H}_2\text{O}]{\text{[1] CH}_3\text{C}\equiv\text{CLi}}$

f. $\xrightarrow[\text{NH}_4\text{OH}]{\text{Ag}_2\text{O}}$ **No reaction**

l. $\xrightarrow[\text{imidazole}]{\text{TBDMSCl}}$ O–TBDMS

20.41

a. $\xrightarrow{\text{Li (2 equiv)}}$ Li   +   LiBr

d. $\xrightarrow{\text{H}_2\text{O}}$ + LiOH

b. $\xrightarrow{\text{Mg}}$ MgBr

e.   MgBr $\xrightarrow{\text{D}_2\text{O}}$ D   + DOMgBr

c.   Br $\xrightarrow[\text{[2] CuI (0.5 equiv)}]{\text{[1] Li (2 equiv)}}$ (CH₃CH₂CH₂CH₂)₂CuLi

f.   Li $\xrightarrow{\text{CH}_3\text{C}\equiv\text{CH}}$ +   LiC≡CCH₃

20.42

a.   MgBr $\xrightarrow{\text{CH}_2=\text{O}\quad\text{H}_2\text{O}}$ OH

g.   MgBr $\xrightarrow{\text{CH}_3\text{COOH}}$ + CH₃COO⁻

b.   MgBr $\xrightarrow{\text{(cyclopentanone)}\quad\text{H}_2\text{O}}$ HO

h.   MgBr $\xrightarrow{\text{HC}\equiv\text{CH}}$ + HC≡C⁻

c.   MgBr $\xrightarrow{\text{CH}_3\text{CH}_2\text{COCl}\quad\text{H}_2\text{O}}$

i.   MgBr $\xrightarrow{\text{CO}_2\quad\text{H}_3\text{O}^+}$

d.   MgBr $\xrightarrow{\text{CH}_3\text{CH}_2\text{COOCH}_3\quad\text{H}_2\text{O}}$

j.   MgBr $\xrightarrow{\text{(epoxide)}\quad\text{H}_2\text{O}}$

e.   MgBr $\xrightarrow{\text{H}_2\text{O}}$ + ⁻OH

k.   MgBr $\xrightarrow{\text{D}_2\text{O}}$ D   + ⁻OD

f.   MgBr $\xrightarrow{\text{CH}_3\text{CH}_2\text{OH}}$ + CH₃CH₂O⁻

l.   MgBr $\xrightarrow{\text{H}_2\text{O}}$

20.43

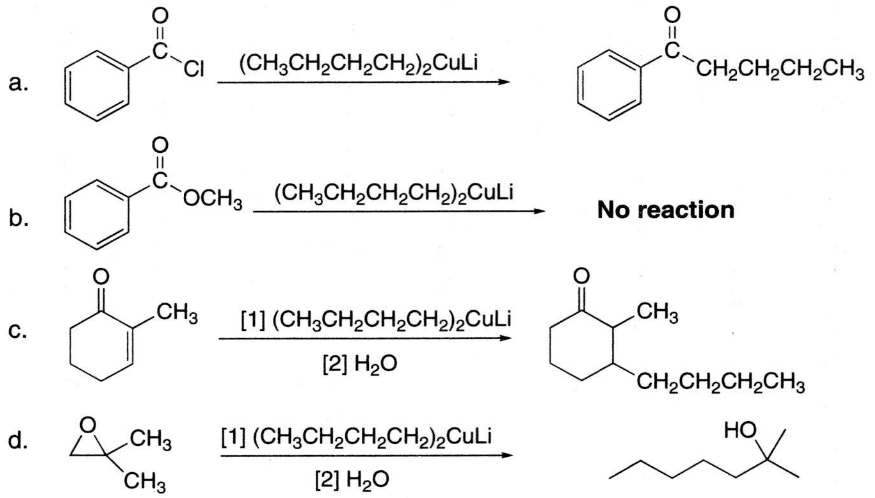

a.

b. **No reaction**

c.

d.

20.44 Arrange the larger group [$(CH_3)_3C–$] on the left side of the carbonyl.

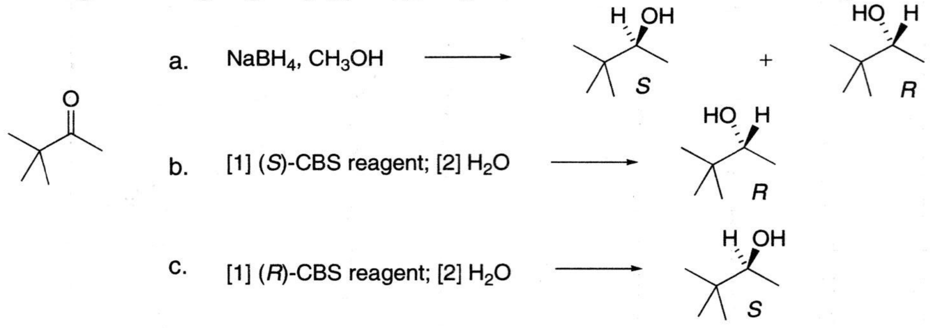

a. NaBH$_4$, CH$_3$OH

b. [1] (*S*)-CBS reagent; [2] H$_2$O

c. [1] (*R*)-CBS reagent; [2] H$_2$O

20.45

a. NaBH$_4$, CH$_3$OH

b. H$_2$ (1 equiv), Pd-C

c. H$_2$ (excess), Pd-C

d. [1] CH$_3$Li; [2] H$_2$O

e. [1] CH$_3$CH$_2$MgBr; [2] H$_2$O

f. [1] (CH$_2$=CH)$_2$CuLi; [2] H$_2$O

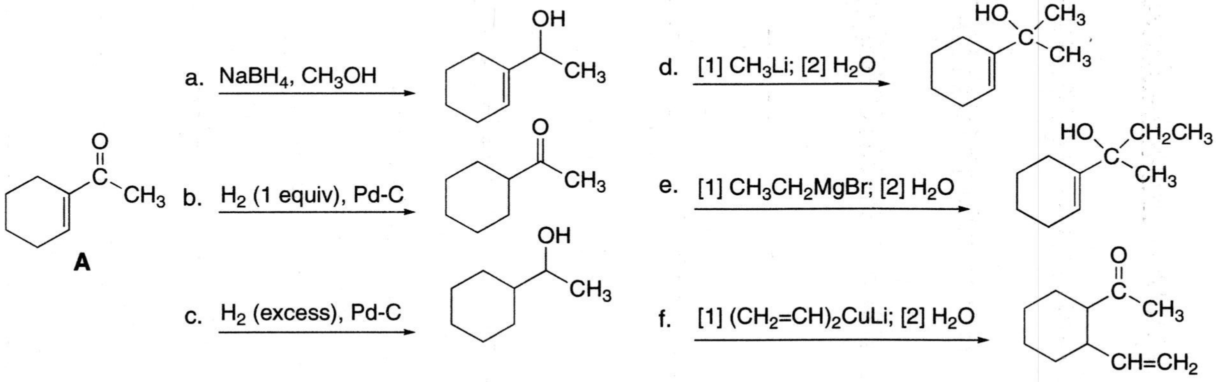

20.46

a. $(CH_3)_2CHCH_2CH_2$—C(=O)—Cl $\xrightarrow[\text{[2] } H_2O]{\text{[1] LiAlH[OC(CH}_3)_3]_3}}$ $(CH_3)_2CHCH_2CH_2$—C(=O)—H

b. $(CH_3)_2CHCH_2CH_2$—C(=O)—Cl $\xrightarrow[\text{[2] } H_2O]{\text{[1] } (CH_2=CH)_2CuLi}}$ $(CH_3)_2CHCH_2CH_2$—C(=O)—CH=CH$_2$

c. $(CH_3)_2CHCH_2CH_2$—C(=O)—Cl $\xrightarrow[\text{[2] } H_2O]{\substack{\text{[1] } C_6H_5MgBr \\ \text{(2 equiv)}}}$ $(CH_3)_2CHCH_2CH_2$—C(OH)(C$_6$H$_5$)—C$_6$H$_5$

d. $(CH_3)_2CHCH_2CH_2$—C(=O)—Cl $\xrightarrow[\text{[2] } H_2O]{\text{[1] LiAlH}_4}$ $(CH_3)_2CHCH_2CH_2$—CH(OH)—H

20.47

a. CH_3CH_2—C(=O)—OCH$_2$CH$_2$CH$_3$ $\xrightarrow[\text{[2] } H_2O]{\text{[1] LiAlH}_4}$ $CH_3CH_2CH_2OH$

b. CH_3CH_2—C(=O)—OCH$_2$CH$_2$CH$_3$ $\xrightarrow[\text{[2] } H_2O]{\substack{\text{[1] } CH_3CH_2CH_2MgCl \\ \text{(2 equiv)}}}$ CH_3CH_2—C(OH)(CH$_2$CH$_2$CH$_3$)—CH$_2$CH$_2$CH$_3$

c. CH_3CH_2—C(=O)—OCH$_2$CH$_2$CH$_3$ $\xrightarrow[\text{[2] } H_2O]{\text{[1] DIBAL-H}}$ CH_3CH_2—C(=O)—H

20.48

a. HO⎯⎯CHO $\xrightarrow[\text{H}_2SO_4,\ H_2O]{\text{CrO}_3}$ HOOC⎯⎯COOH

b. HO⎯⎯CHO $\xrightarrow{\text{PCC}}$ OHC⎯⎯CHO

c. HO⎯⎯CHO $\xrightarrow[\text{NH}_4OH]{\text{Ag}_2O}$ HO⎯⎯COOH

d. HO⎯⎯CHO $\xrightarrow[\text{H}_2SO_4,\ H_2O]{\text{Na}_2Cr_2O_7}$ HOOC⎯⎯COOH

20.49

a. CH$_3$CO⎯⎯COOCH$_3$ $\xrightarrow[\text{CH}_3OH]{\text{NaBH}_4}$ CH$_3$CH(OH)⎯⎯COOCH$_3$

b. CH$_3$CO⎯⎯COOCH$_3$ $\xrightarrow[\text{[2] } H_2O]{\text{[1] LiAlH}_4}$ CH$_3$CH(OH)⎯⎯CH$_2$OH

c. $(CH_3)_2N$—C(=O)⎯⎯COOH $\xrightarrow[\text{[2] } H_2O]{\text{[1] LiAlH}_4}$ $(CH_3)_2N$⎯⎯OH

d. Cl—C(=O)⎯CH$_2$CH$_2$—COOH $\xrightarrow[\text{[2] } H_2O]{\text{[1] LiAlH[OC(CH}_3)_3]_3}$ H—C(=O)⎯CH$_2$CH$_2$—COOH

20.50

a.

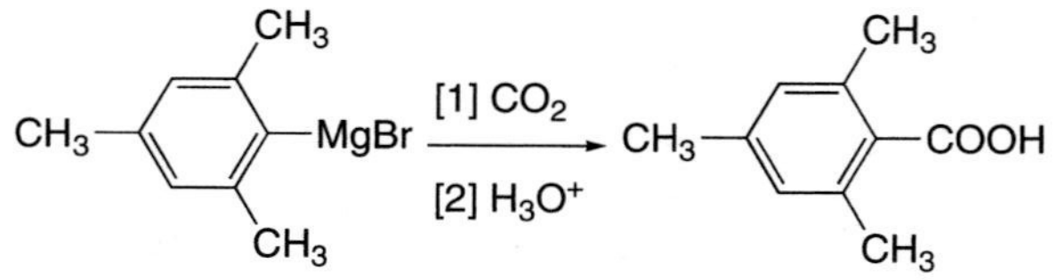

b.

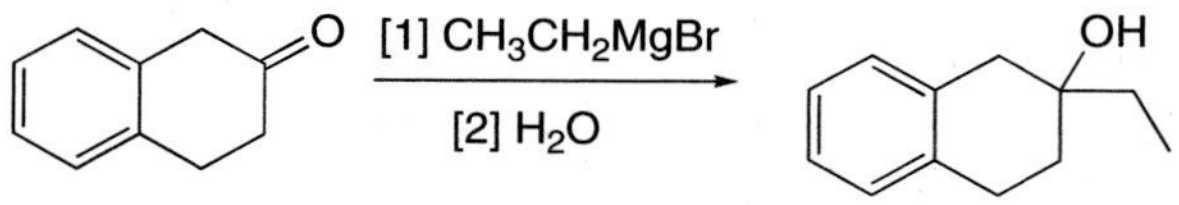

c.

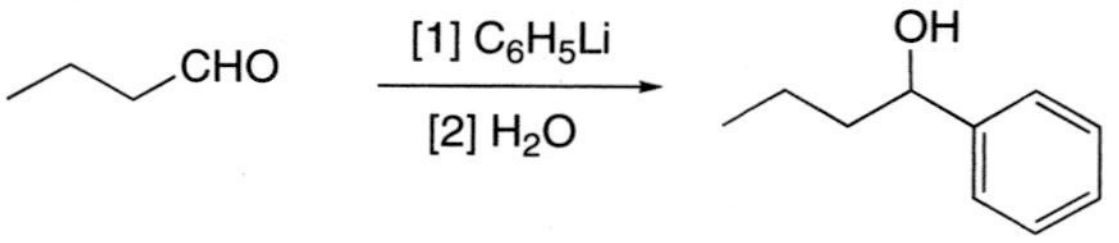

d.

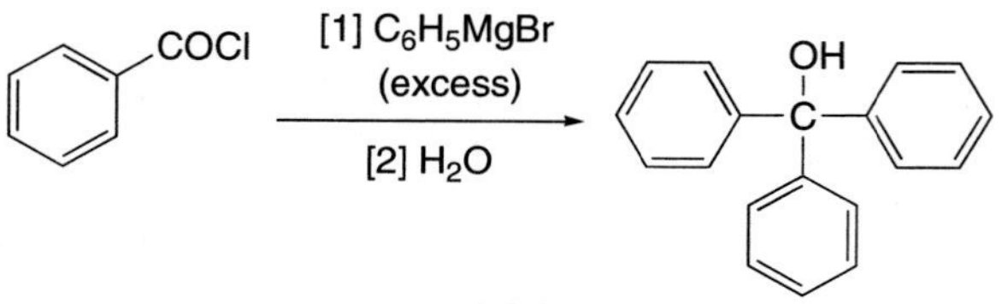

e.

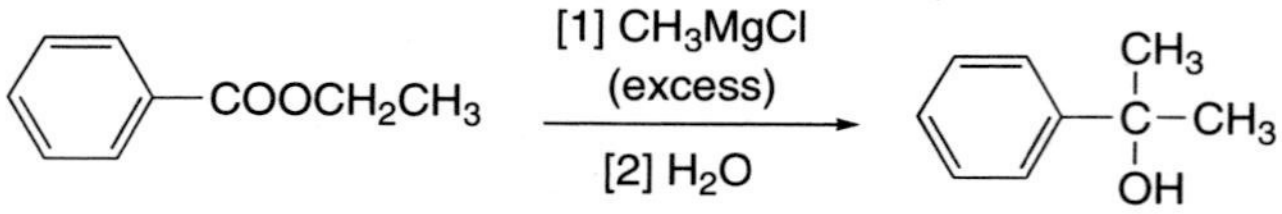

f.

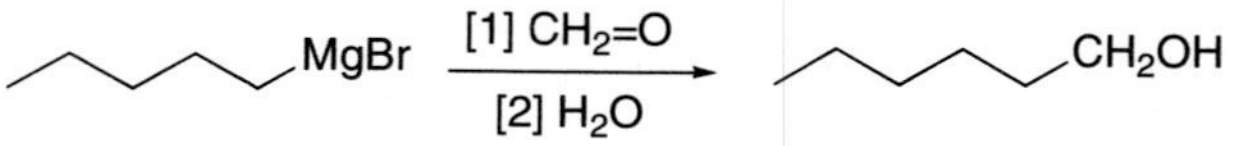

g.

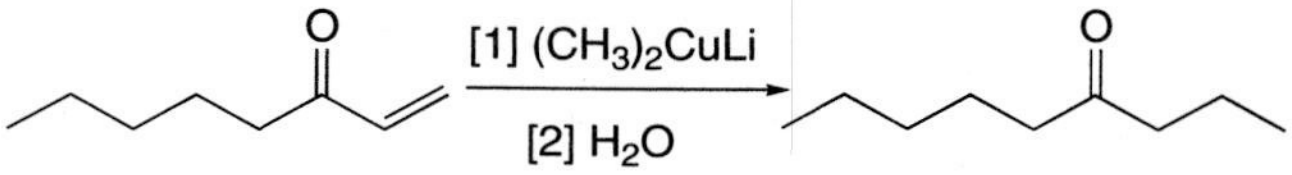

h.

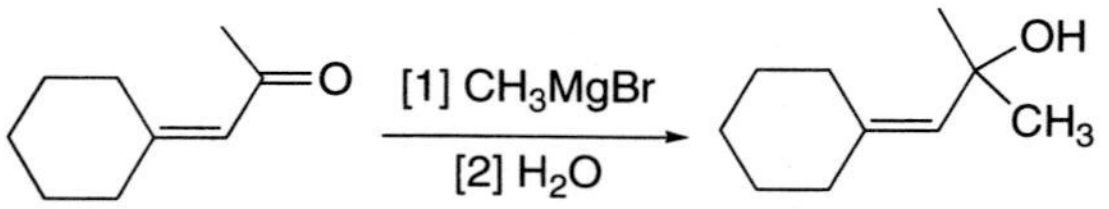

i.

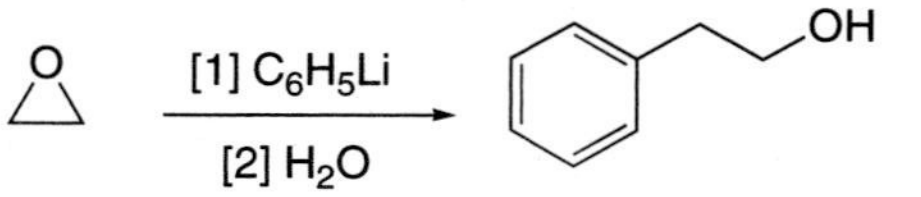

j.

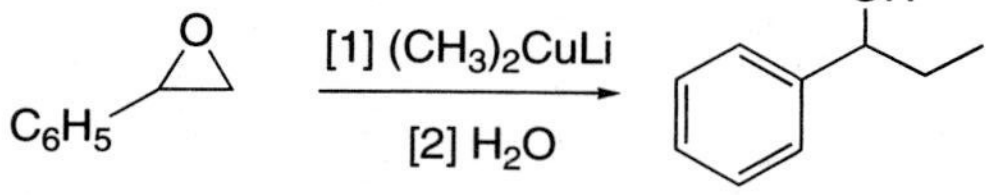

20.51

a.

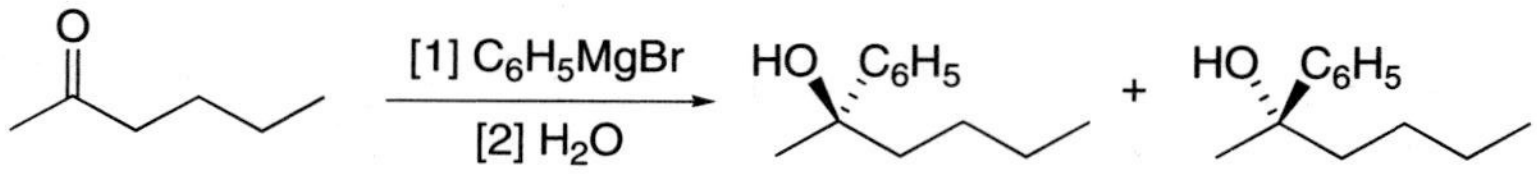

b.

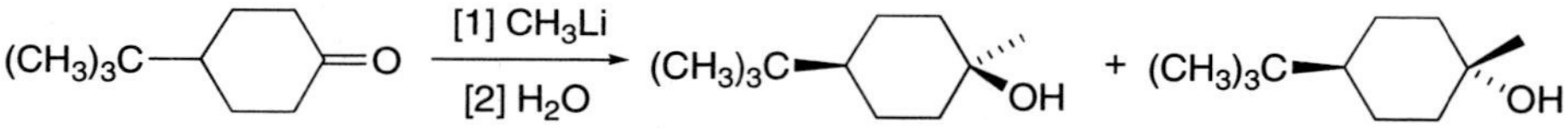

c.

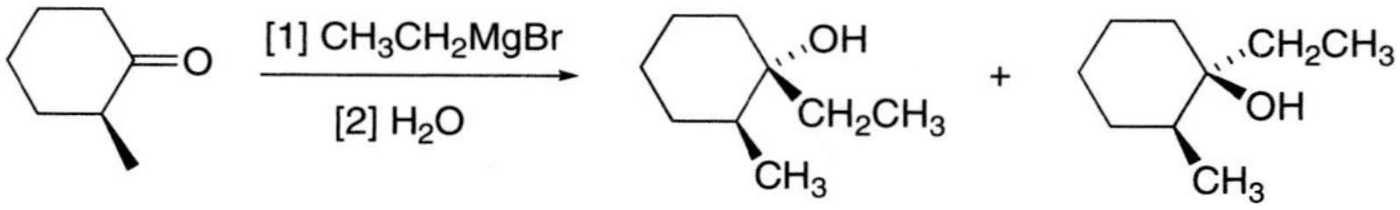

d.

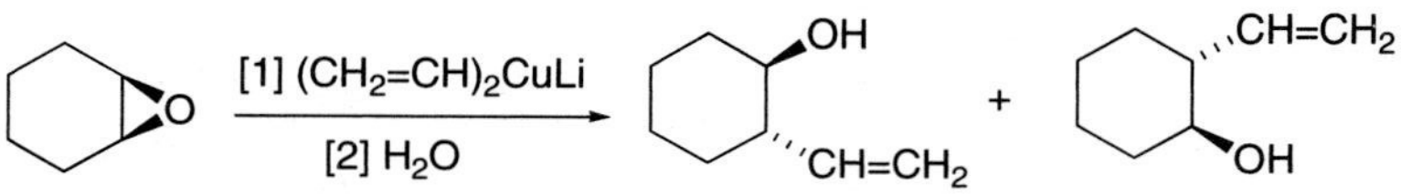

e.

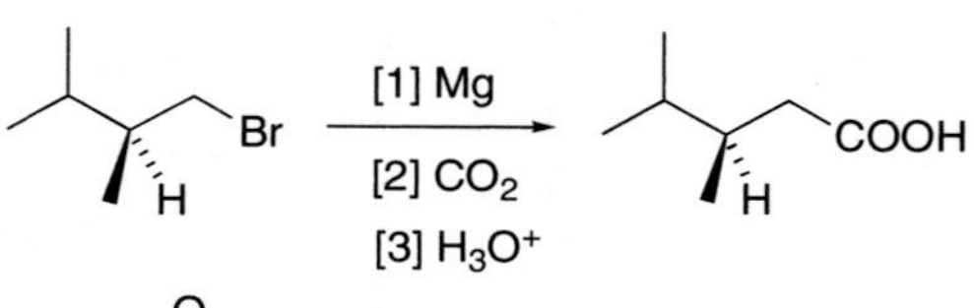

f.

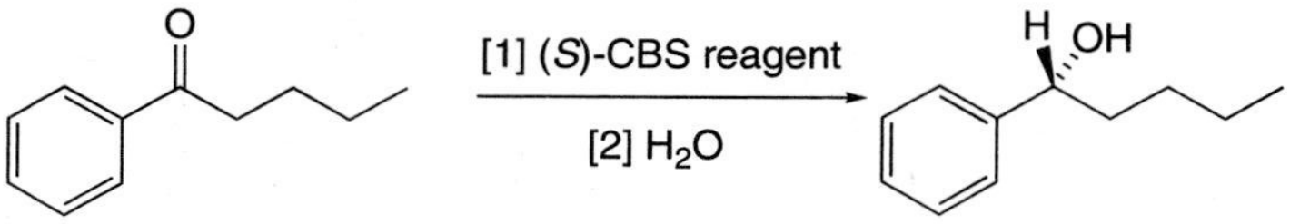

g.

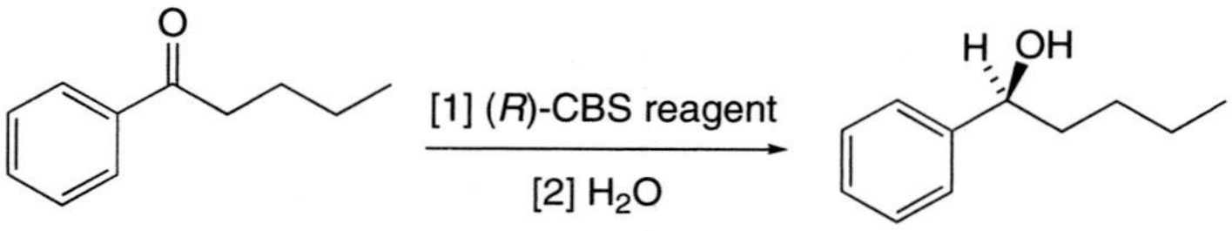

h. 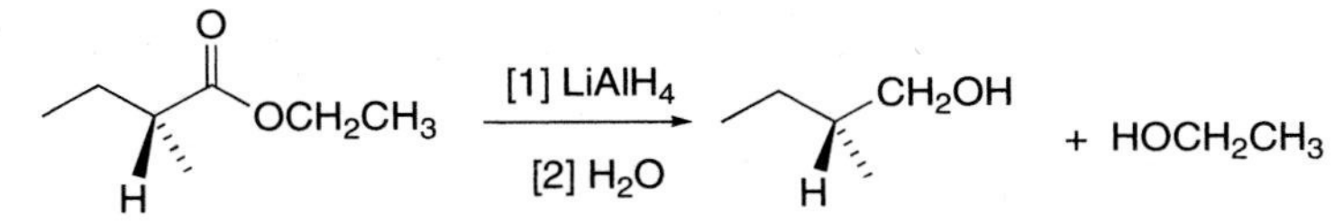
OCH2CH3
[1] LiAlH4
[2] H2O
CH2OH
+ HOCH2CH3
H

20.52 Both ketones are chiral molecules with carbonyl groups that have one side more sterically hindered than the other. In both reductions, hydride approaches from the less hindered side.

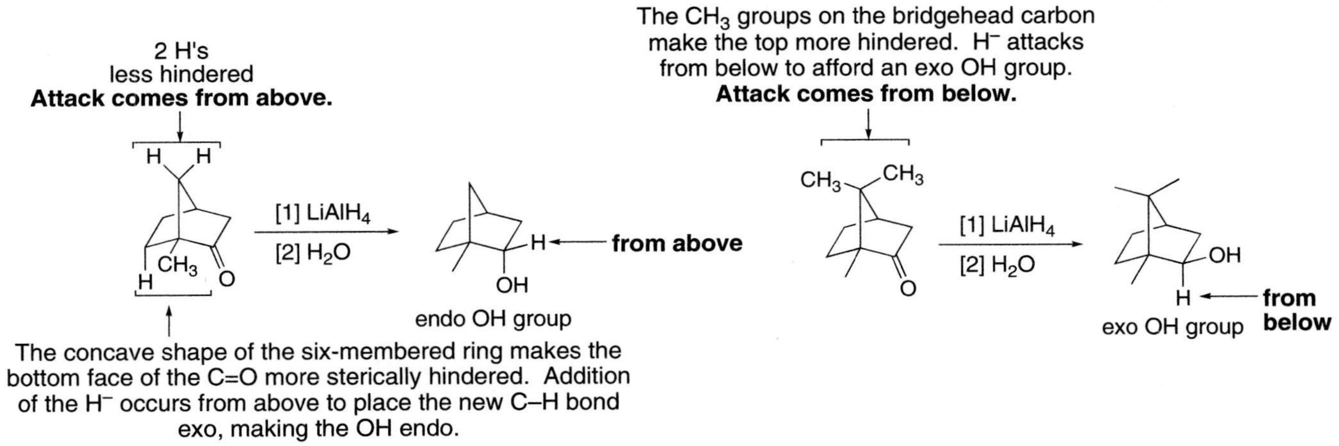
2 H's
less hindered
Attack comes from above.
H H
CH3
O
[1] LiAlH4
[2] H2O
H
OH
endo OH group
from above
The concave shape of the six-membered ring makes the bottom face of the C=O more sterically hindered. Addition of the H⁻ occurs from above to place the new C–H bond exo, making the OH endo.
The CH3 groups on the bridgehead carbon make the top more hindered. H⁻ attacks from below to afford an exo OH group.
Attack comes from below.
CH3 CH3
[1] LiAlH4
[2] H2O
OH
H
exo OH group
from below

20.53 Since a Grignard reagent contains a carbon atom with a partial negative charge, it acts as a base and reacts with the OH of the starting halide, $BrCH_2CH_2CH_2CH_2OH$. This acid–base reaction destroys the Grignard reagent so that addition cannot occur. To get around this problem, the OH group can be protected as a *tert*-butyldimethylsilyl ether, from which a Grignard reagent can be made.

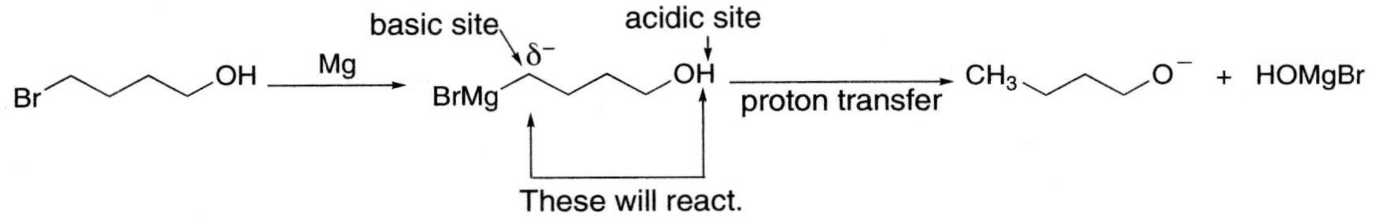
basic site
acidic site
δ⁻
Br
OH
Mg
BrMg
OH
proton transfer
CH3
O⁻
+ HOMgBr
These will react.

INSTEAD: Use a protecting group.

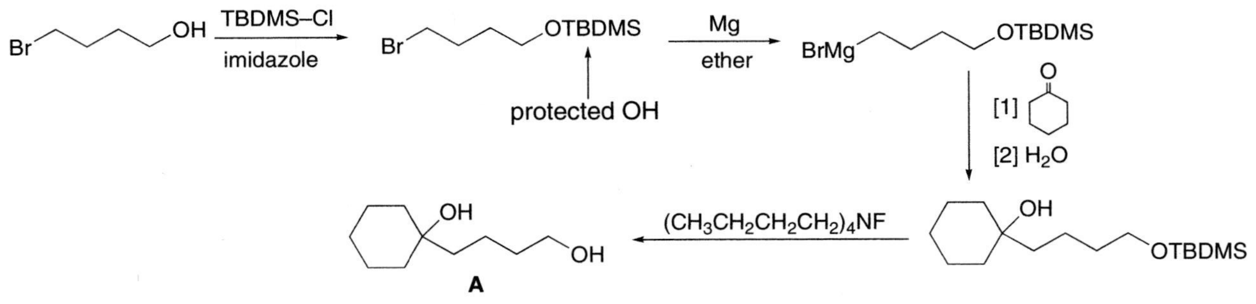
Br
OH
TBDMS–Cl
imidazole
OTBDMS
Mg
ether
BrMg
OTBDMS
protected OH
[1]
[2] H2O
OH
OTBDMS
(CH3CH2CH2CH2)4NF
OH
OH
A

20.54 Compounds **F**, **G**, and **K** are all alcohols with aromatic rings so there will be many similarities in their proton NMR spectra. These compounds will, however, show differences in absorptions due to the CH protons on the carbon bearing the OH group. **F** has a CH$_2$OH group, which will give a singlet in the 3–4 ppm region of the spectrum. **G** is a 3° alcohol that has no protons on the C bonded to the OH group so it will have no peak in the 3–4 ppm region of the spectrum. **K** is a 2° alcohol that will give a doublet in the 3–4 ppm region of the spectrum for the CH proton on the carbon with the OH group.

20.55

b.

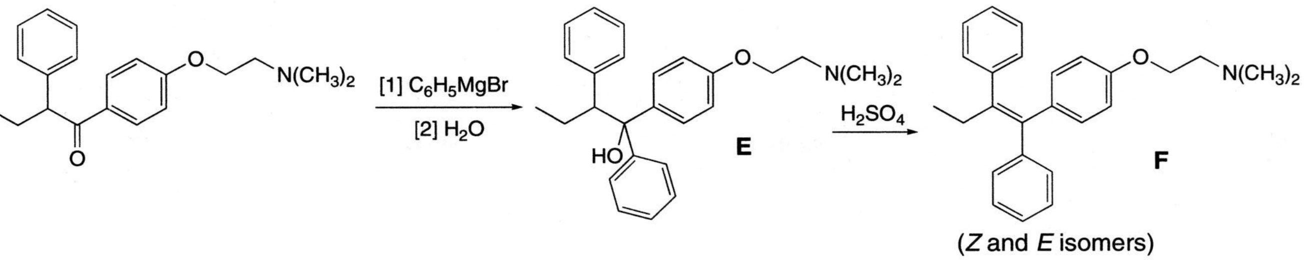

[1] C₆H₅MgBr

[2] H₂O

E

H₂SO₄

F

(Z and E isomers)

c.

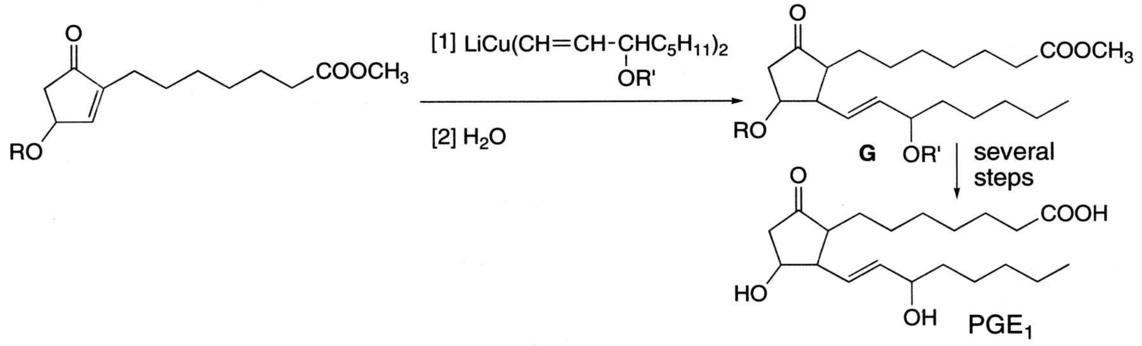

[1] LiCu(CH=CH-CHC₅H₁₁)₂
        OR'

[2] H₂O

G   OR'   several steps

HO   OH   PGE₁

20.56

a.

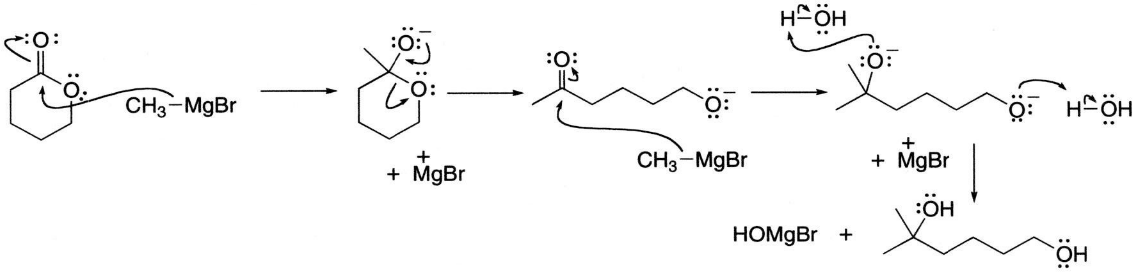

b.

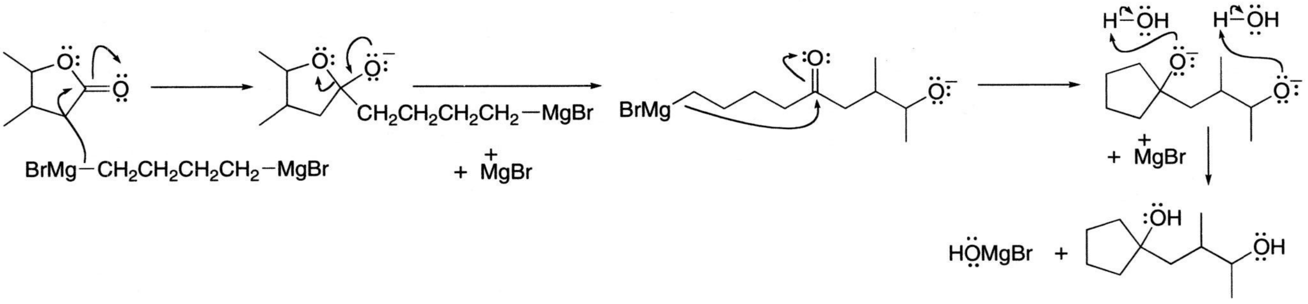

20.57

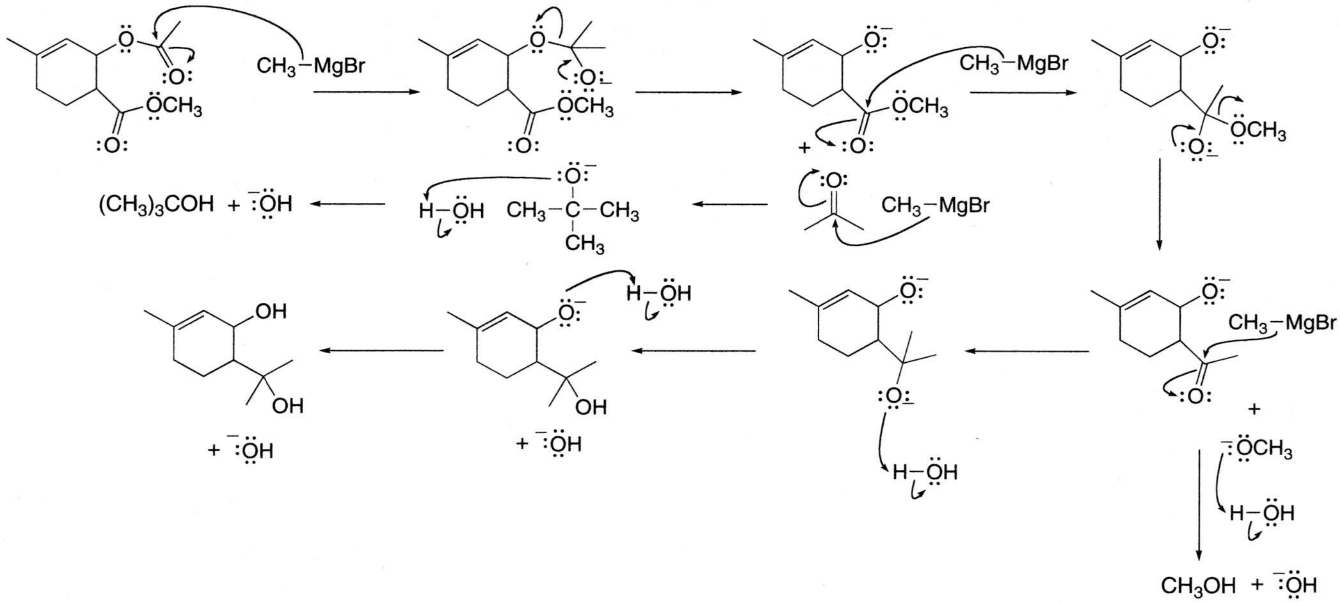

20.58

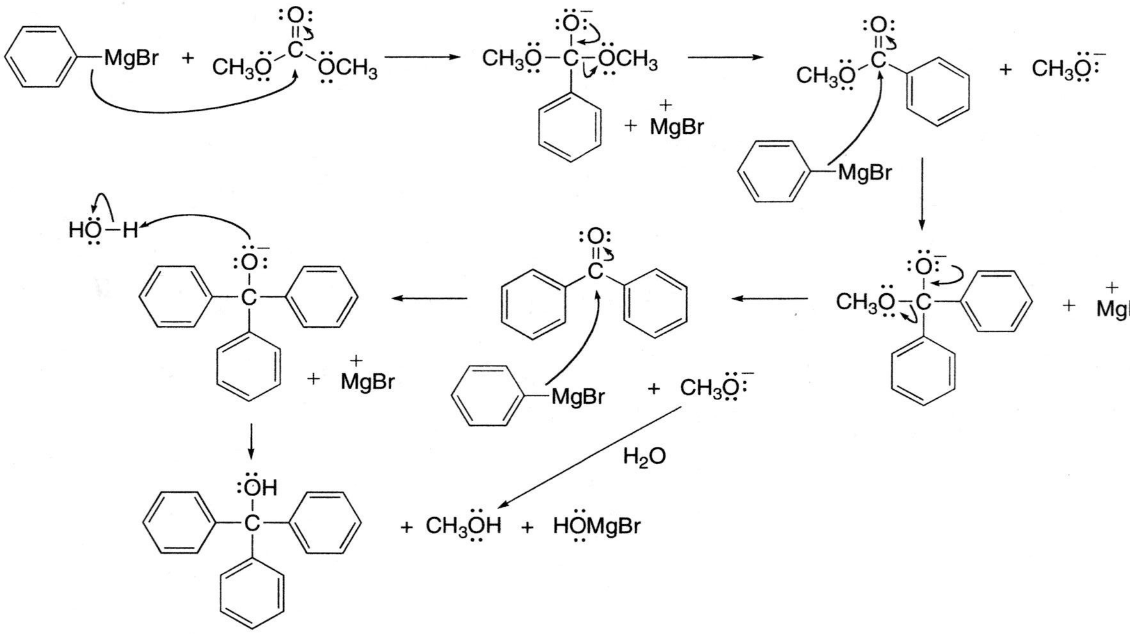

20.59

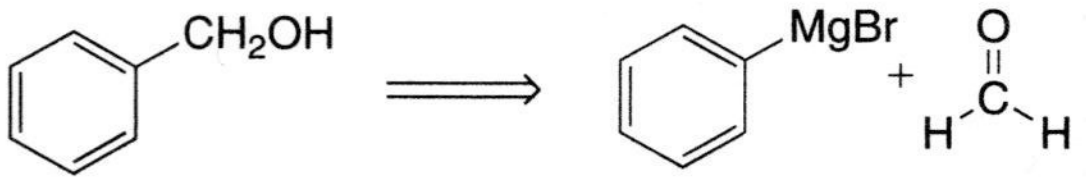

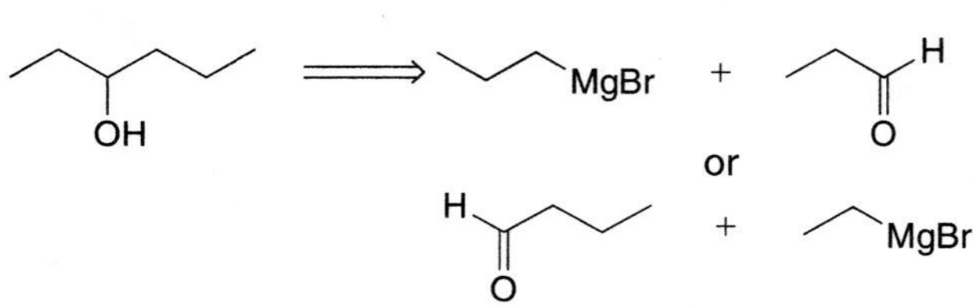

c. $(C_6H_5)_3COH$ $\Longrightarrow$ $(C_6H_5)_2C{=}O$ + BrMg⟨phenyl⟩

e. ⟨1-ethylcyclohexanol⟩ $\Longrightarrow$ ⟨cyclohexanone⟩ + ⟨CH₃CH₂MgBr⟩

d. ⟨2-cyclohexyl-2-butanol⟩ $\Longrightarrow$ ⟨butan-2-one⟩ + BrMg⟨cyclohexyl⟩

or

⟨CH₃CH₂MgBr⟩ + ⟨cyclohexyl methyl ketone⟩

or

CH_3MgBr + ⟨cyclohexyl ethyl ketone⟩

20.60

a. ⟨tricyclohexylcarbinol, OH⟩ $\Longrightarrow$ CH_3O–⟨C(=O)cyclohexyl⟩ + ⟨cyclohexyl⟩–MgBr (2 equiv)

b. $CH_3{-}\overset{OH}{\underset{CH_3}{C}}{-}CH_2CH_2CH(CH_3)_2$ $\Longrightarrow$ CH_3O–C(=O)–$CH_2CH_2CH(CH_3)_2$ + $CH_3{-}MgBr$ (2 equiv)

c. $(CH_3CH_2CH_2CH_2)_2C(OH)CH_3$ $\Longrightarrow$ CH_3O–C(=O)–CH_3 + BrMg⟨butyl⟩ (2 equiv)

20.61

a. ⟨1-cyclohexylhexan-1-ol, OH⟩ (two ways) $\Longrightarrow$ ⟨cyclohexyl⟩–Li + H–C(=O)–⟨pentyl⟩ (hexanal)

or

$\Longrightarrow$ ⟨cyclohexanecarbaldehyde⟩ H + Li⟨pentyl⟩

b. ⟨cyclohexyl phenyl cyclopentyl carbinol, OH⟩ (three ways) $\Longrightarrow$ ⟨cyclohexyl C(=O) cyclopentyl⟩ + ⟨phenyl⟩–Li

or

$\Longrightarrow$ ⟨phenyl C(=O) cyclohexyl⟩ + ⟨cyclopentyl⟩–Li

or

$\Longrightarrow$ ⟨cyclopentyl C(=O) phenyl⟩ + ⟨cyclohexyl⟩–Li

c. ⟨cyclohexyl, cyclohexyl, C(OH)–CH₂CH₃⟩ (three ways)

$\Longrightarrow$

⟨cyclohexyl C(=O) cyclohexyl⟩ + $Li{-}CH_2CH_3$

or

⟨cyclohexyl C(=O) CH₂CH₃⟩ + ⟨cyclohexyl⟩–Li

or

CH_3CH_2–C(=O)–OCH_3 + ⟨cyclohexyl⟩–Li (2 equiv)

20.62

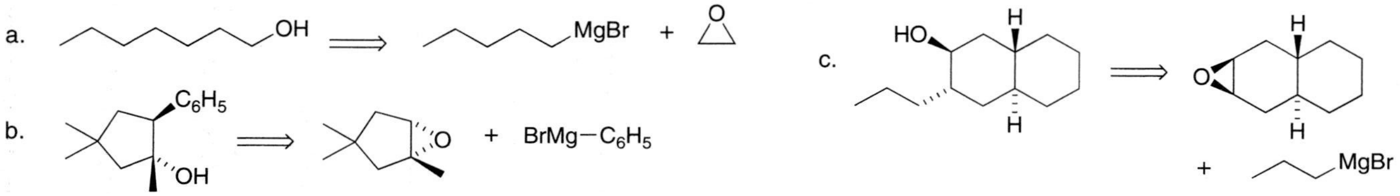

a.

b.

c.

20.63

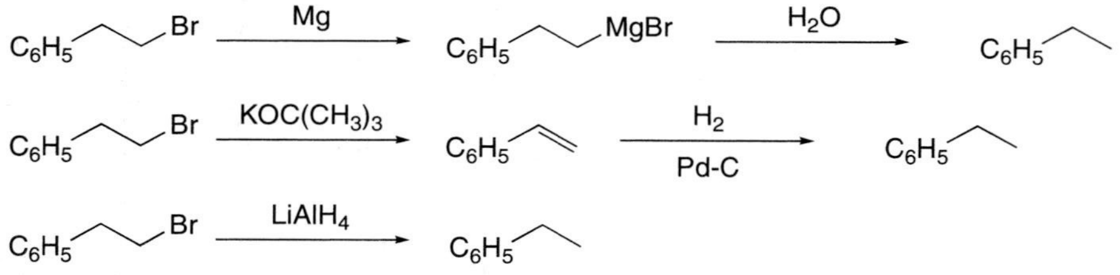

20.64

a.

b.

(from a.)

c.

(from b.)

d.

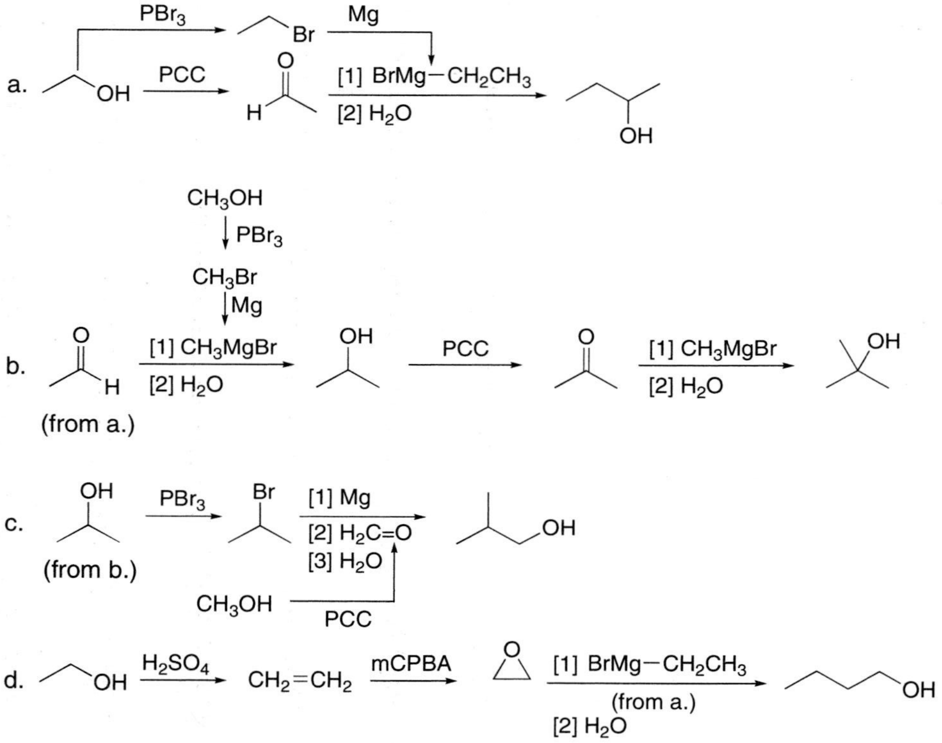

20.65

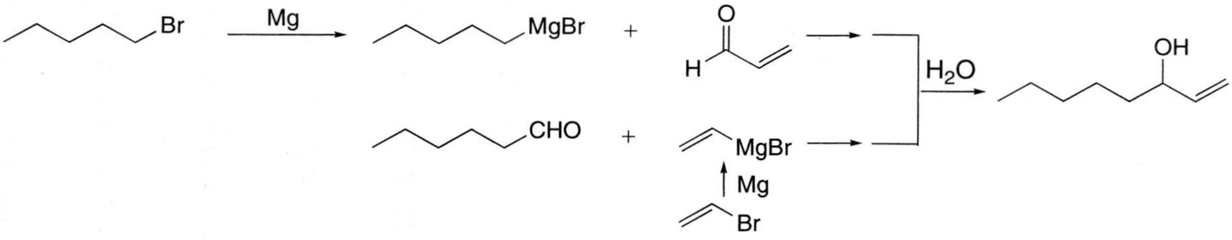

20.66

a.

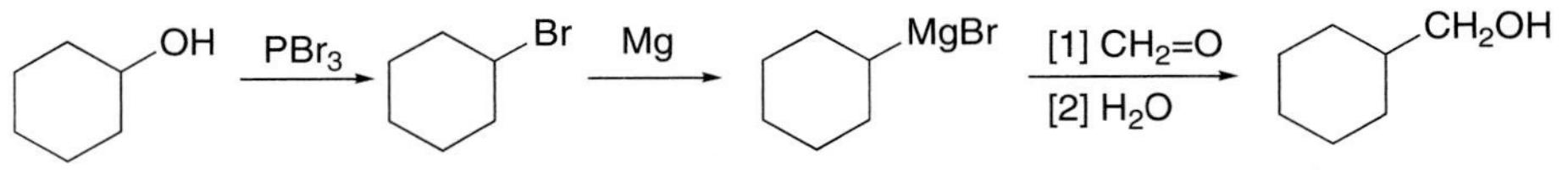

b.

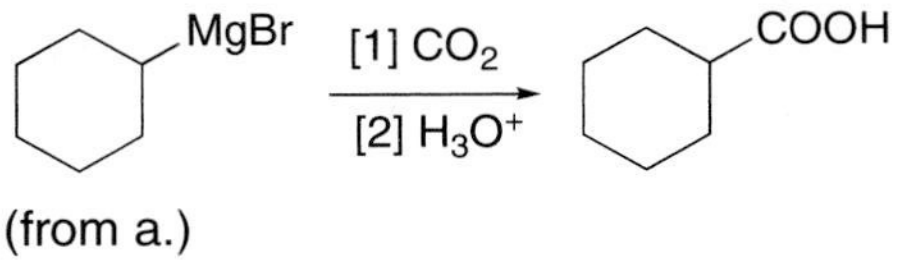

(from a.)

c.

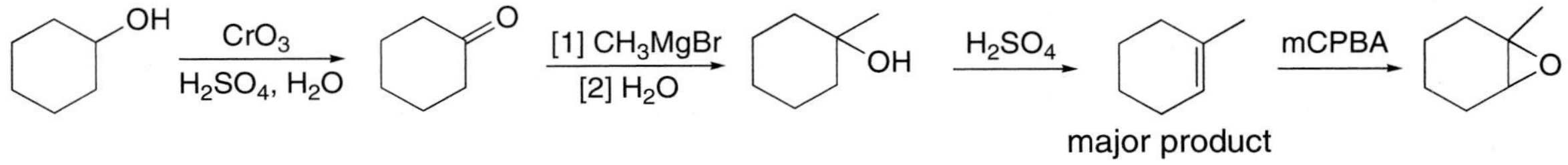

major product

d.

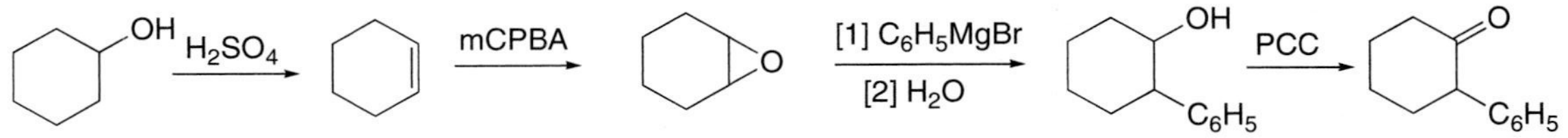

e.

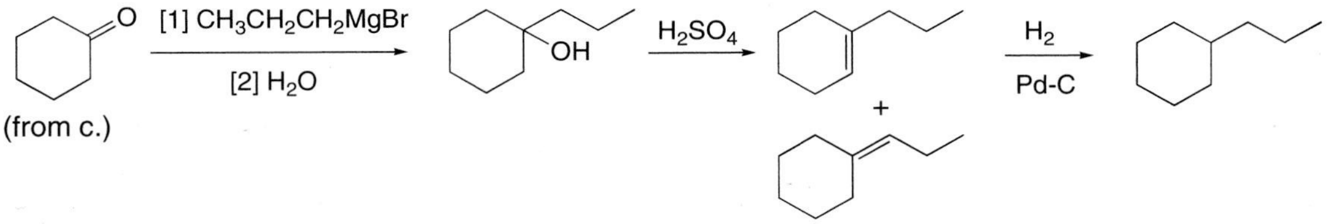

(from c.)

f.

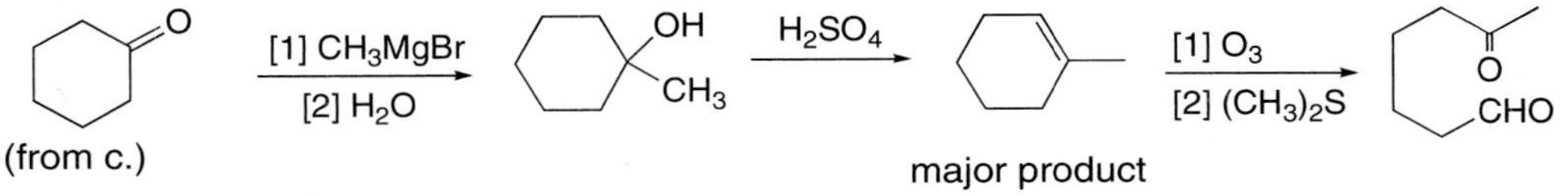

(from c.)

major product

g.

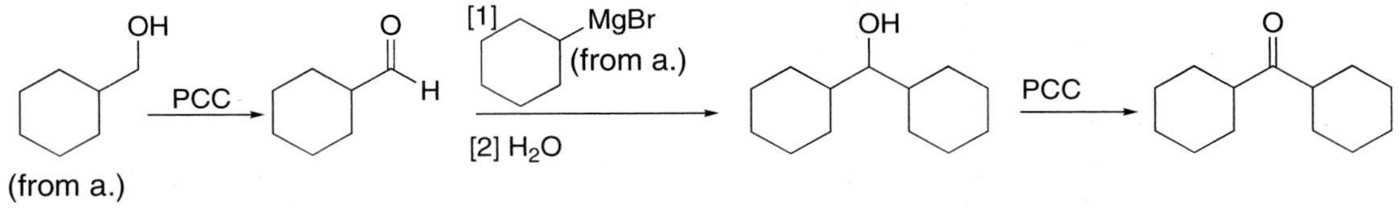

(from a.)

h.

(from c.)

major product

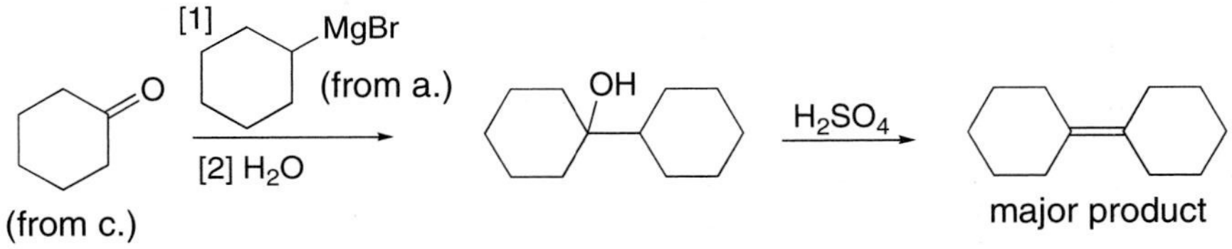

20.67

a.

$$\text{(CH}_3)_2\text{CHOH} \xrightarrow{\text{SOCl}_2} \text{(CH}_3)_2\text{CHCl}$$

b.

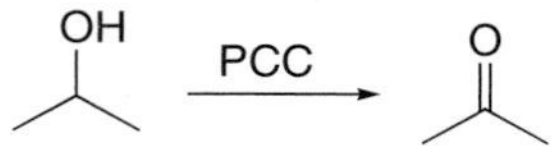

OH → PCC → ketone

c.

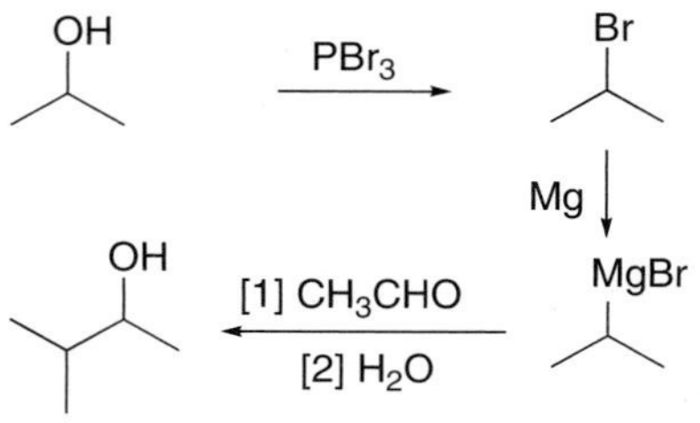

OH → PBr$_3$ → Br → Mg → MgBr → [1] CH$_3$CHO, [2] H$_2$O → OH

d.

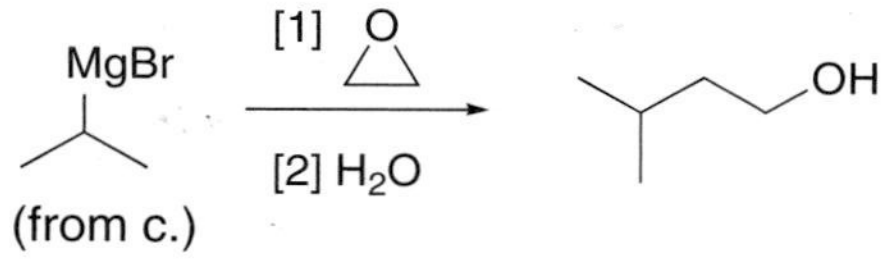

MgBr (from c.) → [1] epoxide, [2] H$_2$O → OH

e.

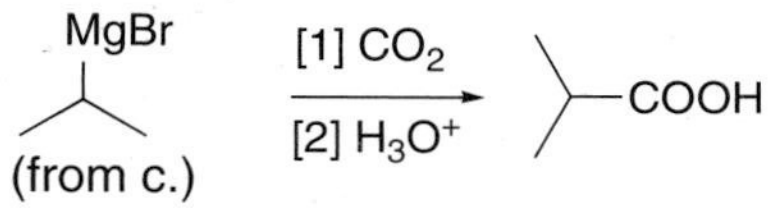

MgBr (from c.) → [1] CO$_2$, [2] H$_3$O$^+$ → COOH

f.

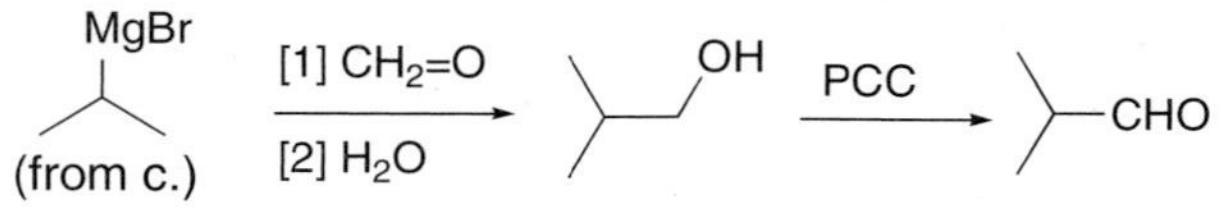

MgBr (from c.) → [1] CH$_2$=O, [2] H$_2$O → OH → PCC → CHO

g.

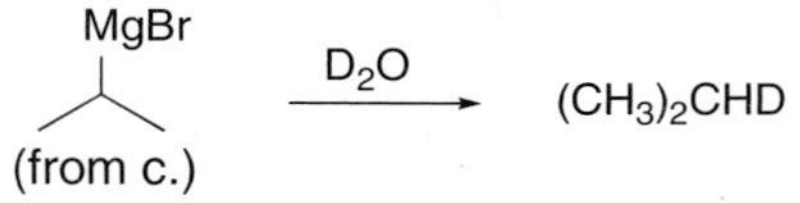

MgBr (from c.) → D$_2$O → (CH$_3$)$_2$CHD

h.

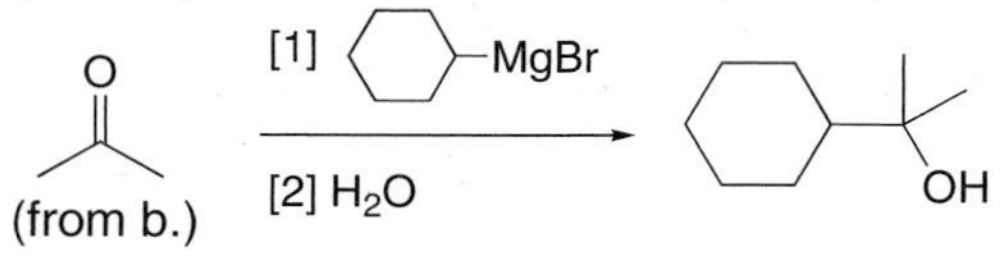

ketone (from b.) → [1] cyclohexyl-MgBr, [2] H$_2$O → product

i.

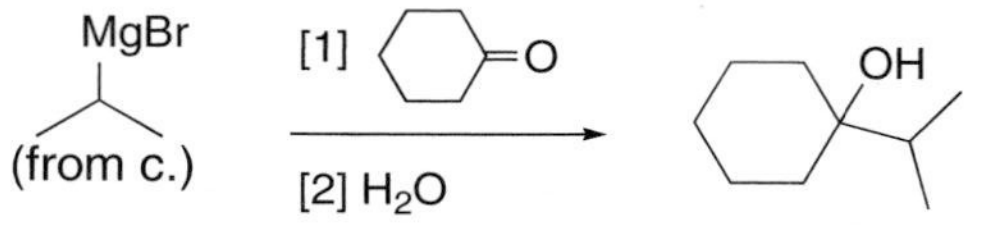

MgBr (from c.) → [1] cyclohexanone, [2] H$_2$O → OH

j.

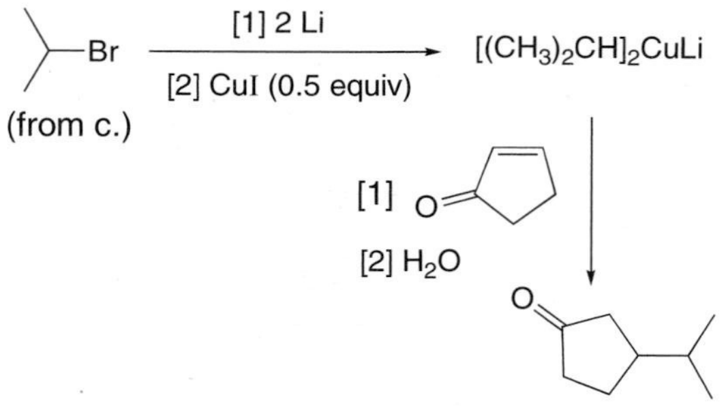

Br (from c.) → [1] 2 Li, [2] CuI (0.5 equiv) → [(CH$_3$)$_2$CH]$_2$CuLi

[1] cyclopentenone
[2] H$_2$O → product

20.68

estradiol → PCC → ketone → [1] NaH, [2] CH$_3$I → CH$_3$O product → [1] HC≡CLi, [2] H$_2$O → mestranol

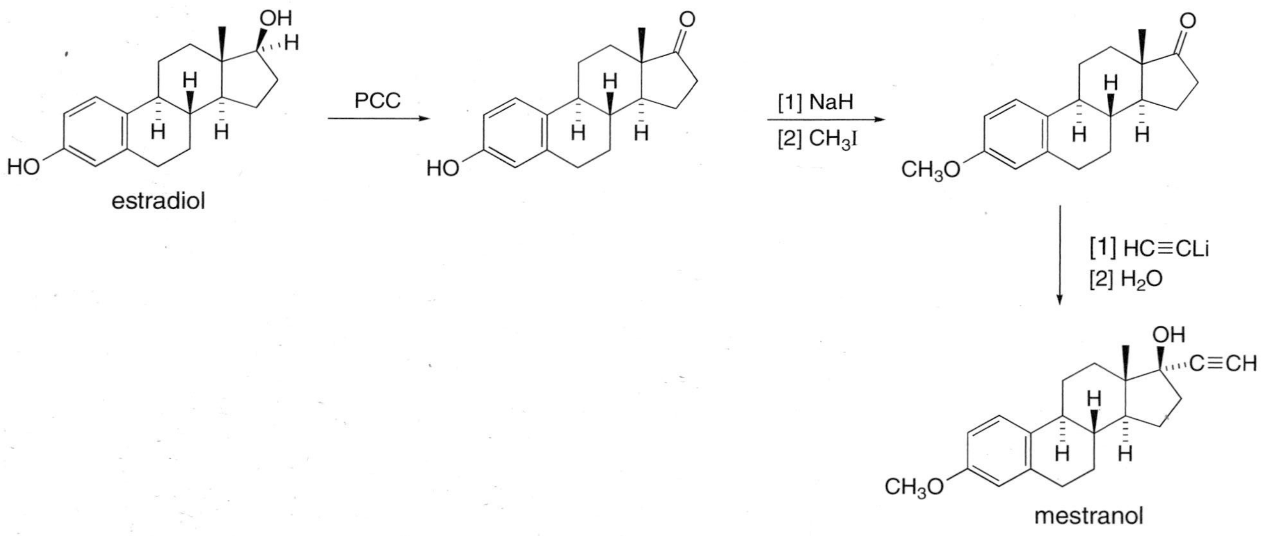

20.69

a.

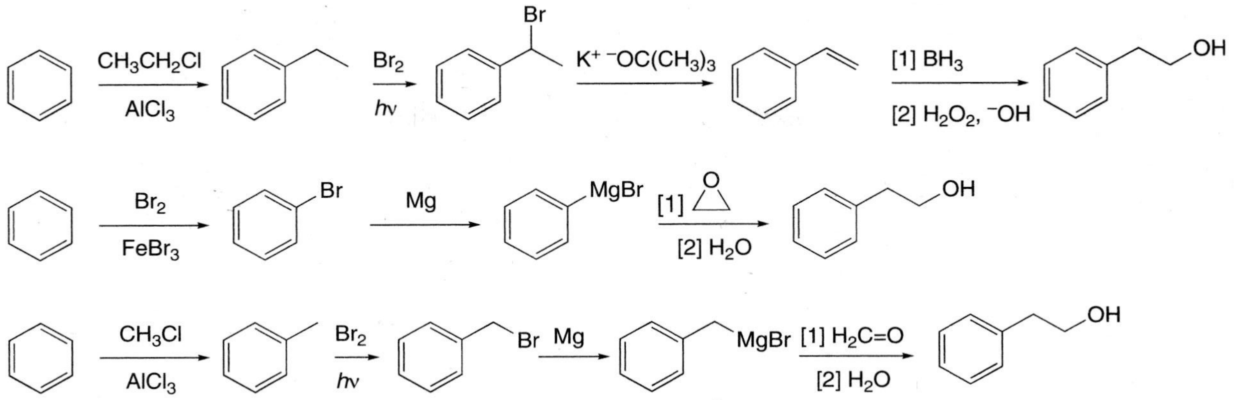

b.

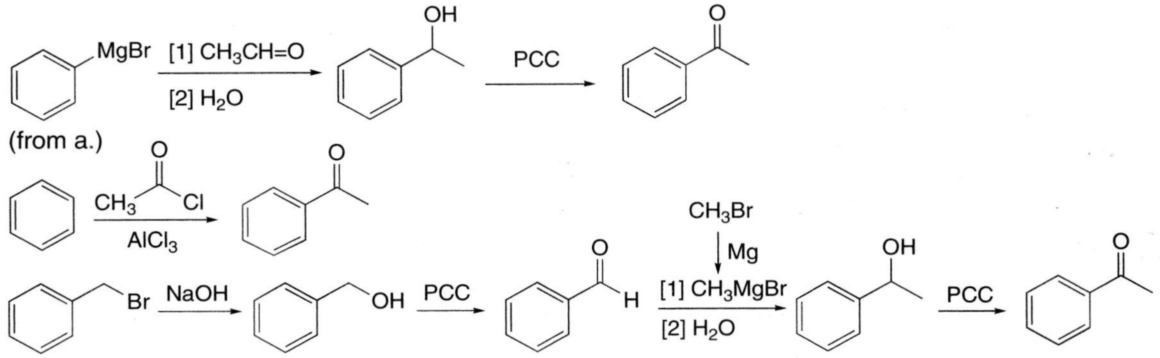

(from a.)

(from a.)

20.70

a.

c.

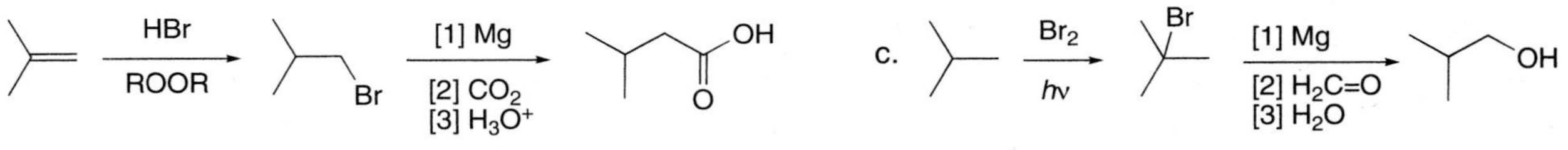

b.

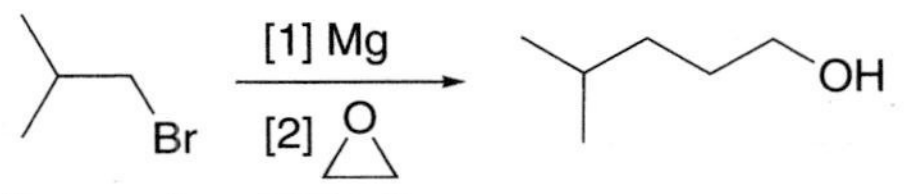

(from a.)

20.71

a.

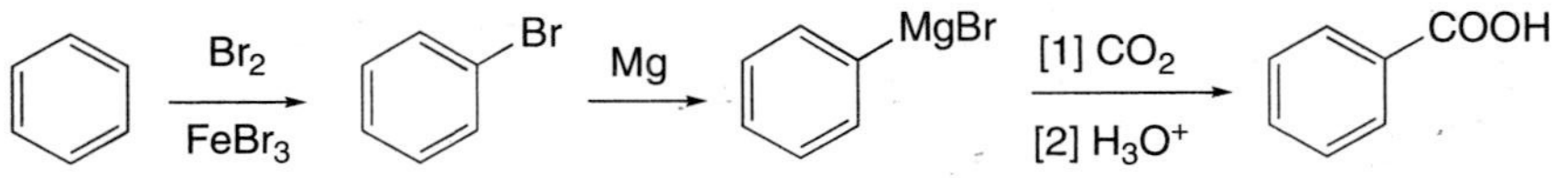

b.

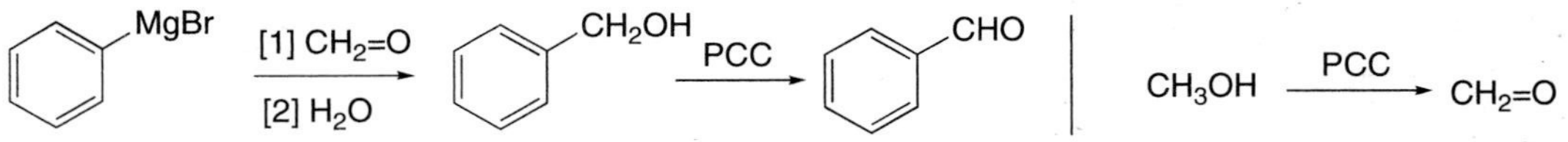

(from a.)

c.

(from b.)

d. $CH_3CH_2CH_2OH$ $\xrightarrow{PBr_3}$ $CH_3CH_2CH_2Br$ $\xrightarrow{Mg}$ $CH_3CH_2CH_2MgBr$

$CH_3CH_2CH_2OH$ $\xrightarrow{PCC}$ CH_3CH_2CHO $\xrightarrow[\text{[2] }H_2O]{\text{[1] }CH_3CH_2CH_2MgBr}}$ $\xrightarrow{PCC}$

$\xrightarrow[\text{[2] }H_2O]{\text{[1] PhMgBr (from a.)}}$

e.

20.72

a. HO $\xrightarrow{PCC}$ $\xrightarrow[\text{[2] }H_2O]{\text{[1] }CH_3CH_2MgBr}}$ $\xrightarrow{PCC}$

CH_3CH_2OH $\xrightarrow{PBr_3}$ CH_3CH_2Br $\xrightarrow{Mg}$ CH_3CH_2MgBr

b. $\xrightarrow{PCC}$ $\xrightarrow[\text{[2] }H_2O]{\text{[1] }CH_3CH_2CH_2MgBr}}$

$CH_3CH_2CH_2OH$ $\xrightarrow{PBr_3}$ $CH_3CH_2CH_2Br$ $\xrightarrow{Mg}$ $CH_3CH_2CH_2MgBr$

c. $\xrightarrow{PCC}$ $\xrightarrow[\text{[2] }H_2O]{\text{[1] }CH_3CH_2CH_2CH_2MgBr}}$ $\xrightarrow{PCC}$

$CH_3CH_2CH_2CH_2OH$ $\xrightarrow{PBr_3}$ $CH_3CH_2CH_2CH_2Br$ $\xrightarrow{Mg}$ $CH_3CH_2CH_2CH_2MgBr$

d. $\xrightarrow[\text{[2] }H_2O]{\text{[1] }(CH_3)_2CHCH_2MgBr}}$ $\xrightarrow{PBr_3}$ $\xrightarrow{Mg}$

(from c.)

$(CH_3)_2CHCH_2OH$ $\xrightarrow{PBr_3}$ $(CH_3)_2CHCH_2Br$ $\xrightarrow{Mg}$ $(CH_3)_2CHCH_2MgBr$

$\xrightarrow[\text{[2] }H_3O^+]{\text{[1] }CO_2}$

e.

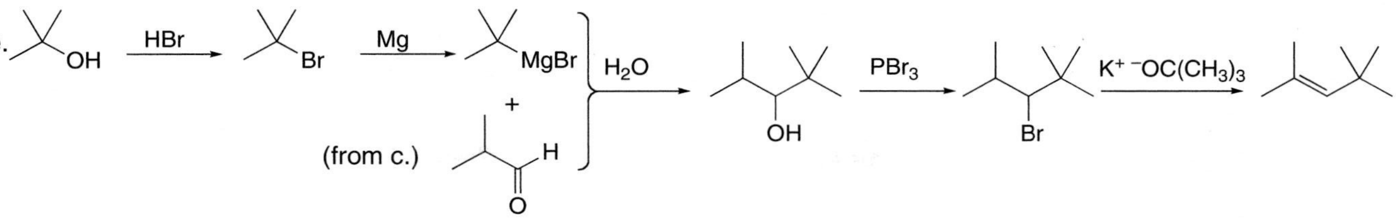

20.73

a.

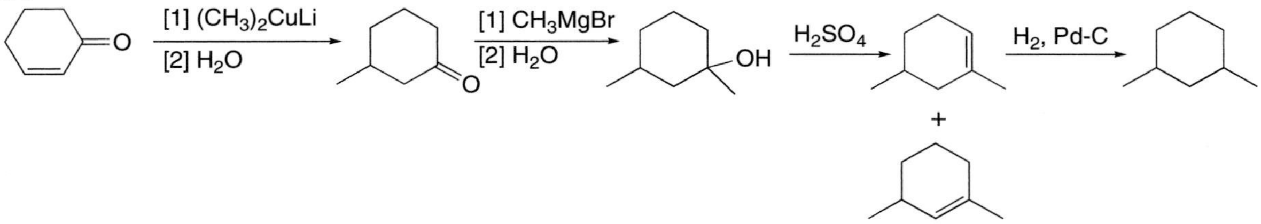

b.

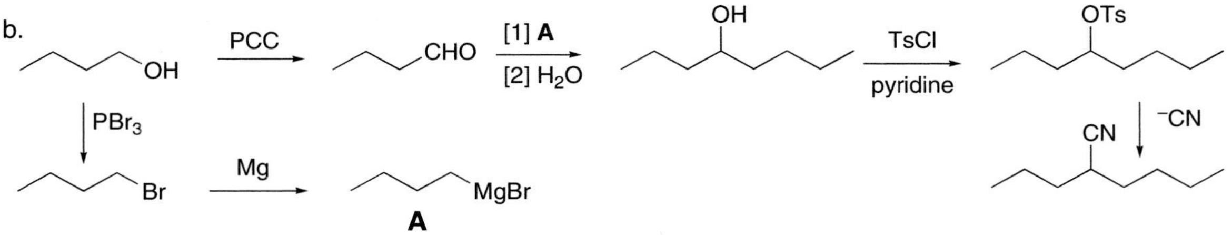

c.

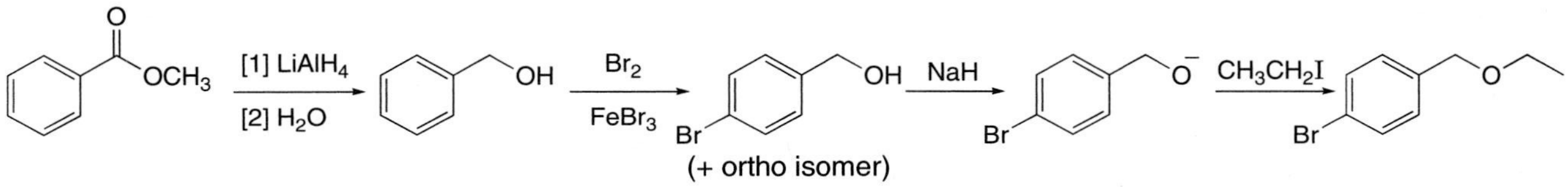

d.

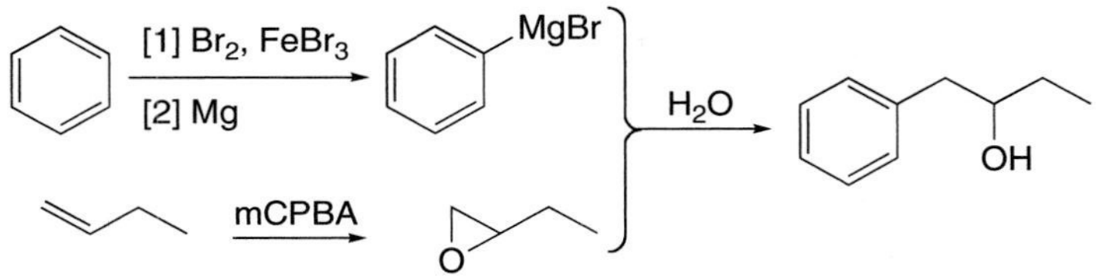

e.

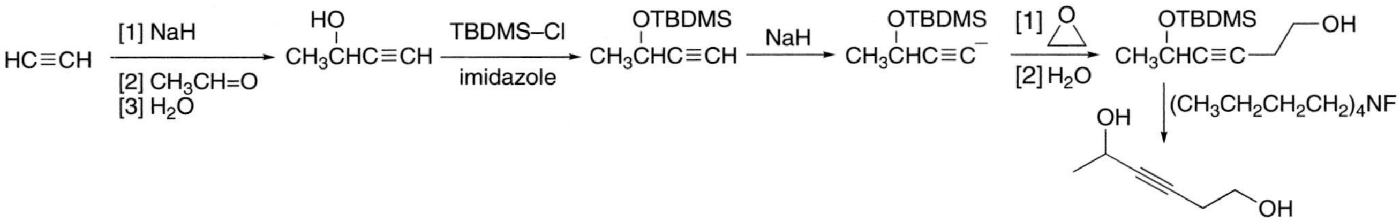

20.74

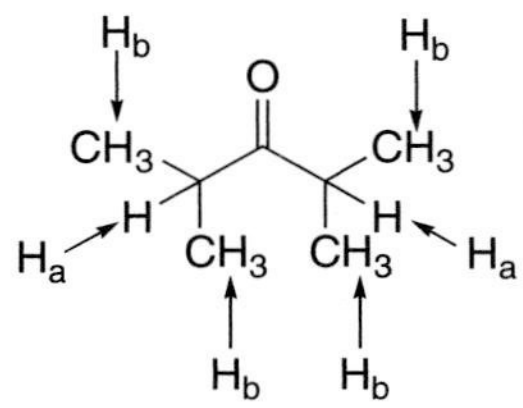

IR peak: 1716 cm^{-1} (C=O)
^{1}H NMR: 2 signals (ppm)
    doublet 1.2 (H$_b$)
    septet 2.7 (H$_a$)

$C_7H_{14}O$
A

NaBH$_4$

CH$_3$OH

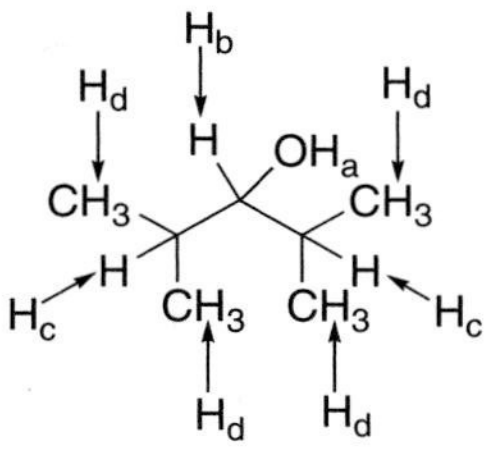

IR peak: 3600–3200 cm^{-1} (OH)
^{1}H NMR: 4 signals (ppm)
    doublet 0.9 (H$_d$)
    singlet 1.5 (H$_a$)
    multiplet 1.7 (H$_c$)
    triplet 3.0 (H$_b$)

$C_7H_{16}O$
B

20.75

C

¹H NMR: 2 signals (ppm)
singlet (6 H) 1.3 (H$_a$)
singlet (2 H) 2.4 (H$_b$)

[1] C$_6$H$_5$MgBr
[2] H$_2$O

C$_{10}$H$_{14}$O
D

IR peak 3600–3200 cm⁻¹ (OH)
¹H NMR: 4 signals (ppm)
singlet (6 H) 1.2 (H$_a$)
singlet (1 H) 1.6 (H$_b$)
singlet (2 H) 2.7 (H$_c$)
multiplet (5 H) 7.2 (benzene ring)

20.76

C$_4$H$_8$O$_2$
E

IR peak 1743 cm⁻¹ (C=O)
¹H NMR: 2 signals (ppm)
triplet (3 H) 1.2 (H$_c$)
singlet (3 H) 2.0 (H$_a$)
quartet (2 H) 4.1 (H$_b$)

[1] CH$_3$CH$_2$MgBr
(excess)
[2] H$_2$O

C$_6$H$_{14}$O
F

IR peak 3600–3200 cm⁻¹ (OH)
¹H NMR: 4 signals (ppm)
triplet (6 H) 0.9 (H$_a$)
singlet (3 H) 1.1 (H$_c$)
quartet (4 H) 1.5 (H$_b$)
singlet (1 H) 1.55 (H$_d$)

20.77 Molecular ion at $m/z = 86$: C$_5$H$_{10}$O (possible molecular formula).

Determine the number of integration units per H:
 Total number of integration units: 25 + 17 + 24 + 17 = 83
 83 units/10 H's = 8.3 units per H
Divide each integration value by 8.3 to determine the number of H's per signal:
 25 units/ 8.3 = 3 H's
 24 units/ 8.3 = 3 H's
 17 units/ 8.3 = 2 H's

CH$_3$CH$_2$CH$_2$C≡N

[1] CH$_3$MgBr
[2] H$_3$O⁺

G

IR peak 1721 cm⁻¹ (C=O)
¹H NMR: 4 signals (ppm)
triplet (3 H) 0.9 (H$_a$)
sextet (2 H) 1.6 (H$_b$)
singlet (3 H) 2.1 (H$_c$)
triplet (2 H) 2.4 (H$_d$)

20.78 Molecular ion at $m/z = 86$: C$_5$H$_{10}$O (possible molecular formula).

[1] (CH$_3$)$_3$CLi
[2] CH$_2$=O
[3] H$_2$O

H

IR peaks: 3600–3200 cm⁻¹ (OH)
1651 cm⁻¹ (C=C)
¹H NMR: 6 signals (ppm)
singlet (1 H) 1.7 (H$_a$)
singlet (3 H) 1.8 (H$_b$)
triplet (2 H) 2.2 (H$_c$)
triplet (2 H) 3.8 (H$_d$)
two signals at 4.8 and 4.9
due to 2 H's: H$_e$ and H$_f$

20.79

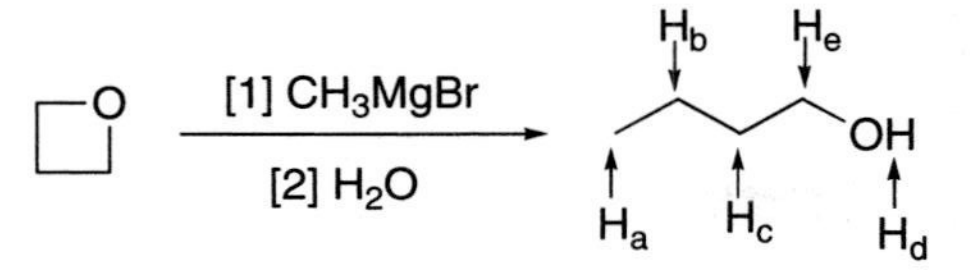

IR peak 3600–3200 cm^{-1} (OH)
^{1}H NMR: 5 signals (ppm)
    triplet (3 H) 0.94 (H$_a$)
    multiplet (2 H) 1.39 (H$_b$)
    multiplet (2 H) 1.53 (H$_c$)
    singlet (1 H) 2.24 (H$_d$)
    triplet (2 H) 3.63 (H$_e$)

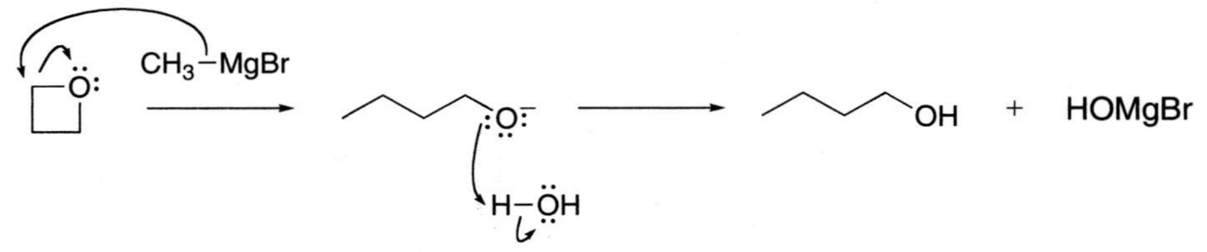

20.80

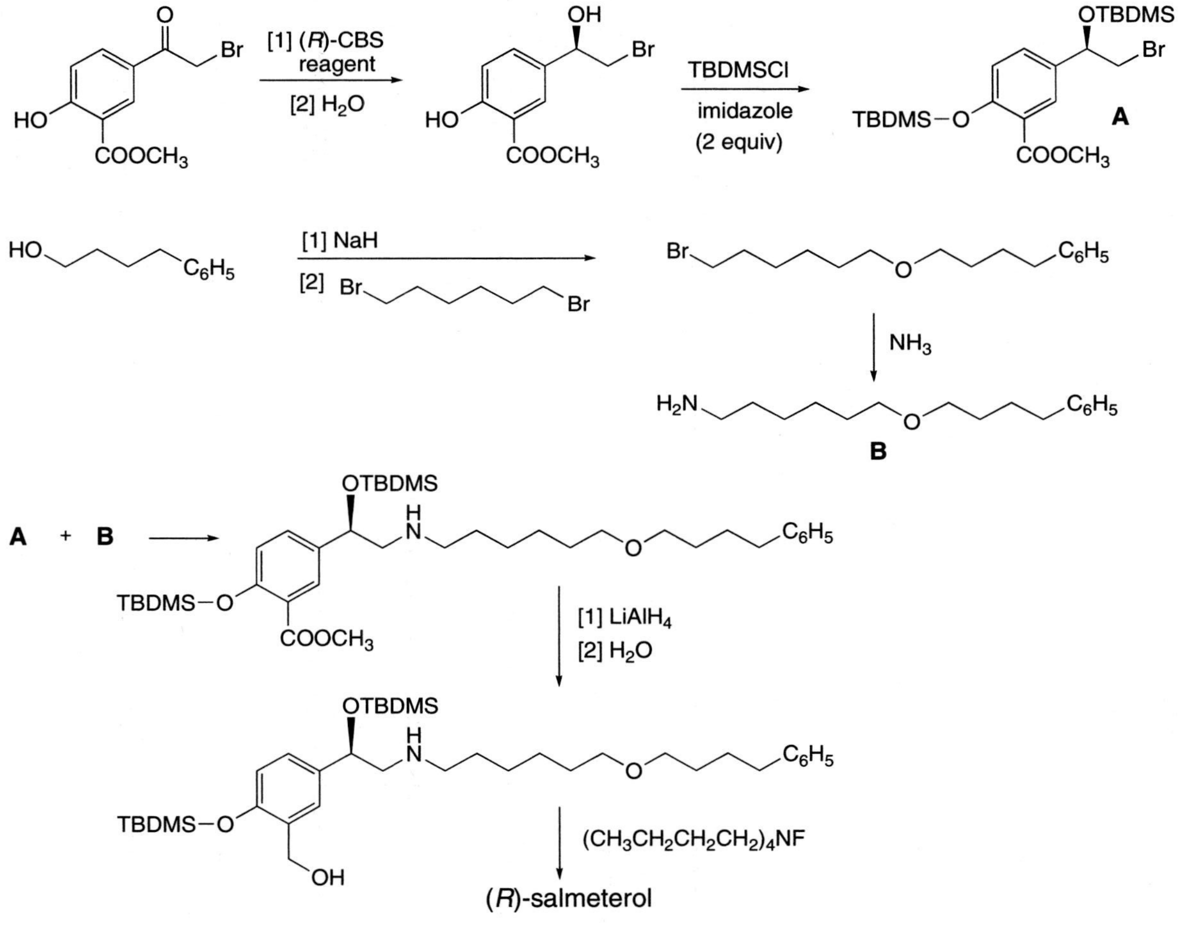

20.81

4-*tert*-butylcyclohexanone → *cis*-4-*tert*-butylcyclohexanol major product

L-Selectride adds H⁻ to a C=O group. There are two possible reduction products—cis and trans isomers—but the cis isomer is favored. The key element is that the three *sec*-butyl groups make L-selectride a large, bulky reducing agent that attacks the carbonyl group from the less hindered direction.

> When H⁻ adds from the equatorial direction, the product has an axial OH and a new equatorial H. Since the equatorial direction is less hindered, this mode of attack is favored with large bulky reducing agents like L-selectride. In this case, the product is cis.

> The axial H's hinder H⁻ attack from the axial direction. As a result, this mode of attack is more difficult with larger reducing agents. In this case the product is trans. This product is not formed to any appreciable extent.

20.82 The β carbon of an α,β-unsaturated carbonyl compound absorbs farther downfield in the ^{13}C NMR spectrum than the α carbon, because the β carbon is deshielded and bears a partial positive charge as a result of resonance. Since three resonance structures can be drawn for an α,β-unsaturated carbonyl compound, one of which places a positive charge on the β carbon, the decrease of electron density at this carbon deshields it, shifting the ^{13}C absorption downfield. This is not the case for the α carbon.

mesityl oxide

20.83

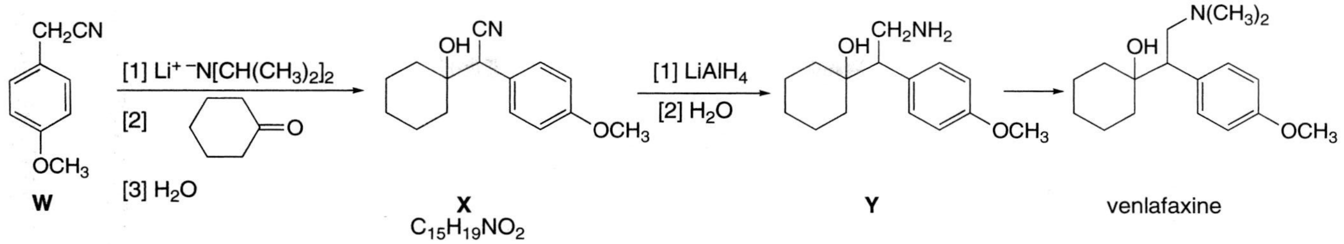

W

[1] Li$^+$ $^-$N[CH(CH$_3$)$_2$]$_2$

[2] (cyclohexanone)

[3] H$_2$O

X
C$_{15}$H$_{19}$NO$_2$

[1] LiAlH$_4$

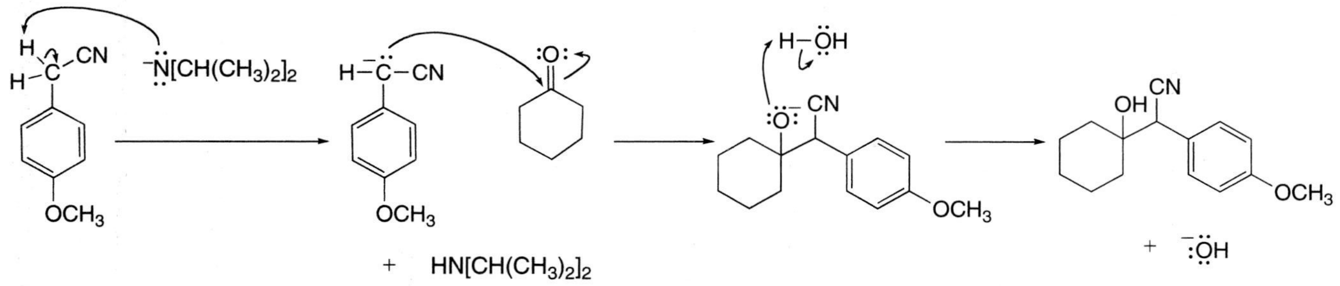

[2] H$_2$O

Y

venlafaxine

+ HN[CH(CH$_3$)$_2$]$_2$

+ $^-$:ÖH

Chapter 21: Aldehydes and Ketones—Nucleophilic Addition

◆ General facts

- Aldehydes and ketones contain a carbonyl group bonded to only H atoms or R groups. The carbonyl carbon is sp^2 hybridized and trigonal planar (21.1).
- Aldehydes are identified by the suffix *-al*, while ketones are identified by the suffix *-one* (21.2).
- Aldehydes and ketones are polar compounds that exhibit dipole–dipole interactions (21.3).

◆ Summary of spectroscopic absorptions of RCHO and R₂CO (21.4)

IR absorptions	C=O	~1715 cm^{-1} for ketones
		• increasing frequency with decreasing ring size
		~1730 cm^{-1} for aldehydes
		• For both RCHO and R₂CO, the frequency decreases with conjugation.
	C_{sp^2}–H of CHO	~2700–2830 cm^{-1} (one or two peaks)
^{1}H NMR absorptions	CHO	9–10 ppm (highly deshielded proton)
	C–H α to C=O	2–2.5 ppm (somewhat deshielded C_{sp^3}–H)
^{13}C NMR absorption	C=O	190–215 ppm

◆ Nucleophilic addition reactions

[1] Addition of hydride (H⁻) (21.8)

- The mechanism has two steps.
- H:⁻ adds to the planar C=O from both sides.

[2] Addition of organometallic reagents (R⁻) (21.8)

- The mechanism has two steps.
- R:⁻ adds to the planar C=O from both sides.

[3] Addition of cyanide (⁻CN) (21.9)

- The mechanism has two steps.
- ⁻CN adds to the planar C=O from both sides.

[4] Wittig reaction (21.10)

- The reaction forms a new C–C σ bond and a new C–C π bond.
- $Ph_3P=O$ is formed as by-product.

[5] Addition of 1° amines (21.11)

- The reaction is fastest at pH 4–5.
- The intermediate carbinolamine is unstable, and loses H_2O to form the C=N.

[6] Addition of 2° amines (21.12)

- The reaction is fastest at pH 4–5.
- The intermediate carbinolamine is unstable, and loses H_2O to form the C=C.

[7] Addition of H₂O—Hydration (21.13)

- The reaction is reversible. Equilibrium favors the product only with less stable carbonyl compounds (e.g., H_2CO and Cl_3CCHO).
- The reaction is catalyzed with either H^+ or ^-OH.

[8] Addition of alcohols (21.14)

- The reaction is reversible.
- The reaction is catalyzed with acid.
- Removal of H_2O drives the equilibrium to favor the products.

◆ Other reactions

[1] Synthesis of Wittig reagents (21.10A)

- Step [1] is best with CH_3X and RCH_2X since the reaction follows an S_N2 mechanism.
- A strong base is needed for proton removal in Step [2].

[2] Conversion of cyanohydrins to aldehydes and ketones (21.9)

$$R-\overset{\overset{\displaystyle OH}{|}}{\underset{\underset{\displaystyle CN}{|}}{C}}-H(R') \xrightarrow{\ ^-OH\ } \boxed{\begin{array}{c} \underset{\displaystyle R}{\overset{\displaystyle O}{\|}}\overset{\displaystyle C}{}H(R') \\ \text{aldehyde or} \\ \text{ketone} \end{array}} \begin{array}{l} + H_2O \\ + \ ^-CN \end{array}$$

- This reaction is the reverse of cyanohydrin formation.

[3] Hydrolysis of nitriles (21.9)

$$R-\overset{\overset{\displaystyle OH}{|}}{\underset{\underset{\displaystyle CN}{|}}{C}}-H(R') \xrightarrow[\substack{H^+ \ or \ ^-OH \\ \Delta}]{H_2O} \boxed{\begin{array}{c} R-\overset{\overset{\displaystyle OH}{|}}{\underset{\underset{\displaystyle COOH}{|}}{C}}-H(R') \\ \alpha\text{-hydroxy} \\ \text{carboxylic acid} \end{array}}$$

[4] Hydrolysis of imines and enamines (21.12)

imine   or   enamine $\xrightarrow{\ H_2O,\ H^+\ }$ aldehyde or ketone $+\ RNH_2\ or\ R_2NH$

[5] Hydrolysis of acetals (21.14)

$$R-\overset{\overset{\displaystyle OR''}{|}}{\underset{\underset{\displaystyle OR''}{|}}{C}}-H(R') + H_2O \ \underset{\longleftarrow}{\overset{H^+}{\longrightarrow}} \ \boxed{\begin{array}{c} \underset{\displaystyle R}{\overset{\displaystyle O}{\|}}\overset{\displaystyle C}{}H(R') \\ \text{aldehyde or} \\ \text{ketone} \end{array}} \begin{array}{l} + \ R''OH \\ (2\ equiv) \end{array}$$

- The reaction is acid catalyzed and is the reverse of acetal synthesis.
- A large excess of H_2O drives the equilibrium to favor the products.

Chapter 21: Answers to Problems

21.1   As the number of R groups bonded to the carbonyl C increases, reactivity towards nucleophilic attack decreases.

a. $(CH_3)_2C{=}O$ $CH_3CH{=}O$ $CH_2{=}O$
     2 R groups   1 R group   0 R groups

Increasing reactivity
decreasing alkyl substitution

b.

Increasing reactivity
decreasing steric hindrance

21.2   More stable aldehydes are less reactive towards nucleophilic attack.

benzaldehyde
Several resonance structures delocalize the partial positive charge on the carbonyl carbon, making it more stable and less reactive towards nucleophilic attack.

cyclohexanecarbaldehyde
This aldehyde has no added resonance stabilization.

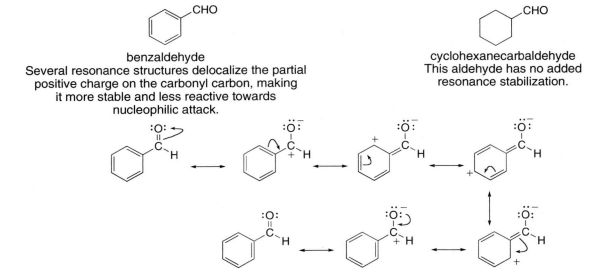

21.3   • To name an aldehyde with a chain of atoms: [1] Find the longest chain with the CHO group and change the *-e* ending to *-al*. [2] Number the carbon chain to put the CHO at C1, but omit this number from the name. Apply all other nomenclature rules.
    • To name an aldehyde with the CHO bonded to a ring: [1] Name the ring and add the suffix *-carbaldehyde*. [2] Number the ring to put the CHO group at C1, but omit this number from the name. Apply all other nomenclature rules.

a. $(CH_3)_3CC(CH_3)_2CH_2CHO$

5 C chain = pentanal   **3,3,4,4-tetramethylpentanal**

b.

8 C chain = octanal   **2,5,6-trimethyloctanal**

c.

4 C ring = cyclobutanecarbaldehyde   **3,3-dichlorocyclobutane-carbaldehyde**

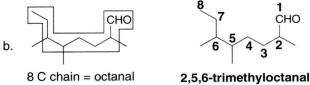

21.4 Work backwards from the name to the structure, referring to the nomenclature rules in Answer 21.3.

a. 2-isobutyl-3-isopropyl**hexanal**

6 C chain

c. 1-methyl**cyclopropanecarbaldehyde**

3 carbon ring

b. *trans*-3-methyl**cyclopentanecarbaldehyde**

5 carbon ring   or

d. 3,6-diethyl**nonanal**

9 C chain

21.5 • To name an acyclic ketone: [1] Find the longest chain with the carbonyl group and change the -*e* ending to -*one*. [2] Number the carbon chain to give the carbonyl C the lower number. Apply all other nomenclature rules.
  • To name a cyclic ketone: [1] Name the ring and change the -*e* ending to -*one*. [2] Number the C's to put the carbonyl C at C1 and give the next substituent the lower number. Apply all other nomenclature rules.

a.

8 C chain = octanone

5-ethyl-4-methyl-3-octanone

c. $(CH_3)_3CCOC(CH_3)_3$

5 C chain = pentanone

2,2,4,4-tetramethyl-3-pentanone

b.

$(CH_3)_3C$ CH_3

5 C ring = cyclopentanone

$(CH_3)_3C$ CH_3

3-*tert*-butyl-2-methylcyclopentanone

21.6 Most common names are formed by naming both alkyl groups on the carbonyl C, arranging them alphabetically, and adding the word ketone.

a. *sec*-butyl ethyl ketone

b. methyl vinyl ketone

c. *p*-ethylacetophenone

d. 3-benzoyl-2-benzylcyclopentanone

benzyl group: benzoyl group: 5 C ketone

$-CH_2-$

e. 6,6-dimethyl-2-cyclohexenone

6 C ketone

f. 3-ethyl-5-hexenal

CHO

21.7 Compounds with both a C–C double bond and an aldehyde are named as enals.

a. (2Z)-3,7-dimethyl-2,6-octadienal

neral

b. (2E,6Z)-2,6-nonadienal

cucumber aldehyde

21.8 Even though both compounds have polar C–O bonds, the electron pairs around the sp^3 hybridized O atom of diethyl ether are more crowded and less able to interact with electron-deficient sites in other diethyl ether molecules. The O atom of the carbonyl group of 2-butanone extends out from the carbon chain making it less crowded. The lone pairs of electrons on the O atom can more readily interact with the electron-deficient sites in the other molecules, resulting in stronger forces.

2-butanone   diethyl ether

21.9 For cyclic ketones, the carbonyl absorption shifts to higher wavenumber as the size of the ring decreases and the ring strain increases. Conjugation of the carbonyl group with a C=C or a benzene ring shifts the absorption to lower wavenumber.

a.   CHO   or   CHO

conjugated C=O
lower wavenumber

higher wavenumber

b.   or

smaller ring
higher wavenumber

21.10 Since a charge-separated resonance structure can be drawn for a carbonyl group, more electron donor R groups stabilize the (+) charge on this resonance form. The two R groups on the ketone C=O thus help to stabilize it.

With an aldehyde, only one electron-donor R:

$CH_3CH_2CH_2CH_2CH_2$ — C — H $CH_3CH_2CH_2CH_2CH_2$ — C — H

The H of RCHO does not stabilize the charge-separated resonance structure, so it contributes less to the hybrid. The C=O has more double bond character.
higher wavenumber

With a ketone, two electron-donor R groups:

CH_3CH_2 — C — $CH_2CH_2CH_3$ CH_3CH_2 — C — $CH_2CH_2CH_3$

The 2 R groups stabilize the charge-separated resonance structure, so it contributes more to the hybrid. The C=O has more single bond character.
weaker bond
lower wavenumber

21.11 The number of lines in their ^{13}C NMR spectra can distinguish the constitutional isomers.

2-pentanone
5 lines

3-pentanone
3 lines

3-methyl-2-butanone
4 lines

21.12

a. $CH_3CH_2CH_2COOCH_3$ $\xrightarrow[\text{[2] } H_2O]{\text{[1] DIBAL-H}}$ $CH_3CH_2CH_2CHO$

b. $CH_3CH_2CH_2CH_2OH$ $\xrightarrow{\text{PCC}}$ $CH_3CH_2CH_2CHO$

c. $HC≡CCH_2CH_3$ $\xrightarrow[\text{[2] } H_2O_2, HO^-]{\text{[1] } BH_3}$ $CH_3CH_2CH_2CHO$

d. $CH_3CH_2CH_2CH=CHCH_2CH_2CH_3$ $\xrightarrow[\text{[2] Zn, } H_2O]{\text{[1] } O_3}$ $CH_3CH_2CH_2CHO$

21.13

a. $\xrightarrow[\text{AlCl}_3]{Cl-CO-CH_3}$

b. $\xrightarrow[\text{[2] } H_2O]{\text{[1] } (CH_3)_2CuLi}$

c. $\xrightarrow[\substack{H_2SO_4 \\ HgSO_4}]{H_2O}$

21.14

Cleave this C=C with O_3.

$\xrightarrow[\text{[2] } (CH_3)_2S]{\text{[1] } O_3}$

21.15

LiAlH$_4$
or
NaBH$_4$

stronger base

weaker base
Equilibrium favors the weaker base.
The H$^-$ nucleophile is a much stronger base than
the alkoxide product.

21.16 Addition of hydride or R–M occurs at a planar carbonyl C, so two different configurations at a new stereogenic center are possible.

a. $\xrightarrow[\text{CH}_3\text{OH}]{\text{NaBH}_4}$

new stereogenic
center

Add
stereochemistry:

b. $\xrightarrow[\text{[2] } H_2O]{\text{[1] } CH_2=CHMgBr}$

Add
stereochemistry:

21.17 Treatment of an aldehyde or ketone with NaCN, HCl adds HCN across the double bond. Cyano groups are hydrolyzed by H_3O^+ to replace the 3 C–N bonds with 3 C–O bonds.

a.

$\text{CHO} \xrightarrow[\text{HCl}]{\text{NaCN}}$ (product: α-hydroxy nitrile on benzene, OH/CHCN)

b.

cyclopentane with OH and CN $\xrightarrow{H_3O^+, \Delta}$ cyclopentane with OH and COOH

21.18

amygdalin $\xrightarrow{\text{enzyme}}$ HO—C(CN)(H)—phenyl $\xrightarrow{\text{enzyme}}$ benzaldehyde + HCN

toxic by-product

21.19

a. $\begin{array}{c}CH_3\\CH_3\end{array}C{=}O \; + \; Ph_3P{=}CH_2 \longrightarrow \begin{array}{c}CH_3\\CH_3\end{array}C{=}CH_2$

b. cyclopentanone ($\bigcirc{=}O$) $+ \; Ph_3P{=}CHCH_2CH_2CH_2CH_3 \longrightarrow$ cyclopentylidene ${=}CHCH_2CH_2CH_2CH_3$

21.20

a. $Ph_3P\!: \; + \; Br{-}CH_2CH_3 \longrightarrow Ph_3\overset{+}{P}{-}CH_2CH_3 \; Br^- \xrightarrow{BuLi} Ph_3P{=}CHCH_3$

b. $Ph_3P\!: \; + \; Br{-}CH(CH_3)_2 \longrightarrow Ph_3\overset{+}{P}{-}CH(CH_3)_2 \; Br^- \xrightarrow{BuLi} Ph_3P{=}C(CH_3)_2$

c. $Ph_3P\!: \; + \; Br{-}CH_2C_6H_5 \longrightarrow Ph_3\overset{+}{P}{-}CH_2C_6H_5 \; Br^- \xrightarrow{BuLi} Ph_3P{=}CHC_6H_5$

21.21

a. benzaldehyde (CHO) $+ \; Ph_3P{=}CHCH_2CH_3 \longrightarrow$ (Z)-alkene with CH_2CH_3 $+$ (E)-alkene with CH_2CH_3

b. benzaldehyde (CHO) $+ \; Ph_3P{=}CHC_6H_5 \longrightarrow$ (E)-stilbene $+$ (Z)-stilbene

c. benzaldehyde (CHO) $+ \; Ph_3P{=}CHCOOCH_3 \longrightarrow$ (E)-$CH{=}CHCOOCH_3$ $+$ (Z)-$CH{=}CHCOOCH_3$

21.22 To draw the starting materials of the Wittig reactions, find the C=C and cleave it. Replace it with a C=O in one half of the molecule and a C=PPh₃ in the other half. The preferred pathway uses a Wittig reagent derived from a less hindered alkyl halide.

a.

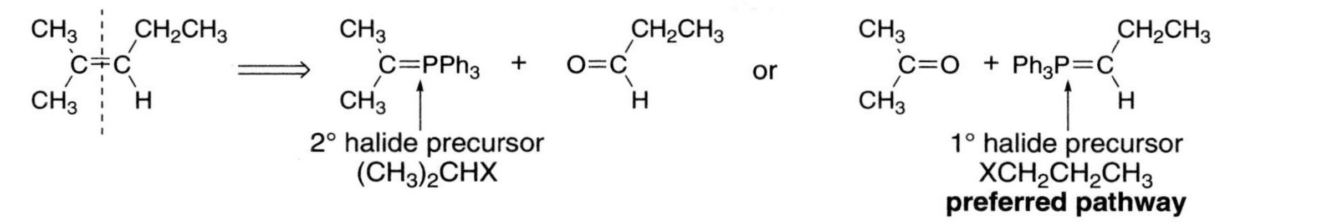

2° halide precursor
(CH₃)₂CHX

1° halide precursor
XCH₂CH₂CH₃
preferred pathway

b. 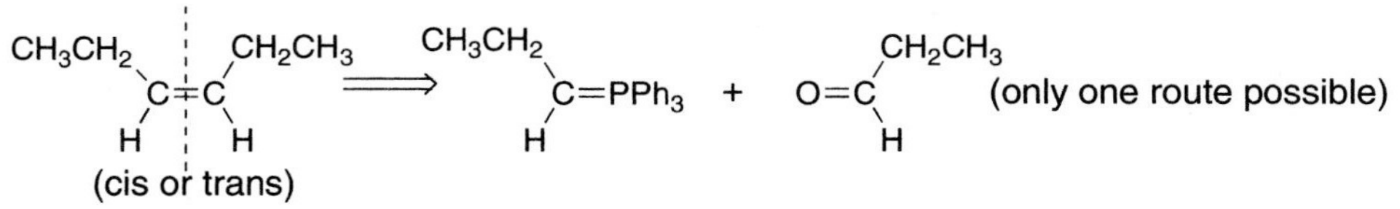
(cis or trans)

(only one route possible)

c. 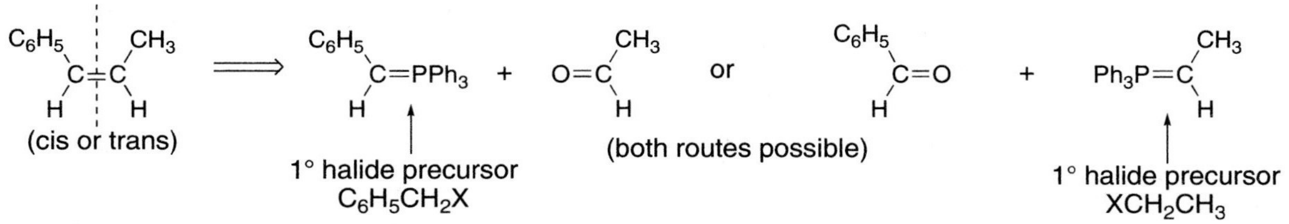
(cis or trans)

1° halide precursor
C₆H₅CH₂X

(both routes possible)

1° halide precursor
XCH₂CH₃

21.23
a.   Two-step sequence:

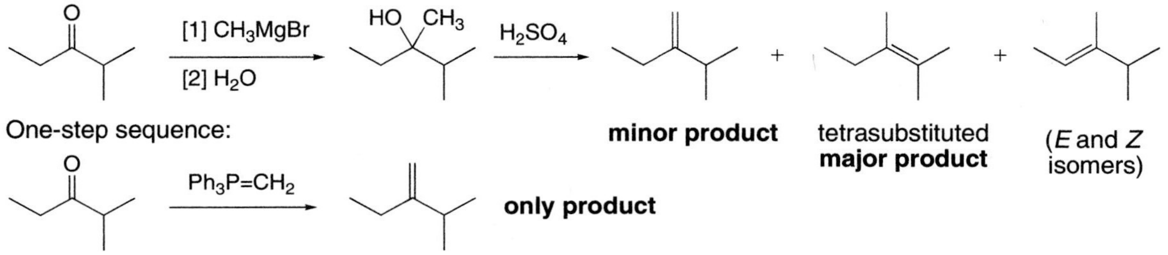

[1] CH₃MgBr
[2] H₂O

H₂SO₄

minor product

tetrasubstituted
major product

(*E* and *Z*
isomers)

One-step sequence:

Ph₃P=CH₂

only product

b.   Two-step sequence:

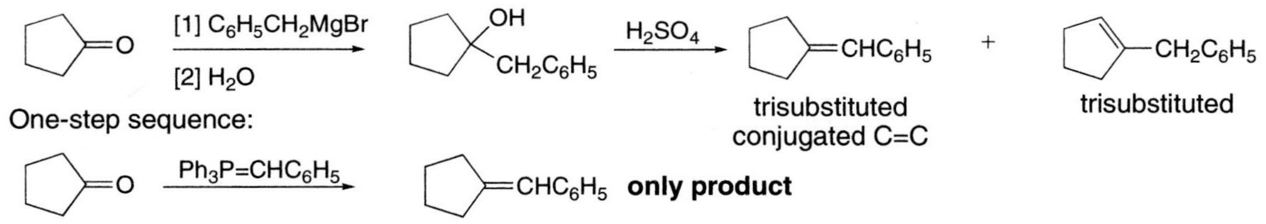

[1] C₆H₅CH₂MgBr
[2] H₂O

H₂SO₄

=CHC₆H₅

trisubstituted
conjugated C=C

—CH₂C₆H₅

trisubstituted

One-step sequence:

Ph₃P=CHC₆H₅

=CHC₆H₅ **only product**

21.24 When a 1° amine reacts with an aldehyde or ketone, the C=O is replaced by C=NR.

a.

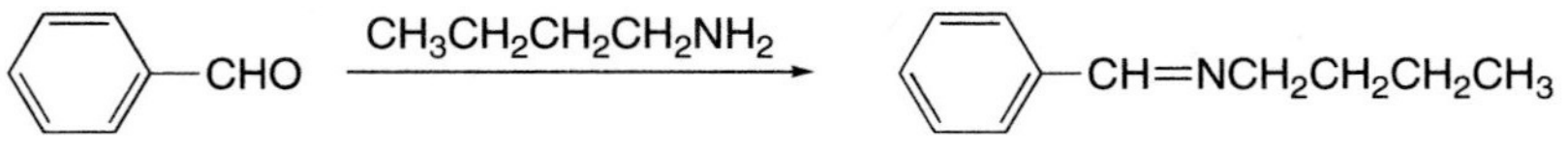

CH₃CH₂CH₂CH₂NH₂

CH=NCH₂CH₂CH₂CH₃

b.

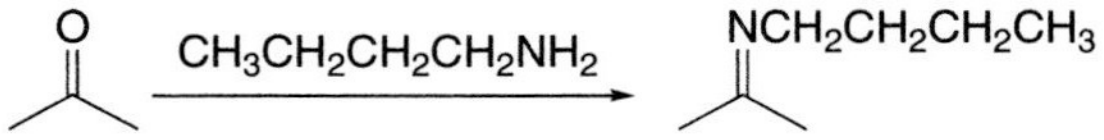

CH₃CH₂CH₂CH₂NH₂

c.

CH₃CH₂CH₂CH₂NH₂

=NCH₂CH₂CH₂CH₃

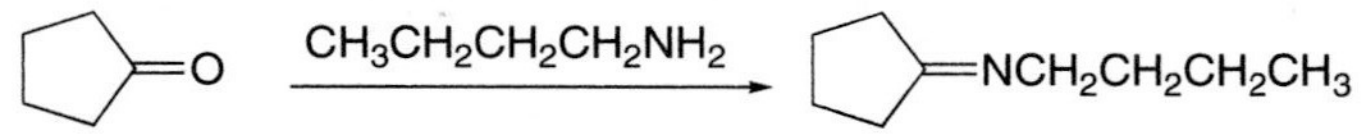

21.25 Remember that the C=NR is formed from a C=O and an NH$_2$ group of a 1° amine.

a.

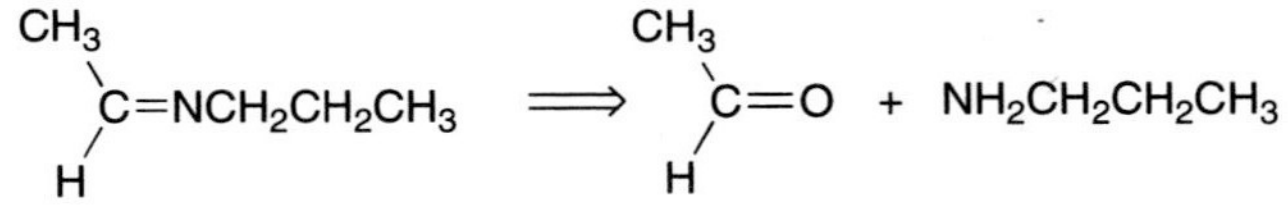

$$CH_3$$

$C=NCH_2CH_2CH_3 \implies CH_3 \quad C=O + NH_2CH_2CH_2CH_3$

b.

$CH_3\text{-}\bigcirc\text{=}N\text{-}\bigcirc \implies CH_3\text{-}\bigcirc\text{=}O + NH_2\text{-}\bigcirc$

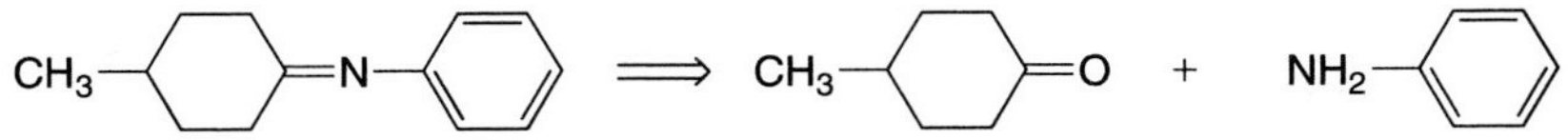

21.26

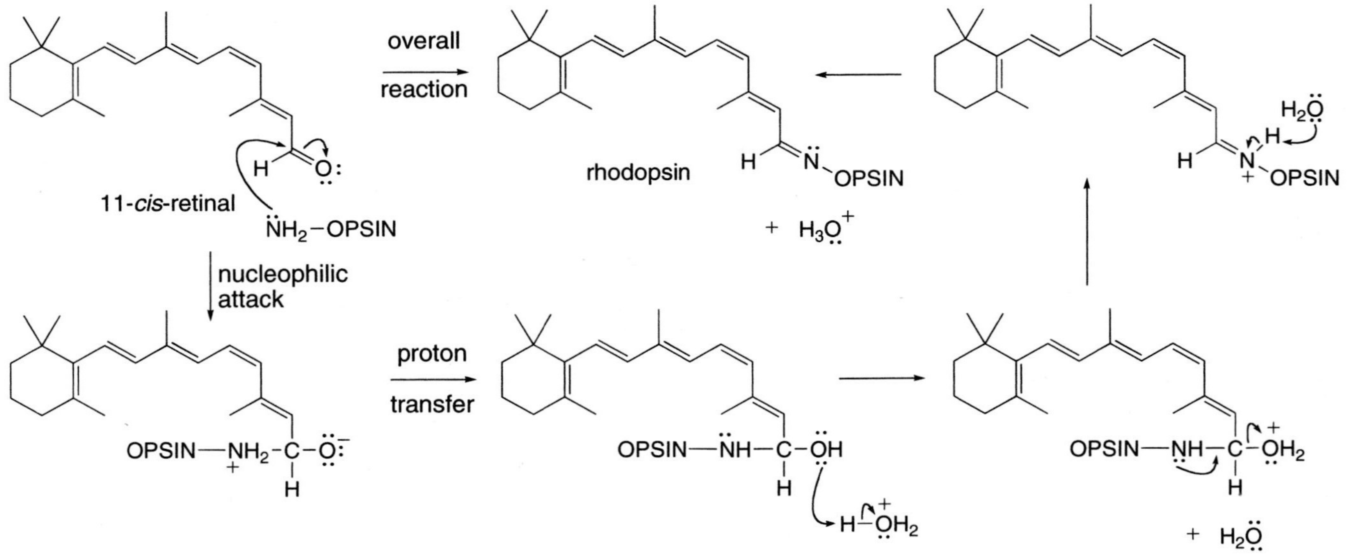

21.27

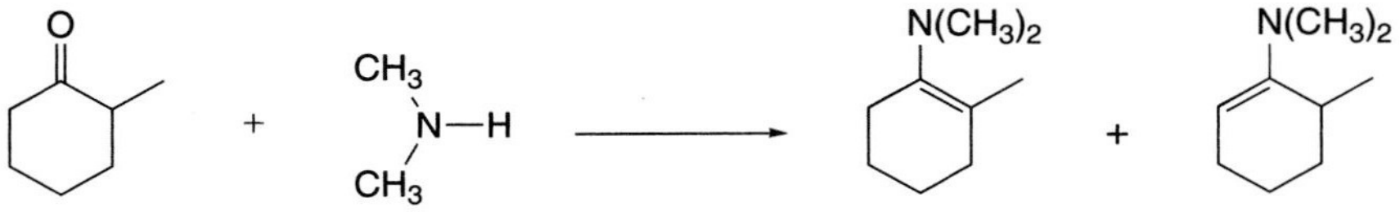

21.28

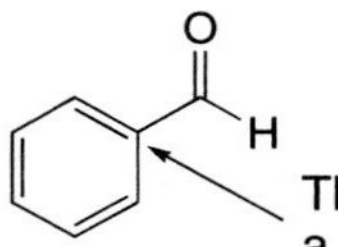

This carbon has four bonds to C's. To make an enamine, it needs a H atom, which is lost as H$_2$O when the enamine is formed.

21.29 • Imines are hydrolyzed to 1° amines and a carbonyl compound.
 • Enamines are hydrolyzed to 2° amines and a carbonyl compound.

a.

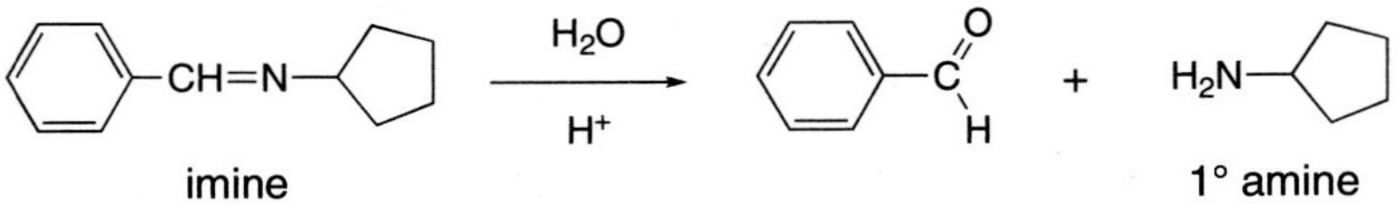

imine   1° amine

b.

enamine   2° amine

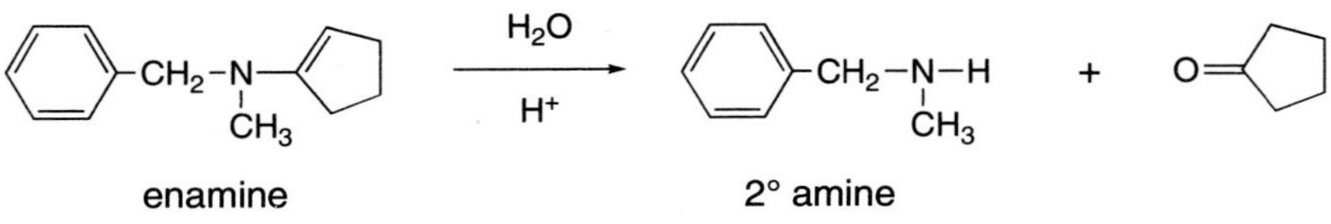

c. $(CH_3)_2NCH=C(CH_3)_2$ $\xrightarrow[H^+]{H_2O}$ $(CH_3)_2NH$ + $\overset{\overset{O}{\|}}{C}-CH(CH_3)_2$

        enamine   2° amine   H

21.30 • A substituent that **donates** electron density to the carbonyl C stabilizes it, **decreasing** the percentage of hydrate at equilibrium.
  • A substituent that **withdraws** electron density from the carbonyl C destabilizes it, **increasing** the percentage of hydrate at equilibrium.

   a. $CH_3CH_2CH_2CHO$ or $CH_3CH_2COCH_3$ b. CH_3CF_2CHO or CH_3CH_2CHO

            one R group on C=O   2 R groups   F atoms are electron withdrawing.
       higher percentage of hydrate   on C=O   higher percentage of hydrate

21.31

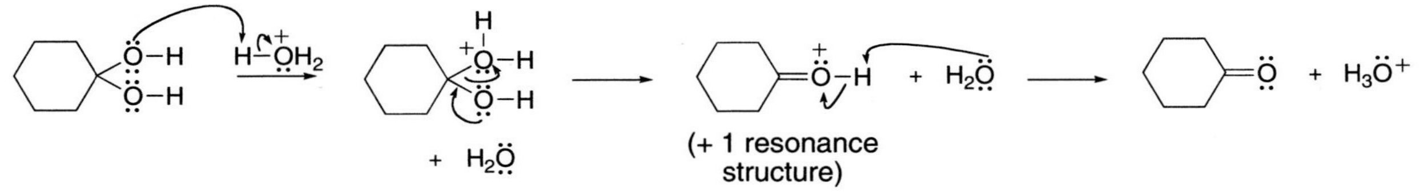

(+ 1 resonance structure)

21.32 Treatment of an aldehyde or ketone with two equivalents of alcohol results in the formation of an acetal (a C bonded to 2 OR groups).

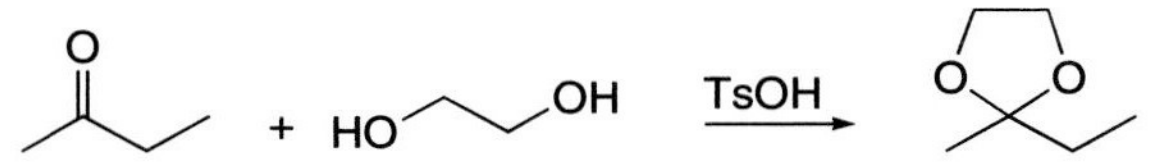

21.33

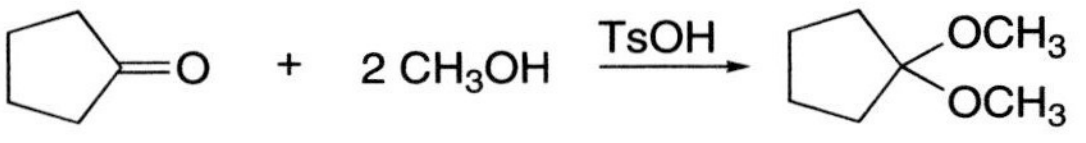

a. 2 OR groups on different C's **2 ethers**

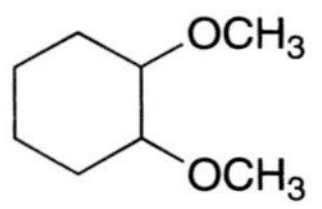

b. 2 OR groups on same C **acetal**

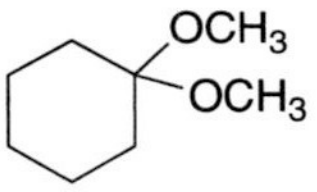

c. 2 OR groups on same C **acetal**

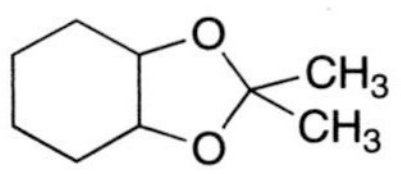

d. 1 OR group and 1 OH group on same C **hemiacetal**

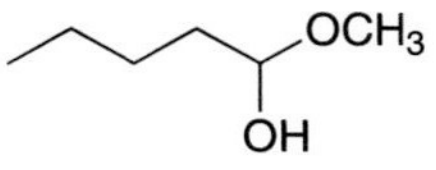

21.34 The mechanism has two parts: [1] nucleophilic addition of ROH to form a hemiacetal; [2] conversion of the hemiacetal to an acetal.

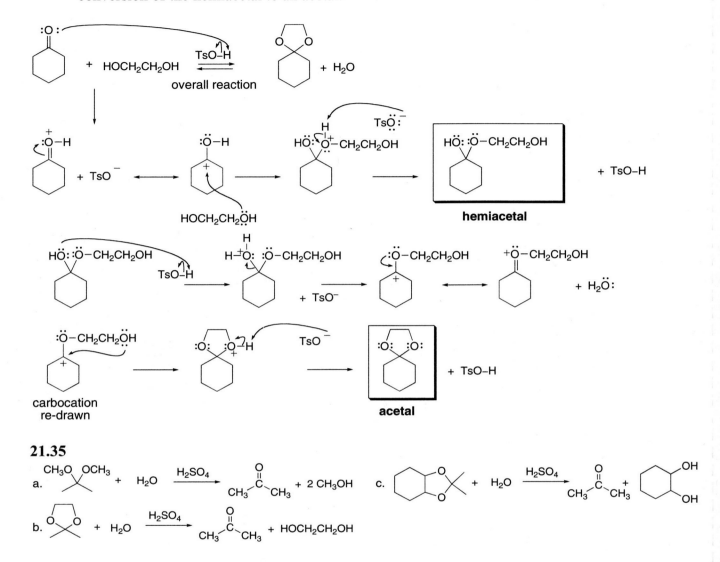

21.35

a.

b.

c.

21.36

safrole   formaldehyde

21.37 Use an acetal protecting group to carry out the reaction.

21.38

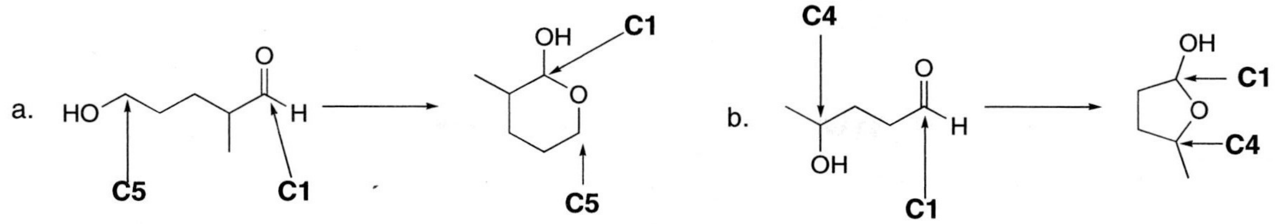

21.39 The hemiacetal OH is replaced by an OR group to form an acetal.

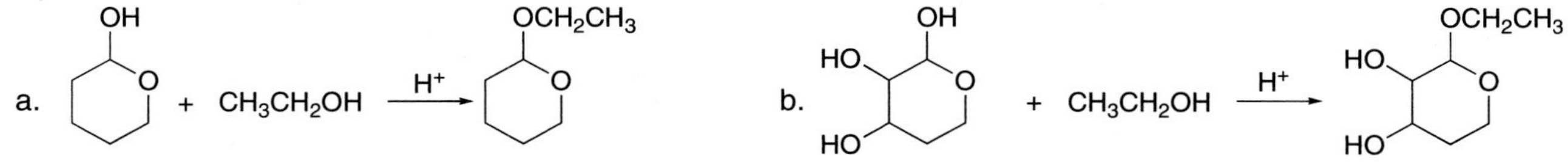

21.40

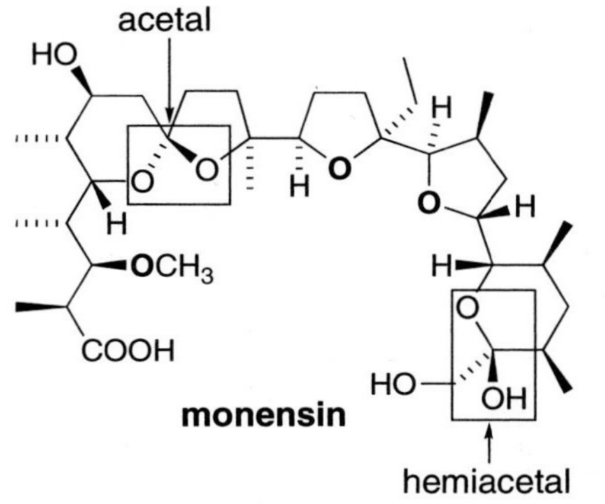

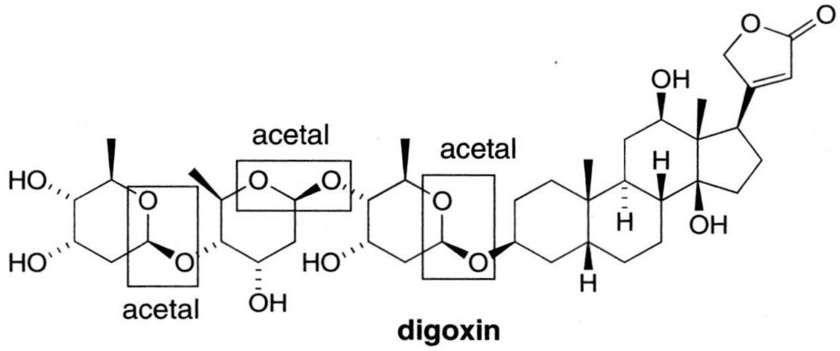

monensin

hemiacetal

digoxin

acetal

Ether **O** atoms are indicated in **bold.**

21.41

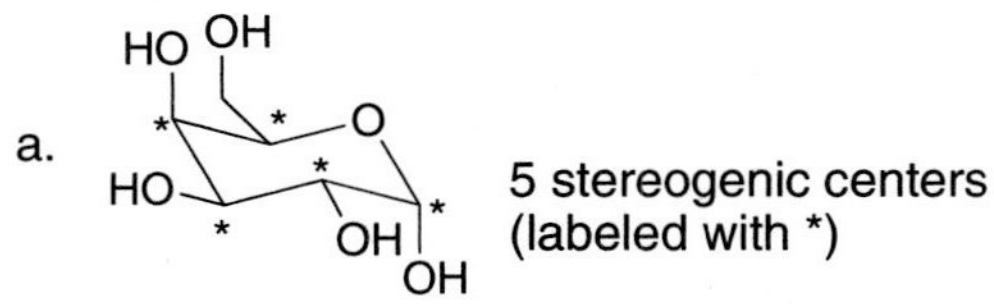

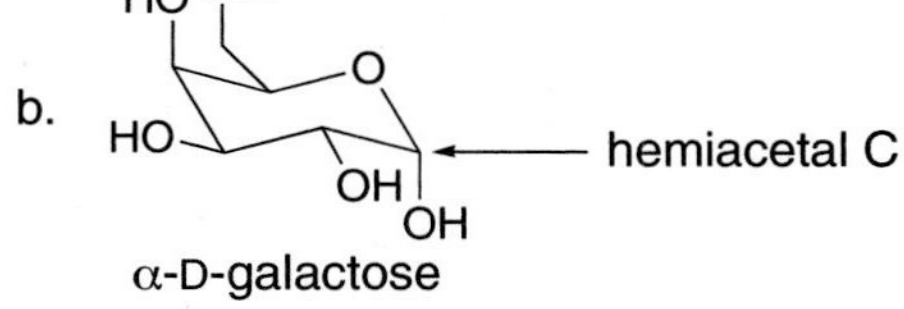

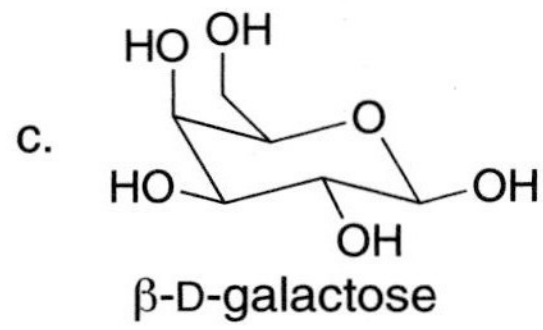

a. 5 stereogenic centers (labeled with *)

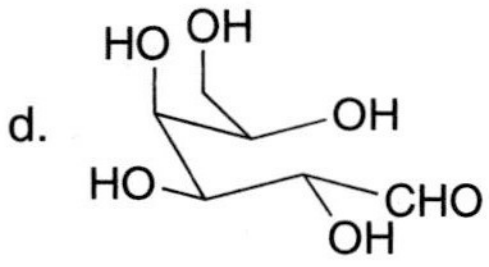

b. hemiacetal C

α-D-galactose

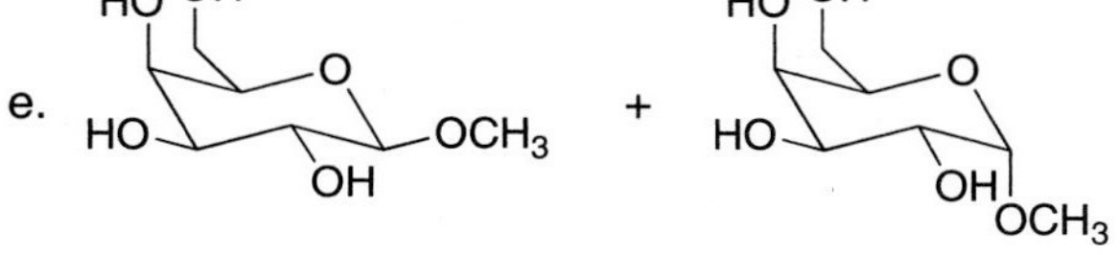

c. β-D-galactose

d.

e.

21.42 Use the rules from Answers 21.3 and 21.5 to name the aldehydes and ketones.

a. $(CH_3)_3CCH_2CHO$ $\xrightarrow{\text{re-draw}}$
4 C = butanal
3,3-dimethylbutanal

h. $(CH_3)_3C$ ... $CH(CH_3)_2$ $\xrightarrow{\text{re-draw}}$
5 C = pentanone
**2,2,4-trimethyl-
3-pentanone**
(common name:
***tert*-butyl isopropyl ketone)**

b.
5 C = pentanone
2-chloro-3-pentanone

c. Ph
8 C = octanone
8-phenyl-3-octanone

d.
5 C ring
**2-methyl-
cyclopentanecarbaldehyde**

e.
6 C ring = cyclohexanone
**5-ethyl-2-methyl-
cyclohexanone**

f. $(CH_3)_2CH$ ... CH_3
6 C ring = cyclohexanone
**5-isopropyl-2-methyl-
cyclohexanone**

g.
***trans*-2-benzylcyclohexanecarbaldehyde**

i. *o*-nitroacetophenone

j. CHO
6 C = hexanal
3,4-diethylhexanal

k.
8 C = octenone
(5*E*)-2,5-dimethyl-5-octen-4-one

l. CHO
6 C = hexenal
3,4-diethyl-2-methyl-3-hexenal

21.43

a. 2-methyl-3-phenylbutanal

b. dipropyl ketone

c. 3,3-dimethylcyclohexanecarbaldehyde

d. α-methoxypropionaldehyde

e. 3-benzoylcyclopentanone

f. 2-formylcyclopentanone

g. (3*R*)-3-methyl-2-heptanone

h. *m*-acetylbenzaldehyde

i. 2-*sec*-butyl-3-cyclopentenone

j. 5,6-dimethyl-1-cyclohexenecarbaldehyde

21.44

hexanal

2-ethylbutanal

2,2-dimethylbutanal

4-methylpentanal

3,3-dimethylbutanal

(2*R*)-2,3-dimethylbutanal

(2*S*)-2,3-dimethylbutanal

(2*R*)-2-methylpentanal

(2*S*)-2-methylpentanal

(3*S*)-3-methylpentanal

(3*R*)-3-methylpentanal

21.45

$C_6H_5CH_2CHO$ phenylacetaldehyde

a.   NaBH$_4$, CH$_3$OH $C_6H_5CH_2CH_2OH$

b.   [1] LiAlH$_4$; [2] H$_2$O $C_6H_5CH_2CH_2OH$

c.   [1] CH$_3$MgBr; [2] H$_2$O $C_6H_5CH_2CH(OH)CH_3$

d.   NaCN, HCl $C_6H_5CH_2CH(OH)CN$

e.   Ph$_3$P=CHCH$_3$ $C_6H_5CH_2CH=CHCH_3$ (*E* and *Z* isomers)

f.   (CH$_3$)$_2$CHNH$_2$, mild acid   [structure: NCH(CH$_3$)$_2$]

g.   (CH$_3$CH$_2$)$_2$NH, mild acid   [structure: N(CH$_2$CH$_3$)$_2$] (*E* and *Z* isomers)

h.   CH$_3$CH$_2$OH (excess), H$^+$ [structure: OCH$_2$CH$_3$, OCH$_2$CH$_3$]

i.   [piperidine] NH, mild acid   (*E* and *Z* isomers)

j.   HO—CH$_2$CH$_2$—OH, H$^+$ [cyclic acetal structure]

21.46

2-heptanone

a.   NaBH$_4$, CH$_3$OH   [OH structure]

b.   [1] LiAlH$_4$; [2] H$_2$O   [OH structure]

c.   [1] CH$_3$MgBr; [2] H$_2$O   [OH structure]

d.   NaCN, HCl   [NC, OH structure]

e.   Ph$_3$P=CHCH$_3$ [CHCH$_3$ structure] (*E* and *Z* isomers)

f.   (CH$_3$)$_2$CHNH$_2$, mild H$^+$ [NCH(CH$_3$)$_2$ structure]

g. $\xrightarrow{\text{(CH}_3\text{CH}_2)_2\text{NH, mild H}^+}$ [structure: N(CH₂CH₃)₂ enamine, (E and Z isomers)] + [structure: N(CH₂CH₃)₂ enamine]

$(E$ and Z isomers)

h. $\xrightarrow{\text{CH}_3\text{CH}_2\text{OH (excess), H}^+}$ [structure: acetal CH₃CH₂O OCH₂CH₃]

i. [piperidine NH] $\xrightarrow{\text{mild H}^+}$ [enamine structures] +

$(E$ and Z isomers)

j. $\xrightarrow[\text{H}^+]{\text{HO—CH}_2\text{CH}_2\text{—OH}}$ [cyclic acetal dioxolane structure]

21.47

a. [cyclopentanone] $\xrightarrow{\text{Ph}_3\text{P}=\text{CHCH}_2\text{CH}_3}$ [cyclopentylidene =CHCH₂CH₃]

b. [cyclopentyl—CHO] $\xrightarrow{\text{Ph}_3\text{P}=\text{cyclohexylidene}}$ [alkene product]

c. [cyclopentyl—CHO] $\xrightarrow{\text{Ph}_3\text{P}=\text{CHCOOCH}_3}$ [two alkene products with COOCH₃, E and Z]

d. [cyclopentanone] $\xrightarrow{\text{Ph}_3\text{P}=\text{CH(CH}_2)_5\text{COOCH}_3}$ [cyclopentylidene =CH(CH₂)₅COOCH₃]

21.48

a. CH₃CH₂Cl $\xrightarrow[\substack{[2]\ \text{BuLi} \\ [3]\ (\text{CH}_3)_2\text{C=O}}]{[1]\ \text{Ph}_3\text{P}}$ CH₃CH=C(CH₃)₂

b. [C₆H₅—CH₂Br] $\xrightarrow[\substack{[2]\ \text{BuLi} \\ [3]\ \text{C}_6\text{H}_5\text{CH}_2\text{CH}_2\text{CHO}}]{[1]\ \text{Ph}_3\text{P}}$ [C₆H₅—CH=CHCH₂CH₂C₆H₅]

$(E$ and Z isomers)

c. [cyclopentyl—CH₂Cl] $\xrightarrow[\substack{[2]\ \text{BuLi} \\ [3]\ \text{CH}_3\text{CH}_2\text{CH}_2\text{CHO}}]{[1]\ \text{Ph}_3\text{P}}$ [cyclopentyl—CH=CHCH₂CH₂CH₃]

$(E$ and Z isomers)

21.49

a. Ph₃P=CHCH₂CH₂CH₃ $\Longrightarrow$ BrCH₂CH₂CH₂CH₃

b. Ph₃P=C(CH₂CH₂CH₃)₂ $\Longrightarrow$ BrCH(CH₂CH₂CH₃)₂

c. Ph₃P=CHCH=CH₂ $\Longrightarrow$ BrCH₂CH=CH₂

21.50

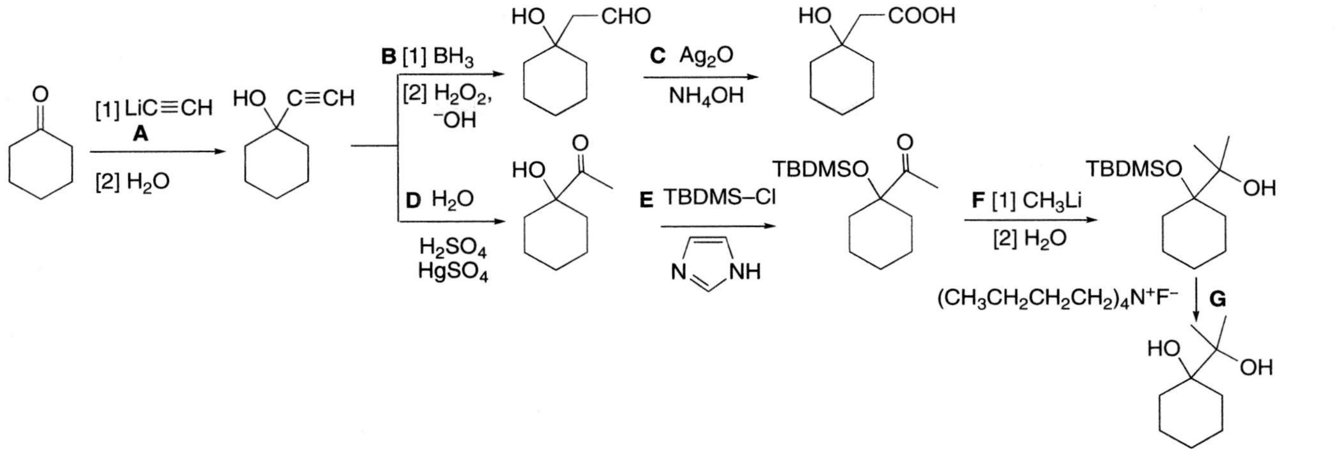

21.51

a. 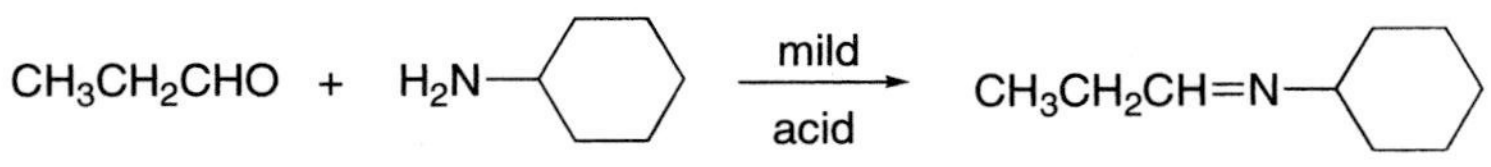 CH_3CH_2CHO + H_2N—(cyclohexyl) $\xrightarrow[\text{acid}]{\text{mild}}$ $CH_3CH_2CH{=}N$—(cyclohexyl)

b. 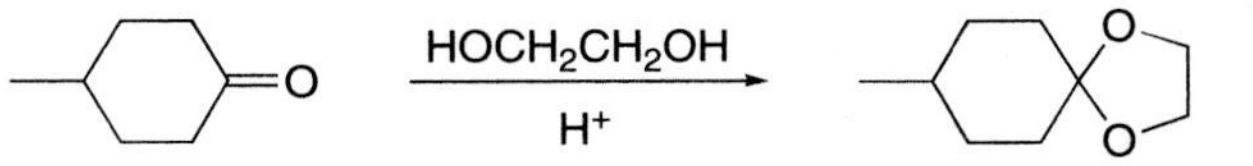 $\xrightarrow[\text{H}^+]{\text{HOCH}_2\text{CH}_2\text{OH}}$

c. 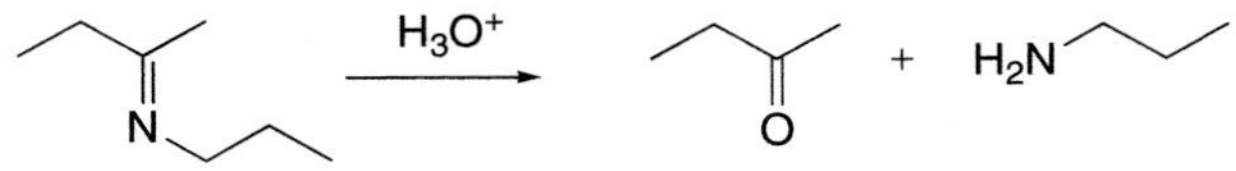 $\xrightarrow{\text{H}_3\text{O}^+}$ + H_2N

d. 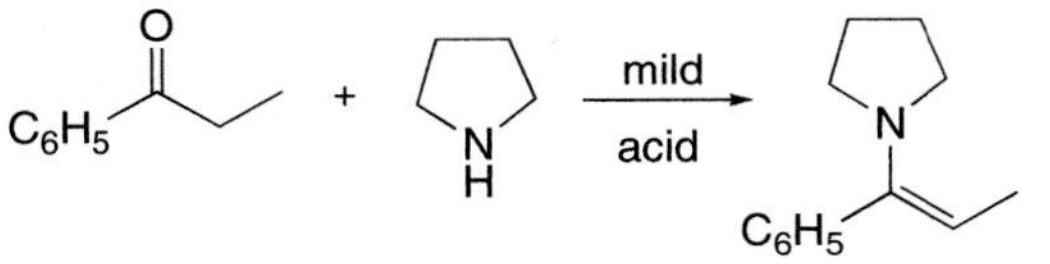 + (pyrrolidine) $\xrightarrow[\text{acid}]{\text{mild}}$ (*E* and *Z* isomers)

e. 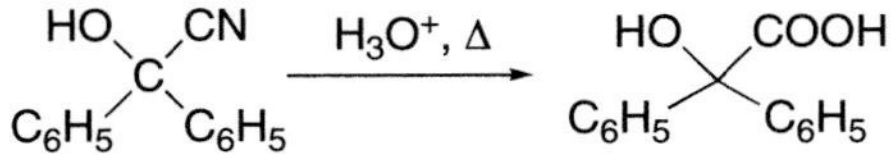 $\xrightarrow{\text{H}_3\text{O}^+, \Delta}$

f. 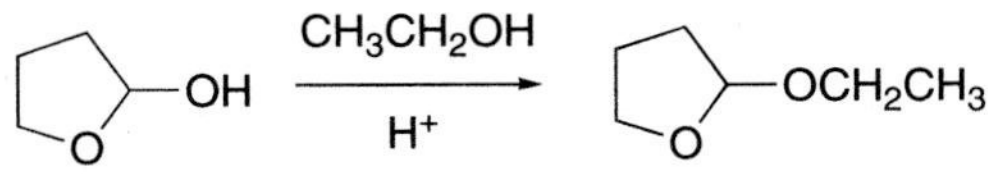 $\xrightarrow[\text{H}^+]{\text{CH}_3\text{CH}_2\text{OH}}$

g. $\xrightarrow{\text{H}_3\text{O}^+}$ + HN(piperidine)

h. CH_3O—$\xrightarrow{\text{H}_3\text{O}^+}$ CH_3O—(cyclohexanone) + $HOCH_3$

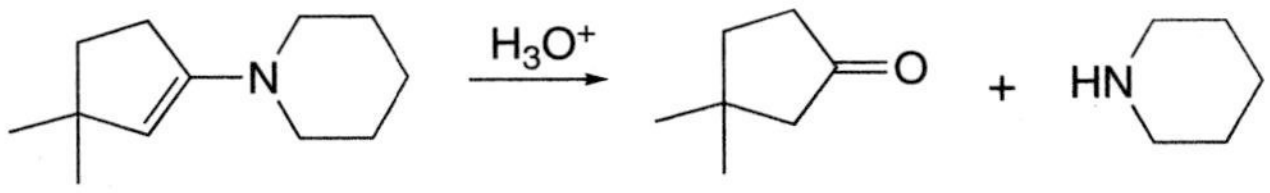

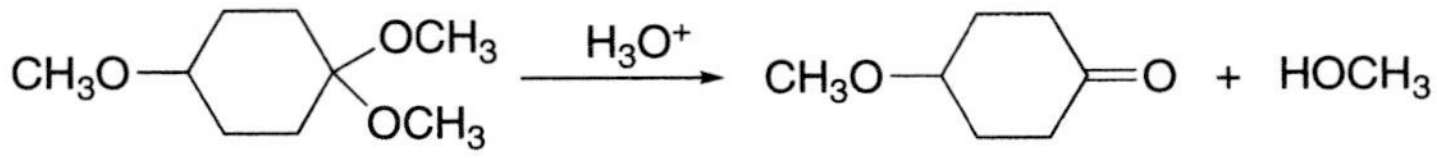

21.52

a. 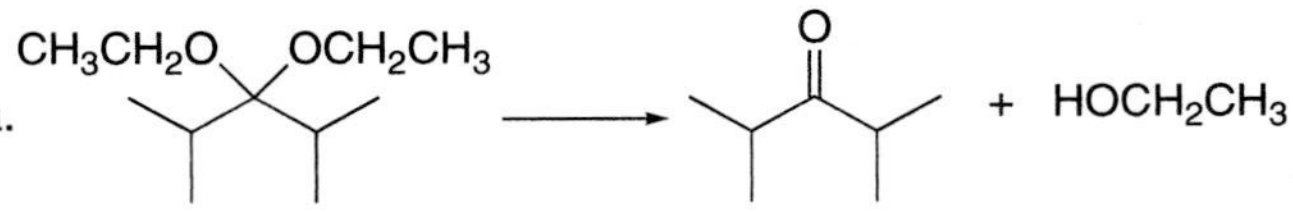 $\longrightarrow$ + $HOCH_2CH_3$

b. 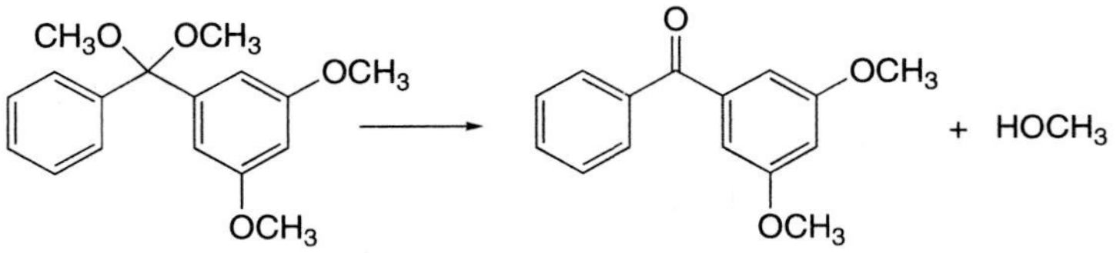 $\longrightarrow$ + $HOCH_3$

c. 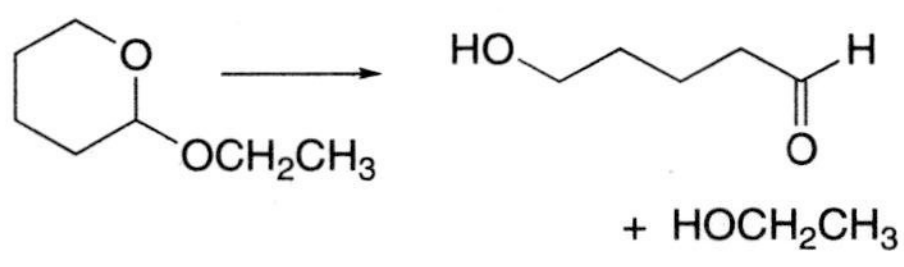 $\longrightarrow$ HO——H + $HOCH_2CH_3$

21.53 Consider para product only, when an ortho, para mixture can result.

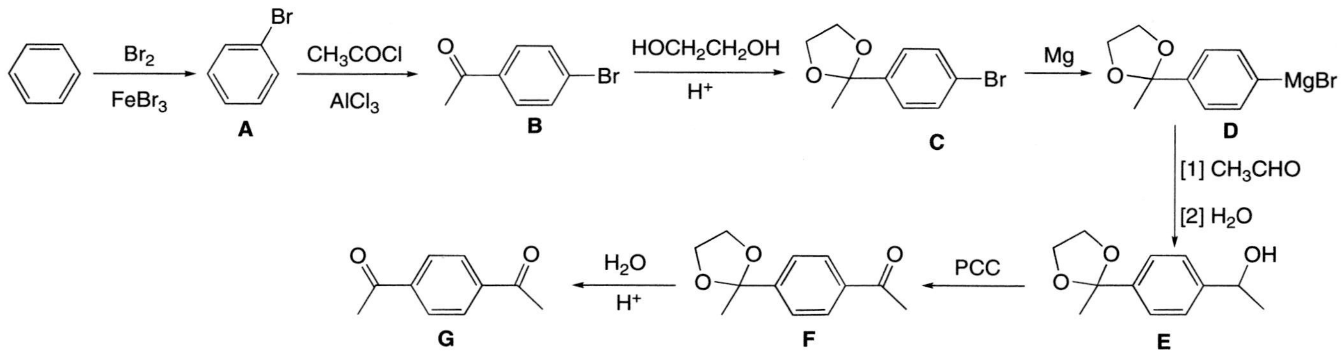

21.54

a. 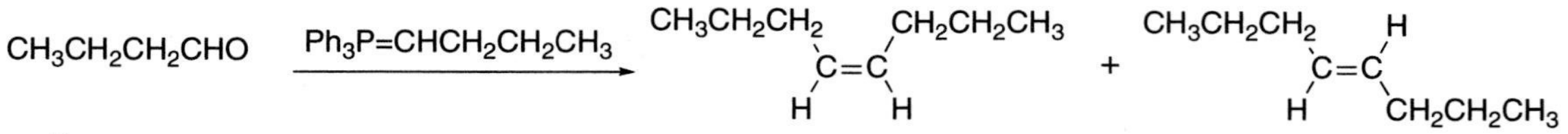 $CH_3CH_2CH_2CHO$ $\xrightarrow{Ph_3P=CHCH_2CH_2CH_3}$

b.

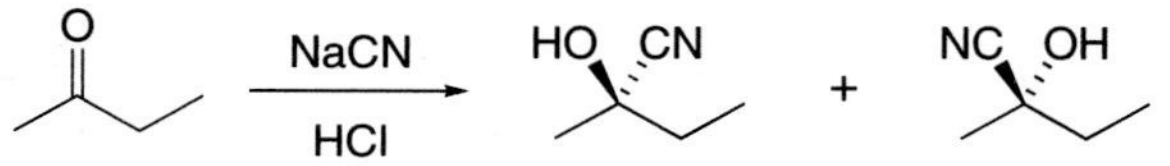

c.

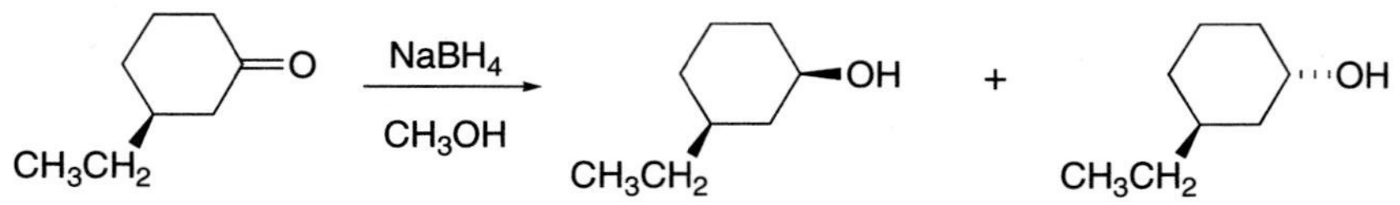

d.

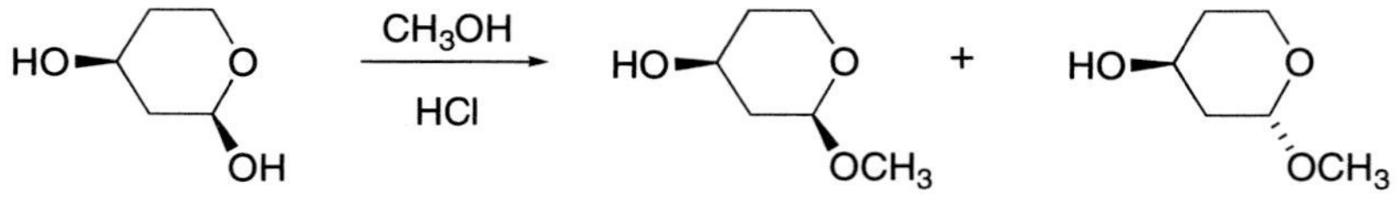

21.55

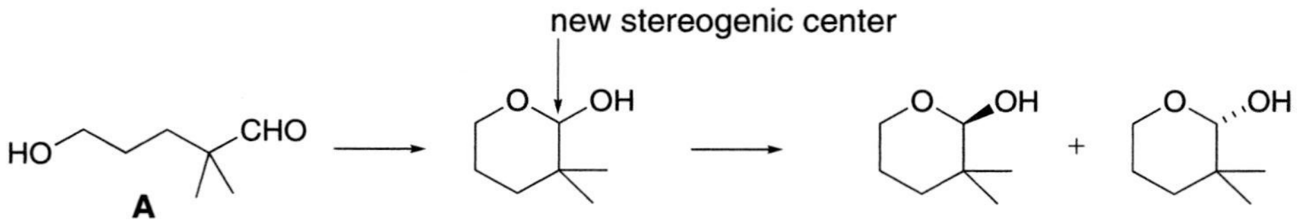

new stereogenic center

An equal mixture of enantiomers results, so the product is optically inactive.

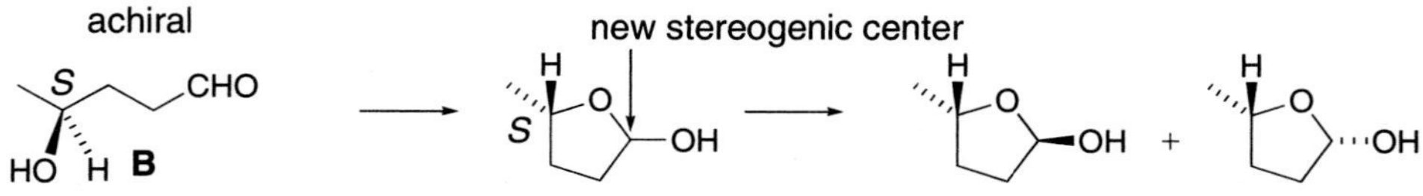

new stereogenic center

A mixture of diastereomers results. Both compounds are chiral and they are not enantiomers, so the mixture is optically active.

21.56

a.

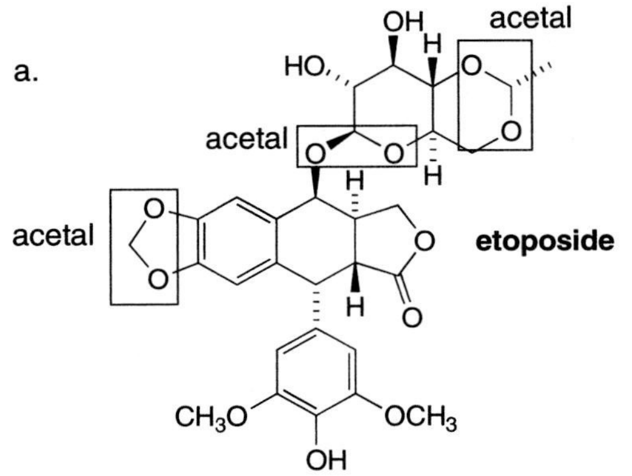

OH
acetal
HO
acetal
acetal
etoposide
CH3O
OCH3
OH

b. Lines of cleavage are drawn in.

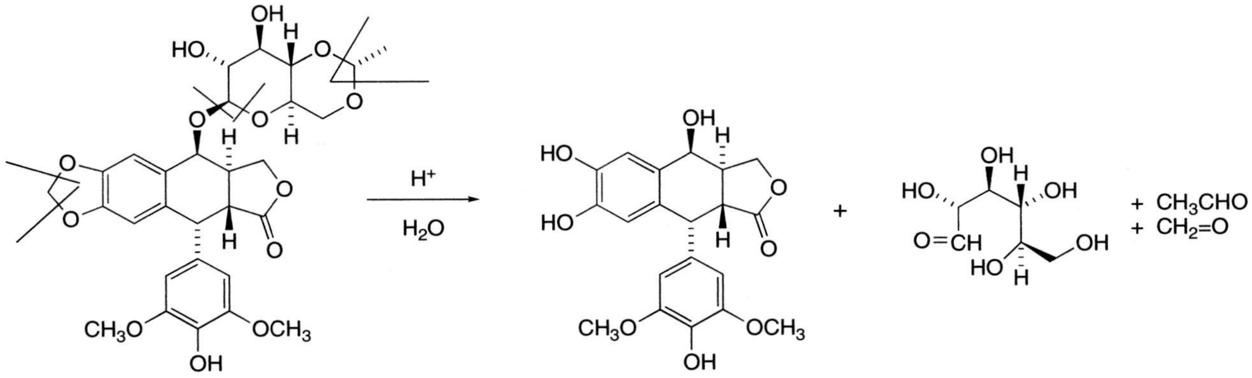

H+
H2O
+
+ CH3CHO
+ CH2=O

21.57 Use the rule from Answer 21.1.

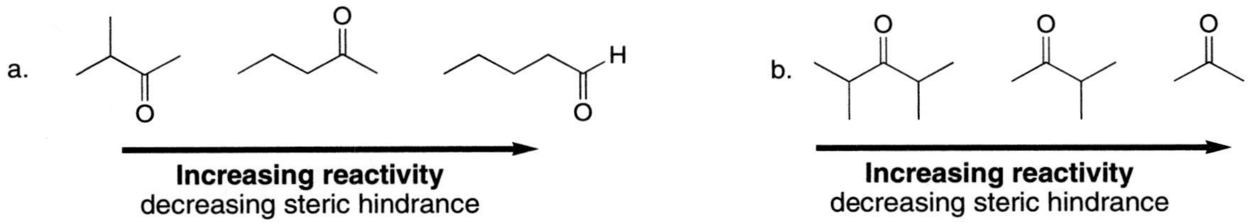

a.
b.
Increasing reactivity
decreasing steric hindrance
Increasing reactivity
decreasing steric hindrance

21.58

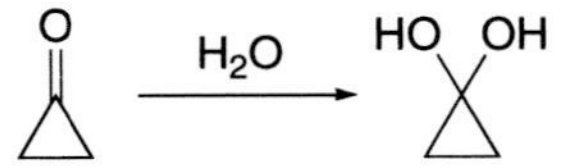

H2O
HO OH

Less stable carbonyl compounds give a higher percentage of hydrate. Cyclopropanone is an unstable carbonyl compound because the bond angles around the carbonyl carbon deviate considerably from the desired angle. Since the carbonyl carbon is sp^2 hybridized, the optimum bond angle is 120°, but the three-membered ring makes the C–C–C bond angles only 60°. This destabilizes the ketone, giving a high concentration of hydrate when dissolved in H_2O.

21.59 Electron-donating groups decrease the amount of hydrate at equilibrium by stabilizing the carbonyl starting material. Electron-withdrawing groups increase the amount of hydrate at equilibrium by destabilizing the carbonyl starting material. Electron-donating groups make the IR absorption of the C=O shift to lower wavenumber because they stabilize the charge-separated resonance form, giving the C=O more single bond character.

O$_2$N— ⬡ —COCH$_3$

p-nitroacetophenone
a. NO$_2$ withdrawing group
less stable

b. **higher percentage of hydrate**

c. **higher wavenumber**

CH$_3$O— ⬡ —COCH$_3$

p-methoxyacetophenone
CH$_3$O donating group
more stable

lower percentage of hydrate

lower wavenumber

21.60 Use the principles from Answer 21.22.

a.

CH$_3$CH$_2$ ⋮ CH$_2$CH$_2$CH$_3$
C⫶C
CH$_3$CH$_2$ ⋮ H
⟹
CH$_3$CH$_2$
C=O + Ph$_3$P=C
CH$_3$CH$_2$ ⋮ H
CH$_2$CH$_2$CH$_3$

or

CH$_3$CH$_2$
C=PPh$_3$ + O=C
CH$_3$CH$_2$
CH$_2$CH$_2$CH$_3$ / H

1° alkyl halide precursor
(XCH$_2$CH$_2$CH$_2$CH$_3$)
preferred pathway

2° alkyl halide precursor
[(CH$_3$CH$_2$)$_2$CHX]

b.

CH$_3$ ⋮ CH$_2$CH$_3$
C⫶C
H ⋮ H
⟹
CH$_3$
C=O + Ph$_3$P=C
H
CH$_2$CH$_3$ / H

or

CH$_3$
C=PPh$_3$ + O=C
H
CH$_2$CH$_3$ / H

1° alkyl halide precursor
(XCH$_2$CH$_2$CH$_3$)

1° alkyl halide precursor
(XCH$_2$CH$_3$)

(Both routes possible.)

c.

⟹

O

+ Ph$_3$P=CH$_2$ **or**

+ CH$_2$=O

methyl halide precursor
(CH$_3$X)
preferred pathway

2° alkyl halide precursor
(CH$_3$CH$_2$CH$_2$CHXCH$_3$)

d.

⬡=CH—C$_6$H$_5$ ⟹ ⬡=O + Ph$_3$P=CH—C$_6$H$_5$ **or**   ⬡=PPh$_3$ + O=CH—C$_6$H$_5$

1° alkyl halide precursor
(C$_6$H$_5$CH$_2$X)
preferred pathway

2° alkyl halide precursor
⬡—X

21.61

a.

⬡—CH$_2$—N=⬠ ⟹ O=⬠
+
⬡—CH$_2$—NH$_2$

b.

⬠—N(piperidine) ⟹ ⬠=O + HN(piperidine)

c.

(dioxolane)—⬡ ⟹ O=CH—⬡ + HOCH$_2$CH$_2$OH

d. CH$_3$O—⟍⟍⟍—OCH$_3$
              |
             OCH$_3$
⟹ CH$_3$O—⟍⟍⟍—H
                    ‖
                    O
+ HOCH$_3$

21.62

a. $C_6H_5-CH_2OH \xrightarrow{\text{PCC}} C_6H_5-CHO$

b. $C_6H_5-COCl \xrightarrow[\text{[2] } H_2O]{\text{[1] LiAlH[OC(CH}_3)_3]_3} C_6H_5-CHO$

c. $C_6H_5-COOCH_3 \xrightarrow[\text{[2] } H_2O]{\text{[1] DIBAL-H}} C_6H_5-CHO$

d. $C_6H_5-COOH \xrightarrow[\text{[2] } H_2O]{\text{[1] LiAlH}_4} C_6H_5-CH_2OH \xrightarrow{\text{PCC}} C_6H_5-CHO$

e. $C_6H_5-CH_3 \xrightarrow{\text{KMnO}_4} C_6H_5-COOH \xrightarrow[\text{[2] } H_2O]{\text{[1] LiAlH}_4} C_6H_5-CH_2OH \xrightarrow{\text{PCC}} C_6H_5-CHO$

f. $C_6H_5-CH=CH_2 \xrightarrow[\text{[2] Zn, } H_2O]{\text{[1] } O_3} C_6H_5-CHO$

g. $C_6H_5-CH=NCH_2CH_2CH_3 \xrightarrow[H^+]{H_2O} C_6H_5-CHO$

h. $C_6H_5-CH(OCH_2CH_3)_2 \xrightarrow[H^+]{H_2O} C_6H_5-CHO$

21.63

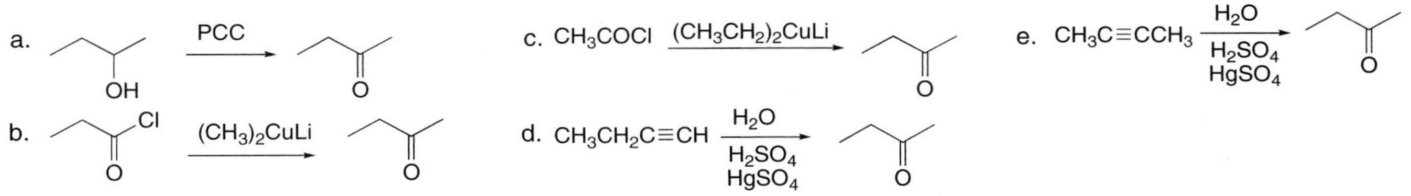

a. $\xrightarrow{\text{PCC}}$

c. $CH_3COCl \xrightarrow{(CH_3CH_2)_2CuLi}$

e. $CH_3C{\equiv}CCH_3 \xrightarrow[\substack{H_2SO_4 \\ HgSO_4}]{H_2O}$

b. $\xrightarrow{(CH_3)_2CuLi}$

d. $CH_3CH_2C{\equiv}CH \xrightarrow[\substack{H_2SO_4 \\ HgSO_4}]{H_2O}$

21.64

a. One-step sequence:

preferred route
only one product formed

or

Two-step sequence:

b. One-step sequence:

preferred route
only one product formed

or

Two-step sequence:

+ other alkenes that result from carbocation rearrangement

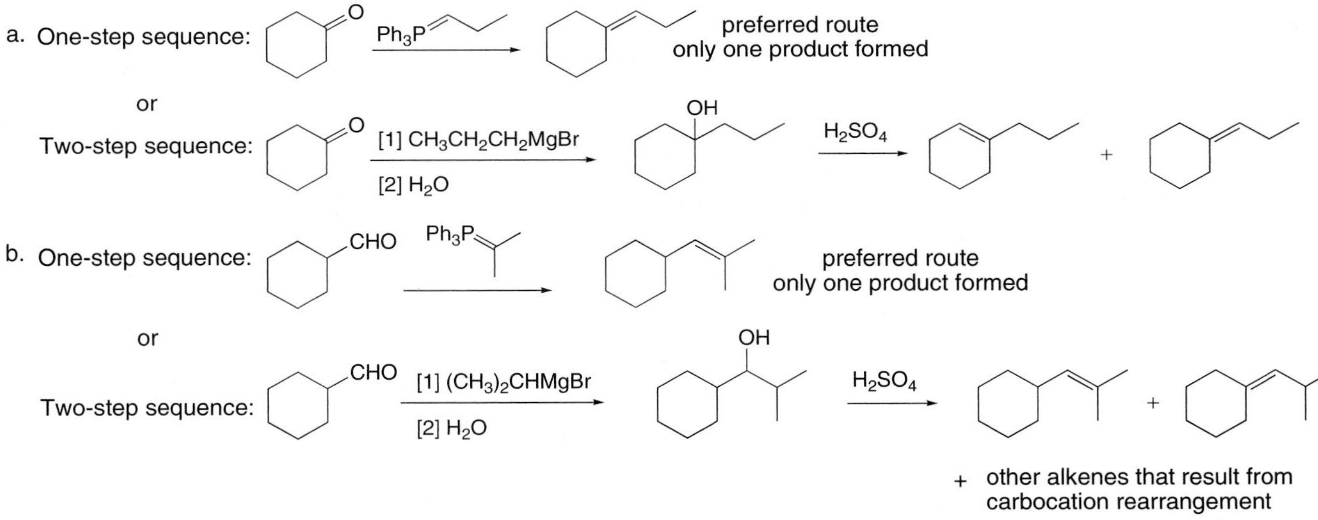

21.65

a. $CH_3CH_2CH_2CH{\doteq}CHCH_3$

One possibility:

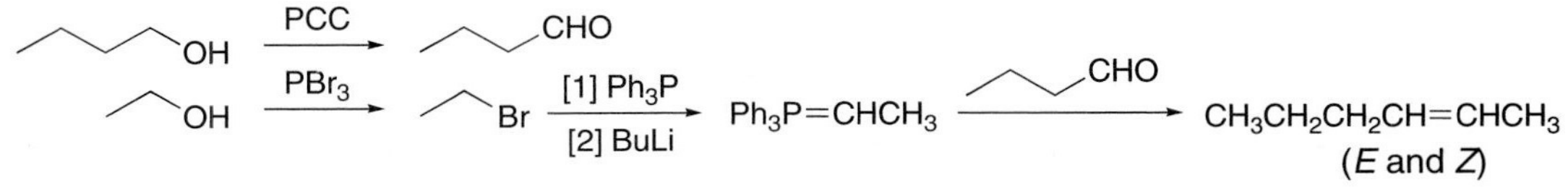

$\xrightarrow{\text{PCC}}$
$\xrightarrow{\text{PBr}_3}$
$\xrightarrow[\text{[2] BuLi}]{\text{[1] Ph}_3P} Ph_3P{=}CHCH_3 \longrightarrow CH_3CH_2CH_2CH{=}CHCH_3$
(*E* and *Z*)

b. $C_6H_5CH \doteq CHCH_2CH_2CH_3$

One possibility:

c.

21.66

a.

b.

c.

d.

e.

21.67

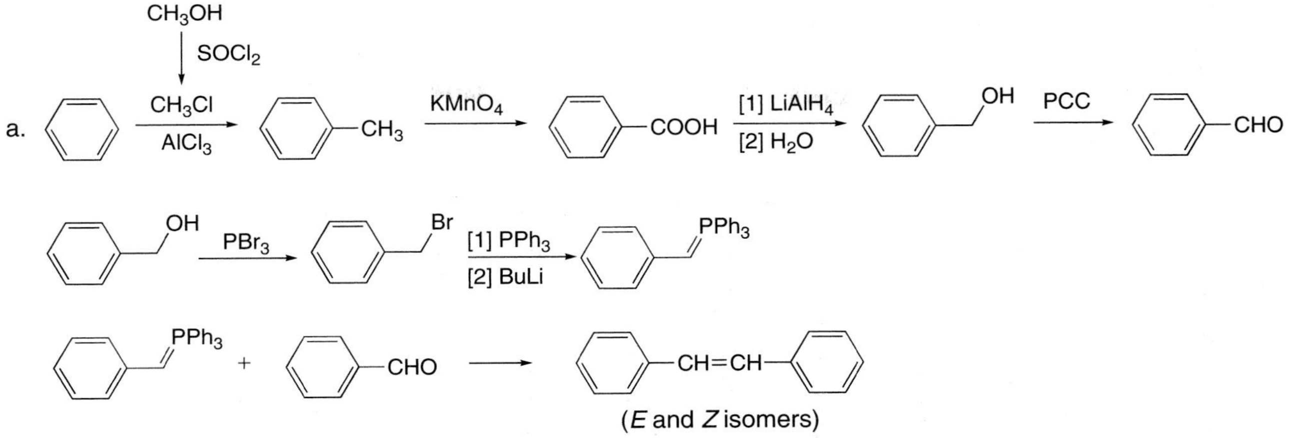
CH3OH
SOCl2
a.
CH3Cl
AlCl3
CH3
KMnO4
COOH
[1] LiAlH4
[2] H2O
OH
PCC
CHO
OH
PBr3
Br
[1] PPh3
[2] BuLi
PPh3
PPh3
+
CHO
CH=CH
(E and Z isomers)

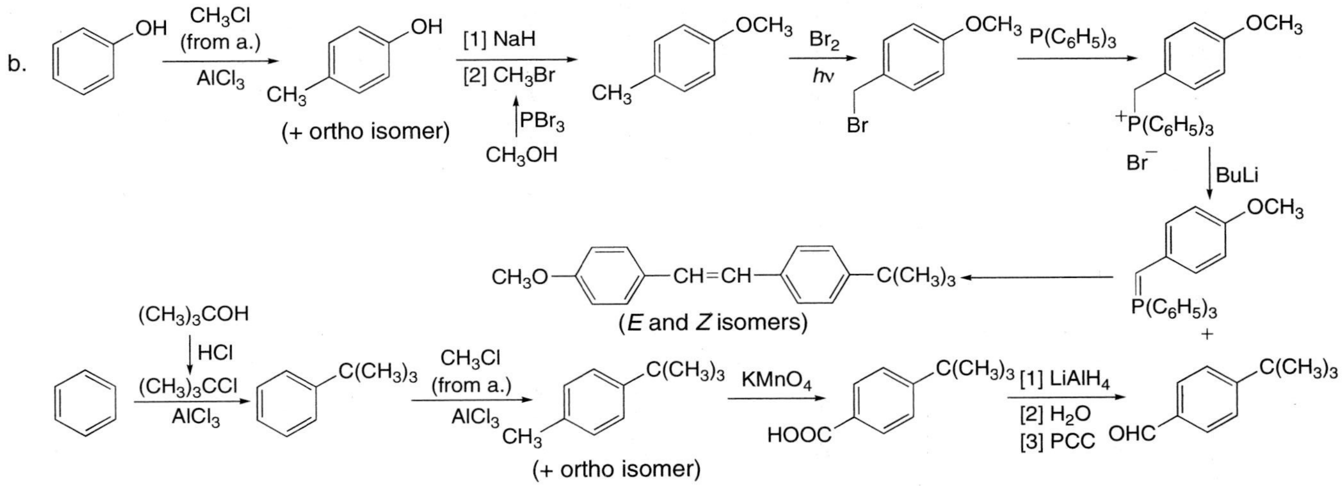
b.
OH
CH3Cl
(from a.)
AlCl3
OH
CH3
(+ ortho isomer)
[1] NaH
[2] CH3Br
PBr3
CH3OH
OCH3
CH3
Br2
hv
OCH3
Br
P(C6H5)3
OCH3
+P(C6H5)3
Br–
BuLi
OCH3
P(C6H5)3
+
CH3O
CH=CH
C(CH3)3
(E and Z isomers)
(CH3)3COH
HCl
(CH3)3CCl
AlCl3
C(CH3)3
CH3Cl
(from a.)
AlCl3
C(CH3)3
CH3
(+ ortho isomer)
KMnO4
C(CH3)3
HOOC
[1] LiAlH4
[2] H2O
[3] PCC
C(CH3)3
OHC

21.68

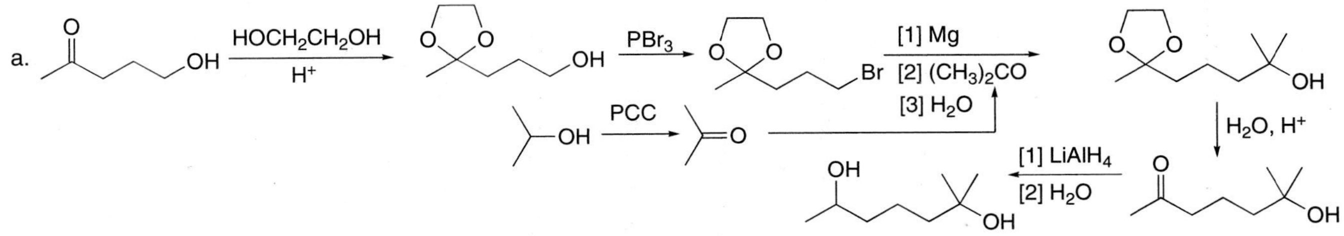
a.
O
OH
HOCH2CH2OH
H+
O
O
OH
PBr3
O
O
Br
[1] Mg
[2] (CH3)2CO
[3] H2O
O
O
OH
OH
PCC
O
H2O, H+
O
OH
[1] LiAlH4
[2] H2O
OH
OH

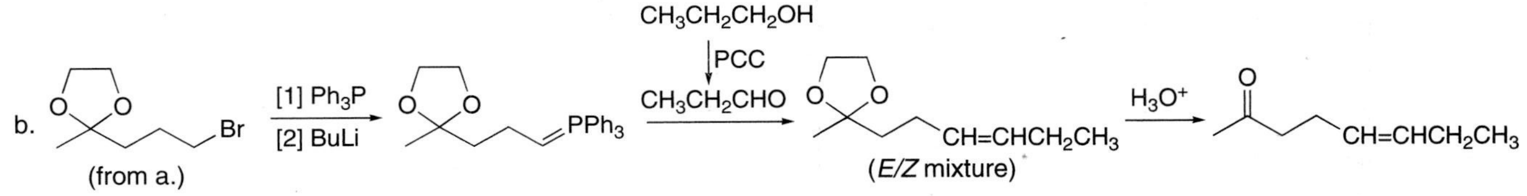
b.
O
O
Br
(from a.)
[1] Ph3P
[2] BuLi
O
O
PPh3
CH3CH2CH2OH
PCC
CH3CH2CHO
O
O
CH=CHCH2CH3
(E/Z mixture)
H3O+
O
CH=CHCH2CH3

21.69

a. $CH_3CH_2OH \xrightarrow{PCC}$ (acetaldehyde)

$CH_3CH_2OH \xrightarrow{PBr_3} CH_3CH_2Br \xrightarrow{Mg} CH_3CH_2MgBr \xrightarrow[{[2] H_2O}]{[1]\ CH_3CHO} $ (2-butanol, $CH_3\underset{CH_2CH_3}{\overset{OH}{C}}H$) $\xrightarrow{PCC}$ (ethyl methyl ketone $CH_3COCH_2CH_3$) $\xrightarrow[TsOH]{CH_3CH_2OH}$ $CH_3-\underset{CH_2CH_3}{\overset{OCH_2CH_3}{\underset{|}{C}}}-OCH_2CH_3$

b. $CH_3CH_2OH \xrightarrow{H_2SO_4} CH_2=CH_2 \xrightarrow{Br_2} Br\text{-}CH_2CH_2\text{-}Br \xrightarrow{2\ NaNH_2} HC\equiv CH \xrightarrow{NaH} HC\equiv C^- Na^+$

$CH_2=CH_2 \xrightarrow[{[2]\ NaHSO_3,\ H_2O}]{[1]\ OsO_4}$ HO–CH2CH2–OH

$HC\equiv C^- Na^+ \xrightarrow[{(from\ a.)}]{[1]\ CH_3CHO\quad [2]\ H_2O}$ (3-butyn-2-ol, OH) $\xrightarrow{PCC}$ (3-butyn-2-one) $\xrightarrow[TsOH]{HO\quad OH}$ (cyclic acetal dioxolane)

21.70

X $\xrightarrow{mCPBA}$ (epoxide) $\xrightarrow{(CH_3)_3CNH_2}$ (alkoxide intermediate, O^-, NHC(CH₃)₃) $\xrightarrow{H_2O}$ (OH, NHC(CH₃)₃) $\xrightarrow{H_3O^+}$ albuterol + $(CH_3)_2C=O$

21.71

Br–CH₂CH₂CH₂–CHO $\xrightarrow[TsOH]{HOCH_2CH_2OH}$ (Br, dioxolane) $\xrightarrow{Mg}$ BrMg (dioxolane) $\xrightarrow[{[2] H_2O}]{[1]\ \text{cyclopentanone}}$ (cyclopentanol, OH, dioxolane)

$\xrightarrow[{H^+}]{H_2O}$ (cyclopentanol OH, CHO) **A**

21.72

a. (cyclohexyl–Ö–H) $\quad Na^+H^- \longrightarrow$ (cyclohexyl–Ö:⁻) + H₂ + Na⁺ $\quad Cl\text{-}CH_2\text{-}O\text{-}CH_3 \longrightarrow$ (cyclohexyl–O–CH₂–O–CH₃) + Cl⁻

methoxy methyl ether

b. (cyclohexyl, O–CH₂–O, CH₃) acetal

c.

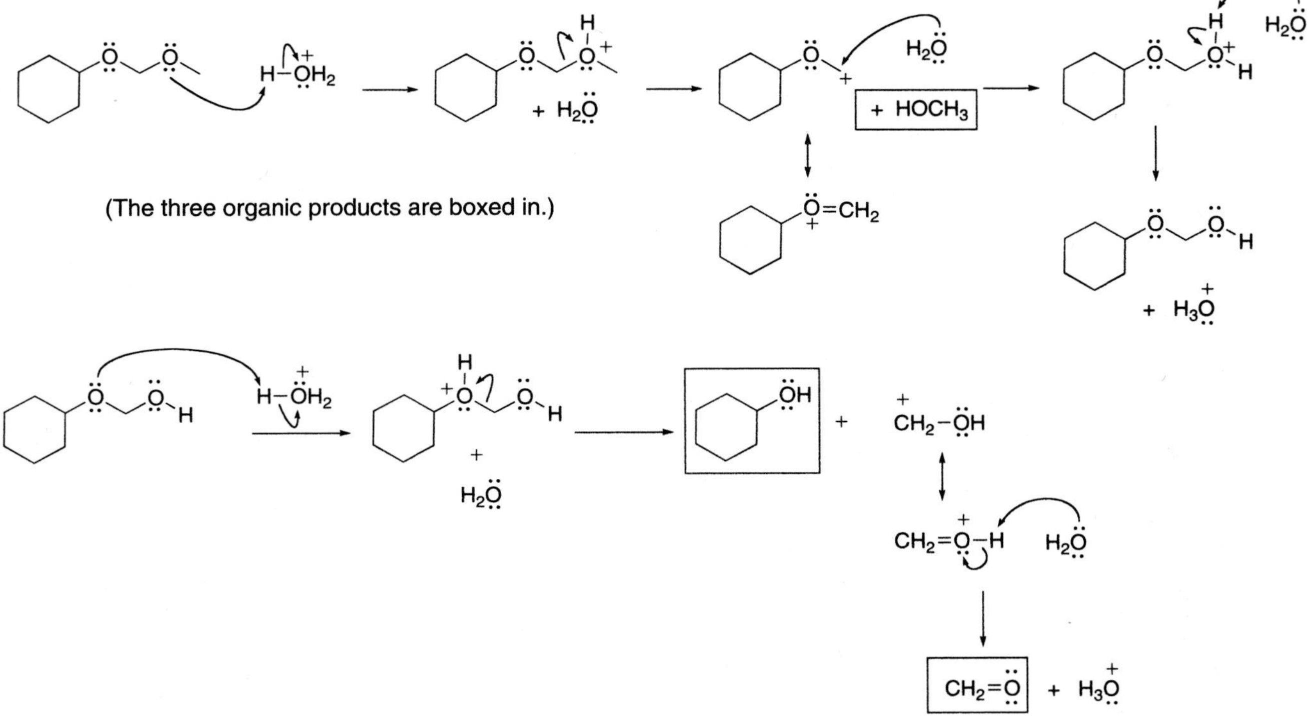
H–OH2
+ H2O
H2O
+ HOCH3
O=CH2
H2O
+ H3O
(The three organic products are boxed in.)
H–OH2
H2O
OH
CH2–OH
CH2=O–H
H2O
CH2=O
+ H3O

21.73

a.

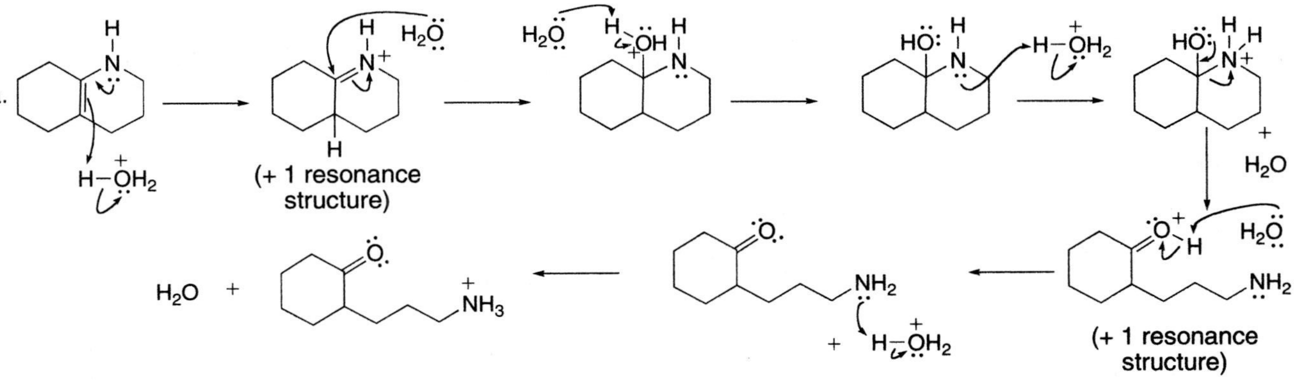
H–OH2
(+ 1 resonance structure)
H2O
H2O
H–OH2
H2O
H2O
NH2
(+ 1 resonance structure)
NH2
H–OH2
H2O
NH3

b.

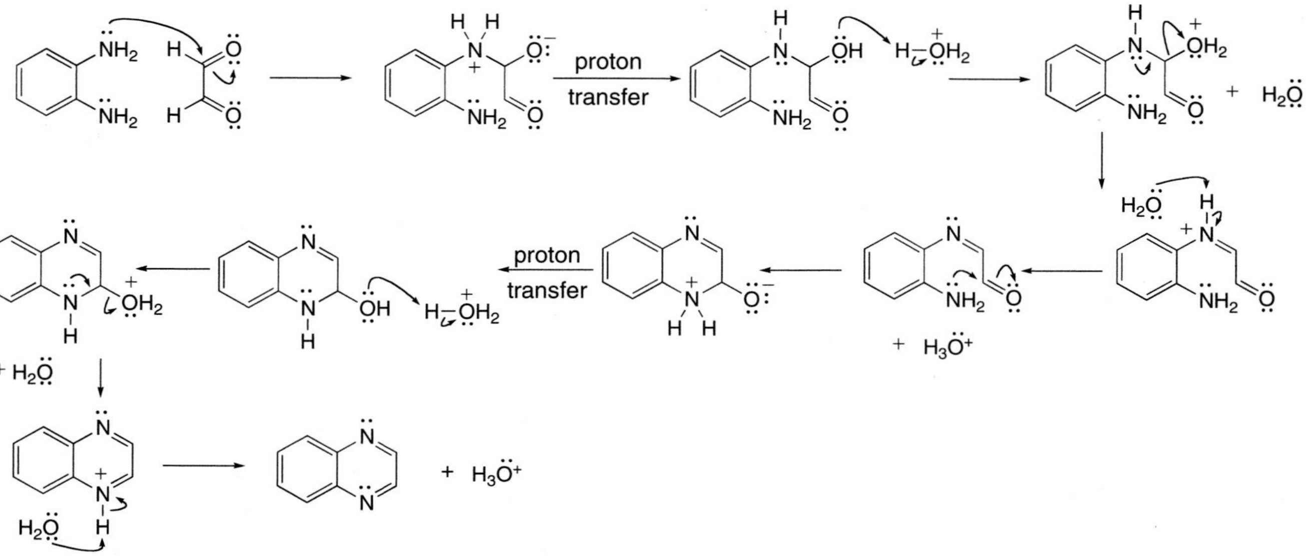

21.74 The OH groups react with the C=O in an intramolecular reaction, first to form a hemiacetal, and then to form an acetal.

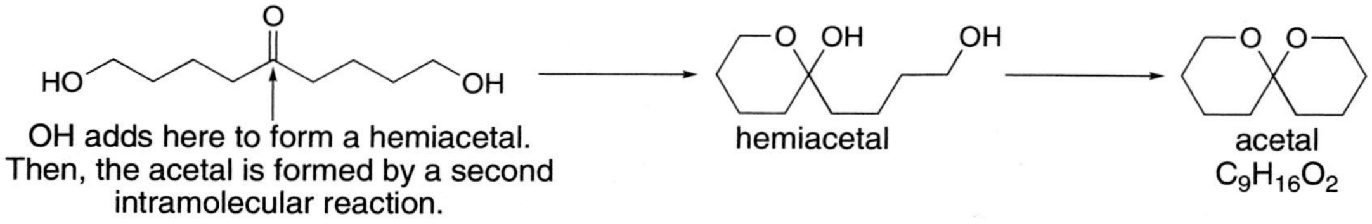

OH adds here to form a hemiacetal.
Then, the acetal is formed by a second
intramolecular reaction.

hemiacetal

acetal
$C_9H_{16}O_2$

21.75

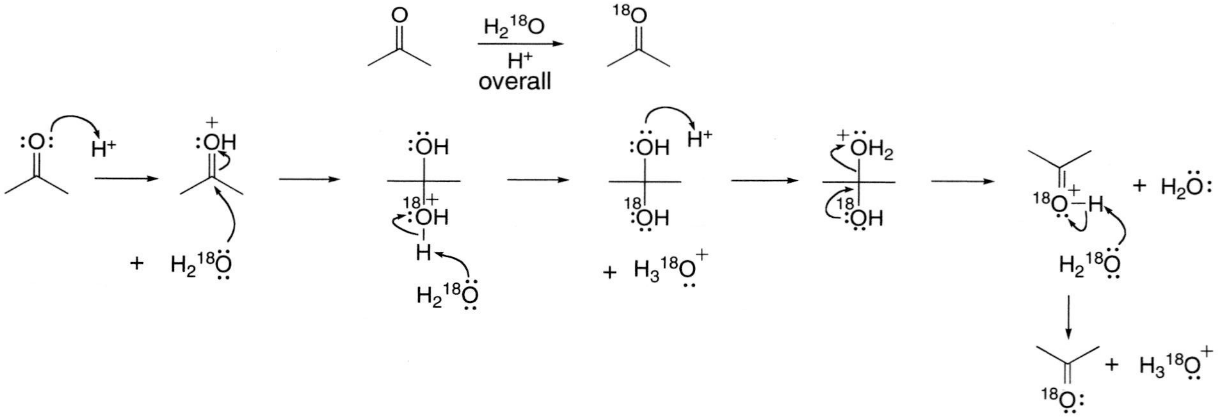

21.76

a.

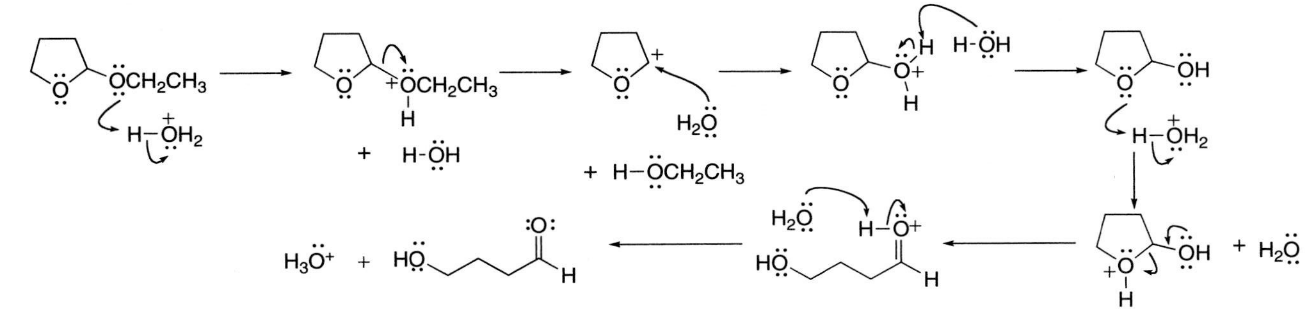

b.

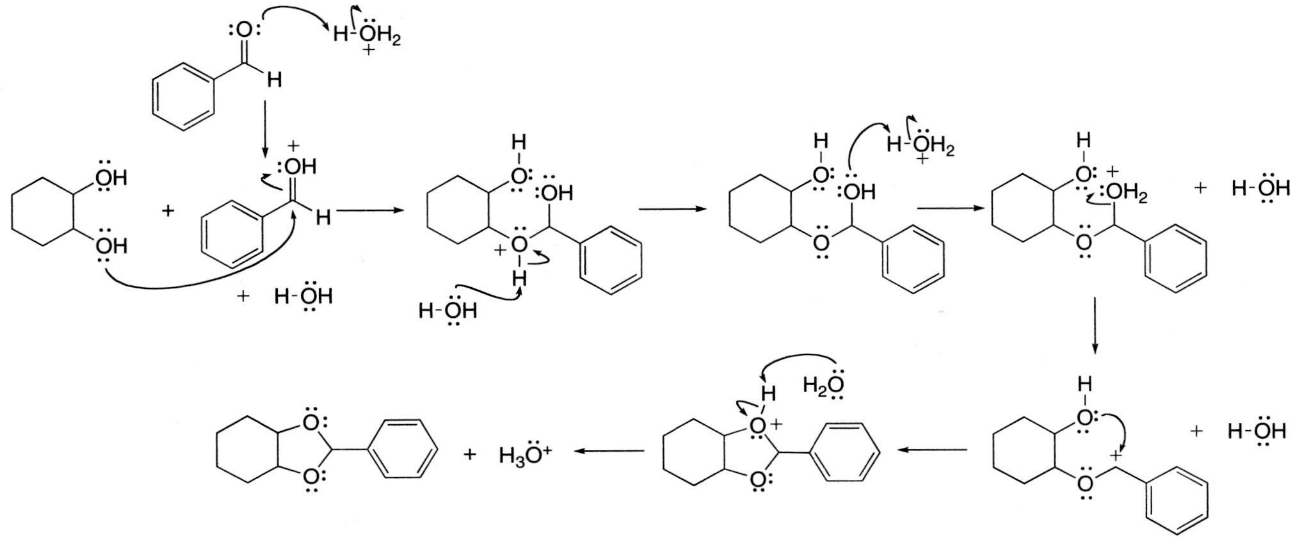

21.77

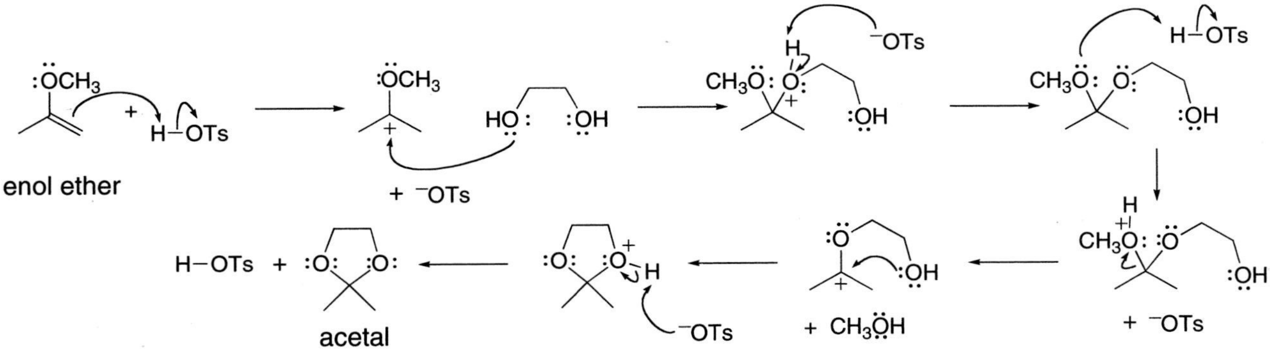

enol ether

acetal

21.78

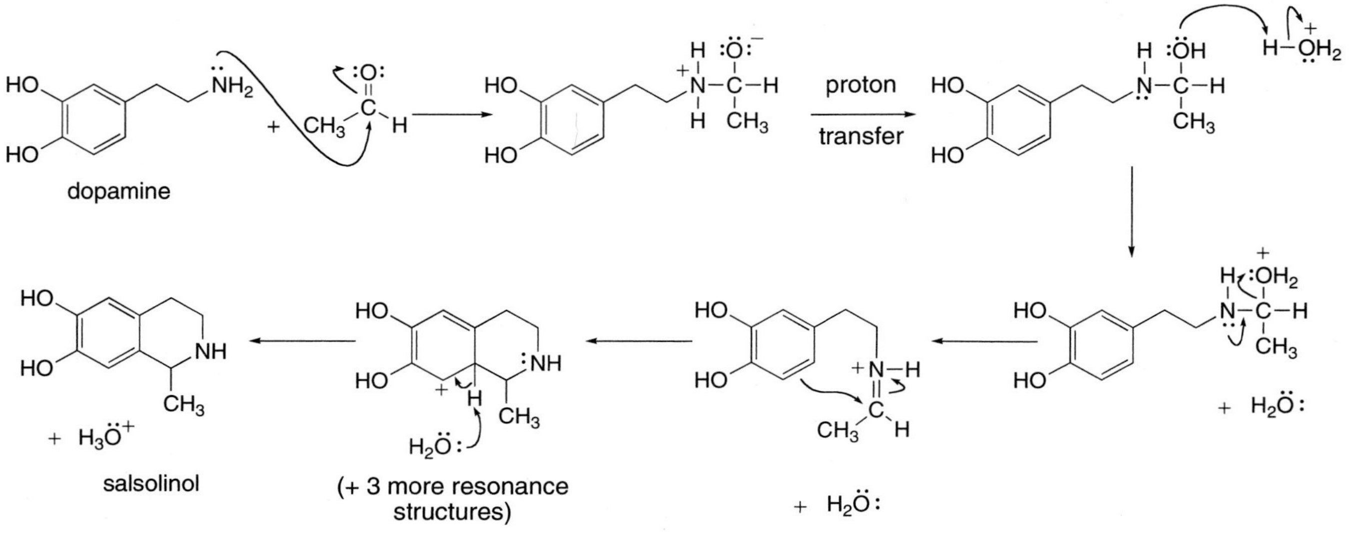

dopamine

salsolinol

+ H₃Ö⁺

(+ 3 more resonance structures)

+ H₂Ö:

21.79

sulfonium salt

sulfur ylide

+ Bu—H + LiX

X

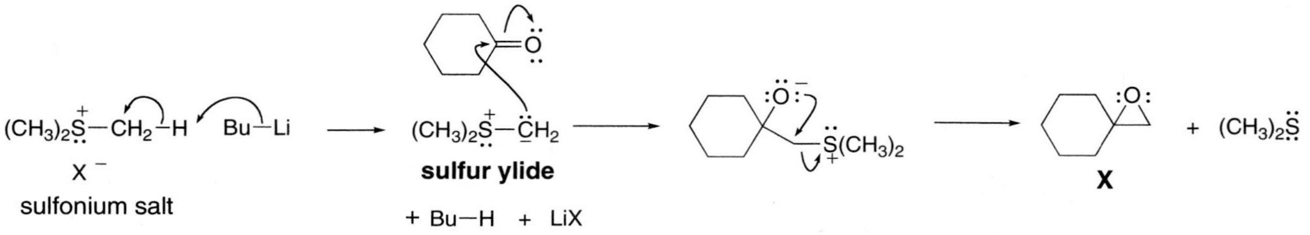

21.80 Hemiacetal **A** is in equilibrium with its acyclic hydroxy aldehyde. The aldehyde can undergo hydride reduction to form 1,4-butanediol and Wittig reaction to form an alkene.

a.

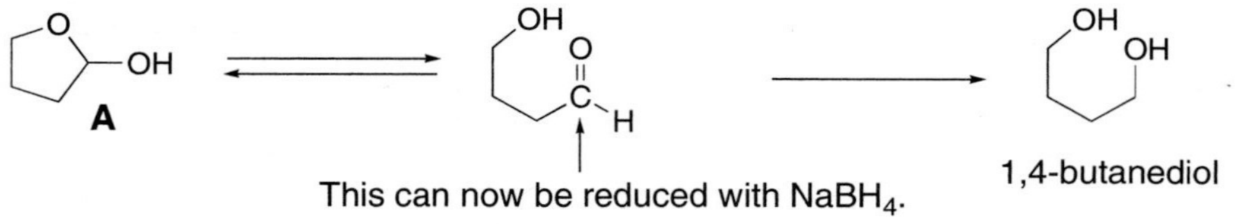

A

This can now be reduced with NaBH₄.

1,4-butanediol

b.

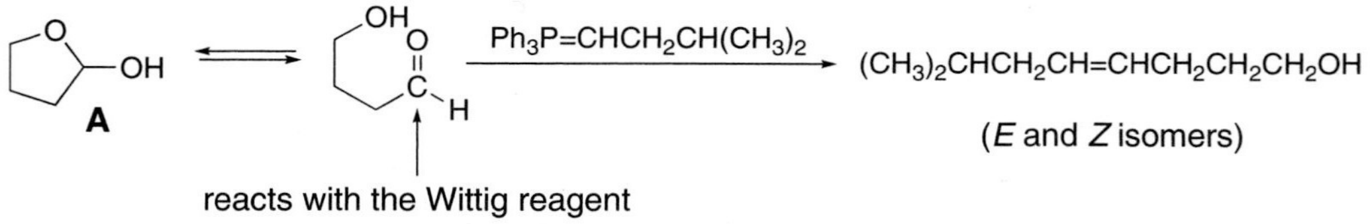

A

$Ph_3P=CHCH_2CH(CH_3)_2$

$(CH_3)_2CHCH_2CH=CHCH_2CH_2CH_2OH$

(*E* and *Z* isomers)

reacts with the Wittig reagent

21.81

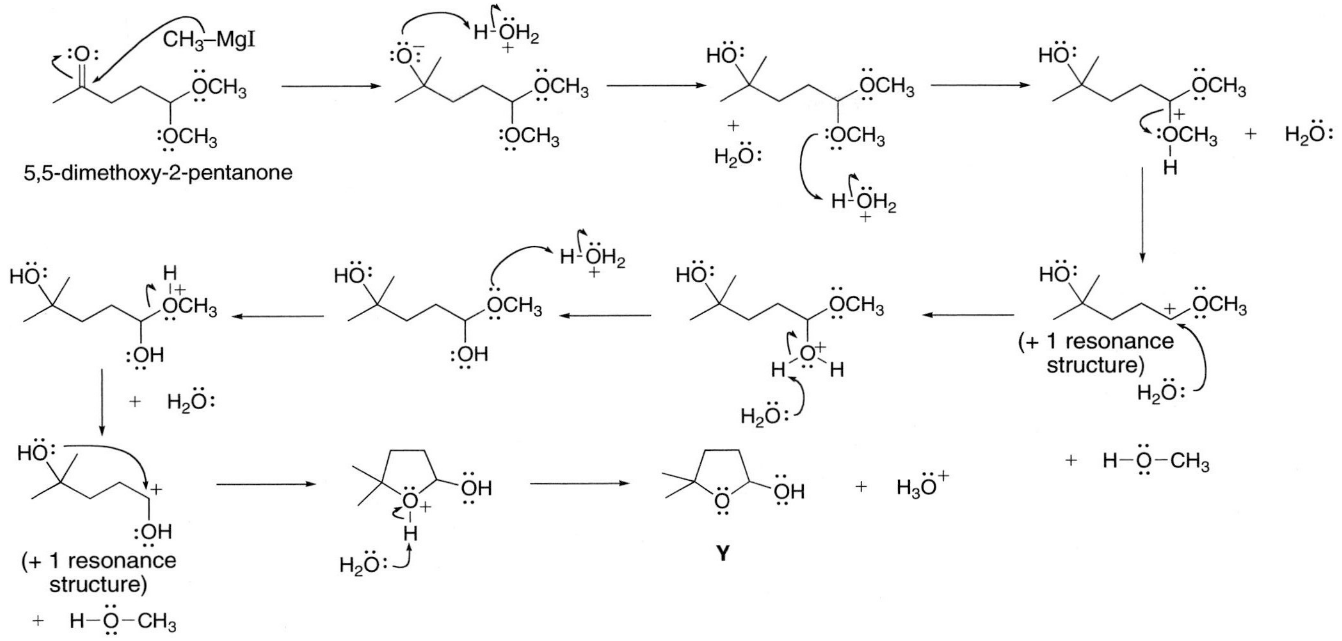

5,5-dimethoxy-2-pentanone

Y

21.82

cyclopropenone
(1640 cm^{-1})

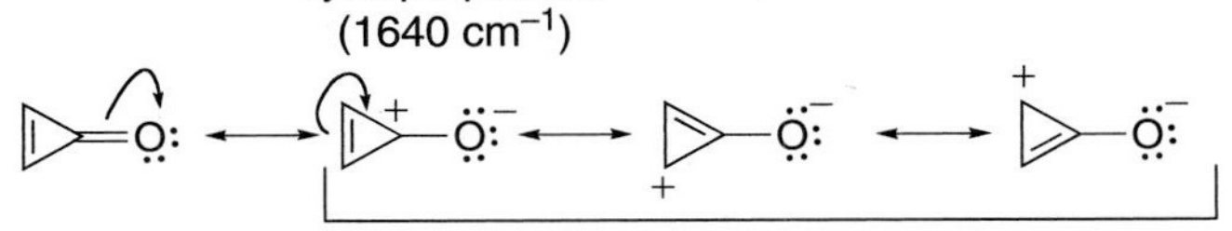

These three resonance structures include an aromatic ring; $4n + 2 = 2\,\pi$ electrons. Although they are charge separated, the stabilized aromatic ring makes these three structures contribute to the hybrid more than usual. Since these three resonance contributors have a C–O single bond, the absorption is shifted to a lower wavenumber.

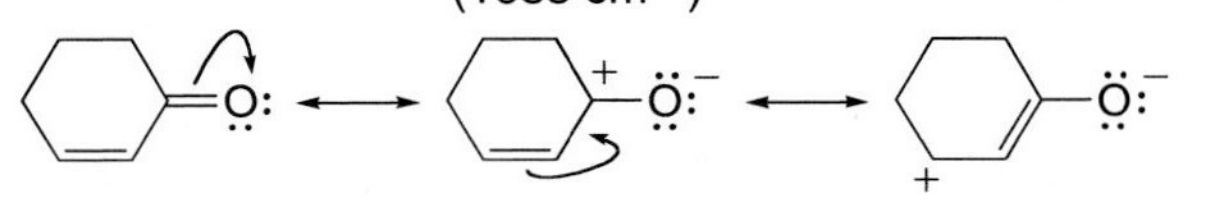

2-cyclohexenone
(1685 cm^{-1})

There are three resonance structures for 2-cyclohexenone, but the charge-separated resonance structures are not aromatic so they contribute less to the resonance hybrid. The C=O absorbs in the usual region for a conjugated carbonyl.

21.83

a.

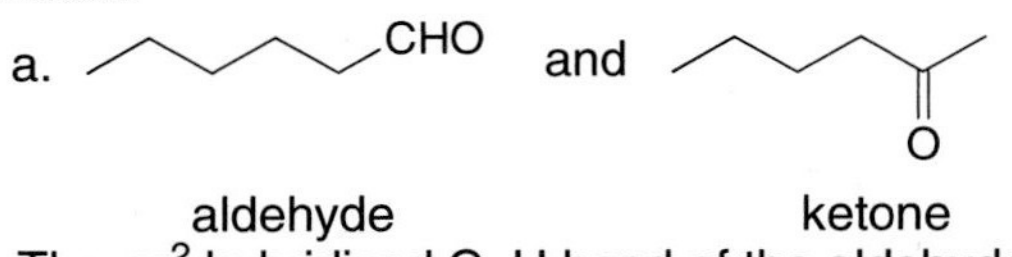

and

aldehyde   ketone

The sp^2 hybridized C–H bond of the aldehyde absorbs at 2700–2830 cm^{-1}.

c.

and

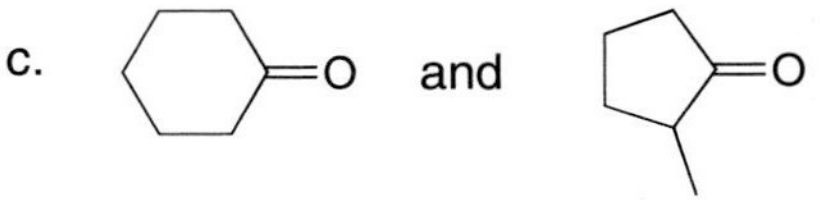

smaller ring
**higher wavenumber
for C=O**

b.

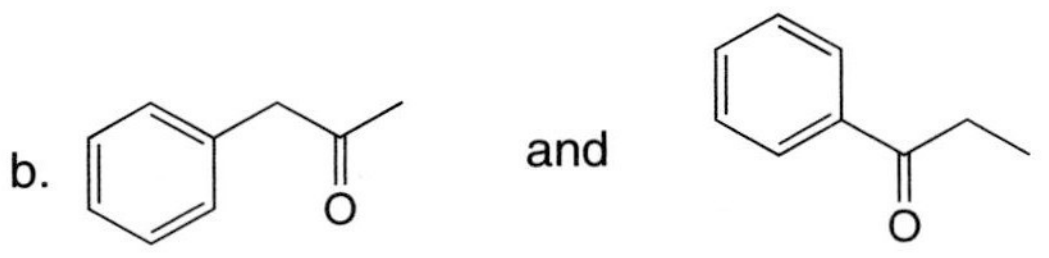

and

**higher wavenumber
for C=O**

conjugated with a
benzene ring
lower wavenumber

21.84

A. Molecular formula $C_5H_{10}O$ ⟶ 1 degree of unsaturation
    IR absorptions at 1728, 2791, 2700 cm^{-1} ⟶ C=O, CHO
    NMR data: singlet at 1.08 (9 H) ⟶ 3 CH_3 groups
            singlet at 9.48 (1 H) ppm ⟶ CHO

B. Molecular formula $C_5H_{10}O$ ⟶ 1 degree of unsaturation
    IR absorption at 1718 cm^{-1} ⟶ C=O
    NMR data: doublet at 1.10 (6 H) ⟶ 2 CH_3's adjacent to H
            singlet at 2.14 (3 H) ⟶ CH_3
            septet at 2.58 (1 H) ppm ⟶ CH adjacent to 2 CH_3's

C. Molecular formula $C_{10}H_{12}O$ ⟶ 5 degrees of unsaturation (4 due to a benzene ring)
    IR absorption at 1686 cm^{-1} ⟶ C=O
    NMR data: triplet at 1.21 (3 H) ⟶ CH_3 adjacent to 2 H's

            singlet at 2.39 (3 H) ⟶ CH_3

            quartet at 2.95 (2 H) ⟶ CH_2 adjacent to 3 H's

            doublet at 7.24 (2 H) ⟶ 2 H's on benzene ring

            doublet at 7.85 (2 H) ppm ⟶ 2 H's on benzene ring

D. Molecular formula $C_{10}H_{12}O$ ⟶ 5 degrees of unsaturation (4 due to a benzene ring)
    IR absorption at 1719 cm^{-1} ⟶ C=O
    NMR data: triplet at 1.02 (3 H) ⟶ CH_3 adjacent 2 H's
          quartet at 2.45 (2 H) ⟶ 2 H's adjacent to 3 H's
          singlet at 3.67 (2 H) ⟶ CH_2
          multiplet at 7.06–7.48 (5 H) ppm ⟶ a monosubstituted benzene ring

21.85

$C_7H_{16}O_2$: 0 degrees of unsaturation
IR: 3000 cm^{-1}: **C–H bonds**
NMR data (ppm):
  H_a: quartet at 3.5 (**4 H**), split by 3 H's
  H_b: singlet at 1.4 (**6 H**)
  H_c: triplet at 1.2 (**6 H**), split by 2 H's

21.86

A. Molecular formula $C_9H_{10}O$
    5 degrees of unsaturation
    IR absorption at 1700 cm^{-1} → C=O
    IR absorption at ~2700 cm^{-1} → CH of RCHO
    NMR data (ppm):
            triplet at 1.2 (2 H's adjacent)
            quartet at 2.7 (3 H's adjacent)
            doublet at 7.3 (2 H's on benzene)
            doublet at 7.7 (2 H's on benzene)
            singlet at 9.9 (CHO)

B. Molecular formula $C_9H_{10}O$
    5 degrees of unsaturation
    IR absorption at 1720 cm^{-1} → C=O
    IR absorption at ~2700 cm^{-1} → CH of RCHO
    NMR data (ppm):
            2 triplets at 2.85 and 2.95 (suggests $-CH_2CH_2-$)
            multiplet at 7.2 (benzene H's)
            signal at 9.8 (CHO)

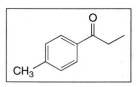

21.87

  C. Molecular formula $C_6H_{12}O_3$

    1 degree of unsaturation

    IR absorption at 1718 cm^{-1} → C=O

    To determine the number of H's that give rise to each signal, first find the number of integration units per H by dividing the total number of integration units (7 + 40 + 14 + 21 = 82) by the number of H's (12); 82/12 = 6.8. Then divide each integration unit by this number (6.8).

    NMR data (ppm):

          singlet at 2.2 (3 H's)

          doublet at 2.7 (2 H's)

          singlet at 3.2 (6 H's – 2 OCH$_3$ groups)

          triplet at 4.8 (1 H)

21.88

  D. Molecular ion at m/z = 150: $C_9H_{10}O_2$ (possible molecular formula)

    5 degrees of unsaturation

    IR absorption at 1692 cm^{-1} → C=O

    NMR data (ppm):

          triplet at 1.5 (3 H's – CH$_3$CH$_2$)

          quartet at 4.1 (2 H's – CH$_3$CH$_2$)

          doublet at 7.0 (2 H's – on benzene ring)

          doublet at 7.8 (2 H's – on benzene ring)

          singlet at 9.9 (1 H – on aldehyde)

21.89

21.90

The carbocation is trigonal planar, so CH$_3$OH attacks from two different directions, and two different acetals are formed.

21.91

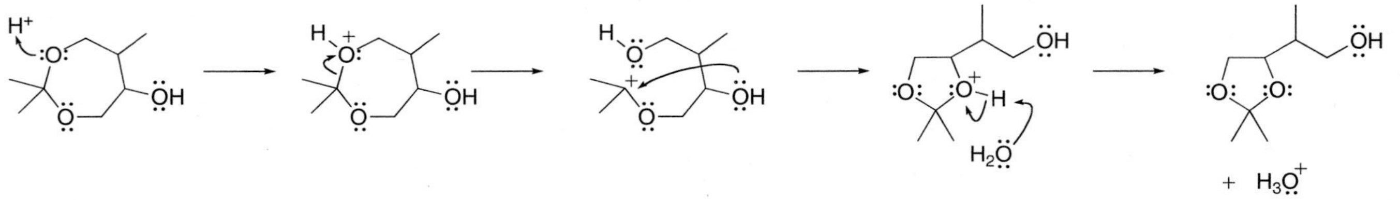
H+
ÖH
+ H3O+

21.92

a.

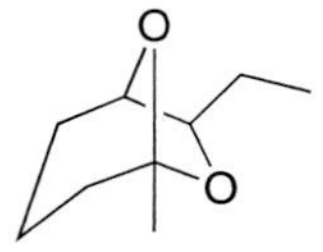

brevicomin

b.

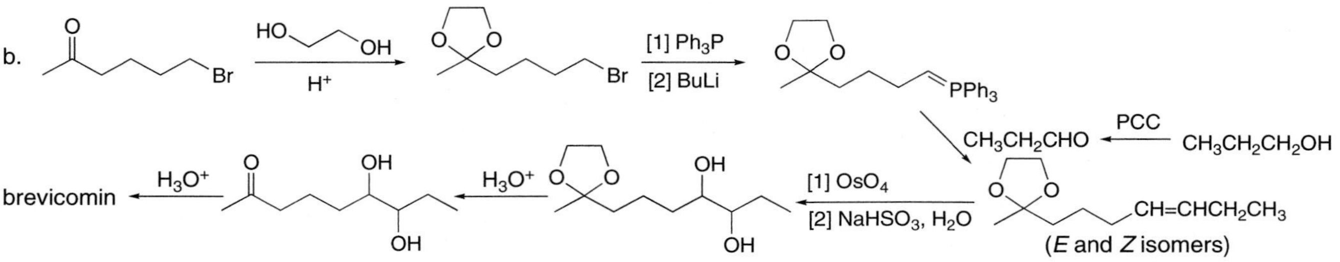
Br
HO OH
H+
[1] Ph3P
[2] BuLi
PPh3
PCC
CH3CH2CHO
CH3CH2CH2OH
CH=CHCH2CH3
(E and Z isomers)
[1] OsO4
[2] NaHSO3, H2O
H3O+
brevicomin

21.93

a.

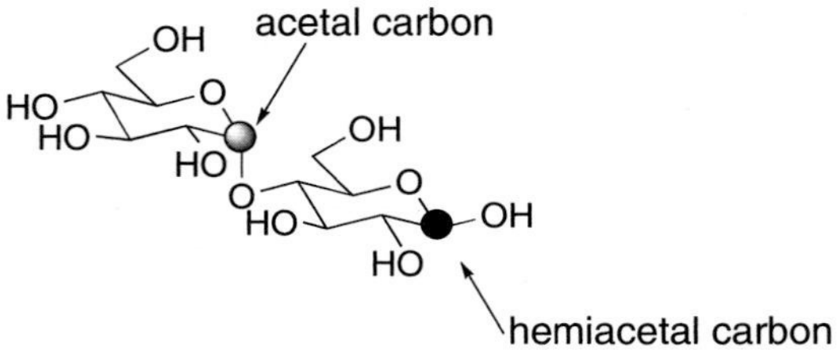
acetal carbon
hemiacetal carbon

c.

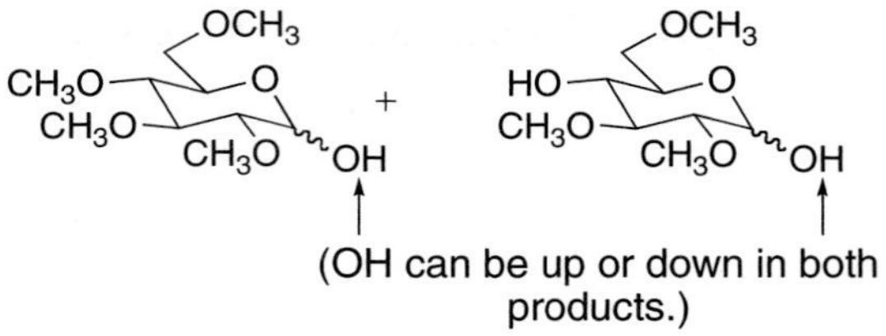
(OH can be up or down in both products.)

b.

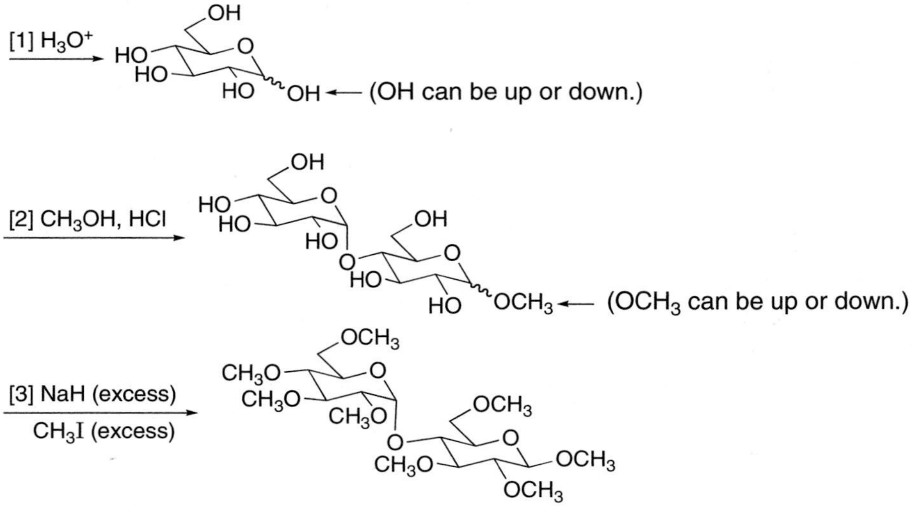
[1] H3O+
(OH can be up or down.)
[2] CH3OH, HCl
(OCH3 can be up or down.)
[3] NaH (excess)
CH3I (excess)

Chapter 22: Carboxylic Acids and Their Derivatives—Nucleophilic Acyl Substitution

♦ Summary of spectroscopic absorptions of RCOZ (22.5)

IR absorptions	• All RCOZ compounds have a C=O absorption in the region 1600–1850 cm^{-1}.
	• RCOCl: 1800 cm^{-1}
	• $(RCO)_2O$: 1820 and 1760 cm^{-1} (two peaks)
	• RCOOR': 1735–1745 cm^{-1}
	• RCONR'$_2$: 1630–1680 cm^{-1}
	• Additional amide absorptions occur at 3200–3400 cm^{-1} (N–H stretch) and 1640 cm^{-1} (N–H bending).
	• Decreasing the ring size of a cyclic lactone, lactam, or anhydride increases the frequency of the C=O absorption.
	• Conjugation shifts the C=O to lower wavenumber.
^{1}H NMR absorptions	• C–H α to the C=O absorbs at 2–2.5 ppm.
	• N–H of an amide absorbs at 7.5–8.5 ppm.
^{13}C NMR absorption	• C=O absorbs at 160–180 ppm.

♦ Summary of spectroscopic absorptions of RCN (22.5)

IR absorption	• C≡N absorption at 2250 cm^{-1}
^{13}C NMR absorption	• C≡N absorbs at 115–120 ppm.

♦ Summary: The relationship between the basicity of Z⁻ and the properties of RCOZ

- **Increasing basicity of the leaving group** (22.2)
- **Increasing resonance stabilization** (22.2)

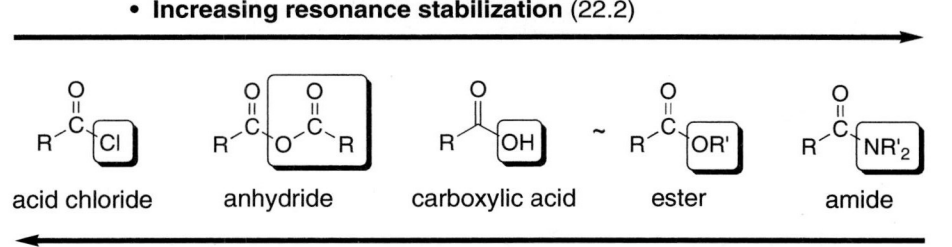

| acid chloride | anhydride | carboxylic acid | ester | amide |

- **Increasing leaving group ability** (22.7B)
- **Increasing reactivity** (22.7B)
- **Increasing frequency of the C=O absorption in the IR** (22.5)

♦ General features of nucleophilic acyl substitution

- The characteristic reaction of compounds having the general structure RCOZ is nucleophilic acyl substitution (22.1).
- The mechanism consists of two basic steps (22.7A):
  [1] Addition of a nucleophile to form a tetrahedral intermediate
  [2] Elimination of a leaving group
- More reactive acyl compounds can be used to prepare less reactive acyl compounds. The reverse is not necessarily true (22.7B).

♦ Nucleophilic acyl substitution reactions

[1] Reactions that synthesize acid chlorides (RCOCl)

[a] From RCOOH (22.10A):

$$R-\overset{\overset{\displaystyle O}{\|}}{C}-OH \ + \ SOCl_2 \ \longrightarrow \ R-\overset{\overset{\displaystyle O}{\|}}{C}-Cl \ + \ SO_2 \ + \ HCl$$

[2] Reactions that synthesize anhydrides [(RCO)₂O]

[a] From RCOCl (22.8):

$$R-\overset{\overset{\displaystyle O}{\|}}{C}-Cl \ + \ {}^{-}O-\overset{\overset{\displaystyle O}{\|}}{C}-R' \ \longrightarrow \ R-\overset{\overset{\displaystyle O}{\|}}{C}-O-\overset{\overset{\displaystyle O}{\|}}{C}-R' \ + \ Cl^{-}$$

[b] From dicarboxylic acids (22.10B):

$$\xrightarrow{\Delta} \quad \text{cyclic anhydride} \ + \ H_2O$$

[3] Reactions that synthesize carboxylic acids (RCOOH)

[a] From RCOCl (22.8):

$$R-\overset{\overset{\displaystyle O}{\|}}{C}-Cl \ + \ H_2O \ \xrightarrow{\text{pyridine}} \ R-\overset{\overset{\displaystyle O}{\|}}{C}-OH \ + \ \text{pyridinium} \ Cl^{-}$$

[b] From (RCO)₂O (22.9):

$$R-\overset{\overset{\displaystyle O}{\|}}{C}-O-\overset{\overset{\displaystyle O}{\|}}{C}-R \ + \ H_2O \ \longrightarrow \ 2\ R-\overset{\overset{\displaystyle O}{\|}}{C}-OH$$

[c] From RCOOR' (22.11):

$$R-\overset{\overset{\displaystyle O}{\|}}{C}-OR' \ + \ H_2O \ \xrightarrow[(H^+ \text{ or } {}^-OH)]{} \ R-\overset{\overset{\displaystyle O}{\|}}{C}-OH \ \text{or} \ R-\overset{\overset{\displaystyle O}{\|}}{C}-O^{-} \ + \ R'OH$$

(with acid)   (with base)

[d] From RCONR'₂ (R' = H or alkyl, 22.13):

$$R-\overset{\overset{\displaystyle O}{\|}}{C}-NR'_2 \quad (R' = H \text{ or alkyl})$$

$$\xrightarrow{H_2O,\ H^+} \ R-\overset{\overset{\displaystyle O}{\|}}{C}-OH \ + \ R'_2\overset{+}{N}H_2$$

$$\xrightarrow{H_2O,\ {}^-OH} \ R-\overset{\overset{\displaystyle O}{\|}}{C}-O^{-} \ + \ R'_2NH$$

[4] Reactions that synthesize esters (RCOOR')

[a] From RCOCl (22.8):

$$R-\overset{\overset{\displaystyle O}{\|}}{C}-Cl \ + \ R'OH \ \xrightarrow{\text{pyridine}} \ R-\overset{\overset{\displaystyle O}{\|}}{C}-OR' \ + \ \text{pyridinium} \ Cl^{-}$$

[b] From (RCO)$_2$O (22.9):

$$R-\overset{\overset{O}{\|}}{C}-O-\overset{\overset{O}{\|}}{C}-R \quad + \quad R'OH \quad \longrightarrow \quad R-\overset{\overset{O}{\|}}{C}-OR' \quad + \quad RCOOH$$

[c] From RCOOH (22.10C):

$$R-\overset{\overset{O}{\|}}{C}-OH \quad + \quad R'OH \quad \xrightarrow{H_2SO_4} \quad R-\overset{\overset{O}{\|}}{C}-OR' \quad + \quad H_2O$$

[5] Reactions that synthesize amides (RCONH$_2$) [The reactions are written with NH$_3$ as the nucleophile to form RCONH$_2$. Similar reactions occur with R'NH$_2$ to form RCONHR', and with R'$_2$NH to form RCONR'$_2$.]

[a] From RCOCl (22.8):

$$R-\overset{\overset{O}{\|}}{C}-Cl \quad + \quad \underset{(2\ equiv)}{NH_3} \quad \longrightarrow \quad R-\overset{\overset{O}{\|}}{C}-NH_2 \quad + \quad NH_4^+Cl^-$$

[b] From (RCO)$_2$O (22.9):

$$R-\overset{\overset{O}{\|}}{C}-O-\overset{\overset{O}{\|}}{C}-R \quad + \quad \underset{(2\ equiv)}{NH_3} \quad \longrightarrow \quad R-\overset{\overset{O}{\|}}{C}-NH_2 \quad + \quad RCOO^-NH_4^+$$

[c] From RCOOH (22.10D):

$$R-\overset{\overset{O}{\|}}{C}-OH \quad \xrightarrow[\text{[2] }\Delta]{\text{[1]}NH_3} \quad R-\overset{\overset{O}{\|}}{C}-NH_2 \quad + \quad H_2O$$

$$R-\overset{\overset{O}{\|}}{C}-OH \quad + \quad R'NH_2 \quad \xrightarrow{DCC} \quad R-\overset{\overset{O}{\|}}{C}-NHR' \quad + \quad H_2O$$

[d] From RCOOR' (22.11):

$$R-\overset{\overset{O}{\|}}{C}-OR' \quad + \quad NH_3 \quad \longrightarrow \quad R-\overset{\overset{O}{\|}}{C}-NH_2 \quad + \quad R'OH$$

◆ Nitrile synthesis (22.18)

Nitriles are prepared by S$_N$2 substitution using unhindered alkyl halides as starting materials.

$$R-X \quad + \quad {}^-CN \quad \xrightarrow{S_N2} \quad R-C\equiv N \quad + \quad X^-$$
$$R = CH_3, 1^\circ$$

◆ Reactions of nitriles

[1] Hydrolysis (22.18A)

$$R-C\equiv N \quad \xrightarrow[(H^+\ or\ {}^-OH)]{H_2O} \quad R-\overset{\overset{O}{\|}}{C}-OH \quad or \quad R-\overset{\overset{O}{\|}}{C}-O^-$$
$$\qquad\qquad\qquad\qquad\qquad (\text{with acid}) \qquad (\text{with base})$$

[2] Reduction (22.18B)

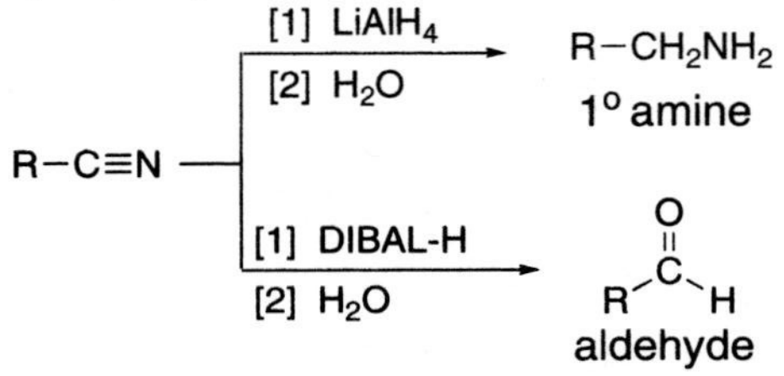

R−C≡N

[1] LiAlH$_4$
[2] H$_2$O
→ R−CH$_2$NH$_2$
1° amine

[1] DIBAL-H
[2] H$_2$O
→ aldehyde

$$R-\underset{H}{\overset{\overset{\displaystyle O}{\|}}{C}}$$

aldehyde

[3] Reaction with organometallic reagents (22.18C)

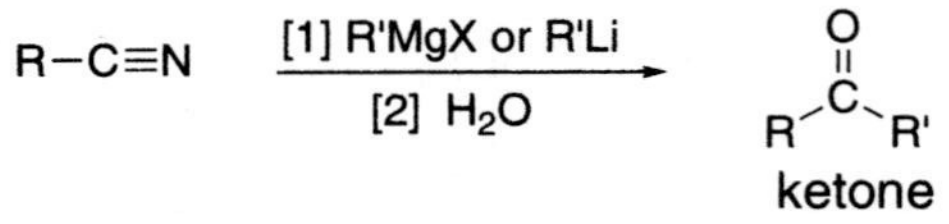

R−C≡N
[1] R'MgX or R'Li
[2] H$_2$O
→ ketone

$$R-\underset{R'}{\overset{\overset{\displaystyle O}{\|}}{C}}$$

ketone

Chapter 22: Answers to Problems

22.1 The number of C–N bonds determines the classification as a 1°, 2°, or 3° amide.

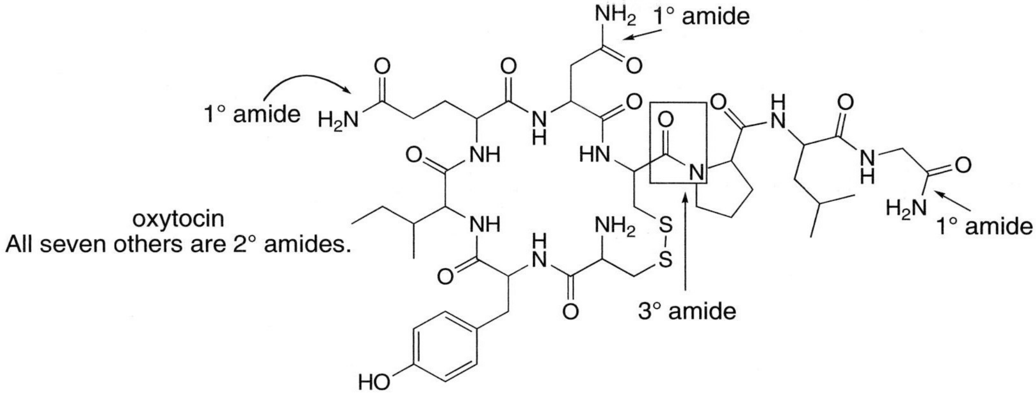
NH2 1° amide
1° amide H2N
oxytocin
All seven others are 2° amides.
H2N 1° amide
3° amide

22.2 As the basicity of Z increases, the stability of RCOZ increases because of added resonance stabilization.

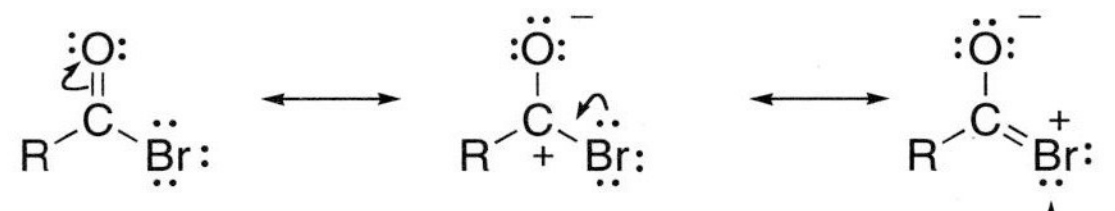

The **basicity of Z** determines how much this structure contributes to the hybrid. Br^- is less basic than ^-OH, so RCOBr is less stable than RCOOH.

22.3

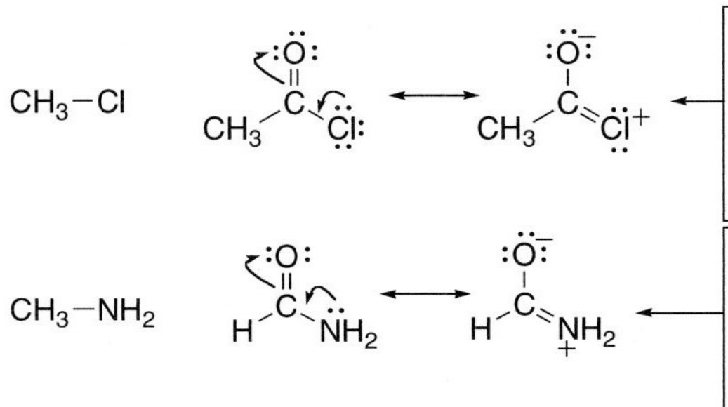

This resonance structure contributes little to the hybrid since Cl^- is a weak base. Thus, the C–Cl bond has little double bond character, making it similar in length to the C–Cl bond in CH_3Cl.

This resonance structure contributes more to the hybrid since $^-NH_2$ is more basic. Thus, the C–N bond in $HCONH_2$ has more double bond character, making it shorter than the C–N bond in CH_3NH_2.

22.4

a. $(CH_3CH_2)_2CH{-}COCl$

re-draw

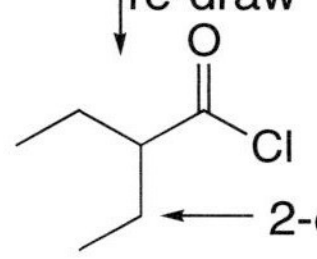

2-ethylbutanoyl chloride

2-ethyl

b. $C_6H_5COOCH_3$

re-draw

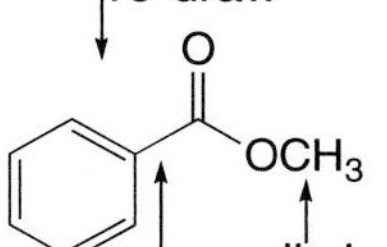

methyl benzoate

alkyl group = methyl

acyl group = benzoate

c. $CH_3CH_2CON(CH_3)CH_2CH_3$

|re-draw

acyl group =
propanamide

N-ethyl-N-methyl

N-ethyl-N-methylpropanamide

e. CH_3CH_2 ... **benzoic propanoic anhydride**

acyl group =   acyl group =
propanoic   benzoic

d. H—C—OCH_2CH_3

alkyl group = ethyl

acyl group =   **ethyl formate**
formate

f. ... CN   **3-ethylhexanenitrile**

6 carbon chain =
hexanenitrile

22.5

a. 5-methylheptanoyl chloride

d. *N*-isobutyl-*N*-methylbutanamide

g. *sec*-butyl 2-methylhexanoate

b. isopropyl propanoate

e. 3-methylpentanenitrile

h. *N*-ethylhexanamide

c. acetic formic anhydride

f. *o*-cyanobenzoic acid

22.6 CH_3CONH_2 has two H's bonded to N that can hydrogen bond. $CH_3CON(CH_3)_2$ does not have any H's capable of hydrogen bonding. This means CH_3CONH_2 has much stronger intermolecular forces, which leads to a higher boiling point.

22.7

a. CH_3—C—OCH_2CH_3 and CH_3—C—$N(CH_2CH_3)_2$

amide: C=O at
lower wavenumber

c. $CH_3CH_2CH_2$—C—$NHCH_3$ and $CH_3CH_2CH_2$—C—NH_2

2° amide: 1 N–H
absorption at
3200–3400 cm^{-1}

1° amide: 2
N–H absorptions

b. ... and ...

smaller ring:
C=O at a higher
wavenumber

d. ... and ...

anhydride:
2 C=O peaks

22.8

$H_b \rightarrow H\ H$ O

H_c signal from the 5 H's on the aromatic ring

A H_a

Molecular formula $C_9H_{10}O_2$
5 degrees of unsaturation
IR: 1743 cm^{-1} from ester C=O
    3091–2895 cm^{-1} from sp^2 and sp^3 C–H
^{1}H NMR: H_a = 2.06 ppm (singlet, 3 H) – CH_3
    H_b = 5.08 ppm (singlet, 2 H) – CH_2
    H_c = 7.33 ppm (broad singlet, 5 H)

$H_a \rightarrow H\ H$ O

H_b

B H_b signal from the 10 H's on the two aromatic rings

Molecular formula $C_{14}H_{12}O_2$
9 degrees of unsaturation
IR: 1718 cm^{-1} from conjugated ester C=O
    3091–2953 cm^{-1} from sp^2 and sp^3 C–H
^{1}H NMR: H_a = 5.35 ppm (singlet, 2 H)
    H_b = 7.26–8.15 ppm (multiplets, 10 H)

22.9

amoxicillin

a. 4 stereogenic centers
b. 2^4 = 16 possible stereoisomers
c. enantiomer

cephalexin
(Trade name: Keflex)

a. 3 stereogenic centers
b. 2^3 = 8 possible stereoisomers
c. enantiomer

22.10 To draw the products of these nucleophilic acyl substitution reactions, find the nucleophile and the leaving group. Then replace the leaving group with the nucleophile and draw a neutral product.

nucleophile

a. CH_3COCl $\xrightarrow{CH_3OH}$ CH_3COOCH_3 + HCl

leaving group

nucleophile

b. $CH_3COOCH_2CH_3$ $\xrightarrow{NH_3}$ CH_3CONH_2 + $HOCH_2CH_3$

leaving group

22.11 More reactive acyl compounds can be converted to less reactive acyl compounds.

a. CH_3COCl $\longrightarrow$ CH_3COOH
    more reactive **YES** less reactive

b. $CH_3CONHCH_3$ $\longrightarrow$ CH_3COOCH_3
    less reactive **NO** more reactive

c. CH_3COOCH_3 $\longrightarrow$ CH_3COCl
    less reactive **NO** more reactive

d. $(CH_3CO)_2O$ $\longrightarrow$ CH_3CONH_2
    more reactive **YES** less reactive

22.12 The better the leaving group is, the more reactive the carboxylic acid derivative. The weakest base is the best leaving group.

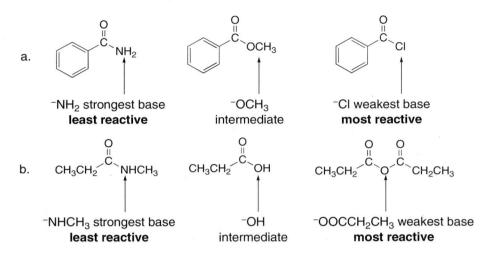

a.

$^-NH_2$ strongest base
least reactive

$^-OCH_3$
intermediate

^-Cl weakest base
most reactive

b. CH_3CH_2 ... $NHCH_3$

CH_3CH_2 ... OH

CH_3CH_2 ... CH_2CH_3

$^-NHCH_3$ strongest base
least reactive

^-OH
intermediate

$^-OOCCH_2CH_3$ weakest base
most reactive

22.13

CH_3 ... O ... CH_3
acetic anhydride

Cl_3C ... O ... CCl_3
trichloroacetic anhydride

The Cl atoms are electron withdrawing, which makes the conjugate base (the leaving group, CCl_3COO^-) weaker and more stable.

22.14

a. $\xrightarrow[\text{pyridine}]{H_2O}$ benzoic acid OH + pyridine $\overset{+}{N}_H$ Cl$^-$

b. $\xrightarrow{CH_3COO^-}$ anhydride O—C—CH₃ + Cl$^-$

c. $\xrightarrow[\text{excess}]{NH_3}$ amide NH₂ + NH₄$^+$Cl$^-$

d. $\xrightarrow[\text{excess}]{(CH_3)_2NH}$ amide N(CH₃)₂ + $(CH_3)_2\overset{+}{N}H_2$ Cl$^-$

22.15 The mechanism has three steps: [1] nucleophilic attack by O; [2] proton transfer; [3] elimination of the Cl$^-$ leaving group to form the product.

OCH₃ ... OH + :Cl: →   OCH₃ ... O⁺—H ... N (pyridine)   →   OCH₃ ... O ... :O::Cl:

+ pyridine $\overset{+}{N}$H

↓

OCH₃ ... O ... :O: + :Cl:$^-$ **A**

OCH₃

22.16

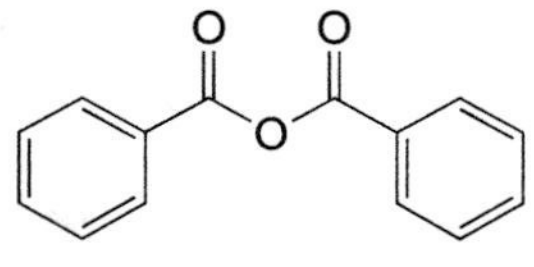

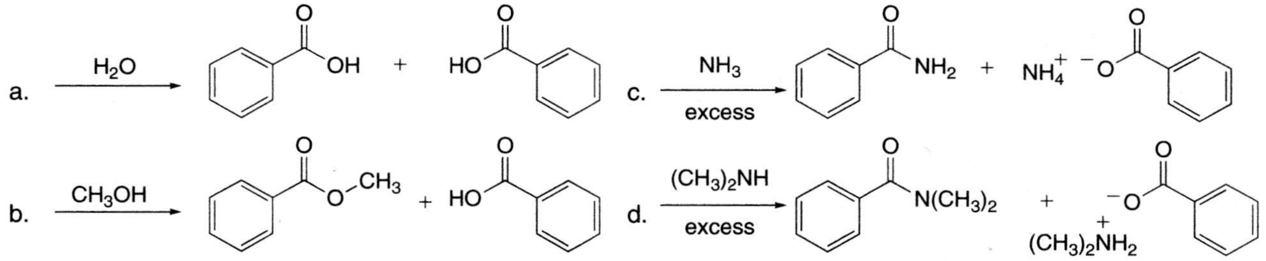

a. $\xrightarrow{H_2O}$ [benzoic acid] OH + HO [benzoic acid]

b. $\xrightarrow{CH_3OH}$ [benzoate] O—CH₃ + HO [benzoic acid]

c. $\xrightarrow[\text{excess}]{NH_3}$ [benzamide] NH₂ + NH_4^+ ^-O [benzoate]

d. $\xrightarrow[\text{excess}]{(CH_3)_2NH}$ [benzamide] N(CH₃)₂ + ^-O [benzoate] $^+_{(CH_3)_2NH_2}$

22.17 Nucleophilicity decreases across a row of the periodic table so an NH_2 group is more nucleophilic than an OH group.

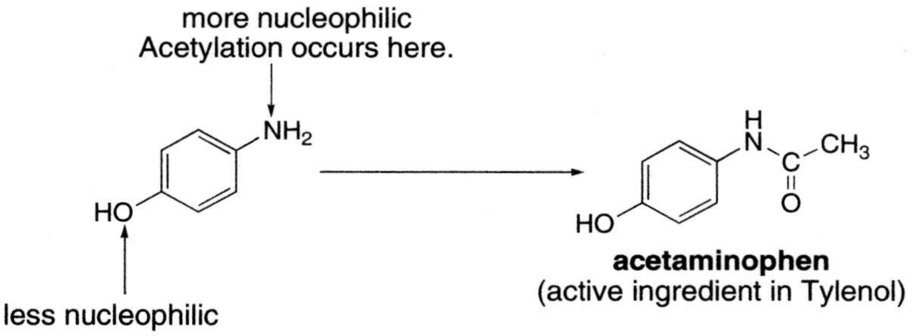

more nucleophilic
Acetylation occurs here.

less nucleophilic

acetaminophen
(active ingredient in Tylenol)

22.18 Reaction of a carboxylic acid with thionyl chloride converts it to an acid chloride.

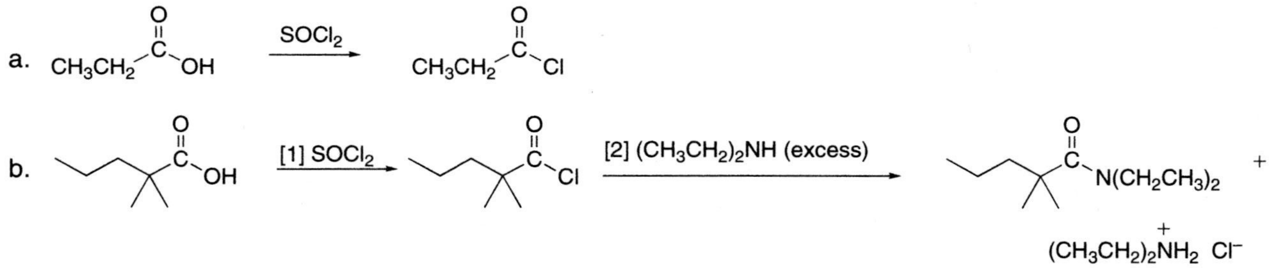

a. CH_3CH_2—C(=O)—OH $\xrightarrow{SOCl_2}$ CH_3CH_2—C(=O)—Cl

b. [acid] C(=O)OH $\xrightarrow{[1]\ SOCl_2}$ [acid chloride] C(=O)Cl $\xrightarrow{[2]\ (CH_3CH_2)_2NH\ (excess)}$ [amide] C(=O)N(CH₂CH₃)₂ + $(CH_3CH_2)_2NH_2\ Cl^-$

22.19

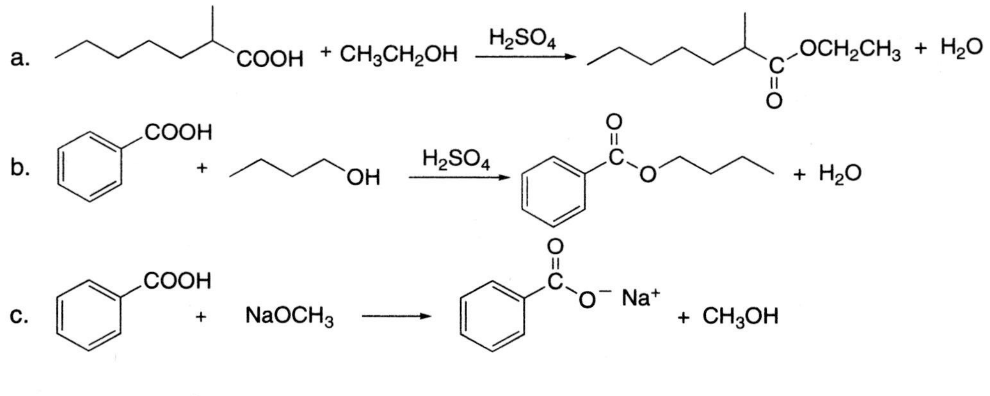

a. [chain]COOH + CH_3CH_2OH $\xrightarrow{H_2SO_4}$ [chain]C(=O)OCH₂CH₃ + H_2O

b. [benzene]COOH + [chain]OH $\xrightarrow{H_2SO_4}$ [benzene]C(=O)O[chain] + H_2O

c. [benzene]COOH + $NaOCH_3$ $\longrightarrow$ [benzene]C(=O)O⁻ Na⁺ + CH_3OH

d.

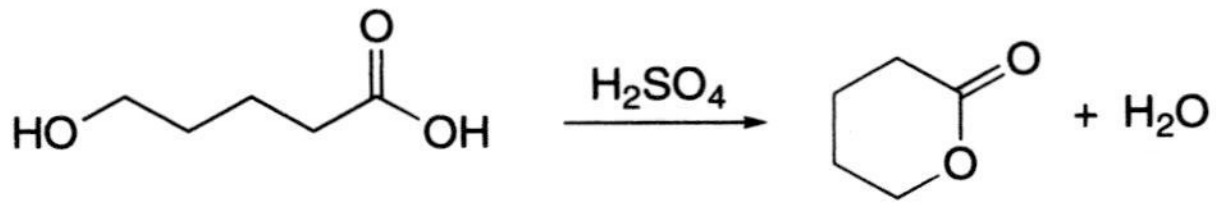

22.20

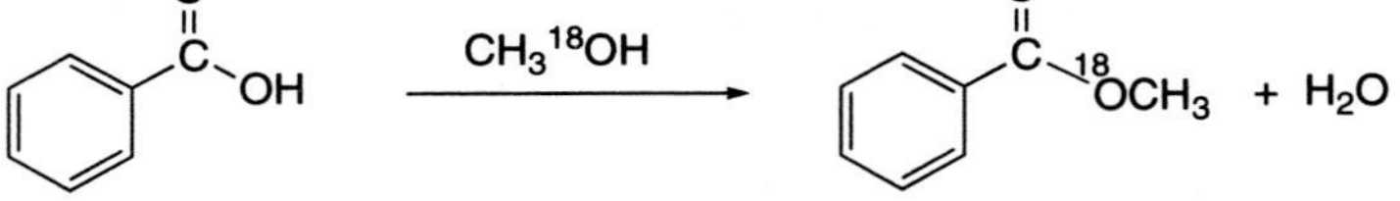

22.21

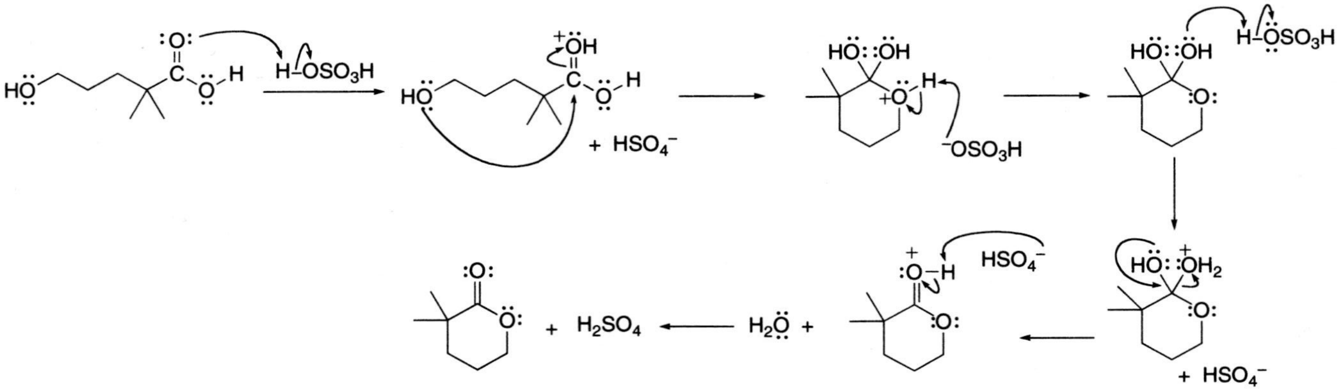

22.22

a. 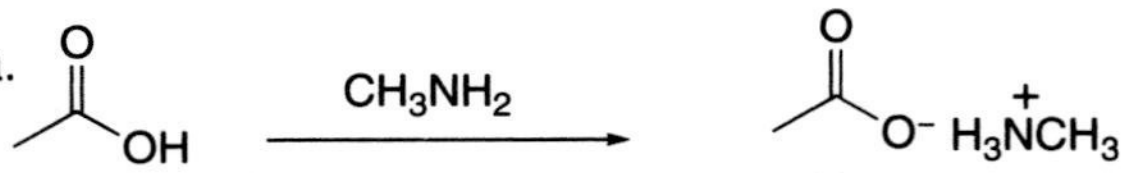 c.

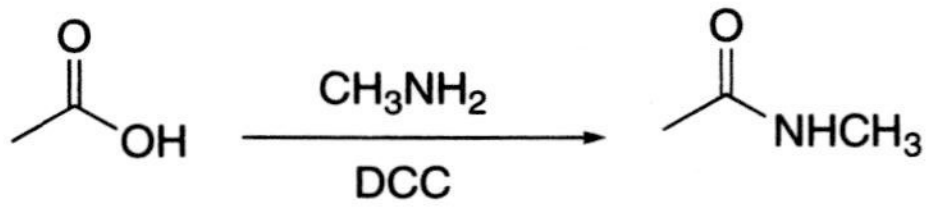

b.

22.23

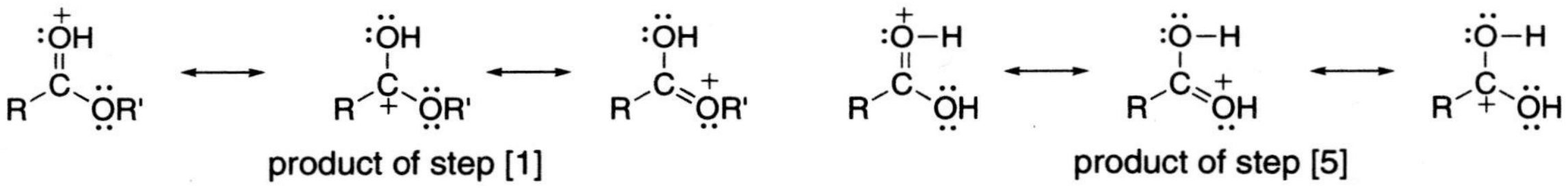

product of step [1] product of step [5]

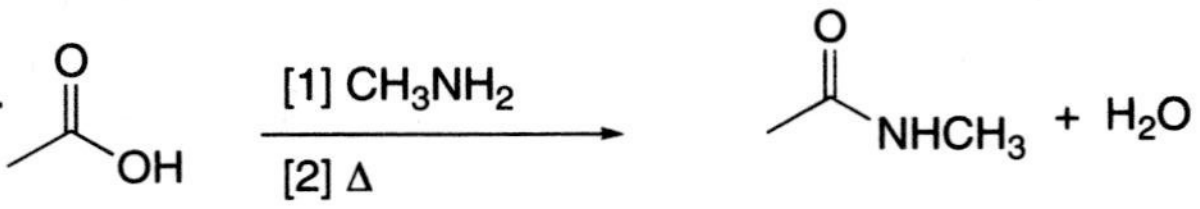

22.24

a.

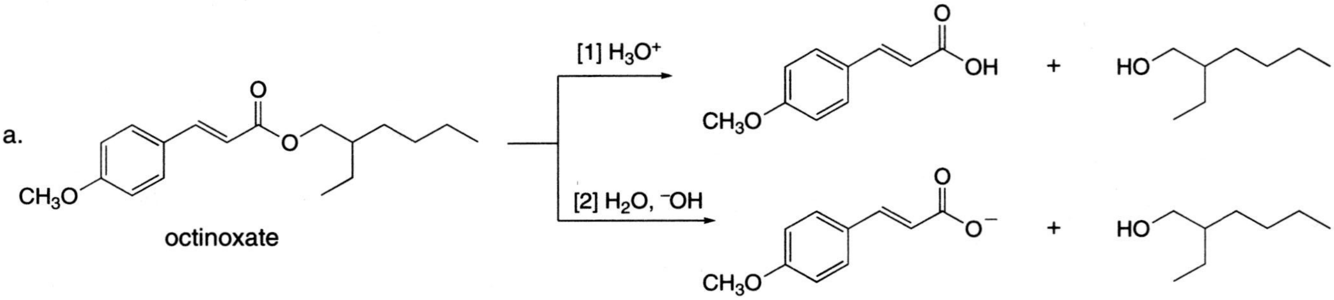

octinoxate

b.

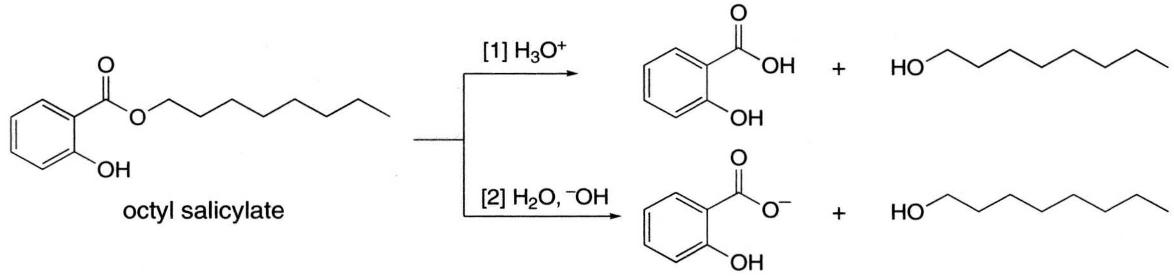

octyl salicylate

[1] H₃O⁺

[2] H₂O, ⁻OH

22.25

a.

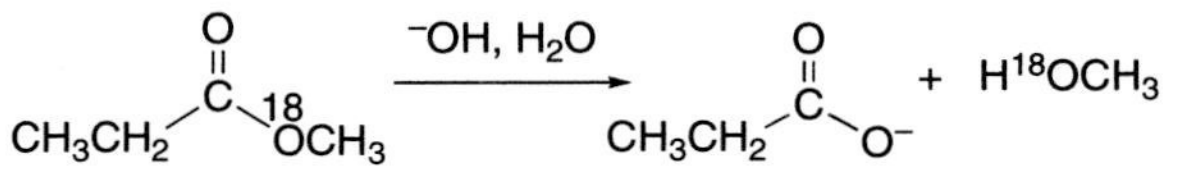

b.

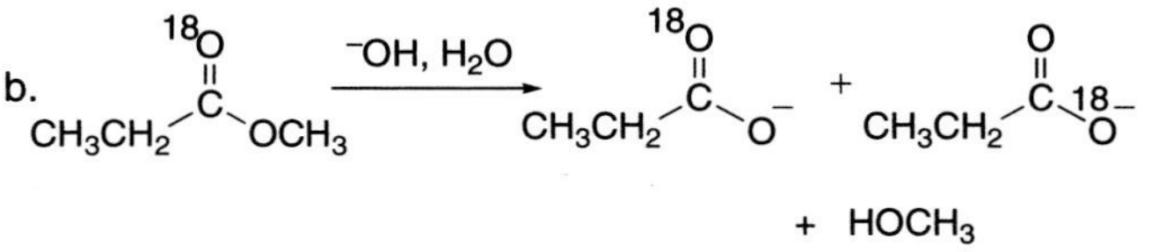

22.26

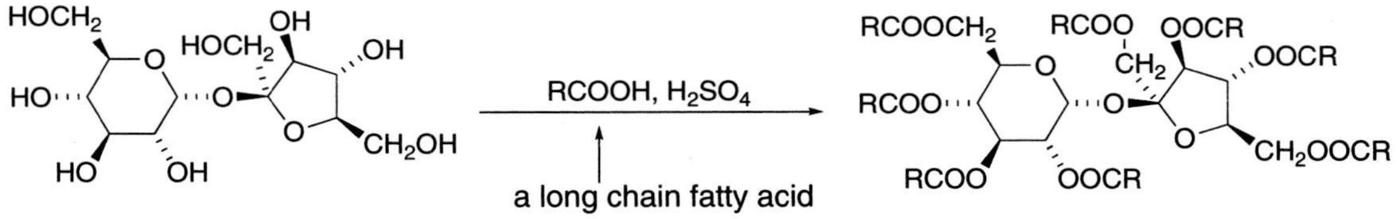

sucrose

RCOOH, H₂SO₄

a long chain fatty acid

olestra

22.27

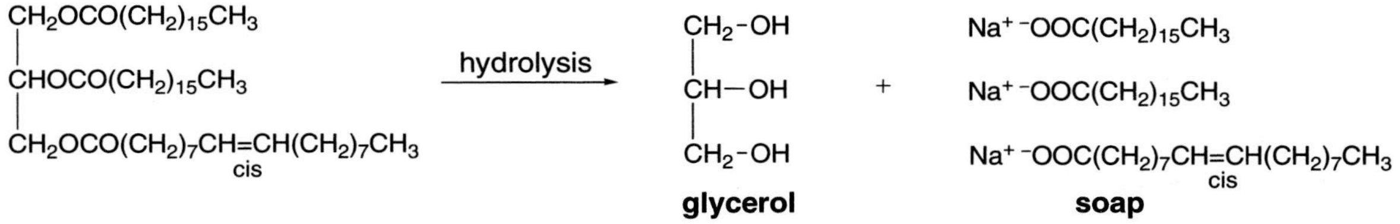

hydrolysis

glycerol + soap

22.28

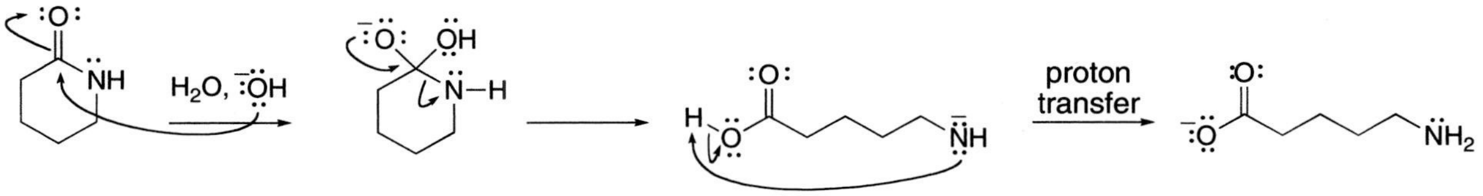

proton transfer

22.29 Aspirin has an ester, a more reactive acyl group, but acetaminophen has an amide, a less reactive acyl group.

a. The ester makes aspirin more easily hydrolyzed with water from the air than acetaminophen. Therefore, Tylenol can be kept for many years, whereas aspirin decomposes.

b. Similarly, aspirin will be hydrolyzed and decompose in the aqueous medium of a liquid medication, but acetaminophen is stable due to the less reactive amide group, allowing it to remain unchanged while dissolved in H_2O.

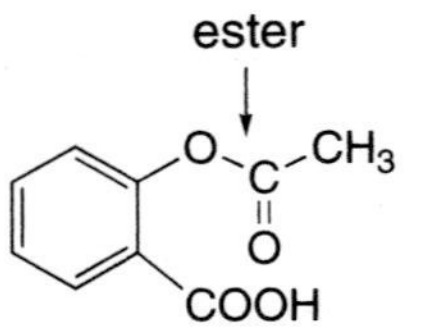
ester
O
C
CH3
O
COOH

acetylsalicylic acid

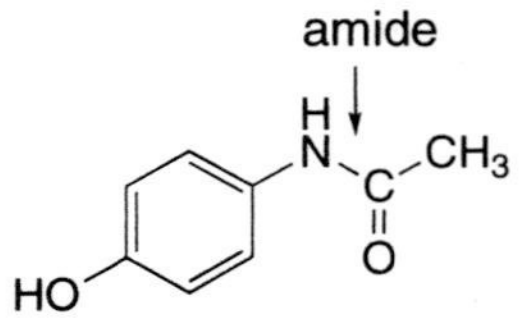
amide
H
N
C
CH3
O
HO

acetaminophen

22.30

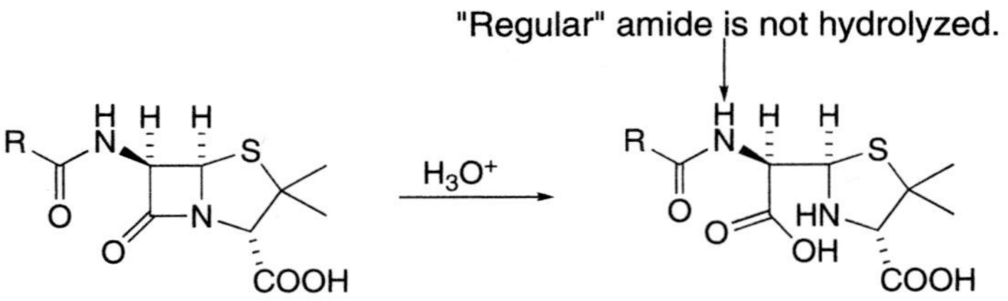
"Regular" amide is not hydrolyzed.
R
N
H H H
S
O
N
COOH
H3O+
HN
OH
COOH

22.31

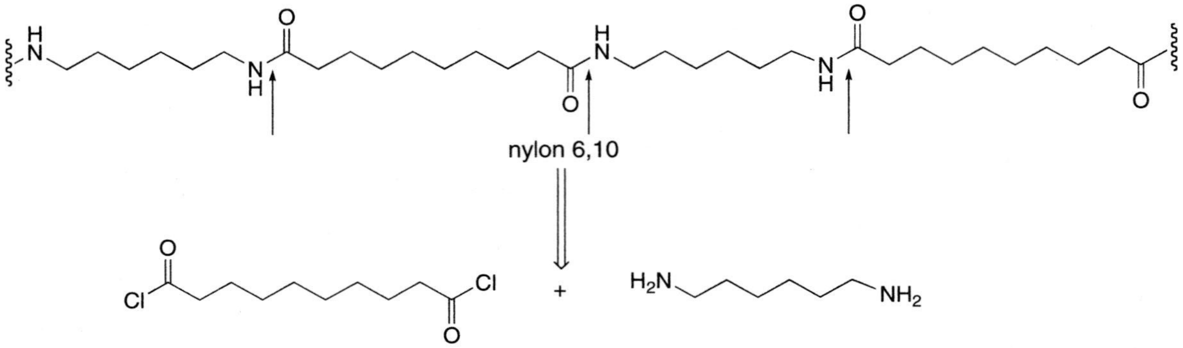
H
N
O
nylon 6,10
Cl
O
Cl
+
H2N
NH2

22.32

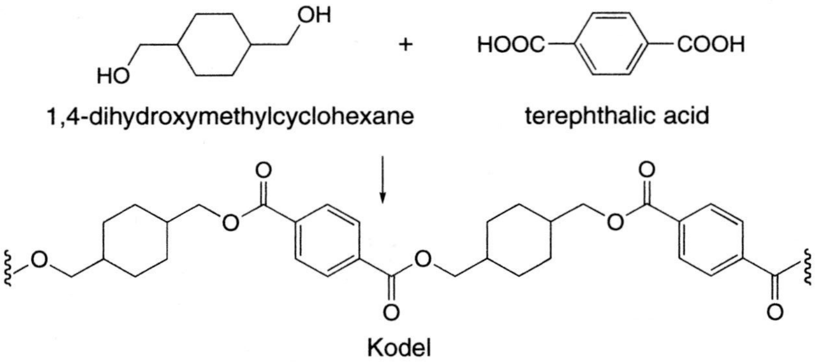
OH
HO
+
HOOC
COOH
1,4-dihydroxymethylcyclohexane
terephthalic acid
O
Kodel

In the polyester Kodel, most of the bonds in the polymer backbone are part of a ring, so there are fewer degrees of freedom. Fabrics made from Kodel are stiff and crease resistant, due to these less flexible polyester fibers.

22.33

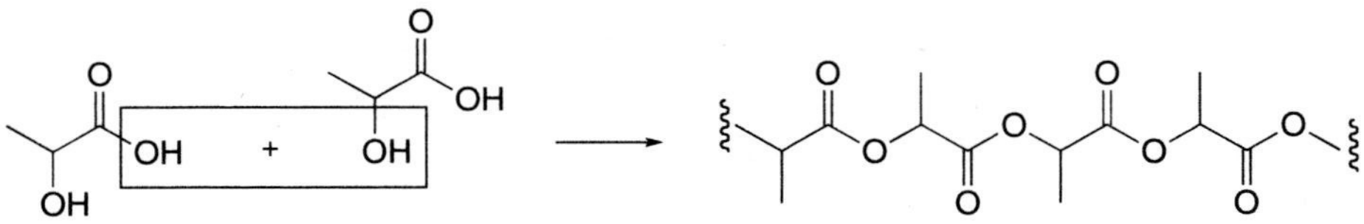
O
OH
+
OH
OH
O

Reaction occurs here (–H_2O).

PLA
polyl(lactic acid)

22.34 Acetyl CoA acetylates the NH_2 group of glucosamine, since the NH_2 group is the most nucleophilic site.

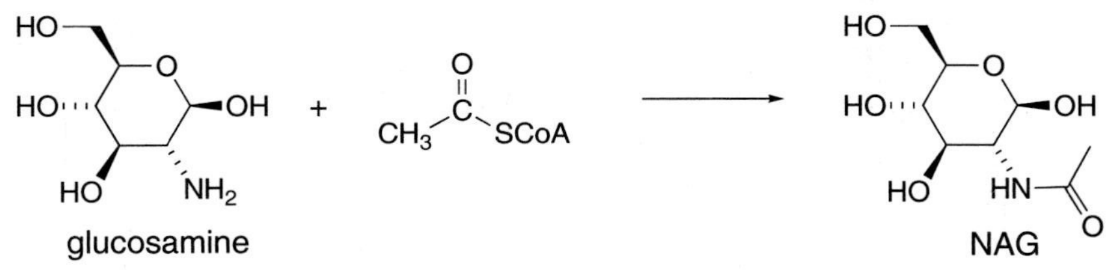

glucosamine + $CH_3\overset{O}{\underset{}{C}}SCoA$ ⟶ NAG

22.35

a. $CH_3CH_2CH_2-Br \xrightarrow{NaCN} CH_3CH_2CH_2-CN$

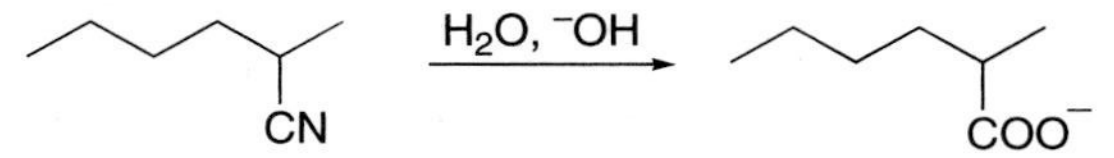

b. 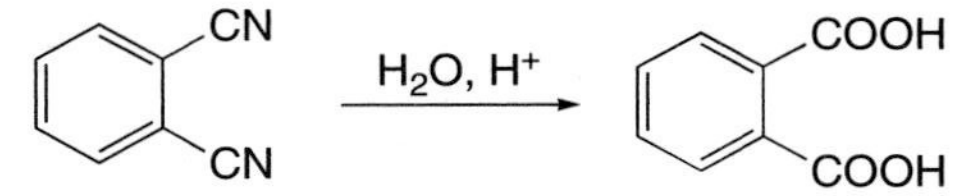 $\xrightarrow{H_2O,\ H^+}$

c. $\xrightarrow{H_2O,\ ^-OH}$

22.36

a.

b.

c.

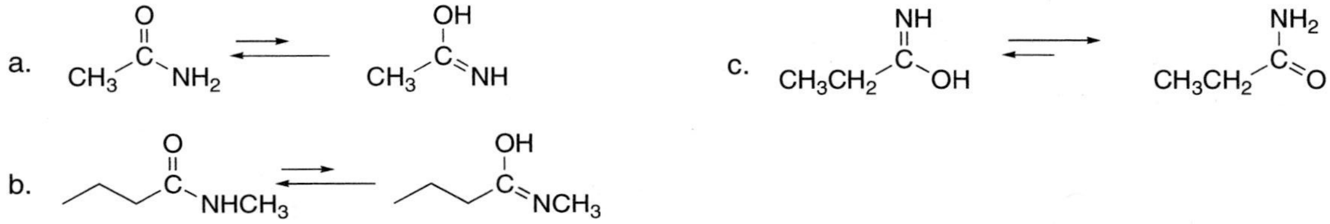

22.37

a. $CH_3CH_2-Br \xrightarrow[\substack{[2]\ LiAlH_4 \\ [3]\ H_2O}]{[1]\ NaCN} CH_3CH_2-CH_2NH_2$

b. 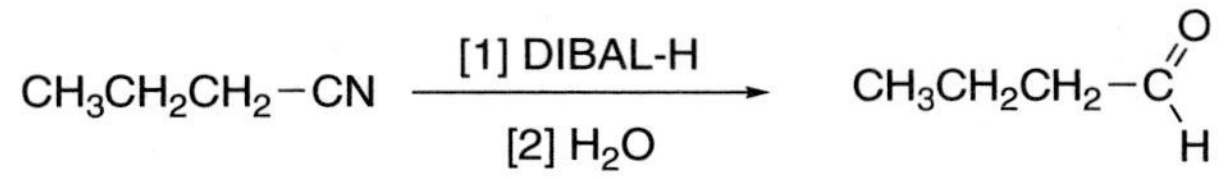 $CH_3CH_2CH_2-CN \xrightarrow[[2]\ H_2O]{[1]\ DIBAL-H} CH_3CH_2CH_2-\overset{O}{\underset{}{C}}H$

22.38

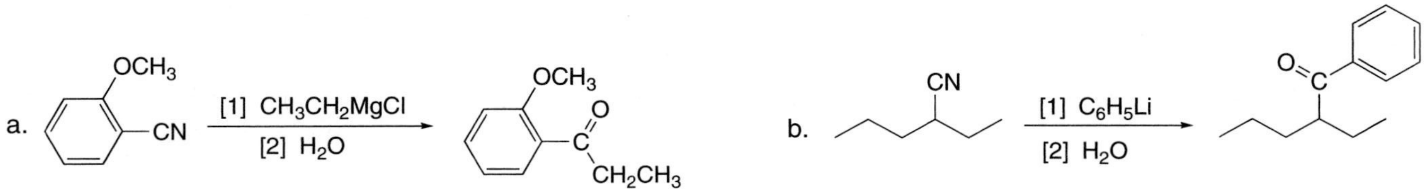

a. $\xrightarrow[[2]\ H_2O]{[1]\ CH_3CH_2MgCl}$

b. $\xrightarrow[[2]\ H_2O]{[1]\ C_6H_5Li}$

22.39

a. $\xrightarrow[[2]\ H_2O]{[1]\ CH_3MgBr}$

b. $\xrightarrow[[2]\ H_2O]{[1]\ (CH_3)_3CMgBr}$

c. $\xrightarrow[[2]\ H_2O]{[1]\ DIBAL-H}$

d. $\xrightarrow{H_3O^+}$

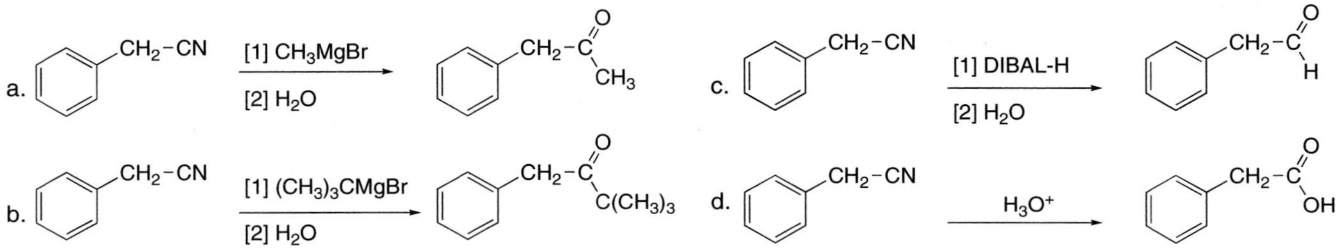

22.40

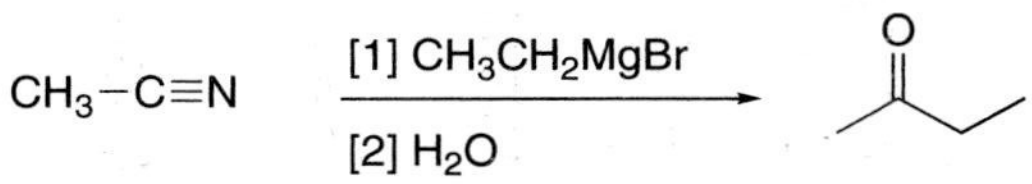

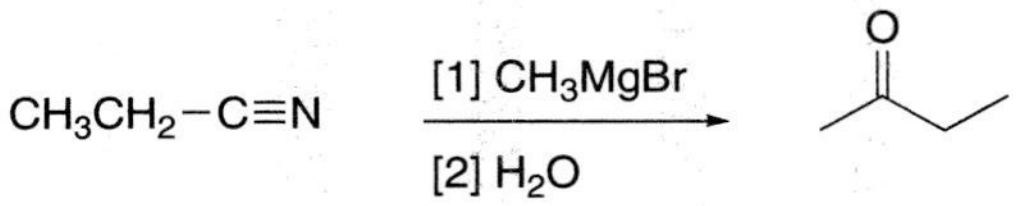

CH₃—C≡N $\xrightarrow[\text{[2] H}_2\text{O}]{\text{[1] CH}_3\text{CH}_2\text{MgBr}}$

CH₃CH₂—C≡N $\xrightarrow[\text{[2] H}_2\text{O}]{\text{[1] CH}_3\text{MgBr}}$

22.41

a. 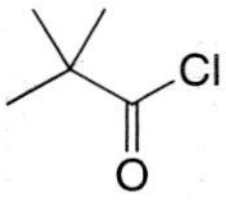 **2,2-dimethylpropanoyl chloride**

b. 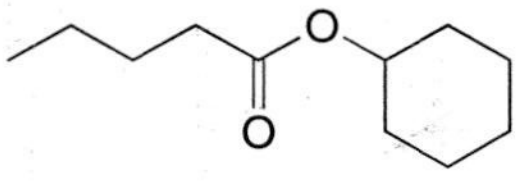 **cyclohexyl pentanoate**

c. 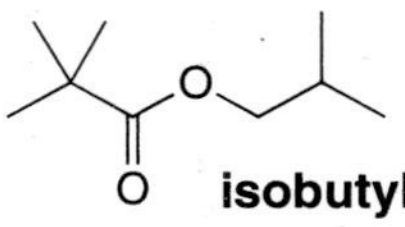 **isobutyl 2,2-dimethylpropanoate**

d. 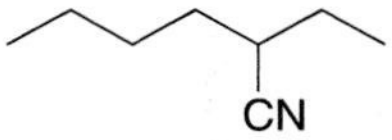 **2-ethylhexanenitrile**

e. 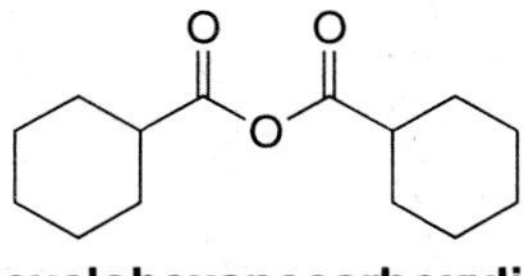 **cyclohexanecarboxylic anhydride**

f. 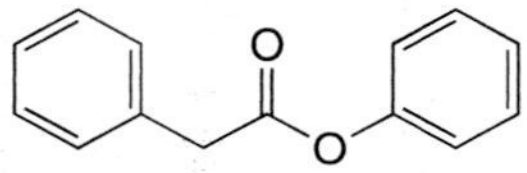 **phenyl phenylacetate**

g. 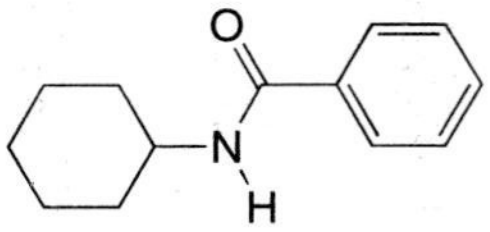 **N-cyclohexylbenzamide**

h. 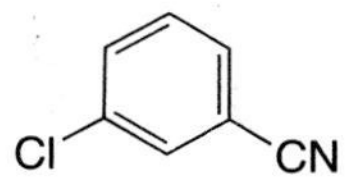 **m-chlorobenzonitrile**

i. 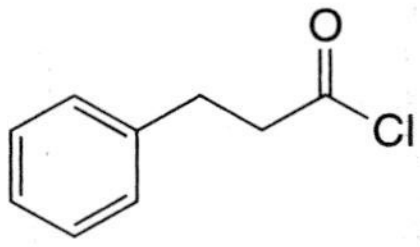 **3-phenylpropanoyl chloride**

j. 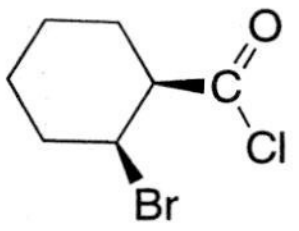 **cis-2-bromocyclohexane-carbonyl chloride**

k. 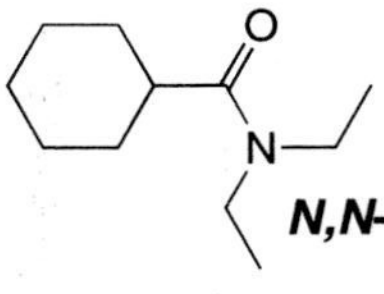 **N,N-diethylcyclohexanecarboxamide**

l. 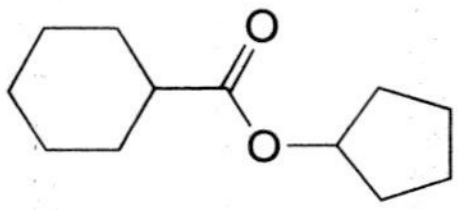 **cyclopentyl cyclohexanecarboxylate**

22.42

a. propanoic anhydride

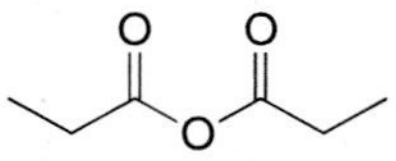

b. α-chlorobutyryl chloride

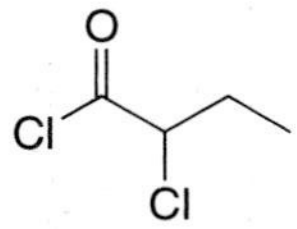

c. cyclohexyl propanoate

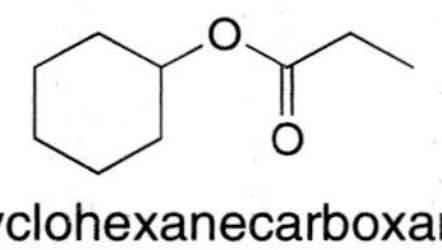

d. cyclohexanecarboxamide

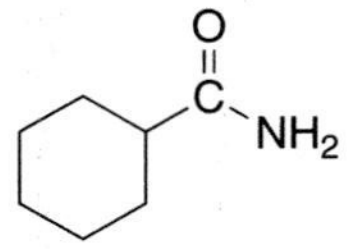

e. isopropyl formate

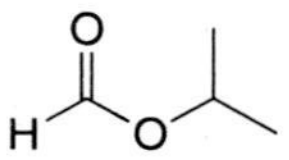

f. N-cyclopentylpentanamide

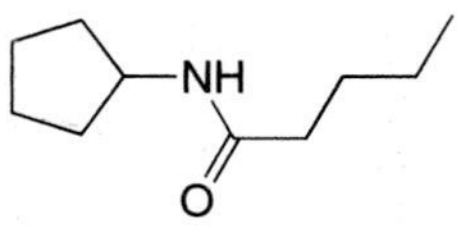

g. 4-methylheptanenitrile

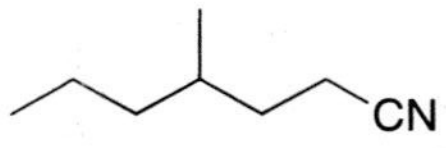

h. vinyl acetate

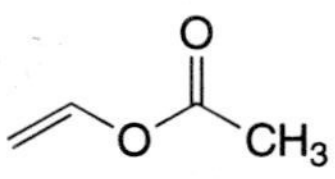

i. benzoic propanoic anhydride

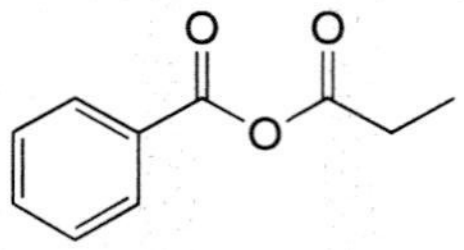

j. 3-methylhexanoyl chloride

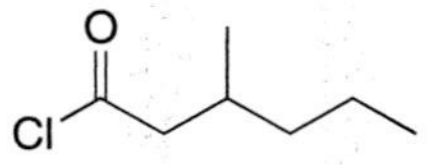

k. octyl butanoate

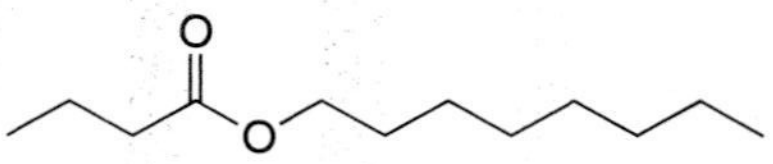

l. N,N-dibenzylformamide

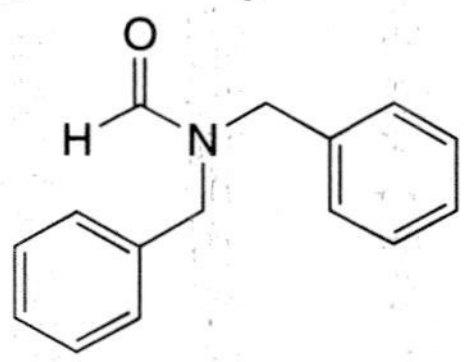

22.43 Rank the compounds using the rules from Answer 22.12.

a.

⁻NH₂ strongest base
least reactive

⁻OCH₂CH₂CH₃
intermediate

⁻Cl weakest base
most reactive

b.

ester
least reactive

anhydride
intermediate

anhydride with electron-
withdrawing F's
most reactive

c.

⁻OH strongest base
least reactive

⁻SH
intermediate

⁻Cl weakest base
most reactive

22.44

a. Better leaving groups make acyl compounds more reactive. **A** has an electron-withdrawing NO_2 group, which stabilizes the negative charge of the leaving group, whereas **B** has an electron-donating OCH_3 group, which destabilizes the leaving group.

A

B

an electron-withdrawing substituent   an electron-donating substituent

leaving group from **A**

one possible
resonance structure

Delocalizing the negative charge on the
NO_2 stabilizes the leaving group
making **A** more reactive than **B**.

leaving group from **B**

one possible
resonance structure

Adjacent negative charges destabilize
the leaving group.

resonance structures for the leaving group

b.

imidazolide

The leaving group is both resonance stabilized and aromatic
(6 π electrons), making it a much better leaving group than
exists in a regular amide.

22.45

Reaction as an acid:

These two resonance structures make the conjugate base more stable, and therefore CH_3CONH_2 a stronger acid.

$CH_3CH_2-\ddot{N}H_2 \xrightarrow{:B} CH_3CH_2-\ddot{N}H$

no resonance stabilization of the conjugate base

Reaction as a base:

This electron pair is delocalized by resonance, making it less available for electron donation. Thus CH_3CONH_2 is a much weaker base.

$CH_3CH_2-\ddot{N}H_2$ ——This electron pair is localized on N.

22.46

$CH_3CH_2CH_2CH_2COCl$

a. $\xrightarrow[\text{pyridine}]{H_2O}$ $CH_3CH_2CH_2CH_2COOH$ +

b. $\xrightarrow[\text{pyridine}]{CH_3CH_2OH}$ $CH_3CH_2CH_2CH_2COOCH_2CH_3$ +

c. $\xrightarrow{CH_3COO^-}$ $CH_3CH_2CH_2CH_2\overset{O}{\underset{}{C}}-O-\overset{O}{\underset{}{C}}CH_3$ + Cl^-

d. $\xrightarrow[\text{excess}]{NH_3}$ $CH_3CH_2CH_2CH_2CONH_2$ + $NH_4^+ Cl^-$

e. $\xrightarrow[\text{excess}]{(CH_3CH_2)_2NH}$ $CH_3CH_2CH_2CH_2CON(CH_2CH_3)_2$
+ $(CH_3CH_2)_2NH_2^+ Cl^-$

f. $\xrightarrow[\text{excess}]{C_6H_5NH_2}$ $CH_3CH_2CH_2CH_2CONHC_6H_5$
+ $C_6H_5NH_3^+ Cl^-$

22.47

a. $\xrightarrow{SOCl_2}$ no reaction

b. $\xrightarrow{H_2O}$ 2

c. $\xrightarrow{CH_3OH}$

d. $\xrightarrow{NaCl}$ no reaction

e. $\xrightarrow[\text{excess}]{(CH_3CH_2)_2NH}$

f. $\xrightarrow[\text{excess}]{CH_3CH_2NH_2}$

22.48

C$_6$H$_5$CH$_2$COOH

a. $\xrightarrow{\text{NaHCO}_3}$ [C$_6$H$_5$CH$_2$CO$_2^-$Na$^+$] + H$_2$CO$_3$

b. $\xrightarrow{\text{NaOH}}$ [C$_6$H$_5$CH$_2$CO$_2^-$Na$^+$] + H$_2$O

c. $\xrightarrow{\text{SOCl}_2}$ [C$_6$H$_5$CH$_2$COCl]

d. $\xrightarrow{\text{NaCl}}$ no reaction

e. $\xrightarrow[\text{(1 equiv)}]{\text{NH}_3}$ [C$_6$H$_5$CH$_2$CO$_2^-$ NH$_4^+$]

f. $\xrightarrow[\text{[2] }\Delta]{\text{[1] NH}_3}$ [C$_6$H$_5$CH$_2$CONH$_2$]

g. $\xrightarrow[\text{H}_2\text{SO}_4]{\text{CH}_3\text{OH}}$ [C$_6$H$_5$CH$_2$CO$_2$CH$_3$]

h. $\xrightarrow[\text{$^-$OH}]{\text{CH}_3\text{OH}}$ [C$_6$H$_5$CH$_2$CO$_2^-$]

i. $\xrightarrow[\text{[2] CH}_3\text{COCl}]{\text{[1] NaOH}}$ [anhydride]

j. $\xrightarrow[\text{DCC}]{\text{CH}_3\text{NH}_2}$ [C$_6$H$_5$CH$_2$CONHCH$_3$]

k. $\xrightarrow[\text{[2] CH}_3\text{CH}_2\text{CH}_2\text{NH}_2]{\text{[1] SOCl}_2}$ [C$_6$H$_5$CH$_2$CONHCH$_2$CH$_2$CH$_3$]

l. $\xrightarrow[\text{[2] (CH}_3)_2\text{CHOH}]{\text{[1] SOCl}_2}$ [C$_6$H$_5$CH$_2$CO$_2$CH(CH$_3$)$_2$]

22.49

CH$_3$CH$_2$CH$_2$CO$_2$CH$_2$CH$_3$

a. $\xrightarrow{\text{SOCl}_2}$ no reaction

b. $\xrightarrow{\text{H}_3\text{O}^+}$ [CH$_3$CH$_2$CH$_2$COOH] + HO—CH$_2$CH$_3$

c. $\xrightarrow{\text{H}_2\text{O, }^-\text{OH}}$ [CH$_3$CH$_2$CH$_2$CO$_2^-$] + HO—CH$_2$CH$_3$

d. $\xrightarrow{\text{NH}_3}$ [CH$_3$CH$_2$CH$_2$CONH$_2$] + HO—CH$_2$CH$_3$

e. $\xrightarrow{\text{CH}_3\text{CH}_2\text{NH}_2}$ [CH$_3$CH$_2$CH$_2$CONHCH$_2$CH$_3$] + HO—CH$_2$CH$_3$

22.50

[C$_6$H$_5$CH$_2$CONH$_2$]

a. $\xrightarrow{\text{H}_3\text{O}^+}$ [C$_6$H$_5$CH$_2$COOH]

b. $\xrightarrow{\text{H}_2\text{O, }^-\text{OH}}$ [C$_6$H$_5$CH$_2$COO$^-$]

22.51

C$_6$H$_5$CH$_2$CN

a. $\xrightarrow{\text{H}_3\text{O}^+}$ [C$_6$H$_5$CH$_2$COOH]

b. $\xrightarrow{\text{H}_2\text{O, }^-\text{OH}}$ [C$_6$H$_5$CH$_2$COO$^-$]

c. $\xrightarrow[\text{[2] H}_2\text{O}]{\text{[1] CH}_3\text{MgBr}}$ [C$_6$H$_5$CH$_2$COCH$_3$]

d. $\xrightarrow[\text{[2] H}_2\text{O}]{\text{[1] CH}_3\text{CH}_2\text{Li}}$ [C$_6$H$_5$CH$_2$COCH$_2$CH$_3$]

e. $\xrightarrow[\text{[2] H}_2\text{O}]{\text{[1] DIBAL-H}}$ [C$_6$H$_5$CH$_2$CHO]

f. $\xrightarrow[\text{[2] H}_2\text{O}]{\text{[1] LiAlH}_4}$ [C$_6$H$_5$CH$_2$CH$_2$NH$_2$]

22.52

a. $\xrightarrow{\text{SOCl}_2}$

g. $\text{CH}_3\text{CH}_2\text{CH}_2\text{CH}_2\text{Br}$ $\xrightarrow[\text{[2] H}_2\text{O, }^-\text{OH}]{\text{[1] NaCN}}$ $\text{CH}_3\text{CH}_2\text{CH}_2\text{CH}_2\overset{\text{O}}{\underset{}{\text{C}}}\text{O}^-$

b. $\text{C}_6\text{H}_5\text{COCl}$ + (excess) ⟶ + Cl^-

h. $\text{C}_6\text{H}_5\text{CH}_2\text{COOH}$ $\xrightarrow[\substack{\text{[2] CH}_3\text{CH}_2\text{CH}_2\text{CH}_2\text{NH}_2 \\ \text{[3] LiAlH}_4 \\ \text{[4] H}_2\text{O}}]{\text{[1] SOCl}_2}$ $\text{C}_6\text{H}_5(\text{CH}_2)_2\text{NH}(\text{CH}_2)_3\text{CH}_3$

c. $\text{C}_6\text{H}_5\text{CN}$ $\xrightarrow[\text{[2] H}_2\text{O}]{\text{[1] CH}_3\text{CH}_2\text{CH}_2\text{MgBr}}$ $\text{C}_6\text{H}_5\overset{\text{O}}{\underset{}{\text{C}}}\text{CH}_2\text{CH}_2\text{CH}_3$

i. $\text{C}_6\text{H}_5\text{CH}_2\text{CH}_2\text{CH}_2\text{CN}$ $\xrightarrow{\text{H}_3\text{O}^+}$ $\text{C}_6\text{H}_5\text{CH}_2\text{CH}_2\text{CH}_2\text{COOH}$

d. $(\text{CH}_3)_2\text{CHCOOH}$ + $\underset{\overset{|}{\text{CH}_3}}{\text{CH}_3\text{CH}_2\text{CHOH}}$ $\xrightarrow{\text{H}_2\text{SO}_4}$ $(\text{CH}_3)_2\text{CH}\overset{\text{O}}{\underset{}{\text{C}}}\text{OCH(CH}_3)\text{CH}_2\text{CH}_3$

j. $\underset{\text{HOOC}}{\overset{}{}}\diagdown\diagup\underset{\text{COOH}}{}$ $\xrightarrow{\Delta}$

k. $(\text{CH}_3\text{CO})_2\text{O}$ + —NH_2 (excess) ⟶ —NHCOCH_3 + CH_3COO^- $\text{H}_3\overset{+}{\text{N}}$—

e. —NHCOCH_3 $\xrightarrow[^-\text{OH}]{\text{H}_2\text{O}}$ $\overset{\text{O}}{\underset{}{\text{C}}}\diagup\text{O}^-$ + —NH_2

f. $\xrightarrow{\text{H}_3\text{O}^+}$

l. $\text{C}_6\text{H}_5\text{CH}_2\text{CH}_2\text{COOCH}_2\text{CH}_3$ $\xrightarrow[^-\text{OH}]{\text{H}_2\text{O}}$ $\text{C}_6\text{H}_5\text{CH}_2\text{CH}_2\text{COO}^-$ + HOCH_2CH_3

22.53 Both lactones and acetals are hydrolyzed with aqueous acid, but only lactones react with aqueous base.

a. **X** $\xrightarrow{\text{H}_3\text{O}^+}$ +

b. **X** $\xrightarrow[\text{H}_2\text{O}]{\text{NaOH}}$

22.54

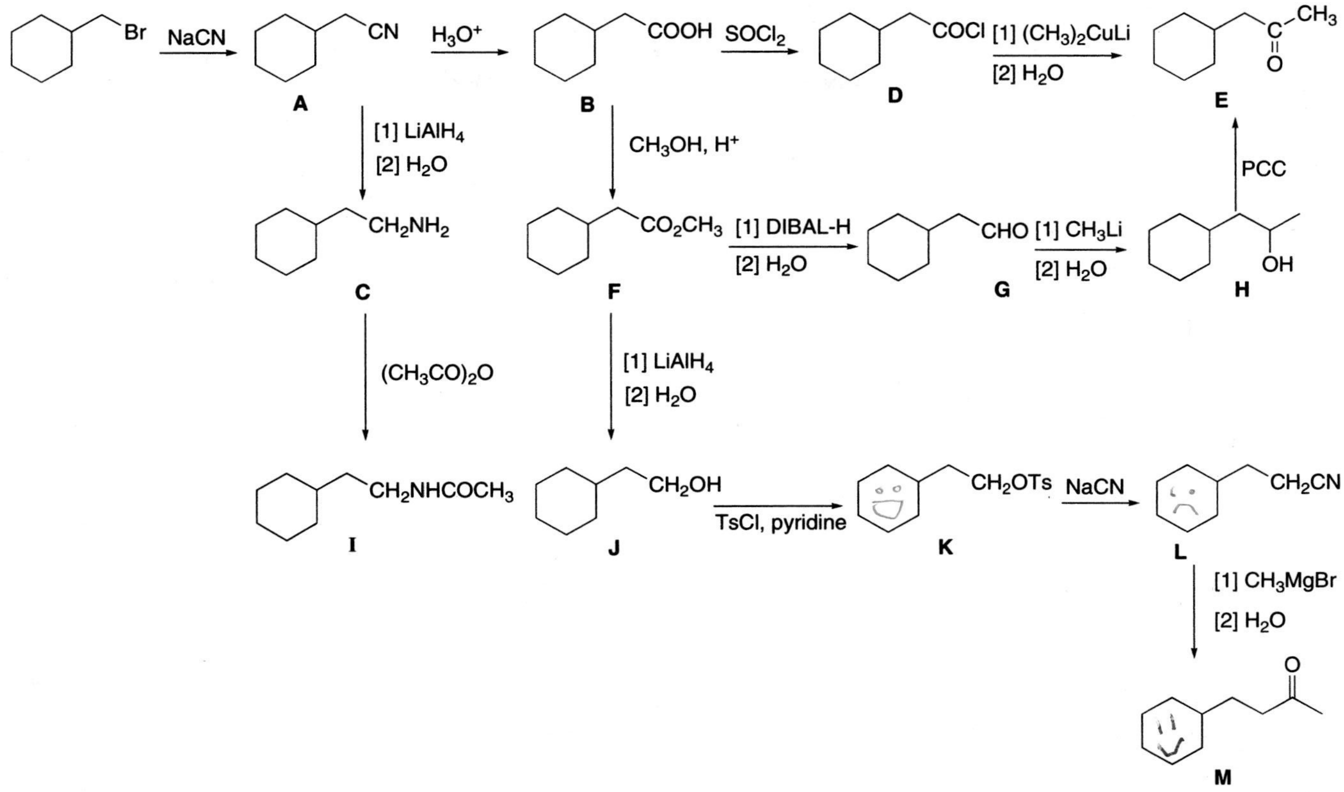

22.55

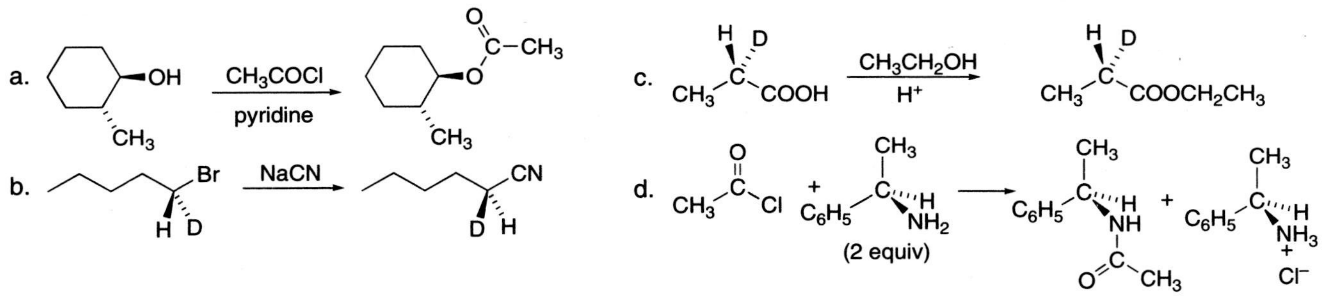

22.56 Hydrolyze the amide and ester bonds in both starting materials to draw the products.

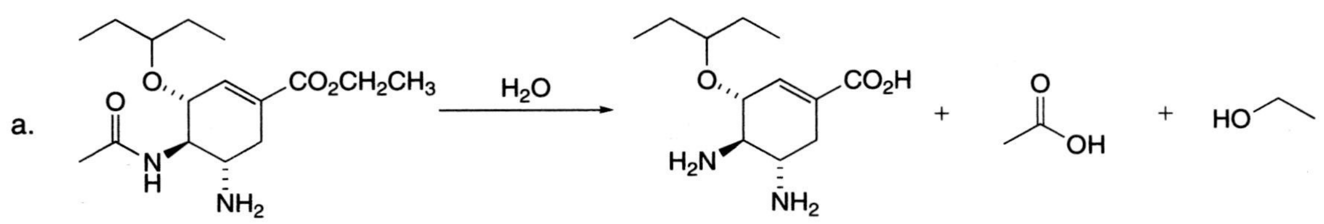

a.

oseltamivir

b.

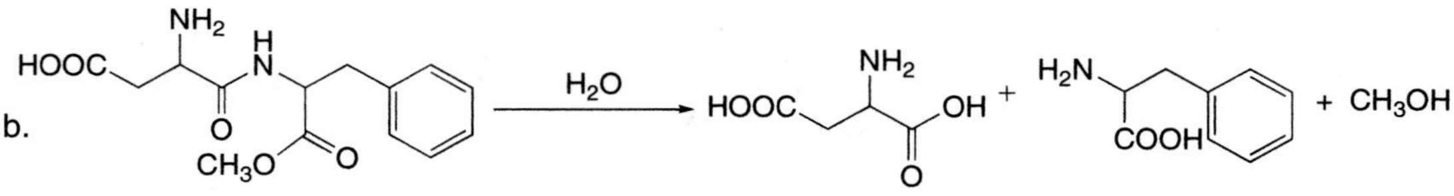

aspartame → phenylalanine + CH₃OH

22.57

a.

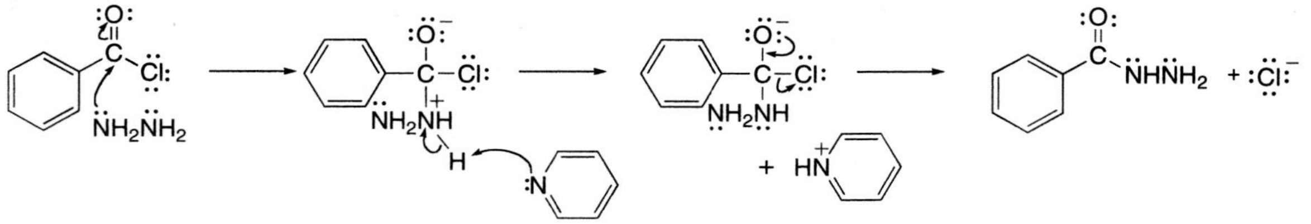

b.

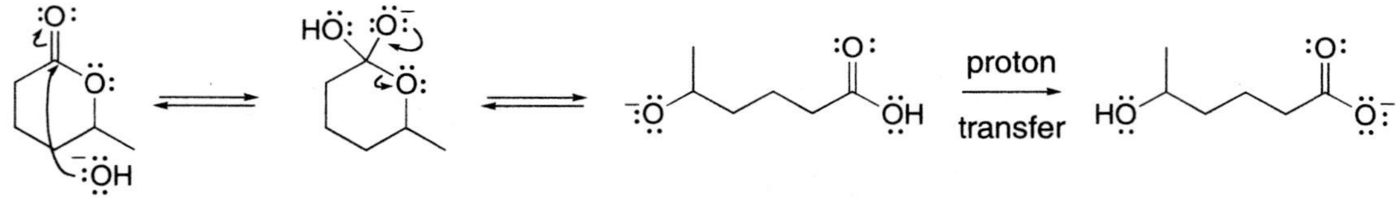

22.58

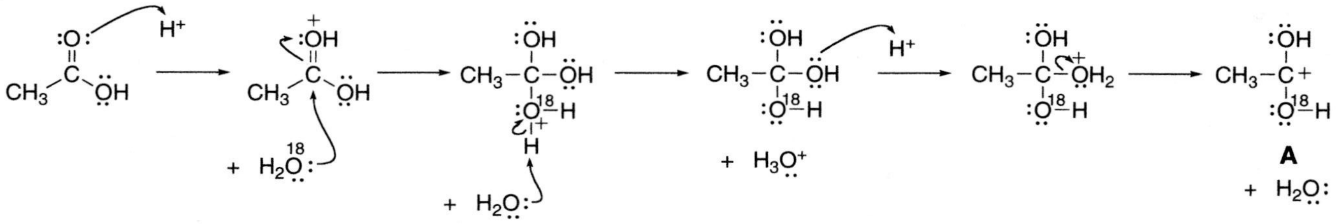

Two possibilities for **A**:

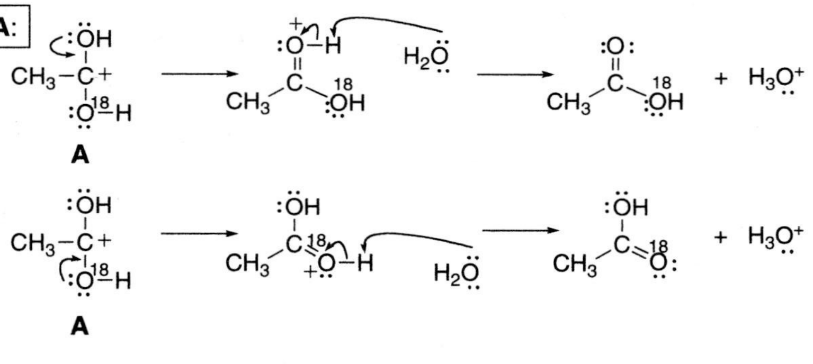

22.59

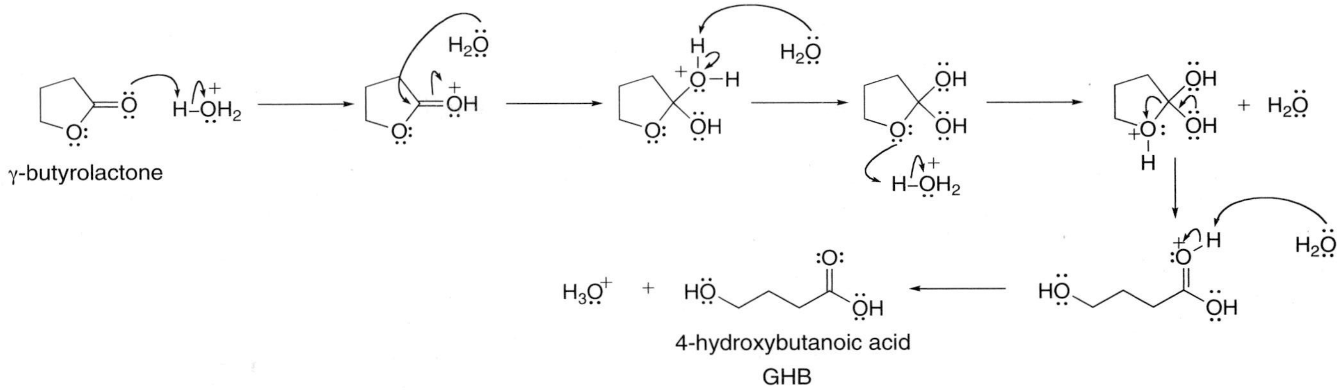

γ-butyrolactone

4-hydroxybutanoic acid
GHB

22.60

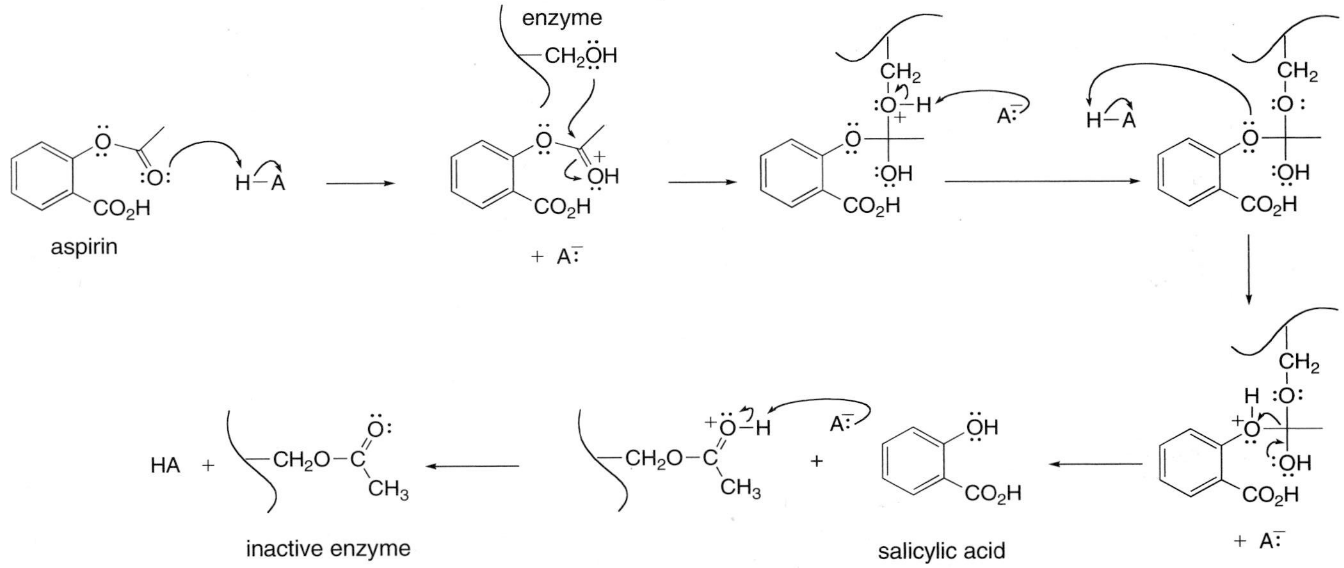

enzyme

aspirin

inactive enzyme

salicylic acid

22.61

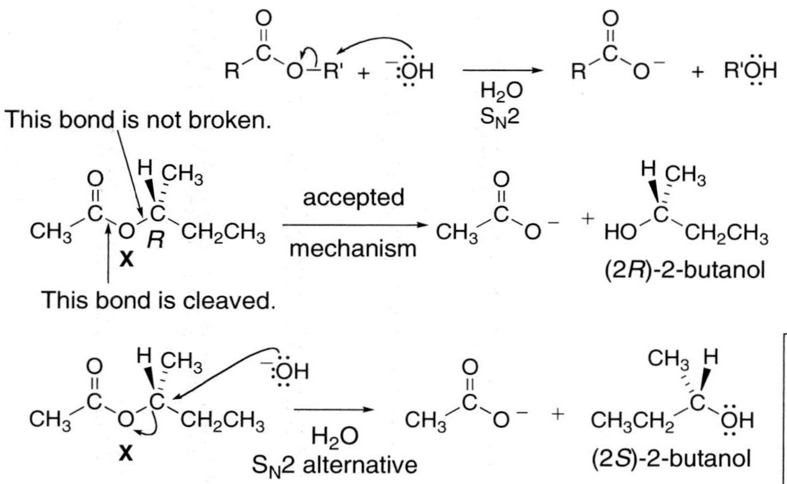
H₂O
SN2
This bond is not broken.
accepted
mechanism
(2R)-2-butanol
This bond is cleaved.
H₂O
SN2 alternative
(2S)-2-butanol

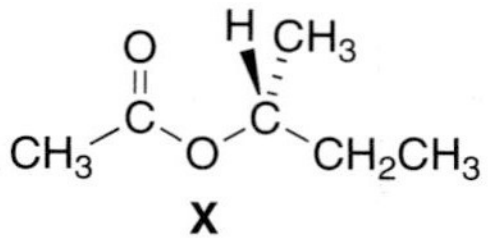
X

According to the accepted mechanism, the stereochemistry around the stereogenic center is retained in the product.

This S_N2 mechanism would form the product of inversion leading to (2*S*)-2-butanol. Since (2*R*)-2-butanol is the only product formed, the S_N2 mechanism does not occur during ester hydrolysis.

22.62

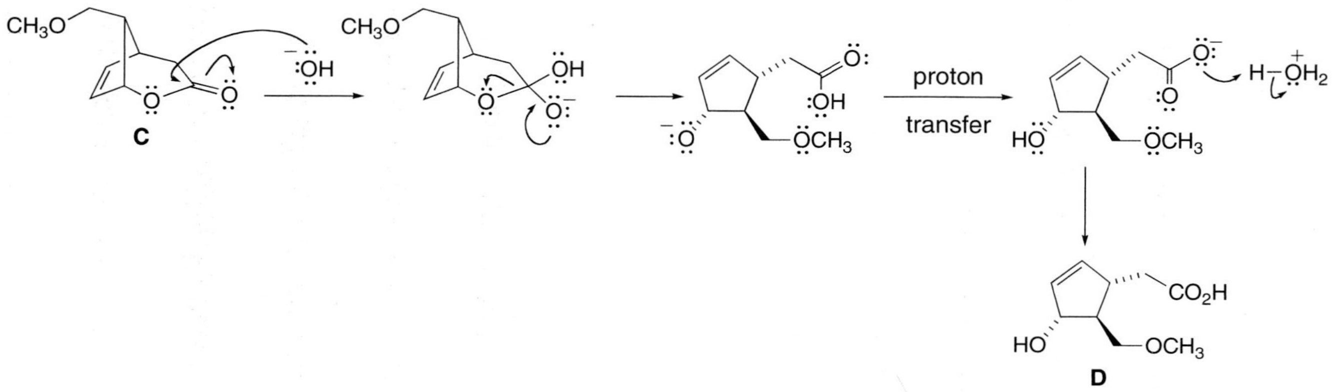
C
proton
transfer
D

22.63

22.64 The mechanism is composed of two parts: hydrolysis of the acetal and intramolecular Fischer esterification of the hydroxy carboxylic acid.

22.65

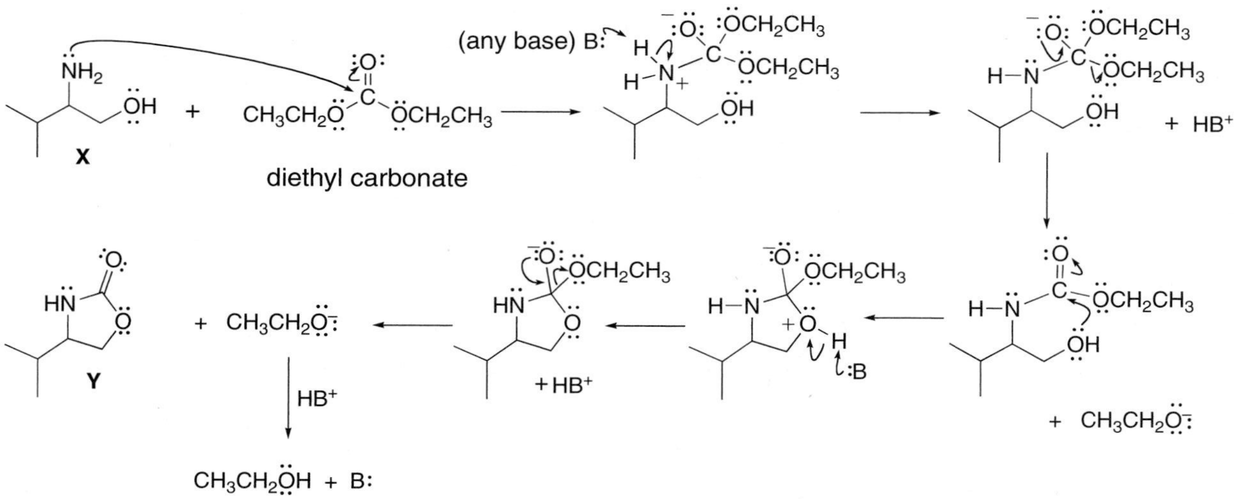

22.66

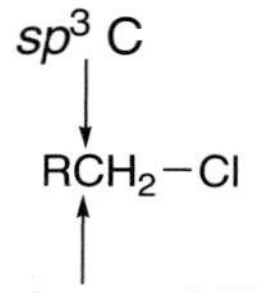

sp^3 C

RCH$_2$—Cl

less electrophilic C
more crowded C since it is
surrounded by four atoms

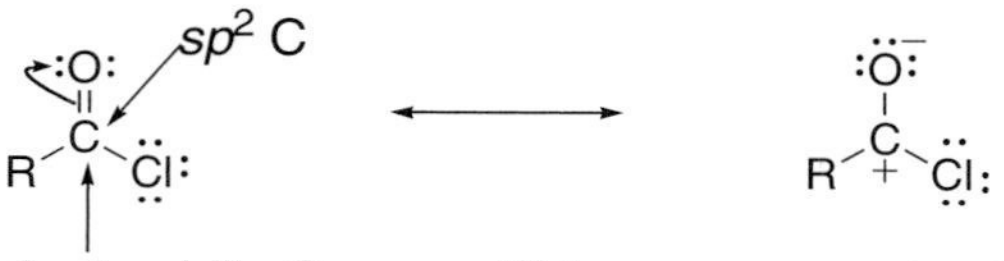

sp^2 C

more electrophilic C
due to electron-withdrawing O
more reactive

This resonance structure illustrates
how the electronegative O atom
withdraws more electron density
from C.

The sp^2 hybridized C of RCOCl is much less crowded,
and this makes nucleophilic attack easier as well.

22.67

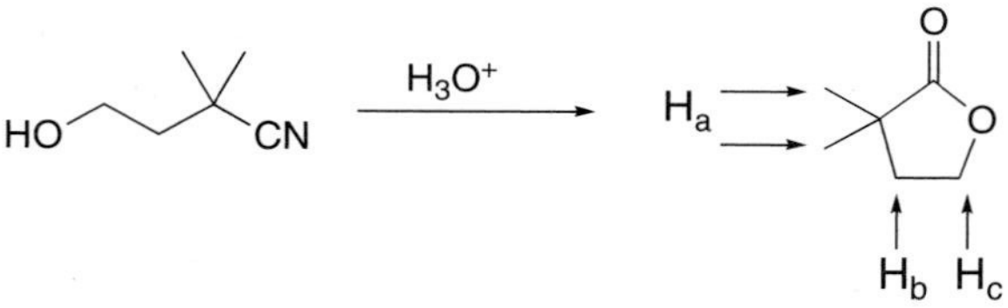

A
$C_6H_{10}O_2$: 2 degrees of unsaturation

IR: 1770 cm^{-1} from ester C=O in a five-membered ring

^{1}H NMR: H_a = 1.27 ppm (singlet, 6 H) – 2 CH$_3$ groups
H_b = 2.12 ppm (triplet, 2 H) – CH$_2$ bonded to CH$_2$
H_c = 4.26 ppm (triplet, 2 H) – CH$_2$ bonded to CH$_2$

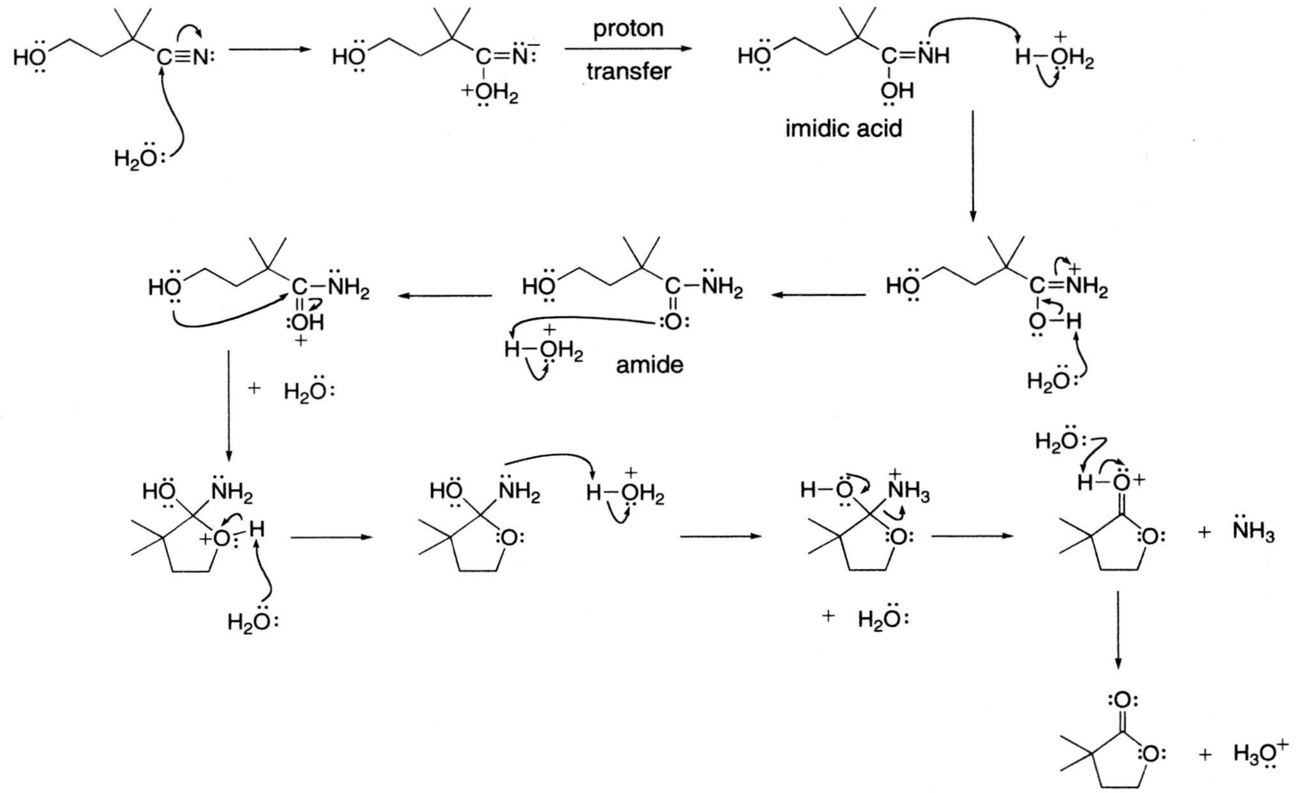

imidic acid

amide

+ $H_2\ddot{O}$:

+ $\ddot{N}H_3$

+ $H_2\ddot{O}$:

+ $H_3\overset{+}{O}$

22.68 Fischer esterification is treatment of a carboxylic acid with an alcohol in the presence of an acid catalyst to form an ester.

a. $(CH_3)_3CCO_2CH_2CH_3 \implies (CH_3)_3CCOOH$
$+ HOCH_2CH_3$

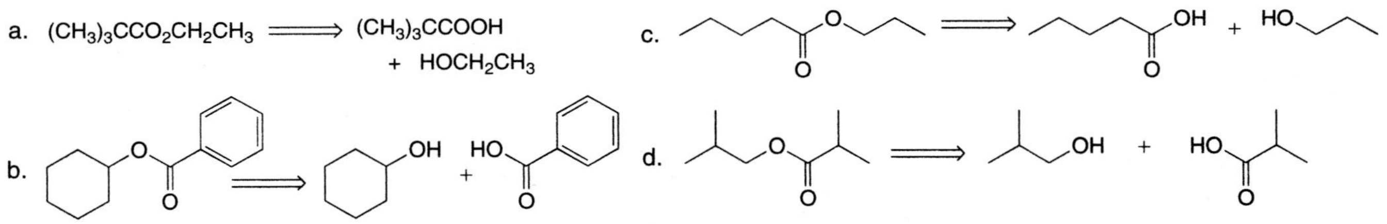

b.

c.

d.

22.69

a.

b.

(from a.)

c.

(from a.)

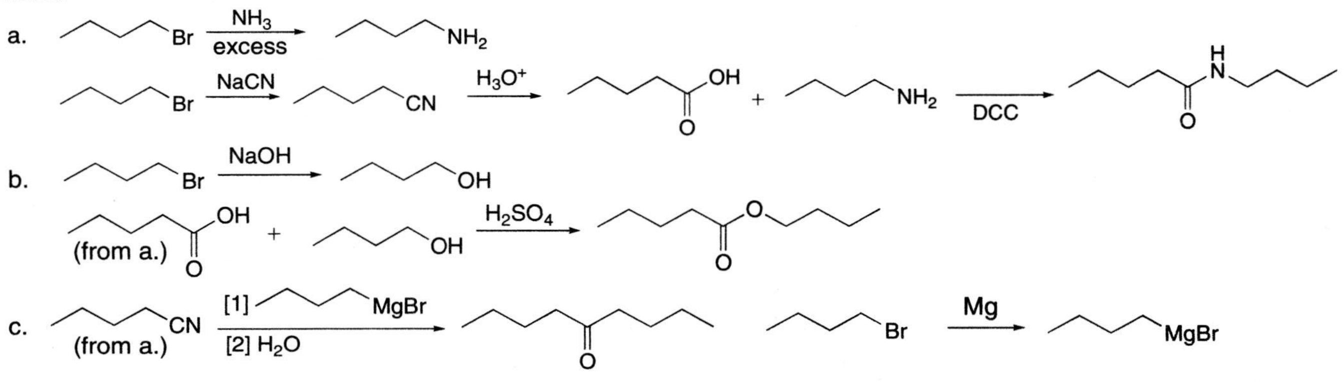

d.

$\overset{CrO_3}{\underset{H_2SO_4, H_2O}{\longrightarrow}}$

(from b.)

+

$\overset{H_2SO_4}{\longrightarrow}$

e.

(from d.)

(from c.)

[1] MgBr
(2 equiv)

[2] H_2O

OH

f.

(from b.)

(from c.)

[1] MgBr
(2 equiv)

[2] H_2O

OH

22.70

a. Br $\overset{^-CN}{\longrightarrow}$ CN

b. CN $\overset{H_2O, H^+}{\longrightarrow}$ COOH

(from a.)

c. COOH $\overset{SOCl_2}{\longrightarrow}$ COCl

(from b.)

d. COOH $\overset{HOCH_2CH_3}{\underset{H_2SO_4}{\longrightarrow}}$ $CO_2CH_2CH_3$

(from b.)

e. CN $\overset{[1]\ CH_3MgBr}{\underset{[2]\ H_2O}{\longrightarrow}}$ $\overset{CH_3}{\underset{O}{C}}$

(from a.)

f. CN $\overset{[1]\ DIBAL-H}{\underset{[2]\ H_2O}{\longrightarrow}}$ CHO

(from a.)

g. CN $\overset{[1]\ LiAlH_4}{\underset{[2]\ H_2O}{\longrightarrow}}$ CH_2NH_2

(from a.)

h.

CH_2NH_2 $\overset{CH_3COCl}{\longrightarrow}$ $CH_2NHCOCH_3$

(from g.)

22.71

a. CH_3Cl + NaCN $\longrightarrow$ CH_3-CN $\overset{H_3O^+}{\longrightarrow}$ CH_3-COOH

CH_3-Cl + Mg $\longrightarrow$ CH_3-MgCl $\overset{[1]\ CO_2}{\underset{[2]\ H_3O^+}{\longrightarrow}}$ CH_3-COOH

b. Br (sp^2) + NaCN $\longrightarrow$ This method can't be used because an S_N2 reaction can't be done on an sp^2 hybridized C.

Br + Mg $\longrightarrow$ MgBr $\overset{[1]\ CO_2}{\underset{[2]\ H_3O^+}{\longrightarrow}}$ COOH

c. $(CH_3)_3CCl$ + NaCN $\longrightarrow$ This method can't be used because an S_N2 reaction can't be done on a 3° C.

$(CH_3)_3C{-}Cl$ + Mg $\longrightarrow$ $(CH_3)_3C{-}MgCl$ $\xrightarrow[\text{[2] } H_3O^+]{\text{[1] } CO_2}$ $(CH_3)_3C{-}COOH$

d. $HOCH_2CH_2CH_2CH_2Br$ + NaCN $\longrightarrow$ $HOCH_2CH_2CH_2CH_2{-}CN$ $\xrightarrow{H_3O^+}$ $HOCH_2CH_2CH_2CH_2{-}COOH$

$HOCH_2CH_2CH_2CH_2{-}Br$ + Mg $\longrightarrow$ This method can't be used because you can't make a Grignard reagent with an acidic OH group.

22.72

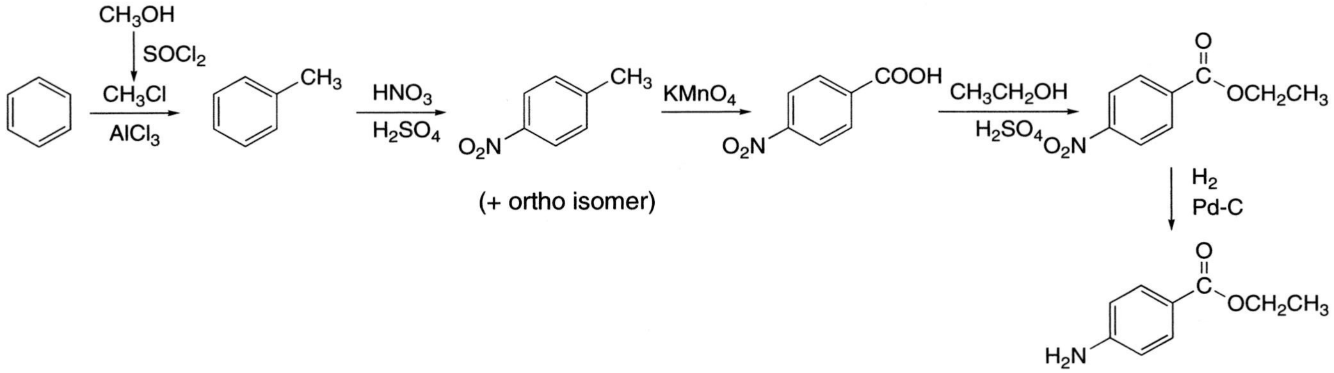

(+ ortho isomer)

22.73

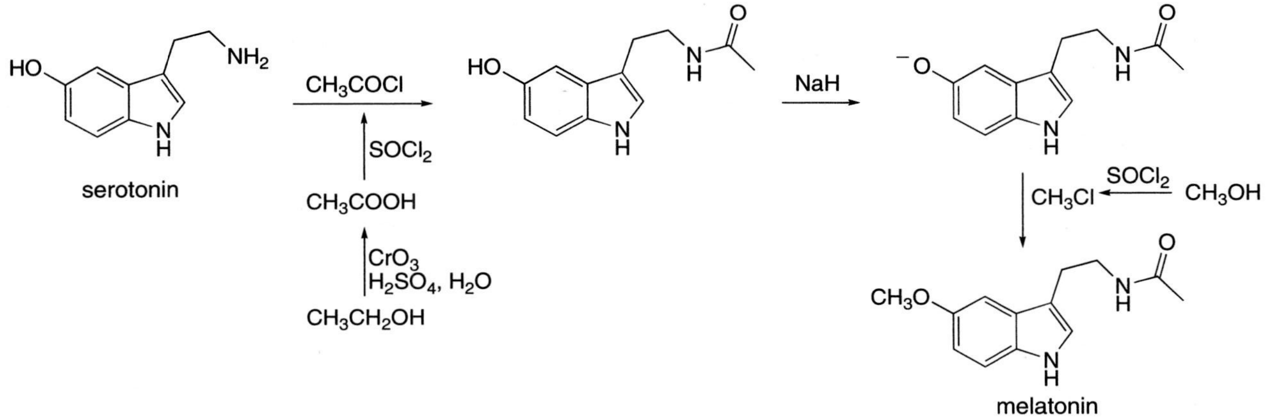

serotonin

melatonin

22.74

a.

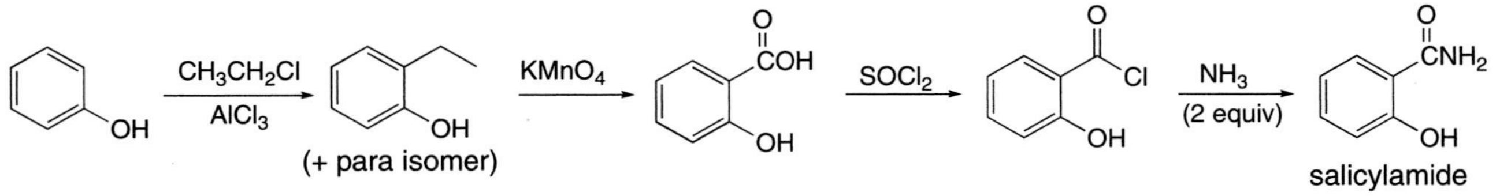

(+ para isomer)

salicylamide

b.

(+ ortho isomer)

(More nucleophilic NH_2 reacts first.)

acetaminophen

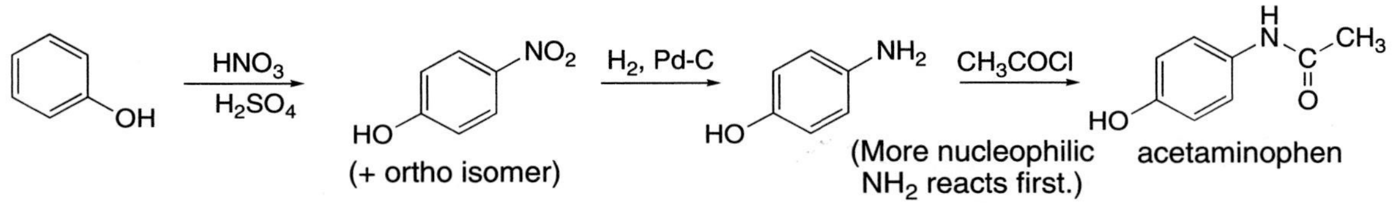

c.

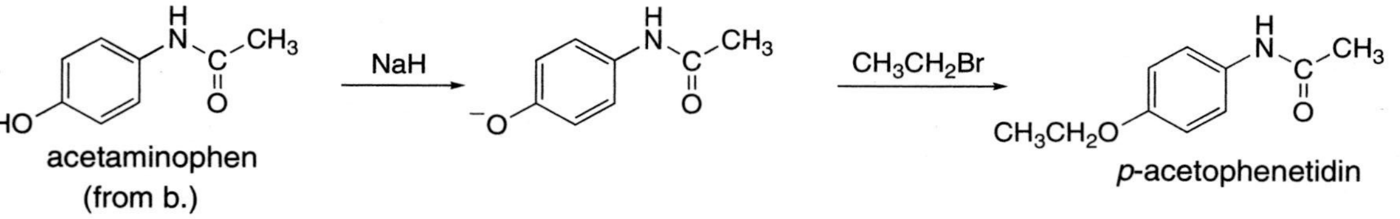
H
N
C
CH3
O
HO
NaH
⁻O
CH3CH2Br
CH3CH2O
acetaminophen
(from b.)
p-acetophenetidin

22.75

a.

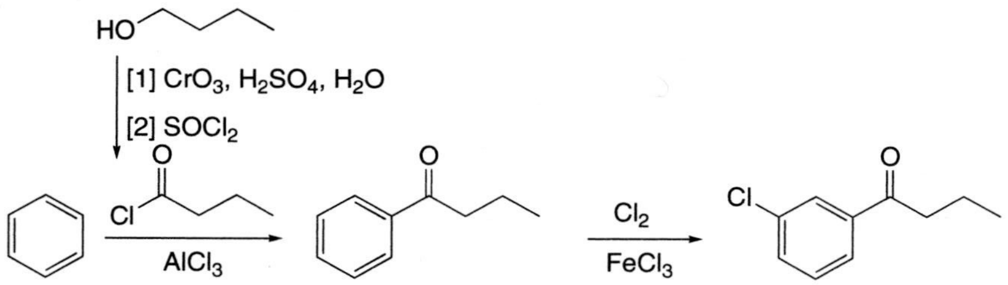
HO
[1] CrO3, H2SO4, H2O
[2] SOCl2
O
Cl
AlCl3
Cl2
FeCl3
Cl

b.

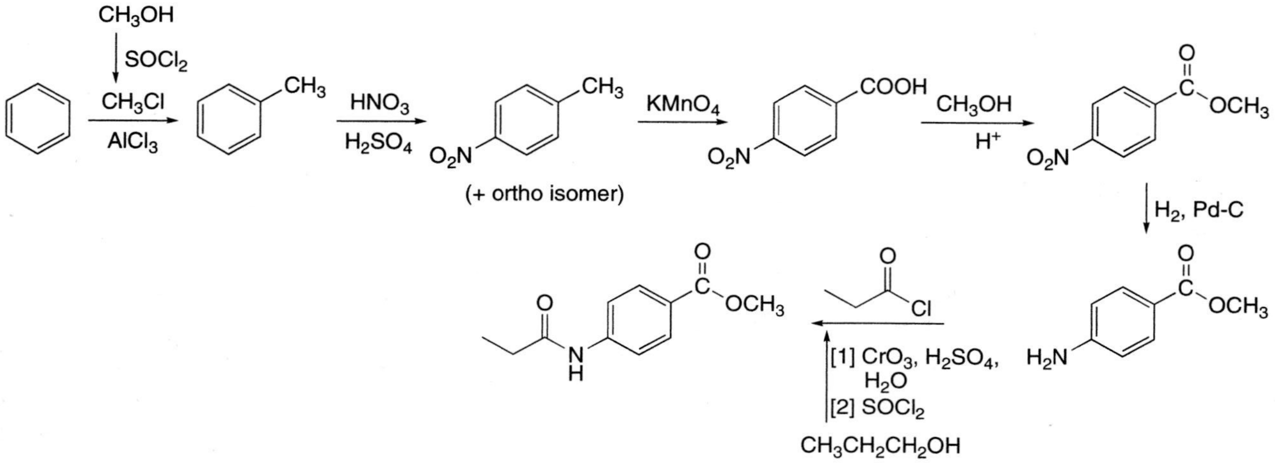
CH3OH
SOCl2
CH3Cl
AlCl3
CH3
HNO3
H2SO4
O2N
(+ ortho isomer)
KMnO4
COOH
CH3OH
H+
OCH3
H2, Pd-C
H2N
Cl
[1] CrO3, H2SO4,
H2O
[2] SOCl2
CH3CH2CH2OH
N
H

c.

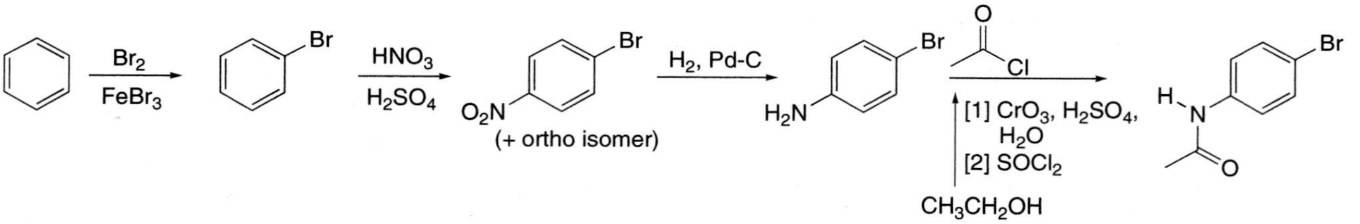
Br2
FeBr3
Br
HNO3
H2SO4
O2N
(+ ortho isomer)
H2, Pd-C
H2N
Cl
[1] CrO3, H2SO4,
H2O
[2] SOCl2
CH3CH2OH
H
N
O

d.

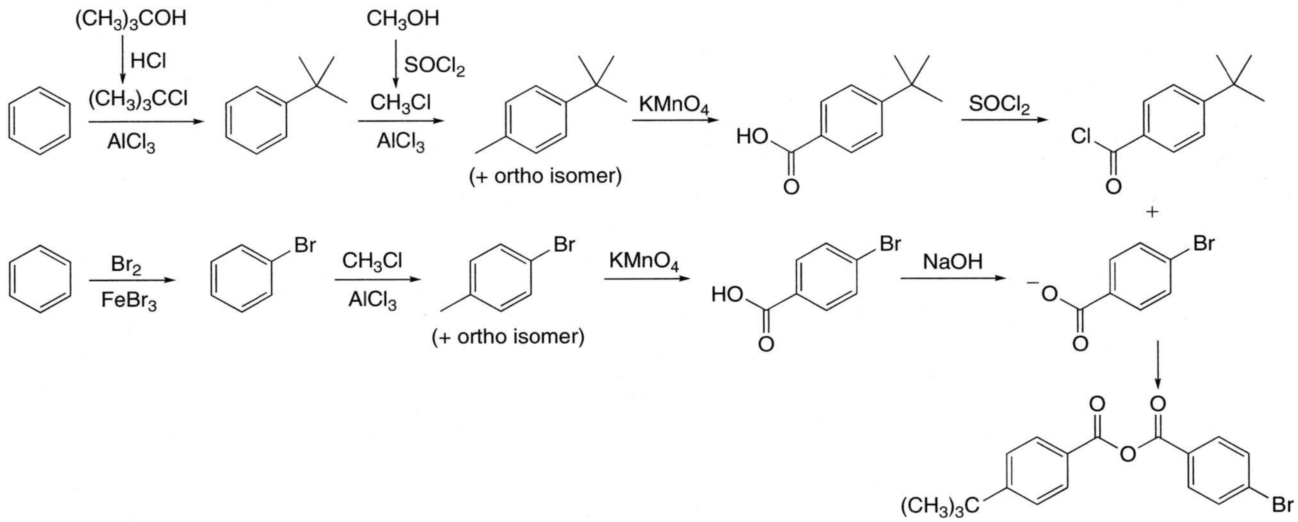
(CH3)3COH
HCl
(CH3)3CCl
AlCl3
CH3OH
SOCl2
CH3Cl
AlCl3
(+ ortho isomer)
KMnO4
HO
SOCl2
Cl
+
Br2
FeBr3
Br
CH3Cl
AlCl3
(+ ortho isomer)
KMnO4
NaOH
(CH3)3C

22.76

a.

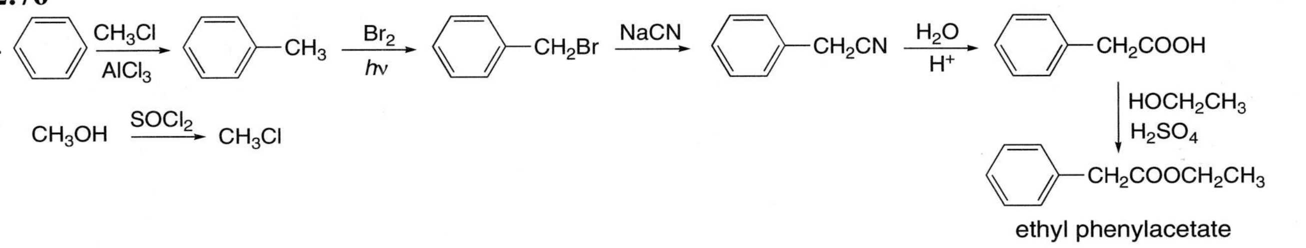
CH3Cl
AlCl3
CH3
Br2
hv
CH2Br
NaCN
CH2CN
H2O
H+
CH2COOH
HOCH2CH3
H2SO4
CH2COOCH2CH3
ethyl phenylacetate
CH3OH
SOCl2
CH3Cl

b.

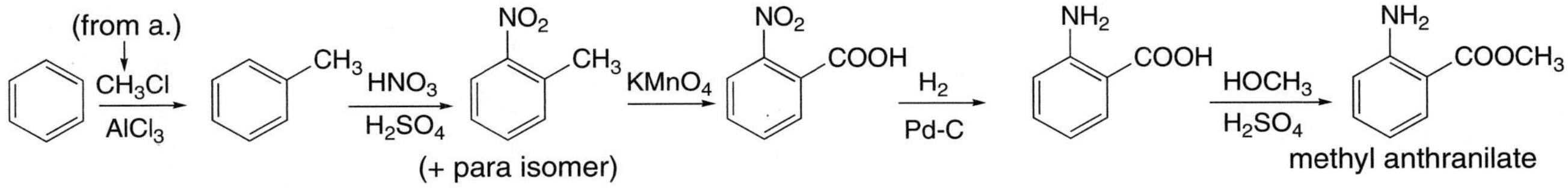
(from a.)
CH3Cl
AlCl3
CH3
HNO3
H2SO4
(+ para isomer)
NO2
KMnO4
COOH
H2
Pd-C
NH2
HOCH3
H2SO4
COOCH3
methyl anthranilate

c.

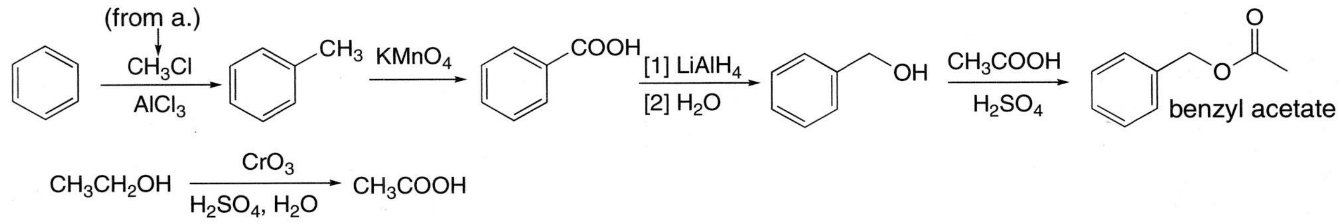
(from a.)
CH3Cl
AlCl3
CH3
KMnO4
COOH
[1] LiAlH4
[2] H2O
OH
CH3COOH
H2SO4
benzyl acetate
CH3CH2OH
CrO3
H2SO4, H2O
CH3COOH

22.77

a.

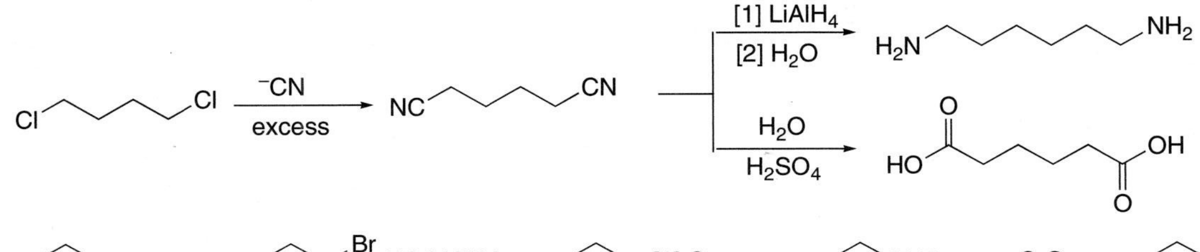
Cl
−CN
excess
NC
CN
[1] LiAlH4
[2] H2O
H2N
NH2
H2O
H2SO4
HO
OH

b.

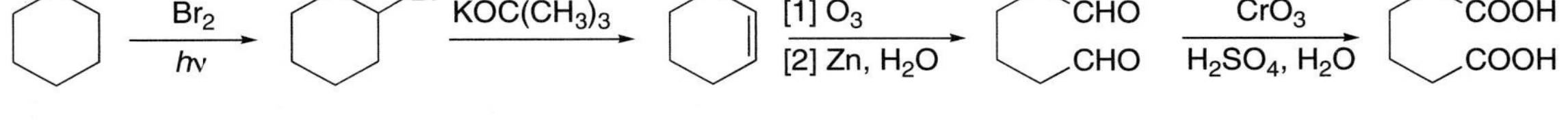
Br2
hv
Br
KOC(CH3)3
[1] O3
[2] Zn, H2O
CHO
CrO3
H2SO4, H2O
COOH

22.78

a.

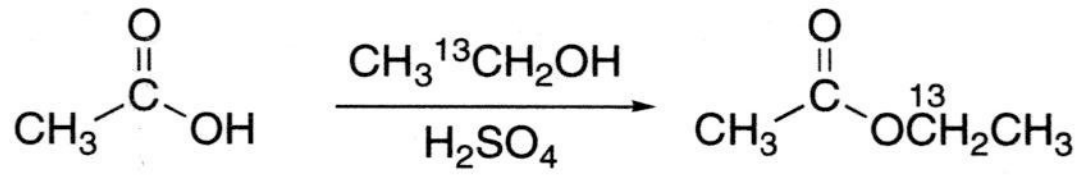

b.

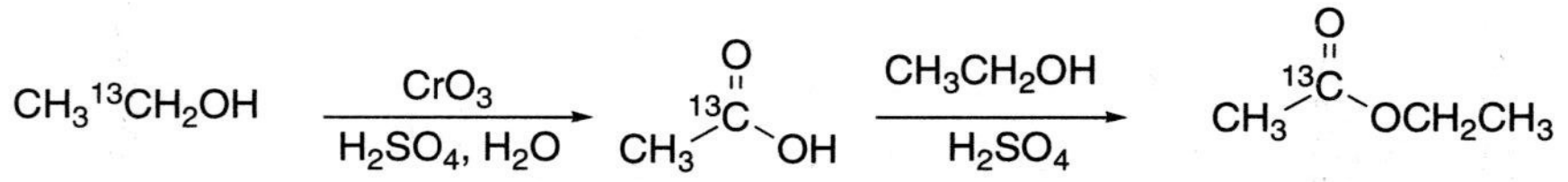

c.

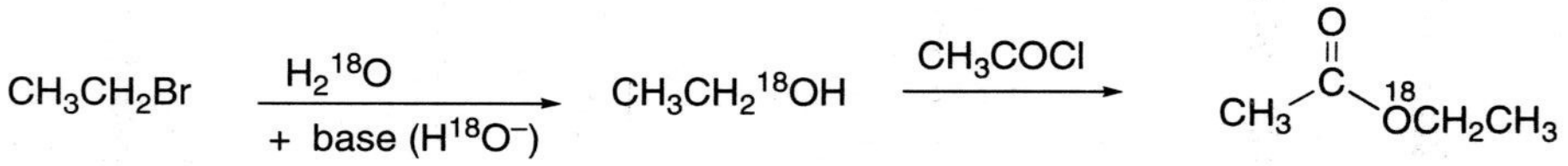

d.

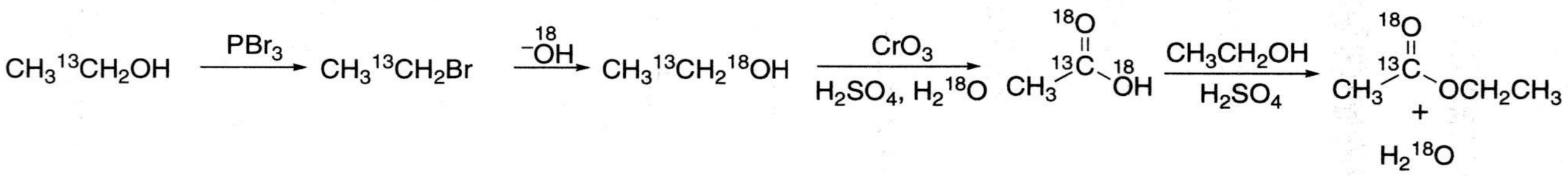

22.79

a.

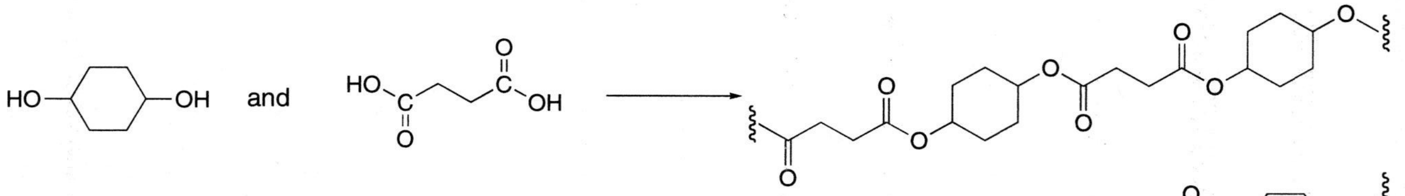

b.

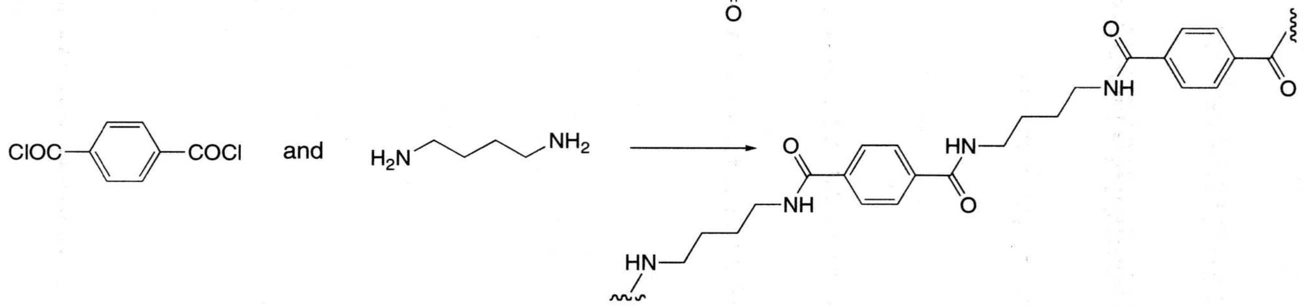

22.80

a.

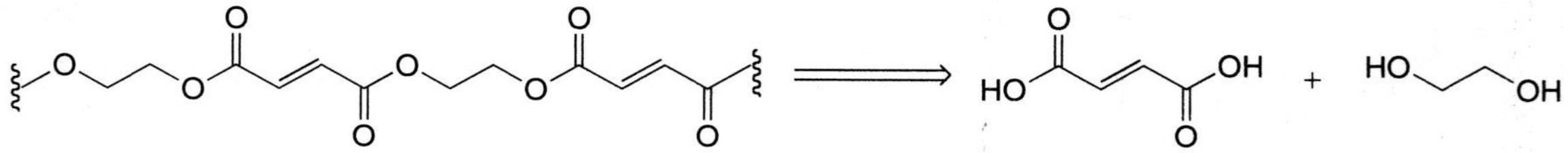

b.

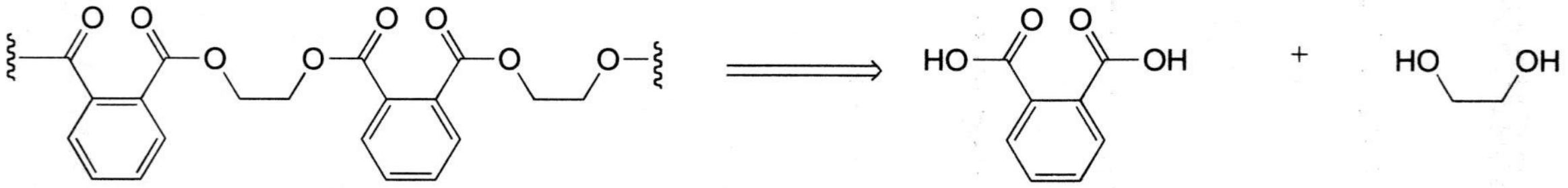

22.81

a. Docetaxel has fewer C's and one more OH group than taxol. This makes docetaxel more water soluble than taxol.

b.

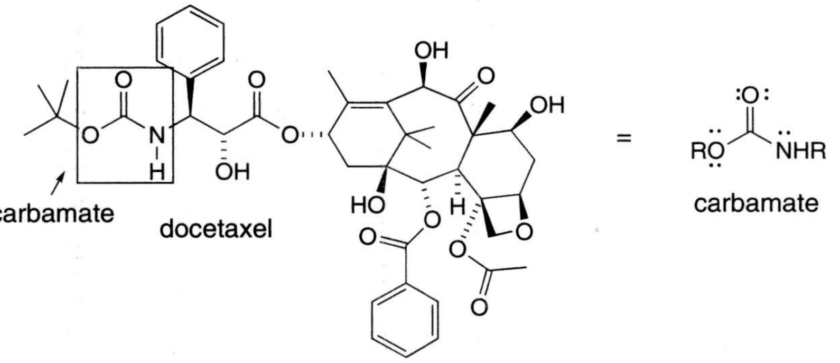

carbamate   docetaxel

=

carbamate

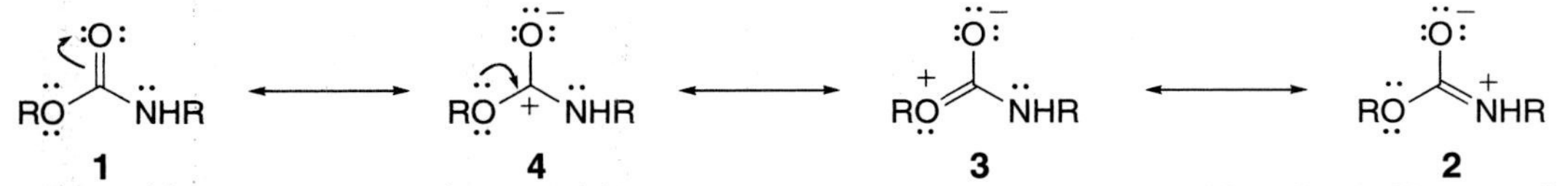

1
most stable

4
least stable

3

2
More basic N atom allows N to donate electron density more than O, so this structure contributes more than **3** to the hybrid.

Increasing stability: 4 < 3 < 2 < 1

c.

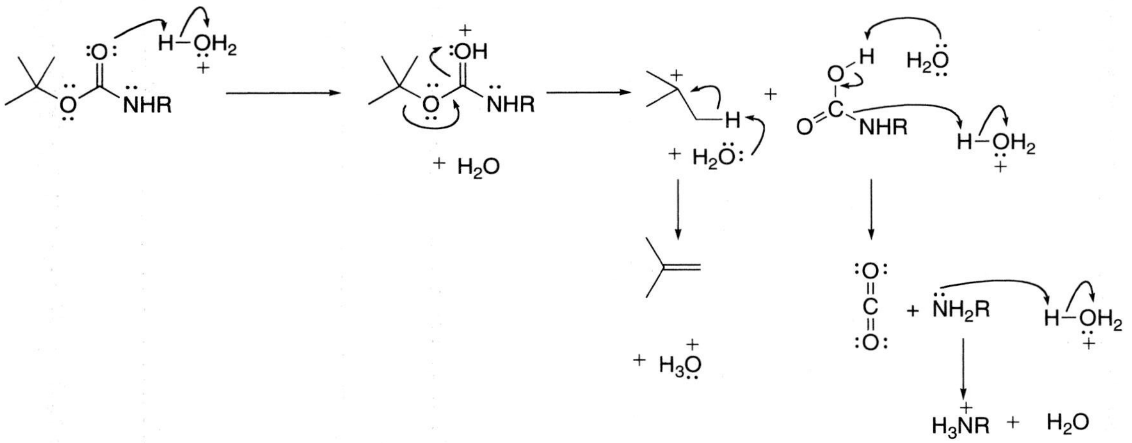

d.

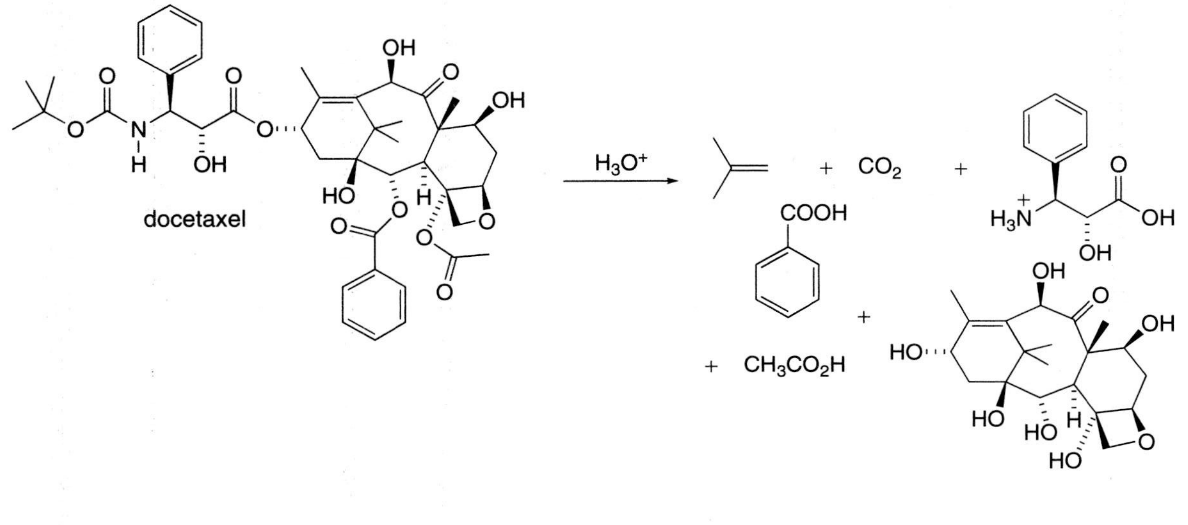

docetaxel

H_3O^+

$+ CO_2 +$

COOH

$+ CH_3CO_2H$

22.82

a.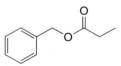

(drawn structures)

CH₃COOCH₃ and CH₃CH₂COOH

contains a broad, strong OH
absorption at 3500–2500 cm⁻¹

c. (drawn structure with N(CH₃)₂) and (drawn structure with NH₂)

C=O at < 1700 cm⁻¹
due to the stabilized amide

2 NH absorptions at 3200–3400 cm⁻¹
C=O absorption higher wavenumber

b. (drawn structure with CI) and (drawn ketone with CI)

Acid chloride CO absorbs at
much higher wavenumber.

ketone

d. O=⟨cyclohexanone⟩–OH and (drawn lactone with CH₃)

OH absorption at
3500–3200 cm⁻¹ + C=O

only C=O

22.83

a. $C_6H_5COOCH_2CH_3$ $CH_3CH_2COOCH_2CH_3$ (drawn lactone structure)

→ Increasing wavenumber

b.

most resonance stabilized		least resonance stabilized
CH_3CONH_2	CH_3COOCH_3	CH_3COCl

→ Increasing wavenumber

22.84

a. $C_6H_{12}O_2$ → one degree of unsaturation
IR: 1738 cm⁻¹ → C=O
NMR: 1.12 (triplet, 3 H), 1.23 (doublet, 6 H),
2.28 (quartet, 2 H), 5.00 (septet, 1 H) ppm

(drawn structure)

b.
C_4H_7N
IR: 2250 cm⁻¹ → triple bond
NMR: 1.08 (triplet, 3 H), 1.70 (multiplet, 2 H),
2.34 (triplet, 2 H) ppm

$CH_3CH_2CH_2C{\equiv}N$

c. C_8H_9NO
IR: 3328 (NH), 1639 (conjugated amide C=O) cm⁻¹
NMR: 2.95 (singlet, 3 H), 6.95 (singlet, 1 H),
7.3–7.7 (multiplet, 5 H) ppm

(drawn structure with N–CH₃)

d. C_4H_7ClO → one degree of unsaturation
IR: 1802 cm⁻¹ → C=O (high wavenumber, RCOCl)
NMR: 0.95 (triplet, 3 H), 1.07 (multiplet, 2 H),
2.90 (triplet, 2 H) ppm

(drawn structure with CI)

e. $C_5H_{10}O_2$ → one degree of unsaturation
IR: 1750 cm⁻¹ → C=O
NMR: 1.20 (doublet, 6 H), 2.00 (singlet, 3 H),
4.95 (septet, 1 H) ppm

(drawn structure)

f. $C_{10}H_{12}O_2$ → five degrees of unsaturation
IR: 1740 cm⁻¹ → C=O
NMR: 1.2 (triplet, 3 H), 2.4 (quartet, 2 H),
5.1 (singlet, 2 H), 7.1–7.5 (multiplet, 5 H) ppm

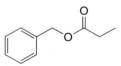

g. $C_8H_{14}O_3$ → two degrees of unsaturation
IR: 1810, 1770 cm⁻¹ → 2 absorptions due to
C=O (anhydride)
NMR: 1.25 (doublet, 12 H), 2.65 (septet, 2 H) ppm

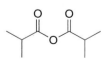

22.85

A. Molecular formula $C_{10}H_{12}O_2$ → five degrees of unsaturation
IR absorption at 1718 cm^{-1} → C=O
NMR data (ppm):

triplet at 1.4 (CH$_3$ adjacent to 2 H's)
singlet at 2.4 (CH$_3$)
quartet at 4.4 (CH$_2$ adjacent to CH$_3$)
doublet at 7.2 (2 H's on benzene ring)
doublet at 7.9 (2 H's on benzene ring)

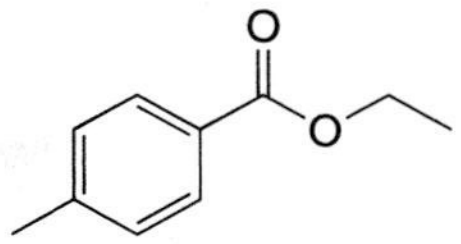

B. IR absorption at 1740 cm^{-1} → C=O
NMR data (ppm):

singlet at 2.0 (CH$_3$)
triplet at 2.9 (CH$_2$ adjacent to CH$_2$)
triplet at 4.4 (CH$_2$ adjacent to CH$_2$)
multiplet at 7.3 (5 H's, monosubstituted benzene)

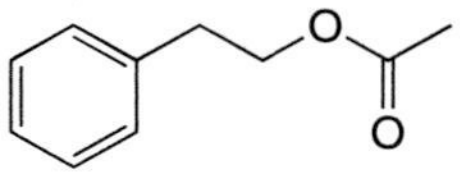

22.86

Molecular formula $C_{10}H_{13}NO_2$ → five degrees of unsaturation
IR absorptions at 3300 (NH) and 1680 (C=O, amide or conjugated) cm^{-1}
NMR data (ppm):

triplet at 1.4 (CH$_3$ adjacent to CH$_2$)
singlet at 2.2 (CH$_3$C=O)
quartet at 3.9 (CH$_2$ adjacent to CH$_3$)
doublet at 6.8 (2 H's on benzene ring)
singlet at 7.2 (NH)
doublet at 7.4 (2 H's on benzene ring)

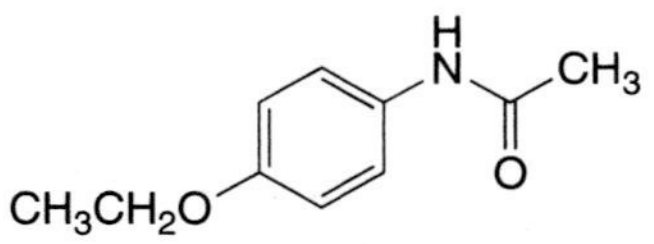

phenacetin

22.87

Molecular formula $C_{11}H_{15}NO_2$ → five degrees of unsaturation
IR absorption 1699 (C=O, amide or conjugated) cm^{-1}
NMR data (ppm):

triplet at 1.3 (3 H) (CH$_3$ adjacent to CH$_2$)
singlet at 3.0 (6 H) (2 CH$_3$ groups on N)
quartet at 4.3 (2 H) (CH$_2$ adjacent to CH$_3$)
doublet at 6.6 (2 H) (2 H's on benzene ring)
doublet at 7.9 (2 H) (2 H's on benzene ring)

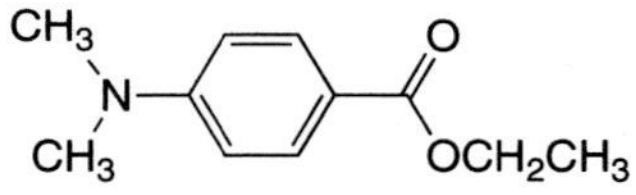

C

22.88

a. Molecular formula $C_6H_{12}O_2$ → one degree of unsaturation
IR absorption at 1743 cm^{-1} → C=O
^{1}H NMR data (ppm):

triplet at 0.9 (3 H) – CH$_3$ adjacent to CH$_2$
multiplet at 1.35 (2 H) – CH$_2$
multiplet at 1.60 (2 H) – CH$_2$
singlet at 2.1 (3 H – from CH$_3$ bonded to C=O)
triplet at 4.1 (2 H) – CH$_2$ adjacent to the electronegative O atom and another CH$_2$

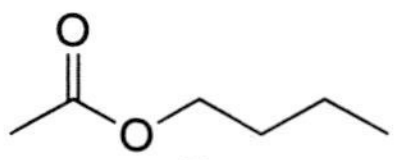

D

b. Molecular formula $C_6H_{12}O_2$ → one degree of unsaturation
IR absorption at 1746 cm^{-1} → C=O
^{1}H NMR data (ppm):

doublet at 0.9 (6 H) – 2 CH$_3$'s adjacent to CH
multiplet at 1.9 (1 H)
singlet at 2.1 (3 H) – CH$_3$ bonded to C=O
doublet at 3.85 (2 H) – CH$_2$ bonded to electronegative O and CH

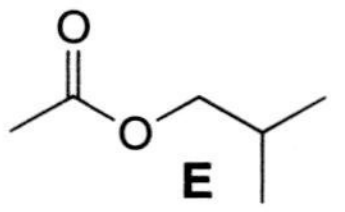

E

22.89

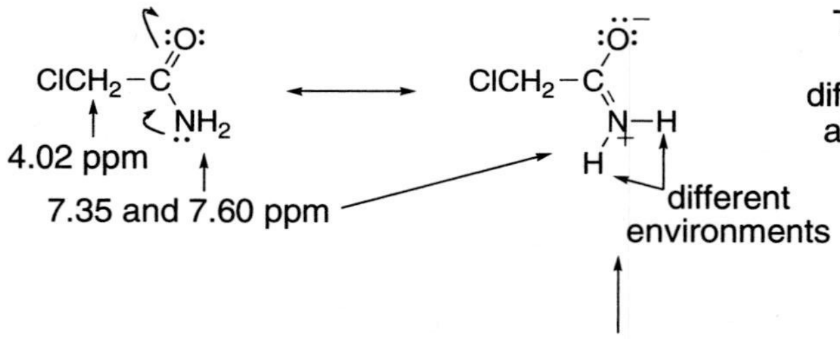

There is restricted rotation around the amide C–N bond. The 2 H's are in different environments (one is cis to an O atom, and one is cis to CH$_2$Cl), so they give different NMR signals.

$ClCH_2-C$ ... 4.02 ppm

7.35 and 7.60 ppm

different environments

This resonance structure gives a significant contribution to the resonance hybrid.

22.90

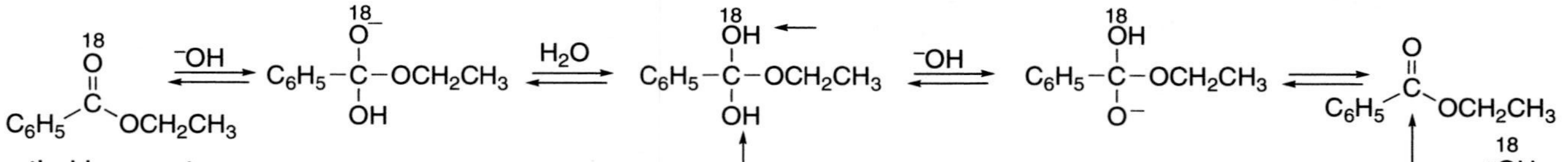

$C_6H_5-C-OCH_2CH_3$ ethyl benzoate

Two OH groups are now equivalent and either can lose H$_2$O to form labeled or unlabeled ethyl benzoate.

Unlabeled starting material was recovered.

22.91

Chapter 23: Substitution Reactions of Carbonyl Compounds at the α Carbon

◆ Kinetic versus thermodynamic enolates (23.4)

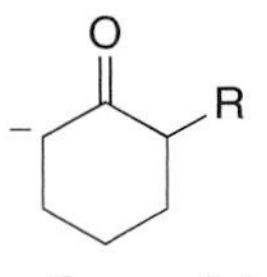

Kinetic enolate
- The less substituted enolate
- Favored by strong base, polar aprotic solvent, low temperature: LDA, THF, −78 °C

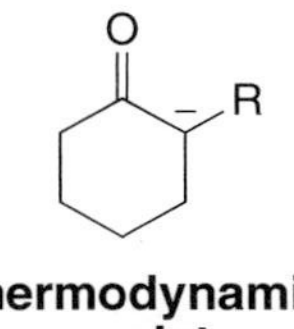

kinetic enolate

Thermodynamic enolate
- The more substituted enolate
- Favored by strong base, protic solvent, higher temperature: $NaOCH_2CH_3$, CH_3CH_2OH, room temperature

thermodynamic enolate

◆ Halogenation at the α carbon

[1] Halogenation in acid (23.7A)

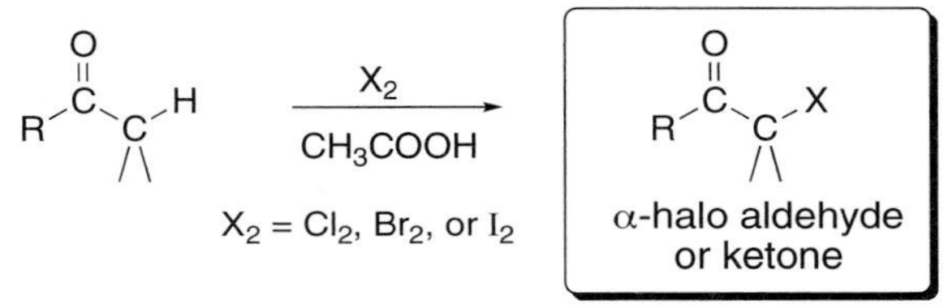

$X_2 = Cl_2, Br_2, or\ I_2$

α-halo aldehyde or ketone

- The reaction occurs via enol intermediates.
- Monosubstitution of X for H occurs on the α carbon.

[2] Halogenation in base (23.7B)

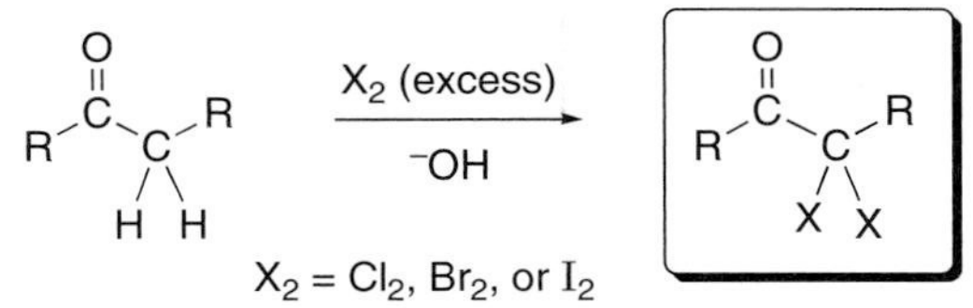

$X_2 = Cl_2, Br_2, or\ I_2$

- The reaction occurs via enolate intermediates.
- Polysubstitution of X for H occurs on the α carbon.

[3] Halogenation of *methyl* ketones in base—The haloform reaction (23.7B)

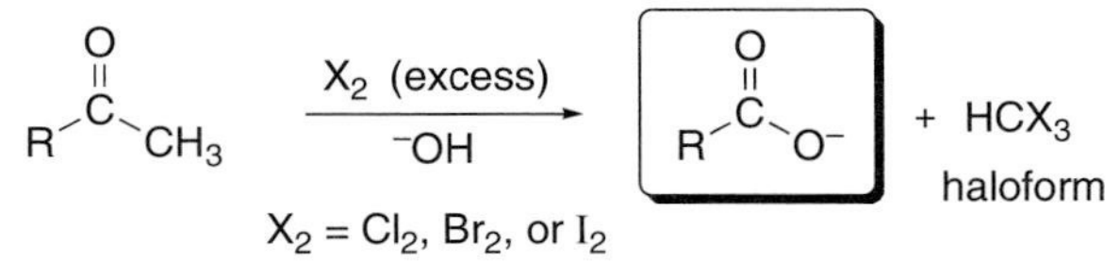

$X_2 = Cl_2, Br_2, or\ I_2$

+ HCX_3

haloform

- The reaction occurs with methyl ketones, and results in cleavage of a carbon–carbon σ bond.

◆ Reactions of α-halo carbonyl compounds (23.7C)

[1] Elimination to form α,β-unsaturated carbonyl compounds

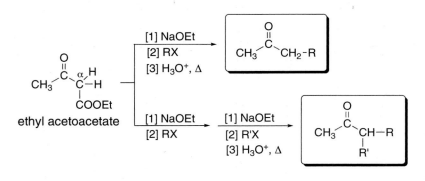

- Elimination of the elements of Br and H forms a new π bond, giving an α,β-unsaturated carbonyl compound.

[2] Nucleophilic substitution

- The reaction follows an S_N2 mechanism, generating an α-substituted carbonyl compound.

◆ Alkylation reactions at the α carbon

[1] Direct alkylation at the α carbon (23.8)

- The reaction forms a new C–C bond to the α carbon.
- LDA is a common base used to form an intermediate enolate.
- The alkylation in Step [2] follows an S_N2 mechanism.

[2] Malonic ester synthesis (23.9)

[1] NaOEt
[2] RX
[3] H_3O^+, Δ

$R-CH_2COOH$
α

H–C–COOEt
   |α
   COOEt

diethyl malonate

[1] NaOEt
[2] RX

[1] NaOEt
[2] R'X
[3] H_3O^+, Δ

R–CHCOOH
   |
   R'
α

- The reaction is used to prepare carboxylic acids with one or two alkyl groups on the α carbon.
- The alkylation in Step [2] follows an S_N2 mechanism.

[3] Acetoacetic ester synthesis (23.10)

[1] NaOEt
[2] RX
[3] H_3O^+, Δ

CH_3–C(=O)–CH_2–R

CH_3–C(=O)–C–H
           |α
           COOEt

ethyl acetoacetate

[1] NaOEt
[2] RX

[1] NaOEt
[2] R'X
[3] H_3O^+, Δ

CH_3–C(=O)–CH–R
              |
              R'

- The reaction is used to prepare ketones with one or two alkyl groups on the α carbon.
- The alkylation in Step [2] follows an S_N2 mechanism.

Chapter 23: Answers to Problems

23.1 • To convert a ketone to its enol tautomer, change the C=O to C–OH, make a new double bond to an α carbon, and remove a proton at the other end of the C=C.

• To convert an enol to its keto form, find the C=C bonded to the OH. Change the C–OH to a C=O, add a proton to the other end of the C=C, and delete the double bond.

[In cases where *E* and *Z* isomers are possible, only one stereoisomer is drawn.]

(Conjugated enols are preferred.)

23.2

2-butanone   C=C has one C bonded to it.   C=C has two C's bonded to it. The more substituted double bond is **more stable.**

23.3 The mechanism has two steps: protonation followed by deprotonation.

23.4

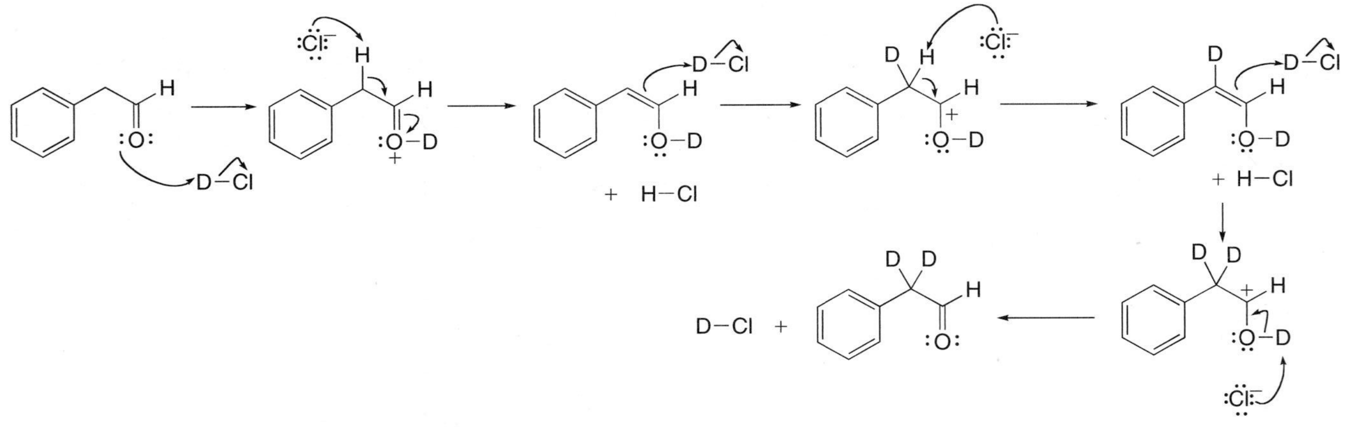

23.5

a.

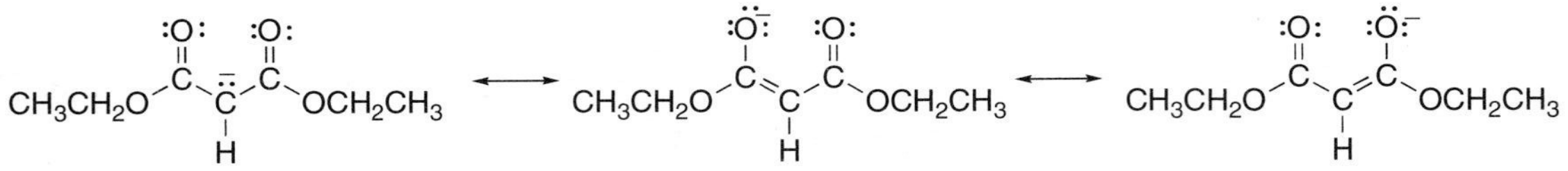

b.

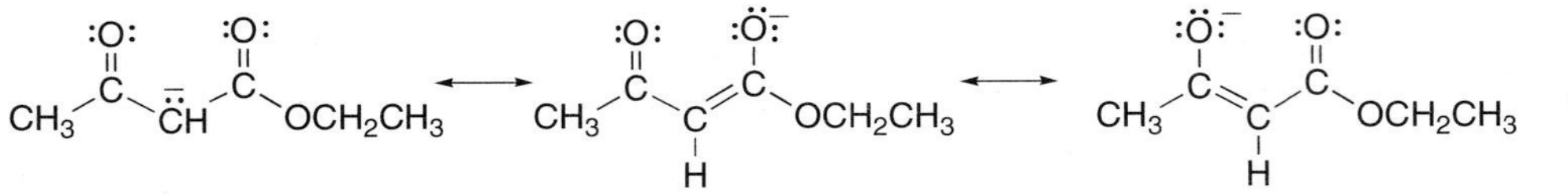

c.

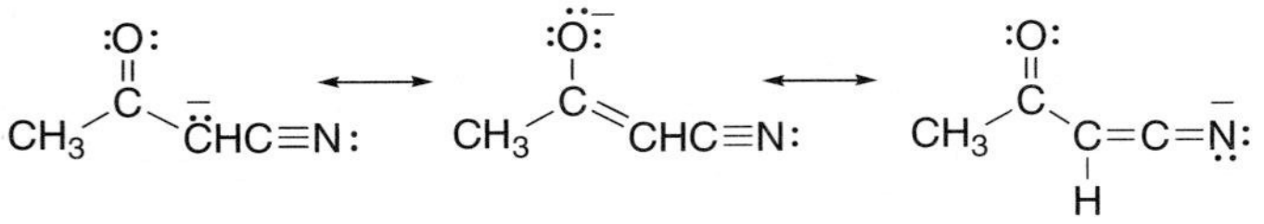

23.6 The indicated H's are α to a C=O or C≡N group, making them more acidic because their removal forms conjugate bases that are resonance stabilized.

a. b. $CH_3CH_2CH_2-CN$ c. 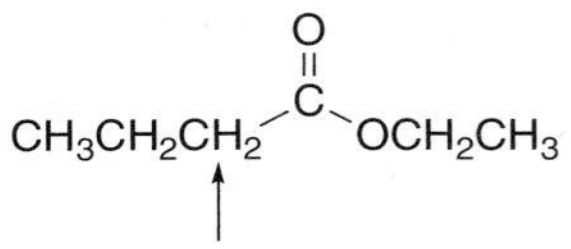 d.

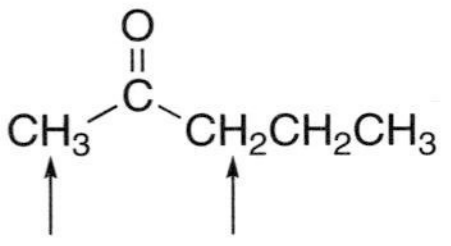

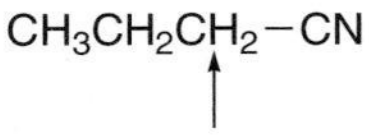

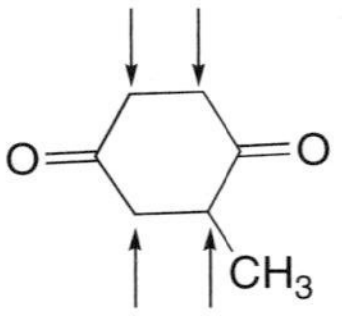

23.7

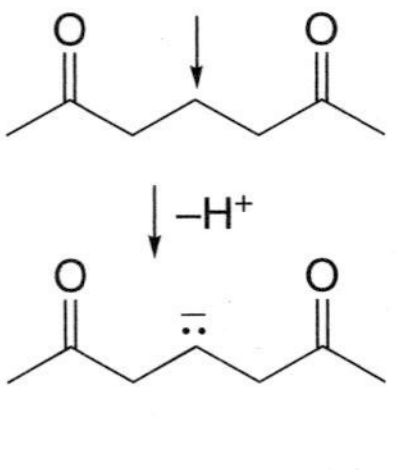

no resonance stabilization
least acidic

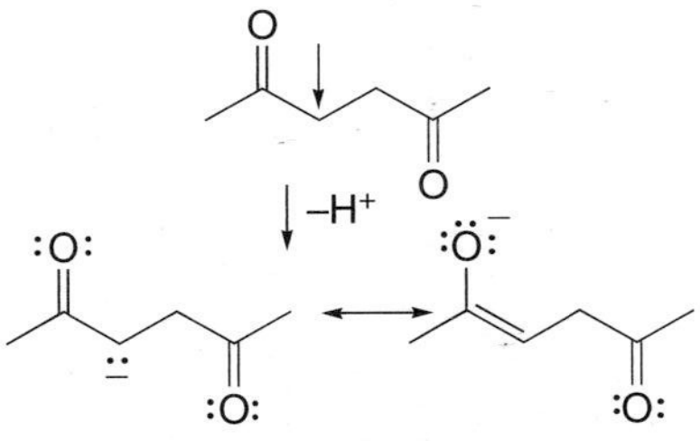

Two resonance structures
stabilize the conjugate base.
intermediate acidity

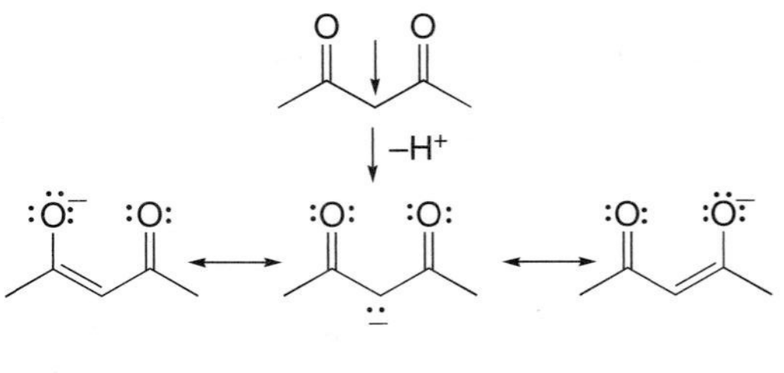

Three resonance structures
stabilize the conjugate base.
most acidic

23.8   In each of the reactions, the LDA pulls off the most acidic proton.

a.

c.

b.

d.

23.9

The CH_2 between the two C=O's contains acidic H's, so CH_3MgBr reacts as a base to remove a proton. Thus, proton transfer (not nucleophilic addition) occurs.

23.10 • LDA, THF forms the kinetic enolate by removing a proton from the less substituted C.
   • Treatment with $NaOCH_3$, CH_3OH forms the thermodynamic enolate by removing a proton from the more substituted C.

a.

c.

b.

23.11
a. This acidic H is removed with base to form an achiral enolate.

(2R)-2-methylcyclohexanone   achiral

Protonation of the planar achiral enolate occurs with equal probability from two sides so a racemic mixture is formed. The racemic mixture is optically inactive.

b.

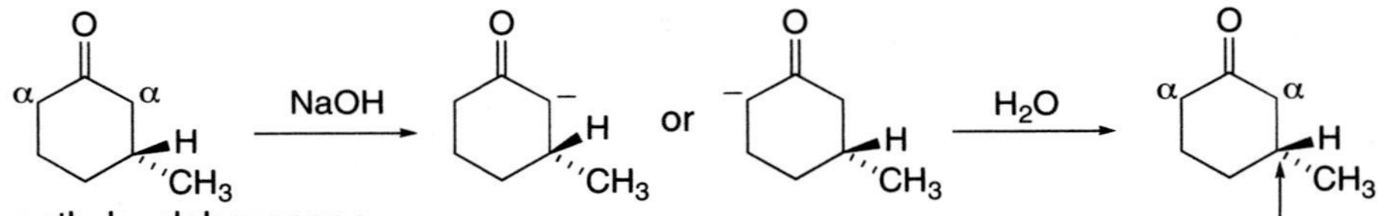

(3R)-3-methylcyclohexanone

NaOH ... or ... H$_2$O

> This stereogenic center is not located at the α carbon, so it is not deprotonated with base. Its configuration is retained in the product, and the product remains optically active.

23.12

a.

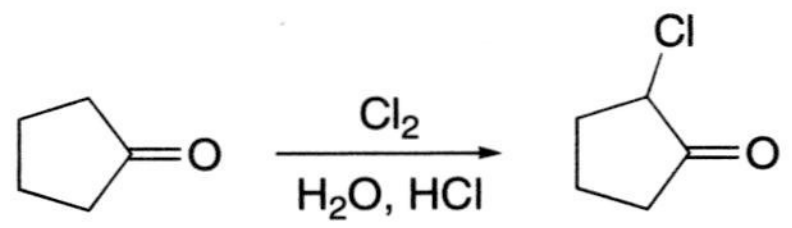

Cl$_2$ / H$_2$O, HCl

c.

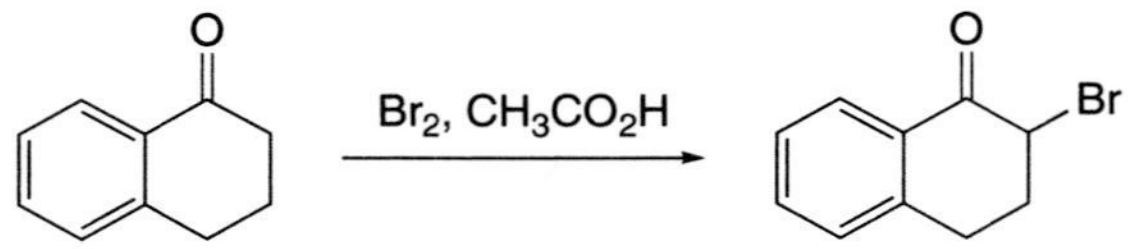

Br$_2$, CH$_3$CO$_2$H

b.

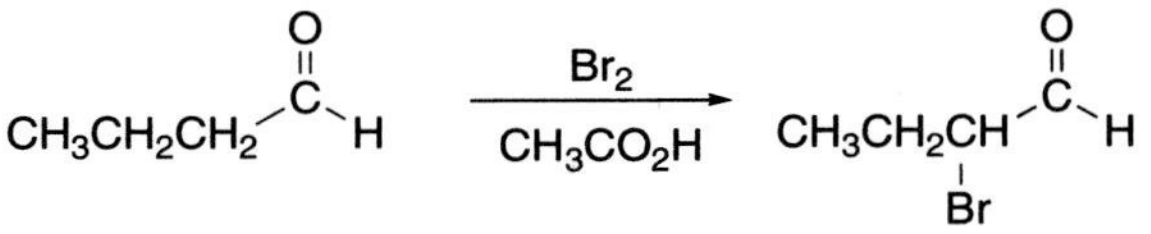

CH$_3$CH$_2$CH$_2$—C(=O)H

Br$_2$ / CH$_3$CO$_2$H

CH$_3$CH$_2$CH(Br)—C(=O)H

23.13

a.

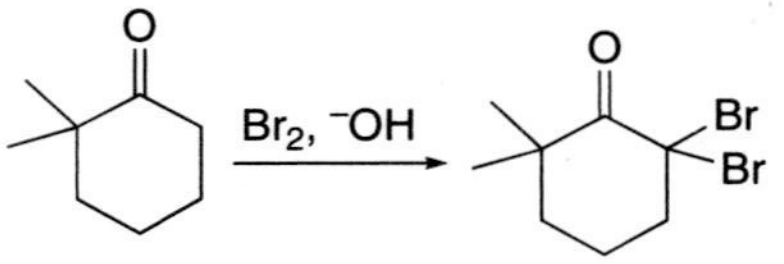

Br$_2$, ⁻OH

b.

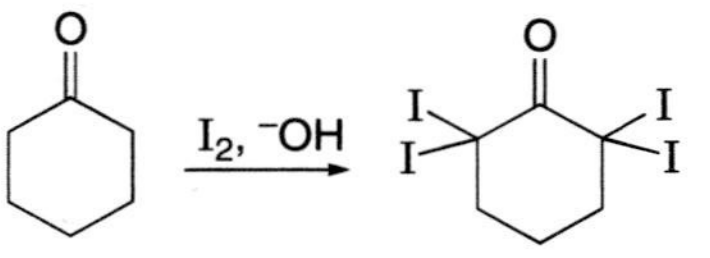

I$_2$, ⁻OH

c.

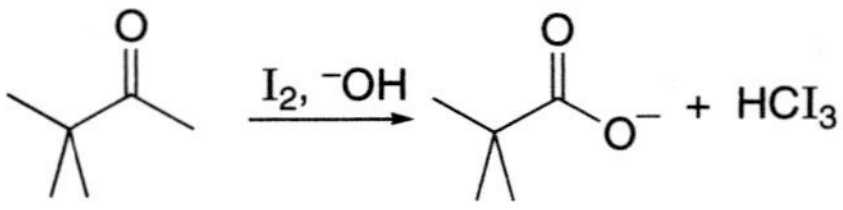

I$_2$, ⁻OH ... + HCI$_3$

23.14

a.

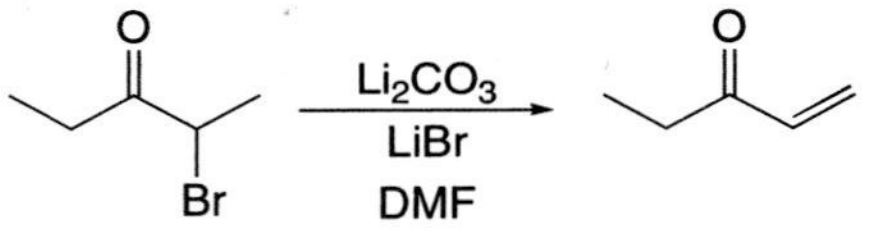

Li$_2$CO$_3$ / LiBr / DMF

c.

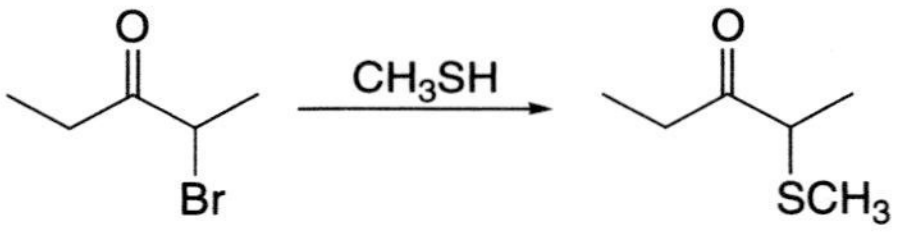

CH$_3$SH

b.

CH$_3$CH$_2$NH$_2$

NHCH$_2$CH$_3$

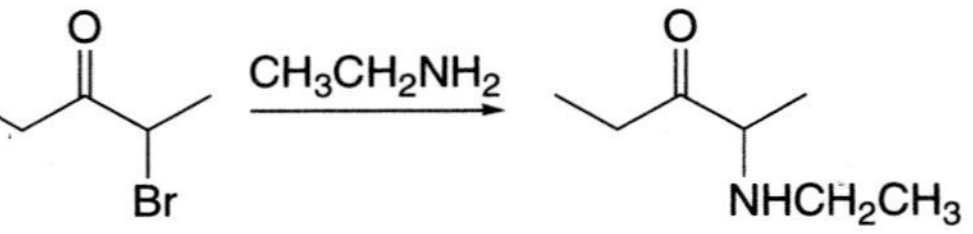

23.15 Bromination takes place on the α carbon to the carbonyl, followed by S$_N$2 reaction with the nitrogen nucleophile.

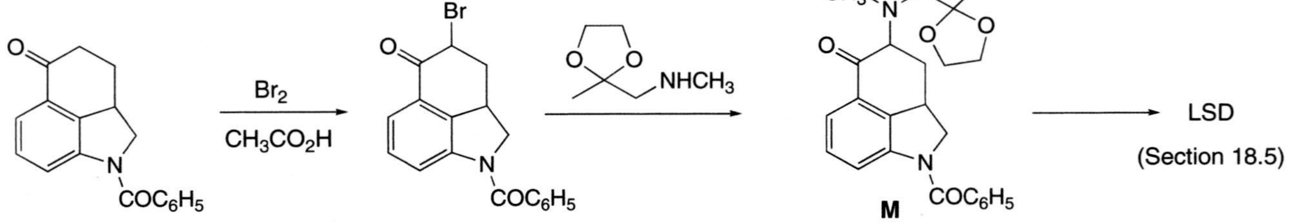

Br$_2$ / CH$_3$CO$_2$H

... NHCH$_3$ ...

LSD

(Section 18.5)

M

23.16

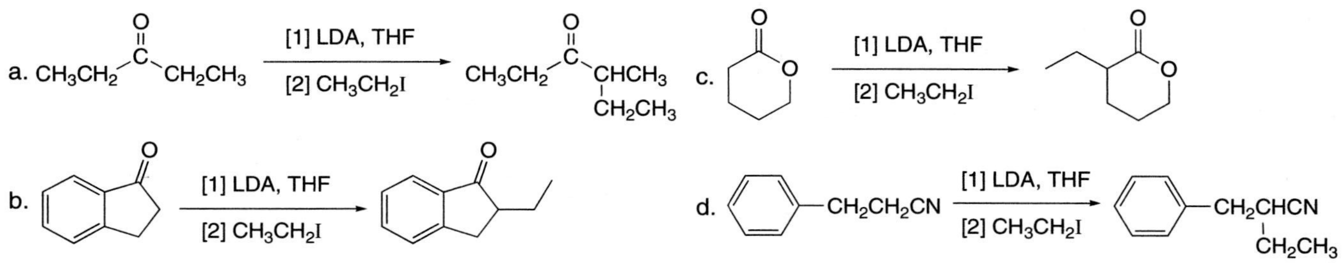

a.

b.

c.

d.

23.17

a.

b.

c.

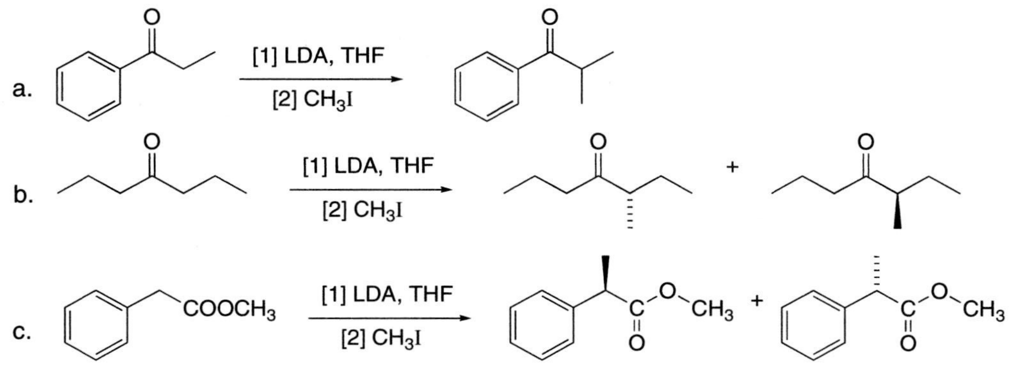

23.18 Three steps are needed: [1] formation of an enolate; [2] alkylation; [3] hydrolysis of the ester.

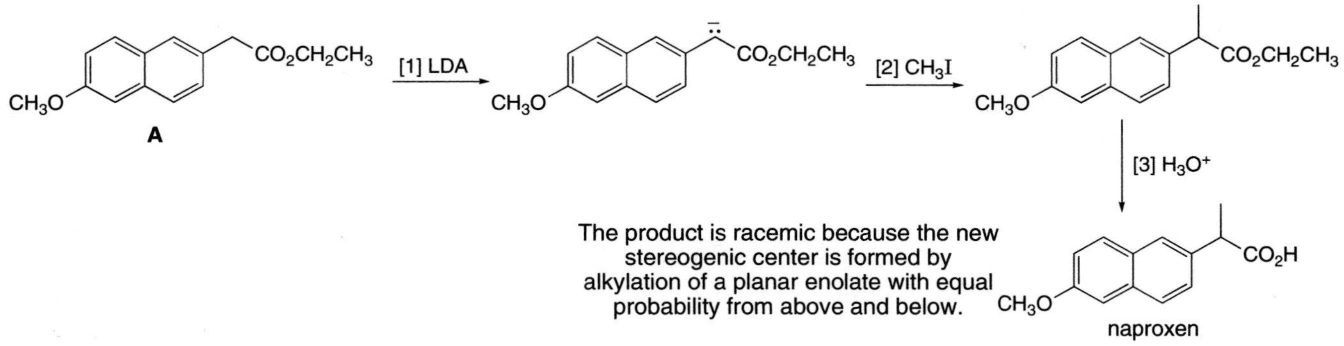

The product is racemic because the new stereogenic center is formed by alkylation of a planar enolate with equal probability from above and below.

naproxen

23.19

a.

b.

c.

(from a.)

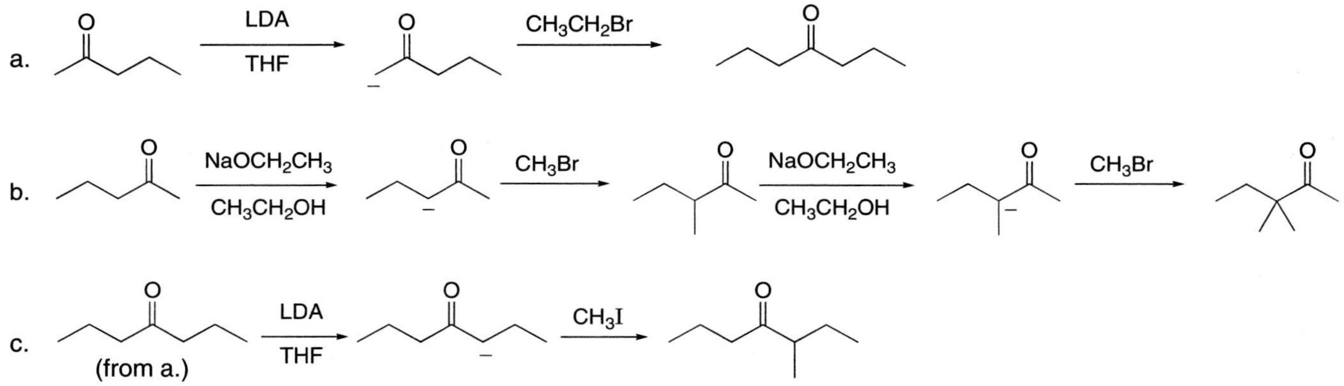

d.

(from b.)

23.20

A   B   C

α-methylene-
γ-butyrolactone

23.21 Decarboxylation occurs only when a carboxy group is bonded to the α C of another carbonyl group.

a.   YES   b.   NO   c.   YES   d.   NO

23.22

a. $CH_2(CO_2Et)_2$ $\xrightarrow[\text{[2]}]{\text{[1] NaOEt}}$ $\xrightarrow[\Delta]{H_3O^+}$ cyclohexyl–CH_2–CH_2COOH

b. $CH_2(CO_2Et)_2$ $\xrightarrow[\text{[2] }CH_3Br]{\text{[1] NaOEt}}$ $\xrightarrow[\text{[2] }CH_3Br]{\text{[1] NaOEt}}$ $\xrightarrow[\Delta]{H_3O^+}$ CH_3–$\overset{CH_3}{\underset{H}{C}}$–$COOH$

23.23

a. $Cl\diagup\diagdown\diagup Cl$ $\longrightarrow$ cyclobutane-COOH

b. $Br\diagup\diagdown O\diagup\diagdown Br$ $\longrightarrow$ tetrahydropyran–COOH

23.24 Locate the α C to the COOH group, and identify all of the alkyl groups bonded to it. These groups are from alkyl halides, and the remainder of the molecule is from diethyl malonate.

a. $(CH_3)_2CHCH_2CH_2CH_2CH_2$ | CH_2COOH

$CH_2(CO_2Et)_2$ $\xrightarrow[\text{[2] }(CH_3)_2CHCH_2CH_2CH_2CH_2Br]{\text{[1] NaOEt}}$ $\xrightarrow[\Delta]{H_3O^+}$ $(CH_3)_2CHCH_2CH_2CH_2CH_2CH_2COOH$

b.

$CH_2(CO_2Et)_2$ $\xrightarrow[\text{[2] }CH_3CH_2CH_2Br]{\text{[1] NaOEt}}$ $\xrightarrow[\text{[2] }CH_3CH_2CH(CH_3)CH_2CH_2Br]{\text{[1] NaOEt}}$ $\xrightarrow[\Delta]{H_3O^+}$

c. $(CH_3CH_2CH_2CH_2)_2$ CHCOOH
α

$CH_2(CO_2Et)_2$ $\xrightarrow[\text{[2] } CH_3CH_2CH_2CH_2Br]{\text{[1] NaOEt}}$ $\xrightarrow[\text{[2] } CH_3CH_2CH_2CH_2Br]{\text{[1] NaOEt}}$ $\xrightarrow[\Delta]{H_3O^+}$ $(CH_3CH_2CH_2CH_2)_2CHCOOH$

23.25 The reaction works best when the alkyl halide is 1° or CH_3X, since this is an S_N2 reaction.

a. $(CH_3)_3C{-}CH_2COOH$

$(CH_3)_3CX$
3° alkyl halide
(too crowded)

b.

COOH

X aryl halide
(leaving group on an
sp^2 hybridized C)

Aryl halides are unreactive
in S_N2 reactions.

c. $(CH_3)_3C{-}COOH$ α

This compound has 3 CH_3 groups on the α
carbon to the COOH. The malonic ester
synthesis can be used to prepare mono- and
disubstituted carboxylic acids only: RCH_2COOH
and $R_2CHCOOH$, but not R_3CCOOH.

23.26

a. $CH_3{-}\underset{O}{\overset{\parallel}{C}}{-}CH_2CO_2Et$ $\xrightarrow[\substack{\text{[2] } CH_3I \\ \text{[3] } H_3O^+,\,\Delta}]{\text{[1] NaOEt}}$ $CH_3{-}\underset{O}{\overset{\parallel}{C}}{-}CH_2CH_3$

b. $CH_3{-}\underset{O}{\overset{\parallel}{C}}{-}CH_2CO_2Et$ $\xrightarrow[\substack{\text{[2] } CH_3CH_2CH_2Br \\ \text{[3] NaOEt} \\ \text{[4] } C_6H_5CH_2I \\ \text{[5] } H_3O^+,\,\Delta}]{\text{[1] NaOEt}}$ $CH_3{-}\underset{O}{\overset{\parallel}{C}}{-}\underset{\underset{CH_2CH_2CH_3}{|}}{CH}{-}CH_2{-}$ [phenyl]

23.27 Locate the α C. All alkyl groups on the α C come from alkyl halides, and the remainder of the molecule comes from ethyl acetoacetate.

a. $CH_3{-}\underset{O}{\overset{\parallel}{C}}{-}CH_2\boxed{CH_2CH_3}$
α

$CH_3{-}\underset{O}{\overset{\parallel}{C}}{-}CH_2{-}COOEt$ $\xrightarrow[\text{[2] } CH_3CH_2Br]{\text{[1] NaOEt}}$ $\xrightarrow[\Delta]{H_3O^+}$ $CH_3{-}\underset{O}{\overset{\parallel}{C}}{-}CH_2CH_2CH_3$

b. $CH_3{-}\underset{O}{\overset{\parallel}{C}}{-}CH\boxed{(CH_2CH_3)_2}$
α

$CH_3{-}\underset{O}{\overset{\parallel}{C}}{-}CH_2{-}COOEt$ $\xrightarrow[\text{[2] } CH_3CH_2Br]{\text{[1] NaOEt}}$ $\xrightarrow[\text{[2] } CH_3CH_2Br]{\text{[1] NaOEt}}$ $\xrightarrow[\Delta]{H_3O^+}$ $CH_3{-}\underset{O}{\overset{\parallel}{C}}{-}CH(CH_2CH_3)_2$

c. $CH_3{-}\underset{O}{\overset{\parallel}{C}}{-}CH_2{-}COOEt$ $\xrightarrow[\text{[2] } CH_3CH_2Br]{\text{[1] NaOEt}}$ $\xrightarrow[\text{[2] } CH_3(CH_2)_3Br]{\text{[1] NaOEt}}$ $\xrightarrow[\Delta]{H_3O^+}$

23.28

$CH_3{-}\underset{O}{\overset{\parallel}{C}}{-}CH_2CO_2Et$ + $Br{-}\diagdown{-}Br$ $\xrightarrow[\text{(2 equiv)}]{\text{NaOEt}}$ $CH_3{-}\underset{O}{\overset{\parallel}{C}}{-}\underset{\mathbf{X}}{\overset{CO_2Et}{\triangleright\!\!\triangleleft}}$

23.29

a.

b.

23.30 Use the directions from Answer 23.1 to draw the enol tautomer(s). In cases where *E* and *Z* isomers can form, only one isomer is drawn.

a.

b.

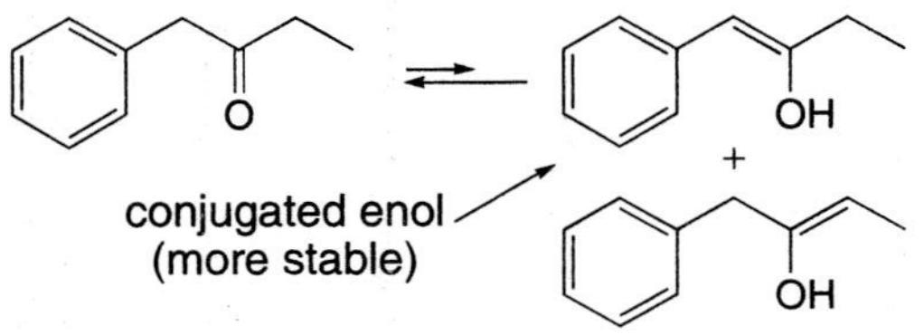

conjugated enol
(more stable)

c.

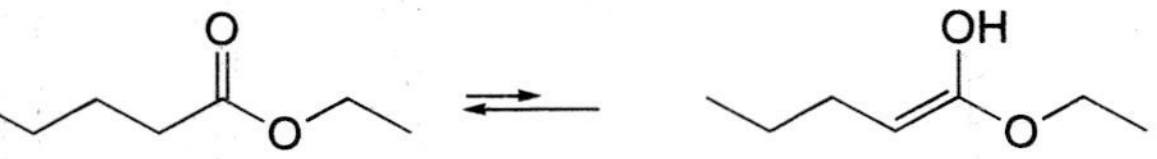

d.

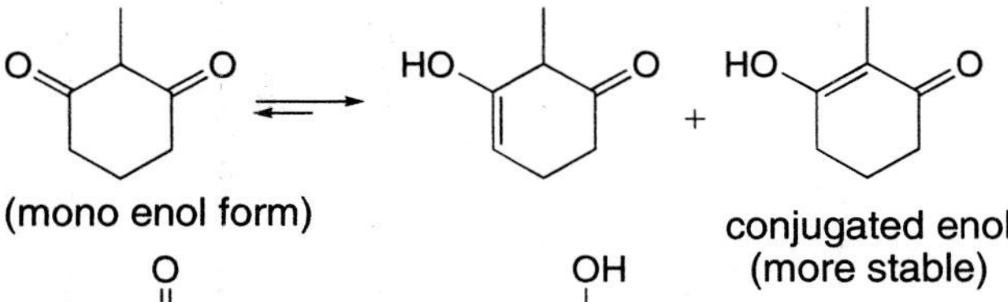

(mono enol form)

conjugated enol
(more stable)

e.

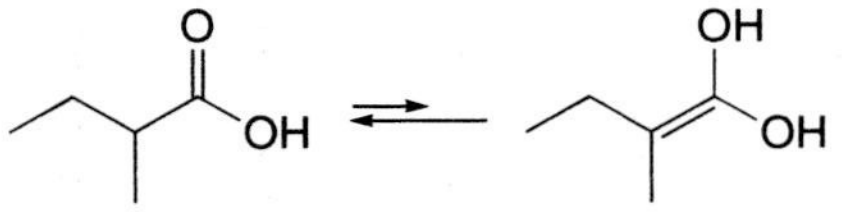

f.

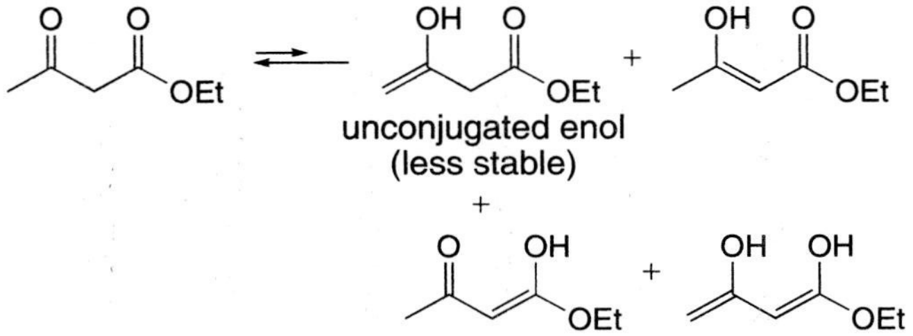

unconjugated enol
(less stable)

23.31

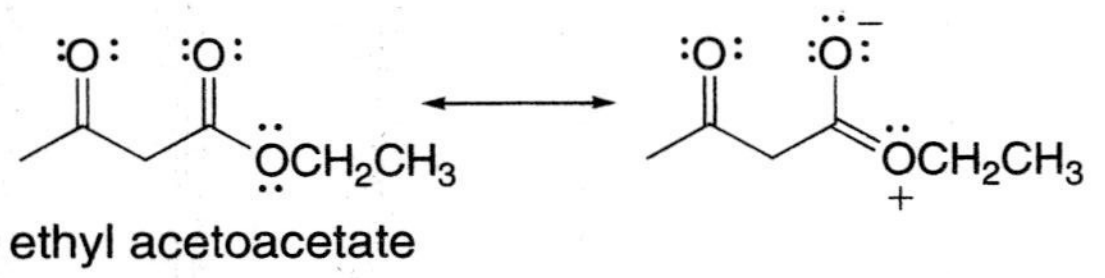

ethyl acetoacetate

The ester C=O is resonance stabilized, and is therefore less available for tautomerization. Since the carbonyl form of the ester group is stabilized by electron delocalization, less enol is present at equilibrium.

23.32

a. 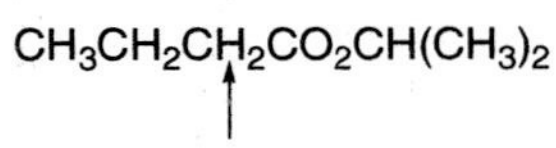 $CH_3CH_2CH_2CO_2CH(CH_3)_2$

b.

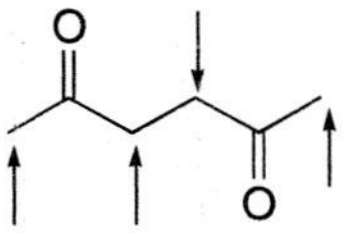

c.

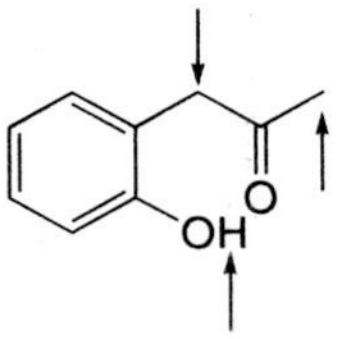

d. CH_3O—⟨ ⟩—CH_2CN

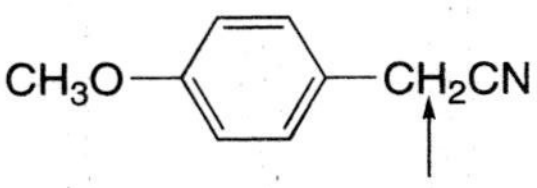

e. NC—⟨ ⟩—$C(=O)CH_2CH_3$

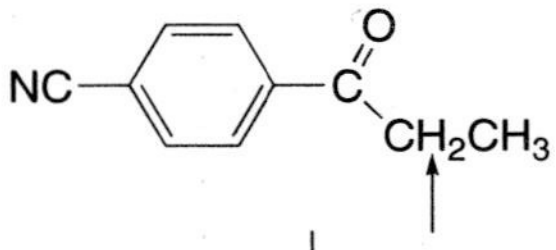

f.

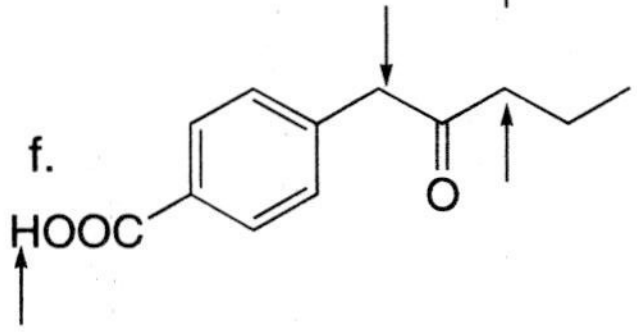

23.33

a. CH_3CH_2—$C(=O)OH$

   H_a H_b H_c

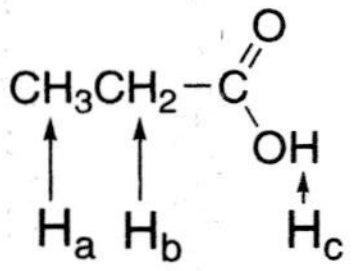

H_a is part of a CH_3 group = **least acidic**.
H_b is bonded to an α C = **intermediate acidity**.
H_c is bonded to O = **most acidic**.

b.

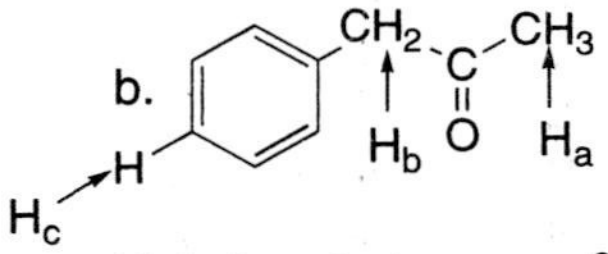

H_c is bonded to an sp^2 hybridized C = **least acidic**.
H_a is bonded to an α C = **intermediate acidity**.
H_b is bonded to an α C, and is adjacent to a benzene ring = **most acidic**.

c.

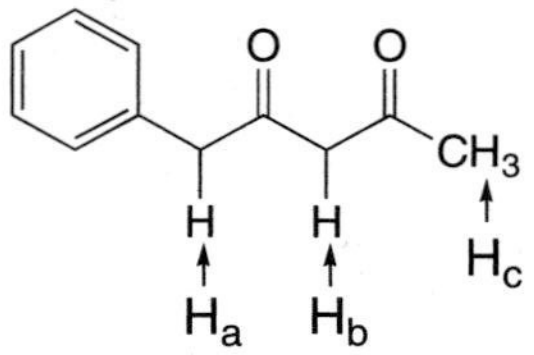

H_c is bonded to an α C = **least acidic**.

H_a is bonded to an α C, and is adjacent to a benzene ring = **intermediate acidity**.

H_b is bonded to an α C between two C=O groups = **most acidic**.

d.

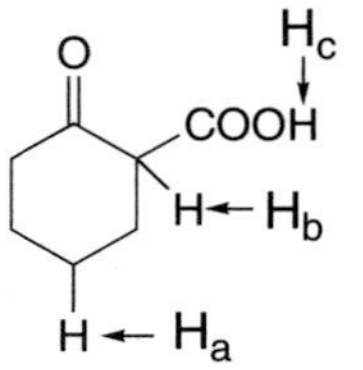

H_a is bonded to an sp^3 hybridized C = **least acidic**.

H_b is bonded to an α C = **intermediate acidity**.

H_c is bonded to O = **most acidic**.

e.

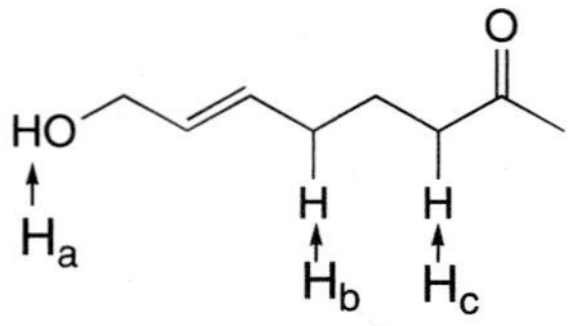

H_b is bonded to an sp^3 hybridized C = **least acidic**.

H_c is bonded to an α C = **intermediate acidity**.

H_a is bonded to O = **most acidic**.

23.34

a.

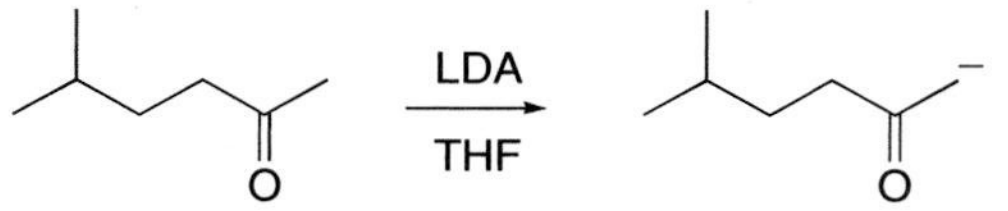

LDA
THF

b.

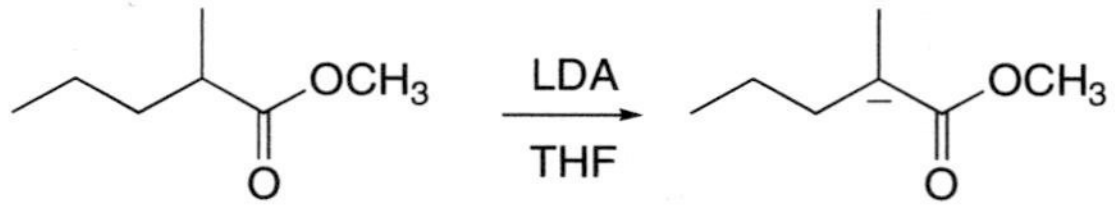

LDA
THF

c.

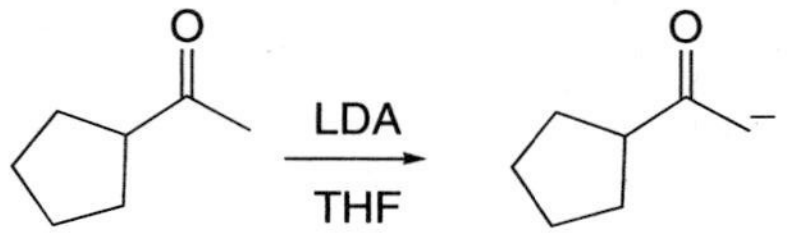

LDA
THF

d.

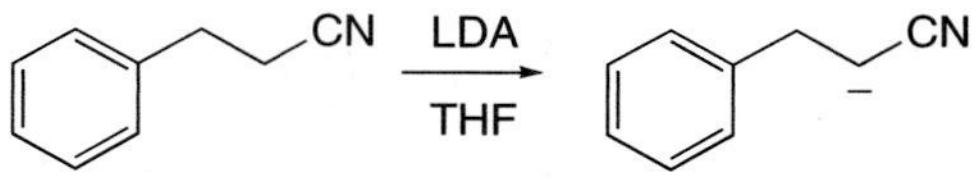

LDA
THF

e.

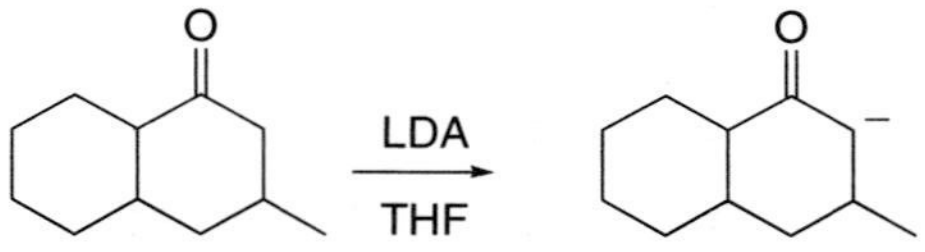

LDA
THF

f.

LDA
THF

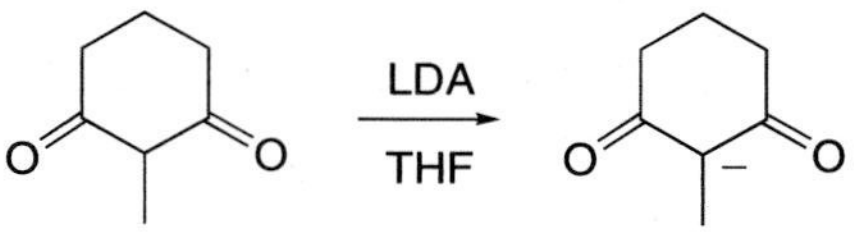

23.35 Enol tautomers have OH groups that give a broad OH absorption at 3600–3200 cm^{-1}, which could be detected readily in the IR.

23.36

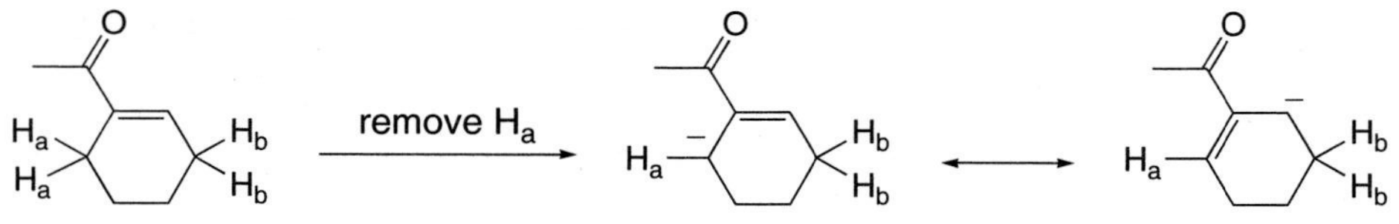

remove H_a

Removal of H_a gives two resonance structures. The negative charge is never on O.

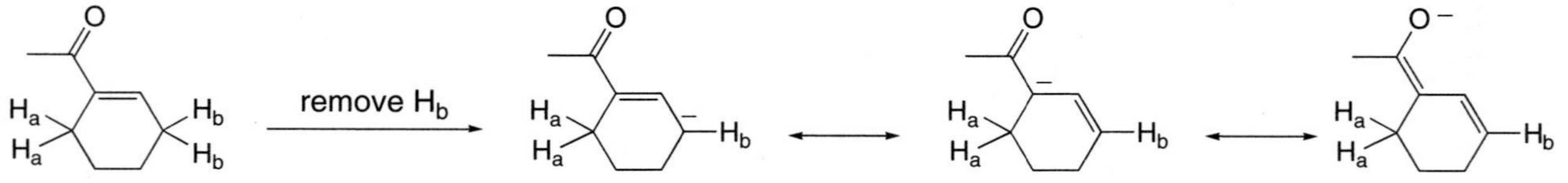

remove H_b

Removal of H_b gives three resonance structures. The negative charge is on O in one resonance structure, making the conjugate base more stable and H_b more acidic (lower pK_a).

23.37

5,5-dimethyl-1,3-cyclohexanedione

5,5-Dimethyl-1,3-cyclohexanedione exists predominantly in its enol form because the C=C of the enol is conjugated with the other C=O of the dicarbonyl compound. Conjugation stabilizes this enol.

2,2-dimethyl-1,3-cyclohexanedione

The enol of 2,2-dimethyl-1,3-cyclohexanedione is not conjugated with the other carbonyl group. In this way it resembles the enol of any other carbonyl compound, and thus it is present in low concentration.

23.38 In the presence of acid, (*R*)-α-methylbutyrophenone enolizes to form an achiral enol. Protonation of the enol from either face forms an equal mixture of two enantiomers, making the solution optically inactive.

(*R*)-α-methylbutyrophenone

achiral
(*E* and *Z* isomers)

In the presence of base, (*R*)-α-methylbutyrophenone is deprotonated to form an achiral enolate, which can then be protonated from either face to form an optically inactive mixture of two enantiomers.

(*R*)-α-methylbutyrophenone

achiral

23.39 Protonation in Step [3] can occur from below (to re-form the *R* isomer) or from above to form the *S* isomer as shown.

R isomer
inactive enantiomer

achiral enol

S isomer
active enantiomer

(+ one more resonance structure)

23.40

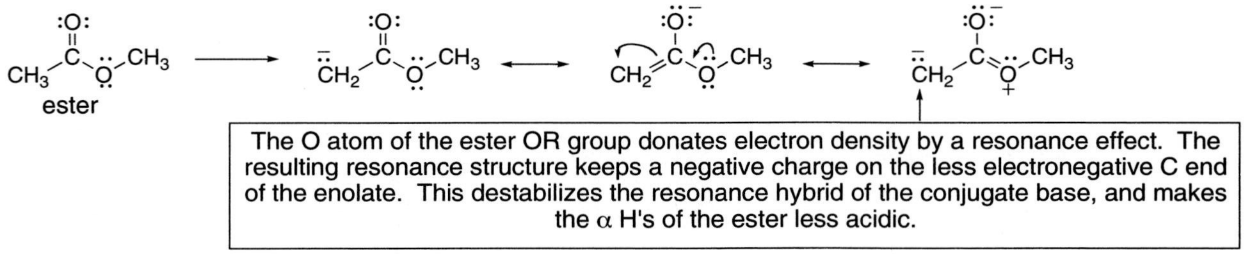
ester
The O atom of the ester OR group donates electron density by a resonance effect. The resulting resonance structure keeps a negative charge on the less electronegative C end of the enolate. This destabilizes the resonance hybrid of the conjugate base, and makes the α H's of the ester less acidic.

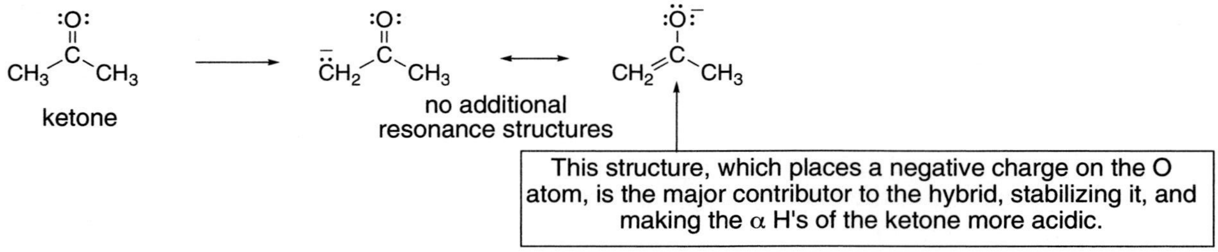
ketone
no additional resonance structures
This structure, which places a negative charge on the O atom, is the major contributor to the hybrid, stabilizing it, and making the α H's of the ketone more acidic.

23.41 LDA reacts with the most acidic proton. If there is any H_2O present, the water would immediately react with the base:

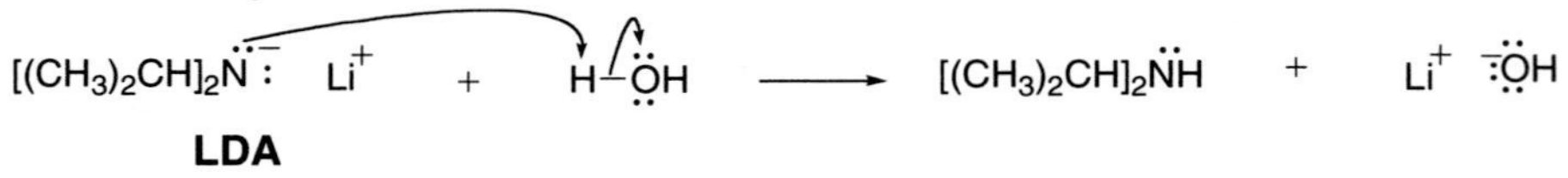
LDA

23.42

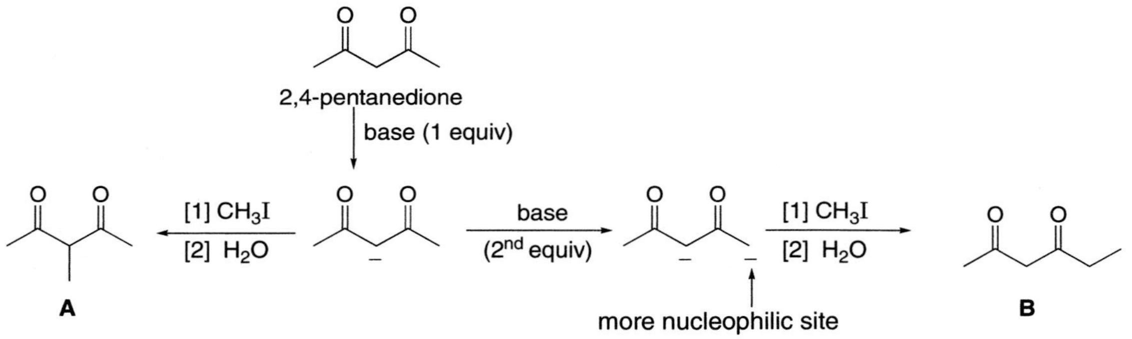
2,4-pentanedione
base (1 equiv)
[1] CH3I
[2] H2O
base
(2nd equiv)
[1] CH3I
[2] H2O
A
more nucleophilic site
B

One equivalent of base removes the most acidic proton between the two C=O's, to form **A** on alkylation with CH_3I.

With a second equivalent of base a dianion is formed. Since the second enolate is less resonance stabilized, it is more nucleophilic and reacts first in an alkylation with CH_3I, forming **B** after protonation with H_2O.

23.43

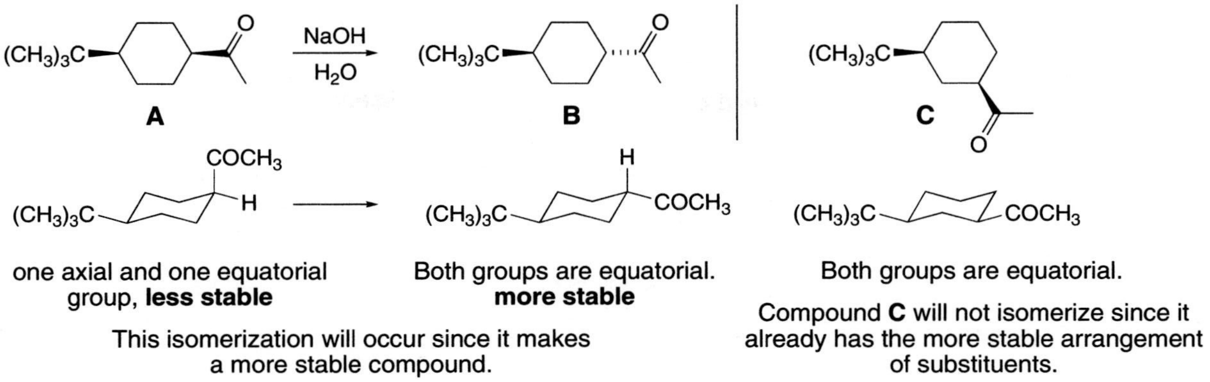

A	**B**	**C**

one axial and one equatorial group, **less stable**

Both groups are equatorial. **more stable**

Both groups are equatorial.

This isomerization will occur since it makes a more stable compound.

Compound **C** will not isomerize since it already has the more stable arrangement of substituents.

Isomerization occurs by way of an intermediate enolate, which can be protonated to either re-form **A**, or give **B**. Since **B** has two large groups equatorial, it is favored at equilibrium.

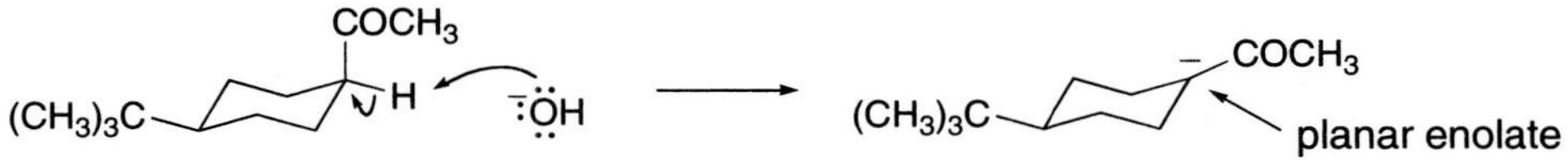

planar enolate

23.44 Protons on the γ carbon of an α,β-unsaturated carbonyl compound are acidic because of resonance.

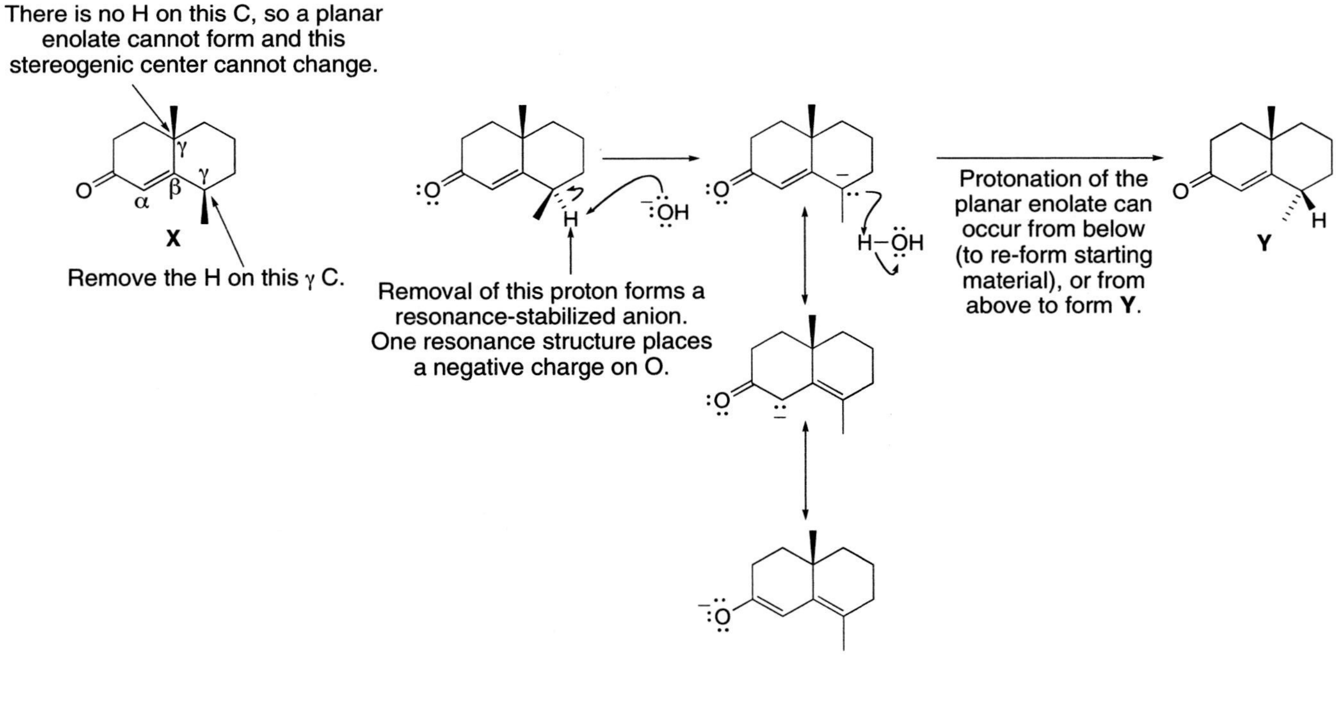

There is no H on this C, so a planar enolate cannot form and this stereogenic center cannot change.

X

Remove the H on this γ C.

Removal of this proton forms a resonance-stabilized anion. One resonance structure places a negative charge on O.

Protonation of the planar enolate can occur from below (to re-form starting material), or from above to form **Y**.

Y

23.45 The mechanism of acid-catalyzed halogenation consists of two parts: **tautomerization** of the carbonyl compound to the enol form, and **reaction of the enol with halogen**.

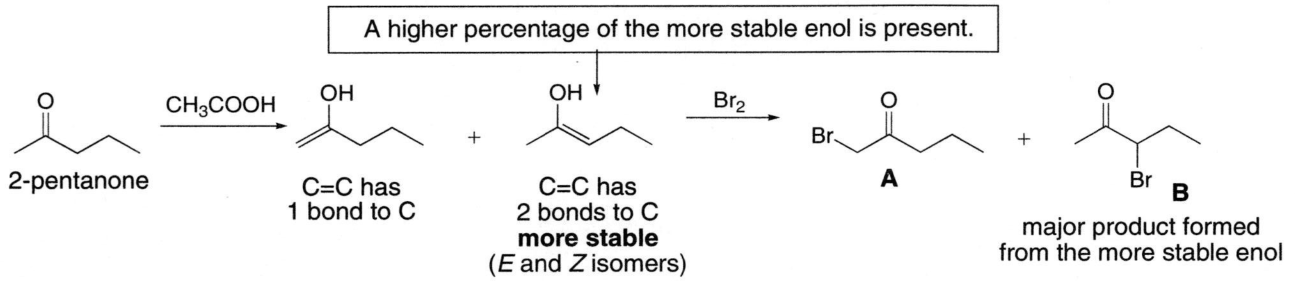
A higher percentage of the more stable enol is present.
CH_3COOH
Br_2
2-pentanone
C=C has
1 bond to C
C=C has
2 bonds to C
more stable
(E and Z isomers)
A
B
major product formed
from the more stable enol

23.46 • The mechanism of acid-catalyzed halogenation [Part (a)] consists of two parts: **tautomerization** of the carbonyl compound to the enol form, and **reaction of the enol with halogen**.

• In the haloform reaction [Part (b)], the three H's of the CH_3 group are successively replaced by X, to form an intermediate that is oxidatively cleaved with base.

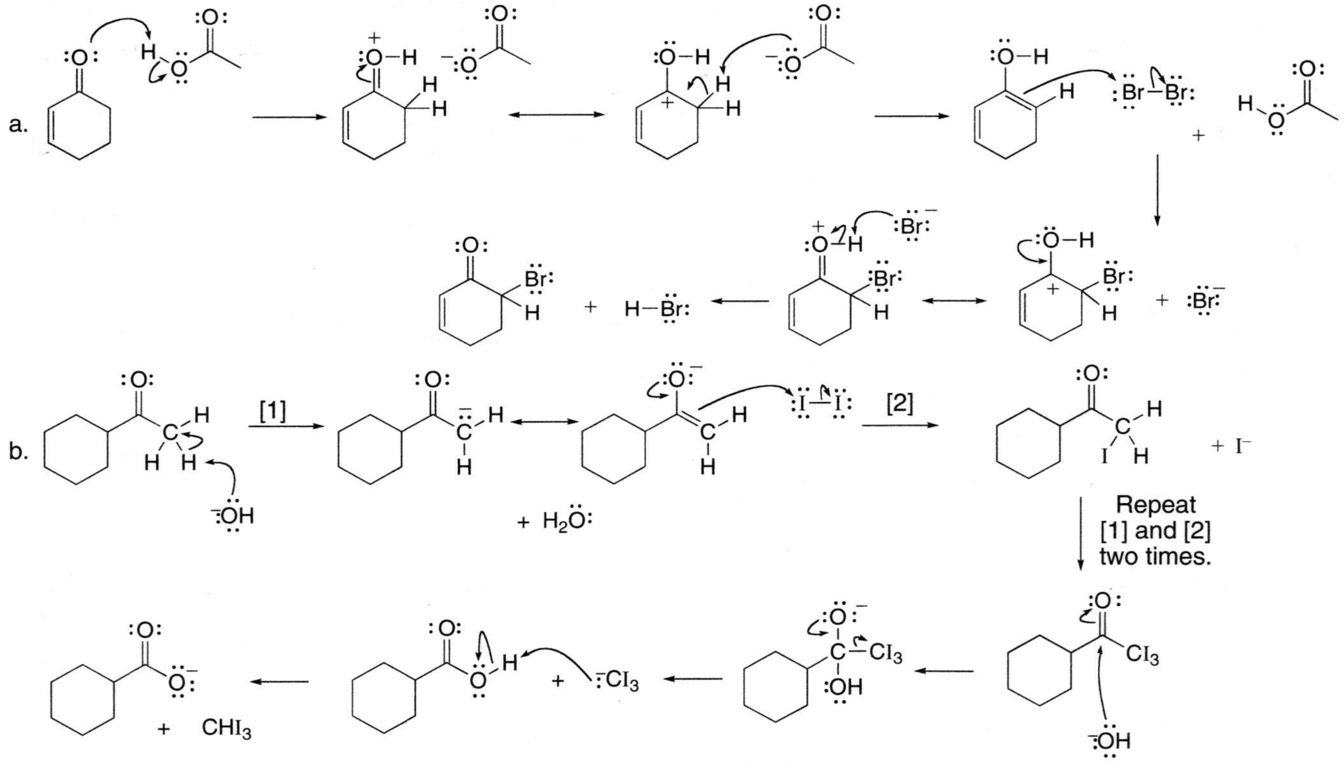
a.
b.
[1]
[2]
Repeat
[1] and [2]
two times.
$+ H_2\ddot{O}$:
$+ CHI_3$

23.47 Use the directions from Answer 23.24.

a. 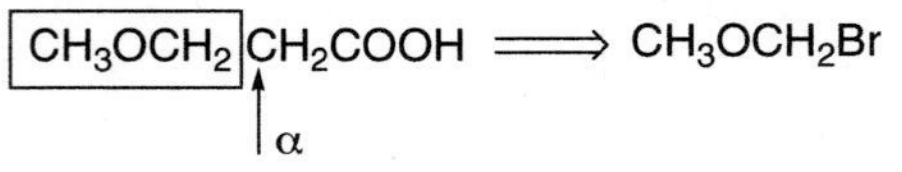
$CH_3OCH_2CH_2COOH$
CH_3OCH_2Br
α

b. 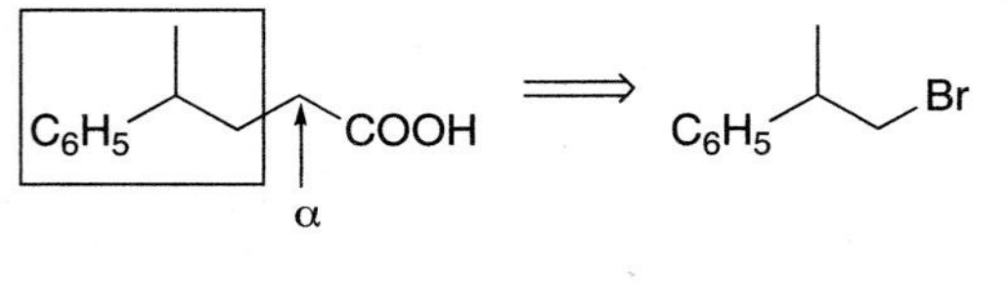
C_6H_5
COOH
α
C_6H_5
Br

c. 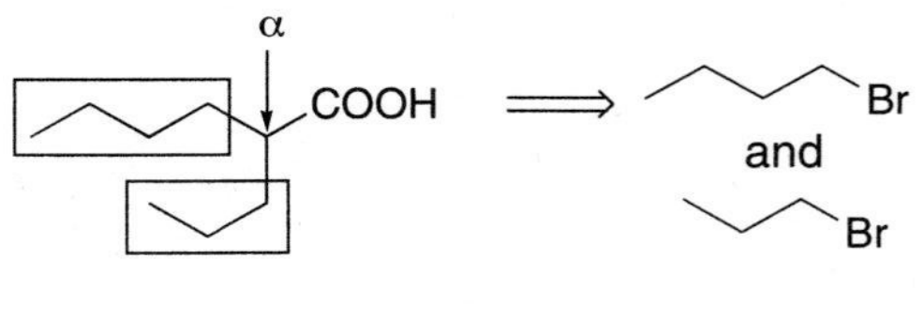
α
COOH
Br
and
Br

23.48

a. $\boxed{CH_3CH_2CH_2CH_2CH_2}\,CH_2COOH$ $CH_2(CO_2Et)_2$ $\xrightarrow[\text{[2] }CH_3CH_2CH_2CH_2CH_2Br]{\text{[1] NaOEt}}$ $\xrightarrow[\Delta]{H_3O^+}$ $CH_3CH_2CH_2CH_2CH_2CH_2COOH$

b.

$CH_2(CO_2Et)_2$ $\xrightarrow[\text{[2] }CH_3CH_2CH_2CH_2CH_2Br]{\text{[1] NaOEt}}$ $\xrightarrow[\text{[2] }CH_3Br]{\text{[1] NaOEt}}$ $\xrightarrow[\Delta]{H_3O^+}$

c.

$CH_2(CO_2Et)_2$ $\xrightarrow[\text{[2] }(CH_3)_2CHCH_2CH_2CH_2CH_2Br]{\text{[1] NaOEt}}$ $\xrightarrow[\text{[2] }CH_3Br]{\text{[1] NaOEt}}$ $\xrightarrow[\Delta]{H_3O^+}$

23.49

$\xrightarrow[\text{[2] }CH_3CH_2CH_2Br]{\text{[1] NaOEt}}$

$\xrightarrow[\text{[2] }CH_3CH_2CH_2Br]{\text{[1] NaOEt}}$

$\xrightarrow[\Delta]{H_3O^+}$

valproic acid

23.50

a. $CH_2(CO_2Et)_2$ $\xrightarrow[\substack{\text{[2] }BrCH_2CH_2CH_2CH_2CH_2Br \\ \text{[3] NaOEt}}]{\text{[1] NaOEt}}$ $\xrightarrow[\Delta]{H_3O^+}$

b.

(from a.) $\xrightarrow[\text{[2] }H_2O]{\text{[1] LiAlH}_4}$

c.

(from a.) $\xrightarrow[H_2SO_4]{CH_3OH}$

$\xrightarrow[\text{[2] }H_2O]{\text{[1] CH}_3\text{MgBr (2 equiv)}}$

d.

(from a.) $\xrightarrow[H_2SO_4]{CH_3CH_2OH}$

$\xrightarrow[\text{[2] }CH_3I]{\text{[1] LDA}}$

23.51

a. epoxide (with CH₃), nucleophilic attack here

$$\text{[1] Na}^+ \ ^-\text{CH(COOEt)}_2 \quad \text{[2] H}_2\text{O}$$

HO
CH₃CHCH₂–CH(COOEt)₂

nucleophilic attack here

c. CH₃COCl

$$\text{[1] Na}^+ \ ^-\text{CH(COOEt)}_2 \quad \text{[2] H}_2\text{O}$$

CH₃COCH(COOEt)₂

b. CH₂=O

$$\text{[1] Na}^+ \ ^-\text{CH(COOEt)}_2 \quad \text{[2] H}_2\text{O}$$

HOCH₂CH(COOEt)₂

d. CH₃CO–O–COCH₃

$$\text{[1] Na}^+ \ ^-\text{CH(COOEt)}_2 \quad \text{[2] H}_2\text{O}$$

CH₃COCH(COOEt)₂

+ CH₃COOH

23.52 Use the directions from Answer 23.27.

a. CH₃COCH₂COOEt

$$\text{[1] NaOEt} \quad \text{[2] Br}\text{-(isohexyl)} \quad \xrightarrow{\text{H}_3\text{O}^+, \ \Delta}$$

b. CH₃COCH₂COOEt

$$\text{[1] NaOEt} \quad \text{[2] CH}_3\text{CH}_2\text{Br} \quad \text{[1] NaOEt} \quad \text{[2] Br-CH}_2\text{-cyclohexyl} \quad \xrightarrow{\text{H}_3\text{O}^+, \ \Delta}$$

c. CH₃COCH₂COOEt

$$\text{[1] NaOEt} \quad \text{[2] Br}\text{-(allyl/butenyl)} \quad \xrightarrow{\text{H}_3\text{O}^+, \ \Delta}$$

d. CH₃COCH₂COOEt

$$\text{[1] NaOEt} \quad \text{[2] Br-CH}_2\text{CH}_2\text{CH(CH}_3)\text{CH}_2\text{-Br} \quad \text{[3] NaOEt} \quad \xrightarrow{\text{H}_3\text{O}^+, \ \Delta}$$

23.53

a. CH₃COCH₂COOEt

$$\text{[1] NaOEt} \quad \text{[2] CH}_3\text{CH}_2\text{Br} \quad \xrightarrow{\text{H}_3\text{O}^+, \ \Delta} \quad \text{CH}_3\text{COCH}_2\text{CH}_2\text{CH}_3$$

b. CH₃COCH₂COOEt

$$\text{[1] NaOEt} \quad \text{[2] CH}_3\text{Br} \quad \text{[1] NaOEt} \quad \text{[2] CH}_3\text{Br} \quad \xrightarrow{\text{H}_3\text{O}^+, \ \Delta} \quad \text{CH}_3\text{COCH(CH}_3)_2$$

c. CH₃COCH(CH₃)₂
(from b.)

$$\xrightarrow{\text{LDA} \ \text{THF}} \quad ^-\text{CH}_2\text{COCH(CH}_3)_2 \quad \xrightarrow{\text{CH}_3\text{I}} \quad \text{CH}_3\text{CH}_2\text{COCH(CH}_3)_2$$

d. CH₃COCH(CH₃)₂
(from b.)

$$\xrightarrow{\text{NaOCH}_3 \ \text{CH}_3\text{OH}} \quad \text{CH}_3\text{CO}\bar{\text{C}}(\text{CH}_3)_2 \quad \xrightarrow{\text{CH}_3\text{I}} \quad \text{CH}_3\text{COC(CH}_3)_3$$

23.54

a.

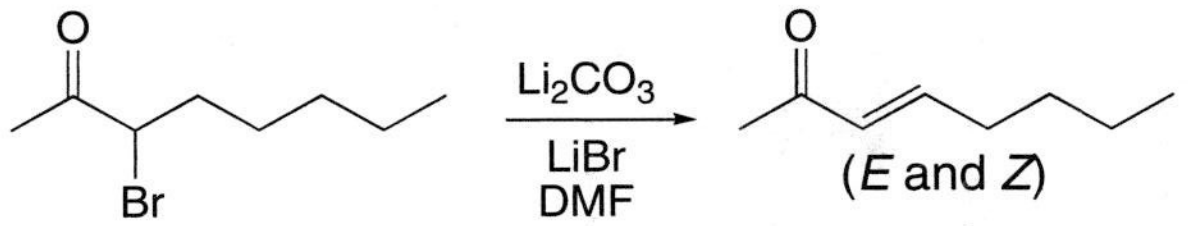

b.

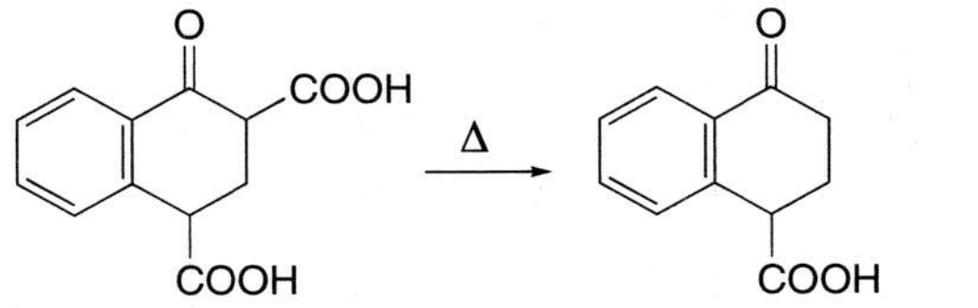

c. 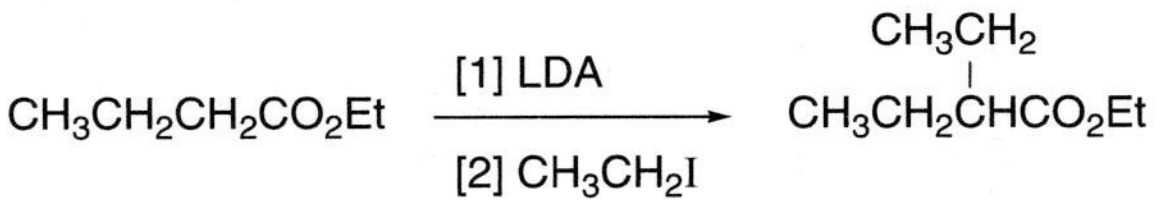 $CH_3CH_2CH_2CO_2Et$ $\xrightarrow[\text{[2] } CH_3CH_2I]{\text{[1] LDA}}$

$$CH_3CH_2\overset{\overset{\displaystyle CH_3CH_2}{|}}{CH}CO_2Et$$

d.

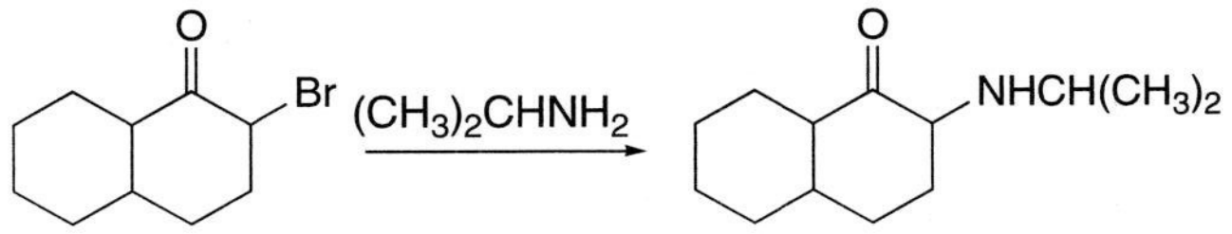

e.

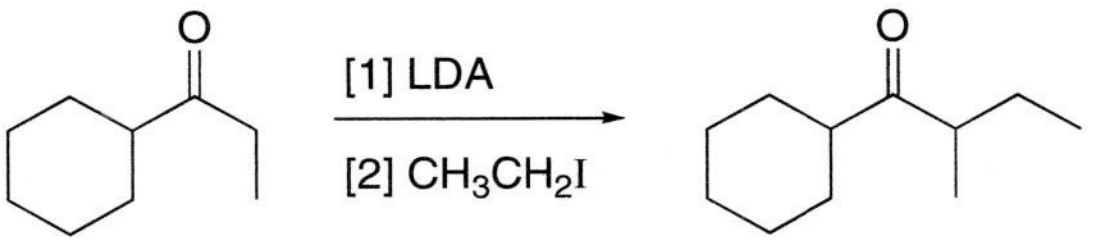

f.

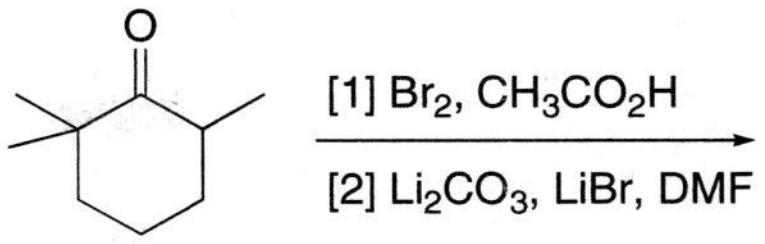

g.

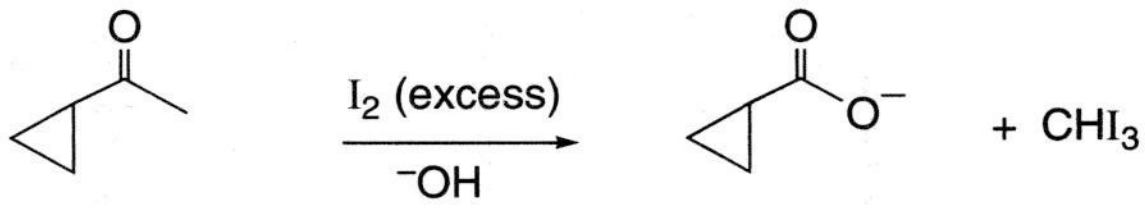

h.

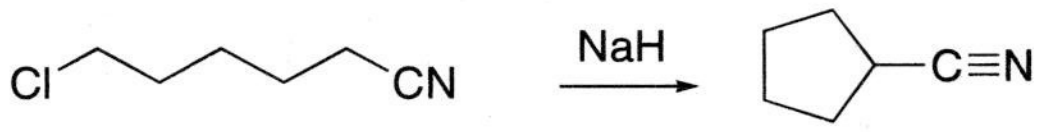

i.

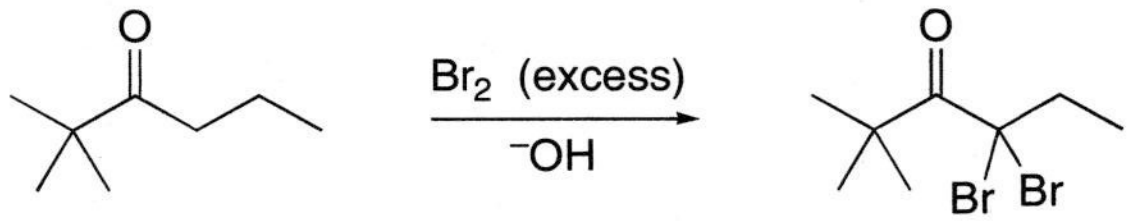

j.

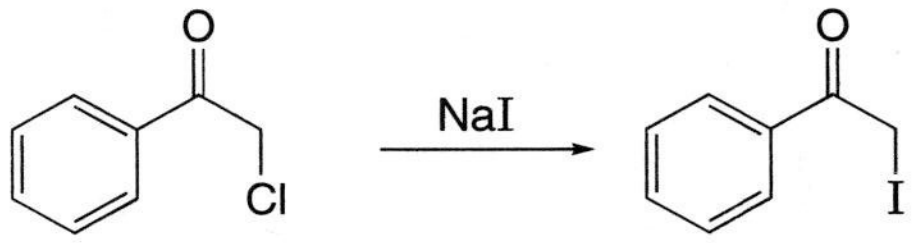

23.55

a.

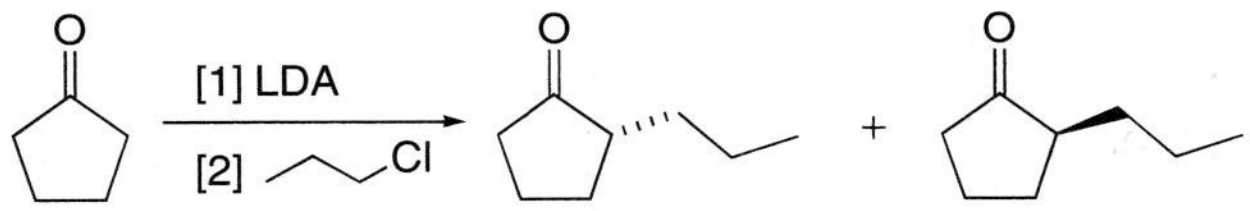

b.

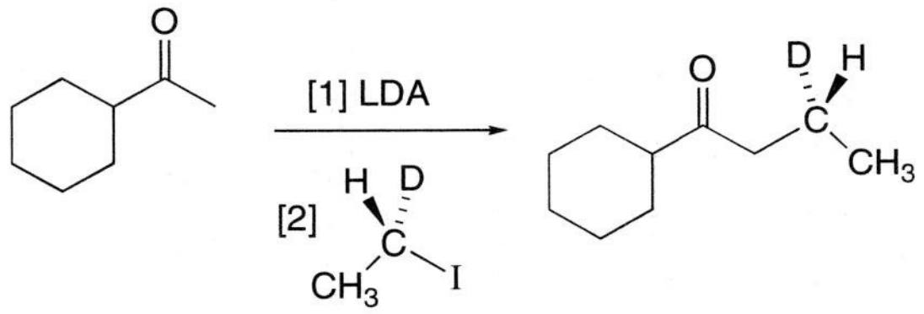

c.

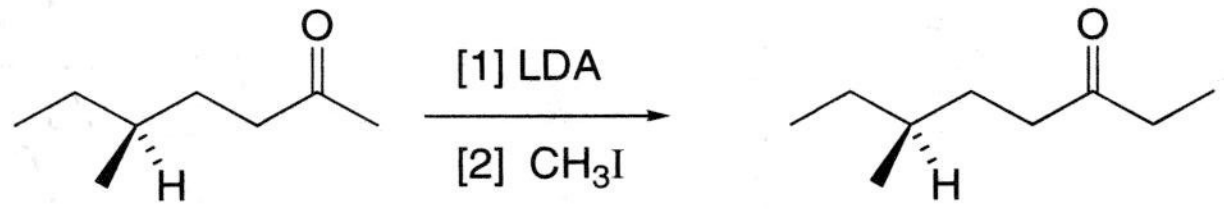

23.56

a.

p-isobutylbenzaldehyde   **A**   **B**   **C**

ibuprofen

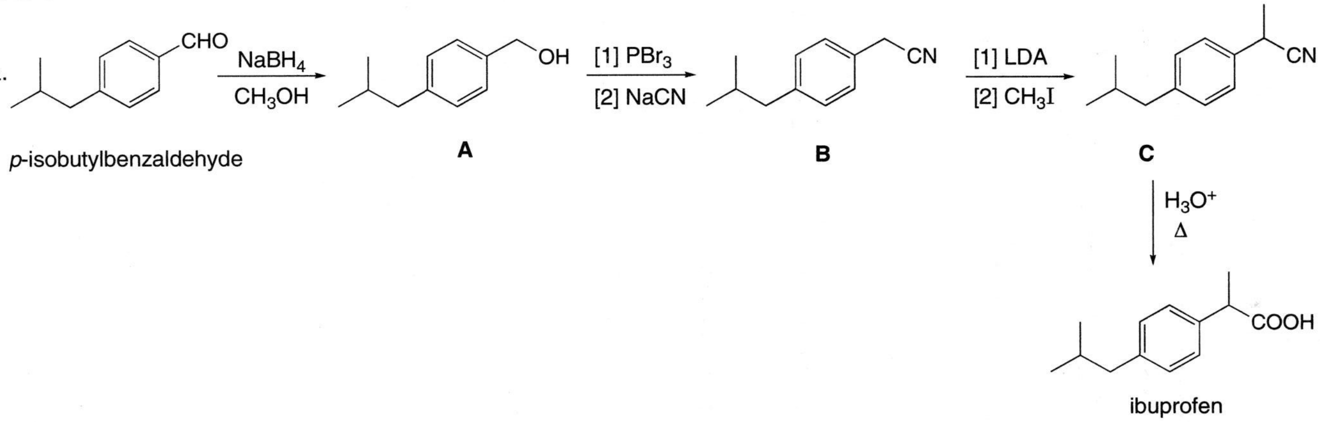

b.

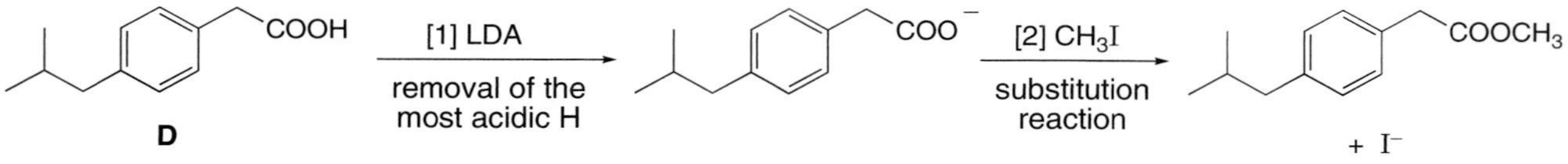
COOH
[1] LDA
removal of the most acidic H
COO⁻
[2] CH₃I
substitution reaction
COOCH₃
D
+ I⁻

Removal of the most acidic proton with LDA forms a carboxylate anion that reacts as a nucleophile with CH_3I to form an ester as substitution product.

23.57

a.

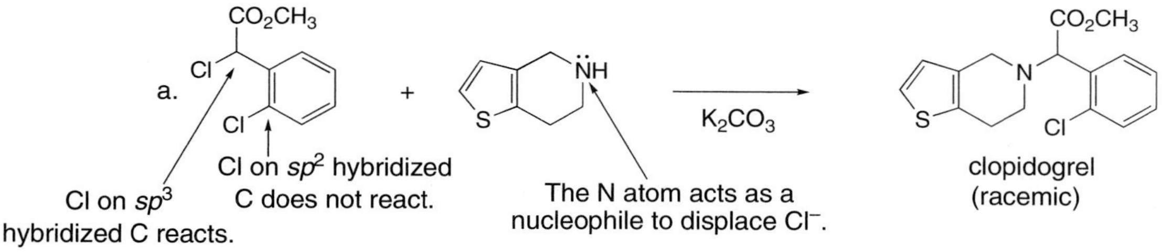
CO₂CH₃
Cl
Cl
+
NH
S
K₂CO₃
N
Cl
Cl on sp² hybridized C does not react.
Cl on sp³ hybridized C reacts.
The N atom acts as a nucleophile to displace Cl⁻.
clopidogrel (racemic)

b.

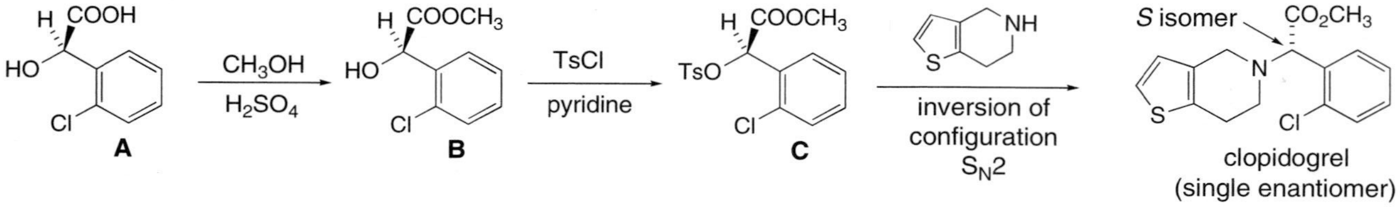
H COOH
HO
Cl
A
CH₃OH
H₂SO₄
H COOCH₃
HO
B
TsCl
pyridine
TsO
C
NH
S
inversion of configuration
SN2
S isomer
CO₂CH₃
N
clopidogrel
(single enantiomer)

23.58

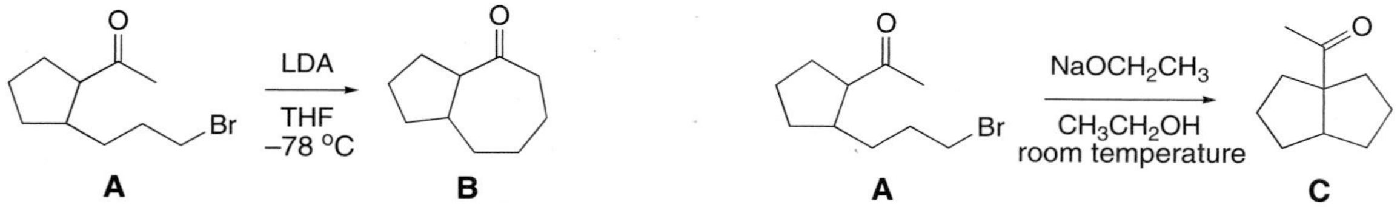
O
Br
A
LDA
THF
–78 °C
B
NaOCH₂CH₃
CH₃CH₂OH
room temperature
C

23.59

a.

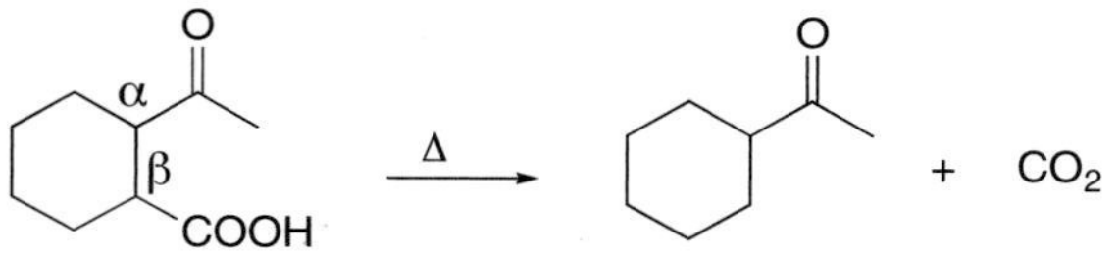
α
β
COOH
Δ
+ CO₂

In order for decarboxylation to occur readily, the COOH group must be bonded to the α C of another carbonyl group. In this case, it is bonded to the β carbon.

b. $CH_2(CO_2Et)_2 \xrightarrow[\text{[2] } (CH_3CH_2)_3CBr]{\text{[1] NaOEt}} (CH_3CH_2)_3CCH(CO_2Et)_2$

The 3° alkyl halide is too crowded to react with the strong nucleophile by an S_N2 mechanism.

c.

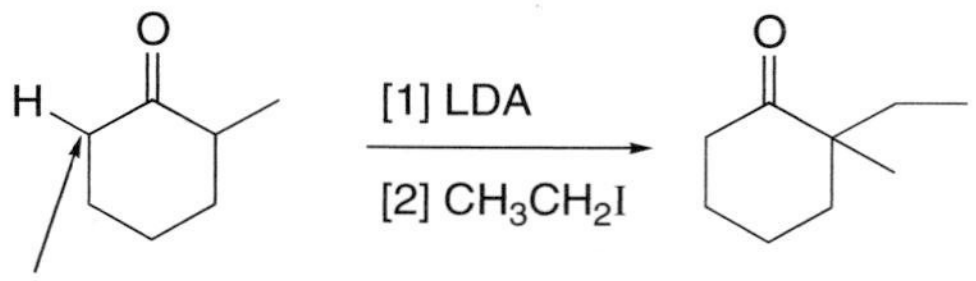
H
O
[1] LDA
[2] CH₃CH₂I

LDA removes a H from the less substituted C, forming the kinetic enolate. This product is from the thermodynamic enolate, which gives substitution on the more substituted α C.

23.60

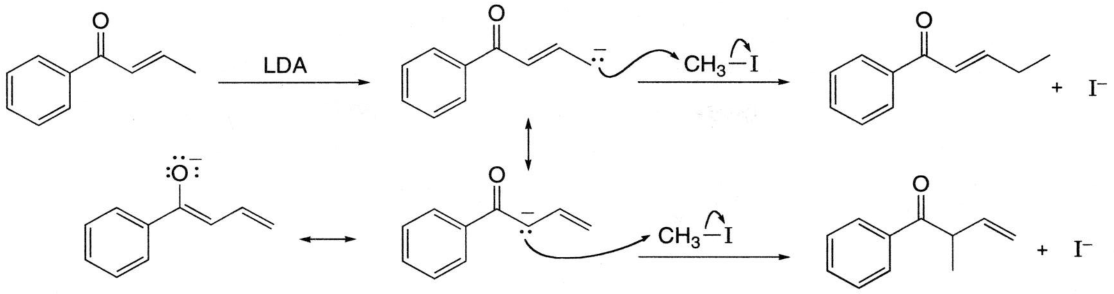

23.61

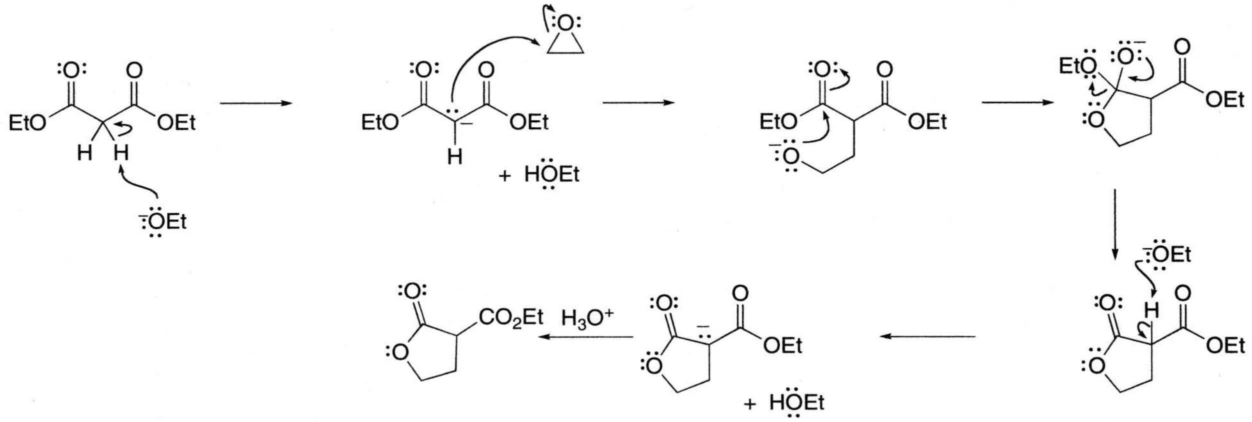

23.62

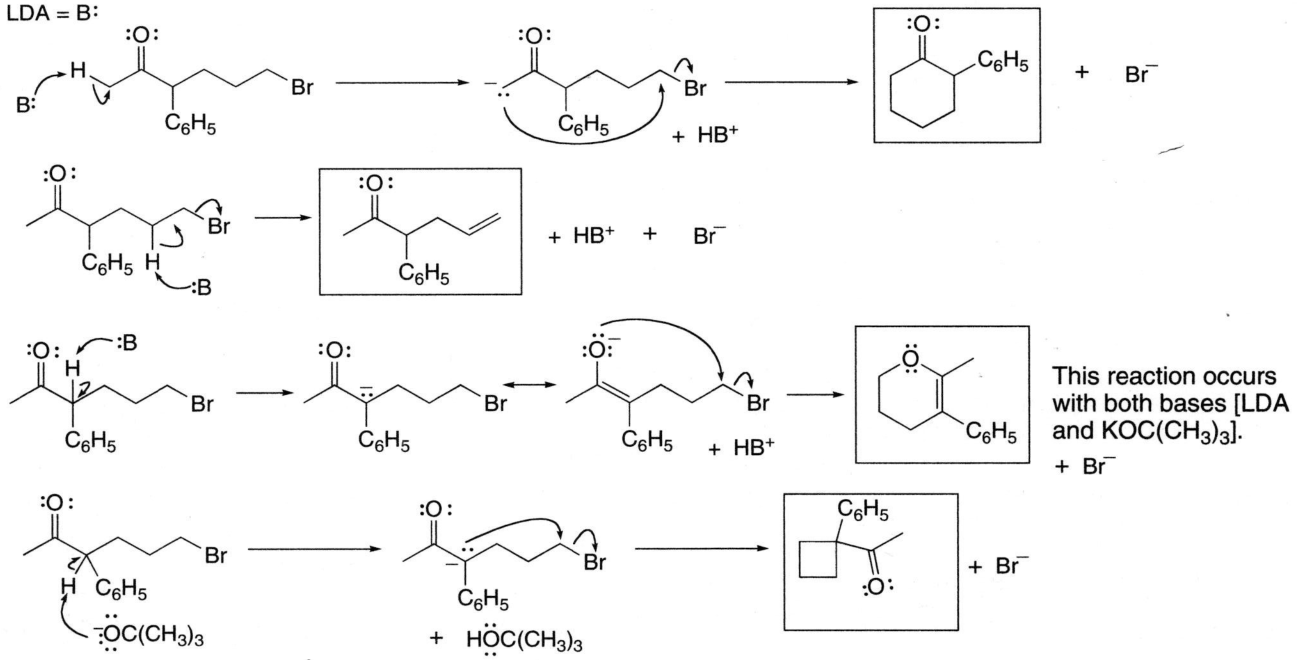

LDA = B:

This reaction occurs with both bases [LDA and KOC(CH₃)₃].

+ Br⁻

23.63

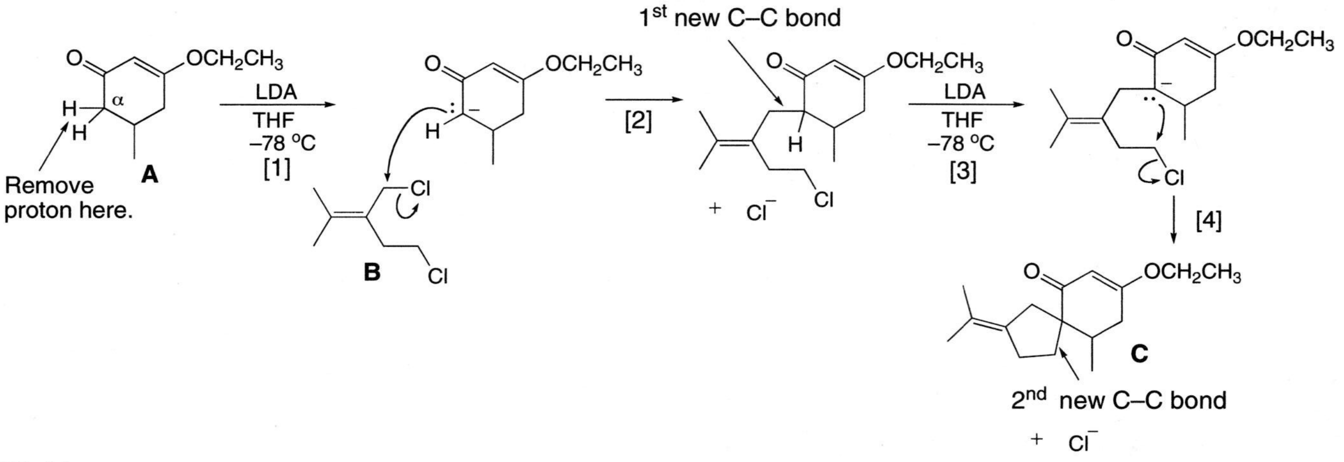

1st new C–C bond

Remove proton here.

A 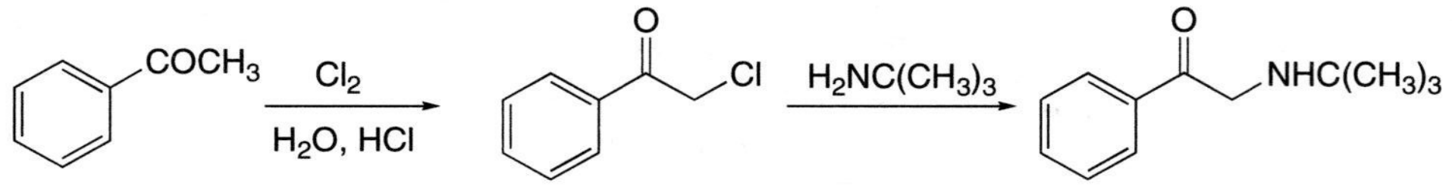   **B** 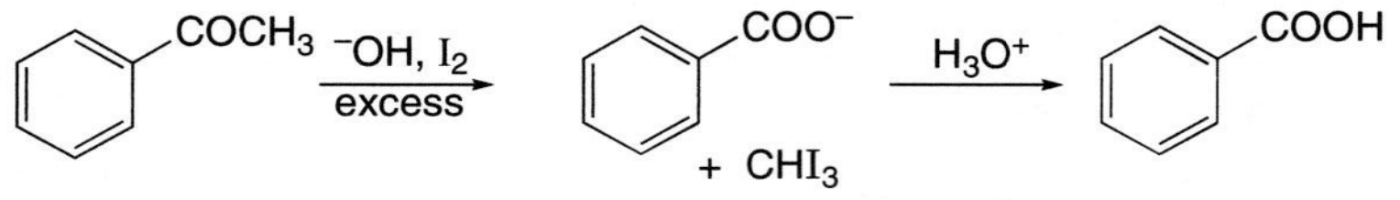   **C**

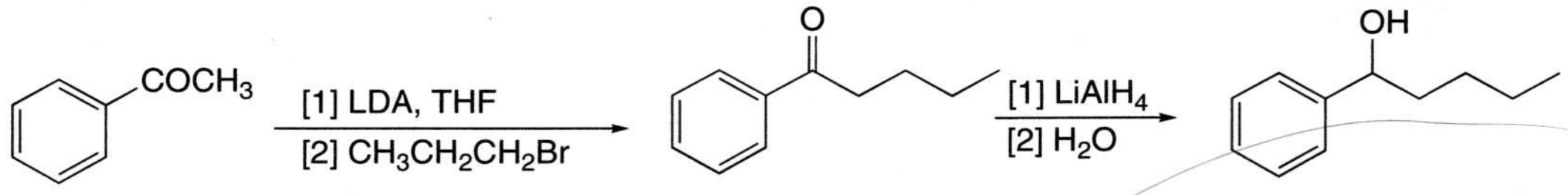

2nd new C–C bond

+ Cl⁻

23.64

a.

$COCH_3$ → Cl_2, H_2O, HCl → CH₂Cl → $H_2NC(CH_3)_3$ → $NHC(CH_3)_3$

b.
$COCH_3$ → ⁻OH, I_2 excess → COO^- + CHI_3 → H_3O^+ → COOH

c.
$COCH_3$ → [1] LDA, THF; [2] $CH_3CH_2CH_2Br$ → [1] $LiAlH_4$; [2] H_2O → OH

d. 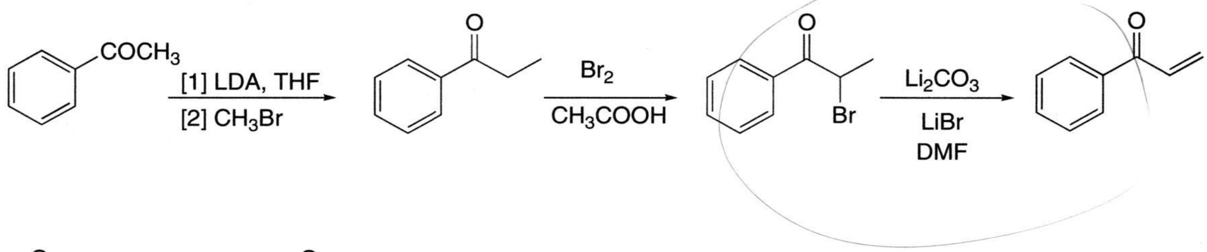
$COCH_3$ → [1] LDA, THF; [2] CH_3Br → Br_2, CH_3COOH → Br → Li_2CO_3, LiBr, DMF →

23.65

a. 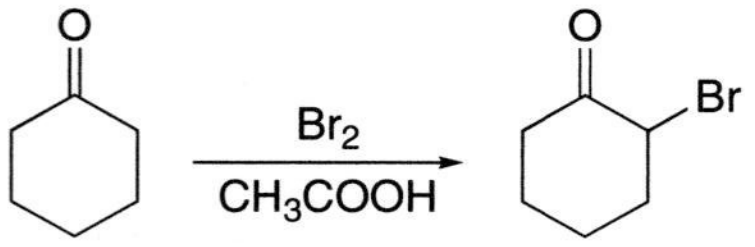
Br_2, CH_3COOH → Br

b. 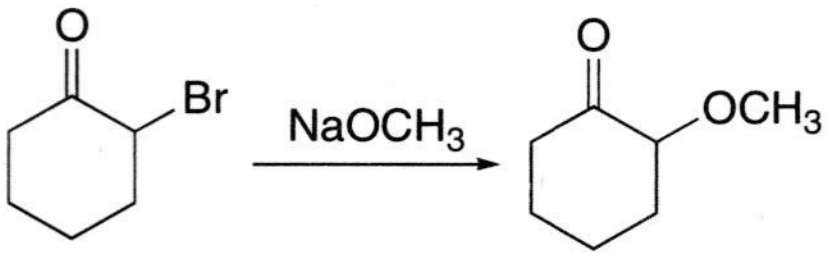
Br (from a.) → $NaOCH_3$ → OCH_3

c. 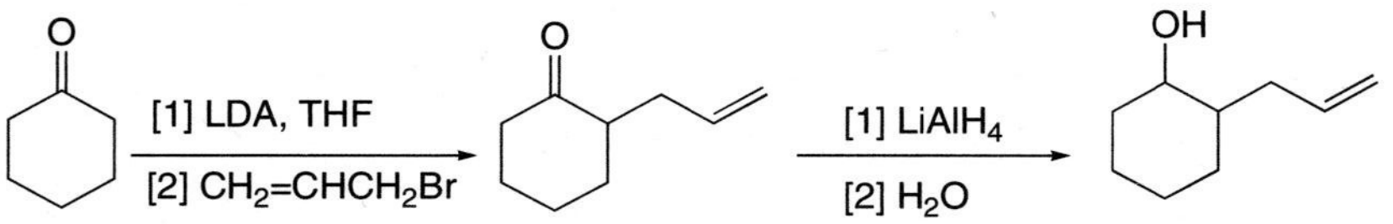
[1] LDA, THF; [2] $CH_2=CHCH_2Br$ → [1] $LiAlH_4$; [2] H_2O → OH

d. 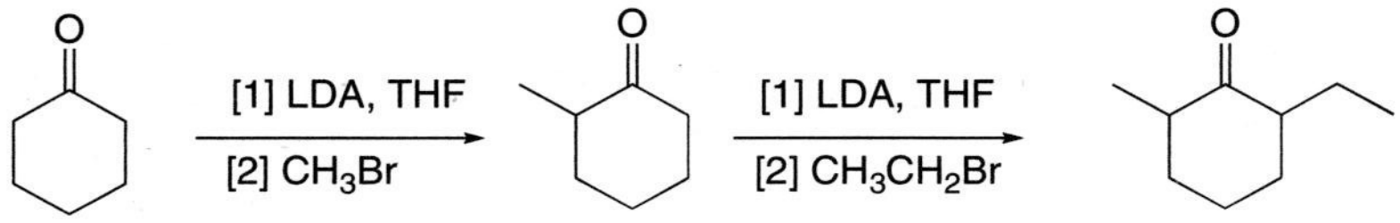
[1] LDA, THF; [2] CH_3Br → [1] LDA, THF; [2] CH_3CH_2Br →

e. (from a.)

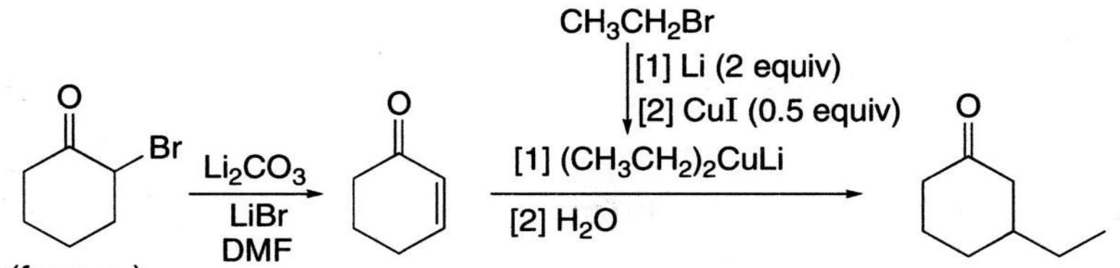

f.

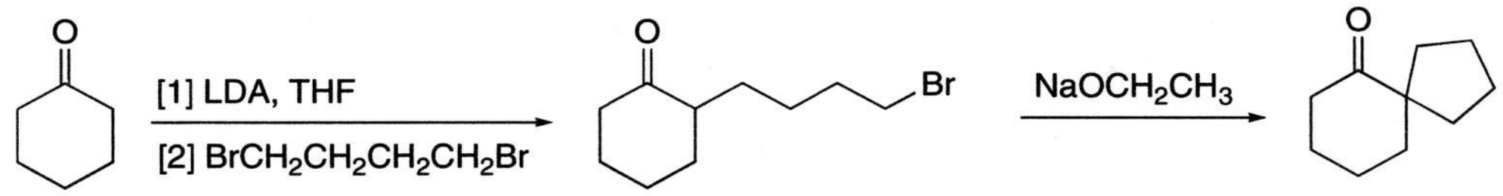

g.

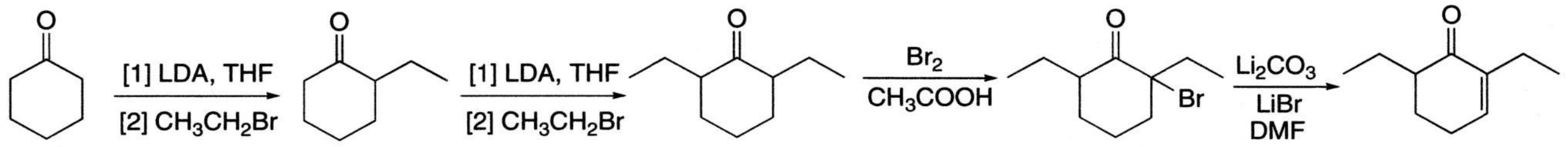

h. (from g.)

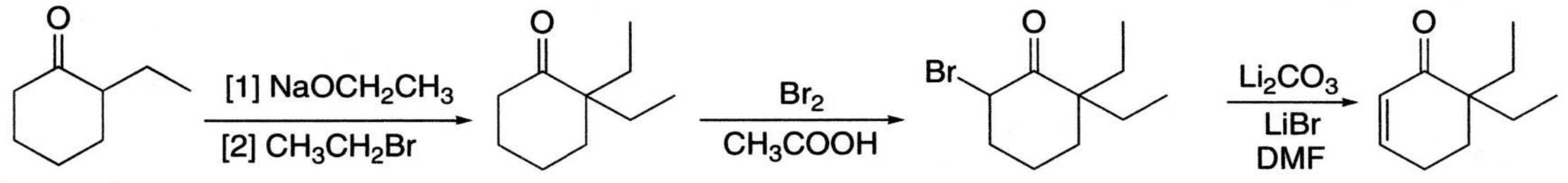

23.66

bupropion

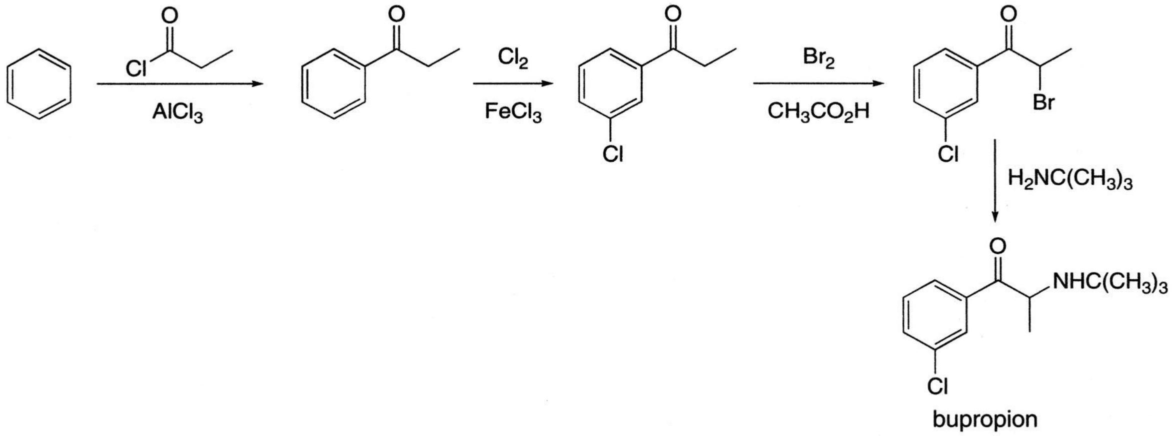

23.67

a.

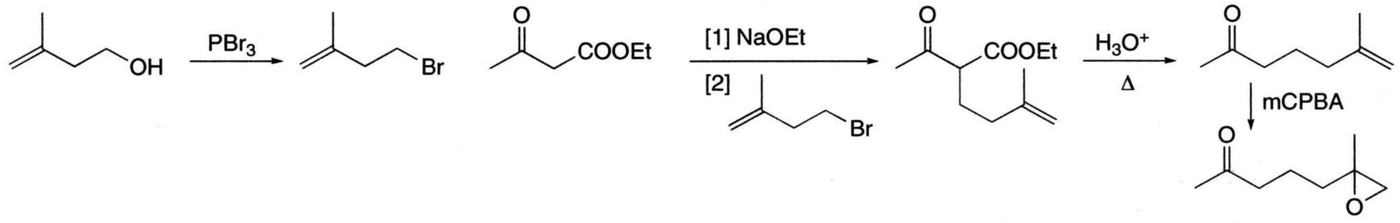

b.

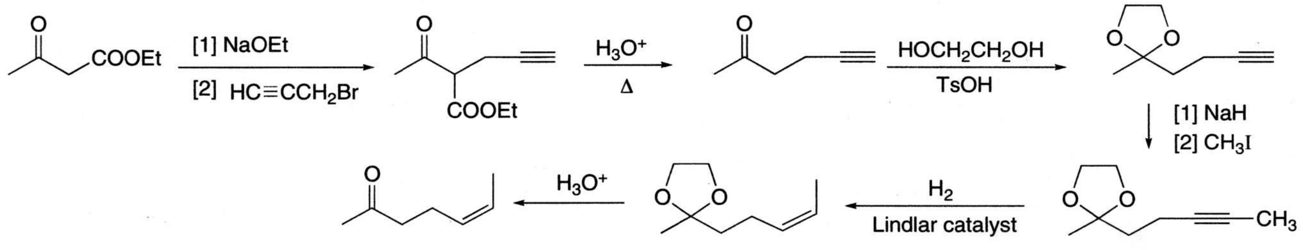

c.

23.68

a.

b.

c.

d.

23.69

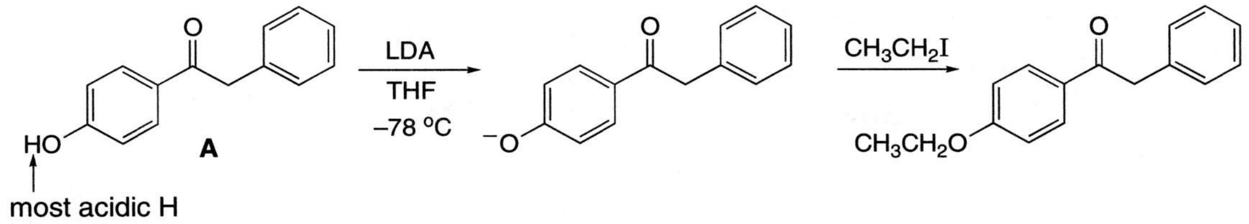

most acidic H

To synthesize the desired product, a protecting group is needed:

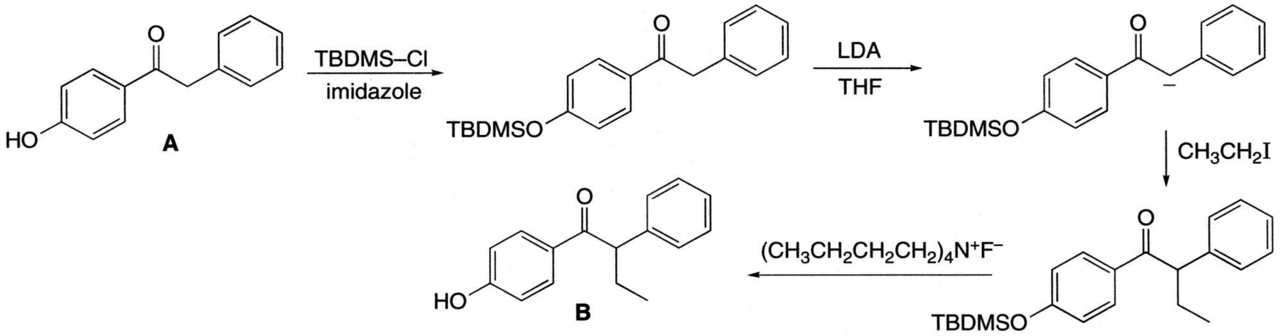

23.70

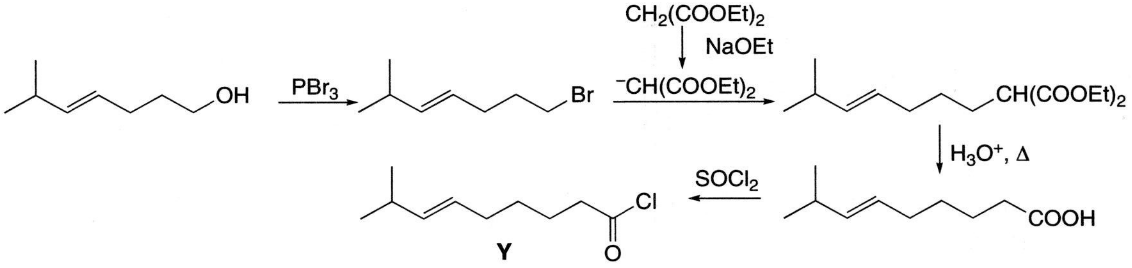

23.71

a.

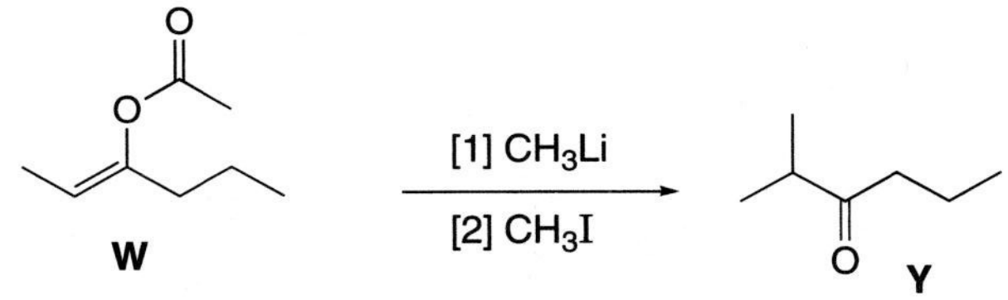

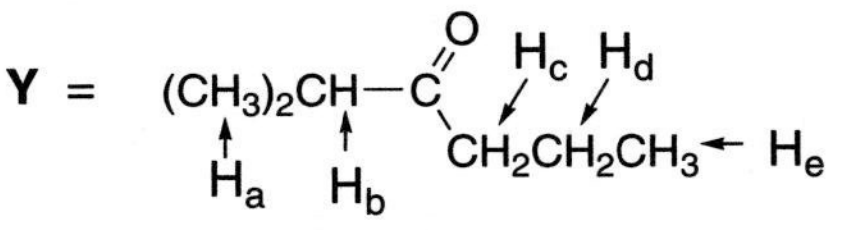

$Y = (CH_3)_2CH-C(=O)-CH_2CH_2CH_3$ with labeled protons H_a, H_b, H_c, H_d, H_e

$C_7H_{14}O \rightarrow$ one degree of unsaturation
IR peak at 1713 cm$^{-1} \rightarrow$ C=O
^{1}H NMR signals at (ppm)
       H_e: triplet at 0.8 (3 H)
       H_a: doublet at 0.9 (6 H)
       H_d: sextet at 1.4 (2 H)
       H_c: triplet at 1.9 (2 H)
       H_b: septet at 2.1 (1 H)

b.

23.72 Removal of H_a with base does not generate an anion that can delocalize onto the carbonyl O atom, whereas removal of H_b generates an enolate that is delocalized on O.

Delocalization of this sort can't occur by removal of H_a, making H_a less acidic.

Removal of H_b gives an anion that is resonance stabilized so H_b is more acidic.

Mechanism:

23.73

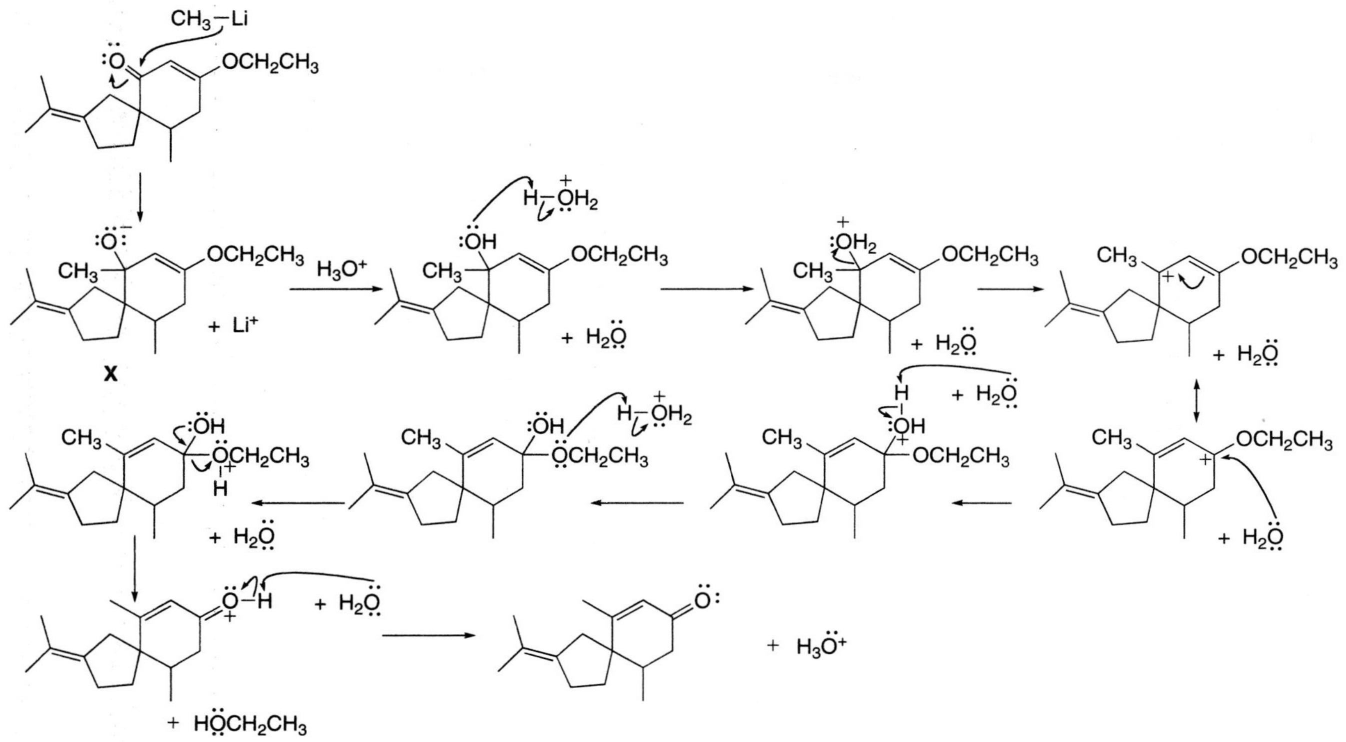

CH3–Li
OCH2CH3
X
+ Li+
H3O+
+ H2O
+ HOCH2CH3
+ H3O+

23.74

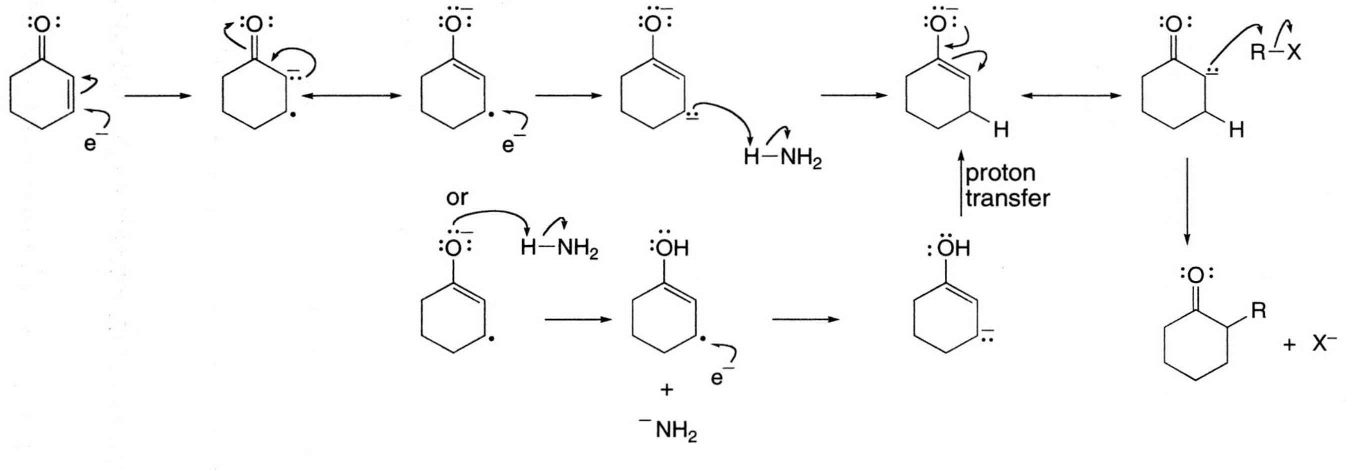

e–
H–NH2
proton transfer
or
+
–NH2
R–X
+ X–

Chapter 24: Carbonyl Condensation Reactions

♦ The four major carbonyl condensation reactions

Reaction type	Reaction
[1] Aldol reaction (24.1)	
[2] Claisen reaction (24.5)	
[3] Michael reaction (24.8)	
[4] Robinson annulation (24.9)	

♦ Useful variations

[1] Directed aldol reaction (24.3)

[2] Intramolecular aldol reaction (24.4)

[a] With 1,4-dicarbonyl compounds:

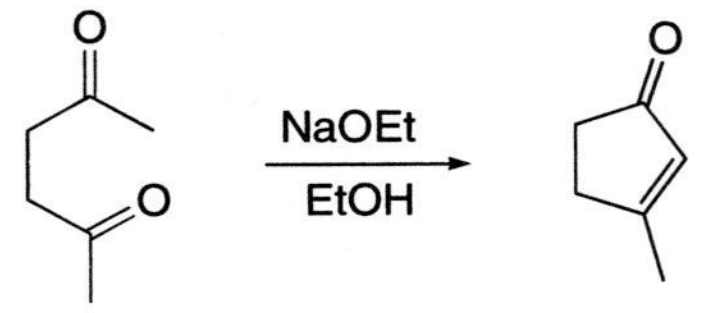

[b] With 1,5-dicarbonyl compounds:

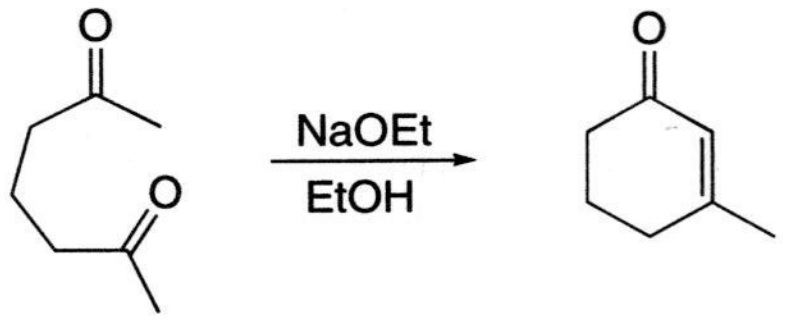

[3] Dieckmann reaction (24.7)

[a] With 1,6-diesters:

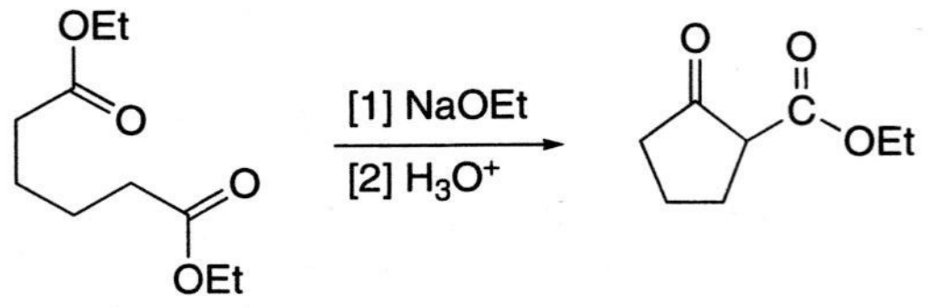

[b] With 1,7-diesters:

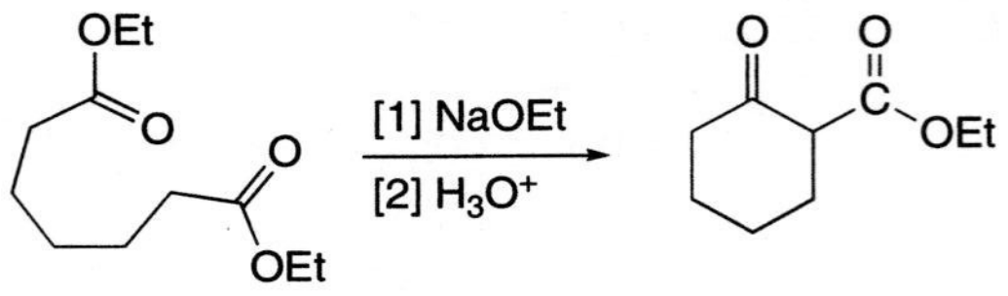

Chapter 24: Answers to Problems

24.1

a.

b. $(CH_3)_3CCH_2CHO \longrightarrow (CH_3)_3CCH_2\overset{HO}{\underset{H}{C}}-\overset{C(CH_3)_3}{\underset{H}{C}}-CHO$

c. $CH_3\overset{O}{\overset{\|}{C}}CH_3 \longrightarrow CH_3-\overset{OH}{\underset{CH_3}{C}}-CH_2-\overset{O}{\overset{\|}{C}}-CH_3$

d.

24.2

a. CHO

no α H
no aldol reaction

b.

α H
yes

c. $(CH_3)_3C-\overset{O}{\overset{\|}{C}}-H$

no α H
no aldol reaction

d. $(CH_3)_3C-\overset{O}{\overset{\|}{C}}-CH_3$

α H
yes

e. CHO
α H
yes

24.3

a. $\xrightarrow{\text{base}}$

b. $\xrightarrow{\text{base}}$

(*E* and *Z* isomers)

c. $\xrightarrow{\text{base}}$

24.4

$CH_3-\overset{H}{\underset{H}{C}}-\overset{HO:H}{\underset{H}{C}}-\overset{O}{\overset{\|}{C}}H \xrightarrow{} CH_3-\overset{H}{\underset{H}{C}}-\overset{+\ddot{O}H_2\ H}{\underset{H}{C}}-\overset{O}{\overset{\|}{C}}H$

$+ HSO_4^-$

$\xrightarrow{} CH_3-\overset{+}{\underset{H}{C}}-\overset{H}{\underset{H}{C}}-\overset{O}{\overset{\|}{C}}H \xrightarrow{HSO_4^-} CH_3CH=CH-\overset{O}{\overset{\|}{C}}H$

$+ H_2\ddot{O}:$

$+ H_2SO_4$

24.5 Locate the α and β C's to the carbonyl group, and break the molecule into two halves at this bond. The α C and all of the atoms bonded to it belong to one carbonyl component. The β C and all of the atoms bonded to it belong to the other carbonyl component.

a.

b. C_6H_5

c.

24.6

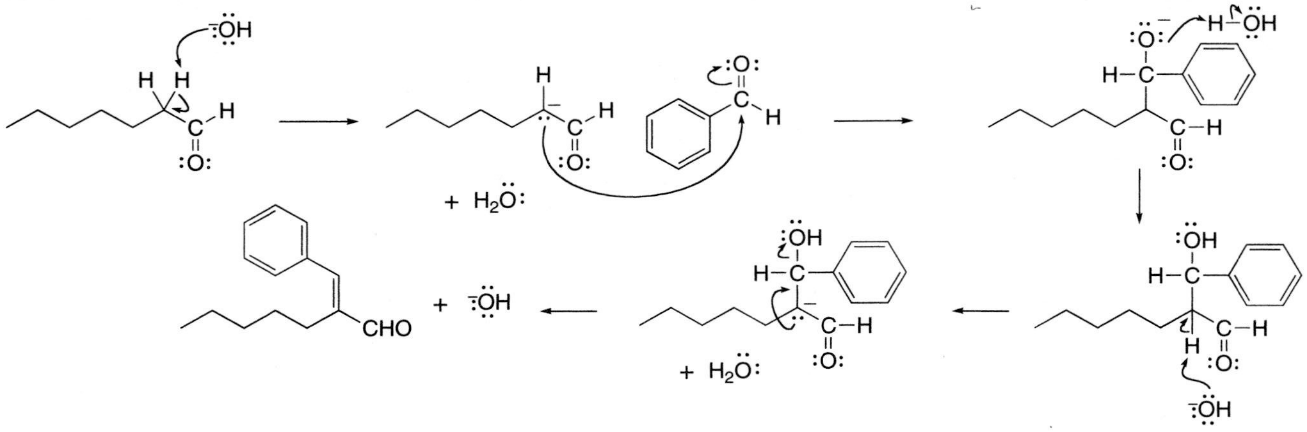

24.7

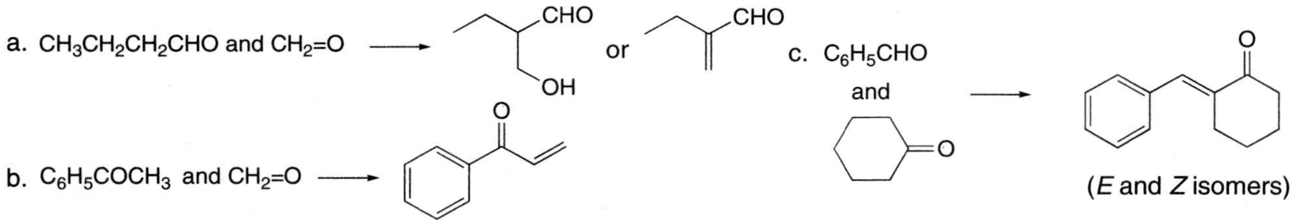

a. CH$_3$CH$_2$CH$_2$CHO and CH$_2$=O ⟶ [structure with CHO and OH] or [structure with CHO]

c. C$_6$H$_5$CHO and [cyclohexanone] ⟶ [benzylidenecyclohexanone] (E and Z isomers)

b. C$_6$H$_5$COCH$_3$ and CH$_2$=O ⟶ [phenyl vinyl ketone]

24.8

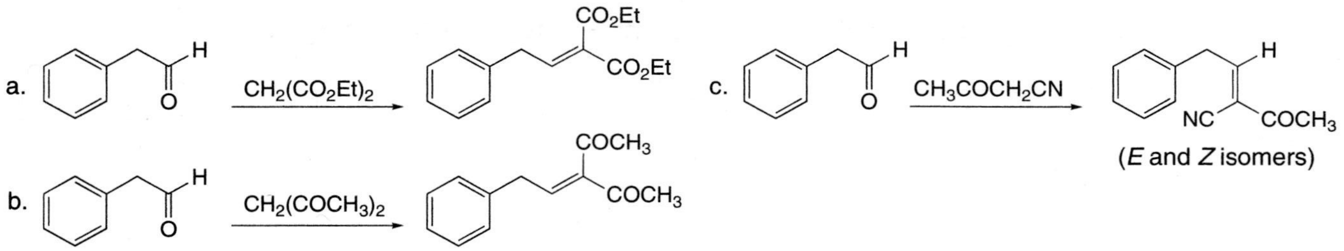

a. [phenylacetaldehyde] $\xrightarrow{CH_2(CO_2Et)_2}$ [product with CO$_2$Et, CO$_2$Et]

b. [phenylacetaldehyde] $\xrightarrow{CH_2(COCH_3)_2}$ [product with COCH$_3$, COCH$_3$]

c. [phenylacetaldehyde] $\xrightarrow{CH_3COCH_2CN}$ [product with NC, COCH$_3$] (E and Z isomers)

24.9

2 [propyl CHO] $\xrightarrow[H_2O]{^-OH}$ [aldol product with CHO, OH]

a. $\xrightarrow[CH_3OH]{NaBH_4}$ [diol product with OH]

b. $\xrightarrow{^-OH}$ [CHO product] (E and Z mixture) $\xrightarrow[\text{[2] H}_2O]{\text{[1] (CH}_3)_2\text{CuLi}}$ [CHO product]

c. [CHO product] (from b.) $\xrightarrow[CH_3OH]{NaBH_4}$ CH$_3$CH$_2$CH$_2$CH=C(CH$_2$CH$_3$)CH$_2$OH

d. [CHO product] (from b.) $\xrightarrow[\text{[2] H}_2O]{\text{[1] CH}_3\text{MgBr}}$ CH$_3$CH$_2$CH$_2$CH=C(CH$_2$CH$_3$)CH(CH$_3$)OH

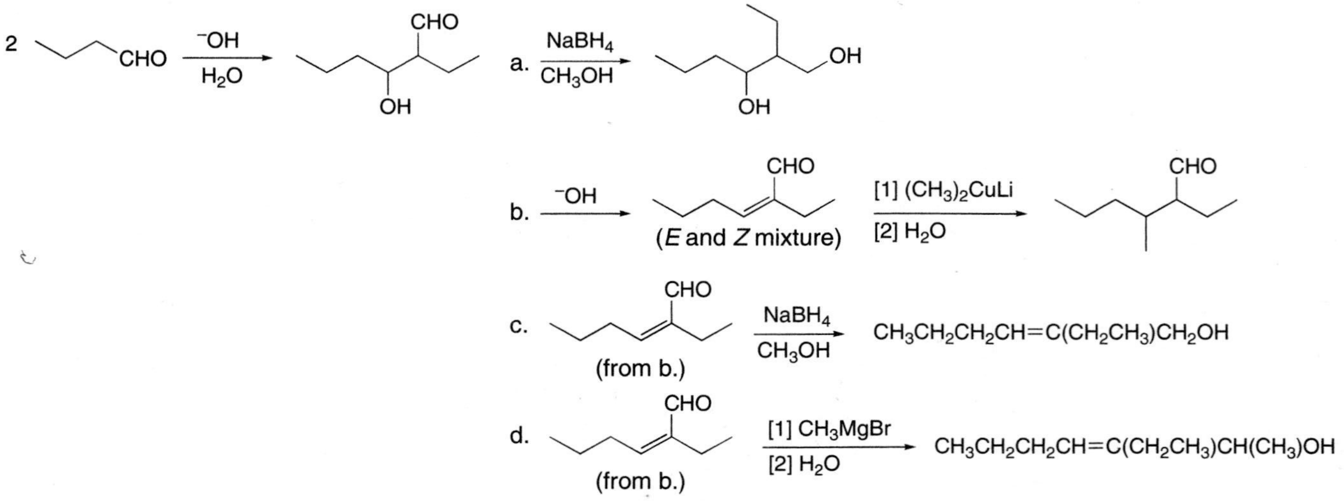

24.10 Find the α and β C's to the carbonyl group and break the bond between them.

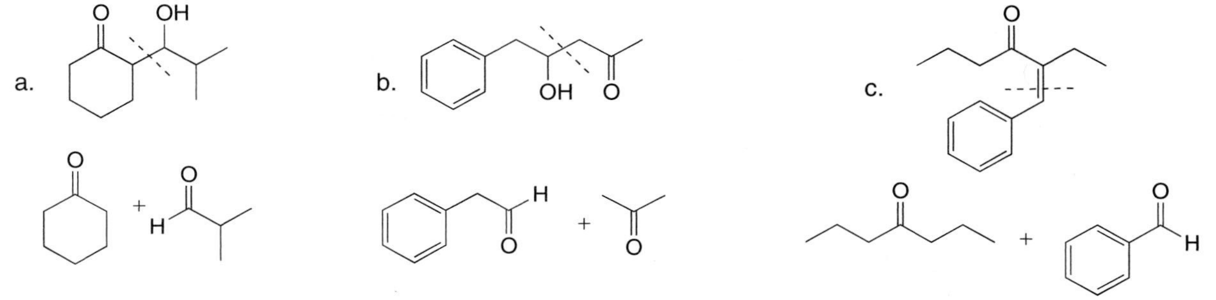

a.

b.

c.

24.11

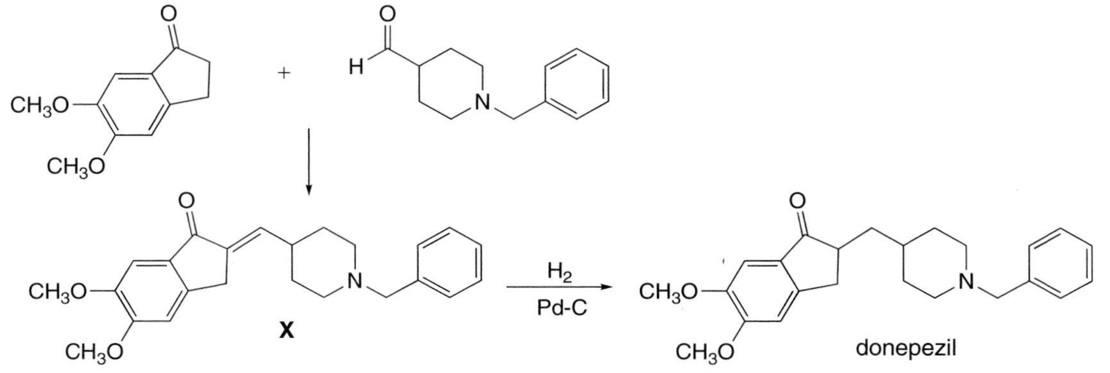

24.12 All enolates have a second resonance structure with a negative charge on O.

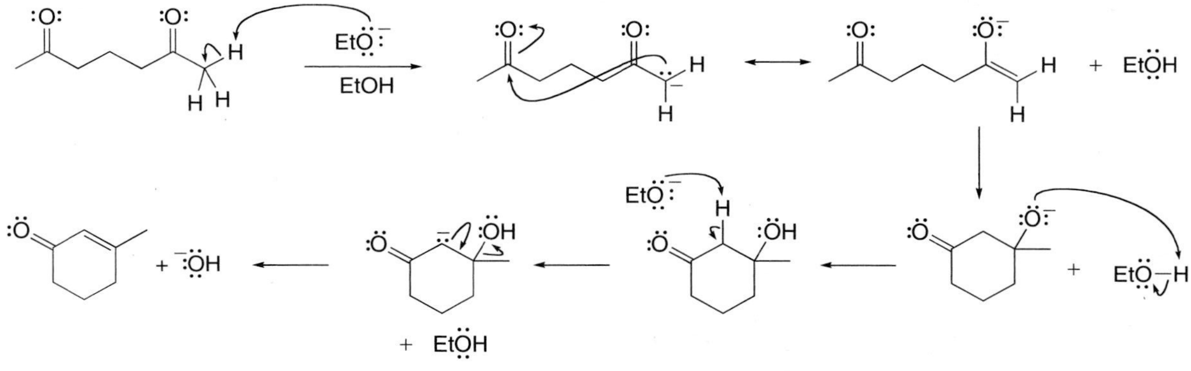

24.13

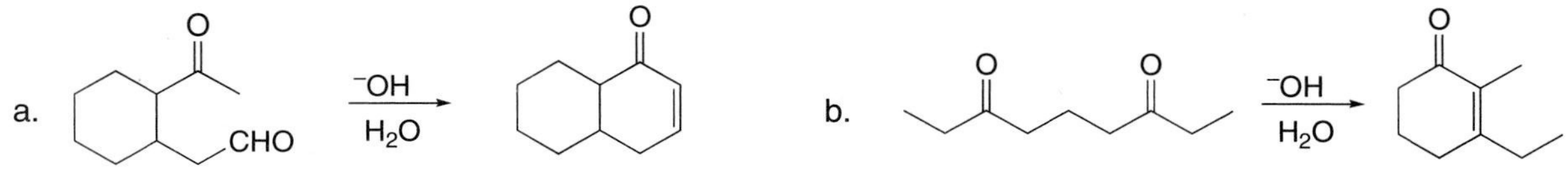

a.

b.

24.14

1-methylcyclopentene

2-cyclohexenone

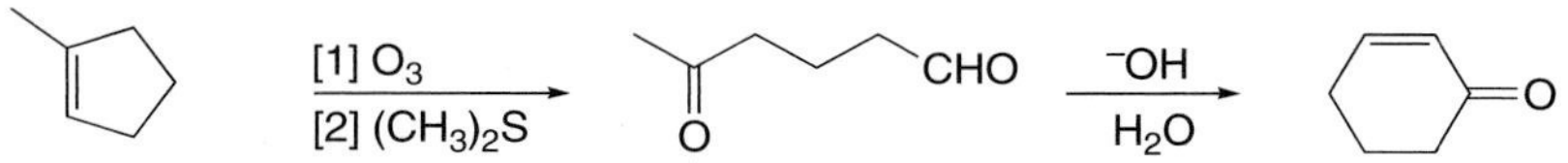

24.15 Join the α C of one ester to the carbonyl C of the other ester to form the β-keto ester.

a.

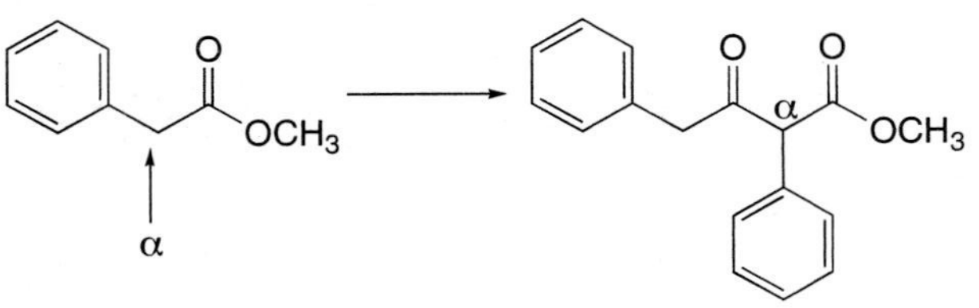

b.

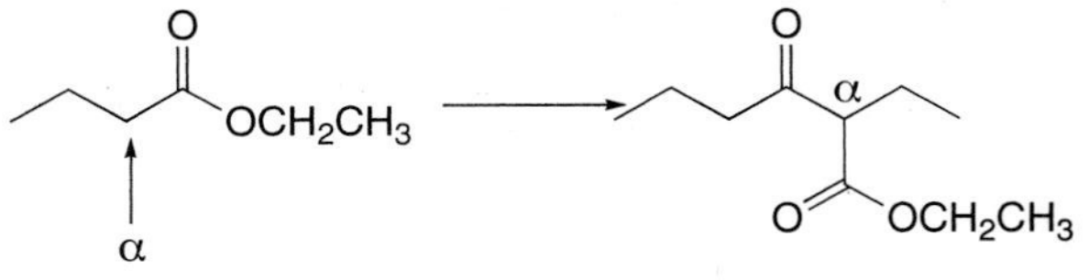

24.16 In a crossed Claisen reaction between an ester and a ketone, the enolate is formed from the ketone, and the product is a β-dicarbonyl compound.

a.   CH₃CH₂CO₂Et and HCO₂Et ⟶ H

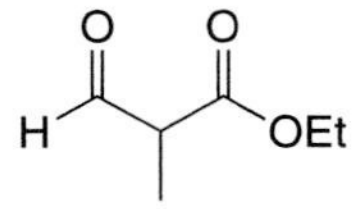

Only this compound
can form an enolate.

b.   CO₂Et   and   HCO₂Et ⟶

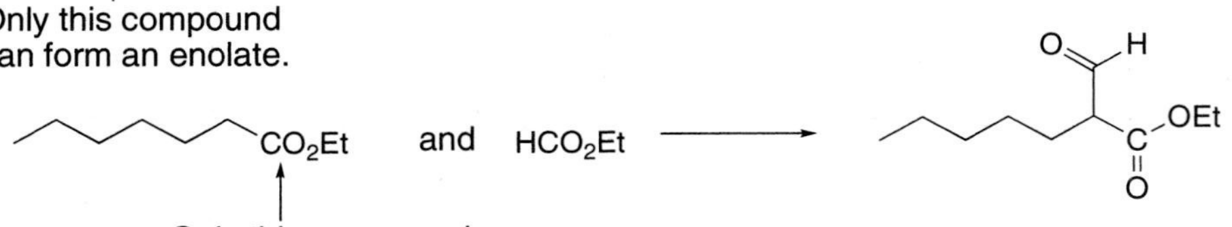

Only this compound
can form an enolate.

c.   CH₃   CH₃   and   CH₃   OEt ⟶

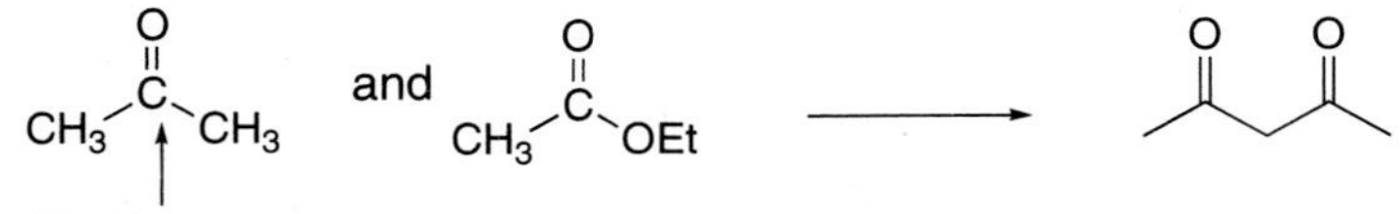

The ketone
forms the enolate.

d.   and   OEt ⟶

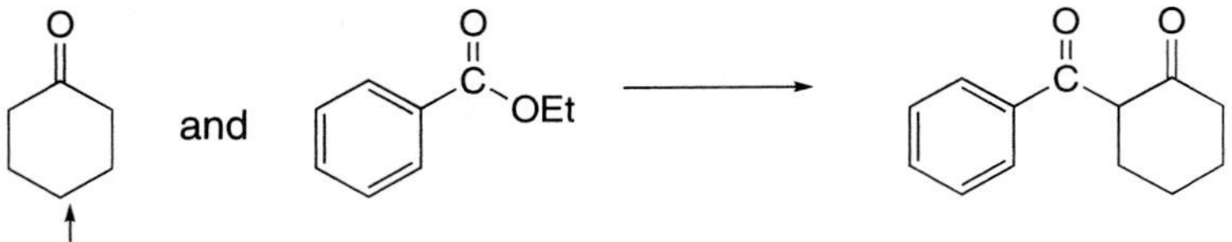

The ketone
forms the enolate.

24.17 A β-dicarbonyl compound like avobenzone is prepared by a crossed Claisen reaction between a ketone and an ester.

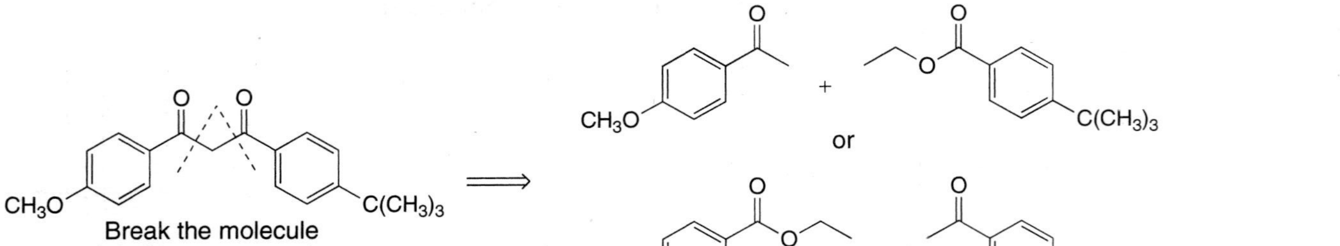

CH₃O   C(CH₃)₃
Break the molecule
into two components
at either dashed line.
avobenzone

CH₃O   +   C(CH₃)₃

or

CH₃O   +   C(CH₃)₃

24.18

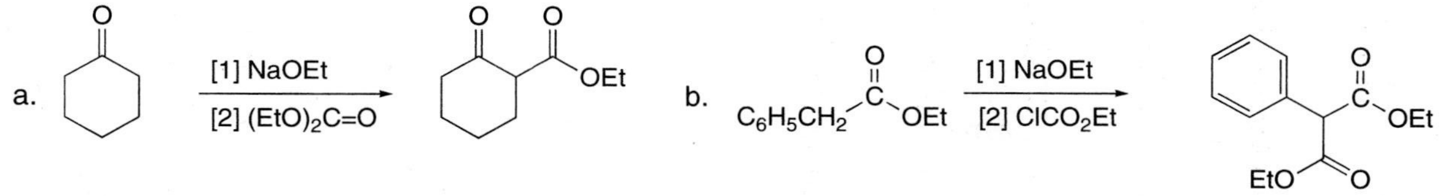

a.

b.

24.19

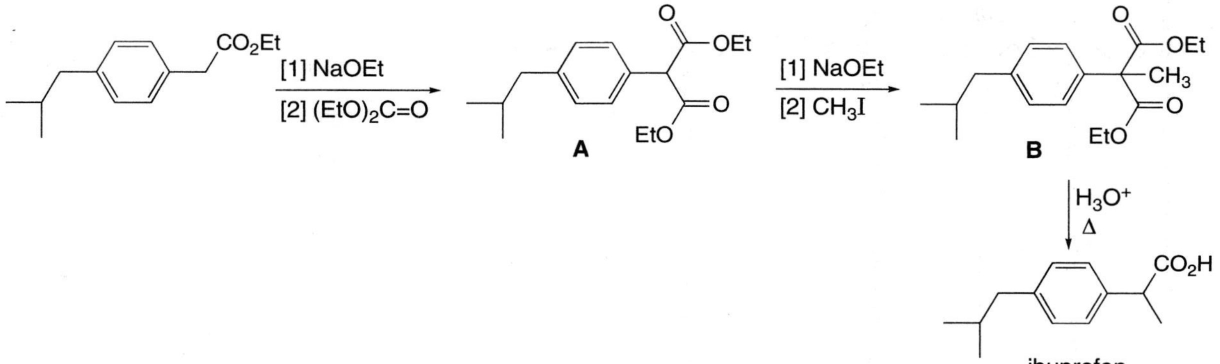

ibuprofen

24.20

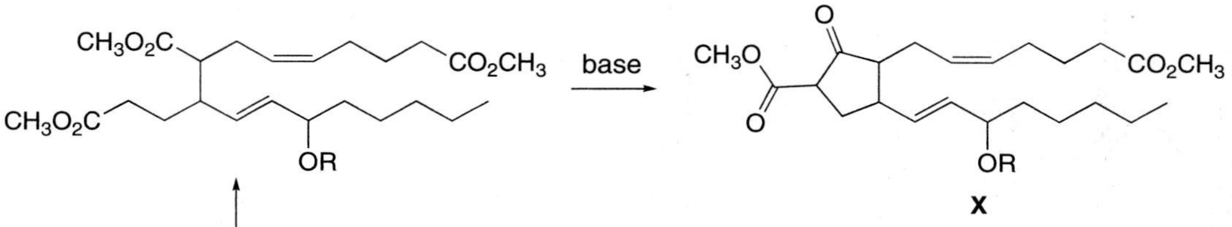

1,6-Diester forms a five-membered ring.

24.21

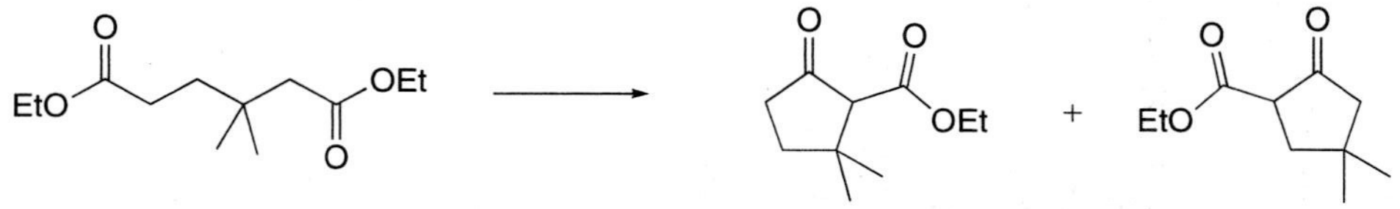

24.22 A Michael acceptor is an α,β-unsaturated carbonyl compound.

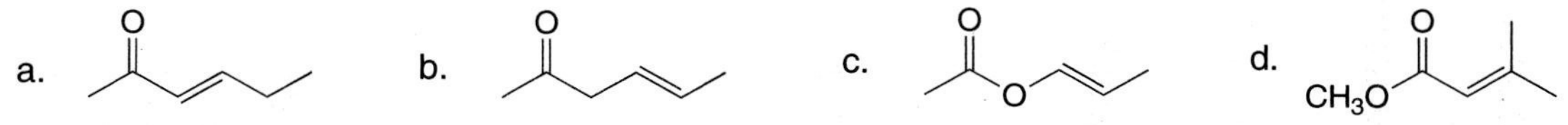

a.
α,β-unsaturated
yes - Michael acceptor

b.
not α,β-unsaturated

c.
not α,β-unsaturated

d.
α,β-unsaturated
yes - Michael acceptor

24.23

a.

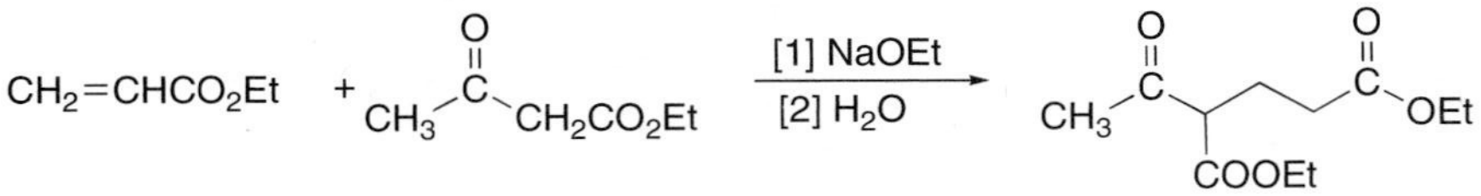

b.

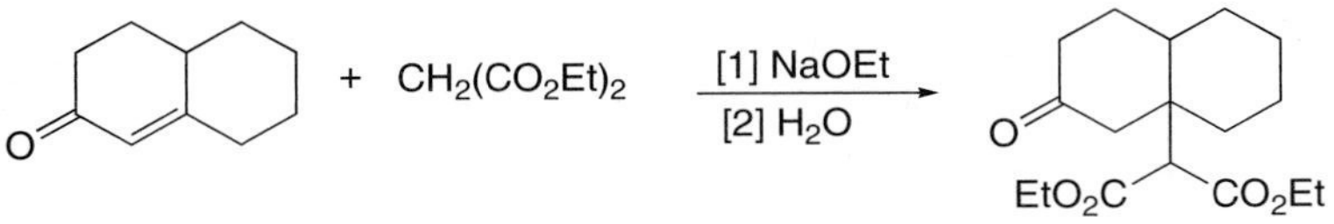

c.

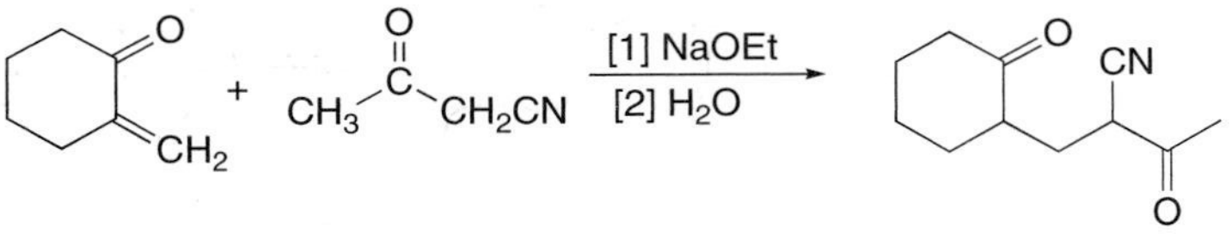

24.24

a.

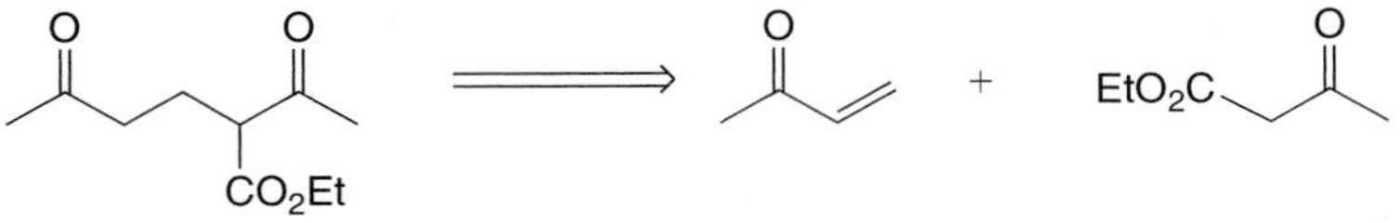

b.

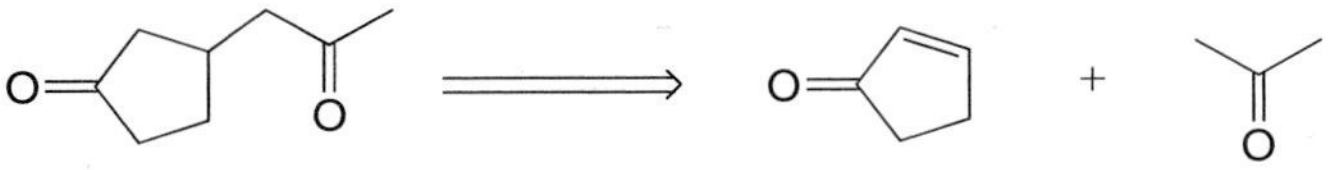

24.25 The Robinson annulation forms a six-membered ring and three new carbon–carbon bonds: two σ bonds and one π bond.

a.

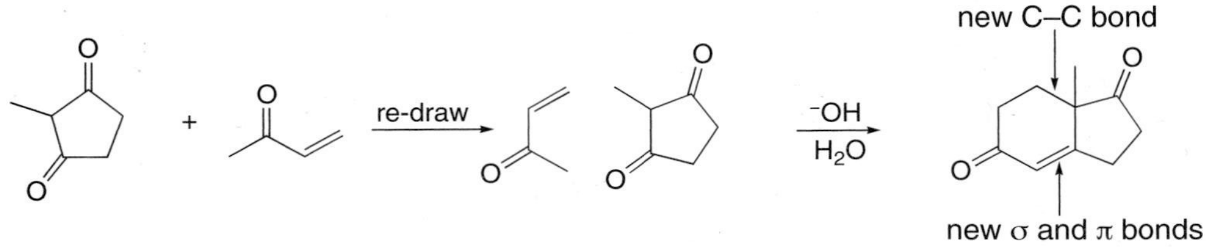

b.

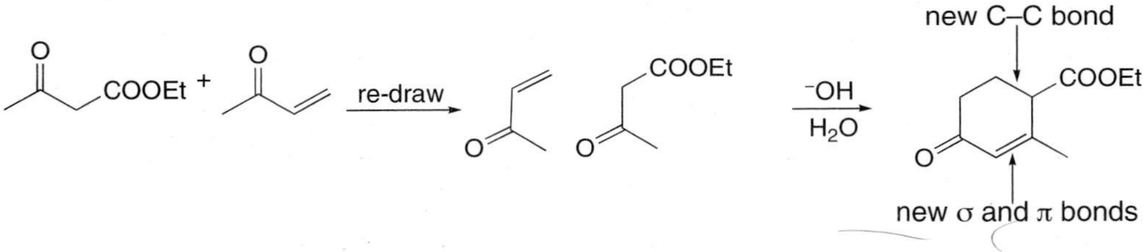

c.

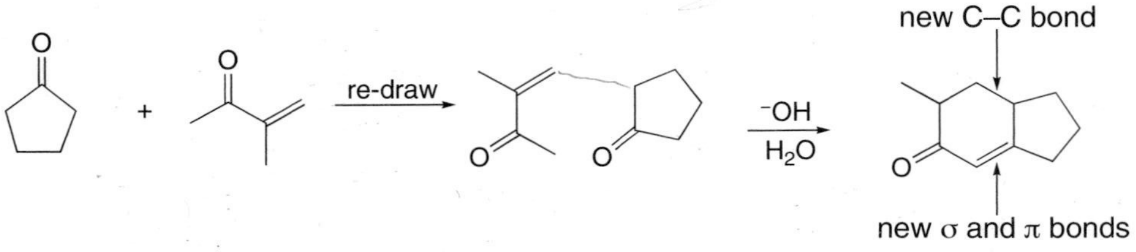

d.

24.26

a.

c.

b.

24.27 The product of an aldol reaction is a β-hydroxy carbonyl compound or an α,β-unsaturated carbonyl compound. The latter type of compound is drawn as product unless elimination of H_2O cannot form a conjugated system.

a. 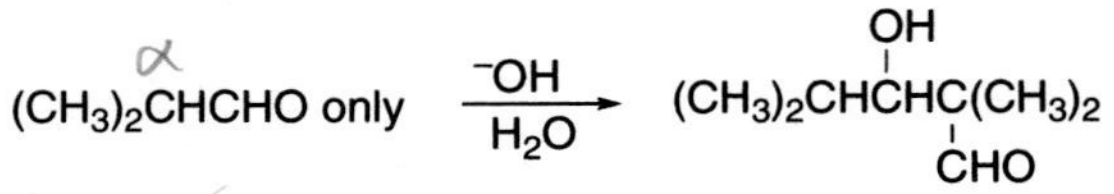 $(CH_3)_2CHCHO$ only $\xrightarrow[H_2O]{^-OH}$ $(CH_3)_2CHCHC(CH_3)_2$ with OH and CHO

d. $(CH_3CH_2)_2C=O$ only $\xrightarrow[H_2O]{^-OH}$

b. 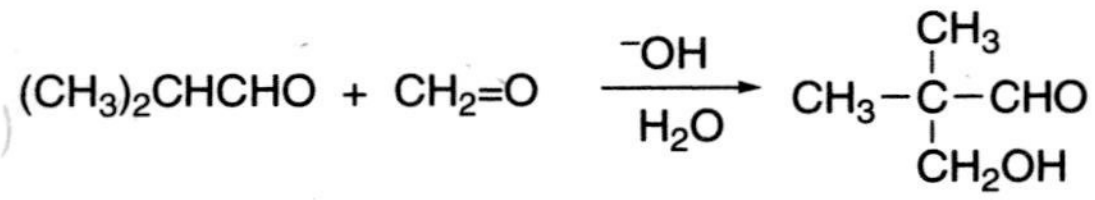 $(CH_3)_2CHCHO + CH_2=O$ $\xrightarrow[H_2O]{^-OH}$ $CH_3-\overset{CH_3}{\underset{CH_2OH}{C}}-CHO$

e. $(CH_3CH_2)_2C=O + CH_2=O$ $\xrightarrow[H_2O]{^-OH}$

c. $C_6H_5CHO + CH_3CH_2CH_2CHO$ $\xrightarrow[H_2O]{^-OH}$

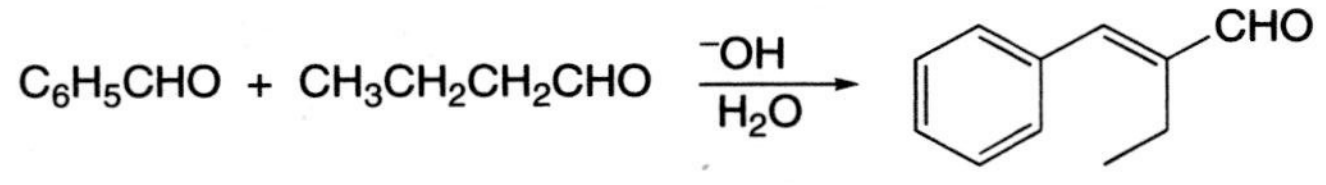

(*E* and *Z* isomers)

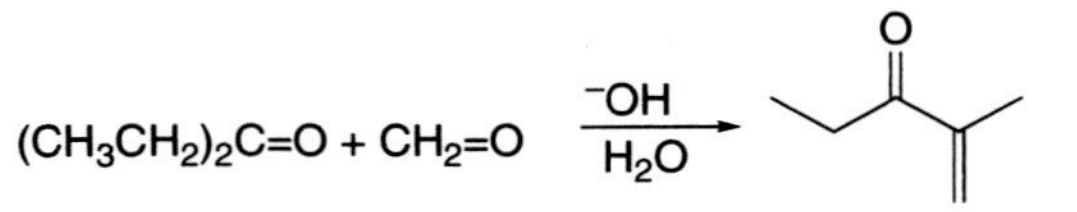

f. 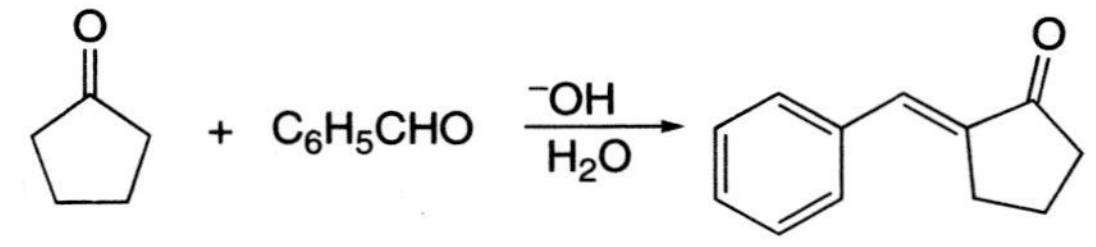 + C_6H_5CHO $\xrightarrow[H_2O]{^-OH}$

(*E* and *Z* isomers)

24.28

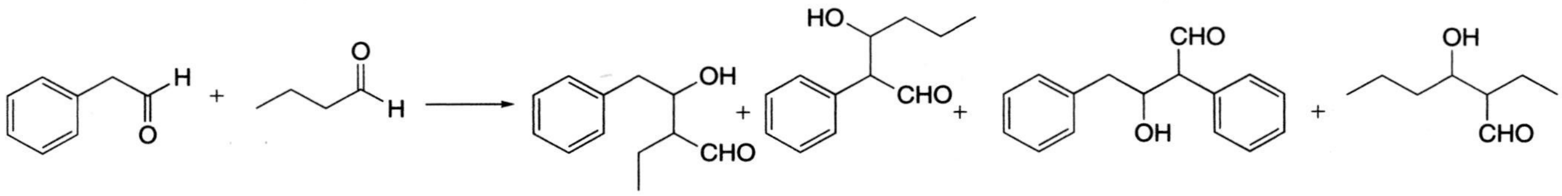

24.29

a. 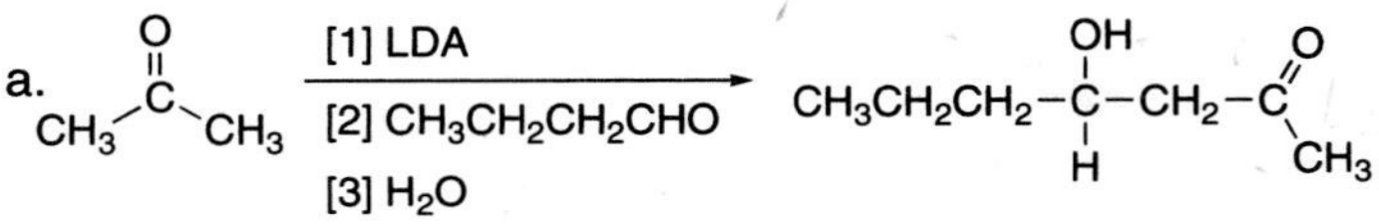 $\xrightarrow[\text{[3] }H_2O]{\text{[1] LDA} \\ \text{[2] }CH_3CH_2CH_2CHO}$ $CH_3CH_2CH_2-\overset{OH}{\underset{H}{C}}-CH_2-\overset{O}{\underset{}{C}}-CH_3$

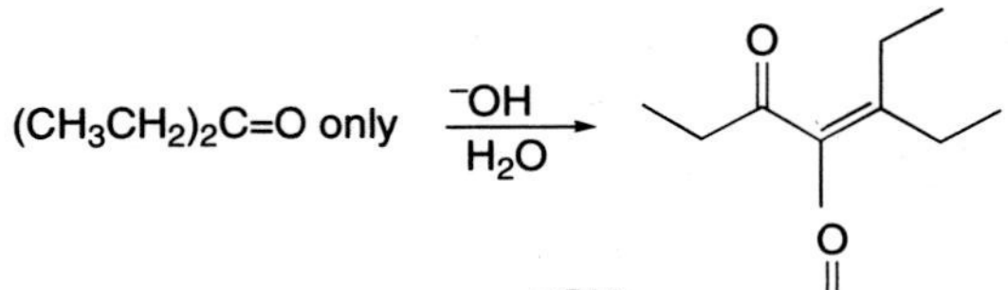

b. 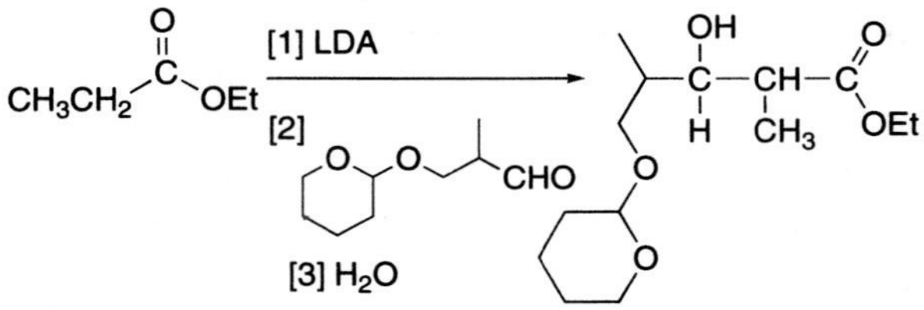 $CH_3CH_2-\overset{O}{\underset{}{C}}-OEt$ $\xrightarrow[\text{[2]}]{\text{[1] LDA}}$ $\xrightarrow{\text{[3] }H_2O}$

24.30

a. 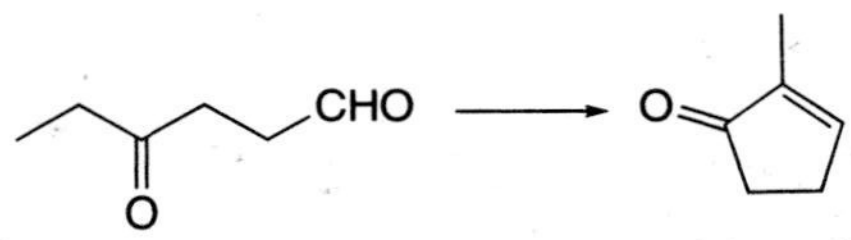 $\longrightarrow$

b. 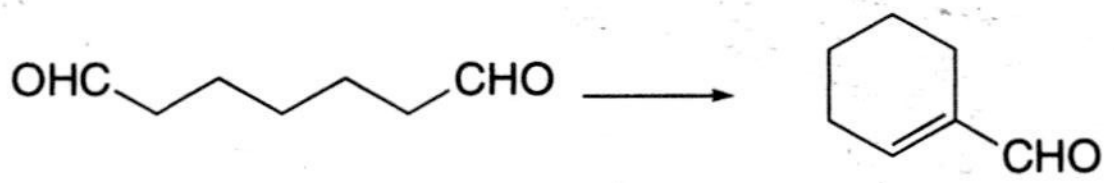 OHC CHO $\longrightarrow$

c. 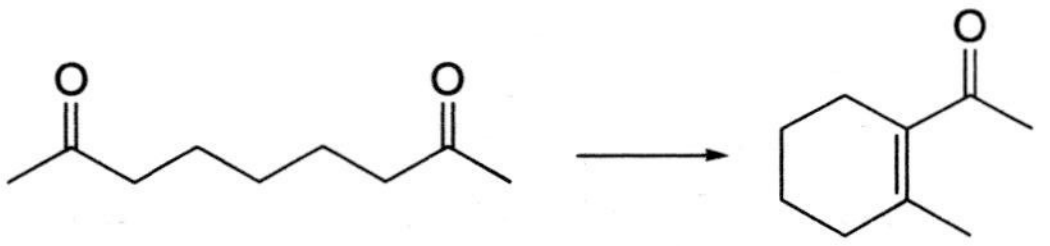 $\longrightarrow$

24.31 Locate the α and β C's to the carbonyl group, and break the molecule into two halves at this bond. The α C and all of the atoms bonded to it belong to one carbonyl component. The β C and all the atoms bonded to it belong to the other carbonyl component.

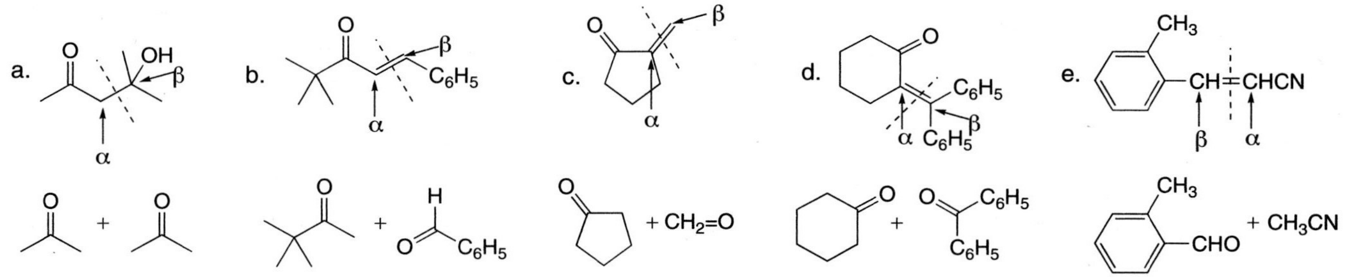

a.   b.   c.   d.   e.

24.32

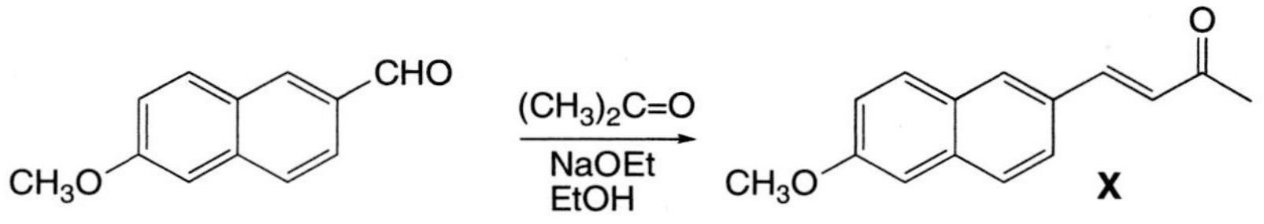

24.33

a.   b.   c.   d.

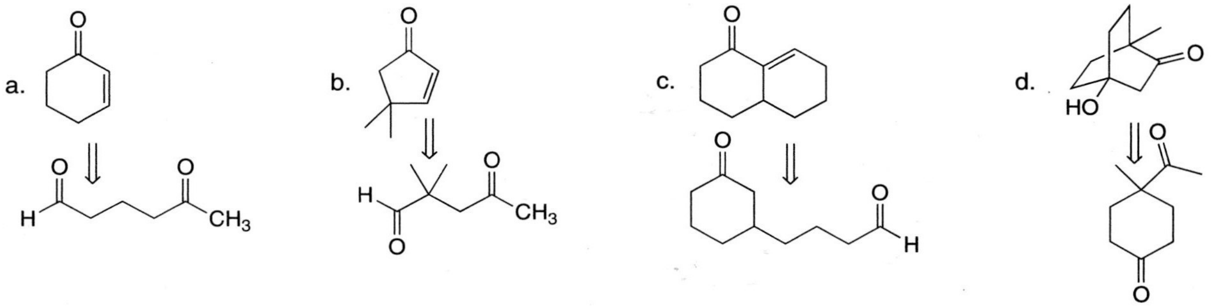

24.34 Ozonolysis cleaves the C=C, and base catalyzes an intramolecular aldol reaction.

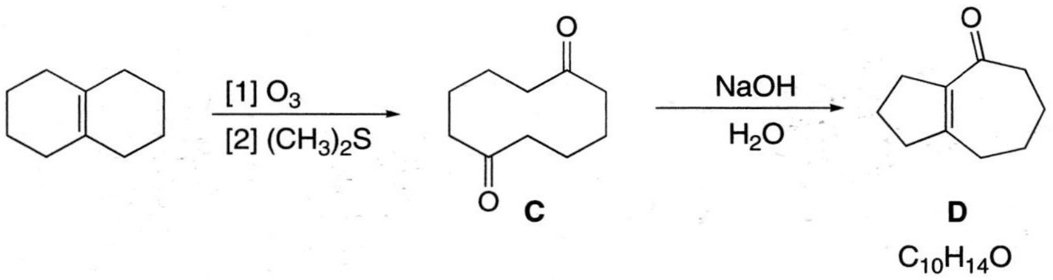

$C_{10}H_{14}O$

24.35

a. $C_6H_5CH_2CH_2CH_2CO_2Et$

b. $(CH_3)_2CHCH_2CH_2CH_2CO_2Et$

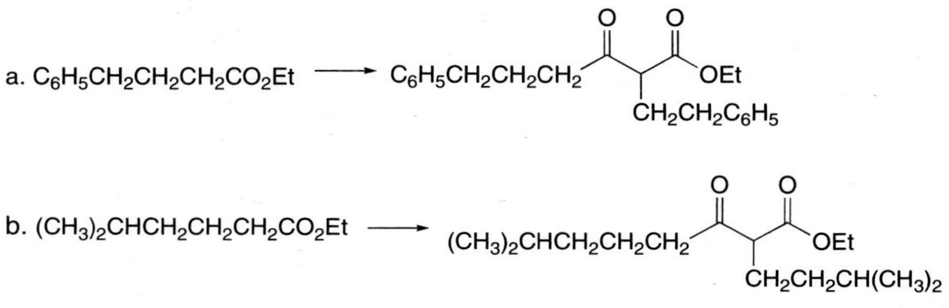

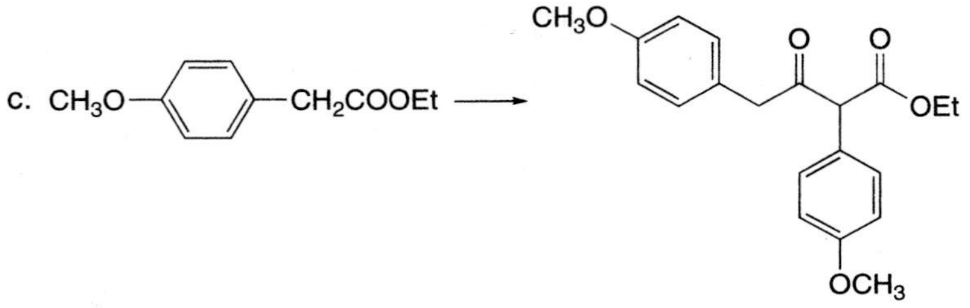

c. CH_3O—⟨aromatic⟩—CH_2COOEt ⟶

24.36

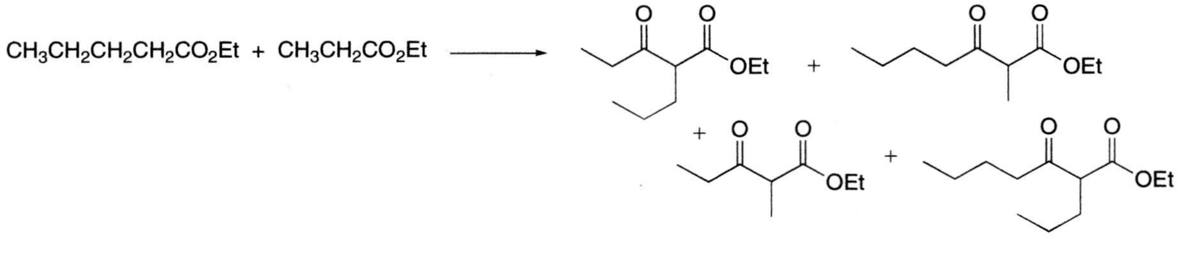

$CH_3CH_2CH_2CH_2CO_2Et$ + $CH_3CH_2CO_2Et$ ⟶

24.37

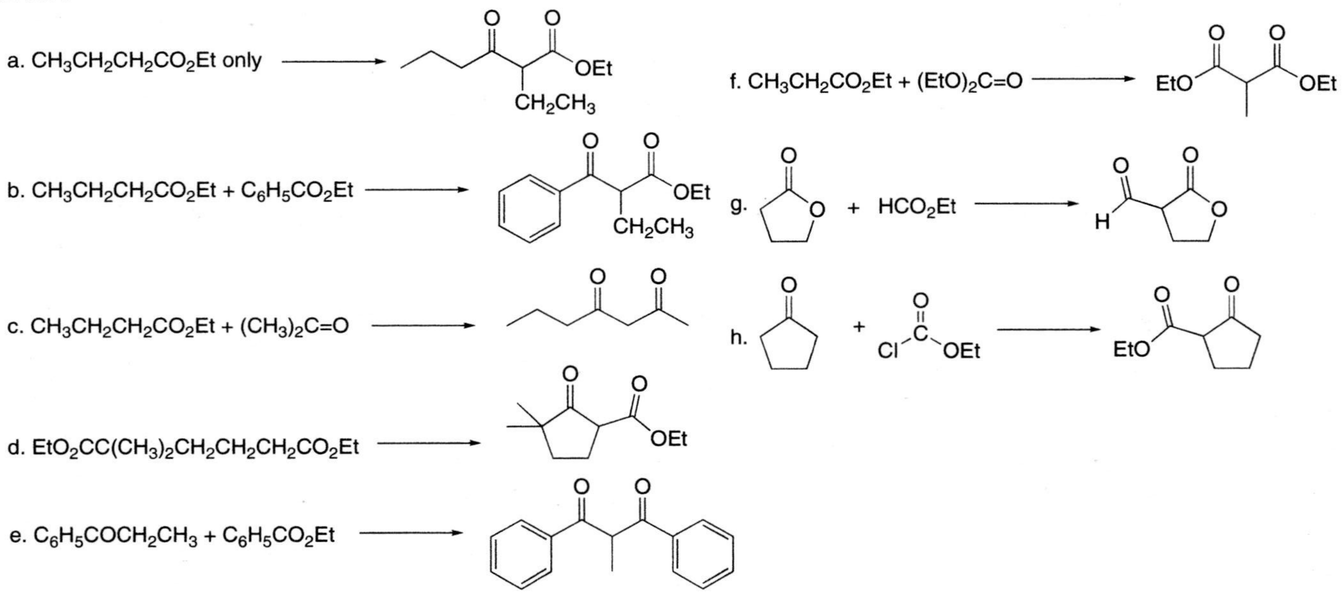

a. $CH_3CH_2CH_2CO_2Et$ only ⟶

b. $CH_3CH_2CH_2CO_2Et$ + $C_6H_5CO_2Et$ ⟶

c. $CH_3CH_2CH_2CO_2Et$ + $(CH_3)_2C=O$ ⟶

d. $EtO_2CC(CH_3)_2CH_2CH_2CH_2CO_2Et$ ⟶

e. $C_6H_5COCH_2CH_3$ + $C_6H_5CO_2Et$ ⟶

f. $CH_3CH_2CO_2Et$ + $(EtO)_2C=O$ ⟶

g. + HCO_2Et ⟶

h. + ⟶

24.38

a. ⟹ + $(EtO)_2C=O$

or

+

b. ⟹ +

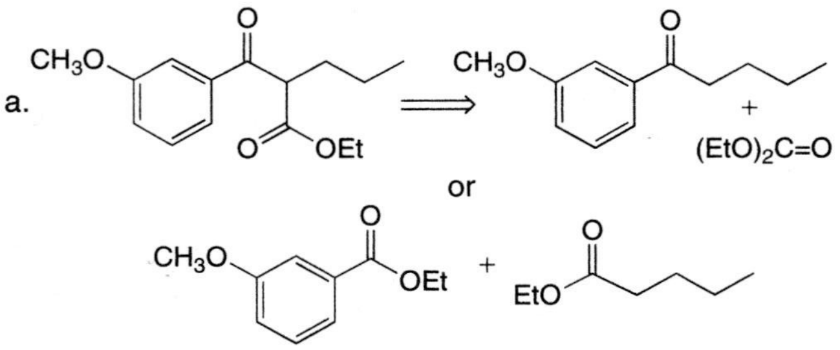

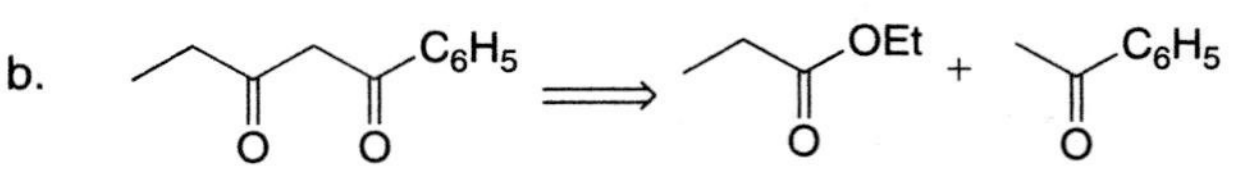

c.

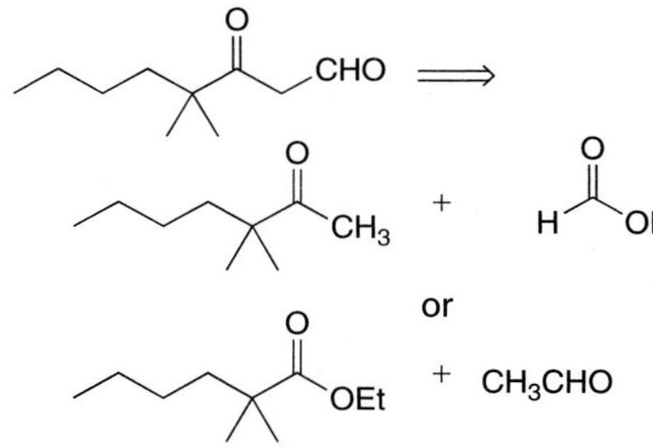

d. $C_6H_5CH(COOEt)_2 \implies$

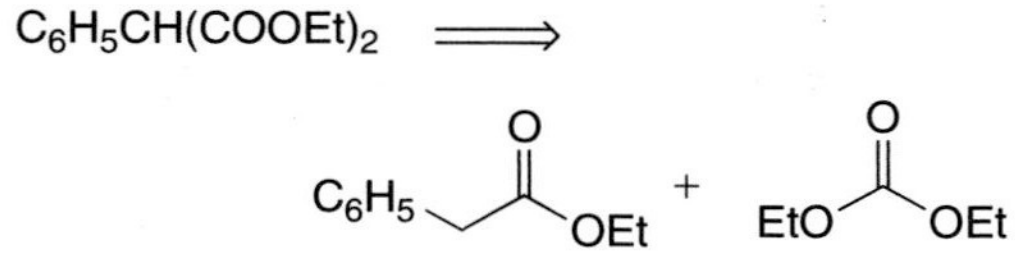

24.39

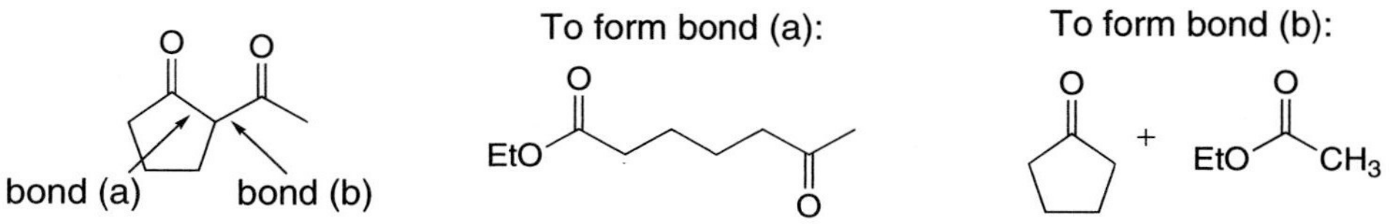

bond (a)   bond (b)

To form bond (a):

To form bond (b):

24.40 Only esters with 2 H's or 3 H's on the α carbon form enolates that undergo Claisen reaction to form resonance-stabilized enolates of the product β-keto ester. Thus, the enolate forms on the CH$_2$ α to one ester carbonyl, and cyclization yields a five-membered ring.

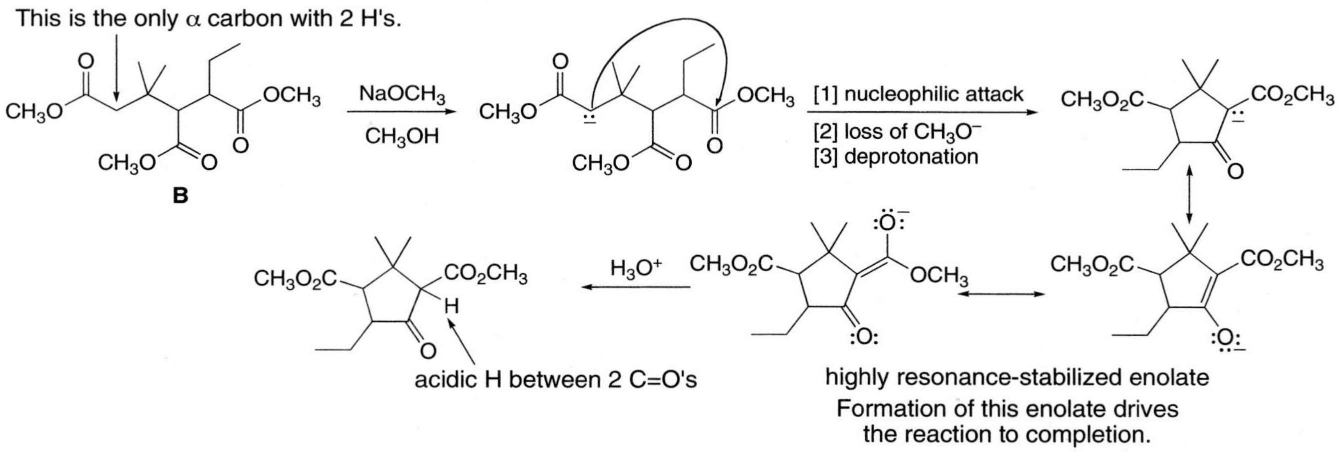

This is the only α carbon with 2 H's.

24.41

a.

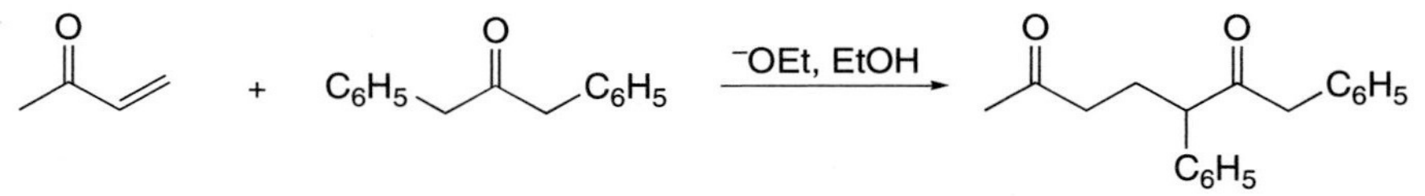

b.

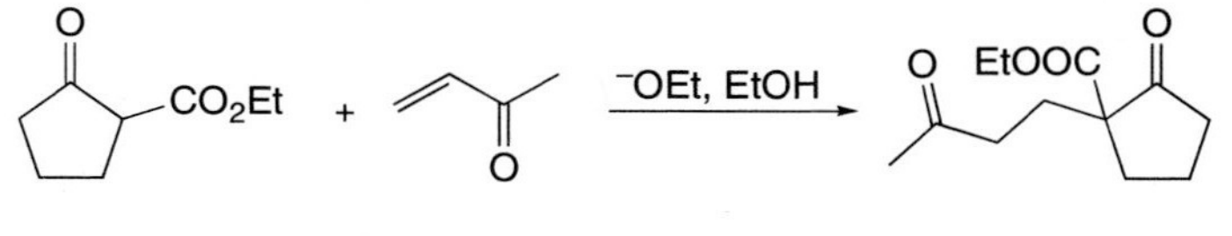

c.

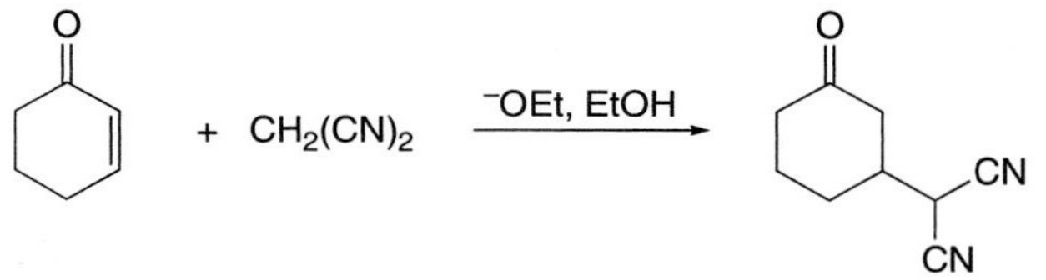

d.

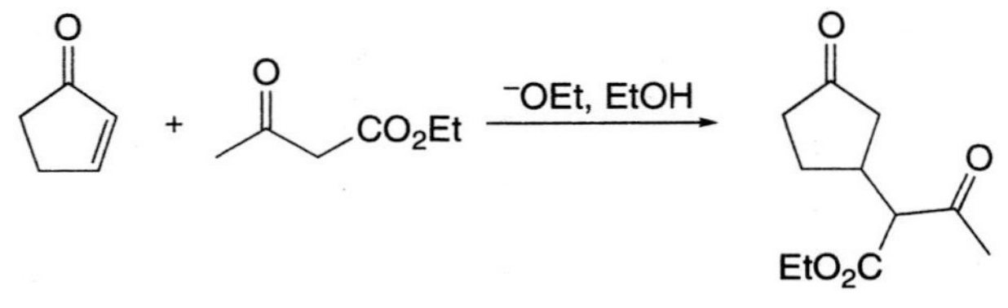

24.42

a.

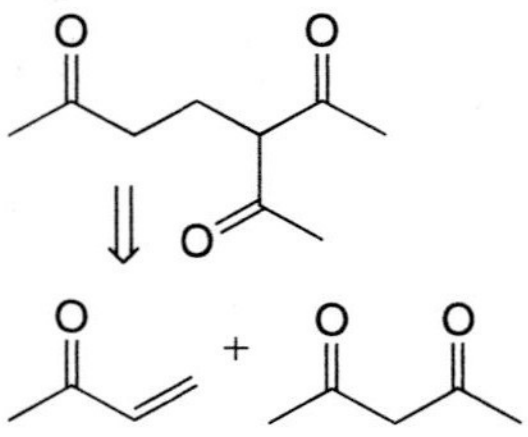

b.

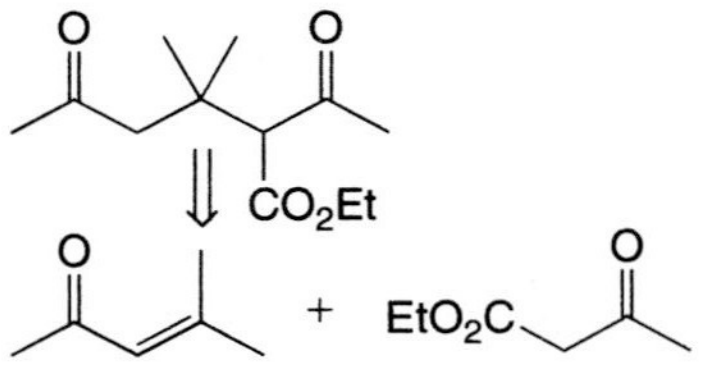

c.

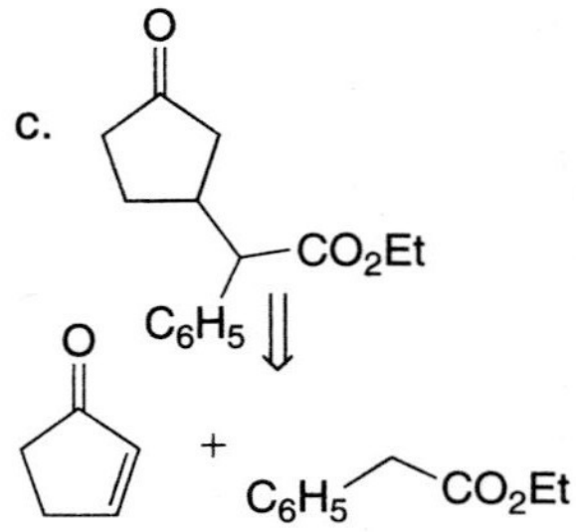

d.

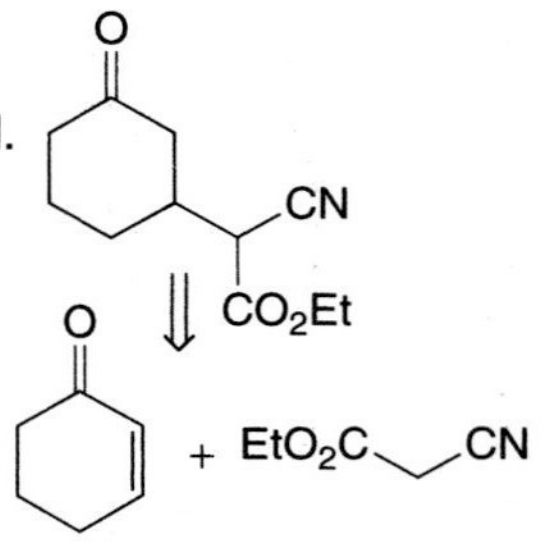

24.43

a.

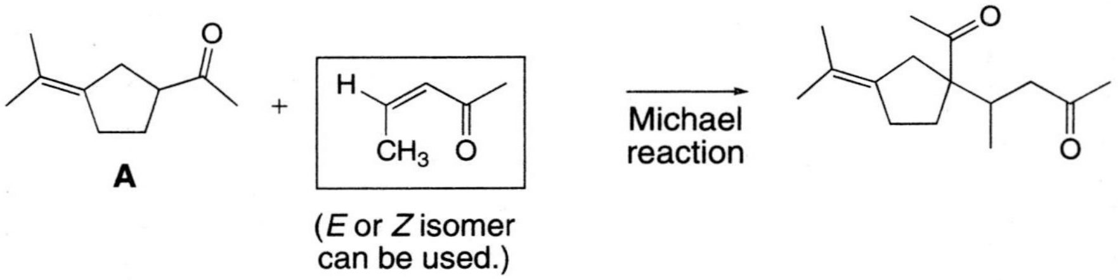

A

+

(*E* or *Z* isomer
can be used.)

$\xrightarrow{\text{Michael reaction}}$

b.

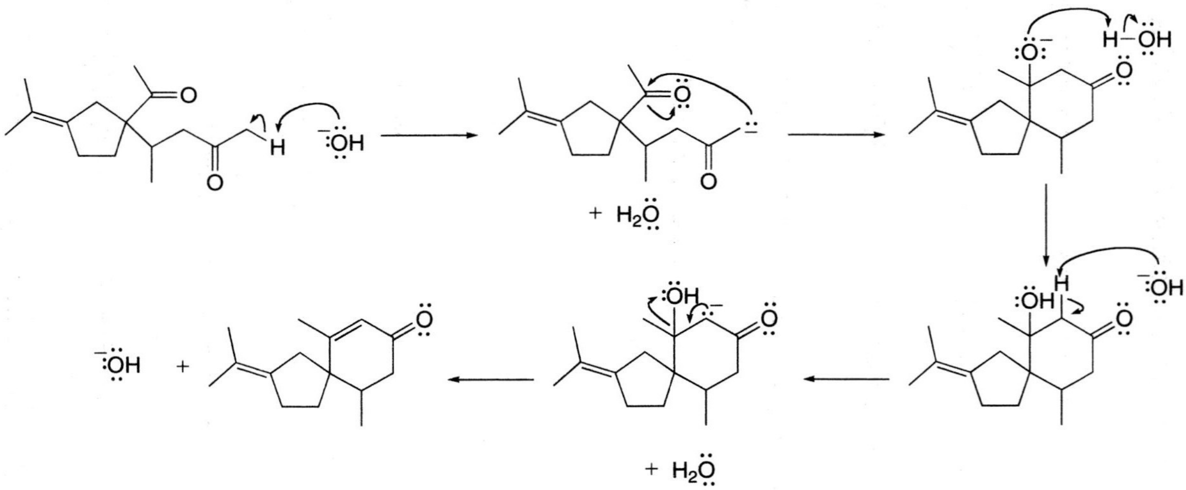

24.44

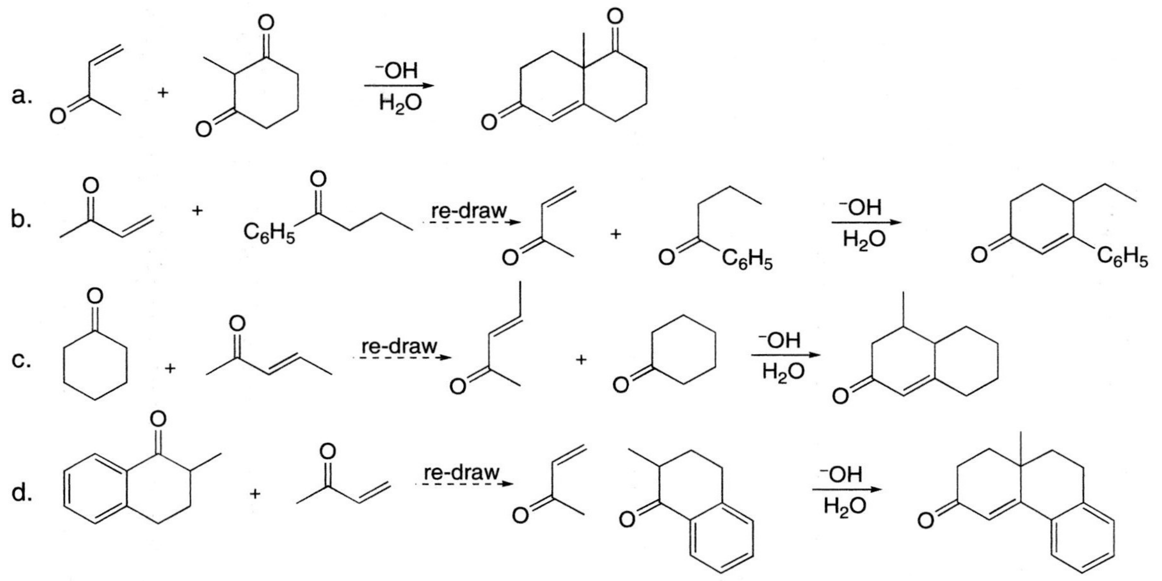

a.

b.

c.

d.

24.45

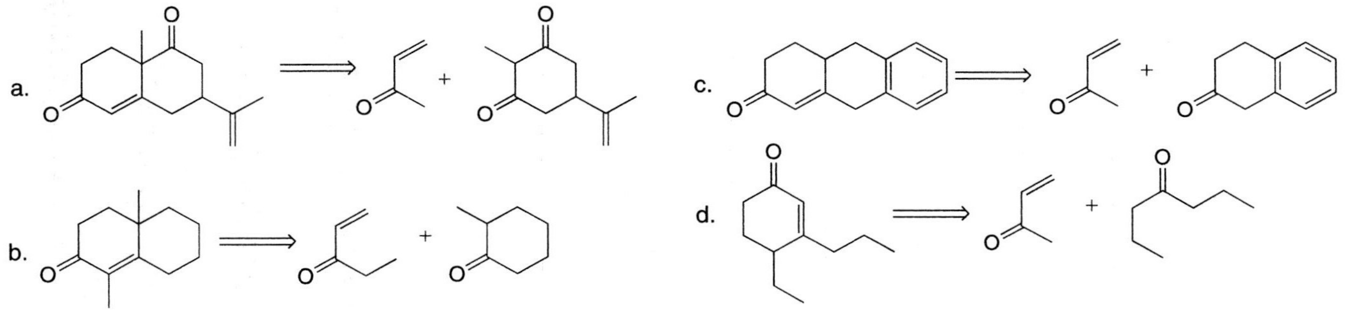

a.

b.

c.

d.

24.46

CH₃CH₂CH₂CHO

a. $\xrightarrow[H_2O]{^-OH}$ (E and Z)

b. $\xrightarrow[CH_2=O, H_2O]{^-OH}$

c. $\xrightarrow{[1] \ LDA \quad [2] \ CH_3CHO; \ [3] \ H_2O}$

d. $\xrightarrow[NaOEt, \ EtOH]{CH_2(CO_2Et)_2}$

e. $\xrightarrow[{[2] \ H_2O}]{[1] \ CH_3Li}$

f. $\xrightarrow[CH_3OH]{NaBH_4}$ CH₃CH₂CH₂CH₂OH

g. $\xrightarrow[Pd-C]{H_2}$ CH₃CH₂CH₂CH₂OH

h. $\xrightarrow[TsOH]{HOCH_2CH_2OH}$

i. $\xrightarrow[mild \ acid]{CH_3NH_2}$

j. $\xrightarrow[mild \ acid]{(CH_3)_2NH}$ CH₃CH₂CH=CHN(CH₃)₂ (E and Z)

k. $\xrightarrow[H_2SO_4]{CrO_3}$ CH₃CH₂CH₂COOH

l. $\xrightarrow[CH_3COOH]{Br_2}$

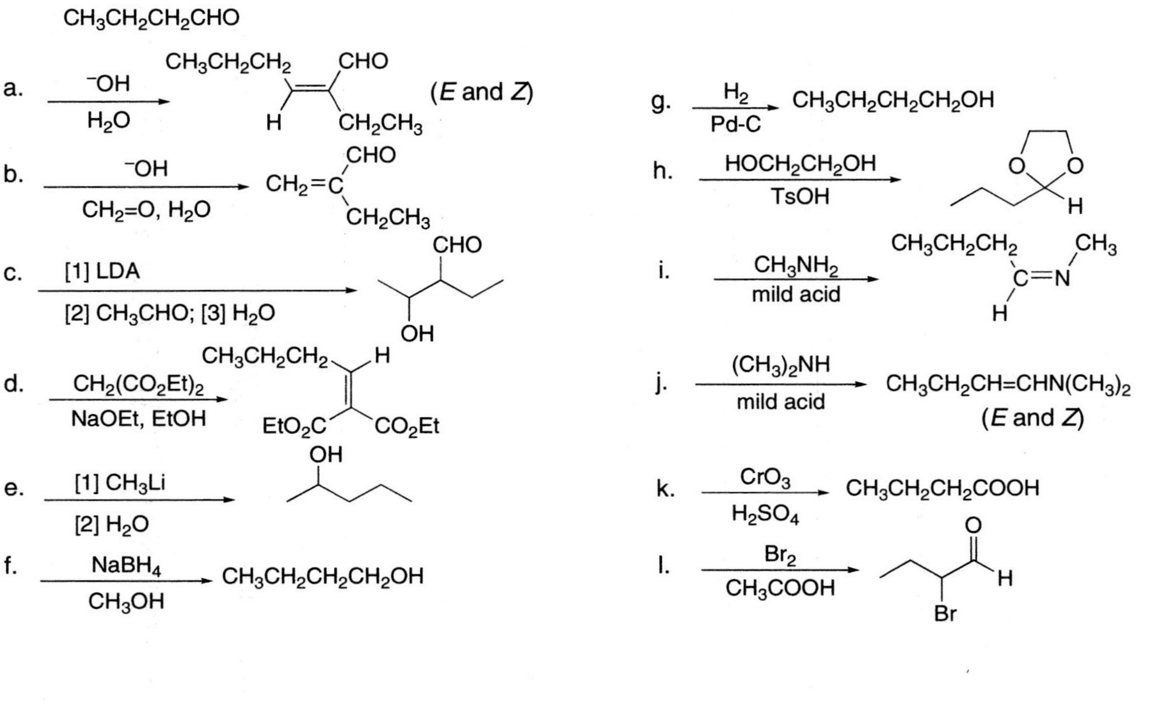

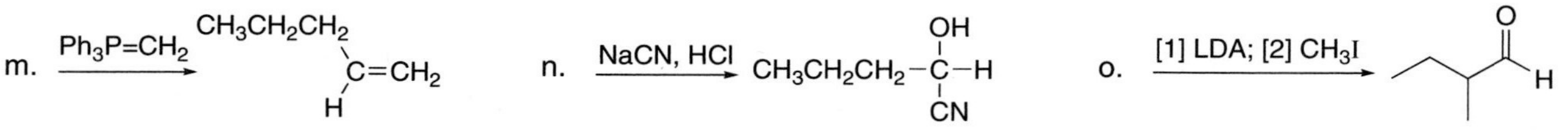

m. $\xrightarrow{Ph_3P=CH_2}$

n. $\xrightarrow{NaCN, HCl}$ $CH_3CH_2CH_2-\underset{CN}{\overset{OH}{C}}-H$

o. $\xrightarrow{[1]\ LDA;\ [2]\ CH_3I}$

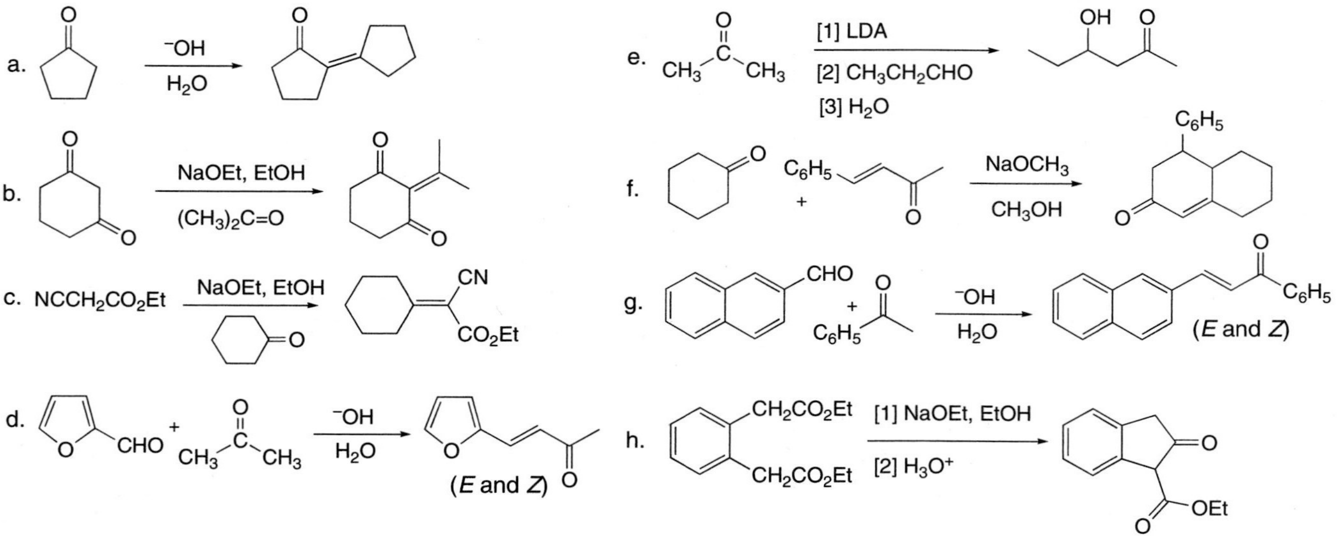

24.47

a. $\xrightarrow[H_2O]{^-OH}$

b. $\xrightarrow[(CH_3)_2C=O]{NaOEt,\ EtOH}$

c. $NCCH_2CO_2Et \xrightarrow[\overset{}{\underset{O}{\bigcirc}}]{NaOEt,\ EtOH}$

d. $\underset{O}{\overset{}{\bigcirc}}-CHO + \underset{CH_3}{\overset{O}{C}}CH_3 \xrightarrow[H_2O]{^-OH}$ (E and Z)

e. $CH_3\overset{O}{C}CH_3 \xrightarrow{\begin{array}{l}[1]\ LDA \\ [2]\ CH_3CH_2CHO \\ [3]\ H_2O\end{array}}$

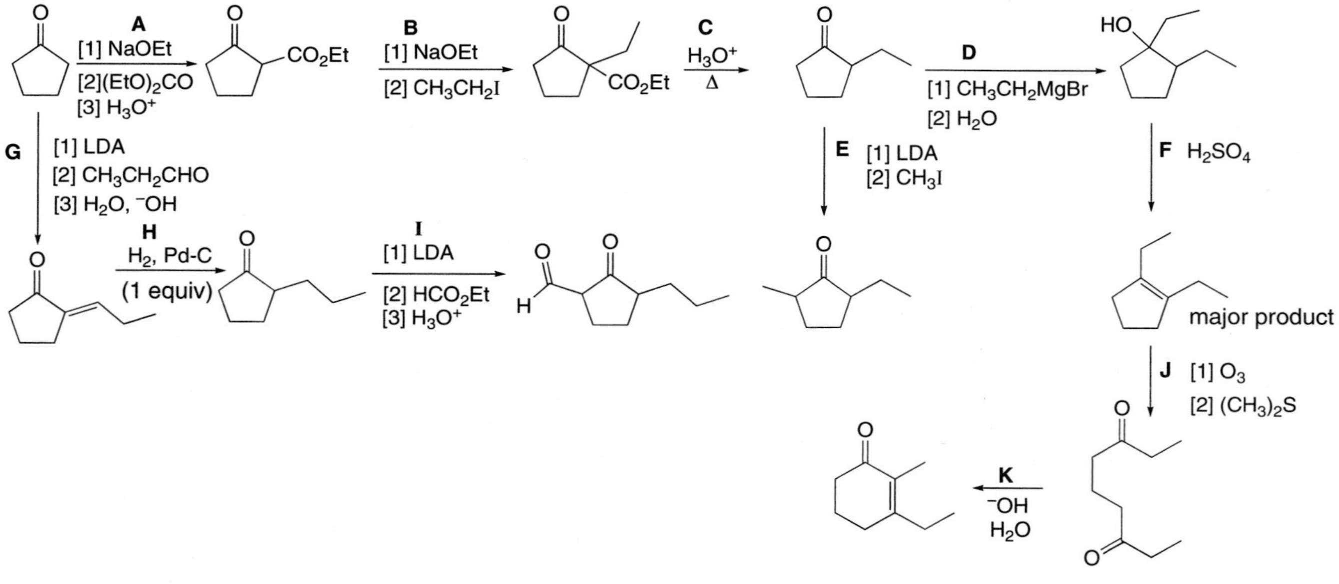

f. $\underset{}{\overset{O}{\bigcirc}} + \underset{}{\overset{C_6H_5}{\overset{}{\underset{O}{\diagup}}}} \xrightarrow[CH_3OH]{NaOCH_3}$

g. $\underset{}{\overset{CHO}{\bigcirc\bigcirc}} + \underset{C_6H_5}{\overset{O}{C}} \xrightarrow[H_2O]{^-OH}$ (E and Z)

h. $\underset{}{\overset{CH_2CO_2Et}{\underset{CH_2CO_2Et}{\bigcirc}}} \xrightarrow[{[2]\ H_3O^+}]{[1]\ NaOEt,\ EtOH}$

24.48

24.49

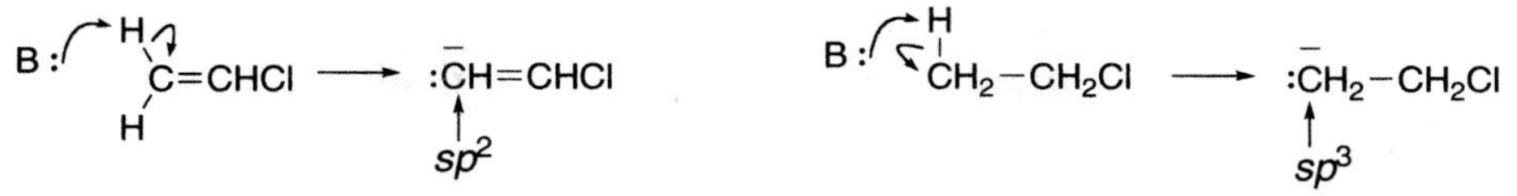

Vinyl halides undergo elimination by an E1cB mechanism more readily than alkyl halides because the carbanion intermediate formed from a vinyl halide has an sp^2 hybridized C, while the carbanion derived from CH_3CH_2Cl is sp^3 hybridized. The higher percent *s*-character of the sp^2 hybridized anion makes it more stable, and therefore, it is formed more readily.

24.50 The final step in the reaction sequence involves an intramolecular crossed Claisen reaction between a ketone and an ester to form a β-dicarbonyl compound.

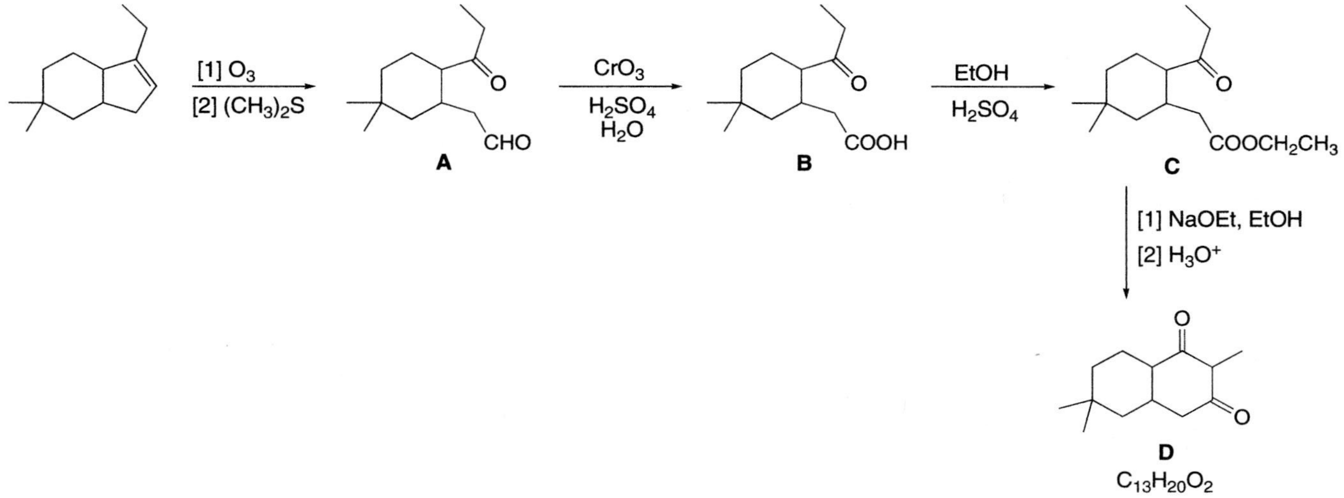

24.51

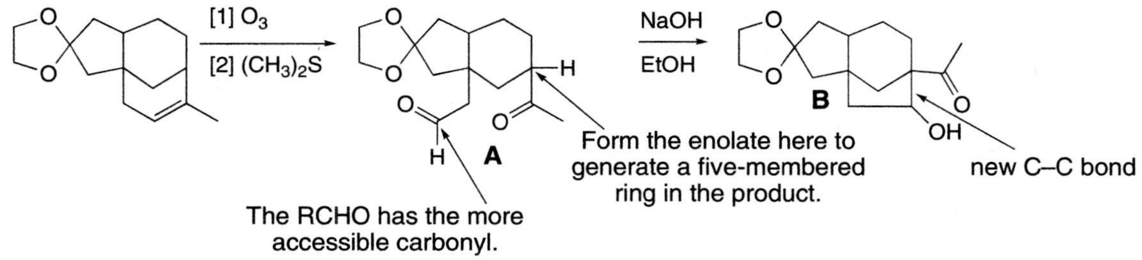

24.52

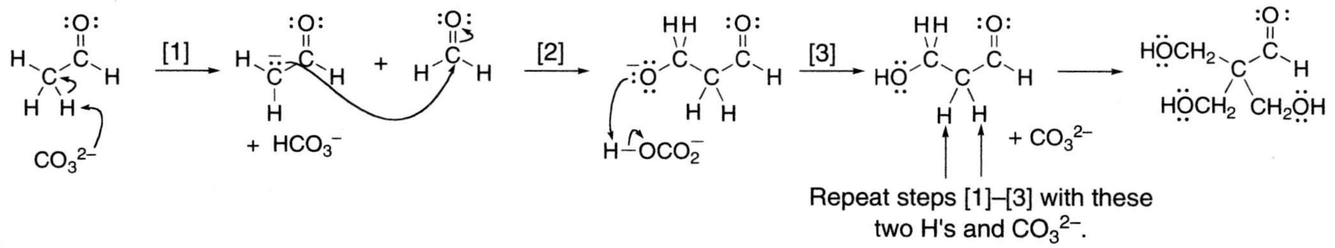

24.53 Enolate **A** is more substituted (and more stable) than either of the other two possible enolates and attacks an aldehyde carbonyl group, which is sterically less hindered than a ketone carbonyl. The resulting ring size (five-membered) is also quite stable. That is why 1-acetylcyclopentene is the major product.

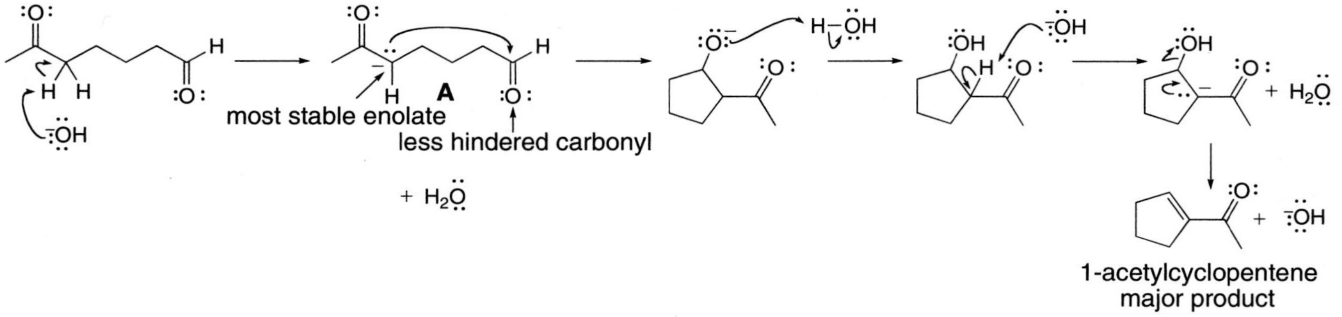

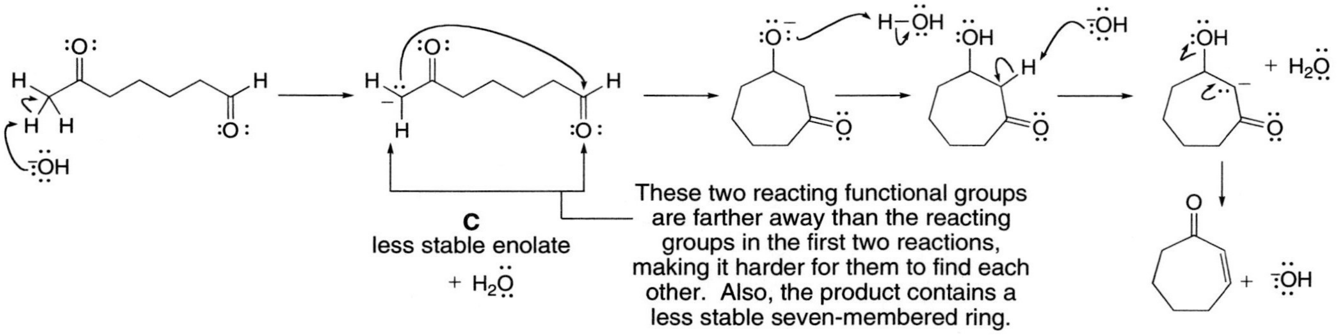

most stable enolate
less hindered carbonyl

+ H₂O

1-acetylcyclopentene
major product

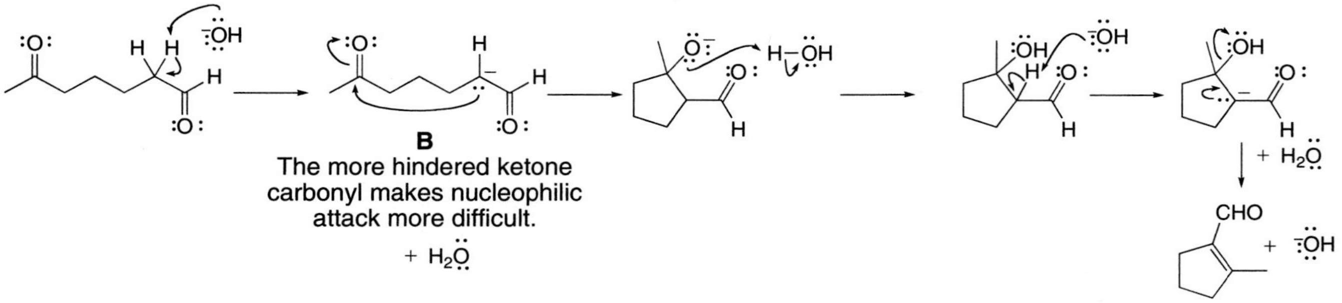

B
The more hindered ketone carbonyl makes nucleophilic attack more difficult.

+ H₂O

C
less stable enolate

+ H₂O

These two reacting functional groups are farther away than the reacting groups in the first two reactions, making it harder for them to find each other. Also, the product contains a less stable seven-membered ring.

24.54 All enolates have a second resonance structure with a negative charge on O.

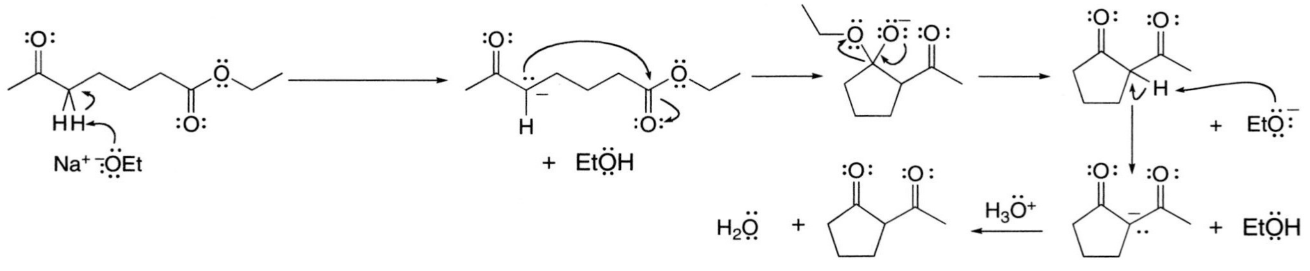

b.

24.55 Removal of a proton from CH_3NO_2 forms an anion for which three resonance structures can be drawn.

24.56 All enolates have a second resonance structure with a negative charge on O.

24.57 Polymerization occurs by repeated Michael reactions.

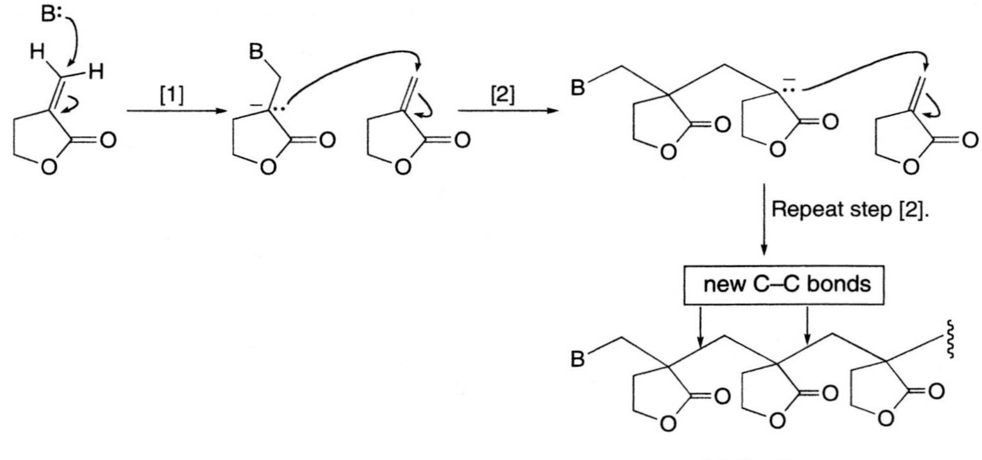

polytulipalin

24.58

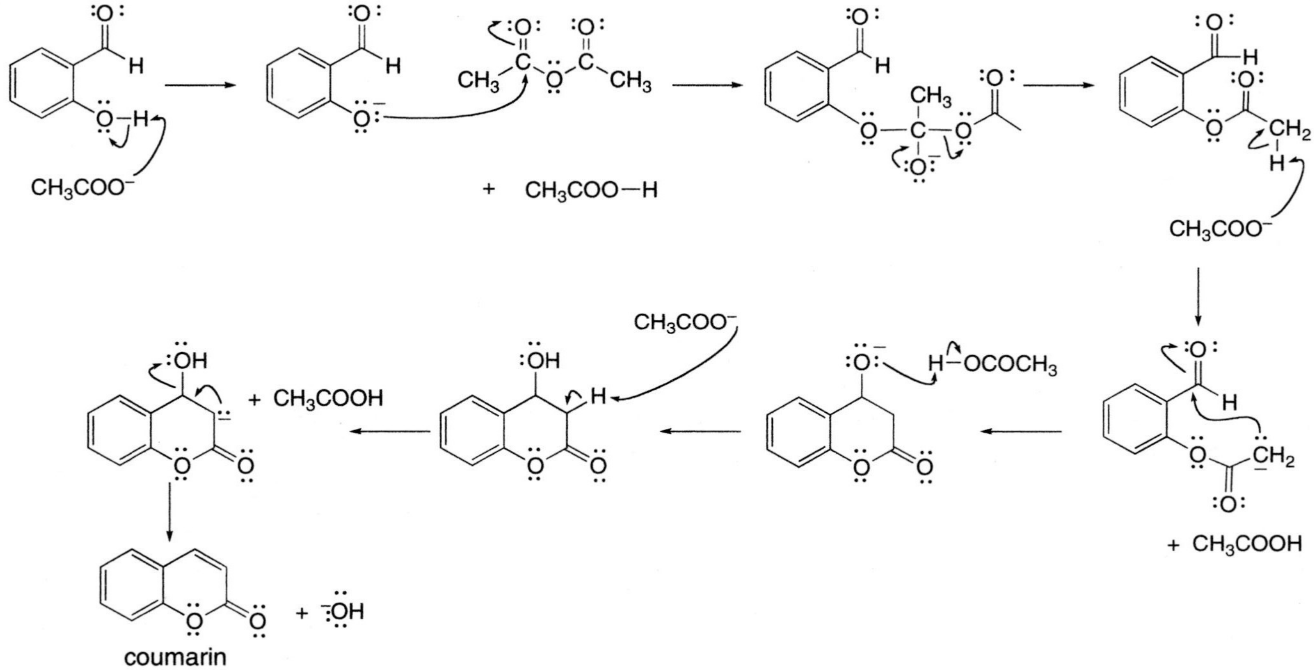

coumarin

24.59

a.

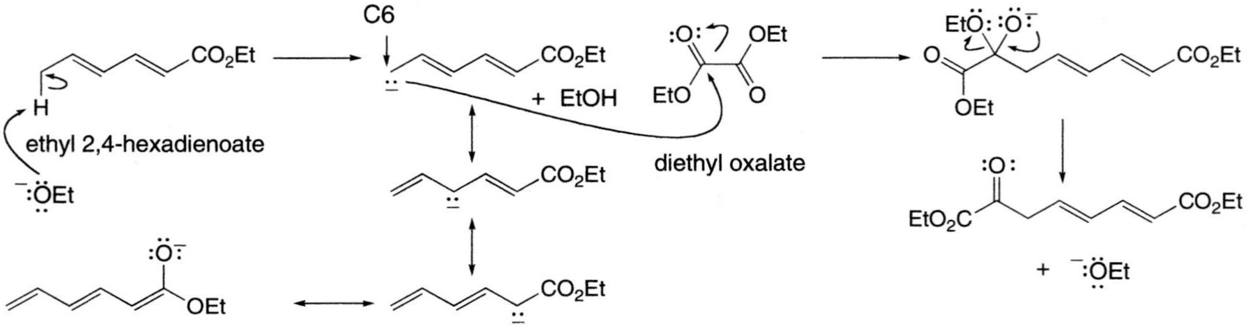

ethyl 2,4-hexadienoate

diethyl oxalate

b. The protons on C6 are more acidic than other sp^3 hybridized C–H bonds because a highly resonance-stabilized carbanion is formed when a proton is removed. One resonance structure places a negative charge on the carbonyl O atom. This makes the protons on C6 similar in acidity to the α H's to a carbonyl.

c. This is a crossed Claisen because it involves the enolate of a conjugated ester reacting with the carbonyl group of a second ester.

24.60

a.

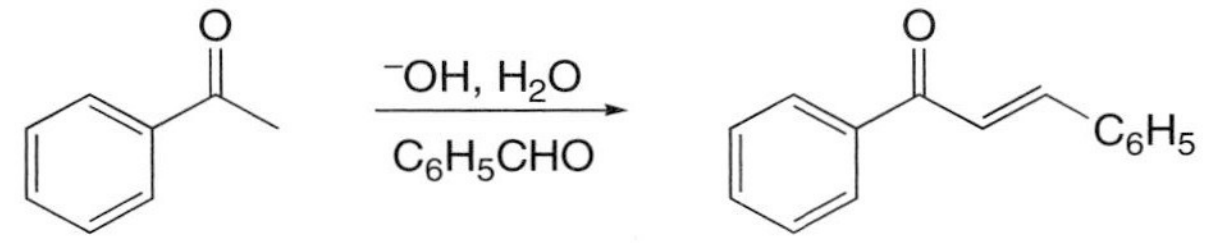

b.

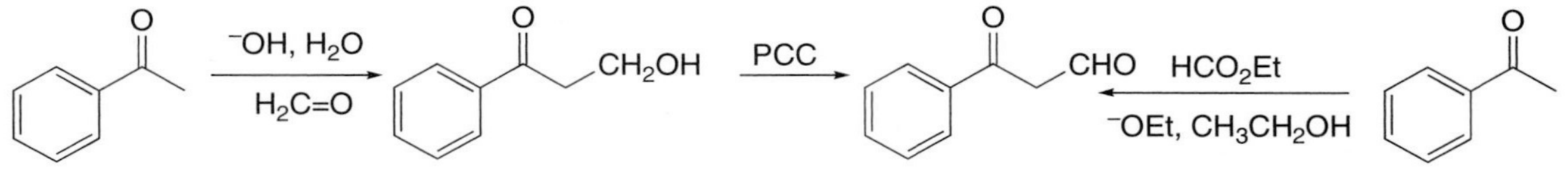

c.

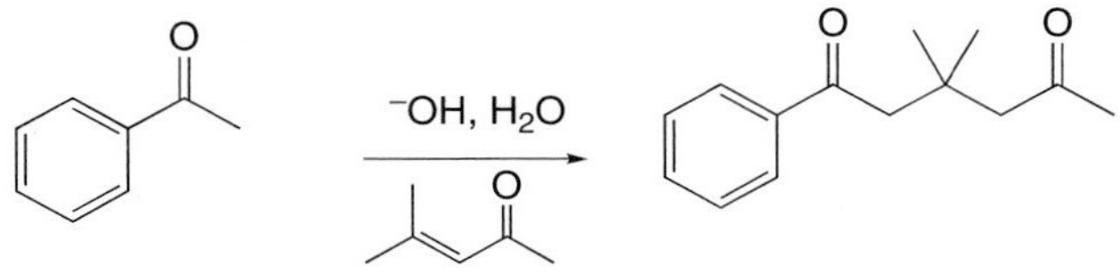

d.

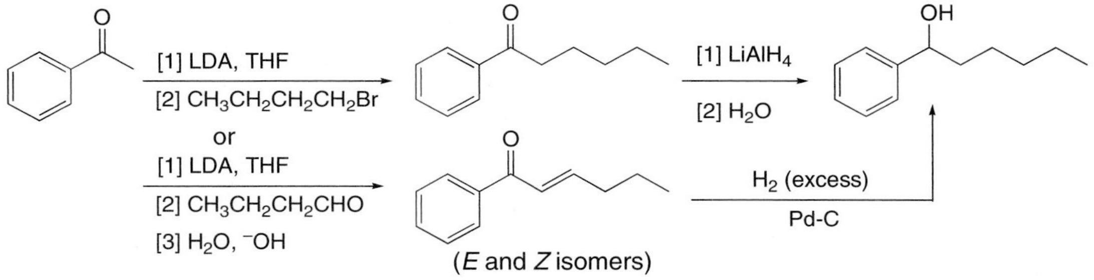

e.

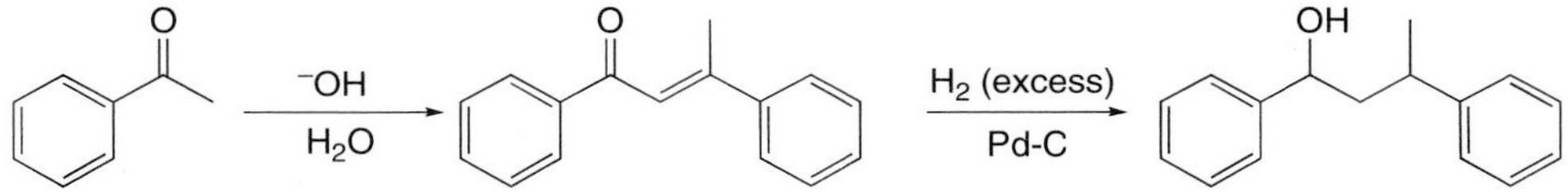

24.61

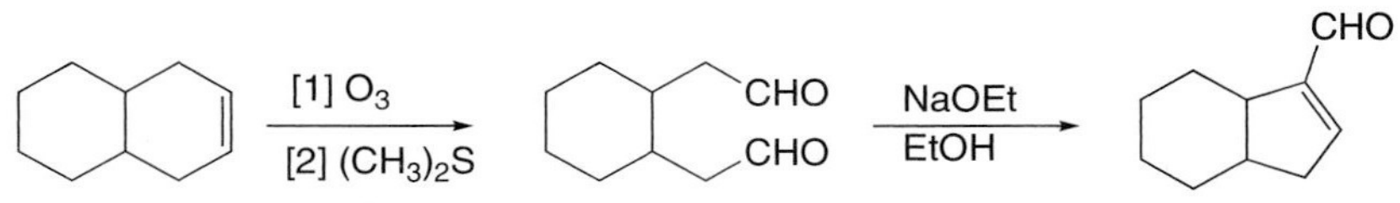

24.62

a.

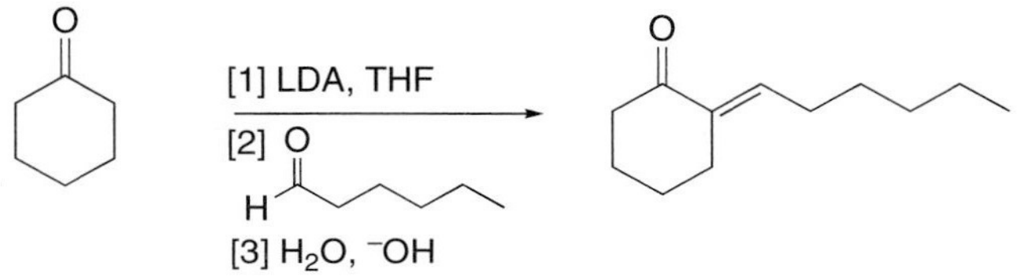

b.

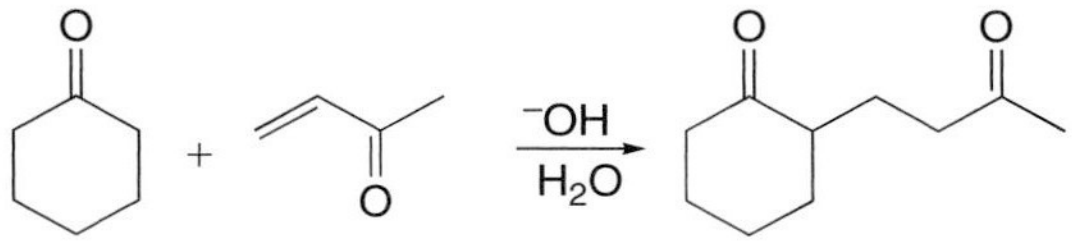

c.

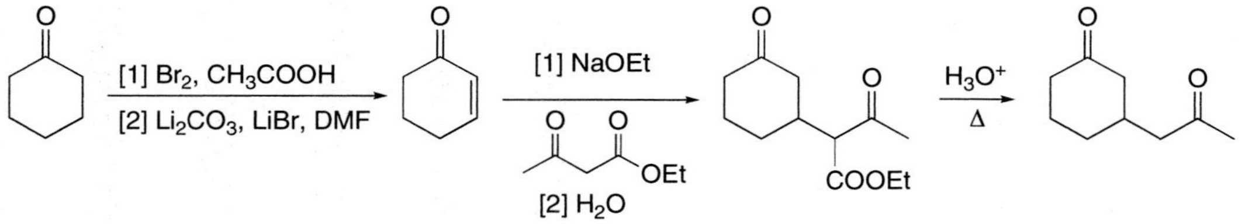

[1] Br₂, CH₃COOH

[2] Li₂CO₃, LiBr, DMF

[1] NaOEt

[2] H₂O

COOEt

H₃O⁺
Δ

d.

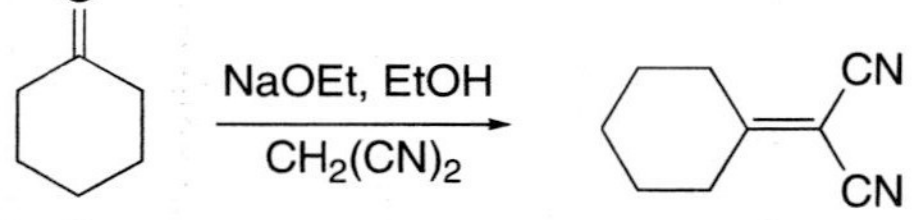

NaOEt, EtOH

CH₂(CN)₂

CN

CN

e.

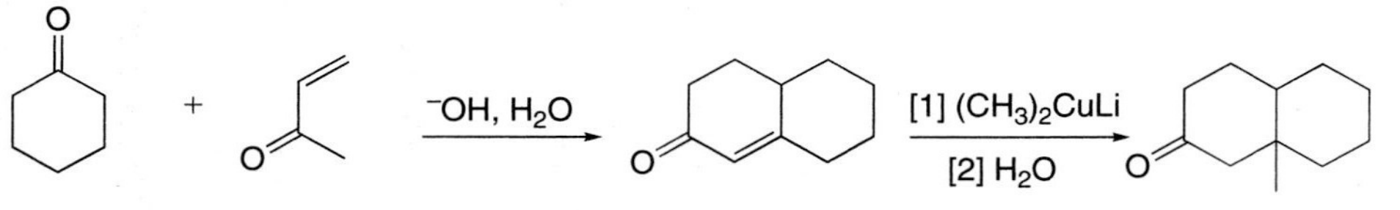

+

⁻OH, H₂O

[1] (CH₃)₂CuLi

[2] H₂O

24.63

a.

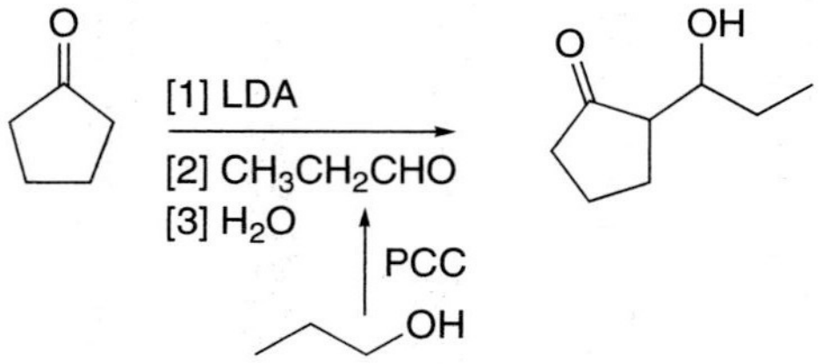

[1] LDA

[2] CH₃CH₂CHO

[3] H₂O

OH

PCC

OH

b.

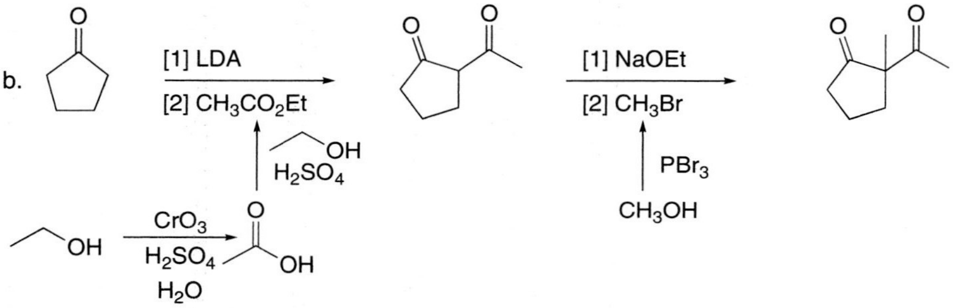

[1] LDA

[2] CH₃CO₂Et

[1] NaOEt

[2] CH₃Br

OH
H₂SO₄

PBr₃

CH₃OH

OH

CrO₃
H₂SO₄
H₂O

O

OH

c.

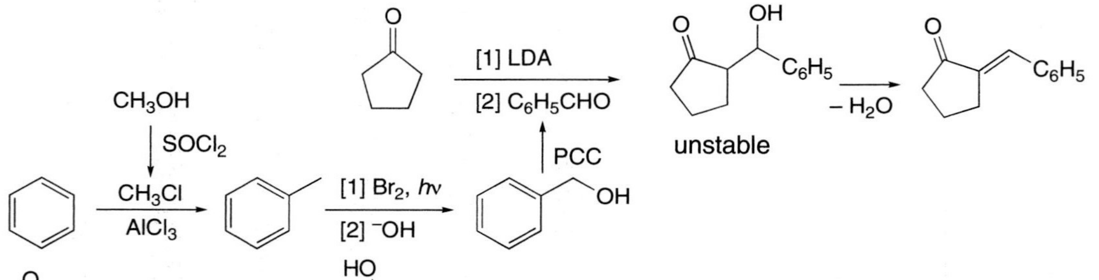

CH₃OH

SOCl₂

CH₃Cl

AlCl₃

[1] Br₂, hv

[2] ⁻OH

OH

PCC

[1] LDA

[2] C₆H₅CHO

OH
C₆H₅

unstable

C₆H₅

– H₂O

d.

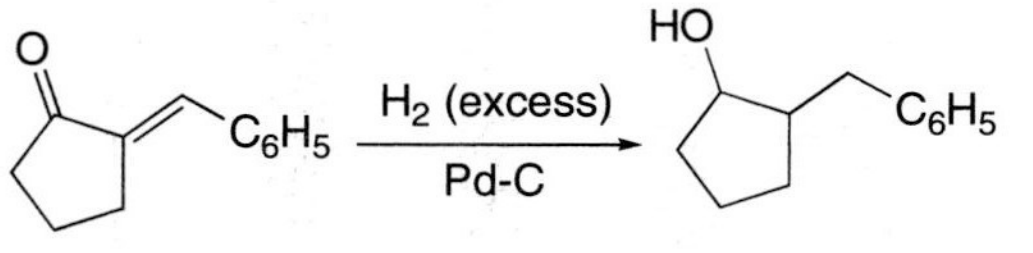

C₆H₅

(from c.)

H₂ (excess)

Pd-C

HO
C₆H₅

e.

[1] CH₃MgBr

[2] H₂O

OH

Mg

H₂SO₄

major
product

[1] O₃

[2] (CH₃)₂S

CHO

⁻OH, H₂O

CH₃OH

PBr₃

CH₃Br

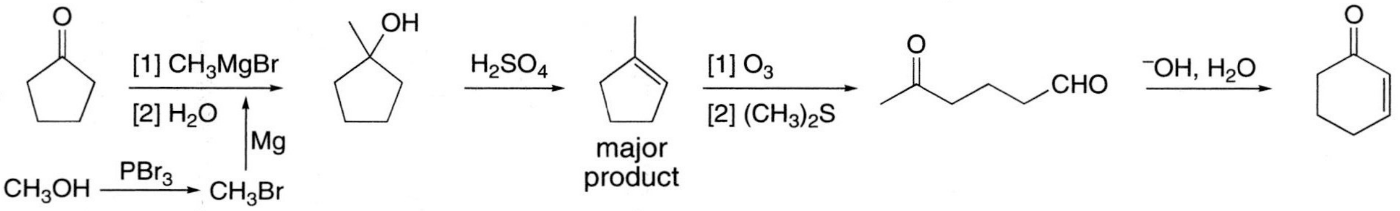

24.64

a.

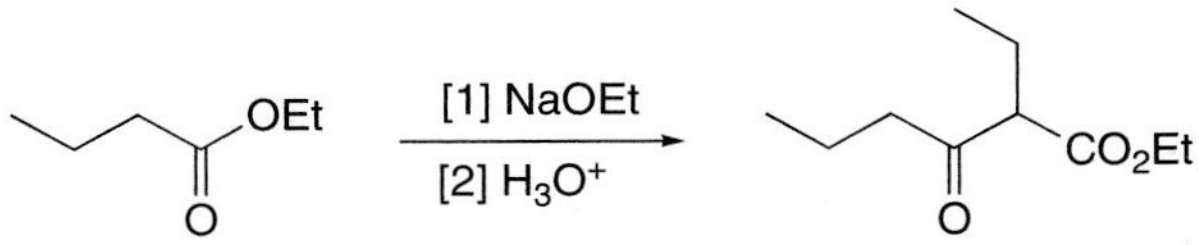

b.

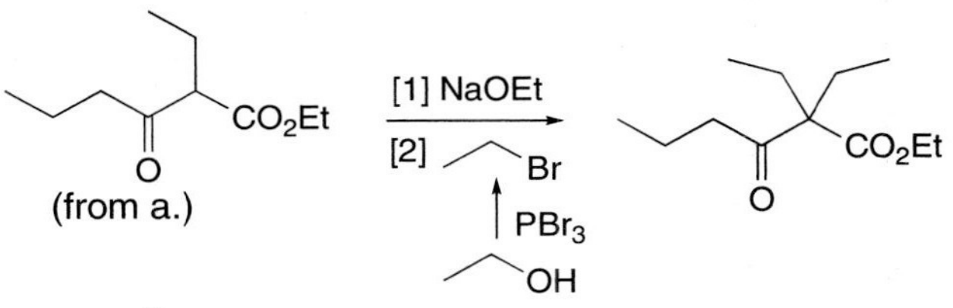

c.

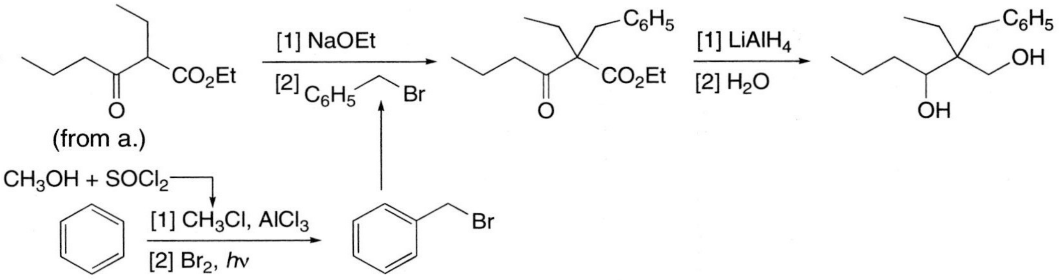

d.

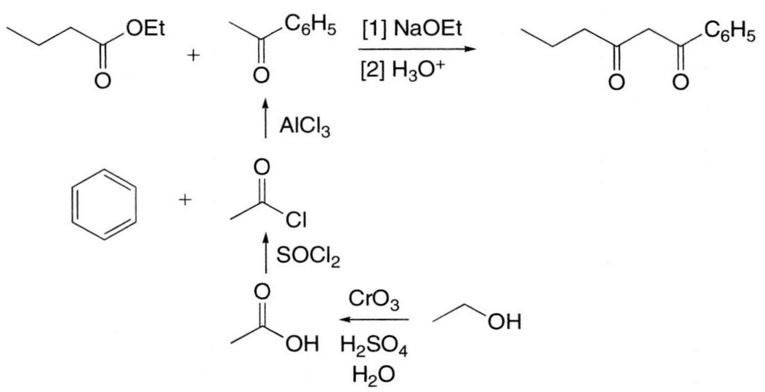

24.65

a.

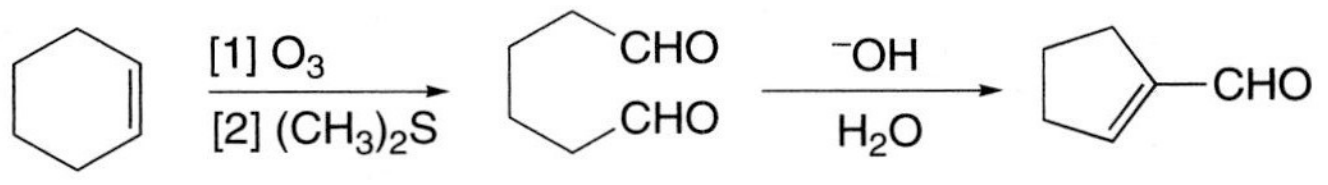

b.

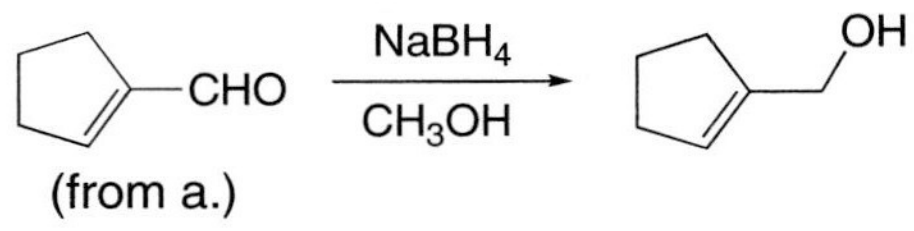

c.

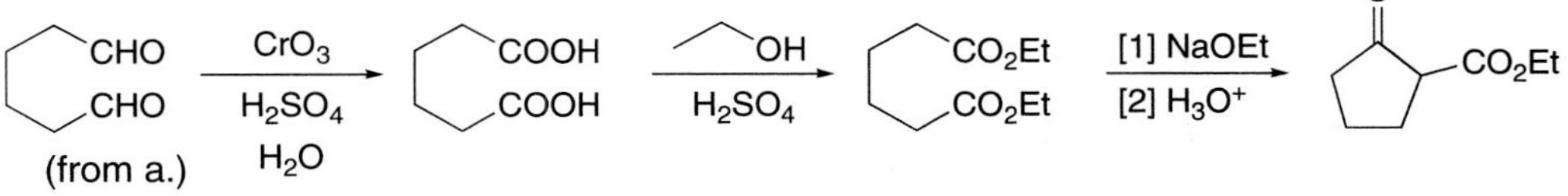

d.

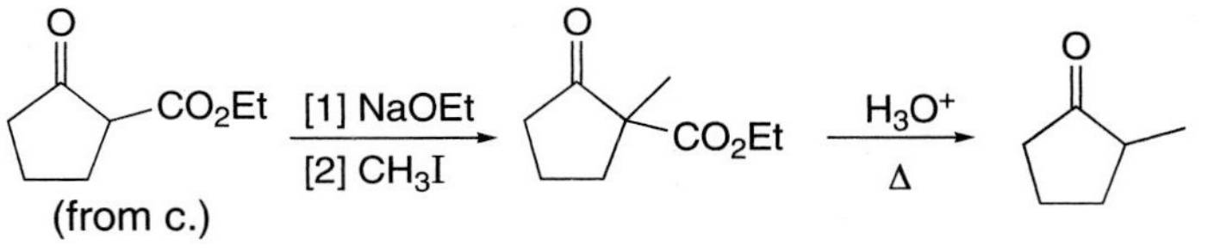

24.66

a.

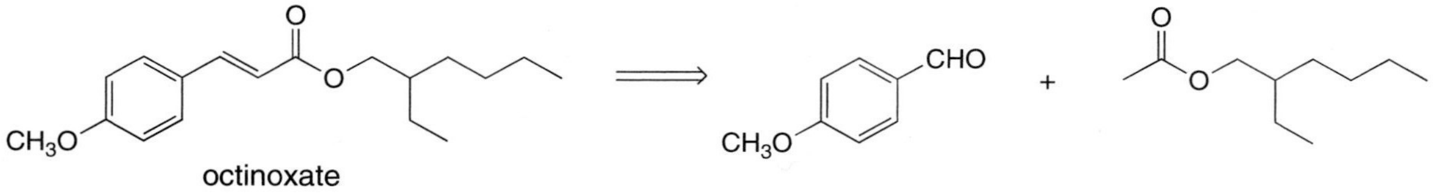

octinoxate

b.

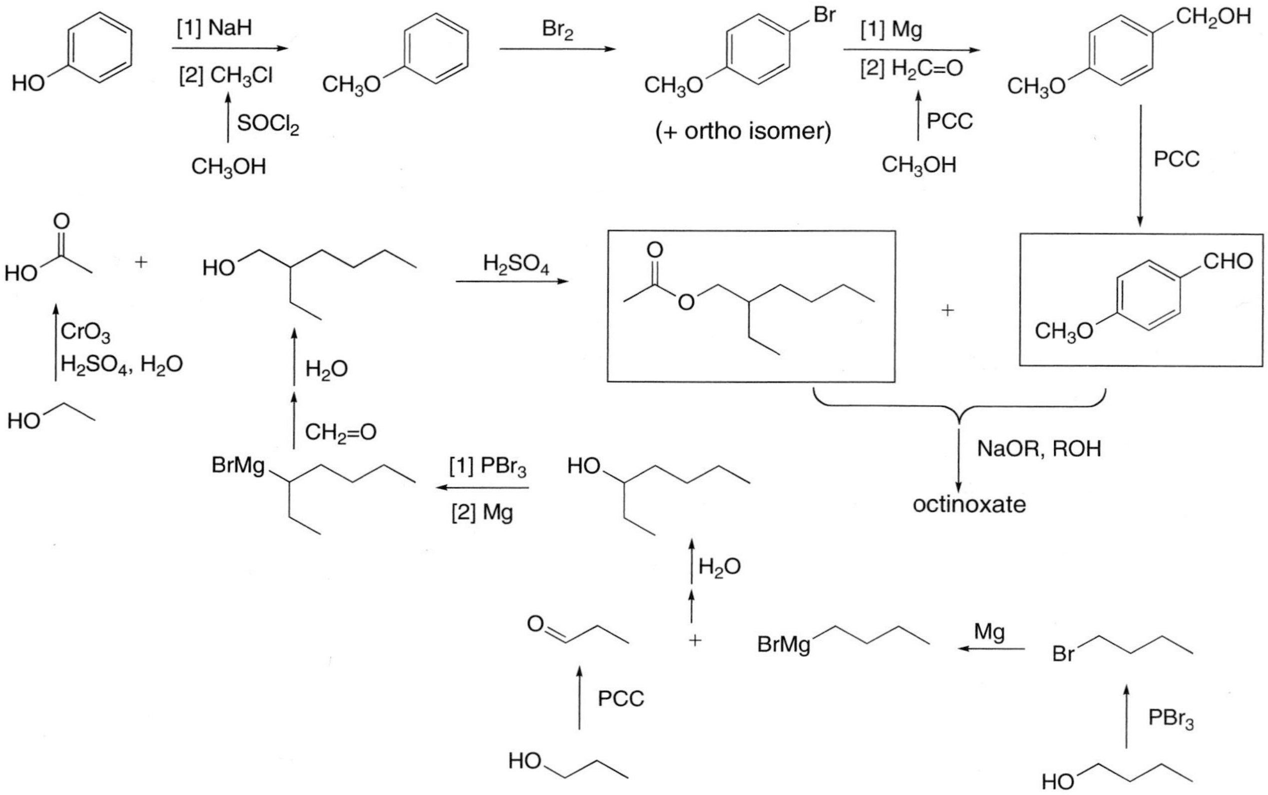

(+ ortho isomer)

24.67

a.

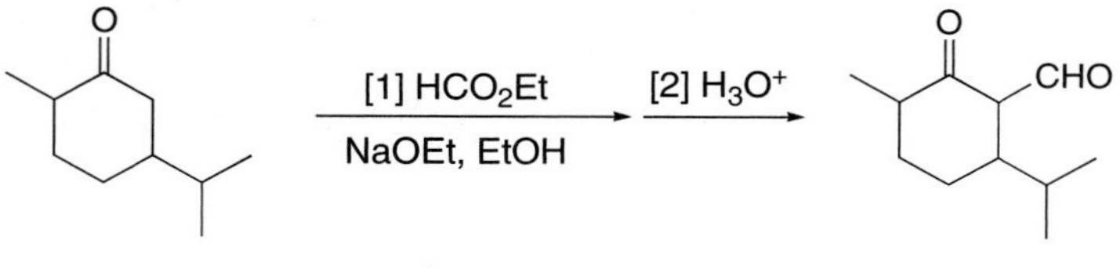

[1] HCO₂Et, NaOEt, EtOH → [2] H₃O⁺

b.

NaOEt, EtOH

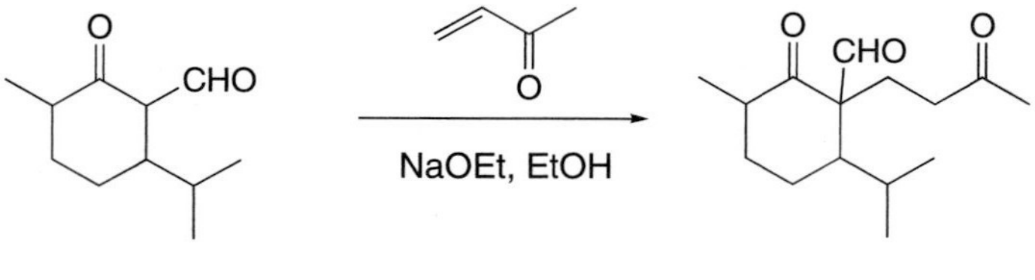

c.

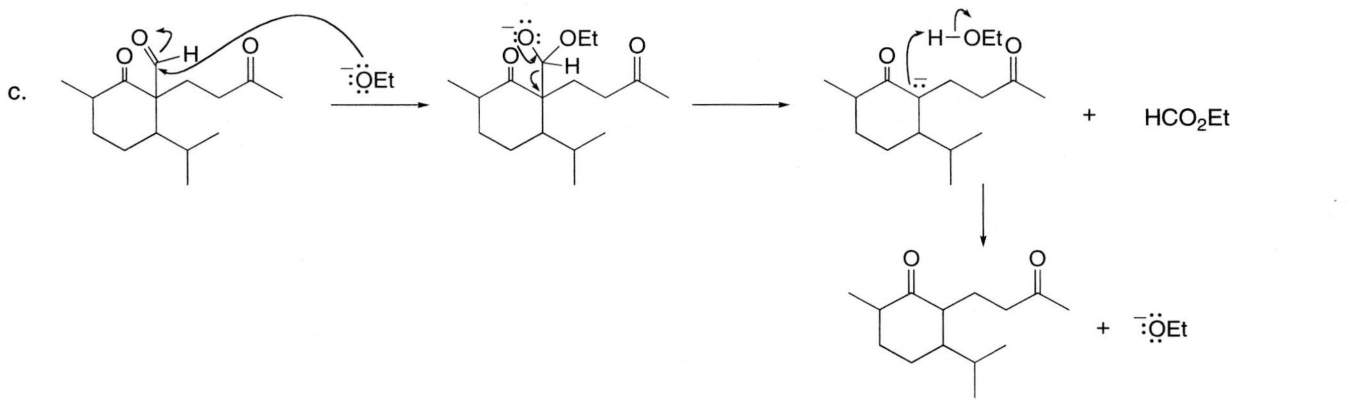

:ÖEt
OEt
H
H–OEt
+ HCO₂Et
+ :ÖEt

d.

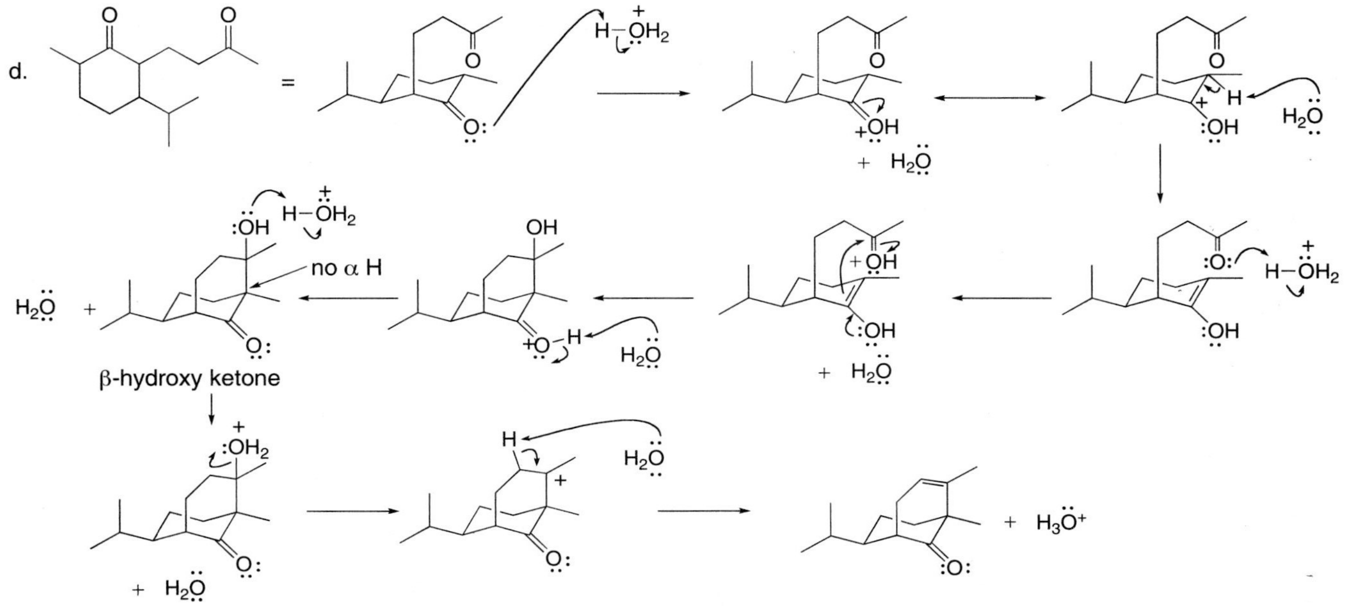

H–OH₂
+OH
+ H₂O
:OH
H₂O
:O:
H–OH₂
+OH
:OH
+ H₂O
OH
H₂O
no α H
H₂O +
β-hydroxy ketone
:OH₂
+ H₂O
H
+ H₃O+

This reaction is an acid-catalyzed aldol that proceeds by way of enols not enolates. The β-hydroxy ketone initially formed cannot dehydrate to form an α,β-unsaturated carbonyl because there is no H on the α carbon. Thus, dehydration occurs but the resulting C=C is not conjugated with the C=O.

e.

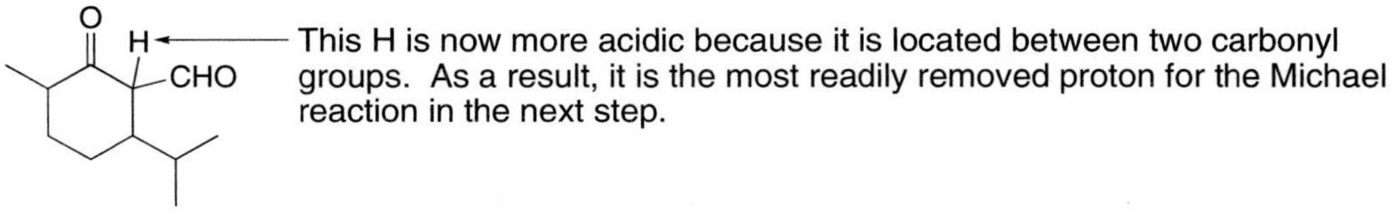

H
CHO
This H is now more acidic because it is located between two carbonyl groups. As a result, it is the most readily removed proton for the Michael reaction in the next step.

24.68 Rearrangement generates a highly resonance-stabilized enolate between two carbonyl groups.

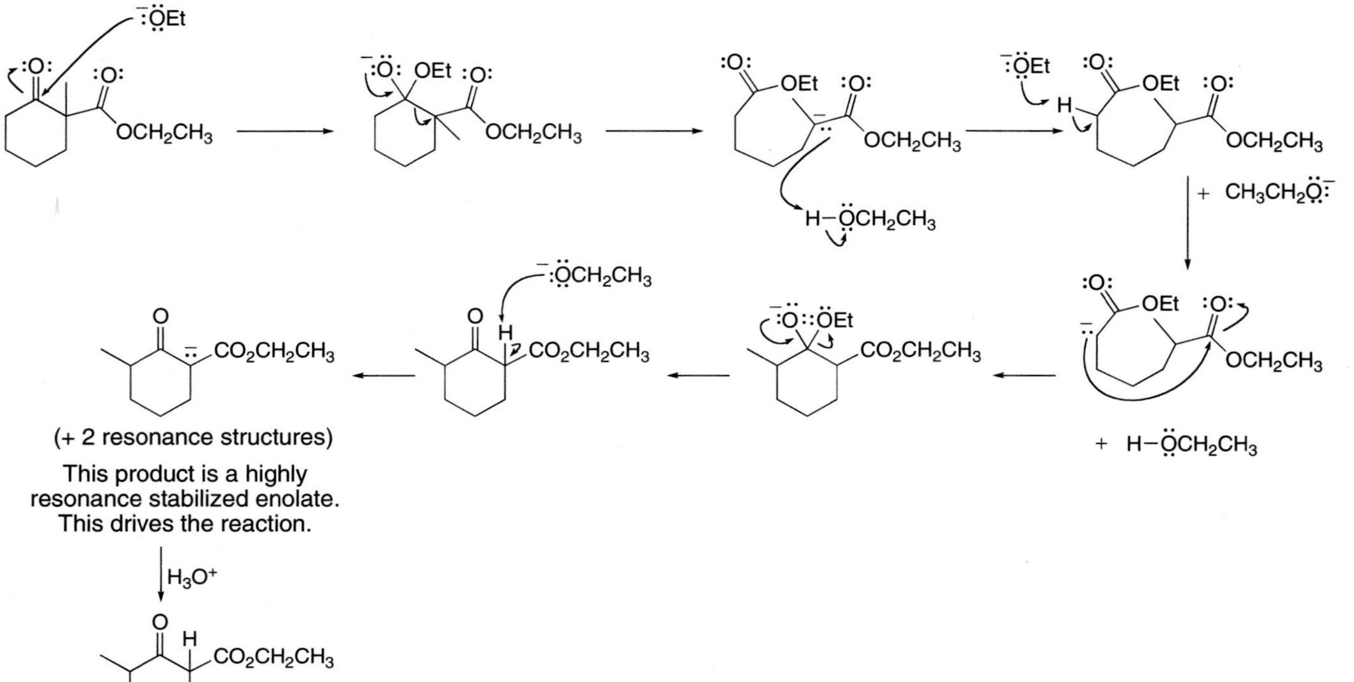

(+ 2 resonance structures)

This product is a highly
resonance stabilized enolate.
This drives the reaction.

24.69 All enolates have a second resonance structure with a negative charge on O.

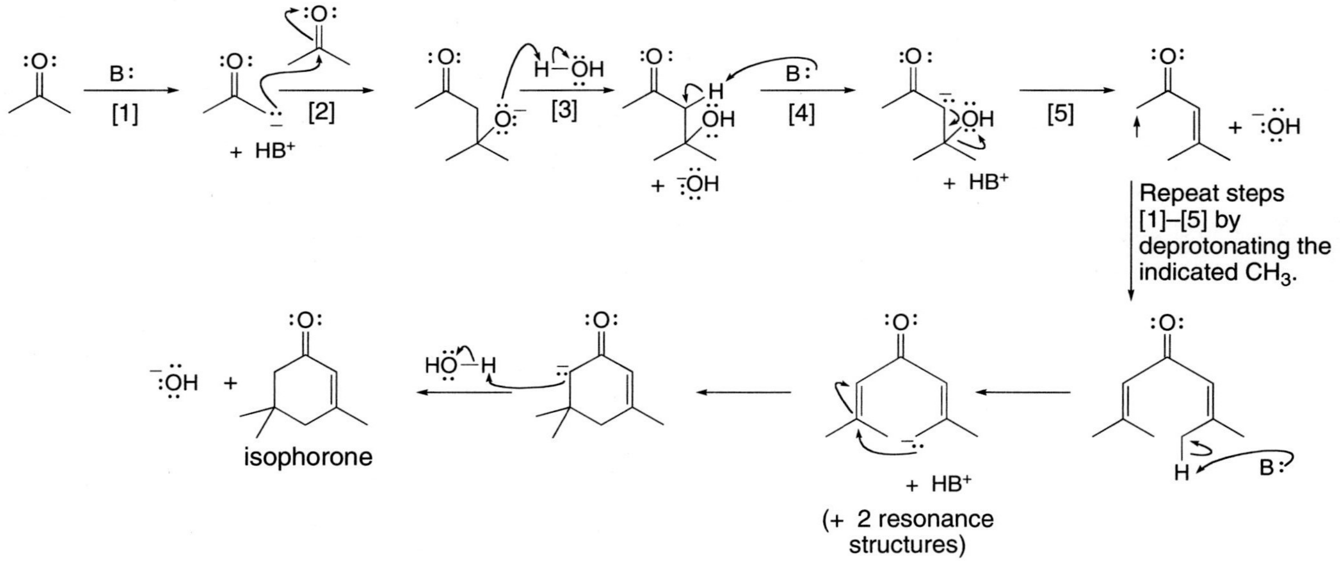

Repeat steps
[1]–[5] by
deprotonating the
indicated CH_3.

isophorone

+ HB+

(+ 2 resonance
structures)

24.70 All enolates have a second resonance structure with a negative charge on O.

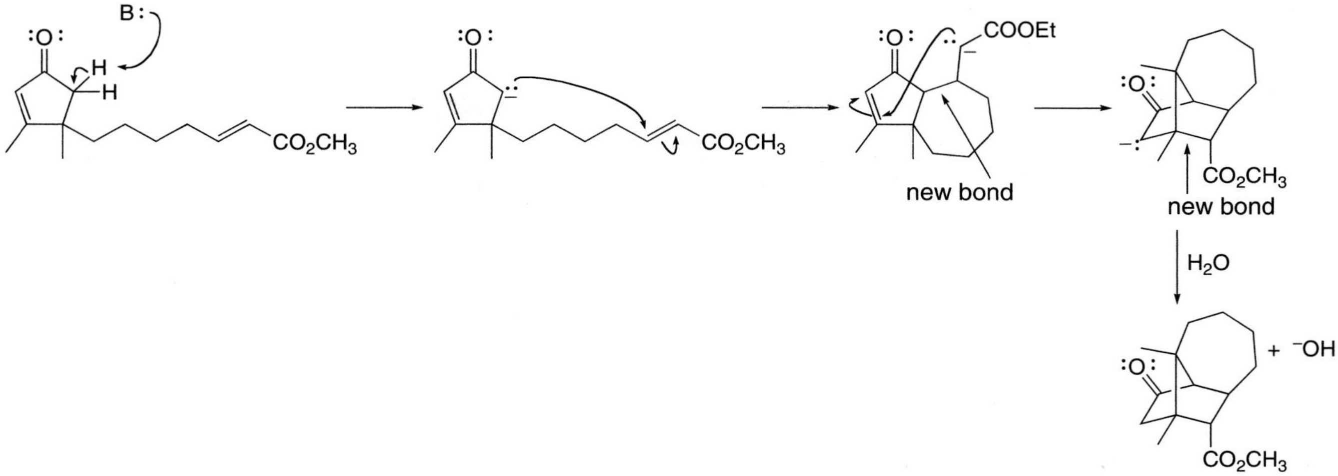

24.71

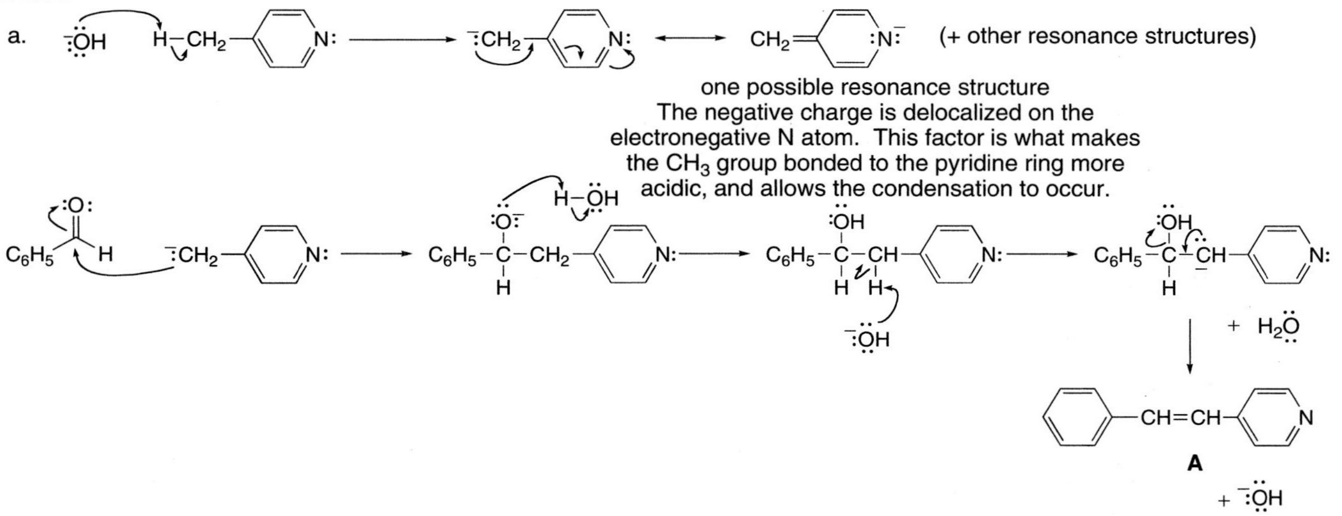

a.

one possible resonance structure
The negative charge is delocalized on the
electronegative N atom. This factor is what makes
the CH$_3$ group bonded to the pyridine ring more
acidic, and allows the condensation to occur.

b. The condensation reaction can occur only if the CH$_3$ group bonded to the pyridine ring has acidic
hydrogens that can be removed with $^-$OH.

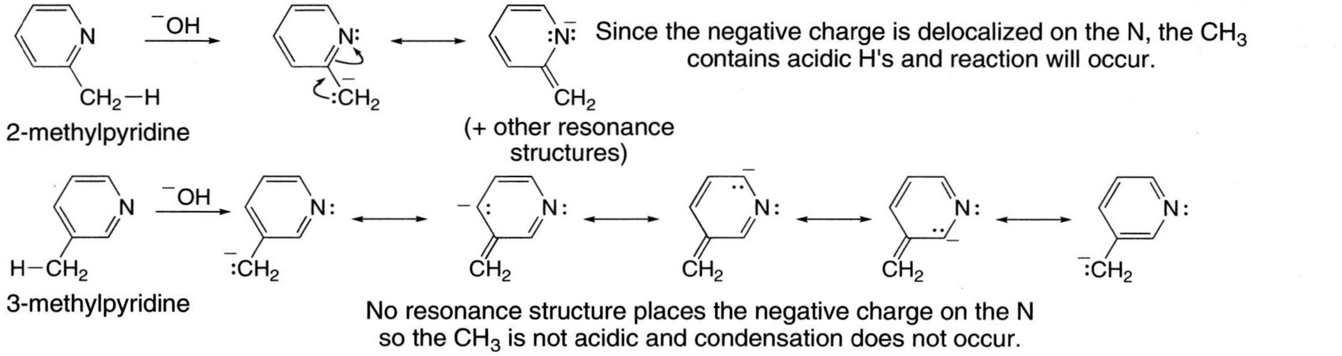

2-methylpyridine

Since the negative charge is delocalized on the N, the CH$_3$
contains acidic H's and reaction will occur.

(+ other resonance
structures)

3-methylpyridine

No resonance structure places the negative charge on the N
so the CH$_3$ is not acidic and condensation does not occur.

Chapter 25: Amines

◆ General facts

- Amines are organic nitrogen compounds having the general structure RNH_2, R_2NH, or R_3N, with a lone pair of electrons on N (25.1).
- Amines are named using the suffix -*amine* (25.3).
- All amines have polar C–N bonds. Primary (1°) and 2° amines have polar N–H bonds and are capable of intermolecular hydrogen bonding (25.4).
- The lone pair on N makes amines strong organic bases and nucleophiles (25.8).

◆ Summary of spectroscopic absorptions (25.5)

Mass spectra	Molecular ion	Amines with an odd number of N atoms give an odd molecular ion.
IR absorptions	N–H	3300–3500 cm^{-1} (two peaks for RNH_2, one peak for R_2NH)
^{1}H NMR absorptions	NH	0.5–5 ppm (no splitting with adjacent protons)
	CH–N	2.3–3.0 ppm (deshielded Csp^3–H)
^{13}C NMR absorption	C–N	30–50 ppm

◆ Comparing the basicity of amines and other compounds (25.10)

- Alkylamines (RNH_2, R_2NH, and R_3N) are more basic than NH_3 because of the electron-donating R groups (25.10A).
- Alkylamines (RNH_2) are more basic than arylamines ($C_6H_5NH_2$), which have a delocalized lone pair from the N atom (25.10B).
- Arylamines with electron-donor groups are more basic than arylamines with electron-withdrawing groups (25.10B).
- Alkylamines (RNH_2) are more basic than amides ($RCONH_2$), which have a delocalized lone pair from the N atom (25.10C).
- Aromatic heterocycles with a localized electron pair on N are more basic than those with a delocalized lone pair from the N atom (25.10D).
- Alkylamines with a lone pair in an sp^3 hybrid orbital are more basic than those with a lone pair in an sp^2 hybrid orbital (25.10E).

◆ Preparation of amines (25.7)

[1] Direct nucleophilic substitution with NH_3 and amines (25.7A)

- The mechanism is S_N2.
- The reaction works best for CH_3X or RCH_2X.
- The reaction works best to prepare 1° amines and ammonium salts.

[2] Gabriel synthesis (25.7A)

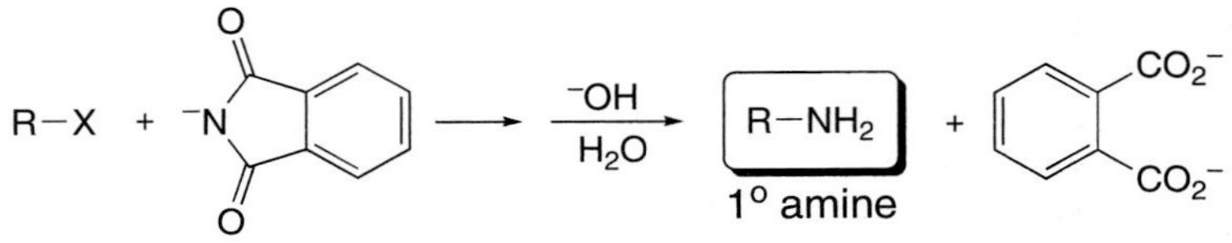

R—X + phthalimide anion $\xrightarrow{\begin{array}{c}^-OH\\ \hline H_2O\end{array}}$ R—NH$_2$ + phthalate

1° amine

- The mechanism is S_N2.
- The reaction works best for CH_3X or RCH_2X.
- Only 1° amines can be prepared.

[3] Reduction methods (25.7B)

[a] From nitro compounds

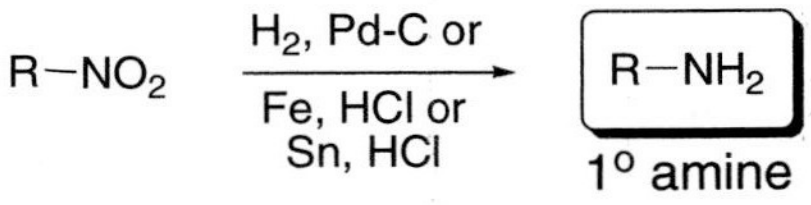

$R-NO_2 \xrightarrow[\begin{array}{c}Fe, HCl\ or\\ Sn, HCl\end{array}]{H_2, Pd-C\ or} R-NH_2$

1° amine

[b] From nitriles

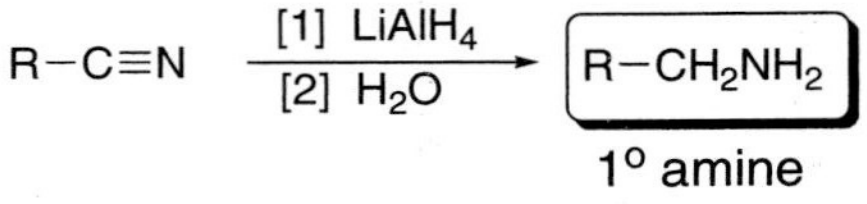

$R-C\equiv N \xrightarrow[\text{[2] } H_2O]{\text{[1] LiAlH}_4} R-CH_2NH_2$

1° amine

[c] From amides

$\underset{R}{\overset{O}{\underset{}{\parallel}}}C-NR'_2 \xrightarrow[\text{[2] } H_2O]{\text{[1] LiAlH}_4} \underset{R'}{RCH_2-N-R'}$

R' = H or alkyl

1°, 2°, and 3° amines

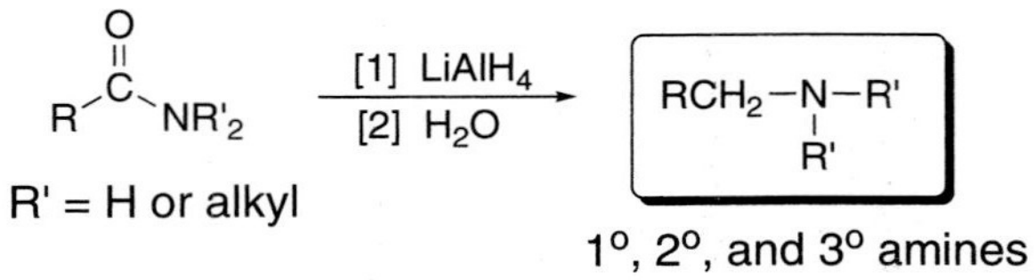

[4] Reductive amination (25.7C)

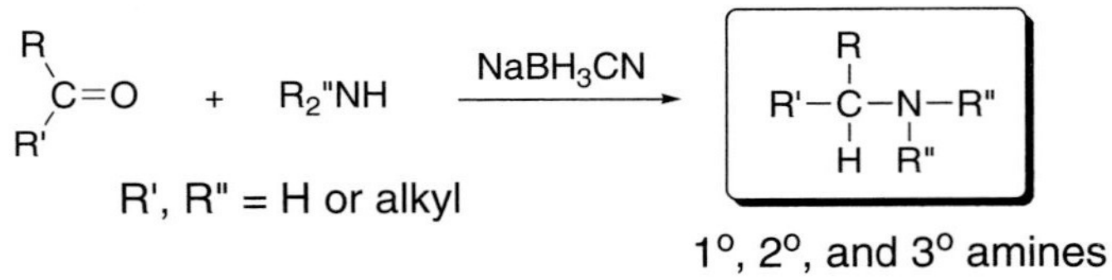

$\underset{R'}{\overset{R}{\underset{}{}}}C=O + R_2''NH \xrightarrow{NaBH_3CN} \underset{H\ R''}{R'-C-N-R''}$

R', R'' = H or alkyl

1°, 2°, and 3° amines

- Reductive amination adds one alkyl group (from an aldehyde or ketone) to a nitrogen nucleophile.
- Primary (1°), 2°, and 3° amines can be prepared.

◆ Reactions of amines

[1] Reaction as a base (25.9)

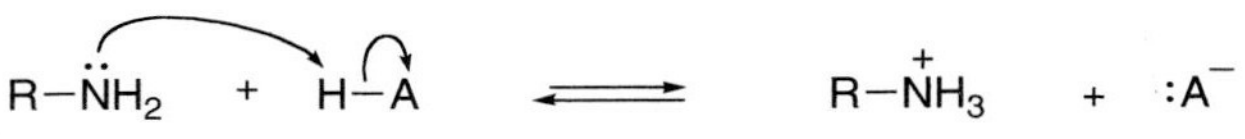

$R-\ddot{N}H_2 + H-A \rightleftharpoons R-\overset{+}{N}H_3 + :A^-$

[2] Nucleophilic addition to aldehydes and ketones (25.11)

With 1° amines:

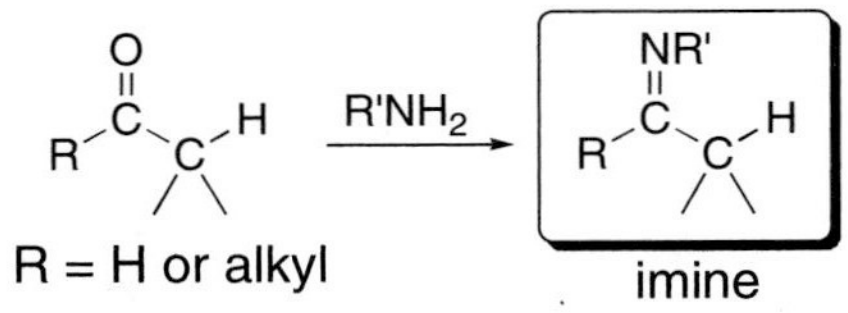

$\underset{R}{\overset{O}{\underset{}{\parallel}}}C\overset{}{\underset{}{C}}H \xrightarrow{R'NH_2} \underset{R}{\overset{NR'}{\underset{}{\parallel}}}C\overset{}{\underset{}{C}}H$

R = H or alkyl

imine

With 2° amines:

$\underset{R}{\overset{O}{\underset{}{\parallel}}}C\overset{}{\underset{}{C}}H \xrightarrow{R'_2NH} \underset{R}{\overset{NR'_2}{\underset{}{\parallel}}}C=C$

R = H or alkyl

enamine

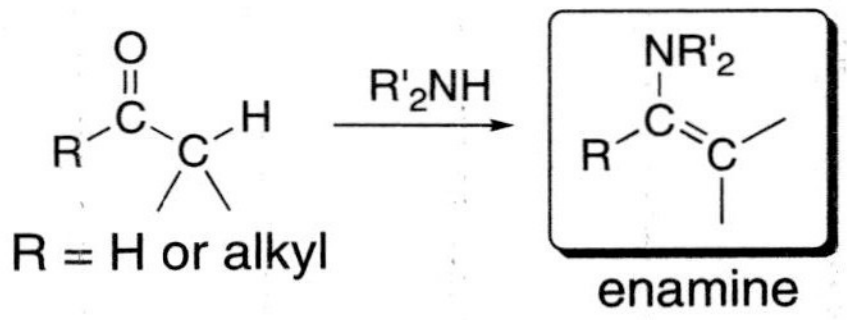

[3] Nucleophilic substitution with acid chlorides and anhydrides (25.11)

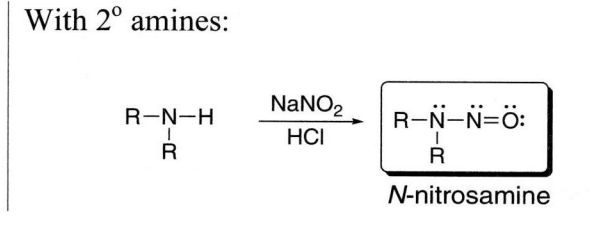

Z = Cl or OCOR
R' = H or alkyl

1°, 2°, and 3° amides

[4] Hofmann elimination (25.12)

- The less substituted alkene is the major product.

[5] Reaction with nitrous acid (25.13)

With 1° amines:

alkyl diazonium salt

With 2° amines:

N-nitrosamine

◆ Reactions of diazonium salts

[1] Substitution reactions (25.14)

With H₂O:

phenol

With CuX:

aryl chloride or
aryl bromide

X = Cl or Br

With HBF₄:

aryl fluoride

With NaI or KI:

aryl iodide

With CuCN:

benzonitrile

With H₃PO₂:

benzene

[2] Coupling to form azo compounds (25.15)

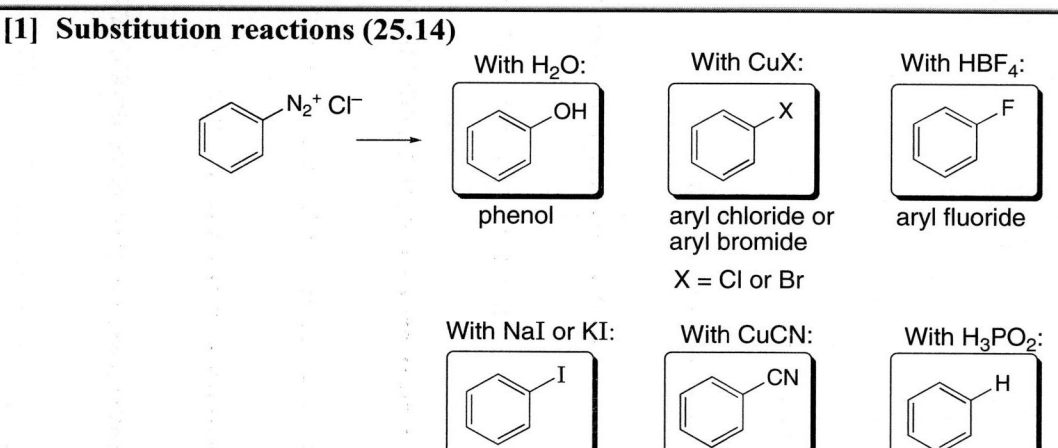

Y = NH₂, NHR, NR₂, OH
(a strong electron-
donor group)

azo compound

+ HCl

Chapter 25: Answers to Problems

25.1 Amines are classified as 1°, 2°, or 3° by the number of alkyl groups bonded to the *nitrogen* atom.

a.
2° amine
1° amine
H_2N —— N —— N —— NH_2
2° amine
1° amine

b. CH_3CH_2O —— C_6H_5 —— N—CH_3 **3° amine**

25.2

a.
$HO-\overset{\underset{|}{CH_3}}{\underset{CH_3}{C}}-CH_2NH_2$
3° alcohol   **1° amine**

b. **3° amine**
$CH_3-\overset{\underset{|}{N}}{\underset{CH_3}{}}-CH_2CH_2OH$
1° alcohol

25.3 The N atom of a quaternary ammonium salt is a stereogenic center when the N is surrounded by four different groups. All stereogenic centers are circled.

a. $CH_3-\overset{+}{\underset{CH_3}{N}}-CH_2CH_2-\overset{+}{N}-CH_2CH_3$
with CH₃ groups
N has 3 similar groups.

b. HO —— HO benzene ring — CH_2CH_2 — N(H) — CH₃ — C_6H_4—OH

25.4

CH_3-NH_2
147 pm
The C–N bond is formed from two *sp³* hybridized atoms and the lone pair is localized on N.

benzene—$\ddot{N}H_2$ ⟷ benzene=$\overset{+}{N}H_2$ + 3 more resonance structures
140 pm
partial double bond character

Because the lone pair on N can be delocalized on the benzene ring, the C–N bond has partial double bond character, making it shorter. Both the C and N atoms must be *sp²* hybridized (+ have a *p* orbital) for delocalization to occur. The higher percent *s*-character of the orbitals of both C and N shortens the bond as well.

25.5

a. $CH_3CH_2CH(NH_2)CH_3$
2-butanamine
or
sec-butylamine

b. $(CH_3CH_2CH_2CH_2)_2NH$
dibutylamine

c. cyclohexane—$N(CH_3)_2$
N,N-dimethylcyclohexanamine

d. chain with NH_2
2-methyl-5-nonanamine

e. $NHCH_2CH_3$
N-ethyl-3-hexanamine

f. CH_3
cyclopentane—$NHCH_2CH_2CH_3$
2-methyl-*N*-propylcyclopentanamine

25.6 An **NH₂** group named as a substituent is called an **amino group**.

a. 2,4-dimethyl-3-hexanamine   c. *N*-isopropyl-*p*-nitroaniline   e. *N,N*-dimethylethylamine   g. *N*-methylaniline

b. *N*-methylpentylamine   d. *N*-methylpiperidine   f. 2-aminocyclohexanone   h. *m*-ethylaniline

25.7 Primary (1°) and 2° amines have higher bp's than similar compounds (like ethers) incapable of hydrogen bonding, but lower bp's than alcohols that have stronger intermolecular hydrogen bonds. Tertiary amines (3°) have lower boiling points than 1° and 2° amines of comparable molecular weight because they have no N–H bonds.

a. $(CH_3)_2CHCH_2CH_3$ $CH_3\overset{\overset{O}{\|}}{C}CH_2CH_3$ $(CH_3)_2CHCH_2NH_2$

alkane **lowest boiling point**	ketone **intermediate boiling point**	amine N–H can hydrogen bond. **highest boiling point**

b.

alkane **lowest boiling point**	ether **intermediate boiling point**	amine N–H can hydrogen bond. **highest boiling point**

25.8 1° Amines show *two* N–H absorptions at 3300–3500 cm⁻¹. 2° Amines show *one* N–H absorption at 3300–3500 cm⁻¹.

molecular weight = 59
one IR peak = 2° amine

$CH_3-\overset{\overset{}{\underset{H}{N}}}{}-CH_2CH_3$

25.9 **The NH signal occurs between 0.5 and 5.0 ppm.** The protons on the carbon bonded to the amine nitrogen are deshielded and typically absorb at 2.3–3.0 ppm. The NH protons are not split.

molecular formula $C_6H_{15}N$
¹H NMR absorptions (ppm):
0.9 (singlet, 1 H) ⟶ NH
1.10 (triplet, 3 H) ⟶ CH_3 adjacent to CH_2
1.15 (singlet, 9 H) ⟶ $(CH_3)_3C$
2.6 (quartet, 2 H) ⟶ CH_2 adjacent to CH_3

25.10 The atoms of 2-phenylethylamine are in bold.

a. $(CH_3CH_2)_2N$ ... **LSD**
lysergic acid diethyl amide

b. CH_3O ... **codeine**

25.11 S_N2 reaction of an alkyl halide with NH_3 or an amine forms an amine or an ammonium salt.

a. (structure) $\xrightarrow[\text{excess}]{NH_3}$ (structure with NH_2)

b. (structure with NH_2) $\xrightarrow[\text{excess}]{CH_3CH_2Br}$ (structure with $\overset{+}{N}(CH_2CH_3)_3$ Br^-)

25.12 The Gabriel synthesis converts an alkyl halide into a 1° amine by a two-step process: nucleophilic substitution followed by hydrolysis.

a. (structure with NH_2) $\Downarrow$ (structure with Br)

b. $(CH_3)_2CHCH_2CH_2NH_2$ $\Downarrow$ $(CH_3)_2CHCH_2CH_2Br$

c.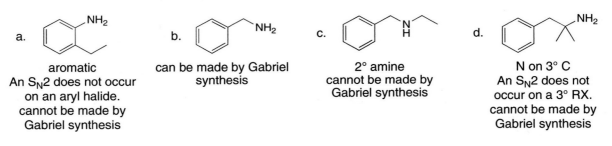

25.13 The Gabriel synthesis prepares 1° amines from alkyl halides. Since the reaction proceeds by an S_N2 mechanism, the halide must be CH_3 or 1°, and X can't be bonded to an sp^2 hybridized C.

a. (structure with NH_2)

aromatic
An S_N2 does not occur
on an aryl halide.
cannot be made by
Gabriel synthesis

b. (structure with NH_2)

can be made by Gabriel
synthesis

c. (structure)

2° amine
cannot be made by
Gabriel synthesis

d. (structure with NH_2)

N on 3° C
An S_N2 does not
occur on a 3° RX.
cannot be made by
Gabriel synthesis

25.14 **Nitriles are reduced to 1° amines with LiAlH₄. Nitro groups are reduced to 1° amines** using a variety of reducing agents. **Primary (1°), 2°, and 3° amides are reduced to 1°, 2°, and 3° amines** respectively, using LiAlH₄.

a. $CH_3CHCH_2NH_2 \Rightarrow CH_3CHCH_2NO_2 \quad CH_3CHC\equiv N \quad CH_3CHCNH_2$
   $\ \ \ \ \ |_{CH_3} \ \ \ \ \ \ \ \ \ \ \ \ \ \ |_{CH_3} \ \ \ \ \ \ \ \ \ \ \ \ |_{CH_3} \ \ \ \ \ \ \ \ \ \ |_{CH_3}$

b. (cyclohexyl)$-CH_2NH_2 \Rightarrow$ (cyclohexyl)$-CH_2NO_2$ (cyclohexyl)$-C\equiv N$ (cyclohexyl amide)

c.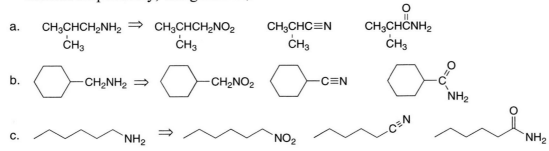

25.15 **Primary (1°), 2°, and 3° amides are reduced to 1°, 2°, and 3° amines** respectively, using LiAlH₄.

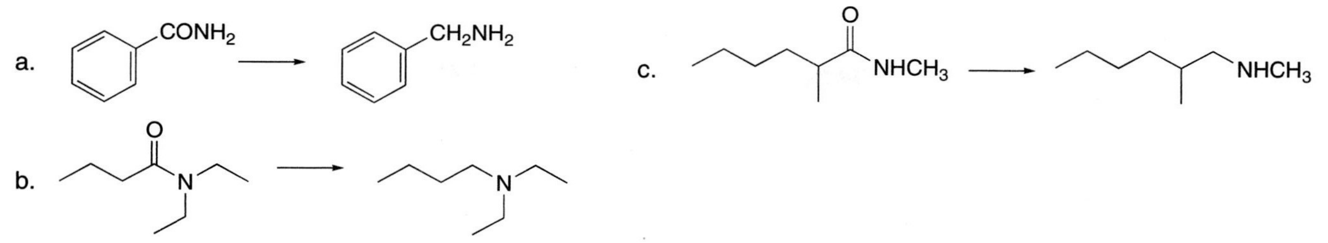

a.

c.

b.

25.16

(CH₃)₂CHNH₂

isopropylamine

General reaction:

$$R-C\equiv N \xrightarrow{[H]} RCH_2NH_2$$

The amine needs 2 H's here.

The C bonded to the N must have 2 H's for the amine to be formed by reduction of a nitrile.

25.17 Only amines with a CH₂ or CH₃ bonded to the N can be made by reduction of an amide.

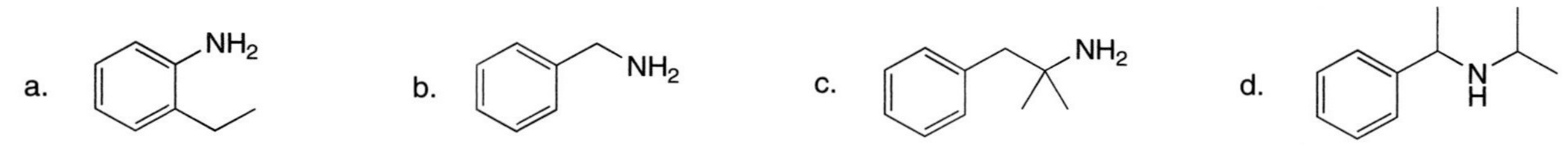

a.

N bonded to benzene cannot be made by reduction of an amide

b.

N bonded to CH₂ can be made by reduction of an amide

c.

N bonded to a 3° C cannot be made by reduction of an amide

d.

N on 2° C on both sides cannot be made by reduction of an amide

25.18 Reductive amination is a two-step method that converts aldehydes and ketones into 1°, 2°, and 3° amines. Reductive amination replaces a C=O by a C–H and C–N bond.

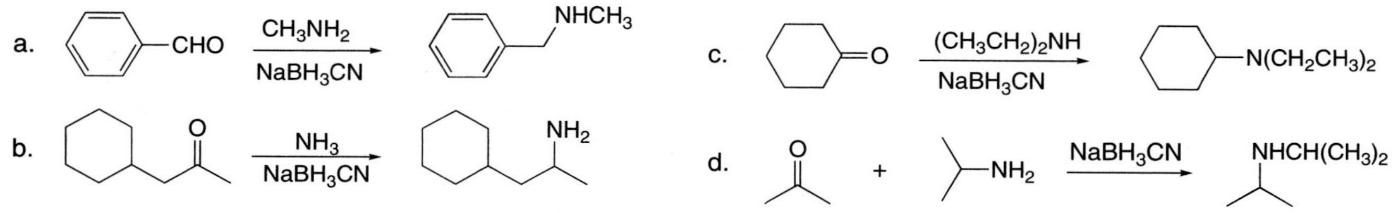

a.

b.

c.

d.

25.19

a.

b.

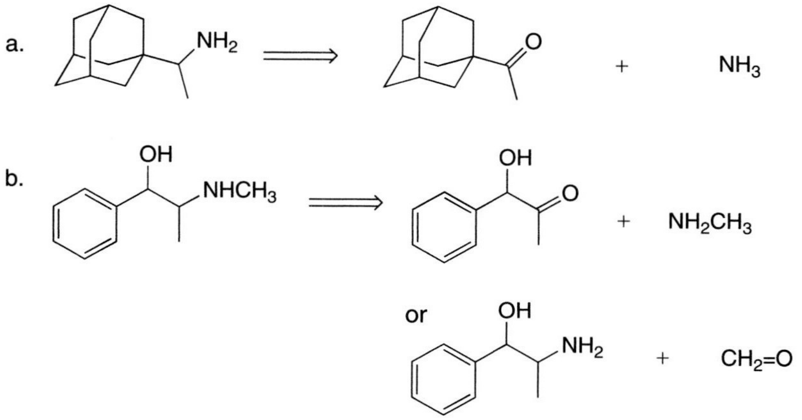

25.20

a.

phentermine

Only amines that have a C bonded to a H and N atom can be made by reductive amination; that is, an amine must have the following structural feature:

$$-\overset{H}{\underset{|}{C}}-\overset{|}{N}-$$

In phentermine, the C bonded to N is not bonded to a H, so it cannot be made by reductive amination.

b. systematic name: 2-methyl-1-phenyl-2-propanamine

25.21 The pK_a of many protonated amines is 10–11, so the pK_a of the starting acid must be **less than 10** for equilibrium to favor the products. Amines are thus readily protonated by strong inorganic acids like HCl and H_2SO_4, and by carboxylic acids as well.

a. $CH_3CH_2CH_2CH_2-NH_2$ + HCl $\rightleftharpoons$ $CH_3CH_2CH_2CH_2-\overset{+}{N}H_3$ + Cl⁻

$pK_a = -7$ $pK_a \approx 10$
weaker acid
products favored

b. C_6H_5COOH + $(CH_3)_2NH$ $\rightleftharpoons$ $(CH_3)_2\overset{+}{N}H_2$ + $C_6H_5COO^-$

$pK_a = 4.2$ $pK_a = 10.7$
weaker acid
products favored

c.

$pK_a = 15.7$ $pK_a \approx 10$
weaker acid
reactants favored

25.22 An amine can be separated from other organic compounds by converting it to a water-soluble ammonium salt by an acid–base reaction. In each case, the extraction procedure would employ the following steps:
- Dissolve the amine and either **X** or **Y** in CH_2Cl_2.
- Add a solution of 10% HCl. The amine will be protonated and dissolve in the aqueous layer, while **X** or **Y** will remain in the organic layer as a neutral compound.
- Separate the layers.

a.

• **soluble in H_2O**
• **insoluble in CH_2Cl_2**

• insoluble in H_2O
• soluble in CH_2Cl_2

b. $(CH_3CH_2CH_2CH_2)_3N$ and $(CH_3CH_2CH_2CH_2)_2O$ $\xrightarrow{H-Cl}$ $(CH_3CH_2CH_2CH_2)_3\overset{+}{N}H$ Cl⁻ + $(CH_3CH_2CH_2CH_2)_2O$

       Y   **Y**

• **soluble in H_2O**   • insoluble in H_2O
• **insoluble in CH_2Cl_2**   • soluble in CH_2Cl_2

25.23 Primary (1°), 2°, and 3° alkylamines are more basic than NH_3 because of the electron-donating inductive effect of the R groups.

a. $(CH_3)_2NH$ and NH_3

2° alkylamine
CH_3 groups are electron donating.
stronger base

b. $CH_3CH_2NH_2$ and $ClCH_2CH_2NH_2$

1° alkylamine   1° alkylamine
stronger base   Cl is electron withdrawing.
                            weaker base

25.24 Arylamines are less basic than alkylamines because the electron pair on N is delocalized. Electron-donor groups add electron density to the benzene ring making the arylamine more basic than aniline. Electron-withdrawing groups remove electron density from the benzene ring, making the arylamine less basic than aniline.

a.

electron-
withdrawing group
least basic

arylamine
**intermediate
basicity**

electron-
donating group
most basic

b.

electron-
withdrawing group
least basic

arylamine
**intermediate
basicity**

alkylamine
most basic

25.25 Amides are much less basic than amines because the electron pair on N is highly delocalized.

amide
least basic

arylamine
intermediate basicity

alkylamine
most basic

25.26

a. Electron pair on N occupies an sp^2 hybrid orbital.

sp^2 hybridized
more basic

This N is also sp^2 hybridized but the electron pair occupies a *p* orpital so it can delocalize onto the aromatic ring. Delocalization makes this N less basic.

DMAP
4-(*N,N*-dimethylamino)pyridine

b.

sp^3 hybridized N
stronger base

nicotine

sp^2 hybridized N
higher percent *s*-character
weaker base

25.27

a.

This electron pair is delocalized, making it a weaker base.

:NH₂

stronger base
This compound is similar to DMAP in Problem 25.26a.

b.

stronger base
sp^3 hybridized N
25% *s*-character

sp^2 hybridized N
33% *s*-character

c.

sp^2 hybridized N
33% *s*-character

stronger base
sp^3 hybridized N
25% *s*-character

25.28 Amines attack carbonyl groups to form products of nucleophilic addition or substitution.

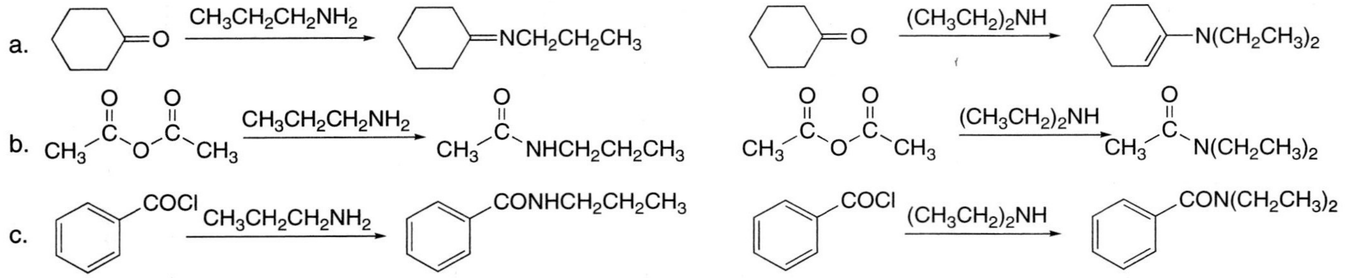

a.

b.

c.

25.29 [1] Convert the amine (aniline) into an amide (acetanilide).
[2] **Carry out the Friedel–Crafts reaction.**
[3] **Hydrolyze the amide** to generate the free amino group.

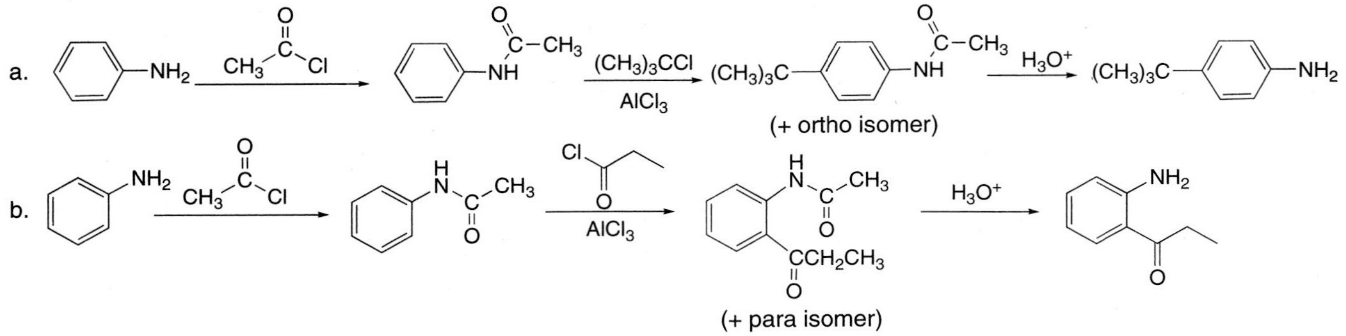

a.

(+ ortho isomer)

b.

(+ para isomer)

25.30

transition state:

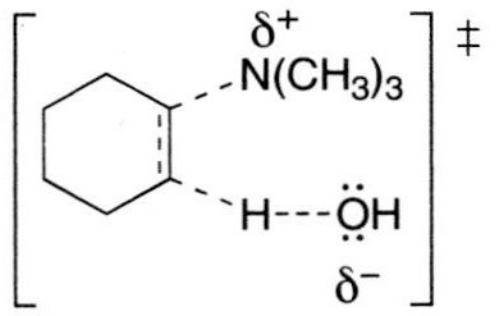

(no 3-D geometry shown here)

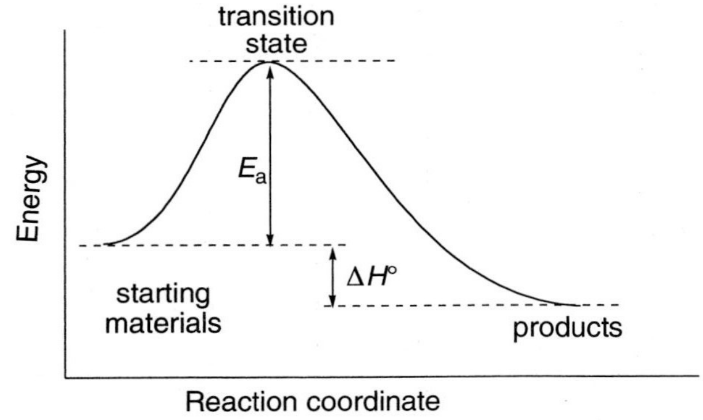

25.31

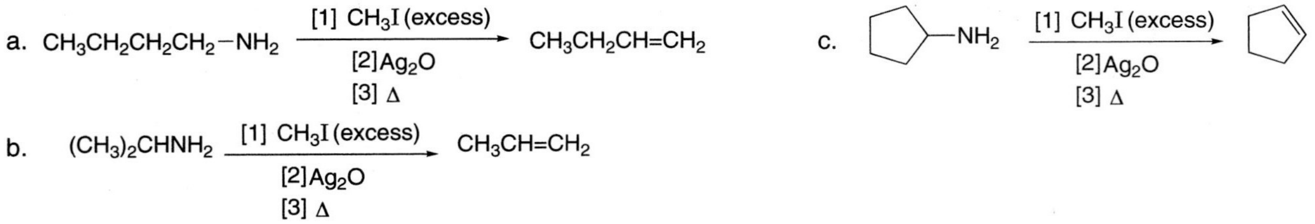

a. $CH_3CH_2CH_2CH_2-NH_2$ $\xrightarrow[\substack{[2]\,Ag_2O \\ [3]\,\Delta}]{[1]\,CH_3I\,(excess)}$ $CH_3CH_2CH=CH_2$

b. $(CH_3)_2CHNH_2$ $\xrightarrow[\substack{[2]\,Ag_2O \\ [3]\,\Delta}]{[1]\,CH_3I\,(excess)}$ $CH_3CH=CH_2$

c.

25.32 In a Hofmann elimination, the base removes a proton from the less substituted, more accessible β carbon atom, because of the bulky leaving group on the nearby α carbon.

a.
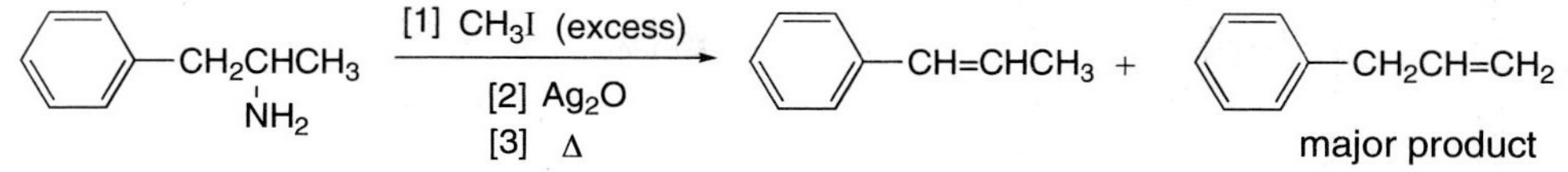
CH2CHCH3
NH2
[1] CH3I (excess)
[2] Ag2O
[3] Δ
CH=CHCH3 +
CH2CH=CH2
major product

b.
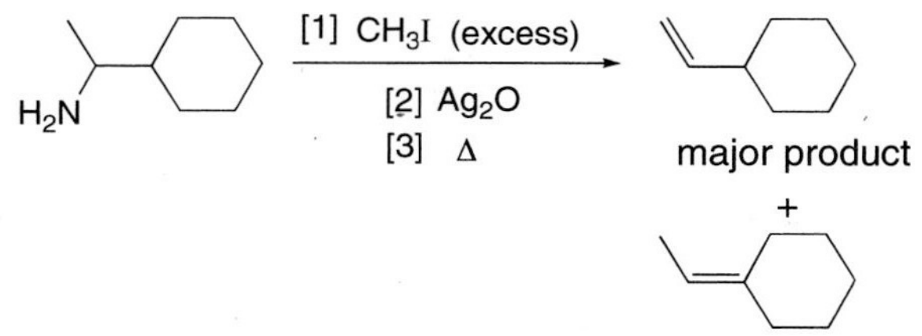
H2N
[1] CH3I (excess)
[2] Ag2O
[3] Δ
major product
+

c.
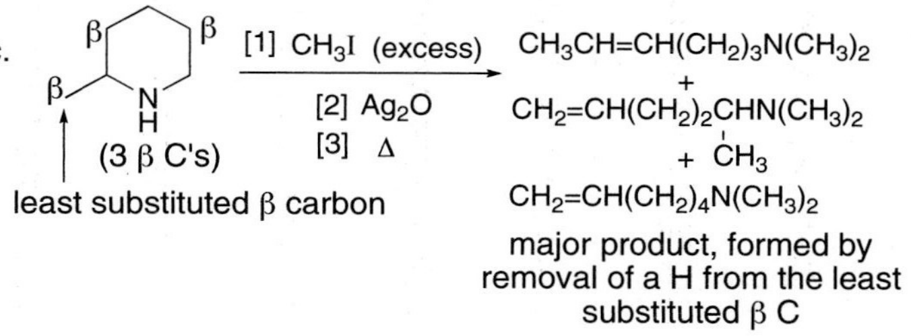
β
β
β
N
H
(3 β C's)
least substituted β carbon
[1] CH3I (excess)
[2] Ag2O
[3] Δ
CH3CH=CH(CH2)3N(CH3)2
+
CH2=CH(CH2)2CHN(CH3)2
+ CH3
CH2=CH(CH2)4N(CH3)2
major product, formed by
removal of a H from the least
substituted β C

25.33

a.
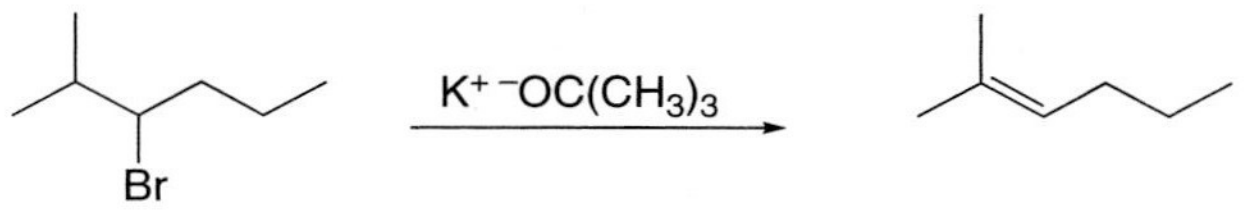
Br
K+ −OC(CH3)3

c.
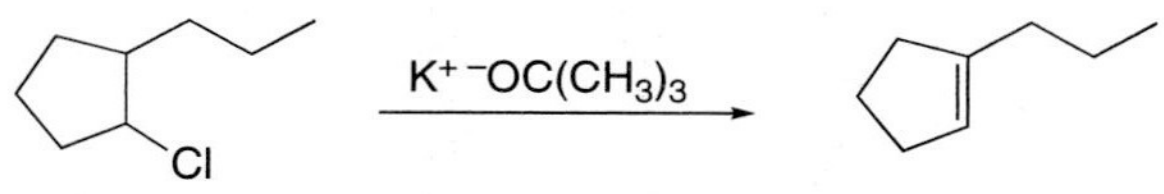
Cl
K+ −OC(CH3)3

b.
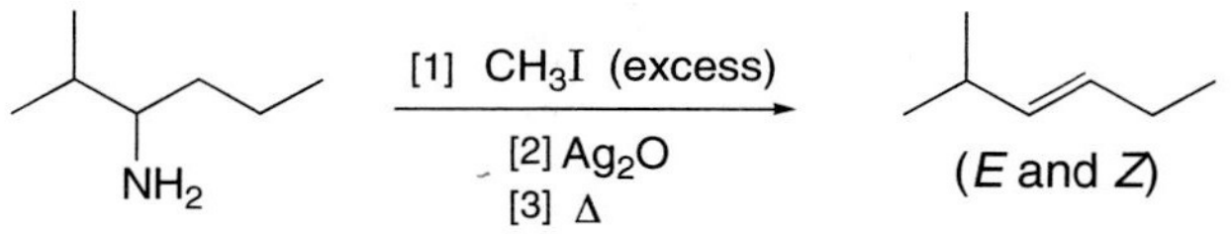
NH2
[1] CH3I (excess)
[2] Ag2O
[3] Δ
(E and Z)

d.
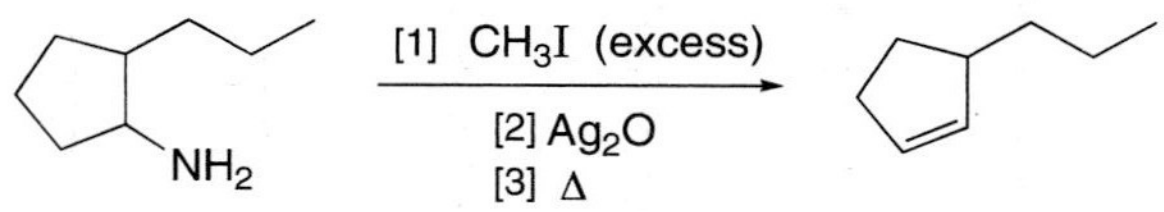
NH2
[1] CH3I (excess)
[2] Ag2O
[3] Δ

25.34

a.
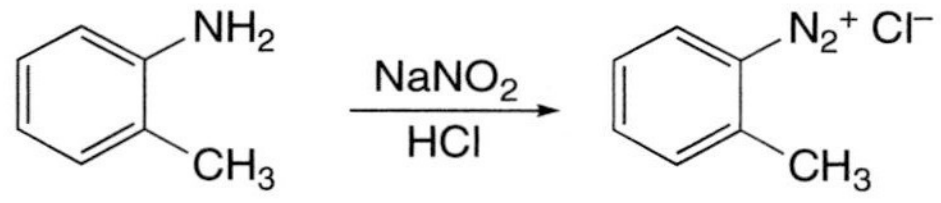
NH2
CH3
NaNO2
HCl
N2+ Cl−
CH3

c.
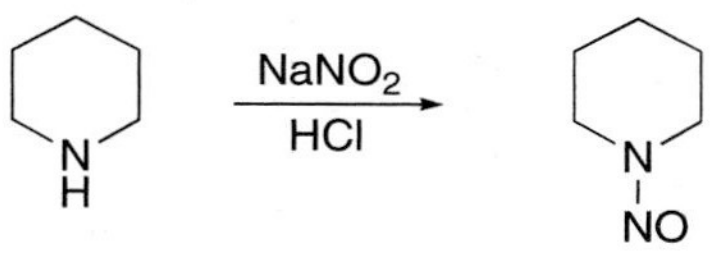
N
H
NaNO2
HCl
N
NO

b.
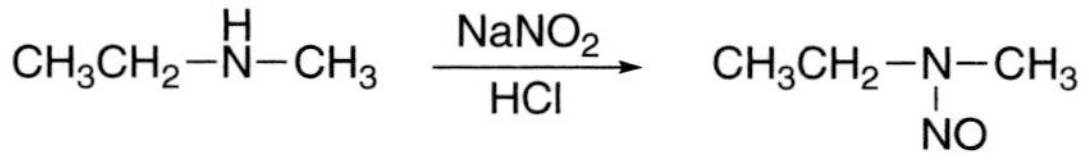
CH3CH2–N(H)–CH3
NaNO2
HCl
CH3CH2–N(NO)–CH3

d.
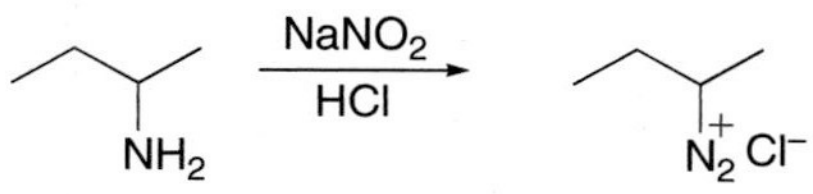
NH2
NaNO2
HCl
N2+ Cl−

25.35

a.
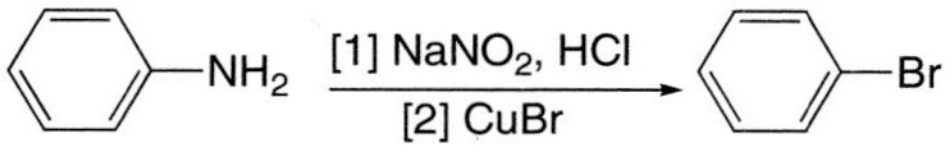
NH2
[1] NaNO2, HCl
[2] CuBr
Br

c.
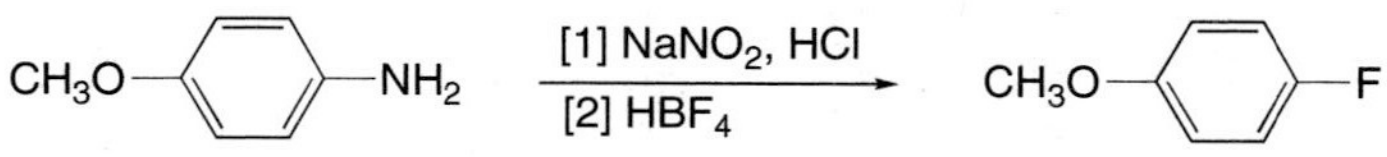
CH3O
NH2
[1] NaNO2, HCl
[2] HBF4
CH3O
F

b.
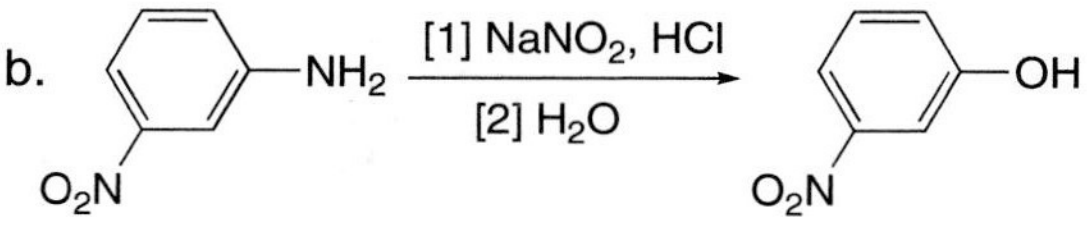
NH2
O2N
[1] NaNO2, HCl
[2] H2O
OH
O2N

d.
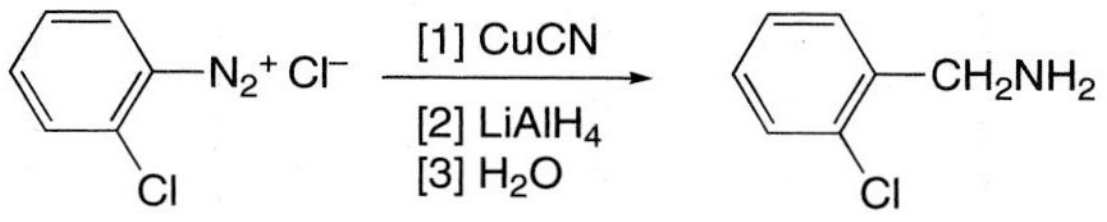
N2+ Cl−
Cl
[1] CuCN
[2] LiAlH4
[3] H2O
CH2NH2
Cl

25.36

a.
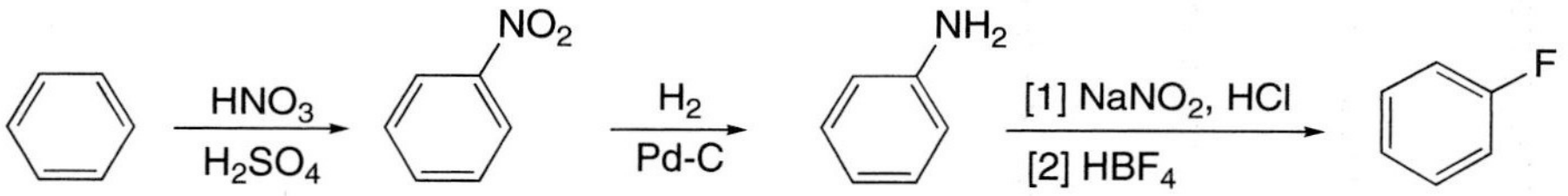
HNO3
H2SO4
NO2
H2
Pd-C
NH2
[1] NaNO2, HCl
[2] HBF4
F

b.

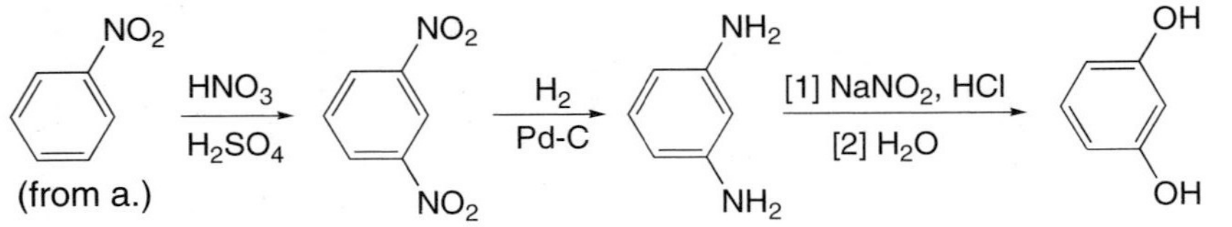

(from a.)

c.

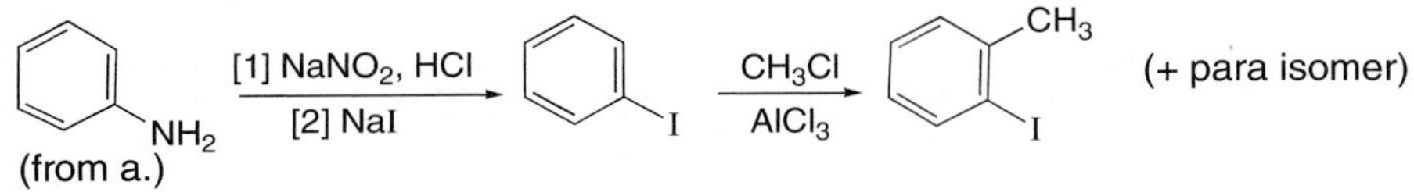

(from a.)

d.

(from a.)

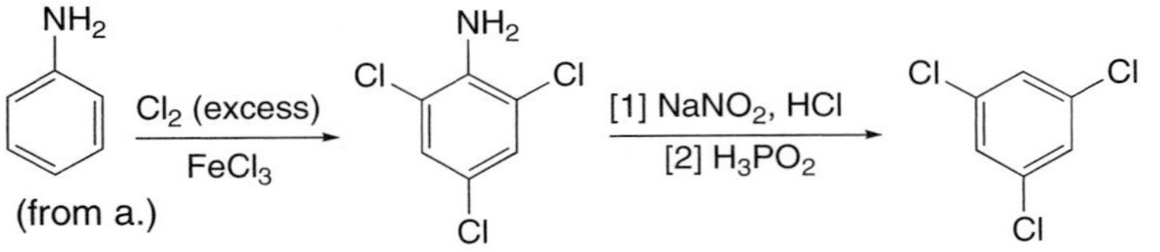

25.37

a.

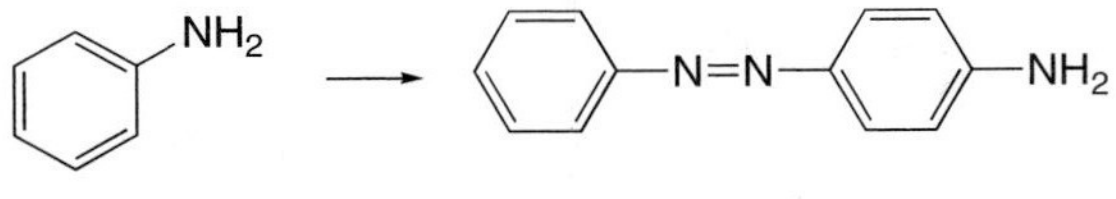

b.

c.

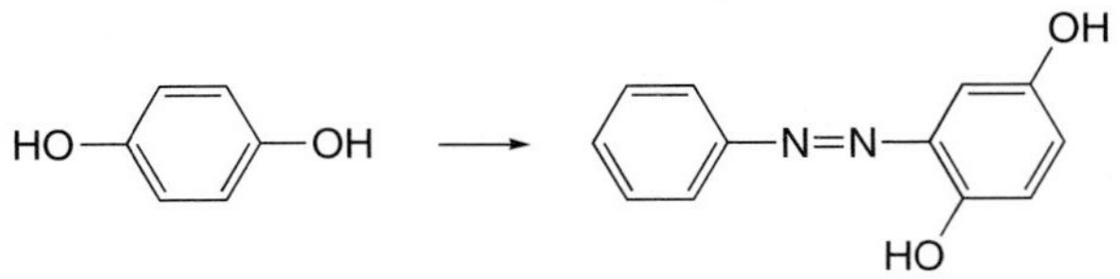

25.38 To determine what starting materials are needed to synthesize a particular azo compound, always divide the molecule into two components: **one has a benzene ring with a diazonium ion, and one has a benzene ring with a very strong electron-donor group.**

a.

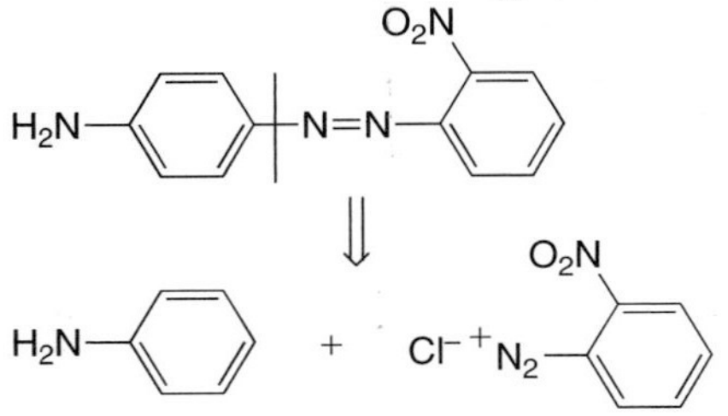

b.

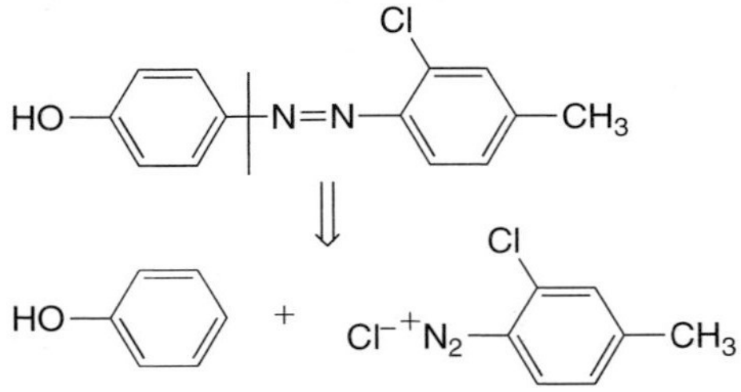

25.39

a.

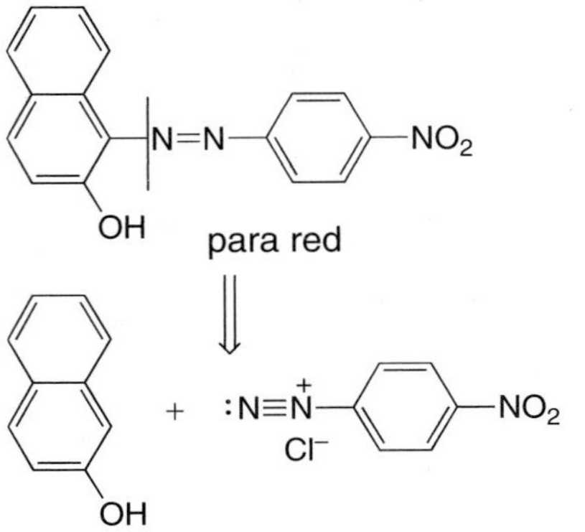

para red

b.

alizarine yellow R

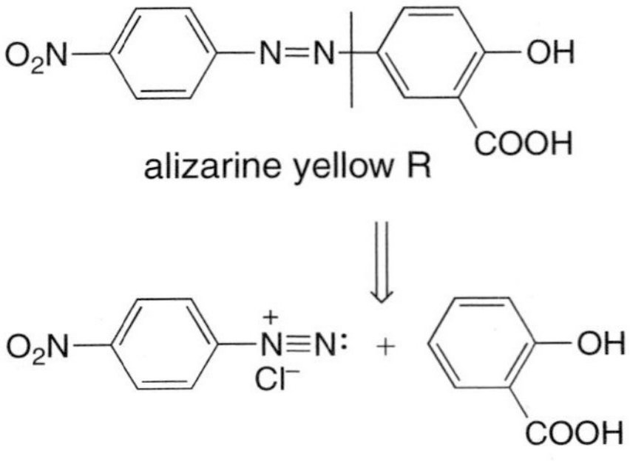

25.40

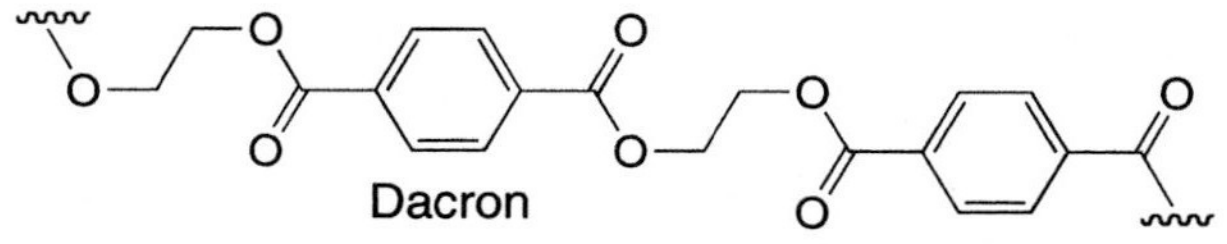

Dacron

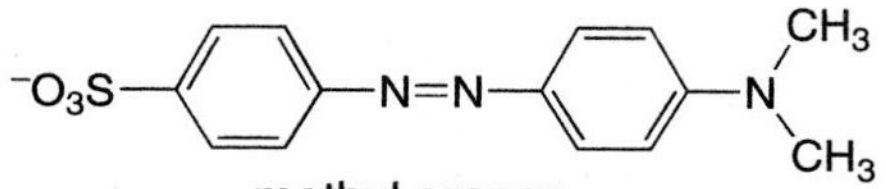

methyl orange

To bind to fabric, methyl orange (an anion) needs to interact with positively charged sites. Since Dacron is a neutral compound with no cationic sites on the chain, it does not bind methyl orange well.

25.41

a. CH₃NHCH₂CH₂CH₂CH₃
   N-methyl-1-butanamine
   (**N-methylbutylamine**)

b. 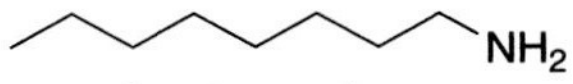
   1-octanamine
   (octylamine)

c. 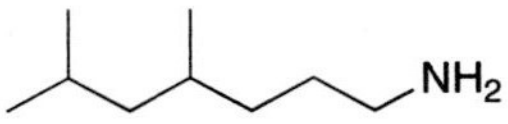
   4,6-dimethyl-1-heptanamine

d. 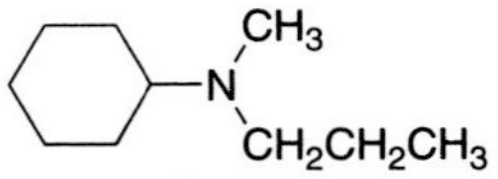
   N-methyl-N-propylcyclohexanamine

e. (CH₃CH₂CH₂)₃N
   tripropylamine

f. (C₆H₅)₂NH
   diphenylamine

g. 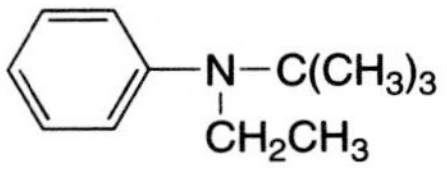
   N-tert-butyl-N-ethylaniline

h. 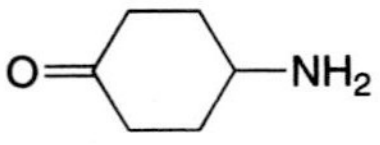
   4-aminocyclohexanone

i. 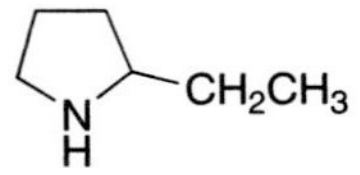
   2-ethylpyrrolidine

j. CH₃CH₂CH₂CH(NH₂)CH(CH₃)₂
   2-methyl-3-hexanamine

k. 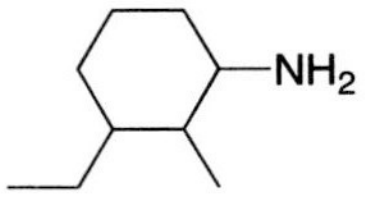
   3-ethyl-2-methylcyclohexanamine

l. 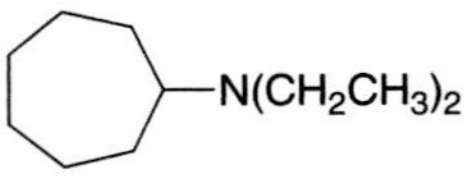
   N,N-diethylcycloheptanamine

25.42

a. cyclobutylamine

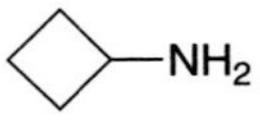

b. **N-isobutylcyclopentylamine**

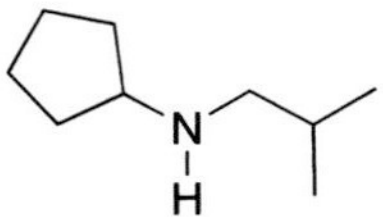

c. tri-tert-butylamine
   N[C(CH₃)₃]₃

d. **N,N-diisopropylaniline**

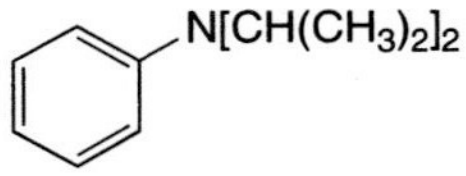

e. **N-methylpyrrole**

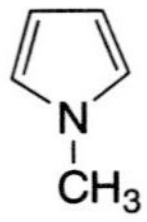

f. **N-methylcyclopentylamine**

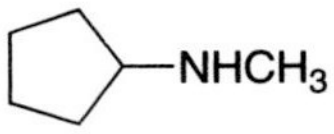

g. cis-2-aminocyclohexanol

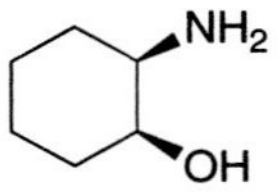

or

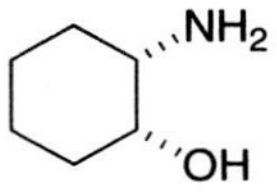

h. 3-methyl-2-hexanamine

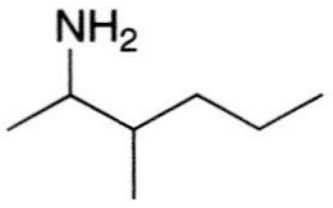

i. 2-sec-butylpiperidine

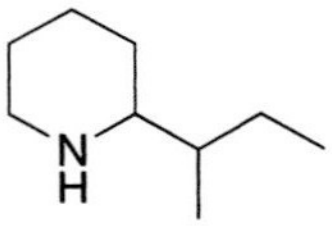

j. (2S)-2-heptanamine

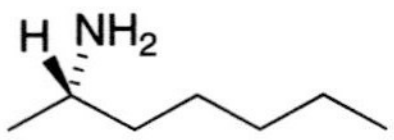

25.43

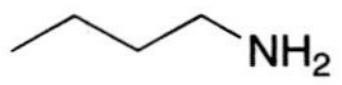

1-butanamine

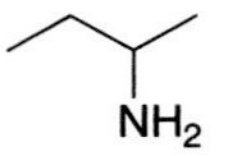

2-butanamine

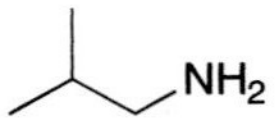

2-methyl-1-propanamine

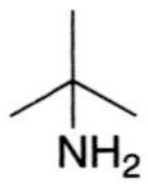

2-methyl-2-propanamine

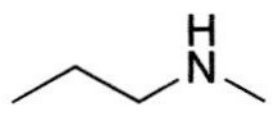

N-methyl-1-propanamine

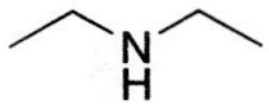

diethylamine

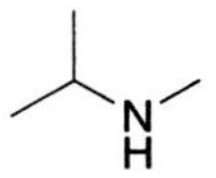

N-methyl-2-propanamine

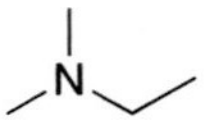

N,N-dimethylethanamine

25.44 [* denotes a stereogenic center.]

a.

1 stereogenic center
2 stereoisomers

b.

$CH_3CH_2\overset{*}{C}HCH_2CH_2CH_2-\overset{+}{\underset{\underset{CH_3}{|}}{N}}\overset{*}{}CH_2CH_2CH_2CH_3$

Cl^-

2 stereogenic centers
4 stereoisomers

25.45

a. $(CH_3CH_2)_2NH$ or

sp^3 hybridized N sp^2 hybridized N
stronger base   **weaker base**

b. $HCON(CH_3)_2$ or $(CH_3)_3N$

amide   alkylamine
weaker base   **stronger base**

c. $(CH_3CH_2)_2NH$ or $(ClCH_2CH_2)_2NH$

2° alkylamine   2° alkylamine
stronger base   Cl is electron withdrawing.
weaker base

d.

or

weaker base   **stronger base**
(delocalized electron
pair on N)

25.46

a.

arylamine   **intermediate**   alkylamine
least basic   **basicity**   **most basic**

b.

delocalized sp^2 hybridized N sp^3 hybridized N
electron pair on N   **intermediate**   **most basic**
least basic   **basicity**

c.

electron-   **intermediate**   electron-
withdrawing group   **basicity**   donating group
least basic   **most basic**

d. $(C_6H_5)_2NH$ $C_6H_5NH_2$

diarylamine   arylamine   alkylamine
least basic   **intermediate**   **most basic**
basicity

25.47 The electron-withdrawing inductive effect of the phenyl group stabilizes benzylamine, making its conjugate acid more acidic than the conjugate acid of cyclohexanamine. The conjugate acid of aniline is more acidic than the conjugate acid of benzylamine, since loss of a proton generates a resonance-stabilized amine, $C_6H_5NH_2$.

$pK_a = 10.7$ pK_a intermediate $pK_a = 4.6$

alkylamine
cyclohexanamine

electron-withdrawing inductive effect of the sp^2 hybridized C's
benzylamine

resonance-stabilized aromatic amine
aniline

25.48 The most basic N atom is protonated on treatment with acid.

a.

benazepril

CH_3CO_2H

more basic

$+ CH_3CO_2^-$

b.

varenicline

CH_3CO_2H

most basic

$+ CH_3CO_2^-$

25.49

a.

$N_b < N_a < N_c$

Order of basicity: $N_b < N_a < N_c$
N_b – The electron pair on this N atom is delocalized on the O atom; least basic.
N_a – The electron pair on this N atom is not delocalized, but is on an sp^2 hybridized atom.
N_c – The electron pair on this N atom is on an sp^3 hybridized N; most basic.

b.

$N_b < N_a < N_c$

Order of basicity: $N_b < N_a < N_c$
N_b – The electron pair on this N atom is delocalized on the aromatic five-membered ring; least basic.
N_a – The electron pair on this N atom is not delocalized, but is on an sp^2 hybridized atom.
N_c – The electron pair on this N atom is on an sp^3 hybridized N; most basic.

25.50 The most basic N atom is protonated on treatment with acid.

a.

a 3° alkylamine with an
sp^3 hybridized N
most basic

aripiprazole

part of an amide
least basic

a delocalized electron
pair on the benzene ring
intermediate

b.

$\xrightarrow{\text{HCl}}$

25.51 The para isomer is the weaker base because the electron pair on its NH_2 group can be delocalized onto the NO_2 group. In the meta isomer, no resonance structure places the electron pair on the NO_2 group, and fewer resonance structures can be drawn:

meta

para

25.52

A

pK_a of the conjugate acid = 5.2
stronger conjugate acid
weaker base
The electron pair of this arylamine
is delocalized on the benzene ring,
decreasing its basicity.

B

This two-carbon bridge makes it difficult for the lone pair on N to delocalize on the aromatic ring.

pK_a of the conjugate acid = 7.29
weaker conjugate acid
stronger base

Resonance structures that place a double bond between the N atom and the benzene ring are destabilized. Since the electron pair is more localized on N, compound **B** is more basic.

B

Geometry makes it difficult
to have a double bond here.

25.53

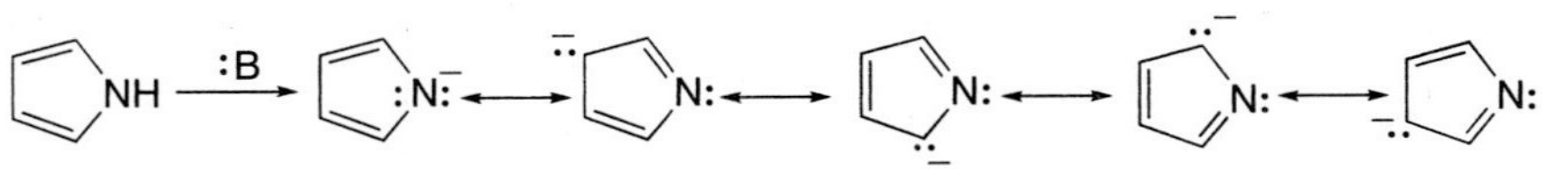

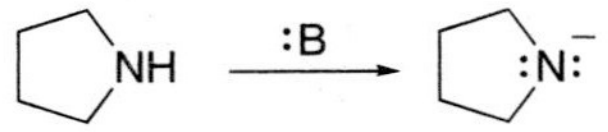

pyrrole
pK_a = 23
stronger acid

weaker conjugate base
The electron pair is delocalized, decreasing the
basicity. The N atom is sp^2 hybridized.

pyrrolidine
pK_a = 44
weaker acid

stronger conjugate base
The electron pair is not
delocalized on the ring.
The N atom is sp^3
hybridized.

25.54

a. $C_6H_5CH_2CH_2CH_2Br \xrightarrow[\text{excess}]{NH_3} C_6H_5CH_2CH_2CH_2NH_2$

b. $C_6H_5CH_2CH_2Br \xrightarrow{NaCN} C_6H_5CH_2CH_2CN \xrightarrow[\text{[2] } H_2O]{\text{[1] LiAlH}_4} C_6H_5CH_2CH_2CH_2NH_2$

c. $C_6H_5CH_2CH_2CH_2NO_2 \xrightarrow[\text{Pd-C}]{H_2} C_6H_5CH_2CH_2CH_2NH_2$

d. $C_6H_5CH_2CH_2CONH_2 \xrightarrow[\text{[2] } H_2O]{\text{[1] LiAlH}_4} C_6H_5CH_2CH_2CH_2NH_2$

e. $C_6H_5CH_2CH_2CHO \xrightarrow[\text{NaBH}_3\text{CN}]{NH_3} C_6H_5CH_2CH_2CH_2NH_2$

25.55

a. $(CH_3CH_2)_2NH$

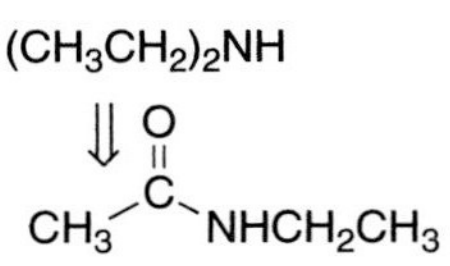

b. (structure)

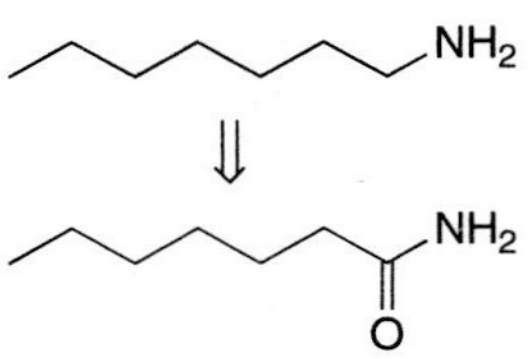

c. (structure)

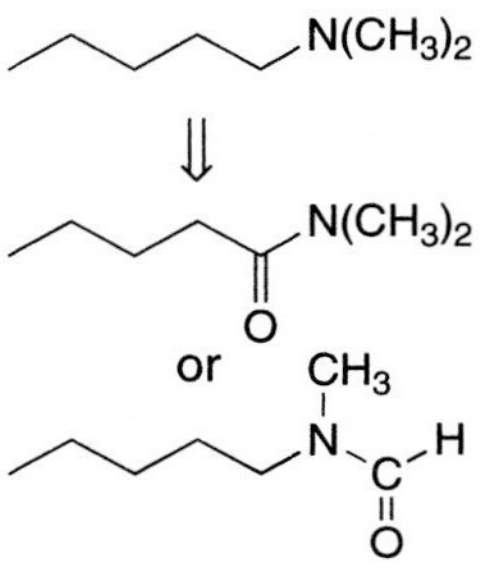

d. (structure)

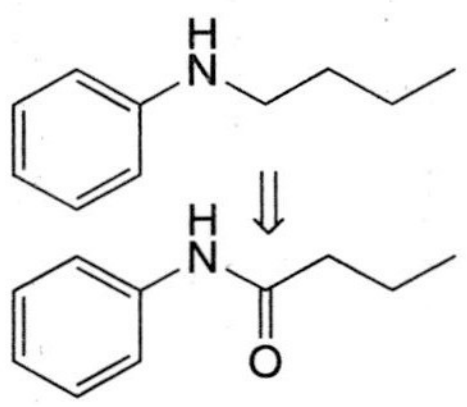

25.56 In reductive amination, one alkyl group on N comes from the carbonyl compound. The remainder of the molecule comes from NH_3 or an amine.

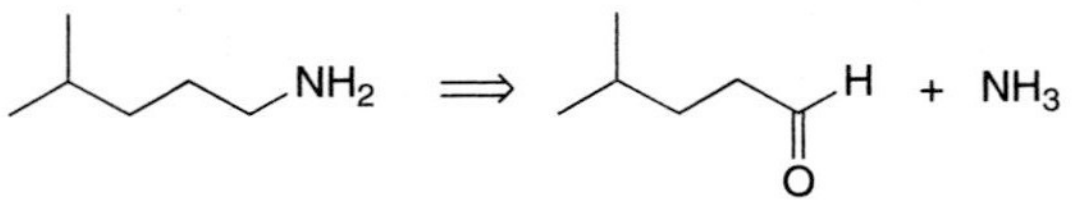

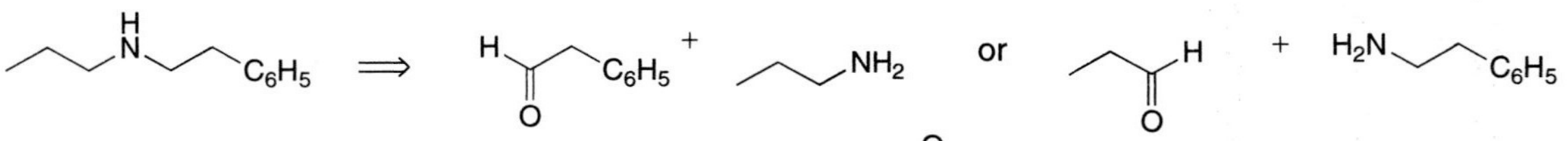

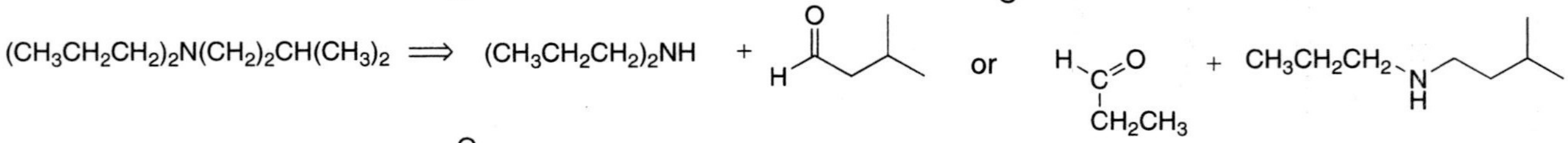

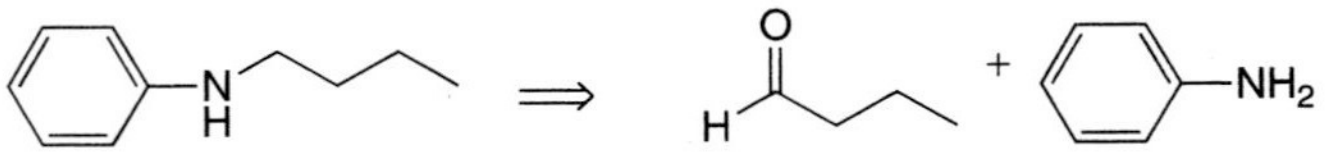

25.57

a. C_6H_5 C=O + $CH_3CH_2CH_2CH_2NH_2$ →(NaBH$_3$CN) product

c. C_6H_5—CHO →(NH$_3$ / NaBH$_3$CN) C_6H_5—CH$_2$NH$_2$

b. cycloheptanone →((CH$_3$)$_2$NH / NaBH$_3$CN) N(CH$_3$)$_2$ product

d. →(cyclohexyl-NH$_2$ / NaBH$_3$CN) product

25.58

a. benzyl bromide →(NH$_3$, excess) CH$_2$NH$_2$

b. C$_6$H$_5$CN →([1] LiAlH$_4$ [2] H$_2$O) CH$_2$NH$_2$

c. C$_6$H$_5$CONH$_2$ →([1] LiAlH$_4$ [2] H$_2$O) CH$_2$NH$_2$

d. C$_6$H$_5$CHO →(NH$_3$ / NaBH$_3$CN) CH$_2$NH$_2$

e. C$_6$H$_5$CH$_3$ →(Br$_2$ / hν) CH$_2$Br →(NH$_3$, excess) CH$_2$NH$_2$

f. C$_6$H$_5$COOH →([1] SOCl$_2$ [2] NH$_3$) CONH$_2$ →([1] LiAlH$_4$ [2] H$_2$O) CH$_2$NH$_2$

g. C$_6$H$_5$NH$_2$ →([1] NaNO$_2$, HCl [2] CuCN) CN →([1] LiAlH$_4$ [2] H$_2$O) CH$_2$NH$_2$

h. benzene →(HNO$_3$ / H$_2$SO$_4$) NO$_2$ →(H$_2$ / Pd-C) NH$_2$ Then as in (g).

25.59 Use the directions from Answer 25.22. Separation can be achieved because benzoic acid reacts with aqueous base and aniline reacts with aqueous acid according to the following equations:

benzoic acid (COOH) + NaOH (10% aqueous) ⇌ COO⁻Na⁺ + H_2O

benzoic acid
• soluble in CH_2Cl_2
• insoluble in H_2O

• soluble in H_2O
• insoluble in CH_2Cl_2

aniline (NH$_2$) + H—Cl (10% aqueous) ⇌ $\overset{+}{N}H_3$ Cl⁻

aniline
• soluble in CH_2Cl_2
• insoluble in H_2O

• soluble in H_2O
• insoluble in CH_2Cl_2

Toluene ($C_6H_5CH_3$), on the other hand, is not protonated or deprotonated in aqueous solution so it is always soluble in CH_2Cl_2 and insoluble in H_2O. The following flow chart illustrates the process.

25.60

25.61

CH$_3$—⟨⟩—NH$_2$ *p*-methylaniline

a. $\xrightarrow{\text{HCl}}$ CH$_3$—⟨⟩—$\overset{+}{\text{N}}$H$_3$ Cl$^-$

b. $\xrightarrow{\text{CH}_3\text{COCl}}$ CH$_3$—⟨⟩—NH—C(=O)—CH$_3$

c. $\xrightarrow{\text{(CH}_3\text{CO)}_2\text{O}}$ CH$_3$—⟨⟩—NH—C(=O)—CH$_3$

d. $\xrightarrow[\text{excess}]{\text{CH}_3\text{I}}$ CH$_3$—⟨⟩—$\overset{+}{\text{N}}$(CH$_3$)$_3$ I$^-$

e. $\xrightarrow{\text{(CH}_3\text{)}_2\text{C=O}}$ CH$_3$—⟨⟩—N=C(CH$_3$)$_2$

f. $\xrightarrow[\text{AlCl}_3]{\text{CH}_3\text{COCl}}$ CH$_3$—⟨⟩—$\overset{+}{\text{N}}$H$_2$ $\overset{-}{|}$AlCl$_3$

g. $\xrightarrow{\text{CH}_3\text{COOH}}$ CH$_3$—⟨⟩—$\overset{+}{\text{N}}$H$_3$ CH$_3$COO$^-$

h. $\xrightarrow[\text{HCl}]{\text{NaNO}_2}$ CH$_3$—⟨⟩—N$_2$$^+Cl^-$

i. Step (b), then $\xrightarrow[\text{AlCl}_3]{\text{CH}_3\text{COCl}}$ CH$_3$—⟨⟩(NH—C(=O)—CH$_3$)(C(=O)CH$_3$)

j. $\xrightarrow[\text{NaBH}_3\text{CN}]{\text{CH}_3\text{CHO}}$ CH$_3$—⟨⟩—NHCH$_2$CH$_3$

25.62

a. CH$_3$CH$_2$CH$_2$CH$_2$NH$_2$ $\xrightarrow{\text{ClCOC}_6\text{H}_5}$ CH$_3$CH$_2$CH$_2$CH$_2$NHCOC$_6$H$_5$

b. CH$_3$CH$_2$CH$_2$CH$_2$NH$_2$ $\xrightarrow{\text{O=C(CH}_2\text{CH}_3\text{)}_2}$ CH$_3$CH$_2$CH$_2$CH$_2$N=C(CH$_2$CH$_3$)$_2$

c. CH$_3$CH$_2$CH$_2$CH$_2$NH$_2$ $\xrightarrow[\text{[2] Ag}_2\text{O} \; \text{[3] }\Delta]{\text{[1] CH}_3\text{I (excess)}}$ CH$_3$CH$_2$CH=CH$_2$

d. CH$_3$CH$_2$CH$_2$CH$_2$NH$_2$ $\xrightarrow[\text{NaBH}_3\text{CN}]{\text{C}_6\text{H}_5\text{CHO}}$ CH$_3$CH$_2$CH$_2$CH$_2$NHCH$_2$C$_6$H$_5$

e. CH$_3$CH$_2$CH$_2$CH$_2$NH$_2$ $\xrightarrow[\text{CH}_3\text{CHO}]{\text{NaBH}_3\text{CN}}$ CH$_3$CH$_2$CH$_2$CH$_2$NHCH$_2$CH$_3$

f. CH$_3$CH$_2$CH$_2$CH$_2$NH$_2$ $\xrightarrow[\text{excess}]{\text{CH}_3\text{I}}$ [CH$_3$CH$_2$CH$_2$CH$_2$N(CH$_3$)$_3$]$^+$I$^-$

25.63

a. CH$_3$(CH$_2$)$_6$NH$_2$ $\xrightarrow[\text{[2]Ag}_2\text{O} \; \text{[3] }\Delta]{\text{[1] CH}_3\text{I (excess)}}$ CH$_3$(CH$_2$)$_4$CH=CH$_2$

b. $\xrightarrow[\text{[2] Ag}_2\text{O} \; \text{[3] }\Delta]{\text{[1] CH}_3\text{I (excess)}}$ (E + Z) / major product

c.

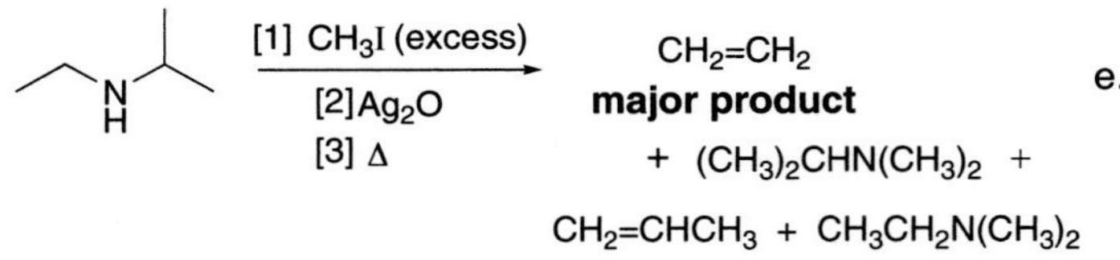

[1] CH$_3$I (excess)
[2] Ag$_2$O
[3] Δ

CH$_2$=CH$_2$
major product
+ (CH$_3$)$_2$CHN(CH$_3$)$_2$ +
CH$_2$=CHCH$_3$ + CH$_3$CH$_2$N(CH$_3$)$_2$

e.

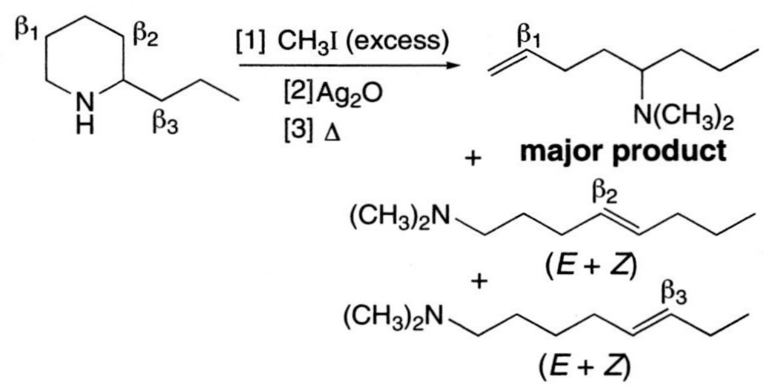

β$_1$ β$_2$
β$_3$

[1] CH$_3$I (excess)
[2] Ag$_2$O
[3] Δ

β$_1$
N(CH$_3$)$_2$
+ **major product**

(CH$_3$)$_2$N
β$_2$
+
(E + Z)

(CH$_3$)$_2$N
β$_3$
(E + Z)

d.

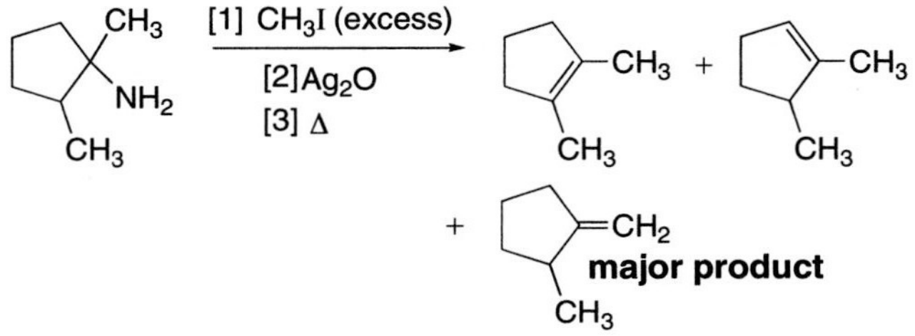

CH$_3$
NH$_2$
CH$_3$

[1] CH$_3$I (excess)
[2] Ag$_2$O
[3] Δ

CH$_3$
+
CH$_3$

CH$_3$
CH$_3$

+

=CH$_2$
major product
CH$_3$

25.64

a.

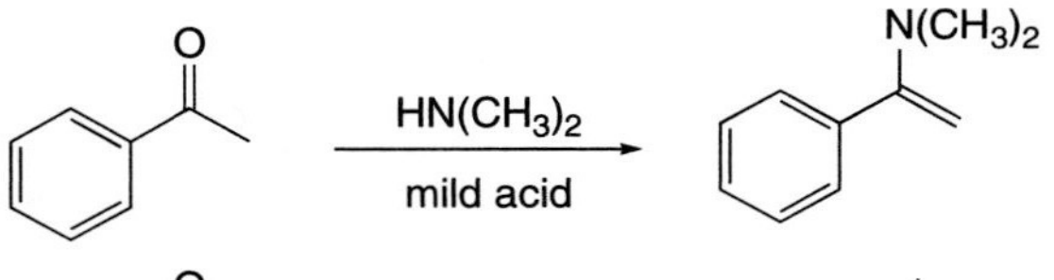

O

HN(CH$_3$)$_2$
mild acid

N(CH$_3$)$_2$

b.

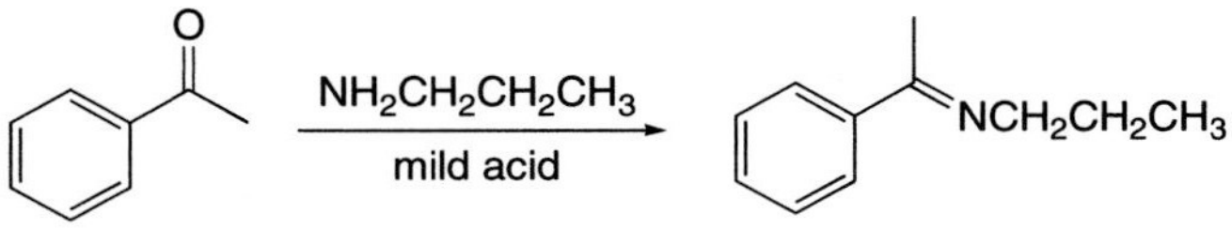

O

NH$_2$CH$_2$CH$_2$CH$_3$
mild acid

NCH$_2$CH$_2$CH$_3$

c.

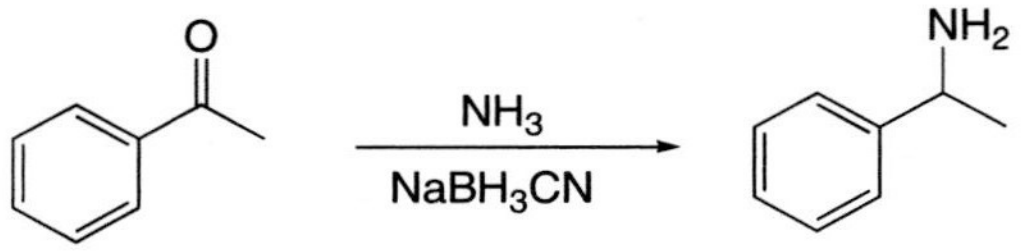

O

NH$_3$
NaBH$_3$CN

NH$_2$

d.

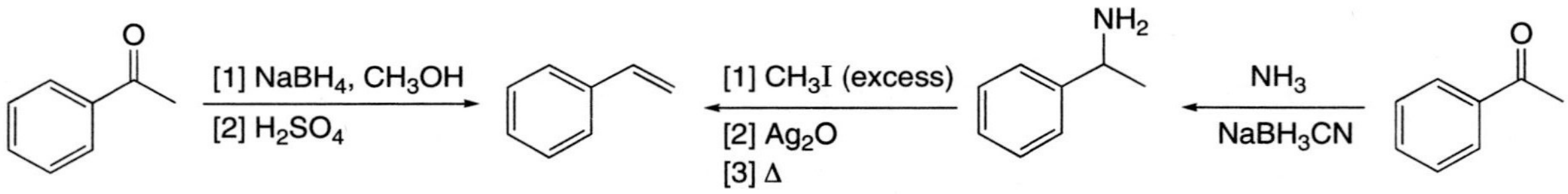

O

[1] NaBH$_4$, CH$_3$OH
[2] H$_2$SO$_4$

[1] CH$_3$I (excess)
[2] Ag$_2$O
[3] Δ

NH$_2$

NH$_3$
NaBH$_3$CN

O

e.

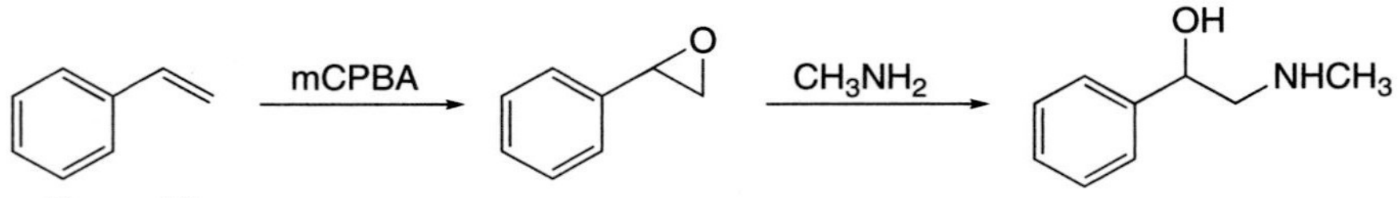

(from d.)

mCPBA

O

CH$_3$NH$_2$

OH
NHCH$_3$

f.

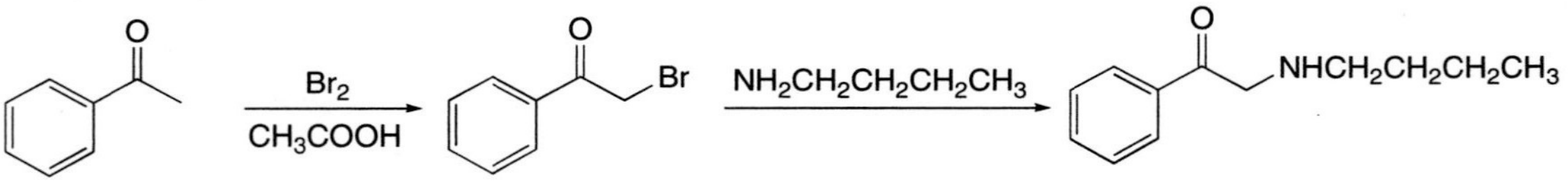

O

Br$_2$
CH$_3$COOH

O

Br

NH$_2$CH$_2$CH$_2$CH$_2$CH$_3$

O

NHCH$_2$CH$_2$CH$_2$CH$_3$

25.65

a.

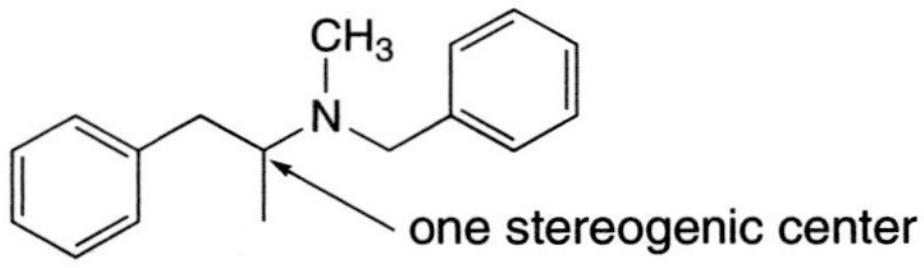

CH$_3$
N
CH$_3$

one stereogenic center

benzphetamine

b. Amides that can be reduced to benzphetamine:

and

c. Amines + carbonyl compounds that form benzphetamine by reductive amination:

NHCH$_3$

+ (benzaldehyde)   or   (phenylacetone)   +

or

+ CH$_2$=O

d.

[1] CH$_3$I
[2] Ag$_2$O
[3] Δ

+   +   + (CH$_3$)$_2$NCH$_2$—

major product
(elimination across α, β$_2$)

elimination across α, β$_1$

25.66

a. —CH$_2$CH$_2$Cl $\xrightarrow[\text{excess}]{\text{NH}_3}$ —CH$_2$CH$_2$NH$_2$

b. $\xrightarrow[\substack{[2]\ (CH_3)_2CHCH_2Cl \\ [3]\ ^-OH,\ H_2O}]{[1]\ KOH}$ NH$_2$ + CO$_2^-$ / CO$_2^-$

c. Br—⟨⟩—NO$_2$ $\xrightarrow[\text{HCl}]{\text{Sn}}$ Br—⟨⟩—NH$_2$

d. (CN) $\xrightarrow[\text{[2] H}_2\text{O}]{\text{[1] LiAlH}_4}$ CH$_2$NH$_2$

e. —CONHCH$_2$CH$_3$ $\xrightarrow[\text{[2] H}_2\text{O}]{\text{[1] LiAlH}_4}$ —CH$_2$NHCH$_2$CH$_3$

f. C$_6$H$_5$CH$_2$CH$_2$NH$_2$ + (C$_6$H$_5$CO)$_2$O ⟶
C$_6$H$_5$CH$_2$CH$_2$NHCOC$_6$H$_5$ + C$_6$H$_5$CH$_2$CH$_2$NH$_3$$^+$ C$_6$H$_5$COO$^-$

g. ⟨NH⟩ $\xrightarrow[\text{HCl}]{\text{NaNO}_2}$ ⟨N—N=O⟩

h. ⟨NH⟩ + C$_6$H$_5$CHO $\xrightarrow{\text{NaBH}_3\text{CN}}$ ⟨N—CH$_2$C$_6$H$_5$⟩

i. (piperidine) + (cyclohexanone)=O ⟶ (enamine)N

j. CH$_3$CH$_2$CH$_2$—N—CH(CH$_3$)$_2$ $\xrightarrow[\substack{[2]\ Ag_2O \\ [3]\ Δ}]{[1]\ CH_3I\ (excess)}$
   |
   H

CH$_3$CH=CH$_2$ + (CH$_3$)$_2$NCH(CH$_3$)$_2$
CH$_3$CH$_2$CH$_2$N(CH$_3$)$_2$

25.67 NH$_2$ and H must be anti for the Hofmann elimination. Rotate around the C–C bond so the NH$_2$ and H are anti.

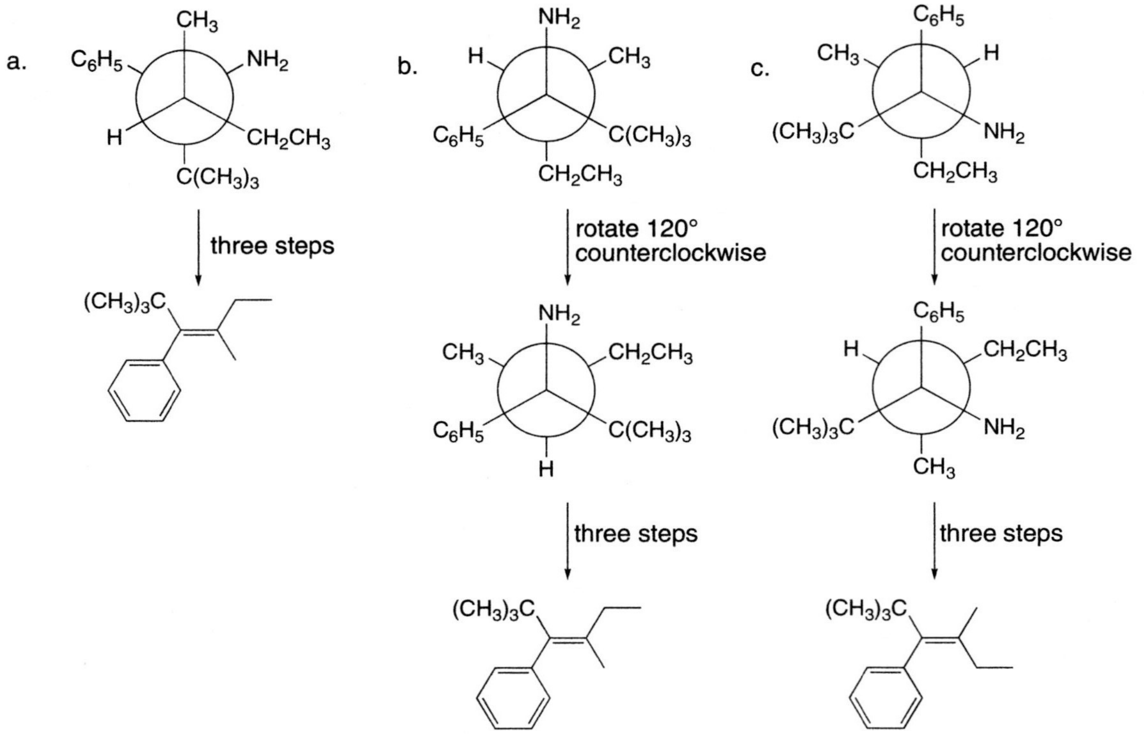

a.

three steps

b.

rotate 120° counterclockwise

three steps

c.

rotate 120° counterclockwise

three steps

25.68

Cl **A**

a. H$_2$O

b. H$_3$PO$_2$

c. CuCl

d. CuBr

e. CuCN

f. HBF$_4$

g. NaI

h. C$_6$H$_5$NH$_2$

i. C$_6$H$_5$OH

j. KI

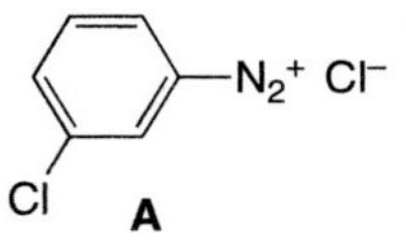

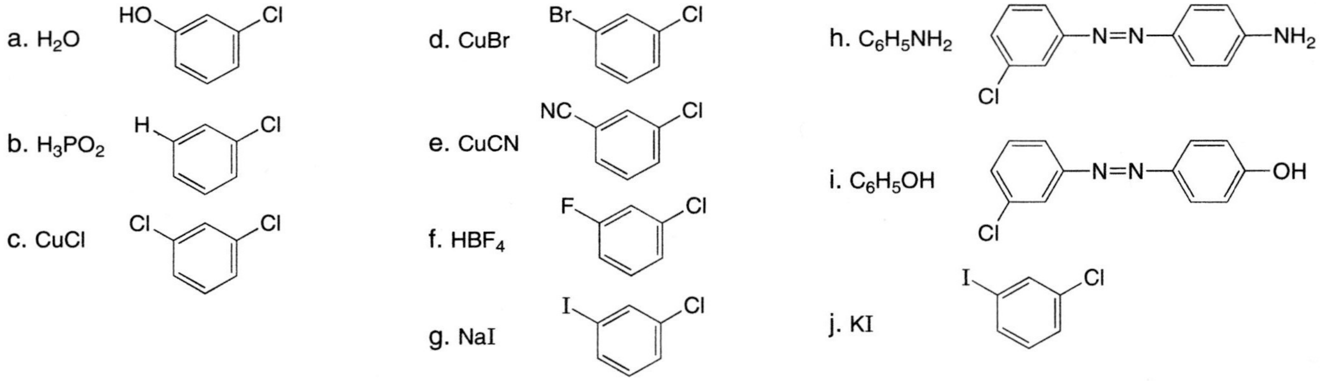

25.69 Under the acidic conditions of the reaction, aniline is first protonated to form an ammonium salt that has a positive charge on the atom bonded to the benzene ring. The –NH$_3^+$ is now an electron-withdrawing meta director, so a significant amount of meta substitution occurs.

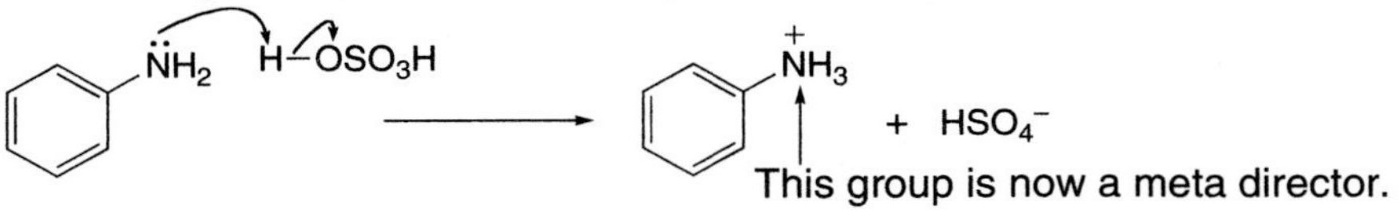

This group is now a meta director.

25.70

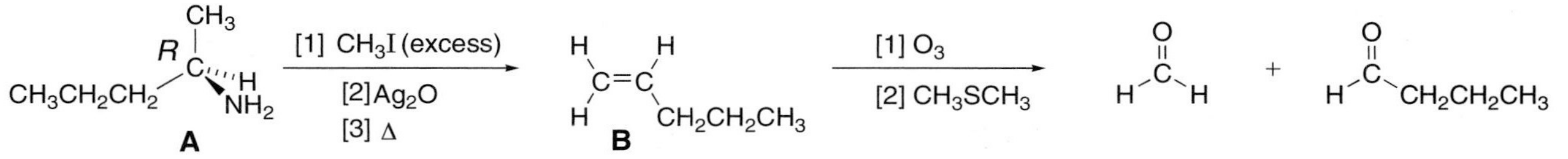

25.71

a.

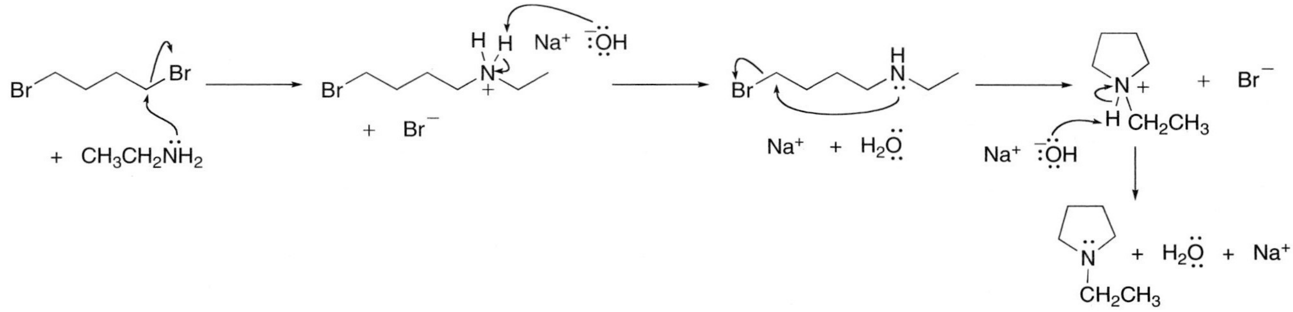

b.

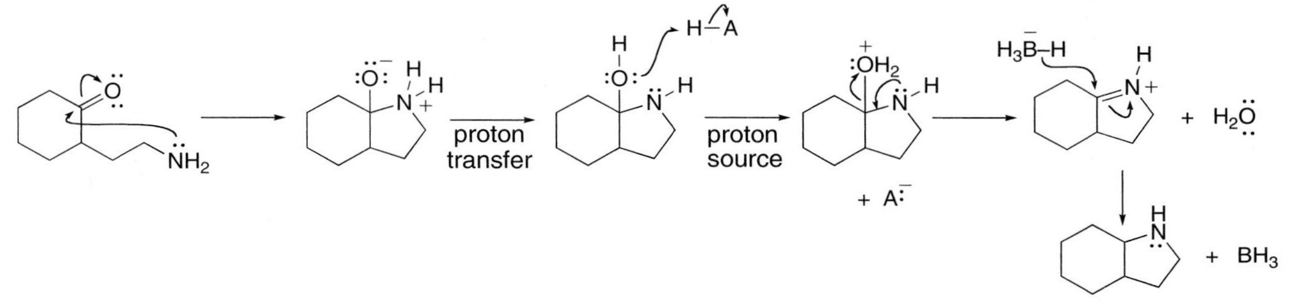

25.72

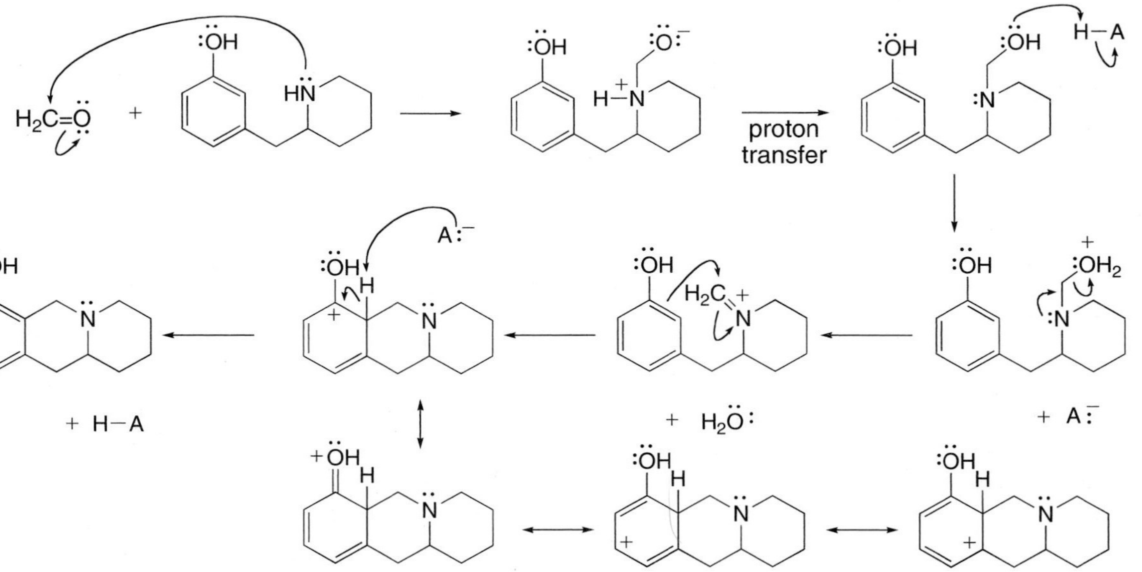

25.73

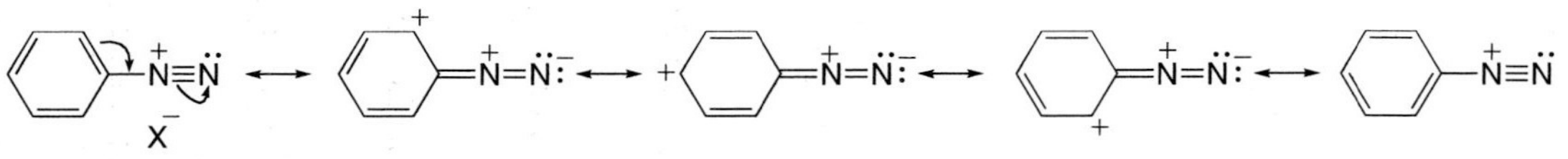

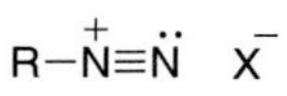

aryl diazonium salt   The N_2^+ group on an aromatic ring is stabilized by
resonance, whereas the alkyl diazonium salt is not.

alkyl diazonium salt

25.74

Overall reaction:

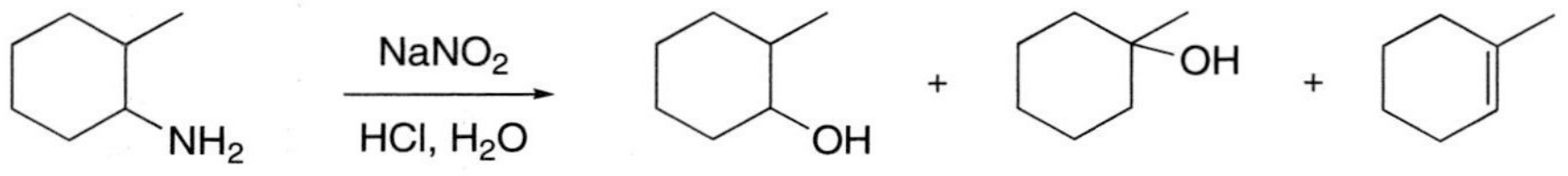

The steps:

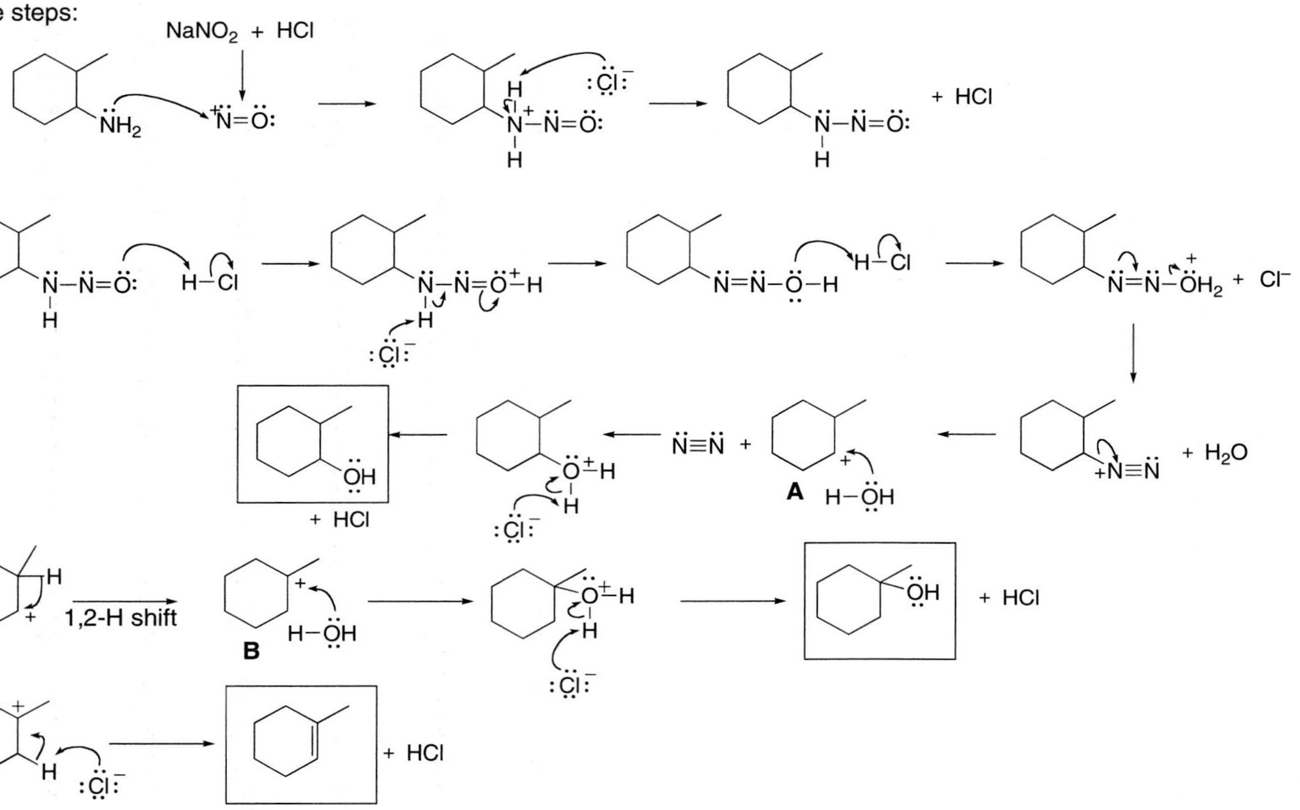

25.75 A nitrosonium ion ($^+$NO) is a weak electrophile so electrophilic aromatic substitution occurs only with a strong electron-donor group that stabilizes the intermediate carbocation.

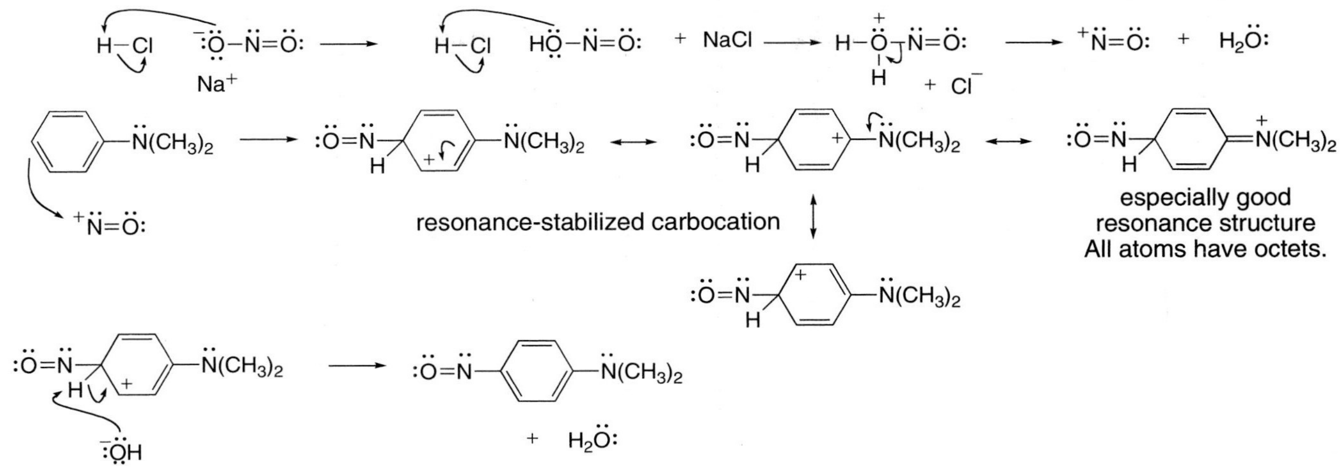

25.76

a.

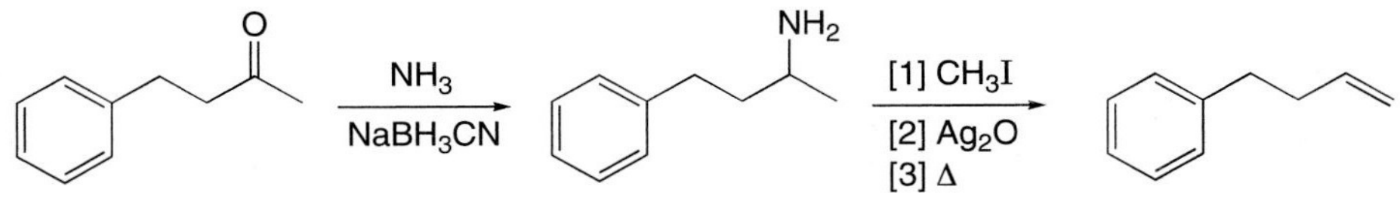

(Hofmann elimination, less substituted C=C favored)

b.

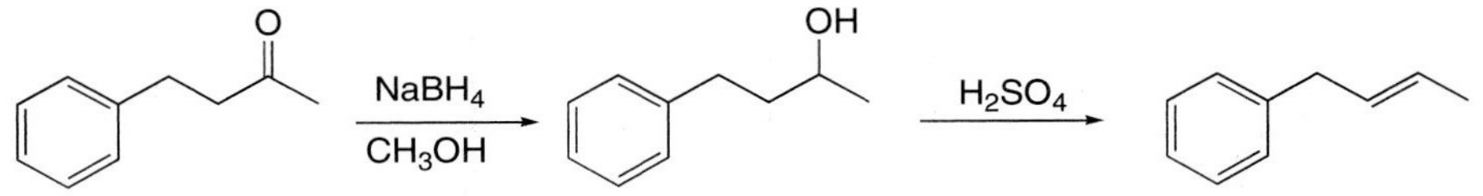

(more substituted C=C favored)

(E + Z isomers formed)

25.77

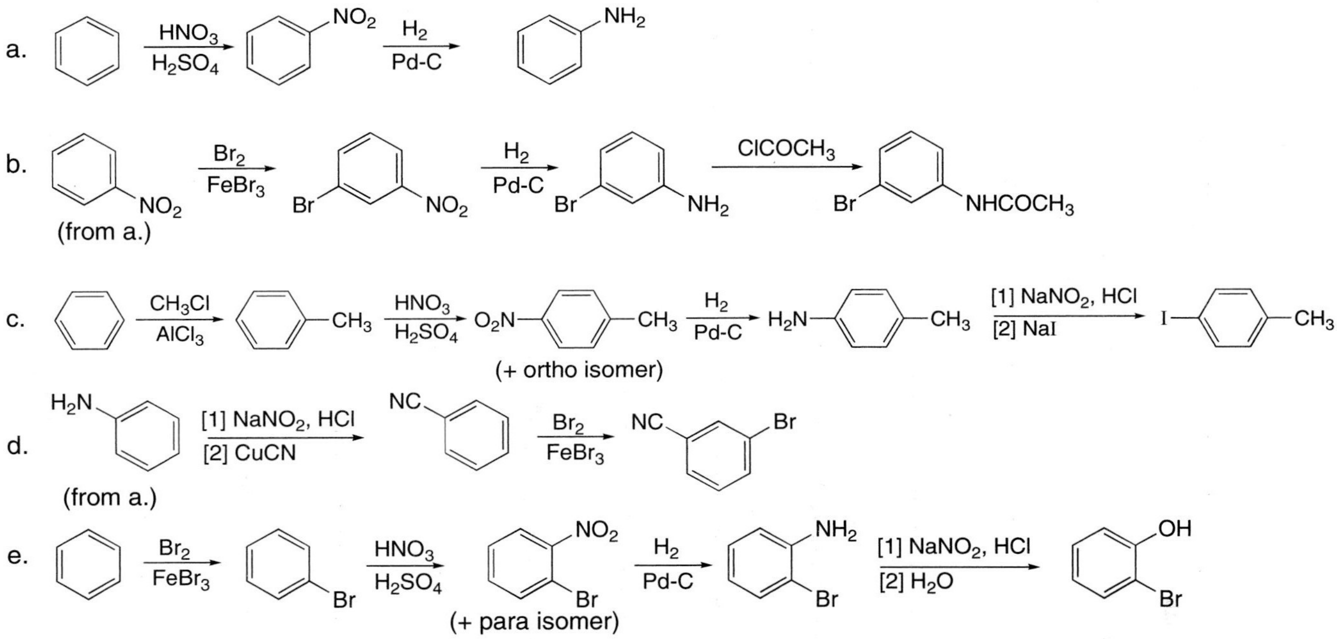

a.

b.

(from a.)

c.

(+ ortho isomer)

d.

(from a.)

e.

(+ para isomer)

f. O_2N—⟨ ⟩—CH_3 $\xrightarrow{KMnO_4}$ O_2N—⟨ ⟩—$COOH$ $\xrightarrow[Pd-C]{H_2}$ H_2N—⟨ ⟩—$COOH$ $\xrightarrow[{[2]\ H_2O}]{[1]\ NaNO_2,\ HCl}$ HO—⟨ ⟩—$COOH$

(from c.)

g. ⟨ ⟩ $\xrightarrow[AlCl_3]{ClCOCH_3}$ ⟨ ⟩ $\xrightarrow[H_2SO_4]{HNO_3}$ O_2N—⟨ ⟩ $\xrightarrow[Pd-C]{H_2}$ H_2N—⟨ ⟩ $\xrightarrow[{[2]\ NaI}]{[1]\ NaNO_2,\ HCl}$ I—⟨ ⟩

h. ⟨ ⟩—NH_2 $\xrightarrow[{[2]\ H_2O}]{[1]\ NaNO_2,\ HCl}$ ⟨ ⟩—OH $\xrightarrow{C_6H_5N_2{}^+Cl^-}$ ⟨ ⟩—$N{=}N$—⟨ ⟩—OH

(from a.)   (from d., Step [1])

25.78

a. ⟨ ⟩—NH_2 $\xrightarrow[{[2]\ NaCN}]{[1]\ NaNO_2,\ HCl}$ ⟨ ⟩—CN $\xrightarrow{H_3O^+}$ ⟨ ⟩—$COOH$ $\xrightarrow[{[2]\ NH_2CH_3}]{[1]\ SOCl_2}$ ⟨ ⟩—$CONHCH_3$

b. ⟨ ⟩—NH_2 $\xrightarrow{CH_3COCl}$ ⟨ ⟩—$NHCOCH_3$ $\xrightarrow[FeBr_3]{Br_2}$ Br—⟨ ⟩—$NHCOCH_3$ $\xrightarrow[AlCl_3]{CH_3Cl}$ Br—⟨ ⟩—$NHCOCH_3$ $\xrightarrow[H_2O]{^-OH}$ Br—⟨ ⟩—NH_2

(+ ortho isomer)

$\xrightarrow[{[2]\ H_3PO_2}]{[1]\ NaNO_2,\ HCl}$ CH_3—⟨ ⟩—Br

c. ⟨ ⟩—$COOH$ $\xrightarrow[H_2SO_4]{HOCH_2CH_3}$ ⟨ ⟩—$COOCH_2CH_3$

(from a.)

d. ⟨ ⟩—$COOH$ $\xrightarrow[{[2]\ H_2O}]{[1]\ LiAlH_4}$ ⟨ ⟩—CH_2OH $\xrightarrow[FeBr_3\ (3x)]{Br_2}$ Br—⟨ ⟩—CH_2OH (with Br, Br)

(from a.)

e. H_2N—⟨ ⟩ $\xrightarrow[{[2]\ H_2O}]{[1]\ NaNO_2,\ HCl}$ HO—⟨ ⟩ $\xrightarrow[AlCl_3]{CH_3Cl}$ HO—⟨ ⟩—CH_3 $\xrightarrow{C_6H_5N_2{}^+Cl^-}$ ⟨ ⟩—$N{=}N$—⟨ ⟩ HO, CH_3

(+ ortho isomer)   (from a., Step [1])

25.79

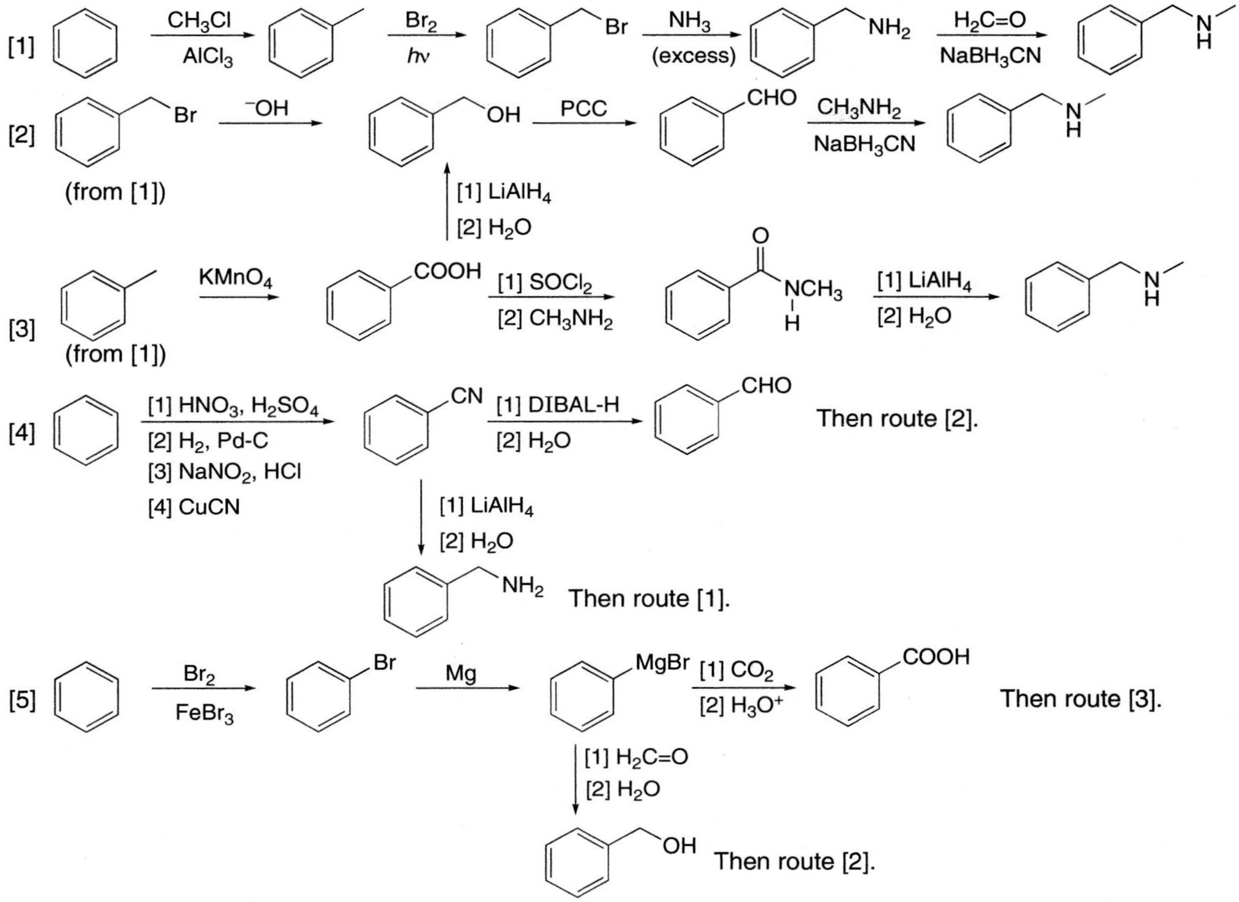

[1]

[2]

(from [1])

[3]

(from [1])

[4]

[5]

25.80

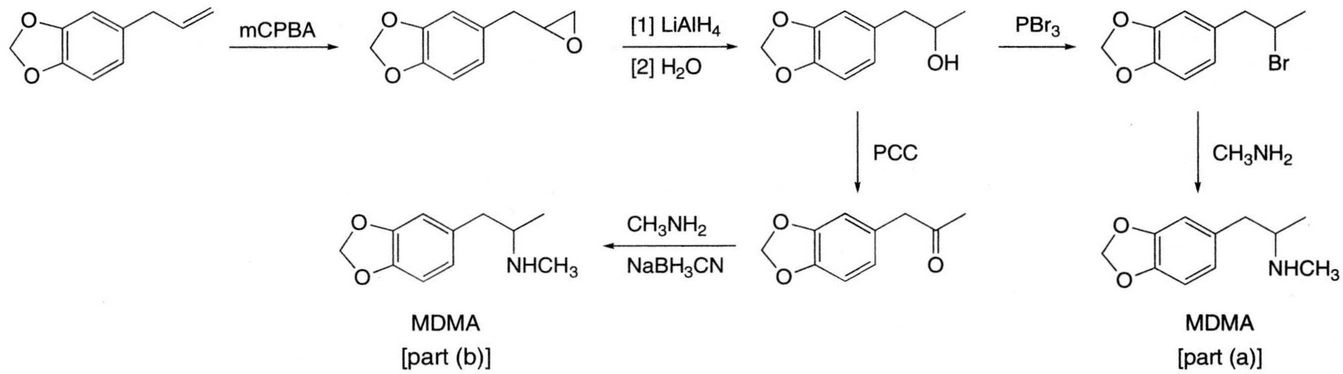

MDMA
[part (b)]

MDMA
[part (a)]

25.81

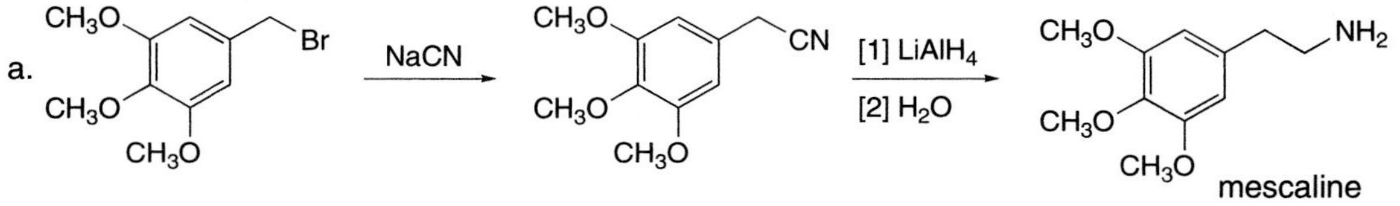

a.

mescaline

b.

c.

25.82

a.

b.

c.

d.

e.

f.

25.83

a.

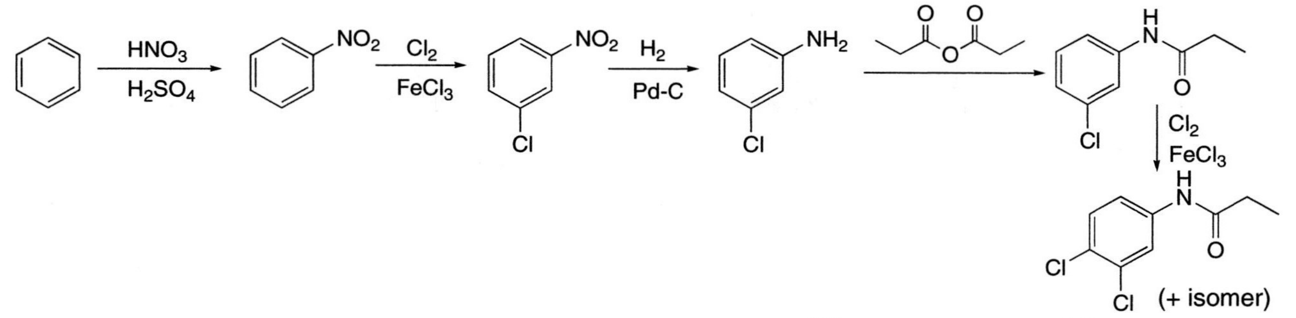

b.

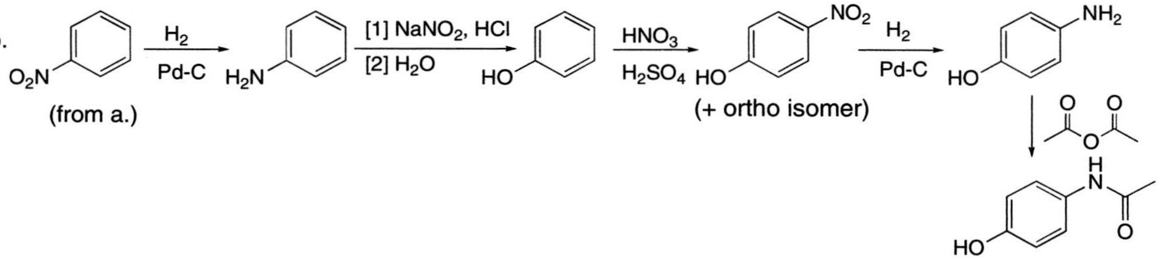

c.

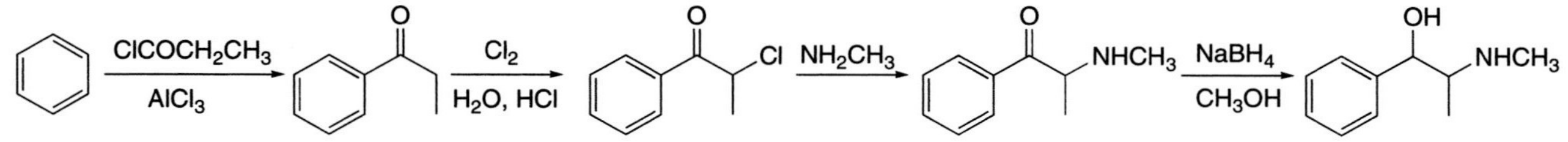

25.84

a.

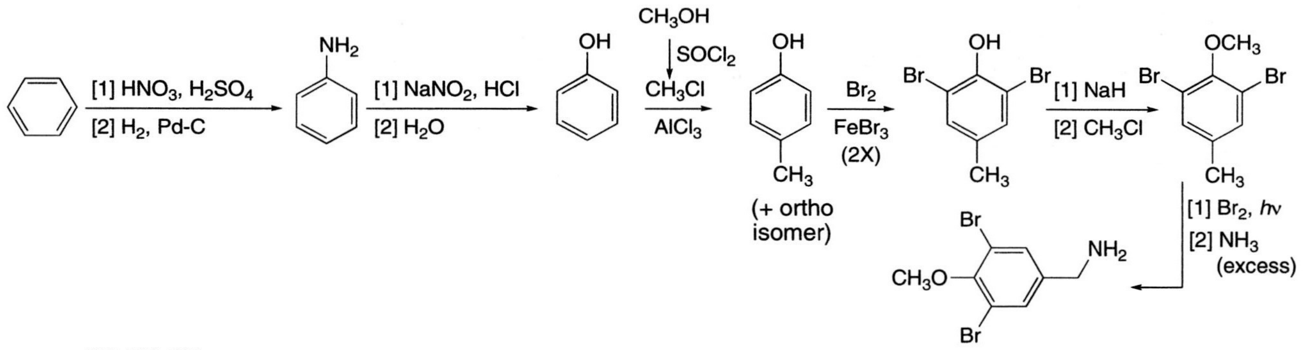

b.

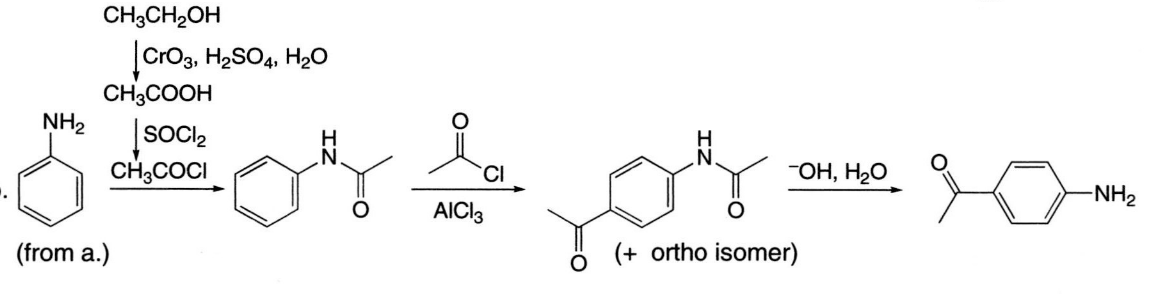

c.

d.

e.

Probably a strong enough activator that the Friedel–Crafts reaction will still occur.

Make two parts:

f.

25.85

molecular weight = 87
$C_5H_{13}N$
two IR peaks = 1° amine

25.86

NHCH_2CH_3

Compound **A**: $C_8H_{11}N$
IR absorption at 3400 cm^{-1}→ 2° amine
^1H NMR signals at (ppm):
 1.3 (triplet, 3 H) CH_3 adjacent to 2 H's
 3.1 (quartet, 2 H) CH_2 adjacent to 3 H's
 3.6 (singlet, 1 H) amine H
 6.8–7.2 (multiplet, 5 H) benzene ring

CH_2NHCH_3

Compound **B**: $C_8H_{11}N$
IR absorption at 3310 cm^{-1} → 2° amine
^1H NMR signals at (ppm):
 1.4 (singlet, 1 H) amine H
 2.4 (singlet, 3 H) CH_3
 3.8 (singlet, 2 H) CH_2
 7.2 (multiplet, 5 H) benzene ring

H_2N——CH_2CH_3

Compound **C**: $C_8H_{11}N$
IR absorption at 3430 and
  3350 cm^{-1} → 1° amine
^1H NMR signals at (ppm):
 1.3 (triplet, 3 H) CH_3 near CH_2
 2.5 (quartet, 2 H) CH_2 near CH_3
 3.6 (singlet, 2 H) amine H's
 6.7 (doublet, 2 H)⎤ para disubstituted
 7.0 (doublet, 2 H)⎦ benzene ring

25.87

HO⎯⎯C≡N   HO⎯⎯NH_2

Compound **D**:
Molecular ion at m/z = 71: C_3H_5NO (possible formula)
IR absorption at 3600–3200 cm^{-1}→ OH
    2263 cm^{-1}→ CN
Use integration values and the molecular formula to
  determine the number of H's that give rise to each
  signal.
^1H NMR signals at (ppm):
 2.6 (triplet, 2 H) CH_2 adjacent to 2 H's
 3.2 (singlet, 1 H) OH
 3.9 (triplet, 2 H) CH_2 adjacent to 2 H's

Compound **E**:
Molecular ion at m/z = 75: C_3H_9NO (possible formula)
IR absorption at 3600–3200 cm^{-1}→ OH
    3636 cm^{-1} → N–H of amine
^1H NMR signals at (ppm):
 1.6 (quintet, 2 H) CH_2 split by 2 CH_2's
 2.5 (singlet, 3 H) NH_2 and OH
 2.8 (triplet, 2 H) CH_2 split by CH_2
 3.7 (triplet, 2 H) CH_2 split by CH_2

25.88 Guanidine is a strong base because its conjugate acid is stabilized by resonance. This resonance delocalization makes guanidine easily donate its electron pair; thus it's a strong base.

guanidine pK_a = 13.6

25.89

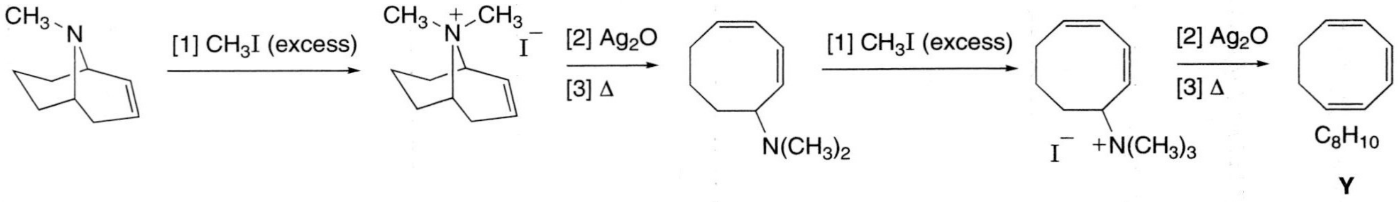

25.90

One possibility:

a.

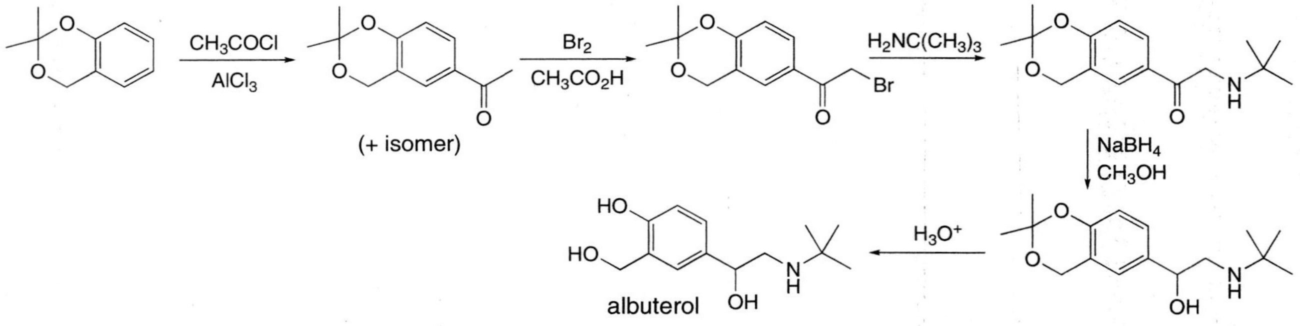

(+ isomer)

albuterol

b.

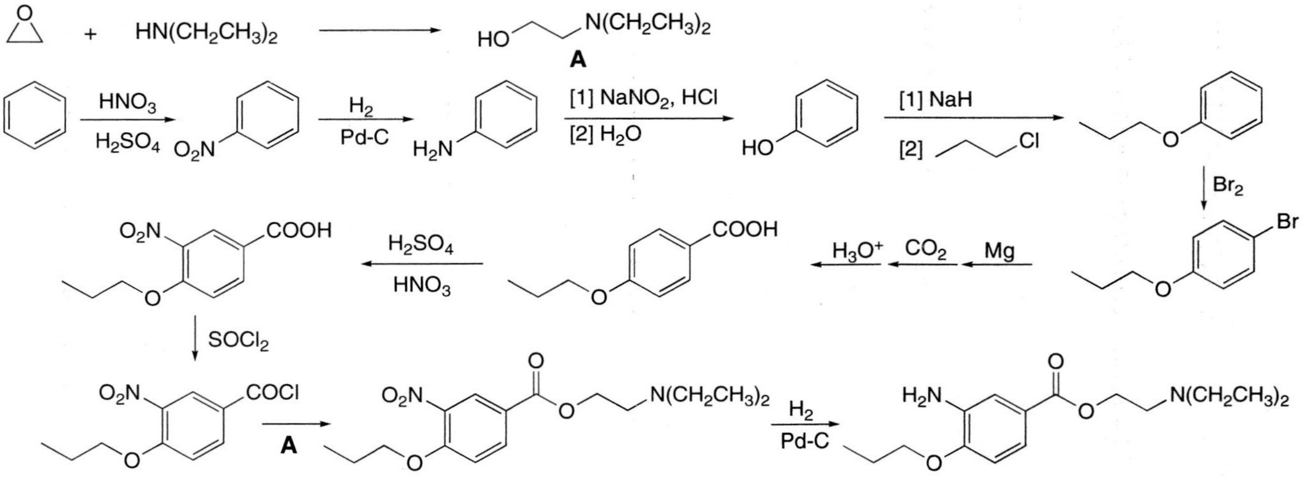

Chapter 26: Carbon–Carbon Bond-Forming Reactions in Organic Synthesis

♦ **Coupling reactions**

[1] Coupling reactions of organocuprate reagents (26.1)

$R'{-}X$ + R_2CuLi $\longrightarrow$ | $R'{-}R$ | + RCu

$X = Cl, Br, I$ + LiX

- R'X can be CH_3X, RCH_2X, 2° cyclic halides, vinyl halides, and aryl halides.
- X may be Cl, Br, or I.
- With vinyl halides, coupling is stereospecific.

[2] Suzuki reaction (26.2)

- R'X is most often a vinyl halide or aryl halide.
- With vinyl halides, coupling is stereospecific.

[3] Heck reaction (26.3)

- R'X is a vinyl halide or aryl halide.
- Z = H, Ph, COOR, or CN
- With vinyl halides, coupling is stereospecific.
- The reaction forms trans alkenes.

♦ **Cyclopropane synthesis**

[1] Addition of dihalocarbenes to alkenes (26.4)

- The reaction occurs with syn addition.
- The position of substituents in the alkene is retained in the cyclopropane.

[2] Simmons–Smith reaction (26.5)

- The reaction occurs with syn addition.
- The position of substituents in the alkene is retained in the cyclopropane.

♦ **Metathesis (26.6)**

- Metathesis works best when $CH_2{=}CH_2$, a gas that escapes from the reaction mixture, is formed as one product.

Chapter 26: Answers to Problems

26.1 A new C–C bond is formed in each coupling reaction.

a. 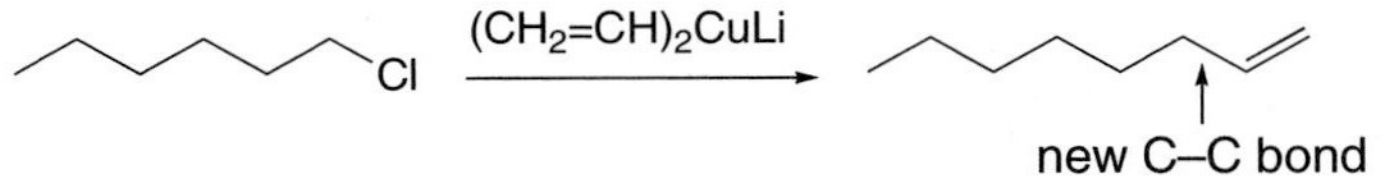
Cl
(CH2=CH)2CuLi
new C–C bond

b. 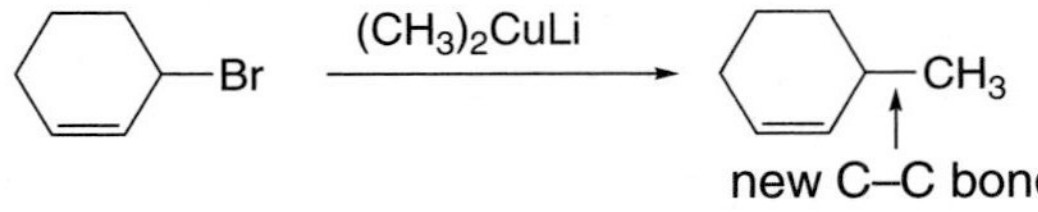
Br
(CH3)2CuLi
CH3
new C–C bond

c. 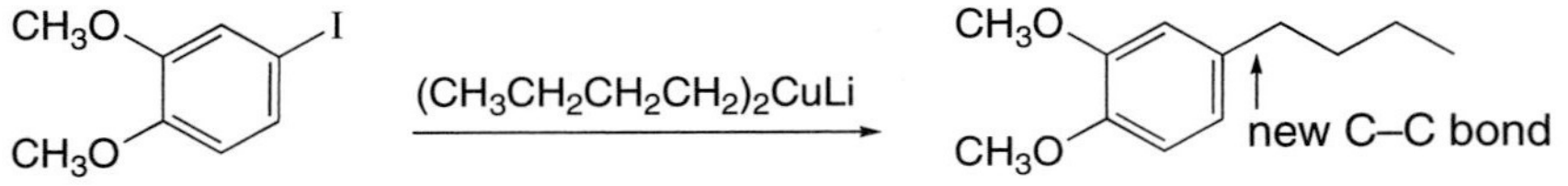
CH3O
CH3O
I
(CH3CH2CH2CH2)2CuLi
new C–C bond

d. 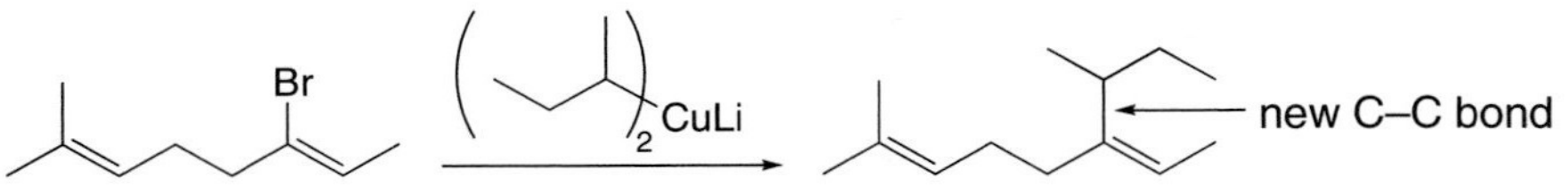
Br
CuLi
2
new C–C bond

26.2

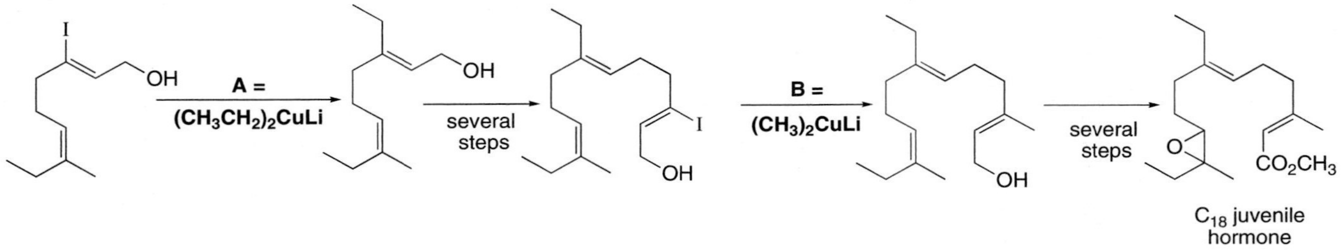
I
OH
A =
(CH3CH2)2CuLi
several steps
B =
(CH3)2CuLi
CO2CH3
C18 juvenile hormone

26.3

a. 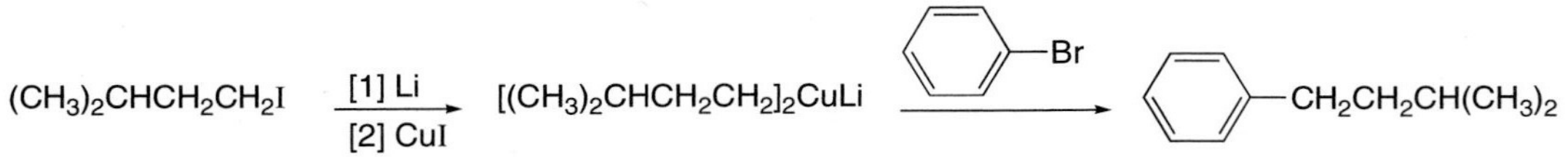
(CH3)2CHCH2CH2I
[1] Li
[2] CuI
[(CH3)2CHCH2CH2]2CuLi
Br
CH2CH2CH(CH3)2

or

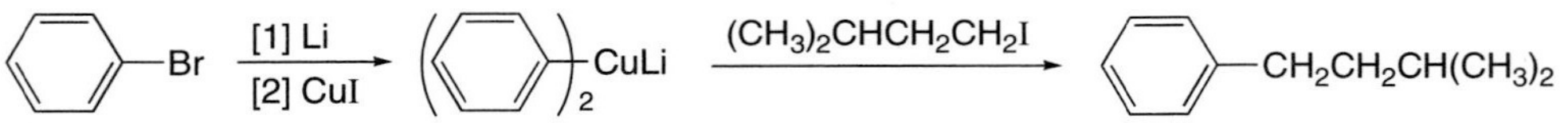
Br
[1] Li
[2] CuI
CuLi
2
(CH3)2CHCH2CH2I
CH2CH2CH(CH3)2

b. 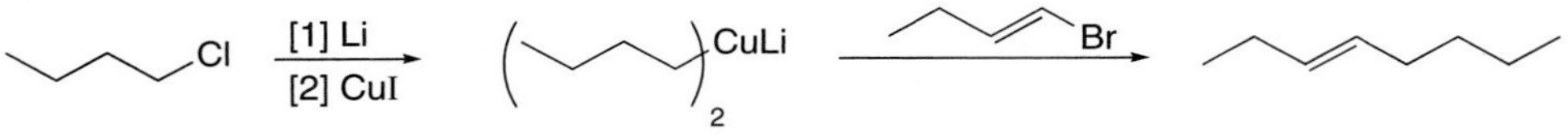
Cl
[1] Li
[2] CuI
CuLi
2
Br

c. 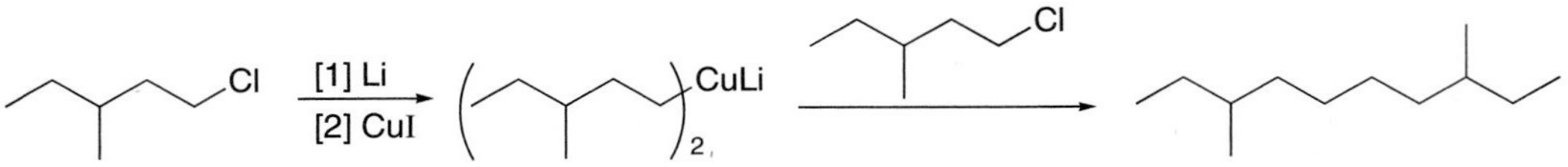
Cl
[1] Li
[2] CuI
CuLi
2

26.4

a.

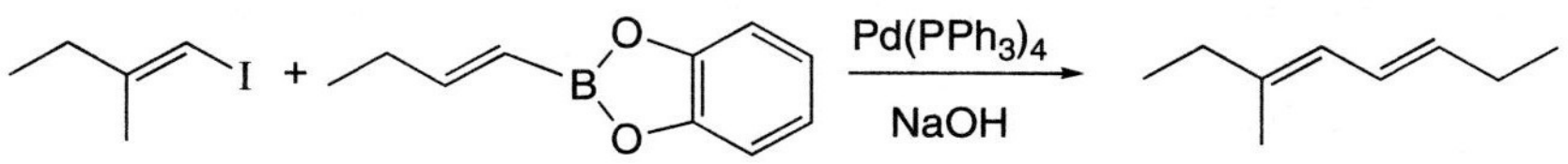

b.

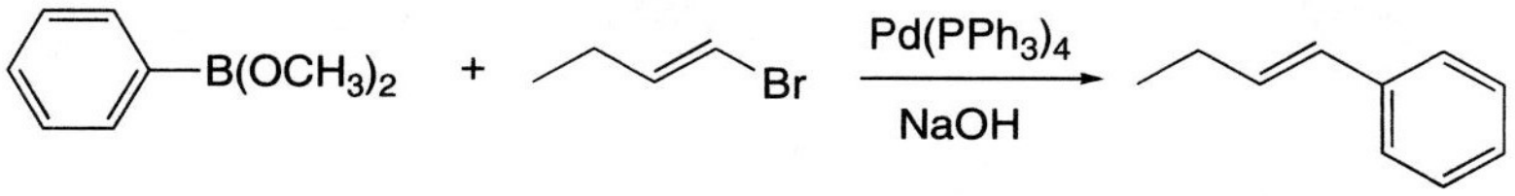

c.

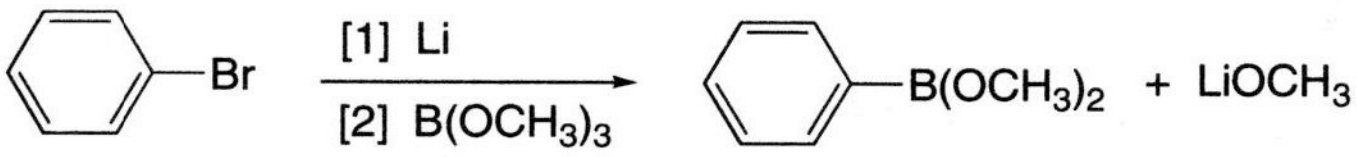

d.

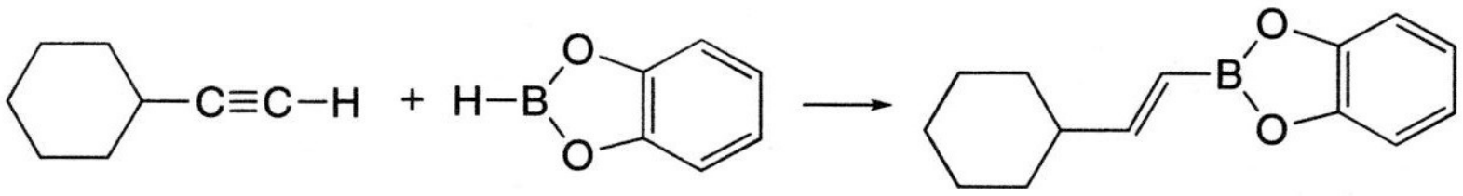

26.5

a. 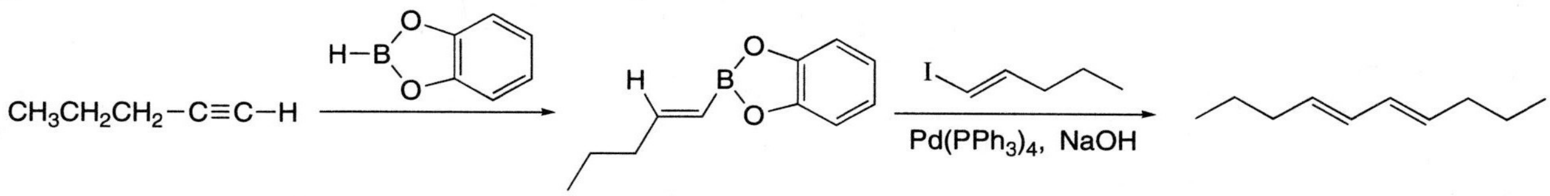 $CH_3CH_2CH_2-C\equiv C-H$

b.

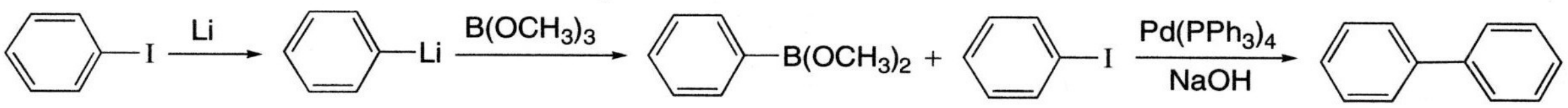

c.

(+ ortho isomer)

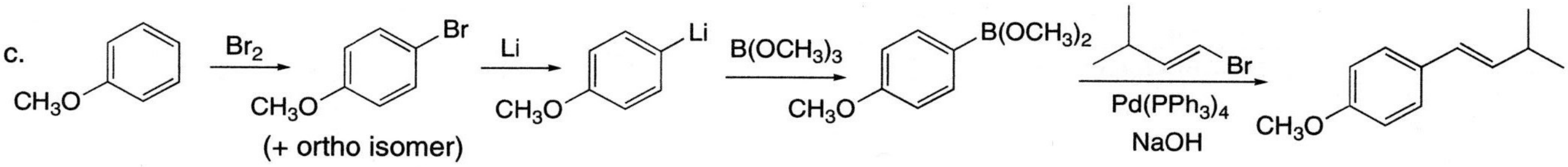

26.6

a.

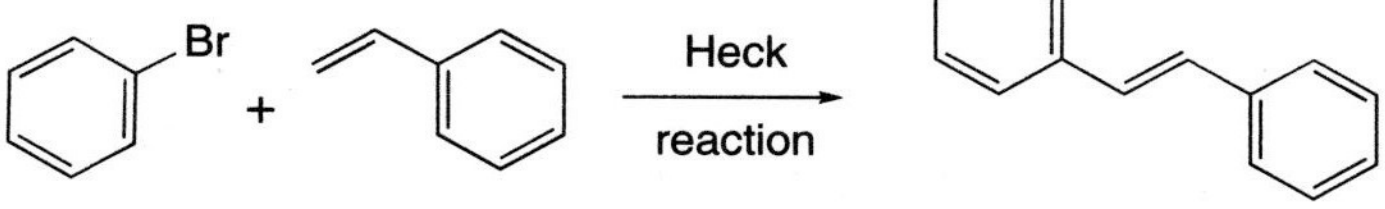

b.

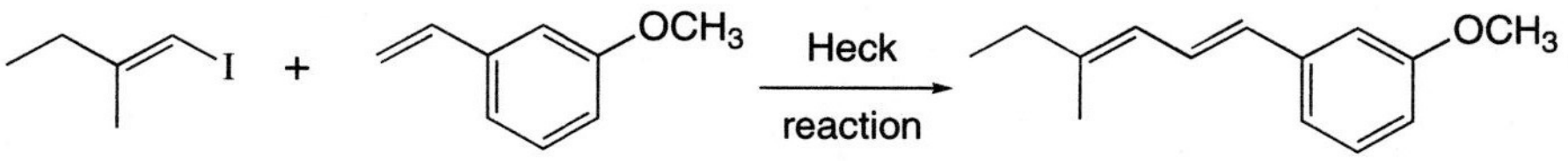

c.

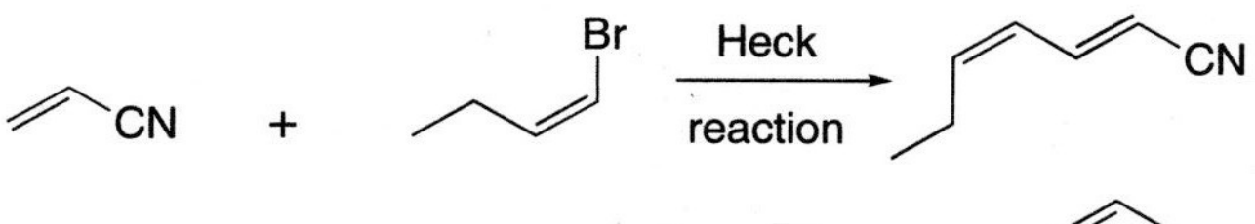

d.

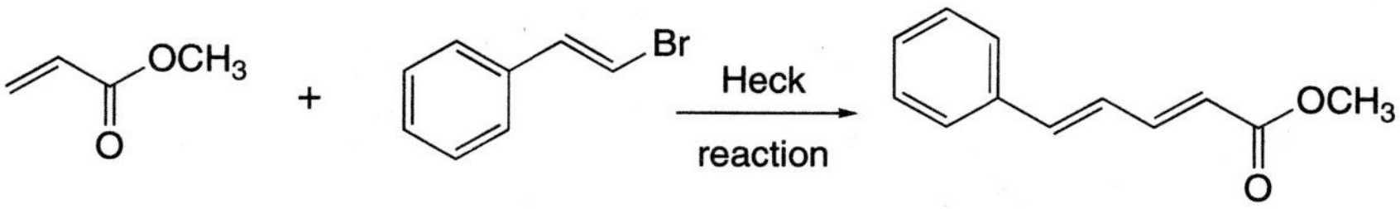

26.7 Locate the double bond with the aryl, COOR, or CN substituent, and break the molecule into two components at the end of the C=C not bonded to one of these substituents.

a.

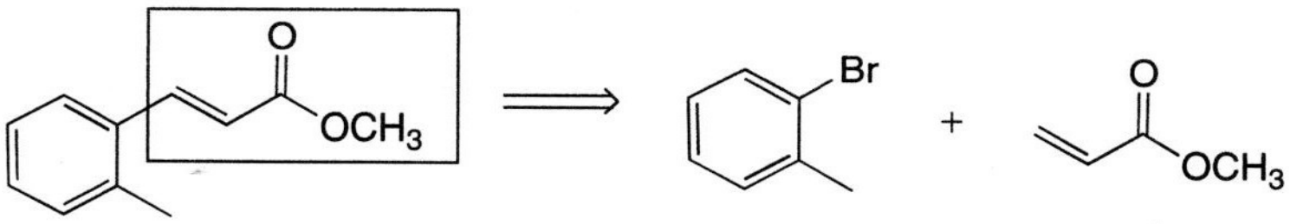

b.

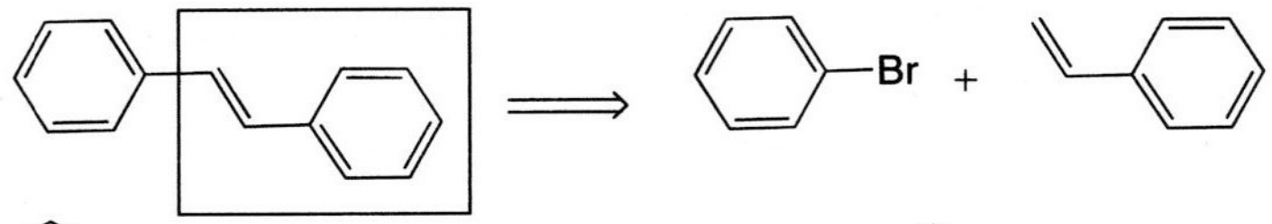

c.

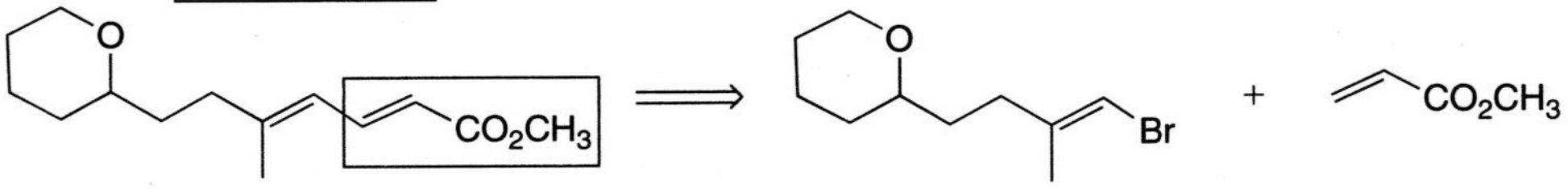

26.8 Add the carbene carbon from either side of the alkene.

a.

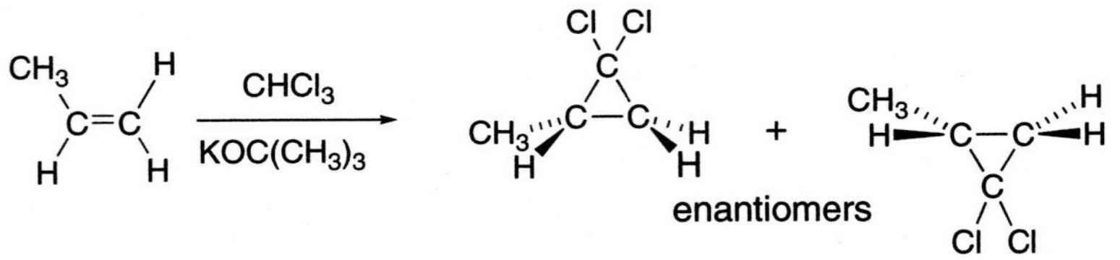

enantiomers

b.

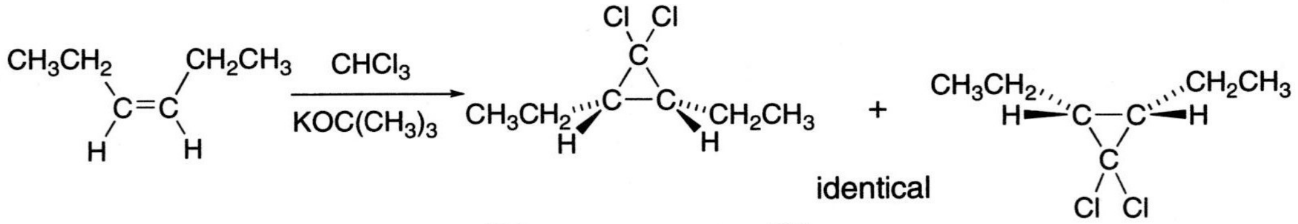

identical

c.

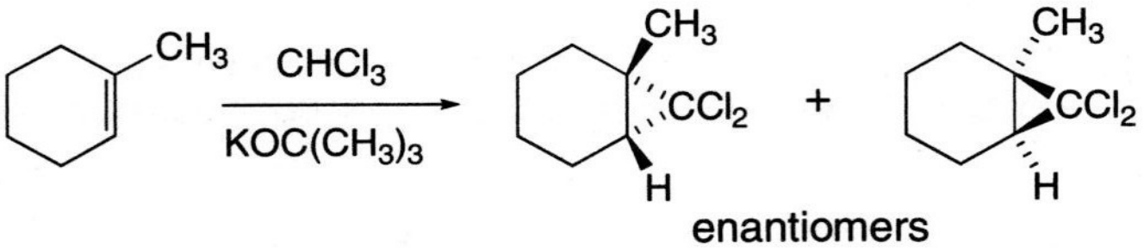

enantiomers

26.9

a.

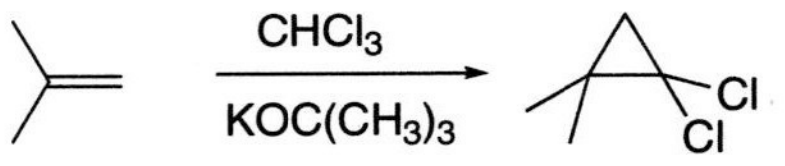

b.

c.

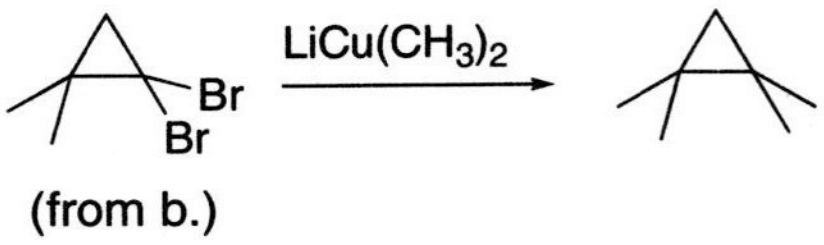

(from b.)

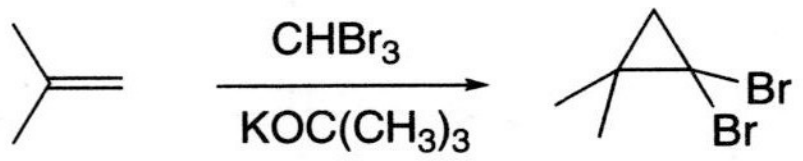

26.10

a.

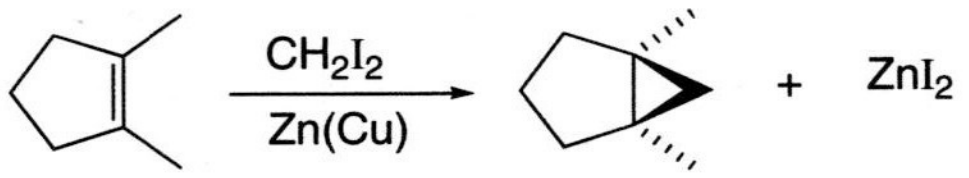

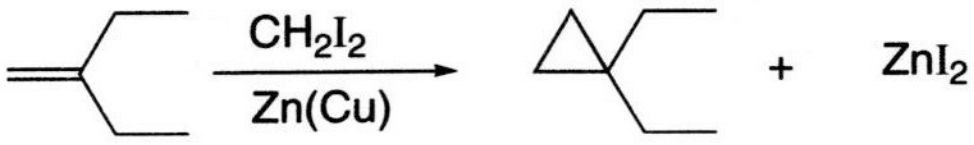

b.

c.

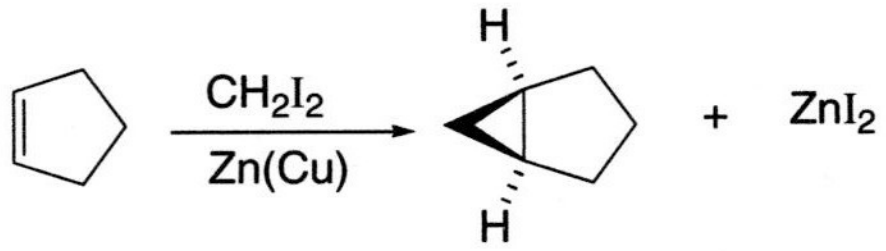

26.11 The relative position of substituents in the reactant is retained in the product.

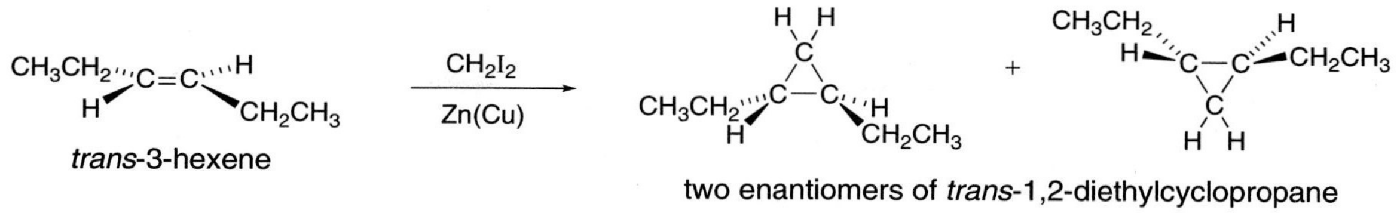

trans-3-hexene
CH_2I_2
Zn(Cu)
two enantiomers of trans-1,2-diethylcyclopropane

26.12

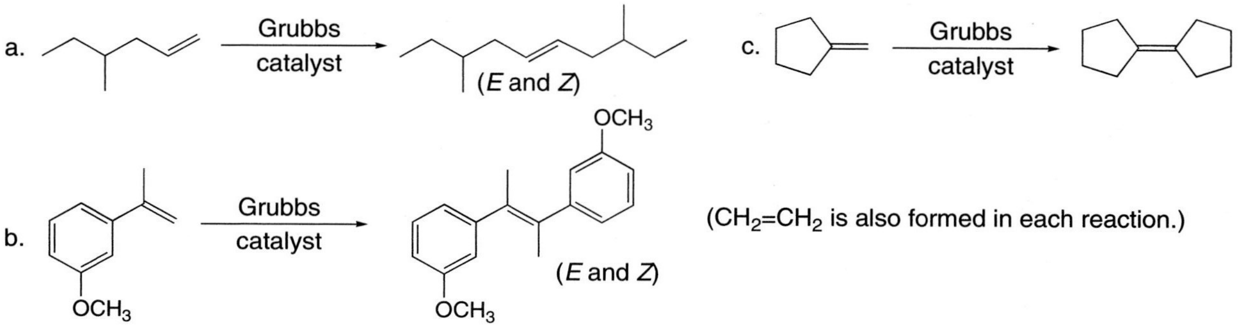

a.
Grubbs
catalyst
(E and Z)
c.
Grubbs
catalyst
b.
Grubbs
catalyst
(E and Z)
OCH_3

($CH_2=CH_2$ is also formed in each reaction.)

26.13

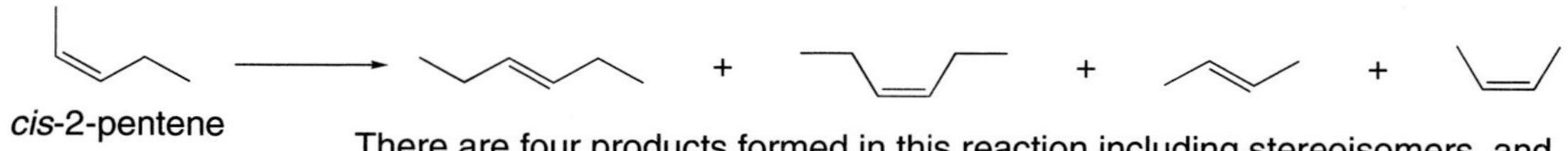

cis-2-pentene

There are four products formed in this reaction including stereoisomers, and therefore, it is not a practical method to synthesize 1,2-disubstituted alkenes.

26.14

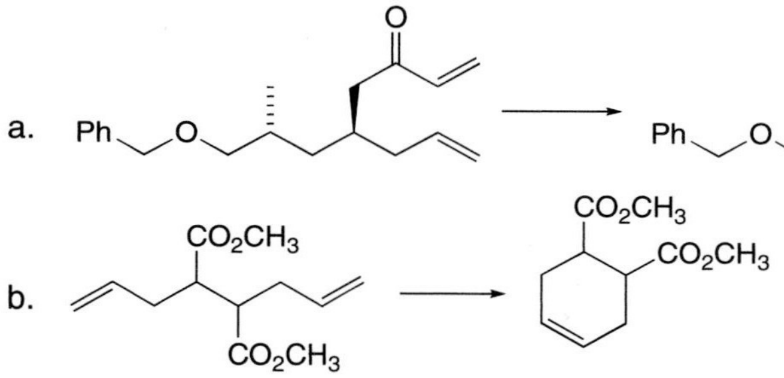

a.
Ph
O
b.
CO_2CH_3

26.15 High dilution conditions favor intramolecular metathesis.

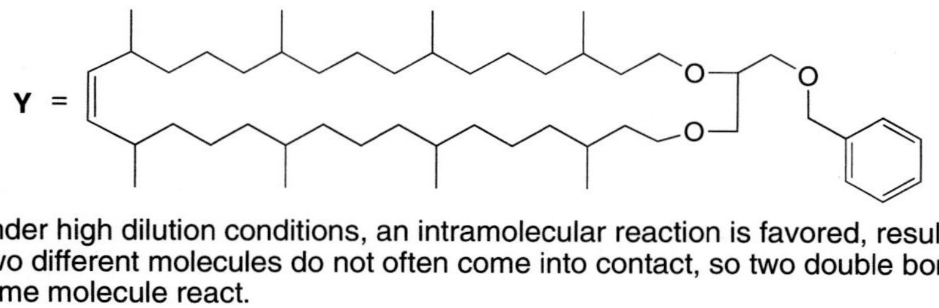

Y =
O

Under high dilution conditions, an intramolecular reaction is favored, resulting in **Y**. Two different molecules do not often come into contact, so two double bonds in the same molecule react.

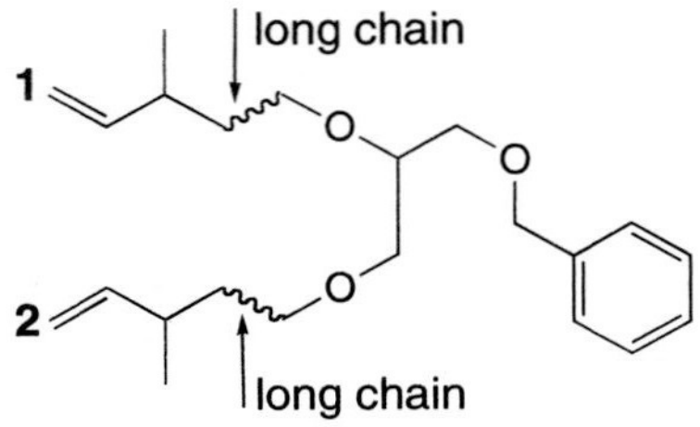

1 long chain

2 long chain

Alkenes **1** and **2** differ in proximity to the ether side chain. Three products are possible because alkene **1** can react with alkene **1** in another molecule (forming **Z**); alkene **1** can react with alkene **2** in another molecule (forming **Z'**); and alkene **2** can react with alkene **2** in another molecule (forming **Z"**). At usual reaction concentrations, the probability of two molecules approaching each other is greater than the probability of two sites (connected by a long chain) in the same molecule reacting together, so the intermolecular reaction is favored.

Z =

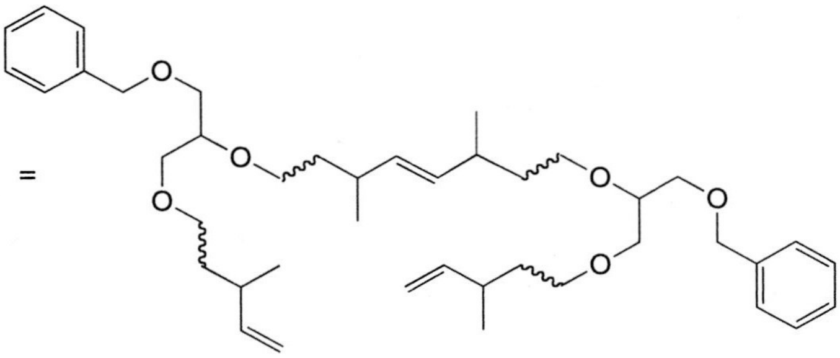

Z' =

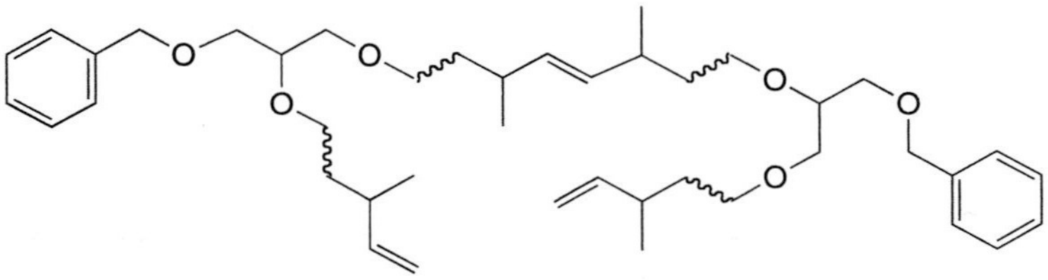

Z" =

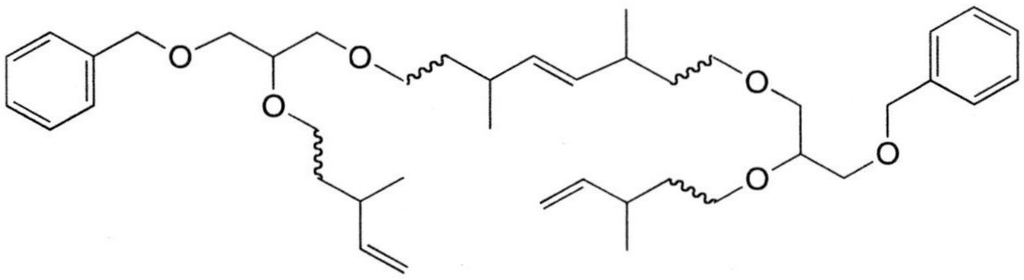

26.16 Cleave the C=C bond in the product, and then bond each carbon of the original alkene to a CH$_2$ group using a double bond.

a.

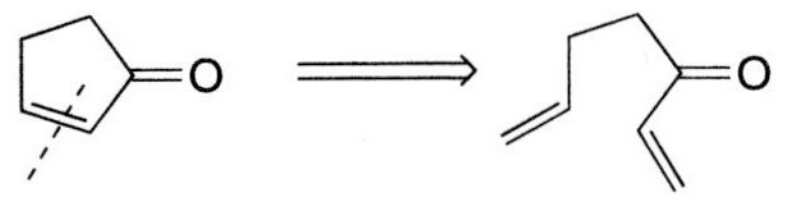

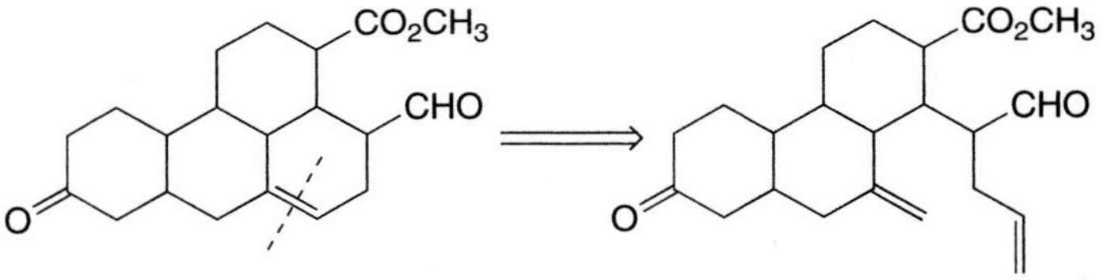

b.

c.

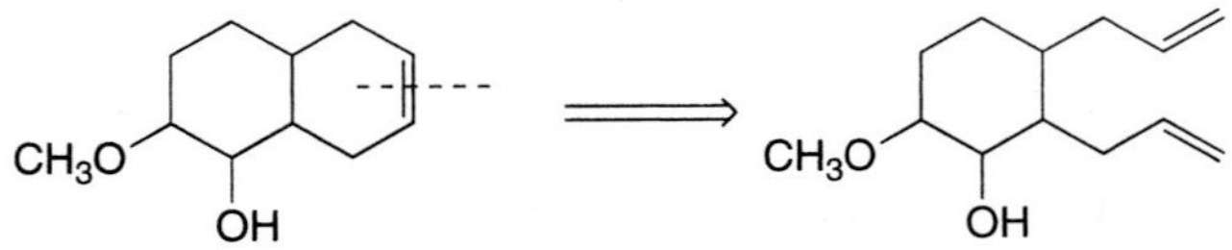

26.17

a.

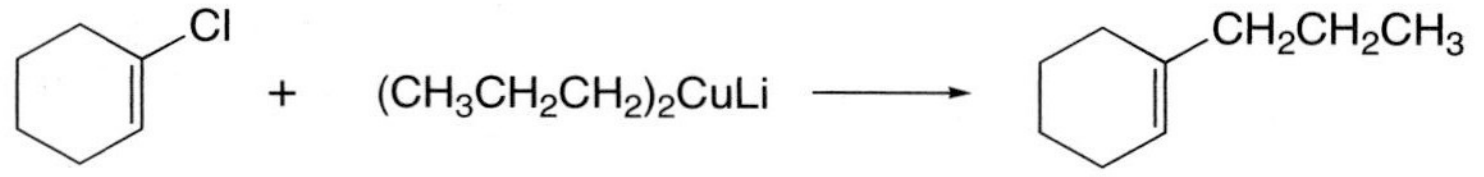

b.

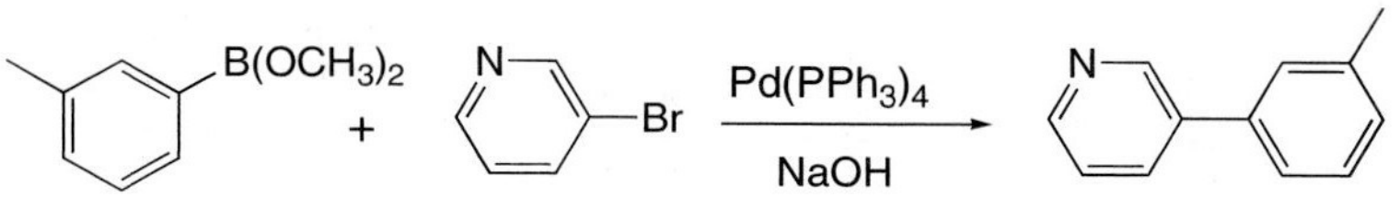

c.

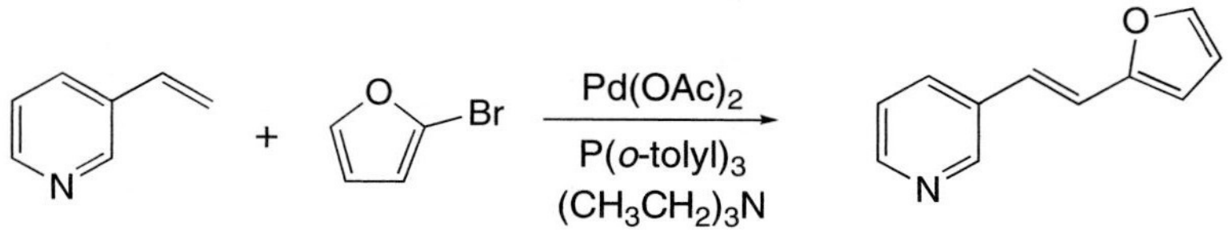

d.

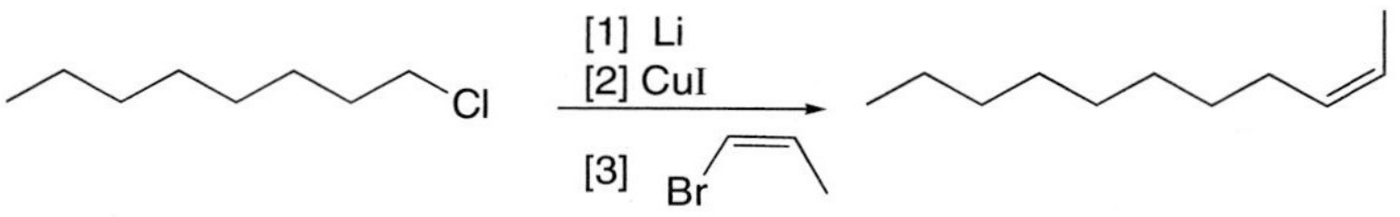

e.

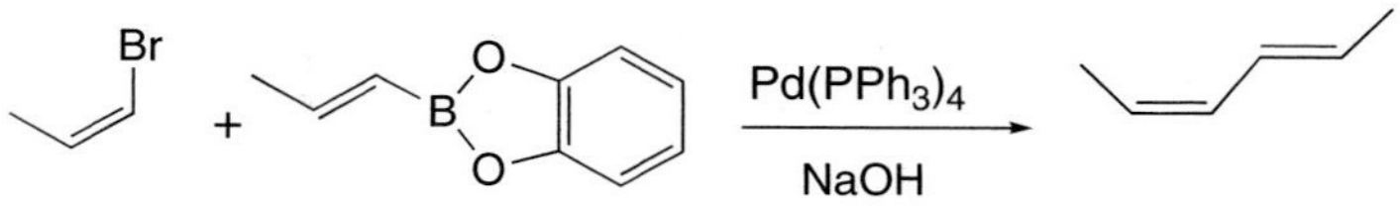

f.

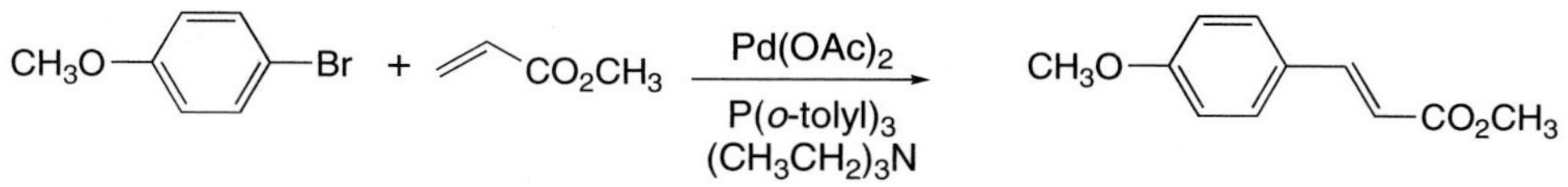

g.

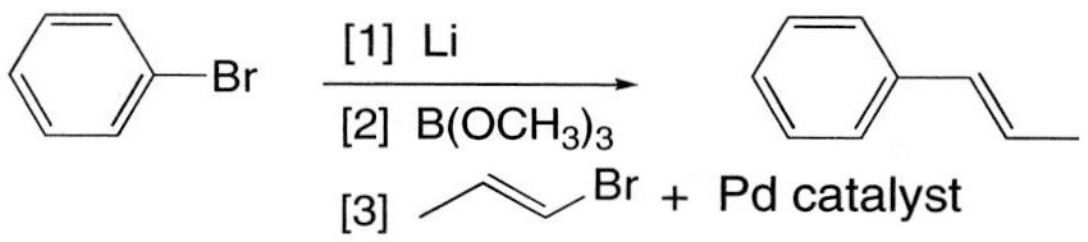

h. $(CH_3)_3C—C≡C—H$

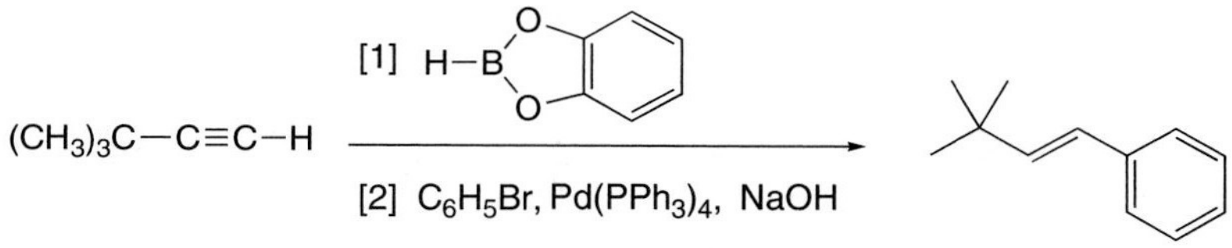

26.18

a.

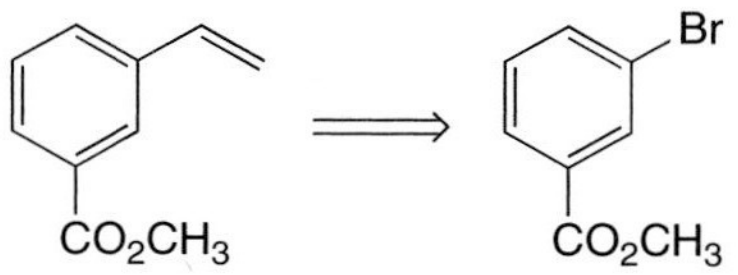

b.

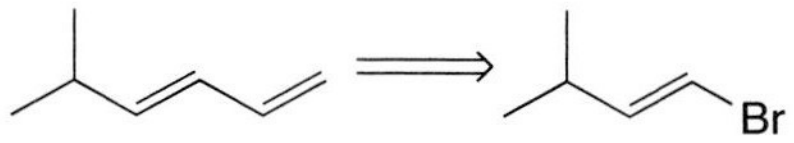

c.

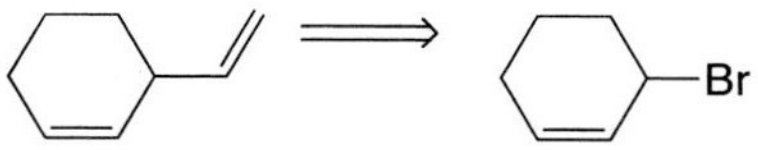

d.

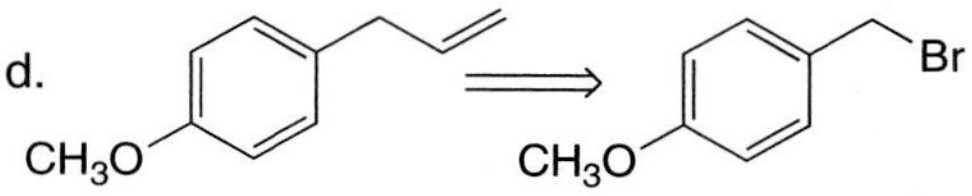

26.19

Each coupling reaction uses Pd(PPh$_3$)$_4$
and NaOH to form the conjugated diene.

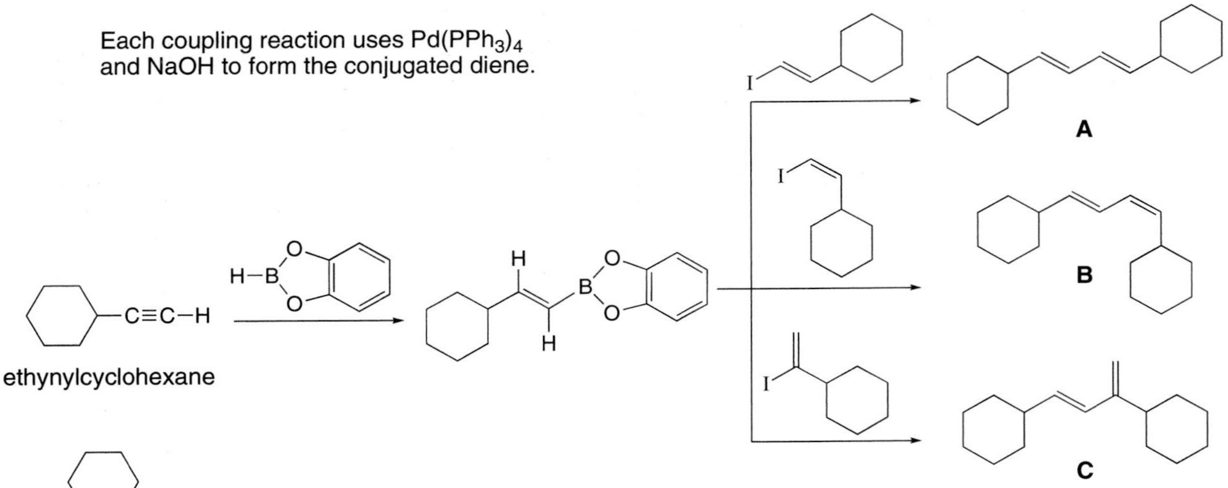

ethynylcyclohexane

A

B

C

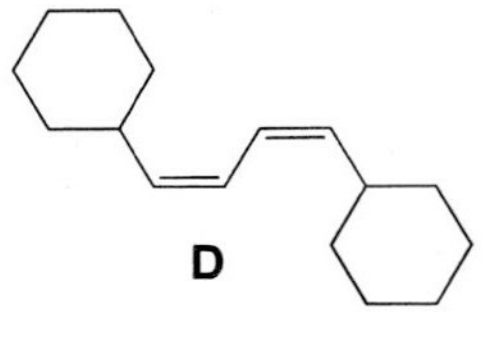

D

It is not possible to synthesize diene **D** using a Suzuki reaction with ethynylcyclohexane as starting material. Hydroboration of ethynylcyclohexane adds the elements of H and B in a syn fashion affording a trans vinylborane. Since the Suzuki reaction is stereospecific, one of the double bonds in the product must therefore be trans.

26.20 Locate the styrene part of the molecule, and break the molecule into two components. The second component in each reaction is styrene, C$_6$H$_5$CH=CH$_2$.

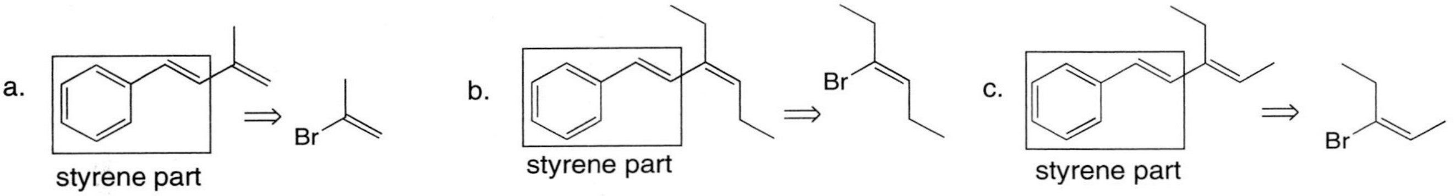

a.

styrene part

b.

styrene part

c.

styrene part

26.21 Inversion of configuration occurs with substitution of the methyl group for the tosylate.

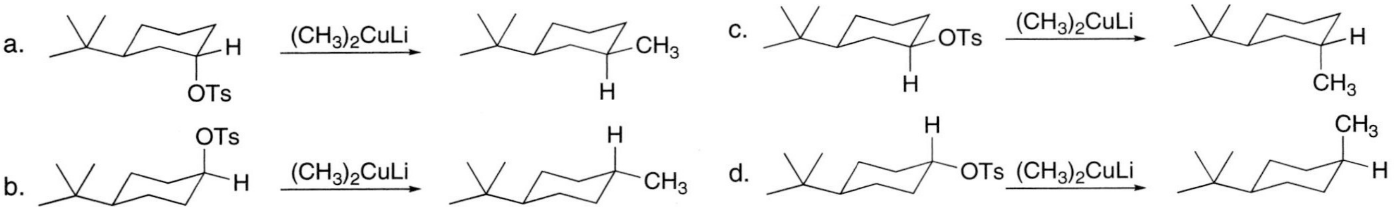

a.

b.

c.

d.

26.22

1-butene

octane

[1] Li
[2] CuI

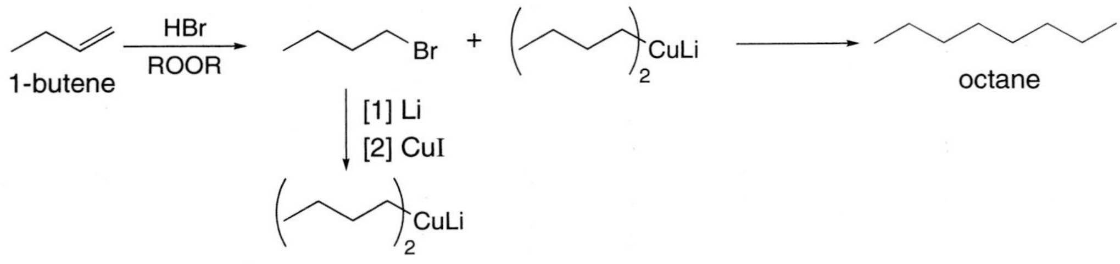

26.23 Add the carbene carbon from either side of the alkene.

a.

b.

c.

d.

e.

f.

26.24 Since the new three-membered ring has a stereogenic center on the C bonded to the phenyl group, the phenyl group can be oriented in two different ways to afford two stereoisomers. These products are diastereomers of each other.

26.25 High dilution conditions favor intramolecular metathesis.

a.

b.

c.

26.26 Retrosynthetically break the double bond in the cyclic compound and add a new =CH$_2$ at each end to find the starting material.

a.

b.

c.

26.27 Alkene metathesis with two different alkenes is synthetically useful only when both alkenes are symmetrically substituted; that is, the two groups on each end of the double bond are identical to the two groups on the other end of the double bond.

a. $CH_3CH_2CH=CH_2$
   +
   $CH_3CH_2CH_2CH=CH_2$ ⟶ $CH_3CH_2CH_2CH=CHCH_2CH_3$ + $CH_3CH_2CH=CHCH_2CH_3$ + $CH_3CH_2CH_2CH=CHCH_2CH_2CH_3$
   + $CH_2=CH_2$ *(Z + E)*   *(Z + E)*   *(Z + E)*

b. 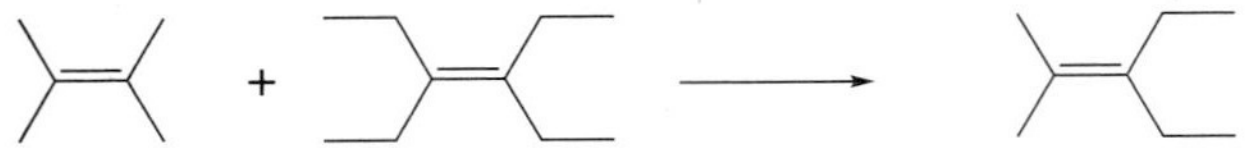 This reaction is synthetically useful since it yields only one product.

c.

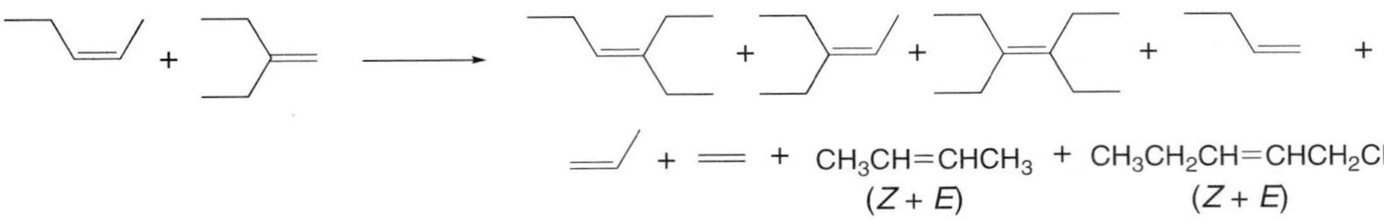

$\diagup\diagdown$ + $=$ + $CH_3CH=CHCH_3$ + $CH_3CH_2CH=CHCH_2CH_3$
                      (Z + E)   *(Z + E)*

26.28

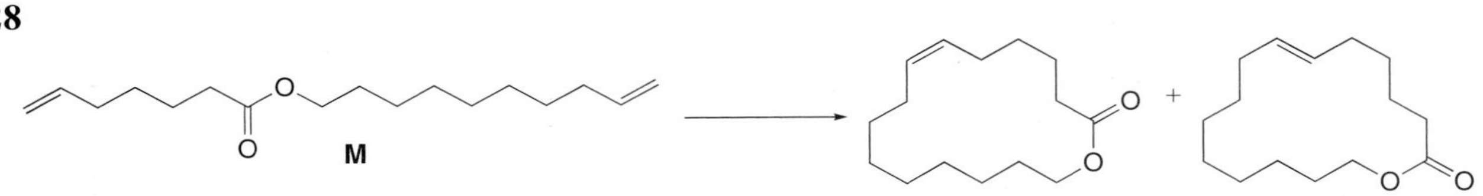

26.29 All double bonds can have either the *E* or *Z* configuration.

a. 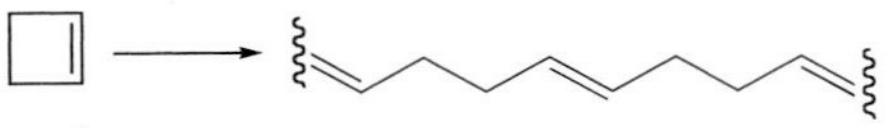   c.

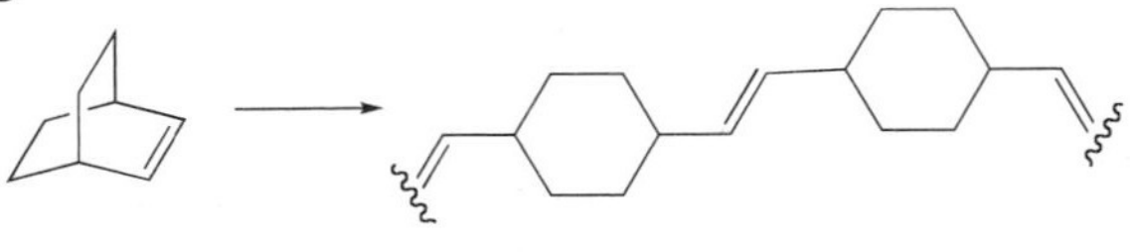

b.

26.30

a. 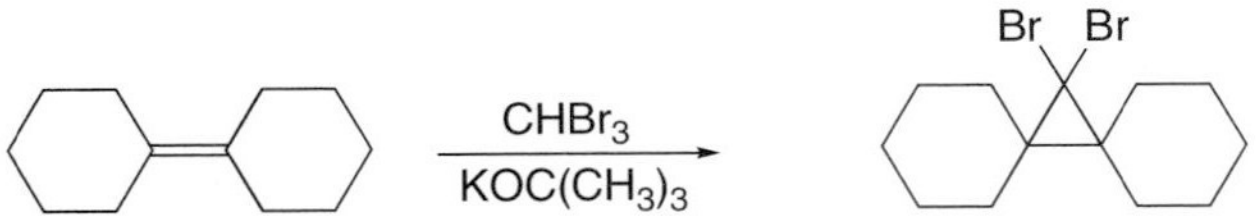 $\xrightarrow{\text{CHBr}_3 \atop \text{KOC(CH}_3)_3}$

b. 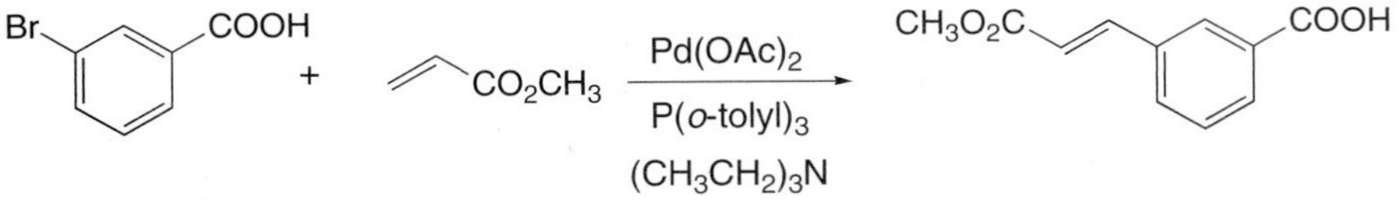 + $\xrightarrow[\text{(CH}_3\text{CH}_2)_3\text{N}]{\text{Pd(OAc)}_2 \atop \text{P(o-tolyl)}_3}$

c. 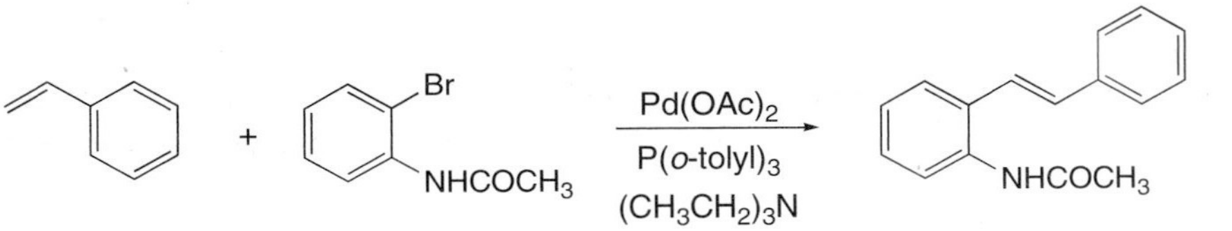 + $\xrightarrow[\text{(CH}_3\text{CH}_2)_3\text{N}]{\text{Pd(OAc)}_2 \atop \text{P(o-tolyl)}_3}$

d. 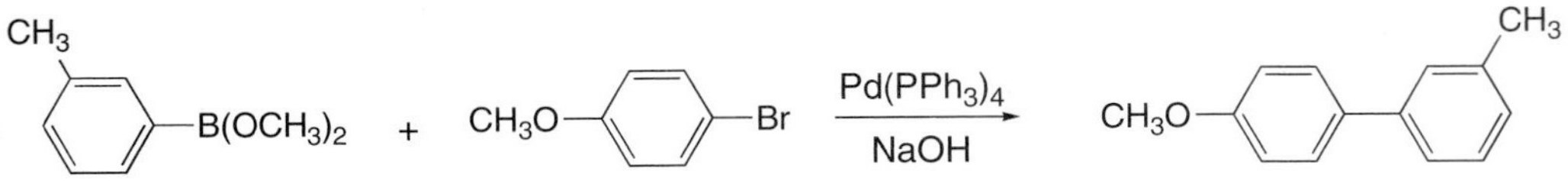 + CH_3O— —Br $\xrightarrow[\text{NaOH}]{\text{Pd(PPh}_3)_4}$

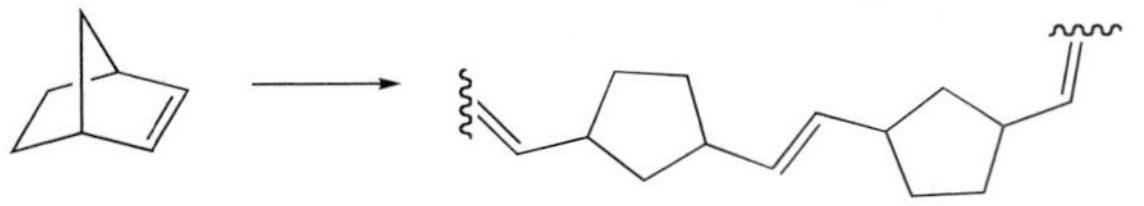

e. 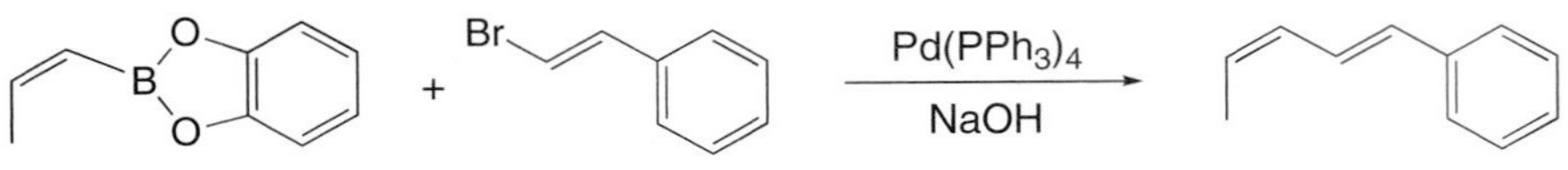 + $\xrightarrow[\text{NaOH}]{\text{Pd(PPh}_3)_4}$

f.

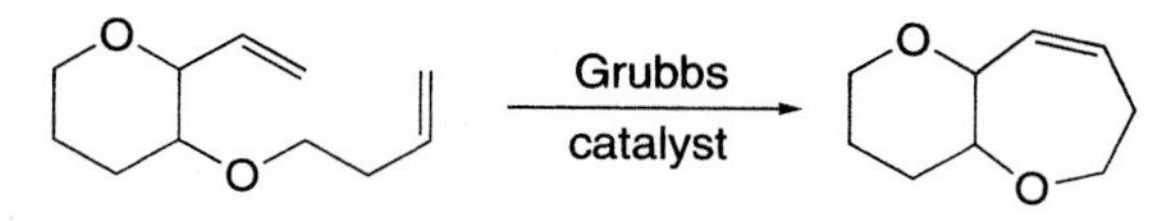

g.

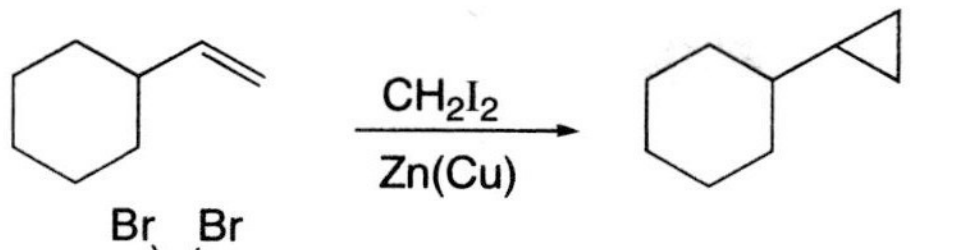

h.

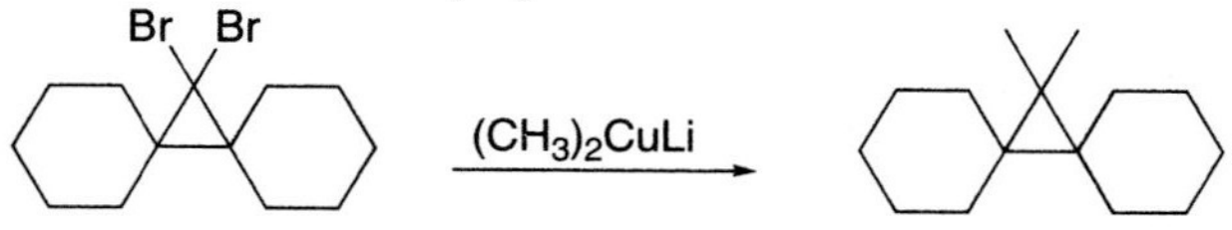

Product in (a)

i.

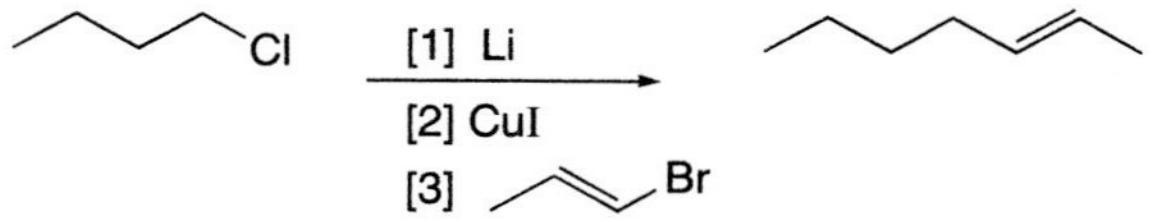

[1] Li
[2] CuI
[3]

j. $CH_3CH_2-C \equiv C-H$

[1] $H-B$ (catecholborane)

[2] (4-isopropylphenyl bromide)

Pd(PPh$_3$)$_4$
NaOH

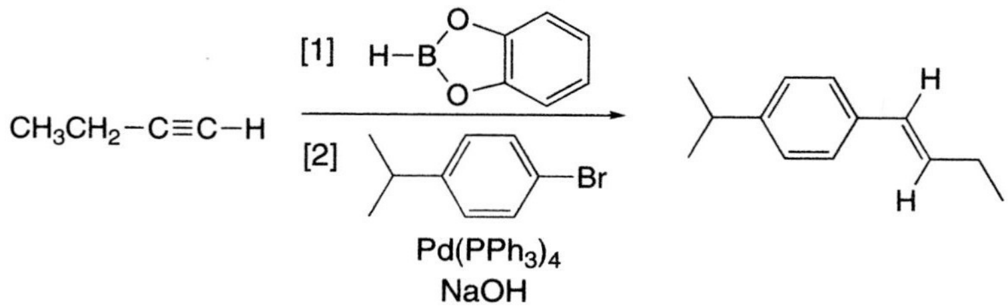

26.31

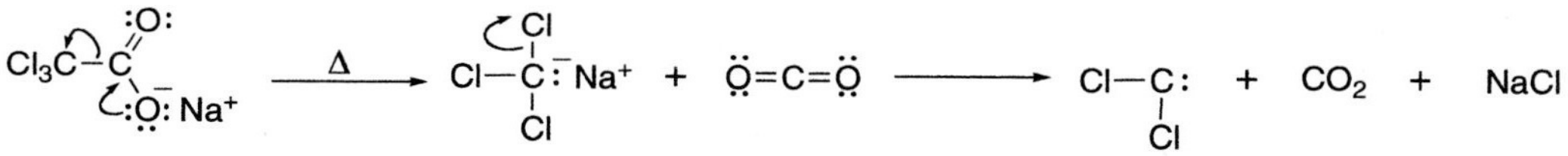

26.32 This reaction follows the Simmons–Smith reaction mechanism illustrated in Mechanism 26.5.

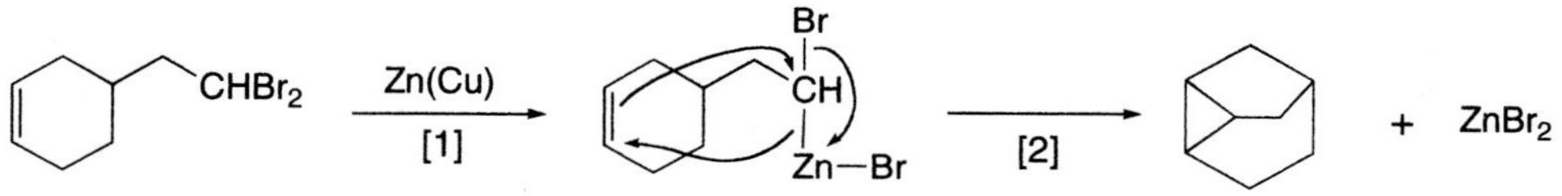

26.33

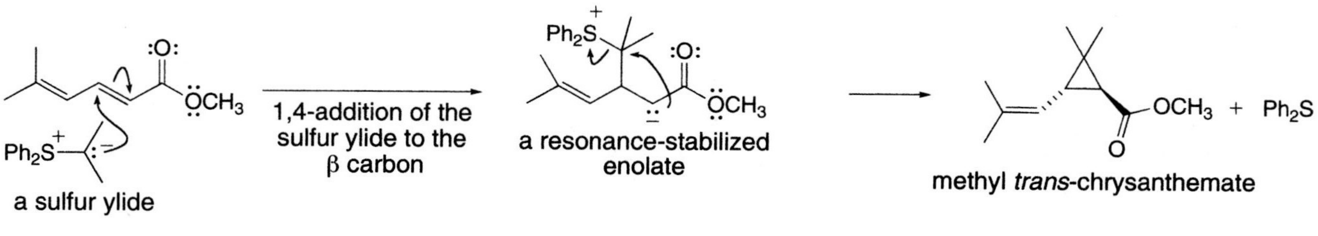

a sulfur ylide

1,4-addition of the
sulfur ylide to the
β carbon

a resonance-stabilized
enolate

methyl *trans*-chrysanthemate

26.34

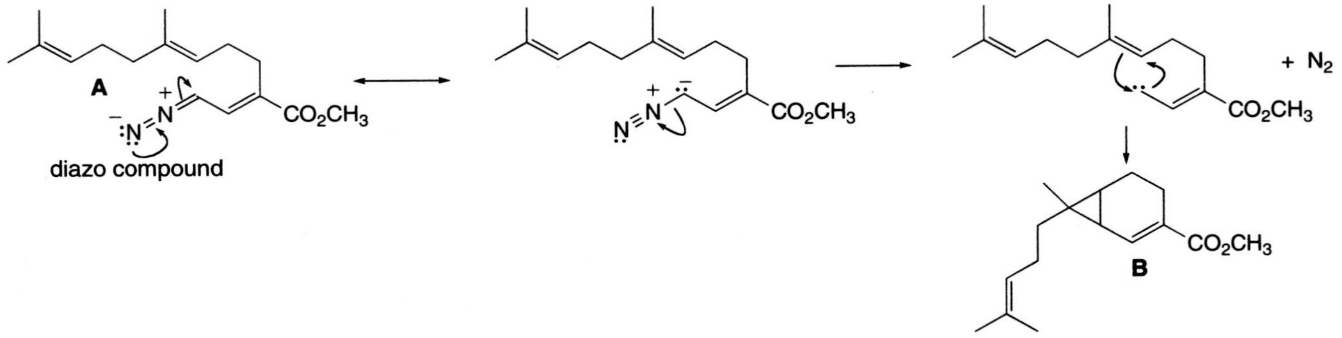

26.35

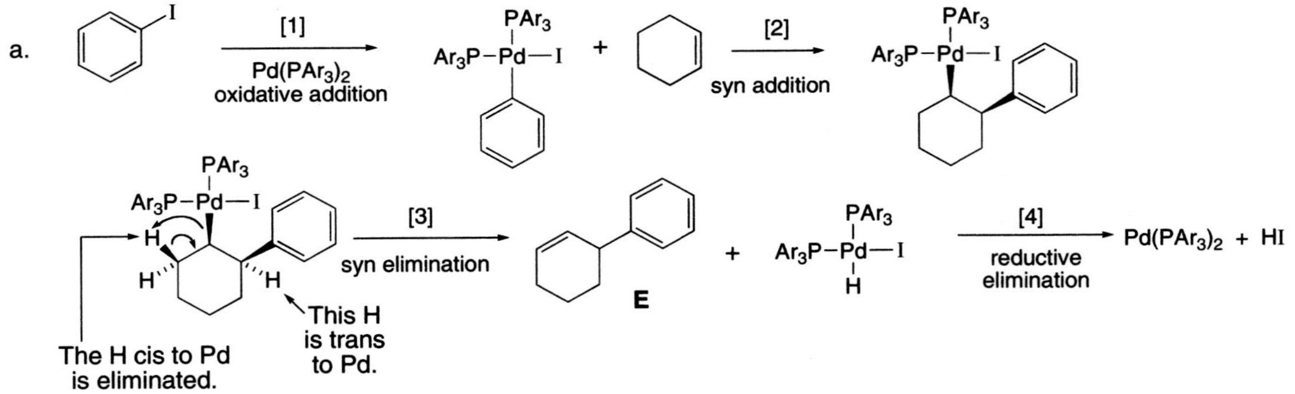

a.

b. This suggests that the stereochemistry in Step [3] must occur with syn elimination of H and Pd to form **E**. Product **F** cannot form because the only H on the C bonded to the benzene ring is trans to the Pd species, and therefore it cannot be removed if elimination occurs in a syn fashion.

26.36

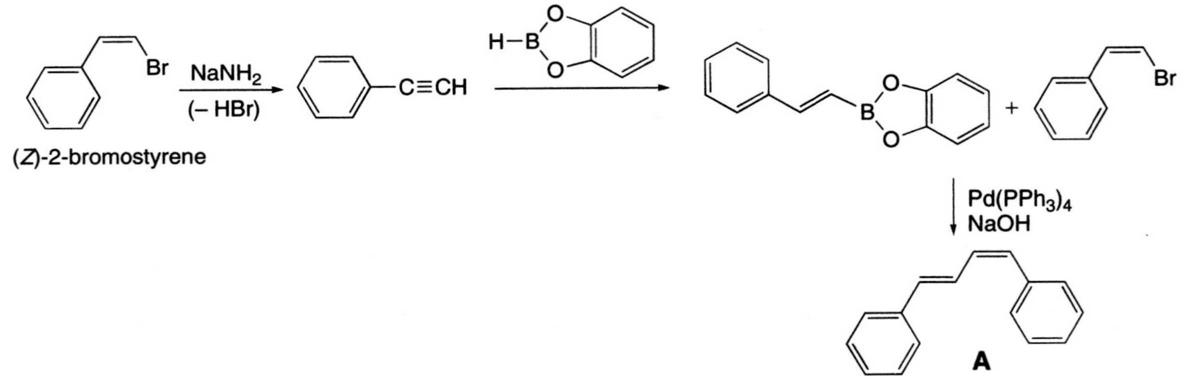

(Z)-2-bromostyrene

26.37

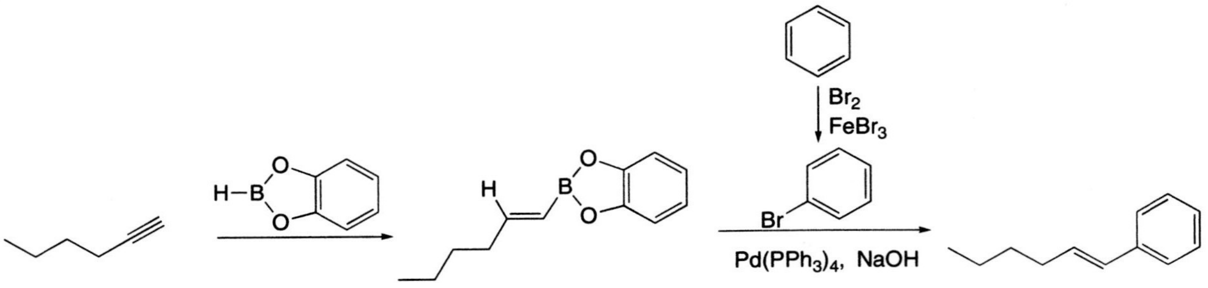

26.38

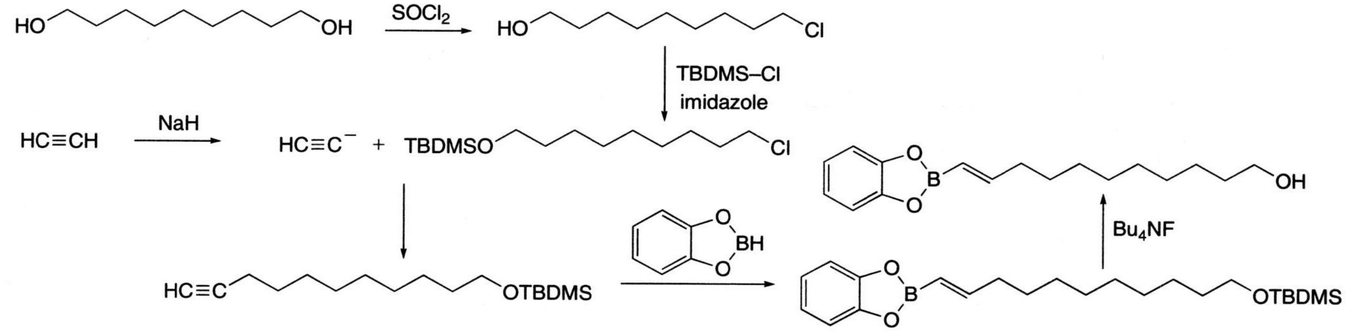

26.39

a.

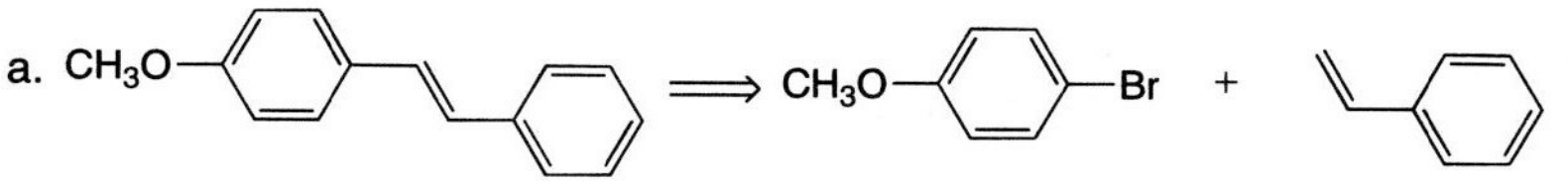

Synthesize these two components, and then use a Heck reaction to synthesize the product.

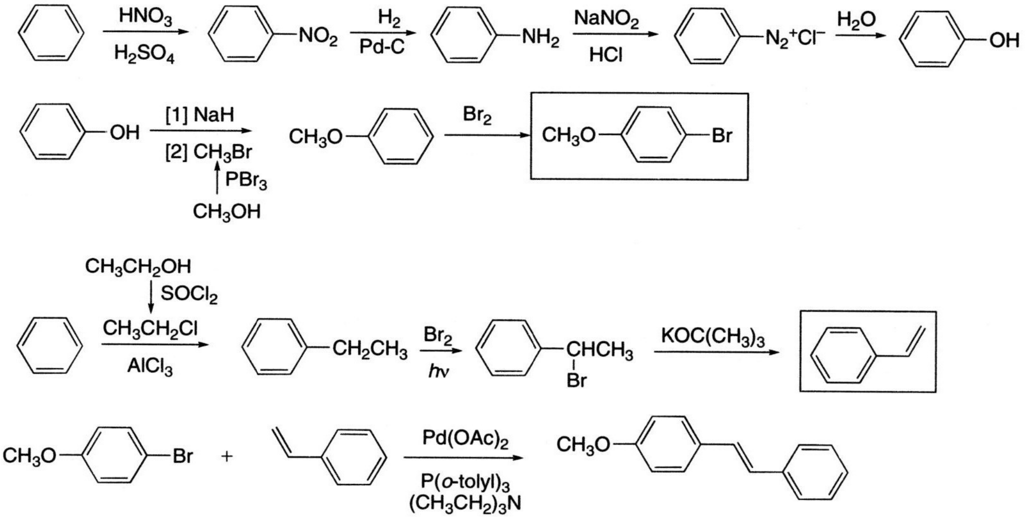

b.

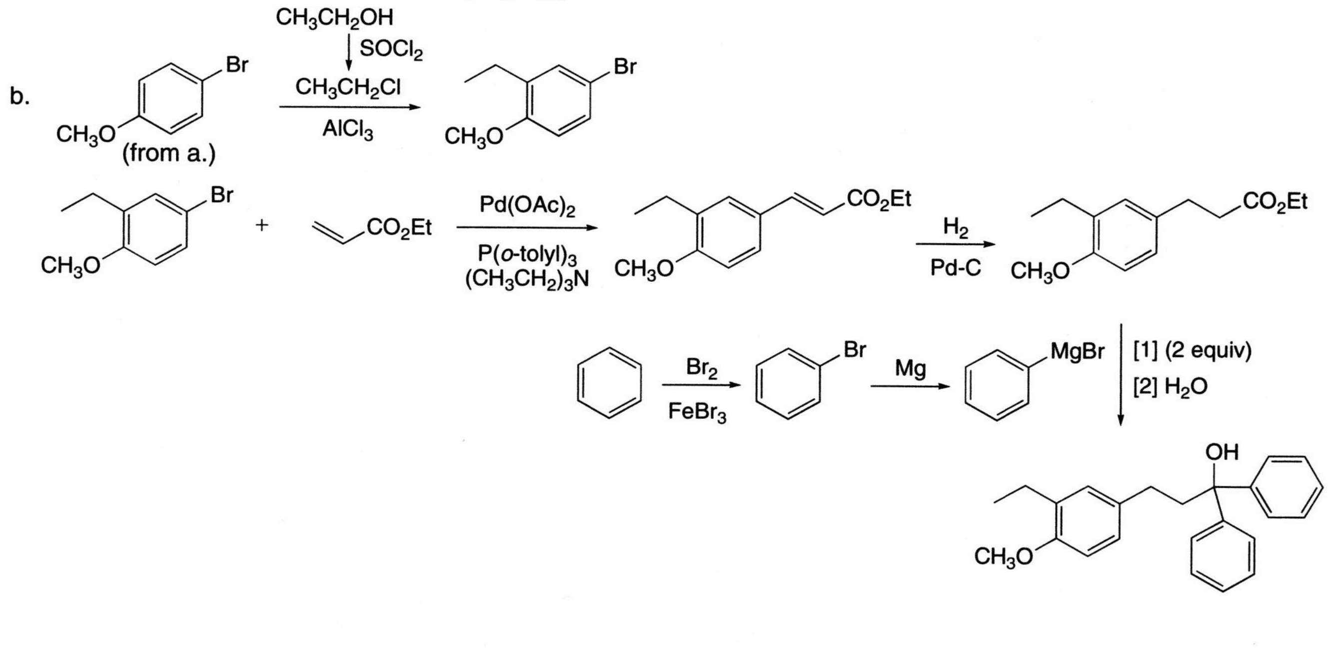

26.40

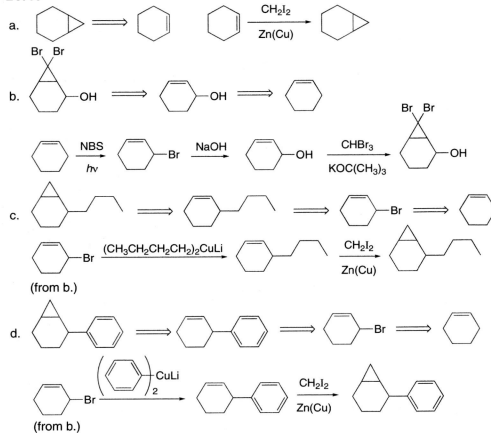

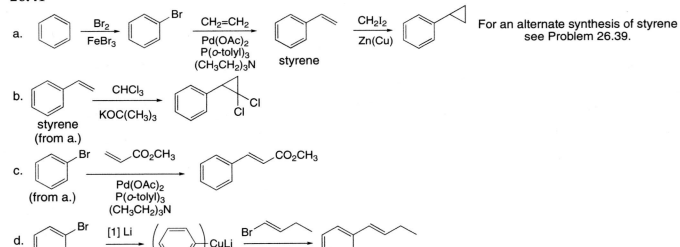

26.41

a.

For an alternate synthesis of styrene
see Problem 26.39.

26.42

a.

b.

c.

d.

26.43

a. CH₃—⟨biphenyl⟩—OCH₃ ⟹ CH₃—⟨⟩—Br + Br—⟨⟩—OCH₃

Either compound can be used to synthesize the organoborane, so two routes are possible.

Possibility [1]:

⟨benzene⟩ →[CH₃Cl / AlCl₃] CH₃—⟨⟩ →[Br₂ / FeBr₃] CH₃—⟨⟩—Br →[[1] Li / [2] B(OCH₃)₃] CH₃—⟨⟩—B(OCH₃)₂

⟨benzene⟩ →[Br₂ / FeBr₃] Br—⟨⟩ →[HNO₃ / H₂SO₄] Br—⟨⟩—NO₂ →[H₂ / Pd-C] Br—⟨⟩—NH₂ →[[1] NaNO₂, HCl / [2] H₂O] Br—⟨⟩—OH →[[1] NaH / [2] CH₃I] Br—⟨⟩—OCH₃

CH₃—⟨⟩—B(OCH₃)₂ + Br—⟨⟩—OCH₃ →[Pd(PPh₃)₄ / NaOH] CH₃—⟨biphenyl⟩—OCH₃

Possibility [2]:

Br—⟨⟩—OCH₃ →[[1] Li / [2] B(OCH₃)₃] (CH₃O)₂B—⟨⟩—OCH₃ + CH₃—⟨⟩—Br →[Pd(PPh₃)₄ / NaOH] CH₃—⟨biphenyl⟩—OCH₃

(from Possibility [1])   (from Possibility [1])

b. HO—⟨biphenyl⟩ ⟹ HO—⟨⟩—Br + Br—⟨⟩

The acidic OH makes it impossible to prepare an organolithium reagent from this aryl halide, so this compound must be used as the aryl halide that couples with the organoborane from bromobenzene.

⟨benzene⟩ →[HNO₃ / H₂SO₄] O₂N—⟨⟩ →[Br₂ / FeBr₃] O₂N—⟨⟩—Br →[H₂ / Pd-C] H₂N—⟨⟩—Br →[[1] NaNO₂, HCl / [2] H₂O] HO—⟨⟩—Br

⟨benzene⟩ →[Br₂ / FeBr₃] Br—⟨⟩ →[[1] Li / [2] B(OCH₃)₃] (CH₃O)₂B—⟨⟩

(CH₃O)₂B—⟨⟩ + HO—⟨⟩—Br →[Pd(PPh₃)₄ / NaOH] HO—⟨biphenyl⟩

c.

This can't be converted to an organoborane reagent via an organolithium reagent.

(as in 26.43b)

26.44

a.

Synthesis of starting material:

b.

Synthesis of starting material:

26.45

a.

Synthesis of starting material:

b.

Synthesis of starting material:

26.46

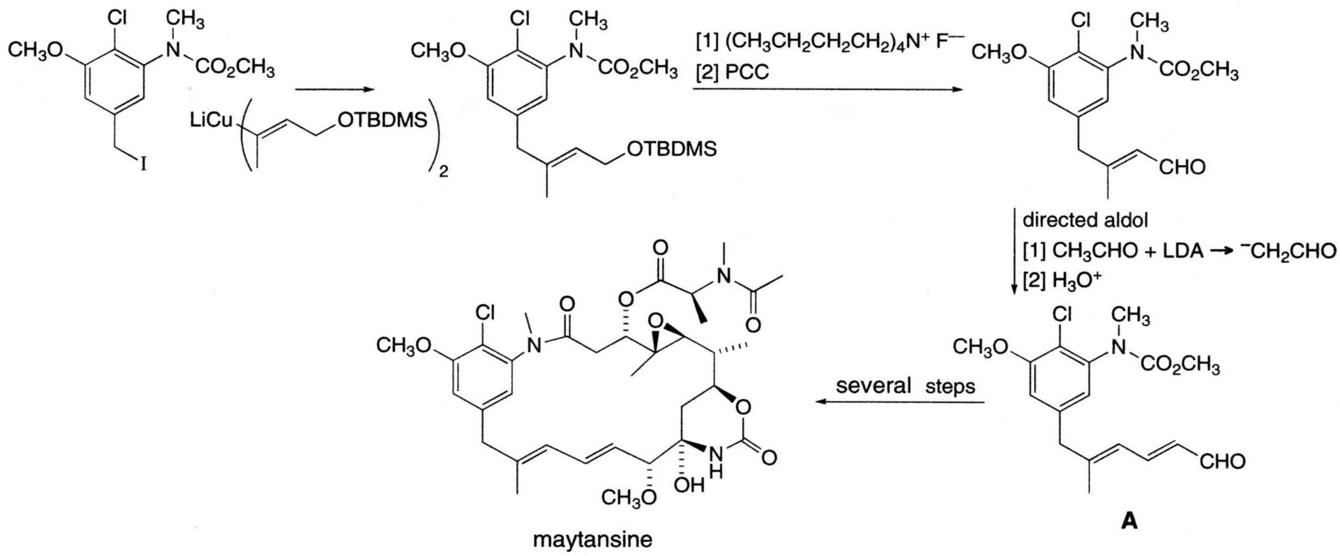

Cl
CH3
CH3O
N
CO2CH3
I
LiCu
OTBDMS
2
[1] (CH3CH2CH2CH2)4N+ F–
[2] PCC
CHO
directed aldol
[1] CH3CHO + LDA → –CH2CHO
[2] H3O+
several steps
OH
maytansine
A

26.47

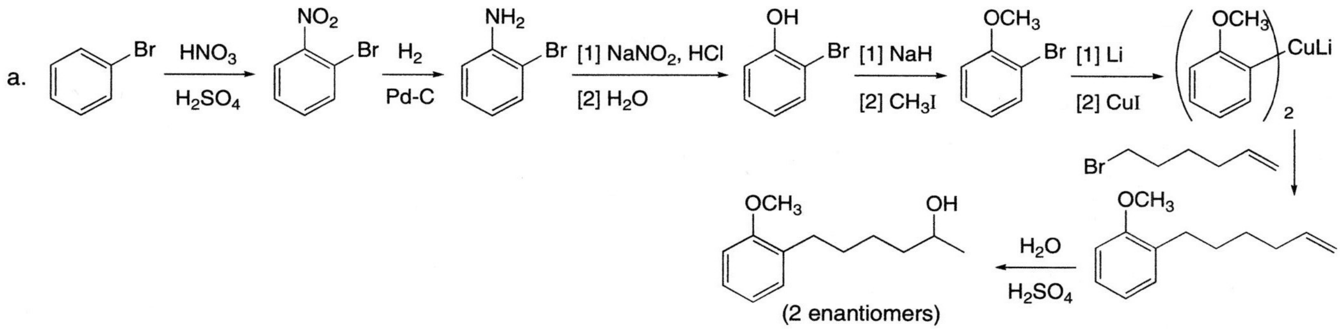

a.
Br
HNO3
H2SO4
NO2
H2
Pd-C
NH2
[1] NaNO2, HCl
[2] H2O
OH
[1] NaH
[2] CH3I
OCH3
[1] Li
[2] CuI
CuLi
2
H2O
H2SO4
(2 enantiomers)

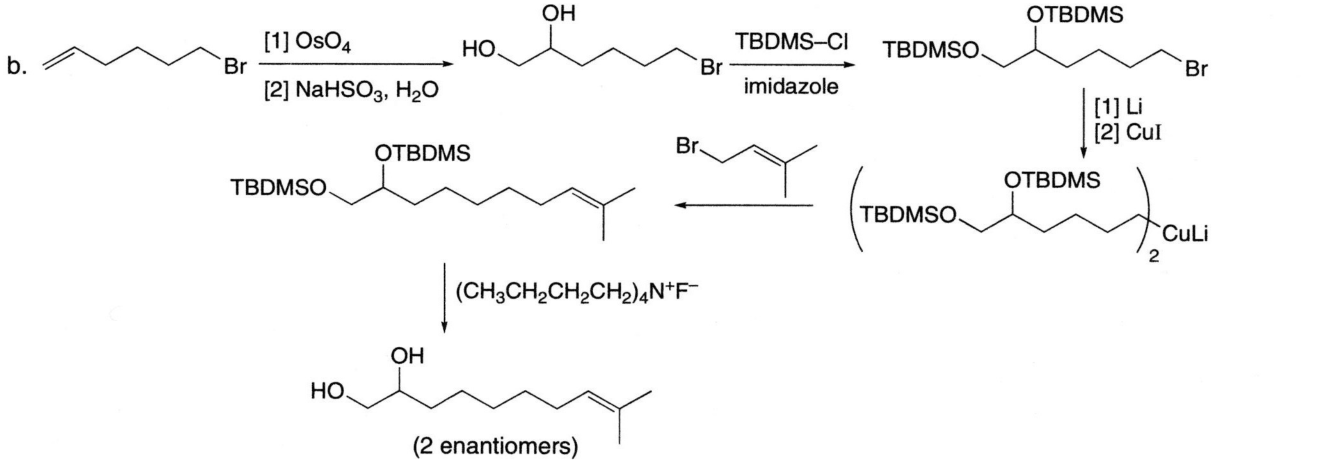

b.
Br
[1] OsO4
[2] NaHSO3, H2O
HO
OH
TBDMS–Cl
imidazole
TBDMSO
OTBDMS
[1] Li
[2] CuI
CuLi
2
(CH3CH2CH2CH2)4N+F–
(2 enantiomers)

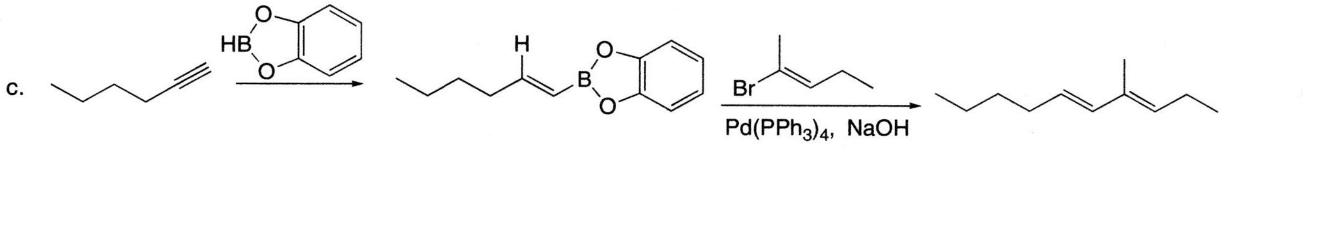

c.
HB
O
H
B
Br
Pd(PPh3)4, NaOH

d. 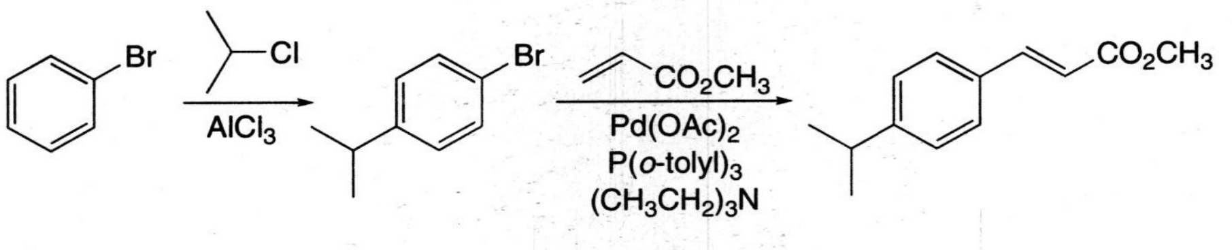
Br
Cl
AlCl3
Br
CO2CH3
Pd(OAc)2
P(o-tolyl)3
(CH3CH2)3N
CO2CH3

e. 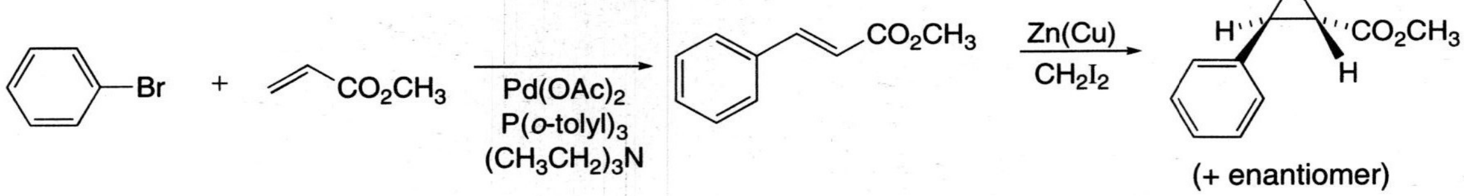
Br
+
CO2CH3
Pd(OAc)2
P(o-tolyl)3
(CH3CH2)3N
CO2CH3
Zn(Cu)
CH2I2
H
CO2CH3
H
(+ enantiomer)

f. 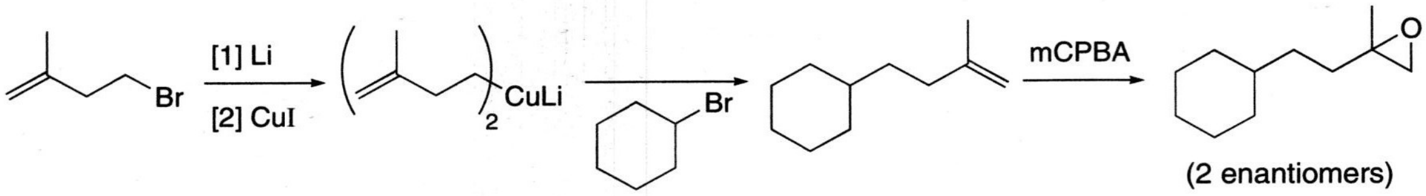
Br
[1] Li
[2] CuI
CuLi
2
Br
mCPBA
O
(2 enantiomers)

26.48

a. 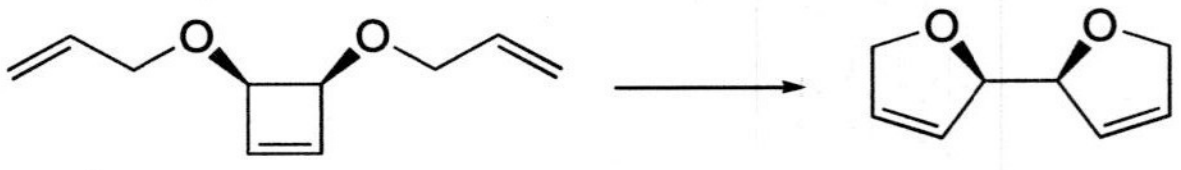
O
O
O
O

b. 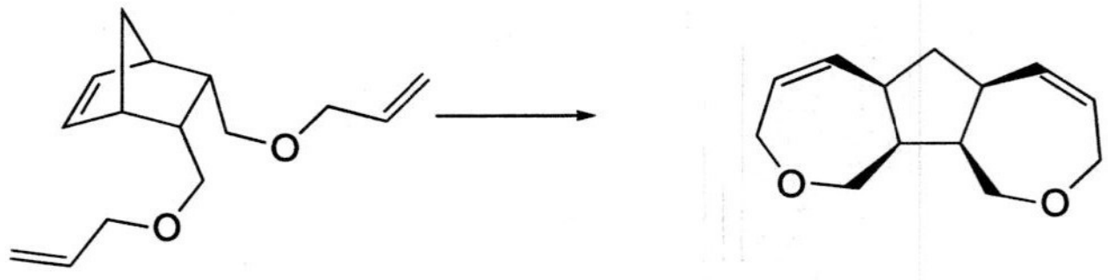
O
O
O
O

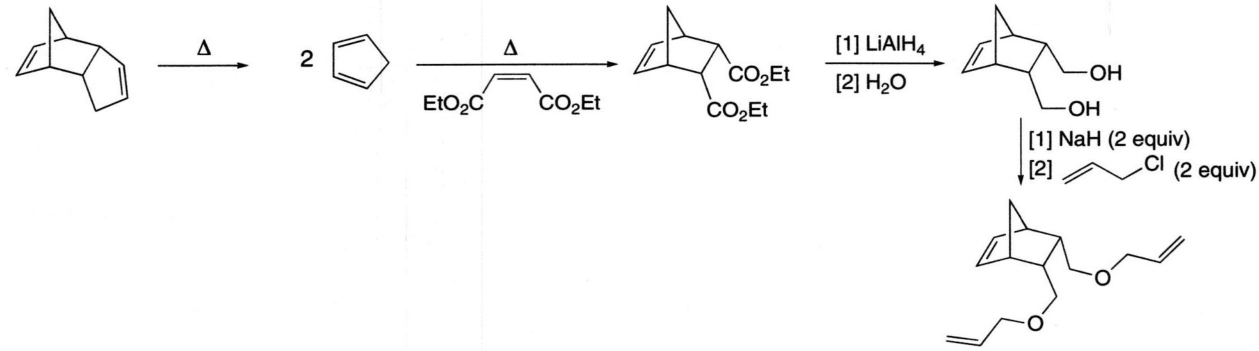
Δ
2
Δ
EtO2C
CO2Et
CO2Et
CO2Et
[1] LiAlH4
[2] H2O
OH
OH
[1] NaH (2 equiv)
[2]
Cl
(2 equiv)
O
O

26.49

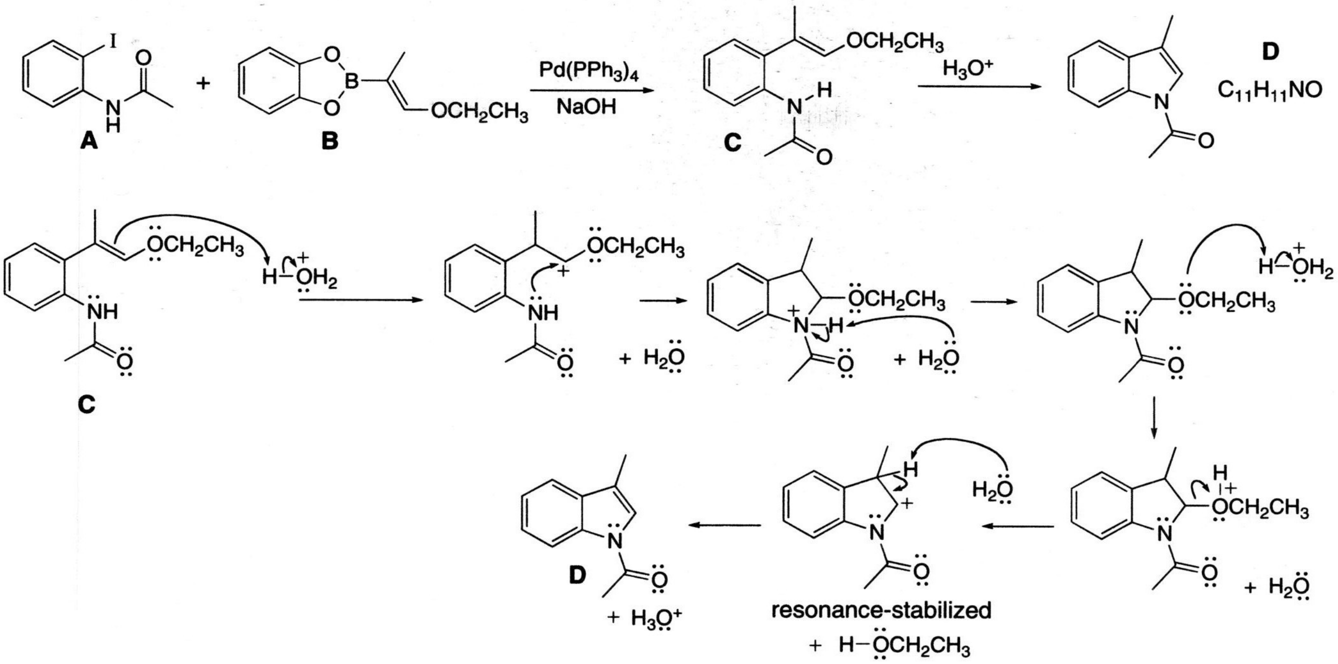

Pd(PPh3)4
NaOH
H3O+
A
B
C
D
C11H11NO
resonance-stabilized
+ H–OCH2CH3
+ H3O+

26.50

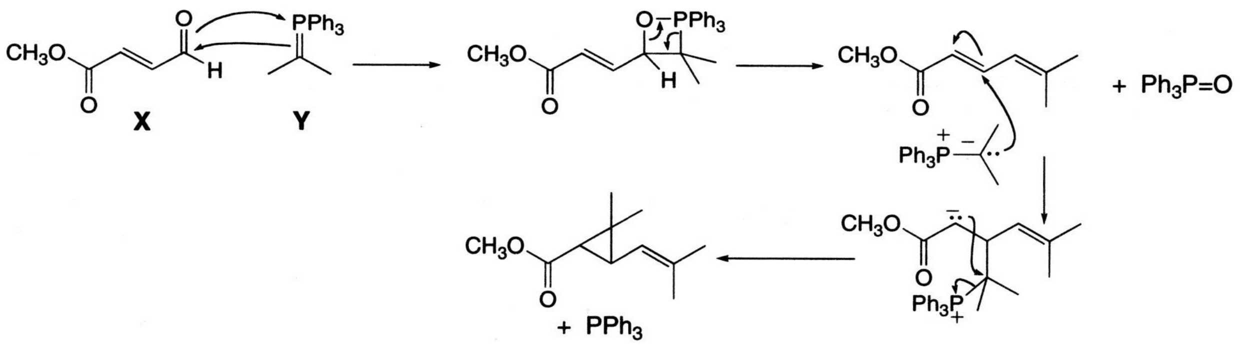

X
Y
+ Ph3P=O
+ PPh3

Chapter 27: Carbohydrates

◆ Important terms

•	**Aldose**	A monosaccharide containing an aldehyde (27.2)
•	**Ketose**	A monosaccharide containing a ketone (27.2)
•	**D-Sugar**	A monosaccharide with the O bonded to the stereogenic center farthest from the carbonyl group drawn on the right in the Fischer projection (27.2C)
•	**Epimers**	Two diastereomers that differ in configuration around one stereogenic center only (27.3)
•	**Anomers**	Monosaccharides that differ in configuration at only the hemiacetal OH group (27.6)
•	**Glycoside**	An acetal derived from a monosaccharide hemiacetal (27.7)

◆ Acyclic, Haworth, and 3-D representations for D-glucose (27.6)

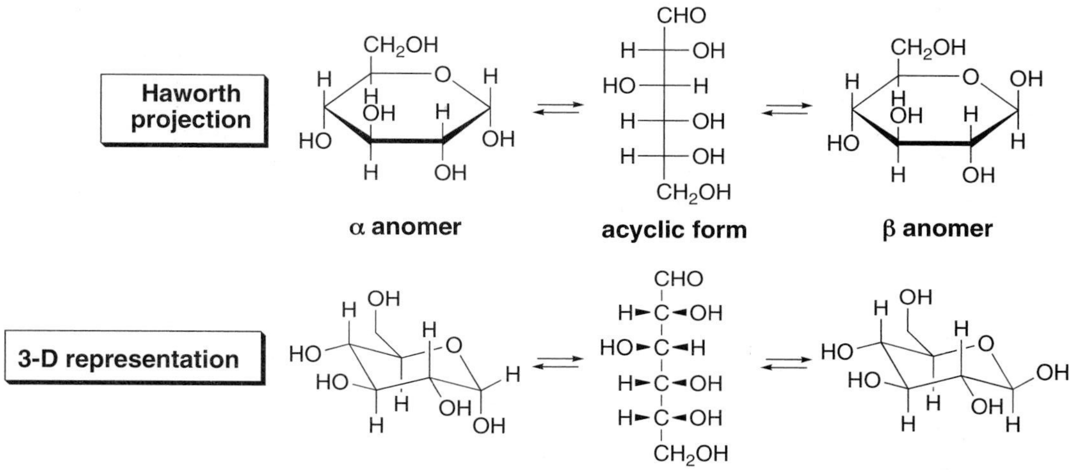

Haworth projection

α anomer   acyclic form   β anomer

3-D representation

◆ Reactions of monosaccharides involving the hemiacetal

[1] Glycoside formation (27.7A)

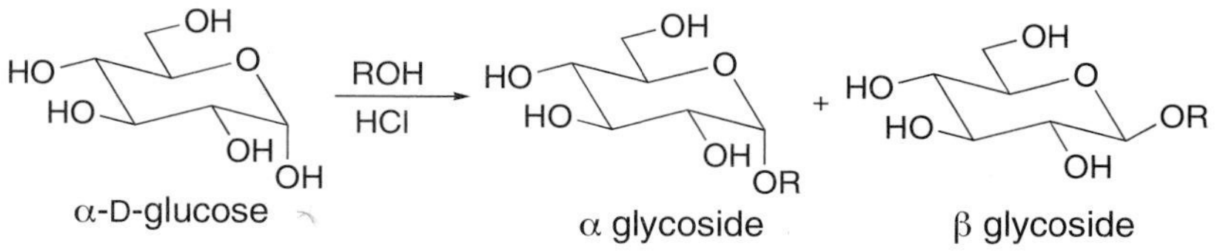

α-D-glucose   α glycoside   β glycoside

• Only the hemiacetal OH reacts.
• A mixture of α and β glycosides forms.

[2] Glycoside hydrolysis (27.7B)

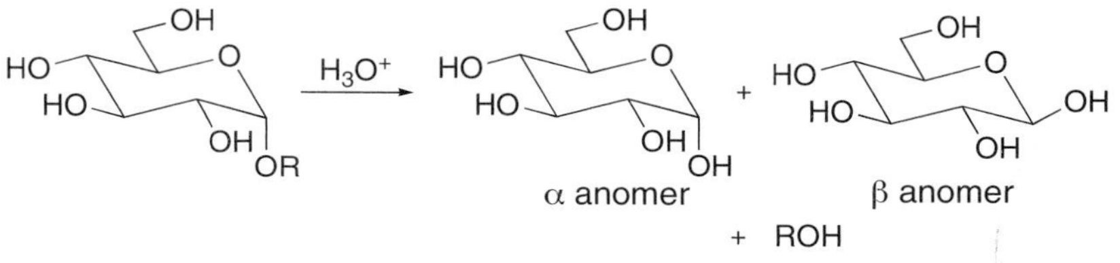

α anomer   β anomer

+ ROH

• A mixture of α and β anomers forms.

♦ **Reactions of monosaccharides at the OH groups**

[1] Ether formation (27.8)

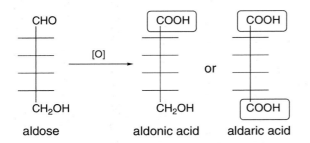

- All OH groups react.
- The stereochemistry at all stereogenic centers is retained.

[2] Ester formation (27.8)

- All OH groups react.
- The stereochemistry at all stereogenic centers is retained.

♦ **Reactions of monosaccharides at the carbonyl group**

[1] Oxidation of aldoses (27.9B)

- Aldonic acids are formed using:
  - Ag_2O, NH_4OH
  - Cu^{2+}
  - Br_2, H_2O
- Aldaric acids are formed with HNO_3, H_2O.

[2] Reduction of aldoses to alditols (27.9A)

[3] Wohl degradation (27.10A)

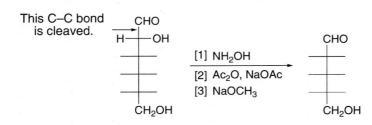

- The C1–C2 bond is cleaved to shorten an aldose chain by one carbon.
- The stereochemistry at all other stereogenic centers is retained.
- Two epimers at C2 form the same product.

[4] Kiliani–Fischer synthesis (27.10B)

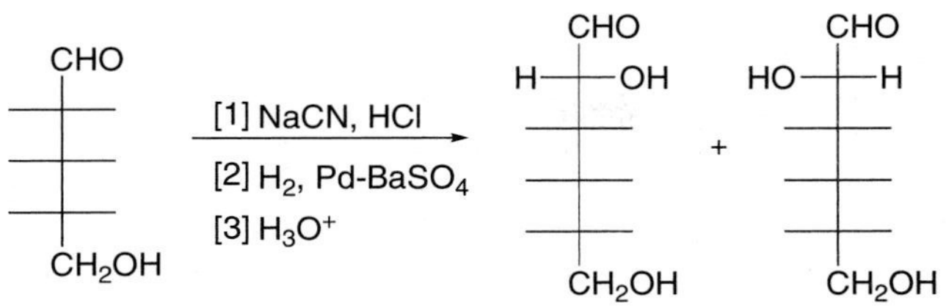

[1] NaCN, HCl
[2] H_2, Pd-BaSO$_4$
[3] H_3O^+

- One carbon is added to the aldehyde end of an aldose.
- Two epimers at C2 are formed.

◆ Other reactions

[1] Hydrolysis of disaccharides (27.12)

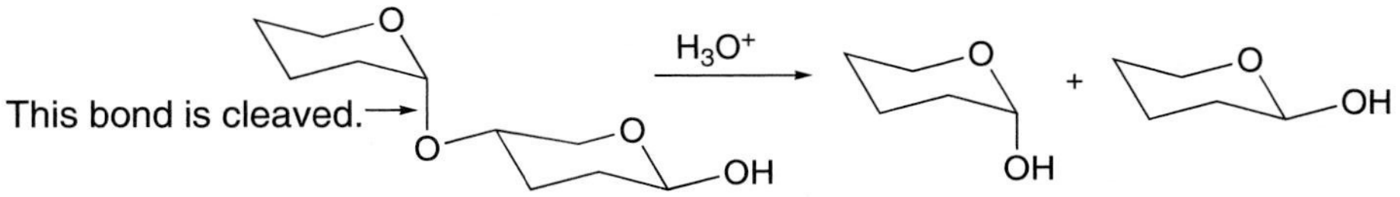

This bond is cleaved.

H_3O^+

A mixture of anomers is formed.

[2] Formation of *N*-glycosides (27.14B)

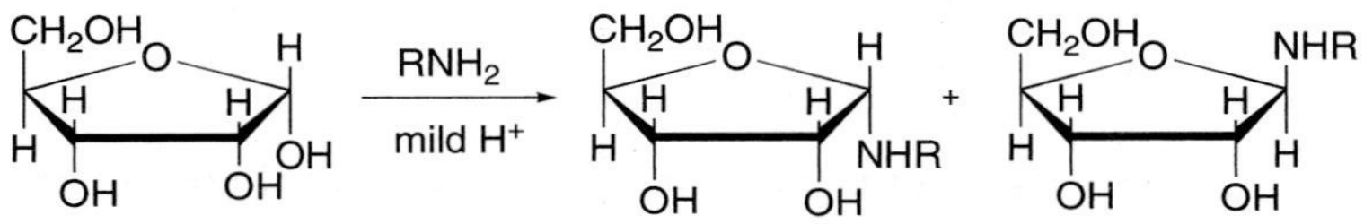

RNH$_2$
mild H$^+$

- Two anomers are formed.

Chapter 27: Answers to Problems

27.1 A *ketose* is a monosaccharide containing a ketone. An *aldose* is a monosaccharide containing an aldehyde. A monosaccharide is called: a *triose* if it has three C's; a *tetrose* if it has four C's; a *pentose* if it has five C's; a *hexose* if it has six C's, and so forth.

a. a ketotetrose

$$CH_2OH$$
$$C=O$$
$$H-C-OH$$
$$CH_2OH$$

b. an aldopentose

$$CHO$$
$$H-C-OH$$
$$H-C-OH$$
$$H-C-OH$$
$$CH_2OH$$

c. an aldotetrose

$$CHO$$
$$H-C-OH$$
$$H-C-OH$$
$$CH_2OH$$

27.2 Rotate and re-draw each molecule to place the horizontal bonds in front of the plane and the vertical bonds behind the plane. Then use a cross to represent the stereogenic center in a Fischer projection formula.

a.
$$COOH$$
$$CH_3-C-OH$$ = $$CH_3—OH$$
$$CH_2CH_2OH$$ $$CH_2CH_2OH$$

c.
$$OHC \quad CH_2CH_3$$
$$C$$
$$HOCH_2 \quad H$$
re-draw
$$CHO$$
$$H-C-CH_2OH$$ = $$H—CH_2OH$$
$$CH_2CH_3$$ $$CH_2CH_3$$

b.
$$CHO$$
$$HO^{\prime\prime}C$$
$$CH_3 \quad H$$
re-draw
$$CHO$$
$$HO-C-CH_3$$ = $$HO—CH_3$$
   $$H$$ $$H$$

d.
$$OH$$
$$H\cdots C\cdots CHO$$
$$CH_3$$
re-draw
$$CHO$$
$$HO-C-CH_3$$ = $$HO—CH_3$$
   $$H$$ $$H$$

27.3 For each molecule:
[1] Convert the Fischer projection formula to a representation with wedges and dashes.
[2] Assign priorities (Section 5.6).
[3] Determine *R* or *S* in the usual manner. Reverse the answer if priority group [4] is oriented forward (on a wedge).

a.
$$CH_2NH_2$$
$$Cl—CH_2Br$$
$$H$$
[1] →
$$CH_2NH_2$$
$$Cl-C-CH_2Br$$
$$H$$
[2] →
1 $$\overset{3}{Cl-C-CH_2Br}$$ 2
$$\underset{4}{H}$$
[3] →
1 $$\overset{3}{Cl-C-CH_2Br}$$ 2
$$H$$
S configuration

b.
$$CHO$$
$$Cl—H$$
$$CH_2NH_2$$
[1] →
$$CHO$$
$$Cl-C-H$$
$$CH_2NH_2$$
[2] →
1 $$\overset{2}{Cl-C-H}$$ 4
$$\underset{3}{CH_2NH_2}$$
[3] →
1 $$\overset{2}{Cl-C-H}$$ — H forward
$$CH_2NH_2$$
S configuration

c.
$$CHO$$
$$Cl—H$$
$$CH_2OH$$
[1] →
$$CHO$$
$$Cl-C-H$$
$$CH_2OH$$
[2] →
1 $$\overset{2}{Cl-C-H}$$ 4
$$\underset{3}{CH_2OH}$$
[3] →
1 $$\overset{2}{Cl-C-H}$$ — H forward
$$CH_2OH$$
S configuration

d.
$$COOH$$
$$Cl—CH_2Br$$
$$H$$
[1] →
$$COOH$$
$$Cl-C-CH_2Br$$
$$H$$
[2] →
1 $$\overset{3}{Cl-C-CH_2Br}$$ 2
$$\underset{4}{H}$$
[3] →
1 $$\overset{3}{Cl-C-CH_2Br}$$ 2
$$H$$
S configuration

27.4

```
        CHO    R
         |    /
   H ► C ◄ OH  S
         |    /
  HO ► C ◄ H   R
         |    /
   H ► C ◄ OH  R
         |    /
   H ► C ◄ OH
         |
       CH₂OH
```

D-glucose

27.5

a. aldotetrose: 2 stereogenic centers   b. a ketohexose: 3 stereogenic centers

```
      CHO
       |
  H — C*— OH
       |
  H — C*— OH
       |
     CH₂OH
```

```
      CH₂OH
       |
      C=O
       |
  H — C*— OH
       |
  H — C*— OH
       |
  H — C*— OH
       |
     CH₂OH
```

[• = stereogenic center]

27.6 A D sugar has the OH group on the stereogenic center farthest from the carbonyl on the right. An L sugar has the OH group on the stereogenic center farthest from the carbonyl on the left.

a.

```
       CHO
    H ——— OH
    H ——— OH
  [HO]——— H
      CH₂OH
        A
```

```
       CHO
   HO ——— H
    H ——— OH
  [HO]——— H
      CH₂OH
        B
```

```
       CHO
   HO ——— H
   HO ——— H
    H ——[OH]
      CH₂OH
        C
```

OH group on the left: **L sugar**   OH group on the left: **L sugar**   OH group on the right: **D sugar**

b. **A** and **B** are diastereomers.
A and **C** are enantiomers.
B and **C** are diastereomers.

27.7 The D- notation signifies the position of the OH group on the stereogenic carbon farthest from the carbonyl group, and does not correlate with dextrorotatory or levorotatory. The latter terms describe a physical phenomenon, the direction of rotation of plane-polarized light.

27.8 There are 32 aldoheptoses; 16 are D sugars.

```
C2 R      CHO
C3 R  H ——— OH
      H ——— OH
      H ——— OH
      H ——— OH
      H ——— OH
        CH₂OH
```

```
       CHO
    H ——— OH
    H ——— OH
   HO ——— H
   HO ——— H
    H ——— OH
      CH₂OH
```

```
       CHO
    H ——— OH
    H ——— OH
   HO ——— H
    H ——— OH
    H ——— OH
      CH₂OH
```

```
       CHO
    H ——— OH
    H ——— OH
    H ——— OH
   HO ——— H
    H ——— OH
      CH₂OH
```

27.9   *Epimers* are two diastereomers that differ in the configuration around only one stereogenic center.

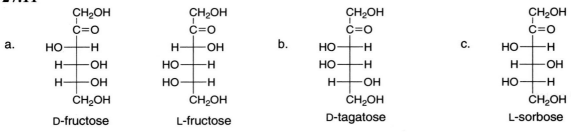

epimers

CHO	CHO		CHO
H——OH	HO——H	and	H——OH
H——OH	H——OH		HO——H
CH₂OH	CH₂OH		CH₂OH
D-erythrose	D-threose		L-threose

27.10   a. D-allose and L-allose: **enantiomers**
b. D-altrose and D-gulose: **diastereomers** but not epimers
c. D-galactose and D-talose: **epimers**
d. D-mannose and D-fructose: **constitutional isomers**
e. D-fructose and D-sorbose: **diastereomers** but not epimers
f. L-sorbose and L-tagatose: **epimers**

27.11

a.

CH₂OH	CH₂OH
C=O	C=O
HO——H	H——OH
H——OH	HO——H
H——OH	HO——H
CH₂OH	CH₂OH
D-fructose	L-fructose

enantiomers

b.

CH₂OH
C=O
HO——H
HO——H
H——OH
CH₂OH
D-tagatose

c.

CH₂OH
C=O
HO——H
H——OH
HO——H
CH₂OH
L-sorbose

27.12

CH₂OH	CH₂OH
S C=O	**S** C=O
HO——H	HO——H
H——OH	HO——H
H——OH	H——OH
CH₂OH	CH₂OH
D-fructose	**D-tagatose**

27.13 Step [1]: Place the O atom in the upper right corner of a hexagon, and add the CH_2OH group on the first carbon counterclockwise from the O atom.

Step [2]: Place the anomeric carbon on the first carbon clockwise from the O atom.

Step [3]: Add the substituents at the three remaining stereogenic centers, clockwise around the ring.

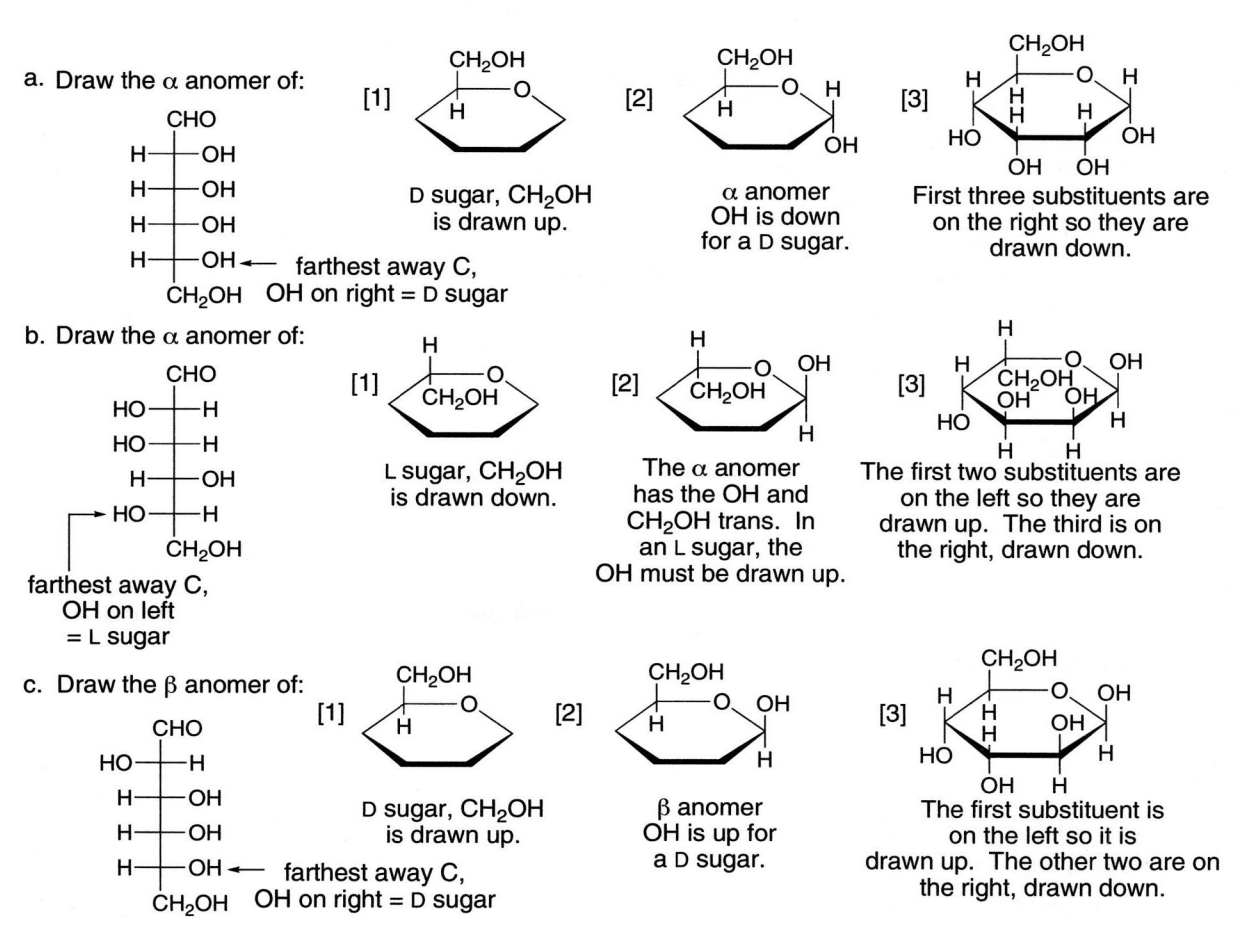

a. Draw the α anomer of:

CHO
H——OH
H——OH
H——OH
H——OH ← farthest away C,
CH_2OH OH on right = D sugar

[1] D sugar, CH_2OH is drawn up.

[2] α anomer OH is down for a D sugar.

[3] First three substituents are on the right so they are drawn down.

b. Draw the α anomer of:

CHO
HO——H
HO——H
H——OH
HO——H
CH_2OH

farthest away C,
OH on left
= L sugar

[1] L sugar, CH_2OH is drawn down.

[2] The α anomer has the OH and CH_2OH trans. In an L sugar, the OH must be drawn up.

[3] The first two substituents are on the left so they are drawn up. The third is on the right, drawn down.

c. Draw the β anomer of:

CHO
HO——H
H——OH
H——OH
H——OH ← farthest away C,
CH_2OH OH on right = D sugar

[1] D sugar, CH_2OH is drawn up.

[2] β anomer OH is up for a D sugar.

[3] The first substituent is on the left so it is drawn up. The other two are on the right, drawn down.

27.14 To convert each Haworth projection into its acyclic form:
  [1] Draw the C skeleton with the CHO on the top and the CH_2OH on the bottom.
  [2] Draw in the OH group farthest from the C=O.
     A CH_2OH group drawn up means a D sugar; a CH_2OH group drawn down means an L sugar.
  [3] Add the three other stereogenic centers, counterclockwise around the ring.
     "Up" groups go on the left, and "down" groups go on the right.

27.15 To convert a Haworth projection into a 3-D representation with a chair cyclohexane:
  [1] Draw the pyranose ring as a chair with the O as an "up" atom.
  [2] Add the substituents around the ring.

With so many axial groups, this is not the more stable conformation of this sugar.

27.16 Cyclization always forms a new stereogenic center at the anomeric carbon, so two different anomers are possible.

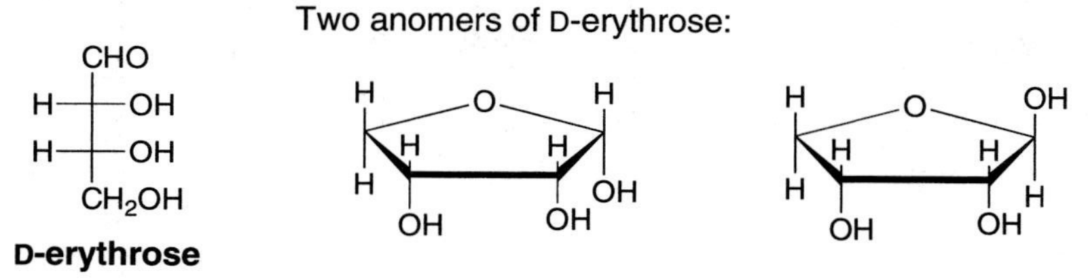
Two anomers of D-erythrose:
D-erythrose

27.17

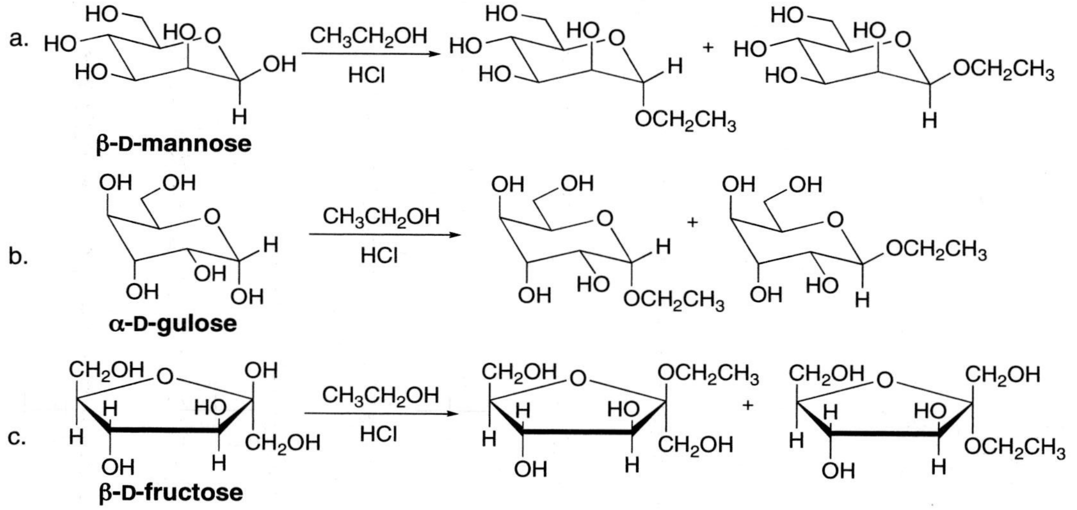
a.
β-D-mannose
CH_3CH_2OH
HCl
b.
α-D-gulose
c.
β-D-fructose

27.18

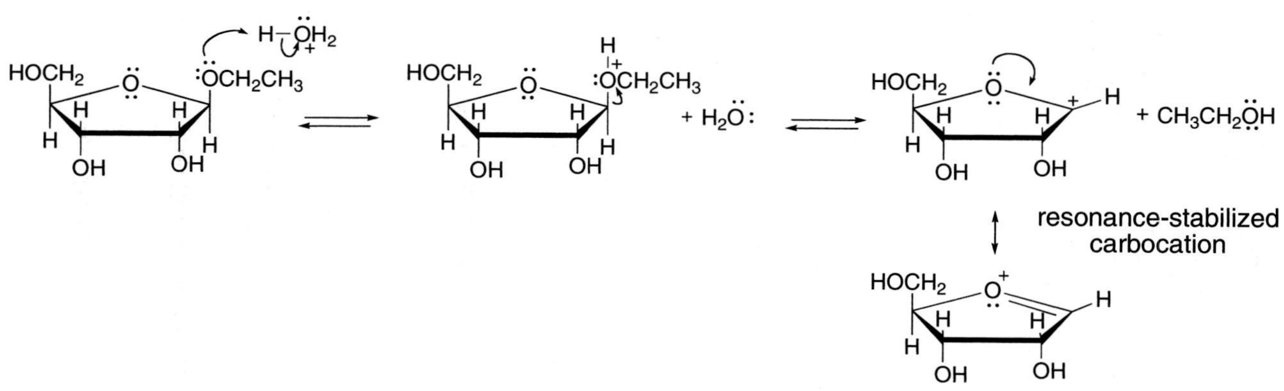
resonance-stabilized carbocation

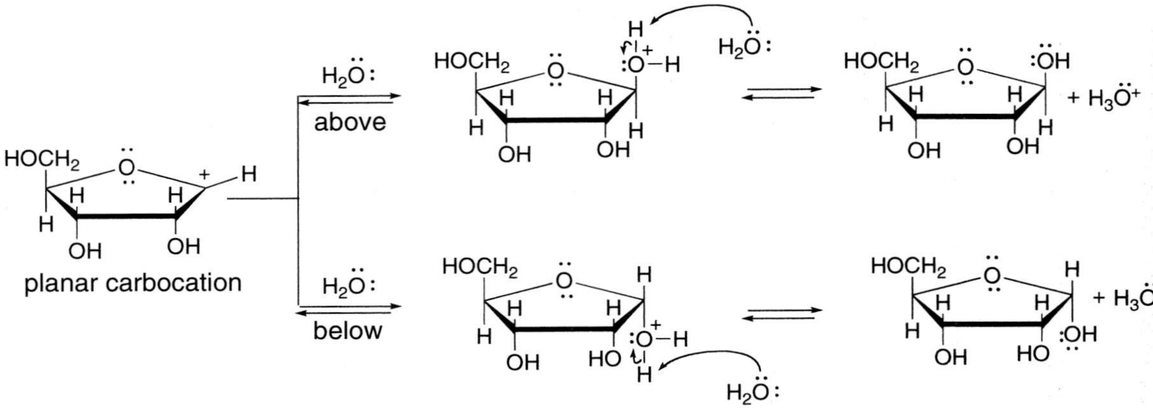

27.19

a. All circled O atoms are part of a glycoside.

b. Hydrolysis of rebaudioside A breaks each bond indicated with a dashed line and forms four molecules of glucose and the aglycon drawn.

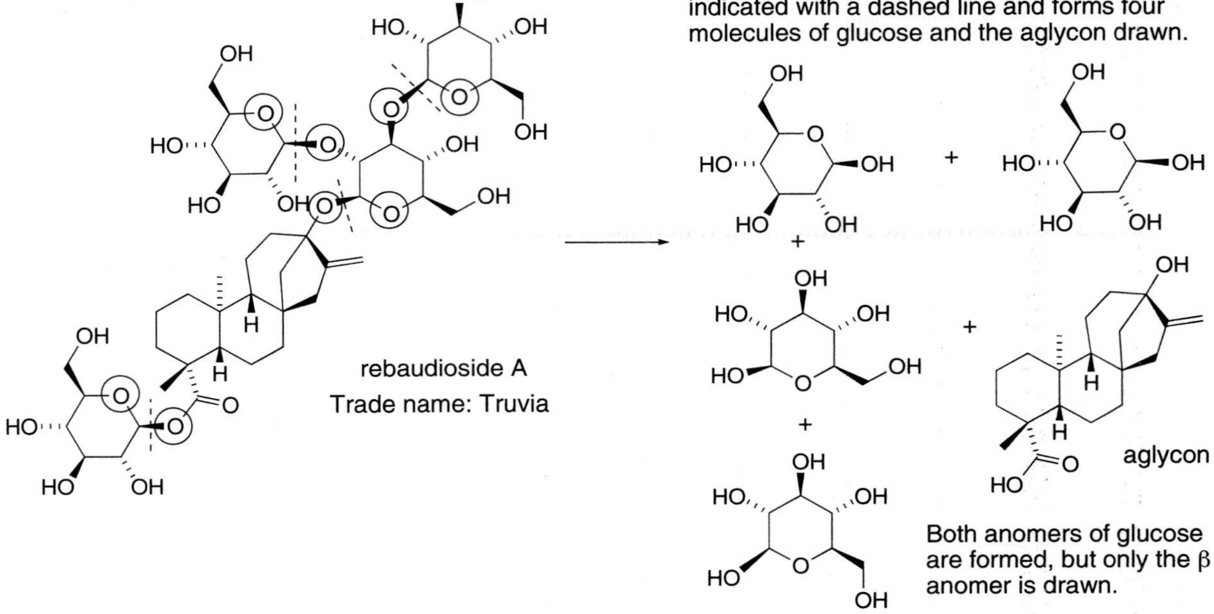

rebaudioside A
Trade name: Truvia

aglycon

Both anomers of glucose are formed, but only the β anomer is drawn.

27.20

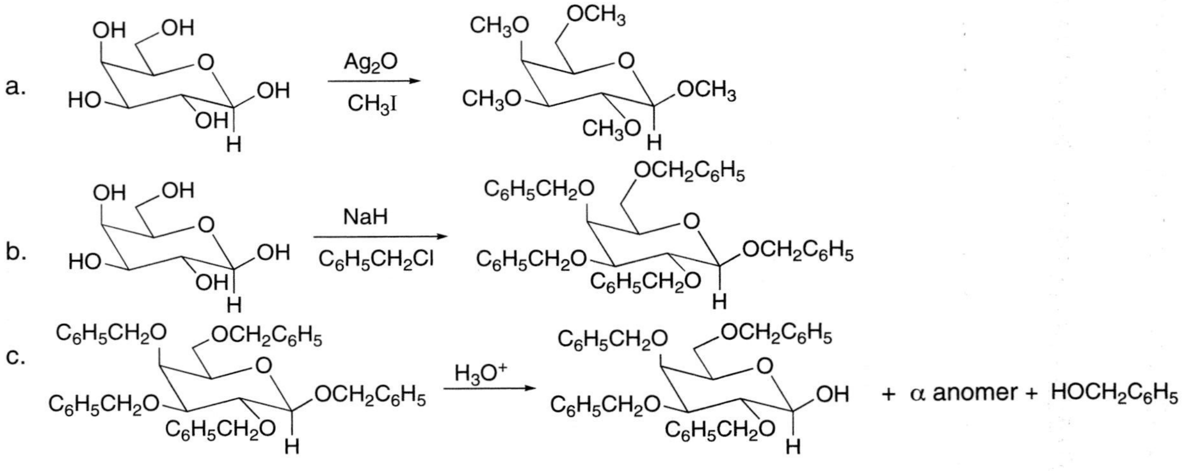

d.

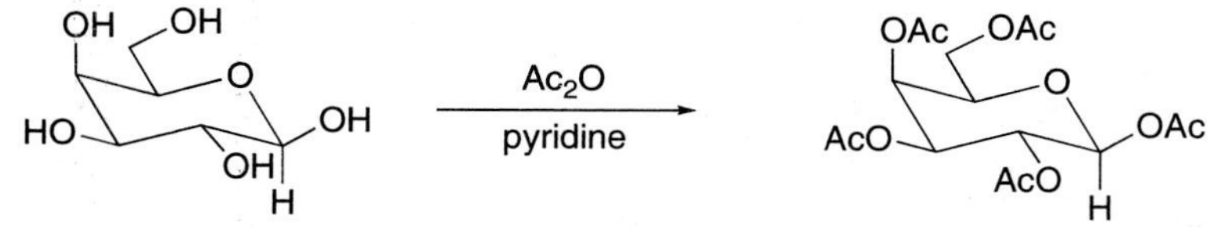

e.

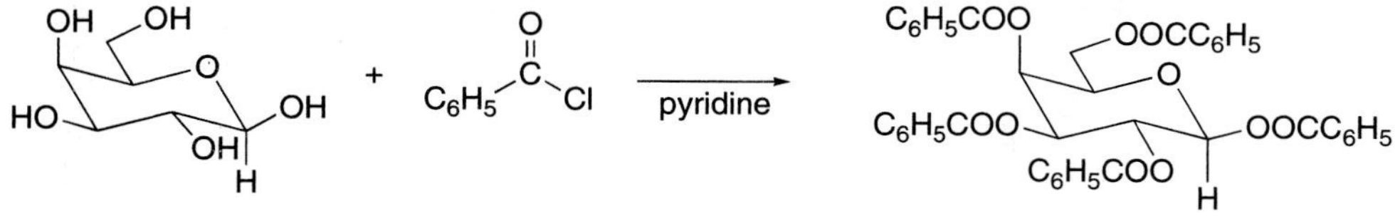

f. product in (c) + + α anomer

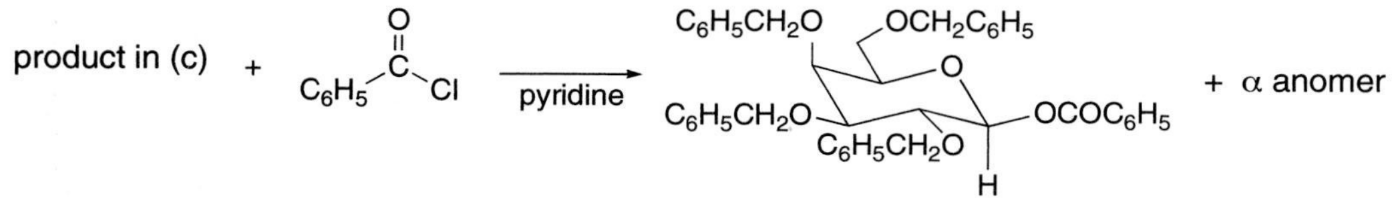

27.21

D-tagatose   **D-galactitol**   **D-talitol**

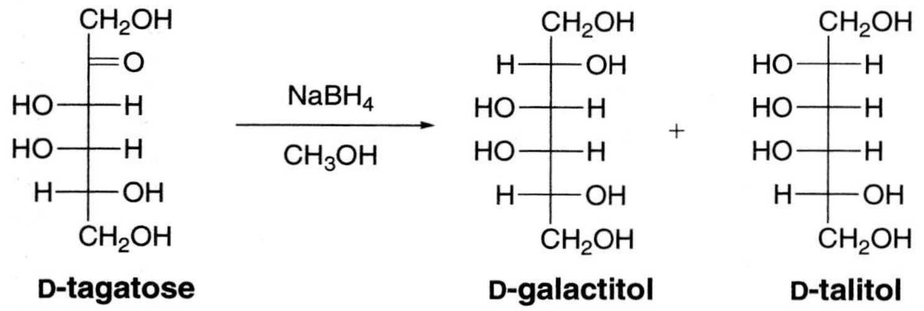

27.22 Carbohydrates containing a hemiacetal are in equilibrium with an acyclic aldehyde, making them reducing sugars. Glycosides are acetals, so they are not in equilibrium with any acyclic aldehyde, making them nonreducing sugars.

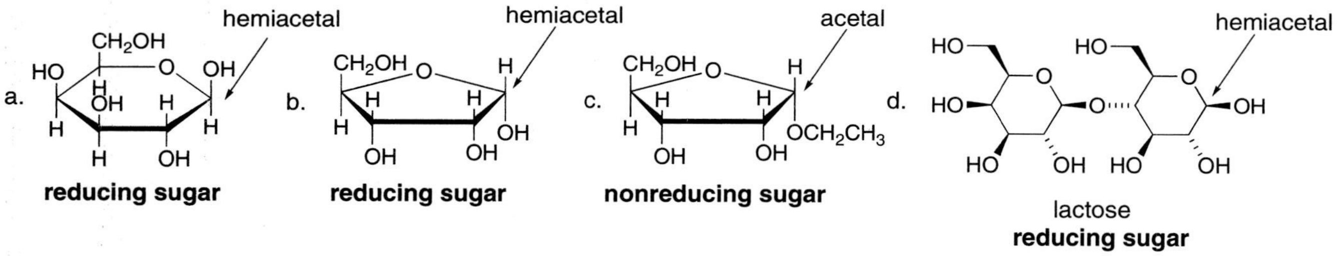

a. reducing sugar   b. reducing sugar   c. nonreducing sugar   d. lactose **reducing sugar**

27.23

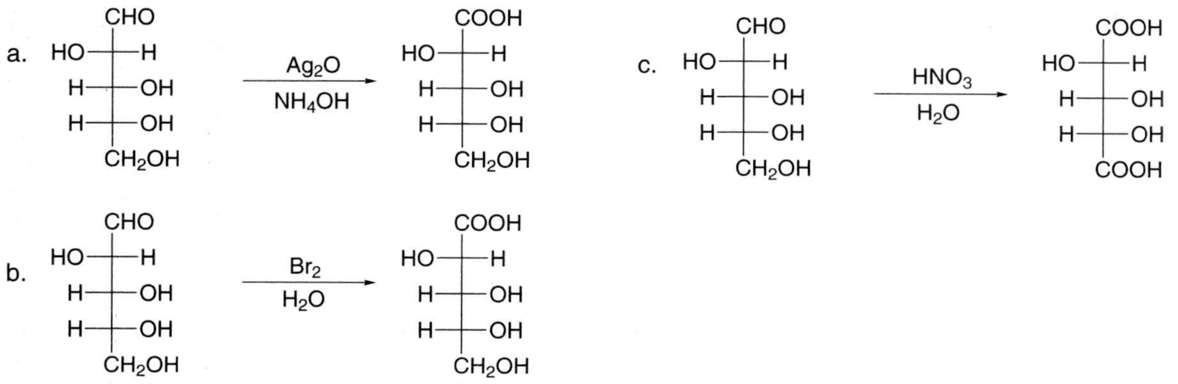

27.24 Molecules with a plane of symmetry are optically inactive.

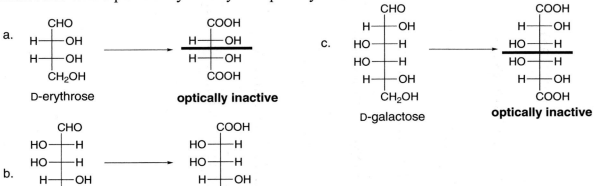

a.

CHO
H——OH
H——OH
CH₂OH
D-erythrose

→

COOH
H——OH
H——OH
COOH
optically inactive

c.

CHO
H——OH
HO——H
HO——H
H——OH
CH₂OH
D-galactose

→

COOH
H——OH
HO——H
HO——H
H——OH
COOH
optically inactive

b.

CHO
HO——H
HO——H
H——OH
CH₂OH
D-lyxose

→

COOH
HO——H
HO——H
H——OH
COOH
optically active

27.25

CHO
HO——H
H——OH
HO——H
H——OH
CH₂OH
D-idose

or

CHO
H——OH
H——OH
HO——H
H——OH
CH₂OH
D-gulose

→

CHO
H——OH
HO——H
H——OH
CH₂OH
D-xylose

27.26

a.

CHO
HO——H
H——OH
CH₂OH
D-threose

→

CHO
HO——H
HO——H
H——OH
CH₂OH

+

CHO
H——OH
HO——H
H——OH
CH₂OH

c.

CHO
H——OH
HO——H
HO——H
H——OH
CH₂OH
D-galactose

→

CHO
HO——H
H——OH
HO——H
HO——H
H——OH
CH₂OH

+

CHO
H——OH
H——OH
HO——H
HO——H
H——OH
CH₂OH

b.

CHO
H——OH
H——OH
H——OH
CH₂OH
D-ribose

→

CHO
HO——H
H——OH
H——OH
H——OH
CH₂OH

+

CHO
H——OH
H——OH
H——OH
H——OH
CH₂OH

27.27

Possible optically inactive D-aldaric acids:

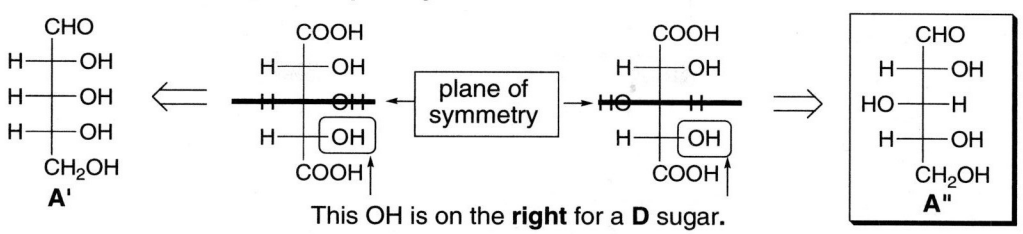

There are two possible structures for the D-aldopentose (**A'** and **A''**), and the Wohl degradation determines which structure corresponds to **A**.

Product of Wohl degradation:

This is **A**.

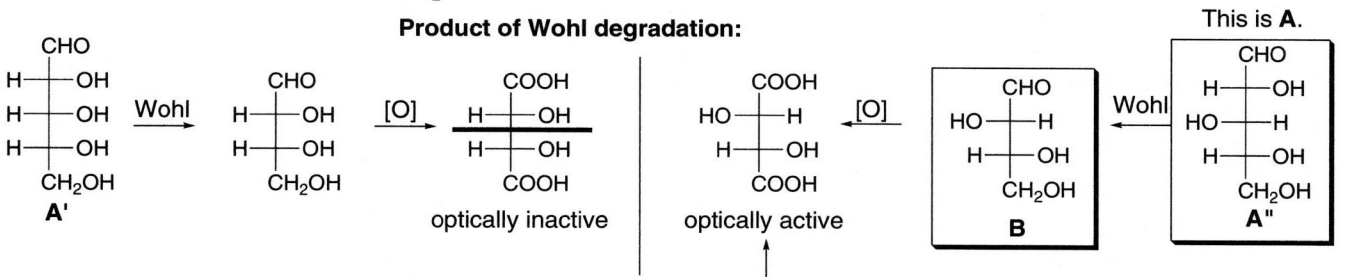

Since this compound has no plane of symmetry, its precursor is **B**, and thus **A''** = **A**.

27.28

CHO — HO—H, H—OH, HO—H, H—OH, CH₂OH → CH₂OH, HO—H, H—OH, HO—H, H—OH, CHO —rotate 180°→ CHO, HO—H, H—OH, HO—H, H—OH, CH₂OH D-idose

identical

27.29

Optically inactive alditols formed from NaBH₄ reduction of a D-aldohexose.

Two D-aldohexoses (**A'** and **A''**) give optically inactive alditols on reduction. **A''** is formed from **B''** by Kiliani–Fischer synthesis. Since **B''** affords an optically active aldaric acid on oxidation, **B''** is **B** and **A''** is **A**. The alternate possibility (**A'**) is formed from an aldopentose **B'** that gives an optically inactive aldaric acid on oxidation.

27.30

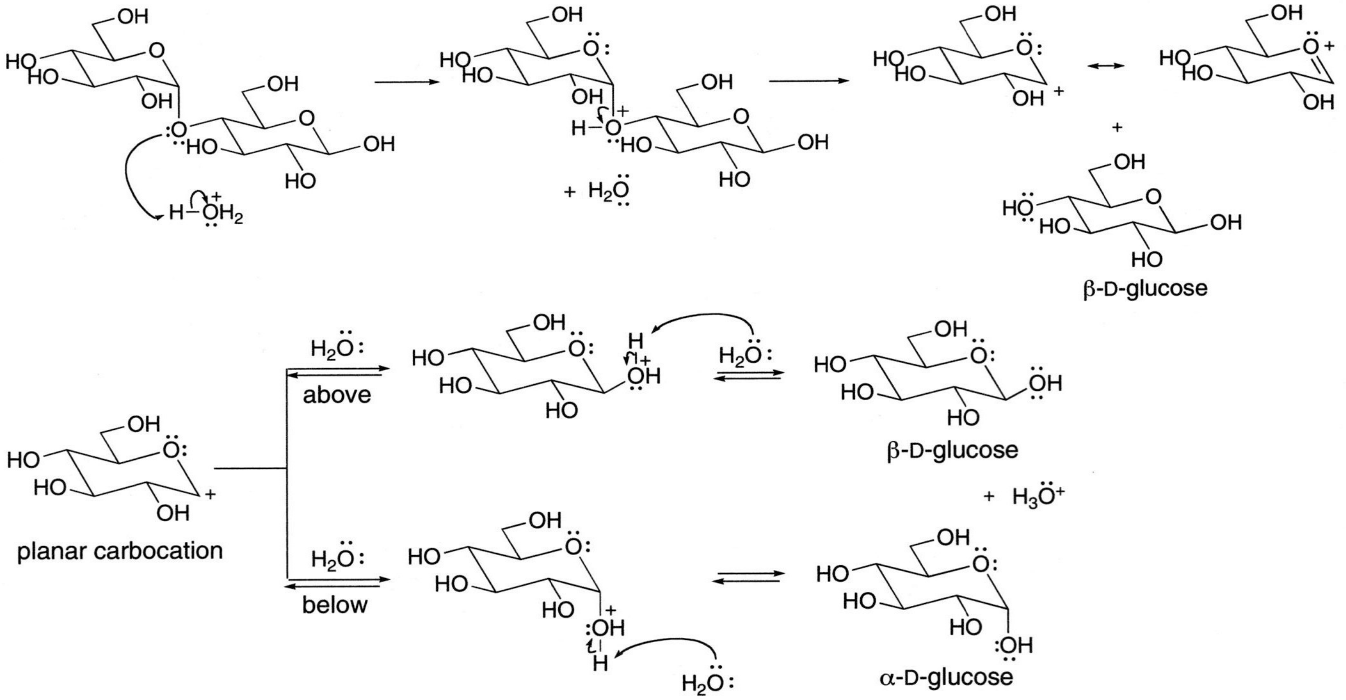

β-D-glucose

planar carbocation

β-D-glucose

+ H$_3$O$^+$

α-D-glucose

27.31

α **anomer**

α-D-glucose   β-D-glucose

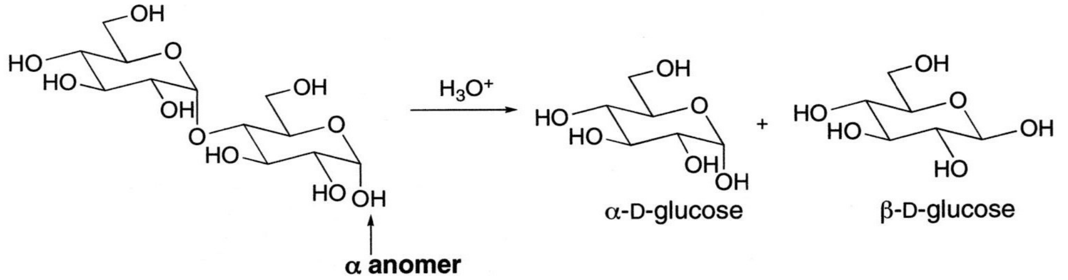

The same products are formed on hydrolysis of the α and β anomers of maltose.

27.32

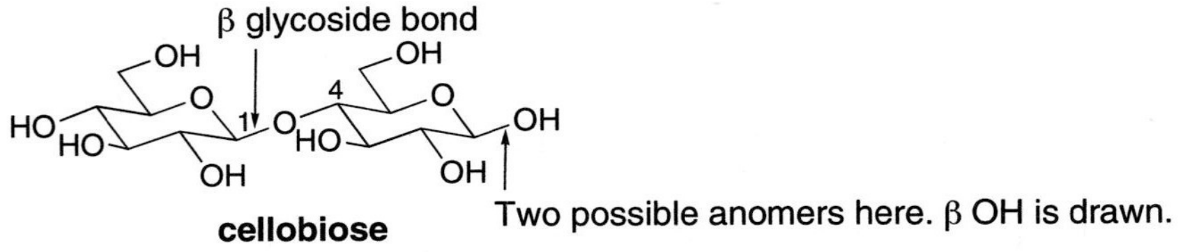

β glycoside bond

cellobiose   Two possible anomers here. β OH is drawn.

27.33

a.

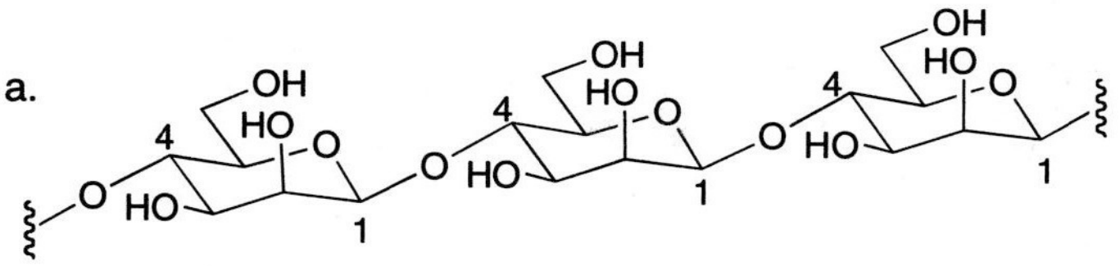

b.

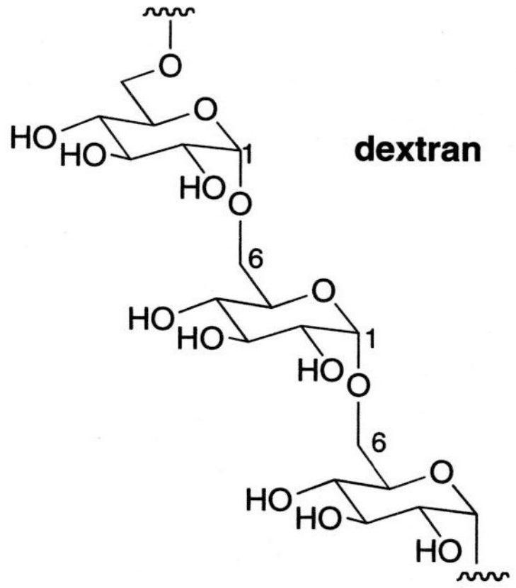

dextran

27.34

chitin—a polysaccharide composed of NAG units

chitosan

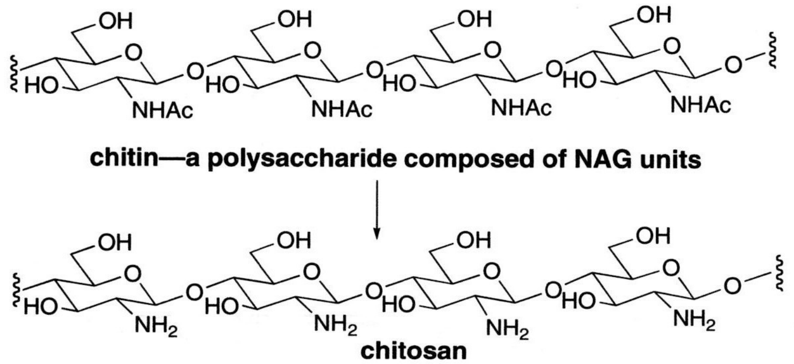

27.35

a.

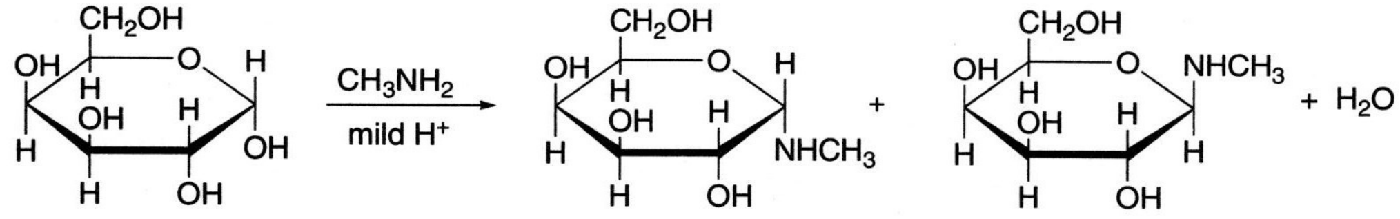

b.

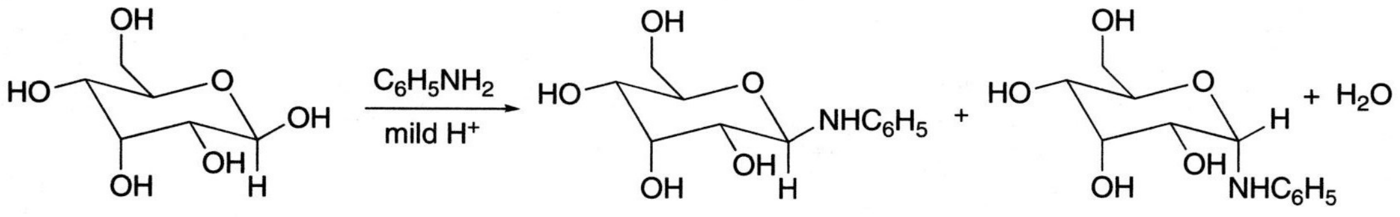

27.36

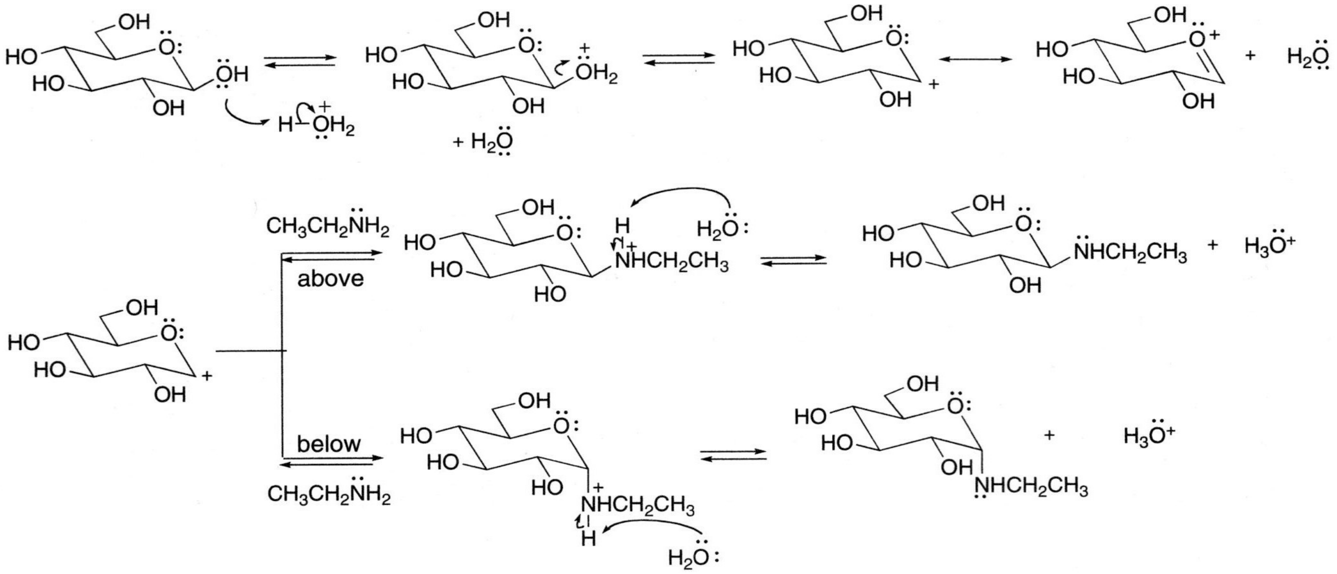

27.37

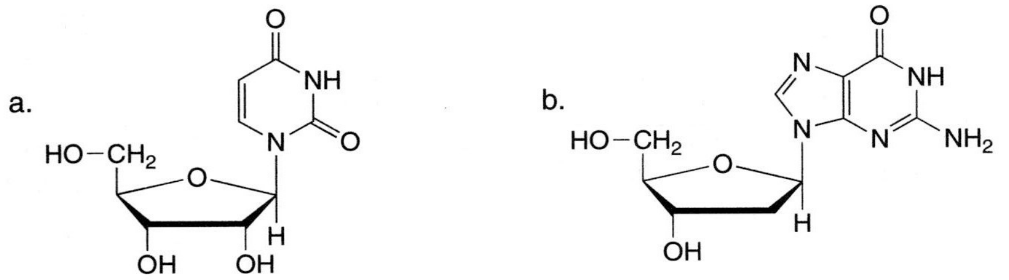

a.

b.

27.38

a. Two purine bases (A and G) are both bicyclic bases. Therefore they are too big to hydrogen bond to each other on the inside of the DNA double helix.

b. Hydrogen bonding between guanine and cytosine has three hydrogen bonds, whereas between guanine and thymine there are only two. This makes hydrogen bonding between guanine and cytosine more favorable.

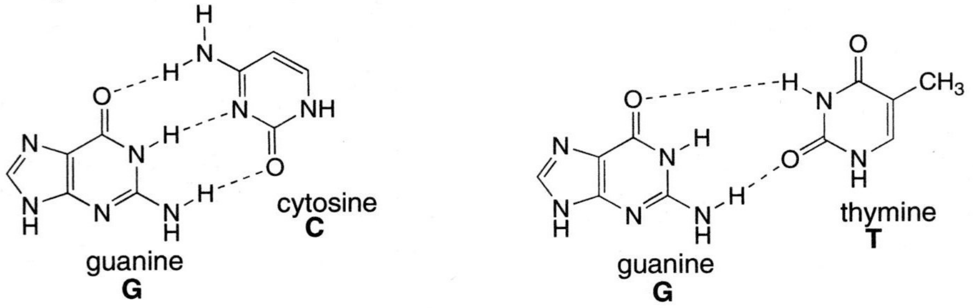

27.39 Label the compounds with *R* or *S* and then classify.

CHO — H—OH — CH₂CH₃ : **A** / **R**

a. CH₃CH₂—C(—OH)(—CHO)(H) : **R** / **identical**

b. (CHO, H, OH) : **R** / **identical**

c. CH₃CH₂—C(H, OH)(CHO) : **S** / **enantiomer**

d. CHO — HO—C—CH₂CH₃ (H) : **S** / **enantiomer**

27.40 Use the directions from Answer 27.2 to draw each Fischer projection.

a. CH₃—C(COOH)(Br)(H) = CH₃—[**S**]—Br (COOH top, H bottom)

b. CH₃—C(Br)(Cl)(H) —re-draw→ H—C—Br (CH₃ top, Cl bottom) = H—[**S**]—Br, Cl bottom

c. (CH₃O)(OCH₂CH₃)C(CH₃)(CH₂CH₃) —re-draw→ CH₃—C(OCH₃)(CH₂CH₃)(OCH₂CH₃) = CH₃—[**S**]—CH₂CH₃ (OCH₃ top, OCH₂CH₃ bottom)

d. (Cl)(H)(Br)C—CH₃CH₂ —re-draw→ CH₃CH₂—C(H)(Cl)(Br) = CH₃CH₂—[**S**]—Cl (H top, Br bottom)

e. CH₃ / H—C—Br / Cl—C—H / CH₂CH₃ = CH₃ / H—[**S**]—Br / Cl—[**S**]—H / CH₂CH₃

f. (H, Cl)(R, S)(Br, Cl)(H, Br) —re-draw→ Br—C—H / Cl—C—Br (**R**, **R**) = Br—[**S**]—H / Cl—[**R**]—Br (Cl top, H bottom)

g. (CH₃)(H, Br)(Br, CH₃) —re-draw→ H—C—Br / Br—C—H (CH₃ top and bottom) = H—[**S**]—Br / Br—[**S**]—H

h. HO—CH₂ ... HO H, H OH, HO H, CHO —re-draw→ HO—C—H / HO—C—H / H—C—OH / CH₂OH = HO—[**S**]—H / HO—[**S**]—H / H—[**R**]—OH / CH₂OH (CHO top)

27.41 *Epimers* are two diastereomers that differ in the configuration around only one stereogenic center.

CHO / H—OH / HO—H / H—OH / CH₂OH —C4— **D-xylose**

CHO / H—OH / HO—H / HO—H / CH₂OH **L-arabinose**

27.42

a. CHO / HO—H / H—OH / H—OH / CH₂OH **D-arabinose**

CHO / H—OH / HO—H / HO—H / CH₂OH **enantiomer**

b. CHO / HO—H / HO—H / H—OH / CH₂OH —C3— **epimer**

c. CHO / H—OH / HO—H / H—OH / CH₂OH **diastereomer (but not epimer)**

d. HO—CH(OH)—C(=O)—CH₂OH with OH groups **constitutional isomer**

27.43

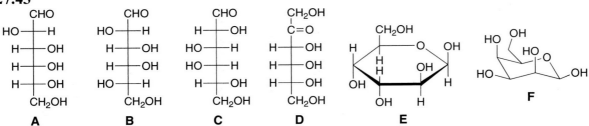

a. **A** and **B** epimers
b. **A** and **C** diastereomers

c. **B** and **C** enantiomers
d. **A** and **D** constitutional isomers

e. **E** and **F** diastereomers

27.44

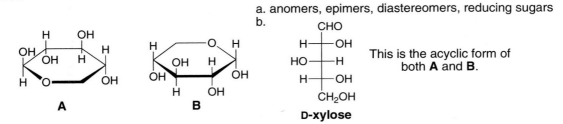

a. anomers, epimers, diastereomers, reducing sugars
b.

D-xylose

This is the acyclic form of
both **A** and **B**.

27.45 Use the directions from Answer 27.13.

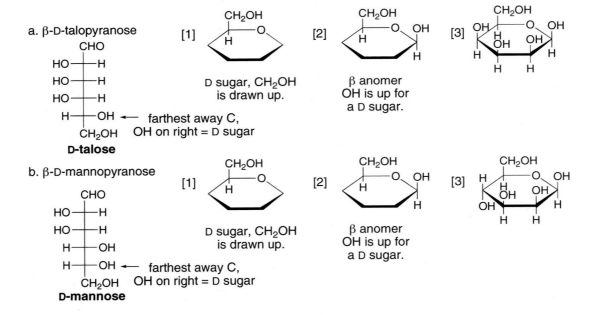

a. β-D-talopyranose

D-talose

[1] D sugar, CH₂OH
is drawn up.

farthest away C,
OH on right = D sugar

[2] β anomer
OH is up for
a D sugar.

[3]

b. β-D-mannopyranose

D-mannose

[1] D sugar, CH₂OH
is drawn up.

farthest away C,
OH on right = D sugar

[2] β anomer
OH is up for
a D sugar.

[3]

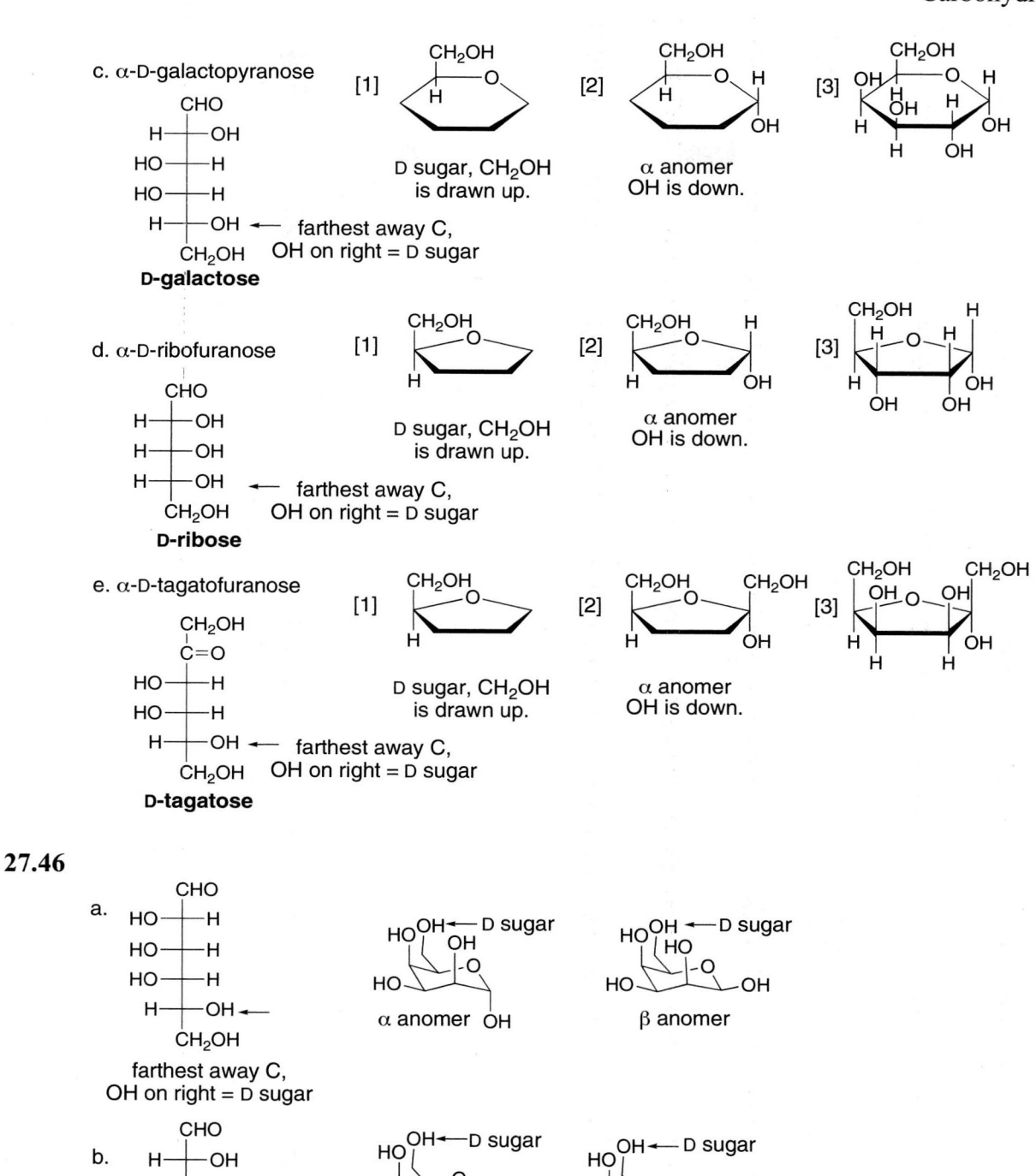

c. α-D-galactopyranose

CHO
H——OH
HO——H
HO——H
H——OH ← farthest away C, OH on right = D sugar
CH₂OH
D-galactose

[1] D sugar, CH₂OH is drawn up.

[2] α anomer OH is down.

[3]

d. α-D-ribofuranose

CHO
H——OH
H——OH
H——OH ← farthest away C, OH on right = D sugar
CH₂OH
D-ribose

[1] D sugar, CH₂OH is drawn up.

[2] α anomer OH is down.

[3]

e. α-D-tagatofuranose

CH₂OH
C=O
HO——H
HO——H
H——OH ← farthest away C, OH on right = D sugar
CH₂OH
D-tagatose

[1] D sugar, CH₂OH is drawn up.

[2] α anomer OH is down.

[3]

27.46

a.

CHO
HO——H
HO——H
HO——H
H——OH ←
CH₂OH
farthest away C,
OH on right = D sugar

α anomer OH

β anomer

b.

CHO
H——OH
H——OH
HO——H
H——OH ←
CH₂OH
farthest away C,
OH on right = D sugar

α anomer

β anomer

c.

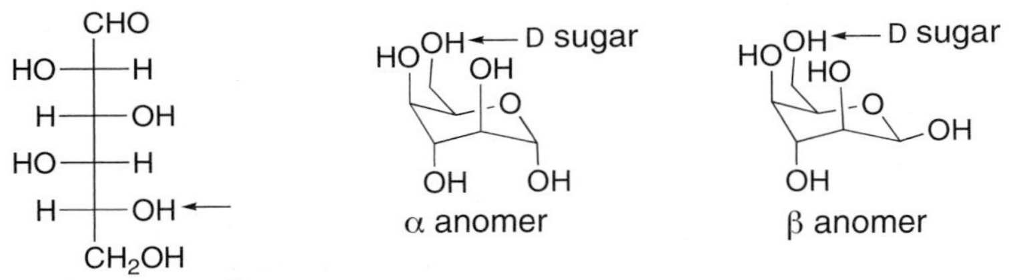

CHO
HO──H
H──OH
HO──H
H──OH ←
CH₂OH

farthest away C,
OH on right = D sugar

HO OH ← D sugar
OH
─O─
OH   OH
α anomer

HO OH ← D sugar
HO
─O─
──OH
OH
β anomer

27.47 Use the directions from Answer 27.14.

a.

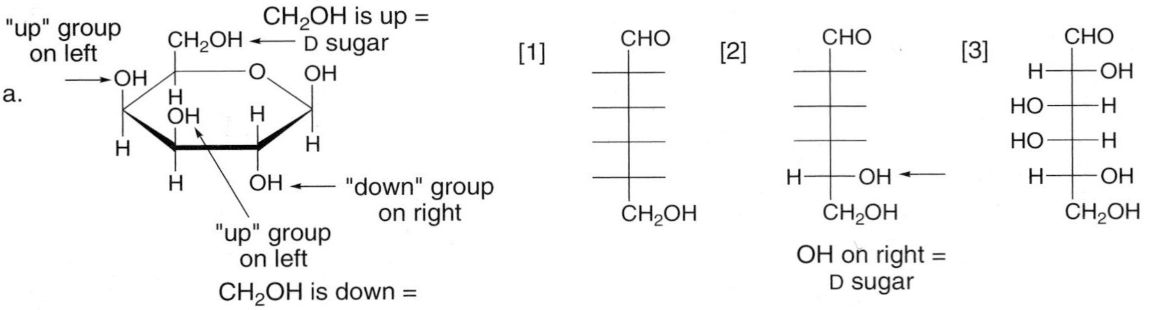

"up" group
on left

CH₂OH ← CH₂OH is up =
D sugar

OH
H
OH   H
H
H   OH ← "down" group
on right

"up" group
on left

[1]

CHO

CH₂OH

[2]

CHO

H──OH ←

CH₂OH

OH on right =
D sugar

[3]

CHO
H──OH
HO──H
HO──H
H──OH
CH₂OH

b.

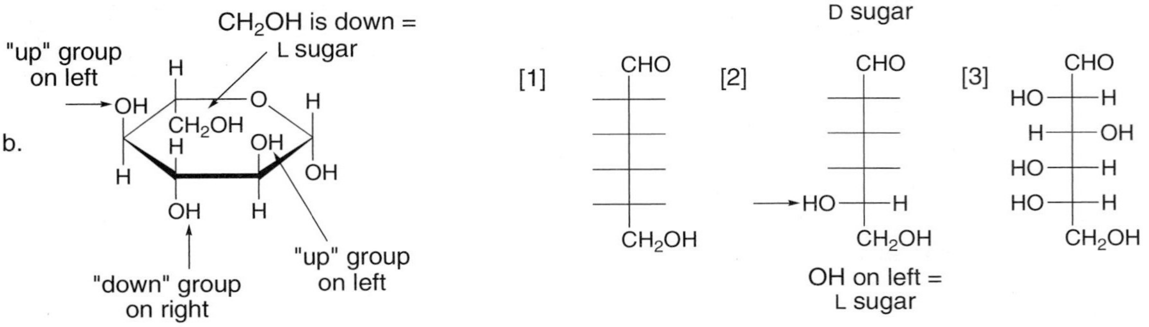

"up" group
on left

CH₂OH is down =
L sugar

H
OH
CH₂OH
H   OH
H
OH   H

OH   H

"up" group
on left

"down" group
on right

[1]

CHO

CH₂OH

[2]

CHO

HO──H

CH₂OH

OH on left =
L sugar

[3]

CHO
HO──H
H──OH
HO──H
HO──H
CH₂OH

c.

OH
HO
HO   O
OH
OH

=

CH₂OH ← CH₂OH is up =
D sugar

H
O   OH
H
H   OH
OH   H

OH   H

[1]

CHO

CH₂OH

[2]

CHO

H──OH ←

CH₂OH

OH on right =
D sugar

[3]

CHO
HO──H
H──OH
H──OH
H──OH
CH₂OH

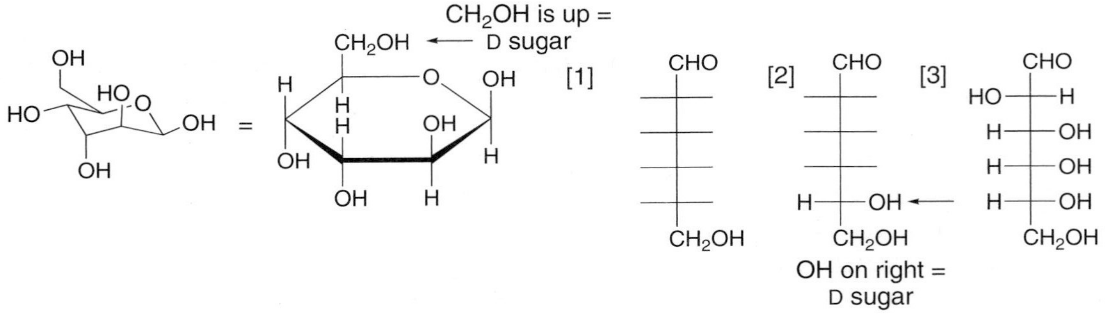

d.

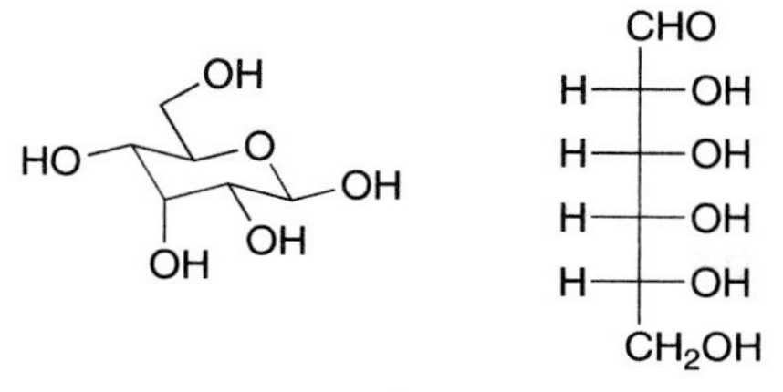

D sugar

e.

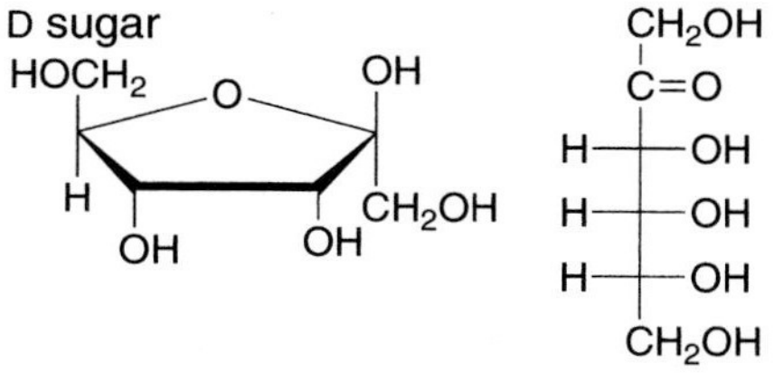

f.

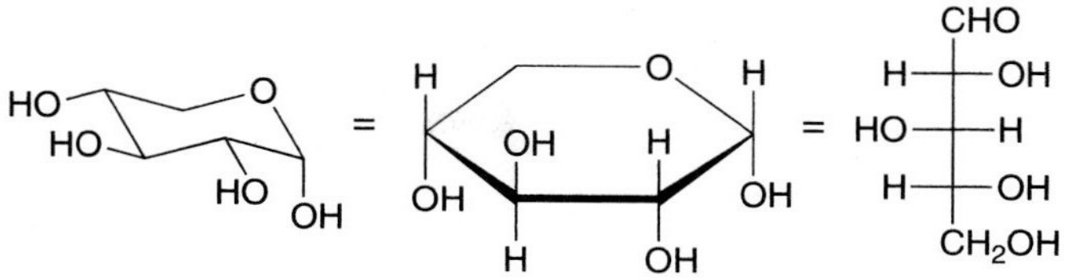

27.48

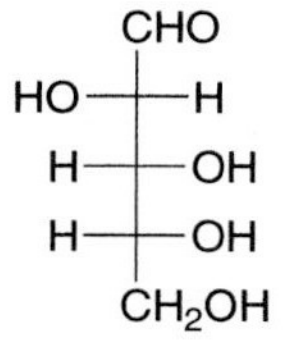

D-arabinose

a.

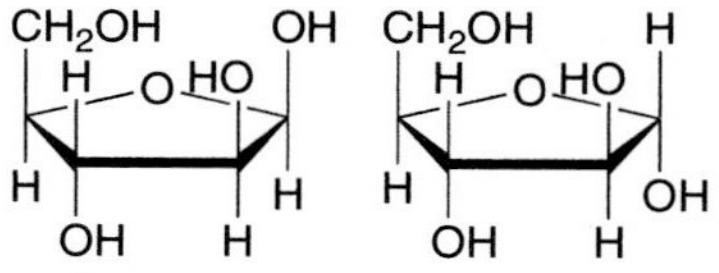

β anomer   α anomer

b.

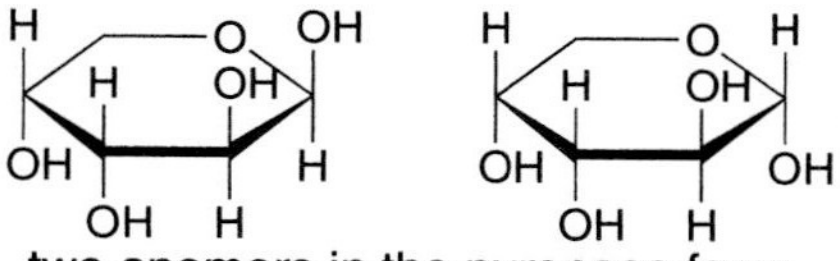

two anomers in the pyranose form

27.49

Two anomers of D-idose, as well as two conformations of each anomer:

α anomer

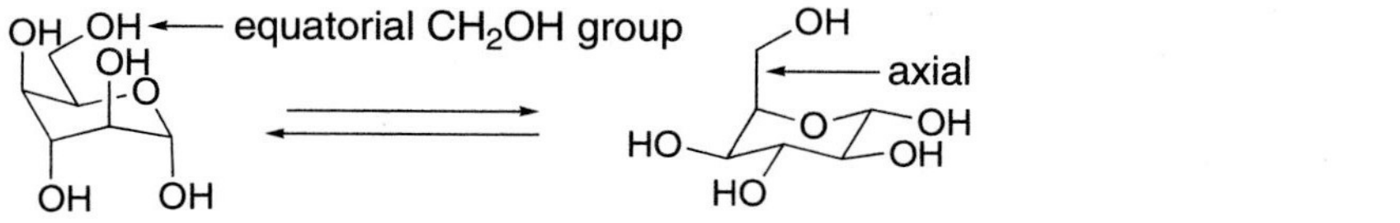

4 axial substituents

4 equatorial OH groups

More stable conformation for the α anomer—the CH₂OH is axial, but all other groups are equatorial.

β anomer

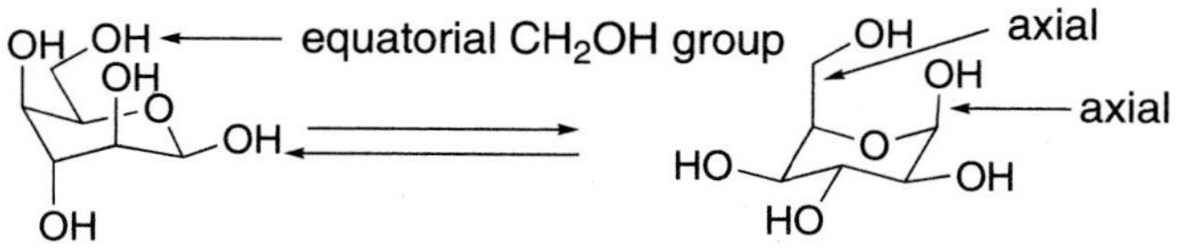

3 axial substituents

3 equatorial OH groups

The more stable conformation for the β anomer—the CH₂OH is axial, as is the anomeric OH, but three other OH groups are equatorial.

27.50

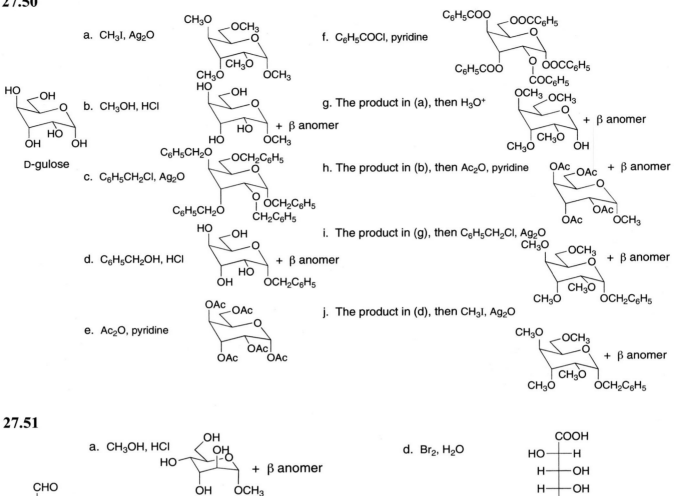

a. CH₃I, Ag₂O

b. CH₃OH, HCl

c. C₆H₅CH₂Cl, Ag₂O

d. C₆H₅CH₂OH, HCl

e. Ac₂O, pyridine

f. C₆H₅COCl, pyridine

g. The product in (a), then H₃O⁺

h. The product in (b), then Ac₂O, pyridine

i. The product in (g), then C₆H₅CH₂Cl, Ag₂O

j. The product in (d), then CH₃I, Ag₂O

27.51

a. CH₃OH, HCl

b. (CH₃)₂CHOH, HCl

c. NaBH₄, CH₃OH

d. Br₂, H₂O

e. HNO₃, H₂O

f. [1] NH$_2$OH
   [2] (CH$_3$CO)$_2$O, NaOCOCH$_3$
   [3] NaOCH$_3$

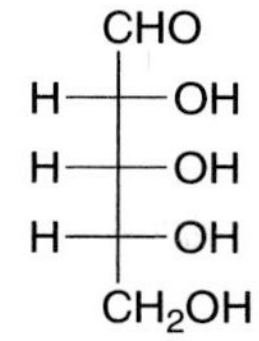

h. CH$_3$I, Ag$_2$O

+ β anomer

g. [1] NaCN, HCl
   [2] H$_2$, Pd-BaSO$_4$
   [3] H$_3$O$^+$

i. Ac$_2$O, pyridine

+ β anomer

j. C$_6$H$_5$CH$_2$NH$_2$, mild H$^+$

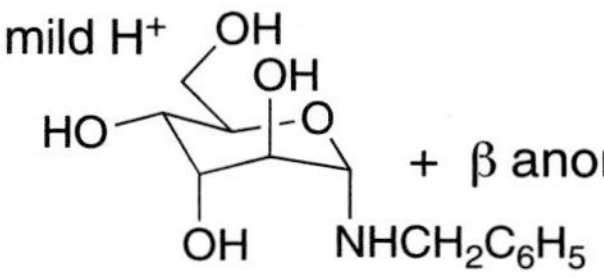

+ β anomer

27.52

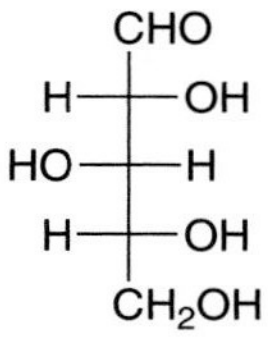

a. CH$_3$OH, HCl

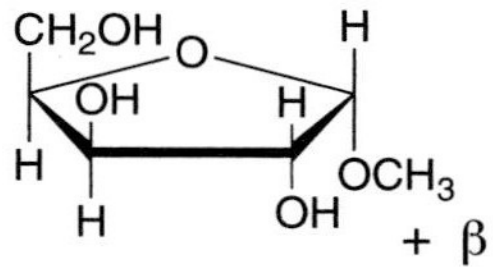

+ β anomer

b. (CH$_3$)$_2$CHOH, HCl

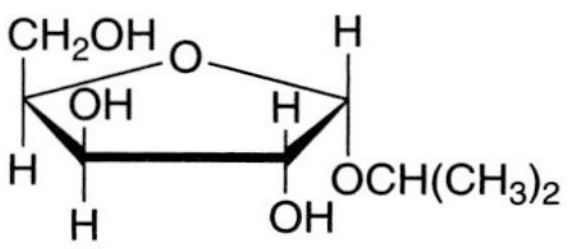

+ β anomer

D-xylose

c. NaBH$_4$, CH$_3$OH

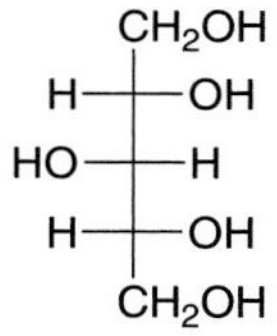

d. Br$_2$, H$_2$O

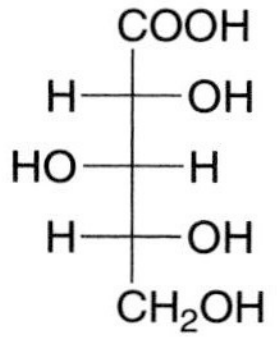

e. HNO$_3$, H$_2$O

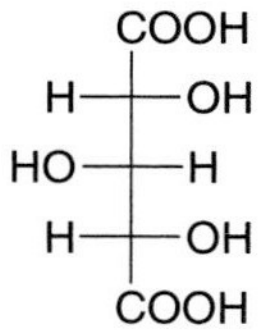

f. [1] NH$_2$OH
   [2] (CH$_3$CO)$_2$O, NaOCOCH$_3$
   [3] NaOCH$_3$

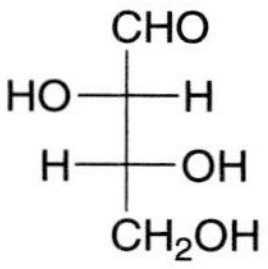

g. [1] NaCN, HCl
   [2] H$_2$, Pd-BaSO$_4$
   [3] H$_3$O$^+$

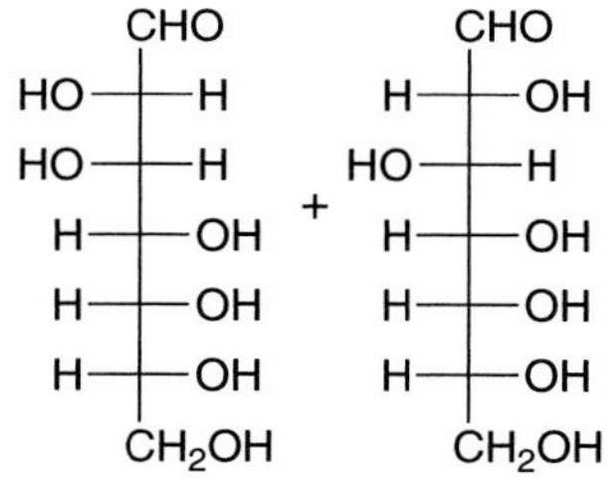

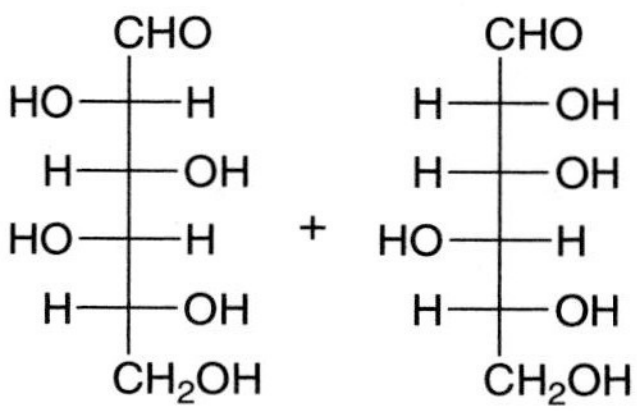

h. CH$_3$I, Ag$_2$O

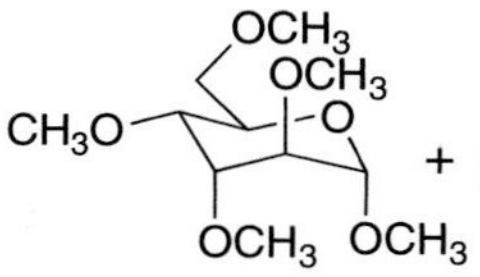

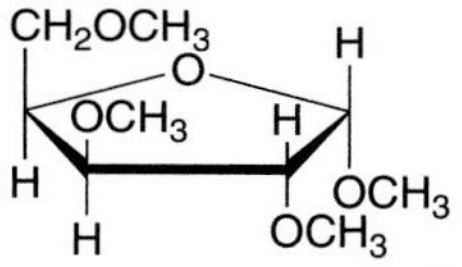

+ β anomer

i. Ac$_2$O, pyridine

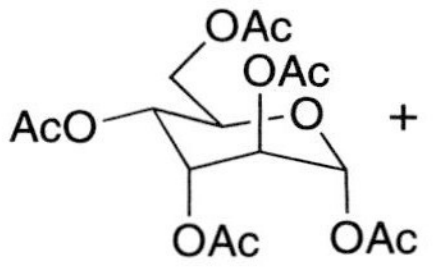

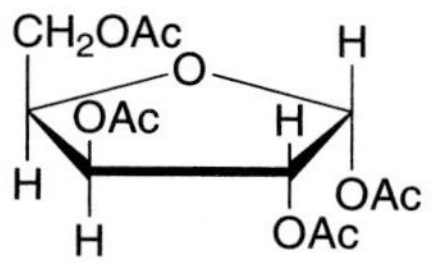

+ β anomer

j. C$_6$H$_5$CH$_2$NH$_2$, mild H$^+$

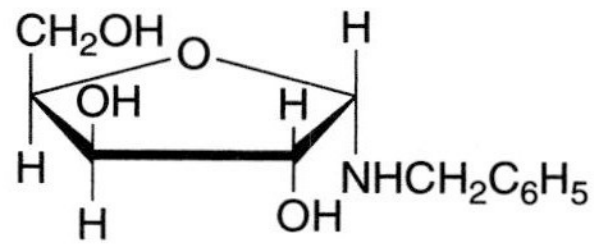

+ β anomer

27.53

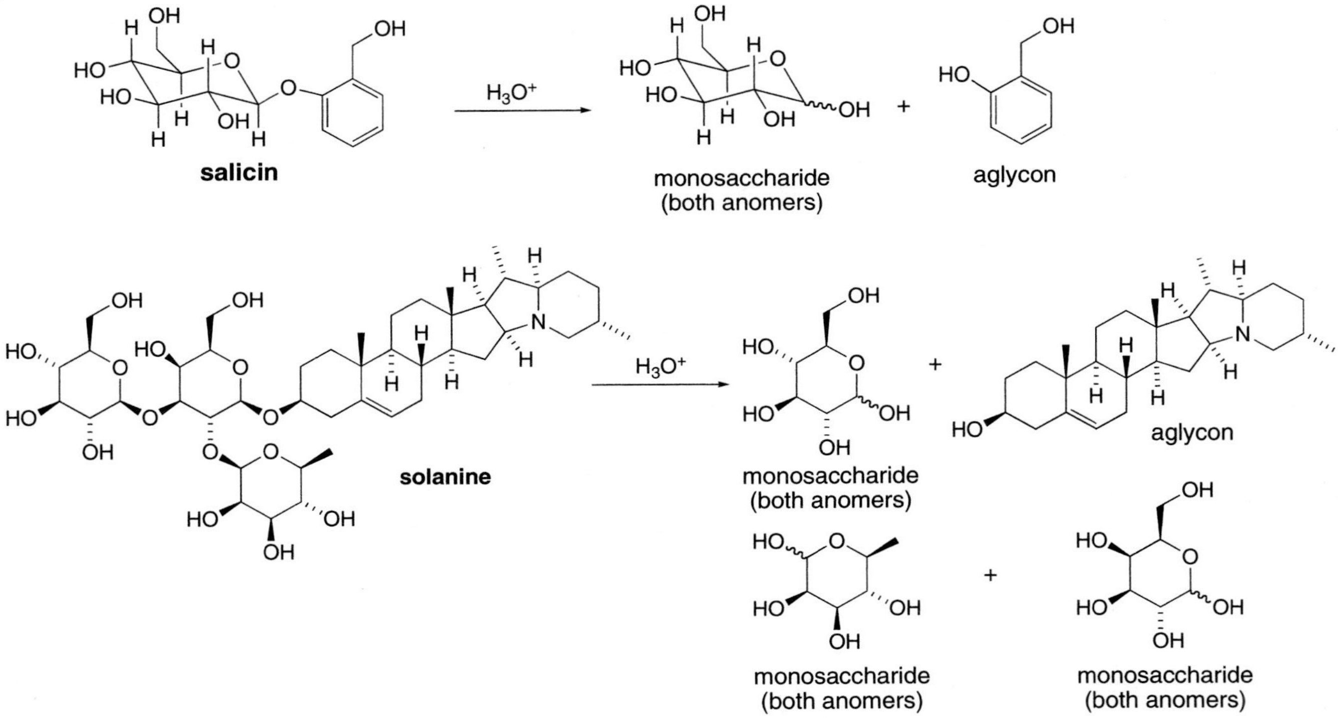

salicin $\xrightarrow{H_3O^+}$ monosaccharide (both anomers) + aglycon

solanine $\xrightarrow{H_3O^+}$ monosaccharide (both anomers) + aglycon

monosaccharide (both anomers) + monosaccharide (both anomers)

27.54

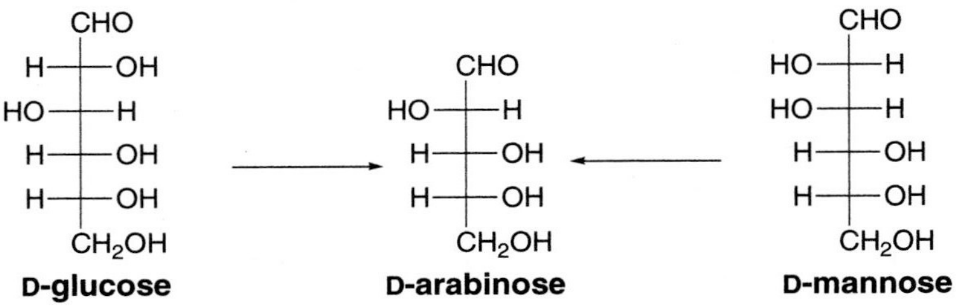

D-glucose → D-arabinose ← D-mannose

27.55

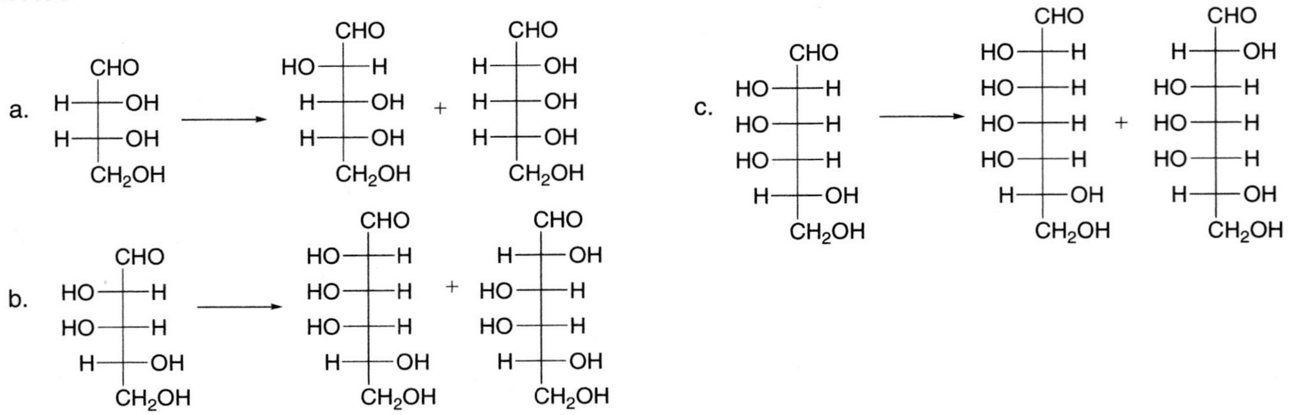

a.

b.

c.

27.56

a.

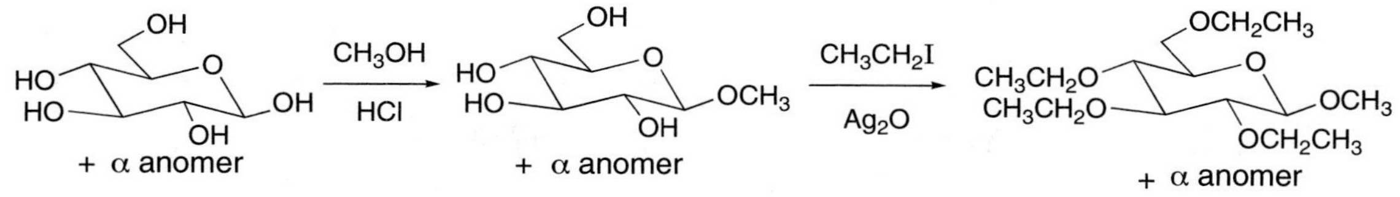

$$\text{+ } \alpha \text{ anomer} \quad\quad \text{+ } \alpha \text{ anomer} \quad\quad \text{+ } \alpha \text{ anomer}$$

CH₃OH / HCl

CH₃CH₂I / Ag₂O

b.

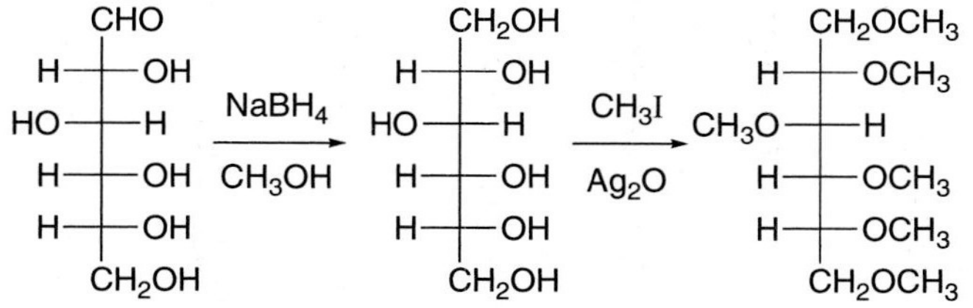

CHO		CH₂OH		CH₂OCH₃
H——OH		H——OH		H——OCH₃
HO——H	NaBH₄	HO——H	CH₃I	CH₃O——H
H——OH	CH₃OH	H——OH	Ag₂O	H——OCH₃
H——OH		H——OH		H——OCH₃
CH₂OH		CH₂OH		CH₂OCH₃

c.

CHO		COOH		COOH
H——OH		H——OH		H——OAc
HO——H	Br₂	HO——H	Ac₂O	AcO——H
H——OH	H₂O	H——OH	pyridine	H——OAc
H——OH		H——OH		H——OAc
CH₂OH		CH₂OH		CH₂OAc

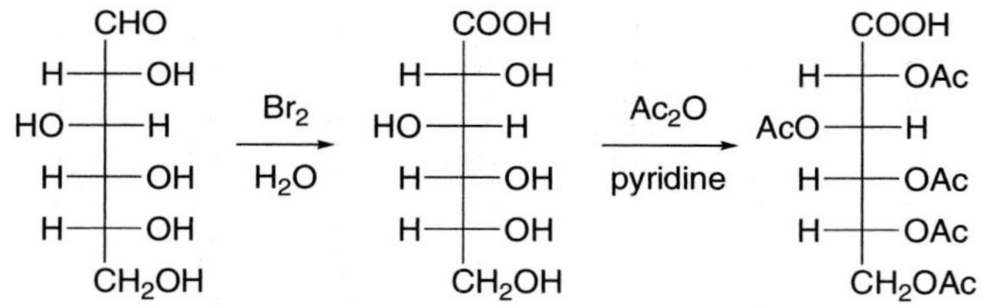

27.57 Molecules with a plane of symmetry are optically inactive.

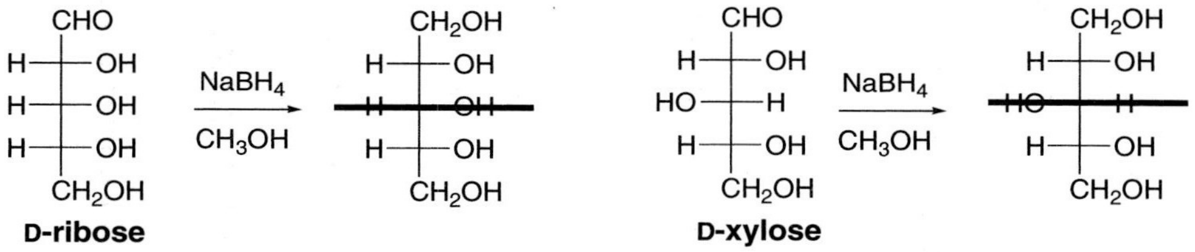

CHO		CH₂OH		CHO		CH₂OH
H——OH		H——OH		H——OH		H——OH
H——OH	NaBH₄	H——OH		HO——H	NaBH₄	HO——H
H——OH	CH₃OH	H——OH		H——OH	CH₃OH	H——OH
CH₂OH		CH₂OH		CH₂OH		CH₂OH
D-ribose				**D-xylose**		

27.58

a.

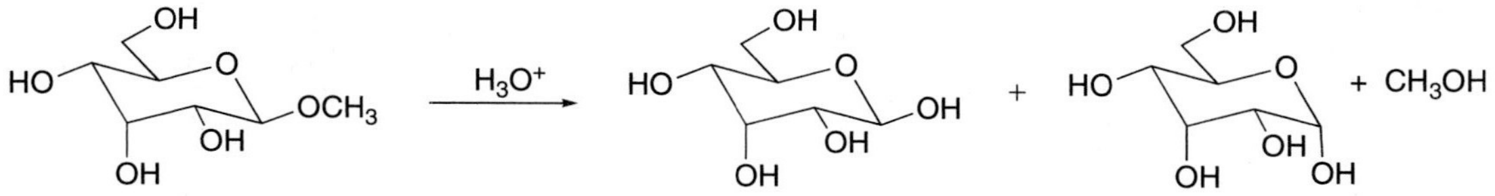

H₃O⁺

+ CH₃OH

b.

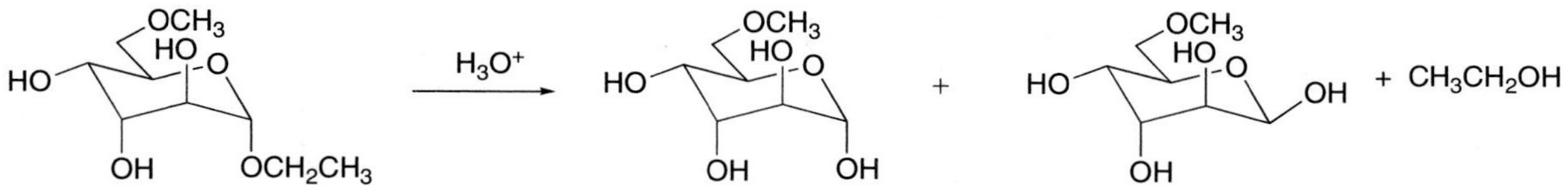

H₃O⁺

+ CH₃CH₂OH

c.

H₃O⁺

+ NH₂CH₂CH₃

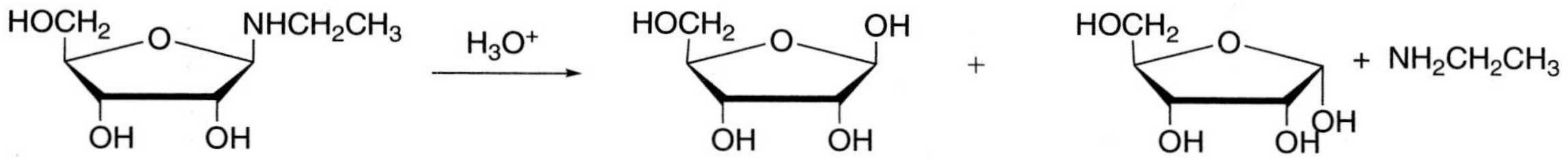

27.59

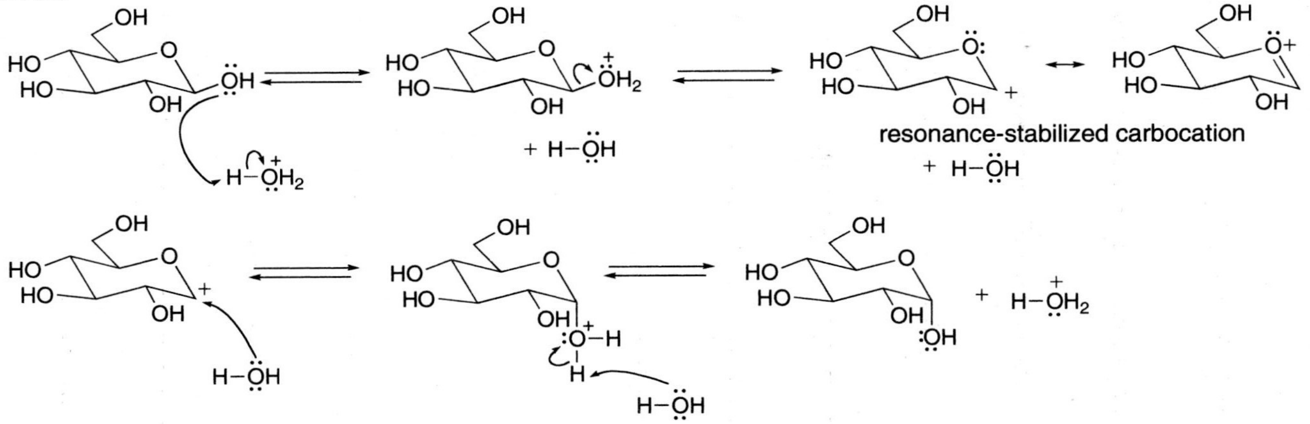
resonance-stabilized carbocation

27.60

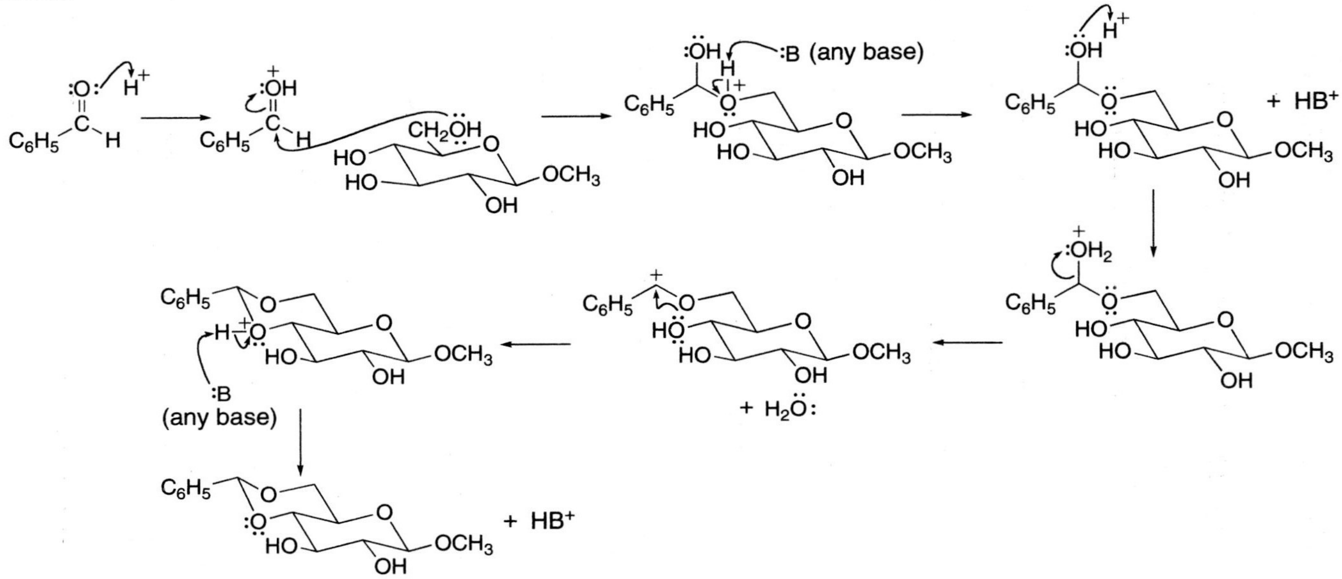
:B (any base)
(any base)

27.61

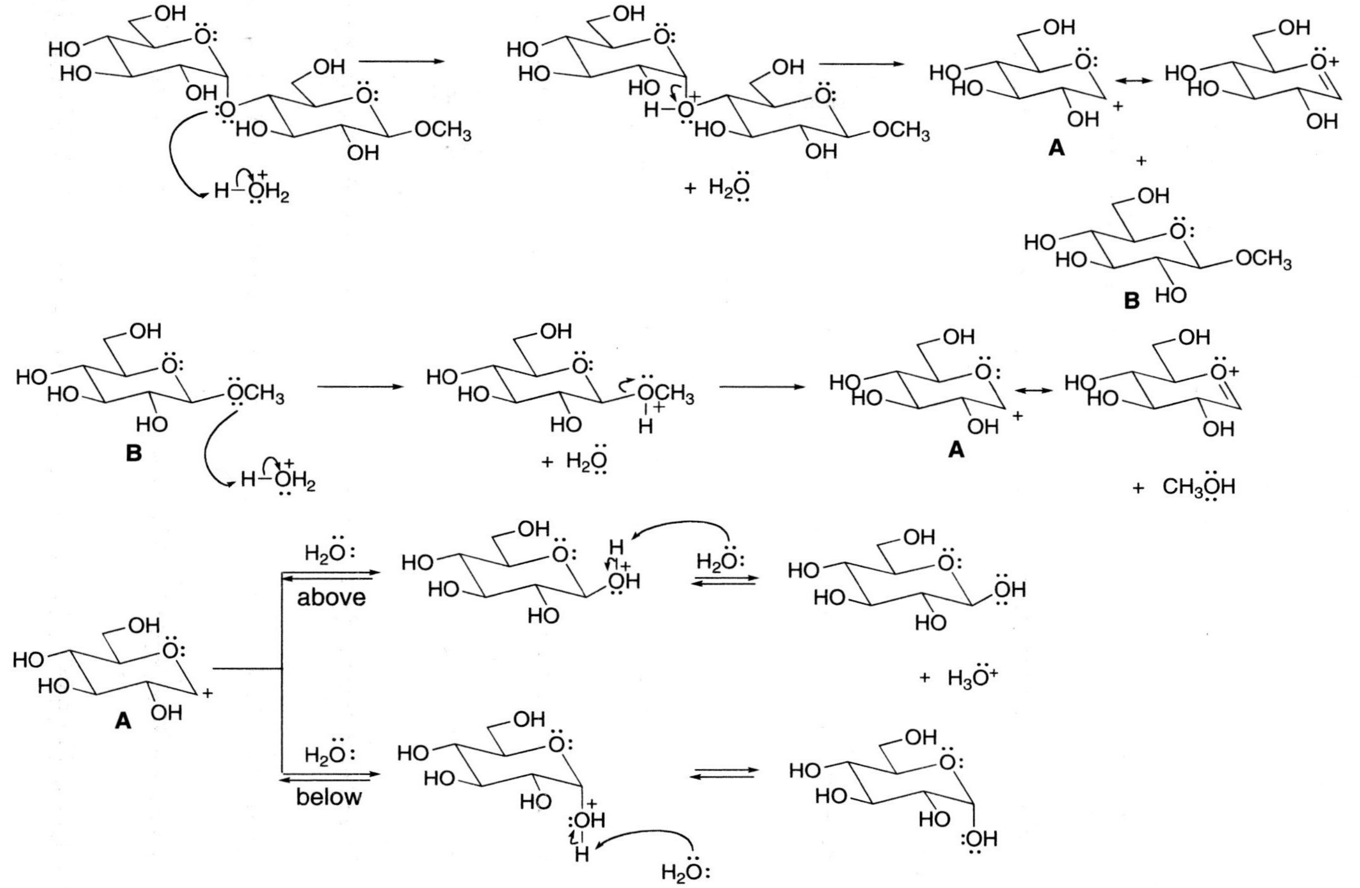

27.62

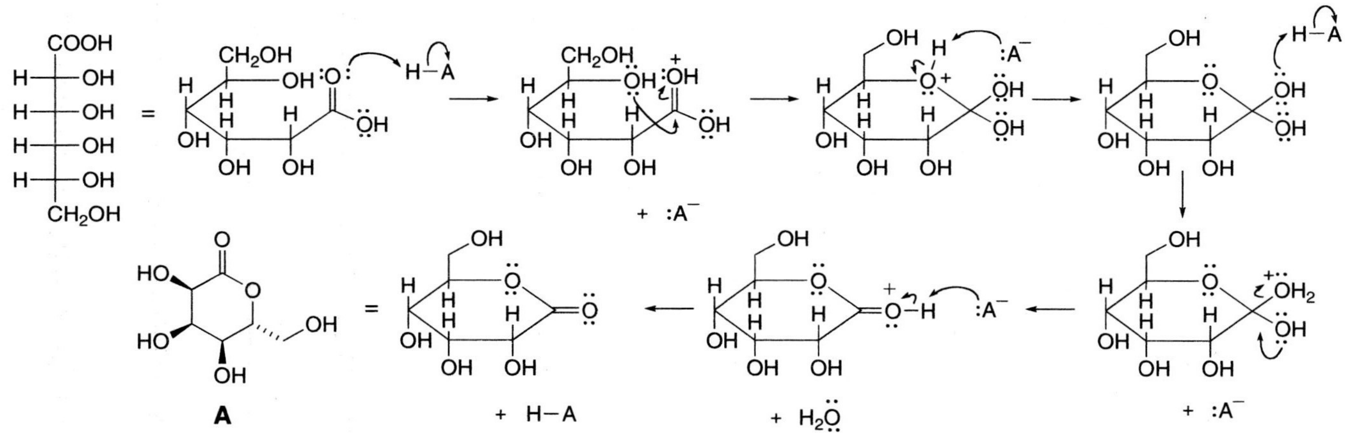

27.63

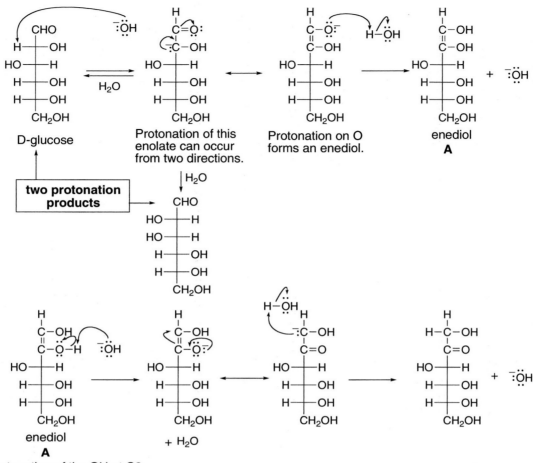

two protonation products →

D-glucose

Protonation of this enolate can occur from two directions.

Protonation on O forms an enediol.

enediol **A**

Deprotonation of the OH at C2 of the enediol forms a new enolate that goes on to form the ketohexose.

enediol **A**

+ H₂O

27.64

Two D-aldopentoses (**A'** and **A''**) yield optically active aldaric acids when oxidized.

Optically active D-aldaric acids:

A' → [O] → optically active

optically active ← [O] → **A''** **D-lyxose**

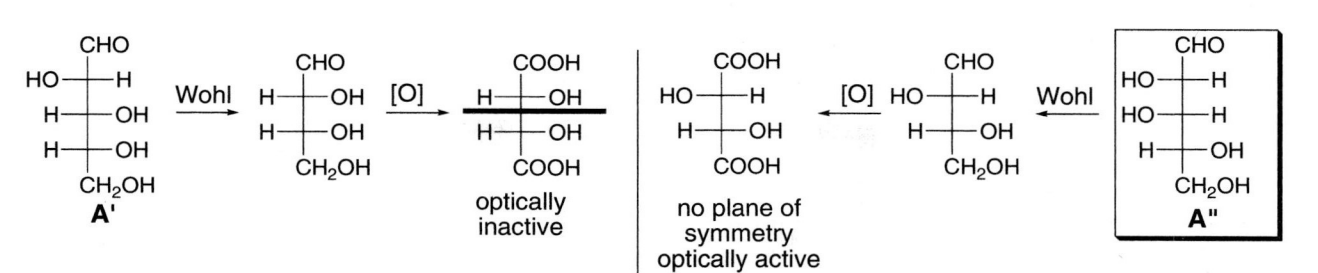

Only **A"** undergoes Wohl degradation to an aldotetrose that is oxidized to an optically active aldaric acid, so **A"** is the structure of the D-aldopentose in question.

27.65

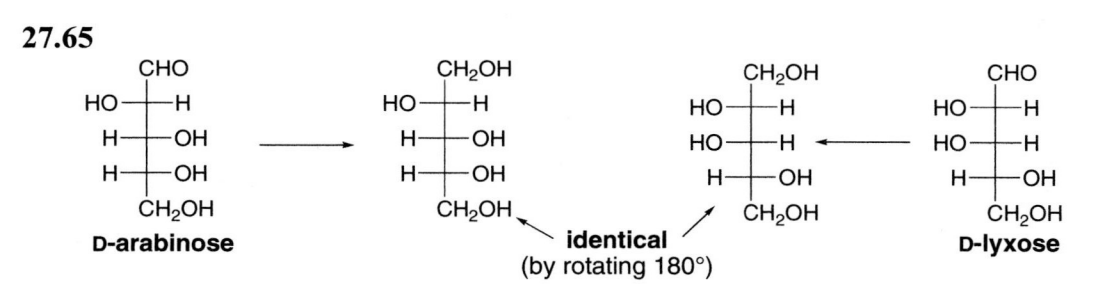

D-arabinose   **identical** (by rotating 180°)   **D-lyxose**

27.66

Only two D-aldopentoses (**A'** and **A"**) yield optically inactive aldaric acids (**B'** and **B"**).

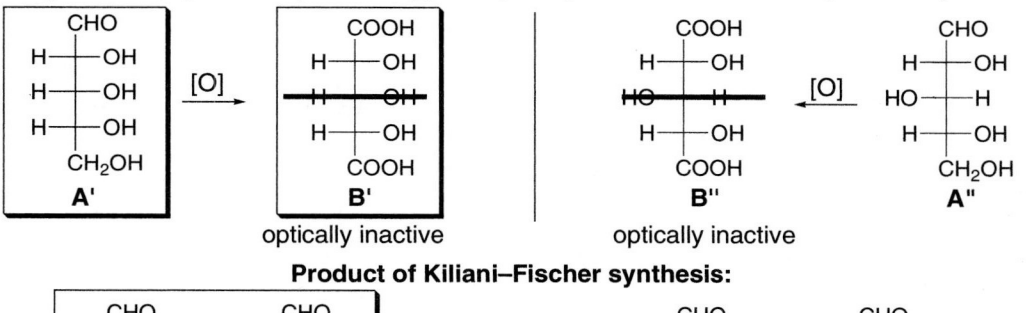

Product of Kiliani–Fischer synthesis:

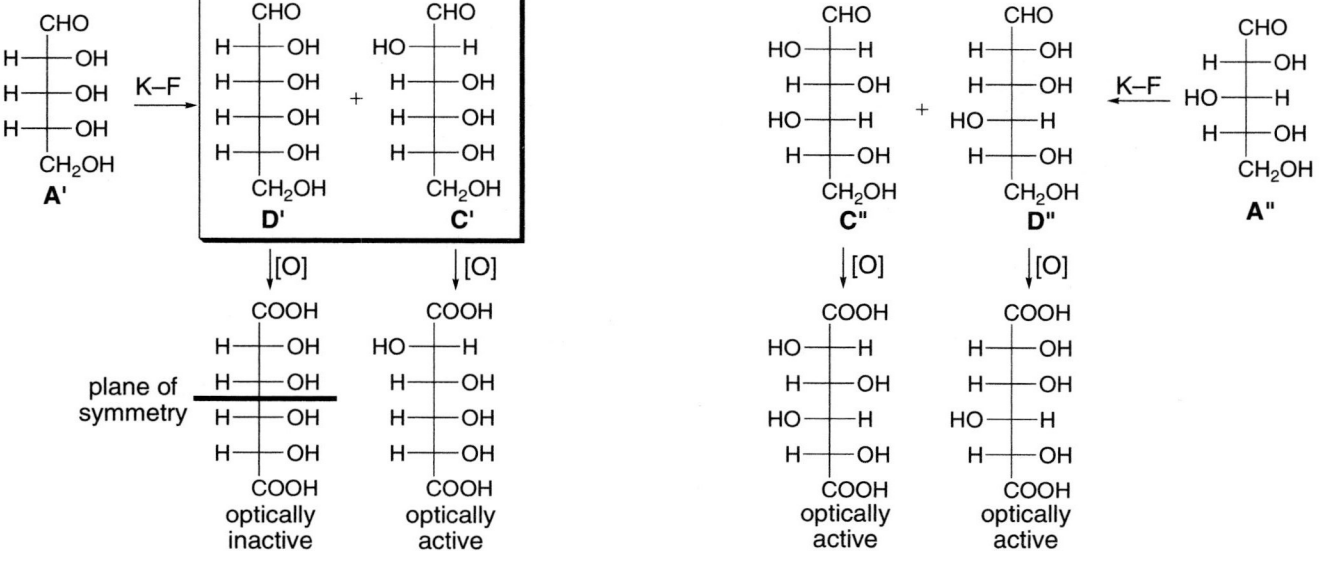

Only **A'** fits the criteria. Kiliani–Fischer synthesis of **A'** forms **C'** and **D'** which are oxidized to one optically active and one optically inactive aldaric acid. A similar procedure with **A''** forms two optically active aldaric acids. Thus, the structures of **A–D** correspond to the structures of **A'–D'**.

27.67

Only two D-aldopentoses (**A'** and **A''**) are reduced to optically active alditols.

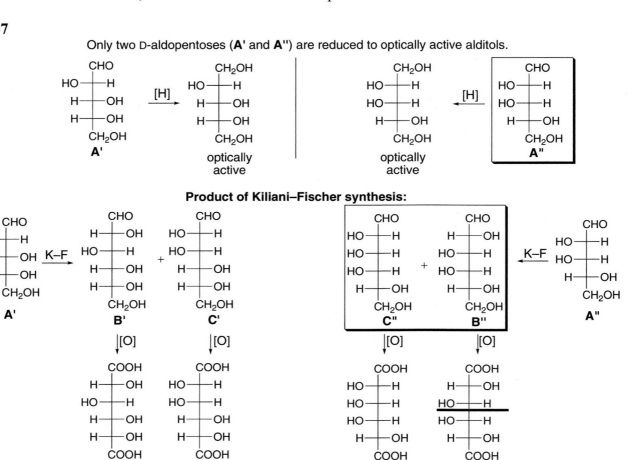

Only **A''** fits the criteria. Kiliani–Fischer synthesis of **A''** forms **B''** and **C''**, which are oxidized to one optically inactive and one optically active diacid. A similar procedure with **A'** forms two optically active diacids. Thus, the structures of **A–C** correspond to **A''–C''**.

27.68

D-gulose

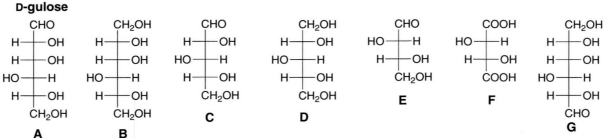

27.69 A disaccharide formed from two galactose units in a 1→4-β-glycosidic linkage:

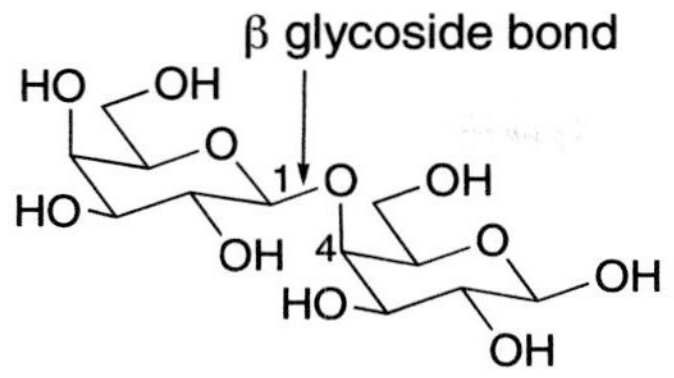

27.70 A disaccharide formed from two mannose units in a 1→4-α-glycosidic linkage:

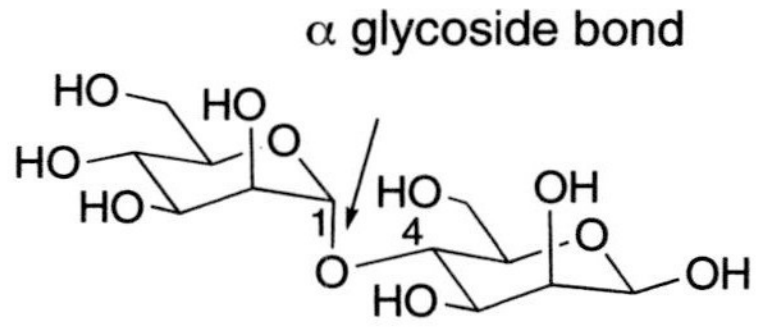

27.71

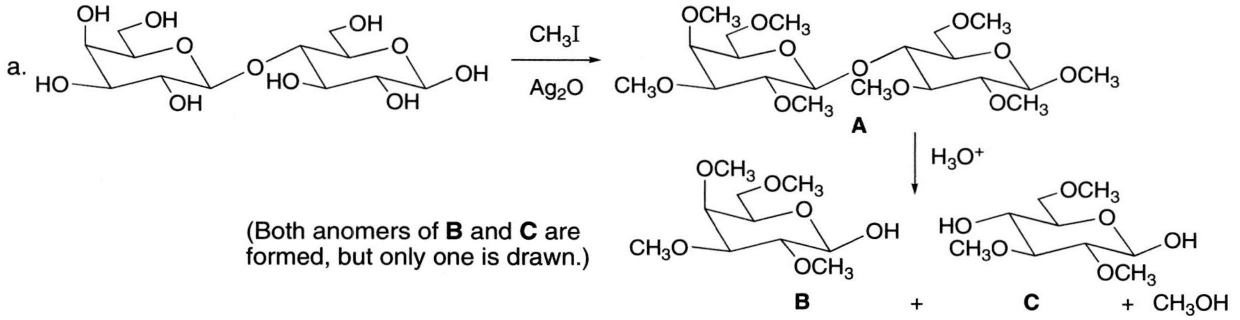

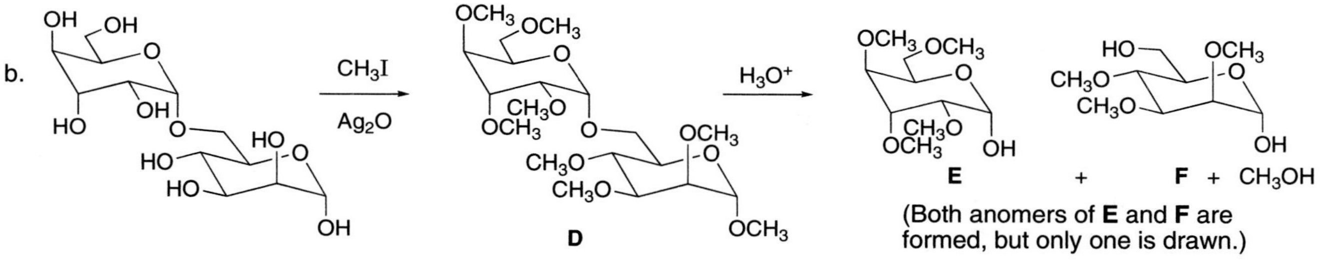

a.

(Both anomers of **B** and **C** are formed, but only one is drawn.)

b.

(Both anomers of **E** and **F** are formed, but only one is drawn.)

27.72

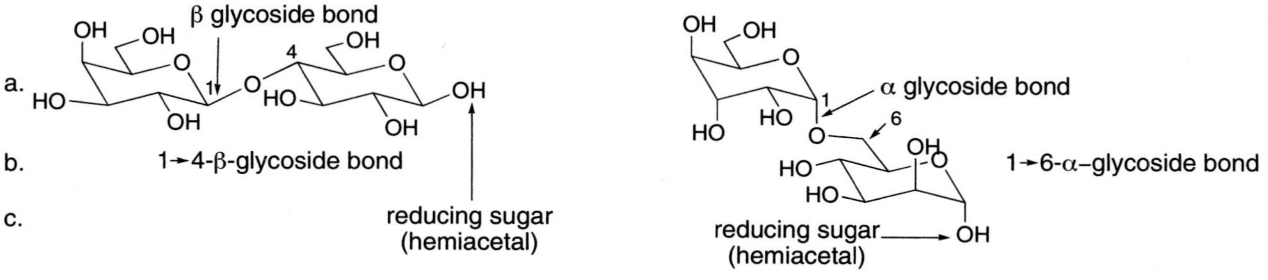

a.

b. 1→4-β-glycoside bond

c.

27.73

a and b.

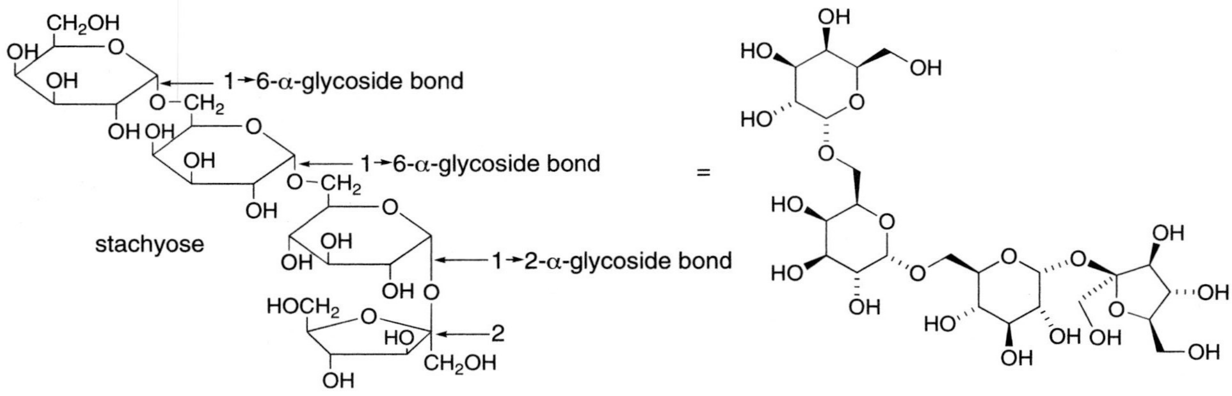

stachyose

c.

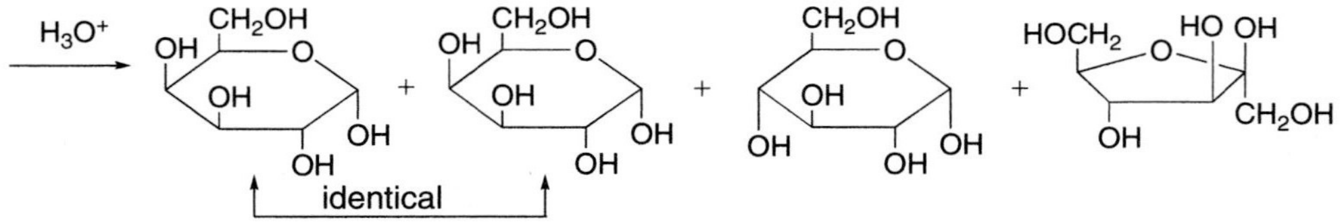

Two anomers of each monosaccharide are formed,
but only one anomer is drawn.

d. Stachyose is not a reducing sugar since it contains no hemiacetal.

e.

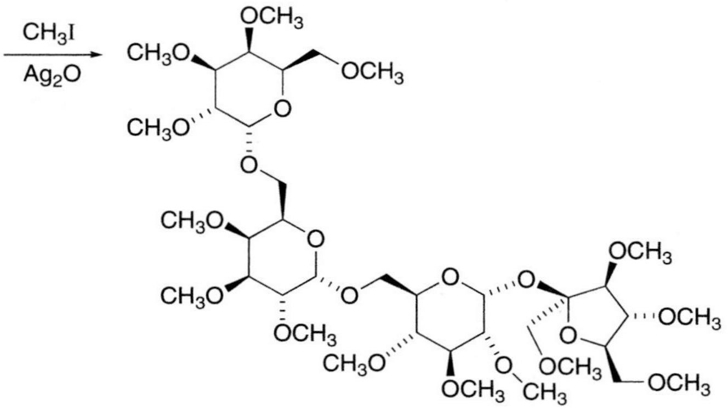

f. product in (e) $\xrightarrow{H_3O^+}$

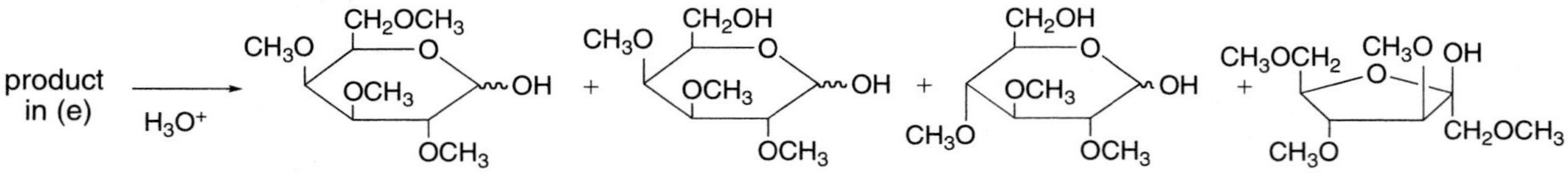

Two anomers of each monosaccharide are formed.

27.74

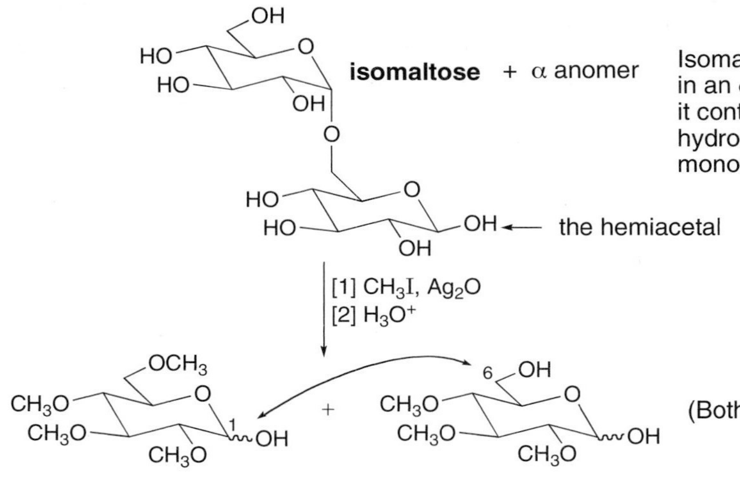

isomaltose + α anomer

the hemiacetal

[1] CH₃I, Ag₂O
[2] H₃O⁺

Isomaltose must be composed of two glucose units in an α-glycosidic linkage. Since it is a reducing sugar it contains a hemiacetal. The free OH groups in the hydrolysis products show where the two monosaccharides are joined.

(Both anomers are present.)

27.75

trehalose

CH₃I
Ag₂O

H₃O⁺

(both anomers)

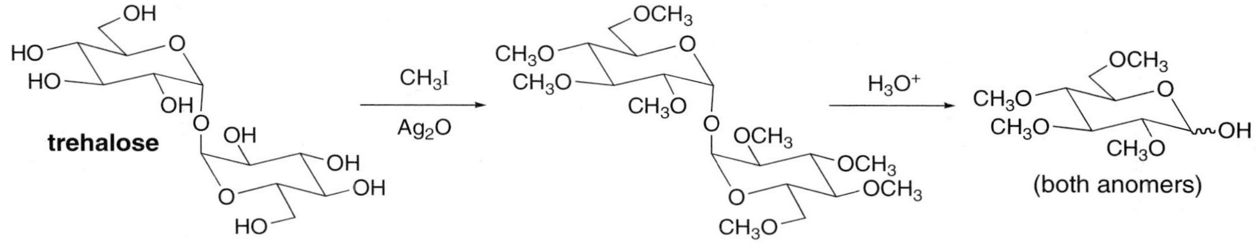

Trehalose must be composed of D-glucose units only, joined in an α-glycosidic linkage. Since trehalose is nonreducing it contains no hemiacetal. Since there is only one product formed after methylation and hydrolysis, the two anomeric C's must be joined.

27.76

a.

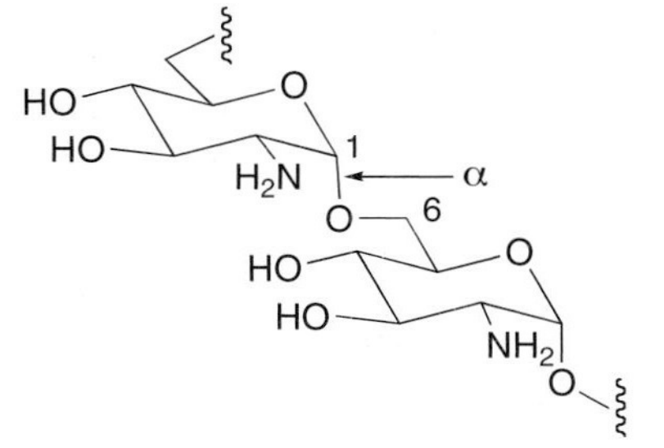

b.

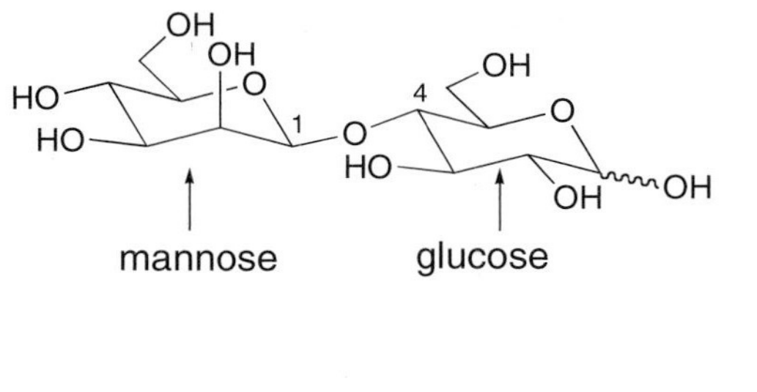

mannose   glucose

c.

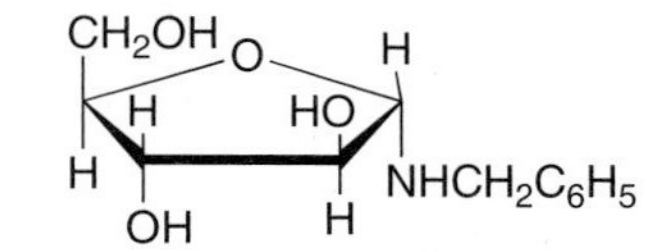

d.

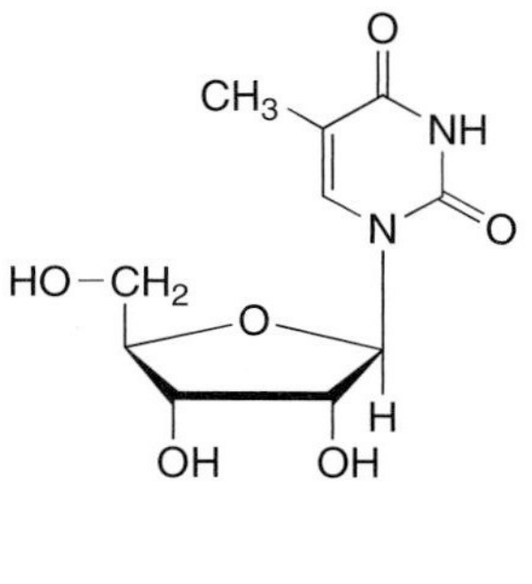

27.77

Ignoring stereochemistry along the way:

27.78 The hydrolysis data suggest that the trisaccharide has D-galactose on one end and D-fructose on the other. D-Galactose must be joined to its adjacent sugar by a β-glycosidic linkage. D-Fructose must be joined to its adjacent sugar by an α-glycosidic linkage.

2,3,4,6-tetra-*O*-methyl-D-galactose

2,3,4-tri-*O*-methyl-D-glucose

1,3,6-tri-*O*-methyl-D-fructose

(Both anomers of each compound are formed.)

Chapter 28: Amino Acids and Proteins

♦ Synthesis of amino acids (28.2)

[1] From α-halo carboxylic acids by S$_N$2 reaction

$$R-\underset{\underset{Br}{|}}{C}HCOOH \xrightarrow[\substack{\text{(large excess)}\\ \textbf{S}_\textbf{N}\textbf{2}}]{NH_3} R-\underset{\underset{NH_2}{|}}{C}HCOO^-NH_4^+ \quad + \quad NH_4^+ \, Br^-$$

[2] By alkylation of diethyl acetamidomalonate

$$\underset{CH_3}{\overset{O}{C}}-\overset{H}{N}-\underset{\underset{COOEt}{|}}{C}-COOEt \xrightarrow[\substack{[2]\ RX\\ [3]\ H_3O^+,\ \Delta}]{[1]\ NaOEt} H_2N-\underset{H}{\overset{R}{C}}-COOH$$

• Alkylation works best with unhindered alkyl halides—that is, CH$_3$X and RCH$_2$X.

[3] Strecker synthesis

$$\underset{R}{\overset{O}{C}}{}^{\diagdown}{}_H \xrightarrow[NaCN]{NH_4Cl} R-\underset{\underset{H}{|}}{\overset{NH_2}{C}}-CN \xrightarrow{H_3O^+} R-\underset{\underset{H}{|}}{\overset{NH_2}{C}}-COOH$$

α-amino nitrile

♦ Preparation of optically active amino acids

[1] Resolution of enantiomers by forming diastereomers (28.3A)
- Convert a racemic mixture of amino acids into a racemic mixture of *N*-acetyl amino acids [(*S*)- and (*R*)-CH$_3$CONHCH(R)COOH].
- React the enantiomers with a chiral amine to form a mixture of diastereomers.
- Separate the diastereomers.
- Regenerate the amino acids by protonation of the carboxylate salt and hydrolysis of the *N*-acetyl group.

[2] Kinetic resolution using enzymes (28.3B)

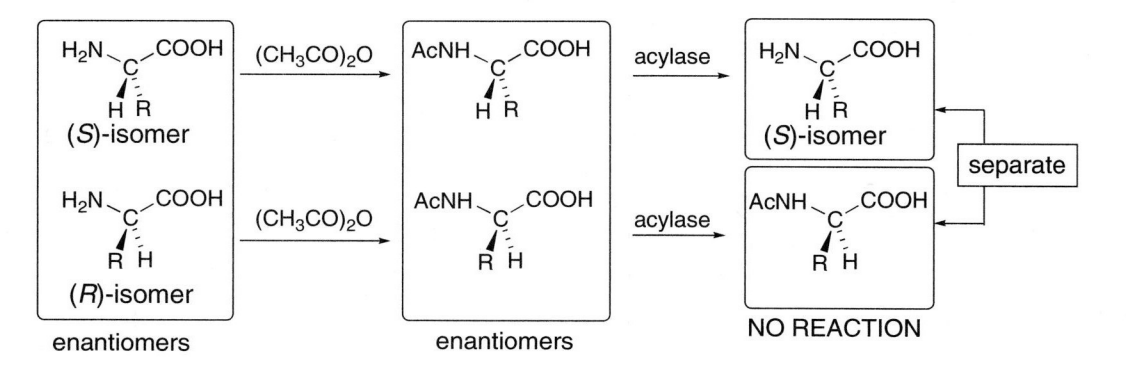

[3] By enantioselective hydrogenation (28.4)

Rh* = chiral Rh hydrogenation catalyst

◆ Summary of methods used for peptide sequencing (28.6)

- Complete hydrolysis of all amide bonds in a peptide gives the identity and amount of the individual amino acids.
- Edman degradation identifies the N-terminal amino acid. Repeated Edman degradations can be used to sequence a peptide from the N-terminal end.
- Cleavage with carboxypeptidase identifies the C-terminal amino acid.
- Partial hydrolysis of a peptide forms smaller fragments that can be sequenced. Amino acid sequences common to smaller fragments can be used to determine the sequence of the complete peptide.
- Selective cleavage of a peptide occurs with trypsin and chymotrypsin to identify the location of specific amino acids (Table 28.2).

◆ Adding and removing protecting groups for amino acids (28.7)

[1] Protection of an amino group as a Boc derivative

[2] Deprotection of a Boc-protected amino acid

[3] Protection of an amino group as an Fmoc derivative

[4] Deprotection of an Fmoc-protected amino acid

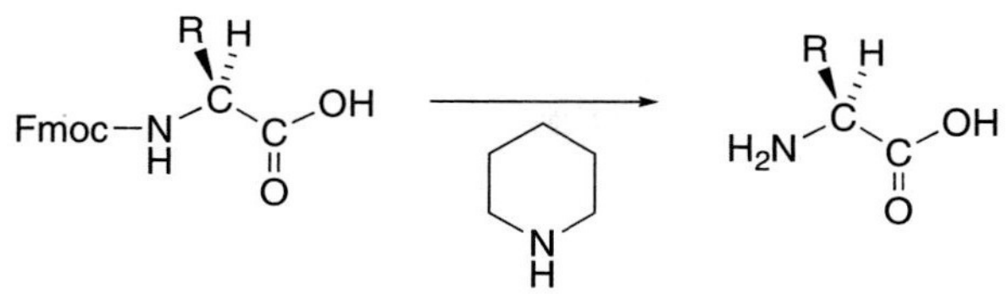

[5] Protection of a carboxy group as an ester

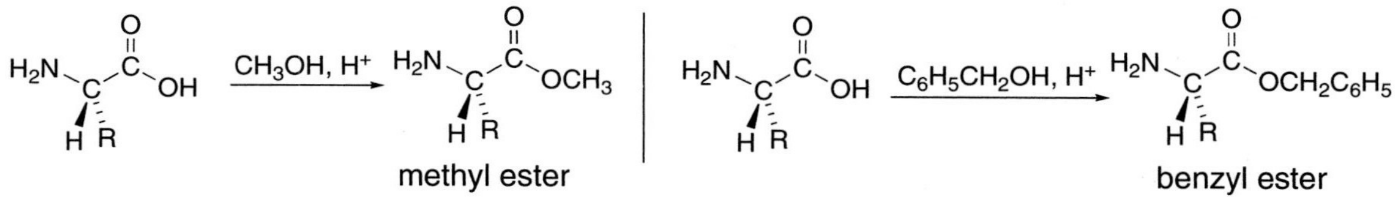

methyl ester

benzyl ester

[6] Deprotection of an ester group

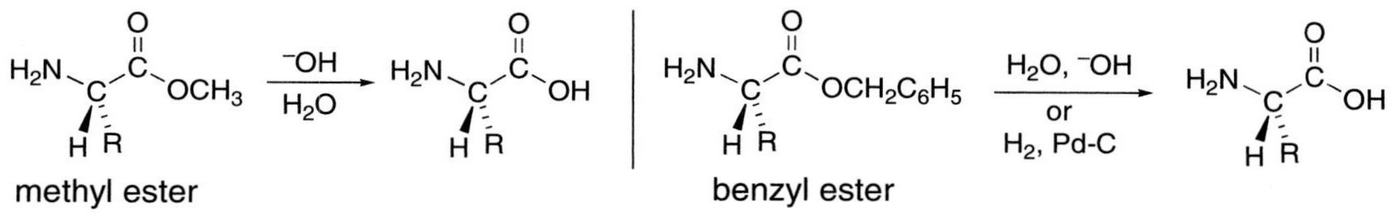

methyl ester

benzyl ester

◆ Synthesis of dipeptides (28.7)

[1] Amide formation with DCC

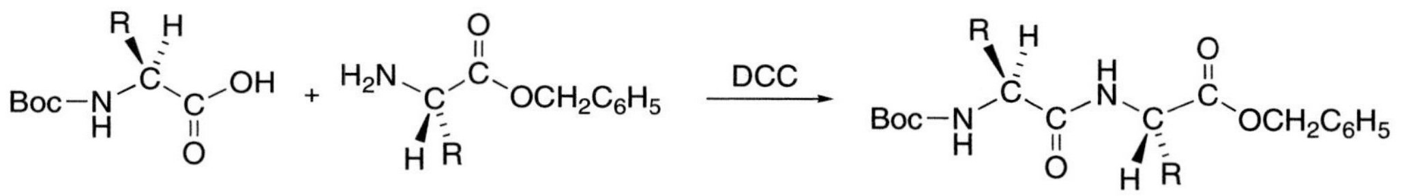

[2] Four steps are needed to synthesize a dipeptide:

a. **Protect** the amino group of one amino acid using a Boc or Fmoc group.
b. **Protect** the carboxy group of the second amino acid using an ester.
c. Form the amide bond with **DCC.**
d. **Remove both protecting groups** in one or two reactions.

◆ Summary of the Merrifield method of peptide synthesis (28.8)

[1] Attach an Fmoc-protected amino acid to a polymer derived from polystyrene.
[2] Remove the Fmoc protecting group.
[3] Form the amide bond with a second Fmoc-protected amino acid using DCC.
[4] Repeat steps [2] and [3].
[5] Remove the protecting group and detach the peptide from the polymer.

28.1

L-isoleucine

28.2

a.

NH_3^+

$(CH_3)_2CH-\overset{\underset{\displaystyle H}{|}}{\overset{\displaystyle |}{C}}-COO^-$

b.

NH_3^+

$(CH_3)_2CHCH_2-\overset{\underset{\displaystyle H}{|}}{\overset{\displaystyle |}{C}}-COO^-$

c.

d.

NH_3^+

$HOOCCH_2CH_2-\overset{\underset{\displaystyle H}{|}}{\overset{\displaystyle |}{C}}-COO^-$

28.3   In an amino acid, the electron-withdrawing carboxy group destabilizes the ammonium ion ($-NH_3^+$), making it more readily donate a proton; that is, it makes it a stronger acid. Also, the electron-withdrawing carboxy group removes electron density from the amino group ($-NH_2$) of the conjugate base, making it a weaker base than a 1° amine, which has no electron-withdrawing group.

28.4   The most direct way to synthesize an α-amino acid is by **S$_N$2 reaction of an α-halo carboxylic acid with a large excess of NH$_3$.**

a. $Br-\overset{\underset{\displaystyle H}{|}}{CH}-COOH$ $\xrightarrow[\text{large excess}]{NH_3}$ $H_2N-\overset{\underset{\displaystyle H}{|}}{CH}-COO^- \, NH_4^+$

glycine

c. $Br-\overset{\underset{\displaystyle CH_2}{|}}{CH}-COOH$ $\xrightarrow[\text{large excess}]{NH_3}$ $H_2N-\overset{\underset{\displaystyle CH_2}{|}}{CH}-COO^- \, NH_4^+$

b. $Br-\overset{\underset{\displaystyle CH-CH_3}{\underset{\underset{\displaystyle CH_3}{|}}{\underset{\displaystyle CH_2}{|}}}}{CH}-COOH$ $\xrightarrow[\text{large excess}]{NH_3}$ $H_2N-\overset{\underset{\displaystyle CH-CH_3}{\underset{\underset{\displaystyle CH_3}{|}}{\underset{\displaystyle CH_2}{|}}}}{CH}-COO^- \, NH_4^+$

isoleucine

phenylalanine

28.5

a. $\xrightarrow{CH_3I}$ $H_2N-\overset{\underset{\displaystyle H}{|}}{\overset{\overset{\displaystyle CH_3}{|}}{C}}-COOH$ alanine

c. $\xrightarrow{CH_3CH_2CH(CH_3)Br}$ $H_2N-\overset{\underset{\displaystyle H}{|}}{\overset{\overset{\displaystyle CH(CH_3)CH_2CH_3}{|}}{C}}-COOH$ isoleucine

b. $\xrightarrow{(CH_3)_2CHCH_2Cl}$ $H_2N-\overset{\underset{\displaystyle H}{|}}{\overset{\overset{\displaystyle CH_2CH(CH_3)_2}{|}}{C}}-COOH$ leucine

28.6

$\xrightarrow[\substack{[2]\ CH_2=O \\ [3]\ H_3O^+,\ \Delta}]{[1]\ NaOEt}$ $H_2N-\overset{\underset{\displaystyle H}{|}}{\overset{\overset{\displaystyle CH_2OH}{|}}{C}}-COOH$

serine

28.7

a. H₂N–CHCOOH ⟹ (valine) ⟹ (CH₃)₂CH–CHO

valine

b. H₂N–CHCOOH ⟹ (leucine) ⟹ (CH₃)₂CHCH₂–CHO

leucine

c. H₂N–CHCOOH ⟹ (phenylalanine) ⟹ C₆H₅CH₂–CHO

phenylalanine

28.8

a. BrCH₂COOH $\xrightarrow[\text{large excess}]{\text{NH}_3}$ NH₂CH₂COO⁻ NH₄⁺

c. CH₃CH₂CH(CH₃)CHO $\xrightarrow[\text{[2] H}_3\text{O}^+]{\text{[1] NH}_4\text{Cl, NaCN}}$ H₂N–CHCOOH | CH(CH₃)CH₂CH₃

b. CH₃CONH–C(H)(COOEt)COOEt $\xrightarrow[\text{[2] (CH}_3)_2\text{CHCl} \;\; \text{[3] H}_3\text{O}^+, \Delta]{\text{[1] NaOEt}}$ H₂N–C(H)(COOH)CH(CH₃)₂

d. CH₃CONH–C(H)(COOEt)COOEt $\xrightarrow[\text{[2] BrCH}_2\text{CO}_2\text{Et} \;\; \text{[3] H}_3\text{O}^+, \Delta]{\text{[1] NaOEt}}$ H₂N–C(H)(COOH)CH₂CO₂H

28.9 A chiral amine must be used to resolve a racemic mixture of amino acids.

a. C₆H₅CH₂CH₂NH₂
achiral

b. (structure)
achiral

c. CH₃CH₂–C(CH₃)(H)–NH₂
chiral
(can be used)

d. (structure)
chiral
(can be used)

28.10

To begin:
Convert the amino acids into *N*-acetyl amino acids (two enantiomers).

NH₂–C(COOH)(H)(CH₂CH(CH₃)₂) **S** + NH₂–C(COOH)((CH₃)₂CHCH₂)(H) **R**   **enantiomers**

↓ Ac₂O

AcNH–C(COOH)(H)(CH₂CH(CH₃)₂) **S** + AcNH–C(COOH)((CH₃)₂CHCH₂)(H) **R**   **enantiomers**

Step [1]:
React both enantiomers with the *R* isomer of the chiral amine.

proton transfer

H₂N–C(C₆H₅)(CH₃)(H)   (*R* isomer only)

AcNH–C(COO⁻)(H)(CH₂CH(CH₃)₂) **S**   H₃N⁺–C(C₆H₅)(CH₃)(H) **R** | AcNH–C(COO⁻)((CH₃)₂CHCH₂)(H) **R**   H₃N⁺–C(C₆H₅)(CH₃)(H) **R**   **diastereomers**

These salts have the *same* configuration around one stereogenic center,
but the *opposite* configuration about the other stereogenic center.

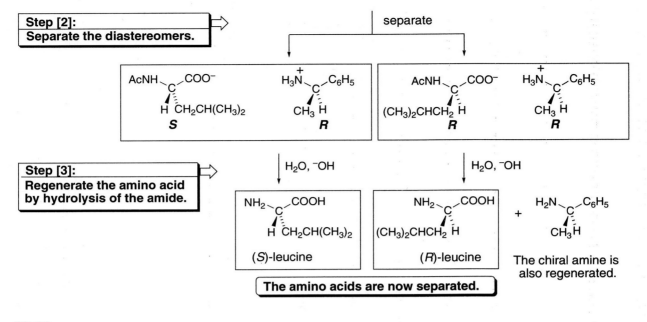

Step [2]:
Separate the diastereomers.

separate

Step [3]:
Regenerate the amino acid by hydrolysis of the amide.

(S)-leucine   (R)-leucine

The chiral amine is also regenerated.

The amino acids are now separated.

28.11

H₂N–C–H with COOH and CH₂CH(CH₃)₂
(mixture of enantiomers)

[1] (CH₃CO)₂O
[2] acylase

→

(S)-leucine   +   N-acetyl-(R)-leucine

28.12

a. H₂N–CHCOOH with CH₃ ⟹ C=C with H, NHAc, H, COOH

b. H₂N–CHCOOH with CH₂, CH-CH₃, CH₃ ⟹ (CH₃)₂CH, C=C, H, NHAc, COOH

c. H₂N–CHCOOH with CH₂, CH₂, CONH₂ ⟹ H₂NCOCH₂, C=C, H, NHAc, COOH

28.13 Draw the peptide by joining adjacent COOH and NH₂ groups in amide bonds.

a. H₂N–CH–C–OH with CH-CH₃ and CH₃ **Val**   H₂N–CH–C–OH with CH₂, CH₂, COOH **Glu** → Val–Glu

amide
N-terminal
C-terminal

b. H₂N–CH–C–OH with H **Gly**   H₂N–CH–C–OH with CH₂, and imidazole (NH, N) **His**   H₂N–CH–C–OH with CH₂, CH-CH₃, CH₃ **Leu** → **Gly–His–Leu**

amide
N-terminal
amide
C-terminal

c.

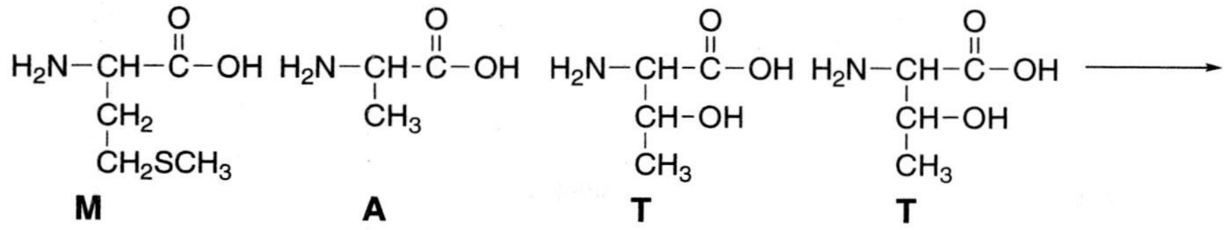

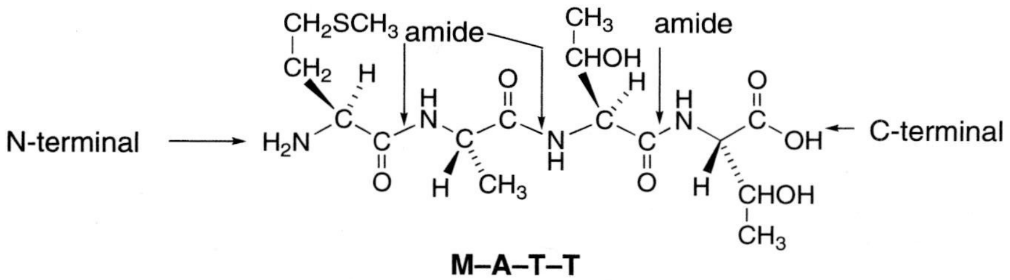

M A T T

M–A–T–T

28.14

a.

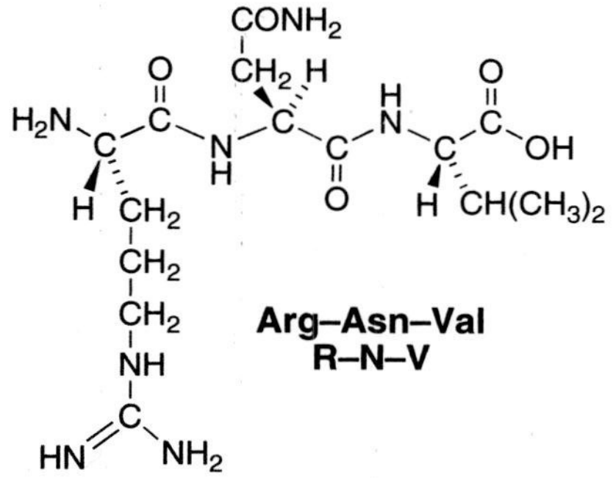

Arg–Asn–Val
R–N–V

b.

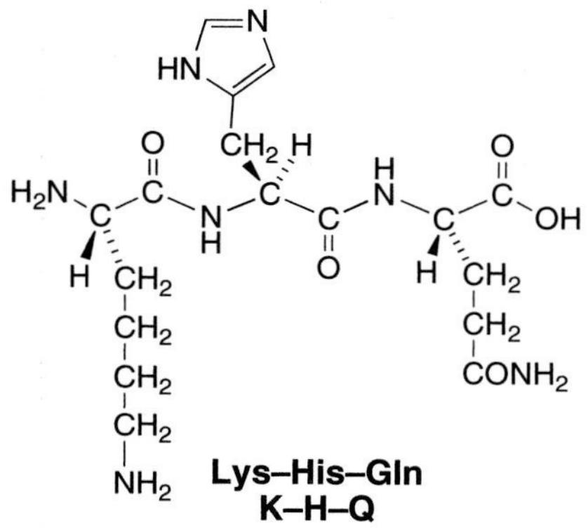

Lys–His–Gln
K–H–Q

28.15 There are six different tripeptides that can be formed from three amino acids (A, B, C): A–B–C, A–C–B, B–A–C, B–C–A, C–A–B, and C–B–A.

28.16 The *s*-trans conformation has the two C's oriented on *opposite* sides of the C–N bond. The *s*-cis conformation has the two C's oriented on the *same* side of the C–N bond.

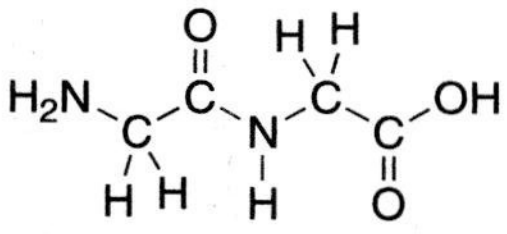

s-trans

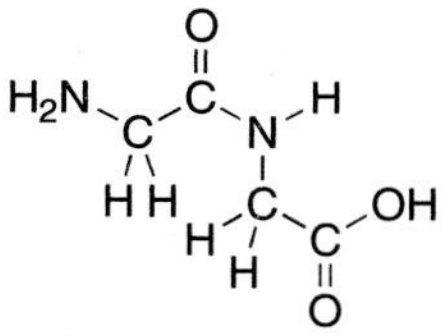

s-cis

28.17

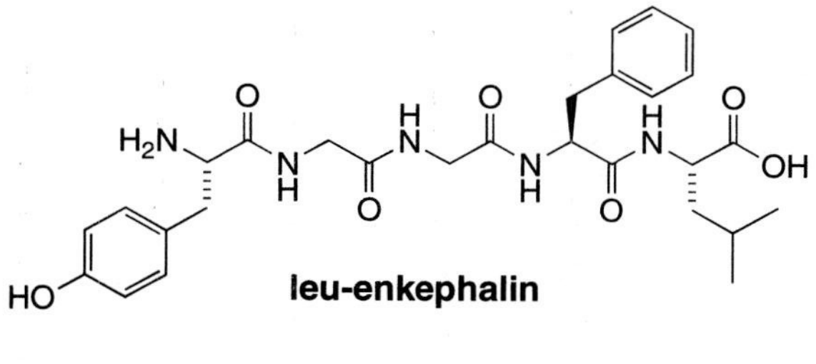

leu-enkephalin

28.18

a. glutathione →

glutathione

b.

The peptide bond beween glutamic acid and its adjacent amino acid (cysteine) is formed from the COOH in the R group of glutamic acid, not the α COOH.

α COOH   This comes from the amino acid glutamic acid.

This carboxy group is used to form the amide bond in the peptide, not the α COOH, as is usual. That's what makes glutathione's structure unusual.

glutamic acid

28.19

a.   from Ala

b.   from Val

28.20 Determine the sequence of the octapeptide as in Sample Problem 28.2. Look for overlapping sequences in the fragments.

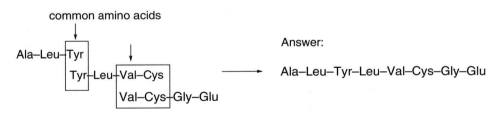

common amino acids

Ala–Leu–Tyr

Tyr–Leu–Val–Cys

Val–Cys–Gly–Glu

Answer:

Ala–Leu–Tyr–Leu–Val–Cys–Gly–Glu

28.21 Trypsin cleaves peptides at amide bonds with a carbonyl group from Arg and Lys. Chymotrypsin cleaves at amide bonds with a carbonyl group from Phe, Tyr, and Trp.

a. [1] Gly–Ala–Phe–Leu–Lys   +   Ala
    [2] Phe–Tyr–Gly–Cys–Arg   +   Ser
    [3] Thr–Pro–Lys   +   Glu–His–Gly–Phe–Cys–Trp–Val–Val–Phe
b. [1] Gly–Ala–Phe   +   Leu–Lys–Ala
    [2] Phe   +   Tyr   +   Gly–Cys–Arg–Ser
    [3] Thr–Pro–Lys–Glu–His–Gly–Phe   +   Cys–Trp   +   Val–Val–Phe

28.22

Edman degradation gives N-terminal amino acid: ⟶ Leu–___–___–___–___–___–___

Carboxypeptidase identifies the C-terminal amino acid: ⟶ Leu–___–___–___–___–___–Glu

Partial hydrolysis
common amino acids
↓

Ala–Ser–Arg

Gly–Ala–Ser

↓

Gly–Ala–Ser–Arg ⟶ Leu–Gly–Ala–Ser–Arg–___–Glu

or

Leu–___–Gly–Ala–Ser–Arg–Glu

Cleavage by trypsin is after Arg and yields a dipeptide; therefore, this must be the peptide: ⟶ Leu–Gly–Ala–Ser–Arg–Phe–Glu

28.23

a.

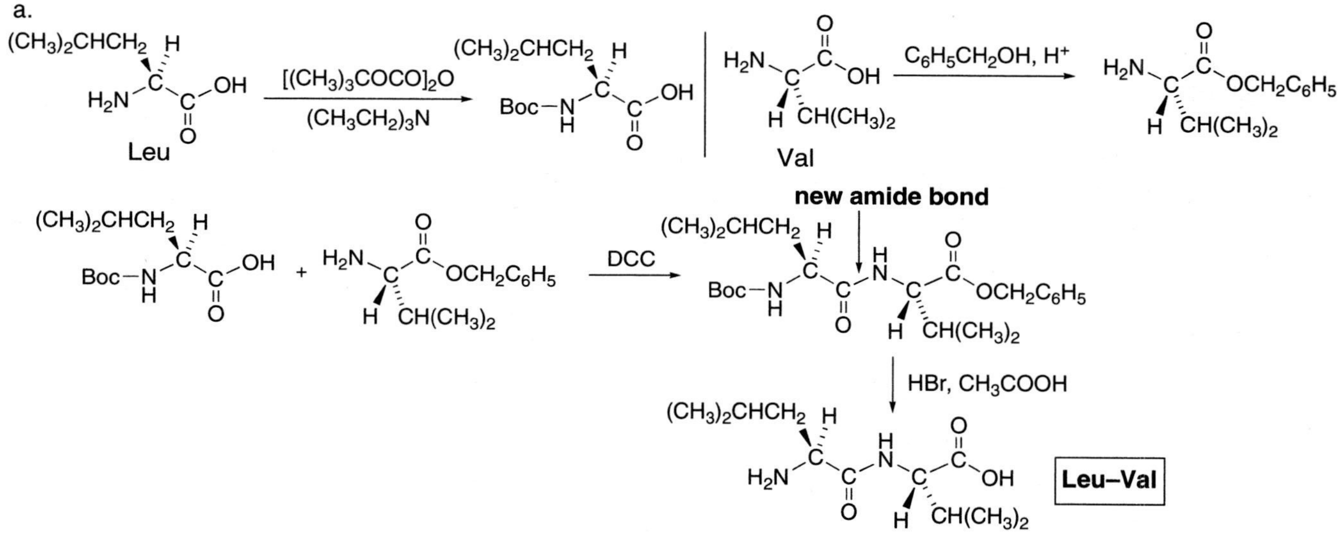

Leu $\xrightarrow[\text{(CH}_3\text{CH}_2)_3\text{N}]{[\text{(CH}_3)_3\text{COCO}]_2\text{O}}$ Boc–

Val $\xrightarrow[\text{H}^+]{\text{C}_6\text{H}_5\text{CH}_2\text{OH}}$

Boc– + H$_2$N– $\xrightarrow{\text{DCC}}$ Boc–

new amide bond

$\xrightarrow{\text{HBr, CH}_3\text{COOH}}$

Leu–Val

b.

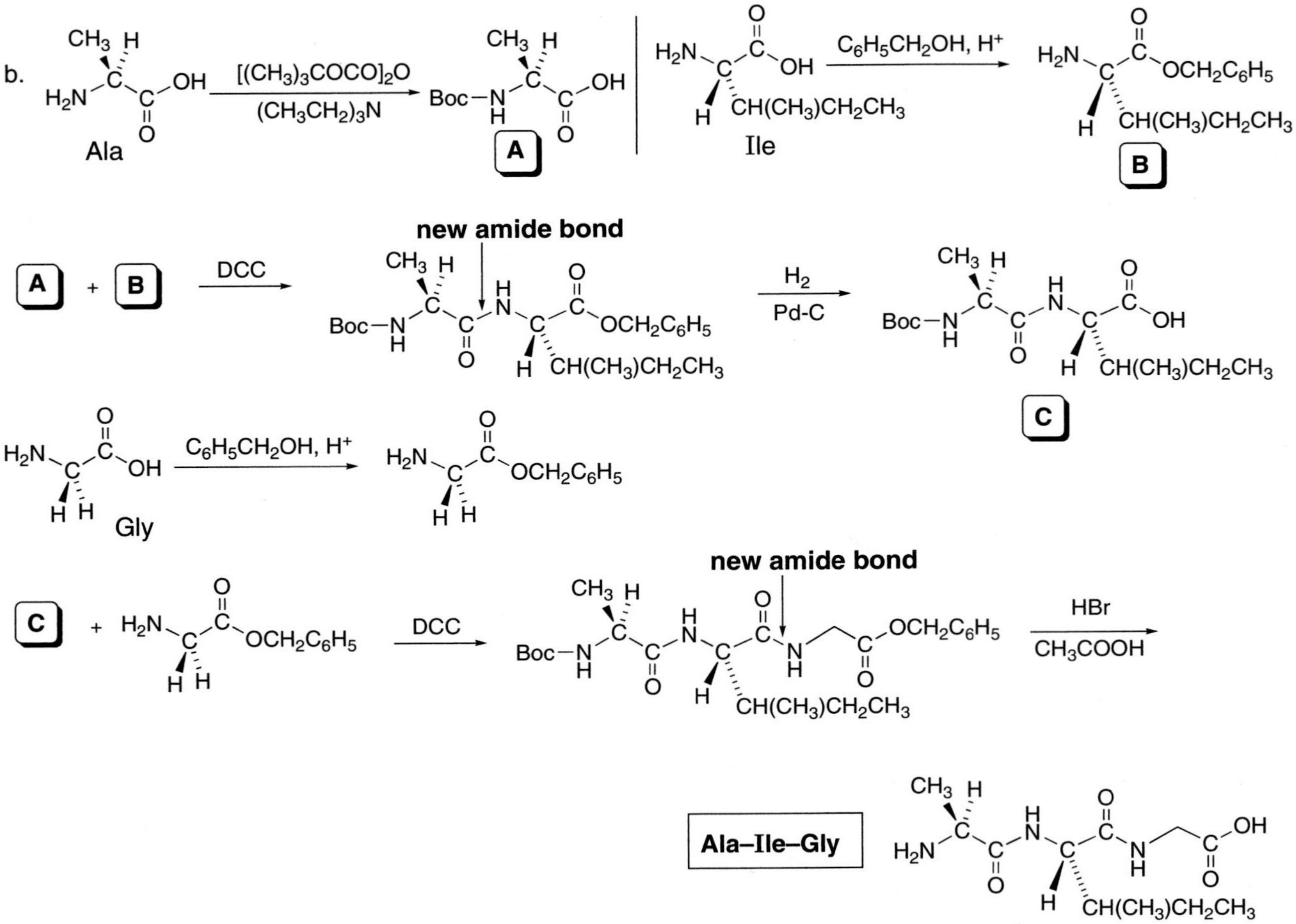

Ala → A

Ile → B

A + B → (DCC) → **new amide bond** → C (H₂/Pd-C)

Gly → benzyl ester

C + glycine benzyl ester → (DCC) → **new amide bond** → (HBr/CH₃COOH)

Ala–Ile–Gly

c.

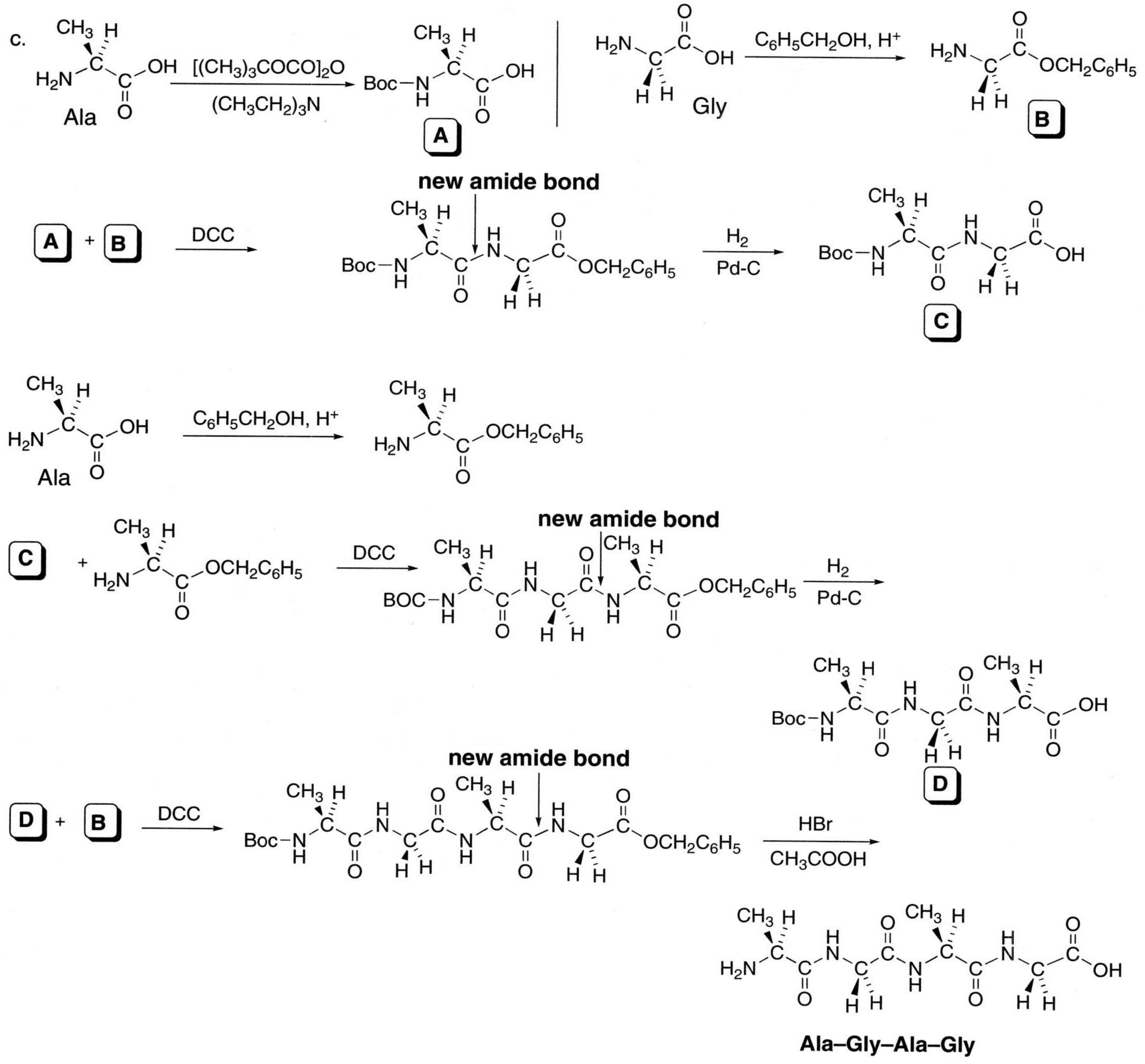
Ala
[(CH3)3COCO]2O
(CH3CH2)3N
A
Gly
C6H5CH2OH, H+
B
new amide bond
A + B
DCC
H2
Pd-C
C
Ala
C6H5CH2OH, H+
C
DCC
new amide bond
H2
Pd-C
D
D + B
DCC
new amide bond
HBr
CH3COOH
Ala–Gly–Ala–Gly

28.24

All Fmoc-protected amino acids are made by the following general reaction:

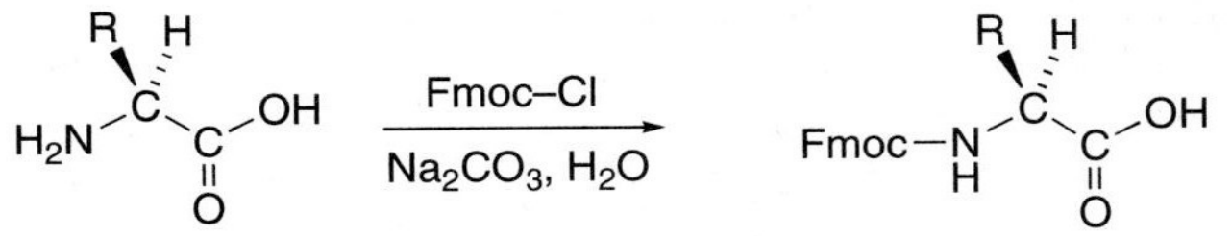

The steps:

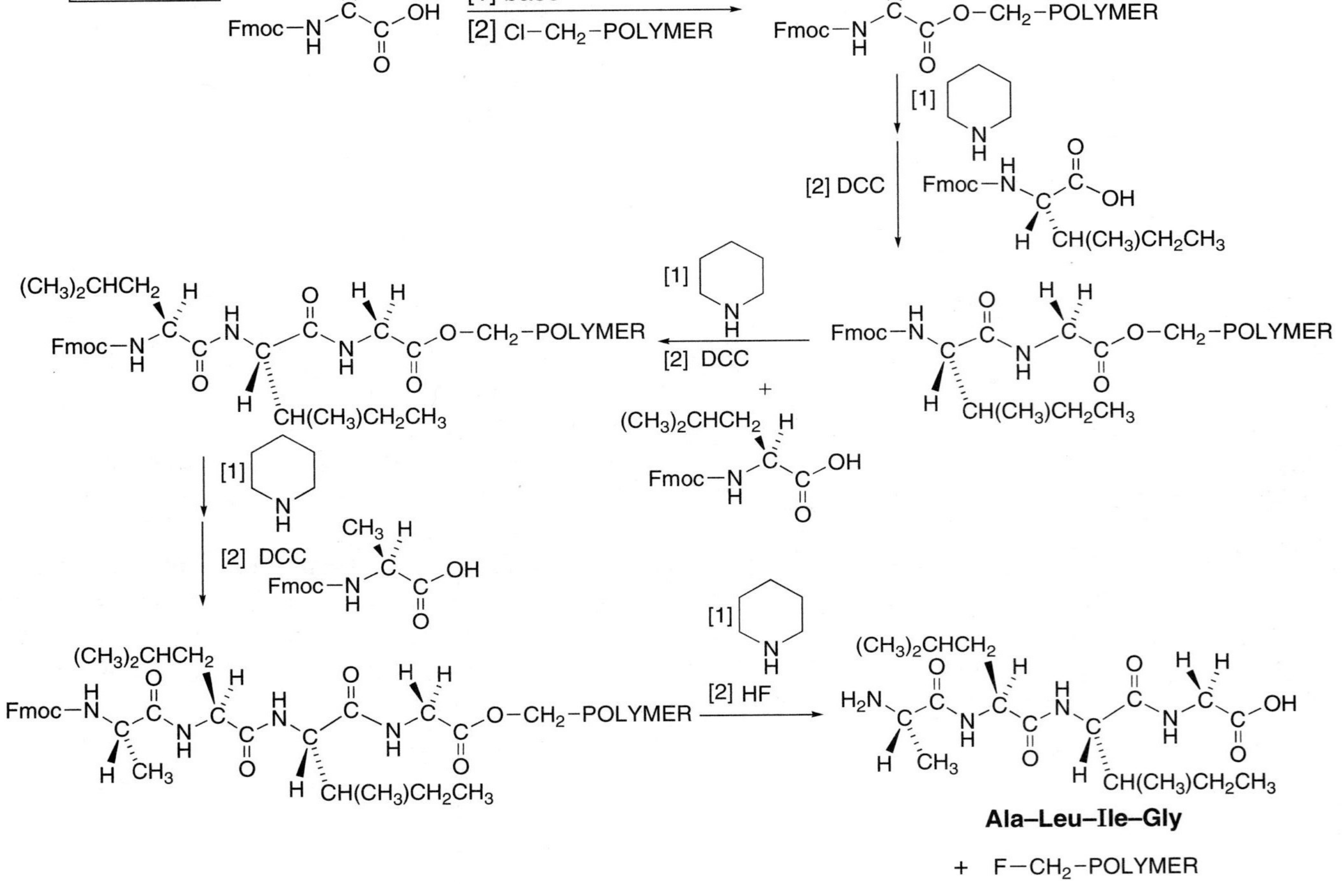

Ala–Leu–Ile–Gly

+ F–CH$_2$–POLYMER

28.25 Antiparallel β-pleated sheets are more stable then parallel β-pleated sheets because of geometry. The N–H and C=O of one chain are directly aligned with the N–H and C=O of an adjacent chain in the antiparallel β-pleated sheet, whereas they are not in the parallel β-pleated sheet. This makes the latter set of hydrogen bonds weaker.

28.26 In a *parallel* β-pleated sheet, the strands run in the *same* direction from the N- to C-terminal amino acid. In an *antiparallel* β-pleated sheet, the strands run in the *opposite* direction.

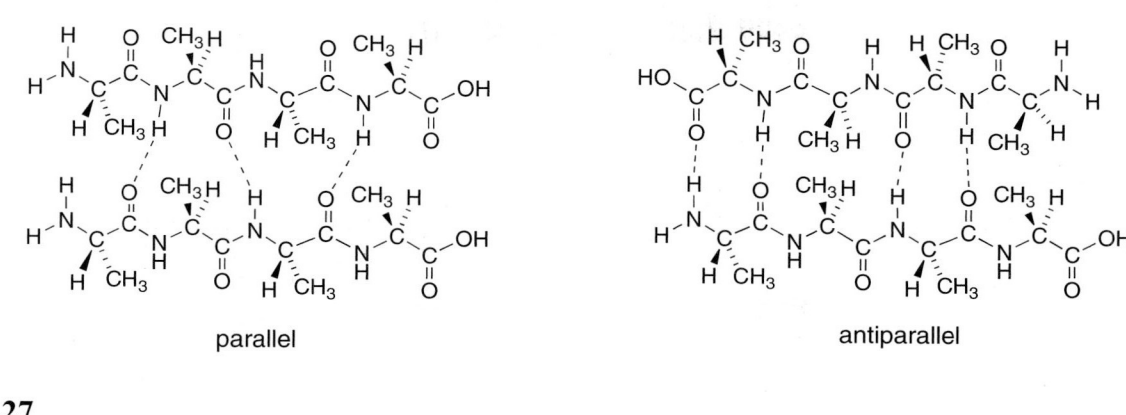

parallel   antiparallel

28.27

a. Ser and Tyr

H₂N–CH–COOH   H₂N–CH–COOH
     | |
    CH₂   CH₂
     |
    OH

side chains with
OH groups

hydrogen bonding

b. Val and Leu

H₂N–CH–COOH   H₂N–CH–COOH
     | |
   CH–CH₃   CH₂
     | |
    CH₃   CH–CH₃
                          |
                         CH₃

side chains with only
C–C and C–H bonds

van der Waals forces

c. 2 Phe residues

H₂N–CH–COOH   H₂N–CH–COOH
     | |
    CH₂   CH₂

van der Waals forces

28.28 a. The R group for glycine is a hydrogen. The R groups must be small to allow the β-pleated sheets to stack on top of each other. With large R groups, steric hindrance prevents stacking.

b. Silk fibers are water insoluble because most of the polar functional groups are in the interior of the stacked sheets. The β-pleated sheets are stacked one on top of another so few polar functional groups are available for hydrogen bonding to water.

28.29 All L-amino acids except cysteine have the **S configuration**. L-Cysteine has the *R* configuration because the R group contains a sulfur atom, which has higher priority.

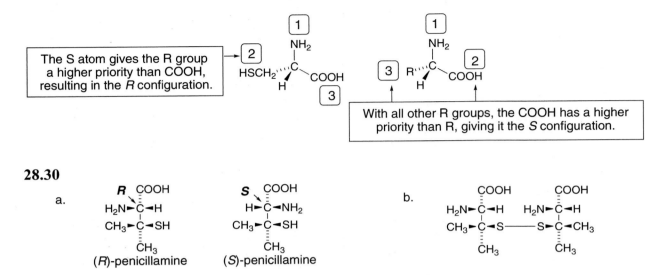

The S atom gives the R group a higher priority than COOH, resulting in the *R* configuration.

With all other R groups, the COOH has a higher priority than R, giving it the *S* configuration.

28.30

a. (*R*)-penicillamine   (*S*)-penicillamine

b.

28.31 Amino acids are insoluble in diethyl ether because amino acids are highly polar; they exist as salts in their neutral form. Diethyl ether is weakly polar, so amino acids are not soluble in it. *N*-Acetyl amino acids are soluble because they are polar but not salts.

amino acid, a salt
H_2O soluble and ether insoluble

N-acetyl amino acid
ether soluble

28.32 The electron pair on the N atom not part of a double bond is delocalized on the five-membered ring, making it less basic.

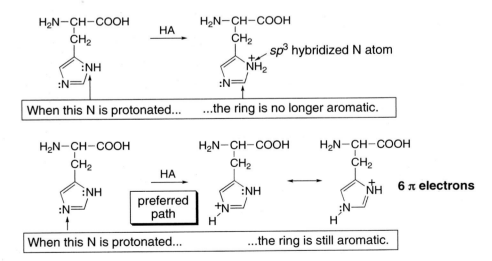

sp^3 hybridized N atom

When this N is protonated...   ...the ring is no longer aromatic.

preferred path

6π **electrons**

When this N is protonated...   ...the ring is still aromatic.

28.33

The ring structure on tryptophan is aromatic since each atom contains a *p* orbital. Protonation of the N atom would disrupt the aromaticity, making this a less favorable reaction.

no *p* orbital on N

This electron pair is delocalized on the bicyclic ring system (giving it 10 π electrons), making it less available for donation, and thus less basic.

28.34 At its isoelectric point, each amino acid is neutral.

a.

$$H_3N^+-\overset{\displaystyle COO^-}{\underset{\displaystyle CH_3}{\overset{|}{\underset{|}{C}}}-H}$$

alanine

b.

$$H_3N^+-\overset{\displaystyle COO^-}{\underset{\displaystyle CH_2CH_2SCH_3}{\overset{|}{\underset{|}{C}}}-H}$$

methionine

c.

$$H_3N^+-\overset{\displaystyle COO^-}{\underset{\displaystyle CH_2COOH}{\overset{|}{\underset{|}{C}}}-H}$$

aspartic acid

d.

$$H_2N-\overset{\displaystyle COO^-}{\underset{\displaystyle CH_2CH_2CH_2CH_2NH_3^+}{\overset{|}{\underset{|}{C}}}-H}$$

lysine

28.35

a. [1] glutamic acid: use the pK_a's $2.10 + 4.07$
[2] lysine: use the pK_a's $8.95 + 10.53$
[3] arginine: use the pK_a's $9.04 + 12.48$

b. In general, the p*I* of an acidic amino acid is lower than that of a neutral amino acid.

c. In general, the p*I* of a basic amino acid is higher than that of a neutral amino acid.

28.36

a. threonine
p*I* = 5.06
(+1) charge at pH = 1

$$H_3N^+-\overset{|}{\underset{|}{CH}}-COOH$$
$$\overset{|}{\underset{|}{CH}}-OH$$
$$CH_3$$

b. methionine
p*I* = 5.74
(+1) charge at pH = 1

$$H_3N^+-\overset{|}{\underset{|}{CH}}-COOH$$
$$CH_2$$
$$CH_2$$
$$S$$
$$CH_3$$

c. aspartic acid
p*I* = 2.98
(+1) charge at pH = 1

$$H_3N^+-\overset{|}{\underset{|}{CH}}-COOH$$
$$CH_2$$
$$COOH$$

d. arginine
p*I* = 5.41
(+2) charge at pH = 1

$$H_3N^+-\overset{|}{\underset{|}{CH}}-COOH$$
$$CH_2$$
$$CH_2$$
$$CH_2$$
$$NH$$
$$C=NH_2^+$$
$$NH_2$$

28.37

a. **valine**	b. **proline**	c. **glutamic acid**	d. **lysine**
$pI = 6.00$	$pI = 6.30$	$pI = 3.08$	$pI = 9.74$
(–1) charge	(–1) charge	(–2) charge	(–1) charge
at pH = 11	at pH = 11	at pH = 11	at pH = 11

a. valine structure:
$H_2N-CH-COO^-$
$\quad CH-CH_3$
$\quad CH_3$

b. proline structure:
COO^- (ring with HN)

c. glutamic acid structure:
$H_2N-CH-COO^-$
$\quad CH_2$
$\quad CH_2$
$\quad COO^-$

d. lysine structure:
$H_2N-CH-COO^-$
$\quad CH_2$
$\quad CH_2$
$\quad CH_2$
$\quad CH_2$
$\quad NH_2$

28.38 The terminal NH_2 and COOH groups are ionizable functional groups, so they can gain or lose protons in aqueous solution.

a. $H_2N-CH-C-OH$ (Ala) with CH_3 substituent and C=O; arrow to A–A–A tripeptide structure with H_2N, CH_3, H, amide linkages, and terminal OH.

Ala

A–A–A

b. At pH = 1

tripeptide structure with H_3N^+ terminal, CH_3, H substituents, amide linkages, terminal OH.

c. The pK_a of the COOH of the tripeptide is higher than the pK_a of the COOH group of alanine, making it less acidic. This occurs because the COOH group in the tripeptide is farther away from the $-NH_3^+$ group. The positively charged $-NH_3^+$ group stabilizes the negatively charged carboxylate anion of alanine more than the carboxylate anion of the tripeptide because it is so much closer in alanine. The opposite effect is observed with the ionization of the $-NH_3^+$ group. In alanine, the $-NH_3^+$ is closer to the COO^- group, so it is more difficult to lose a proton, resulting in a higher pK_a. In the tripeptide, the $-NH_3^+$ is farther away from the COO^-, so it is less affected by its presence.

28.39

leucine

H_2N — C — COOH, with H and $CH_2CH(CH_3)_2$ on the α-carbon

a. $\xrightarrow{CH_3OH, \ H^+}$ H_2N — C — $COOCH_3$ (H, $CH_2CH(CH_3)_2$)

b. $\xrightarrow{CH_3COCl, \ pyridine}$ CH_3 — C(=O) — NH — C — C(=O) — OH (H, $CH_2CH(CH_3)_2$)

c. $\xrightarrow{C_6H_5CH_2OH, \ H^+}$ H_2N — C — $COOCH_2C_6H_5$ (H, $CH_2CH(CH_3)_2$)

d. $\xrightarrow{Ac_2O, \ pyridine}$ AcNH — C — COOH (H, $CH_2CH(CH_3)_2$)

e. $\xrightarrow{HCl \ (1 \ equiv)}$ $H_3\overset{+}{N}$ — C — COOH (H, $CH_2CH(CH_3)_2$)

f. $\xrightarrow{NaOH \ (1 \ equiv)}$ H_2N — C — COO^- (H, $CH_2CH(CH_3)_2$)

g. $\xrightarrow{C_6H_5COCl, \ pyridine}$ C_6H_5 — C(=O) — NH — C — C(=O) — OH (H, $CH_2CH(CH_3)_2$)

h. $\xrightarrow[(CH_3CH_2)_3N]{[(CH_3)_3COCO]_2O}$ Boc — NH — C — COOH (H, $CH_2CH(CH_3)_2$)

i. The product in (d), then $NH_2CH_2COOCH_3$ + DCC

AcNH — C — C(=O) — NH — C — $COOCH_3$ (H, $CH_2CH(CH_3)_2$; H, H)

j. The product in (h), then $NH_2CH_2COOCH_3$ + DCC

Boc — NH — C — C(=O) — NH — C — $COOCH_3$ (H, $CH_2CH(CH_3)_2$; H, H)

k. $\xrightarrow{Fmoc-Cl, \ Na_2CO_3, \ H_2O}$ Fmoc — NH — C — COOH (H, $CH_2CH(CH_3)_2$)

l. $\xrightarrow{C_6H_5N=C=S}$ phenylthiohydantoin ring with C_6H_5 on N, =O, =S, NH, and $CH_2CH(CH_3)_2$ substituent

28.40

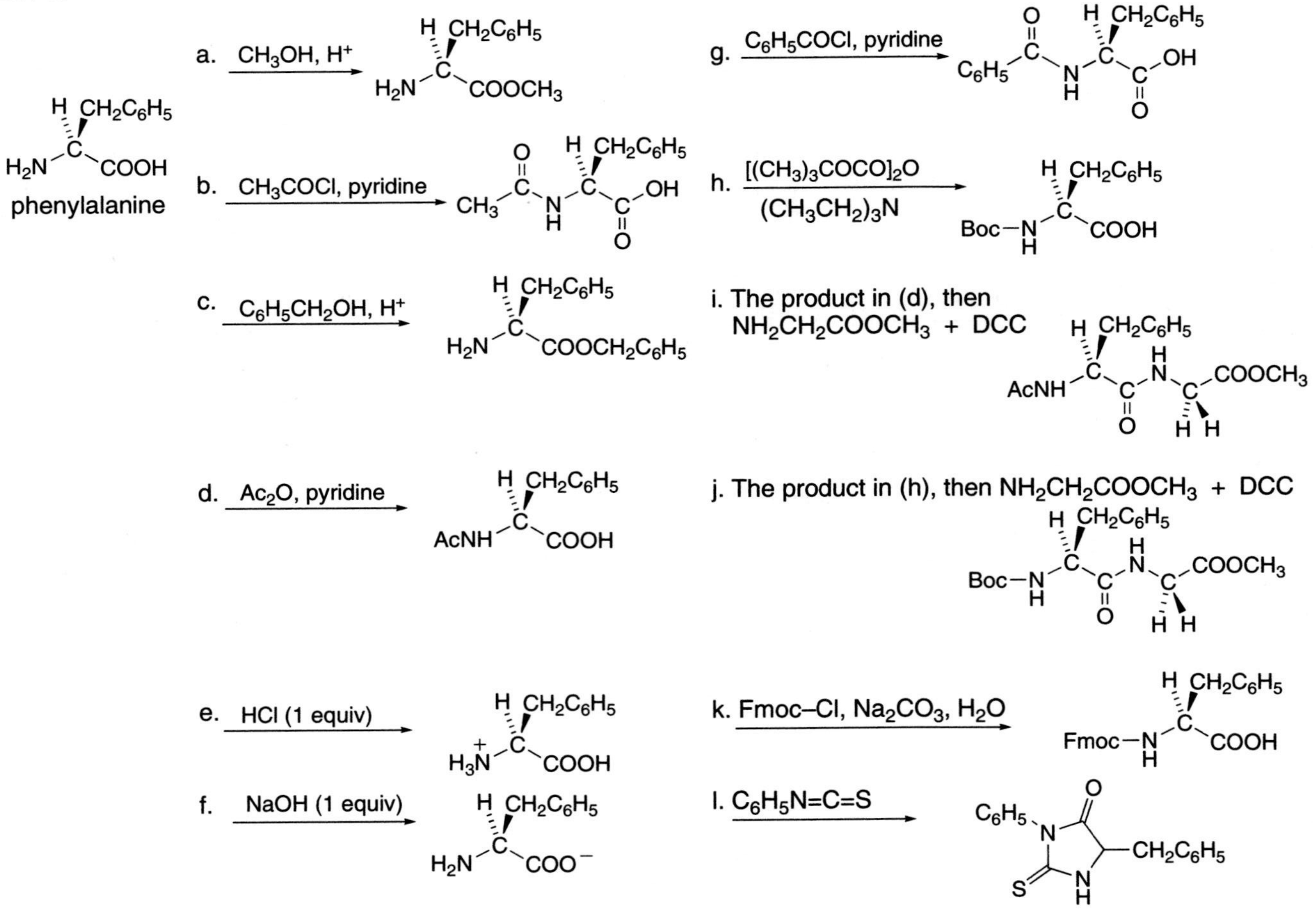

a. CH$_3$OH, H$^+$

b. CH$_3$COCl, pyridine

c. C$_6$H$_5$CH$_2$OH, H$^+$

d. Ac$_2$O, pyridine

e. HCl (1 equiv)

f. NaOH (1 equiv)

g. C$_6$H$_5$COCl, pyridine

h. [(CH$_3$)$_3$COCO]$_2$O / (CH$_3$CH$_2$)$_3$N

i. The product in (d), then NH$_2$CH$_2$COOCH$_3$ + DCC

j. The product in (h), then NH$_2$CH$_2$COOCH$_3$ + DCC

k. Fmoc–Cl, Na$_2$CO$_3$, H$_2$O

l. C$_6$H$_5$N=C=S

phenylalanine

28.41

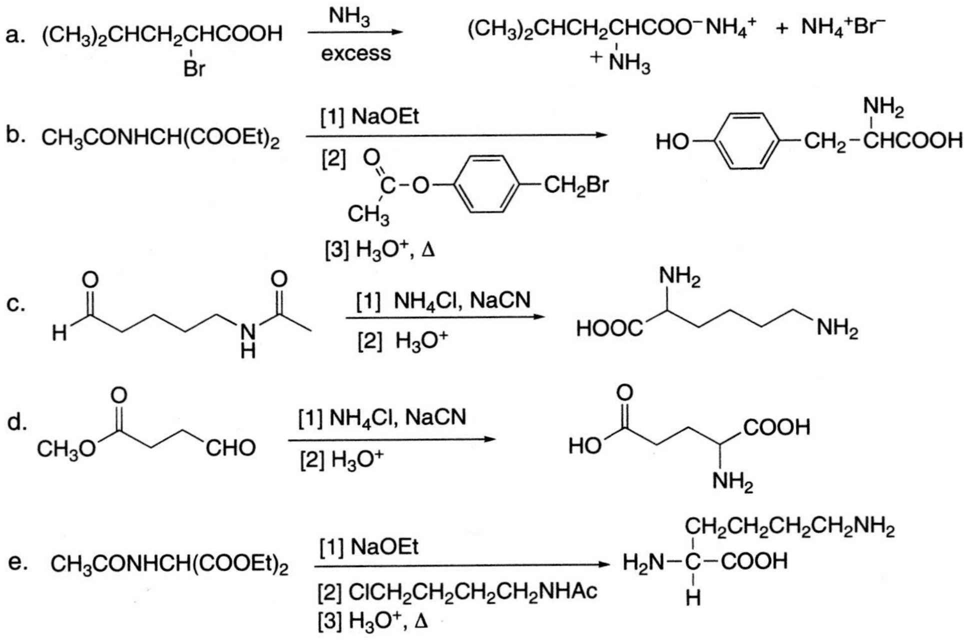

a. (CH$_3$)$_2$CHCH$_2$CHCOOH $\xrightarrow[\text{excess}]{\text{NH}_3}$ (CH$_3$)$_2$CHCH$_2$CHCOO$^-$NH$_4^+$ + NH$_4^+$Br$^-$
  |Br |$^+$NH$_3$

b. CH$_3$CONHCH(COOEt)$_2$ [1] NaOEt   [2] (image: acetate benzyl bromide)   [3] H$_3$O$^+$, Δ

c. (image: aldehyde amide)   [1] NH$_4$Cl, NaCN   [2] H$_3$O$^+$

d. (image: CH$_3$O ester CHO)   [1] NH$_4$Cl, NaCN   [2] H$_3$O$^+$

e. CH$_3$CONHCH(COOEt)$_2$ [1] NaOEt   [2] ClCH$_2$CH$_2$CH$_2$CH$_2$NHAc   [3] H$_3$O$^+$, Δ

28.42

a. Asn

$H_2N-CH-C(=O)-OH$ with side chain CH_2, $C=O$, NH_2 $\Longrightarrow$ CH_2Br, $C=O$, NH_2

b. His

$H_2N-CH-C(=O)-OH$ with side chain CH_2 attached to imidazole ring $\Longrightarrow$ CH_2Br attached to imidazole ring

c. Trp

$H_2N-CH-C(=O)-OH$ with side chain CH_2 attached to indole ring $\Longrightarrow$ CH_2Br attached to indole ring

28.43

$CH_3-C(=O)-N(H)-C(H)(COOEt)(COOEt)$

$\xrightarrow{\text{[1] NaOEt} \quad \text{[2] CH}_3\text{CHO} \quad \text{[3] H}_3\text{O}^+,\ \Delta}$

$H_2N-C(H)(CH(OH)CH_3)-COOH$
threonine

28.44

a. $CH_3CHO \xrightarrow[\text{CH}_3\text{COOH}]{\text{Br}_2} BrCH_2CHO \xrightarrow[\text{H}_2\text{SO}_4,\ \text{H}_2\text{O}]{\text{CrO}_3} BrCH_2COOH \xrightarrow[\text{excess}]{\text{NH}_3} \overset{+}{H_3N}CH_2COO^-$
glycine

b. $CH_3-C(=O)-H \xrightarrow[\text{[2] H}_3\text{O}^+]{\text{[1] NH}_4\text{Cl, NaCN}} CH_3-C(NH_2)(H)-COOH$
alanine

28.45

$CH_2(COOEt)_2 \xrightarrow[\text{CH}_3\text{COOH}]{\text{Br}_2} Br-CH(COOEt)_2$
A

$\xrightarrow{\text{potassium phthalimide}}$ phthalimide$-N-CH(COOEt)_2$
B

$\xrightarrow{\text{[1] NaOEt} \quad \text{[2] ClCH}_2\text{CH}_2\text{SCH}_3}$

phthalimide$-N-C(COOEt)(COOEt)(CH_2CH_2SCH_3)$
C

$\xrightarrow[\text{[2] H}_3\text{O}^+,\ \Delta]{\text{[1] NaOH, H}_2\text{O}}$

$H_2N-C(H)(CH_2CH_2SCH_3)-COOH$
D

28.46

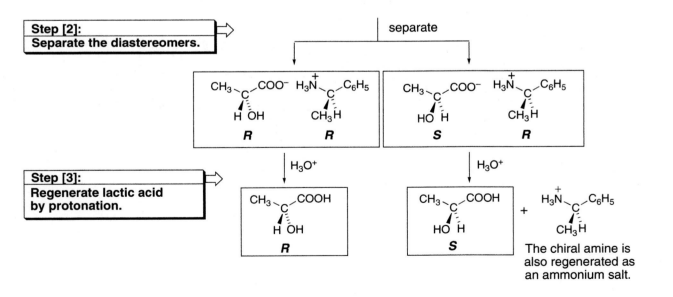

28.47

CH₃ ─ C ─ COOH (R)   CH₃ ─ C ─ COOH (S)   **enantiomers**

Step [1]:
React both enantiomers with the R isomer of the chiral amine.

proton transfer

H₂N ─ C ─ C₆H₅   (R isomer only)

diastereomers

These salts have the *same* configuration around one stereogenic center, but the *opposite* configuration about the other stereogenic center.

Step [2]:
Separate the diastereomers.

separate

Step [3]:
Regenerate lactic acid by protonation.

The chiral amine is also regenerated as an ammonium salt.

28.48

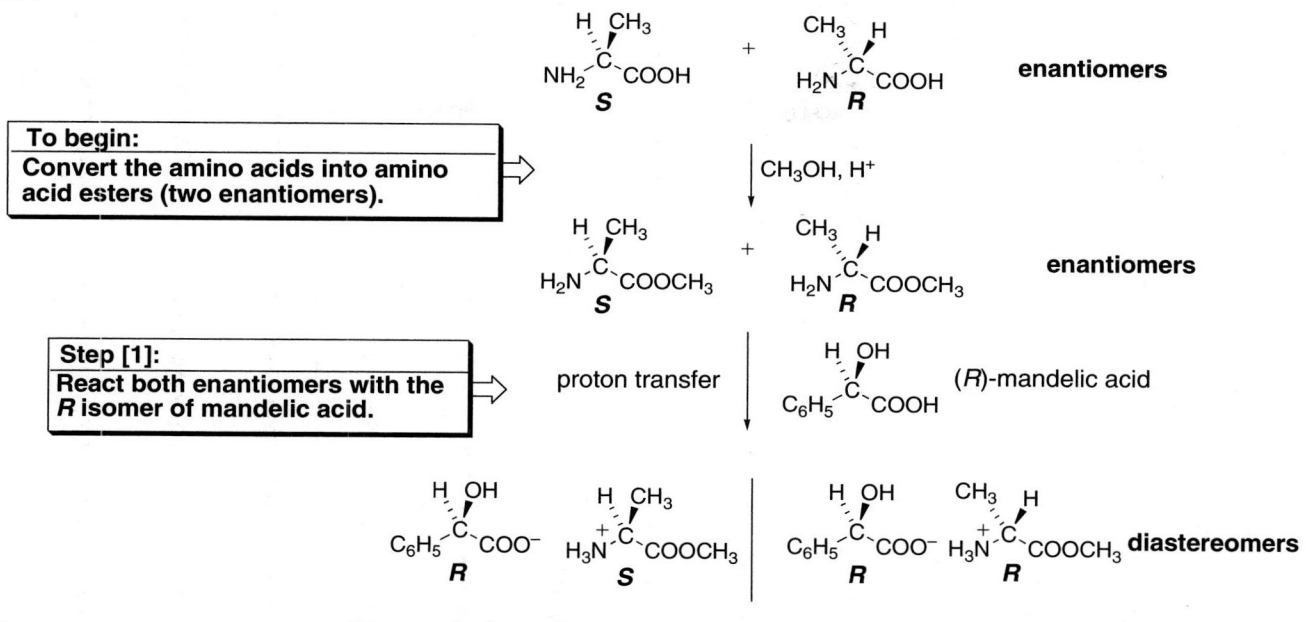

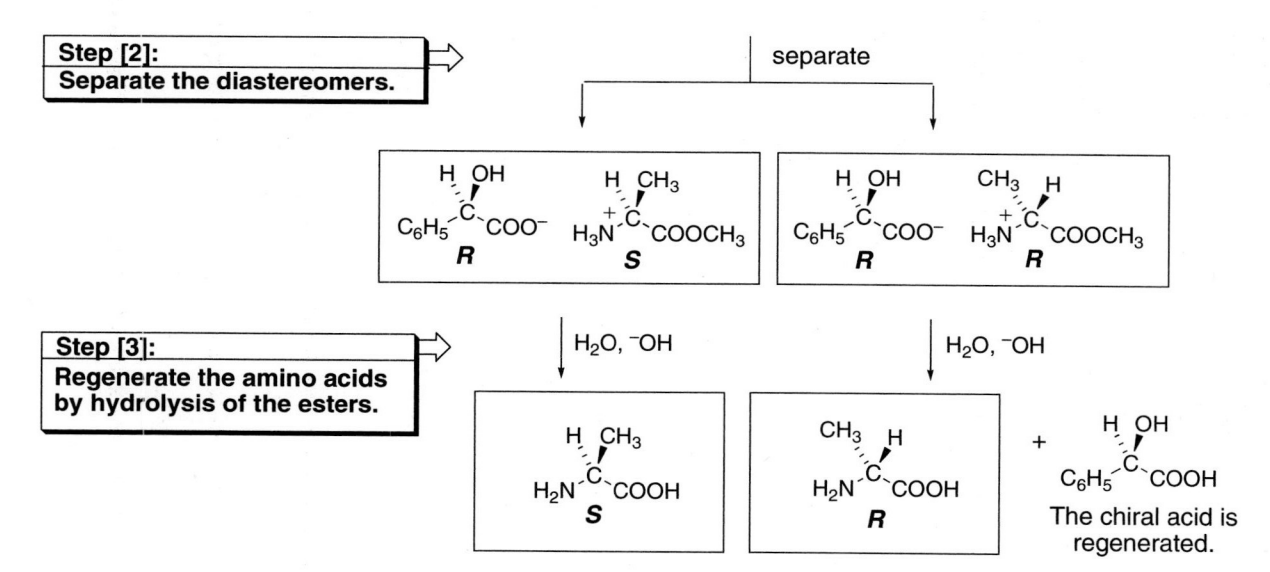

28.49

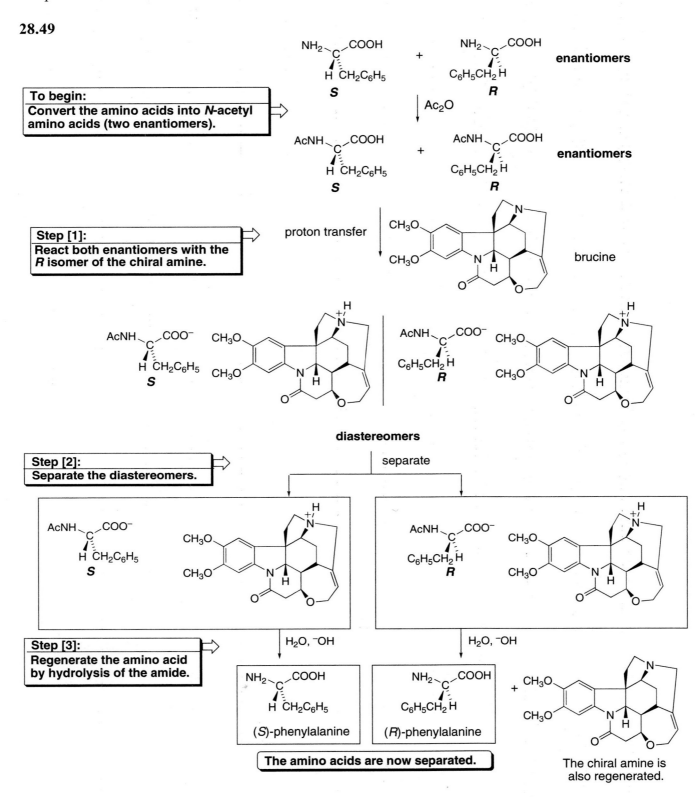

To begin:
Convert the amino acids into *N*-acetyl amino acids (two enantiomers).

enantiomers

enantiomers

Step [1]:
React both enantiomers with the *R* isomer of the chiral amine.

proton transfer

brucine

diastereomers

Step [2]:
Separate the diastereomers.

separate

Step [3]:
Regenerate the amino acid by hydrolysis of the amide.

(*S*)-phenylalanine

(*R*)-phenylalanine

The amino acids are now separated.

The chiral amine is also regenerated.

28.50

a.

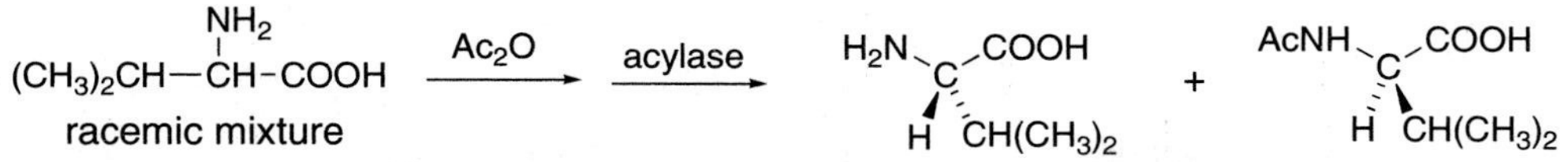

b.

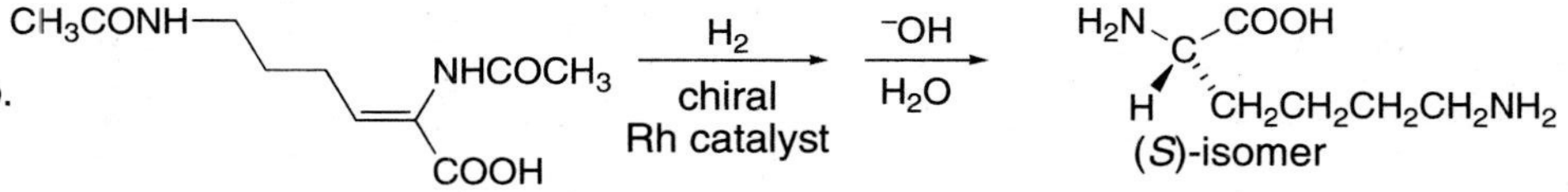

c.

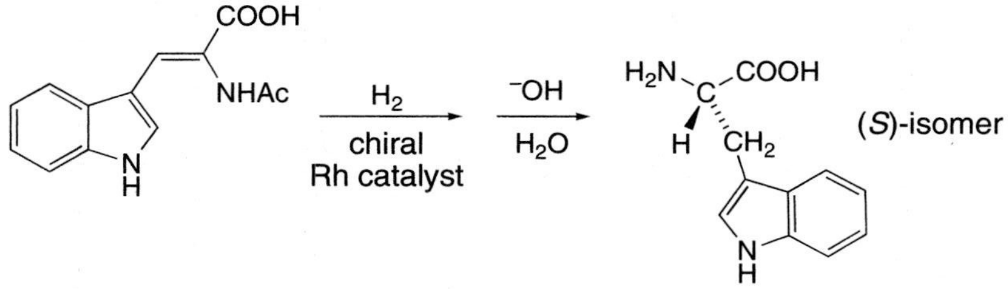

28.51

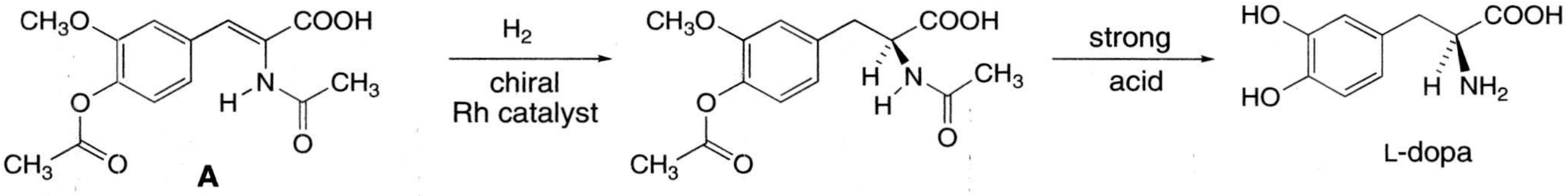

28.52

a.

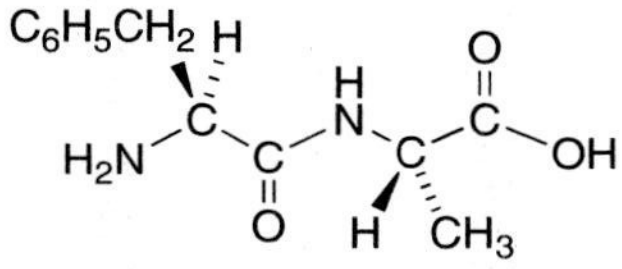

Phe–Ala

b.

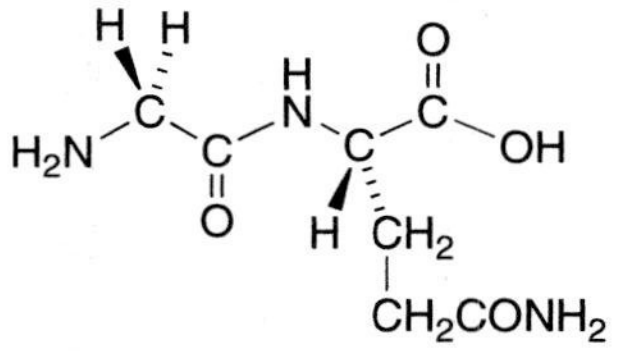

Gly–Gln

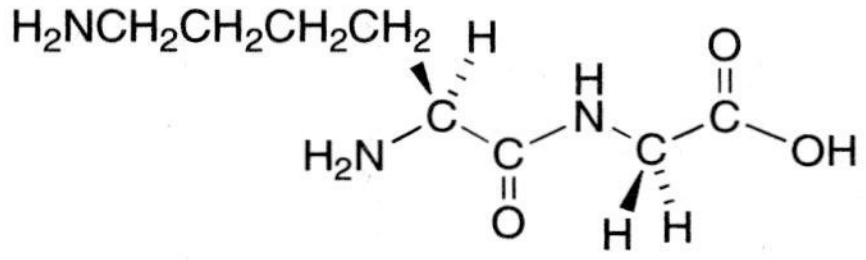

c.

Lys–Gly

d.

R–H

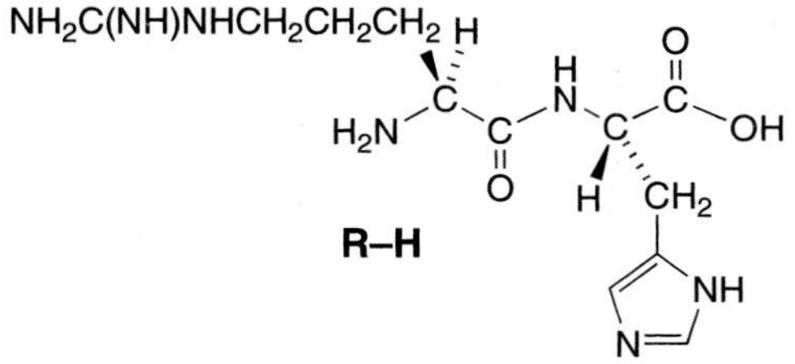

28.53 Amide bonds are bold lines (not wedges).

[1]

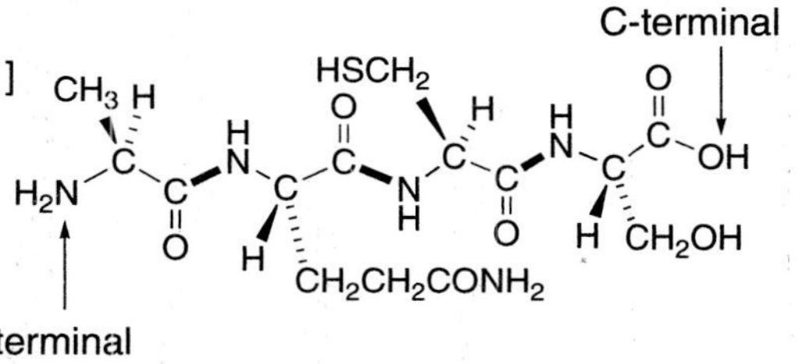

Ala–Gln–Cys–Ser
A–Q–C–S

[2]

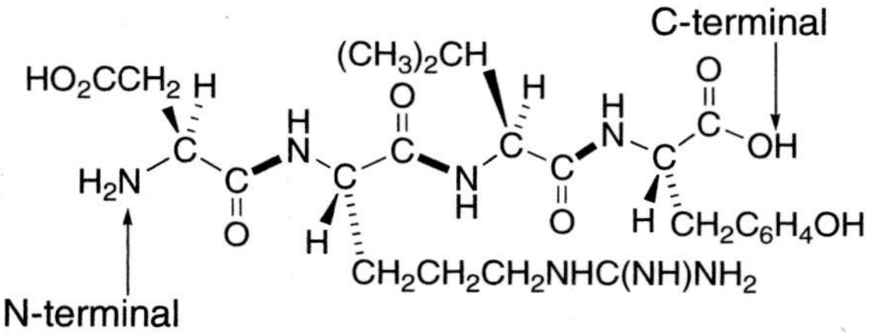

Asp–Arg–Val–Tyr
D–R–V–Y

28.54 Name a peptide from the N-terminal to the C-terminal end.

a.

CH₂COOH

Gly–Asp–Glu
G–D–E

b.

C-terminal   N-terminal

HOOC

Ala–Gly–Arg
A–G–R

28.55 A peptide C–N bond is stronger than an ester C–O bond because the C–N bond has more double bond character due to resonance. Since N is more basic than O, an amide C–N bond is more stabilized by delocalization of the lone pair on N.

A   B

Structure **B** contributes greatly to the resonance hybrid and this shortens and strengthens the C–N bond.

28.56 Use the principles from Answer 28.16.

s-trans

s-cis

28.57

a. A–P–F + L–K–W + S–G–R–G
b. A–P–F–L–K + W–S–G–R + G
c. A–P–F–L–K–W–S–G–R + G
d. A + P–F–L–K–W–S–G–R–G

28.58

a.

common amino acids

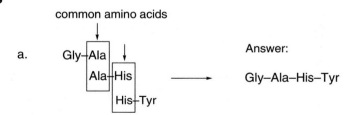

Answer:

Gly–Ala–His–Tyr

common amino acids

b.   Lys┼His
         His┼Gly–Glu
             Gly–Glu┼Phe ⟶

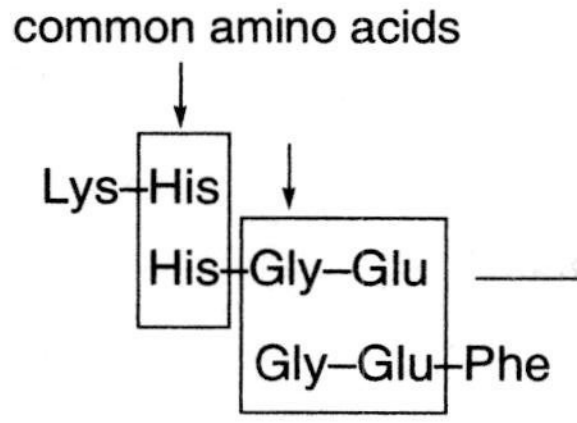

Answer:

Lys–His–Gly–Glu–Phe

28.59

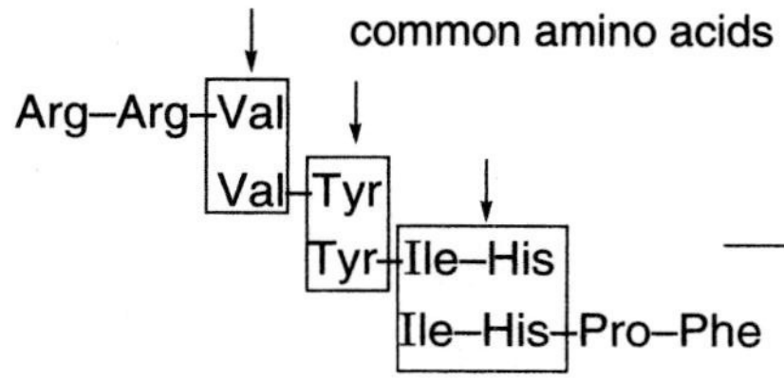

common amino acids

Arg–Arg┼Val
            Val┼Tyr
               Tyr┼Ile–His
                   Ile–His┼Pro–Phe ⟶

Answer:

Arg–Arg–Val–Tyr–Ile–His–Pro–Phe

28.60 Gly is the N-terminal amino acid (from Edman degradation), and Leu is the C-terminal amino acid (from treatment with carboxypeptidase). Partial hydrolysis gives the rest of the sequence.

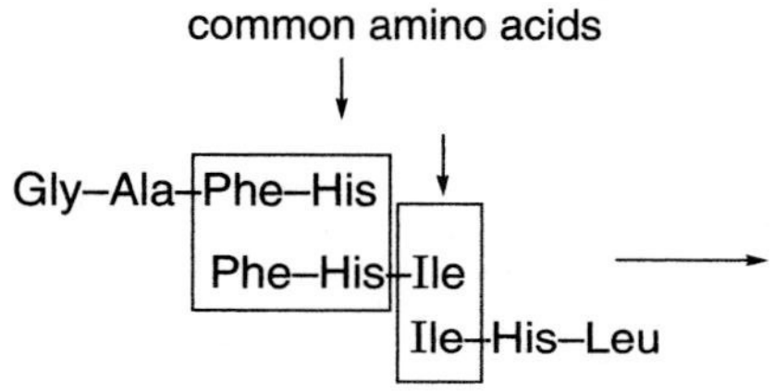

common amino acids

Gly–Ala┼Phe–His
            Phe–His┼Ile
                   Ile┼His–Leu ⟶

Answer:

Gly–Gly–Ala–Phe–His–Ile–His–Leu

28.61 Edman degradation data give the N-terminal amino acid for the octapeptide and all smaller peptides.

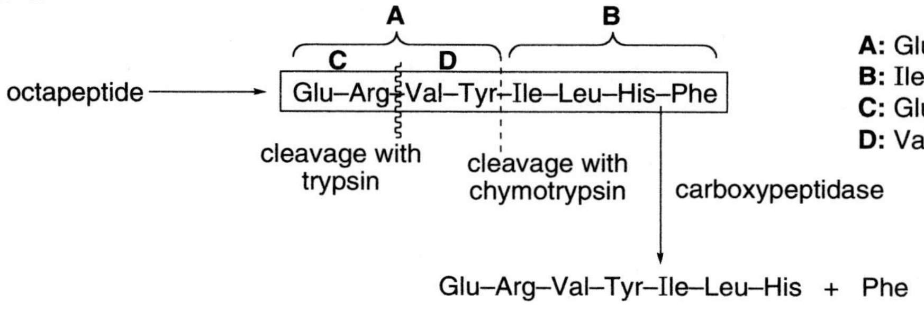

octapeptide ⟶ Glu–Arg⌇Val–Tyr┊Ile–Leu–His–Phe

A: Glu–Arg–Val–Tyr
B: Ile–Leu–His–Phe
C: Glu–Arg
D: Val–Tyr

cleavage with trypsin

cleavage with chymotrypsin

carboxypeptidase

Glu–Arg–Val–Tyr–Ile–Leu–His + Phe

28.62 A and B can react to form an amide, or two molecules of B can form an amide.

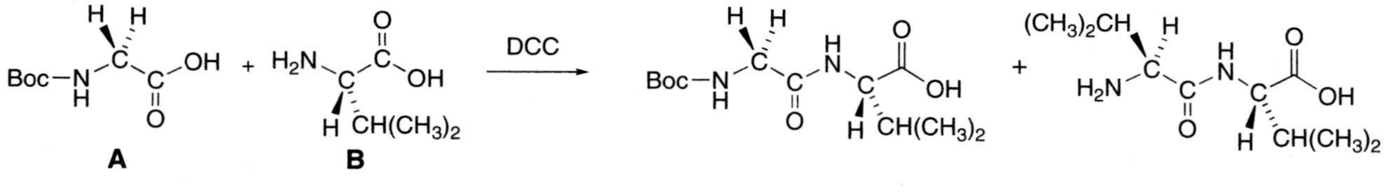

28.63

a.

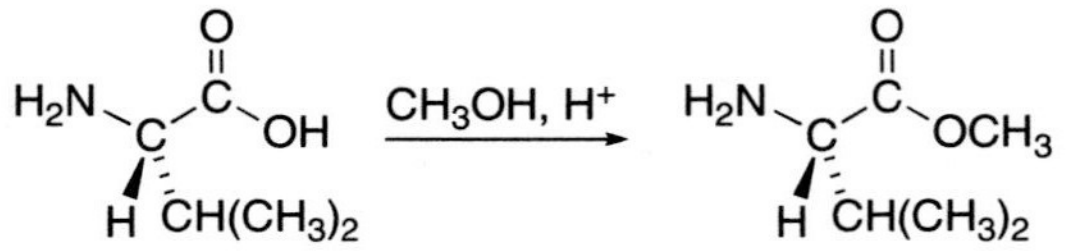

b.

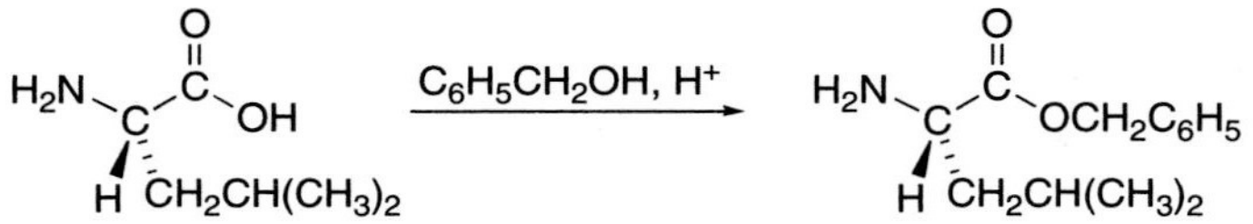

c.

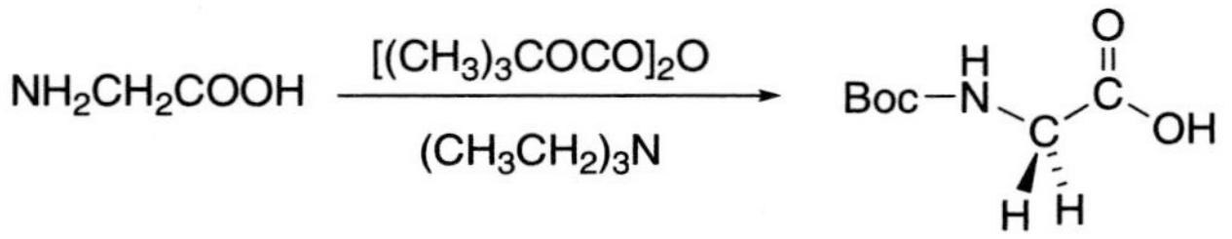

d. 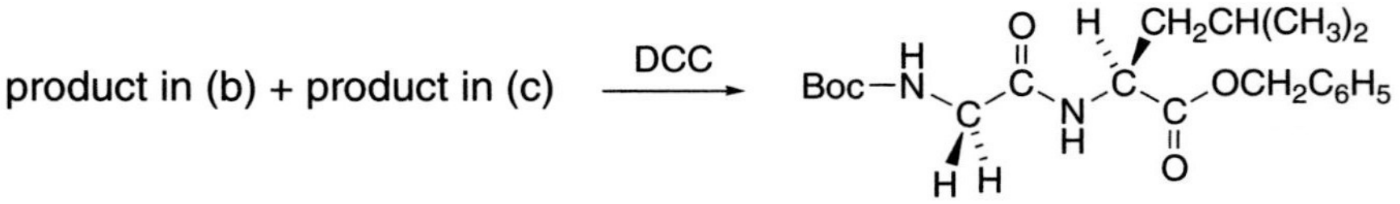 product in (b) + product in (c) $\xrightarrow{\text{DCC}}$

e.

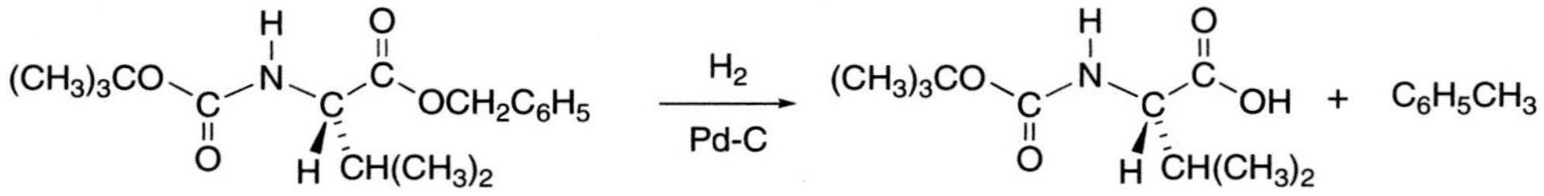

f. 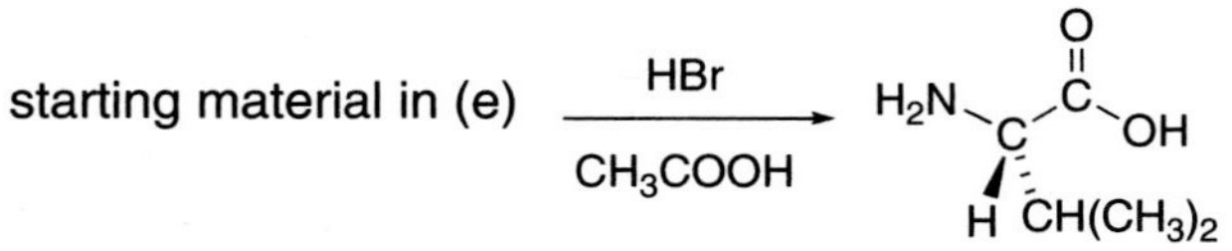 starting material in (e) $\xrightarrow[\text{CH}_3\text{COOH}]{\text{HBr}}$

g. 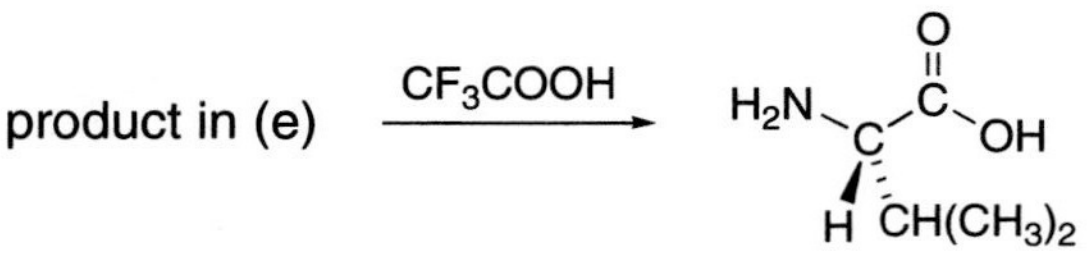 product in (e) $\xrightarrow{\text{CF}_3\text{COOH}}$

h.

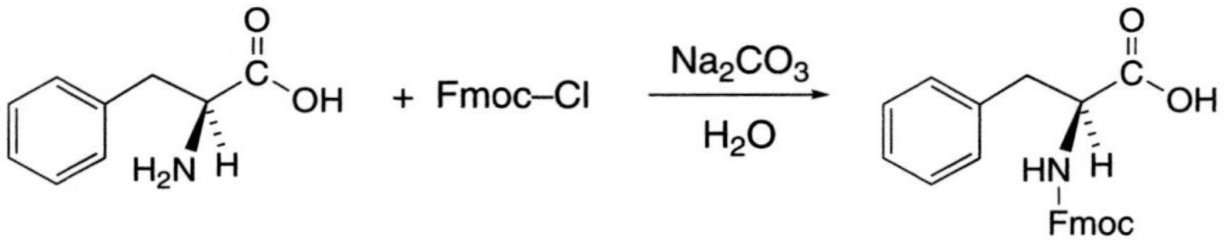

28.64

a.

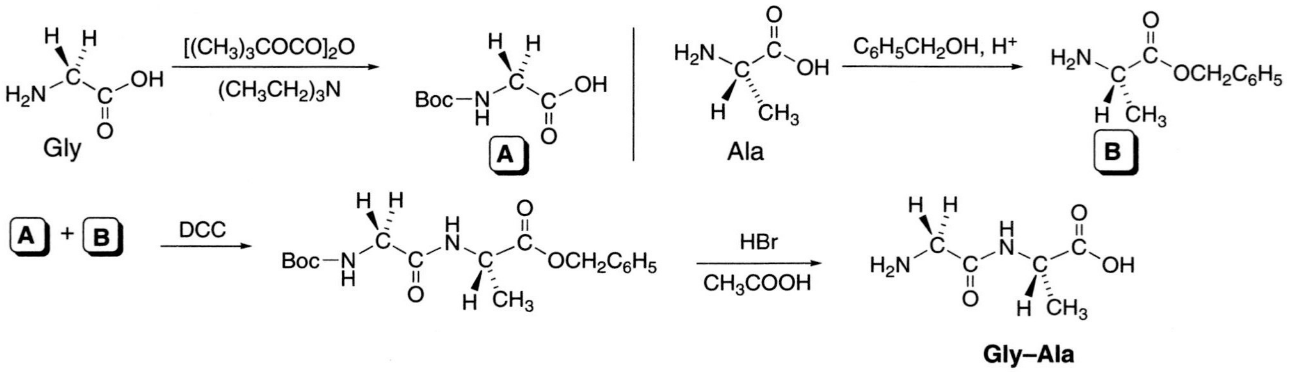

b.

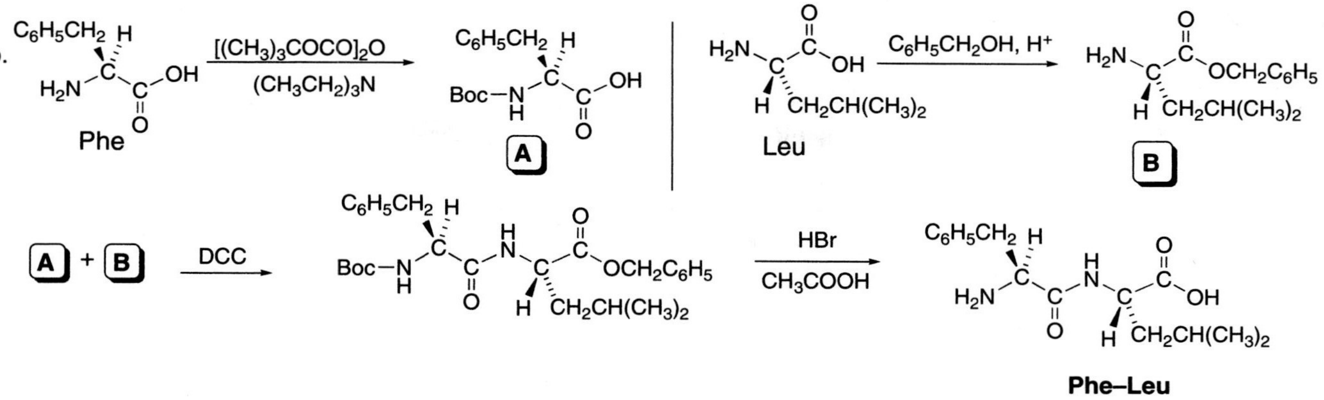

Phe $\xrightarrow{[(CH_3)_3COCO]_2O,\ (CH_3CH_2)_3N}$ **A** (Boc–Phe)

Leu $\xrightarrow{C_6H_5CH_2OH,\ H^+}$ **B**

A + **B** $\xrightarrow{DCC}$ Boc–Phe–Leu–OCH$_2$C$_6$H$_5$ $\xrightarrow{HBr,\ CH_3COOH}$ **Phe–Leu**

c.

Ile $\xrightarrow{[(CH_3)_3COCO]_2O,\ (CH_3CH_2)_3N}$ **A**

Ala $\xrightarrow{C_6H_5CH_2OH,\ H^+}$ **B**

A + **B** $\xrightarrow{DCC}$ Boc–Ile–Ala–OCH$_2$C$_6$H$_5$ $\xrightarrow{H_2,\ Pd\text{-}C}$ **C**

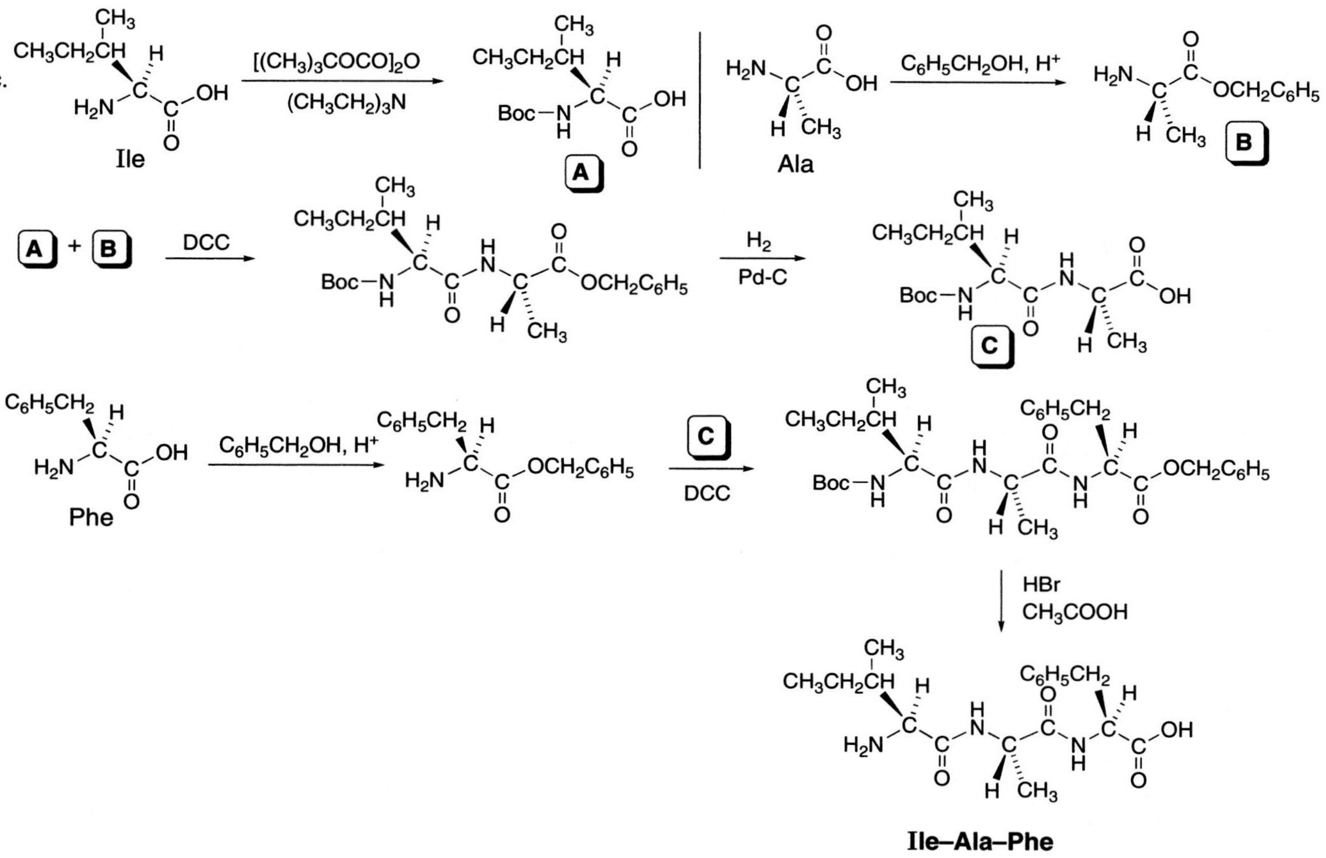

Phe $\xrightarrow{C_6H_5CH_2OH,\ H^+}$ H$_2$N–CH(CH$_2$C$_6$H$_5$)–OCH$_2$C$_6$H$_5$ $\xrightarrow{C,\ DCC}$ Boc–Ile–Ala–Phe–OCH$_2$C$_6$H$_5$ $\xrightarrow{HBr,\ CH_3COOH}$ **Ile–Ala–Phe**

28.65 Make all the Fmoc derivatives as described in Problem 28.24.

a.

b.

28.66 An acetyl group on the NH$_2$ forms an amide. Although this amide does block an amino group from reaction, this amide is no different in reactivity than any of the peptide amide bonds. To remove the acetyl group after the peptide bond is formed would require harsh reaction conditions that would also cleave the amide bonds of the peptide.

N-acetyl amino acid

28.67

a. A *p*-nitrophenyl ester activates the carboxy group of the first amino acid to amide formation by converting the OH group into a good leaving group, the *p*-nitrophenoxide group, which is highly resonance stabilized. In this case the electron-withdrawing NO_2 group further stabilizes the leaving group.

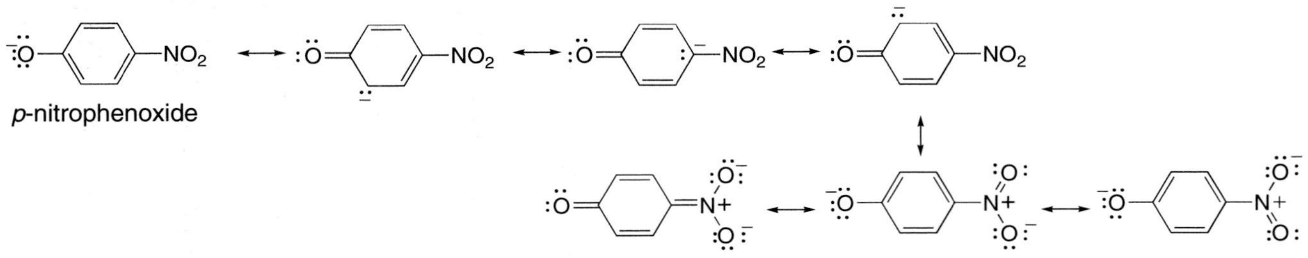

p-nitrophenoxide

The negative charge is delocalized
on the O atom of the NO_2 group.

b. The *p*-methoxyphenyl ester contains an electron-donating OCH_3 group, making $CH_3OC_6H_4O^-$ a poorer leaving group than $NO_2C_6H_4O^-$, so this ester does not activate the amino acid to amide formation as much.

28.68

a.

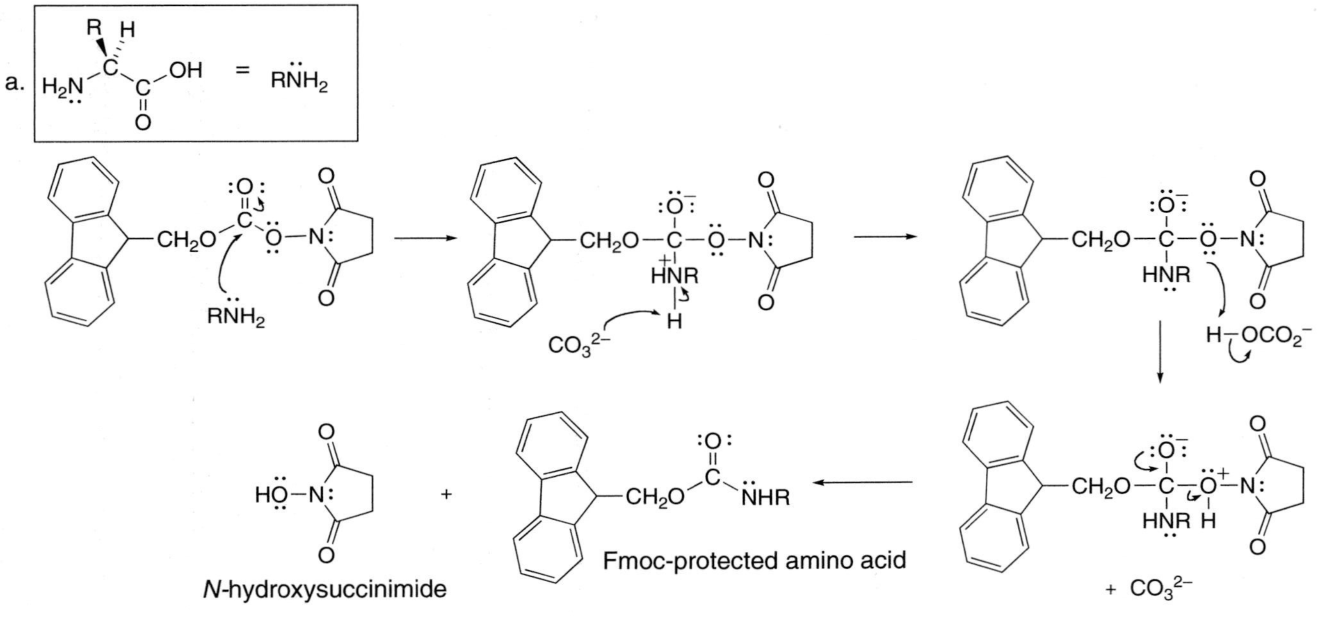

$$H_2N-\underset{O}{\underset{\|}{C}}\text{...}\quad = \quad R\ddot{N}H_2$$

N-hydroxysuccinimide

Fmoc-protected amino acid

Fmoc-protected amino acid

28.69 Reaction of the OH groups of the Wang resin with the COOH group of the Fmoc-protected amino acids would form esters by Fischer esterification. After the peptide has been synthesized, the esters can be hydrolyzed with aqueous acid or base, but the conditions cannot be too harsh to break the amide bond or cause epimerization.

28.70 Amino acids commonly found in the interior of a globular protein have nonpolar or weakly polar side chains: isoleucine and phenylalanine. Amino acids commonly found on the surface have COOH, NH_2, and other groups that can hydrogen bond to water: aspartic acid, lysine, arginine, and glutamic acid.

28.71 The proline residues on collagen are hydroxylated to increase hydrogen bonding interactions.

The new OH group allows more hydrogen bonding interactions between the chains of the triple helix, thus stabilizing it.

28.72

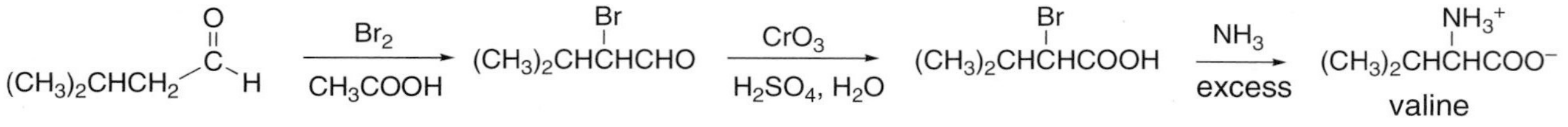

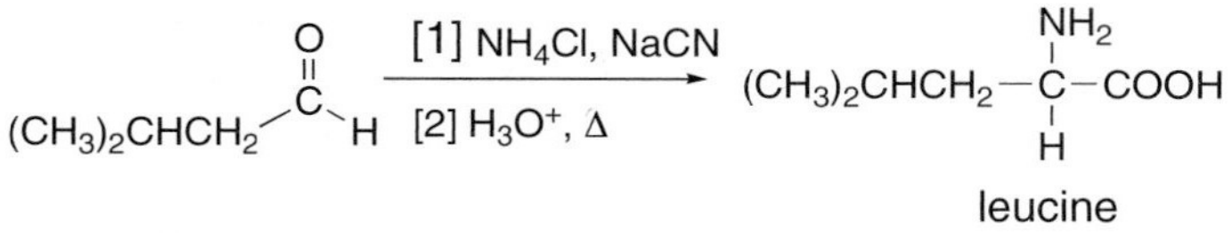

(Racemic valine and leucine are formed as products, but the synthesis of the tripeptide is drawn with one enantiomer only.)

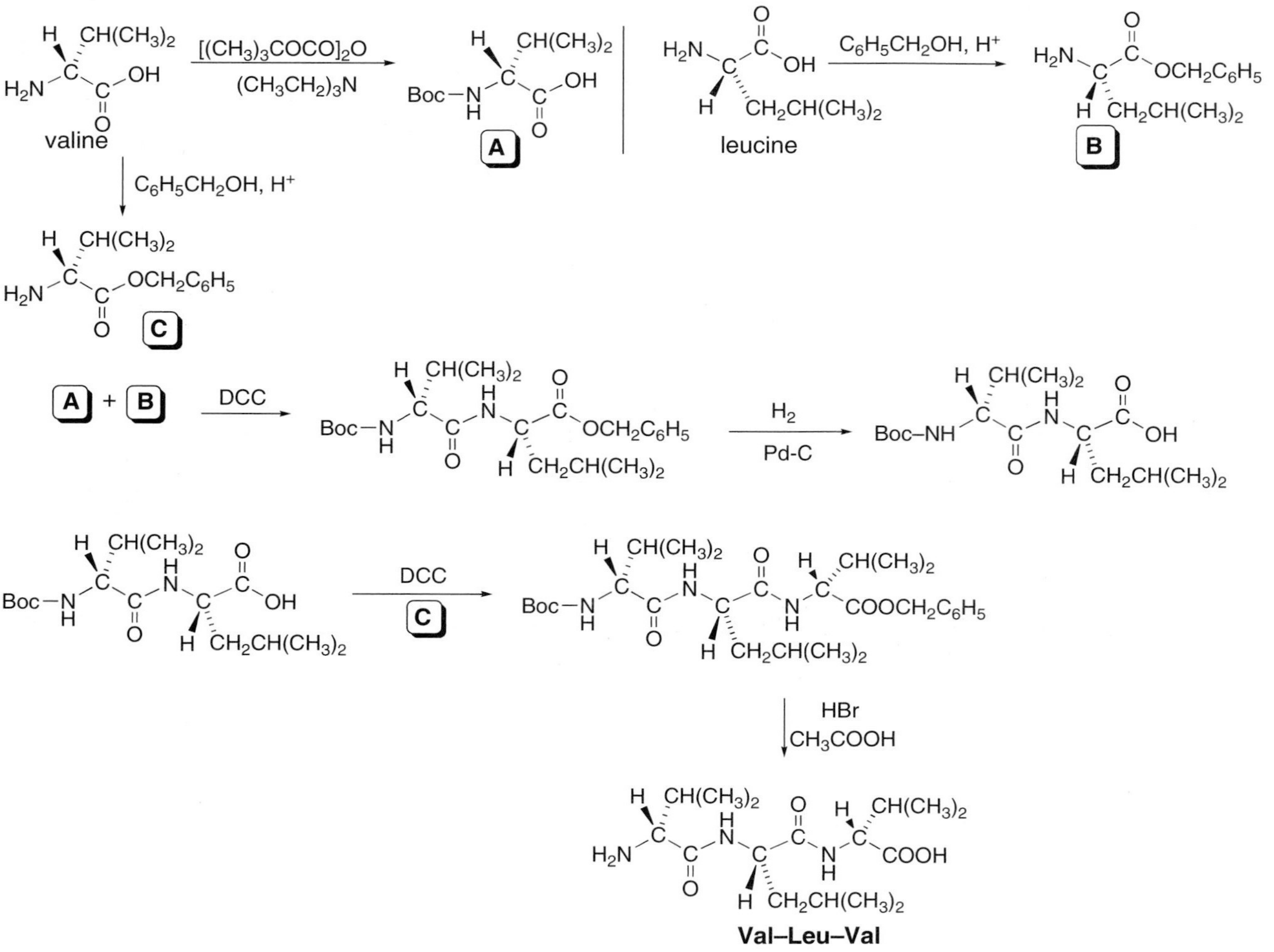

Val–Leu–Val

28.73 Perhaps using a chiral amine R*NH₂ (or related chiral nitrogen-containing compound) to make a chiral imine, will now favor formation of one of the amino nitriles in the Strecker synthesis. Hydrolysis of the CN group and removal of R* would then form the amino acid.

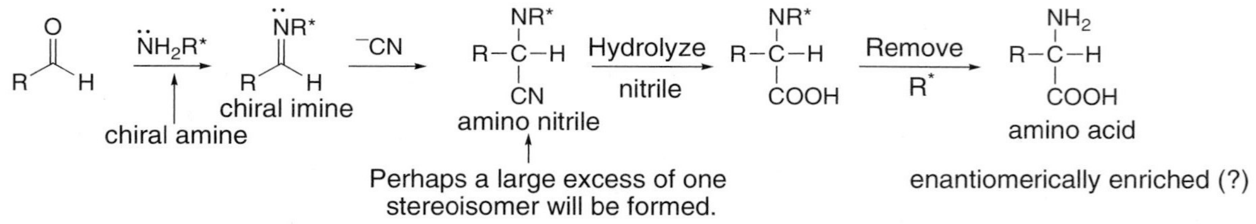

28.74

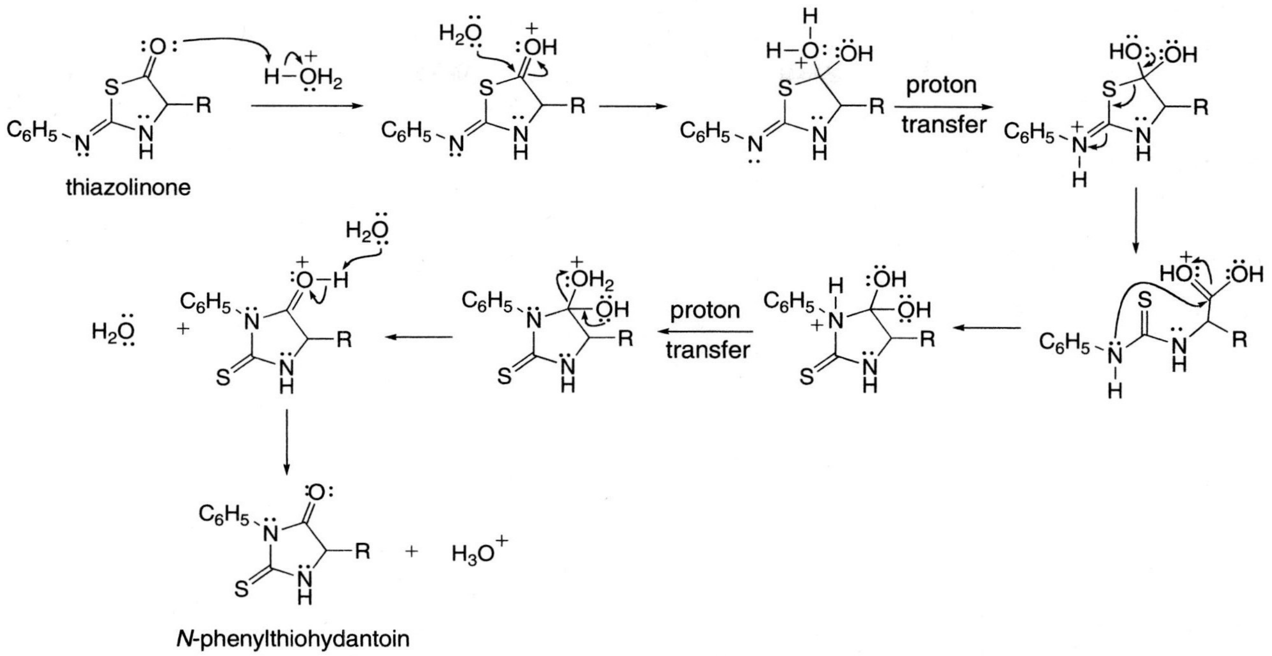

thiazolinone
proton
transfer
proton
transfer
N-phenylthiohydantoin

Chapter 29: Lipids

♦ Hydrolyzable lipids

[1] **Waxes (29.2)**—Esters formed from a long-chain alcohol and a long-chain carboxylic acid

R, R' = long chains of C's

[2] **Triacylglycerols (29.3)**—Triesters of glycerol with three fatty acids

R, R', R" = alkyl groups with 11–19 C's

[3] **Phospholipids (29.4)**

[a] Phosphatidylethanolamine (cephalin)

$O-P-O-CH_2CH_2\overset{+}{N}H_3$

R, R' = long carbon chain

[b] Phosphatidylcholine (lecithin)

$O-P-O-CH_2CH_2\overset{+}{N}(CH_3)_3$

R, R' = long carbon chain

[c] Sphingomyelin

HO $(CH_2)_{12}CH_3$

$O-P-O-CH_2CH_2\overset{+}{N}R'_3$

R = long carbon chain
R' = H or CH_3

♦ Nonhydrolyzable lipids

[1] **Fat-soluble vitamins (29.5)**—Vitamins A, D, E, and K

[2] **Eicosanoids (29.6)**—Compounds containing 20 carbons derived from arachidonic acid. There are four types: prostaglandins, thromboxanes, prostacyclins, and leukotrienes.

[3] **Terpenes (29.7)**—Lipids composed of repeating five-carbon units called isoprene units

Isoprene unit	Types of terpenes			
	[1] monoterpene	10 C's	[4] sesterterpene	25 C's
	[2] sesquiterpene	15 C's	[5] triterpene	30 C's
	[3] diterpene	20 C's	[6] tetraterpene	40 C's

[4] Steroids (29.8)—Tetracyclic lipids composed of three six-membered and one five-membered ring

Chapter 29: Answers to Problems

29.1 Waxes are esters (RCOOR') formed from a high molecular weight alcohol (R'OH) and a fatty acid (RCOOH).

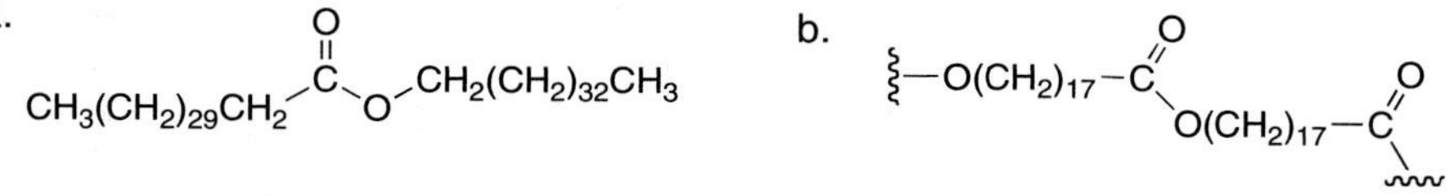

a.

$$CH_3(CH_2)_{29}CH_2-C(=O)-O-CH_2(CH_2)_{32}CH_3$$

b.

$$-O(CH_2)_{17}-C(=O)-O(CH_2)_{17}-C(=O)$$

29.2 Eicosapentaenoic acid has 20 C's and 5 C=C's. Since an increasing number of double bonds decreases the melting point, eicosapentaenoic acid should have a melting point lower than arachidonic acid; that is, $< 49°C$.

29.3

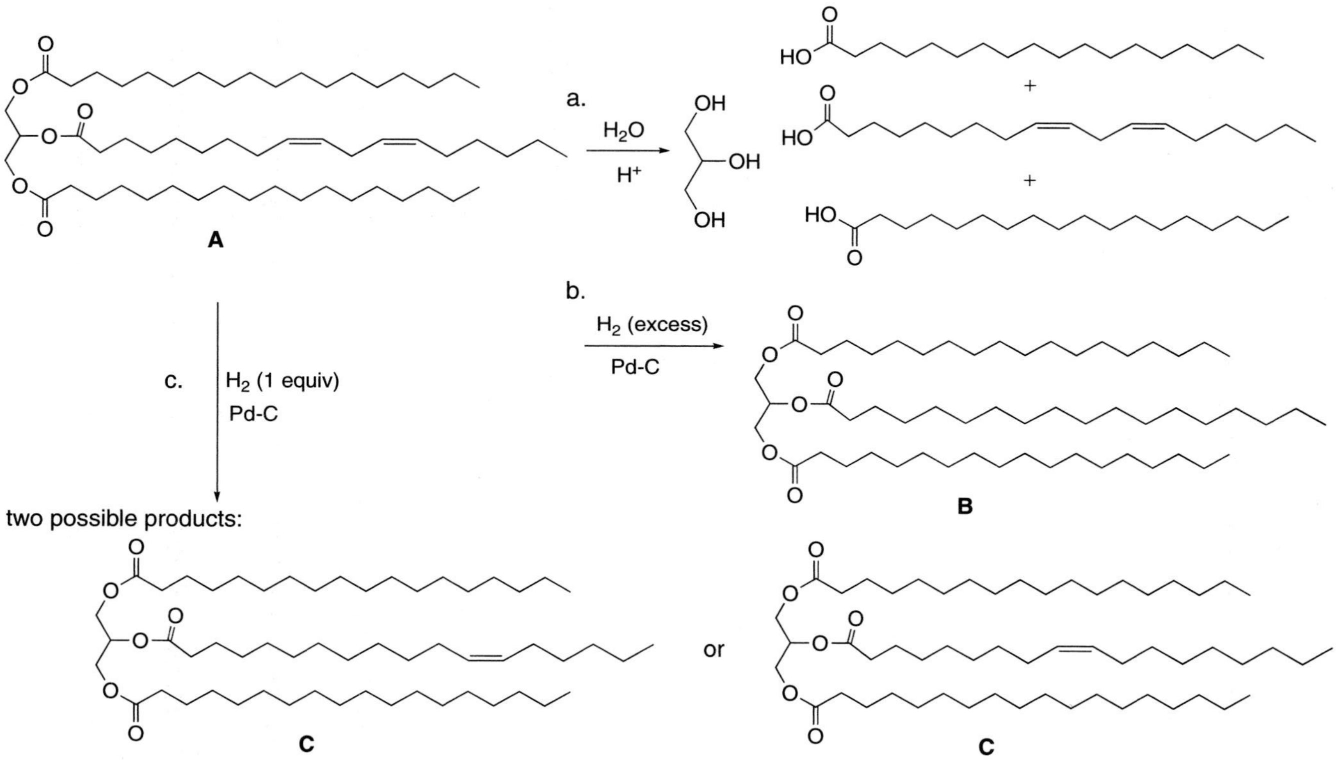

a. H_2O, H^+

b. H_2 (excess), Pd-C → **B**

c. H_2 (1 equiv), Pd-C

two possible products:

C or **C**

A	<	**C**	<	**B**
2 double bonds		1 double bond		0 double bonds
lowest melting point		**intermediate melting point**		**highest melting point**

29.4

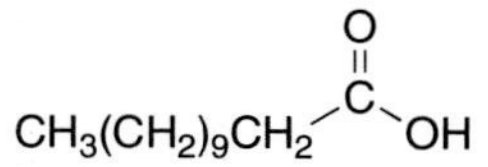

$$CH_3(CH_2)_9CH_2-C(=O)-OH$$
lauric acid

Lauric acid is a saturated fatty acid but has only 12 C's. The carbon chain is much shorter than palmitic acid (16 C's) and stearic acid (18 C's), making coconut oil a liquid at room temperature.

29.5

29.6 A lecithin is a type of phosphoacylglycerol. Two of the hydroxy groups of glycerol are esterified with fatty acids. The third OH group is part of a phosphodiester, which is also bonded to another low molecular weight alcohol.

general structure
of a lecithin

29.7 Soaps and phosphoacylglycerols have hydrophilic and hydrophobic components. Both compounds have an ionic "head" that is attracted to polar solvents like H_2O. This head is small in size compared to the hydrophobic region, which consists of one or two long hydrocarbon chains. These nonpolar chains consist of only C–C and C–H bonds and exhibit only van der Waals forces.

29.8 Phospholipids have a polar (ionic) head and two nonpolar tails. These two regions, which exhibit very different forces of attraction, allow the phospholipids to form a bilayer with a central hydrophobic region that serves as a barrier to agents crossing a cell membrane, while still possessing an ionic head to interact with the aqueous environment inside and outside the cell. Two different regions are needed in the molecule. Triacylglycerols have three polar, uncharged ester groups, but they are not nearly as polar as phospholipids. They do not have an ionic head with nonpolar tails and so they do not form bilayers. They are largely nonpolar C–C and C–H bonds so they are not attracted to an aqueous medium, making them H_2O insoluble.

29.9 Fat-soluble vitamins are hydrophobic and therefore are readily stored in the fatty tissues of the body. Water-soluble vitamins, on the other hand, are readily excreted in the urine and large concentrations cannot build up in the body.

29.10

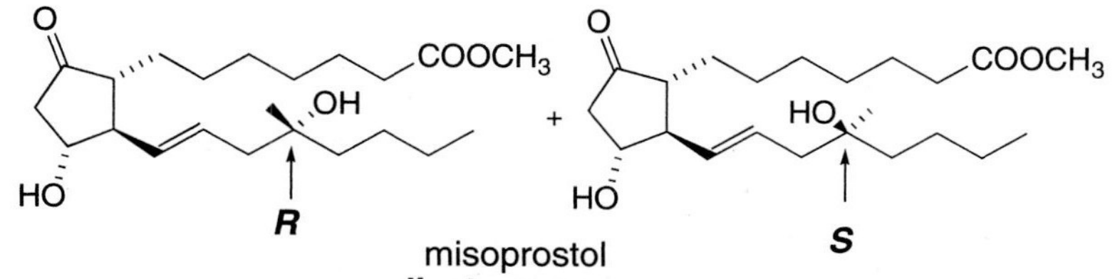
O
COOCH3
OH
HO
R
+
O
COOCH3
HO
HO
S

misoprostol
diastereomers
Only one tetrahedral stereogenic center is different in these two compounds.

29.11 Isoprene units are shown in bold.

a. 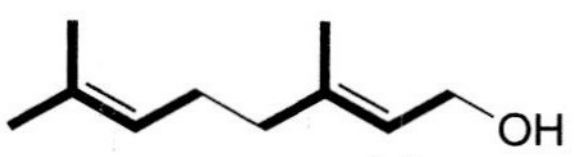
OH

geraniol

c. 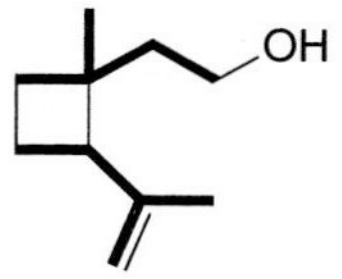
OH

grandisol

b. 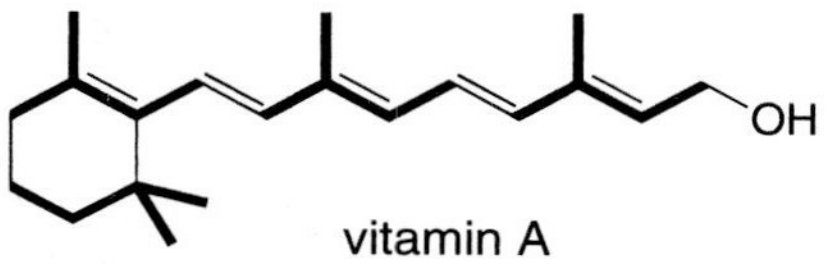
OH

vitamin A

d. 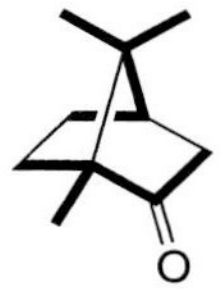
O

camphor

29.12

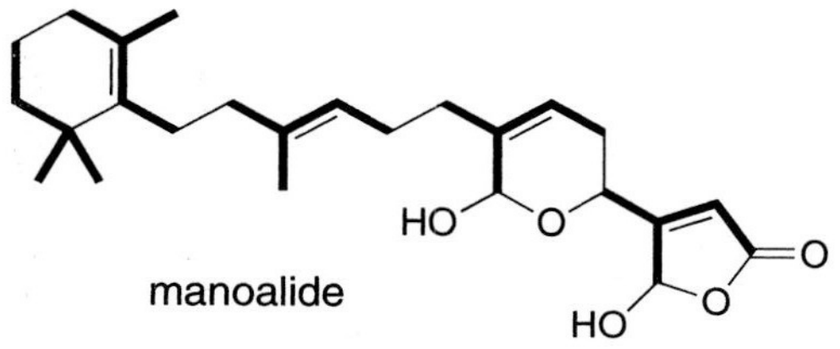
HO
O
O
O
HO

manoalide

29.13

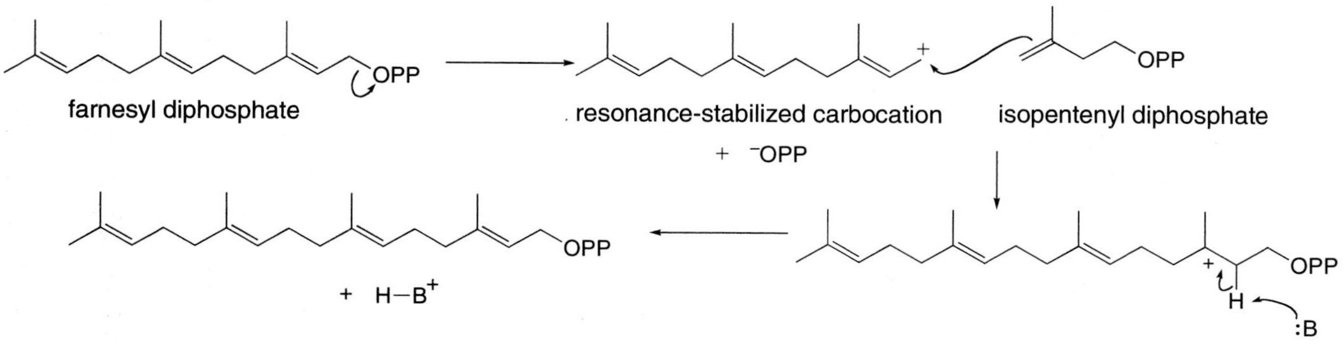
OPP
farnesyl diphosphate
+
OPP
resonance-stabilized carbocation
isopentenyl diphosphate
+ ⁻OPP
OPP
+
H
:B
OPP
+ H–B+

29.14

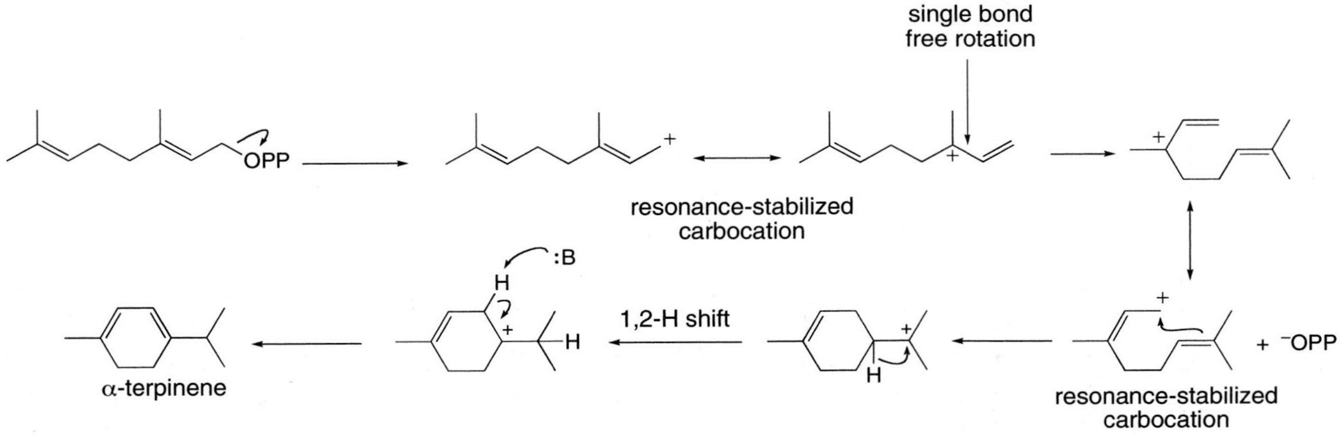

29.15

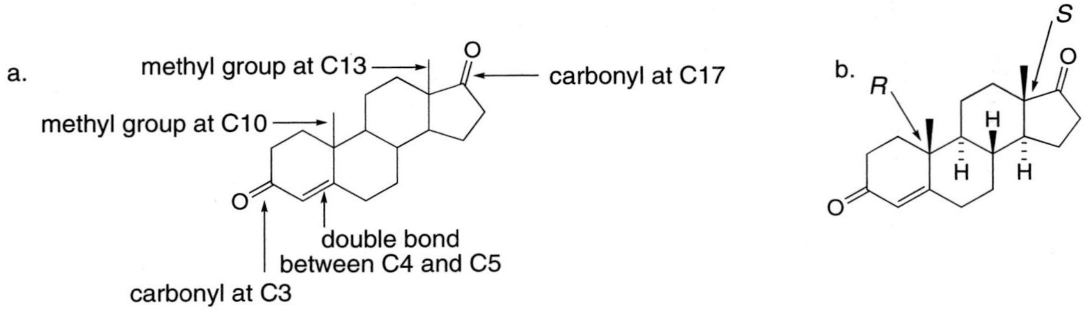

a.

methyl group at C13 ⟶
carbonyl at C17
methyl group at C10 ⟶
double bond between C4 and C5
carbonyl at C3

b.

S
R

29.16

cholesterol
equatorial OH

enantiomer

different here
diastereomer

different here
diastereomer

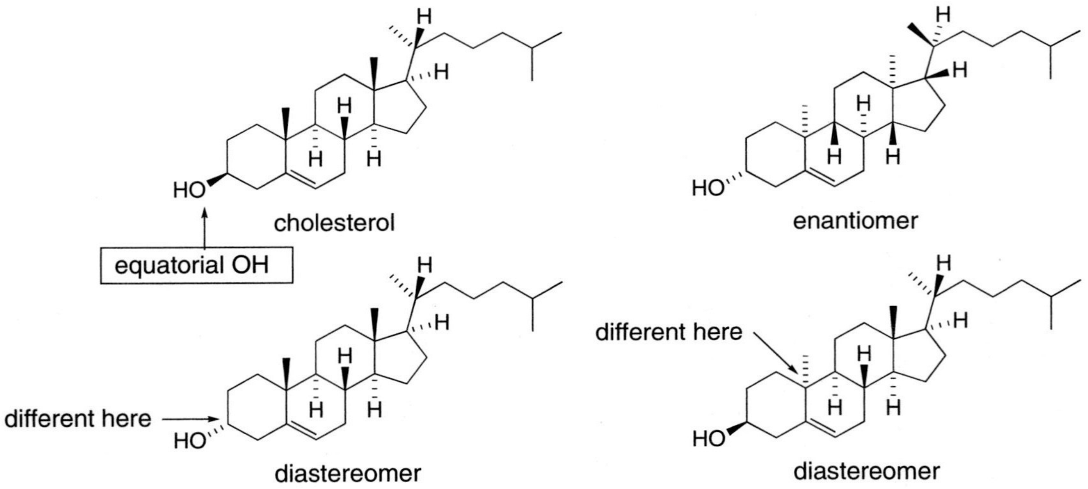

29.17

All four rings are in the same plane. The bulky CH$_3$ groups (arrows) are located above the plane. Epoxide **A** is favored, because it results from epoxidation below the plane, on the opposite side from the CH$_3$ groups that shield the top of the molecule somewhat to attack by reagents. In **B**, the epoxide ring is above the plane on the same side as the CH$_3$ groups. Formation of **B** would require epoxidation of the planar C=C from the less accessible, more sterically hindered side of the double bond. This path is thus disfavored.

29.18

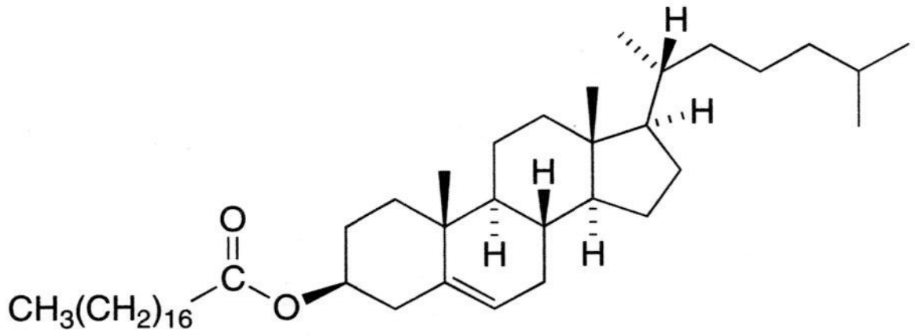
H
H
H
H
H
O
$CH_3(CH_2)_{16}$
C
O

29.19 Each compound has one tetrahedral stereogenic center (circled), so there are two stereoisomers (two enantiomers) possible. All C=C's have the *Z* configuration.

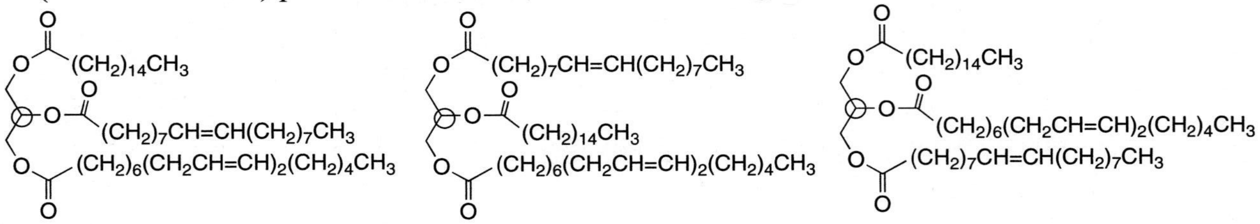
$(CH_2)_{14}CH_3$
$(CH_2)_7CH{=}CH(CH_2)_7CH_3$
$(CH_2)_6(CH_2CH{=}CH)_2(CH_2)_4CH_3$
$(CH_2)_7CH{=}CH(CH_2)_7CH_3$
$(CH_2)_{14}CH_3$
$(CH_2)_6(CH_2CH{=}CH)_2(CH_2)_4CH_3$
$(CH_2)_{14}CH_3$
$(CH_2)_6(CH_2CH{=}CH)_2(CH_2)_4CH_3$
$(CH_2)_7CH{=}CH(CH_2)_7CH_3$

29.20

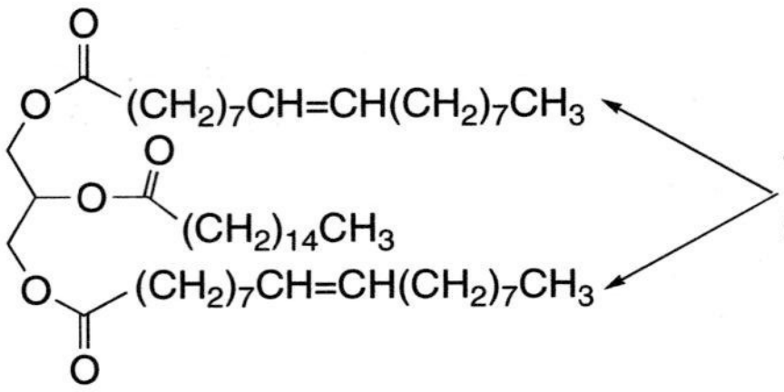
$(CH_2)_7CH{=}CH(CH_2)_7CH_3$
$(CH_2)_{14}CH_3$
$(CH_2)_7CH{=}CH(CH_2)_7CH_3$

There is no stereogenic center in this triacylglycerol since these R groups are identical, making this triacylglycerol optically inactive.

29.21

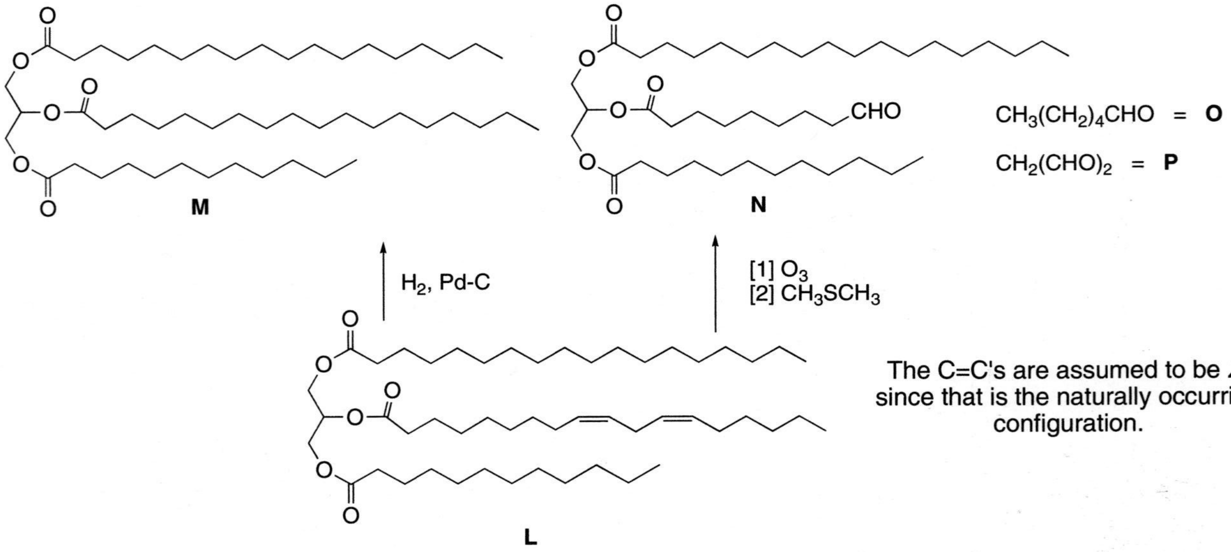
M
CHO
N
$CH_3(CH_2)_4CHO$ = O
$CH_2(CHO)_2$ = P
H_2, Pd-C
[1] O_3
[2] CH_3SCH_3
L

The C=C's are assumed to be *Z*, since that is the naturally occurring configuration.

29.22 When R" = $CH_2CH_2NH_3^+$, the compound is called a **phosphatidylethanolamine** or **cephalin.**

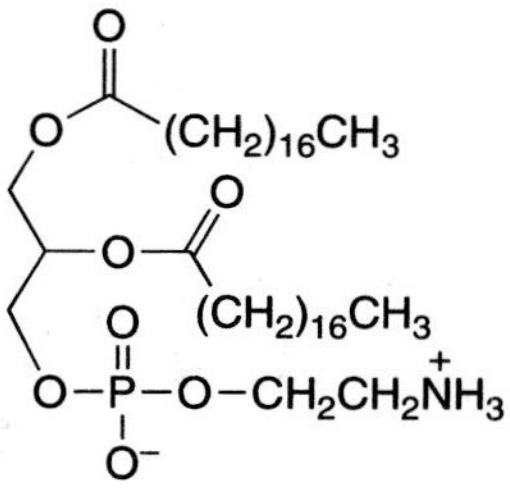
(CH2)16CH3
(CH2)16CH3
O–P–O–CH2CH2NH3

cephalin

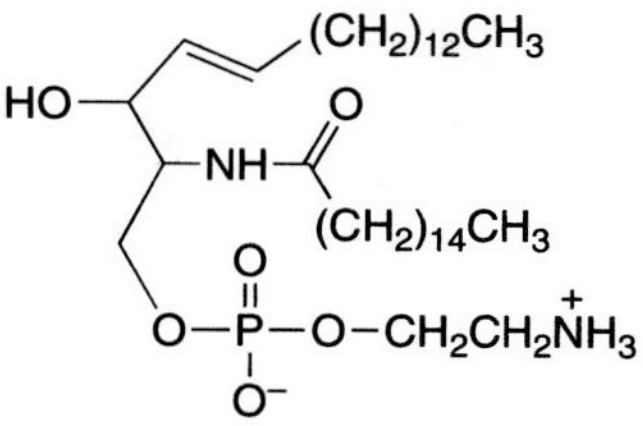
(CH2)12CH3
HO
NH
(CH2)14CH3
O–P–O–CH2CH2NH3

sphingomyelin

29.23

a.

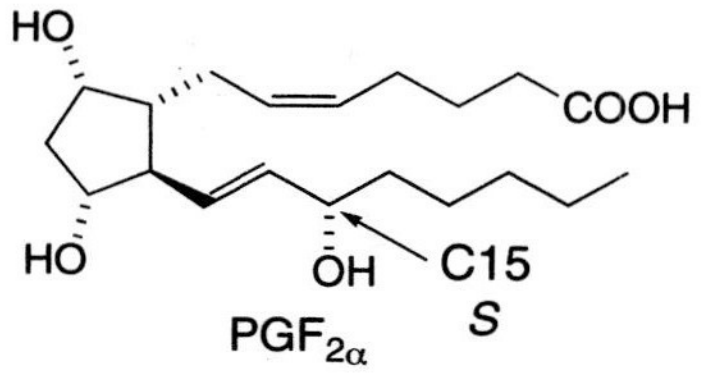
HO
COOH
HO
OH
C15
S
PGF2α

b.

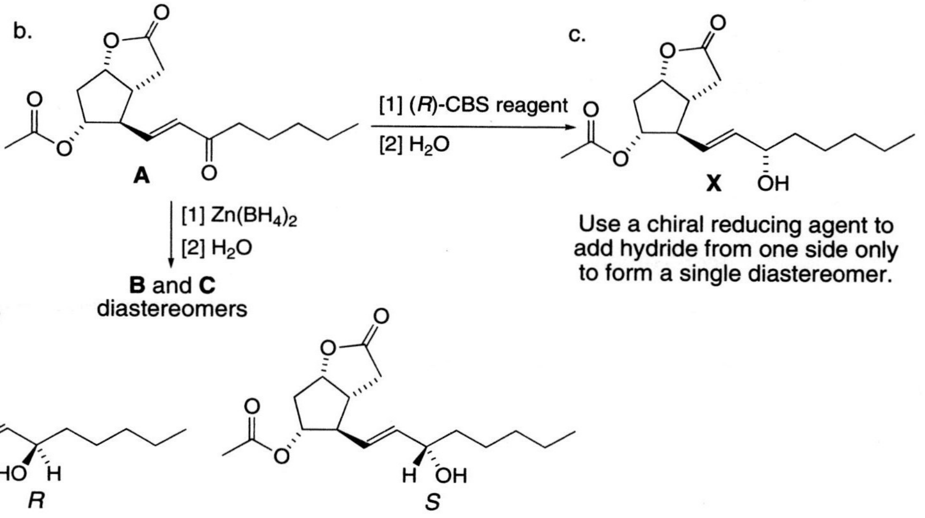
A
[1] (R)-CBS reagent
[2] H2O
c.
X
OH
Use a chiral reducing agent to
add hydride from one side only
to form a single diastereomer.
[1] Zn(BH4)2
[2] H2O
B and C
diastereomers
HO H
R
H OH
S

29.24

a.

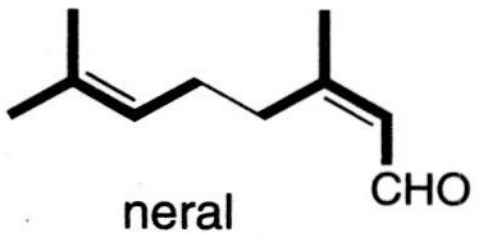
CHO

neral

b.

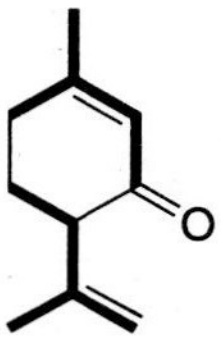

carvone

c.

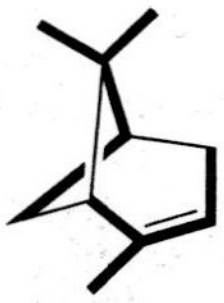

α-pinene

d.

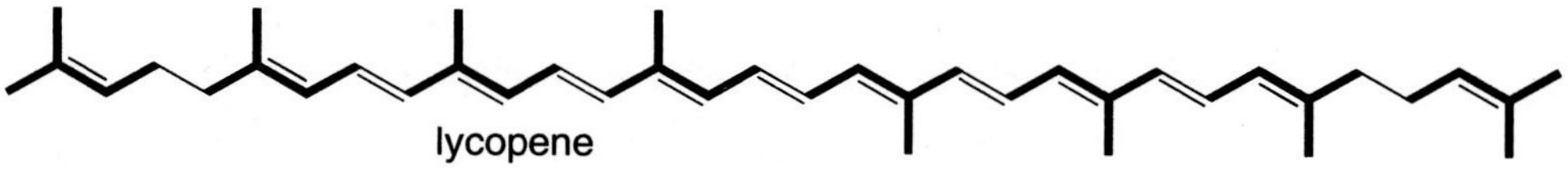

lycopene

e.

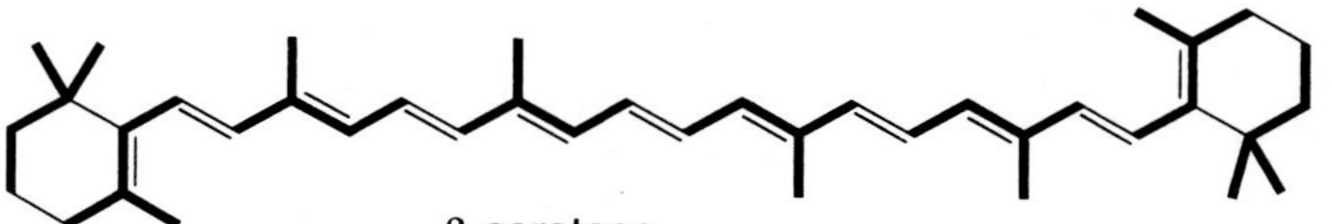

β-carotene

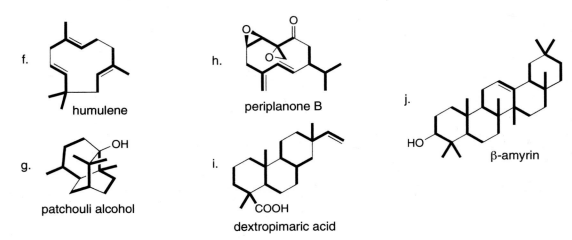

f. humulene

h. periplanone B

g. patchouli alcohol

i. dextropimaric acid
COOH

j. β-amyrin
HO

29.25 A *monoterpene* **contains 10 carbons** and two isoprene units; a *sesquiterpene* **contains 15 carbons** and three isoprene units, etc. See Table 29.5.

a. monoterpene CHO

b. monoterpene
O

c. monoterpene

d. tetraterpene

e. tetraterpene

f. sesquiterpene

g. sesquiterpene
OH

h. sesquiterpene
O
O

i. diterpene
COOH

j. triterpene
HO

29.26

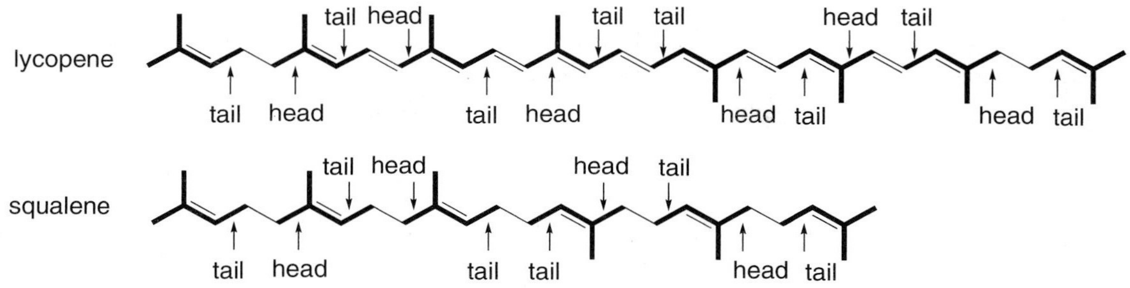

lycopene

squalene

29.27

resonance-stabilized carbocation

α-pinene

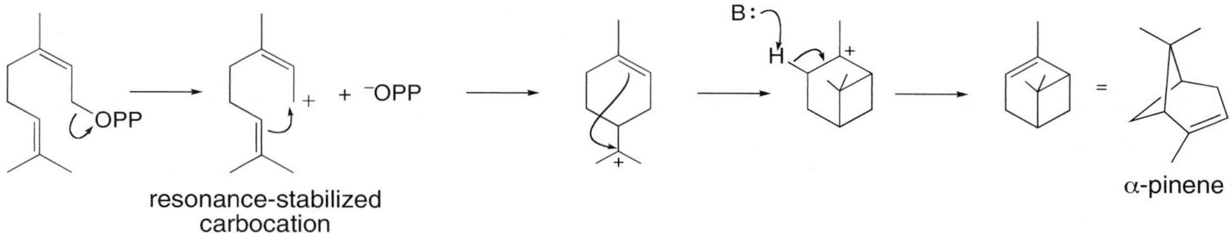

29.28 The unusual feature in the cyclization that forms flexibilene is that a 2° carbocation rather than a 3° carbocation is generated. Cyclization at the other end of the C=C would have given a 3° carbocation and formed a 14-membered ring. In addition, the 2° carbocation does not rearrange to form a 3° carbocation.

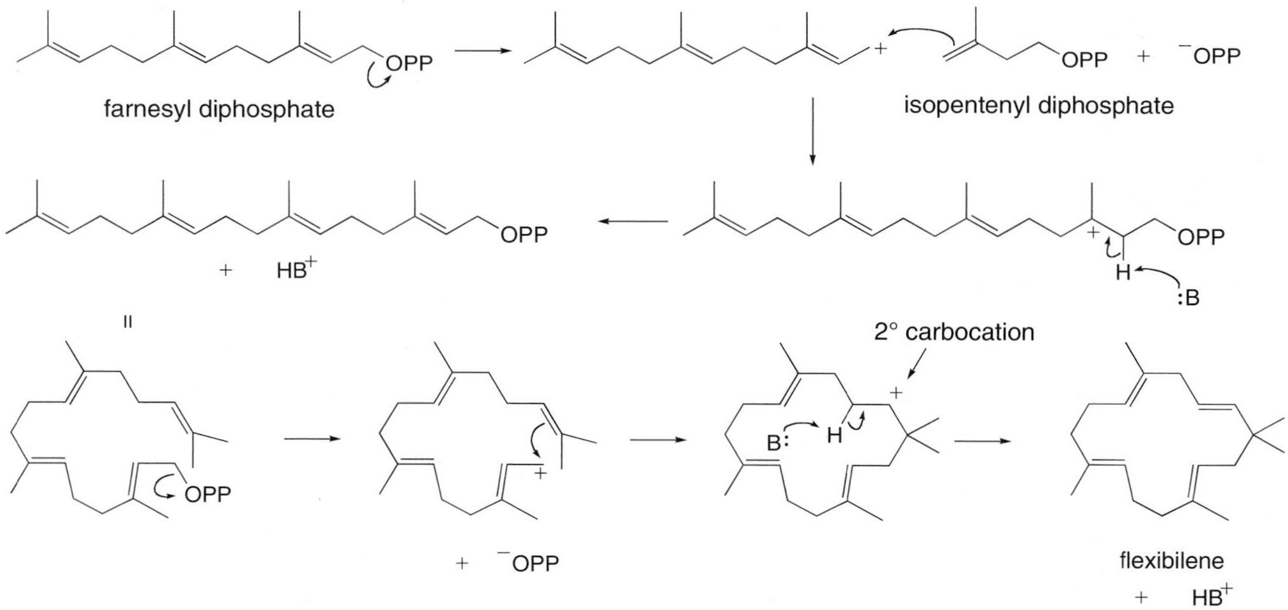

farnesyl diphosphate

isopentenyl diphosphate

+ HB⁺

2° carbocation

+ ⁻OPP

flexibilene
+ HB⁺

29.29

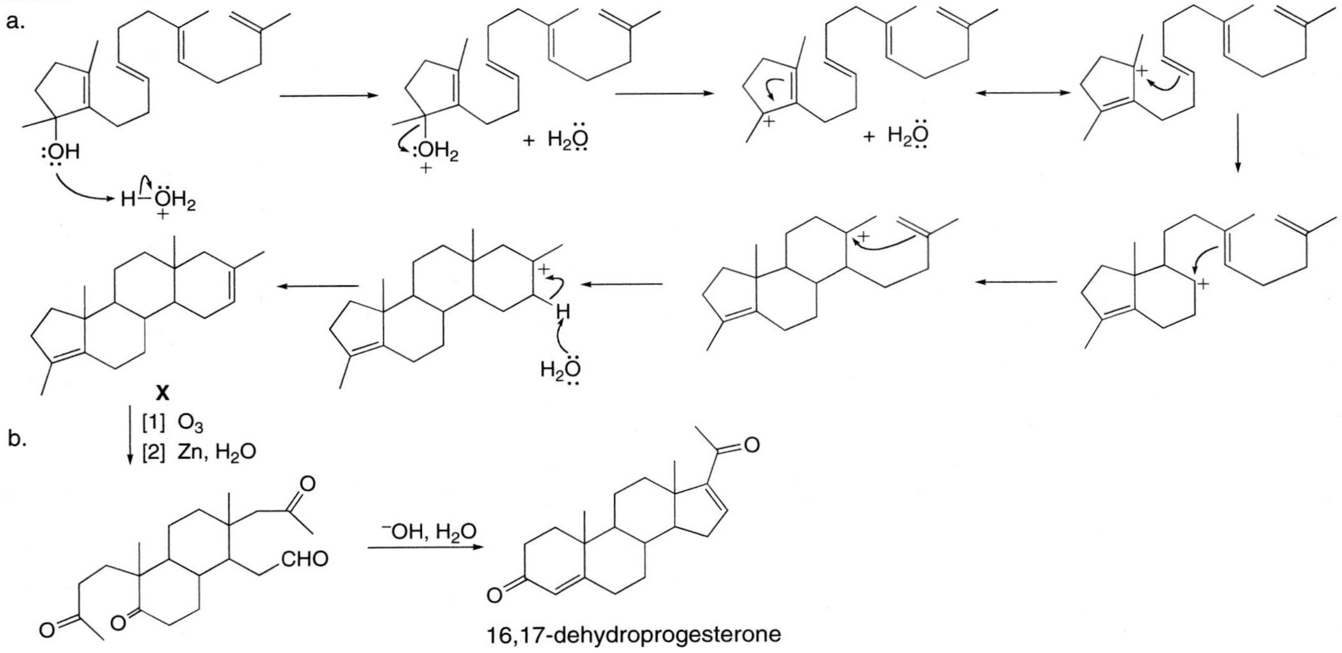
a.
:ÖH
H–ÖH2
+
:ÖH2
+
+ H2Ö
+ H2Ö
+
X
H
H2Ö
b.
[1] O3
[2] Zn, H2O
O
CHO
O
O
−OH, H2O
O
O
16,17-dehydroprogesterone

29.30

a.
H
CH3
H
b.
H
CH3
H

29.31

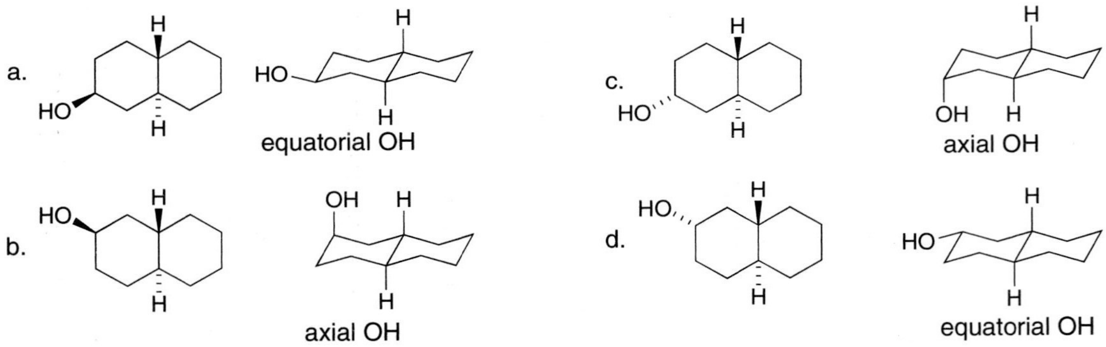
a.
H
HO
H
H
HO
H
equatorial OH
b.
HO
H
H
OH
H
H
axial OH
c.
H
HO
H
H
OH
H
axial OH
d.
HO
H
H
H
HO
H
equatorial OH

29.32

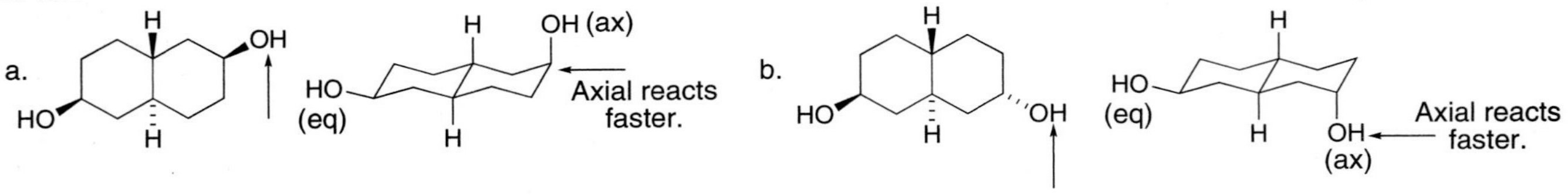
a.
H
OH
HO
H
H
OH (ax)
HO
(eq)
H
Axial reacts
faster.
b.
H
HO
H
OH
H
HO
(eq)
H
OH
(ax)
Axial reacts
faster.

29.33

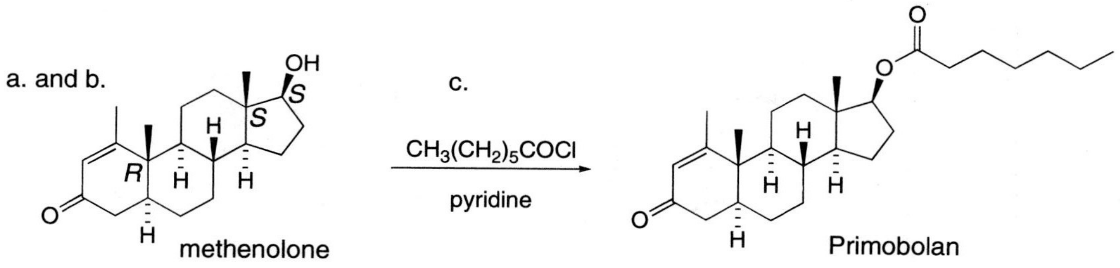
a. and b.
OH
S
S
R
H
methenolone
c.
$CH_3(CH_2)_5COCl$
pyridine
O
Primobolan

29.34

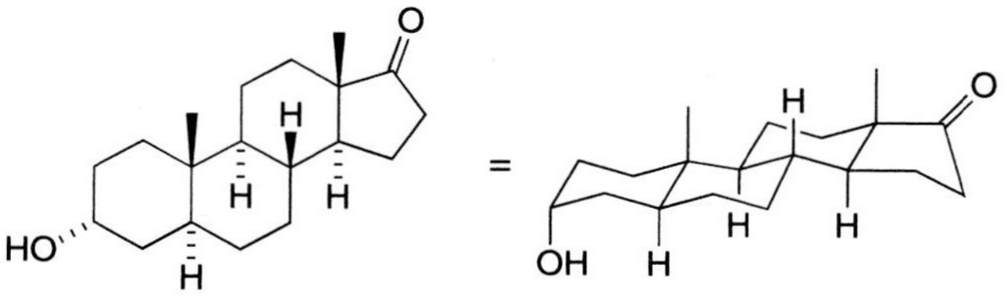
HO
O
=
OH
H

29.35

a.

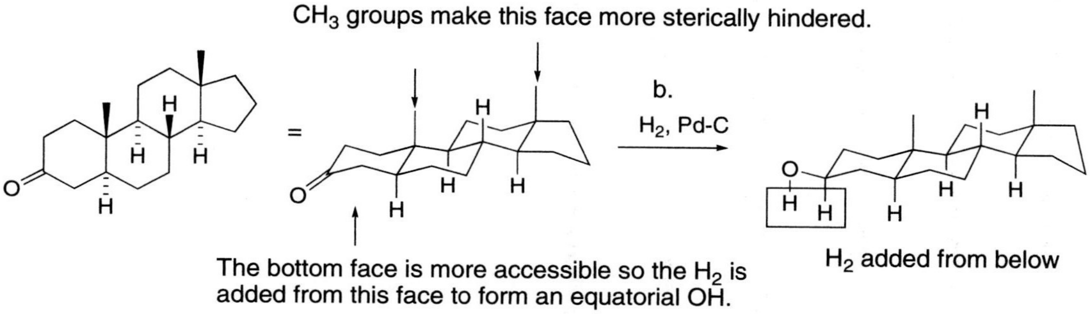
CH_3 groups make this face more sterically hindered.
=
b.
H_2, Pd-C
The bottom face is more accessible so the H_2 is added from this face to form an equatorial OH.
H_2 added from below

29.36

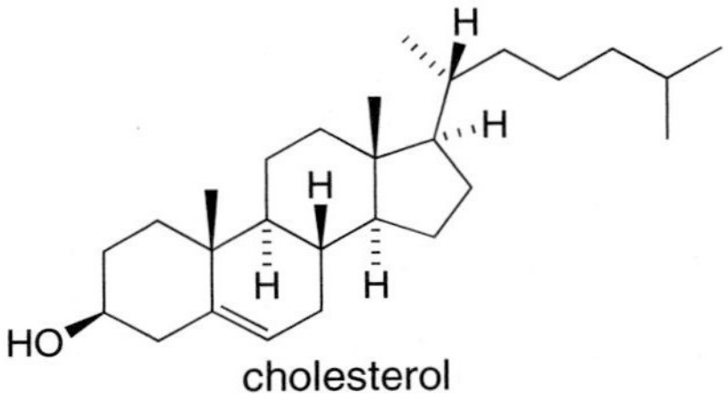

cholesterol

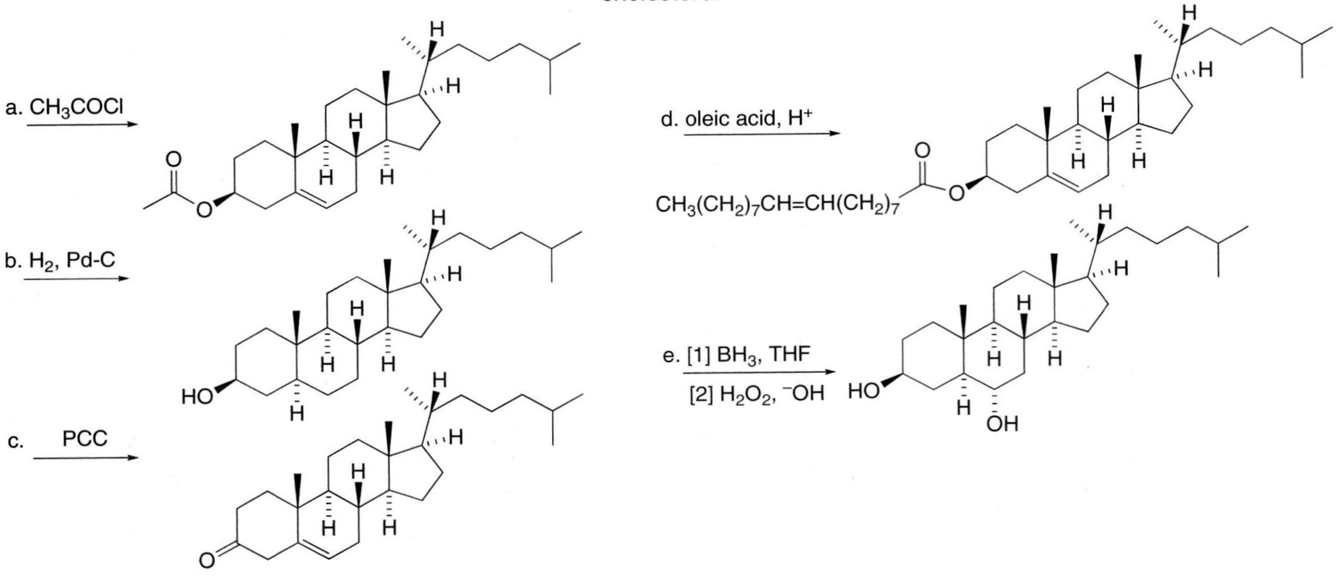

a. CH$_3$COCl

b. H$_2$, Pd-C

c.   PCC

d. oleic acid, H$^+$

CH$_3$(CH$_2$)$_7$CH=CH(CH$_2$)$_7$

e. [1] BH$_3$, THF
   [2] H$_2$O$_2$, $^-$OH

29.37

+ $^-$OPP
resonance-
stabilized
carbocation

1,2-shift   1,2-shift

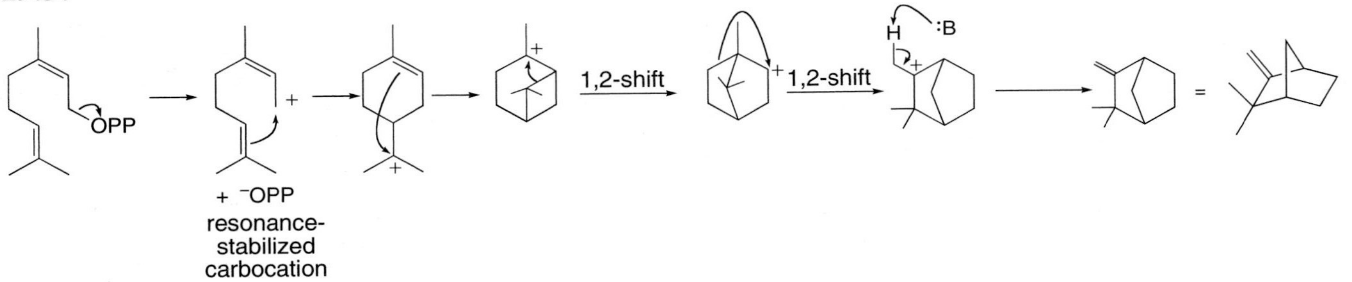

29.38 Re-draw the starting material in a conformation that suggests the structure of the product.

29.39

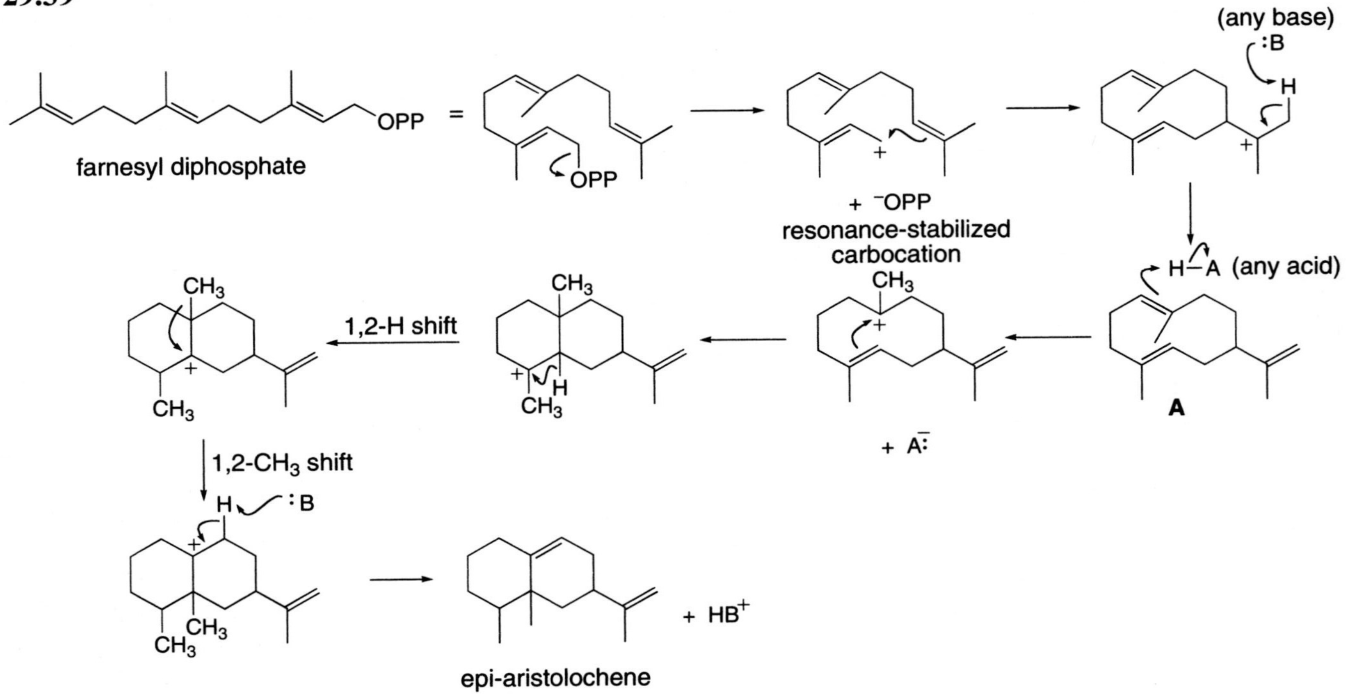

(any base)
:B
H
OPP
=
farnesyl diphosphate
OPP
+
+ ⁻OPP
resonance-stabilized
carbocation
+
H–A (any acid)
A
CH3
+
+ A:
CH3
+
H
CH3
1,2-H shift
CH3
+
CH3
1,2-CH3 shift
H
:B
+
CH3
CH3
+ HB+
epi-aristolochene

Chapter 30: Synthetic Polymers

♦ Chain-growth polymers—Addition polymers

[1] Chain-growth polymers with alkene starting materials (30.2)

- General reaction:

- Mechanism—three possibilities, depending on the identity of Z:

Type	Identity of Z	Initiator	Comments
[1] radical polymerization	Z stabilizes a radical. Z = R, Ph, Cl, etc.	A source of radicals (ROOR)	Termination occurs by radical coupling or disproportionation. Chain branching occurs.
[2] cationic polymerization	Z stabilizes a carbocation. Z = R, Ph, OR, etc.	H–A or a Lewis acid (BF$_3$ + H$_2$O)	Termination occurs by loss of a proton.
[3] anionic polymerization	Z stabilizes a carbanion. Z = Ph, COOR, COR, CN, etc.	An organolithium reagent (R–Li)	Termination occurs only when an acid or other electrophile is added.

[2] Chain-growth polymers with epoxide starting materials (30.3)

- The mechanism is S$_N$2.
- Ring opening occurs at the less substituted carbon of the epoxide.

◆ **Examples of step-growth polymers—Condensation polymers (30.6)**

Polyamides	Polyesters
nylon 6	polyethylene terephthalate
Kevlar	copolymer of glycolic and lactic acids
Polyurethanes	**Polycarbonates**
a polyurethane	Lexan

◆ **Structure and properties**

- Polymers prepared from monomers having the general structure $CH_2=CHZ$ can be **isotactic**, **syndiotactic**, or **atactic** depending on the identity of Z and the method of preparation (30.4).
- **Ziegler–Natta catalysts** form polymers without significant branching. Polymers can be isotactic, syndiotactic, or atactic depending on the catalyst. Polymers prepared from 1,3-dienes have the *E* or *Z* configuration depending on the monomer (30.4, 30.5).
- Most polymers contain ordered crystalline regions and less ordered amorphous regions (30.7). The greater the crystallinity, the harder the polymer.
- **Elastomers** are polymers that stretch and can return to their original shape (30.5).
- **Thermoplastics** are polymers that can be molded, shaped, and cooled such that the new form is preserved (30.7).
- **Thermosetting polymers** are composed of complex networks of covalent bonds so they cannot be melted to form a liquid phase (30.7).

Chapter 30: Answers to Problems

30.1   Place brackets around the repeating unit that creates the polymer.

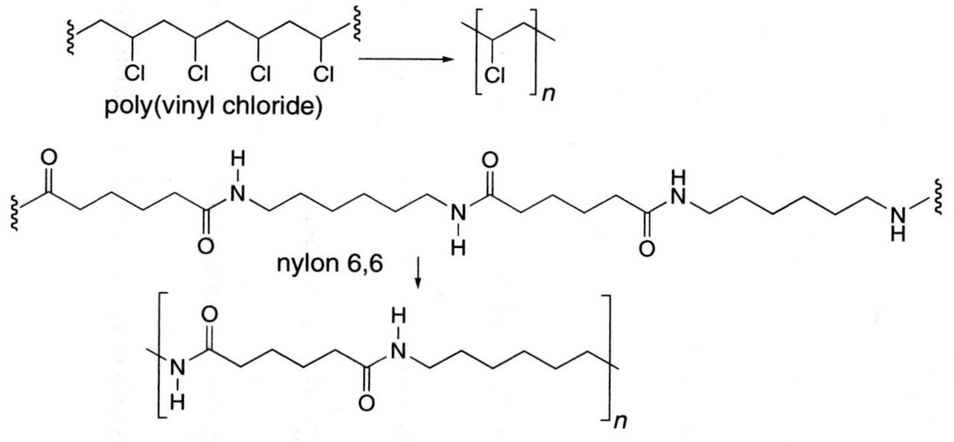

poly(vinyl chloride)

nylon 6,6

30.2   Draw each polymer formed by chain-growth polymerization.

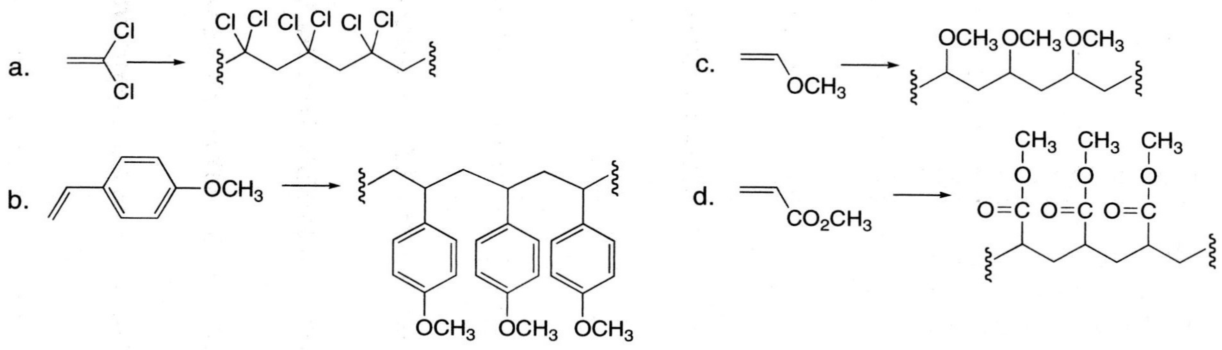

a.

b.

c.

d.

30.3   Draw each polymer formed by radical polymerization.

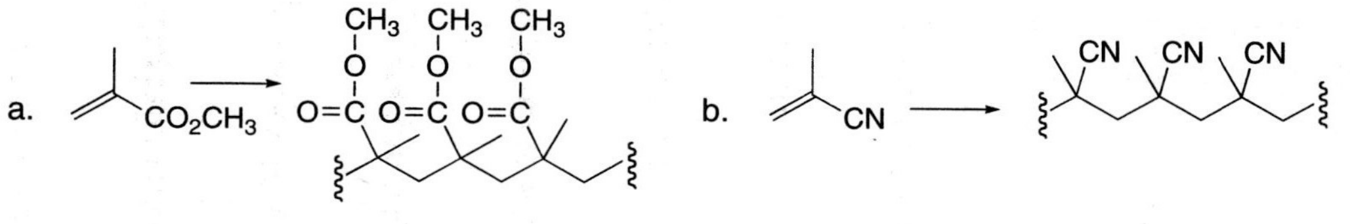

a.

b.

30.4   Use Mechanism 30.1 as a model of radical polymerization.

Initiation:

Propagation:

Repeat Step [3] over and over.

Termination:

(Ac = CH₃CO–)

30.5   Radical polymerization forms a long chain of polystyrene with phenyl groups bonded to every other carbon. To form branches on this polystyrene chain, a radical on a second polymer chain abstracts a H atom. Abstraction of H_a forms a resonance-stabilized radical **A'**. The 2° radical **B'** (without added resonance stabilization) is formed by abstraction of H_b. Abstraction of H_a is favored, therefore, and this radical goes on to form products with 4° C's (**A**).

30.6   Cationic polymerization proceeds via a carbocation intermediate. Substrates that form more stable 3° carbocations react more readily in these polymerization reactions than substrates that form less stable 1° carbocations. $CH_2=C(CH_3)_2$ will form a more substituted carbocation than $CH_2=CH_2$.

30.7 Cationic polymerization occurs with alkene monomers having substituents that can stabilize carbocations, such as alkyl groups and other electron-donor groups. Anionic polymerization occurs with alkene monomers having substituents that can stabilize a negative charge, such as COR, COOR, or CN.

a. $CH_2=C(CH_3)COOCH_3$ b. $CH_2=CHCH_3$ c. $CH_2=CHOC(CH_3)_3$ d. $CH_2=CHCOCH_3$

electron-withdrawing group
anion polymerization

alkyl group
cationic polymerization

an electron-donating
resonance effect
cationic polymerization

electron-withdrawing group
anion polymerization

30.8 Use Mechanism 30.4 as a model of anion polymerization.

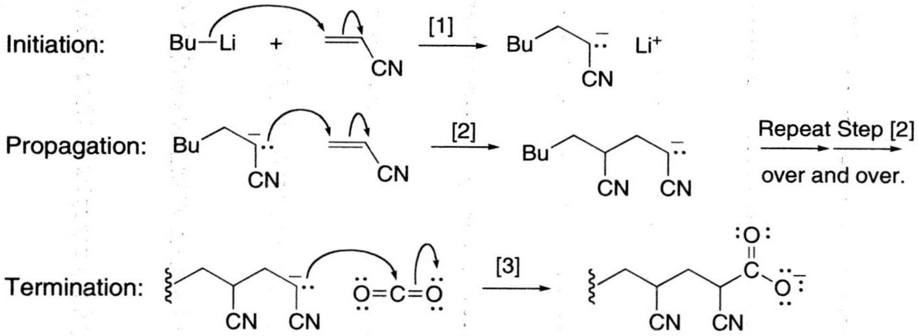

30.9 Styrene ($CH_2=CHPh$) can by polymerized by all three methods of chain-growth polymerization because a benzene ring can stabilize a radical, a carbocation, and also a carbanion by resonance delocalization.

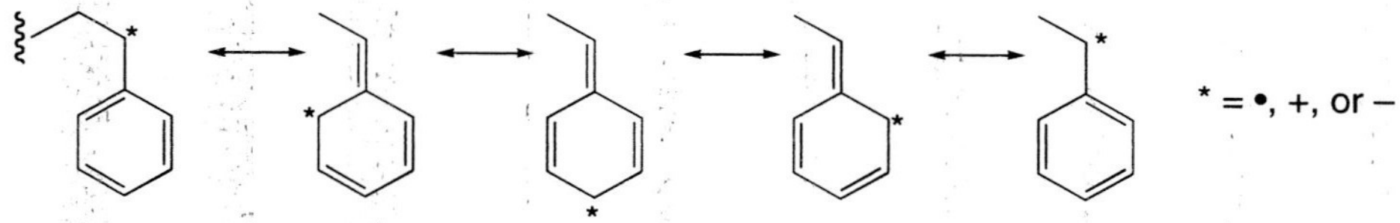

$* = •, +, or −$

30.10 Draw the copolymers formed in each reaction.

30.11

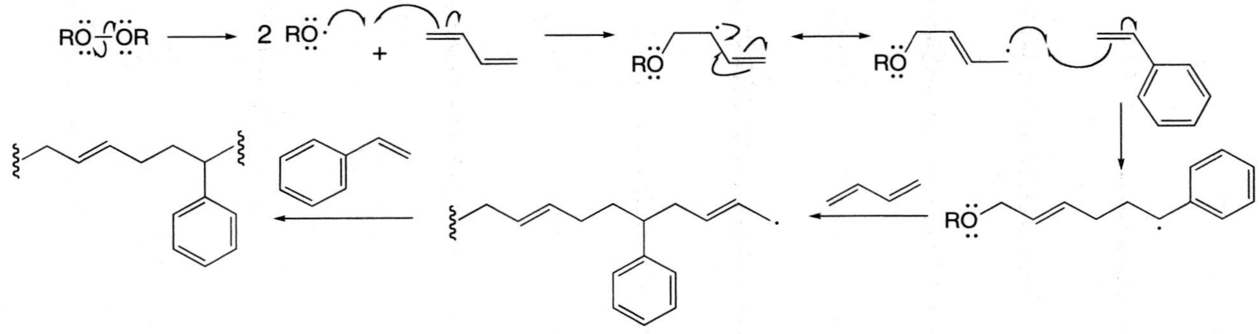

30.12

a.

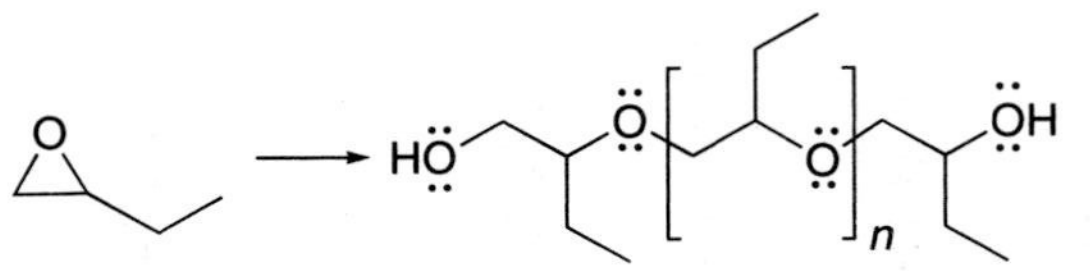

c.

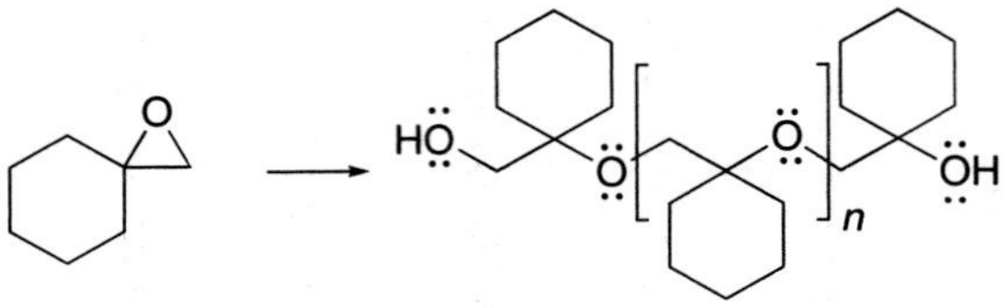

b.

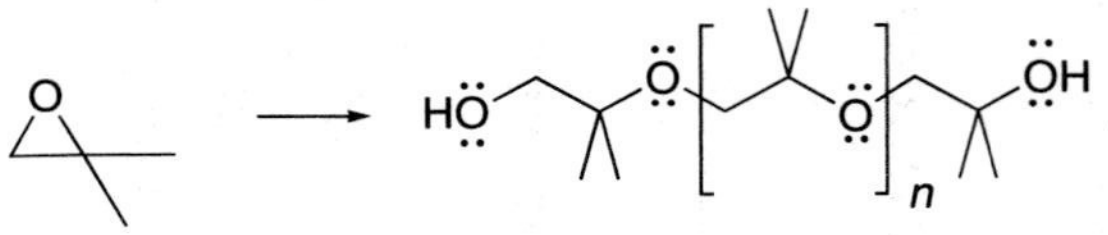

30.13

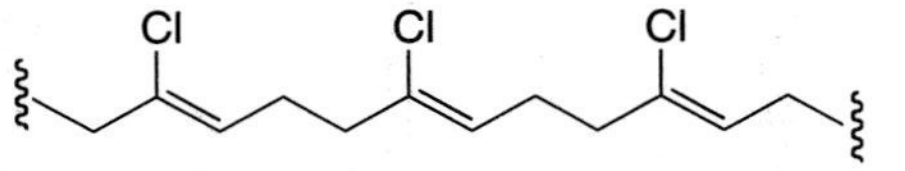

neoprene

All double bonds have the *Z* configuration.

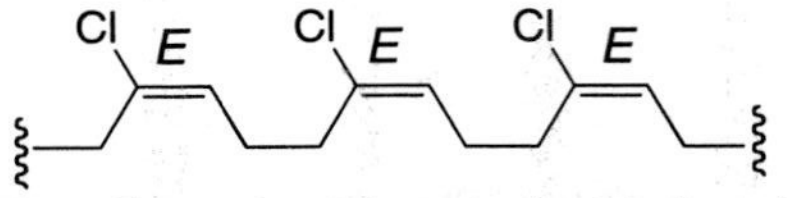

E configuration of each double bond

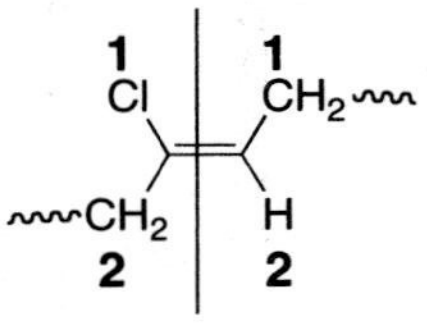
1
Cl
1
CH2
CH2
2
H
2

Two higher priority groups (1's) are on the same side of the double bond - - - > *Z* configuration.

30.14

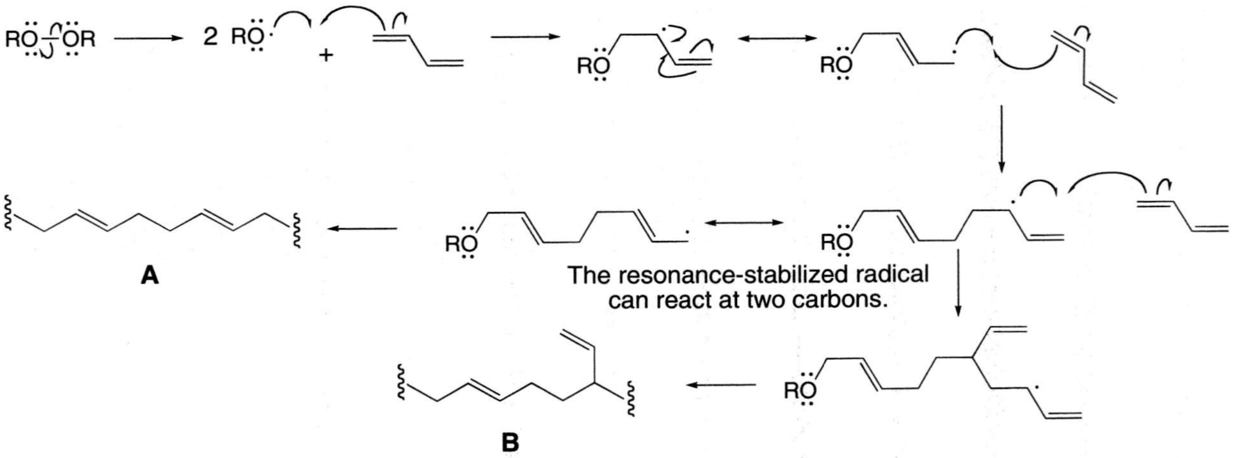
A
The resonance-stabilized radical
can react at two carbons.
B

30.15

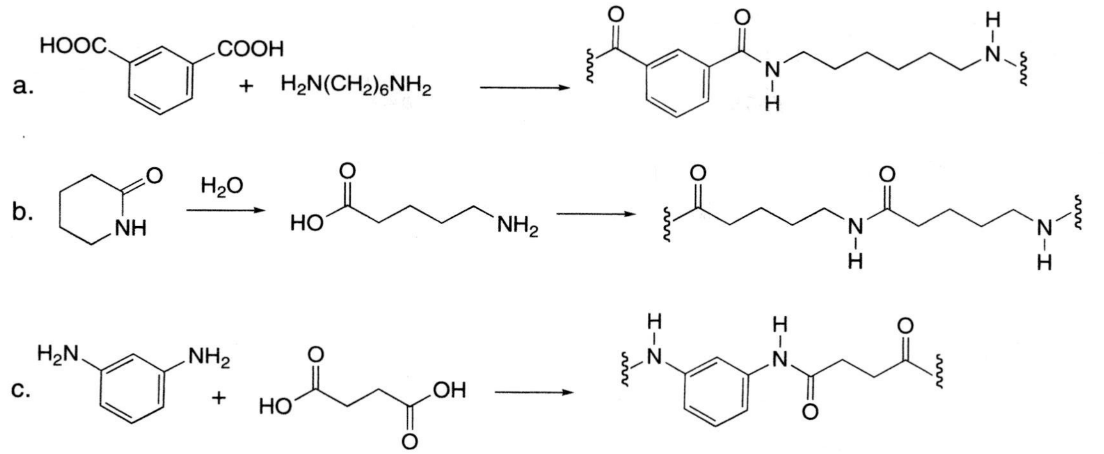

a.

HOOC — COOH + H₂N(CH₂)₆NH₂ ⟶ [amide polymer structure]

b.

[lactam] + H₂O ⟶ [amino acid] ⟶ [amide polymer structure]

c.

H₂N — NH₂ + [succinic acid] ⟶ [amide polymer structure]

30.16

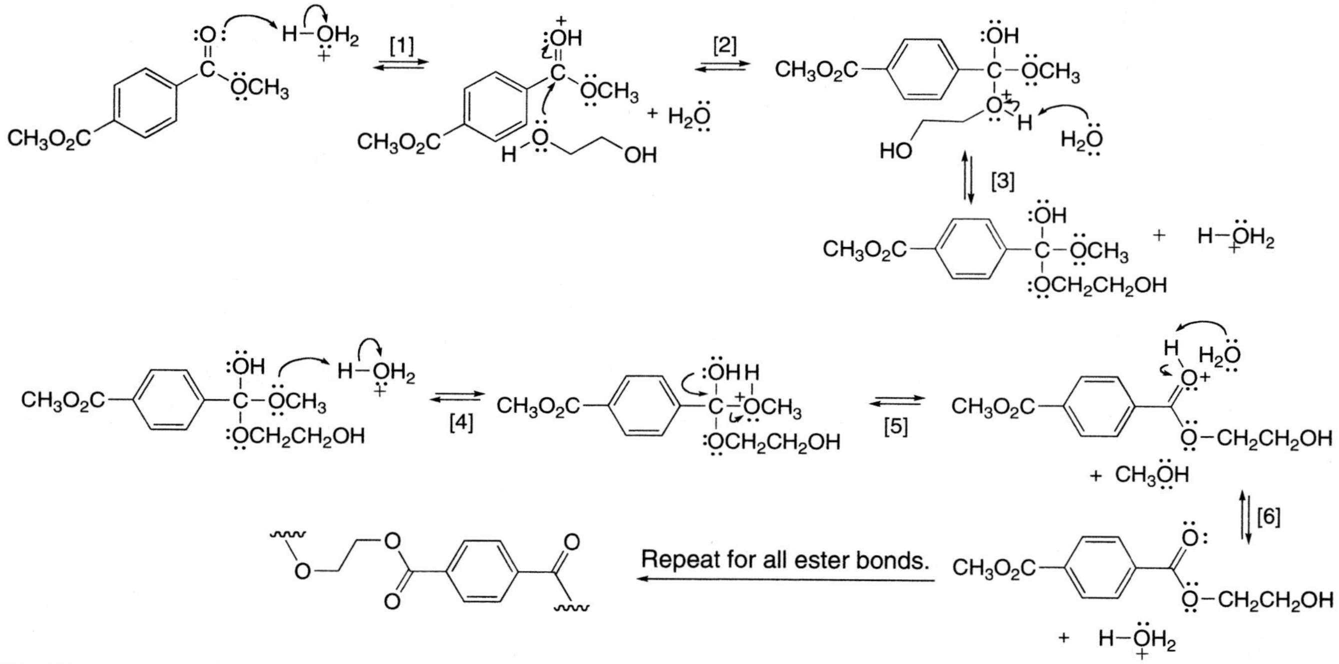

[1] [2] [3] [4] [5] [6]

Repeat for all ester bonds.

+ CH₃ÖH

+ H—ÖH₂

30.17

[diacid] + HO — OH ⟶ [polyester structure]

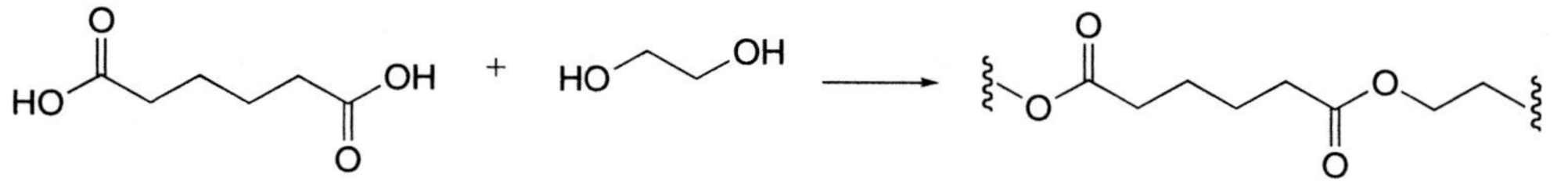

This compound is less suitable than either nylon 6,6 or PET for use in consumer products because esters are more easily hydrolyzed than amides, so this polyester is less stable than the polyamide nylon. This polyester has more flexible chains than PET, and this translates into a less strong fiber.

30.18

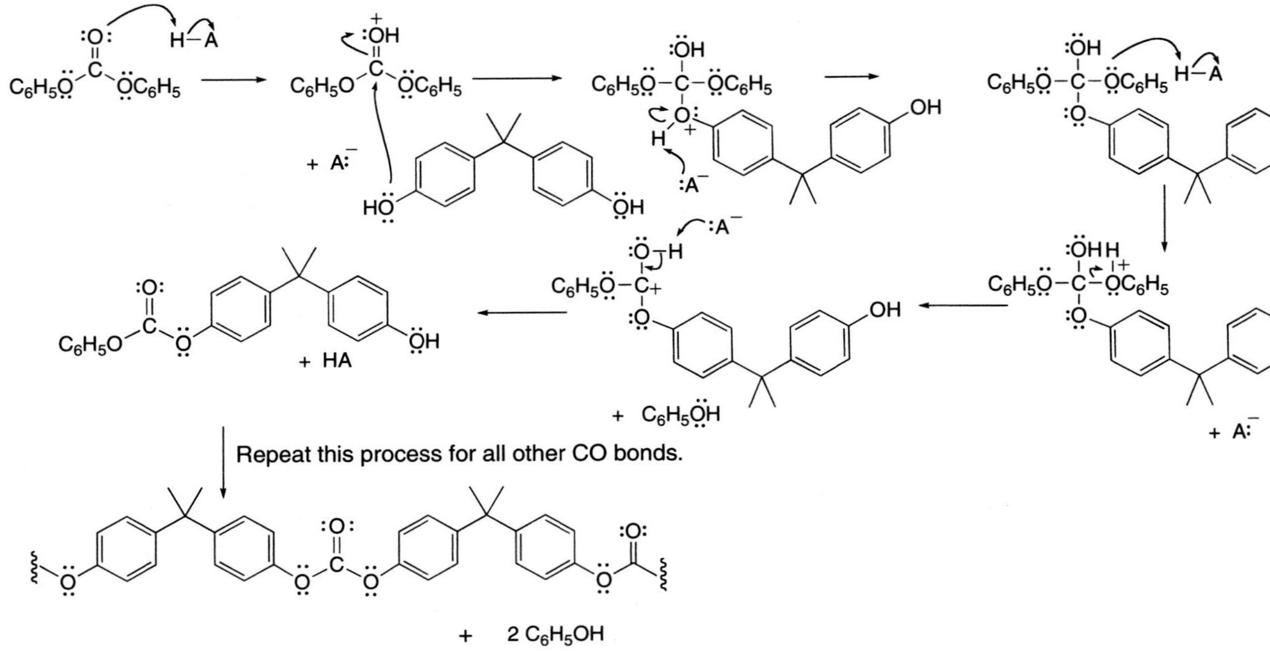

Repeat this process for all other CO bonds.

30.19

HO⟨benzene⟩OH + epichlorohydrin (excess)

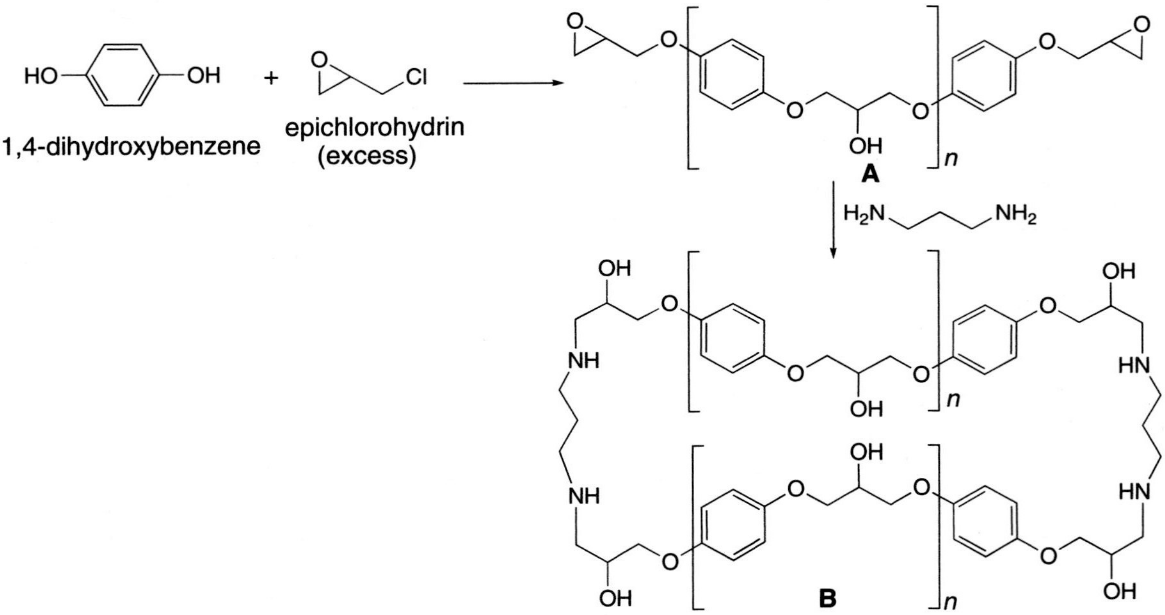

1,4-dihydroxybenzene

epichlorohydrin (excess)

A

H_2N—⟨⟩—NH_2

B

30.20

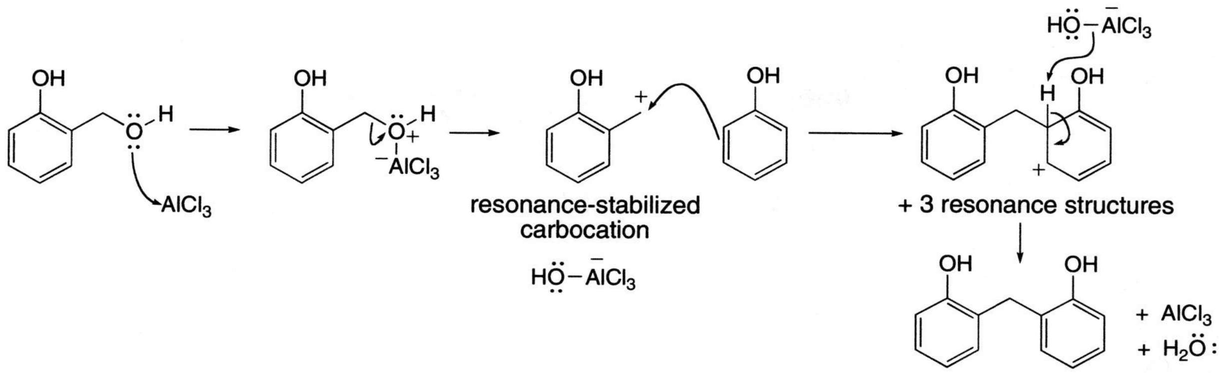

30.21

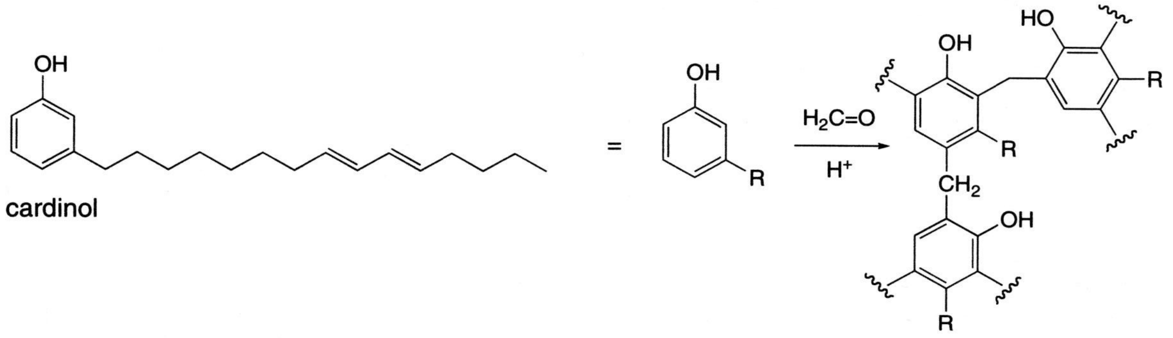

cardinol

30.22 Chemical recycling of HDPE and LDPE is not easily done because these polymers are both long chains of CH_2 groups joined together in a linear fashion. Since there are only C–C bonds and no functional groups in the polymer chain, there are no easy methods to convert the polymers to their monomers. This process is readily accomplished only when the polymer backbone contains hydrolyzable functional groups.

30.23 a. Combustion of polyethylene forms $CO_2 + H_2O$.
  b. Combustion of polyethylene terephthalate forms $CO_2 + H_2O$.
  c. These reactions are exothermic.
  d. HDPE and PET must be separated from poly(vinyl chloride) prior to incineration because combustion of hydrocarbons (like HDPE) and oxygen-containing organics (like PET) releases only $CO_2 + H_2O$ into the atmosphere. Poly(vinyl chloride) also contains Cl atoms bonded to a hydrocarbon chain. On combustion this forms HCl, which cannot be released directly into the atmosphere, making incineration of halogen-containing polymers more laborious and more expensive.

30.24

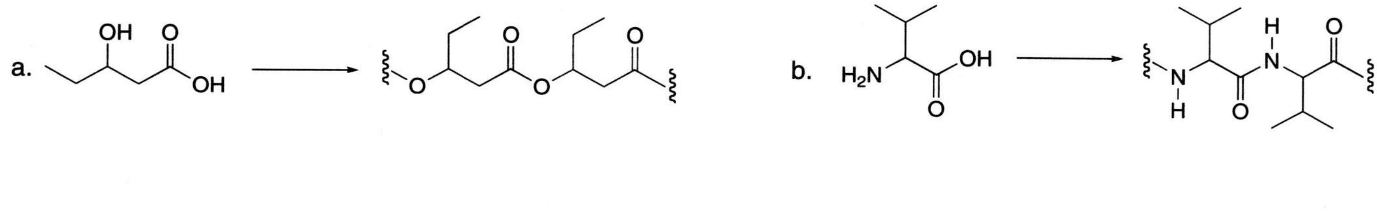

30.25 Draw the polymer formed by chain-growth polymerization as in Answer 30.2.

a.

c.

b.

d.

30.26 Draw the copolymers.

a.

d.

b.

e.

c.

30.27

a.

b.

c.

d.

e.

f.

30.28

a.

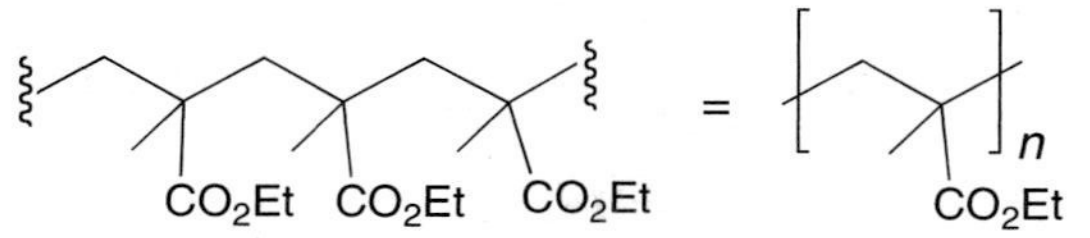

b.

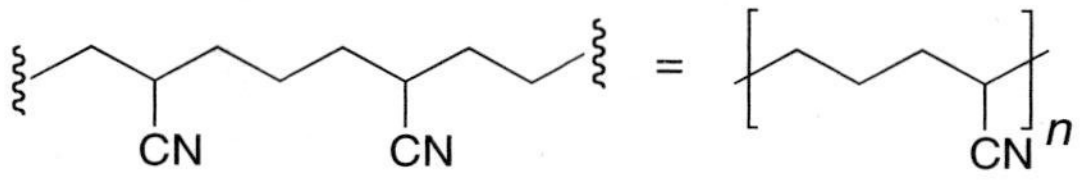

c.

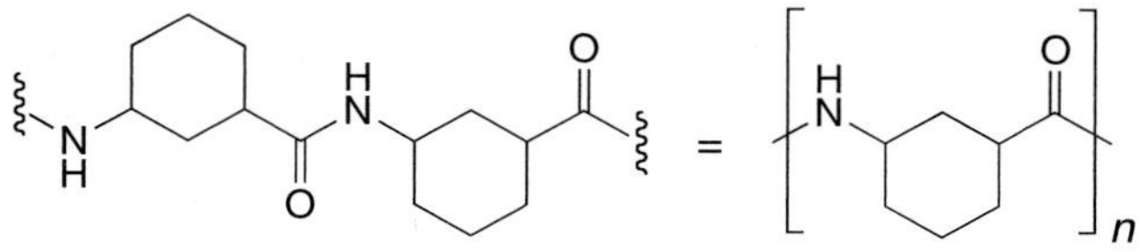

d.

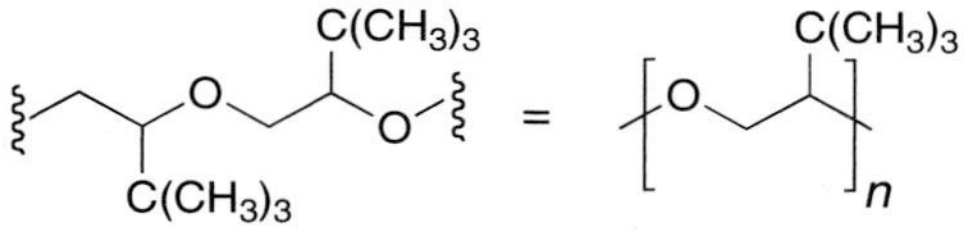

e.

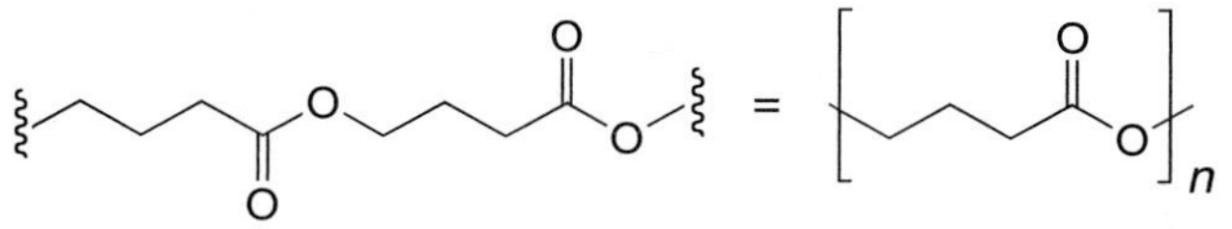

f.

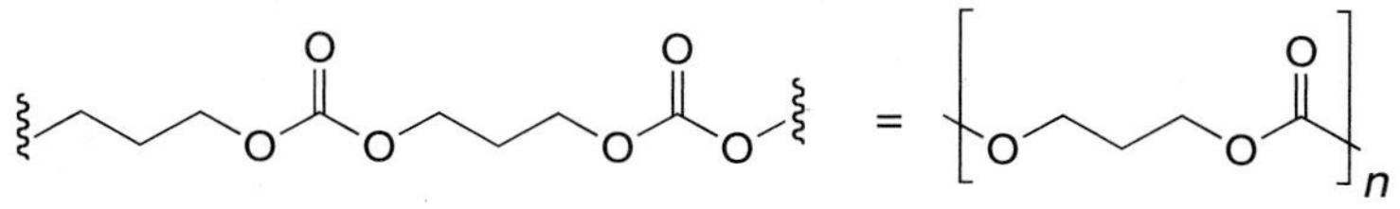

30.29 An **isotactic polymer** has all Z groups on the same side of the carbon backbone. A **syndiotactic polymer** has the Z groups alternating from one side of the carbon chain to the other. An **atactic polymer** has the Z groups oriented randomly along the polymer chain.

a. 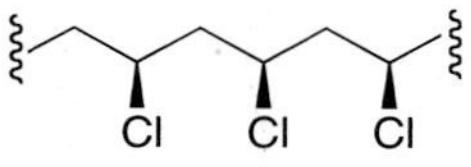   b. 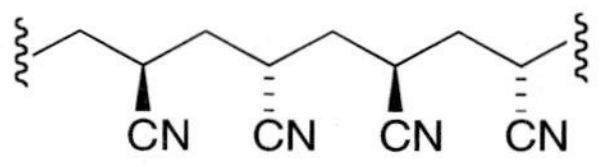   c.

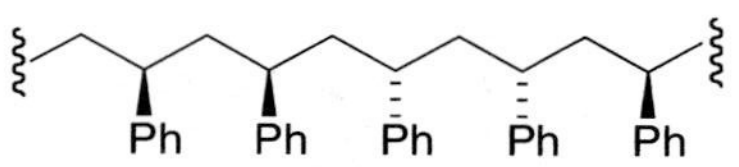

30.30

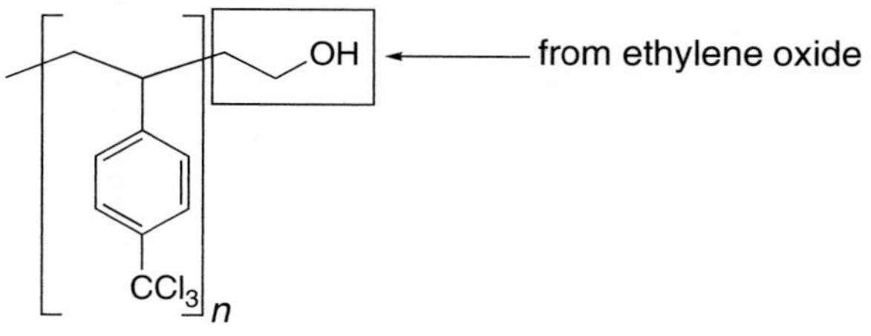

from ethylene oxide

30.31

a.

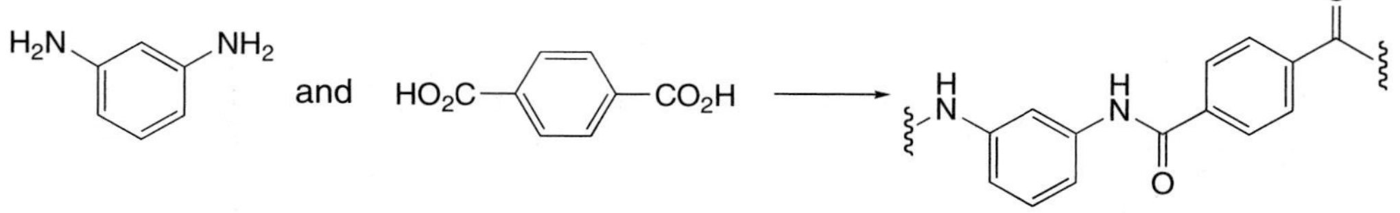

b. 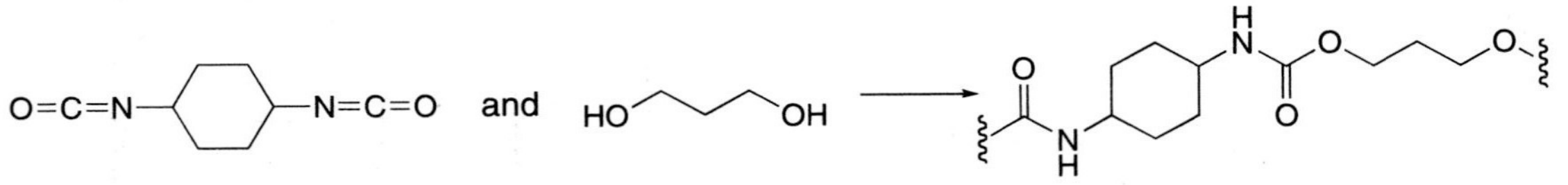   O=C=N—[cyclohexane]—N=C=O   and   HO—[chain]—OH   →   [structure with amide and carbamate groups]

c.   [benzene with COCl, COCl]   and   HO—[cyclohexane]—OH   →   [phthalate diester structure]

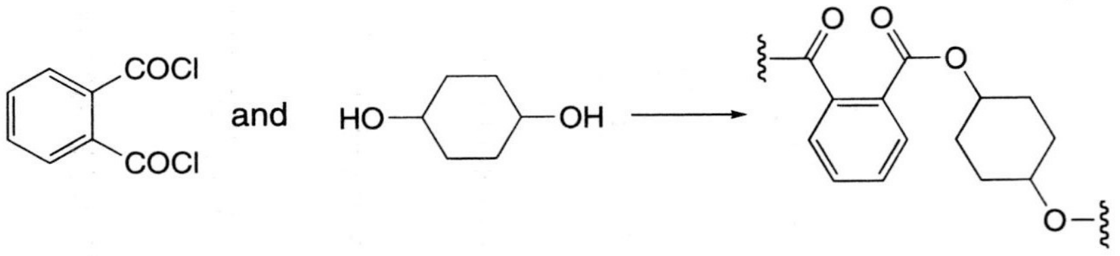

d. 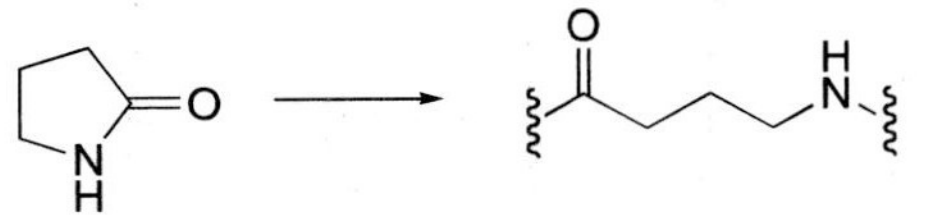   [pyrrolidinone] →   [ring-opened amide structure]

e. 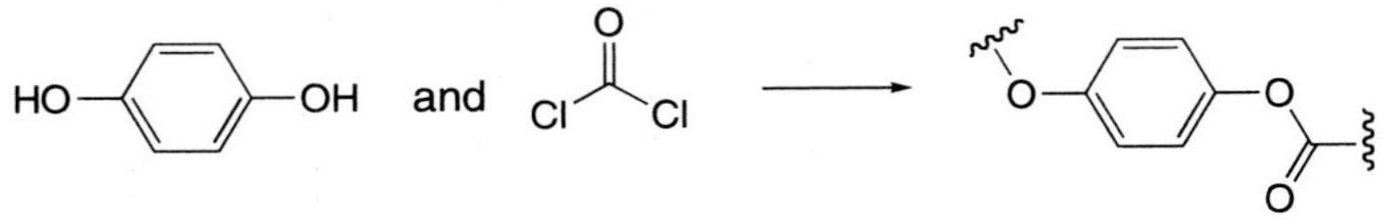   HO—[benzene]—OH   and   Cl—C(=O)—Cl   →   [polycarbonate structure]

f. 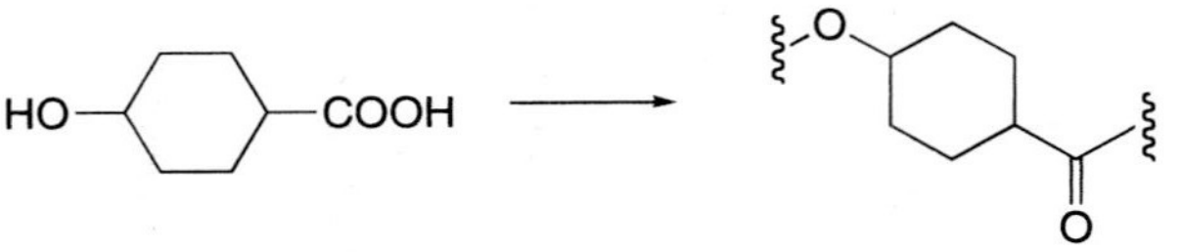   HO—[cyclohexane]—COOH   →   [ester structure]

30.32

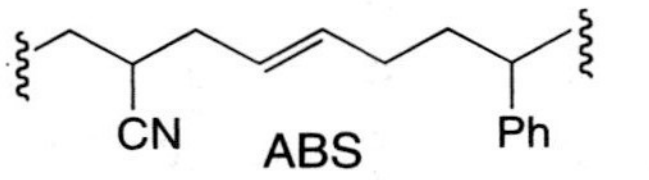

[polymer chain structure]

CN   ABS   Ph

30.33

a. 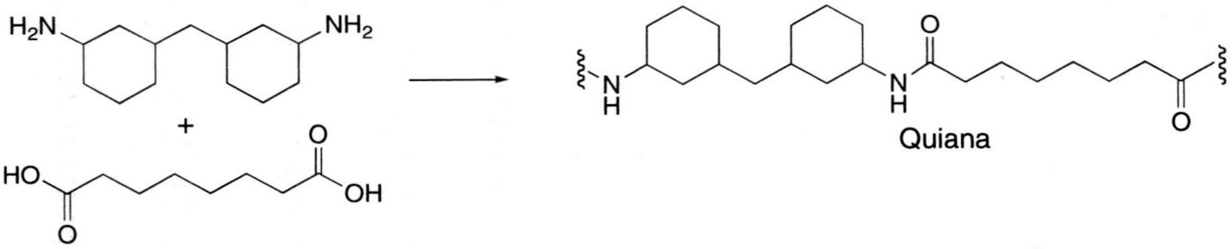   H₂N—[cyclohexane]—CH₂—[cyclohexane]—NH₂

   +

   HO—[chain]—OH (diacid)

   →   [polyamide structure]   Quiana

b. 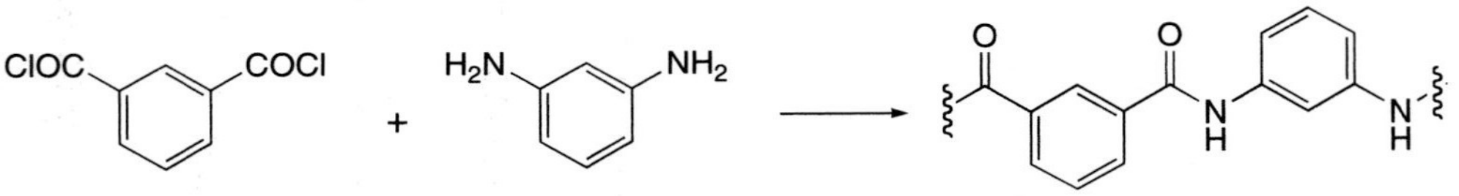   ClOC—[benzene]—COCl   +   H₂N—[benzene]—NH₂   →   [aramid structure]   Nomex

30.34

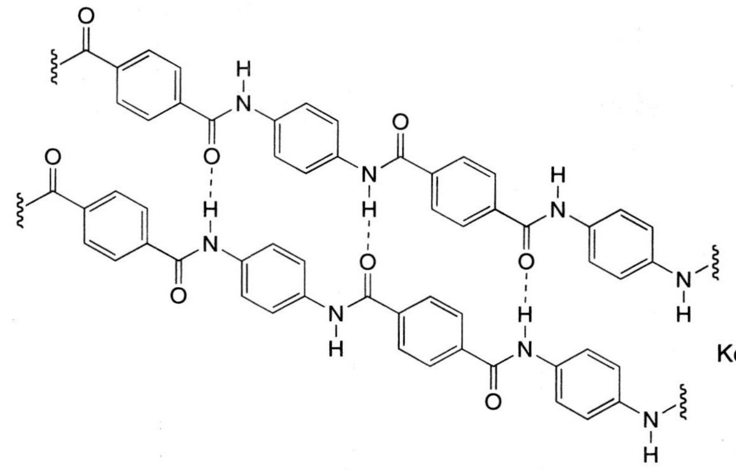

Kevlar

30.35

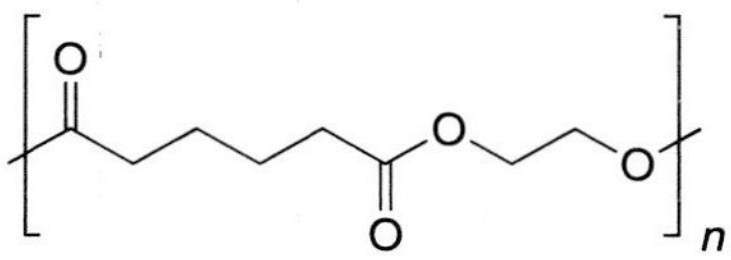

polyester **A**	PET	nylon 6,6
$T_g = < 0\ ^\circ C$	$T_g = 70\ ^\circ C$	$T_g = 53\ ^\circ C$
$T_m = 50\ ^\circ C$	$T_m = 265\ ^\circ C$	$T_m = 265\ ^\circ C$

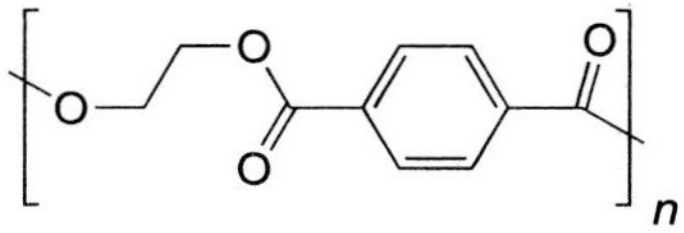

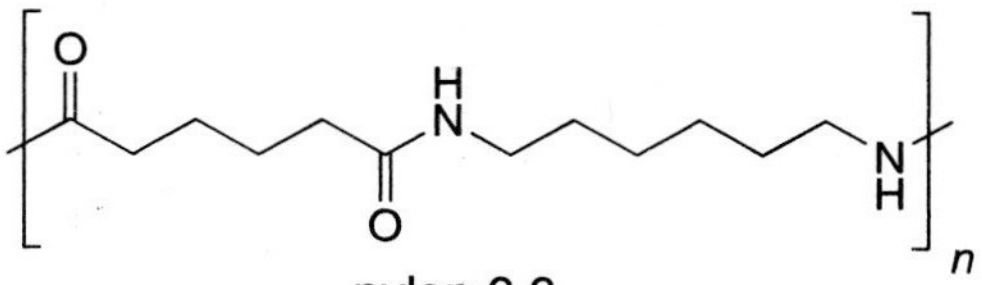

a. Polyester **A** has a lower T_g and T_m than PET because its polymer chain is more flexible. There are no rigid benzene rings so the polymer is less ordered.

b. Polyester **A** has a lower T_g and T_m than nylon 6,6 because the N–H bonds of nylon 6,6 allow chains to hydrogen bond to each other, which makes the polymer more ordered.

c. The T_m for Kevlar would be higher than that of nylon 6,6 because in addition to extensive hydrogen bonding between chains, each chain contains rigid benzene rings. This results in a more ordered polymer.

30.36

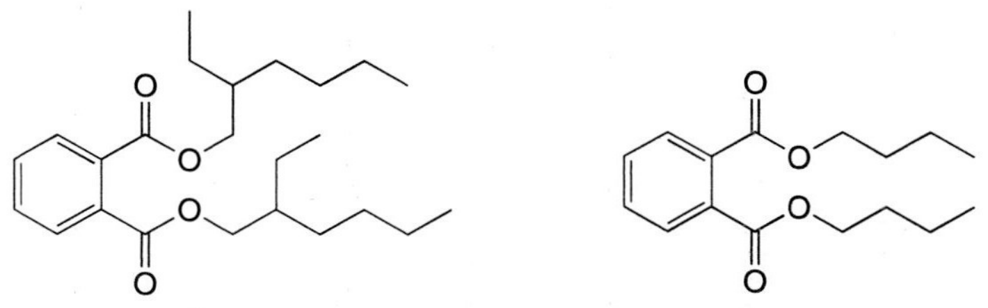

A   dibutyl phthalate

Diester **A** is often used as a plasticizer in place of dibutyl phthalate because it has a higher molecular weight, giving it a higher boiling point. **A** should therefore be less volatile than dibutyl phthalate, so it should evaporate from a polymer less readily.

30.37

Initiation:

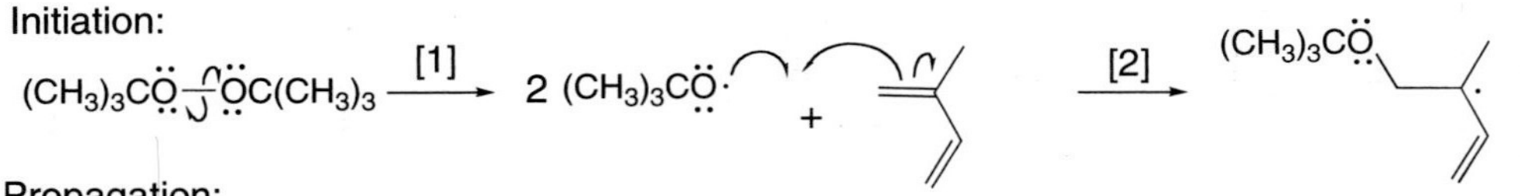

Propagation:

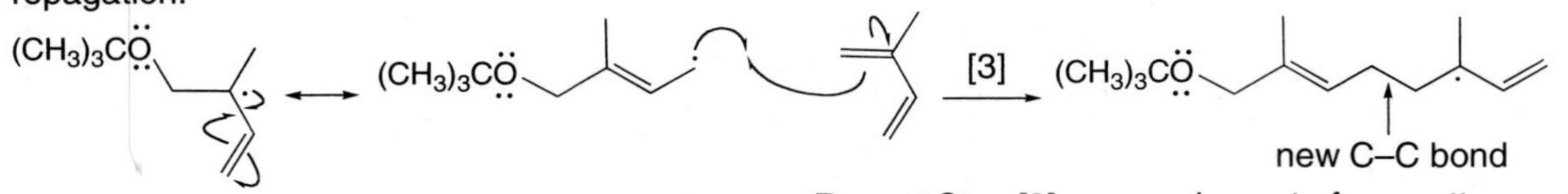

new C–C bond

Repeat Step [3] over and over to form gutta-percha.

Termination:

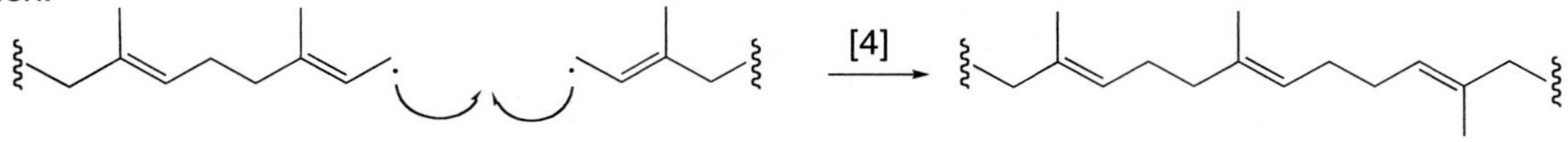

30.38

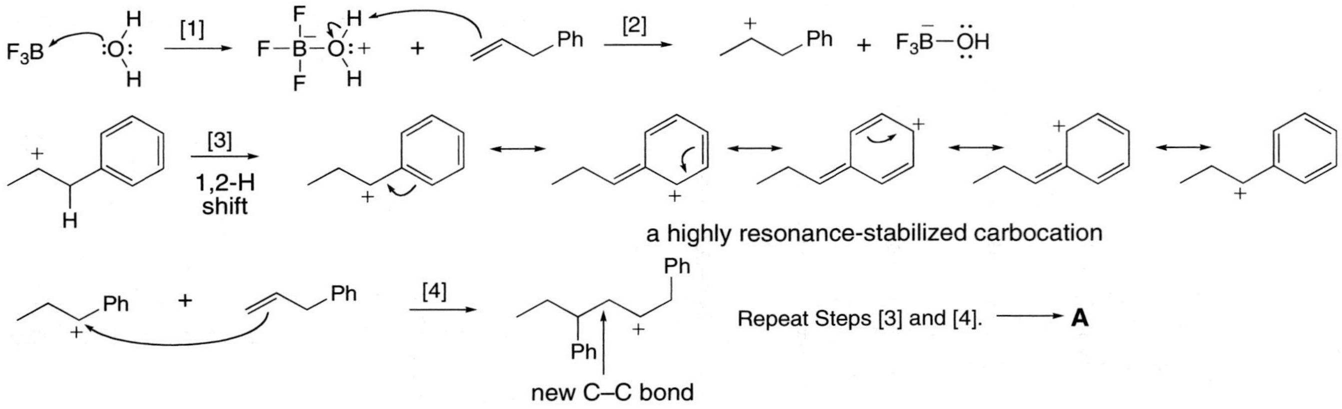

a highly resonance-stabilized carbocation

new C–C bond

Repeat Steps [3] and [4]. ⟶ **A**

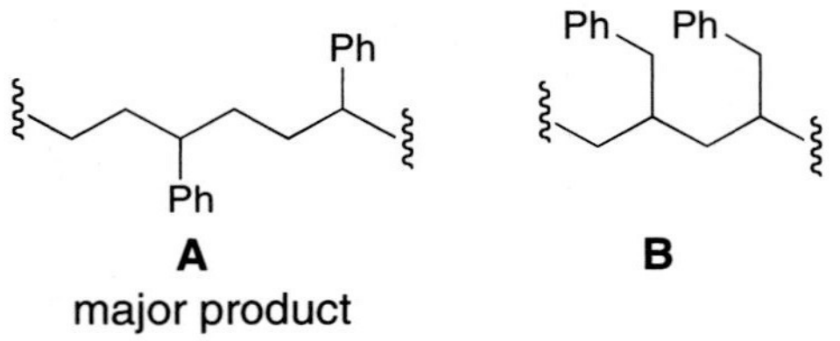

A
major product

B

A is the major product formed due to the 1,2-H shift (Step [3]) that occurs to form a resonance-stabilized carbocation.
B is the product that would form without this shift.

30.39

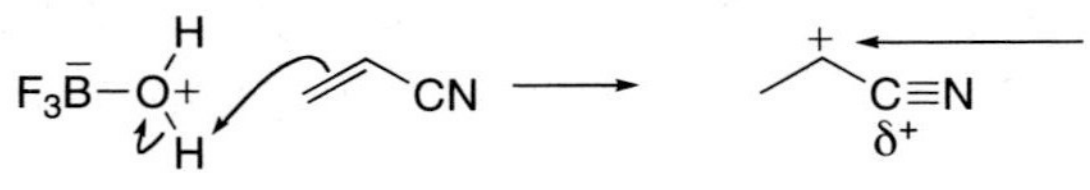

This carbocation is unstable because it is located next to an electron-withdrawing CN group that bears a δ^+ on its C atom. This carbocation is difficult to form, so $CH_2=CHCN$ is only slowly polymerized under cationic conditions.

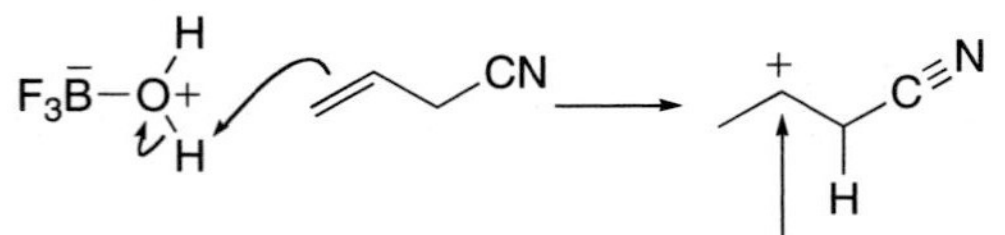

This 2° carbocation is more stable because it is not directly bonded to the electron-withdrawing CN group. As a result, it is more readily formed. Thus, cationic polymerization can occur more readily.

30.40

30.41 The substituent on styrene determines whether cationic or anionic polymerization is preferred. When the substituent stabilizes a carbocation, cationic polymerization will occur. When the substituent stabilizes a carbanion, anionic polymerization will occur.

a. cationic polymerization   b. anionic polymerization   c. anionic polymerization   d. cationic polymerization

30.42 The rate of anionic polymerization depends on the ability of the substituents on the alkene to stabilize an intermediate carbanion: the better a substituent stabilizes a carbanion, the faster anionic polymerization occurs.

least   most

increasing ability to undergo anionic polymerization

30.43 The reason for this selectivity is explained in Figure 9.9. In the ring opening of an unsymmetrical epoxide under acidic conditions, nucleophilic attack occurs at the carbon atom that is more able to accept a δ^+ in the transition state; that is, nucleophilic attack occurs at the more substituted carbon. The transition state having a δ^+ on a C with an electron-donating CH_3 group is more stabilized (lower in energy), permitting a faster reaction.

Repeat Steps [4] and [5] over and over.

30.44

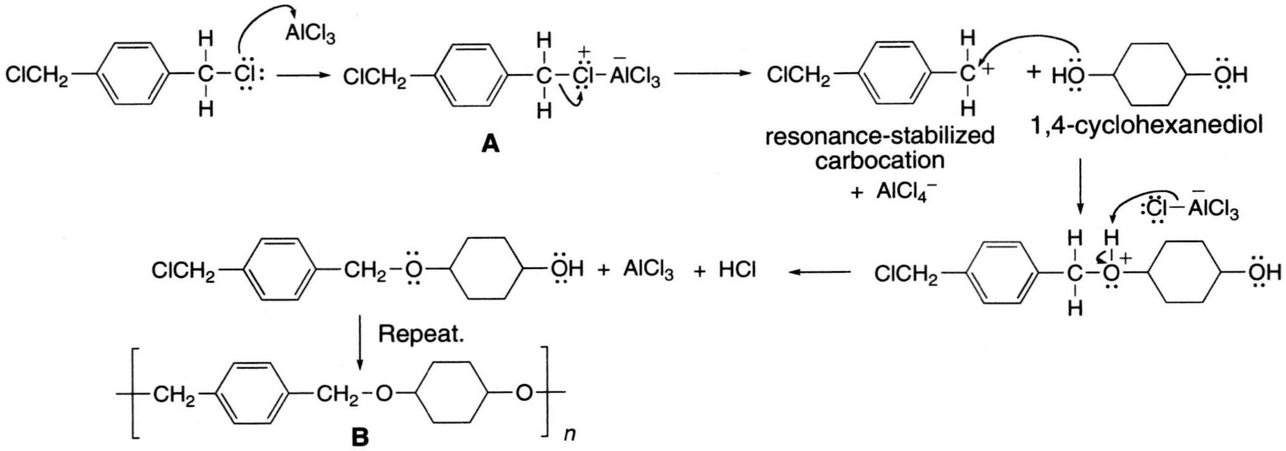

30.45

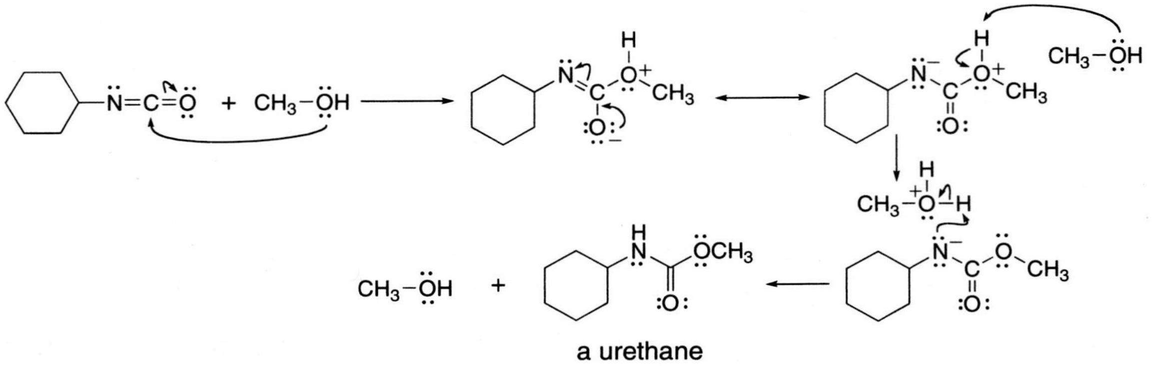

a urethane

30.46

a.

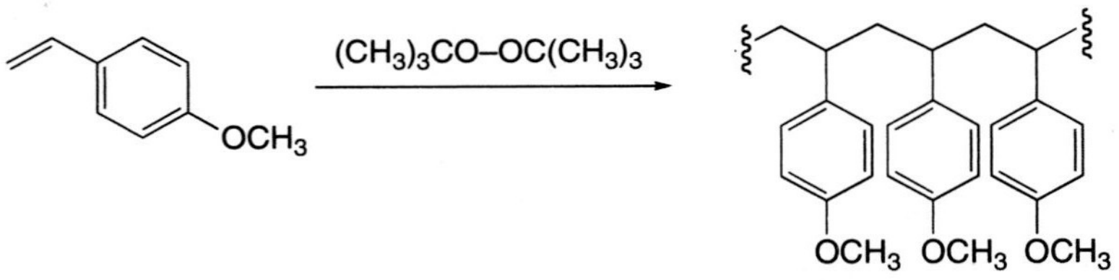

b.

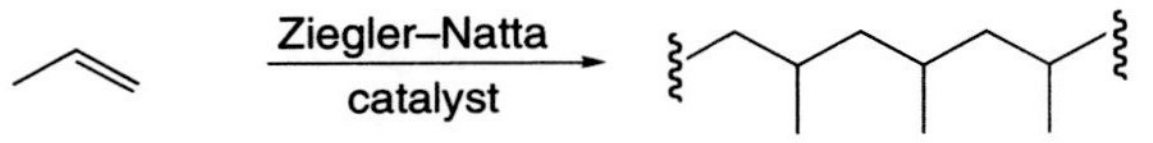

c.

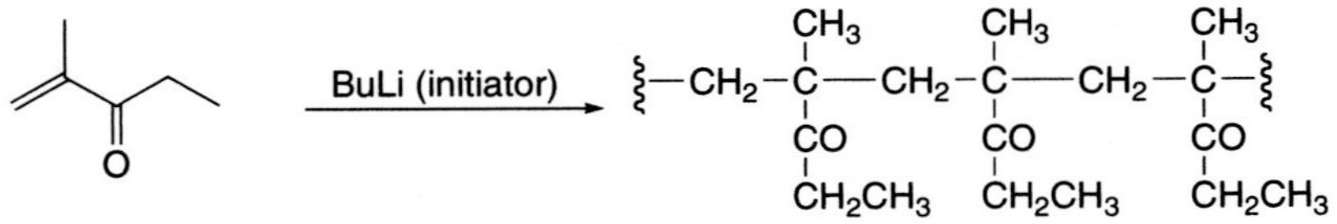

d.

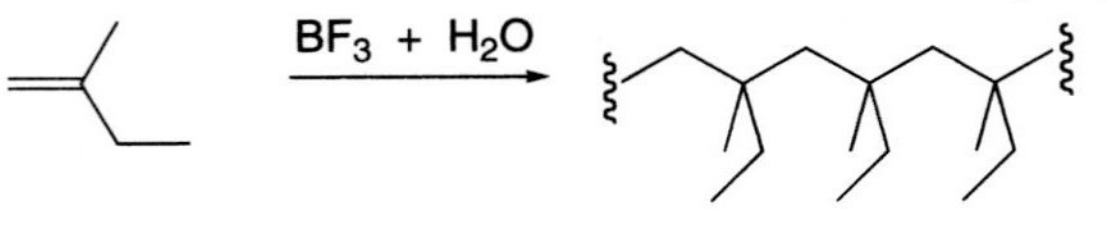

e.

f.

g.

h.

i.

j.

30.47 Polyethylene bottles are resistant to NaOH because they are hydrocarbons with no reactive sites. Polyester shirts and nylon stockings both contain functional groups. Nylon contains amides and polyester contains esters, two functional groups that are susceptible to hydrolysis with aqueous NaOH. Thus, the polymers are converted to their monomer starting materials, creating a hole in the garment.

30.48

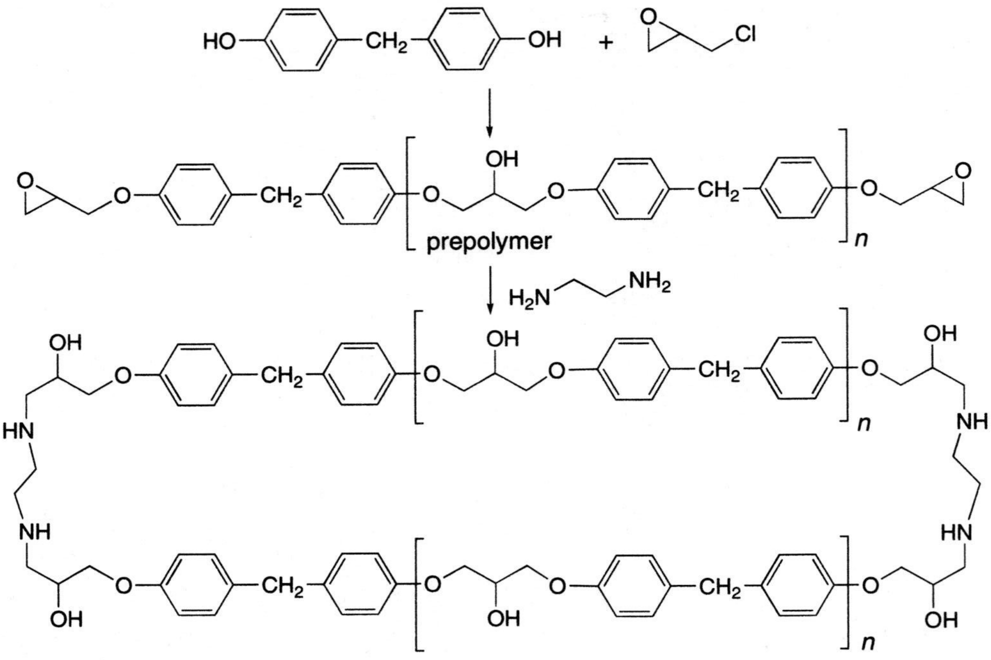

30.49

a.

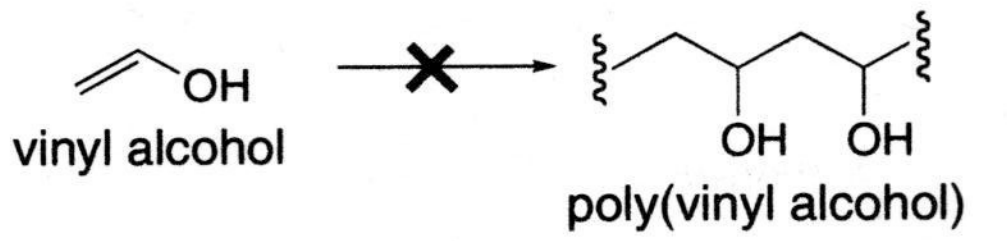

vinyl alcohol

poly(vinyl alcohol)

Poly(vinyl alcohol) cannot be prepared from vinyl alcohol because vinyl alcohol is not a stable monomer. It is the enol of acetaldehyde (CH_3CHO), and thus it can't be converted to poly(vinyl alcohol).

b.

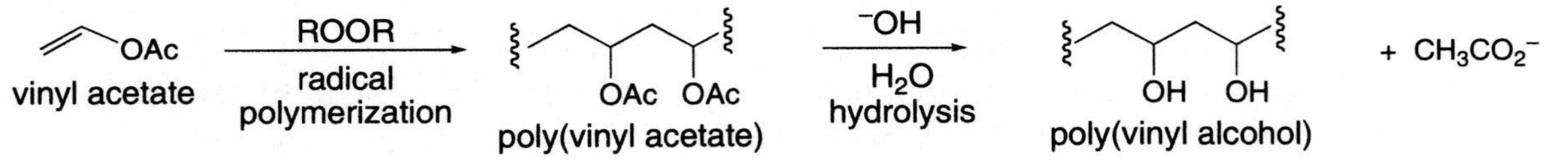

vinyl acetate

poly(vinyl acetate)

poly(vinyl alcohol)

$+ CH_3CO_2^-$

c.

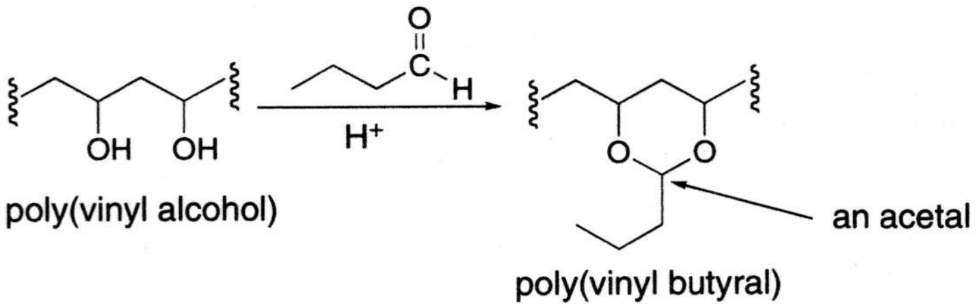

poly(vinyl alcohol)

an acetal

poly(vinyl butyral)

30.50

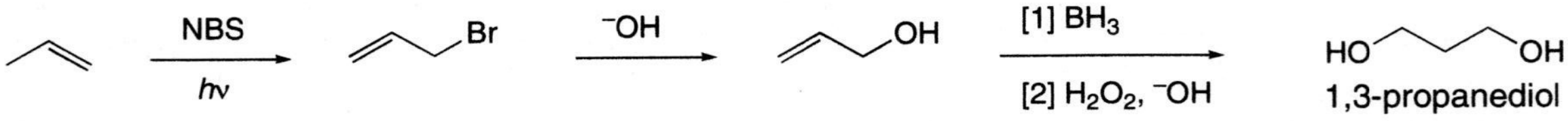

1,3-propanediol

30.51

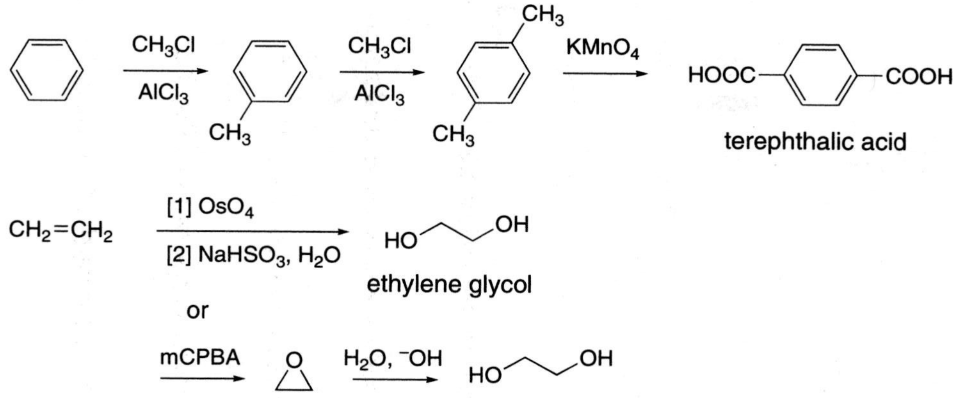

terephthalic acid

ethylene glycol

or

30.52

phenol

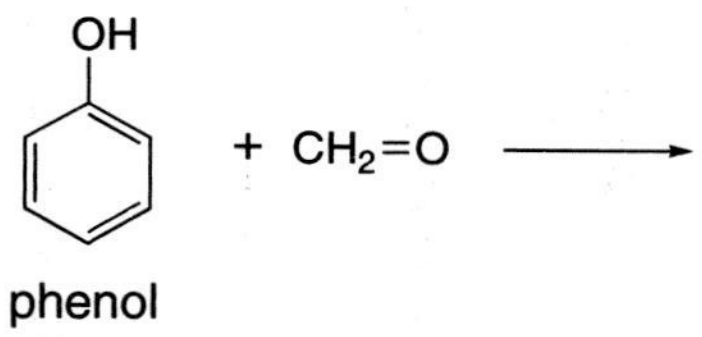

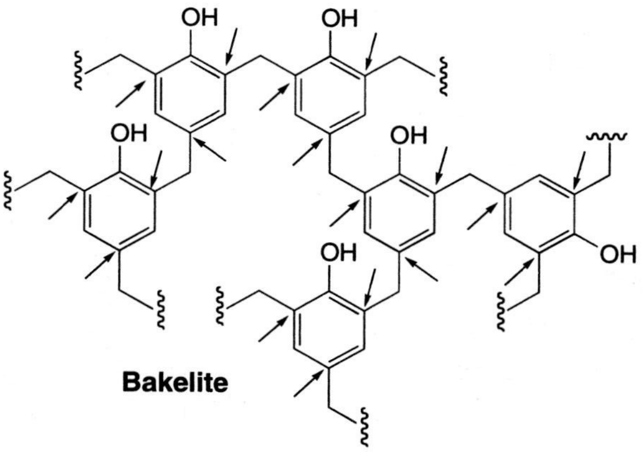

Since phenol has no substituents at any ortho or para position, an extensive network of covalent bonds can join the benzene rings together at all ortho and para positions to the OH groups.

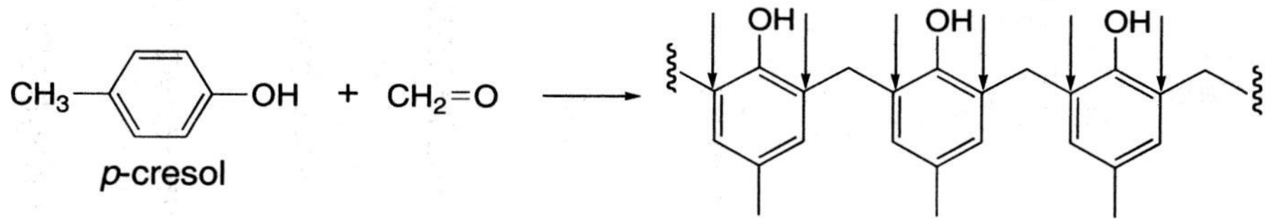

Bakelite

p-cresol

Since *p*-cresol has a CH₃ group at the para position to the OH group, new bonds can be formed only at two ortho positions so that a less extensive three-dimensional network can form.

30.53

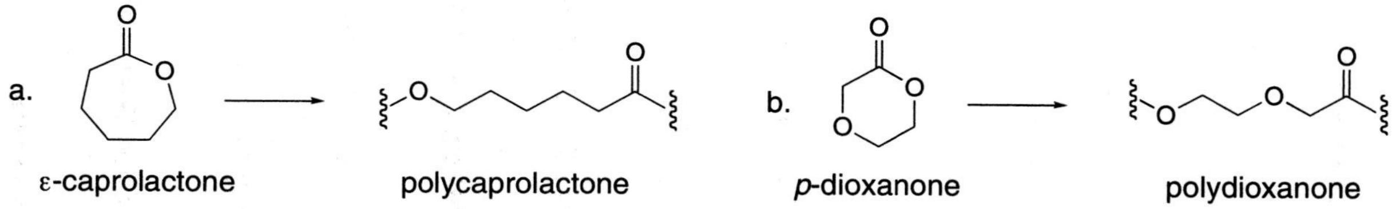

a.

ε-caprolactone → polycaprolactone

b.

p-dioxanone → polydioxanone

30.54

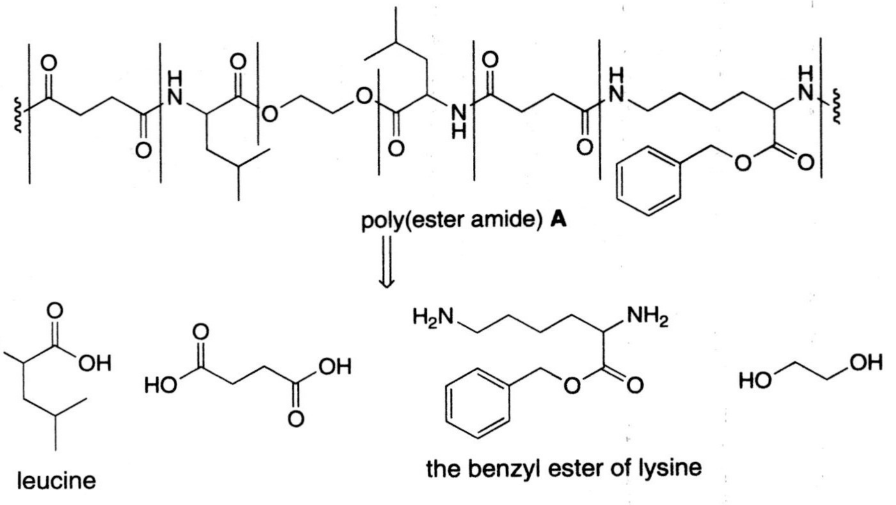

poly(ester amide) **A**

leucine

the benzyl ester of lysine

30.55

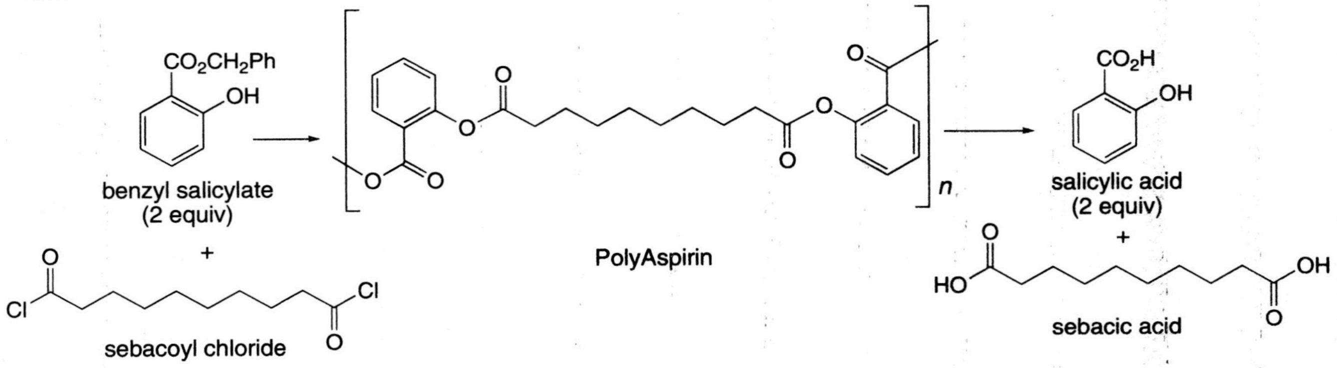

benzyl salicylate
(2 equiv)

+

sebacoyl chloride

PolyAspirin

salicylic acid
(2 equiv)

+

sebacic acid

30.56

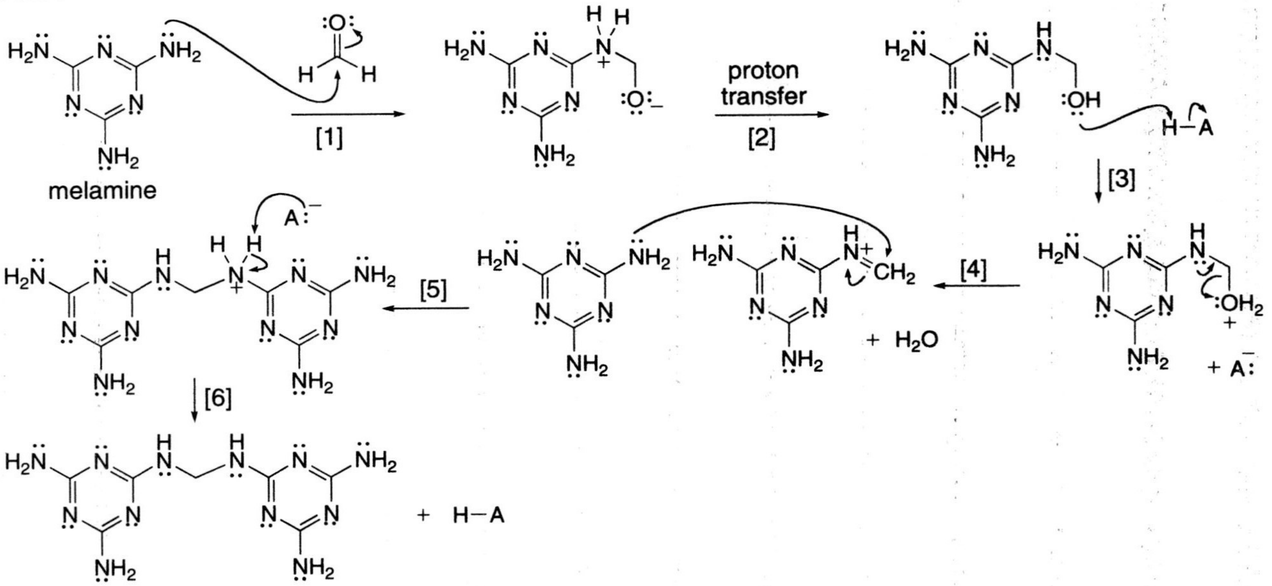

melamine

proton
transfer
[2]

[1]

[3]

[4]

+ H₂O

[5]

[6]

+ H—A

30.57

a.

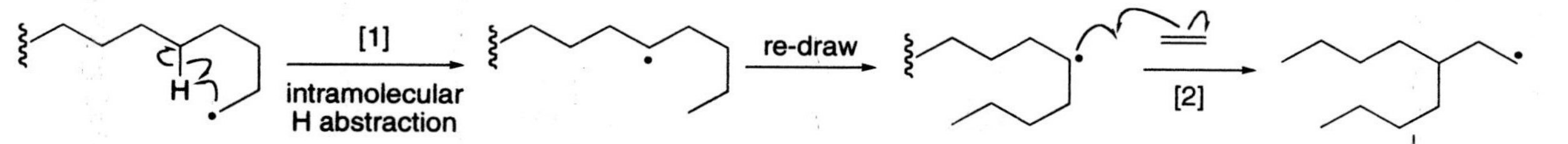

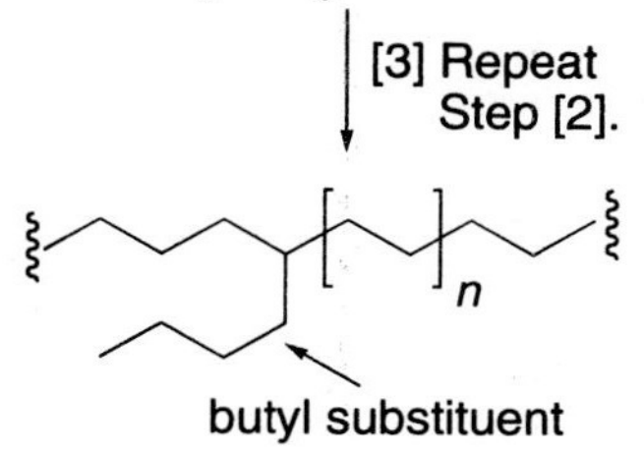

b. Abstraction of the H is more facile than abstraction of the other H's because the H atom that is removed is six atoms from the radical. The transition state for this intramolecular reaction is cyclic, and resembles a six-membered ring, the most stable ring size. Other H's are too far away or the transition state would resemble a smaller, less stable ring.

30.58

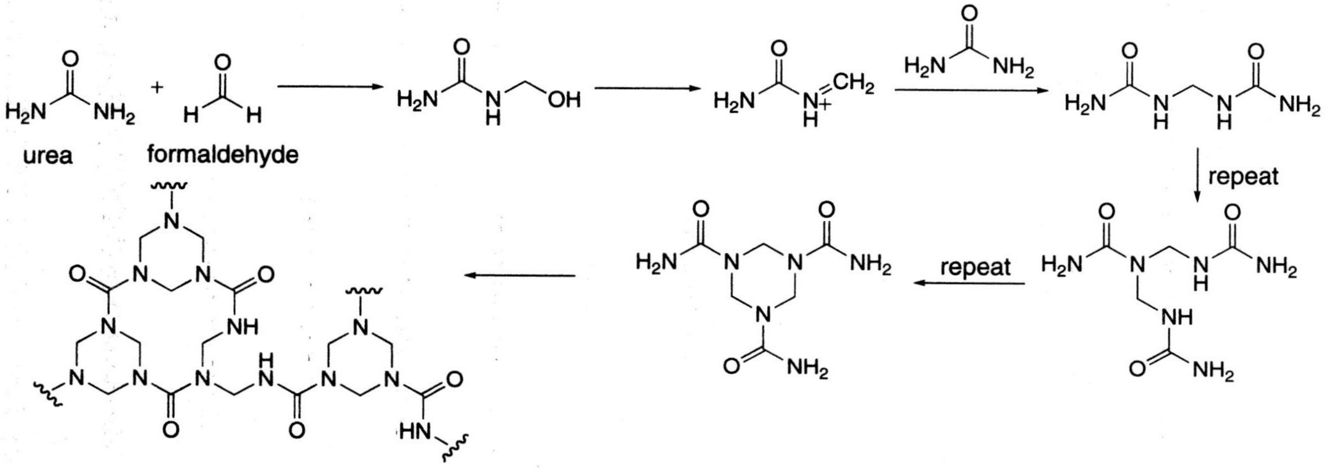

urea   formaldehyde